107. *Area of a Window* A window is in the shape of a square with a side of length s, with a semicircle of diameter s adjoining the top of the square. Write the total area of the window W as a function of s.

HINT Start with the formulas for the area of a square and the area of a circle.

■ **Figure for Exercises 107 and 108**

Hints:

Within the exercise sets, there are now Hints for starting approximately five selected application problems within each set. These Hints suggest ways of approaching the problem (visual, algebraic) and give a starting point to solve the application problem. The process of problem-solving is the focus of those Hints.

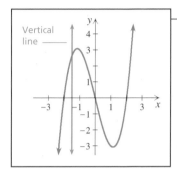

Art Annotations:

This innovative feature highlights key ideas within the mathematical art, making each illustration a more useful learning tool.

Linking Concepts:

This is a multi-part exercise or exploration that uses a concept from the current section along with concepts from preceding sections in the book to solve problems that illustrate the links among various concepts. This is an excellent check for understanding of major or multiple concepts in the course. Success at these likely means success in the course.

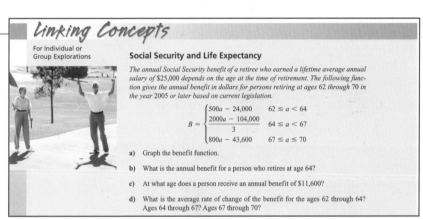

Linking Concepts

For Individual or Group Explorations

Social Security and Life Expectancy

The annual Social Security benefit of a retiree who earned a lifetime average annual salary of $25,000 depends on the age at the time of retirement. The following function gives the annual benefit in dollars for persons retiring at ages 62 through 70 in the year 2005 or later based on current legislation.

$$B = \begin{cases} 500a - 24{,}000 & 62 \le a < 64 \\ \dfrac{2000a - 104{,}000}{3} & 64 \le a < 67 \\ 800a - 43{,}600 & 67 \le a \le 70 \end{cases}$$

a) Graph the benefit function.

b) What is the annual benefit for a person who retires at age 64?

c) At what age does a person receive an annual benefit of $11,600?

d) What is the average rate of change of the benefit for the ages 62 through 64? Ages 64 through 67? Ages 67 through 70?

Function Gallery: Some Inverse Functions

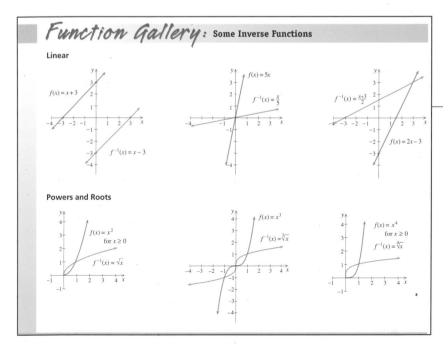

Linear

$f(x) = x + 3$
$f^{-1}(x) = x - 3$

$f(x) = 5x$
$f^{-1}(x) = \dfrac{x}{5}$

$f^{-1}(x) = \dfrac{x+3}{2}$
$f(x) = 2x - 3$

Powers and Roots

$f(x) = x^2$ for $x \ge 0$
$f^{-1}(x) = \sqrt{x}$

$f(x) = x^3$
$f^{-1}(x) = \sqrt[3]{x}$

$f(x) = x^4$ for $x \ge 0$
$f^{-1}(x) = \sqrt[4]{x}$

Function Galleries:

These graphical galleries are designed to help students link the visual aspects of various families of functions to the properties of the functions.

COLLEGE ALGEBRA AND TRIGONOMETRY

4th Edition

Mark Dugopolski

Southeastern Louisiana University

PEARSON

Addison
Wesley

Boston San Francisco New York
London Toronto Sydney Tokyo Singapore Madrid
Mexico City Munich Paris Cape Town Hong Kong Montreal

Publisher: Greg Tobin
Executive Editor: Anne Kelly
Project Editor: Joanne Ha
Assistant Project Editors: Marcia Emerson and Elizabeth Bernardi
Editorial Assistant: Ashley O'Shaughnessy
Managing Editor: Karen Wernholm
Senior Production Supervisor: Peggy McMahon
Senior Designer: Barbara T. Atkinson
Cover Design: Minko Images
Photo Researcher: Beth Anderson
Media Producer: Michelle Murray
Software Development: Mary Durnwald & John O'Brien
Senior Marketing Manager: Becky Anderson
Marketing Coordinator: Maureen McLaughlin
Senior Author Support/Technology Specialist: Joe Vetere
Rights and Permissions Advisor: Dana Weightman
Manufacturing Buyer: Carol Melville
Text Design, Production Coordination, Composition, and Illustrations: Nesbitt Graphics

Cover photo: © Orange County Choppers, Inc. All rights reserved.

For permission to use copyrighted material, grateful acknowledgment has been made to the copyright holders on page C-1 in the back of the book, which is hereby made part of this copyright page.

Many of the designations used by manufacturers and sellers to distinguish their products are claimed as trademarks. Where those designations appear in this book, and Pearson Education was aware of a trademark claim, the designations have been printed in initial caps or all caps.

The Library of Congress has already cataloged the Student Edition as follows:

Library of Congress Cataloging-in-Publication Data

Dugopolski, Mark.
    College algebra and trigonometry / Mark Dugopolski.--4th ed.
       p. cm.
    Includes index.
    ISBN 0-321-35692-6
       1. Algebra--Textbooks. 2. Trigonometry--Textbooks. I. Title.

QA154.3.D84 2007
512'.13--dc22

2005045805

ISBN 0-321-35692-6

2 3 4 5 6 7 8 9 10—QWT—09 08 07

CONTENTS

Haselbach, Lv

P

Prerequisites

1

Equations, Inequalities, and Modeling

2

Functions and Graphs

3

Polynomial and Rational Functions

4

Exponential and Logarithmic Functions

5

The Trigonometric Functions

6

Trigonometric Identities and Conditional Equations

7

Applications of Trigonometry

8

Systems of Equations and Inequalities

9

Matrices and Determinants

10

The Conic Sections

11

Sequences, Series, and Probability

College Algebra and Trigonometry is designed for a variety of students with different mathematical needs. For those students who will take additional mathematics, this text will provide the proper foundation of skills, understanding, and insights for success in subsequent courses. For those students who will not pursue further mathematics, the extensive emphasis on applications and modeling will demonstrate the usefulness and applicability of algebra and trigonometry in the everyday world. I am always on the lookout for real-life applications of mathematics and I have included many real problems that people actually encounter on the job. With an emphasis on problem solving, this text provides students with an excellent opportunity to sharpen their reasoning and thinking skills. With increased problem-solving capabilities, students will have confidence to tackle problems that they encounter outside the classroom.

Content Changes for the Fourth Edition

For this fourth edition of *College Algebra and Trigonometry,* I have extensively updated explanations, examples, exercises, and art in response to comments from the users of the third edition. In tackling the exercise sets, over 1000 additional exercises have been added to this edition, and many of the existing exercises have been improved, updated, and reorganized with the goal of providing students a smoother path to success. Using a pedagogical approach to color, more color has been added to key examples to emphasize certain computations. The art package has been revamped with new photographs and situational art, which accompanies a fresh new design that will help students visualize the mathematics being discussed. The section on complex numbers has been moved to Chapter P (Prerequisites) so that Chapter 3 (Polynomial and Rational Functions) would contain fewer sections and be easier to teach.

New or Enhanced Features

I have included several new features and revised some existing features. These improvements are designed to make the book easier to use and to support student and instructor success. I believe students and instructors will welcome the following new and enhanced features:

For students:

- **Try This** (page 92) Occurring immediately after every example in the text is an exercise that is very similar to the example. These problems give students the opportunity to immediately try a problem that is just like the example and check their work. Solutions to all *Try This* problems are in the appendix of the *Student Edition.*
- **Pop Quizzes** (page 66) Included in every section of the text, the *Pop Quizzes* give instructors convenient quizzes that can be used in the classroom. They are short and cover just the basic facts. The answers appear in the *Annotated Instructor's Edition* only.
- **Thinking Outside the Box** (page 696) Found throughout the text, these problems are designed to get students (and instructors) to do some mathematics just for fun. Yes, fun. I enjoyed solving these problems and hope that you will also. The problems can be used for individual or group work. They may or may not have anything to do with the sections in which they are located and might not

even require any techniques discussed in this text. So be creative on these problems and try *Thinking Outside the Box.* The answers are given in the *Annotated Instructor's Edition* only and complete solutions can be found in the *Instructor's Solutions Manual.*

■ **Function Galleries** (page 415) Located throughout the text, the function summaries have been improved with a new design and are now also gathered together at the beginning of the text. These graphical galleries are designed to help the students link the visual aspects of various families of functions to the properties of the functions.

■ **Hints** (page 107) are provided as a starting point for selected application problems and encourage students to think through the problems. A *Hint* logo HINT is used where a hint is given.

■ **Highlights** (page 409) This end-of-chapter feature has been completely rewritten with the student in mind. Each highlight contains an overview of all of the concepts presented in the chapter along with brief examples to illustrate the concept.

■ **Art Annotations** (page 281) have been added in order to make the mathematical art a better learning tool and easier for students to use.

■ **Summaries** of important concepts are included to help students clarify ideas that have multiple parts. For example, see the *Summary* of the Types of Symmetry in Section 3.5 on page 316.

■ **Strategies** contain general guidelines for accomplishing tasks. They have been given a more visible design so that students will be more aware of them and use them to sharpen their problem-solving skills. For example, see the *Strategy* for Solving Exponential and Logarithmic Equations in Section 4.4 on page 399.

■ **Procedures** are similar to *Strategies,* but are more specific and more algorithmic. They have also been highlighted in this edition and are designed to give students a step-by-step approach. For an example, see the *Procedure* for Finding an Inverse Function by the Switch and Solve Method in Section 2.5 on page 243.

For instructors:

■ **Annotated Instructor's Edition** New to this edition, this text is available with answers beside the exercises for most exercises, plus a complete answer section at the back.

■ **Example Numbers** (page 276) Occurring at the end of the direction line for nearly all groups of exercises, the *Example Numbers* indicate which examples correspond to the exercises. This keying of exercises to examples appears only in the *Instructor's Edition* and will make it easier for instructors to assign homework problems.

■ **Insider's Guide, 4/e** This new supplement includes resources to help faculty with course preparation and classroom management. Included are helpful teaching tips correlated to each section of the text as well as additional resources for classroom enrichment.

Continuing Features

■ **Chapter Opener** Each chapter begins with chapter opener text that discusses a real-world situation in which the mathematics of the chapter is used. Examples and exercises that relate back to the opener are included in the chapter.

■ **Graphing Calculator Discussions** Optional graphing calculator discussions have been included in the text. They are clearly marked by graphing calculator icons so that they can be easily skipped if desired. While use of this technology is optional, students who do not use a graphing calculator can still benefit from the technology discussions, as well as from the calculator generated graphs that occur in the text. In this text, the graphing calculator is used as a tool to support and enhance algebraic conclusions, not to make conclusions.

- **For Thought** Each exercise set is preceded by a set of ten true/false questions that review the basic concepts in the section, help check student understanding before beginning the exercises, and offer opportunities for writing and/or discussion. The answers to all *For Thought* exercises are included in the back of the student edition of this text.

- **Writing/Discussion and Cooperative Learning Exercises** These exercises deepen students' understanding by giving them the opportunity to express mathematical ideas both in writing and to their classmates during small group or team discussions.

- **Linking Concepts** This feature is located at the end of nearly every exercise set. It is a multi-part exercise or exploration that can be used for individual or group work. The idea of this feature is to use concepts from the current section along with concepts from preceding sections (or chapters) to solve problems that illustrate the links among various concepts. Some parts of these questions are open-ended, and require somewhat more thought than standard skill-building exercises. Answers are given in the instructor's edition only and full solutions can be found in the *Instructor's Solutions Manual*.

- **Chapter Review Exercises** These exercises are designed to give students a comprehensive review of the chapter without reference to individual sections, and prepare students for a chapter test.

- **Chapter Test** The problems in the *Chapter Test* are designed to measure the student's readiness for a typical one-hour classroom test. Instructors may also use them as a model for their own end-of-chapter tests. Students should be aware that their in-class test could vary from the *Chapter Test* due to different emphasis placed on the topics by individual instructors.

- **Tying It All Together** Found at the end of most chapters in the text, these exercises help students review selected concepts from the present and prior chapters, and require students to integrate multiple concepts and skills.

- **Index of Applications** The many applications contained within the text are listed in the *Index of Applications* that appears at the end of the text. The applications are page referenced and grouped by subject matter.

Acknowledgments

Thanks to the many professors and students who have used this text in previous editions. I am always glad to hear from users of my texts. You can email me at mdugopolski@bellsouth.net. Thanks to the following reviewers whose comments and suggestions were invaluable in preparing this fourth edition:

Sharon Christensen, *Cameron University*
Randy Combs, *Western Texas A&M University*
Steven M. Davis, *Macon State College*
Daniel Fahringer, *Harrisburg Area Community College*
Brian Garant, *Morton College*
Joe Howe, *St. Charles Community College*
Richard Mason, *Indian Hills Community College*
Timothy Miller, *Missouri Western State College*

Tonie Niblett, *Northeast Alabama Community College*
Eugene Robkin, *University of Wisconsin—Baraboo*
Ali Saadat, *University of California—Riverside*
Ed Slaminka, *Auburn University*
Nesan Sriskanda, *Claflin University*
Jennie Thompson, *Leeward Community College*
Gary Zielinski, *Texas State University*

Thanks to Edgar N. Reyes, Southeastern Louisiana University, for working all of the exercises and writing the *Solutions Manuals;* Rebecca W. Muller, Southeastern Louisiana University, for writing the *Instructor's Testing Manual,* Darryl Nester, Bluffton University, for writing the *Graphing Calculator Manual;* and Abby Tanenbaum for editing the *Insider's Guide to Teaching with College Algebra and Trigonometry.* I wish to express my thanks to Perian Herring, Steve Ouellette, John Samons, and Lauri Semarne for accuracy checking this text. A special thanks also goes out to Nesbitt Graphics, the compositor, for the superb work they did on this book.

Finally, it has been another wonderful experience working with the talented and dedicated Addison-Wesley team. I wish to thank them for their assistance, encouragement, and direction throughout this project: Greg Tobin, VP/Publisher, for his superior guidance; Executive Editor Anne Kelly, whose ideas were central in making this revision a success. Becky Anderson, Senior Marketing Manager, for creatively tying all the ideas together to get the message out; Joanne Ha, Project Editor, for keeping it all on track with poise; Peggy McMahon, Senior Production Supervisor, whose knowledge in both production and sports were integral to the production of this text; Barbara Atkinson, Senior Designer, her eye for design helped improve the book's look dramatically; Michelle Murray, Media Producer, who put together a top-notch media package; Elizabeth Bernardi and Marcia Emerson, Assistant Editors, Ashley O'Shaughnessy, Editorial Assistant, and Maureen McLaughlin, Marketing Coordinator, who ran the behind-the-scenes work. And thanks also to Joseph Vetere, Author Support/Technical Specialist, for his expertise in the production of this book.

As always, thanks to my wife, Cheryl, whose love, encouragement, understanding, support, and patience is invaluable.

Mark Dugopolski
Ponchatoula, Louisiana

SUPPLEMENTS LIST

▪▪▪ Student Supplements

Student's Solutions Manual
- By Edgar N. Reyes, *Southeastern Louisiana University*.
- Provides detailed solutions to all odd-numbered text exercises.
- ISBN: 0-321-36875-4

Graphing Calculator Manual
- By Darryl Nester, *Bluffton University*.
- Provides instructions and keystroke operations for the TI-83/83 Plus, TI-84 Plus, TI-86, and TI-89.
- Also contains worked-out examples taken directly from the text.
- ISBN: 0-321-36876-2

Videotape Series
- Provides comprehensive coverage of each section and topic in the text in an engaging format on VHS that stresses student interaction.
- Includes examples and problems from the text.
- ISBN: 0-321-36872-X

Digital Video Tutor
- Complete set of digitized videos on CD for student use at home or on campus.
- Ideal for distance learning or supplemental instruction.
- ISBN: 0-321-36877-0

A Review of Algebra
- By Heidi Howard, *Florida Community College at Jacksonville*.
- Provides additional support for those students needing further algebra review.
- ISBN: 0-201-77347-3

Addison-Wesley Math Tutor Center
- Provides tutoring through a registration number that can be packaged with a new textbook or purchased separately.
- Staffed by qualified college mathematics instructors.
- Accessible via toll-free telephone, toll-free fax, e-mail, and the Internet.
- www.aw-bc.com/tutorcenter

▪▪▪ Instructor Supplements

Annotated Instructor's Edition
- NEW! Special edition of the text.
- Provides answers beside the text exercises for most exercises, plus a full answer section at the back.
- ISBN: 0-321-35771-X

Instructor's Solutions Manual
- By Edgar N. Reyes, *Southeastern Louisiana University*.
- Provides complete solutions to all text exercises, including the *For Thought* and *Linking Concepts* exercises.
- ISBN: 0-321-36871-1

Instructor's Testing Manual
- By Rebecca W. Muller, *Southeastern Louisiana University*.
- Contains six alternative forms of tests per chapter (two are multiple choice).
- Answer keys are included.
- ISBN: 0-321-36868-1

TestGen®
- Enables instructors to build, edit, print, and administer tests.
- Features a computerized bank of questions developed to cover all text objectives.
- Available on a dual-platform Windows/Macintosh CD-ROM.
- ISBN: 0-321-36873-8

NEW! Insider's Guide to Teaching with College Algebra and Trigonometry, 4/e
- Provides helpful teaching tips correlated to each section of the text as well as general teaching advice.
- ISBN: 0-321-36870-3

PowerPoint Lecture Presentation
- Classroom presentation slides are geared specifically to sequence this textbook.
- Available within MyMathLab or at www.aw-bc.com/irc

NEW! Adjunct Support Center
- Offers consultation on suggested syllabi, helpful tips on using the textbook support package, assistance with content, and advice on classroom strategies.
- Available Sunday–Thursday evenings from 5 P.M. to midnight: telephone: 1-800-435-4084; e-mail: AdjunctSupport@aw.com; fax: 1-877-262-9774

MathXL®

Math XL® is a powerful online homework, tutorial, and assessment system that accompanies your Addison-Wesley textbook in mathematics or statistics. With MathXL, instructors can create, edit, and assign online homework and tests using algorithmically generated exercises correlated at the objective level to your textbook. They can also create and assign their own online exercises and import TestGen tests for added flexibility. All student work is tracked in MathXL's online gradebook. Students can take chapter tests in MathXL and receive personalized study plans based on their test results. The study plan diagnoses weaknesses and links students directly to tutorial exercises for the objectives they need to study and retest. Students can also access supplemental animations and video clips directly from selected exercises. MathXL is available to qualified adopters. For more information, visit our Web site at www.mathxl.com, or contact your Addison-Wesley sales representative.

MyMathLab

MyMathLab is a series of text-specific, easily customizable online courses for Addison-Wesley textbooks in mathematics and statistics. Powered by CourseCompass™ (Pearson Education's online teaching and learning environment) and by MathXL® (our online homework, tutorial, and assessment system), MyMathLab gives you the tools you need to deliver all or a portion of your course online, whether your students are in a lab setting or working from home. MyMathLab provides a rich and flexible set of course materials, featuring free-response exercises that are algorithmically generated for unlimited practice and mastery. Students can also use online tools, such as video lectures, animations, and a multimedia textbook, to independently improve their understanding and performance. Instructors can use MyMathLab's homework and test managers to select and assign online exercises correlated directly to the textbook, and they can also create and assign their own online exercises and import TestGen tests into MyMathLab for added flexibility. MyMathLab's online gradebook—designed specifically for mathematics and statistics—automatically tracks students' homework and test results and gives the instructor control over how to calculate final grades. Instructors can also add offline (paper-and-pencil) grades to the gradebook.

MyMathLab is available to qualified adopters. For more information, visit our Web site at www.mymathlab.com, or contact your Addison-Wesley sales representative.

MathXL® Tutorials on CD (ISBN: 0-321-36878-9)

This interactive tutorial CD-ROM provides algorithmically generated practice exercises that are correlated at the objective level to the exercises in the textbook. Every practice exercise is accompanied by an example and a guided solution designed to involve students in the solution process. Selected exercises may also include a video clip to help students visualize concepts. The software provides helpful feedback for incorrect answers and can generate printed summaries of students' progress.

InterAct Math Tutorial Web Site: www.interactmath.com

Get practice and tutorial help online! This interactive tutorial Web site provides algorithmically generated practice exercises that correlate directly to the exercises in the textbook. Students can retry an exercise as many times as they like with new values each time for unlimited practice and mastery. Every exercise is accompanied by an interactive guided solution that provides helpful feedback for incorrect answers, and students can also view a worked-out sample problem that steps them through an exercise similar to the one they're working on.

Function Gallery: Some Basic Functions and Their Properties

Constant Function

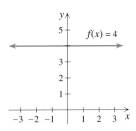

Domain $(-\infty, \infty)$
Range $\{4\}$
Constant on $(-\infty, \infty)$
Symmetric about y-axis

Identity Function

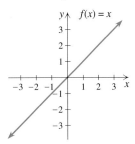

Domain $(-\infty, \infty)$
Range $(-\infty, \infty)$
Increasing on $(-\infty, \infty)$
Symmetric about origin

Linear Function

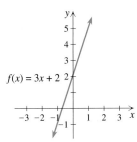

Domain $(-\infty, \infty)$
Range $(-\infty, \infty)$
Increasing on $(-\infty, \infty)$

Absolute-Value Function

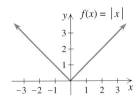

Domain $(-\infty, \infty)$
Range $[0, \infty)$
Increasing on $(0, \infty)$
Decreasing on $(-\infty, 0)$
Symmetric about y-axis

Square Function

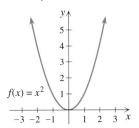

Domain $(-\infty, \infty)$
Range $[0, \infty)$
Increasing on $(0, \infty)$
Decreasing on $(-\infty, 0)$
Symmetric about y-axis

Square-Root Function

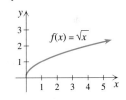

Domain $[0, \infty)$
Range $[0, \infty)$
Increasing on $(0, \infty)$

Cube Function

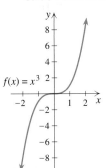

Domain $(-\infty, \infty)$
Range $(-\infty, \infty)$
Increasing on $(-\infty, \infty)$
Symmetric about origin

Cube-Root Function

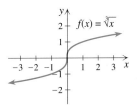

Domain $(-\infty, \infty)$
Range $(-\infty, \infty)$
Increasing on $(-\infty, \infty)$
Symmetric about origin

Greatest Integer Function

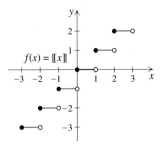

Domain $(-\infty, \infty)$
Range $\{n \mid n \text{ is an integer}\}$
Constant on $[n, n + 1)$
for every integer n

Linear

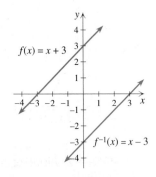

$f(x) = x + 3$

$f^{-1}(x) = x - 3$

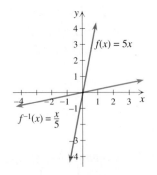

$f(x) = 5x$

$f^{-1}(x) = \dfrac{x}{5}$

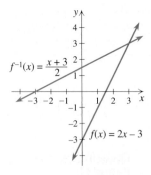

$f^{-1}(x) = \dfrac{x + 3}{2}$

$f(x) = 2x - 3$

Powers and Roots

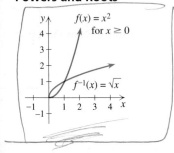

$f(x) = x^2$
for $x \geq 0$

$f^{-1}(x) = \sqrt{x}$

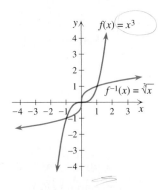

$f(x) = x^3$

$f^{-1}(x) = \sqrt[3]{x}$

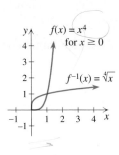

$f(x) = x^4$
for $x \geq 0$

$f^{-1}(x) = \sqrt[4]{x}$

Function Gallery: **Polynomial Functions**

Linear: $f(x) = mx + b$, domain $(-\infty, \infty)$, range $(-\infty, \infty)$ if $m \neq 0$, slope m, y-intercept $(0, b)$

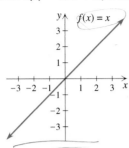

Slope 1, y-intercept $(0, 0)$

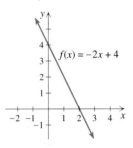

Slope 3, y-intercept $(0, -2)$

Slope -2, y-intercept $(0, 4)$

Quadratic: $f(x) = ax^2 + bx + c$ or $f(x) = a(x - h)^2 + k$, domain $(-\infty, \infty)$, vertex (h, k) or $x = -b/(2a)$

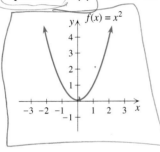

Vertex $(0, 0)$
Range $[0, \infty)$

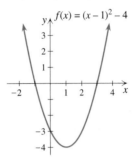

Vertex $(1, -4)$
Range $[-4, \infty)$

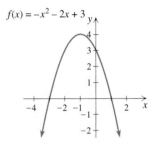

Vertex $(-1, 4)$
Range $(-\infty, 4]$

Cubic: $f(x) = ax^3 + bx^2 + cx + d$, domain $(-\infty, \infty)$, range $(-\infty, \infty)$

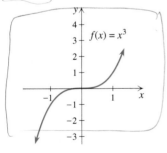

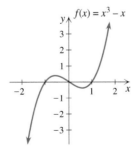

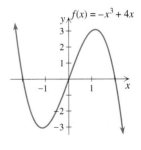

Quartic or Fourth-Degree: $f(x) = ax^4 + bx^3 + cx^2 + dx + e$, domain $(-\infty, \infty)$

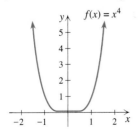

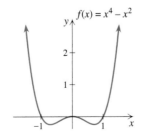

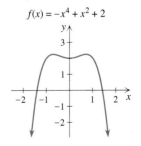

xvii

Function Gallery: Some Basic Rational Functions

Horizontal Asymptote *x*-axis and Vertical Asymptote *y*-axis

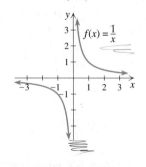

$$f(x) = \frac{1}{x}$$

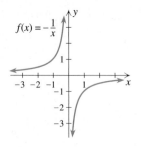

$$f(x) = -\frac{1}{x}$$

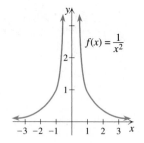

$$f(x) = \frac{1}{x^2}$$

Various Asymptotes

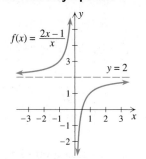

$$f(x) = \frac{2x-1}{x}$$

$$y = 2$$

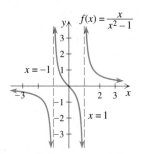

$$f(x) = \frac{x}{x^2-1}$$

$$x = -1$$

$$x = 1$$

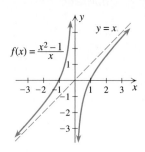

$$f(x) = \frac{x^2-1}{x}$$

$$y = x$$

Exponential: $f(x) = a^x$, domain $(-\infty, \infty)$, range $(0, \infty)$

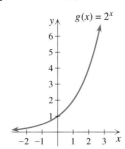

$g(x) = 2^x$

Increasing on $(-\infty, \infty)$
y-intercept $(0, 1)$

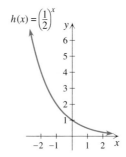

$h(x) = \left(\frac{1}{2}\right)^x$

Decreasing on $(-\infty, \infty)$
y-intercept $(0, 1)$

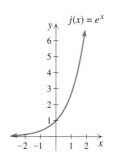

$j(x) = e^x$

Increasing on $(-\infty, \infty)$
y-intercept $(0, 1)$

Logarithmic: $f^{-1}(x) = \log_a(x)$, domain $(0, \infty)$, range $(-\infty, \infty)$

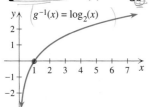

$g^{-1}(x) = \log_2(x)$

Increasing on $(0, \infty)$
x-intercept $(1, 0)$

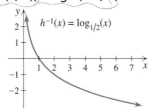

$h^{-1}(x) = \log_{1/2}(x)$

Decreasing on $(0, \infty)$
x-intercept $(1, 0)$

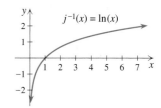

$j^{-1}(x) = \ln(x)$

Increasing on $(0, \infty)$
x-intercept $(1, 0)$

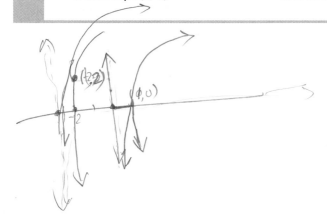

Function Gallery: Some Basic Functions of Algebra with Transformations

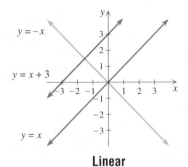

Linear

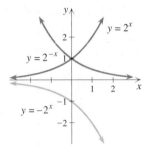

Quadratic

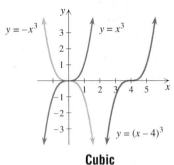

Cubic

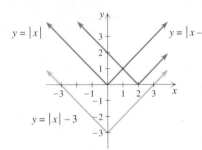

Absolute value

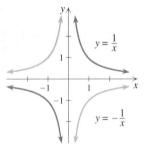

Exponential

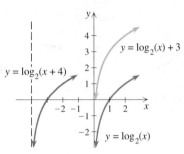

Logarithmic

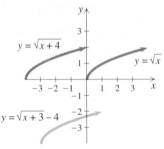

Square root

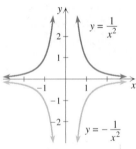

Reciprocal

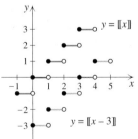

Rational

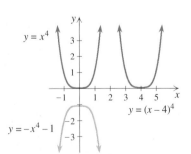

Fourth degree

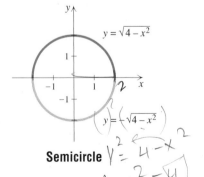

Semicircle

Greatest integer

Function Gallery: The Sine and Cosine Functions

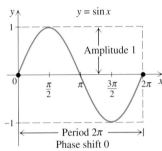

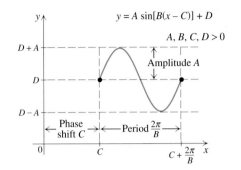

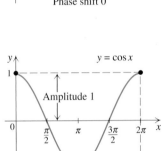

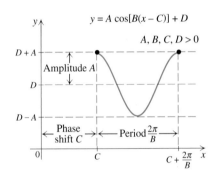

Function Gallery: Periods of Sine, Cosine, and Tangent ($B > 1$)

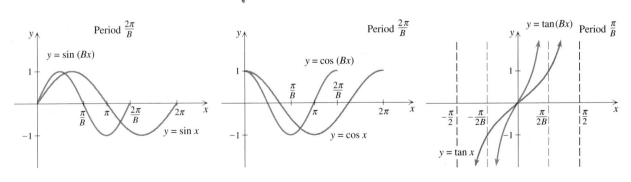

Function Gallery: Trigonometric Functions

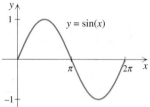

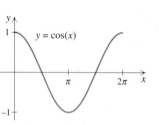

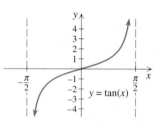

Domain (k any integer)	$(-\infty, \infty)$	$(-\infty, \infty)$	$x \neq \dfrac{\pi}{2} + k\pi$
Range	$[-1, 1]$	$[-1, 1]$	$(-\infty, \infty)$
Period	2π	2π	π
Fundamental cycle	$[0, 2\pi]$	$[0, 2\pi]$	$\left[-\dfrac{\pi}{2}, \dfrac{\pi}{2}\right]$

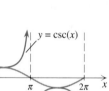

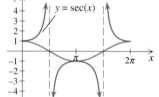

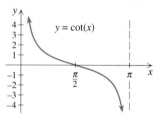

Domain (k any integer)	$x \neq k\pi$	$x \neq \dfrac{\pi}{2} + k\pi$	$x \neq k\pi$
Range	$(-\infty, -1] \cup [1, \infty)$	$(-\infty, -1] \cup [1, \infty)$	$(-\infty, \infty)$
Period	2π	2π	π
Fundamental cycle	$[0, 2\pi]$	$[0, 2\pi]$	$[0, \pi]$

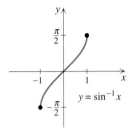

Domain $[-1, 1]$
Range $\left[-\frac{\pi}{2}, \frac{\pi}{2}\right]$

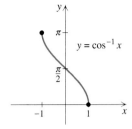

Domain $[-1, 1]$
Range $[0, \pi]$

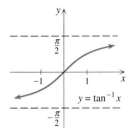

Domain $(-\infty, \infty)$
Range $\left(-\frac{\pi}{2}, \frac{\pi}{2}\right)$

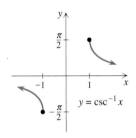

Domain $(-\infty, -1] \cup [1, \infty)$
Range $\left[-\frac{\pi}{2}, 0\right) \cup \left(0, \frac{\pi}{2}\right]$

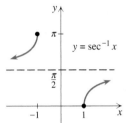

Domain $(-\infty, -1] \cup [1, \infty)$
Range $\left[0, \frac{\pi}{2}\right) \cup \left(\frac{\pi}{2}, \pi\right]$

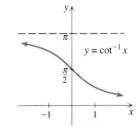

Domain $(-\infty, \infty)$
Range $(0, \pi]$

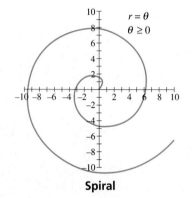

$r = \theta$
$\theta \geq 0$

Spiral

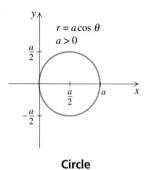

$r = a\cos\theta$
$a > 0$

Circle

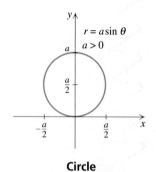

$r = a\sin\theta$
$a > 0$

Circle

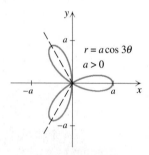

$r = a\cos 3\theta$
$a > 0$

Three-Leaf Rose

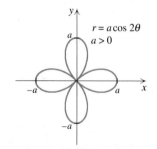

$r = a\cos 2\theta$
$a > 0$

Four-Leaf Rose

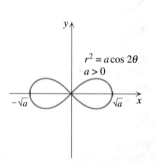

$r^2 = a\cos 2\theta$
$a > 0$

Lemniscate

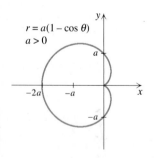

$r = a(1 - \cos\theta)$
$a > 0$

Cardioid

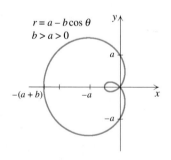

$r = a - b\cos\theta$
$b > a > 0$

Limaçon

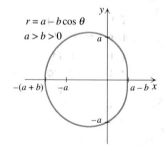

$r = a - b\cos\theta$
$a > b > 0$

Limaçon

P Prerequisites

GONE are the days when raw sailing ability and stamina won races like the America's Cup. In the modern sailing world, technology plays an ever increasing role in determining outcomes. All yachts now have a crewman on board to interpret computer data on sailing conditions.

Today mathematical formulas are used in designing the boats and in stating the rules that govern the race. Formulas are used to keep the race competitive by establishing strict boundaries for yacht sail area, hull shape, length, and displacement. Yachtsmen must do their homework and prepare for this race for years in advance. Preparation is just as important in algebra.

WHAT YOU WILL LEARN In this chapter we will review the basic concepts that are necessary for success in algebra. Throughout this chapter you will see that even basic concepts have applications in business, science, engineering, and sailing.

P.1

Real Numbers and Their Properties

In arithmetic we learn facts about the real numbers and how to perform operations with them. Since algebra is an extension of arithmetic, we begin our study of algebra with a discussion of the real numbers and their properties.

The Real Numbers

A **set** is a collection of objects or **elements.** The set containing the numbers 1, 2, and 3 is written as {1, 2, 3}. To indicate a continuing pattern, we use three dots as in {1, 2, 3, . . .}. The set of real numbers is a collection of many types of numbers. To better understand the real numbers we recall some of the basic subsets of the real numbers:

Subset	Name (symbol)
{1, 2, 3, . . .}	**Counting** or **natural numbers** (N)
{0, 1, 2, 3, . . .}	**Whole numbers** (W)
{. . . , −3, −2, −1, 0, 1, 2, 3, . . .}	**Integers** (J)

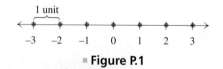

1 unit

■ **Figure P.1**

Numbers can be pictured as points on a line, the **number line.** To draw a number line, draw a line and label any convenient point with the number 0. Now choose a convenient length, one **unit,** and use it to locate evenly spaced points as shown in Fig. P.1. The positive integers are located to the right of zero and the negative integers to the left of zero. The numbers corresponding to the points on the line are called the **coordinates** of the points.

The integers and their ratios form the set of **rational numbers,** Q. The rational numbers also correspond to points on the number line. For example, the rational number 1/2 is found halfway between 0 and 1 on the number line. In set notation, the set of rational numbers is written as

$$\left\{ \frac{a}{b} \,\middle|\, a \text{ and } b \text{ are integers with } b \neq 0 \right\}.$$

This notation is read "The set of all numbers of the form a/b such that a and b are integers with b not equal to zero." In our set notation we used letters to represent integers. A letter that is used to represent a number is called a **variable.**

There are infinitely many rational numbers located between each pair of consecutive integers, yet there are infinitely many points on the number line that do not correspond to rational numbers. The numbers that correspond to those points are called **irrational** numbers. In decimal notation, the rational numbers are the numbers that are repeating or terminating decimals, and the irrational numbers are the nonrepeating nonterminating decimals. For example, the number 0.595959 . . . is a rational number because the pair 59 repeats indefinitely. By contrast, notice that in the number 5.010010001 . . . , each group of zeros contains one more zero than the previous group. Because no group of digits repeats, 5.010010001 . . . is an irrational number.

Numbers such as $\sqrt{2}$ or π are also irrational. We can visualize $\sqrt{2}$ as the length of the diagonal of a square whose sides are one unit in length. See Fig. P.2. In any circle, the ratio of the circumference c to the diameter d is π ($\pi = c/d$).

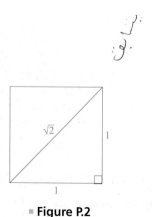

$\sqrt{2}$

1

1

■ **Figure P.2**

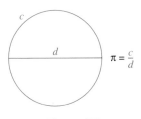

$$\pi = \frac{c}{d}$$

■ **Figure P.3**

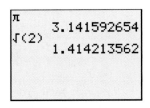

■ **Figure P.4**

See Fig. P.3. It is difficult to see that numbers like $\sqrt{2}$ and π are irrational because their decimal representations are not apparent. However, the irrationality of π was proven in 1767 by Johann Heinrich Lambert, and it can be shown that the square root of any positive integer that is not a perfect square is irrational.

Since a calculator operates with a fixed number of decimal places, it gives us a *rational approximation* for an irrational number such as $\sqrt{2}$ or π. See Fig. P.4. □

The set of rational numbers, Q, together with the set of irrational numbers, I, is called the set of **real numbers, R**. The following are examples of real numbers:

$$-3, \quad -0.025, \quad 0, \quad \frac{1}{3}, \quad 0.595959\ldots, \quad \sqrt{2}, \quad \pi, \quad 5.010010001\ldots$$

These numbers are **graphed** on a number line in Fig. P.5.

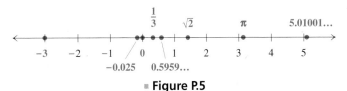

■ **Figure P.5**

Since there is a one-to-one correspondence between the points of the number line and the real numbers, we often refer to a real number as a point. Figure P.6 shows how the various subsets of the real numbers are related to one another.

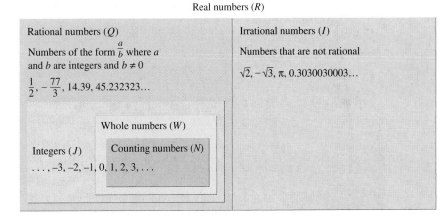

■ **Figure P.6**

To indicate that a number is a member of a set, we write $a \in A$, which is read "a is a member of set A." We write $a \notin A$ for "a is not a member of set A." Set A is a subset of set B ($A \subseteq B$) means that every member of set A is also a member of set B, and A is not a subset of B ($A \nsubseteq B$) means that there is at least one member of A that is not a member of B.

Example **1**   **Classifying numbers and sets of numbers**

Determine whether each statement is true or false and explain. See Fig. P.6.

a. $0 \in R$ **b.** $\pi \in Q$ **c.** $R \subseteq Q$ **d.** $I \nsubseteq Q$ **e.** $\sqrt{5} \in Q$

Solution

a. True, because 0 is a member of the set of whole numbers, a subset of the set of real numbers.
b. False, because π is irrational.
c. False, because every irrational number is a member of R but not Q.
d. True, because the irrational numbers and the rational numbers have no numbers in common.
e. False, because the square root of any integer that is not a perfect square is irrational.

Try This. True or false? **a.** $0 \in I$ **b.** $I \subseteq R$ **c.** $\sqrt{5} \in I$ ■

Properties of the Real Numbers

In arithmetic we can observe that $3 + 4 = 4 + 3$, $6 + 9 = 9 + 6$, etc. We get the same sum when we add two real numbers in either order. This property of addition of real numbers is the **commutative property.** Using variables, the commutative property of addition is stated as $a + b = b + a$ for any real numbers a and b. There is also a commutative property of multiplication, which is written as $a \cdot b = b \cdot a$ or $ab = ba$. There are many properties concerning the operations of addition and multiplication on the real numbers that are useful in algebra.

Properties of the Real Numbers

For any real numbers a, b, and c:	
$a + b$ and ab are real numbers	**Closure property**
$a + b = b + a$ and $ab = ba$	**Commutative properties**
$a + (b + c) = (a + b) + c$ and $a(bc) = (ab)c$	**Associative properties**
$a(b + c) = ab + ac$	**Distributive property**
$0 + a = a$ and $1 \cdot a = a$ (Zero is the **additive identity,** and 1 is the **multiplicative identity.**)	**Identity properties**
$0 \cdot a = 0$	**Multiplication property of zero**
For each real number a, there is a unique real number $-a$ such that $a + (-a) = 0$. ($-a$ is the **additive inverse** of a.)	**Additive inverse property**
For each nonzero real number a, there is a unique real number $1/a$ such that $a \cdot 1/a = 1$. ($1/a$ is the **multiplicative inverse** or **reciprocal** of a.)	**Multiplicative inverse property**

The closure property indicates that the sum and product of any pair of real numbers is a real number. The commutative properties indicate that we can add or multiply in either order and get the same result. Since we can add or multiply only a pair of numbers, the associative properties indicate two different ways to obtain the result when adding or multiplying three numbers. The operations within parentheses are performed first. Because of the commutative property, the distributive property can be used also in the form $(b + c)a = ab + ac$.

Note that the properties stated here involve only addition and multiplication, considered the basic operations of the real numbers. Subtraction and division are defined in terms of addition and multiplication. By definition $a - b = a + (-b)$ and $a \div b = a \cdot 1/b$ for $b \neq 0$. Note that $a - b$ is called the **difference** of a and b and $a \div b$ is called the **quotient** of a and b.

Example 2 Using the properties

Complete each statement using the property named.

a. $a7 =$ _____, commutative
b. $2x + 4 =$ _____, distributive
c. $8(\text{_____}) = 1$, multiplicative inverse
d. $\dfrac{1}{3}(3x) =$ _____, associative

Solution

a. $a7 = 7a$ **b.** $2x + 4 = 2(x + 2)$

c. $8\left(\dfrac{1}{8}\right) = 1$ **d.** $\dfrac{1}{3}(3x) = \left(\dfrac{1}{3} \cdot 3\right)x$

Try This. Complete the statement $x \cdot 3 =$ _____ using the commutative property. ■

Additive Inverses

The negative sign is used to indicate negative numbers as in -7 (negative seven). If the negative sign precedes a variable as in $-b$ it is read as "additive inverse" or "opposite" because $-b$ could be positive or negative. If b is positive then $-b$ is negative and if b is negative then $-b$ is positive.

Using two "opposite" signs has a cancellation effect. For example, $-(-5) = 5$ and $-(-(-3)) = -3$. Note that the additive inverse of a number can be obtained by multiplying the number by -1. For example, $-1 \cdot 3 = -3$.

Calculators usually use the negative sign (-) to indicate opposite or negative and the subtraction sign $(-)$ for subtraction as shown in Fig. P.7. □

We know that $a + b = b + a$ for any real numbers a and b, but is $a - b = b - a$ for any real numbers a and b? In general, $a - b$ is not equal to $b - a$. For example, $7 - 3 = 4$ and $3 - 7 = -4$. So subtraction is not commutative. Since $a - b + b - a = 0$, we can conclude that $a - b$ and $b - a$ are opposites or additive inverses of each other. We summarize these properties of opposites as follows.

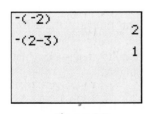

■ **Figure P.7**

Properties of Opposites

For any real numbers a and b:

1. $-1 \cdot a = -a$ (The product of -1 and a is the opposite of a.)
2. $-(-a) = a$ (The opposite of the opposite of a is a.)
3. $-(a - b) = b - a$ (The opposite of $a - b$ is $b - a$.)

Example **3** **Using properties of opposites**

Use the properties of opposites to complete each equation.

a. $-(-\pi) =$ _____   **b.** $-1(-2) =$ _____   **c.** $-1(x - h) =$ _____

Solution

a. $-(-\pi) = \pi$
b. $-1(-2) = -(-2) = 2$
c. $-1(x - h) = -(x - h) = h - x$

Try This. Complete the equation $-(1 - w) =$ _____ using the properties of opposites.   ■

Relations

Symbols such as $<, >, =, \leq$, and $\geq$ are called **relations** because they indicate how numbers are related. We can visualize these relations by using a number line. For example, $\sqrt{2}$ is located to the right of 0 in Fig. P.5, so $\sqrt{2} > 0$. Since $\sqrt{2}$ is to the left of π in Fig. P.5, $\sqrt{2} < \pi$. In fact, if a and b are any two real numbers, we say that a is less than b (written $a < b$) provided that a is to the left of b on the number line. We say that a is greater than b (written $a > b$) if a is to the right of b on the number line. We say $a = b$ if a and b correspond to the same point on the number line. The fact that there are only three possibilities for ordering a pair of real numbers is called the **trichotomy property.**

Trichotomy Property

> For any two real numbers a and b, exactly one of the following is true: $a < b, a = b$, or $a > b$.

The trichotomy property is very natural to use. For example, if we know that $r = t$ is false, then we can conclude (using the trichotomy property) that either $r > t$ or $r < t$ is true. If we know that $w + 6 > z$ is false, then we can conclude that $w + 6 \leq z$ is true. The following four properties of equality are also very natural to use, and we often use them without even thinking about them.

Properties of Equality

> For any real numbers a, b, and c:
>
> **1.** $a = a$ Reflexive property
> **2.** If $a = b$, then $b = a$.   Symmetric property
> **3.** If $a = b$ and $b = c$, then $a = c$.   Transitive property
> **4.** If $a = b$, then a and b may be substituted   Substitution property
>   for one another in any expression involving
> a or b.

Absolute Value

The **absolute value** of a (in symbols, $|a|$) can be thought of as the distance from a to 0 on a number line. Since both 3 and -3 are three units from 0 on a number line as shown in Fig. P.8 on the next page, $|3| = 3$ and $|-3| = 3$:

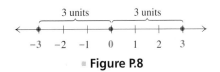

■ **Figure P.8**

A symbolic definition of absolute value is written as follows.

Definition: Absolute Value

For any real number a,

$$|a| = \begin{cases} a & \text{if } a \geq 0 \\ -a & \text{if } a < 0. \end{cases}$$

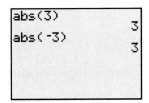

■ **Figure P.9**

A calculator typically uses **abs** for absolute value as shown in Fig. P.9. □

The symbolic definition of absolute value indicates that for $a \geq 0$ we use the equation $|a| = a$ (the absolute value of a is just a). For $a < 0$ we use the equation $|a| = -a$ (the absolute value of a is the opposite of a, a positive number).

Example **4** **Using the definition of absolute value**

Use the symbolic definition of absolute value to simplify each expression.

a. $|5.6|$ **b.** $|0|$ **c.** $|-3|$

Solution

a. Since $5.6 \geq 0$, we use the equation $|a| = a$ to get $|5.6| = 5.6$.
b. Since $0 \geq 0$, we use the equation $|a| = a$ to get $|0| = 0$.
c. Since $-3 < 0$, we use the equation $|a| = -a$ to get $|-3| = -(-3) = 3$.

Try This. Use the definition of absolute value to simplify $|-9|$.   ■

The definition of absolute value guarantees that the absolute value of any number is nonnegative. The definition also implies that additive inverses (or opposites) have the same absolute value. These properties of absolute value and two others are stated as follows.

Properties of Absolute Value

For any real numbers a and b:

1. $|a| \geq 0$ (The absolute value of any number is nonnegative.)
2. $|-a| = |a|$ (Additive inverses have the same absolute value.)
3. $|a \cdot b| = |a| \cdot |b|$ (The absolute value of a product is the product of the absolute values.)
4. $\left|\dfrac{a}{b}\right| = \dfrac{|a|}{|b|}, b \neq 0$ (The absolute value of a quotient is the quotient of the absolute values.)

Absolute value is used in finding the distance between points on a number line. Since 9 lies four units to the right of 5, the distance between 5 and 9 is 4. In symbols,

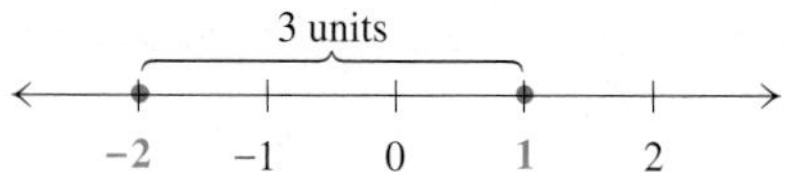

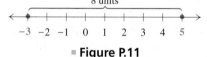

3 units

▪ **Figure P.10**

$d(5, 9) = 4$. We can obtain 4 by $9 - 5 = 4$ or $|5 - 9| = 4$. In general, $|a - b|$ gives the distance between a and b for any values of a and b. For example, the distance between -2 and 1 in Fig. P.10 is three units and

$$d(-2, 1) = |-2 - 1| = |-3| = 3.$$

Distance Between Two Points on the Number Line

> If a and b are any two points on the number line, then the distance between a and b is $|a - b|$. In symbols, $d(a, b) = |a - b|$.

Note that $d(a, 0) = |a - 0| = |a|$, which is consistent with the definition of absolute value of a as the distance between a and 0 on the number line.

Example **5**   **Distance between two points on a number line**

Find the distance between -3 and 5 on the number line.

Solution

The points corresponding to -3 and 5 are shown on the number line in Fig. P.11. The distances between these points is found as follows:

$$d(-3, 5) = |-3 - 5| = |-8| = 8$$

Notice that $d(-3, 5) = d(5, -3)$:

$$d(5, -3) = |5 - (-3)| = |8| = 8$$

⌨ When you use a calculator to find the absolute value of a difference or a sum, you must use parentheses as shown in Fig. P.12.

Try This. Find the distance between -5 and -9 on the number line.   ▪

8 units

▪ **Figure P.11**

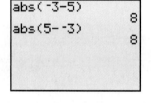

▪ **Figure P.12**

Exponential Expressions

We use positive integral exponents to indicate the number of times a number occurs in a product. For example, $2 \cdot 2 \cdot 2 \cdot 2$ is written as 2^4. We read 2^4 as "the fourth power of 2" or "2 to the fourth power."

Definition: Positive Integral Exponents

> For any positive integer n
> $$a^n = \underbrace{a \cdot a \cdot a \cdot \cdots \cdot a}_{n \text{ factors of } a}.$$
> We call a the **base,** n the **exponent** or **power,** and a^n an **exponential expression.**

We read a^n as "a to the nth power." For a^1 we usually omit the exponent and write a. We refer to the exponents 2 and 3 as squares and cubes. For example, 3^2 is read "3 squared," 2^3 is read "2 cubed," x^4 is read "x to the fourth," b^5 is read "b to the fifth," and so on. To evaluate an expression such as -3^2 we square 3 first, then take the opposite. So $-3^2 = -9$ and $(-3)^2 = (-3)(-3) = 9$.

Example **6** **Evaluating exponential expressions**

Evaluate.

a. 4^3 **b.** $(-2)^4$ **c.** -2^4

Solution

a. $4^3 = 4 \cdot 4 \cdot 4 = 16 \cdot 4 = 64$
b. $(-2)^4 = (-2)(-2)(-2)(-2) = 16$
c. $-2^4 = -(2 \cdot 2 \cdot 2 \cdot 2) = -16$

Try This. Evaluate.   **a.** 5^2 **b.** -5^2 ■

Arithmetic Expressions

The result of writing numbers in a meaningful combination with the ordinary operations of arithmetic is called an **arithmetic expression** or simply an **expression.** The **value** of an arithmetic expression is the real number obtained when all operations are performed. Symbols such as parentheses, brackets, braces, absolute value bars, and fraction bars are called **grouping symbols.** Operations within grouping symbols are performed first.

Example **7** **Evaluating an arithmetic expression with grouping symbols**

Evaluate each expression.

a. $(-7 \cdot 3) + (5 \cdot 8)$ **b.** $\dfrac{3 - 9}{-2 - (-5)}$ **c.** $3 - |5 - (2 \cdot 9)|$

Solution

a. Perform the operations within the parentheses first and remove the parentheses:

$$(-7 \cdot 3) + (5 \cdot 8) = -21 + 40 = 19$$

b. Since the fraction bar acts as a grouping symbol, we evaluate the numerator and denominator before dividing.

$$\frac{3 - 9}{-2 - (-5)} = \frac{-6}{3} = -2$$

c. First evaluate within the innermost grouping symbols:

$$
\begin{aligned}
3 - |5 - (2 \cdot 9)| &= 3 - |5 - 18| && \text{Innermost grouping symbols} \\
&= 3 - |-13| && \text{Innermost grouping symbols} \\
&= 3 - 13 && \text{Evaluate the absolute value.} \\
&= -10 && \text{Subtract.}
\end{aligned}
$$

Try This. Evaluate.   **a.** $(-1 + 3)(5 - 6)$ **b.** $2 - |3 - 9|$ ■

The Order of Operations

When some or all grouping symbols are omitted in an expression we evaluate the expression using the following order of operations. Any operations contained within grouping symbols are performed first, using the order of operations.

Order of Operations

> **1.** Evaluate exponential expressions.
> **2.** Perform multiplication and division in order from left to right.
> **3.** Perform addition and subtraction in order from left to right.

Example **8** Using the order of operations to evaluate an expression

Evaluate each expression.

a. $3 - 4 \cdot 2^3$ **b.** $5 \cdot 8 \div 4 \cdot 2$ **c.** $3 - 4 + 9 - 2$ **d.** $5 - 2(3 - 4 \cdot 2)^2$

Solution

a. By the order of operations evaluate 2^3, then multiply, and then subtract:

$$3 - 4 \cdot 2^3 = 3 - 4 \cdot 8 = 3 - 32 = -29$$

b. In an expression with only multiplication and division, the operations are performed from left to right:

$$5 \cdot 8 \div 4 \cdot 2 = 40 \div 4 \cdot 2 = 10 \cdot 2 = 20$$

c. In an expression with only addition and subtraction, the operations are performed from left to right:

$$3 - 4 + 9 - 2 = -1 + 9 - 2 = 8 - 2 = 6$$

d. Perform operations within parentheses first, using the order of operations:

$$5 - 2(3 - 4 \cdot 2)^2 = 5 - 2(-5)^2 = 5 - 2 \cdot 25 = -45$$

Try This. Evaluate.   **a.** $3 - 6 \cdot 2$ **b.** $4 - 5 \cdot 2^3$ ■

Algebraic Expressions

When we write numbers and one or more variables in a meaningful combination with the ordinary operations of arithmetic, the result is called an **algebraic expression,** or simply an expression. The **value of an algebraic expression** is the value of the arithmetic expression that is obtained when the variables are replaced by real numbers.

Example **9** Evaluating an algebraic expression

Find the value of $b^2 - 4ac$ when $a = -1$, $b = -2$, and $c = 3$.

Solution

Replace the variables by the appropriate numbers:

$$b^2 - 4ac = (-2)^2 - 4(-1)(3) = 16$$

Try This. Evaluate $a^2 - b^2$ if $a = -2$ and $b = -3$.   ■

The **domain** of an algebraic expression in one variable is the set of all real numbers that can be used for the variable. For example, the domain of $1/x$ is the set of nonzero real numbers, because division by 0 is undefined. Two algebraic expressions in one variable are **equivalent** if they have the same domain and if they have the same value for each member of the domain. The expressions $1/x$ and x/x^2 are equivalent.

A **term** is the product of a number and one or more variables raised to powers. Expressions such as $3x$, $2kab^2$, and πr^2 are terms. Numbers or expressions that are multiplied are called **factors.** For example, 3 and x are factors of the term $3x$. The **coefficient** of any variable part of a term is the product of the remaining factors in the term. For example, the coefficient of x in $3x$ is 3. The coefficient of ab^3 in $2kab^3$ is $2k$ and the coefficient of b^3 is $2ka$. If two terms contain the same variables with the same exponents, then they are called **like terms.** The distributive property allows us to **combine like terms.** For example, $3x + 2x = (3 + 2)x = 5x$.

To **simplify** an expression means to find a simpler-looking equivalent expression. The properties of the real numbers are used to simplify expressions.

Example **10** **Using properties to simplify an expression**

Simplify each expression.

a. $-4x - (6 - 7x)$ **b.** $\dfrac{1}{2}x - \dfrac{3}{4}x$ **c.** $-6(x - 3) - 3(5 - 7x)$

Solution

a.
$$\begin{aligned}
-4x - (6 - 7x) &= -4x + [-(6 + (-7x))] && \text{Definition of subtraction} \\
&= -4x + [-1(6 + (-7x))] && \text{First property of opposites} \\
&= -4x + [(-6) + 7x] && \text{Distributive property} \\
&= [-4x + 7x] + (-6) && \text{Commutative and associative properties} \\
&= 3x - 6 && \text{Combine like terms.}
\end{aligned}$$

b.
$$\begin{aligned}
\frac{1}{2}x - \frac{3}{4}x &= \frac{2}{4}x - \frac{3}{4}x && \text{Write } \frac{1}{2} \text{ as } \frac{2}{4} \text{ to obtain a common denominator.} \\
&= -\frac{1}{4}x && \text{Combine like terms.}
\end{aligned}$$

c.
$$\begin{aligned}
-6(x - 3) - 3(5 - 7x) &= -6x + 18 - 15 + 21x && \text{Distributive property} \\
&= 15x + 3 && \text{Combine like terms.}
\end{aligned}$$

Try This. Simplify $-2(x - 3) - 3(1 - x)$.   ■

For Thought

True or False? Explain.

1. Zero is the only number that is both rational and irrational.

2. Between any two distinct rational numbers there is another rational number.

3. Between any two distinct real numbers there is an irrational number.

4. Every real number has a multiplicative inverse.

5. If a is not less than and not equal to 3, then a is greater than 3.

6. If $a \le w$ and $w \le z$, then $a < z$.

7. For any real numbers a, b, and c, $a - (b - c) = (a - b) - c$.

8. If a and b are any two real numbers, then the distance between a and b on the number line is $a - b$.

9. Calculators give only rational answers.

10. For any real numbers a and b, the opposite of $a + b$ is $a - b$.

P.1 Exercises

Match each given statement with its symbolic form and determine whether the statement is true or false. If the statement is false, correct it.

1. The number $\sqrt{2}$ is a real number.

2. The number $\sqrt{3}$ is rational.

3. The number 0 is not an irrational number.

4. The number -6 is not an integer.

5. The set of integers is a subset of the real numbers.

6. The set of irrational numbers is a subset of the rationals.

7. The set of real numbers is not a subset of the rational numbers.

8. The set of natural numbers is not a subset of the whole numbers.
   - **a.** $\sqrt{3} \in Q$
   - **b.** $-6 \notin J$
   - **c.** $R \not\subseteq Q$
   - **d.** $I \subseteq Q$
   - **e.** $\sqrt{2} \in R$
   - **f.** $N \not\subseteq W$
   - **g.** $J \subseteq R$
   - **h.** $0 \notin I$

Determine which elements of the set $\{-3.5, -\sqrt{2}, -1, 0, 1, \sqrt{3}, 3.14, \pi, 4.3535\ldots, 5.090090009\ldots\}$ are members of the following sets.

9. Real numbers

10. Rational numbers

11. Irrational numbers

12. Integers

13. Whole numbers

14. Natural numbers

Complete each statement using the property named.

15. $7 + x =$ _____, commutative

16. $5(4y) =$ _____, associative

17. $5(x + 3) =$ _____, distributive

18. $-3(x - 4) =$ _____, distributive

19. $5x + 5 =$ _____, distributive

20. $-5x + 10 =$ _____, distributive

21. $-13 + (4 + x) =$ _____, associative

22. $yx =$ _____, commutative

23. $0.125($_____$) = 1$, multiplicative inverse

24. $-3 + ($_____$) = 0$, additive inverse

Use the properties of opposites to complete each equation.

25. $-(-\sqrt{3}) =$ _____

26. $-1(-6.4) =$ _____

27. $-1(x^2 - y^2) =$ _____

28. $-(1 - a^2) =$ _____

Use the symbolic definition of absolute value to simplify each expression.

29. $|7.2|$

30. $|0/3|$

31. $\left|-\sqrt{5}\right|$

32. $|-3/4|$

Find the distance on the number line between each pair of numbers.

33. 8, 13 **34.** 1, 99 **35.** −5, 17

36. 22, −9 **37.** −6, −18 **38.** −3, −14

39. $-\dfrac{1}{2}, \dfrac{1}{4}$ **40.** $-\dfrac{1}{2}, -\dfrac{3}{4}$

Evaluate each exponential expression.

41. 2^3 **42.** 3^4 **43.** -7^2 **44.** -9^2

45. $(-4)^2$ **46.** $(-10)^2$ **47.** $\left(-\dfrac{1}{4}\right)^3$ **48.** $\left(-\dfrac{3}{4}\right)^4$

Evaluate each expression.

49. $(2 \cdot 5) - (3 \cdot 6)$ **50.** $(5 - 3)(2 - 6)$

51. $|3 - (4 \cdot 5)| - 5$ **52.** $5 - |4 - (2 \cdot 3)|$

53. $|-4 \cdot 3| - |-3 \cdot 5|$ **54.** $(-8 \cdot 3) - |-3 \cdot 7|$

55. $\dfrac{-2 - (-6)}{-5 - (-9)}$ **56.** $\dfrac{4 - (-3)}{-3 - (-1)}$

Use the order of operations to evaluate each expression.

57. $4 - 5 \cdot 3^2$ **58.** $4 + 2(-6)^2$

59. $3 - 4 + 5 - 7 - 4$ **60.** $4 - 3 + 2 - 5 + 6$

61. $3 \cdot 6 + 2 \cdot 4$ **62.** $-2 \cdot 9 + 3 \cdot 5$

63. $26 \cdot \dfrac{1}{5} \div \dfrac{1}{2} \cdot 5$ **64.** $\dfrac{4}{3} \cdot 50(0.75) \div 2$

65. $(3 \cdot 4 - 1)(1 + 2 \cdot 4)$ **66.** $-2 - 3(5 - 2 \cdot 8)$

67. $2 - 3|3 - 4 \cdot 6|$ **68.** $1 - (3 - |1 - 2 \cdot 3|)$

69. $7^2 - 2(-3)(-6)$ **70.** $(-3)^2 - 4(-2)(-5)$

71. $3 - 4(5 - 3 \cdot 2)^2$ **72.** $1 - 3(6 \cdot 5 - 4 \cdot 8)^2$

73. $\dfrac{2(5 - 2)^2}{5^2 - 4^2}$ **74.** $\dfrac{(2 - 3 \cdot 4)^2}{3^2 + 4^2}$

Evaluate each expression if $a = -2$, $b = 3$, and $c = 4$.

75. $b^2 - 4ac$ **76.** $(b - 4ac)^2$

77. $\dfrac{a - c}{b - c}$ **78.** $\dfrac{a^2 - c}{b^3 + c^4}$

79. $a^2 - b^2$ **80.** $a^2 + b^2$

81. $(a - b)(a + b)$ **82.** $(a + b)^2$

83. $(a - b)(a^2 + ab + b^2)$ **84.** $(a + b)(a^2 - ab + b^2)$

85. $a^b + c^b$ **86.** $(a + c)^b$

Use the properties of the real numbers to simplify each expression.

87. $-5x + 3x$ **88.** $-5x - (-8x)$

89. $x - 0.15x$ **90.** $x + 3 - 0.9x$

91. $-3(2xy)$ **92.** $\dfrac{1}{2}(8wz)$

93. $\dfrac{1}{2}(6 - 4x)$ **94.** $\dfrac{1}{4}(8x - 4)$

95. $\dfrac{6x - 2y}{2}$ **96.** $\dfrac{-9 - 6x}{-3}$

97. $(3 - 4x) + (x - 9)$ **98.** $(9x - 3) + (4 - 6x)$

99. $-2(4 - x) - 3(3 - 3x)$ **100.** $5(4 - 2x) - 2(x - 5)$

Solve each problem.

101. *Target Heart Rate* The expression $0.60(220 - a - r) + r$ is used to obtain the target heart rate for a cardiovascular workout for a nonathletic male with age a and resting heart rate r (Stevens Creek Software, www.stevenscreek.com).
   a. Simplify the expression.

   b. Find the target heart rate for a 20-year-old nonathletic male with resting heart rate of 70 beats per minute.

   c. Simplify the expression $0.60\left(205 - \dfrac{a}{2} - r\right) + r$, which is used for an athletic male.

102. *Target Heart Rate* The expression $0.60(226 - a - r) + r$ is used to obtain the target heart rate for a cardiovascular workout for a nonathletic female with age a and resting heart rate r (Stevens Creek Software, www.stevenscreek.com).
   a. Simplify the expression.

   b. The accompanying table shows the target heart rate for a 22-year-old nonathletic female for various resting heart rates. Find the missing entries.

■ **Table for Exercise 102**

Resting Heart Rate	Target Heart Rate	
55	144.4	
60	146.4	
65		
70	150.4	
75		

For Writing/Discussion

103. Graph the numbers $\frac{1}{2}$, $-\frac{1}{2}$, $\frac{1}{3}$, $-\frac{1}{3}$, 0, $\frac{5}{12}$, and $-\frac{5}{12}$ on a number line. Explain how you decided where to put the numbers. Arrange these same numbers in order from smallest to largest. Explain your method. Did you use a calculator? If so, explain how it could be done without one.

104. Use a calculator to arrange the numbers $\frac{10}{3}$, $\sqrt{10}$, $\frac{22}{7}$, π, and $\frac{157}{50}$ in order from smallest to largest. Explain what you did to make your decisions on the order of these numbers. Could these numbers be arranged without using a calculator? How do these numbers differ from those in the previous exercise?

Thinking Outside the Box I

Paying Up A king agreed to pay his gardener one dollar's worth of titanium per day for seven days of work on the castle grounds. The king has a seven-dollar bar of titanium that is segmented so that it can be broken into seven one-dollar pieces, but it is bad luck to break a seven-dollar bar of titanium more than twice. How can the king make two breaks in the bar and pay the gardener exactly one dollar's worth of titanium per day for seven days?

■ **Figure for Thinking Outside the Box I**

P.1 Pop Quiz

1. Is 0 an irrational number?

2. Simplify $|-2|$.   **3.** Simplify $-(1 - y)$.

4. Find the distance between -3 and 9.

5. Evaluate $3 \cdot 4 - 5 \cdot 2^3$.

6. Evaluate $5 - |3 - 2 \cdot 4|$.

7. Evaluate $\dfrac{a^2 - b^2}{a - b}$ if $a = 2$ and $b = -3$.

8. Simplify $3(x - 5) - 2(5 - x)$.

Linking Concepts

For Individual or Group Explorations

Number Puzzles

Puzzles concerning numbers are as old as numbers themselves. The best number puzzles can be solved without a lot of mathematics, but that does not necessarily make them easy.

a) Think of a number, and add $\frac{2}{3}$ of this number to itself. From this sum, subtract $\frac{1}{3}$ of its value and say what your answer is. From your answer subtract $\frac{1}{10}$ of your answer. You will now have your original number (Rhind papyrus, 1849 B.C.).

Explain why this works.

b) Think of a number between 1 and 10. Think of the product of your number and 9. Think of the sum of the digits in your answer. Think of that number minus 5. Think of the letter in the alphabet that corresponds to the number you are thinking about. Think of a state that begins with the letter. Think of the second letter in the state. Think of a big animal that begins with that letter. Think of the color of that animal. The color is gray. Explain.

c) Sandy has one quart of grass seed and one quart of sand, each stored in one gallon containers. Sandy pours a little seed into the sand and shakes well. She then pours the same amount of the mix back into the container of seed so that both containers again contain exactly one quart. Is there more sand in Sandy's seed or more seed in Sandy's sand? Explain.

d) In each cell of the second row of the following table put one of the digits 0 through 9. You may use a digit more than once, but each digit in the second row must indicate the number of times that the digit above it appears in the second row.

0	1	2	3	4	5	6	7	8	9

P.2

Integral Exponents and Scientific Notation

We defined positive integral exponents in Section P.1. In this section we will define negative integral exponents and review the rules for exponents. Then we will see how exponents are used in scientific notation to indicate very large and very small numbers.

Negative Integral Exponents

We use a negative sign in an exponent to represent multiplicative inverses or reciprocals. For negative exponents we do not allow the base to be zero because zero does not have a reciprocal.

Definition: Negative Integral Exponents

> If a is a nonzero real number and n is a positive integer, then $a^{-n} = \dfrac{1}{a^n}$.

Example **1** **Evaluating expressions that have negative exponents**

Simplify each expression without using a calculator, then check with a calculator.

a. $3^{-1} \cdot 5^{-2} \cdot 10^2$ **b.** $\left(\dfrac{2}{3}\right)^{-3}$ **c.** $\dfrac{6^{-2}}{2^{-3}}$

Solution

a. $3^{-1} \cdot 5^{-2} \cdot 10^2 = \dfrac{1}{3} \cdot \dfrac{1}{5^2} \cdot 100 = \dfrac{1}{3} \cdot \dfrac{1}{25} \cdot 100 = \dfrac{100}{75} = \dfrac{4}{3}$

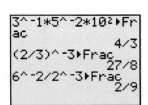

■ **Figure P.13**

b. $\left(\dfrac{2}{3}\right)^{-3} = \dfrac{1}{\left(\dfrac{2}{3}\right)^{3}} = \dfrac{1}{\dfrac{2}{3} \cdot \dfrac{2}{3} \cdot \dfrac{2}{3}} = \dfrac{1}{\dfrac{8}{27}} = \dfrac{27}{8}$ Note that $\left(\frac{2}{3}\right)^{-3} = \left(\frac{3}{2}\right)^{3}$.

c. $\dfrac{6^{-2}}{2^{-3}} = \dfrac{\dfrac{1}{6^{2}}}{\dfrac{1}{2^{3}}} = \dfrac{1}{6^{2}} \cdot \dfrac{2^{3}}{1} = \dfrac{8}{36} = \dfrac{2}{9}$ Note that $\frac{6^{-2}}{2^{-3}} = \frac{2^{3}}{6^{2}}$.

 These three expressions are shown on a graphing calculator in Fig. P.13. Note that the fractional base must be in parentheses. The fraction feature was used to get fractional answers.

Try This. Simplify.   **a.** $2^{-2} \cdot 4^{3}$ **b.** $\left(\dfrac{1}{2}\right)^{-3} \cdot 12^{-2}$ ■

Example 1(b) illustrates the fact that a fractional base can be inverted, if the sign of the exponent is changed. Example 1(c) illustrates the fact that a factor of the numerator or denominator can be moved from the numerator to the denominator or vice versa as long as we change the sign of the exponent. These rules follow from the definition of negative exponents.

Rules for Negative Exponents and Fractions

> If a and b are nonzero real numbers and m and n are integers, then
>
> $$\left(\dfrac{a}{b}\right)^{-m} = \left(\dfrac{b}{a}\right)^{m} \quad \text{and} \quad \dfrac{a^{-m}}{b^{-n}} = \dfrac{b^{n}}{a^{m}}.$$

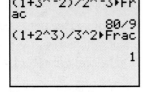

■ **Figure P.14**

Using this rule, we could shorten Examples 1(b) and (c) as follows:

$$\left(\dfrac{2}{3}\right)^{-3} = \left(\dfrac{3}{2}\right)^{3} = \dfrac{27}{8} \quad \text{and} \quad \dfrac{6^{-2}}{2^{-3}} = \dfrac{2^{3}}{6^{2}} = \dfrac{8}{36} = \dfrac{2}{9}.$$

Note that we cannot apply these rules when addition or subtraction is involved.

$$\dfrac{1 + 3^{-2}}{2^{-3}} \neq \dfrac{1 + 2^{3}}{3^{2}}$$

 Figure P.14 shows these expressions on a graphing calculator.   □

Rules of Exponents

Consider the product $a^{2} \cdot a^{3}$. Using the definition of exponents, we can simplify this product as follows:

$$a^{2} \cdot a^{3} = (a \cdot a)(a \cdot a \cdot a) = a^{5}$$

Similarly, if m and n are any positive integers we have

$$a^{m} \cdot a^{n} = \underbrace{\overbrace{a \cdot a \cdot \cdots \cdot a}^{m \text{ factors}} \cdot \overbrace{a \cdot a \cdot a \cdot \cdots \cdot a}^{n \text{ factors}}}_{m + n \text{ factors}} = a^{m+n}.$$

This equation indicates that the product of exponential expressions *with the same base* is obtained by adding the exponents. This fact is called the **product rule.**

Example **2** **Using the product rule**

Simplify each expression.

a. $(3x^8y^2)(-2xy^4)$ **b.** $2^3 \cdot 3^2$

Solution

a. Use the product rule to add the exponents when bases are identical:

$$(3x^8y^2)(-2xy^4) = -6x^9y^6$$

b. Since the bases are different we cannot use the product rule, but we can simplify the expression using the definition of exponents:

$$2^3 \cdot 3^2 = 8 \cdot 9 = 72$$

Try This. Simplify $-2a^4b^3(-3a^5b^6)$.   ■

So far we have defined positive and negative integral exponents. The definition of zero as an exponent is given in the following box. Note that the zero power of zero is not defined.

Definition: Zero Exponent

If a is a nonzero real number, then $a^0 = 1$.

The definition of zero exponent allows us to extend the product rule to any integral exponents. For example, using the definition of negative exponents, we get

$$2^{-3} \cdot 2^3 = \frac{1}{2^3} \cdot 2^3 = 1.$$

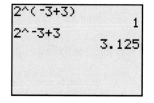

■ **Figure P.15**

Adding exponents, we get $2^{-3} \cdot 2^3 = 2^{-3+3} = 2^0$. The answer is the same because 2^0 is defined to be 1.

To evaluate 2^{-3+3} on a calculator, the expression $-3 + 3$ must be in parentheses as in Fig. P.15.   □

Using the definitions of positive, negative, and zero exponents, we can show that the product rule and several other rules hold for any integral exponents. We list these rules in the following box.

Rules for Integral Exponents

If a and b are nonzero real numbers and m and n are integers, then

1. $a^m a^n = a^{m+n}$ **Product rule**

2. $\dfrac{a^m}{a^n} = a^{m-n}$ **Quotient rule**

3. $(a^m)^n = a^{mn}$ **Power of a power rule**

4. $(ab)^n = a^n b^n$ **Power of a product rule**

5. $\left(\dfrac{a}{b}\right)^n = \dfrac{a^n}{b^n}$ **Power of a quotient rule**

The rules for integral exponents are used to simplify expressions.

Example **3** Simplifying expressions with integral exponents

Simplify each expression. Write your answer without negative exponents. Assume that all variables represent nonzero real numbers.

a. $(3x^2y^3)(-4x^{-2}y^{-5})$ **b.** $\dfrac{-6a^5b^{-1}}{2a^7b^{-3}}$

Solution

a. $(3x^2y^3)(-4x^{-2}y^{-5}) = -12x^{2+(-2)}y^{3+(-5)}$ Product rule

$\qquad\qquad\qquad\qquad\quad = -12x^0y^{-2}$ Simplify the exponents.

$\qquad\qquad\qquad\qquad\quad = -\dfrac{12}{y^2}$ Definition of negative and zero exponents

b. $\dfrac{-6a^5b^{-1}}{2a^7b^{-3}} = -3a^{5-7}b^{-1-(-3)}$ Quotient rule

$\qquad\qquad\quad = -3a^{-2}b^2$ Simplify the exponents.

$\qquad\qquad\quad = -\dfrac{3b^2}{a^2}$ Definition of negative exponents.

Try This. Simplify $-5a^{-7}b^{-5}(9a^{-2}b^8)$.   ■

In the next example, we use the rules of exponents to simplify expressions that have variables in the exponents.

Example **4** Simplifying expressions with variable exponents

Simplify each expression. Assume that all bases are nonzero real numbers and all exponents are integers.

a. $(-3x^{a-5}y^{-3})^4$ **b.** $\left(\dfrac{3a^{2m-1}}{2a^{-3m}}\right)^{-3}$

Solution

a. $(-3x^{a-5}y^{-3})^4 = (-3)^4(x^{a-5})^4(y^{-3})^4$ Power of a product rule

$\qquad\qquad\qquad\quad = 81x^{4a-20}y^{-12}$ Power of a power rule

$\qquad\qquad\qquad\quad = \dfrac{81x^{4a-20}}{y^{12}}$ Definition of negative exponents

b. $\left(\dfrac{3a^{2m-1}}{2a^{-3m}}\right)^{-3} = \dfrac{3^{-3}(a^{2m-1})^{-3}}{2^{-3}(a^{-3m})^{-3}}$ Power of a quotient rule

$\qquad\qquad\qquad = \dfrac{2^3a^{-6m+3}}{3^3a^{9m}}$ Power of a power rule and definition of negative exponents

$\qquad\qquad\qquad = \dfrac{8a^{-15m+3}}{27}$ Quotient rule

Try This. Simplify $(2a^{m-2})^2(-2a^{4m})^3$.   ■

Scientific Notation

Archimedes (287–212 B.C.) was a brilliant Greek inventor and mathematician who studied at the Egyptian city of Alexandria, then the center of the scientific world. Archimedes used his knowledge of mathematics to calculate for King Gelon the number of grains of sand in the universe: 1 followed by 63 zeros. Of course, Archimedes' universe was different from our universe, and he performed his computations with Greek letter numerals, since the modern number system and scientific notation had not yet been invented. Although it is impossible to calculate the number of grains of sand in the universe, this story illustrates how long scientists have been interested in quantities ranging in size from the diameter of our galaxy to the diameter of an atom. Scientific notation offers a convenient way of expressing very large or very small numbers.

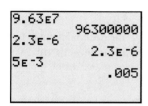

▪ **Figure P.16**

In scientific notation, a positive number is written as a product of a number between 1 and 10 and a power of 10. For example, 9.63×10^7 and 2.3×10^{-6} are numbers written in scientific notation.

These numbers are shown on a graphing calculator in scientific notation in Fig. P.16. Note that only the power of 10 shows in scientific notation on a calculator. The graphing calculator shown in Fig. P.16 converts to standard notation when the ENTER key is pressed (provided the number is neither too large nor too small). ☐

Conversion of numbers from scientific notation to standard notation is actually just multiplication.

Example **5** **Scientific notation to standard notation**

Convert each number to standard notation.

a. 9.63×10^7 **b.** 2.3×10^{-6}

Solution

a. $9.63 \times 10^7 = 9.63 \times 10,000,000$ Evaluate 10^7.

$\qquad\qquad\quad = 96,300,000$ Move decimal point seven places to the right.

b. $2.3 \times 10^{-6} = 2.3 \times \dfrac{1}{1,000,000}$ Evaluate 10^{-6}.

$\qquad\qquad\quad = 2.3 \times 0.000001$

$\qquad\qquad\quad = 0.0000023$ Move decimal point six places to the left.

Try This. Convert 3.78×10^{-2} to standard notation.   ▪

Observe how the decimal point is relocated in Example 5. Converting from scientific notation to standard notation is simply a matter of moving the decimal point. To convert a number from scientific to standard notation, we move the decimal point the number of places indicated by the power of 10. Move the decimal point to the right for a positive power and to the left for a negative power. Note that in scientific notation a number greater than 10 is written with a positive power of 10 and a number less than 1 is written with a negative power of 10. Numbers between 1 and 10 are not written in scientific notation.

In the next example, we convert from standard notation to scientific notation by reversing the process used in Example 5.

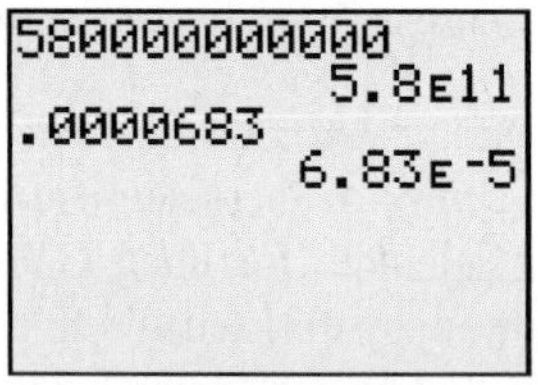

```
580000000000
              5.8E11
.0000683
              6.83E-5
```

■ **Figure P.17**

⬚ A graphing calculator converts to scientific notation when you press ENTER as shown in Fig. P.17. □

Example **6** **Standard notation to scientific notation**

Convert each number to scientific notation.

a. 580,000,000,000 **b.** 0.0000683

Solution

a. Determine the power of 10 by counting the number of places that the decimal must move so that there is a single nonzero digit to the left of the decimal point (11 places). Since 580,000,000,000 is larger than 10, we use a positive power of 10:

$$580,000,000,000 = 5.8 \times 10^{11}$$

b. Determine the power of 10 by counting the number of places the decimal must move so that there is a single nonzero digit to the left of the decimal point (five places). Since 0.0000683 is smaller than 1, we use a negative power of 10:

$$0.0000683 = 6.83 \times 10^{-5}$$

Try This. Convert 5,480,000 to scientific notation. ■

One advantage of scientific notation is that the rules of exponents can be used when performing certain computations involving scientific notation. Calculators can be used to perform computations with scientific notation, but it is good to practice some computation without a calculator.

Example **7** **Using scientific notation in computations**

Perform the indicated operations without a calculator. Write your answers in scientific notation. Check your answers with a calculator.

a. $(4 \times 10^{13})(5 \times 10^{-9})$ **b.** $\dfrac{1.2 \times 10^{-9}}{4 \times 10^{-7}}$ **c.** $\dfrac{(2,000,000,000)^3(0.00009)}{600,000,000}$

Solution

a. $(4 \times 10^{13})(5 \times 10^{-9}) = 20 \times 10^{13+(-9)}$ Product rule for exponents

$\qquad\qquad\qquad\qquad = 20 \times 10^4$ Simplify the exponent.

$\qquad\qquad\qquad\qquad = 2 \times 10^1 \times 10^4$ Write 20 in scientific notation.

$\qquad\qquad\qquad\qquad = 2 \times 10^5$ Product rule for exponents

b. $\dfrac{1.2 \times 10^{-9}}{4 \times 10^{-7}} = \dfrac{1.2}{4} \times 10^{-9-(-7)}$ Quotient rule for exponents

$\qquad\qquad\qquad = 0.3 \times 10^{-2}$ Simplify the exponent.

$\qquad\qquad\qquad = 3 \times 10^{-3}$ Use $0.3 = 3 \times 10^{-1}$.

c. First convert each number to scientific notation, then use the rules of exponents to simplify:

$$\frac{(2,000,000,000)^3(0.00009)}{600,000,000} = \frac{(2 \times 10^9)^3(9 \times 10^{-5})}{6 \times 10^8}$$

$$= \frac{(8 \times 10^{27})(9 \times 10^{-5})}{6 \times 10^8}$$

$$= \frac{72 \times 10^{22}}{6 \times 10^8} = 12 \times 10^{14} = 1.2 \times 10^{15}$$

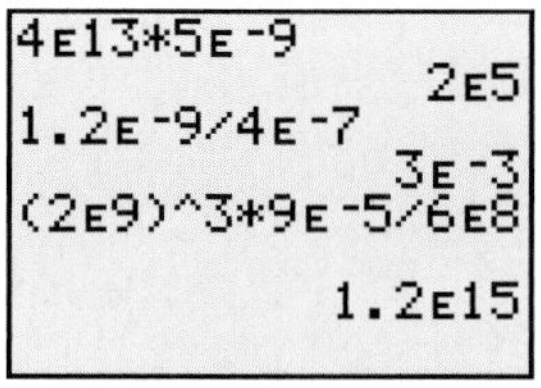

These three computations are done on a calculator in Fig. P.18. Set the mode to scientific to get the answers in scientific notation.

Try This. Find the product $(7 \times 10^{14})(5 \times 10^{-3})$. ■

In the next example we use scientific notation to perform the type of computation performed by Archimedes when he attempted to determine the number of grains of sand in the universe.

Example **8** **The number of grains of sand in Archimedes' earth**

If the radius of the earth is approximately 6.38×10^3 kilometers and the radius of a grain of sand is approximately 1×10^{-3} meters, then what number of grains of sand have a volume equal to the volume of the earth?

Solution
Since the volume of a sphere is given by $V = \frac{4}{3}\pi r^3$, the volume of the earth is

$$\frac{4}{3}\pi(6.38 \times 10^3)^3 \text{ km}^3 \approx 1.09 \times 10^{12} \text{ km}^3.$$

Since $1 \text{ km} = 10^3$ m, the radius of a grain of sand is 1×10^{-6} km and its volume is

$$\frac{4}{3}\pi(1 \times 10^{-6})^3 \text{ km}^3 \approx 4.19 \times 10^{-18} \text{ km}^3.$$

To get the number of grains of sand, divide the volume of the earth by the volume of a grain of sand:

$$\frac{1.09 \times 10^{12} \text{ km}^3}{4.19 \times 10^{-18} \text{ km}^3} \approx 2.60 \times 10^{29}$$

See Fig. P.19 for the computations.

Try This. Find the volume of a sphere that has radius 2.4×10^{-3} in. ■

■ **Figure P.18**

■ **Figure P.19**

For Thought

True or False? Explain. Do Not Use a Calculator.

1. $2^{-1} + 2^{-1} = 1$ **2.** $2^{100} = 4^{50}$

3. $9^8 \cdot 9^8 = 81^8$ **4.** $(0.25)^{-1} = 4$

5. $\dfrac{5^{10}}{5^{-12}} = 5^{-2}$ **6.** $2 \cdot 2 \cdot 2 \cdot 2^{-1} = \dfrac{1}{16}$

7. $-3^{-3} = -\dfrac{1}{27}$ **8.** $\left(\dfrac{3}{4}\right)^{-2} = \left(\dfrac{4}{3}\right)^2$

9. $10^{-4} = 0.00001$

10. $98.6 \times 10^8 = 9.86 \times 10^7$

P.2 Exercises

Evaluate each expression without using a calculator. Check your answers with a calculator.

1. 3^{-1} **2.** 2^{-1} **3.** -4^{-2}

4. -5^{-3} **5.** $\dfrac{1}{2^{-3}}$ **6.** $\dfrac{1}{10^{-3}}$

7. $\left(\dfrac{3}{2}\right)^{-3}$ **8.** $\left(\dfrac{2}{3}\right)^{-2}$ **9.** $\left(-\dfrac{1}{2}\right)^{-2}$

10. $\left(-\dfrac{1}{3}\right)^{-4}$ **11.** $2^{-1} \cdot 4^2 \cdot 10^{-1}$

12. $4 \cdot 4 \cdot 4 \cdot 4^{-1}$ **13.** $\dfrac{3^{-2}}{6^{-3}}$

14. $\dfrac{3^{-1}}{2^3}$ **15.** $2^0 + 2^{-1}$

16. $6^{-1} + 5^{-1}$ **17.** $-2 \cdot 10^{-3}$

18. $-1^{-1} \cdot (-2)^{-2}$

Simplify each expression.

19. $(-3x^2y^3)(2x^9y^8)$ **20.** $(-6a^7b^4)(3a^3b^5)$

21. $x^2x^4 + x^3x^3$ **22.** $a^7a^5 + a^3a^9$

23. $(-2b^2)(3b^3) + (5b^3)(-3b^2)$

24. $(4y^6)(-xy^4) + (3xy^9)(5y)$

25. $(-2a^2)(a^9) - (a^5)(-4a^6)$

26. $(w^2)(-3w) - (-5w)(-7w^2)$

27. $(-m^2)(-m) - m(-m) + m(3m^2)$

28. $(z^2)(-z) - (-z) - z(-z^2) + z(2z)$

29. $5^2 \cdot 3^2$ **30.** $2^3 \cdot 4^2$

Simplify each expression. Write answers without negative exponents. Assume that all variables represent nonzero real numbers.

31. $\left(\dfrac{1}{2}x^{-4}y^3\right)\left(\dfrac{1}{3}x^4y^{-6}\right)$ **32.** $\left(\dfrac{1}{3}a^{-5}b\right)(a^4b^{-1})$

33. $\dfrac{6x^7}{2x^3}$ **34.** $\dfrac{-9x^2y}{3xy^2}$

35. $\dfrac{-3m^{-1}n}{-6m^{-1}n^{-1}}$ **36.** $\dfrac{-p^{-1}q^{-1}}{-3pq^{-3}}$

37. $(2a^2)^3 + (-3a^3)^2$ **38.** $(b^{-4})^2 - (-b^{-2})^4$

39. $-1(2x^3)^2$ **40.** $(-3y^{-1})^{-1}$

41. $\left(\dfrac{-2x^2}{3}\right)^3$ **42.** $\left(\dfrac{-1}{2a}\right)^{-2}$

43. $\left(\dfrac{y^2}{5}\right)^{-2}$ **44.** $\left(-\dfrac{y^2}{2a}\right)^4$

45. $\left(\dfrac{6xy^2}{8x^{-4}y^3}\right)^{-3}$ **46.** $\left(-\dfrac{15x^{-2}y^9}{18x^2y^3}\right)^{-2}$

Simplify each expression. Assume that all bases are nonzero real numbers and all exponents are integers.

47. $(x^{b-1})^3(x^{b-4})^{-2}$ **48.** $(a^2)^{m+2}(a^3)^{4m}$

49. $(-5a^{2t}b^{-3t})^3$ **50.** $(-2x^{-5v}y^3)^2$

51. $\dfrac{-9x^{3w}y^{9v}}{6x^{8w}y^{3v}}$ **52.** $\dfrac{6c^{9s}d^{4t}}{-9c^{3s}d^{8t}}$

53. $\left(\dfrac{a^{s+2}}{a^{2s-3}}\right)^4$ **54.** $\left(\dfrac{x^{2a-3}}{x^{-4a+1}}\right)^{-4}$

Convert each number given in standard notation to scientific notation and each number given in scientific notation to standard notation.

55. 4.3×10^4

56. 5.98×10^5

57. 3.56×10^{-5}

58. 9.333×10^{-9}

59. 5,000,000

60. 16,587,000

61. 0.0000672

62. 0.000000981

63. 7×10^{-9}

64. 6×10^{-3}

65. 20,000,000,000

66. 0.00000000004

Perform the indicated operations without a calculator. Write your answers in scientific notation. Use a calculator to check.

67. $(5 \times 10^8)(4 \times 10^7)$

68. $(5 \times 10^{-10})(6 \times 10^5)$

69. $\dfrac{8.2 \times 10^{-6}}{4.1 \times 10^{-3}}$

70. $\dfrac{9.3 \times 10^{12}}{3.1 \times 10^{-3}}$

71. $5(2 \times 10^{-10})^3$

72. $2(2 \times 10^8)^{-4}$

73. $\dfrac{(2,000,000)^3(0.000005)}{(0.00002)^2}$

74. $\dfrac{(6,000,000)^2(0.000003)^{-3}}{(2000)^3(1,000,000)}$

Use a calculator to perform the indicated operations. Give your answers in scientific notation.

75. $(4.32 \times 10^{-9})(2.3 \times 10^4)$

76. $(2.33 \times 10^{23})(3.98 \times 10^{-9})$

77. $\dfrac{(5.63 \times 10^{-6})^3(3.5 \times 10^7)^{-4}}{\pi(8.9 \times 10^{-4})^2}$

78. $\dfrac{\pi(2.39 \times 10^{-12})^2}{(6.75 \times 10^{-8})^3}$

Solve each problem. Use your calculator.

79. *Body-Mass Index* The body mass index (*BMI*) is used to assess the relative amount of fat in a person's body (Shape Up America, www.shapeup.org). If *w* is weight in pounds and *h* is height in inches, then

$$BMI = 703wh^{-2}.$$

Find the *BMI* for Dallas Cowboys' player Flozell Adams at 6'7" and 335 pounds.

80. *Abnormal Fat* The accompanying table shows the height and weight of five Dallas Cowboys players in training camp (ESPN, www.espn.com). Some physicians consider your body fat

abnormal if your *BMI* (from the previous exercise) satisfies

$$|BMI - 23| > 3.$$

Which of the five players does not have abnormal body fat?

■ **Table for Exercise 80**

Player	Ht	Wt	
Adams	6-7	335	
Allen	6-3	326	
Anderson	6-2	215	
Robinson	6-4	250	
Smith	5-11	175	

81. *Displacement-Length Ratio* The displacement-length ratio *D* indicates whether a sailboat is relatively heavy or relatively light (Ted Brewer Yacht Design, www.tedbrewer.com). *D* is given by the formula

$$D = \frac{dL^{-3} \cdot 10^6}{2240},$$

where *L* is the length at the water line in feet, and *d* is the displacement in pounds. If $D > 300$, then a boat is considered relatively heavy. Find *D* for the USS Constitution, the nation's oldest floating ship, which has a displacement of 2200 tons and a length of 175 feet.

82. *A Lighter Boat* If $D < 150$ (from the previous exercise), then a boat is considered relatively light.
   a. Use the accompanying graph to estimate the length for which $D < 150$ for a boat with a displacement of 11,500 pounds.

   b. Use the formula to find *D* for a 50-foot boat with a displacement of 11,500 pounds.

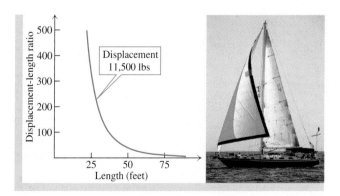

■ **Figure for Exercise 82**

83. *National Debt* In February 2005, the national debt was 7.634×10^{12} dollars and the U.S. population was 2.956×10^8 people (U.S. Treasury, www.treas.gov). What was each person's share of the debt at that time?

84. *Increasing Debt* In the three years prior to February of 2005, the national debt was increasing at an average rate of 1.44×10^9 dollars per day. If the debt keeps increasing at that rate, then what will be the amount of the debt in February of 2011? See the previous exercise.

85. *Energy from the Sun* Merely an average star, our sun is a swirling mass of dense gases powered by thermonuclear reactions. The great solar furnace transforms 5 million tons of mass into energy every second. How many tons of mass will be transformed into energy during the sun's 10 billion-year lifetime?

> HINT Find the number of seconds in 10 billion years.

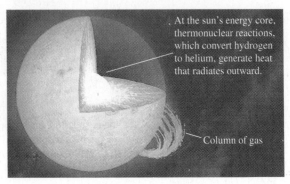

At the sun's energy core, thermonuclear reactions, which convert hydrogen to helium, generate heat that radiates outward.

Column of gas

■ **Figure for Exercise 85**

86. *Orbit of the Earth* The earth orbits the sun in an approximately circular orbit with a radius of 1.495979×10^8 km. What is the area of the circle?

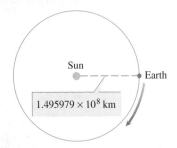

Sun

Earth

1.495979×10^8 km

■ **Figure for Exercises 86 and 87**

87. *Radius of the Earth* The radius of the sun is 6.9599×10^5 km, which is 109.1 times the earth's radius. What is the radius of the earth?

88. *Distance to the Sun* The distance from the sun to the earth is 1.495979×10^8 km. Use the fact that 1 km = 0.621 mi to find the distance in miles from the earth to the sun.

89. *Mass of the Sun* The mass of the sun is 1.989×10^{30} kg and the mass of the earth is 5.976×10^{24} kg. How many times larger in mass is the sun than the earth?

90. *Speed of Light* If the speed of light is 3×10^8 m/sec, then how long does it take light from the sun to reach the earth? (See Exercise 88.)

> HINT $T = D/R$

For Writing/Discussion

91. The integral powers of 10 are important in the decimal number system.

a. Use a calculator to evaluate

$$5 \cdot 10^3 + 3 \cdot 10^2 + 6 \cdot 10^1 + 5 \cdot 10^0 + 4 \cdot 10^{-1}.$$

b. Without using a calculator, evaluate

$$9 \cdot 10^3 + 7 \cdot 10^2 + 2 \cdot 10^0 + 3 \cdot 10^{-1} + 8 \cdot 10^{-2}.$$

c. Write a brief explanation of how you did part (b).

d. Write 9063.241 as a sum of integral multiples of powers of 10. Is your answer unique? Write a detailed description of how you did it.

e. Try to write 43,002.19 as a sum of integral multiples of powers of 10 using the description that you wrote for part (d).

92. Many calculators can handle scientific notation for powers of 10 between -99 and 99. How can you be expected to compute $(4 \times 10^{220})^2$?

Thinking Outside the Box II & III

Powers of Three   What is x if

$$\frac{1}{27} \cdot 3^{100} \cdot \frac{1}{81} \cdot 9^x = \frac{1}{3} \cdot 3^x?$$

Common Remainders   The numbers

$$1{,}576{,}231, \quad 4{,}080{,}602, \quad \text{and} \quad 2{,}690{,}422$$

all yield the same remainder when divided by a certain natural number that is greater than one. What is the divisor and the remainder?

P.2 Pop Quiz

1. Evaluate $2^{-1} + 4^{-1}$.

2. Evaluate $(-1/2)^{-3}$.

3. Simplify $(-2xy^2)(3x^3y)$.

4. Simplify $(-2a^{-3}b^3)(-a^4b^{-1})$.

5. Simplify $a^m(a^{m-2})^3$.

6. Write 3.6×10^3 in standard notation.

7. Write 0.005 in scientific notation.

8. Evaluate $\dfrac{3 \times 10^{-7}}{6 \times 10^{-8}}$.

Linking Concepts

For Individual or Group Explorations

Financial Expressions

What's my monthly payment? How much should I be saving monthly for retirement? Most people will ask those questions and other financial questions at some point in their lives. Three expressions that are used to answer basic financial questions are given in the following table, where n is the number of periods in a year, r is the annual percentage rate (APR), t is the number of years, and i is the interest rate per period (i = r/n) (www.interest.com).

What $P left at compound interest will grow to	What $R deposited periodically will grow to	Periodic payment necessary to pay off a loan of $P
$P(1 + i)^{nt}$	$R\dfrac{(1 + i)^{nt} - 1}{i}$	$P\dfrac{i}{1 - (1 + i)^{-nt}}$

For periodic payments or deposits, these expressions apply only if the compounding period equals the payment period. For example, if the payments are made monthly then the interest must be compounded monthly.

a) Jaime inherited $12,000 and left it in a bank to grow at 6% APR compounded monthly. What was the inheritance worth after 8 years?

b) When Herman quit smoking at age 21 he began depositing the $100 per month that he saved into an account paying 8% APR compounded monthly. How much money will he have in the account when he is 65 years old?

c) Beatrice plans to pay off her $30,000 car loan in 6 years at $9\frac{3}{4}$% APR compounded monthly. What is the monthly payment?

d) Find a newspaper ad where a car dealer lists the monthly payment necessary to buy a car and cut out the ad. Calculate the monthly payment with the expression given here and compare.

P.3

Rational Exponents and Radicals

Raising a number to a power is reversed by finding the root of a number. We indicate roots by using rational exponents or radicals. In this section we will review definitions and rules concerning rational exponents and radicals.

Roots

Since $2^4 = 16$ and $(-2)^4 = 16$, both 2 and -2 are fourth roots of 16. The nth root of a number is defined in terms of the nth power.

Definition: *n*th Roots

> If n is a positive integer and $a^n = b$, then a is called an **nth root** of b. If $a^2 = b$, then a is a **square root** of b. If $a^3 = b$, then a is the **cube root** of b.

We also describe roots as even or odd, depending on whether the positive integer is even or odd. For example, if n is even (or odd) and a is an nth root of b, then a is called an **even** (or **odd**) **root** of b. Every positive real number has *two* real even roots, a positive root and a negative root. For example, both 5 and -5 are square roots of 25 because $5^2 = 25$ and $(-5)^2 = 25$. Moreover, every real number has exactly *one* real odd root. For example, because $2^3 = 8$ and 3 is odd, 2 is the only real cube root of 8. Because $(-2)^3 = -8$ and 3 is odd, -2 is the only real cube root of -8.

Finding an nth root is the reverse of finding an nth power, so we use the notation $a^{1/n}$ for the nth root of a. For example, since the positive square root of 25 is 5, we write $25^{1/2} = 5$.

Definition: Exponent 1/*n*

> If n is a positive even integer and a is positive, then $a^{1/n}$ denotes the **positive real nth root of a** and is called the **principal nth root of a.**
>
> If n is a positive odd integer and a is any real number, then $a^{1/n}$ denotes the real nth root of a.
>
> If n is a positive integer, then $0^{1/n} = 0$.

Example 1 Evaluating expressions involving exponent 1/*n*

Evaluate each expression.

a. $4^{1/2}$ **b.** $8^{1/3}$ **c.** $(-8)^{1/3}$ **d.** $(-4)^{1/2}$

Solution

a. The expression $4^{1/2}$ represents the positive real square root of 4. So $4^{1/2} = 2$.
b. $8^{1/3} = 2$
c. $(-8)^{1/3} = -2$

d. Since the definition of nth root does not include an even root of a negative number, $(-4)^{1/2}$ has not yet been defined. Even roots of negative numbers do exist in the complex number system, which we define in Section P.4. So an even root of a negative number is not a real number.

Try This. Evaluate.   **a.** $9^{1/2}$ **b.** $16^{1/4}$ ■

Rational Exponents

We have defined $a^{1/n}$ as the nth root of a. We extend this definition to $a^{m/n}$, which is defined as the mth power of the nth root of a. A rational exponent indicates both a root and a power.

Definition:
Rational Exponents

> If m and n are positive integers, then
>
> $$a^{m/n} = (a^{1/n})^m$$
>
> provided that $a^{1/n}$ is a real number.

Note that $a^{1/n}$ is not real when a is negative and n is even. According to the definition of rational exponents, expressions such as $(-25)^{-3/2}$, $(-43)^{1/4}$, and $(-1)^{2/2}$ are not defined because each of them involves an even root of a negative number. Note that some authors define $a^{m/n}$ only for m/n in lowest terms. In that case the fourth power of the square root of 3 could *not* be written as $3^{4/2}$. This author prefers the more general definition given above.

The root and the power indicated in a rational exponent can be evaluated in either order. That is, $(a^{1/n})^m = (a^m)^{1/n}$ provided $a^{1/n}$ is real. For example,

$$8^{2/3} = (8^{1/3})^2 = 2^2 = 4 \qquad \text{or} \qquad 8^{2/3} = (8^2)^{1/3} = 64^{1/3} = 4.$$

A negative rational exponent indicates reciprocal just as a negative integral exponent does. So

$$8^{-2/3} = \frac{1}{8^{2/3}} = \frac{1}{(8^{1/3})^2} = \frac{1}{2^2} = \frac{1}{4}.$$

Evaluate $8^{-2/3}$ mentally as follows: The cube root of 8 is 2, 2 squared is 4, and the reciprocal of 4 is $\frac{1}{4}$. The three operations indicated by a negative rational exponent can be performed in any order, but the simplest procedure for mental evaluations is summarized as follows.

P R O C E D U R E   **Evaluating $a^{-m/n}$**

To evaluate $a^{-m/n}$ mentally,
1. find the nth root of a,
2. raise it to the m power,
3. find the reciprocal.

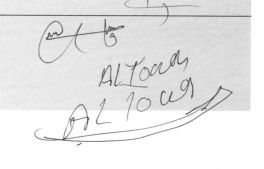

Rational exponents can be reduced to lowest terms. For example, we can evaluate $2^{6/2}$ by first reducing the exponent:

$$2^{6/2} = 2^3 = 8$$

Exponents can be reduced only on expressions that are real numbers. For example, $(-1)^{2/2} \neq (-1)^1$ because $(-1)^{2/2}$ is not a real number, while $(-1)^1$ is a real number. ⊡ Your graphing calculator will probably evaluate $(-1)^{2/2}$ as -1, because it is not using our definition. Moreover, some calculators will not evaluate an expression with a negative base such as $(-8)^{2/3}$, but will evaluate the equivalent expression $((-8)^2)^{1/3}$. To use your calculator effectively, you must get to know it well.   □

Example **2** **Evaluating expressions with rational exponents**

Evaluate each expression.

a. $(-8)^{2/3}$ **b.** $27^{-2/3}$ **c.** $100^{6/4}$

Solution

a. Mentally, the cube root of -8 is -2 and the square of -2 is 4. In symbols:

$$(-8)^{2/3} = ((-8)^{1/3})^2 = (-2)^2 = 4$$

b. Mentally, the cube root of 27 is 3, the square of 3 is 9, and the reciprocal of 9 is $\frac{1}{9}$. In symbols:

$$27^{-2/3} = \frac{1}{(27^{1/3})^2} = \frac{1}{3^2} = \frac{1}{9}$$

c. $100^{6/4} = 100^{3/2} = 10^3 = 1000$

⊡ The expressions are evaluated with a graphing calculator in Fig. P.20.

Try This. Evaluate.   **a.** $9^{3/2}$ **b.** $16^{-5/4}$ ■

```
(-8)^(2/3)
                    4
27^(-2/3)▶Frac
                  1/9
100^(6/4)
                 1000
```

■ **Figure P.20**

Rules for Rational Exponents

The rules for integral exponents from Section P.2 also hold for rational exponents.

Rules for Rational Exponents

The following rules are valid for all real numbers a and b and rational numbers r and s, provided that all indicated powers are real and no denominator is zero.

1. $a^r a^s = a^{r+s}$ **2.** $\dfrac{a^r}{a^s} = a^{r-s}$ **3.** $(a^r)^s = a^{rs}$

4. $(ab)^r = a^r b^r$ **5.** $\left(\dfrac{a}{b}\right)^r = \dfrac{a^r}{b^r}$ **6.** $\left(\dfrac{a}{b}\right)^{-r} = \left(\dfrac{b}{a}\right)^r$ **7.** $\dfrac{a^{-r}}{b^{-s}} = \dfrac{b^s}{a^r}$

When variable expressions involve even roots, we must be careful with signs. For example, $(x^2)^{1/2} = x$ is not correct for all values of x, because $((-5)^2)^{1/2} = 25^{1/2} = 5$. However, using absolute value we can write

$$(x^2)^{1/2} = |x| \qquad \text{for every real number } x.$$

When finding an even root of an expression involving variables, remember that if n is even, $a^{1/n}$ is the *positive* nth root of a.

Example **3**   Using absolute value with rational exponents

Simplify each expression, using absolute value when necessary. Assume that the variables can represent any real numbers.

a. $(64a^6)^{1/6}$ **b.** $(x^9)^{1/3}$ **c.** $(a^8)^{1/4}$ **d.** $(y^{12})^{1/4}$

Solution

a. For any nonnegative real number a, we have $(64a^6)^{1/6} = 2a$. If a is negative, $(64a^6)^{1/6}$ is positive and $2a$ is negative. So we write

$$(64a^6)^{1/6} = |2a| = 2|a| \qquad \text{for every real number } a.$$

b. For any nonnegative x, we have $(x^9)^{1/3} = x^{9/3} = x^3$. If x is negative, $(x^9)^{1/3}$ and x^3 are both negative. So we have

$$(x^9)^{1/3} = x^3 \qquad \text{for every real number } x.$$

c. For nonnegative a, we have $(a^8)^{1/4} = a^2$. Since $(a^8)^{1/4}$ and a^2 are both positive if a is negative, no absolute value sign is needed. So

$$(a^8)^{1/4} = a^2 \qquad \text{for every real number } a.$$

d. For nonnegative y, we have $(y^{12})^{1/4} = y^3$. If y is negative, $(y^{12})^{1/4}$ is positive but y^3 is negative. So

$$(y^{12})^{1/4} = |y^3| \qquad \text{for every real number } y.$$

Try This. Let w represent any real number. Simplify $(w^4)^{1/4}$.   ■

When simplifying expressions we often assume that the variables represent positive real numbers so that we do not have to be concerned about undefined expressions or absolute value. In the following example we make that assumption as we use the rules of exponents to simplify expressions involving rational exponents.

Example **4**   Simplifying expressions with rational exponents

Use the rules of exponents to simplify each expression. Assume that the variables represent positive real numbers. Write answers without negative exponents.

a. $x^{2/3}x^{4/3}$ **b.** $(x^4y^{1/2})^{1/4}$ **c.** $\left(\dfrac{a^{3/2}b^{2/3}}{a^2}\right)^3$

Solution

a. $x^{2/3}x^{4/3} = x^{6/3}$ 　　　　　　Product rule

　　　　　 $= x^2$ 　　　　　　　　Simplify the exponent.

b. $(x^4y^{1/2})^{1/4} = (x^4)^{1/4}(y^{1/2})^{1/4}$ 　　Power of a product rule

　　　　　　　 $= xy^{1/8}$ 　　　　　Power of a power rule

c. $\left(\dfrac{a^{3/2}b^{2/3}}{a^2}\right)^3 = \dfrac{(a^{3/2})^3(b^{2/3})^3}{(a^2)^3}$ Power of a quotient rule

$= \dfrac{a^{9/2}b^2}{a^6}$ Power of a power rule

$= a^{-3/2}b^2$ Quotient rule $\left(\dfrac{9}{2} - 6 = -\dfrac{3}{2}\right)$

$= \dfrac{b^2}{a^{3/2}}$ Definition of negative exponents

Try This. Simplify $(a^{1/3}a^{1/2})^{12}$.   ■

Radical Notation

The exponent $1/n$ and the **radical sign** $\sqrt[n]{}$ are both used to indicate the nth root.

Definition: Radical

> If n is a positive integer and a is a number for which $a^{1/n}$ is defined, then the expression $\sqrt[n]{a}$ is called a **radical,** and
> $$\sqrt[n]{a} = a^{1/n}.$$
> If $n = 2$, we write $\sqrt{a}$ rather than $\sqrt[2]{a}$.

The number a is called the **radicand** and n is the **index** of the radical. Expressions such as $\sqrt{-3}$, $\sqrt[4]{-81}$, and $\sqrt[6]{-1}$ do not represent real numbers because each is an even root of a negative number.

Example **5** **Evaluating radicals**

Evaluate each expression and check with a calculator.

a. $\sqrt{49}$ **b.** $\sqrt[3]{-1000}$ **c.** $\sqrt[4]{\dfrac{16}{81}}$

Solution

a. The symbol $\sqrt{49}$ indicates the positive square root of 49. So $\sqrt{49} = 49^{1/2} = 7$. Writing $\sqrt{49} = \pm 7$ is incorrect.

b. $\sqrt[3]{-1000} = (-1000)^{1/3} = -10$ Check that $(-10)^3 = -1000$.

c. $\sqrt[4]{\dfrac{16}{81}} = \left(\dfrac{16}{81}\right)^{1/4} = \dfrac{2}{3}$ Check that $\left(\dfrac{2}{3}\right)^4 = \dfrac{16}{81}$.

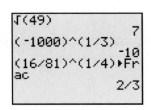

■ **Figure P.21**

These expressions are evaluated with a calculator in Fig. P.21.

Try This. Evaluate. **a.** $\sqrt{100}$ **b.** $\sqrt[3]{-27}$ ■

Since $a^{1/n} = \sqrt[n]{a}$, expressions involving rational exponents can be written with radicals.

Rule: Converting $a^{m/n}$ to Radical Notation

If a is a real number and m and n are integers for which $\sqrt[n]{a}$ is real, then

$$a^{m/n} = \left(\sqrt[n]{a}\right)^m = \sqrt[n]{a^m}.$$

Example **6** **Writing rational exponents as radicals**

Write each expression in radical notation. Assume that all variables represent positive real numbers. Simplify the radicand if possible.

a. $2^{2/3}$ **b.** $(3x)^{3/4}$ **c.** $2(x^2 + 3)^{-1/2}$

Solution

a. $2^{2/3} = \sqrt[3]{2^2} = \sqrt[3]{4}$

b. $(3x)^{3/4} = \sqrt[4]{(3x)^3} = \sqrt[4]{27x^3}$

c. $2(x^2 + 3)^{-1/2} = 2 \cdot \dfrac{1}{(x^2 + 3)^{1/2}} = \dfrac{2}{\sqrt{x^2 + 3}}$

Try This. Write $5^{2/3}$ in radical notation.   ■

The Product and Quotient Rules for Radicals

Using rational exponents we can write

$$(ab)^{1/n} = a^{1/n}b^{1/n} \qquad \text{and} \qquad \left(\frac{a}{b}\right)^{1/n} = \frac{a^{1/n}}{b^{1/n}}.$$

These equations say that the nth root of a product (or quotient) is the product (or quotient) of the nth roots. Using radical notation these rules are written as follows.

Rule: Product and Quotient

For any positive integer n and real numbers a and b $(b \neq 0)$,

1. $\sqrt[n]{ab} = \sqrt[n]{a} \cdot \sqrt[n]{b}$ **Product rule for radicals**

2. $\sqrt[n]{\dfrac{a}{b}} = \dfrac{\sqrt[n]{a}}{\sqrt[n]{b}}$ **Quotient rule for radicals**

provided that all of the roots are real.

An expression that is the square of a term that is free of radicals is called a **perfect square.** For example, $9x^6$ is a perfect square because $9x^6 = (3x^3)^2$. Likewise, $27y^{12}$ is a **perfect cube.** In general, an expression that is the nth power of an expression free of radicals is a **perfect nth power.** In the next example, the product and quotient rules for radicals are used to simplify radicals containing perfect squares, cubes, and so on.

Example **7** **Using the product and quotient rules for radicals**

Simplify each radical expression. Assume that all variables represent positive real numbers.

a. $\sqrt[3]{125a^6}$ **b.** $\sqrt{\dfrac{3}{16}}$ **c.** $\sqrt[5]{\dfrac{-32y^5}{x^{20}}}$

Solution

a. Both 125 and a^6 are perfect cubes. So use the product rule to simplify:

$$\sqrt[3]{125a^6} = \sqrt[3]{125} \cdot \sqrt[3]{a^6} = 5a^2 \quad \text{Since } \sqrt[3]{a^6} = a^{6/3} = a^2$$

b. Since 16 is a perfect square, use the quotient rule to simplify the radical:

$$\sqrt{\dfrac{3}{16}} = \dfrac{\sqrt{3}}{\sqrt{16}} = \dfrac{\sqrt{3}}{4}$$

```
√(3/16)
        .4330127019
√(3)/4
        .4330127019
```

■ **Figure P.22**

We can check this answer by using a calculator as shown in Fig. P.22. Note that agreement in the first 10 decimal places supports our belief that the two expressions are equal, but does not prove it. The expressions are equal because of the quotient rule. □

c. $\sqrt[5]{\dfrac{-32y^5}{x^{20}}} = \dfrac{\sqrt[5]{-32y^5}}{\sqrt[5]{x^{20}}} = \dfrac{-2y}{x^4} \quad \text{Since } \sqrt[5]{x^{20}} = x^{20/5} = x^4$

Try This. Simplify $\sqrt[3]{-8m^9}$. ■

Simplified Form and Rationalizing the Denominator

We have been simplifying radical expressions by just making them look simpler. However, a radical expression is in *simplified form* only if it satisfies the following three specific conditions. (You should check that the simplified expressions of Example 7 satisfy these conditions.)

Definition: Simplified Form for Radicals of Index *n*

A radical of index *n* in **simplified form** has

1. *no* perfect *n*th powers as factors of the radicand,
2. *no* fractions inside the radical, and
3. *no* radicals in a denominator.

The product rule is used to remove the perfect *n*th powers that are factors of the radicand, and the quotient rule is used when fractions occur inside the radical. The process of removing radicals from a denominator is called **rationalizing the denominator.** Radicals can be removed from the numerator by using the same type of procedure.

Example **8** **Simplified form and rationalizing the denominator**

Write each radical expression in simplified form. Assume that all variables represent positive real numbers.

a. $\sqrt{20}$ **b.** $\sqrt{24x^8y^9}$ **c.** $\dfrac{9}{\sqrt{3}}$ **d.** $\sqrt[3]{\dfrac{3}{5a^4}}$

Solution

a. Since 4 is a factor of 20, $\sqrt{20}$ is not in its simplified form. Use the product rule for radicals to simplify it:

$$\sqrt{20} = \sqrt{4} \cdot \sqrt{5} = 2\sqrt{5}$$

b. Use the product rule to factor the radical, putting all perfect squares in the first factor:

$$\sqrt{24x^8y^9} = \sqrt{4x^8y^8} \cdot \sqrt{6y} \qquad \text{Product rule}$$
$$= 2x^4y^4\sqrt{6y} \qquad \text{Simplify the first radical.}$$

c. Since $\sqrt{3}$ appears in the denominator, we multiply the numerator and denominator by $\sqrt{3}$ to rationalize the denominator. Note that multiplying by $\dfrac{\sqrt{3}}{\sqrt{3}}$ is equivalent to multiplying the expression by 1. So its appearance is changed, but not its value. The following display illustrates this point.

$$\frac{9}{\sqrt{3}} = \frac{9}{\sqrt{3}} \cdot 1 = \frac{9}{\sqrt{3}} \cdot \frac{\sqrt{3}}{\sqrt{3}} = \frac{9\sqrt{3}}{3} = 3\sqrt{3}$$

d. To rationalize this denominator, we must get a perfect cube in the denominator. The radicand $5a^4$ can be made into the perfect cube $125a^6$ by multiplying by $25a^2$:

$$\sqrt[3]{\frac{3}{5a^4}} = \frac{\sqrt[3]{3}}{\sqrt[3]{5a^4}} \qquad \text{Quotient rule for radicals}$$

$$= \frac{\sqrt[3]{3} \cdot \sqrt[3]{25a^2}}{\sqrt[3]{5a^4} \cdot \sqrt[3]{25a^2}} \qquad \text{Multiply numerator and denominator by } \sqrt[3]{25a^2}.$$

$$= \frac{\sqrt[3]{75a^2}}{\sqrt[3]{125a^6}} \qquad \text{Product rule for radicals}$$

$$= \frac{\sqrt[3]{75a^2}}{5a^2} \qquad \text{Since } (5a^2)^3 = 125a^6$$

Try This. Write $\sqrt{8y^7}$ in simplified form.   ■

Operations with Radical Expressions

Radical expressions with the same index can be added, subtracted, multiplied, or divided. For example, $2\sqrt{7} + 3\sqrt{7} = 5\sqrt{7}$ because $2x + 3x = 5x$ is true for any value of x. Because $2\sqrt{7}$ and $3\sqrt{7}$ are added in the same manner as like terms, they are called **like terms** or **like radicals.** Note that sums such as $\sqrt{3} + \sqrt{5}$ or $\sqrt[3]{2y} + \sqrt{2y}$ cannot be written as a single radical because the terms are not like terms. The next example further illustrates the basic operations with radicals.

Example 9 Operations with radicals of the same index

Perform each operation and simplify each answer. Assume that each variable represents a positive real number.

a. $\sqrt{20} + \sqrt{5}$ **b.** $\sqrt[3]{24x} - \sqrt[3]{81x}$ **c.** $\sqrt[4]{4y^3} \cdot \sqrt[4]{12y^2}$ **d.** $\sqrt{40} \div \sqrt{5}$

Solution

a. $\sqrt{20} + \sqrt{5} = \sqrt{4} \cdot \sqrt{5} + \sqrt{5}$ Product rule for radicals

$\qquad\qquad = 2\sqrt{5} + \sqrt{5} = 3\sqrt{5}$ Simplify. Add like terms.

b. $\sqrt[3]{24x} - \sqrt[3]{81x} = \sqrt[3]{8} \cdot \sqrt[3]{3x} - \sqrt[3]{27} \cdot \sqrt[3]{3x}$ Product rule for radicals

$\qquad\qquad = 2\sqrt[3]{3x} - 3\sqrt[3]{3x} = -\sqrt[3]{3x}$ Simplify. Subtract like terms.

c. $\sqrt[4]{4y^3} \cdot \sqrt[4]{12y^2} = \sqrt[4]{48y^5}$ Product rule for radicals

$\qquad\qquad = \sqrt[4]{16y^4} \cdot \sqrt[4]{3y} = 2y\sqrt[4]{3y}$ Factor out the perfect fourth powers. Simplify.

d. $\sqrt{40} \div \sqrt{5} = \sqrt{\dfrac{40}{5}} = \sqrt{8}$ Quotient rule for radicals; divide.

$\qquad\qquad = \sqrt{4} \cdot \sqrt{2} = 2\sqrt{2}$ Product rule; simplify.

Try This. Subtract and simplify $\sqrt{50} - \sqrt{8}$. ■

Radicals with different indices are not usually added or subtracted, but they can be combined in certain cases as shown in the next example.

Example 10 Combining radicals with different indices

Write each expression using a single radical symbol. Assume that each variable represents a positive real number.

a. $\sqrt[3]{2} \cdot \sqrt{3}$ **b.** $\sqrt[3]{y} \cdot \sqrt[4]{2y}$ **c.** $\sqrt{\sqrt[3]{2}}$

Solution

a. $\sqrt[3]{2} \cdot \sqrt{3} = 2^{1/3} \cdot 3^{1/2}$ Rewrite radicals as rational exponents.

$\qquad\qquad = 2^{2/6} \cdot 3^{3/6}$ Write exponents with the least common denominator.

$\qquad\qquad = \sqrt[6]{2^2 \cdot 3^3}$ Rewrite in radical notation using the product rule.

$\qquad\qquad = \sqrt[6]{108}$ Simplify inside the radical.

b. $\sqrt[3]{y} \cdot \sqrt[4]{2y} = y^{1/3}(2y)^{1/4}$ Rewrite radicals as rational exponents.

$$= y^{4/12}(2y)^{3/12}$$ Write exponents with the LCD.

$$= \sqrt[12]{y^4(2y)^3}$$ Rewrite in radical notation using the product rule.

$$= \sqrt[12]{8y^7}$$ Simplify inside the radical.

c. $\sqrt{\sqrt[3]{2}} = (2^{1/3})^{1/2} = 2^{1/6} = \sqrt[6]{2}$

Try This. Write $\sqrt[3]{5} \cdot \sqrt{2}$ using a single radical symbol.   ■

In Example 10(c) we found that the square root of a cube root is a sixth root. In general, an *m*th root of an *n*th root is an *mn*th root.

Theorem: *m*th Root of an *n*th Root

> If *m* and *n* are positive integers for which all of the following roots are real, then
> $$\sqrt[m]{\sqrt[n]{a}} = \sqrt[mn]{a}.$$

For Thought

True or False? Explain. Do Not Use a Calculator.

1. $8^{-1/3} = -2$

2. $16^{1/4} = 4^{1/2}$

3. $\sqrt{\dfrac{4}{6}} = \dfrac{2}{3}$

4. $(\sqrt{3})^3 = 3\sqrt{3}$

5. $(-1)^{2/2} = -1$

6. $\sqrt[3]{7^2} = 7^{3/2}$

7. $9^{1/2} = \sqrt{3}$

8. $\dfrac{1}{\sqrt{3}} = \dfrac{\sqrt{3}}{3}$

9. $\dfrac{2^{1/2}}{2^{1/3}} = \sqrt[6]{2}$

10. $\sqrt[3]{7^5} = 7\sqrt[3]{49}$

P.3 Exercises

Use the procedure for evaluating $a^{-m/n}$ on page 27 to evaluate each expression. Use a calculator to check.

1. $-9^{1/2}$ **2.** $27^{1/3}$ **3.** $64^{1/2}$ **4.** $-144^{1/2}$

5. $(-64)^{1/3}$ **6.** $81^{1/4}$ **7.** $(-27)^{4/3}$ **8.** $125^{-2/3}$

9. $8^{-4/3}$ **10.** $4^{-3/2}$ **11.** $\left(\dfrac{1}{4}\right)^{1/2}$ **12.** $\left(\dfrac{1}{32}\right)^{1/5}$

13. $\left(\dfrac{4}{9}\right)^{3/2}$ **14.** $\left(-\dfrac{8}{27}\right)^{2/3}$

Simplify each expression. Use absolute value when necessary.

15. $(x^6)^{1/6}$ **16.** $(x^{10})^{1/5}$

17. $(a^{15})^{1/5}$ **18.** $(y^2)^{1/2}$

19. $(a^8)^{1/4}$ **20.** $(z^{12})^{1/4}$

21. $(x^3y^6)^{1/3}$ **22.** $(16x^4y^8)^{1/4}$

Simplify each expression. Assume that all variables represent positive real numbers. Write your answers without negative exponents.

23. $y^{2/3} \cdot y^{7/3}$ **24.** $a^{3/5} \cdot a^{7/5}$

25. $(x^4y)^{1/2}$ **26.** $(a^{1/2}b^{1/3})^2$

27. $(2a^{1/2})(3a)$ **28.** $(-3y^{1/3})(-2y^{1/2})$

29. $\dfrac{6a^{1/2}}{2a^{1/3}}$ **30.** $\dfrac{-4y}{2y^{2/3}}$

31. $(a^2 b^{1/2})(a^{1/3} b^{-1/2})$

32. $(4^{3/4} a^2 b^3)(4^{3/4} a^{-2} b^{-5})$

33. $\left(\dfrac{x^6 y^3}{z^9} \right)^{1/3}$

34. $\left(\dfrac{x^{1/2} y}{y^{1/2}} \right)^3$

Evaluate each radical expression. Use a calculator to check.

35. $\sqrt{900}$ **36.** $\sqrt{400}$ **37.** $\sqrt[3]{-8}$

38. $\sqrt[3]{64}$ **39.** $\sqrt[5]{-32}$ **40.** $\sqrt[6]{64}$

41. $\sqrt{\dfrac{4}{9}}$ **42.** $\sqrt{\dfrac{9}{16}}$ **43.** $\sqrt{0.01}$

44. $\sqrt{0.25}$ **45.** $\sqrt[3]{-\dfrac{8}{1000}}$ **46.** $\sqrt[4]{\dfrac{1}{625}}$

47. $\sqrt[4]{16^3}$ **48.** $\sqrt[3]{8^5}$

Write each expression involving rational exponents in radical notation and each expression involving radicals in exponential notation.

49. $10^{2/3}$ **50.** $-2^{3/4}$ **51.** $3y^{-3/5}$

52. $a(b^4 + 1)^{-1/2}$ **53.** $\dfrac{1}{\sqrt{x}}$ **54.** $-4\sqrt{x^3}$

55. $\sqrt[5]{x^3}$ **56.** $\sqrt[3]{x^3 + y^3}$

Simplify each radical expression. Assume that all variables represent positive real numbers. (

57. $\sqrt{16x^2}$ **58.** $\sqrt{121y^4}$

59. $\sqrt[3]{8y^9}$ **60.** $\sqrt[3]{125x^{18}}$

61. $\sqrt{\dfrac{xy}{100}}$ **62.** $\sqrt{\dfrac{t}{81}}$

63. $\sqrt[3]{\dfrac{-8a^3}{b^{15}}}$ **64.** $\sqrt[4]{\dfrac{16t^4}{y^8}}$

Write each radical expression in simplified form. Assume that all variables represent positive real numbers. (

65. $\sqrt{28}$ **66.** $\sqrt{50}$ **67.** $\dfrac{1}{\sqrt{5}}$ **68.** $\dfrac{7}{\sqrt{7}}$

69. $\sqrt{\dfrac{x}{8}}$ **70.** $\sqrt{\dfrac{3y}{20}}$ **71.** $\sqrt[3]{40}$ **72.** $\sqrt[3]{54}$

73. $\sqrt[3]{-250x^4}$ **74.** $\sqrt[3]{-24a^5}$ **75.** $\sqrt[3]{\dfrac{1}{2}}$ **76.** $\sqrt[3]{\dfrac{3x}{25}}$

Perform the indicated operations and simplify your answer. Assume that all variables represent positive real numbers. When possible use a calculator to verify your answer.

77. $\sqrt{8} + \sqrt{20} - \sqrt{12}$

78. $\sqrt{18} - \sqrt{50} + \sqrt{12} - \sqrt{75}$

79. $(-2\sqrt{3})(5\sqrt{6})$ **80.** $(-3\sqrt{2})(-2\sqrt{3})$

81. $(3\sqrt{5a})(4\sqrt{5a})$ **82.** $(-2\sqrt{6})(3\sqrt{6})$

83. $(-5\sqrt{3})^2$ **84.** $(3\sqrt{5})^2$

85. $\sqrt{18a} \div \sqrt{2a^4}$ **86.** $\sqrt{21x^7} \div \sqrt{3x^2}$

87. $5 \div \sqrt{x}$ **88.** $a \div \sqrt{b}$

89. $\sqrt{20x^3} + \sqrt{45x^3}$ **90.** $\sqrt[3]{16a^4} + \sqrt[3]{54a^4}$

Write each expression using a single radical sign. Assume that all variables represent positive real numbers. Simplify the radicand where possible.

91. $\sqrt[3]{3} \cdot \sqrt{2}$ **92.** $\sqrt{5} \cdot \sqrt[3]{4}$

93. $\sqrt[3]{3} \cdot \sqrt[4]{x}$ **94.** $\sqrt[3]{3} \cdot \sqrt[4]{4}$

95. $\sqrt{xy} \cdot \sqrt[3]{2xy}$ **96.** $\sqrt[3]{2a} \cdot \sqrt{2a}$

97. $\sqrt[3]{\sqrt{7}}$ **98.** $\sqrt[3]{\sqrt[3]{2a}}$

Solve each problem.

99. *Economic Order Quantity* Purchasing managers use the formula

$$E = \sqrt{\dfrac{2AS}{I}}$$

to determine the most economic order quantity E for parts used in production. A is the quantity that the plant will use in one year, S is the cost of setup for making the part, and I is the cost of holding one unit in stock for one year. Find E if $S = \$6000$, $A = 25$, and $I = \$140$.

100. *Piano Tuning* The note middle C on a piano is tuned so that the string vibrates at 262 cycles per second, or 262 Hz (Hertz). The C note that is one octave higher is tuned to 524 Hz. Tuning for the 11 notes in between using the method of *equal temperament* is $262 \cdot 2^{n/12}$, where n takes the values 1 through 11. Find the tuning rounded to the nearest whole Hertz for those 11 notes.

101. *Sail Area-Displacement Ratio*   The sail area-displacement ratio S is given by

$$S = \frac{16A}{d^{2/3}},$$

where A is the sail area (ft^2) and d is the displacement (lbs). S measures the amount of power available to drive a sailboat (Ted Brewer Yacht Design, www.tedbrewer.com). Ratios typically range from 15 to 25, with a high ratio indicating a powerful boat. Find S for the USS Constitution, which has a displacement of 2200 tons, a sail area of 42,700 ft^2, and 44 guns.

102. *A Less Powerful Boat*   Find S (from the previous exercise) for the Ted Hood 51. It has a sail area of 1302 ft^2, a displacement of 49,400 pounds, a length of 51 feet, and no guns.

103. *Depreciation Rate*   If the cost of an item is C and after n years its value is S, then the annual depreciation rate r is given by $r = 1 - (S/C)^{1/n}$. After a useful life of n years a computer with an original cost of $5000 has a salvage value of $200.
 a. Use the accompanying graph to estimate r if $n = 5$ and if $n = 10$.

 b. Use the formula to find r if $n = 5$ and if $n = 10$.

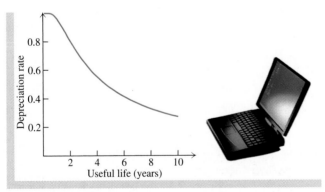

■ **Figure for Exercise 103**

104. *BMW Depreciation*   A new BMW Z3 convertible sells for $30,193 while a five-year-old model sells for $17,095 (Edmund's, www.edmunds.com). Use the formula from the previous exercise to find the annual depreciation rate.

105. *Longest Screwdriver*   A toolbox has length L, width W, and height H. The length D of the longest screwdriver that will fit inside the box is given by

$$D = (L^2 + W^2 + H^2)^{1/2}.$$

Find the length of the longest screwdriver that will fit in a 4 in. by 6 in. by 12 in. box.

106. *Changing Radius*   The radius of a sphere r is given in terms of its volume V by the formula

$$r = \left(\frac{0.75V}{\pi}\right)^{1/3}.$$

By how many inches has the radius of a spherical balloon increased when the amount of air in the balloon is increased from 4.2 ft^3 to 4.3 ft^3?

107. *Heron's Formula*   If the lengths of the sides of a triangle are a, b, and c, and $s = (a + b + c)/2$, then the area A is given by the formula

$$A = \sqrt{s(s - a)(s - b)(s - c)}.$$

Find the area of a triangle whose sides are 6 ft, 7 ft, and 11 ft (see figure for Exercises 107 and 108).

108. *Area of an Equilateral Triangle*   Use Heron's formula from the previous exercise to find a formula for the area of an equilateral triangle with sides of length a and simplify it.

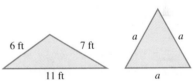

■ **Figure for Exercises 107 and 108**

For Writing/Discussion

109. *Roots or Powers*   Which one of the following expressions is not equivalent to the others? Explain in writing how you arrived at your decision.
 a. $\left(\sqrt[5]{t}\right)^4$ **b.** $\sqrt[4]{t^5}$ **c.** $\sqrt[5]{t^4}$
 d. $t^{4/5}$ **e.** $(t^{1/5})^4$

110. Which one of the following expressions is not equivalent to $\sqrt{a^2b^4}$? Write your reasoning in a paragraph.
 a. $|a| \cdot b^2$ **b.** $|ab^2|$ **c.** ab^2
 d. $(a^2b^4)^{1/2}$ **e.** $b^2\sqrt{a^2}$

111. *The Lost Rule?*   Is it true that the square root of a sum is equal to the sum of the square roots? Explain. Give examples.

112. *Technicalities*   If m and n are real numbers and $m^2 = n$, then m is a square root of n, but if $m^3 = n$ then m is *the* cube root of n. How do we know when to use "a" or "the"?

Thinking Outside the Box IV & V

Perfect Squares The square of an integer is called a *perfect square*. How many perfect squares divide evenly into the number 349,272,000?

Door Prizes One thousand tickets numbered 1 through 1000 are placed in a bowl. The tickets are drawn out for door prizes one at a time without replacement. How many tickets must be drawn from the bowl to be certain that one of the numbers on a selected ticket is twice as large as another selected number?

P.3 Pop Quiz

1. Evaluate $(-8)^{5/3}$.

2. Simplify $(8a^3b^{27})^{1/3}$.

3. Simplify $(x^{1/2}x^{1/3})^2$.

4. Evaluate $\sqrt[3]{1000}$.

5. Simplify $\sqrt{60}$.

6. Simplify $\sqrt{40} + \sqrt{90}$.

7. Simplify $\dfrac{6}{\sqrt{20}}$.

8. Write $\sqrt{3} \cdot \sqrt[3]{2}$ as single radical.

Linking Concepts

For Individual or Group Explorations

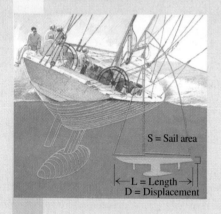

S = Sail area

←— L = Length —→
D = Displacement

Yachting Rules

The new International America's Cup Class rules, which took effect in 1989, establish the boundaries of a new class of longer, lighter, and faster yachts. In addition to satisfying 200 pages of other rules, the basic dimensions of any yacht competing for the silver trophy must satisfy the inequality

$$L + 1.25S^{1/2} - 9.8D^{1/3} \le 16.296$$

where L is the length in meters, S is the sail area in square meters, and D is the displacement in cubic meters (Americas Cup, www.americascup.org).

a) Determine whether The New Zealand Challenge (considered a short, light boat) with a length of 20.95 m, a sail area of 277.3 m², and a displacement of 17.56 m³ satisfies the inequality.

b) Verify that the published size of The Challenge Australia (considered a long, heavy boat) with a length of 21.87 m, a sail area of 311.78 m², and a displacement of 22.44 m³ does not satisfy the inequality.

c) Is the inequality satisfied if the displacement of The Challenge Australia is increased by 0.01 m³?

d) What does it mean to say that the displacement of a boat is 22.44 m³?

e) What do you think could be done to The Challenge Australia to increase its displacement?

Complex Numbers

Our system of numbers developed as the need arose. Numbers were first used for counting. As society advanced, the rational numbers were formed to express fractional parts and ratios. Negative numbers were invented to express losses or debts. When it was discovered that the exact size of some very real objects could not be expressed with rational numbers, the irrational numbers were added to the system, forming the set of real numbers. Later still, there was a need for another expansion to the number system. In this section we study that expansion, the set of complex numbers.

Definitions

In Section P.3 we saw that there are no even roots of negative numbers in the set of real numbers. So, the real numbers are inadequate or incomplete in this regard. The imaginary numbers were invented to complete the set of real numbers. Using real and imaginary numbers, every nonzero real number has *two* square roots, *three* cube roots, *four* fourth roots, and so on. (Actually finding all of the roots of any real number is done in trigonometry.)

The imaginary numbers are based on the symbol $\sqrt{-1}$. Since there is no real number whose square is -1, a new number called i is defined such that $i^2 = -1$.

Definition:
Imaginary Number *i*

> The **imaginary number** i is defined by
> $$i^2 = -1.$$
> We may also write $i = \sqrt{-1}$.

A complex number is formed as a real number plus a real multiple of i.

Definition:
Complex Numbers

> The set of **complex numbers** is the set of all numbers of the form $a + bi$, where a and b are real numbers.

In $a + bi$, a is called the **real part** and b is called the **imaginary part.** Two complex numbers $a + bi$ and $c + di$ are **equal** if and only if their real parts are equal ($a = c$) and their imaginary parts are equal ($b = d$). If $b = 0$, then $a + bi$ is a **real number.** If $b \neq 0$, then $a + bi$ is an imaginary number.

The form $a + bi$ is the **standard form** of a complex number, but for convenience we use a few variations of that form. If either the real or imaginary part of a complex number is 0, then that part is omitted. For example,

$$0 + 3i = 3i, \qquad 2 + 0i = 2, \qquad \text{and} \qquad 0 + 0i = 0.$$

If b is a radical, then i is usually written before b. For example, we write $2 + i\sqrt{3}$ rather than $2 + \sqrt{3}i$, which could be confused with $2 + \sqrt{3i}$. If b is negative, a subtraction symbol can be used to separate the real and imaginary parts as in $3 + (-2)i = 3 - 2i$. A complex number with fractions, such as $\frac{1}{3} - \frac{2}{3}i$, may be written as $\frac{1 - 2i}{3}$.

Example **1** Standard form of a complex number

Determine whether each complex number is real or imaginary and write it in the standard form $a + bi$.

a. $3i$ **b.** 87 **c.** $4 - 5i$ **d.** 0 **e.** $\dfrac{1 + \pi i}{2}$

Solution

a. The complex number $3i$ is imaginary, and $3i = 0 + 3i$.
b. The complex number 87 is a real number, and $87 = 87 + 0i$.
c. The complex number $4 - 5i$ is imaginary, and $4 - 5i = 4 + (-5)i$.
d. The complex number 0 is real, and $0 = 0 + 0i$.
e. The complex number $\dfrac{1 + \pi i}{2}$ is imaginary, and $\dfrac{1 + \pi i}{2} = \dfrac{1}{2} + \dfrac{\pi}{2}i$.

Try This. Determine whether $i - 5$ is real or imaginary and write it in standard form. ■

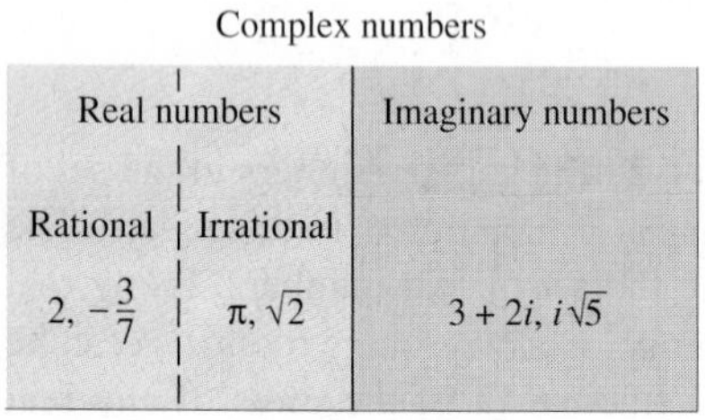

The real numbers can be classified as rational or irrational. The complex numbers can be classified as real or imaginary. The relationship between these sets of numbers is shown in Fig. P.23.

Complex numbers

Real numbers		Imaginary numbers
Rational	Irrational	
$2, -\frac{3}{7}$	$\pi, \sqrt{2}$	$3 + 2i, i\sqrt{5}$

■ **Figure P.23**

Addition, Subtraction, and Multiplication

Now that we have defined complex numbers, we define the operations of arithmetic with them.

Definition: Addition, Subtraction, and Multiplication

If $a + bi$ and $c + di$ are complex numbers, we define their sum, difference, and product as follows.

$$(a + bi) + (c + di) = (a + c) + (b + d)i$$

$$(a + bi) - (c + di) = (a - c) + (b - d)i$$

$$(a + bi)(c + di) = (ac - bd) + (bc + ad)i$$

It is not necessary to memorize these definitions, because the results can be obtained by performing the operations as if the complex numbers were binomials with i being a variable, replacing i^2 with -1 wherever it occurs.

Example **2** Operations with complex numbers

Perform the indicated operations with the complex numbers.

a. $(-2 + 3i) + (-4 - 9i)$ **b.** $(-1 - 5i) - (3 - 2i)$ **c.** $2i(3 + i)$
d. $(3i)^2$ **e.** $(-3i)^2$ **f.** $(5 - 2i)(5 + 2i)$

Solution

a. $(-2 + 3i) + (-4 - 9i) = -6 - 6i$
b. $(-1 - 5i) - (3 - 2i) = -1 - 5i - 3 + 2i = -4 - 3i$
c. $2i(3 + i) = 6i + 2i^2 = 6i + 2(-1) = -2 + 6i$
d. $(3i)^2 = 3^2 i^2 = 9(-1) = -9$

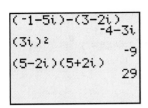

■ **Figure P.24**

e. $(-3i)^2 = (-3)^2 i^2 = 9(-1) = -9$

f. $(5 - 2i)(5 + 2i) = 25 - 4i^2 = 25 - 4(-1) = 29$

◰ Check these results with a calculator that handles complex numbers, as in Fig. P.24.

Try This. Find the product $(4 - 3i)(1 + 2i)$. ■

We can find whole-number powers of i by using the definition of multiplication. Since $i^1 = i$ and $i^2 = -1$, we have

$$i^3 = i^1 \cdot i^2 = i(-1) = -i \qquad \text{and} \qquad i^4 = i^1 \cdot i^3 = i(-i) = -i^2 = 1.$$

The first eight powers of i are listed here.

$$i^1 = i \qquad i^2 = -1 \qquad i^3 = -i \qquad i^4 = 1$$
$$i^5 = i \qquad i^6 = -1 \qquad i^7 = -i \qquad i^8 = 1$$

This list could be continued in this pattern, but any other whole-number power of i can be obtained from knowing the first four powers. We can simplify a power of i by using the fact that $i^4 = 1$ and $(i^4)^n = 1$ for any integer n.

Example **3** **Simplifying a power of** *i*

Simplify.

a. i^{83} **b.** i^{-46}

Solution

a. Divide 83 by 4 and write $83 = 4 \cdot 20 + 3$. So

$$i^{83} = (i^4)^{20} \cdot i^3 = 1^{20} \cdot i^3 = 1 \cdot i^3 = -i.$$

b. Since $-46 = 4(-12) + 2$, we have

$$i^{-46} = (i^4)^{-12} \cdot i^2 = 1^{-12} \cdot i^2 = 1(-1) = -1.$$

Try This. Simplify i^{35}. ■

Division of Complex Numbers

The complex numbers $a + bi$ and $a - bi$ are called **complex conjugates** of each other.

Example **4** **Complex conjugates**

Find the product of the given complex number and its conjugate.

a. $3 - i$ **b.** $4 + 2i$ **c.** $-i$

Solution

a. The conjugate of $3 - i$ is $3 + i$, and $(3 - i)(3 + i) = 9 - i^2 = 10$.

b. The conjugate of $4 + 2i$ is $4 - 2i$, and $(4 + 2i)(4 - 2i) = 16 - 4i^2 = 20$.

c. The conjugate of $-i$ is i, and $-i \cdot i = -i^2 = 1$.

Try This. Find the product of $3 - 5i$ and its conjugate. ■

In general we have the following theorem about complex conjugates.

Theorem:
Complex Conjugates

> If a and b are real numbers, then the product of $a + bi$ and its conjugate $a - bi$ is the real number $a^2 + b^2$. In symbols,
>
> $$(a + bi)(a - bi) = a^2 + b^2.$$

We use the theorem about complex conjugates to divide imaginary numbers, in a process that is similar to rationalizing a denominator.

Example **5** **Dividing imaginary numbers**

Write each quotient in the form $a + bi$.

a. $\dfrac{8 - i}{2 + i}$ **b.** $\dfrac{1}{5 - 4i}$ **c.** $\dfrac{3 - 2i}{i}$

Solution

a. Multiply the numerator and denominator by $2 - i$, the conjugate of $2 + i$:

$$\frac{8 - i}{2 + i} = \frac{(8 - i)(2 - i)}{(2 + i)(2 - i)} = \frac{16 - 10i + i^2}{4 - i^2} = \frac{15 - 10i}{5} = 3 - 2i$$

Check division using multiplication: $(3 - 2i)(2 + i) = 8 - i$.

b. $\dfrac{1}{5 - 4i} = \dfrac{1(5 + 4i)}{(5 - 4i)(5 + 4i)} = \dfrac{5 + 4i}{25 + 16} = \dfrac{5 + 4i}{41} = \dfrac{5}{41} + \dfrac{4}{41}i$

Check: $\left(\dfrac{5}{41} + \dfrac{4}{41}i\right)(5 - 4i) = \dfrac{25}{41} + \dfrac{20}{41}i - \dfrac{20}{41}i - \dfrac{16}{41}i^2$

$$= \frac{25}{41} + \frac{16}{41} = 1.$$

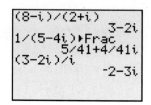

You can also check with a calculator that handles complex numbers, as in Fig. P.25. □

■ **Figure P.25**

c. $\dfrac{3 - 2i}{i} = \dfrac{(3 - 2i)(-i)}{i(-i)} = \dfrac{-3i + 2i^2}{-i^2} = \dfrac{-2 - 3i}{1} = -2 - 3i$

Check: $(-2 - 3i)(i) = -3i^2 - 2i = 3 - 2i.$

Try This. Write $\dfrac{4}{1 + i}$ in the form $a + bi$.   ■

Roots of Negative Numbers

In Examples 2(d) and 2(e), we saw that both $(3i)^2 = -9$ and $(-3i)^2 = -9$. This means that in the complex number system there are two square roots of -9, $3i$ and $-3i$. For any positive real number b, we have $(i\sqrt{b})^2 = -b$ and $(-i\sqrt{b})^2 = -b$. So there are two square roots of $-b$, $i\sqrt{b}$ and $-i\sqrt{b}$. We call $i\sqrt{b}$ the **principal square root** of $-b$ and make the following definition.

Definition: Square Root of a Negative Number

> For any positive real number b, $\sqrt{-b} = i\sqrt{b}$.

In the real number system, $\sqrt{-2}$ and $\sqrt{-8}$ are undefined, but in the complex number system they are defined as $\sqrt{-2} = i\sqrt{2}$ and $\sqrt{-8} = i\sqrt{8}$. Even though we now have meaning for a symbol such as $\sqrt{-2}$, *all operations with complex numbers must be performed after converting to the $a + bi$ form*. If we perform operations with roots of negative numbers using properties of the real numbers, we can get contradictory results:

$$\sqrt{-2} \cdot \sqrt{-8} = \sqrt{(-2)(-8)} = \sqrt{16} = 4 \quad \text{Incorrect.}$$

$$i\sqrt{2} \cdot i\sqrt{8} = i^2 \cdot \sqrt{16} = -4 \quad \text{Correct.}$$

The product rule $\sqrt{a} \cdot \sqrt{b} = \sqrt{ab}$ is used *only* for nonnegative numbers a and b.

Example 6 Square roots of negative numbers

Write each expression in the form $a + bi$, where a and b are real numbers.

a. $\sqrt{-8} + \sqrt{-18}$ **b.** $\dfrac{-4 + \sqrt{-50}}{4}$ **c.** $\sqrt{-27}\left(\sqrt{9} - \sqrt{-2}\right)$

Solution

The first step in each case is to replace the square roots of negative numbers by expressions with i.

a. $\sqrt{-8} + \sqrt{-18} = i\sqrt{8} + i\sqrt{18} = 2i\sqrt{2} + 3i\sqrt{2}$
$$= 5i\sqrt{2}$$

b. $\dfrac{-4 + \sqrt{-50}}{4} = \dfrac{-4 + i\sqrt{50}}{4} = \dfrac{-4 + 5i\sqrt{2}}{4}$
$$= -1 + \frac{5}{4}i\sqrt{2}$$

c. $\sqrt{-27}\left(\sqrt{9} - \sqrt{-2}\right) = 3i\sqrt{3}\left(3 - i\sqrt{2}\right) = 9i\sqrt{3} - 3i^2\sqrt{6}$
$$= 3\sqrt{6} + 9i\sqrt{3}$$

Try This. Write $\dfrac{2 - \sqrt{-12}}{2}$ in the form $a + bi$. ■

For Thought

True or False? Explain.

1. The multiplicative inverse of i is $-i$.

2. The conjugate of i is $-i$.

3. The set of complex numbers is a subset of the set of real numbers.

4. $\left(\sqrt{3} - i\sqrt{2}\right)\left(\sqrt{3} + i\sqrt{2}\right) = 5$

5. $(2 + 5i)(2 + 5i) = 4 + 25$

6. $5 - \sqrt{-9} = 5 - 9i$

7. $(3i)^2 + 9 = 0$

8. $(-3i)^2 + 9 = 0$

9. $i^4 = 1$

10. $i^{18} = 1$

P.4 Exercises

Determine whether each complex number is real or imaginary and write it in the standard form a + bi.

1. $6i$

2. $-3i + \sqrt{6}$

3. $\dfrac{1 + i}{3}$

4. -72

5. $\sqrt{7}$

6. $-i\sqrt{5}$

7. $\dfrac{\pi}{2}$

8. 0

Perform the indicated operations and write your answers in the form a + bi, where a and b are real numbers.

9. $(3 - 3i) + (4 + 5i)$

10. $(-3 + 2i) + (5 - 6i)$

11. $(1 - i) - (3 + 2i)$

12. $(6 - 7i) - (3 - 4i)$

13. $(1 - i\sqrt{2}) + (3 + 2i\sqrt{2})$

14. $(5 + 3i\sqrt{5}) + (-4 - 5i\sqrt{5})$

15. $\left(5 + \dfrac{1}{3}i\right) - \left(\dfrac{1}{2} - \dfrac{1}{2}i\right)$

16. $\left(\dfrac{1}{2} - \dfrac{2}{3}i\right) - \left(3 - \dfrac{1}{4}i\right)$

17. $-6i(3 - 2i)$

18. $-3i(5 + 2i)$

19. $(2 - 3i)(4 + 6i)$

20. $(3 - i)(5 - 2i)$

21. $(4 - 5i)(6 + 2i)$

22. $(3 + 7i)(2 + 5i)$

23. $(5 - 2i)(5 + 2i)$

24. $(4 + 3i)(4 - 3i)$

25. $(\sqrt{3} - i)(\sqrt{3} + i)$

26. $(\sqrt{2} + i\sqrt{3})(\sqrt{2} - i\sqrt{3})$

27. $(3 + 4i)^2$

28. $(-6 - 2i)^2$

29. $(\sqrt{5} - 2i)^2$

30. $(\sqrt{6} + i\sqrt{3})^2$

31. i^{17}

32. i^{24}

33. i^{98}

34. i^{19}

35. i^{-1}

36. i^{-2}

37. i^{-3}

38. i^{-4}

39. i^{-13}

40. i^{-27}

41. i^{-38}

42. i^{-66}

Find the product of the given complex number and its conjugate.

43. $3 - 9i$

44. $4 + 3i$

45. $\dfrac{1}{2} + 2i$

46. $\dfrac{1}{3} - i$

47. i

48. $-i\sqrt{5}$

49. $3 - i\sqrt{3}$

50. $\dfrac{5}{2} + i\dfrac{\sqrt{2}}{2}$

Write each quotient in the form a + bi.

51. $\dfrac{1}{2 - i}$

52. $\dfrac{1}{5 + 2i}$

53. $\dfrac{-3i}{1 - i}$

54. $\dfrac{3i}{-2 + i}$

55. $\dfrac{-3 + 3i}{i}$

56. $\dfrac{-2 - 4i}{-i}$

57. $\dfrac{1 - i}{3 + 2i}$

58. $\dfrac{4 + 2i}{2 - 3i}$

59. $\dfrac{2 - i}{3 + 5i}$

60. $\dfrac{4 + 2i}{5 - 3i}$

Write each expression in the form a + bi, where a and b are real numbers.

61. $\sqrt{-4} - \sqrt{-9}$

62. $\sqrt{-16} + \sqrt{-25}$

63. $\sqrt{-4} - \sqrt{16}$

64. $\sqrt{-3} \cdot \sqrt{-3}$

65. $(\sqrt{-6})^2$

66. $(\sqrt{-5})^3$

67. $\sqrt{-2} \cdot \sqrt{-50}$

68. $\dfrac{-6 + \sqrt{-3}}{3}$

69. $\dfrac{-2 + \sqrt{-20}}{2}$

70. $\dfrac{9 - \sqrt{-18}}{-6}$

71. $-3 + \sqrt{3^2 - 4(1)(5)}$

72. $1 - \sqrt{(-1)^2 - 4(1)(1)}$

73. $\sqrt{-8}(\sqrt{-2} + \sqrt{8})$

74. $\sqrt{-6}(\sqrt{2} - \sqrt{-3})$

Evaluate the expression $\dfrac{-b + \sqrt{b^2 - 4ac}}{2a}$ for each choice of a, b, and c.

75. $a = 1, b = 2, c = 5$

76. $a = 5, b = -4, c = 1$

77. $a = 2, b = 4, c = 3$

78. $a = 2, b = -4, c = 5$

Evaluate the expression $\dfrac{-b - \sqrt{b^2 - 4ac}}{2a}$ for each choice of a, b, and c.

79. $a = 1, b = 6, c = 17$

80. $a = 1, b = -12, c = 84$

81. $a = -2, b = 6, c = 6$

82. $a = 3, b = 6, c = 8$

Perform the indicated operations. Write the answers in the form a + bi where a and b are real numbers.

83. $(3 - 5i)(3 + 5i)$

84. $(2 - 4i)(2 + 4i)$

85. $(3 - 5i) + (3 + 5i)$

86. $(2 - 4i) + (2 + 4i)$

87. $\dfrac{3 - 5i}{3 + 5i}$

88. $\dfrac{2 - 4i}{2 + 4i}$

89. $(6 - 2i) - (7 - 3i)$

90. $(5 - 6i) - (8 - 9i)$

91. $i^5(i^2 - 3i)$

92. $3i^7(i - 5i^3)$

For Writing/Discussion

93. Explain in detail how to find i^n for any positive integer n.

94. Find a number $a + bi$ such that $a^2 + b^2$ is irrational.

95. Let $w = a + bi$ and $\overline{w} = a - bi$, where a and b are real numbers. Show that $w + \overline{w}$ is real and that $w - \overline{w}$ is imaginary. Write sentences (containing no mathematical symbols) stating these results.

96. Is it true that the product of a complex number and its conjugate is a real number? Explain.

97. Prove that the reciprocal of $a + bi$, where a and b are not both zero, is $\dfrac{a}{a^2 + b^2} - \dfrac{b}{a^2 + b^2}i$.

98. *Cooperative Learning* Work in a small group to find the two square roots of 1 and two square roots of -1 in the complex number system. How many fourth roots of 1 are there in the complex number system and what are they? Explain how to find all of the fourth roots in the complex number system for any positive real number.

99. Evaluate $i^{0!} + i^{1!} + i^{2!} + \cdots + i^{100!}$, where $n!$ (read "n factorial") is the product of the integers from 1 through n if $n \geq 1$ and $0! = 1$.

Thinking Outside the Box VI & VII

Reversing the Digits Find a four-digit integer x such that $4x$ is another four-digit integer whose digits are in the reverse order of the digits of x.

Summing Reciprocals There is only one way to write 1 as a sum of the reciprocals of three different positive integers:

$$\frac{1}{2} + \frac{1}{3} + \frac{1}{6} = 1$$

Find all possible ways to write 1 as a sum of the reciprocals of four different positive integers.

P.4 Pop Quiz

1. Find the sum of $3 + 2i$ and $4 - i$.

2. Find the product of $4 - 3i$ and $2 + i$.

3. Find the product of $2 - 3i$ and its conjugate.

4. Write $\dfrac{5}{2 - 3i}$ in the form $a + bi$.

5. Find i^{27}.

6. What are the two square roots of -16?

P.5

Polynomials

In this section we will review some basic facts about polynomials.

Definitions

A term was defined in Section P.1 as the product of a number and one or more variables raised to powers. A **polynomial** is simply a single term or a finite sum of terms in which the powers of the variables are whole numbers. A polynomial in one variable is defined as follows.

Definition: Polynomial in One Variable x

If n is a nonnegative integer and $a_0, a_1, a_2, \ldots, a_n$ are real numbers, then

$$a_n x^n + a_{n-1} x^{n-1} + a_{n-2} x^{n-2} + \cdots + a_1 x + a_0$$

is a **polynomial** in one variable x.

In algebra a single number is often referred to as a **constant.** The last term a_0 is called the **constant term.**

Polynomials with one, two, and three terms are called **monomials, binomials,** and **trinomials,** respectively. We usually write the terms of a polynomial in a single variable so that the exponents are in descending order from left to right. When a polynomial is written in this manner, the coefficient of the first term is the **leading coefficient.** The **degree** of a polynomial in one variable is the highest power of the variable in the polynomial. A constant such as 5 is a monomial with zero degree because $5 = 5x^0$. First-, second-, and third-degree polynomials are called **linear, quadratic,** and **cubic polynomials,** respectively.

Example 1 Using the definitions

Find the degree and leading coefficient of each polynomial and determine whether the polynomial is a monomial, binomial, or trinomial.

a. $\dfrac{x^3}{2} - \dfrac{1}{8}$ **b.** $5y^2 + y - 9$ **c.** $3w$

Solution

Polynomial (a) is a third-degree binomial, (b) is a second-degree trinomial, and (c) is a first-degree monomial. The leading coefficients are $1/2$, 5, and 3, respectively. We can also describe (a) as a cubic polynomial, (b) as a quadratic polynomial, and (c) as a linear polynomial.

Try This. Find the degree and leading coefficient of $-x^3 + 6x^2$.   ■

Naming and Evaluating Polynomials

Polynomials are often used to model quantities such as profit, revenue, and cost. A profit polynomial might be named P. For example, if the expression $3x - 10$ is a profit polynomial, we write $P = 3x - 10$ or $P(x) = 3x - 10$. P and $P(x)$ (read "P of x") both represent the profit when x units are sold. If $x = 6$, then the value of the polynomial is $3 \cdot 6 - 10$ or 8. If $x = 7$ the value is 11. Using the $P(x)$ notation we write $P(6) = 8$ and $P(7) = 11$. With the $P(x)$ notation it is clear that the profit for 6 units is 8 and the profit for 7 units is 11. The $P(x)$ notation is called **function notation.** Functions are discussed in great detail in Chapter 2.

Example 2 Evaluating a polynomial

Let $P(x) = x^2 - 5$ and $C(x) = -x^3 + 5x - 3$. Find the following.

a. $P(3)$ **b.** $P(-50)$ **c.** $C(10)$

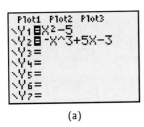

(a)

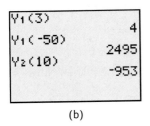

(b)

■ **Figure P.26**

Solution

a. $P(3) = 3^2 - 5 = 4$

b. $P(-50) = (-50)^2 - 5 = 2500 - 5 = 2495$

c. $C(10) = -10^3 + 5(10) - 3 = -1000 + 50 - 3 = -953$

To evaluate the polynomials in Example 2 on a calculator, define the polynomials with the $Y =$ key as in Fig. P.26(a), then use the y-variables to indicate which polynomials are to be evaluated as in Fig. P.26(b).

Try This. Find $P(-2)$ if $P(x) = -x^2 - 4x + 9$. ■

We will mainly study polynomials in one variable, but you will also see polynomials in more than one variable. The degree of a term in more than one variable is the sum of the powers of the variables. For example, the degree of $5x^3y^4z$ is 8. The degree of a polynomial in more than one variable is equal to the highest degree of any of its terms. For example, the degree of $3x^2y^2 + x^2y - 5y$ is 4.

Addition and Subtraction of Polynomials

Since the coefficients of a polynomial are real numbers and the variables represent real numbers, the properties of the real numbers are valid for polynomials. We add or subtract polynomials by adding or subtracting the like terms. The results are the same whether you arrange your work horizontally, as in Example 3, or vertically, as in Example 4.

Example **3** Adding and subtracting polynomials horizontally

Find each sum or difference.

a. $(3x^3 - x + 5) + (-8x^3 + 3x - 9)$ **b.** $(x^2 - 5x) - (3x^2 - 4x - 1)$

Solution

a. We use the commutative and associative properties of addition to rearrange the terms.

$$(3x^3 - x + 5) + (-8x^3 + 3x - 9) = (3x^3 - 8x^3) + (-x + 3x) + (5 - 9)$$
$$= -5x^3 + 2x - 4 \quad \text{Combine like terms.}$$

b. The first step is to remove the parentheses and change the sign of every term in the second polynomial.

$$(x^2 - 5x) - (3x^2 - 4x - 1) = x^2 - 5x - 3x^2 + 4x + 1 \quad \text{Distributive property}$$
$$= -2x^2 - x + 1 \quad \text{Combine like terms.}$$

Try This. Find the difference $(-x^2 + 3x) - (x^2 - 5x + 1)$. ■

Example **4** Adding and subtracting polynomials vertically

Find each sum or difference.

a. $(3x^3 - x + 5) + (-8x^3 + 3x - 9)$ **b.** $(x^2 - 5x) - (3x^2 - 4x - 1)$

Solution

a. Add:
$$\begin{array}{r} 3x^3 - x + 5 \\ -8x^3 + 3x - 9 \\ \hline -5x^3 + 2x - 4 \end{array}$$

b. Subtract:
$$\begin{array}{r} x^2 - 5x \\ 3x^2 - 4x - 1 \\ \hline -2x^2 - x + 1 \end{array}$$

Try This. Find the difference $(-3x^2 - x + 1) - (x^2 - 9)$ vertically.   ■

Multiplication of Polynomials

To multiply polynomials, we use the product rule and the distributive property as shown in the following example.

Example **5** Multiplying polynomials

Find each product.

a. $(3x^2)(4x^5)$ **b.** $-3x(2x - 3)$
c. $(3x + 1)(2x + 5)$ **d.** $(x^2 - 3x + 4)(2x - 3)$

Solution

a. $(3x^2)(4x^5) = 3 \cdot 4 \cdot x^2 \cdot x^5 = 12x^7$ Product rule for exponents

b. $-3x(2x - 3) = -6x^2 + 9x$ Distributive property

c. $(3x + 1)(2x + 5) = 3x(2x + 5) + 1(2x + 5)$ Distributive property
$$= 6x^2 + 15x + 2x + 5$$ Distributive property
$$= 6x^2 + 17x + 5$$ Combine like terms.

Note that we can also multiply polynomials vertically:

$$\begin{array}{r} 2x + 5 \\ 3x + 1 \\ \hline 2x + 5 \\ 6x^2 + 15x \\ \hline 6x^2 + 17x + 5 \end{array}$$
$1(2x + 5) = 2x + 5$
$3x(2x + 5) = 6x^2 + 15x$
Add.

d. $(x^2 - 3x + 4)(2x - 3) = x^2(2x - 3) - 3x(2x - 3) + 4(2x - 3)$
$$= 2x^3 - 3x^2 - 6x^2 + 9x + 8x - 12$$
$$= 2x^3 - 9x^2 + 17x - 12$$

Try This. Find the product $(x^2 - 2)(x^2 - 3)$.   ■

Using FOIL

The product of two binomials (as in Example 5c) results in four terms. Such products can be found quickly if we memorize where those four terms come from. The word FOIL is used as a memory aid. Consider the following product:

$$(a + b)(c + d) = a(c + d) + b(c + d) = \overset{F}{ac} + \overset{O}{ad} + \overset{I}{bc} + \overset{L}{bd}$$

The product of the two binomials consists of four terms:

the product of the *First* term of each (ac),

the product of the *Outer* terms (ad),

the product of the *Inner* terms (bc), and

the product of the *Last* term of each (bd).

Using FOIL to multiply two binomials is simply a way of speeding up the distributive property.

Example 6 Multiplying binomials using FOIL

Find each product using FOIL.

a. $(x + 4)(2x - 3)$ **b.** $(x^2 - 3)(2x - 3)$

c. $(a^2b + 5)(a^2b - 5)$ **d.** $(a + b)^2$

Solution

a. $(x + 4)(2x - 3) = \overset{F}{2x^2} - \overset{O}{3x} + \overset{I}{8x} - \overset{L}{12} = 2x^2 + 5x - 12$

b. $(x^2 - 3)(2x - 3) = 2x^3 - 3x^2 - 6x + 9$

c. $(a^2b + 5)(a^2b - 5) = a^4b^2 - 5a^2b + 5a^2b - 25 = a^4b^2 - 25$

d. $(a + b)^2 = (a + b)(a + b) = a^2 + ab + ab + b^2 = a^2 + 2ab + b^2$

Try This. Use FOIL to find the product $(x - 2)(x + 9)$.   ■

Example 7 Multiplying radicals using FOIL

Use FOIL to find the product $\left(\sqrt{2} - 2\sqrt{3}\right)\left(\sqrt{2} + 4\sqrt{3}\right)$.

Solution

$$\left(\sqrt{2} - 2\sqrt{3}\right)\left(\sqrt{2} + 4\sqrt{3}\right) = \sqrt{2}\sqrt{2} + 4\sqrt{6} - 2\sqrt{6} - 2\sqrt{3} \cdot 4\sqrt{3}$$
$$= 2 + 2\sqrt{6} - 24 \quad {\scriptstyle 2\sqrt{3} \cdot 4\sqrt{3} = 8 \cdot 3 = 24}$$
$$= -22 + 2\sqrt{6}$$

Try This. Use FOIL to find the product $\left(2\sqrt{3} - 1\right)\left(\sqrt{3} + 2\right)$.   ■

Special Products

The products found in Examples 6(c) and 6(d) are called **special products** because of the way that the result is simplified. We used FOIL to find those special products, but it is better to memorize the rules given here so that the products can be found quickly.

The Special Products

$$(a + b)^2 = a^2 + 2ab + b^2 \qquad \textbf{The square of a sum}$$
$$(a - b)^2 = a^2 - 2ab + b^2 \qquad \textbf{The square of a difference}$$
$$(a + b)(a - b) = a^2 - b^2 \qquad \textbf{The product of a sum and a difference}$$

It is helpful to memorize the verbal statements of the special products. For example, the first special product would be: "The square of a sum is equal to the square of the first term, plus twice the product of the first and last terms, plus the square of the last term." Or, for the third special product: "The product of a sum and a difference (of the same terms) is equal to the difference of their squares."

Example 8 Finding special products

Find each product by using the special product rules.

a. $(2x + 3)^2$ **b.** $(x^3 - 9)^2$ **c.** $(3x^m + 5)(3x^m - 5)$
d. $(3 - \sqrt{6})(3 + \sqrt{6})$

Solution

a. To find $(2x + 3)^2$ substitute $2x$ for a and 3 for b in $(a + b)^2 = a^2 + 2ab + b^2$:

$$(2x + 3)^2 = (2x)^2 + 2(2x)(3) + 3^2 = 4x^2 + 12x + 9$$

b. To find $(x^3 - 9)^2$ substitute x^3 for a and 9 for b in $(a - b)^2 = a^2 - 2ab + b^2$:

$$(x^3 - 9)^2 = (x^3)^2 - 2(x^3)(9) + 9^2 = x^6 - 18x^3 + 81$$

c. Substitute $3x^m$ for a and 5 for b in $(a + b)(a - b) = a^2 - b^2$:

$$(3x^m + 5)(3x^m - 5) = (3x^m)^2 - 5^2 = 9x^{2m} - 25$$

d. Substitute 3 for a and $\sqrt{6}$ for b in $(a - b)(a + b) = a^2 - b^2$:

$$(3 - \sqrt{6})(3 + \sqrt{6}) = 3^2 - (\sqrt{6})^2 = 9 - 6 = 3$$

Try This. Find $(2a - 5)^2$.   ■

In Example 8(d) the expressions $3 - \sqrt{6}$ and $3 + \sqrt{6}$ are called **conjugates.** Their product is a rational number. This fact is used to rationalize a denominator in the next example.

Example **9** **Using conjugates to rationalize a denominator**

Simplify the expression and check with a calculator: $\dfrac{3}{3 - \sqrt{6}}$

Solution

Multiply the numerator and denominator by $3 + \sqrt{6}$, the conjugate of $3 - \sqrt{6}$:

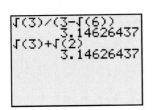

■ **Figure P.27**

$$\frac{\sqrt{3}}{3 - \sqrt{6}} = \frac{\sqrt{3}(3 + \sqrt{6})}{(3 - \sqrt{6})(3 + \sqrt{6})} = \frac{3\sqrt{3} + \sqrt{18}}{3}$$

$$= \frac{3\sqrt{3} + 3\sqrt{2}}{3} = \sqrt{3} + \sqrt{2}$$

Checking the approximate values for the original expression and the answer shown in Fig. P.27 gives us confidence that our exact answer is correct.

Try This. Simplify $\dfrac{2}{2 - \sqrt{2}}$.   ■

Division of Polynomials

For two given whole numbers a (the dividend) and d (the divisor, $d \neq 0$), there are unique whole numbers q (the quotient) and r (the remainder, $r < d$) such that $a = qd + r$. For example, if the dividend is 23 and the divisor is 7, then the unique quotient is 3, the unique remainder is 2, and $23 = 3 \cdot 7 + 2$. The same situation is true for polynomials.

Theorem:
Division of Polynomials

> If the **dividend** $P(x)$ and the **divisor** $D(x)$ are polynomials such that $D(x) \neq 0$, then there exist unique polynomials, the **quotient** $Q(x)$ and the **remainder** $R(x)$, such that
>
> $$P(x) = Q(x)D(x) + R(x),$$
>
> where $R(x) = 0$ or the degree of $R(x)$ is less than the degree of $D(x)$.

In words, we have

$$\text{dividend} = (\text{quotient})(\text{divisor}) + \text{remainder}.$$

We can also express the relationship between these quantities as

$$\frac{\text{dividend}}{\text{divisor}} = \text{quotient} + \frac{\text{remainder}}{\text{divisor}}.$$

If the dividend is 23 and the divisor is 7, then $\frac{23}{7} = 3 + \frac{2}{7}$.

To find the quotient when dividing by a monomial we use the quotient rule as illustrated in Example 10(a) and (b). If the divisor is a polynomial with more than one term, we use an algorithm that is similar to the algorithm for dividing whole numbers as shown in Example 10(c) and (d).

Example **10** Dividing polynomials

Find the quotient and remainder when the first polynomial is divided by the second.

a. $(8x^5) \div (2x^2)$ **b.** $(6x^3 - 9x^2 + 12x) \div (3x)$

c. $(x^2 - 3x - 10) \div (x - 5)$ **d.** $(x^3 - 8) \div (x - 2)$

Solution

a. $(8x^5) \div (2x^2) = \dfrac{8x^5}{2x^2} = 4x^{5-2} = 4x^3$

The quotient is $4x^3$ and the remainder is 0. Check: $4x^3 \cdot 2x^2 = 8x^5$.

b. $(6x^3 - 9x^2 + 12x) \div (3x) = \dfrac{6x^3}{3x} - \dfrac{9x^2}{3x} + \dfrac{12x}{3x} = 2x^2 - 3x + 4$

The quotient is $2x^2 - 3x + 4$ and the remainder is 0. Check as follows:

$$(2x^2 - 3x + 4)(3x) = 6x^3 - 9x^2 + 12x$$

c. Find the quotient and remainder for $(x^2 - 3x - 10) \div (x - 5)$ using the long division algorithm:

$$
\begin{array}{r}
x + 2 \\
x - 5 \overline{)\, x^2 - 3x - 9} \\
\underline{x^2 - 5x} \\
2x - 9 \\
\underline{2x - 10} \\
1
\end{array}
$$

  $x^2 \div x = x$

  $x(x - 5) = x^2 - 5x$

  Subtract: $-3x - (-5x) = 2x$. Bring down -9.

  $2x \div x = 2, 2(x - 5) = 2x - 10$

  Subtract: $-9 - (-10) = 1$.

The quotient is $x + 2$ and the remainder is 1. Check that $(x + 2)(x - 5) + 1 = x^2 - 3x - 9$. Note that we also have

$$\frac{x^2 - 3x - 9}{x - 5} = x + 2 + \frac{1}{x - 5}.$$

d. To keep the division organized, insert $0x^2$ and $0x$ for the missing x^2- and x-terms.

$$
\begin{array}{r}
x^2 + 2x + 4 \\
x - 2 \overline{)\, x^3 + 0x^2 + 0x - 8} \\
\underline{x^3 - 2x^2} \\
2x^2 + 0x \\
\underline{2x^2 - 4x} \\
4x - 8 \\
\underline{4x - 8} \\
0
\end{array}
$$

  $x^3 \div x = x^2$

  $x^2(x - 2) = x^3 - 2x^2$

  $0x^2 - (-2x^2) = 2x^2$

  $2x(x - 2) = 2x^2 - 4x$

  $4(x - 2) = 4x - 8$

The quotient is $x^2 + 2x + 4$ and the remainder is 0. Check:

$$(x - 2)(x^2 + 2x + 4) = x^3 - 8.$$

Try This. Find the quotient $(x^2 - 3x + 2) \div (x - 1)$. ■

For Thought

True or False? Explain. Do Not Use a Calculator.

1. The expression $x^{-2} + 3x^{-1} + 9$ is a trinomial.

2. The degree of the polynomial $x^3 + 5x^2 - 6x^2y^2$ is 3.

3. $(3 + 5)^2 = 3^2 + 5^2$

4. $(50 + 1)(50 - 1) = 2499$

5. If the side of a square is $x + 3$ ft, then $x^2 + 9$ ft^2 is its area.

6. $(2x - 1) - (3x + 5) = -x + 4$ for any real number x.

7. $(a + b)^3 = a^3 + b^3$ for any real numbers a and b.

8. The dividend times the quotient plus the remainder equals the divisor.

9. $\dfrac{x^2 + 5x + 7}{x + 2} = x + 3 + \dfrac{1}{x + 2}$ for any real number x.

10. If $P(x) = 3x^2 + 7$, $Q(x) = 5x^2 - 9$, and $S(x) = 8x^2 - 2$, then $P(17) + Q(17) = S(17)$.

P.5 Exercises

Find the degree and leading coefficient of each polynomial. Determine whether the polynomial is a monomial, binomial, or trinomial.

1. $x^3 - 4x^2 + \sqrt{5}$ **2.** $-x^7 - 6x^4$ **3.** $x - 3x^2$

4. $x + 5 + x^2$ **5.** 79 **6.** $-\dfrac{x}{\sqrt{2}}$

Let $P(x) = x^2 - 3x + 2$ and $M(x) = -x^3 + 5x^2 - x + 2$. Find the following.

7. $P(-2)$ **8.** $P(1)$ **9.** $M(-3)$ **10.** $M(50)$

Find each sum or difference.

11. $(3x^2 - 4x) + (5x^2 + 7x - 1)$

12. $(-3x^2 - 4x + 2) + (5x^2 - 8x - 7)$

13. $(4x^2 - 3x) - (9x^2 - 4x + 3)$

14. $(x^2 + 2x + 4) - (x^2 + 4x + 4)$

15. $(4ax^3 - a^2x) - (5a^2x^3 - 3a^2x + 3)$

16. $(x^2y^2 - 3xy + 2x) - (6x^2y^2 + 4y - 6x)$

Find the sum or difference.

17. Add: $\begin{aligned}3x - 4 \\ -x + 3\end{aligned}$ **18.** Add: $\begin{aligned}-2x^2 - 5 \\ 3x^2 - 6\end{aligned}$

19. Subtract: $\begin{aligned}x^2 \quad\;\; - 8 \\ -2x^2 + 3x - 2\end{aligned}$ **20.** Subtract: $\begin{aligned}-2x^2 - 5x + 9 \\ 4x^2 - 7x\end{aligned}$

Use the distributive property to find each product.

21. $-3a^3(6a^2 - 5a + 2)$ **22.** $-2m(m^2 - 3m + 9)$

23. $(3b^2 - 5b + 2)(b - 3)$ **24.** $(-w^2 - 5w + 6)(w + 5)$

25. $(2x - 1)(4x^2 + 2x + 1)$ **26.** $(3x - 2)(9x^2 + 6x + 4)$

27. $(x - 4)(z + 3)$ **28.** $(a - 3)(b + c)$

29. $(a - b)(a^2 + ab + b^2)$ **30.** $(a + b)(a^2 - ab + b^2)$

Find each product using FOIL.

31. $(a + 9)(a - 2)$ **32.** $(z - 3)(z - 4)$

33. $(2y - 3)(y + 9)$ **34.** $(2y - 1)(3y + 4)$

35. $(2x - 9)(2x + 9)$ **36.** $(4x - 6y)(4x + 6y)$

37. $(2x + 5)^2$ **38.** $(5x - 3)^2$

39. $(2x^2 + 4)(3x^2 + 5)$ **40.** $(3x^3 - 2)(5x^3 + 6)$

Find each product using FOIL.

41. $\left(1 + \sqrt{2}\right)\left(3 + \sqrt{2}\right)$ **42.** $\left(5 + \sqrt{6}\right)\left(2 + \sqrt{6}\right)$

43. $\left(5 + \sqrt{2}\right)\left(4 - 3\sqrt{2}\right)$ **44.** $\left(2 - 3\sqrt{7}\right)\left(5 - \sqrt{7}\right)$

45. $\left(3\sqrt{2} + \sqrt{3}\right)\left(2\sqrt{2} - \sqrt{3}\right)$

46. $\left(2\sqrt{6} - \sqrt{3}\right)\left(4\sqrt{6} - \sqrt{3}\right)$

47. $\left(\sqrt{5} + \sqrt{3}\right)^2$ **48.** $\left(\sqrt{8} - \sqrt{3}\right)^2$

Find each product using the special product rules.

49. $(3x + 5)^2$

50. $(x^3 - 2)^2$

51. $(x^n - 3)(x^n + 3)$

52. $(2z^b + 1)(2z^b - 1)$

53. $(\sqrt{2} - 5)(\sqrt{2} + 5)$

54. $(6 - \sqrt{3})(6 + \sqrt{3})$

55. $(3\sqrt{6} - 1)^2$

56. $(2\sqrt{5} + 2)^2$

57. $(3x^3 - 4)^2$

58. $(2x^2y^3 + 1)^2$

Simplify each expression by rationalizing the denominator.

59. $\dfrac{5}{1 - \sqrt{7}}$

60. $\dfrac{2}{3 + \sqrt{5}}$

61. $\dfrac{\sqrt{10}}{\sqrt{5} - 2}$

62. $\dfrac{\sqrt{3}}{\sqrt{2} - \sqrt{3}}$

63. $\dfrac{\sqrt{6}}{6 + \sqrt{3}}$

64. $\dfrac{\sqrt{2}}{\sqrt{8} + \sqrt{3}}$

65. $\dfrac{1 + \sqrt{2}}{2 - \sqrt{3}}$

66. $\dfrac{\sqrt{2} + 5}{3 + \sqrt{8}}$

67. $\dfrac{\sqrt{2} + \sqrt{3}}{\sqrt{6}}$

68. $\dfrac{2 + \sqrt{10}}{\sqrt{2}}$

Find each quotient.

69. $(36x^6) \div (-4x^3)$

70. $(-24y^{12}) \div (8y^3)$

71. $(3x^2 - 6x) \div (-3x)$

72. $(6x^3 - 9x^2) \div (3x^2)$

73. $(x^2 + 6x + 9) \div (x + 3)$

74. $(x^2 - 3x - 54) \div (x - 9)$

75. $(a^3 - 1) \div (a - 1)$

76. $(b^6 + 8) \div (b^2 + 2)$

Find the quotient and remainder when the first polynomial is divided by the second.

77. $x^2 + 3x + 3, \ x - 2$

78. $3x^2 - x + 4, \ x + 2$

79. $2x^2 - 5, \ x + 3$

80. $-4x^2 + 1, \ x - 1$

81. $x^3 - 2x^2 - 2x - 3, \ x - 3$

82. $x^3 + 3x^2 - 3x + 4, \ x + 4$

Use division to express each fraction in the form

$quotient + \dfrac{remainder}{divisor}.$

83. $\dfrac{x^2 - 2}{x - 1}$

84. $\dfrac{x^2 + 4x + 5}{x + 1}$

85. $\dfrac{2x^2 - 3x + 1}{x}$

86. $\dfrac{2x^2 + 1}{x^2}$

87. $\dfrac{x^2 + x + 1}{x}$

88. $\dfrac{-x^2 + x - 1}{x}$

89. $\dfrac{x^2 - 2x + 1}{x - 2}$

90. $\dfrac{x^2 - x - 9}{x - 3}$

Perform the indicated operations mentally. Write down only the answer.

91. $(x - 4)(x + 6)$

92. $(z^4 + 5)(z^4 - 4)$

93. $(2a^5 - 9)(a^5 + 3)$

94. $(3b^2 + 1)(2b^2 + 5)$

95. $(y - 3) - (2y + 6)$

96. $(a^2 + 3) - (a^2 - 6)$

97. $(7a - 9) + (3 - a)$

98. $(3m - 8) + (7 - m)$

99. $(w + 4)^2$

100. $(t - 2)^2$

101. $(4x - 9)(4x + 9)$

102. $(3x + 1)(3x - 1)$

103. $3y^2(y^3 - 3x)$

104. $a^4b(a^2b^2 + 1)$

105. $(6b^3 - 3b^2) \div (3b^2)$

106. $(3w - 8) \div (8 - 3w)$

107. $(x^3 + 8)(x^3 - 8)$

108. $(x^2 - 4)(x^2 + 4)$

109. $(3w^2 - 2n)^2$

110. $(7y^2 - 3x)^2$

111. $(\sqrt{5} - 2)(\sqrt{5} + 2)$

112. $(\sqrt{2} - \sqrt{3})(\sqrt{2} + \sqrt{3})$

Solve each problem.

113. *Area of a Rectangle* If the length of a field is $x + 3$ m and the width is $2x - 1$ m, then what trinomial represents the area of the field?

   HINT $A = LW$.

114. *Area of a Triangle* If the base of a triangular sign is $2x - 3$ cm and the height is $2x + 6$ cm, then what trinomial represents the area of the sign?

   HINT $A = \frac{1}{2}bh$.

115. *Volume of a Rectangular Box* If the volume of a box is $x^3 + 9x^2 + 26x + 24$ cm^3 and the height is $x + 4$ cm, then what polynomial represents the area of the bottom?

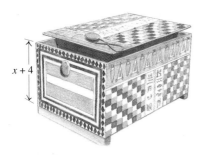

■ **Figure for Exercise 115**

116. *Area of a Triangle* If the area of a triangle is $x^2 - x - 6$ m^2 and the base is $x - 3$ m, then what polynomial represents the height?

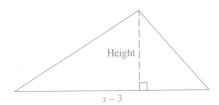

■ **Figure for Exercise 116**

117. *Available Habitat* A wild animal generally stays at least x mi from the edge of a forest. For a rectangular forest preserve that is 6 mi long and 4 mi wide, write a polynomial that represents the area of the available habitat for the wild animal.

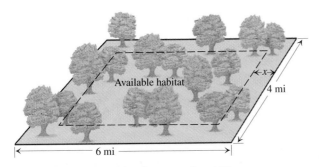

■ **Figure for Exercise 117**

118. *Red Fox* What percent of the forest in Exercise 117 is available habitat for the red fox if $x = 0.5$ mi for this wild animal?

119. *Comparing Investments* If an investment of $1 grows at an annual rate r, then the amount in dollars after three years is given by the polynomial

$$A(r) = r^3 + 3r^2 + 3r + 1.$$

From 1995 to 2005, Fidelity's Short-Term Bond Fund has averaged 5.69% annual growth while Fidelity's Equity-Income Fund has averaged 11.60% annual growth (Fidelity Investments, www.fidelity.com). Assume that these growth rates continue. Determine how much more an investment of $1 is worth in the higher yielding fund after three years

a. by estimating from the accompanying graph.

b. by using the polynomial.

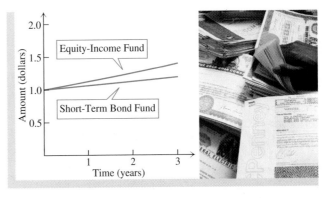

■ **Figure for Exercise 119**

120. *Four-Year Growth* If an investment of $1 grows at an annual rate r, then the amount in dollars after four years is given by the polynomial

$$A(r) = r^4 + 4r^3 + 6r^2 + 4r + 1.$$

From 1995 to 2005 Fidelity's Magellan®. Fund, the world's largest mutual fund, has averaged 10.06% annual growth (Fidelity Investments, www.fidelity.com). How much would $1 amount to in four years at 10.06%?

121. *Water Pollution* The polynomial $F(x) = 500 + 200x$ is used to determine the fine in dollars for violating water pollution standards, where x is the number of days that the violation persists after the citation. If a chemical plant persists in violating water pollution standards for 25 days after being cited, then how much is the fine?

122. *Profit for Computer Sales* The polynomial $P(n) = -2n^2 + 800n$ gives the profit in dollars on the sale of n computers in a single order. Find the profit for the sale of 200 computers.

For Writing/Discussion

123. *Area of a Rectangle* Wilson thought that his lot was in the shape of a square. After he had it surveyed he discovered that it was rectangular in shape with the length 10 ft longer than he thought and the width 10 ft shorter than he thought. Does he have more or less area than he thought? How much? Explain.

124. *Egyptian Area Formula* Surrounded by thousands of square miles of arid desert, ancient Egypt was sustained by the green strips of land bordering the Nile. Villagers cooperated to control the annual flooding so that all might reap abundant harvests from a network of fields and orchards divided by irrigation canals. The Egyptians used the expression

$$\frac{(a + b)(c + d)}{4}$$

to approximate the area of a four-sided field, where a and b represent the lengths of two opposite sides and c and d represent the lengths of the other two opposite sides. Is the formula correct if the field is rectangular? Is it correct for every four-sided figure? Explain why the Egyptians used this formula.

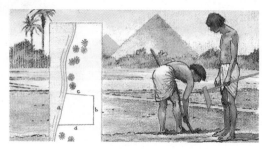

■ **Figure for Exercise 124**

125. *Cooperative Learning*   Each student in your small group should write a polynomial $P(x)$ of a different degree. Use a calculator to evaluate each polynomial for $x = 5000$ and $x = -5000$. Repeat this procedure until you can make a conjecture about the relationship between the sign of x, the sign of the leading coefficient, the degree of $P(x)$, and the value of $P(x)$.

Thinking Outside the Box VIII

Maximizing the Sum   Consider the following sum:

$$\frac{1}{1} + \frac{2}{2} + \frac{3}{3} + \frac{4}{4} + \frac{5}{5} + \cdots + \frac{2006}{2006}$$

You are allowed to rearrange the numerators of the fractions in any way that you choose, but keep the denominators as they are given. What arrangement would give the largest sum for the 2006 fractions?

P.5 Pop Quiz

1. What is the degree of $3x^2 - x^4$?

2. Find $P(-1)$ if $P(x) = 3x^2 - x^4$.

3. Find $(x - 2)(x^2 + 2x + 4)$.   **4.** Find $(a + 3)(2a - 1)$.

5. Simplify $\left(\sqrt{5} + 3\right)^2$.   **6.** Simplify $\dfrac{\sqrt{6}}{\sqrt{2} - 1}$.

7. What is the remainder when $x^2 + 7x + 14$ is divided by $x + 3$?

Linking Concepts

For Individual or Group Explorations

Maximum Sail Area

According to the International America's Cup Class rules, the polynomial

$$S(L) = \left(\frac{16.296 + 9.8\sqrt[3]{18} - L}{1.25}\right)^2$$

gives the maximum sail area S (in square meters) in terms of the length L (in meters) for a boat with a fixed displacement of 18 m^3 and lengths less than 40 m.

a)   What do you think is the purpose of rules such as this one?

b)   Find the maximum sail area (to the nearest hundredth of a square meter) for lengths of 19.82 m, 20.56 m, and 21.24 m.

c) Is the maximum sail area increasing or decreasing as the length increases?

d) How does your answer to part (c) relate to your answer to part (a)?

e) Write the polynomial in the form $S(L) = aL^2 + bL + c$ where a, b, and c are real numbers rounded to the nearest hundredth.

f) Repeat part (b) using the polynomial that you found in part (e) and compare your answers to those found in part (b).

Factoring Polynomials

In Section P.5 we studied multiplication of polynomials. In this section we will factor polynomials. **Factoring** "reverses" multiplication. By factoring, we can express a complicated polynomial as a product of several simpler expressions, which are often easier to study.

Factoring Out the Greatest Common Factor

To factor $6x^2 - 3x$, notice that $3x$ is a monomial that can be divided evenly into each term. We can use the distributive property to write

$$6x^2 - 3x = 3x(2x - 1).$$

We call this process **factoring out** $3x$. Both $3x$ and $2x - 1$ are **factors** of $6x^2 - 3x$. Since 3 is a factor of $6x^2$ and $3x$, 3 is a **common factor** of the terms of the polynomial. The **greatest common factor (GCF)** is a monomial that includes every number and variable that is a factor of all terms of the polynomial. The monomial $3x$ is the greatest common factor of $6x^2 - 3x$. Usually the common factor has a positive coefficient, but at times it is useful to factor out a common factor with a negative coefficient. For example, we could factor out $-3x$ from $6x^2 - 3x$:

$$6x^2 - 3x = -3x(-2x + 1)$$

Example **1** **Factoring out the greatest common factor**

Factor out the greatest common factor from each polynomial, first using the GCF with a positive coefficient and then using a negative coefficient.

a. $9x^4 - 6x^3 + 12x^2$ **b.** $x^2y + 10xy + 25y$ **c.** $-z - w$

Solution

a. $9x^4 - 6x^3 + 12x^2 = 3x^2(3x^2 - 2x + 4)$

$$= -3x^2(-3x^2 + 2x - 4)$$

b. $x^2y + 10xy + 25y = y(x^2 + 10x + 25)$

$$= -y(-x^2 - 10x - 25)$$

c. $-z - w = 1(-z - w)$

$$= -1(z + w)$$

Try This. Factor out the GCF in $-12a^3 + 8a^2 - 16a$.   ■

Factoring by Grouping

Some four-term polynomials can be factored by **grouping** the terms in pairs and factoring out a common factor from each pair of terms. We show this technique in the next example. As we will see later, grouping can also be used to factor trinomials.

Example **2** **Factoring four-term polynomials**

Factor each polynomial by grouping.

a. $x^3 + x^2 + 3x + 3$ **b.** $aw + bc - bw - ac$

Solution

a. Factor the common factor x^2 out of the first two terms and the common factor 3 out of the last two terms:

$$x^3 + x^2 + 3x + 3 = x^2(x + 1) + 3(x + 1) \quad \text{Factor out common factors.}$$

$$= (x + 1)(x^2 + 3) \quad \text{Factor out the common factor } (x + 1).$$

b. We must first arrange the polynomial so that the first group of two terms has a common factor and the last group of two terms also has a common factor:

$$aw + bc - bw - ac = aw - bw - ac + bc \quad \text{Rearrange.}$$

$$= w(a - b) - c(a - b) \quad \text{Factor out common factors.}$$

$$= (w - c)(a - b) \quad \text{Factor out } (a - b).$$

Try This. Factor $w^3 + w^2 - 3w - 3$.   ■

Factoring $ax^2 + bx + c$

Factoring by grouping was used in Example 2 to factor polynomials with four terms. Factoring by grouping can be used also to factor a trinomial that is the product of two binomials, because there are four terms in the product of two binomials (FOIL). If we examine the multiplication of two binomials using the distributive property, we can develop a strategy for factoring trinomials.

$$(x + 3)(x + 5) = (x + 3)x + (x + 3)5 \quad \text{Distributive property}$$

$$= x^2 + 3x + 5x + 15 \quad \text{Distributive property}$$

$$= x^2 + 8x + 15 \quad \text{Combine like terms.}$$

To factor $x^2 + 8x + 15$ we simply reverse the steps in this multiplication. First write $8x$ as $3x + 5x$, and then factor by grouping. We could write $8x$ as $x + 7x$, $2x + 6x$, and so on, but we choose $3x + 5x$ because 3 and 5 have a product of 15. So to factor $x^2 + 8x + 15$ proceed as follows:

$$x^2 + 8x + 15 = x^2 + 3x + 5x + 15 \qquad \text{Replace } 8x \text{ by } 3x + 5x.$$
$$= (x + 3)x + (x + 3)5 \qquad \text{Factor out common factors.}$$
$$= (x + 3)(x + 5) \qquad \text{Factor out } x + 3.$$

The key to factoring $ax^2 + bx + c$ with $a = 1$ is to find two numbers that have a product equal to c and a sum equal to b, then proceed as above. After you have practiced this procedure, you should skip the middle two steps shown here and just write the answer.

Example 3 Factoring $ax^2 + bx + c$ with $a = 1$

Factor each trinomial.

a. $x^2 - 5x - 14$ **b.** $x^2 + 4x - 21$ **c.** $x^2 - 5x + 6$

Solution

a. Two numbers that have a product of -14 and a sum of -5 are -7 and 2.

$$x^2 - 5x - 14 = x^2 - 7x + 2x - 14 \qquad \text{Replace } -5x \text{ by } -7x + 2x.$$
$$= (x - 7)x + (x - 7)2 \qquad \text{Factor out common factors.}$$
$$= (x - 7)(x + 2) \qquad \text{Factor out } (x - 7).$$

Check by using FOIL.

b. Two numbers that have a product of -21 and a sum of 4 are 7 and -3.

$$x^2 + 4x - 21 = (x + 7)(x - 3)$$

Check by using FOIL.

c. Two numbers that have a product of 6 and a sum of -5 are -2 and -3.

$$x^2 - 5x + 6 = (x - 2)(x - 3)$$

Check by using FOIL.

Try This. Factor $x^2 - 3x - 70$.   ▪

Before trying to factor $ax^2 + bx + c$ with $a \neq 1$, consider the product of two linear factors:

$$(Mx + N)(Px + Q) = MPx^2 + (MQ + NP)x + NQ$$

Observe that the product of MQ and NP (from the coefficient of x) is $MNPQ$. The product of MP (the leading coefficient) and NQ (the constant term) is also $MNPQ$. So if the trinomial $ax^2 + bx + c$ can be factored, there are two numbers that have a sum equal to b (the coefficient of x) and a product equal to ac (the product of the leading coefficient and the constant term). This fact is the key to the **ac-method** for factoring $ax^2 + bx + c$.

P R O C E D U R E	**The *ac*-Method for Factoring**

To factor $ax^2 + bx + c$ with $a \neq 1$:

1. Find two numbers whose sum is b and whose product is ac.
2. Replace b by the sum of these two numbers.
3. Factor the resulting four-term polynomial by grouping.

Example 4 Factoring $ax^2 + bx + c$ with $a \neq 1$

Factor each trinomial.

a. $2x^2 + 5x + 2$ **b.** $6x^2 - x - 12$ **c.** $15x^2 - 14x + 3$

Solution

a. Since $ac = 2 \cdot 2 = 4$ and $b = 5$, we need two numbers that have a product of 4 and a sum of 5. The numbers are 4 and 1.

$$2x^2 + 5x + 2 = 2x^2 + 4x + x + 2 \qquad \text{Replace } 5x \text{ by } 4x + x.$$
$$= (x + 2)2x + (x + 2)1 \qquad \text{Factor by grouping.}$$
$$= (x + 2)(2x + 1) \qquad \text{Check using FOIL.}$$

b. Since $ac = 6(-12) = -72$ and $b = -1$, we need two numbers that have a product of -72 and a sum of -1. The numbers are 8 and -9.

$$6x^2 - x - 12 = 6x^2 - 9x + 8x - 12 \qquad \text{Replace } -x \text{ by } -9x + 8x.$$
$$= (2x - 3)3x + (2x - 3)4 \qquad \text{Factor by grouping.}$$
$$= (2x - 3)(3x + 4) \qquad \text{Check using FOIL.}$$

c. Since $ac = 15 \cdot 3 = 45$ and $b = -14$, we need two numbers that have a product of 45 and a sum of -14. The numbers are -5 and -9.

$$15x^2 - 14x + 3 = 15x^2 - 5x - 9x + 3 \qquad \text{Replace } -14x \text{ by } -5x - 9x.$$
$$= (3x - 1)5x + (3x - 1)(-3) \qquad \text{Factor by grouping.}$$
$$= (3x - 1)(5x - 3) \qquad \text{Check using FOIL.}$$

Try This. Factor $12x^2 + 5x - 3$. ■

After factoring some polynomials by the grouping method of Example 3 and the *ac*-method of Example 4, you will find that you can often guess the factors without going through all of the steps in those methods. Guessing the factors is an acceptable method that is called the **trial-and-error** method. For this method you simply write down any factors that you think might work, use FOIL to check if they are correct, then try again if they are wrong.

Example 5 Factoring by trial and error

Use trial and error to factor $2x^2 + 11x - 6$.

Solution

Since $2x^2$ is $2x \cdot x$ and 6 is either $2 \cdot 3$ or $1 \cdot 6$, there are several possibilities to try for the factors:

$$(2x \quad 1)(x \quad 6) \qquad (2x \quad 2)(x \quad 3)$$
$$(2x \quad 6)(x \quad 1) \qquad (2x \quad 3)(x \quad 2)$$

The factors must have opposite signs to get the negative sign in the original polynomial. You can rule out a binomial that has a common factor such as $2x + 6$ because there is no common factor in the original trinomial. We now use FOIL to multiply some possible factors until we find the correct factors:

$$(2x + 1)(x - 6) = 2x^2 - 11x - 6 \quad \text{Error.}$$
$$(2x - 1)(x + 6) = 2x^2 + 11x - 6 \quad \text{Correct.}$$

So $2x^2 + 11x - 6$ is factored as $(2x - 1)(x + 6)$.

Try This. Factor $3x^2 - 20x - 7$.   ■

Factoring the Special Products

In Section P.5 we learned how to find the special products: the square of a sum, the square of a difference, and the product of a sum and a difference. The trinomial that results from squaring a sum or a difference is called a **perfect square trinomial.** We can write a factoring rule for each special product.

Factoring the Special Products

$a^2 - b^2 = (a + b)(a - b)$	**Difference of two squares**
$a^2 + 2ab + b^2 = (a + b)^2$	**Perfect square trinomial**
$a^2 - 2ab + b^2 = (a - b)^2$	**Perfect square trinomial**

Example **6** **Factoring special products**

Factor each polynomial.

a. $4x^2 - 1$ **b.** $x^2 - 6x + 9$ **c.** $9y^2 + 30y + 25$ **d.** $x^{2t} - 9$

Solution

a. $4x^2 - 1 = (2x)^2 - 1^2$ ⟶ Recognize the difference of two squares.

$\qquad\qquad = (2x + 1)(2x - 1)$

b. $x^2 - 6x + 9 = x^2 - 2(3x) + 3^2$ ⟶ Recognize the perfect square trinomial.

$\qquad\qquad = (x - 3)^2$

c. $9y^2 + 30y + 25 = (3y)^2 + 2(3y)(5) + 5^2$ ⟶ Recognize the perfect square trinomial.

$\qquad\qquad = (3y + 5)^2$

d. $x^{2t} - 9 = (x^t - 3)(x^t + 3)$ ⟶ Recognize the difference of two squares.

Try This. Factor $4x^2 - 28x + 49$.   ■

Factoring the Difference and Sum of Two Cubes

The following formulas are used to factor the difference of two cubes and the sum of two cubes. You should verify these formulas using multiplication.

Factoring the Difference and Sum of Two Cubes

$$a^3 - b^3 = (a - b)(a^2 + ab + b^2) \qquad \text{Difference of two cubes}$$
$$a^3 + b^3 = (a + b)(a^2 - ab + b^2) \qquad \text{Sum of two cubes}$$

Example **7** Factoring differences and sums of cubes

Factor each polynomial.

a. $x^3 - 27$ **b.** $8w^6 + 125z^3$ **c.** $y^{3m} - 1$

Solution

a. Since $x^3 - 27 = x^3 - 3^3$, we use $a = x$ and $b = 3$ in the formula for factoring the difference of two cubes:

$$x^3 - 27 = (x - 3)(x^2 + 3x + 9)$$

b. Since $8w^6 + 125z^3 = (2w^2)^3 + (5z)^3$, we use $a = 2w^2$ and $b = 5z$ in the formula for factoring the sum of two cubes:

$$8w^6 + 125z^3 = (2w^2 + 5z)(4w^4 - 10w^2z + 25z^2)$$

c. Since $y^{3m} = (y^m)^3$, we can use the formula for the difference of two cubes with $a = y^m$ and $b = 1$:

$$y^{3m} - 1 = (y^m - 1)(y^{2m} + y^m + 1)$$

Try This. Factor $8x^3 + 125$. ■

Factoring by Substitution

When a polynomial involves a complicated expression, we can use two substitutions to help us factor. First we replace the complicated expression by a single variable and factor the simpler-looking polynomial. Then we replace the single variable by the complicated expression. This method is called **substitution.**

Example **8** Factoring higher-degree polynomials

Use substitution to factor each polynomial.

a. $w^4 - 6w^2 - 16$ **b.** $(a^2 - 3)^2 + 7(a^2 - 3) + 12$

Solution

a. Replace w^2 by x and w^4 by x^2.

$$w^4 - 6w^2 - 16 = x^2 - 6x - 16$$
$$= (x - 8)(x + 2) \qquad \text{Factor the trinomial.}$$
$$= (w^2 - 8)(w^2 + 2) \qquad \text{Replace } x \text{ by } w^2.$$

b. Replace $a^2 - 3$ by b in the polynomial.

$$(a^2 - 3)^2 + 7(a^2 - 3) + 12 = b^2 + 7b + 12$$
$$= (b + 3)(b + 4) \qquad \text{Factor the trinomial.}$$
$$= (a^2 - 3 + 3)(a^2 - 3 + 4) \qquad \text{Replace } b \text{ by } a^2 - 3.$$
$$= a^2(a^2 + 1) \qquad \text{Simplify.}$$

Try This. Factor $(w^2 + 1)^2 - 3(w^2 + 1) - 18$. ■

Factoring Completely

A polynomial is factored when it is written as a product. Unless noted otherwise, our discussion of factoring will continue to be limited to polynomials whose factors have integral coefficients. Polynomials that cannot be factored using integral coefficients are called **prime** or **irreducible over the integers.** For example, $a^2 + 1$, $b^2 + b + 1$, and $x + 5$ are prime polynomials because they cannot be expressed as a product (in a nontrivial manner). A polynomial is said to be **factored completely** when it is written as a product of prime polynomials. When factoring polynomials, we usually do not factor integers that are common factors. For example, $4x^2(2x - 3)$ is factored completely even though the coefficient 4 could be factored.

Example **9** **Factoring completely**

Factor each polynomial completely.

a. $2w^4 - 32$ **b.** $-6x^7 + 6x$

Solution

a. $2w^4 - 32 = 2(w^4 - 16)$ $\qquad$ Factor out the greatest common factor.

$\qquad\qquad\quad = 2(w^2 - 4)(w^2 + 4)$ $\qquad$ Difference of two squares

$\qquad\qquad\quad = 2(w - 2)(w + 2)(w^2 + 4)$ $\qquad$ Difference of two squares

The polynomial is now factored completely because $w^2 + 4$ is prime.

b. $-6x^7 + 6x = -6x(x^6 - 1)$ $\qquad$ Greatest common factor

$\qquad\qquad\quad = -6x(x^3 - 1)(x^3 + 1)$ $\qquad$ Difference of two squares

$\qquad\qquad\quad = -6x(x - 1)(x^2 + x + 1)(x + 1)(x^2 - x + 1)$

$\qquad\qquad\qquad\qquad\qquad$ Difference of two cubes; sum of two cubes

The polynomial is factored completely because all of the factors are prime.

Try This. Factor $2y^4 - 162$ completely. ■

If 7 is a factor of 147, then 7 is a divisor of 147. If we divide 147 by 7 to get 21 (with no remainder), then we have $147 = 7 \cdot 21$. By factoring 21, we factor 147 completely as $147 = 3 \cdot 7^2$. The same ideas hold for polynomials. If we know one factor of a polynomial, we can use division to find the other factor and then factor the polynomial completely.

Example **10** **Using division to factor completely**

Factor $x^3 - 7x + 6$ completely, given that $x - 2$ is a factor.

Solution

Use long division to find the other factor:

$$
\begin{array}{r}
x^2 + 2x\ - 3 \\
x - 2{\overline{\smash{\big)}\,x^3 + 0x^2 - 7x + 6}} \\
\underline{x^3 - 2x^2} \\
2x^2 - 7x \\
\underline{2x^2 - 4x} \\
-3x + 6 \\
\underline{-3x + 6} \\
0
\end{array}
$$

Because the remainder is 0, we have $x^3 - 7x + 6 = (x - 2)(x^2 + 2x - 3)$. Factoring completely gives

$$x^3 - 7x + 6 = (x - 2)(x + 3)(x - 1).$$

Try This. Factor $y^3 - 6y^2 + 11y - 6$ given $y - 1$ is a factor. ■

For Thought

True or False? Mark true if the polynomial is factored correctly and false if it is not. Explain.

1. $x^2 + 6x - 16 = x(x + 6) - 16$

2. $2x^4 - 5x^2 = -x^2(5 - 2x^2)$

3. $2x^4 - 5x^2 - 3 = (x^2 - 3)(2x^2 + 1)$

4. $a^2 - 1 = (a - 1)^2$

5. $a^3 - 1 = (a - 1)(a^2 + 1)$

6. $x^3 - y^3 = (x - y)(x^2 + 2xy + y^2)$

7. $8a^3 + 27b^3 = (2a + 3b)(4a^2 + 6ab + 9b^2)$

8. $x^2 + 8x - 12 = (x + 2)(x - 6)$

9. $(a + 3)6 - (a + 3)x = (a + 3)(6 - x)$

10. $a - b = -1(b - a)$

P.6 Exercises

Factor out the greatest common factor from each polynomial, first using a positive coefficient on the GCF and then using a negative coefficient.

1. $6x^3 - 12x^2$

2. $12x^2 + 18x^3$

3. $4a - 8ab$

4. $3wm + 15wm^2$

5. $-ax^3 + 5ax^2 - 5ax$

6. $-sa^3 + sb^3 - sb$

7. $m - n$

8. $y - x$

Factor each polynomial by grouping.

9. $x^3 + 2x^2 + 5x + 10$

10. $2w^3 - 2w^2 + 3w - 3$

11. $y^3 - y^2 - 3y + 3$

12. $x^3 + x^2 - 7x - 7$

13. $ady - w + d - awy$

14. $xy + ab + by + ax$

15. $x^2y^2 + ab - ay^2 - bx^2$ **16.** $6yz - 3y - 10z + 5$

58. $(t^3 + 5)^2 + 5(t^3 + 5) + 4$

Factor each trinomial. See the procedure for factoring by the ac-method on page 60.

Factor each polynomial completely.

17. $x^2 + 10x + 16$ **18.** $x^2 + 7x + 12$

59. $-3x^3 + 27x$ **60.** $a^4b^2 - 16b^2$

19. $x^2 - 4x - 12$ **20.** $y^2 + 3y - 18$

61. $16t^4 + 54w^3t$ **62.** $8a^6 - a^3b^3$

21. $m^2 - 12m + 20$ **22.** $n^2 - 8n + 7$

63. $a^3 + a^2 - 4a - 4$ **64.** $2b^3 + 3b^2 - 18b - 27$

23. $t^2 + 5t - 84$ **24.** $s^2 - 6s - 27$

65. $x^4 - 2x^3 - 8x + 16$ **66.** $a^4 - a^3 + a - 1$

25. $2x^2 - 7x - 4$ **26.** $3x^2 + 5x - 2$

67. $-36x^3 + 18x^2 + 4x$ **68.** $-6a^4 - a^3 + 15a^2$

27. $8x^2 - 10x - 3$ **28.** $18x^2 - 15x + 2$

69. $a^7 - a^6 - 64a + 64$

29. $6y^2 + 7y - 5$ **30.** $15x^2 - 14x - 8$

70. $a^5 - 4a^4 - 4a + 16$

71. $-6x^2 - x + 15$ **72.** $-6x^2 - 9x + 42$

Factor each special product.

73. $(a^2 + 2)^2 - 4(a^2 + 2) + 3$

31. $t^2 - u^2$ **32.** $9t^2 - v^2$

74. $(z^3 + 5)^2 - 10(z^3 + 5) + 24$

33. $t^2 + 2t + 1$ **34.** $m^2 + 10m + 25$

35. $4w^2 - 4w + 1$ **36.** $9x^2 - 12xy + 4y^2$

For each pair of polynomials, use long division to determine whether the first polynomial is a factor of the second.

37. $y^{4t} - 25$ **38.** $121w^{2t} - 1$

75. $x + 3, x^3 + 4x^2 + 4x + 3$

39. $9z^2x^2 + 24zx + 16$ **40.** $25t^2 - 20tw^3 + 4w^6$

76. $x - 2, x^3 - 3x^2 + 5x - 6$

77. $x - 1, 3x^3 + 5x^2 - 12x - 9$

78. $x + 1, 2x^3 + 4x^2 - 9x + 2$

Factor each sum or difference of two cubes.

41. $t^3 - u^3$ **42.** $m^3 + n^3$

43. $a^3 - 8$ **44.** $b^3 + 1$

In each pair of polynomials, the second polynomial is a factor of the first. Factor the first polynomial completely.

45. $27y^3 + 8$ **46.** $1 - 8a^6$

79. $x^3 + 4x^2 + x - 6, x - 1$

47. $27x^3y^6 - 8z^9$ **48.** $8t^3h^3 + n^9$

80. $x^3 - 13x - 12, x + 1$

49. $x^{3n} - 8$ **50.** $y^{3w} + 8$

81. $x^3 - x^2 - 4x - 6, x - 3$

82. $2x^3 + 7x^2 + 5x + 1, 2x + 1$

Use substitution to factor each polynomial.

83. $x^4 + 5x^3 + 5x^2 - 5x - 6, x + 2$

51. $y^6 + 10y^3 + 25$ **52.** $y^8 + 8y^4 + 12$

84. $4x^4 + 4x^3 - 25x^2 - x + 6, x - 2$

53. $4a^4b^8 - 8a^2b^4 - 5$ **54.** $3c^2m^{14} - 22cm^7 + 7$

55. $(2a + 1)^2 + 2(2a + 1) - 24$

56. $2(w - 3)^2 + 3(w - 3) - 2$

57. $(b^2 + 2)^2 - 5(b^2 + 2) + 4$

Solve each problem.

85. *Volume of a Rectangular Box* If the volume of a box is $x^3 - 1$ ft^3 and the height is $x - 1$ ft, then what polynomial represents the area of the bottom?

HINT $V = LWH$.

▪ Figure for Exercise 85

86. *Area of a Trapezoid* If the area of a trapezoid is $2x^2 + 5x + 2$ square meters and the two parallel sides are x meters and $x + 1$ meters, then what polynomial represents the height of the trapezoid?

HINT $A = \frac{1}{2}h(b_1 + b_2)$.

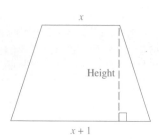

▪ Figure for Exercise 86

87. *Making a Box* An open-top puzzle box is to be constructed from a 6 in. by 7 in. piece of cardboard by cutting out squares from each corner and folding up the sides as shown in the figure. Find the volume of the box for $x = 0.5$ in., 1 in., and 2 in. Which of these values for x produces the largest volume?

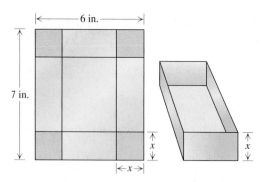

▪ Figure for Exercises 87 and 88

88. *Area and Volume* Find a polynomial that represents the area of the bottom of the box in the accompanying figure. Find a polynomial that represents the volume of the box.

For Writing/Discussion

89. Which of the following is not a perfect square trinomial? Explain your reasoning in writing.
  a. $4x^8 - 20x^4y^3 + 25y^6$ **b.** $1000a^2 - 200ab + b^2$

  c. $400w^4 - 40w^2 + 1$ **d.** $36a^{14} - 36a^7 + 9$

90. Which of the following is not a difference of two squares? Explain your reasoning in writing.
  a. $196a^6b^4 - 289w^8$ **b.** $100x^{16} - 16h^{100}$

  c. $25w^9 - 9y^{25}$ **d.** $1 - 9y^{36}$

91. *Another Lost Rule?* Is it true that the sum of two cubes factors as the cube of a sum? Explain. Give examples.

92. *Cooperative Learning* Work in a small group to write a summary of the techniques used for factoring polynomials. When given a polynomial to factor completely, what should you look for first, second, third, and so on? Evaluate your summary by using it as a guide to factor some polynomials selected from the exercises.

Thinking Outside the Box IX

Wrong Division Is it possible to divide the integers from 1 through 9 into three sets of any size so that the product of the integers in each of the three sets is less than 72?

P.6 Pop Quiz

Factor each polynomial completely.

1. $12a^2 - 8a$

2. $a^3 + a^2 - 3a - 3$

3. $x^2 - 2x - 35$

4. $m^2 - 12m + 36$

5. $n^2 - 9m^2$

6. $w^3 - 1000$

7. $m^4 - n^4$

8. $-6a^3 + 54ab^2$

Linking Concepts

For Individual or
Group Explorations

Pyramid Builders

The Egyptians knew that the formula $V = \frac{1}{3}ha^2$ gives the volume of a pyramid that has a base with area a^2 and a height h. A pyramid missing its top (as it is during construction) is called a truncated pyramid. The ancient Egyptians used the formula $V = \frac{H}{3}(a^2 + ab + b^2)$ for the volume of a truncated pyramid with a square base of area a^2, a square top of area b^2, and height H as shown in the drawing. The Great Pyramid of Egypt was built with a square base of 755 feet on each side and a height of 480 feet.

a) Find the volume of the Great Pyramid.

b) When the Great Pyramid was half as tall as its planned height (like a truncated pyramid), what volume of stone had been set in place?

c) Was the pyramid half finished when it was half as high as planned?

d) Assuming that the formula $V = \frac{1}{3}ha^2$ is correct, prove that the formula for the volume of a truncated pyramid is correct.

> **HINT** Subtract the volume of the missing top from the whole pyramid to get the volume of the truncated pyramid. You will need the fact that $a^2 + ab + b^2$ is a factor of $a^3 - b^3$ to get the given formula.

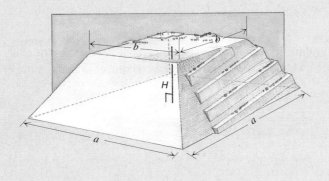
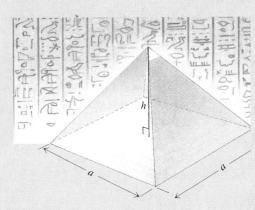

P.7

Rational Expressions

Rational expressions are to algebra what fractions are to arithmetic. In this section we will learn to reduce or build up rational expressions, and to perform basic operations with rational expressions.

Reducing

A **rational expression** is a ratio of two polynomials in which the denominator is not the zero polynomial. For example,

$$\frac{7}{2}, \quad \frac{2x - 1}{x + 3}, \quad 2a - 9, \quad \text{and} \quad \frac{2x + 4}{x^2 + 5x + 6}$$

are rational expressions. The **domain** of a rational expression is the set of all real numbers that can be used in place of the variable. Since division by zero is undefined, the only numbers that *cannot* be used in place of the variable are those that cause the denominator to have a value of zero.

✓ *Example* **1** Domain of a rational expression

Find the domain of each rational expression.

a. $\dfrac{2x - 1}{x + 3}$ **b.** $\dfrac{2x + 4}{x^2 + 5x + 6}$ **c.** $\dfrac{1}{x^2 + 8}$

Solution

a. The domain is the set of all real numbers except those that cause $x + 3$ to have a value of 0. So -3 is excluded from the domain, because $x + 3$ has a value of 0 for $x = -3$. We write the domain in set notation as $\{x \mid x \neq -3\}$.

b. First factor $x^2 + 5x + 6$ as $(x + 2)(x + 3)$. The domain is the set of all real numbers except -2 and -3, because replacing x by either of these numbers would cause the denominator to be 0. The domain is written in set notation as $\{x \mid x \neq -2 \text{ and } x \neq -3\}$.

c. The value of $x^2 + 8$ is positive for any real number x. So the domain is the set of all real numbers, R.

Try This. Find the domain of $\dfrac{x - 2}{x^2 - 9}$. ■

In arithmetic we learned that each rational number has infinitely many equivalent forms. This fact is due to the **basic principle of rational numbers.**

Basic Principle of Rational Numbers

If a, b, and c are integers with $b \neq 0$ and $c \neq 0$, then

$$\frac{ac}{bc} = \frac{a}{b}.$$

■ **Figure P.28**

We use the basic principle to reduce fractions. For example, we reduce $\frac{3}{6}$ by removing or dividing out the common factor 3 from the numerator and denominator. Since $\frac{3}{3} = 1$ and 1 is the multiplicative identity, the value of the fraction is unchanged:

$$\frac{3}{6} = \frac{1 \cdot 3}{2 \cdot 3} = \frac{1}{2} \cdot \frac{3}{3} = \frac{1}{2} \cdot 1 = \frac{1}{2}$$

Your calculator reduces fractions as shown in Fig. P.28. □

To reduce a rational expression we factor the numerator and denominator completely, then *divide out* the common factors. A rational expression is in *lowest terms* when all common factors have been divided out.

Example 2 Reducing to lowest terms

Reduce each rational expression to lowest terms.

a. $\dfrac{2x + 4}{x^2 + 5x + 6}$ **b.** $\dfrac{b - a}{a^3 - b^3}$ **c.** $\dfrac{x^2 z^3}{x^5 z}$

Solution

a. $\dfrac{2x + 4}{x^2 + 5x + 6} = \dfrac{2(x + 2)}{(x + 2)(x + 3)}$ Factor the numerator and denominator.

$\qquad\qquad = \dfrac{2}{x + 3}$ Divide out the common factor $x + 2$.

b. $\dfrac{b - a}{a^3 - b^3} = \dfrac{-1(a - b)}{(a - b)(a^2 + ab + b^2)}$ Factor -1 out of $b - a$.

$\qquad\qquad = \dfrac{-1}{a^2 + ab + b^2}$

c. $\dfrac{x^2 z^3}{x^5 z} = \dfrac{(x^2 z)(z^2)}{(x^2 z)(x^3)} = \dfrac{z^2}{x^3}$ (The quotient rule yields the same result.)

Try This. Reduce $\dfrac{3x - 9}{x^2 - 9}$ to lowest terms. ■

Be careful when reducing. The *only* way to reduce rational expressions is to factor and divide out the common *factors*. Identical terms that are not factors cannot be eliminated from a rational expression. For example,

$$\frac{x + 3}{3} \neq x$$

for all real numbers, because 3 is not a factor of the numerator.

Multiplication and Division

We multiply two rational numbers by multiplying their numerators and their denominators. For example,

$$\frac{2}{3} \cdot \frac{5}{7} = \frac{10}{21}.$$

Definition: Multiplication of Rational Numbers

If a/b and c/d are rational numbers, then

$$\frac{a}{b} \cdot \frac{c}{d} = \frac{ac}{bd}.$$

We multiply rational expressions in the same manner as rational numbers. Of course, any common factor can be divided out as we do when reducing rational expressions.

Example **3** Multiplying rational expressions

Multiply the rational expressions.

a. $\dfrac{2a - 2b}{6} \cdot \dfrac{9a}{a^2 - b^2}$ **b.** $\dfrac{x - 1}{x^2 + 4x + 4} \cdot \dfrac{x + 2}{x^2 + 2x - 3}$

Solution

a. $\dfrac{2a - 2b}{6} \cdot \dfrac{9a}{a^2 - b^2} = \dfrac{2(a - b)}{2 \cdot 3} \cdot \dfrac{3 \cdot 3a}{(a - b)(a + b)}$ Factor completely.

$\qquad\qquad = \dfrac{3a}{a + b}$ Divide out common factors.

b. $\dfrac{x - 1}{x^2 + 4x + 4} \cdot \dfrac{x + 2}{x^2 + 2x - 3} = \dfrac{x - 1}{(x + 2)^2} \cdot \dfrac{x + 2}{(x + 3)(x - 1)}$ Factor completely.

$\qquad\qquad = \dfrac{1}{x^2 + 5x + 6}$ Divide out common factors.

Try This. Find the product $\dfrac{3x + 3}{x^2 - 1} \cdot \dfrac{2x - 2}{9}$. ■

We divide rational numbers by multiplying by the reciprocal of the divisor, or *invert and multiply*. For example, $6 \div \frac{1}{2} = 6 \cdot 2 = 12$.

Definition: Division of Rational Numbers

If a/b and c/d are rational numbers with $c \neq 0$, then

$$\frac{a}{b} \div \frac{c}{d} = \frac{a}{b} \cdot \frac{d}{c}.$$

Rational expressions are divided in the same manner as rational numbers.

Example **4** Dividing rational expressions

Perform the indicated operations.

a. $\dfrac{9}{2x} \div \dfrac{3}{x}$ **b.** $\dfrac{4 - x^2}{6} \div \dfrac{x - 2}{2}$

Solution

a. $\dfrac{9}{2x} \div \dfrac{3}{x} = \dfrac{9}{2x} \cdot \dfrac{x}{3}$ Invert and multiply.

        $= \dfrac{3 \cdot 3}{2x} \cdot \dfrac{x}{3}$ Factor completely.

        $= \dfrac{3}{2}$ Divide out common factors.

b. $\dfrac{4 - x^2}{6} \div \dfrac{x - 2}{2} = \dfrac{-1(x - 2)(x + 2)}{2 \cdot 3} \cdot \dfrac{2}{x - 2}$ Invert and multiply.

        $= \dfrac{-x - 2}{3}$ Divide out common factors.

Try This.   Find the quotient $\dfrac{25 - a^2}{6} \div \dfrac{2a - 10}{8}$.   ■

Note that the division in Example 4(a) is valid only if $x \neq 0$. The division in Example 4(b) is valid only if $x \neq 2$ because $x - 2$ appears in the denominator after the rational expression is inverted.

Building Up the Denominator

The addition of fractions can be carried out only when their denominators are identical. To get a required denominator, we may **build up** the denominator of a fraction. To build up the denominator we use the basic principle of rational numbers in the reverse of the way we use it for reducing. We multiply the numerator and denominator of a fraction by the same nonzero number to get an equivalent fraction. For example, to get a fraction equivalent to $\frac{1}{3}$ with a denominator of 12, we multiply the numerator and denominator by 4:

$$\frac{1}{3} = \frac{1 \cdot 4}{3 \cdot 4} = \frac{4}{12}$$

To build up the denominator of a rational expression, we use the same procedure.

Example **5**   Writing equivalent rational expressions

Convert the first rational expression into an equivalent one that has the indicated denominator.

a. $\dfrac{3}{2a}, \dfrac{?}{6ab}$ **b.** $\dfrac{x - 1}{x + 2}, \dfrac{?}{x^2 + 6x + 8}$ **c.** $\dfrac{a}{3b - a}, \dfrac{?}{a^2 - 9b^2}$

Solution

a. Compare the two denominators. Since $6ab = 2a(3b)$, we multiply the numerator and denominator of the first expression by $3b$:

$$\frac{3}{2a} = \frac{3 \cdot 3b}{2a \cdot 3b} = \frac{9b}{6ab}$$

b. Factor the second denominator and compare it to the first. Since $x^2 + 6x + 8 = (x + 2)(x + 4)$, we multiply the numerator and denominator by $x + 4$:

$$\frac{x - 1}{x + 2} = \frac{(x - 1)(x + 4)}{(x + 2)(x + 4)} = \frac{x^2 + 3x - 4}{x^2 + 6x + 8}$$

c. Factor the second denominator as

$$a^2 - 9b^2 = (a - 3b)(a + 3b) = -1(3b - a)(a + 3b).$$

Since $3b - a$ is a factor of $a^2 - 9b^2$, we multiply the numerator and denominator by $-1(a + 3b)$:

$$\frac{a}{3b - a} = \frac{a(-1)(a + 3b)}{(3b - a)(-1)(a + 3b)} = \frac{-a^2 - 3ab}{a^2 - 9b^2}$$

Try This. Complete the equation $\dfrac{y}{y + 2} = \dfrac{?}{y^2 + 3y + 2}$. ■

Addition and Subtraction

Fractions can be added or subtracted only if their denominators are identical. For example, $\frac{1}{3} + \frac{1}{3} = \frac{2}{3}$ and $\frac{7}{12} - \frac{2}{12} = \frac{5}{12}$.

Definition: Addition and Subtraction of Rational Numbers

> If a/b and c/d are rational numbers, then
>
> $$\frac{a}{b} + \frac{c}{b} = \frac{a + c}{b} \qquad \text{and} \qquad \frac{a}{b} - \frac{c}{b} = \frac{a - c}{b}.$$

For fractions with different denominators, we build up one or both denominators to get denominators that are equal to the least common multiple (LCM) of the denominators. The **least common denominator (LCD)** is the smallest number that is a multiple of all the denominators. Use the following steps to find the LCD.

P R O C E D U R E   **Finding the LCD**

1. Factor each denominator completely.
2. Write a product using each factor that appears in a denominator.
3. For each factor, use the highest power of that factor that occurs in the denominators.

For example, to find the LCD for 10 and 12, we write $10 = 2 \cdot 5$ and $12 = 2^2 \cdot 3$. The LCD contains the factors 2, 3, and 5. Using the highest power of each, we get $2^2 \cdot 3 \cdot 5 = 60$ for the LCD. So, to add fractions with denominators of 10 and 12, we build up each fraction to a denominator of 60:

$$\frac{1}{10} + \frac{1}{12} = \frac{1 \cdot 6}{10 \cdot 6} + \frac{1 \cdot 5}{12 \cdot 5} = \frac{6}{60} + \frac{5}{60} = \frac{11}{60}$$

We use the same method to add or subtract rational expressions.

 *Example* Adding and subtracting
rational expressions

Perform the indicated operations.

a. $\dfrac{x}{x-1} + \dfrac{2x+3}{x^2-1}$ **b.** $\dfrac{x}{x^2+6x+9} - \dfrac{x-3}{x^2+5x+6}$

Solution

a. $\dfrac{x}{x-1} + \dfrac{2x+3}{x^2-1} = \dfrac{x}{x-1} + \dfrac{2x+3}{(x-1)(x+1)}$ Factor denominators completely.

$= \dfrac{x(x+1)}{(x-1)(x+1)} + \dfrac{2x+3}{(x-1)(x+1)}$ Build up using the LCD, $(x-1)(x+1)$.

$= \dfrac{x^2+x+2x+3}{(x-1)(x+1)}$ Add the fractions.

$= \dfrac{x^2+3x+3}{(x-1)(x+1)}$ Simplify the numerator.

b. $\dfrac{x}{x^2+6x+9} - \dfrac{x-3}{x^2+5x+6} = \dfrac{x}{(x+3)^2} - \dfrac{x-3}{(x+2)(x+3)}$

$= \dfrac{x(x+2)}{(x+3)^2(x+2)} - \dfrac{(x-3)(x+3)}{(x+2)(x+3)(x+3)}$

$= \dfrac{x^2+2x}{(x+3)^2(x+2)} - \dfrac{x^2-9}{(x+3)^2(x+2)}$

$= \dfrac{x^2+2x-(x^2-9)}{(x+3)^2(x+2)}$ Subtract numerators.

$= \dfrac{x^2+2x-x^2+9}{(x+3)^2(x+2)}$ Remove parentheses and change signs.

$= \dfrac{2x+9}{(x+3)^2(x+2)}$ Simplify the numerator.

Note that the numerator $x^2 - 9$ must be in parentheses when it is subtracted.

Try This. Find the sum $\dfrac{y}{y+2} + \dfrac{2}{y+1}$.   ■

Complex Fractions

A **complex fraction** is a fraction having rational expressions in the numerator, denominator, or both. Complex fractions can be simplified quickly by multiplying the numerator and denominator by the LCD of all of the denominators.

Example **7** Simplifying a complex fraction

Simplify each complex fraction.

a. $\dfrac{4 - \dfrac{3}{4x}}{\dfrac{1}{x^2} - \dfrac{1}{6}}$ **b.** $\dfrac{\dfrac{x + 6}{x^2 - 9}}{\dfrac{x}{x - 3} - \dfrac{x + 4}{x + 3}}$

Solution

a. The LCD for the denominators 6, $4x$, and x^2 is $12x^2$. Multiply the numerator and denominator of the complex fraction by $12x^2$.

$$\frac{4 - \dfrac{3}{4x}}{\dfrac{1}{x^2} - \dfrac{1}{6}} = \frac{\left(4 - \dfrac{3}{4x}\right)12x^2}{\left(\dfrac{1}{x^2} - \dfrac{1}{6}\right)12x^2} = \frac{48x^2 - 9x}{12 - 2x^2}$$

b. The LCD for the denominators $x^2 - 9$, $x - 3$, and $x + 3$ is $x^2 - 9$, because $x^2 - 9 = (x - 3)(x + 3)$. Multiply the numerator and denominator by $(x - 3)(x + 3)$, or $x^2 - 9$:

$$\frac{\dfrac{x + 6}{x^2 - 9}}{\dfrac{x}{x - 3} - \dfrac{x + 4}{x + 3}} = \frac{\left(\dfrac{x + 6}{x^2 - 9}\right)(x - 3)(x + 3)}{\left(\dfrac{x}{x - 3} - \dfrac{x + 4}{x + 3}\right)(x - 3)(x + 3)}$$

$$= \frac{x + 6}{x(x + 3) - (x + 4)(x - 3)}$$

$$= \frac{x + 6}{x^2 + 3x - (x^2 + x - 12)}$$

$$= \frac{x + 6}{2x + 12} = \frac{x + 6}{2(x + 6)} = \frac{1}{2}$$

Try This. Simplify $\dfrac{\dfrac{1}{2a} - \dfrac{1}{4}}{\dfrac{3}{a} + \dfrac{1}{6}}$. ■

The fractions in a complex fraction can be written with negative exponents. To simplify complex fractions with negative exponents we still multiply the numerator and denominator by the LCD.

Example **8** A complex fraction with negative exponents

Simplify each complex fraction. Write answers with positive exponents only.

a. $\dfrac{a^{-1} + b^{-3}}{ab^{-2} + ba^{-4}}$ **b.** $pq + p^{-1}q^{-2}$

Solution

a. If we use the definition of negative exponent to rewrite each term, then the denominators would be a, b^3, b^2, and a^4. The LCD for these denominators is a^4b^3. Multiply the numerator and denominator by a^4b^3:

$$\frac{a^{-1} + b^{-3}}{ab^{-2} + ba^{-4}} = \frac{(a^{-1} + b^{-3})a^4b^3}{(ab^{-2} + ba^{-4})a^4b^3}$$

$$= \frac{a^3b^3 + a^4}{a^5b + b^4}$$

Note that the exponents in a^4b^3 are just large enough to eliminate all negative exponents in the multiplication. Note also that we could have rewritten the complex fraction without negative exponents before multiplying by a^4b^3, but it is not necessary to do so.

b. Although this expression is not exactly a complex fraction, we can use the same technique to eliminate the negative exponents. Multiply the numerator and denominator by pq^2:

$$pq + p^{-1}q^{-2} = \frac{(pq + p^{-1}q^{-2})pq^2}{1 \cdot pq^2}$$

$$= \frac{p^2q^3 + 1}{pq^2}$$

Try This. Simplify $\dfrac{x^{-2}}{y^{-3} - x^{-3}}$. ■

For Thought

True or False? Explain.

1. The rational expression $\dfrac{2x + 5}{2y}$ reduces to $\dfrac{x + 5}{y}$.

2. The rational expression $\dfrac{-3}{a - 5}$ is equivalent to $\dfrac{3}{5 - a}$.

3. The expressions $\dfrac{x(x + 2)}{x(x + 3)}$ and $\dfrac{x + 2}{x + 3}$ are equivalent.

4. The expression $\dfrac{a^2 - b^2}{a - b}$ reduced to lowest terms is $a - b$.

5. The LCD for the rational expressions $\dfrac{1}{x}$ and $\dfrac{1}{x + 1}$ is $x + 1$.

6. $\dfrac{2x - 1}{x - 3} + \dfrac{x + 5}{x - 3} = \dfrac{3x + 4}{x - 3}$ provided that $x \neq 3$.

7. $\dfrac{x}{2} = \dfrac{1}{2}x$ for all nonzero values of x.

8. $\dfrac{x}{3} - 1 = \dfrac{x - 3}{3}$ for any real number x.

9. If $x = 500$, then the approximate value of $\dfrac{2x + 1}{x - 3}$ is 2.

10. If $|x|$ is very large, then $\dfrac{5x + 1}{x}$ has an approximate value of 5.

P.7 Exercises

Find the domain of each rational expression.

1. $\dfrac{x-3}{x+2}$

2. $\dfrac{x^2-1}{x-5}$

3. $\dfrac{x^2-9}{(x-4)(x+2)}$

4. $\dfrac{2x-3}{(x+1)(x-3)}$

5. $\dfrac{x+1}{x^2-9}$

6. $\dfrac{x+2}{x^2+3x+2}$

7. $\dfrac{3x^2-2x+1}{x^2+3}$

8. $\dfrac{-2x^2-7}{3x^2+8}$

Reduce each rational expression to lowest terms.

9. $\dfrac{3x-9}{x^2-x-6}$

10. $\dfrac{-2x-4}{x^2-3x-10}$

11. $\dfrac{10a-8b}{12b-15a}$

12. $\dfrac{a^2-b^2}{b-a}$

13. $\dfrac{a^3b^6}{a^2b^3-a^4b^2}$

14. $\dfrac{18u^6v^5+24u^3v^3}{42u^2v^5}$

15. $\dfrac{x^4y^5z^2}{x^7y^3z}$

16. $\dfrac{t^3u^7}{-t^8u^5}$

17. $\dfrac{a^3-b^3}{a^2-b^2}$

18. $\dfrac{a^3+b^3}{a^2+b^2}$

19. $\dfrac{9y^2-6y+1}{1-9y^2}$

20. $\dfrac{6-y-y^2}{y^2-4y+4}$

Find the products or quotients.

21. $\dfrac{2a}{3b^2}\cdot\dfrac{9b}{14a^2}$

22. $\dfrac{14w}{51y}\cdot\dfrac{3w}{7y}$

23. $\dfrac{12a}{7}\div\dfrac{2a^3}{49}$

24. $\dfrac{20x}{y^3}\div\dfrac{30}{y^5}$

25. $\dfrac{a^2-9}{3a-6}\cdot\dfrac{a^2-4}{a^2-a-6}$

26. $\dfrac{6x^2+x-1}{6x+3}\cdot\dfrac{15}{9x^2-1}$

27. $\dfrac{x^2-y^2}{9}\div\dfrac{x^2+2xy+y^2}{18}$

28. $\dfrac{a^3-b^3}{a^2-2ab+b^2}\div\dfrac{2a^2+2ab+2b^2}{9a^2-9b^2}$

29. $\dfrac{x^2-y^2}{-3xy}\cdot\dfrac{6x^2y^3}{2y-2x}$

30. $\dfrac{a^2-a-2}{2}\cdot\dfrac{1}{4-a^2}$

31. $\dfrac{b^2-b-6}{9-b^2}\div\dfrac{b^2+6b+8}{6+2b}$

32. $\dfrac{3a^2+a-4}{15-a-2a^2}\div\dfrac{6a^2-7a-20}{4a^2-20a+25}$

Convert the first rational expression into an equivalent one that has the indicated denominator.

33. $\dfrac{4}{3a},\dfrac{?}{12a^2}$

34. $\dfrac{a+2}{4a^2},\dfrac{?}{20a^3b}$

35. $\dfrac{x-5}{x+3},\dfrac{?}{x^2-9}$

36. $\dfrac{x+2}{x-8},\dfrac{?}{16-2x}$

37. $\dfrac{x}{x+5},\dfrac{?}{x^2+6x+5}$

38. $\dfrac{3a-b}{a+b},\dfrac{?}{9b^2-9a^2}$

Find the least common denominator (LCD) for each given pair of rational expressions. See the procedure for finding the LCD on page 72.

39. $\dfrac{1}{4ab^2},\dfrac{7}{6a^2b^3}$

40. $\dfrac{3}{2x^2y},\dfrac{a}{5xy}$

41. $\dfrac{-7a}{3a+3b},\dfrac{5b}{2a+2b}$

42. $\dfrac{1}{3a-3b},\dfrac{2}{a^2-b^2}$

43. $\dfrac{2x}{x^2+5x+6},\dfrac{3x}{x^2-x-6}$

44. $\dfrac{x+7}{2x^2+7x-15},\dfrac{x-5}{2x^2-5x+3}$

Perform the indicated operations.

45. $\dfrac{3}{2x}+\dfrac{1}{6}$

46. $\dfrac{-7}{3a^2b}+\dfrac{4}{6ab^2}$

47. $\dfrac{x+3}{x-1}-\dfrac{x+4}{x+1}$

48. $\dfrac{x+2}{x-3}-\dfrac{x^2+3x-2}{x^2-9}$

49. $3+\dfrac{1}{a}$

50. $-1-\dfrac{3}{c}$

51. $t-1-\dfrac{1}{t+1}$

52. $w+\dfrac{1}{w-1}$

53. $\dfrac{x}{x^2 + 3x + 2} + \dfrac{x - 1}{x^2 + 5x + 6}$

54. $\dfrac{x - 1}{x^2 + x - 6} - \dfrac{x - 2}{x^2 + 4x + 3}$

55. $\dfrac{1}{x - 3} - \dfrac{5}{6 - 2x}$

56. $\dfrac{5}{4 - x^2} - \dfrac{2x}{x - 2}$

57. $\dfrac{y^2}{x^3 - y^3} + \dfrac{x + y}{x^2 + xy + y^2}$

58. $\dfrac{ab}{a^3 + b^3} + \dfrac{a}{2a^2 - 2ab + 2b^2}$

59. $\dfrac{1}{x} + \dfrac{1}{x - 1} - \dfrac{1}{x + 1}$

60. $\dfrac{3}{x} - \dfrac{x - 1}{x^2 - 9} + \dfrac{1}{x - 3}$

61. $\dfrac{5}{x} + \dfrac{2}{x + 1} - \dfrac{6}{x - 3}$

62. $\dfrac{3x - 2}{4x} - \dfrac{2x + 1}{3x} + \dfrac{x - 5}{2x}$

Simplify each complex fraction.

63. $\dfrac{\dfrac{25}{36a}}{\dfrac{10a}{27}}$

64. $\dfrac{\dfrac{9a}{35}}{\dfrac{6}{25a}}$

65. $\dfrac{\dfrac{4}{a} - \dfrac{3}{b}}{\dfrac{1}{ab} + \dfrac{2}{b^2}}$

66. $\dfrac{\dfrac{2}{6xy} - \dfrac{1}{4x}}{\dfrac{1}{3y^2} + \dfrac{1}{2x}}$

67. $\dfrac{\dfrac{a}{ab^2} - \dfrac{b}{ab^3}}{\dfrac{3}{a^2} + \dfrac{a}{a^3b}}$

68. $\dfrac{\dfrac{1}{a^2b^3c}}{\dfrac{c}{ab^2} + \dfrac{a}{b^2c}}$

69. $\dfrac{a + \dfrac{4}{a + 4}}{a - \dfrac{4a + 4}{a + 4}}$

70. $\dfrac{y - \dfrac{y + 6}{y + 2}}{y - \dfrac{4y + 15}{y + 2}}$

71. $\dfrac{\dfrac{t + 2}{t - 1} - \dfrac{t - 3}{t}}{\dfrac{t + 4}{t} + \dfrac{t - 2}{t - 1}}$

72. $\dfrac{\dfrac{3}{2 + x} - \dfrac{4}{2 - x}}{\dfrac{1}{x + 2} - \dfrac{3}{x - 2}}$

Simplify. Write answers with positive exponents only.

73. $\dfrac{x^{-1} + 1}{x^{-1} - 1}$

74. $\dfrac{x^{-2} - 4}{x^{-1} - 2}$

75. $a^2 + a^{-1}b^{-3}$

76. $m + m^{-1} + m^{-2}$

77. $\dfrac{x^2 - y^2}{x^{-1} - y^{-1}}$

78. $\dfrac{x^{-2} - y^{-2}}{(x - y)^2}$

79. $(m^{-1} - n^{-1})^{-2}$

80. $(a^{-1} + y^{-1})^{-1}$

Let $R(x) = \dfrac{3}{x + 6}$, $S(x) = \dfrac{2x - 5}{2x - 9}$, and $T(x) = \dfrac{9x^2 - 1}{3x^2 - 2}$. Find the following values. (The notation $R(x)$ was introduced in Section P.5.)

81. $R(1)$ **82.** $R(-1)$ **83.** $R(500)$

84. $R(-1000)$ **85.** $S(2)$ **86.** $S(-2)$

87. $S(600)$ **88.** $S(-600)$ **89.** $T(-4)$

90. $T(7)$ **91.** $T(-400)$ **92.** $T(500)$

Solve each problem.

93. *Incinerating Garbage* High Tech Incineration (HTI) charges $50, plus $20 per ton to incinerate a truckload of garbage. The average cost per ton (in dollars) for incinerating n tons of garbage is given by

$$A(n) = \dfrac{20n + 50}{n}.$$

a. Use the accompanying graph to determine whether the average cost per ton is increasing or decreasing as the size of the load increases?

b. A company uses trucks with capacities of 7, 12, and 22 tons. What is the average cost per ton for each truck?

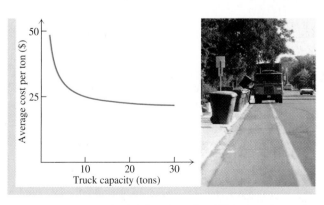

■ **Figure for Exercise 93**

94. *Water Pollution*   High Tech Incineration (HTI) has been told by the EPA to stop polluting the Red River immediately or it must pay $5000 plus $400 per day until it stops. Write a rational expression like that in the previous exercise for calculating the average fine per day *F*. What is the average fine per day if HTI stops polluting in 10 days, 20 days, or 30 days? Is the average fine per day increasing or decreasing as the number of days gets larger?

95. *Costly Cleanup*   The cost of cleaning up the ConChem hazardous waste site has been modeled by the rational expression

$$C(p) = \frac{6{,}000{,}000p}{100 - p}$$

where $C(p)$ is the cost in dollars for cleaning up $p\%$ of the pollution.

  a. Find the costs for cleaning up 50%, 75%, and 99% of the pollution.

  b. Use the accompanying graph to determine what happens to the cost as the percentage approaches 100%. What is the cost for a 100% clean up?

  c. What is the domain of the rational expression?

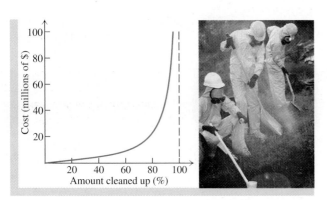

■ **Figure for Exercise 95**

96. *Costly Campaign*   A political strategist used the rational expression

$$C(p) = \frac{2{,}000{,}000 + 1{,}000{,}000p}{50 - p}$$

to estimate the cost in dollars of a campaign that would get her candidate for governor $p\%$ of the votes. Find the costs for 30%, 40%, and 49%. Could this candidate raise enough money to get at least 50% of the vote?

97. *Filing Invoices*   Gina can file all of the daily invoices in 4 hr and Bert can do the same job in 6 hr. If they work together, then what portion of the invoices can they file in 1 hr?

98. *Painting a House*   Melanie can paint the entire house in *x* hours and Timothy can do the same job in *y* hours. Write a rational expression that represents the portion of the house that they can paint in 2 hr working together.

99. *Average Speed*   Wally West, alias the Flash, finishes a meal at the Golden Buddha restaurant only to find he's left his wallet at home. Not wanting to reveal his secret identity to his date, he excuses himself and slips outside. Dashing home at 250 mph, he snatches his wallet from his nightstand and races back to the restaurant at 300 mph. Discounting the time it took to find his wallet, what was his average speed for the trip?
  HINT Average speed is total distance divided by total time for the trip.

100. *Average Cost*   Every day Denise spends the same amount on eggs for Denise's Diner. On Monday eggs were 50 cents per dozen, on Tuesday they were 60 cents per dozen, and on Wednesday they were 70 cents per dozen. What was her average cost for a dozen eggs for the three-day period?
  HINT The average cost is the total cost divided by the total number of dozens purchased.

For Writing/Discussion

101. *Domain*   Why is it important to know the domain of a rational expression?

102. *Cooperative Learning*   Each student in your small group should write down a rational expression in which the degree of the denominator is greater than or equal to the degree of the numerator. Evaluate your rational expression for several "very large" values of *x*, using a calculator. Discuss your results. Can you predict the approximate value by looking at the rational expression?

103. Find the exact value of each expression. Explain how you did it.

  a. $\dfrac{1}{1 + \dfrac{1}{1 + \dfrac{1}{1 + \dfrac{1}{1 + \dfrac{1}{2}}}}}$

  b. $\dfrac{1}{1 - \dfrac{1}{1 - \dfrac{1}{1 - \dfrac{1}{1 - \dfrac{1}{3}}}}}$

104. Find the domain of each expression. Explain your reasoning in writing.

a. $\dfrac{1}{1 + \dfrac{1}{1 + \dfrac{1}{x}}}$

b. $\dfrac{1}{1 + \dfrac{1}{1 + \dfrac{1}{1 + \dfrac{1}{x - 1}}}}$

Thinking Outside the Box X

Minimizing a Sum If x and N are real numbers such that $N = \sqrt{5x + 6} + \sqrt{7x + 8}$, then what is the smallest possible value of N? The answer must be exact.

P.7 Pop Quiz

1. What is the domain for $\dfrac{x}{x - 1}$?

2. Reduce $\dfrac{x^2 - y^2}{y - x}$ to lowest terms.

3. Simplify $\dfrac{b^{-1} + a^{-3}}{a^{-3}b^{-2}}$.

Perform the indicated operations.

4. $\dfrac{x^2 y^3}{8z^4} \cdot \dfrac{12z^3 y^2}{x^5}$

5. $\dfrac{8}{w - p} \div \dfrac{4}{w^2 - p^2}$

6. $\dfrac{1}{m - n} + \dfrac{4}{m + n}$

7. $\dfrac{5}{a - b} - \dfrac{5a}{a^2 - b^2}$

Linking Concepts

For Individual or Group Explorations

Average Cost Per Drink

To save the environment, the Lion's Den Cafeteria sells an empty reusable 24-ounce plastic mug to students for $2.99. Using the mug, a student can get a 24-ounce soft drink for a reduced price of 69 cents.

a) If a student buys the mug and uses it only once, then what is the cost for that one drink?

b) If a student buys the mug and uses it twice, then what is the average cost per drink?

c) Does the average cost per drink increase or decrease as the mug is used more and more?

d) If the regular price of a 24-ounce soft drink is 89 cents, then when does the student start saving money?

e) Write a rational expression for which the value of the expression is the average cost per drink and n is the number of times the mug is used.

f) If a student buys the mug and uses it every day for a semester, then what is the approximate average cost per drink?

g) If a professional student buys the mug and uses it every day for many years, then what value is the average cost per drink approaching?

h) Use a computer to create a graph showing the number of drinks purchased and the average cost per drink. Make sure that the graph illustrates your answer to part (g).

■■■Highlights

P.1 Real Numbers and Their Properties

Rationals and Irrationals

Every real number is either rational (a ratio of integers) or irrational (not a ratio of integers).

Rational: $1, 2/3, -44.7$
Irrational: $\sqrt{2}, \sqrt{7}, \pi$

Commutative Property

Addition: $a + b = b + a$
Multiplication: $ab = ba$

$3 + 4 = 4 + 3$
$5 \cdot 7 = 7 \cdot 5$

Associative Property

Addition: $a + (b + c) = (a + b) + c$
Multiplication: $a(bc) = (ab)c$

$3 + (4 + 5) = (3 + 4) + 5$
$3(4 \cdot 5) = (3 \cdot 4)5$

Distributive Property

$a(b + c) = ab + ac$

$5(6 + 7) = 5 \cdot 6 + 5 \cdot 7$

Absolute Value

A number's distance from 0 on the number line

$|5| = 5, |-5| = 5, |0| = 0$

Exponents

a^n is the product of n a's for a positive integer n

$4^3 = 4 \cdot 4 \cdot 4 = 64$

Order of Operations

When some or all grouping symbols are omitted evaluate exponents, multiplication and division, and addition and subtraction in this order.

$3 - 5 \cdot 6 = 3 - 30 = -27$
$2 + 3^2 = 2 + 9 = 11$
$4 + 6/3 = 4 + 2 = 6$

P.2 Integral Exponents and Scientific Notation

Negative and Zero Exponents

$a^{-n} = 1/a^n$ provided n is a positive integer and $a \neq 0$. If $a \neq 0$, then $a^0 = 1$.

$2^{-3} = 1/2^3, 2^0 = 1$

Rules

Product rule: $a^m a^n = a^{m+n}$
Quotient rule: $a^m/a^n = a^{m-n}$
Power rule: $(a^m)^n = a^{mn}$
Power of a product: $(ab)^m = a^m b^m$
Power of a quotient: $(a/b)^m = a^m/b^m$

$x^2 x^3 = x^5$
$w^4/w^5 = w^{-1}$
$(s^3)^2 = s^6$
$(3x)^2 = 9x^2$
$(w/2)^3 = w^3/8$

Scientific Notation

A positive number less than 1 or greater than 10 is written as a product of a number between 1 and 10 and a power of 10.

$0.005 = 5 \times 10^{-3}$
$340 = 3.4 \times 10^2$

P.3 Rational Exponents and Radicals

nth Root

If n is a positive integer and $a^n = b$, then a is an nth root of b. If n is even, then the positive nth root is represented as $\sqrt[n]{a}$ or $a^{1/n}$. If n is odd, $\sqrt[n]{a}$ or $a^{1/n}$ represent the real nth root of a. An even root of a negative number is not real.

2 and -2 are fourth roots of 16.

$\sqrt[4]{16} = 16^{1/4} = 2$
$\sqrt[3]{-8} = (-8)^{1/3} = -2$
$\sqrt{-2}$ is not real.

Rules

Product rule for radicals: $\sqrt[n]{ab} = \sqrt[n]{a} \cdot \sqrt[n]{b}$
Quotient rule for radicals: $\sqrt[n]{a/b} = \sqrt[n]{a}/\sqrt[n]{b}$

$\sqrt[3]{6} = \sqrt[3]{2} \cdot \sqrt[3]{3}$
$\sqrt{5/2} = \sqrt{5}/\sqrt{2}$

Simplified Radical Expression

1. No perfect nth powers as factors of the radicand.
2. No fractions inside the radical, and
3. No radicals in a denominator.

$\sqrt[3]{8x} = \sqrt[3]{8} \cdot \sqrt[3]{x} = 2\sqrt[3]{x}$

$\sqrt{1/2} = \sqrt{1}/\sqrt{2} = 1/\sqrt{2}$

$\dfrac{1}{\sqrt{2}} = \dfrac{1 \cdot \sqrt{2}}{\sqrt{2} \cdot \sqrt{2}} = \dfrac{\sqrt{2}}{2}$

Exponent m/n

$a^{m/n} = \sqrt[n]{a^m} = \left(\sqrt[n]{a}\right)^m$
All rules of exponents apply to rational exponents.

$8^{2/3} = \left(\sqrt[3]{8}\right)^2 = 2^2 = 4$

P.4 Complex Numbers

Standard Form	Numbers of the form $a + bi$ where a and b are real numbers, $i = \sqrt{-1}$, and $i^2 = -1$	$2 + 3i,\ -\pi + i\sqrt{2},\ 6,\ 0,\ \frac{1}{2}i$
Add, Subtract, Multiply	Add, subtract, and multiply like binomials with variable i, using $i^2 = -1$ to simplify.	$(3 - 2i)(4 + 5i)$ $= 12 + 7i - 10i^2$ $= 22 + 7i$
Divide	Divide by multiplying the numerator and denominator by the complex conjugate of the denominator.	$6/(1 + i)$ $= \dfrac{6(1 - i)}{(1 + i)(1 - i)} = 3 - 3i$
Square Roots of Negative Numbers	Square roots of negative numbers must be converted to standard form using $\sqrt{-b} = i\sqrt{b}$, for $b > 0$, before doing computations.	$\sqrt{-4} \cdot \sqrt{-9} = 2i \cdot 3i = -6$

P.5 Polynomials

Polynomial	$a_n x^n + a_{n-1}x^{n-1} + \cdots + a_1 x + a_0$ where n is a nonnegative integer and $a_0, a_1, a_2, \ldots, a_n$ are real numbers	$x^4 - 5x^3 + 6x^2 - 7x + 9$
Operations	Polynomials can be added, subtracted, multiplied, and divided.	$(x^2 + 2x) - (x^2 - x) = 3x$ $(x^2 + 2x)/x = x + 2$
FOIL	A rule for finding the product of two binomials quickly (First, Outer, Inner, Last)	$(x + 3)(x + 2) = x^2 + 5x + 6$
Value of a Polynomial	The value of the arithmetic expression obtained when the variable is replaced by a real number	$P(x) = x^3 - x + 5$ $P(2) = 11$

P.6 Factoring Polynomials

Factor	To write as a product of two or more polynomials	$x^2 + x = x(x + 1)$
Perfect Square Trinomial	$a^2 + 2ab + b^2 = (a + b)^2$ $a^2 - 2ab + b^2 = (a - b)^2$	$x^2 + 6x + 9 = (x + 3)^2$ $4a^2 - 4ab + b^2 = (2a - b)^2$
Difference of Two Squares	$a^2 - b^2 = (a + b)(a - b)$	$9x^2 - y^2 = (3x + y)(3x - y)$
Difference or Sum of Two Cubes	$a^3 - b^3 = (a - b)(a^2 + ab + b^2)$ $a^3 + b^3 = (a + b)(a^2 - ab + b^2)$	$x^3 - 8 = (x - 2)(x^2 + 2x + 4)$ $w^3 + 1 = (w + 1)(w^2 - w + 1)$
***ac*-Method**	To factor $ax^2 + bx + c$ with $a \neq 1$ 1. Find two numbers whose sum is ac and whose product is b. 2. Replace b by the sum of these two numbers. 3. Factor by grouping.	$2x^2 - x - 3$ $= 2x^2 - 3x + 2x - 3$ $= x(2x - 3) + 1(2x - 3)$ $= (2x - 3)(x + 1)$

P.7 Rational Expressions

Rational Expression	A ratio of two polynomials, in which the denominator is not the zero polynomial	$\dfrac{x + 2}{x - 1}$

Domain   The set of all real numbers for which the   Domain of $\dfrac{x+2}{x-1}$ is $\{x|x \neq 1\}$.
                                    denominator is not zero

Operations   Performed just like the operations with $\dfrac{x+2}{x-1} + \dfrac{x-2}{x-1} = \dfrac{2x}{x-1}$
                                    fractions

Complex Fraction   A rational expression with rational expressions $\dfrac{\frac{1}{x}+2}{x-\frac{2}{x}} = \dfrac{\left(\frac{1}{x}+2\right)x}{\left(x-\frac{2}{x}\right)x} = \dfrac{1+2x}{x^2-2}$
                                    in the numerator and/or denominator
                                    Simplify by multiplying by the LCD of all
                                    denominators.

■ ■ ■ Chapter P Review Exercises

Determine whether each statement is true or false and explain your answer.

1. Every real number is a rational number.

2. Zero is neither rational nor irrational.

3. There are no negative integers.

4. Every repeating decimal number is a rational number.

5. The terminating decimal numbers are irrational numbers.

6. The number $\sqrt{289}$ is a rational number.

7. Zero is a natural number.

8. The multiplicative inverse of 8 is 0.125.

9. The reciprocal of 0.333 is 3.

10. The real number π is irrational.

11. The additive inverse of 0.5 is 0.

12. The distributive property is used in adding like terms.

Simplify each expression.

13. $-3x - 4(3 - 5x)$

14. $x - 0.02(x - 9)$

15. $\dfrac{x}{5} + \dfrac{x}{10}$

16. $\dfrac{1}{3}x - \dfrac{1}{8}x$

17. $\dfrac{3x-6}{9}$

18. $\dfrac{1}{2}(4x - 6)$

19. $\dfrac{-7-(-1)}{3-(-5)}$

20. $\dfrac{6-(3-x)}{2-(-1)}$

21. $|-3| - |-5|$

22. $|5 - (-2)|$

23. $|3 - 7|$

24. $|-3 - (-4)|$

25. $8 - 9 \cdot 2 \div 3 + 5$

26. $3 - 4(2 - 3 \cdot 5^2)$

27. $12 \div 4 \cdot 3 \div 6 + 3^3$

28. $8 \cdot 3^2 - 3\sqrt{3^2 + 4^2}$

Simplify each expression. Assume that all variables represent positive real numbers.

29. 5^4

30. 2^{-4}

31. $(-2)^2 - 4(-2)(5)$

32. $6^2 - 4(-1)(-3)$

33. $2^{-1} + 2^0$

34. $\dfrac{3^{-1}}{-3^2}$

35. $\dfrac{-3^{-1}2^3}{2^{-1}}$

36. $\dfrac{-1}{-1^{-1}}$

37. $8^{-2/3}$

38. $-16^{-3/4}$

39. $(125x^6)^{1/3}$

40. $\dfrac{1}{(27t^{12})^{-1/3}}$

41. $\sqrt{121}$

42. $\sqrt[3]{-1000}$

43. $\sqrt{28s^3}$

44. $\sqrt{75a^2b^9}$

45. $\sqrt[3]{-2000}$

46. $\sqrt[3]{56w^4}$

47. $\sqrt{\dfrac{5}{2a}}$

48. $\sqrt{\dfrac{1}{18z^3}}$

49. $\sqrt[3]{\dfrac{2}{5}}$

50. $\sqrt[3]{\dfrac{3}{4y}}$

51. $\sqrt{18n^3} + \sqrt{50n^3}$

52. $\sqrt[3]{24} - \sqrt[3]{81}$

53. $\dfrac{2\sqrt{3}}{\sqrt{3} - 1}$

54. $\dfrac{2}{\sqrt{6} - 2}$

55. $\dfrac{\sqrt{6}}{\sqrt{8} + \sqrt{18}}$

56. $\dfrac{\sqrt{15}}{\sqrt{75} + \sqrt{20}}$

Convert each number given in scientific notation to standard notation and each number given in standard notation to scientific notation.

57. 3.2×10^8

58. 4.543×10^9

59. 1.85×10^{-4}

60. 9.44×10^{-5}

61. 0.000056

62. 0.000341

63. 2,340,000

64. 88,300,000,000

Perform the indicated operations. Write the answer in scientific notation.

65. $(5 \times 10^6)^3$

66. $\dfrac{(0.00000046)(3000)}{2,300,000}$

67. $\dfrac{(800)^2(0.00001)^{-3}}{(2,000,000)^3(0.00002)}$

68. $\dfrac{(5.1 \times 10^8)(2 \times 10^{-3})}{1.7 \times 10^{-6}}$

Write each expression in the form $a + bi$, where a and b are real numbers.

69. $(3 - 7i) + (-4 + 6i)$

70. $(-6 - 3i) - (3 - 2i)$

71. $(4 - 5i)^2$

72. $7 - i(2 - 3i)^2$

73. $(1 - 3i)(2 + 6i)$

74. $(0.3 + 2i)(0.3 - 2i)$

75. $(2 - 3i) \div i$

76. $(-2 + 4i) \div (-i)$

77. $(1 - i) \div (2 + i)$

78. $(3 + 6i) \div (4 - i)$

79. $\dfrac{1 + i}{2 - 3i}$

80. $\dfrac{3 - i}{4 - 3i}$

81. $\dfrac{6 + \sqrt{-8}}{2}$

82. $\dfrac{-2 - \sqrt{-18}}{2}$

83. $\dfrac{-6 + \sqrt{(-2)^2 - 4(-1)(-6)}}{-8}$

84. $\dfrac{-9 - \sqrt{(-9)^2 - 4(-3)(-9)}}{-6}$

85. $i^{34} + i^{19}$

86. $\sqrt{6} + \sqrt{-3}\sqrt{-2}$

Perform the indicated operations.

87. $(3x^2 - x - 2) + (-x^2 + 2x - 5)$

88. $(4y^3 - y^2 + 5y - 9) - (y^3 - 6y^2 + 3y - 2)$

89. $(-4x^4 - 3x^3 + x) - (x^4 - 6x^3 - 2x)$

90. $(3y^4 - 4y^2 - 6) + (-y^4 - 8y + 7)$

91. $(3a^2 - 2a + 5)(a - 2)$

92. $(w - 5)(w^2 + 5w + 25)$

93. $(b - 3y)^2$

94. $(x - 1)^3$

95. $(t - 3)(3t + 2)$

96. $(5y - 9)(5y + 9)$

97. $(-35y^5) \div (7y^2)$

98. $(3x^3 - 6x^2) \div (3x)$

99. $(3 + \sqrt{2})(3 - \sqrt{2})$

100. $(2\sqrt{3} - 1)(3\sqrt{3} + 2)$

101. $(1 + \sqrt{3})^2$

102. $(\sqrt{2} - 1)^2$

103. $(2\sqrt{5} + \sqrt{3})^2$

104. $(3\sqrt{2} - \sqrt{7})^2$

Find the quotient and remainder when the first polynomial is divided by the second.

105. $x^3 + 2x^2 - 9x + 3, x - 2$

106. $x^3 - 6x^2 + 3x - 9, x + 3$

107. $6x^2 + x + 2, 2x - 1$

108. $12x^2 - x - 21, 3x - 4$

Use division to express each fraction in the form

quotient $+ \dfrac{\text{remainder}}{\text{divisor}}$.

109. $\dfrac{x^2 - 3}{x + 2}$

110. $\dfrac{a^2 + 5a + 2}{a}$

111. $\dfrac{2x + 3}{x - 5}$

112. $\dfrac{-3x + 2}{5x - 4}$

Factor each polynomial completely.

113. $6x^3 - 6x$

114. $4u^2 - 9v^2$

115. $9h^2 + 24ht + 16t^2$

116. $b^2 + 6b - 16$

117. $t^3 + y^3$

118. $8a^3 - 27$

119. $x^3 + 3x^2 - 9x - 27$

120. $3x - by + bx - 3y$

121. $t^6 - 1$

122. $y^4 - 625$

123. $18x^2 - 9x - 20$

124. $(x - 1)^2 - (x - 1) - 2$

125. $a^3b + 3a^2b - 18ab$

126. $x^3y^4 + 4x^2y^2 - 12x$

127. $2x^3 + x^2y - y - 2x$

128. $12x^3 - 2x^2y - 24xy^2$

Perform the indicated operations with the rational expressions.

129. $\dfrac{x-2}{x+3} + \dfrac{x+8}{x+3}$

130. $\dfrac{3x-5}{x^2-4} - \dfrac{3-x}{x^2-4}$

131. $\dfrac{x-1}{x-2} - \dfrac{x+3}{x+4}$

132. $\dfrac{1-x}{x} + \dfrac{3-x}{x-2}$

133. $\dfrac{x^2-9}{x+3} \cdot \dfrac{1}{6-2x}$

134. $\dfrac{x^3-8}{6} \cdot \dfrac{3x+6}{x^2-4}$

135. $\dfrac{a^3bc^8}{a^9b^3c} \cdot \dfrac{(ab^3c^5)^2}{a^4b^3}$

136. $\dfrac{(x^3z^2)^5}{xz^4} \cdot \left(\dfrac{xz^2}{x^3}\right)^2$

137. $\dfrac{1}{x^2-4} + \dfrac{3}{x-2}$

138. $\dfrac{3}{2x-4} + \dfrac{5}{3x-6}$

139. $\dfrac{1}{6x} - \dfrac{7}{10x^2}$

140. $\dfrac{1}{3y^2} - \dfrac{2}{9y}$

141. $\dfrac{a^2-25}{a^2-4a-5} \div \dfrac{2a+10}{a^2-1}$

142. $\dfrac{y^4-16}{y^2+y-2} \div \dfrac{y^3+4y}{y^3-y}$

143. $\dfrac{x^2-16}{x^2+5x+4} \div \dfrac{8-2x}{x^3+1}$

144. $\dfrac{x^2+ax+bx+ab}{x^2+2bx+b^2} \div \dfrac{x^2+2ax+a^2}{x^3+b^3}$

145. $\dfrac{a-2}{a^2+6a+5} + \dfrac{2a+1}{a^2-1}$

146. $\dfrac{y-1}{y^2-2y-24} - \dfrac{y-3}{y^2+2y-8}$

Simplify.

147. $\dfrac{\dfrac{5}{2x} - \dfrac{3}{4x}}{\dfrac{1}{2} - \dfrac{2}{x}}$

148. $\dfrac{\dfrac{4}{4-y^2} - \dfrac{5}{y-2}}{\dfrac{1}{2-y} - \dfrac{3}{y+2}}$

149. $\dfrac{\dfrac{1}{y^2-2} - 3}{\dfrac{5}{y^2-2} + 4}$

150. $\dfrac{\dfrac{1}{6a^2b^3}}{\dfrac{5}{8a^3b} - \dfrac{3}{10a^3b^4}}$

151. $\dfrac{a^{-2} - b^{-3}}{a^{-1}b^{-1}}$

152. $\dfrac{x^{-1} - y^{-1}}{x^{-3} - y^{-3}}$

153. $p^{-1} + pq^{-3}$

154. $a^{-1} + x^{-1}$

Given that

$$P(x) = x^3 - 3x^2 + x - 9 \quad \text{and} \quad R(x) = \dfrac{3x-1}{2x-9},$$

find each of the following.

155. $P(2)$

156. $P(-1)$

157. $P(0)$

158. $P\left(\dfrac{1}{2}\right)$

159. $R(-1)$

160. $R(3)$

161. $R(50)$

162. $R(-40)$

Solve each problem.

163. *Home Run* If a baseball is hit at a 45° angle with an initial velocity of v_0 ft/sec then it travels d feet, where

$$d = \dfrac{(v_0)^2}{32}.$$

a. Use the accompanying graph to estimate the distance if the ball is hit at 100 ft/sec.

b. If a major league power hitter increases his initial velocity from 125 ft/sec to 130 ft/sec, then how much distance will he gain?

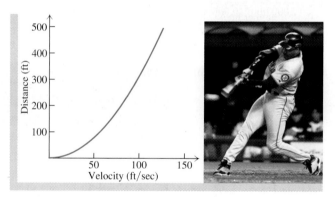

■ **Figure for Exercise 163**

164. *Thirty-Year Bonds* An investment of P dollars amounts to $P(1 + r)^n$ dollars after n years with interest rate r compounded annually. In February of 2005, the thirty-year treasury-bond yield was 5.375% (CNNFN, www.cnnfn.com). If you had invested $10,000 in treasury-bonds at that time, then what would it amount to in February of 2035 at 5.375% compounded annually?

165. *Weight of a Hydrogen Atom* One hydrogen atom weighs 1.7×10^{-24} g. Find the number of hydrogen atoms in 1 kg of hydrogen.

166. *Moon Talk* Radio waves travel at 3.0×10^8 m/sec (the same as the speed of light) and the distance from the earth to the moon is 3.84×10^8 m. How many seconds did it take for a radio wave to travel from mission control in Houston to the astronauts on the surface of the moon and then back to Houston? (A good demonstration of the speed of light actually occurred when Houston controllers heard an echo to their words that traveled to the moon and back.)

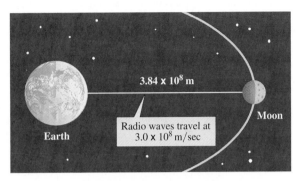

■ Figure for Exercise 166

167. *Distance*   What is the distance between -2.35 and 8.77 on the number line?

168. *Absolute Value*   Write an expression involving absolute value that gives the distance between the points a and -5 on the number line.

169. *Mowing a Lawn*   Howard and Will get summer jobs doing yard work. If Howard can mow an entire lawn in 6 hr and Will can mow it in 4 hr, then what portion of it can they mow in 2 hr working together?

170. *Cost of Watermelons*   Write a polynomial that expresses the total cost of $x + 3$ watermelons at $2 apiece and $x + 9$ watermelons at $3 apiece.

Thinking Outside the Box XI

Renumbering Elm Street   The city council in Perfect City has changed the numbering scheme for the 200 houses on Elm Street. The houses will be numbered with the natural numbers from 1 through 200. A city worker is given a box containing 1000 metal numbers, 100 of each digit, and told to distribute new house numbers in order of the addresses starting with 1 Elm Street. What address is the first one for which she will not have the correct digits?

■ Figure for Thinking Outside the Box XI

■■■ Chapter P Test

Determine which elements of the set

$$\{-\pi, -\sqrt{3}, -1.22, -1, 0, 2, \sqrt{5}, 10/3, 6.020020002\ldots\}$$

are members of the following sets.

  1. Real numbers

  2. Rational numbers

  3. Irrational numbers

  4. Nonnegative integers

Evaluate each expression.

  5. $|2 \cdot 3 - 5^2| - 6$

  6. $\dfrac{1}{8^{-1/3}}$

  7. $-27^{-2/3}$

  8. $\dfrac{(-3)^2 - 4(-3) + 9}{(-3 - 2)(5 - (-1))}$

Simplify each expression. Assume that all variables represent positive numbers.

  9. $(-2x^4y^2)(-3xy^5)$

  10. $(4x^4)^{1/2} + (-8x^6)^{1/3}$

  11. $\dfrac{(ab^2 + a^2b)^2}{a^2b^2}$

  12. $\dfrac{(-2a^{-1}b^6)^3}{(8a^9b^{-12})^{-2/3}}$

Simplify each expression. Assume that all variables represent positive real numbers.

  13. $\sqrt{27} - \sqrt{8} + \sqrt{32}$

  14. $\dfrac{\sqrt{8}}{\sqrt{6} - \sqrt{2}}$

  15. $\sqrt[3]{\dfrac{1}{4x^4}}$

  16. $\sqrt{12x^3y^9z^0}$

Perform the indicated operations and write the answer in the form $a + bi$, where a and b are real.

  17. $(4 - 3i)^2$

  18. $\dfrac{2 - i}{3 + i}$

  19. $i^6 - i^{35}$

  20. $\sqrt{-8}\left(\sqrt{-2} + \sqrt{6}\right)$

Perform the indicated operations.

  21. $(-x^3 - 5x) + (4x^3 + 3x^2 - 7x)$

  22. $(-x^2 + 3x - 4) - (4x^2 - 6x + 9)$

  23. $(x + 3)(x^2 - 2x - 1)$

  24. $(8h^3 - 1) \div (2h - 1)$

25. $(x - 9y)(x + 3y)$ **26.** $(x^3 - 2x - 4) \div (x - 2)$

27. $(3x - 8)^2$ **28.** $(2t^4 - 1)(2t^4 + 1)$

Perform each operation.

29. $\dfrac{x^3 - 5x^2 + 6x}{2x^2 - 6x} \cdot \dfrac{4x^3 + 32}{2x^3 - 8x}$

30. $\dfrac{x + 5}{x^2 - 4x + 3} + \dfrac{x - 1}{x^2 + x - 12}$

31. $\dfrac{a - 1}{4a^2 - 9} - \dfrac{a - 2}{3 - 2a}$ **32.** $\dfrac{\dfrac{1}{2a^2b} - 2a}{\dfrac{1}{4ab^3} + \dfrac{1}{3b}}$

Factor completely.

33. $ax^2 - 11ax + 18a$ **34.** $m^5 - m$

35. $3x^2 + 14x - 5$ **36.** $bx^2 - 3bx + wx - 3w$

Solve each problem.

37. In 2005 the total income for the 2.956×10^8 people in the United States was 9.777×10^{12} dollars (U.S. Census Bureau, www.census.gov). What was the per capita income?

38. If an investment of $P appreciates to $A after n years, then the value of the expression $(A/P)^{1/n} - 1$ is the annual appreciation rate. A couple paid $1.3 million for a house with a view of the Golden Gate Bridge and sold the house three years later for $2.5 million. Find the annual appreciation rate.

39. The altitude in feet of an arrow t seconds after it is shot straight upward is calculated from the polynomial $A(t) = -16t^2 + 120t$. Find the altitude of the arrow 2 sec after it is shot upward.

1 Equations, Inequalities, and Modeling

EVEN infamous Heartbreak Hill couldn't break the winning spirit of Kenyan runner Catherine Ndereba as she focused on first place. It was the 109th running of the Boston Marathon, a 26.2-mile ordeal that one runner called "14 miles of fun, 8 miles of sweat, and 4 miles of hell!"

Sporting events like the 2005 Marathon have come a long way since the first Olympic Games were held over 2500 years ago. Today, sports and science go hand in hand. Modern athletes often use mathematics to analyze the variables that help them increase aerobic capacity, reduce air resistance, or strengthen muscles.

WHAT YOU WILL LEARN In this chapter you will see examples of sports applications while you learn to solve equations and inequalities. By the time you reach the finish line, you will be using algebra to model and solve problems.

1.1

Equations in One Variable

One of our main goals in algebra is to develop techniques for solving a wide variety of equations. In this section we will solve linear equations and other similar equations.

Definitions

An **equation** is a statement (or sentence) indicating that two algebraic expressions are equal. The verb in an equation is the equality symbol. For example, $2x + 8 = 0$ is an equation. If we replace x by -4, we get $2 \cdot (-4) + 8 = 0$, a true statement. So we say that -4 is a **solution** or **root** to the equation or -4 **satisfies** the equation. If we replace x by 3, we get $2 \cdot 3 + 8 = 0$, a false statement. So 3 is not a solution.

Whether the equation $2x + 8 = 0$ is true or false depends on the value of x, and so it is called an **open sentence.** The equation is neither true nor false until we choose a value for x. The set of all solutions to an equation is called the **solution set** to the equation. To **solve** an equation means to find the solution set. The solution set for $2x + 8 = 0$ is $\{-4\}$. The equation $2x + 8 = 0$ is an example of a linear equation.

Definition: Linear Equation in One Variable

> A **linear equation in one variable** is an equation of the form $ax + b = 0$, where a and b are real numbers, with $a \neq 0$.

Note that other letters can be used in place of x. For example, $3t + 5 = 0$, $2w - 6 = 0$, and $-2u + 7 = 0$ are linear equations.

Solving Linear Equations

The equations $2x + 8 = 0$ and $2x = -8$ both have the solution set $\{-4\}$. Two equations with the same solution set are called **equivalent** equations. Adding the same real number to or subtracting the same real number from each side of an equation results in an equivalent equation. Multiplying or dividing each side of an equation by the same nonzero real number also results in an equivalent equation. These **properties of equality** are stated in symbols in the following box.

Properties of Equality

> If A and B are algebraic expressions and C is a real number, then the following equations are equivalent to $A = B$:
>
> | | |
> |---|---|
> | $A + C = B + C$ | **Addition property of equality** |
> | $A - C = B - C$ | **Subtraction property of equality** |
> | $CA = CB \ (C \neq 0)$ | **Multiplication property of equality** |
> | $\dfrac{A}{C} = \dfrac{B}{C} \ (C \neq 0)$ | **Division property of equality** |

We can use an algebraic expression for C in the properties of equality, because the value of an algebraic expression is a real number. However, this can produce nonequivalent equations. For example,

$$x = 0 \qquad \text{and} \qquad x + \frac{1}{x} = 0 + \frac{1}{x}$$

appear to be equivalent by the addition property of equality. But the first is satisfied by 0 and the second is not. When an equation contains expressions that are undefined for some real number(s) then we must check all solutions carefully.

Any linear equation, $ax + b = 0$, can be solved in two steps. Subtract b from each side and then divide each side by a ($a \neq 0$), to get $x = -b/a$. Although the equations in our first example are not exactly in the form $ax + b = 0$, they are often called linear equations because they are equivalent to linear equations.

Example 1 Using the properties of equality

Solve each equation.

a. $3x - 4 = 8$ **b.** $\frac{1}{2}x - 6 = \frac{3}{4}x - 9$ **c.** $3(4x - 1) = 4 - 6(x - 3)$

Solution

a. $3x - 4 = 8$

$3x - 4 + 4 = 8 + 4$ Add 4 to each side.

$3x = 12$ Simplify.

$\dfrac{3x}{3} = \dfrac{12}{3}$ Divide each side by 3.

$x = 4$ Simplify.

Since the last equation is equivalent to the original, the solution set to the original equation is $\{4\}$. We can check by replacing x by 4 in $3x - 4 = 8$. Since $3 \cdot 4 - 4 = 8$ is correct, we are confident that the solution set is $\{4\}$.

b. Multiplying each side of the equation by the least common denominator, LCD, will eliminate all of the fractions:

$$\frac{1}{2}x - 6 = \frac{3}{4}x - 9$$

$$4\left(\frac{1}{2}x - 6\right) = 4\left(\frac{3}{4}x - 9\right) \qquad \text{Multiply each side by 4, the LCD.}$$

$$2x - 24 = 3x - 36 \qquad \text{Distributive property}$$

$$2x - 24 - 3x = 3x - 36 - 3x \qquad \text{Subtract } 3x \text{ from each side.}$$

$$-x - 24 = -36 \qquad \text{Simplify.}$$

$$-x = -12 \qquad \text{Add 24 to each side.}$$

$$(-1)(-x) = (-1)(-12) \qquad \text{Multiply each side by } -1.$$

$$x = 12 \qquad \text{Simplify.}$$

Check 12 in the original equation. The solution set is $\{12\}$.

c. $3(4x - 1) = 4 - 6(x - 3)$

$\quad 12x - 3 = 4 - 6x + 18$ Distributive property

$\quad 12x - 3 = 22 - 6x$ Simplify.

$\quad\quad 18x - 3 = 22$ Add $6x$ to each side.

$\quad\quad\quad 18x = 25$ Add 3 to each side.

$\quad\quad\quad\quad x = \dfrac{25}{18}$ Divide each side by 18.

Check $25/18$ in the original equation. The solution set is $\left\{\dfrac{25}{18}\right\}$.

You can use a graphing calculator to calculate the value of each side of the equation when x is $25/18$ as shown in Fig. 1.1.

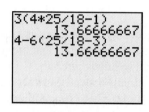

■ **Figure 1.1**

Try This. Solve $5(3x - 2) = 5 - 7(x - 1)$.   ■

Note that checking the equation in Example 1(c) with a calculator did not prove that $25/18$ is the correct solution. The properties of equality that were applied correctly in each step guarantee that we have the correct solution. The values of the two sides of the equation could agree for the 10 digits shown on the calculator and disagree for the digits not shown. Since that possibility is extremely unlikely, the calculator check does support our belief that we have the correct solution.   □

Identities, Conditional Equations, and Inconsistent Equations

An equation that is satisfied by every real number for which both sides are defined is an **identity.** Some examples of identities are

$$3x - 1 = 3x - 1, \quad\quad 2x + 5x = 7x, \quad \text{and} \quad \frac{x}{x} = 1.$$

The solution set to the first two identities is the set of all real numbers, R. Since $0/0$ is undefined, the solution set to $x/x = 1$ is the set of nonzero real numbers, $\{x \mid x \neq 0\}$.

A **conditional equation** is an equation that is satisfied by at least one real number but is not an identity. The equation $3x - 4 = 8$ is true only on condition that $x = 4$, and it is a conditional equation. The equations of Example 1 are conditional equations.

An **inconsistent equation** is an equation that has no solution. Some inconsistent equations are

$$0 \cdot x + 1 = 2, \quad\quad x + 3 = x + 5, \quad \text{and} \quad 9x - 9x = 8.$$

Note that each of these inconsistent equations is equivalent to a false statement: $1 = 2, 3 = 5,$ and $0 = 8,$ respectively.

Example **2** Classifying an equation

Determine whether the equation $3(x - 1) - 2x(4 - x) = (2x + 1)(x - 3)$ is an identity, an inconsistent equation, or a conditional equation.

Solution

$$3(x - 1) - 2x(4 - x) = (2x + 1)(x - 3)$$

$$3x - 3 - 8x + 2x^2 = 2x^2 - 5x - 3 \qquad \text{Simplify each side.}$$

$$2x^2 - 5x - 3 = 2x^2 - 5x - 3$$

Since the last equation is equivalent to the original and the last equation is an identity, the original equation is an identity.

Try This. Determine whether $x(x - 1) - 6 = (x - 3)(x + 2)$ is an identity, an inconsistent equation, or a conditional equation. ■

Equations Involving Rational Expressions

In Example 1(b) we solved an equation involving fractions. To simplify that equation, the first step was to multiply by the LCD of the fractions. The equations in the next example all involve rational expressions. Notice that multiplying each side of these equations by the LCD greatly simplifies the equations.

Example **3** **Equations involving rational expressions**

Solve each equation and identify each as an identity, an inconsistent equation, or a conditional equation.

a. $\dfrac{y}{y - 3} + 3 = \dfrac{3}{y - 3}$ **b.** $\dfrac{1}{x - 1} - \dfrac{1}{x + 1} = \dfrac{2}{x^2 - 1}$ **c.** $\dfrac{1}{2} + \dfrac{1}{x - 1} = 1$

Solution

a. Since $y - 3$ is the denominator in each rational expression, $y - 3$ is the LCD. Note that using 3 in place of y in the equation would cause 0 to appear in the denominators.

$$(y - 3)\left(\frac{y}{y - 3} + 3\right) = (y - 3)\frac{3}{y - 3} \qquad \text{Multiply each side by the LCD.}$$

$$(y - 3)\frac{y}{y - 3} + (y - 3)3 = 3 \qquad \text{Distributive property}$$

$$y + 3y - 9 = 3$$

$$4y - 9 = 3$$

$$4y = 12 \qquad \text{Add 9 to each side.}$$

$$y = 3 \qquad \text{Divide each side by 4.}$$

If we replace y by 3 in the original equation, then we get two undefined expressions. So 3 is not a solution to the original equation. The original equation has no solution. The equation is inconsistent.

b. Since $x^2 - 1 = (x - 1)(x + 1)$, the LCD is $(x - 1)(x + 1)$. Note that using 1 or -1 for x in the equation would cause 0 to appear in a denominator.

$$\frac{1}{x - 1} - \frac{1}{x + 1} = \frac{2}{x^2 - 1}$$

$$(x - 1)(x + 1)\left(\frac{1}{x - 1} - \frac{1}{x + 1}\right) = (x - 1)(x + 1)\frac{2}{x^2 - 1} \quad \text{Multiple by the LCD.}$$

$$(x - 1)(x + 1)\frac{1}{x - 1} - (x - 1)(x + 1)\frac{1}{x + 1} = 2 \qquad \text{Distributive property}$$

$$x + 1 - (x - 1) = 2$$

$$2 = 2$$

Since the last equation is an identity, the original equation is also an identity. The solution set is $\{x \mid x \neq 1 \text{ and } x \neq -1\}$, because 1 and -1 cannot be used for x in the original equation.

c. Note that we cannot use 1 for x in the original equation. To solve the equation multiply each side by the LCD:

$$\frac{1}{2} + \frac{1}{x - 1} = 1$$

$$2(x - 1)\left(\frac{1}{2} + \frac{1}{x - 1}\right) = 2(x - 1)1 \quad \text{Multiply by the LCD.}$$

$$x - 1 + 2 = 2x - 2$$

$$x + 1 = 2x - 2$$

$$3 = x$$

Check 3 in the original equation. The solution set is $\{3\}$, and the equation is a conditional equation.

Try This. Solve $\dfrac{2}{x - 3} - \dfrac{3}{x + 3} = \dfrac{4}{x^2 - 9}$. ▪

In Example 3(a) the final equation had a root that did not satisfy the original equation, because the domain of the rational expression excluded the root. Such a root is called an **extraneous root.** If an equation has no solution, then its solution set is the **empty set** (the set with no members). The symbol $\varnothing$ is used to represent the empty set.

Example **4** **Using a calculator in solving an equation**

Solve

$$\frac{7}{2.4x} + \frac{3}{5.9} = \frac{1}{8.2}$$

with the aid of a calculator. Round the answer to three decimal places.

Solution

We could multiply each side by the LCD, but since we are using a calculator, we can subtract 3/5.9 from each side to isolate x.

$$\frac{7}{2.4x} + \frac{3}{5.9} = \frac{1}{8.2}$$

$$\frac{7}{2.4x} = \frac{1}{8.2} - \frac{3}{5.9}$$

$$\frac{7}{2.4x} \approx -0.38652336 \qquad \text{Use a calculator to simplify.}$$

$$\frac{7}{2.4} \approx -0.38652336x \qquad \text{Multiply each side by } x.$$

$$\frac{7}{2.4(-0.38652336)} \approx x \qquad \text{Divide each side by } -0.38652336.$$

$$x \approx -7.546 \qquad \text{Round to three decimal places.}$$

The solution set is $\{-7.546\}$. Since -0.38652336 is an approximate value, the sign $\approx$ (for "approximately equal to") is used instead of the equal sign. To get three-decimal-place accuracy in the final answer, use as many digits as your calculator allows until you get to the final computation.

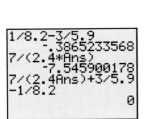

Figure 1.2 shows the computations and the check. Note how the ANS key is used.

Try This. Solve $\dfrac{2}{3.4x} - \dfrac{1}{8.9} = \dfrac{4}{4.7}$ with the aid of a calculator.   ■

■ **Figure 1.2**

Equations Involving Absolute Value

To solve equations involving absolute value, remember that $|x| = x$ if $x \geq 0$, and $|x| = -x$ if $x < 0$. The absolute value of x is greater than or equal to 0 for any real number x. So an equation such as $|x| = -6$ has no solution. Since a number and its opposite have the same absolute value, $|x| = 4$ is equivalent to $x = 4$ or $x = -4$. The only number that has 0 absolute value is 0. These ideas are summarized as follows.

S U M M A R Y   **Basic Absolute Value Equations**

Absolute value equation	Equivalent statement	Solution set
$\|x\| = k$ $(k > 0)$	$x = -k$ or $x = k$	$\{-k, k\}$
$\|x\| = 0$	$x = 0$	$\{0\}$
$\|x\| = k$ $(k < 0)$		$\varnothing$

Example **5** Equations involving absolute value

Solve each equation.

a. $|x - 5| = 4$ **b.** $2|x + 8| - 6 = 0$

Solution

a. First write an equivalent statement without using absolute value symbols.

$$|x - 5| = 4$$

$$x - 5 = -4 \quad \text{or} \quad x - 5 = 4$$

$$x = 1 \quad \text{or} \quad x = 9$$

The solution set is $\{1, 9\}$. Check that $|1 - 5| = 4$ and $|9 - 5| = 4$.

b. First isolate $|x + 8|$.

$$2|x + 8| - 6 = 0$$

$$2|x + 8| = 6$$

$$|x + 8| = 3$$

Now write an equivalent statement without absolute value symbols.

$$x + 8 = -3 \quad \text{or} \quad x + 8 = 3$$

$$x = -11 \quad \text{or} \quad x = -5$$

The solution set is $\{-11, -5\}$. Check that $2|-11 + 8| - 6 = 0$ and $2|-5 + 8| - 6 = 0$.

Try This. Solve $|2x - 3| = 5$. ■

S T R A T E G Y   Equation-Solving Overview

In solving equations, there are usually many different sequences of correct steps that lead to the correct solution. If the approach that you try first does not work, *try another approach*, but learn from your failures as well as your successes. Solving equations successfully takes patience and practice.

Equations Used as Models

Equations are often used to describe or **model** real situations. In the coming sections we will learn how to write equations that model real situations. In the next example you are simply given an equation and asked to solve it.

Example **6** An equation used as a model

If x is the number of years after 1990 and y is the median income in dollars for working women in the United States, then the equation $y = 355.9x + 11,075.3$ models the real data (www.infoplease.com). In what year (to the nearest year) is the median income $20,000?

Solution

Since y is 20,000, we solve the following equation to find x:

$$20,000 = 355.9x + 11,075.3$$

$$8924.7 = 355.9x \qquad \text{Subtract 11,075.3 from each side.}$$

$$25.08 \approx x \qquad \text{Divide each side by 355.9.}$$

So to the nearest year the median income reaches $20,000 in 25 years after 1990, or 2015.

Try This. In what year (to the nearest year) will the median income be $25,000? ■

For Thought

True or False? Explain.

1. The number 3 is in the solution set to
$5(4 - x) = 2x - 1$.

2. The equation $3x - 1 = 8$ is equivalent to $3x - 2 = 7$.

3. The equation $x + \sqrt{x} = -2 + \sqrt{x}$ is equivalent to
$x = -2$.

4. The solution set to $x - x = 7$ is the empty set.

5. The equation $12x = 0$ is an inconsistent equation.

6. The equation $x - 0.02x = 0.98x$ is an identity.

7. The equation $|x| = -8$ is equivalent to $x = 8$ or
$x = -8$.

8. The equations $\dfrac{x}{x - 5} = \dfrac{5}{x - 5}$ and $x = 5$ are equivalent.

9. To solve $-\dfrac{2}{3}x = \dfrac{3}{4}$, we should multiply each side by $-\dfrac{2}{3}$.

10. If a and b are real numbers, then $ax + b = 0$ has a
solution.

1.1 Exercises

Determine whether each given number is a solution to the equation following it.

1. $3, 2x - 4 = 9$

2. $-2, \dfrac{1}{x} - \dfrac{1}{2} = -1$

3. $-3, (x - 1)^2 = 16$

4. $4, \sqrt{3x + 4} = -4$

Solve each equation and check your answer.

5. $3x - 5 = 0$

6. $-2x + 3 = 0$

7. $-3x + 6 = 12$

8. $5x - 3 = -13$

9. $8x - 6 = 1 - 6x$

10. $4x - 3 = 6x - 1$

11. $7 + 3x = 4(x - 1)$

12. $-3(x - 5) = 4 - 2x$

13. $-\dfrac{3}{4}x = 18$

14. $\dfrac{2}{3}x = -9$

15. $\dfrac{x}{2} - 5 = -12 - \dfrac{2x}{3}$

16. $\dfrac{x}{4} - 3 = \dfrac{x}{2} + 3$

17. $\dfrac{3}{2}x + \dfrac{1}{3} = \dfrac{1}{4}x - \dfrac{1}{6}$

18. $\dfrac{x}{2} + \dfrac{x}{5} = \dfrac{x}{6} - \dfrac{1}{3}$

Solve each equation. Identify each equation as an identity, an inconsistent equation, or a conditional equation.

19. $3(x - 6) = 3x - 18$

20. $2a + 3a = 6a$

21. $2x + 3x = 4x$

22. $4(y - 1) = 4y - 4$

23. $2(x + 3) = 3(x - 1)$

24. $2(x + 1) = 3x + 2$

25. $3(x - 6) = 3x + 18$

26. $2x + 3x = 5x + 1$

27. $\dfrac{3x}{x} = 3$

28. $\dfrac{x(x + 2)}{x + 2} = x$

Solve each equation involving rational expressions. Identify each equation as an identity, an inconsistent equation, or a conditional equation.

29. $\dfrac{1}{w - 1} - \dfrac{1}{2w - 2} = \dfrac{1}{2w - 2}$

30. $\dfrac{1}{x} + \dfrac{1}{x - 3} = \dfrac{9}{x^2 - 3x}$

31. $\dfrac{1}{x} - \dfrac{1}{3x} = \dfrac{1}{2x} + \dfrac{1}{6x}$

32. $\dfrac{1}{5x} - \dfrac{1}{4x} + \dfrac{1}{3x} = -\dfrac{17}{60}$

33. $\dfrac{z + 2}{z - 3} = \dfrac{5}{-3}$

34. $\dfrac{2x - 3}{x - 4} = \dfrac{5}{x - 4}$

35. $\dfrac{1}{x - 3} - \dfrac{1}{x + 3} = \dfrac{6}{x^2 - 9}$

36. $\dfrac{4}{x - 1} - \dfrac{9}{x + 1} = \dfrac{3}{x^2 - 1}$

37. $4 + \dfrac{6}{y - 3} = \dfrac{2y}{y - 3}$

38. $\dfrac{x}{x + 6} - 3 = 1 - \dfrac{6}{x + 6}$

39. $\dfrac{t}{t + 3} + 4 = \dfrac{2}{t + 3}$

40. $\dfrac{3x}{x + 1} - 5 = \dfrac{x - 11}{x + 1}$

Use a calculator to help you solve each equation. Round each approximate answer to three decimal places.

41. $0.27x - 3.9 = 0.48x + 0.29$

42. $x - 2.4 = 0.08x + 3.5$

43. $0.06(x - 3.78) = 1.95$

44. $0.86(3.7 - 2.3x) = 4.9$

45. $2a + 1 = -\sqrt{17}$ **46.** $3c + 4 = \sqrt{38}$

47. $\dfrac{0.001}{y - 0.333} = 3$ **48.** $1 + \dfrac{0.001}{t - 1} = 0$

49. $\dfrac{x}{0.376} + \dfrac{x}{0.135} = 2$ **50.** $\dfrac{1}{x} + \dfrac{5}{6.72} = 10.379$

51. $(x + 3.25)^2 = (x - 4.1)^2$

52. $0.25(2x - 1.6)^2 = (x - 0.9)^2$

53. $(2.3 \times 10^6)x + 8.9 \times 10^5 = 1.63 \times 10^4$

54. $(-3.4 \times 10^{-9})x + 3.45 \times 10^{-8} = 1.63 \times 10^4$

Solve each absolute value equation. Use the summary on page 93.

55. $|x| = 8$ **56.** $|x| = 2.6$

57. $|x - 4| = 8$ **58.** $|x - 5| = 3.6$

59. $|x - 6| = 0$ **60.** $|x - 7| = 0$

61. $|x + 8| = -3$ **62.** $|x + 9| = -6$

63. $|2x - 3| = 7$ **64.** $|3x + 4| = 12$

65. $\dfrac{1}{2}|x - 9| = 16$ **66.** $\dfrac{2}{3}|x + 4| = 8$

67. $2|x + 5| - 10 = 0$ **68.** $6 - 4|x + 3| = -2$

69. $8|3x - 2| = 0$ **70.** $5|6 - 3x| = 0$

71. $2|x| + 7 = 6$ **72.** $5 + 3|x - 4| = 0$

Solve each equation.

73. $x - 0.05x = 190$ **74.** $x + 0.1x = 121$

75. $0.1x - 0.05(x - 20) = 1.2$

76. $0.03x - 0.2 = 0.2(x + 0.03)$

77. $(x + 2)^2 = x^2 + 4$ **78.** $(x - 3)^2 = x^2 - 9$

79. $(2x - 3)^2 = (2x + 5)^2$

80. $(3x - 4)^2 + (4x + 1)^2 = (5x + 2)^2$

81. $\dfrac{x}{2} + 1 = \dfrac{1}{4}(x - 6)$

82. $-\dfrac{1}{6}(x + 3) = \dfrac{1}{4}(3 - x)$

83. $\dfrac{y - 3}{2} + \dfrac{y}{5} = 3 - \dfrac{y + 1}{6}$

84. $\dfrac{y - 3}{5} - \dfrac{y - 4}{2} = 5$

85. $5 + 7|x + 6| = 19$ **86.** $9 - |2x - 3| = 6$

87. $9 - 4|2x - 3| = 9$

88. $-7 - |3x + 1| = |3x + 1| - 7$

89. $8 - 5|5x + 1| = 12$

90. $5|7 - 3x| + 2 = 4|7 - 3x| - 1$

91. $\dfrac{3}{x - 2} + \dfrac{4}{x + 2} = \dfrac{7x - 2}{x^2 - 4}$

92. $\dfrac{2}{x - 1} - \dfrac{3}{x + 2} = \dfrac{8 - x}{x^2 + x - 2}$

93. $\dfrac{4}{x + 3} - \dfrac{3}{2 - x} = \dfrac{7x + 1}{x^2 + x - 6}$

94. $\dfrac{3}{x} - \dfrac{4}{1 - x} = \dfrac{7x - 3}{x^2 - x}$ **95.** $\dfrac{x - 2}{x - 3} = \dfrac{x - 3}{x - 4}$

96. $\dfrac{y - 1}{y + 4} = \dfrac{y + 1}{y - 2}$

Solve each problem.

97. *Working Mothers* The percentage of working mothers y can be modeled by the equation

$$y = 0.0102x + 0.644$$

where x is the number of years since 1990 (U.S. Census Bureau, www.census.gov).

a. Use the accompanying graph to estimate the year in which 70% of mothers were in the work force.

b. Is the percentage of mothers in the labor force increasing or decreasing?

c. Use the equation to find the year in which 90% of mothers will be in the labor force.

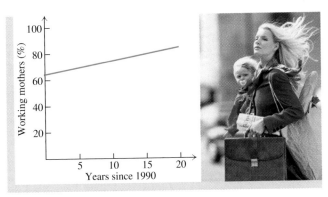

■ **Figure for Exercises 97 and 98**

98. *Working Mothers*   Use the equation in the previous exercise to find the year in which 64.4% of mothers were in the labor force.

99. *Cost Accounting*   An accountant has been told to distribute a bonus to the employees that is 15% of the company's net income. Since the bonus is an expense to the accountant, it must be subtracted from the income to determine the net income. If the company has an income of $140,000 before the bonus, then the accountant must solve

$$B = 0.15(140,000 - B)$$

to find the bonus B. Find B.

100. *Corporate Taxes*   For a class C corporation in Louisiana, the amount of state income tax S is deductible on the federal return and the amount of federal income tax F is deductible on the state return. With $200,000 taxable income and a 30% federal tax rate, the federal tax is $0.30(200,000 - S)$. If the state tax rate is 6% then the state tax satisfies

$$S = 0.06(200,000 - 0.30(200,000 - S)).$$

Find the state tax S and the federal tax F.

101. *Production Cost*   An automobile manufacturer, who spent $500 million to develop a new line of cars, wants the cost of development and production to be $12,000 per vehicle. If the production costs are $10,000 per vehicle, then the cost per vehicle for development and production of x vehicles is $(10,000x + 500,000,000)/x$ dollars. Solve the equation

$$\frac{10,000x + 500,000,000}{x} = 12,000$$

to find the number of vehicles that must be sold so that the cost of development and production is $12,000 per vehicle.

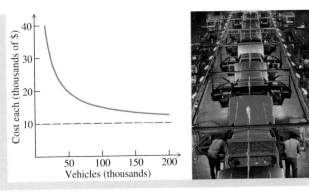

■ **Figure for Exercise 101**

102. *Harmonic Mean*   The *harmonic mean* of the numbers $x_1, x_2, \ldots, x_n$ is defined as

$$HM = \frac{n}{\dfrac{1}{x_1} + \dfrac{1}{x_2} + \cdots + \dfrac{1}{x_n}}.$$

The accompanying table shows the number of vehicles (in millions) exported from Japan for the years 1999 through 2004. (Japanese Automobile Manufacturers Association, www.jama.org.)

■ **Table for Exercise 102**

Year	Vehicles (millions)
1999	4.409
2000	4.455
2001	4.166
2002	4.698
2003	4.756
2004	4.832

a. What is the harmonic mean for the number of vehicles exported for those years?

b. If the harmonic mean for 1999 through 2005 is 4.627 million vehicles per year, then what is the number of vehicles exported in 2005?

For Writing/Discussion

103. *Definitions*   Without looking back in the text, write the definitions of linear equation, identity, and inconsistent equation. Use complete sentences.

104. *Cooperative Learning*   Each student in a small group should write a linear equation, an identity, an inconsistent equation, and an equation that has an extraneous root. The group should solve each equation and determine whether each equation is of the required type.

Thinking Outside the Box XII

Nines   Nine people applying for credit at the Highway 99 Loan Company listed nine different incomes each containing a different number of digits. Each of the nine incomes was a whole number of dollars and the maximum income was a nine-digit number. The loan officer found that the arithmetic mean of the nine incomes was $123,456,789. What are the nine incomes?

1.1 Pop Quiz

Solve each equation and identify each as an identity, inconsistent equation, or a conditional equation.

1. $7x - 6 = 0$

2. $\dfrac{1}{4}x - \dfrac{1}{6} = \dfrac{1}{3}$

3. $3(x - 9) = 3x - 27$

4. $|w - 1| = 6$

5. $2(x + 6) = 2x + 6$

6. $(x + 1)^2 = x^2 + 1$

Linking Concepts

For Individual or Group Explorations

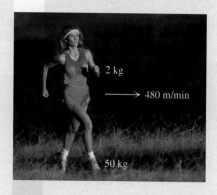

Modeling Oxygen Uptake in Runners

In a study of oxygen uptake rate for marathon runners (Costill and Fox, Medicine and Science in Sports, Vol. 1) it was found that the power expended in kilocalories per minute is given by the formula $P = M(av - b)$ where M is the mass of the runner in kilograms and v is the speed in meters per minute. The constants a and b have values

$$a = 1.02 \times 10^{-3} \; kcal/kg \, m \quad and \quad b = 2.62 \times 10^{-2} \; kcal/kg \, min.$$

a)   Runners with masses of 60 kg, 65 kg, and 70 kg are running together at 400 m/min. Find the power expenditure for each runner.

b)   With a constant velocity, does the power expended increase or decrease as the mass of the runner increases?

c)   Runners of 80 kg, 84 kg, and 90 kg, are all expending power at the rate of 38.7 kcal/min. Find the velocity at which each is running.

d)   With a constant power expenditure, does the velocity increase or decrease as the mass of the runner increases?

e)   A 50 kg runner in training has a velocity of 480 m/min while carrying weights of 2 kg. Assume that her power expenditure remains constant when the weights are removed, and find her velocity without the weights.

f)   Why do runners in training carry weights?

g)   Use a computer to create graphs showing power expenditure versus mass with a velocity fixed at 400 m/min and velocity versus mass with a fixed power expenditure of 40 kcal/min.

1.2

Constructing Models to Solve Problems

In Section 1.1 we solved equations. In this section we will use those skills in problem solving. To solve a problem we construct a mathematical model of the problem. Sometimes we use well-known formulas to model real situations, but we must often construct our own models.

Modeling with Formulas

A **formula** is an equation that involves two or more variables. Formulas are used to **model** real-life phenomena. For example, the formula $D = RT$ models the relationship between distance, rate, and time in uniform motion. The formula $P = 2L + 2W$ models the relationship between the perimeter, length, and width of a rectangle. A list of formulas commonly used as mathematical models is given inside the back cover of this text.

The formula $D = RT$ is said to be **solved for** D. Dividing each side of $D = RT$ by T yields $R = D/T$ and we have solved the formula for R. When a formula is solved for a specified variable, that variable is isolated on one side of the equal sign and must not occur on the other side.

Example **1** Solving a formula for a specified variable

Solve the formula $S = P + Prt$ for P.

Solution

$$P + Prt = S \qquad \text{Write the formula with } P \text{ on the left.}$$

$$P(1 + rt) = S \qquad \text{Factor out } P.$$

$$P = \frac{S}{1 + rt} \qquad \text{Divide each side by } 1 + rt.$$

Try This. Solve $A = \frac{1}{2}hb_1 + \frac{1}{2}hb_2$ for h.   ■

In some situations we know the values of all variables except one. After we substitute values for those variables, the formula is an equation in one variable. We can then solve the equation to find the value of the remaining variable.

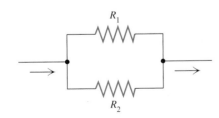

■ **Figure 1.3**

Example **2** Finding the value of a variable in a formula

The total resistance R (in ohms) in the parallel circuit shown in Fig. 1.3 is modeled by the formula

$$\frac{1}{R} = \frac{1}{R_1} + \frac{1}{R_2}.$$

The subscripts 1 and 2 indicate that R_1 and R_2 represent the resistance for two different receivers. If $R = 7$ ohms and $R_1 = 10$ ohms, then what is the value of R_2?

Solution

Substitute the values for R and R_1 and solve for R_2.

$$\frac{1}{7} = \frac{1}{10} + \frac{1}{R_2}$$
$\qquad$ The LCD of 7, 10, and R_2 is $70R_2$.

$$70R_2 \cdot \frac{1}{7} = 70R_2\left(\frac{1}{10} + \frac{1}{R_2}\right)$$
$\qquad$ Multiply each side by $70R_2$.

$$10R_2 = 7R_2 + 70$$

$$3R_2 = 70$$

$$R_2 = \frac{70}{3}$$

Check the solution in the original formula. The resistance R_2 is 70/3 ohms or about 23.3 ohms.

Try This. Find b_1 if $A = 20$, $h = 2$, $b_2 = 3$, and $A = \frac{1}{2}h(b_1 + b_2)$.   ■

Constructing Your Own Models

Applied problems in mathematics often involve solving an equation. The equations for some problems come from known formulas, as in Example 2, while in other problems we must write an equation that models a particular problem situation.

The best way to learn to solve problems is to study a few examples and then solve lots of problems. We will first look at an example of problem solving, then give a problem-solving strategy.

Example **3** Solving a problem involving sales tax

Jeannie Fung bought a Ford Mustang GT for a total cost of $18,966, including sales tax. If the sales tax rate is 9%, then what amount of tax did she pay?

Solution

There are two unknown quantities here, the price of the car and the amount of the tax. Represent the unknown quantities as follows:

$$x = \text{the price of the car}$$
$$0.09x = \text{amount of sales tax}$$

The price of the car plus the amount of sales tax is the total cost of the car. We can model this relationship with an equation and solve it for x:

$$x + 0.09x = 18{,}966$$

$$1.09x = 18{,}966$$

$$x = \frac{18{,}966}{1.09} = 17{,}400$$

$$0.09x = 1566$$

You can check this answer by adding $17,400 and $1566 to get $18,966, the total cost. The amount of tax that Jeannie Fung paid was $1566.

Try This. Joe bought a new computer for $1506.75, including sales tax at 5%. What amount of tax did he pay?
$\qquad$ ■

No two problems are exactly alike, but there are similarities. The following strategy will assist you in solving problems on your own.

S T R A T E G Y   Problem Solving

1. Read the problem as many times as necessary to get an understanding of the problem.

2. If possible, draw a diagram to illustrate the problem.

3. Choose a variable, write down what it represents, and if possible, represent any other unknown quantities in terms of that variable.

4. Write an equation that models the situation. You may be able to use a known formula, or you may have to write an equation that models only that particular problem.

5. Solve the equation.

6. Check your answer by using it to solve the original problem (not just the equation).

7. Answer the question posed in the original problem.

In the next example we have a geometric situation. Notice how we are following the strategy for problem solving.

Example **4** Solving a geometric problem

In 1974, Chinese workers found three pits that contained life-size sculptures of warriors, which were created to guard the tomb of an emperor (www.chinatour.com). The largest rectangular pit has a length that is 40 yards longer than three times the width. If its perimeter is 640 yards, what are the length and width?

Solution

First draw a diagram as shown in Fig. 1.4. Use the fact that the length is 40 yards longer than three times the width to represent the width and length as follows:

$$x = \text{the width in yards}$$
$$3x + 40 = \text{the length in yards}$$

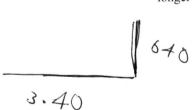

■ Figure 1.4

The formula for the perimeter of a rectangle is $2L + 2W = P$. Replace W by x, L by $3x + 40$, and P by 640.

$$2L + 2W = P \qquad \text{Perimeter formula}$$
$$2(3x + 40) + 2x = 640 \qquad \text{Substitution}$$
$$6x + 80 + 2x = 640$$
$$8x + 80 = 640$$
$$8x = 560$$
$$x = 70$$

If $x = 70$, then $3x + 40 = 250$. Check that the dimensions of 70 yards and 250 yards give a perimeter of 640 yards. We conclude that the length of the pit is 250 yards, and its width is 70 yards.

Try This. The length of a rectangle is 20 cm shorter than five times its width. If the perimeter is 800 cm, then what is the length?   ■

The next problem is called a **uniform-motion** problem because it involves motion at a constant rate. Of course, people do not usually move at a constant rate, but their average speed can be assumed to be constant over some time interval. The problem also illustrates how a table can be used as an effective technique for organizing information.

Example 5 Solving a uniform-motion problem

A group of hikers from Tulsa hiked down into the Grand Canyon in 3 hours 30 minutes. Coming back up on a trail that was 4 miles shorter, they hiked 2 mph slower and it took them 1 hour longer. What was their rate going down?

Solution

Let x represent the rate going down into the canyon. Make a table to show the distance, rate, and time for both the trip down and the trip back up. Once we fill in any two entries in a row of the table, we can use $D = RT$ to obtain an expression for the third entry:

	Rate	Time	Distance
Down	x mi/hr	3.5 hr	$3.5x$ mi
Up	$x - 2$ mi/hr	4.5 hr	$4.5(x - 2)$ mi

Using the fact that the distance up was 4 miles shorter, we can write the following equation:

$$4.5(x - 2) = 3.5x - 4$$
$$4.5x - 9 = 3.5x - 4$$
$$x - 9 = -4$$
$$x = 5$$

After checking that 5 mph satisfies the conditions given in the problem, we conclude that the hikers traveled at 5 mph going down into the canyon.

Try This. Bea hiked uphill from her car to a waterfall in 6 hours. Hiking back to her car over the same route, she averaged 2 mph more and made the return trip in half the time. How far did she hike?   ■

Average speed is not necessarily the average of your speeds. For example, if you drive 70 mph for 3 hours and then 30 mph for 1 hour, you will travel 240 miles in 4 hours. Your average speed for the trip is 60 mph, which is not the average of 70 and 30.

Example 6 Average speed

Shelly drove 40 miles from Peoria to Bloomington at 20 mph. She then drove back to Peoria at a higher rate of speed so that she averaged 30 mph for the whole trip. What was her speed on the return trip?

Solution

Let x represent the speed on the return trip. Make a table as follows:

	Rate	Time	Distance
Going	20 mph	2 hr	40 mi
Returning	x mph	$40/x$ hr	40 mi
Round trip	30 mph	$2 + 40/x$ hr	80 mi

Use $D = RT$ to write an equation for the round trip and solve it.

$$80 = 30\left(2 + \frac{40}{x}\right)$$
$$80 = 60 + \frac{1200}{x}$$
$$20 = \frac{1200}{x}$$
$$20x = 1200$$
$$x = 60$$

So the speed on the return trip was 60 mph. It is interesting to note that the distance between the cities does not affect the solution. You should repeat this example using d as the distance between the cities.

Try This. Dee drove 20 miles to work at 60 mph. She then drove back home at a lower rate of speed, averaging 50 mph for the round trip. What was her speed on the return trip?   ■

The next example involves mixing beverages with two different concentrations of orange juice. In other **mixture problems,** we may mix chemical solutions, candy, or even people.

Example 7 Solving a mixture problem

A beverage producer makes two products, Orange Drink, containing 10% orange juice, and Orange Delight, containing 50% orange juice. How many gallons of Orange Delight must be mixed with 300 gallons of Orange Drink to create a new product containing 40% orange juice?

Solution

300 gallons x gallons $x + 300$ gallons

■ **Figure 1.5**

Let x represent the number of gallons of Orange Delight and make a sketch as in Fig. 1.5. Next, we make a table that shows three pertinent expressions for each product: the amount of the product, the percent of orange juice in the product, and the actual amount of orange juice in that product.

	Amount of Product	**Percent Orange Juice**	**Amount Orange Juice**
Drink	300 gal	10%	0.10(300) gal
Delight	x gal	50%	0.50x gal
Mixture	x + 300 gal	40%	0.40(x + 300) gal

We can now write an equation expressing the fact that the actual amount of orange juice in the mixture is the sum of the amounts of orange juice in the Orange Drink and in the Orange Delight:

$$0.40(x + 300) = 0.10(300) + 0.50x$$
$$0.4x + 120 = 30 + 0.5x$$
$$90 = 0.1x$$
$$900 = x$$

Mix 900 gallons of Orange Delight with the 300 gallons of Orange Drink to obtain the proper mixture. Check.

Try This. How many gallons of a 40% acid solution must be mixed with 30 gallons of a 20% acid solution to obtain a mixture that is 35% acid? ■

Work problems are problems in which people or machines are working together to accomplish a task. A typical situation might have two people painting a house at different rates. Suppose Joe and Frank are painting a house together for 2 hours and Joe paints at the rate of one-sixth of the house per hour while Frank paints at the rate of one-third of the house per hour. Note that

$$\left(\frac{1}{6} \text{ of house per hour}\right)(2 \text{ hr}) = \frac{1}{3} \text{ of house}$$

and

$$\left(\frac{1}{3} \text{ of house per hour}\right)(2 \text{ hr}) = \frac{2}{3} \text{ of house}$$

The product of the rate and the time gives the fraction of the house completed by each person, and these fractions have a sum of 1 because the entire job is completed. Note how similar this situation is to a uniform-motion problem where $RT = D$.

Example **8** Solving a work problem

Aboard the starship *Nostromo*, the human technician, Brett, can process the crew's medical history in 36 minutes. However, the android Science Officer, Ash, can process the same records in 24 minutes. After Brett worked on the records for 1 minute, Ash joined in and both crew members worked until the job was done. How long did Ash work on the records?

Solution

Let x represent the number of minutes that Ash worked and $x + 1$ represent the number of minutes that Brett worked. Ash works at the rate of 1/24 of the job per minute, while Brett works at the rate of 1/36 of the job per minute. The following table shows all of the pertinent quantities.

	Rate	Time	Work Completed
Ash	$\dfrac{1}{24}$ job/min	x min	$\dfrac{1}{24}x$ job
Brett	$\dfrac{1}{36}$ job/min	$x + 1$ min	$\dfrac{1}{36}(x + 1)$ job

The following equation expresses the fact that the work completed together is the sum of the work completed by each worker alone.

$$\frac{1}{24}x + \frac{1}{36}(x + 1) = 1$$

$$72\left[\frac{1}{24}x + \frac{1}{36}(x + 1)\right] = 72 \cdot 1 \quad \text{Multiply by the LCD 72.}$$

$$3x + 2x + 2 = 72$$

$$5x = 70$$

$$x = 14$$

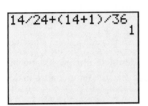

■ **Figure 1.6**

Ash worked for 14 minutes.

Check 14 in the original equation as shown in Fig. 1.6.

Try This. A tank can be filled by a small pipe in 12 hours or a large pipe in 8 hours. How long will it take to fill the tank if the small pipe is used alone for 2 hours and then both pipes are used until the tank is full? ■

For Thought

True or False? Explain.

1. If we solve $P + Prt = S$ for P, we get $P = S - Prt$.

2. The perimeter of any rectangle is the product of its length and width.

3. If n is an odd integer, then $n + 1$ and $n + 3$ represent odd integers.

4. Solving $x - y = 1$ for y gives us $y = x - 1$.

5. Two numbers that have a sum of -3 can be represented by x and $-3 - x$.

6. If P is the number of professors and S is the number of students at the play, and there are twice as many professors as students, then $2P = S$.

7. If you need \$100,000 for your house, and the agent gets 9% of the selling price, then the agent gets \$9000, and the house sells for \$109,000.

8. If John can mow the lawn in x hours, then he mows the lawn at the rate of $1/x$ of the lawn per hour.

9. If George hiked $3x$ miles and Anita hiked $4(x - 2)$ miles, and George hiked 5 more miles than Anita, then $3x + 5 = 4(x - 2)$.

10. Two numbers that differ by 9 can be represented as 9 and $x + 9$.

1.2 Exercises

Solve each formula for the specified variable. The use of the formula is indicated in parentheses.

1. $I = Prt$ for r (simple interest)

2. $D = RT$ for R (uniform motion)

3. $F = \frac{9}{5}C + 32$ for C (temperature)

4. $C = \frac{5}{9}(F - 32)$ for F (temperature)

5. $A = \frac{1}{2}bh$ for b (area of a triangle)

6. $A = \frac{1}{2}bh$ for h (area of a triangle)

7. $Ax + By = C$ for y (equation of a line)

8. $Ax + By = C$ for x (equation of a line)

9. $\frac{1}{R} = \frac{1}{R_1} + \frac{1}{R_2} + \frac{1}{R_3}$ for R_1 (resistance)

10. $\frac{1}{R} = \frac{1}{R_1} + \frac{1}{R_2} + \frac{1}{R_3}$ for R_2 (resistance)

11. $a_n = a_1 + (n - 1)d$ for n (arithmetic sequence)

12. $S_n = \frac{n}{2}(a_1 + a_n)$ for a_1 (arithmetic series)

13. $S = \frac{a_1 - a_1 r^n}{1 - r}$ for a_1 (geometric series)

14. $S = 2LW + 2LH + 2HW$ for H (surface area)

15. *The 2.4-Meter Rule*   A 2.4-meter sailboat is a one-person boat that is about 13 ft in length, has a displacement of about 550 lb, and a sail area of about 81 ft². To compete in the 2.4-meter class, a boat must satisfy the formula

$$2.4 = \frac{L + 2D - F\sqrt{S}}{2.37},$$

where L = length, F = freeboard, D = girth, and S = sail area. Solve the formula for D.

16. *Finding the Freeboard*   Solve the formula in the previous exercise for F.

Use the appropriate formula to solve each problem.

17. *Simple Interest*   If $51.30 in interest is earned on a deposit of $950 in one year, then what is the simple interest rate?

18. *Simple Interest*   If you borrow $100 and pay back $105 at the end of one month, then what is the simple annual interest rate?

19. *Uniform Motion*   How long does it take an SR-71 Blackbird, one of the fastest U.S. jets, to make a surveillance run of 5570 mi if it travels at an average speed of Mach 3 (2228 mph)?

20. *Circumference of a Circle*   If the circumference of a circular sign is 72π in., then what is the radius?

21. *Fahrenheit Temperature*   If the temperature at 11 P.M. on New Year's Eve in Times Square was 23°F, then what was the temperature in degrees Celsius?

22. *Celsius Temperature*   If the temperature at 1 P.M. on July 9 in Toronto was 30°C, then what was the temperature in degrees Fahrenheit?

Solve each problem. See the strategy for problem solving on page 101.

23. *Cost of a Car*   Jeff knows that his neighbor Sarah paid $40,230, including sales tax, for a new Buick Park Avenue. If the sales tax rate is 8%, then what is the cost of the car before the tax?

24. *Real Estate Commission*   To be able to afford the house of their dreams, Dave and Leslie must clear $128,000 from the sale of their first house. If they must pay $780 in closing costs and 6% of the selling price for the sales commission, then what is the minimum selling price for which they will get $128,000?

25. *Adjusting the Saddle*   The saddle height on a bicycle should be 109% of the inside leg measurement of the rider (www.harriscyclery.com). See the figure. If the saddle height is 37 in., then what is the inside leg measurement?

109% of the inside leg measurement

■ **Figure for Exercise 25**

26. *Target Heart Rate* For a cardiovascular workout, fitness experts recommend that you reach your target heart rate and stay at that rate for at least 20 minutes (www.healthstatus.com). To find your target heart rate find the sum of your age and your resting heart rate, then subtract that sum from 220. Find 60% of that result and add it to your resting heart rate. If the target heart rate for a 30-year-old person is 144, then what is that person's resting heart rate?

27. *Investment Income* Tara paid one-half of her game-show winnings to the government for taxes. She invested one-third of her winnings in Jeff's copy shop at 14% interest and one-sixth of her winnings in Kaiser's German Bakery at 12% interest. If she earned a total of $4000 on the investments in one year, then how much did she win on the game show?

28. *Construction Penalties* Gonzales Construction contracted Kentwood High and Memorial Stadium for a total cost of $4.7 million. Because the construction was not completed on time, Gonzales paid 5% of the amount of the high school contract in penalties and 4% of the amount of the stadium contract in penalties. If the total penalty was $223,000, then what was the amount of each contract?

29. *Trimming a Garage Door* A carpenter used 30 ft of molding in three pieces to trim a garage door. If the long piece was 2 ft longer than twice the length of each shorter piece, then how long was each piece?

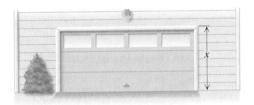

■ **Figure for Exercise 29**

30. *Increasing Area of a Field* Julia's soybean field is 3 m longer than it is wide. To increase her production, she plans to increase both the length and width by 2 m. If the new field is 46 m^2 larger than the old field, then what are the dimensions of the old field?

31. *Fencing a Feed Lot* Peter plans to fence off a square feed lot and then cross-fence to divide the feed lot into four smaller square feed lots. If he uses 480 ft of fencing, then how much area will be fenced in?

■ **Figure for Exercise 31**

32. *Fencing Dog Pens* Clint is constructing two adjacent rectangular dog pens. Each pen will be three times as long as it is wide, and the pens will share a common long side. If Clint has 65 ft of fencing, what are the dimensions of each pen?

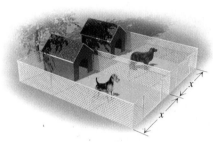

■ **Figure for Exercise 32**

33. *Racing Speed* Bobby and Rick are in a 10-lap race on a one-mile oval track. Bobby, averaging 90 mph, has completed two laps just as Rick is getting his car onto the track. What speed does Rick have to average to be even with Bobby at the end of the tenth lap?

HINT Bobby does 8 miles in the same time as Rick does 10 miles.

34. *Rowing a Boat* Boudreaux rowed his pirogue from his camp on the bayou to his crab traps. Going down the bayou, he caught a falling tide that increased his normal speed by 2 mph, but coming back it decreased his normal speed by 2 mph. Going with the tide, the trip took only 10 min; going against the tide, the trip took 30 min. How far is it from Boudreaux's camp to his crab traps?

HINT With the tide his rate is $x + 2$ mph and against the tide it is $x - 2$ mph.

35. *Average Speed* Junior drove his rig on Interstate 10 from San Antonio to El Paso. At the halfway point he noticed that he had been averaging 80 mph, while his company requires his average speed to be 60 mph. What must be his speed for the last half of the trip so that he will average 60 mph for the entire trip?

HINT The distance from San Antonio to El Paso is irrelevant. Use d or simply make up a distance.

36. *Basketball Stats* Two-thirds of the way through the basketball season, Tina Thompson of the Houston Comets has an average of 10 points per game. What must her point average be for the remaining games to average 20 points per game for the season?

37. *Start-Up Capital* Norma invested the start-up capital for her Internet business in two hedge funds. After one year one of the funds returned 5% and the other returned 6%, for a total return of $5880. If the amount on which she made 6% was $10,000 larger than the amount on which she made 5%, then what was the original amount of her start-up capital?

38. *Combining Investments* Brent lent his brother Bob some money at 8% simple interest, and he lent his sister Betty half as much money at twice the interest rate. Both loans were for one year. If Brent made a total of 24 cents in interest, then how much did he lend to each one?

39. *Percentage of Minority Workers* At the Northside assembly plant, 5% of the workers were classified as minority, while at the Southside assembly plant, 80% of the workers were classified as minority. When Northside and Southside were closed, all workers transferred to the new Eastside plant to make up its entire work force. If 50% of the 1500 employees at Eastside are minority, then how many employees did Northside and Southside have originally?

40. *Mixing Alcohol Solutions* A pharmacist needs to obtain a 70% alcohol solution. How many ounces of a 30% alcohol solution must be mixed with 40 ounces of an 80% alcohol solution to obtain a 70% alcohol solution?

> **HINT** Add x ounces of 30% solution to 40 ounces of 80% solution to get $x + 40$ ounces of 70% solution.

41. *Harvesting Wheat* With the old combine, Nikita's entire wheat crop can be harvested in 72 hr, but a new combine can do the same job in 48 hr. How many hours would it take to harvest the crop with both combines operating?

> **HINT** The rate for the old combine is $1/72$ crop/hr and for the new one it is $1/48$ crop/hr. Together the rate is $1/x$ crop/hr.

42. *Processing Forms* Rita can process a batch of insurance claims in 4 hr working alone. Eduardo can process a batch of insurance claims in 2 hr working alone. How long would it take them to process a batch of claims if they worked together?

43. *Batman and Robin* Batman can clean up all of the crime in Gotham City in 8 hr working alone. Robin can do the same job alone in 12 hr. If Robin starts crime fighting at 8 A.M. and Batman joins him at 10 A.M., then at what time will they have all of the crime cleaned up?

44. *Scraping Barnacles* Della can scrape the barnacles from a 70-ft yacht in 10 hr using an electric barnacle scraper. Don can do the same job in 15 hr using a manual barnacle scraper. If Don starts scraping at noon and Della joins him at 3 P.M., then at what time will they finish the job?

45. *Planning a Race Track* If Mario plans to develop a circular race track one mile in circumference on a square plot of land, then what is the minimum number of acres that he needs? (One acre is equal to 43,560 ft^2.)

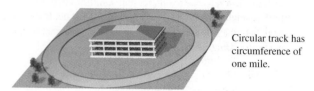

Circular track has circumference of one mile.

■ **Figure for Exercise 45**

46. *Volume of a Can of Coke* If a can of Coke contains 12 fluid ounces and the diameter of the can is 2.375 in., then what is the height of the can? (One fluid ounce equals approximately 1.8 in.^{3}.)

47. *Area of a Lot* Julio owns a four-sided lot that lies between two parallel streets. If his 90,000-ft^2 lot has 500 ft frontage on one street and 300 ft frontage on the other, then how far apart are the streets?

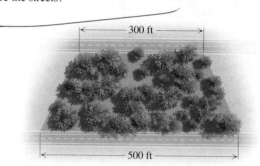

300 ft

500 ft

■ **Figure for Exercise 47**

48. *Width of a Football Field* If the perimeter of a football field in the NFL including the end zones is 1040 ft and the field is 120 yd long, then what is the width of the field in feet?

49. *Depth of a Swimming Pool* A circular swimming pool with a diameter of 30 ft and a horizontal bottom contains 22,000 gal of water. What is the depth of the water in the pool? (One cubic foot contains approximately 7.5 gal of water.)

30 ft

x

■ **Figure for Exercise 49**

50. *Depth of a Reflecting Pool*   A rectangular reflecting pool with a horizontal bottom is 100 ft by 150 ft and contains 200,000 gal of water. How deep is the water in the pool?

51. *Olympic Track*   To host the Summer Olympics, a city plans to build an eight-lane track. The track will consist of parallel 100-m straightaways with semicircular turns on either end as shown in the figure. The distance around the outside edge of the oval track is 514 m. If the track is built on a rectangular lot as shown in the drawing, then how many hectares (1 hectare = 10,000 m²) of land are needed in the rectangular lot?

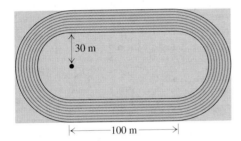

30 m

|←——— 100 m ———→|

■ **Figure for Exercises 51 and 52**

52. *Green Space*   If the inside radius of the turns is 30 m and grass is to be planted inside and outside the track of Exercise 51, then how many square meters of the rectangular lot will be planted in grass?

53. *Taxable Income*   According to the IRS (Internal Revenue Service, www.irs.gov), a single taxpayer with a taxable income between $70,350 and $146,750 in 2004 paid $14,325, plus 28% of the amount over $70,350. If Lorinda paid $17,167 in federal income tax for 2004, then what was her taxable income in 2004?

54. *Higher Bracket*   According to the IRS (Internal Revenue Service, www.irs.gov), a single taxpayer with a taxable income over $319,100 in 2004 paid $92,592.50 plus 35% of the amount over $319,100. If Glen paid $255,836 in federal income tax for 2004, then what was his taxable income for 2004?

55. *Diluting Baneberry*   How much water must Poison Ivy add to a 4-liter solution that contains 5% extract of baneberry to get a solution that contains 3% extract of baneberry?

56. *Diluting Antifreeze*   A mechanic is working on a car with a 20-qt radiator containing a 60% antifreeze solution. How much of the solution should he drain and replace with pure water to get a solution that is 50% antifreeze?

57. *Mixing Dried Fruit*   The owner of a health-food store sells dried apples for $1.20 per quarter-pound, and dried apricots for $1.80 per quarter-pound. How many pounds of each must he mix together to get 20 lb of a mixture that sells for $1.68 per quarter-pound?

58. *Mixing Breakfast Cereal*   Raisins sell for $4.50/lb, and bran flakes sell for $2.80/lb. How many pounds of raisins should be mixed with 12 lb of bran flakes to get a mixture that sells for $3.14/lb?

59. *Coins in a Vending Machine*   Dana inserted eight coins, consisting of dimes and nickels, into a vending machine to purchase a Snickers bar for 55 cents. How many coins of each type did she use?

60. *Cost of a Newspaper*   Ravi took eight coins from his pocket, which contained only dimes, nickels, and quarters, and bought the Sunday Edition of *The Daily Star* for 75 cents. If the number of nickels he used was one more than the number of dimes, then how many of each type of coin did he use?

61. *Active Ingredients*   A pharmacist has 200 milliliters of a solution that is 40% active ingredient. How much pure water should she add to the solution to get a solution that is 25% active ingredient?

62. *Mixed Nuts*   A manager bought 12 pounds of peanuts for $30. He wants to mix $5 per pound cashews with the peanuts to get a batch of mixed nuts that is worth $4 per pound. How many pounds of cashews are needed?

63. *Salt Solution I*   A chemist has 5 gallons of salt solution with a concentration of 0.2 pounds per gallon and another solution with a concentration of 0.5 pound per gallon. How many gallons of the stronger solution must be added to the weaker solution to get a solution that contains 0.3 pounds per gallon?

64. *Salt Solution II*   Suppose the 5-gallon solution in the previous problem is contained in a 5-gallon container. The chemist plans to remove some amount of the 0.2 pound per gallon solution and replace it with 0.5 pound per gallon solution so that he ends up with 5 gallons of a 0.3 pound per gallon solution. What amount should he remove?

65. *Draining a Pool I*   A small pump can drain a pool in 8 hours. A large pump could drain the same pool in 5 hours. How long (to the nearest minute) will it take to drain the pool if both pumps are used simultaneously?

66. *Draining a Pool II*   Suppose that the pumps in the previous exercise could not be used simultaneously and that the pool was drained in exactly 6 hours. How long was each pump used?

67. *Dining In* Revenue for restaurants and supermarkets can be modeled by the equations $R = 13.5n + 190$ and $S = 7.5n + 225$, where n is the number of years since 1986 (*Forbes*, www.forbes.com).

a. Use the accompanying graph to estimate the year in which restaurant and supermarket revenue were equal.

b. Use the equations to find the year in which restaurant and supermarket revenue were equal.

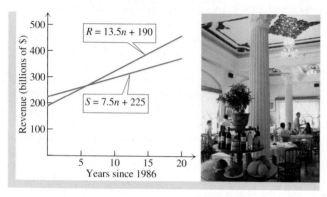

■ **Figure for Exercises 67 and 68**

68. *Dining Out* Use the equations in the previous exercise to find the year in which restaurant revenue will double supermarket revenue.

For Writing/Discussion

69. *Working Together* If one lost hiker can pick a gallon of wild berries in 2 hr, then how long should it take two lost hikers working together to pick one gallon of wild berries? If one

mechanic at Spee-Dee Oil Change can change the oil in a Saturn in 6 min, then how long should it take two mechanics working together to change the oil in a Saturn? How long would it take 60 mechanics? Are your answers reasonable? Explain.

70. *Average Speed* Imagine that you are cruising down I-75 headed to spring break in Florida.

a. Suppose that you cruise at 70 mph for 3 hours and then at 60 mph for 1 hour. What is your average speed for those 4 hours? When is the average speed over two time intervals actually the average of the two speeds? Explain your answer.

b. Suppose that you cruise at 60 mph for 180 miles and then cruise at 40 mph for 160 miles. What is your average speed? When is the average speed over two distance intervals actually the average of the two speeds? Explain your answer.

Thinking Outside the Box XIII

Roughing It Milo and Bernard are planning a three-day canoe trip on the Roaring Fork River. Their friend Vince will drop them off at the Highway 14 bridge. From there they will paddle upstream for 12 hours on the first day and 9 hours on the second day. They have been on this river before and know that their average paddling rate is twice the rate of the current in the river. At what time will they have to start heading downstream on the third day to meet Vince at the Highway 14 bridge at 5 P.M?

1.2 Pop Quiz

1. Solve $dx + dy = w$ for y.

2. If the length of a rectangle is 3 feet longer than the width and the perimeter is 62 feet, then what is the width?

3. How many liters of water should be added to 3 liters of a 70% alcohol solution to obtain a 50% alcohol solution?

Linking Concepts

For Individual or Group Explorations

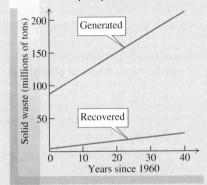

A Recycling Model

In 1960 the United States generated 87.1 million tons of municipal solid waste and recovered (or recycled) only 4.3% of it (U.S. EPA, www.epa.gov). The amount of municipal solid waste generated in the United States can be modeled by the formula $w = 3.14n + 87.1$, while the amount recovered can be modeled by the formula $w = 0.576n + 3.78$, where w is in millions of tons and n is the number of years since 1960.

a)   Use the graph to estimate the first year in which the United States generated over 100 million tons of municipal solid waste.

b)   Use the formulas to determine the year in which the United States generated 150 million more tons than it recovered.

c)   Find the year in which 13% of the municipal solid waste generated will be recovered.

d)   Find the years in which the recovery rates will be 14%, 15%, and 16%.

e)   Will the recovery rate ever reach 25%?

f)   According to this model, what is the maximum percentage of solid waste that will ever be recovered?

1.3

Equations and Graphs in Two Variables

In Section 1.1 we studied equations in one variable. In this section we study pairs of variables. For example, p might represent the price of gasoline and n the number of gallons that you consume in a month; x might be the number of toppings on a medium pizza and y the cost of that pizza; or h might be the height of a two-year-old child and w the weight. To study relationships between pairs of variables we use a two-dimensional coordinate system.

■ **Table 1.1**

Toppings x	Cost y	
0	$ 5	
1	7	
2	9	
3	11	
4	13	

The Cartesian Coordinate System

If x and y are real numbers then (x, y) is called an **ordered pair** of real numbers. The numbers x and y are the **coordinates** of the ordered pair, with x being the **first coordinate** or **abscissa**, and y being the **second coordinate** or **ordinate**. For example, Table 1.1 shows the number of toppings on a medium pizza and the corresponding cost. The ordered pair (3, 11) indicates that a three-topping pizza costs $11. The order of the numbers matters. In this context, (11, 3) would indicate that an 11-topping pizza costs $3.

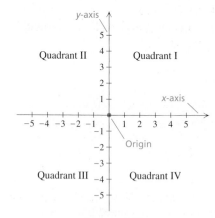

■ **Figure 1.7**

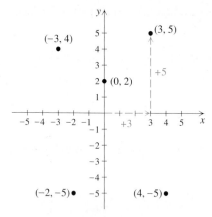

■ **Figure 1.8**

To picture ordered pairs of real numbers we use the **rectangular coordinate system** or **Cartesian coordinate system,** named after the French mathematician René Descartes (1596–1650). The Cartesian coordinate system consists of two number lines drawn perpendicular to one another, intersecting at zero on each number line as shown in Fig. 1.7. The point of intersection of the number lines is called the **origin.** The horizontal number line is the **x-axis** and its positive numbers are to the right of the origin. The vertical number line is the **y-axis** and its positive numbers are above the origin. The two number lines divide the plane into four regions called **quadrants,** numbered as shown in Fig. 1.7. The quadrants do not include any points on the axes. We call a plane with a rectangular coordinate system the **coordinate plane** or the **xy-plane.**

Just as every real number corresponds to a point on the number line, every ordered pair of real numbers (a, b) corresponds to a point P in the xy-plane. For this reason, ordered pairs of numbers are often called **points.** So a and b are the coordinates of (a, b) or the coordinates of the point P. Locating the point P that corresponds to (a, b) in the xy-plane is referred to as **plotting** or **graphing** the point, and P is called the *graph* of (a, b). In general, a **graph** is a set of points in the rectangular coordinate system.

Example 1 Plotting points

Plot the points $(3, 5)$, $(4, -5)$, $(-3, 4)$, $(-2, -5)$, and $(0, 2)$ in the xy-plane.

Solution

The point $(3, 5)$ is located three units to the right of the origin and five units above the x-axis as shown in Fig. 1.8. The point $(4, -5)$ is located four units to the right of the origin and five units below the x-axis. The point $(-3, 4)$ is located three units to the left of the origin and four units above the x-axis. The point $(-2, -5)$ is located two units to the left of the origin and five units below the x-axis. The point $(0, 2)$ is on the y-axis because its first coordinate is zero.

Try This. Plot $(-3, -2)$, $(-1, 3)$, $(5/2, 0)$, and $(2, -3)$. ■

Note that for points in quadrant I, both coordinates are positive. In quadrant II the first coordinate is negative and the second is positive, while in quadrant III, both coordinates are negative. In quadrant IV the first coordinate is positive and the second is negative.

The Distance Formula

Consider the points $A(x_1, y_1)$ and $B(x_2, y_2)$ shown in Fig. 1.9. Let AB represent the length of line segment $\overline{AB}$. Now $\overline{AB}$ is the hypotenuse of the right triangle in Fig. 1.9. Since A and C lie on a horizontal line, the distance between them is $|x_2 - x_1|$. Likewise $CB = |y_2 - y_1|$. Since the sum of the squares of the legs of a right triangle is equal to the square of the hypotenuse (the Pythagorean theorem) we have

$$d^2 = |x_2 - x_1|^2 + |y_2 - y_1|^2.$$

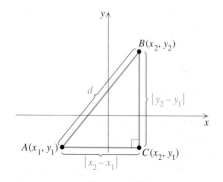

■ **Figure 1.9**

Since the distance between two points is a nonnegative real number, we have $d = \sqrt{(x_2 - x_1)^2 + (y_2 - y_1)^2}$. The absolute value symbols are replaced with parentheses, because $|a - b|^2 = (a - b)^2$ for any real numbers a and b.

The Distance Formula

The distance d between the points (x_1, y_1) and (x_2, y_2) is given by the formula

$$d = \sqrt{(x_2 - x_1)^2 + (y_2 - y_1)^2}.$$

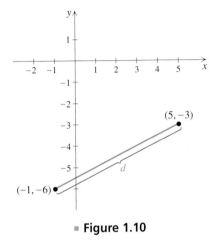

■ **Figure 1.10**

Example **2** **Finding the distance between two points**

Find the distance between $(5, -3)$ and $(-1, -6)$.

Solution

Let $(x_1, y_1) = (5, -3)$ and $(x_2, y_2) = (-1, -6)$. These points are shown on the graph in Fig. 1.10. Substitute these values into the distance formula:

$$d = \sqrt{(-1 - 5)^2 + (-6 - (-3))^2}$$
$$= \sqrt{(-6)^2 + (-3)^2}$$
$$= \sqrt{36 + 9} = \sqrt{45} = 3\sqrt{5}$$

The exact distance between the points is $3\sqrt{5}$.

Try This. Find the distance between $(-3, -2)$ and $(-1, 4)$.   ■

Note that the distance between two points is the same regardless of which point is chosen as (x_1, y_1) or (x_2, y_2).

The Midpoint Formula

When you average two test scores (by finding their sum and dividing by 2), you are finding a number midway between the two scores. Likewise, the midpoint of the line segment with endpoints -1 and 7 in Fig. 1.11 is $(-1 + 7)/2$ or 3. In general, $(a + b)/2$ is the midpoint of the line segment with endpoints a and b shown in Fig. 1.12. (See Exercises 95 and 96.) We can find the midpoint of a line segment in the xy-plane in the same manner.

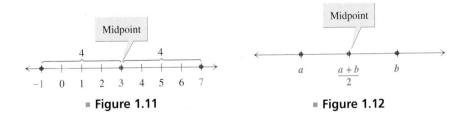

■ **Figure 1.11**   ■ **Figure 1.12**

Theorem:
The Midpoint Formula

The midpoint of the line segment with endpoints (x_1, y_1) and (x_2, y_2) is

$$\left(\frac{x_1 + x_2}{2}, \frac{y_1 + y_2}{2} \right).$$

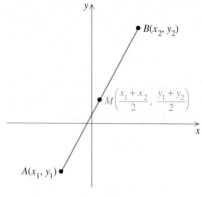

■ **Figure 1.13**

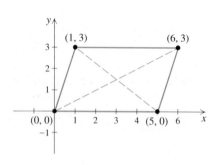

■ **Figure 1.14**

■ **Proof**   To prove that M is the midpoint of $\overline{AB}$ as shown in Fig. 1.13 we will show that $AM = MB$ and that A, M, and B are colinear (on the same line). To prove that they are colinear, it is sufficient to show that $AM + MB = AB$. (If $AM + MB > AB$, then they form a triangle.) First we use the distance formula to find AM and MB:

$$AM = \sqrt{\left(\frac{x_1 + x_2}{2} - x_1\right)^2 + \left(\frac{y_1 + y_2}{2} - y_1\right)^2} = \sqrt{\left(\frac{x_2 - x_1}{2}\right)^2 + \left(\frac{y_2 - y_1}{2}\right)^2}$$

$$MB = \sqrt{\left(x_2 - \frac{x_1 + x_2}{2}\right)^2 + \left(y_2 - \frac{y_1 + y_2}{2}\right)^2} = \sqrt{\left(\frac{x_2 - x_1}{2}\right)^2 + \left(\frac{y_2 - y_1}{2}\right)^2}$$

Since $AM = MB$, we can conclude that M is equidistant from A and B. The expressions for AM and MB can be further simplified by factoring 2^2 out of each denominator:

$$AM = MB = \frac{1}{2}\sqrt{(x_2 - x_1)^2 + (y_2 - y_1)^2}$$

Using the distance formula $AB = \sqrt{(x_2 - x_1)^2 + (y_2 - y_1)^2}$. Since

$$AM + MB = \frac{1}{2}\sqrt{(x_2 - x_1)^2 + (y_2 - y_1)^2} + \frac{1}{2}\sqrt{(x_2 - x_1)^2 + (y_2 - y_1)^2}$$

$$= \sqrt{(x_2 - x_1)^2 + (y_2 - y_1)^2},$$

we have $AM + MB = AB$. So A, M, and B are colinear.   ■

Example ▮3▮ Using the midpoint formula

Prove that the diagonals of the parallelogram with vertices $(0, 0)$, $(1, 3)$, $(5, 0)$, and $(6, 3)$ bisect each other.

Solution

The parallelogram is shown in Fig. 1.14. The midpoint of the diagonal from $(0, 0)$ to $(6, 3)$ is

$$\left(\frac{0 + 6}{2}, \frac{0 + 3}{2}\right),$$

or $(3, 1.5)$. The midpoint of the diagonal from $(1, 3)$ to $(5, 0)$ is

$$\left(\frac{1 + 5}{2}, \frac{3 + 0}{2}\right),$$

or $(3, 1.5)$. Since the diagonals have the same midpoint, they bisect each other.

Try This.   Prove that the diagonals of the square with vertices $(0, 0)$, $(4, 1)$, $(3, 5)$, and $(-1, 4)$ bisect each other.   ■

The Circle

An ordered pair is a **solution to** or **satisfies** an equation in two variables if the equation is correct when the variables are replaced by the coordinates of the ordered pair. For example, $(3, 11)$ satisfies $y = 2x + 5$ because $11 = 2(3) + 5$ is correct. The

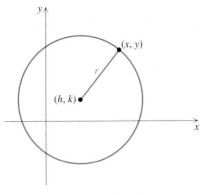

solution set to an equation in two variables is the set of all ordered pairs that satisfy the equation. The graph of (the solution set to) an equation is a geometric object that gives us a visual image of an algebraic object. Circles provide a nice example of this relationship between algebra and geometry.

A **circle** is the set of all points in a plane that lie a fixed distance from a given point in the plane. The fixed distance is called the **radius,** and the given point is the **center.** The distance formula can be used to write an equation for the circle shown in Fig. 1.15 with center (h, k) and radius r for $r > 0$. A point (x, y) is on the circle if and only if it satisfies the equation

$$\sqrt{(x - h)^2 + (y - k)^2} = r.$$

Since both sides of $\sqrt{(x - h)^2 + (y - k)^2} = r$ are positive, we can square each side to get the following **standard form** for the equation of a circle.

Theorem: Equation for a Circle in Standard Form

The equation for a circle with center (h, k) and radius r for $r > 0$ is

$$(x - h)^2 + (y - k)^2 = r^2.$$

A circle centered at the origin has equation $x^2 + y^2 = r^2$.

Note that squaring both sides of an equation produces an equivalent equation only when both sides are positive. If we square both sides of $\sqrt{x} = -3$, we get $x = 9$. But $\sqrt{9} \neq -3$ since the square root symbol always represents the nonnegative square root. Raising both sides of an equation to a power is discussed further in Section 3.4.

Example **4** **Graphing a circle**

Sketch the graph of the equation $(x - 1)^2 + (y + 2)^2 = 9$.

Solution

Since $(x - h)^2 + (y - k)^2 = r^2$ is a circle with center (h, k) and radius r (for $r > 0$), the graph of $(x - 1)^2 + (y + 2)^2 = 9$ is a circle with center $(1, -2)$ and radius 3. You can draw the circle as in Fig. 1.16 with a compass. To draw a circle by hand, locate the points that lie 3 units above, below, right, and left of the center, as shown in Fig. 1.16. Then sketch a circle through these points.

To support these results with a graphing calculator you must first solve the equation for y:

$$(x - 1)^2 + (y + 2)^2 = 9$$
$$(y + 2)^2 = 9 - (x - 1)^2$$
$$y + 2 = \pm\sqrt{9 - (x - 1)^2}$$
$$y = -2 \pm \sqrt{9 - (x - 1)^2}$$

Now enter y_1 and y_2 as in Fig. 1.17(a) on the next page. Set the viewing window as in Fig. 1.17(b). The graph in Fig. 1.17(c) supports our previous conclusion. A circle looks round only if the same unit distance is used on both axes. Some calculators

Figure 1.16 caption area:

$(x - 1)^2 + (y + 2)^2 = 9$

$(1, -2)$

■ **Figure 1.16**

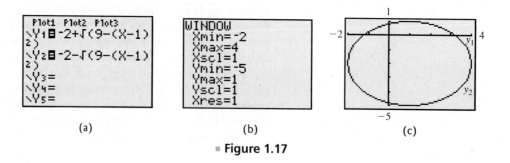

(a)　　　　　　　(b)　　　　　　(c)

■ **Figure 1.17**

automatically draw the graph with the same unit distance on both axes given the correct command (Zsquare on a TI-83).

Try This. Sketch the graph of $(x + 2)^2 + (y - 4)^2 = 25$. ■

Note that an equation such as $(x - 1)^2 + (y + 2)^2 = -9$ is not satisfied by any pair of real numbers, because the left-hand side is a nonnegative real number, while the right-hand side is negative. The equation $(x - 1)^2 + (y + 2)^2 = 0$ is satisfied only by $(1, -2)$. Since only one point satisfies $(x - 1)^2 + (y + 2)^2 = 0$, its graph is sometimes called a degenerate circle with radius zero. We study circles again later in this text when we study the conic sections.

In the next example we start with a description of a circle and write its equation.

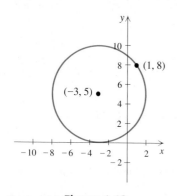

■ **Figure 1.18**

Example **5** **Writing an equation of a circle**

Write the standard equation for the circle with the center $(-3, 5)$ and passing through $(1, 8)$ as shown in Fig. 1.18.

Solution

The radius of this circle is the distance between $(-3, 5)$ and $(1, 8)$:

$$r = \sqrt{(8 - 5)^2 + (1 - (-3))^2} = \sqrt{9 + 16} = 5$$

Now use $h = -3, k = 5$, and $r = 5$ in the standard equation of the circle $(x - h)^2 + (y - k)^2 = r^2$.

$$(x - (-3))^2 + (y - 5)^2 = 5^2$$

So the equation of the circle is $(x + 3)^2 + (y - 5)^2 = 25$.

Try This. Write the standard equation for the circle with center $(2, -1)$ and passing through $(3, 6)$. ■

If the equation of a circle is not given in the standard form, we can convert it to standard form by **completing the square**. Completing the square means finding the third term of a perfect square trinomial when given the first two. For example, we can recognize $x^2 + 6x$ as the first two terms of $x^2 + 6x + 9$, which is $(x + 3)^2$. Note that one-half of 6 is 3 and 3^2 is 9. We can recognize $x^2 + 10x$ as the first two terms of the perfect square trinomial $x^2 + 10x + 25$, which is $(x + 5)^2$. Note that one-half of 10 is 5, and 5^2 is 25. These examples suggest the following rule.

Rule for Completing the Square of $x^2 + bx + ?$

The last term of a perfect square trinomial (with $a = 1$) is the square of one-half of the coefficient of the middle term. In symbols, the perfect square trinomial whose first two terms are $x^2 + bx$ is

$$x^2 + bx + \left(\frac{b}{2}\right)^2.$$

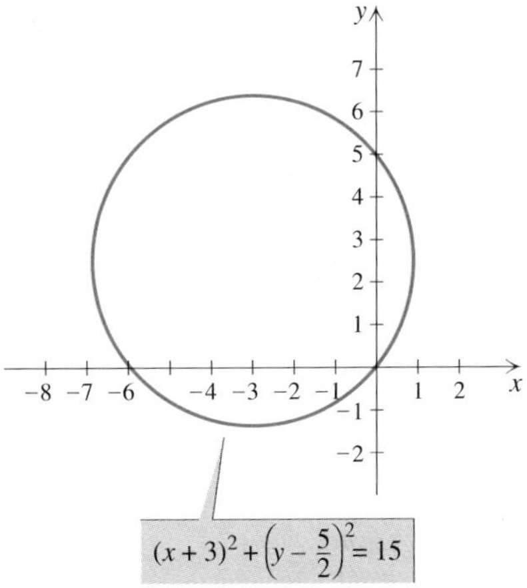

$(x + 3)^2 + \left(y - \frac{5}{2}\right)^2 = 15$

■ **Figure 1.19**

Example **6** **Changing an equation of a circle to standard form**

Graph the equation $x^2 + 6x + y^2 - 5y = -\frac{1}{4}$.

Solution

Complete the square for both x and y to get the standard form.

$$x^2 + 6x + 9 + y^2 - 5y + \frac{25}{4} = -\frac{1}{4} + 9 + \frac{25}{4} \qquad \left(\frac{1}{2} \cdot 6\right)^2 = 9, \left(\frac{1}{2} \cdot 5\right)^2 = \frac{25}{4}$$

$$(x + 3)^2 + \left(y - \frac{5}{2}\right)^2 = 15 \qquad \text{Factor the trinomials on the left side.}$$

The graph is a circle with center $\left(-3, \frac{5}{2}\right)$ and radius $\sqrt{15}$. See Fig. 1.19.

Try This. Graph the equation $x^2 + 3x + y^2 - 2y = 0$.   ■

Whether an equation of the form $x^2 + y^2 + Ax + By = C$ is a circle depends on the value of C. We can always complete the squares for $x^2 + Ax$ and $y^2 + By$ by adding $(A/2)^2$ and $(B/2)^2$ to each side of the equation. Since a circle must have a positive radius we will have a circle only if $C + (A/2)^2 + (B/2)^2 > 0$.

The Line

Like the circle, the line is another geometric object that can be described algebraically. The **standard form** of the equation of a line is $Ax + By = C$.

Theorem: Equation of a Line in Standard Form

If A, B, and C are real numbers, then the graph of the equation

$$Ax + By = C$$

is a straight line, provided A and B are not both zero.

An equation of the form $Ax + By = C$ is called a **linear equation in two variables.** The equations,

$$2x + 3y = 5, \qquad x = 4, \qquad \text{and} \qquad y = 5,$$

are linear equations in standard form. An equation such as $y = 3x - 1$ that can be rewritten in standard form is also called a linear equation.

There is only one line containing any two distinct points. So to graph a linear equation we simply find two points that satisfy the equation and draw a line through them. We often use the point where the line crosses the x-axis, the **x-intercept,** and the point where the line crosses the y-axis, the **y-intercept.** Since every point on the x-axis has y-coordinate 0, we find the x-intercept by replacing y with 0 and then solving the equation for x. Since every point on the y-axis has x-coordinate 0, we find the y-intercept by replacing x with 0 and then solving for y. If the x- and y-intercepts are both at the origin, then you must find another point that satisfies the equation.

Example 7 Graphing lines and showing the intercepts

Graph each equation. Be sure to find and show the intercepts.

a. $2x - 3y = 9$ b. $y = 40 - x$

Solution

a. Since the y-coordinate of the x-intercept is 0, we replace y by 0 in the equation:

$$2x - 3(0) = 9$$
$$2x = 9$$
$$x = 4.5$$

To find the y-intercept, we replace x by 0 in the equation:

$$2(0) - 3y = 9$$
$$-3y = 9$$
$$y = -3$$

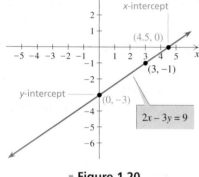

■ **Figure 1.20**

The x-intercept is $(4.5, 0)$ and the y-intercept is $(0, -3)$. Locate the intercepts and draw the line as shown in Fig. 1.20. To check, locate a point such as $(3, -1)$, which also satisfies the equation, and see if the line goes through it.

b. If $x = 0$, then $y = 40 - 0 = 40$ and the y-intercept is $(0, 40)$. If $y = 0$, then $0 = 40 - x$ or $x = 40$. The x-intercept is $(40, 0)$. Draw a line through these points as shown in Fig. 1.21. Check that $(10, 30)$ and $(20, 20)$ also satisfy $y = 40 - x$ and the line goes through these points.

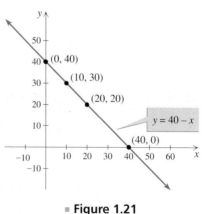

■ **Figure 1.21**

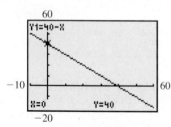

■ **Figure 1.22**

The calculator graph shown in Fig. 1.22 is consistent with the graph in Fig. 1.21. Note that the viewing window is set to show the intercepts.

Try This. Graph $2x + 5y = 10$ and determine the intercepts.   ■

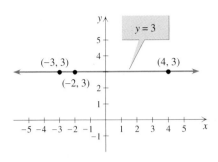

■ **Figure 1.23**

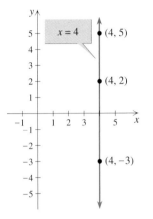

■ **Figure 1.24**

Example **8** Graphing horizontal and vertical lines

Sketch the graph of each equation in the rectangular coordinate system.

a. $y = 3$ **b.** $x = 4$

Solution

a. The equation $y = 3$ is equivalent to $0 \cdot x + y = 3$. Because x is multiplied by 0, we can choose any value for x as long as we choose 3 for y. So ordered pairs such as $(-3, 3)$, $(-2, 3)$, and $(4, 3)$ satisfy the equation $y = 3$. The graph of $y = 3$ is the horizontal line shown in Fig. 1.23.

b. The equation $x = 4$ is equivalent to $x + 0 \cdot y = 4$. Because y is multiplied by 0, we can choose any value for y as long as we choose 4 for x. So ordered pairs such as $(4, -3)$, $(4, 2)$, and $(4, 5)$ satisfy the equation $x = 4$. The graph of $x = 4$ is the vertical line shown in Fig. 1.24.

   Note that you cannot graph $x = 4$ using the Y= key on your calculator. You can graph it on a calculator using polar coordinates or parametric equations (discussed later in this text).

Try This. Graph $y = 5$ in the rectangular coordinate system.   ■

   In the context of two variables the equation $x = 4$ has infinitely many solutions. Every ordered pair on the vertical line in Fig. 1.24 satisfies $x = 4$. In the context of one variable $x = 4$ has only one solution, 4.

Using a Graph to Solve an Equation

Graphing and solving equations go hand in hand. For example, the graph of $y = 2x - 6$ in Fig. 1.25 has x-intercept $(3, 0)$, because if $x = 3$ then $y = 0$. Of course 3 is also the solution to the corresponding equation $2x - 6 = 0$ (where y is replaced by 0). For this reason, the solution to an equation is also called a **zero** or **root** of the equation. Every x-intercept on a graph provides us a solution to the corresponding equation. However, an x-intercept on a graph may not be easy to identify. In the next example we see how a graphing calculator identifies an x-intercept and thus gives us the approximate solution to an equation.

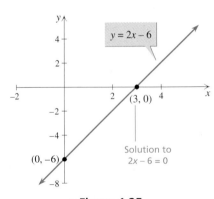

■ **Figure 1.25**

Example **9** Using a graph to solve an equation

Use a graphing calculator to solve $0.55(x - 3.45) + 13.98 = 0$.

Solution

First graph $y = 0.55(x - 3.45) + 13.98$ using a viewing window that shows the x-intercept as in Fig. 1.26(a) on the next page. Next press ZERO or ROOT on the CALC menu. The calculator can find a zero between a *left bound* and a *right bound*, which you must enter. The calculator also asks you to make a guess. See Fig. 1.26(b). The more accurate the guess, the faster the calculator will find the zero. The solution to the equation rounded to two decimal places is -21.97. See Fig. 1.26(c). As always, consult your calulator manual if you are having difficulty.

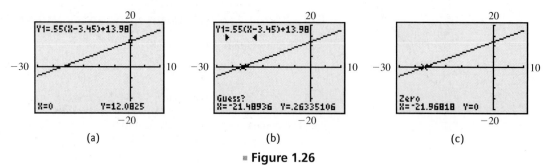

(a) (b) (c)

▪ **Figure 1.26**

Try This. Use a graphing calculator to solve $0.34(x - 2.3) + 4.5 = 0$. ▪

For Thought

True or False? Explain.

1. The point $(2, -3)$ is in quadrant II.

2. The point $(4, 0)$ is in quadrant I.

3. The distance between (a, b) and (c, d) is

$$\sqrt{(a - b)^2 + (c - d)^2}.$$

4. The equation $3x^2 + y = 5$ is a linear equation.

5. The solution to $7x - 9 = 0$ is the x-coordinate of the x-intercept of $y = 7x - 9$.

6. $\sqrt{7^2 + 9^2} = 7 + 9$

7. The origin lies midway between $(1, 3)$ and $(-1, -3)$.

8. The distance between $(3, -7)$ and $(3, 3)$ is 10.

9. The x-intercept for the graph of $3x - 2y = 7$ is $(7/3, 0)$.

10. The graph of $(x + 2)^2 + (y - 1)^2 = 5$ is a circle centered at $(-2, 1)$ with radius 5.

1.3 Exercises

In Exercises 1–10, for each point shown in the xy-plane, write the corresponding ordered pair and name the quadrant in which it lies or the axis on which it lies.

1. A **2.** B **3.** C **4.** D

5. E **6.** F **7.** G **8.** H

9. I **10.** J

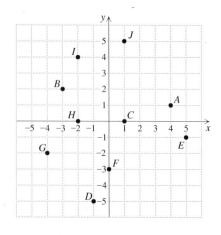

For each pair of points find the distance between them and the midpoint of the line segment joining them.

11. $(1, 3), (4, 7)$ **12.** $(-3, -2), (9, 3)$

13. $(-1, -2), (1, 0)$ **14.** $(-1, 0), (1, 2)$

15. $(12, -11), (5, 13)$ **16.** $(-4, -7), (4, 8)$

17. $(-1, 1), \left(-1 + 3\sqrt{3}, 4\right)$

18. $\left(1 + \sqrt{2}, -2\right), \left(1 - \sqrt{2}, 2\right)$

19. $(1.2, 4.8), (-3.8, -2.2)$ **20.** $(-2.3, 1.5), (4.7, -7.5)$

21. $(a, 0), (b, 0)$

22. $(a, 0), \left(\dfrac{a + b}{2}, 0\right)$

23. $(\pi, 0), (\pi/2, 1)$

24. $(0, 0), (\pi/2, 1)$

Determine the center and radius of each circle and sketch the graph.

25. $x^2 + y^2 = 16$

26. $x^2 + y^2 = 1$

27. $(x + 6)^2 + y^2 = 36$

28. $x^2 + (y - 2)^2 = 16$

29. $y^2 = 25 - (x + 1)^2$

30. $x^2 = 9 - (y - 3)^2$

31. $(x - 2)^2 = 8 - (y + 2)^2$

32. $(y + 2)^2 = 20 - (x - 4)^2$

Write the standard equation for each circle.

33. Center at $(0, 0)$ with radius 7

34. Center at $(0, 0)$ with radius 5

35. Center at $(-2, 5)$ with radius 1/2

36. Center at $(-1, -6)$ with radius 1/3

37. Center at $(3, 5)$ and passing through the origin

38. Center at $(-3, 9)$ and passing through the origin

39. Center at $(5, -1)$ and passing through $(1, 3)$

40. Center at $(-2, -3)$ and passing through $(2, 5)$

Determine the center and radius of each circle and sketch the graph. See the rule for completing the square on page 117.

41. $x^2 + y^2 = 9$

42. $x^2 + y^2 = 100$

43. $x^2 + y^2 + 6y = 0$

44. $x^2 + y^2 = 4x$

45. $x^2 + 6x + y^2 + 8y = 0$

46. $x^2 - 8x + y^2 - 10y = -5$

47. $x^2 - 3x + y^2 + 2y = \dfrac{3}{4}$

48. $x^2 + 5x + y^2 - y = \dfrac{5}{2}$

49. $x^2 - 6x + y^2 - 8y = 0$

50. $x^2 + 10x + y^2 - 8y = -40$

51. $x^2 + y^2 = 4x + 3y$

52. $x^2 + y^2 = 5x - 6y$

53. $x^2 + y^2 = \dfrac{x}{2} - \dfrac{y}{3} - \dfrac{1}{16}$

54. $x^2 + y^2 = x - y + \dfrac{1}{2}$

Write the standard equation for each of the following circles.

55. a.

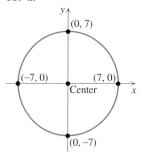

b.

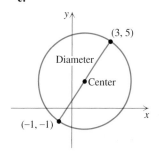

c.

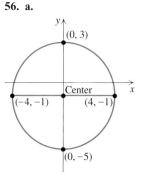

56. a.

b.

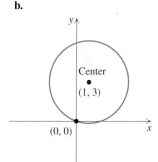

c.

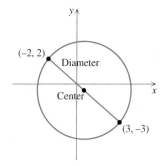

Write the equation of each circle in standard form. The coordinates of the center and the radius for each circle are integers.

57. a.

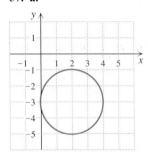

b.

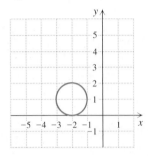

c.

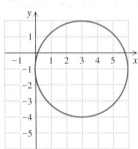

d.

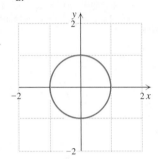

58. a.

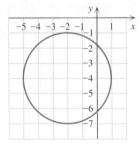

b.

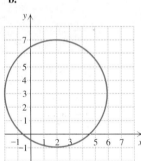

c.

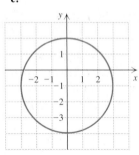

d.

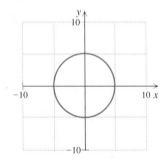

Sketch the graph of each linear equation. Be sure to find and show the x- and y-intercepts.

59. $y = 3x - 4$

60. $y = 5x - 5$

61. $3x - y = 6$

62. $5x - 2y = 10$

63. $x = 3y - 90$

64. $x = 80 - 2y$

65. $\frac{2}{3}y - \frac{1}{2}x = 400$

66. $\frac{1}{2}x - \frac{1}{3}y = 600$

67. $2x + 4y = 0.01$

68. $3x - 5y = 1.5$

69. $0.03x + 0.06y = 150$

70. $0.09x - 0.06y = 54$

Graph each equation in the rectangular coordinate system.

71. $x = 5$ **72.** $y = -2$ **73.** $y = 4$

74. $x = -3$ **75.** $x = -4$ **76.** $y = 5$

77. $y - 1 = 0$ **78.** $5 - x = 4$

Find the solution to each equation by reading the accompanying graph.

79. $2.4x - 8.64 = 0$

80. $8.84 - 1.3x = 0$

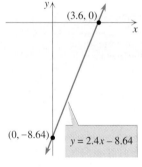

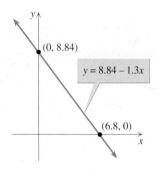

81. $-\frac{3}{7}x + 6 = 0$

82. $\frac{5}{6}x + 30 = 0$

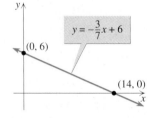

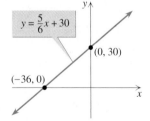

Use a graphing calculator to estimate the solution to each equation to two decimal places. Then find the solution algebraically and compare it with your estimate.

83. $1.2x + 3.4 = 0$

84. $3.2x - 4.5 = 0$

85. $1.23x - 687 = 0$

86. $-2.46x + 1500 = 0$

87. $0.03x - 3497 = 0$

88. $0.09x + 2000 = 0$

89. $4.3 - 3.1(2.3x - 9.9) = 0$

90. $9.4x - 4.37(3.5x - 9.76) = 0$

Solve each problem.

91. *First Marriage*   The median age at first marriage for women went from 20.8 in 1970 to 25.1 in 2000 as shown in the accompanying figure (U.S. Census Bureau, www.census.gov).
   a. Find the midpoint of the line segment in the figure and interpret your result.

   b. Find the distance between the two points shown in the figure and interpret your result.

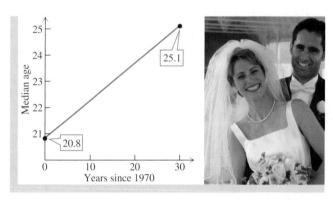

■ **Figure for Exercise 91**

92. *Unmarried Couples*   The number of unmarried-couple households h (in millions), can be modeled using the equation $h = 0.171n + 2.913$, where n is the number of years since 1990 (U.S. Census Bureau, www.census.gov).
   a. Find and interpret the n-intercept for the line. Does it make sense?

   b. Find and interpret the h-intercept for the line.

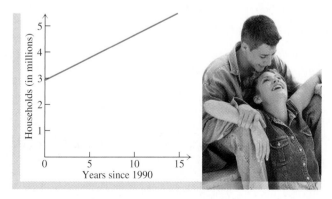

■ **Figure for Exercise 92**

93. *Capsize Control*   The capsize screening value C is an indicator of a sailboat's suitability for extended offshore sailing. C is determined by the formula

$$C = 4D^{-1/3}B,$$

where D is the displacement of the boat in pounds and B is its beam (or width) in feet. Sketch the graph of this equation for B ranging from 0 to 20 ft assuming that D is fixed at 22,800 lbs. Find C for the Island Packet 40, which has a displacement of 22,800 pounds and a beam of 12 ft 11 in. (Island Packet Yachts, www.ipy.com).

94. *Limiting the Beam*   The International Offshore Rules require that the capsize screening value C (from the previous exercise) be less than or equal to 2 for safety. What is the maximum allowable beam (to the nearest inch) for a boat with a displacement of 22,800 lbs? For a fixed displacement, is a boat more or less likely to capsize as its beam gets larger?

■ **Figure for Exercises 93 and 94**

95. Prove that if a and b are real numbers such that $a < b$, then $a < (a + b)/2 < b$.

■ **Figure for Exercises 95 and 96**

96. Prove that the point $(a + b)/2$ is equidistant from a and b on the number line for any real numbers a and b. The proof of this statement together with the previous exercise proves that $(a + b)/2$ is the midpoint of the segment with endpoints a and b on a number line.

97. Show that the points $A(-4, -5)$, $B(1, 1)$, and $C(6, 7)$ are colinear.
   HINT   Points A, B, and C lie on a straight line if $AB + BC = AC$.

98. Show that the midpoint of the hypotenuse of any right triangle is equidistant from all three vertices.

For Writing/Discussion

99. *Finding Points*   Can you find two points such that their co-ordinates are integers and the distance between them is 10? $\sqrt{10}$? $\sqrt{19}$? Explain.

100. *Plotting Points*   Plot at least five points in the *xy*-plane that satisfy the inequality $y > 2x$. Give a verbal description of the solution set to $y > 2x$.

101. *Cooperative Learning*   Work in a small group to plot the points $(-1, 3)$ and $(4, 1)$ on graph paper. Assuming that these two points are adjacent vertices of a square, find the other two vertices. Now select your own pair of adjacent vertices and "complete the square." Now generalize your results. Start

with the points (x_1, y_1) and (x_2, y_2) as adjacent vertices of a square and write expressions for the coordinates of the other two vertices. Repeat this exercise assuming that the first two points are opposite vertices of a square.

102. *Distance to the Origin*   Let *m* and *n* be any real numbers. What is the distance between $(0, 0)$ and $(2m, m^2 - 1)$? What is the distance between $(0, 0)$ and $(2mn, m^2 - n^2)$?

Thinking Outside the Box XIV

Methodical Mower   Eugene is mowing a rectangular lawn that is 300 ft by 400 ft. He starts at one corner and mows a swath of uni-form width around the outside edge in a clockwise direction. He continues going clockwise, widening the swath that is mowed and shrinking the rectangular section that is yet to be mowed. When he is half done with the lawn, how wide is the swath?

1.3 Pop Quiz

1. Find the distance between $(-1, 3)$ and $(3, 5)$.

2. Find the center and radius for the circle
$$(x - 3)^2 + (y + 5)^2 = 81.$$

3. Find the center and radius for the circle
$$x^2 + 4x + y^2 - 10y = -28.$$

4. Find the equation of the circle that passes through the origin and has center at $(3, 4)$.

5. Find all intercepts for $2x - 3y = 12$.

6. Which point is on both of the lines $x = 5$ and $y = -1$?

Linking Concepts

For Individual or Group Explorations

Modeling Energy Requirements

Clinical dietitians must design diets for patients that meet their basic energy re-quirements and are suitable for the condition of their health. The basic energy re-quirement B (in calories) for a male is given by the formula

$$B = 655.096 + 9.563W + 1.85H - 4.676A,$$

where W is the patient's weight in kilograms, H is the height in centimeters, and A is the age in years (www.eatwell.com). For a female the formula is

$$B = 66.473 + 13.752W + 5.003H - 6.755A.$$

a)   Find your basic energy requirement.

b)   Replace *H* and *A* with your actual height and age. Now draw a graph showing how *B* depends on weight for suitable values of *W*.

c)   For a fixed height and age, does the basic energy requirement increase or decrease as weight increases?

d) Replace W and H with your actual weight and height. Now draw a graph showing how B depends on age for suitable values of A.

e) For a fixed weight and height, does the basic energy requirement increase or decrease as a person gets older?

f) Replace W and A with your actual weight and height. Now draw a graph showing how B depends on height for suitable values of H.

g) For a fixed weight and age, does the basic energy requirement increase or decrease for taller persons?

h) For the three graphs that you have drawn, explain how to determine whether the graph is increasing or decreasing from the formula for the graph.

1.4

Linear Equations in Two Variables

In Section 1.3 we graphed lines, including horizontal and vertical lines. We learned that every line has an equation in standard form $Ax + By = C$. In this section we will continue to study lines.

Slope of a Line

A road that has a 5% grade rises 5 feet for every horizontal run of 100 feet. A roof that has a 5–12 pitch rises 5 feet for every horizontal run of 12 feet. The grade of a road and the pitch of a roof are measurements of *steepness*. The steepness or *slope* of a line in the xy-coordinate system is the ratio of the **rise** (the change in y-coordinates) to the **run** (the change in x-coordinates) between two points on the line:

$$\text{slope} = \frac{\text{change in } y\text{-coordinates}}{\text{change in } x\text{-coordinates}} = \frac{\text{rise}}{\text{run}}$$

If (x_1, y_1) and (x_2, y_2) are the coordinates of the two points in Fig. 1.27, then the rise is $y_2 - y_1$ and the run is $x_2 - x_1$:

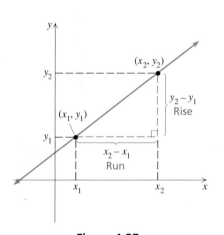

■ **Figure 1.27**

Definition: Slope

> The **slope** of the line through (x_1, y_1) and (x_2, y_2) with $x_1 \neq x_2$ is
>
> $$\frac{y_2 - y_1}{x_2 - x_1}.$$

$Y = 2x - 9$

$Y + 2 = -2(x + 1)$

$Y = -2x - 4$

Note that if (x_1, y_1) and (x_2, y_2) are two points for which $x_1 = x_2$ then the line through them is a vertical line. Since this case is not included in the definition of slope, a vertical line does not have a slope. We also say that the slope of a vertical line is undefined. If we choose two points on a horizontal line then $y_1 = y_2$ and $y_2 - y_1 = 0$. For any horizontal line the rise between two points is 0 and the slope is 0.

Example 1 Finding the slope

In each case find the slope of the line that contains the two given points.

a. $(-3, 4), (-1, -2)$ **b.** $(-3, 7), (5, 7)$ **c.** $(-3, 5), (-3, 8)$

Solution

a. Use $(x_1, y_1) = (-3, 4)$ and $(x_2, y_2) = (-1, -2)$ in the formula:

$$\text{slope} = \frac{y_2 - y_1}{x_2 - x_1} = \frac{-2 - 4}{-1 - (-3)} = \frac{-6}{2} = -3$$

The slope of the line is -3.

b. Use $(x_1, y_1) = (-3, 7)$ and $(x_2, y_2) = (5, 7)$ in the formula:

$$\text{slope} = \frac{y_2 - y_1}{x_2 - x_1} = \frac{7 - 7}{5 - (-3)} = \frac{0}{8} = 0$$

The slope of this horizontal line is 0.

c. The line through $(-3, 5)$ and $(-3, 8)$ is a vertical line and so it does not have a slope.

Try This. Find the slope of the line that contains $(-2, 5)$ and $(-1, -3)$.   ■

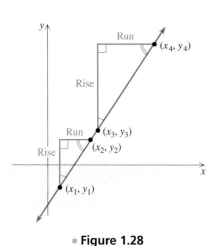

■ **Figure 1.28**

The slope of a line is the same number regardless of which two points on the line are used in the calculation of the slope. To understand why, consider the two triangles shown in Fig. 1.28. These triangles have the same shape and are called similar triangles. Because the ratios of corresponding sides of similar triangles are equal, the ratio of rise to run is the same for either triangle.

Point-Slope Form

Suppose that a line through (x_1, y_1) has slope m. Every other point (x, y) on the line must satisfy the equation

$$\frac{y - y_1}{x - x_1} = m$$

because any two points can be used to find the slope. Multiply both sides by $x - x_1$ to get $y - y_1 = m(x - x_1)$, which is the **point-slope form** of the equation of a line.

$3x + 3y = -1$

$3y = -3x - 1$

$y = -x - \frac{1}{3}$

$(y - 3) = \frac{1}{1}(x + 3)$

$y = -x + 3 + 3$

$y = x + 6$

Theorem:
Point-Slope Form

> The equation of the line (in point-slope form) through (x_1, y_1) with slope m is
>
> $$y - y_1 = m(x - x_1).$$

In Section 1.3 we started with the equation of a line and graphed the line. Using the point-slope form, we can start with a graph of a line or a description of the line and write the equation for the line.

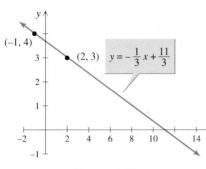

■ **Figure 1.29**

Example 2 The equation of a line given two points

In each case graph the line through the given pair of points. Then find the equation of the line and solve it for y if possible.

a. $(-1, 4), (2, 3)$ **b.** $(2, 5), (-6, 5)$ **c.** $(3, -1), (3, 9)$

Solution

a. Find the slope of the line shown in Fig. 1.29 as follows:

$$m = \frac{y_2 - y_1}{x_2 - x_1} = \frac{3 - 4}{2 - (-1)} = \frac{-1}{3} = -\frac{1}{3}$$

Now use a point, say $(2, 3)$, and $m = -\frac{1}{3}$ in the point-slope form:

$$y - y_1 = m(x - x_1)$$

$$y - 3 = -\frac{1}{3}(x - 2) \qquad \text{The equation in point-slope form}$$

$$y - 3 = -\frac{1}{3}x + \frac{2}{3}$$

$$y = -\frac{1}{3}x + \frac{11}{3} \qquad \text{The equation solved for } y$$

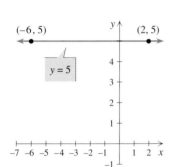

■ **Figure 1.30**

b. The slope of the line through $(2, 5)$ and $(-6, 5)$ shown in Fig. 1.30 is 0. The equation of this horizontal line is $y = 5$.

c. The line through $(3, -1)$ and $(3, 9)$ shown in Fig. 1.31 is vertical and it does not have slope. Its equation is $x = 3$.

Try This. Find the equation of the line that contains $(-2, 5)$ and $(-1, -3)$ and solve the equation for y. ■

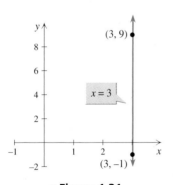

■ **Figure 1.31**

Slope-Intercept Form

In Example 2, notice that the slope $-\frac{1}{3}$ appears as the coefficient of x in the equation $y = -\frac{1}{3}x + \frac{11}{3}$. Also, notice that the y-intercept $\left(0, \frac{11}{3}\right)$ can be determined by simply examining the equation. So the form $y = mx + b$ is called **slope-intercept form:**

Theorem:
Slope-Intercept Form

> The equation of the line (in slope-intercept form) with slope m and y-intercept $(0, b)$ is
>
> $$y = mx + b.$$

If you know the slope and y-intercept for a line then you can use slope-intercept form to write its equation. For example, the equation of the line through $(0, 9)$ with slope 4 is $y = 4x + 9$. In the next example we use the slope-intercept form to determine the slope and y-intercept for a line.

Example **3** **Find the slope and y-intercept**

Identify the slope and y-intercept for the line $2x - 3y = 6$.

Solution

First solve the equation for y to get it in slope-intercept form:

$$2x - 3y = 6$$
$$-3y = -2x + 6$$
$$y = \frac{2}{3}x - 2$$

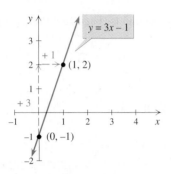

■ **Figure 1.32**

So the slope is $\frac{2}{3}$ and the y-intercept is $(0, -2)$.

Try This. Identify the slope and y-intercept for $3x + 5y = 15$.   ■

Using Slope to Graph a Line

Slope is the ratio $\frac{\text{rise}}{\text{run}}$ that results from moving from one point to another on a line. A positive rise indicates a motion upward and a negative rise indicates a motion downward. A positive run indicates a motion to the right and a negative run indicates a motion to the left. If you start at any point on a line with slope $\frac{1}{2}$, then moving up 1 unit and 2 units to the right will bring you back to the line. On a line with slope -3 or $\frac{-3}{1}$, moving down 3 units and 1 unit to the right will bring you back to the line.

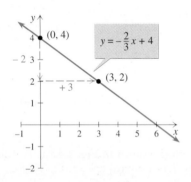

■ **Figure 1.33**

Example **4** **Graphing a line using its slope and y-intercept**

Graph each line.

a. $y = 3x - 1$ **b.** $y = -\frac{2}{3}x + 4$

Solution

a. The line $y = 3x - 1$ has y-intercept $(0, -1)$ and slope 3 or $\frac{3}{1}$. Starting at $(0, -1)$ we obtain a second point on the line by moving up 3 units and 1 unit to the right. So the line goes through $(0, -1)$ and $(1, 2)$ as shown in Fig. 1.32.

b. The line $y = -\frac{2}{3}x + 4$ has y-intercept $(0, 4)$ and slope $-\frac{2}{3}$ or $\frac{-2}{3}$. Starting at $(0, 4)$ we obtain a second point on the line by moving down 2 units and then 3 units to the right. So the line goes through $(0, 4)$ and $(3, 2)$ as shown in Fig. 1.33.

Try This. Use the slope and y-intercept to graph $y = -\frac{3}{2}x + 1$.   ■

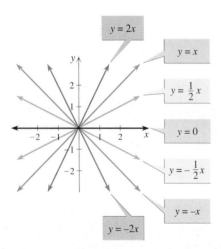

Graphs of $y = mx$

■ **Figure 1.34**

As the x-coordinate increases on a line with positive slope, the y-coordinate increases also. As the x-coordinate increases on a line with negative slope, the y-coordinate decreases. Figure 1.34 shows some lines of the form $y = mx$ with positive slopes and negative slopes. Observe the effect that the slope has on the position of the line.

The Three Forms for the Equation of a Line

There are three forms for the equation of a line. The following strategy will help you decide when and how to use these forms.

STRATEGY   **Finding the Equation of a Line**

1. Since vertical lines have no slope they cannot be written in slope-intercept form $y = mx + b$ or point-slope form $y - y_1 = m(x - x_1)$. All lines can be described by an equation in standard form $Ax + By = C$.

2. For any constant k, $y = k$ is a horizontal line and $x = k$ is a vertical line.

3. If you know two points on a line, then find the slope.

4. If you know the slope and point on the line, use point-slope form. If the point is the y-intercept, then use slope-intercept form.

5. Final answers are usually written in slope-intercept or standard form. Standard form is often simplified by using only integers for the coefficients.

Example **5** **Standard form using integers**

Find the equation of the line through $\left(0, \frac{1}{3}\right)$ with slope $\frac{1}{2}$. Write the equation in standard form using only integers.

Solution

Since we know the slope and y-intercept, start with slope-intercept form:

$$y = \frac{1}{2}x + \frac{1}{3} \qquad \text{Slope-intercept form}$$

$$-\frac{1}{2}x + y = \frac{1}{3}$$

$$-6\left(-\frac{1}{2}x + y\right) = -6 \cdot \frac{1}{3} \qquad \text{Multiply by } -6 \text{ to get integers.}$$

$$3x - 6y = -2 \qquad \text{Standard form with integers}$$

Any integral multiple of $3x - 6y = -2$ would also be standard form, but we usually use the smallest possible positive coefficient for x.

Try This. Find the equation of the line through $\left(0, \frac{1}{2}\right)$ with slope $\frac{3}{4}$ and write the equation in standard form using only integers.   ■

Parallel Lines

Two lines in a plane are said to be **parallel** if they have no points in common. Any two vertical lines are parallel, and slope can be used to determine whether nonvertical lines are parallel. For example, the lines $y = 3x - 4$ and $y = 3x + 1$ are parallel because their slopes are equal.

Theorem: Parallel Lines

> Two nonvertical lines in the coordinate plane are parallel if and only if their slopes are equal.

A proof to this theorem is outlined in Exercises 97 and 98.

Example **6** **Writing equations of parallel lines**

Find the equation in slope-intercept form of the line through $(1, -4)$ that is parallel to $y = 3x + 2$.

Solution

Since $y = 3x + 2$ has slope 3, any line parallel to it also has slope 3. Write the equation of the line through $(1, -4)$ with slope 3 in point-slope form:

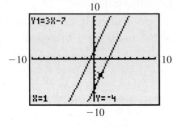

■ **Figure 1.35**

$$y - (-4) = 3(x - 1) \quad \text{Point-slope form}$$
$$y + 4 = 3x - 3$$
$$y = 3x - 7 \quad \text{Slope-intercept form}$$

The line $y = 3x - 7$ goes through $(1, -4)$ and is parallel to $y = 3x + 2$.

The graphs of $y_1 = 3x - 7$ and $y_2 = 3x + 2$ in Fig. 1.35 support the answer.

Try This. Find the equation in slope-intercept form of the line through $(2, 4)$ that is parallel to $y = -\frac{1}{2}x + 9$.   ■

Perpendicular Lines

Two lines are **perpendicular** if they intersect at a right angle. Slope can be used to determine whether lines are perpendicular. For example, lines with slopes such as $2/3$ and $-3/2$ are perpendicular. The slope $-3/2$ is the opposite of the reciprocal of $2/3$. In the following theorem we use the equivalent condition that the product of the slopes of two perpendicular lines is -1, provided they both have slopes.

**Theorem:
Perpendicular Lines**

> Two lines with slopes m_1 and m_2 are perpendicular if and only if $m_1 m_2 = -1$.

■ **PROOF** The phrase "if and only if" means that there are two statements to prove. First we prove that if two lines are perpendicular, then the product of their slopes is -1. Suppose $y = m_1 x + b_1$ and $y = m_2 x + b_2$ are perpendicular as shown in Fig. 1.36. Assuming that $m_1 > 0$ and $m_2 < 0$, start at the intersection of the lines and form a triangle by a rise of m_1 and a run of 1 to indicate a slope of m_1. Use the slope m_2 as a negative rise of m_2 and a run of 1. By the Pythagorean theorem, $a = \sqrt{1 + m_1^2}$ and $b = \sqrt{1 + m_2^2}$. Applying the Pythagorean theorem to

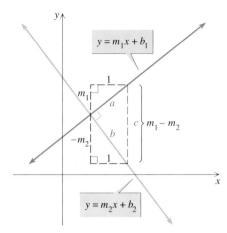

■ **Figure 1.36**

the third right triangle with $c = m_1 - m_2$, we get

$$(m_1 - m_2)^2 = \left(\sqrt{1 + m_1^2}\right)^2 + \left(\sqrt{1 + m_2^2}\right)^2 \quad \text{Since } c^2 = a^2 + b^2$$

$$(m_1 - m_2)^2 = 1 + m_1^2 + 1 + m_2^2 \qquad \text{Simplify.}$$

$$m_1^2 - 2m_1m_2 + m_2^2 = 2 + m_1^2 + m_2^2$$

$$-2m_1m_2 = 2$$

$$m_1m_2 = -1$$

For the second half of the proof, we show that if $m_1m_2 = -1$, then the lines are perpendicular. If $m_1m_2 = -1$, we can reverse the order of the preceding equations to conclude that the triangle with sides a, b, and c is a right triangle and that the lines are perpendicular. Thus, we have proved the theorem. ■

Example **7** **Writing equations of perpendicular lines**

Find the equation of the line perpendicular to the line $3x - 4y = 8$ and containing the point $(-2, 1)$. Write the answer in slope-intercept form.

Solution

Rewrite $3x - 4y = 8$ in slope-intercept form:

$$-4y = -3x + 8$$

$$y = \frac{3}{4}x - 2 \qquad \text{Slope of this line is 3/4.}$$

Since the product of the slopes of perpendicular lines is -1, the slope of the line that we seek is $-4/3$. Use the slope $-4/3$ and the point $(-2, 1)$ in the point-slope form:

$$y - 1 = -\frac{4}{3}(x - (-2))$$

$$y - 1 = -\frac{4}{3}x - \frac{8}{3}$$

$$y = -\frac{4}{3}x - \frac{5}{3}$$

■ **Figure 1.37**

The last equation is the required equation in slope-intercept form. The graphs of these two equations should look perpendicular.

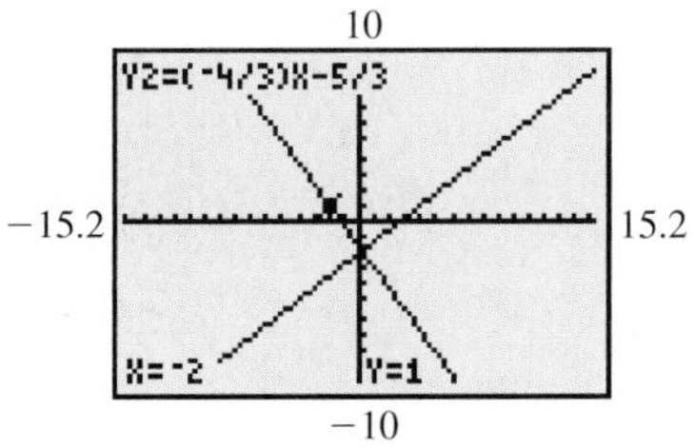

If you graph $y_1 = \frac{3}{4}x - 2$ and $y_2 = -\frac{4}{3}x - \frac{5}{3}$ with a graphing calculator, the graphs will not appear perpendicular in the standard viewing window because each axis has a different unit length. The graphs appear perpendicular in Fig. 1.37 because the unit lengths were made equal with the ZSquare feature of the TI-83.

Try This. Find the equation in slope-intercept form of the line through $(-2, 1)$ that is perpendicular to $2x - y = 8$. ■

Applications

In Section 1.3 the distance formula was used to establish some facts about geometric figures in the coordinate plane. We can also use slope to prove that lines are parallel or perpendicular in geometric figures.

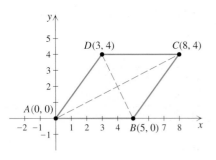

■ **Figure 1.38**

Example **8** **The diagonals of a rhombus are perpendicular**

Given a rhombus with vertices (0, 0), (5, 0), (3, 4), and (8, 4), prove that the diagonals of this rhombus are perpendicular. (A rhombus is a four-sided figure in which the sides have equal length.)

Solution

Plot the four points *A*, *B*, *C*, and *D* as shown in Fig. 1.38. You should show that each side of this figure has length 5, verifying that the figure is a rhombus. Find the slopes of the diagonals and their product:

$$m_{AC} = \frac{4 - 0}{8 - 0} = \frac{1}{2} \qquad m_{BD} = \frac{0 - 4}{5 - 3} = -2$$

$$m_{AC} \cdot m_{BD} = \frac{1}{2}(-2) = -1$$

Since the product of the slopes of the diagonals is −1, the diagonals are perpendicular.

Try This. Prove that the triangle with vertices (0, 0), (5, 2), and (1, 12) is a right triangle. ■

If the value of one variable can be determined from the value of another variable, then we say that the first variable **is a function of** the second variable. Because the area of a circle can be determined from the radius by the formula $A = \pi r^2$, we say that *A* is a function of *r*. If *y* is determined from *x* by using the slope-intercept form of the equation of a line $y = mx + b$, then *y* **is a linear function of x.** The formula $F = \frac{9}{5}C + 32$ expresses *F* as a linear function of *C*. We will discuss functions in depth in Chapter 2.

Example **9** **A linear function**

From ABC Wireless the monthly cost for a cell phone with 100 minutes per month is $35, or 200 minutes per month for $45. See Fig. 1.39. The cost in dollars is a linear function of the time in minutes.

a. Find the formula for *C*.
b. What is the cost for 400 minutes per month?

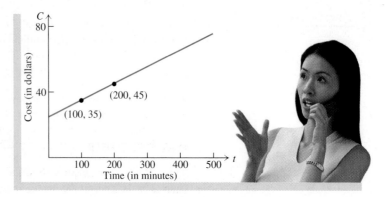

■ **Figure 1.39**

Solution

a. First find the slope:

$$m = \frac{C_2 - C_1}{t_2 - t_1} = \frac{45 - 35}{200 - 100} = \frac{10}{100} = 0.10$$

The slope is \$0.10 per minute. Now find b by using $C = 35, t = 100$, and $m = 0.10$ in the slope-intercept form $C = mt + b$:

$$35 = 0.10(100) + b$$
$$35 = 10 + b$$
$$25 = b$$

So the formula is $C = 0.10t + 25$.

b. Use $t = 400$ in the formula $C = 0.10t + 25$:

$$C = 0.10(400) + 25 = 65$$

The cost for 400 minutes per month is \$65.

Try This. The cost of a truck rental is \$70 for 100 miles and \$90 for 200 miles. Write the cost as a linear function of the number of miles.   ■

Note that in Example 9 we could have used the point-slope form $C - C_1 = m(t - t_1)$ to get the formula $C = 0.10t + 25$. Try it.

The situation in the next example leads naturally to an equation in standard form.

Example 10 Interpreting slope

A manager for a country market will spend a total of \$80 on apples at \$0.25 each and pears at \$0.50 each. Write the number of apples she can buy as a linear function of the number of pears. Find the slope and interpret your answer.

Solution

Let a represent the number of apples and p represent the number of pears. Write an equation in standard form about the total amount spent:

$$0.25a + 0.50p = 80$$
$$0.25a = 80 - 0.50p$$
$$a = 320 - 2p \qquad \text{Solve for } a.$$

The equation $a = 320 - 2p$ or $a = -2p + 320$ expresses the number of apples as a function of the number of pears. Since p is the first coordinate and a is the second, the slope is -2 apples per pear. So if the number of apples is decreased by 2, then the number of pears can be increased by 1 and the total is still \$80. This makes sense because the pears cost twice as much as the apples.

Try This. A manager will spend \$3000 on file cabinets at \$100 each and book-shelves at \$150 each. Write the number of file cabinets as a function of the number of bookshelves and interpret the slope.   ■

For Thought

True or False? Explain.

1. The slope of the line through $(2, 2)$ and $(3, 3)$ is $3/2$.

2. The slope of the line through $(-3, 1)$ and $(-3, 5)$ is 0.

3. Any two distinct parallel lines have equal slopes.

4. The graph of $x = 3$ in the coordinate plane is the single point $(3, 0)$.

5. Two lines with slopes m_1 and m_2 are perpendicular if $m_1 = -1/m_2$.

6. Every line in the coordinate plane has an equation in slope-intercept form.

7. The slope of the line $y = 3 - 2x$ is 3.

8. Every line in the coordinate plane has an equation in standard form.

9. The line $y = 3x$ is parallel to the line $y = -3x$.

10. The line $x - 3y = 4$ contains the point $(1, -1)$ and has slope $1/3$.

1.4 Exercises

Find the slope of the line containing each pair of points.

1. $(-2, 3), (4, 5)$

2. $(-1, 2), (3, 6)$

3. $(1, 3), (3, -5)$

4. $(2, -1), (5, -3)$

5. $(5, 2), (-3, 2)$

6. $(0, 0), (5, 0)$

7. $\left(\dfrac{1}{8}, \dfrac{1}{4}\right), \left(\dfrac{1}{4}, \dfrac{1}{2}\right)$

8. $\left(-\dfrac{1}{3}, \dfrac{1}{2}\right), \left(\dfrac{1}{6}, \dfrac{1}{3}\right)$

9. $(5, -1), (5, 3)$

10. $(-7, 2), (-7, -6)$

Find the equation of the line through the given pair of points. Solve it for y if possible.

11. $(-1, -1), (3, 4)$

12. $(-2, 1), (3, 5)$

13. $(-2, 6), (4, -1)$

14. $(-3, 5), (2, 1)$

15. $(3, 5), (-3, 5)$

16. $(-6, 4), (2, 4)$

17. $(4, -3), (4, 12)$

18. $(-5, 6), (-5, 4)$

Write an equation in slope-intercept form for each of the lines shown.

19.

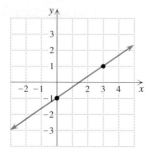

20.

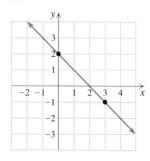

21.

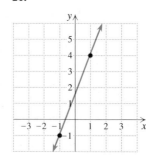

22.

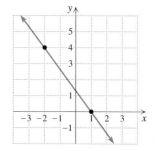

23.   **24.**

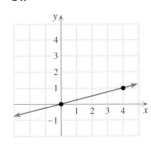

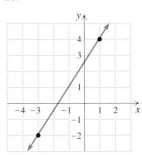   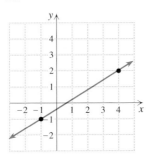

Find the equation of the line through the given pair of points in standard form using only integers. See the strategy for finding the equation of a line on page 129.

47. (3, 0) and (0, −4)   **48.** (−2, 0) and (0, 3)

49. (2, 3) and (−3, −1)   **50.** (4, −1) and (−2, −6)

51. (−4, 2) and (−4, 5)   **52.** (−3, 6) and (9, 6)

53. $\left(2, \frac{2}{3}\right), \left(-\frac{1}{2}, -2\right)$ **54.** $\left(\frac{3}{4}, -3\right), \left(-5, \frac{1}{8}\right)$

55. $\left(\frac{1}{2}, \frac{1}{4}\right), \left(-\frac{1}{3}, \frac{1}{5}\right)$ **56.** $\left(-\frac{3}{8}, \frac{1}{4}\right), \left(\frac{1}{2}, -\frac{1}{6}\right)$

25.   **26.**

Find the slope of each line described.

57. A line parallel to $y = 0.5x - 9$

58. A line parallel to $3x - 9y = 4$

59. A line perpendicular to $3y - 3x = 7$

60. A line perpendicular to $2x - 3y = 8$

61. A line perpendicular to the line $x = 4$

62. A line parallel to $y = 5$

Write each equation in slope-intercept form and identify the slope and y-intercept of the line.

27. $3x - 5y = 10$ **28.** $2x - 2y = 1$

29. $y - 3 = 2(x - 4)$ **30.** $y + 5 = -3(x - (-1))$

31. $y + 1 = \frac{1}{2}(x - (-3))$ **32.** $y - 2 = -\frac{3}{2}(x + 5)$

33. $y - 4 = 0$ **34.** $-y + 5 = 0$

35. $y - 0.4 = 0.03(x - 100)$

36. $y + 0.2 = 0.02(x - 3)$

Write an equation in standard form using only integers for each of the lines described. In each case make a sketch.

63. The line with slope 2, going through $(1, -2)$

64. The line with slope -3, going through $(-3, 4)$

65. The line through $(1, 4)$, parallel to $y = -3x$

66. The line through $(-2, 3)$ parallel to $y = \frac{1}{2}x + 6$

67. The line parallel to $5x - 7y = 35$ and containing $(6, 1)$

68. The line parallel to $4x + 9y = 5$ and containing $(-4, 2)$

Use the y-intercept and slope to sketch the graph of each equation.

37. $y = \frac{1}{2}x - 2$ **38.** $y = \frac{2}{3}x + 1$

39. $y = -3x + 1$ **40.** $y = -x + 3$

41. $y = -\frac{3}{4}x - 1$ **42.** $y = -\frac{3}{2}x$

43. $x - y = 3$ **44.** $2x - 3y = 6$

45. $y - 5 = 0$ **46.** $6 - y = 0$

69. The line perpendicular to $y = \frac{2}{3}x + 5$ and containing $(2, -3)$

70. The line perpendicular to $y = 9x + 5$ and containing $(5, 4)$

71. The line perpendicular to $x - 2y = 3$ and containing $(-3, 1)$

72. The line perpendicular to $3x - y = 9$ and containing $(0, 0)$

73. The line perpendicular to $x = 4$ and containing $(2, 5)$

74. The line perpendicular to $y = 9$ and containing $(-1, 3)$

Find the value of a in each case.

75. The line through $(-2, 3)$ and $(8, 5)$ is perpendicular to $y = ax + 2$.

76. The line through $(3, 4)$ and $(7, a)$ has slope $2/3$.

77. The line through $(-2, a)$ and $(a, 3)$ has slope $-1/2$.

78. The line through $(-1, a)$ and $(3, -4)$ is parallel to $y = ax$.

Either prove or disprove each statement. Use a graph only as a guide. Your proof should rely on algebraic calculations.

79. The points $(-1, 2)$, $(2, -1)$, $(3, 3)$, and $(-2, -2)$ are the vertices of a parallelogram.

80. The points $(-1, 1)$, $(-2, -5)$, $(2, -4)$, and $(3, 2)$ are the vertices of a parallelogram.

81. The points $(-5, -1)$, $(-3, -4)$, $(3, 0)$, and $(1, 3)$ are the vertices of a rectangle.

82. The points $(-5, -1)$, $(1, -4)$, $(4, 2)$, and $(-1, 5)$ are the vertices of a square.

83. The points $(-5, 1)$, $(-2, -3)$, and $(4, 2)$ are the vertices of a right triangle.

84. The points $(-4, -3)$, $(1, -2)$, $(2, 3)$, and $(-3, 2)$ are the vertices of a rhombus.

Use a graphing calculator to solve each problem.

85. Graph $y_1 = (x - 5)/3$ and $y_2 = x - 0.67(x + 4.2)$. Do the lines appear to be parallel? Are the lines parallel?

86. Graph $y_1 = 99x$ and $y_2 = -x/99$. Do the lines appear to be perpendicular? Should they appear perpendicular?

87. Graph $y = (x^3 - 8)/(x^2 + 2x + 4)$. Use TRACE to examine points on the graph. Write a linear function for the graph. Explain why this function is linear.

88. Graph $y = (x^3 + 2x^2 - 5x - 6)/(x^2 + x - 6)$. Use TRACE to examine points on the graph. Write a linear function for the graph. Factor $x^3 + 2x^2 - 5x - 6$ completely.

Solve each problem.

89. *Celsius to Fahrenheit Formula*   Fahrenheit temperature F is a linear function of Celsius temperature C. The ordered pair $(0, 32)$ is an ordered pair of this function because $0°C$ is equivalent to $32°F$, the freezing point of water. The ordered pair $(100, 212)$ is also an ordered pair of this function because $100°C$ is equivalent to $212°F$, the boiling point of water. Use

the two given points and the point-slope formula to write F as a function of C. Find the Fahrenheit temperature of an oven at $150°C$.

90. *Cost of Business Cards*   Speedy Printing charges $23 for 200 deluxe business cards and $35 for 500 deluxe business cards. Given that the cost is a linear function of the number of cards printed, find a formula for that function and find the cost of 700 business cards.

91. *Volume Discount*   Mona Kalini gives a walking tour of Honolulu to one person for $49. To increase her business, she advertised at the National Orthodontist Convention that she would lower the price by $1 per person for each additional person. Write the cost per person c as a function of the number of people n on the tour. How much does she make for a tour with 40 people?

HINT Find the equation of the line through $(1, 49)$, $(2, 48)$, $(3, 47)$, etc.

92. *Ticket Pricing*   At $10 per ticket, Willie Williams and the Wranglers will fill all 8000 seats in the Assembly Center. The manager knows that for every $1 increase in the price, 500 tickets will go unsold. Write the number of tickets sold, n, as a function of the ticket price, p. How much money will be taken in if the tickets are $20 each?

HINT Find the equation of the line through $(10, 8000)$, $(11, 7500)$, $(12, 7000)$, etc.

93. *Lindbergh's Air Speed*   Charles Lindbergh estimated that at the start of his historic flight the practical economical air speed was 95 mph and at 4000 statute miles from the starting point it was 75 mph (www.charleslindbergh.com). Assume that the practical economical air speed S is a linear function of the distance D from the starting point as shown in the accompanying figure. Find a formula for that function.

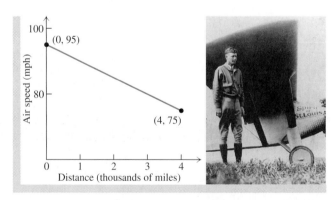

■ **Figure for Exercise 93**

94. *Speed Over Newfoundland*   In Lindbergh's flying log he recorded his air speed over Newfoundland as 98 mph, 1100 miles into his flight. According to the formula from Exercise 93, what should have been his air speed over Newfoundland?

■ **Figure for Exercise 94**

95. *Computers and Printers*   An office manager will spend a total of $60,000 on computers at $2000 each and printers at $1500 each. Write the number of computers purchased as a function of the number of printers purchased. Find and interpret the slope.

    HINT Start with standard form.

96. *Carpenters and Helpers*   Because a job was finished early, a contractor will distribute a total of $2400 in bonuses to 9 carpenters and 3 helpers. The carpenters all get the same amount and the helpers all get the same amount. Write the amount of a helper's bonus as a function of the amount of a carpenter's bonus. Find and interpret the slope.

For Writing/Discussion

97. *Equal Slopes*   Show that if $y = mx + b_1$ and $y = mx + b_2$ are equations of lines with equal slopes, but $b_1 \neq b_2$, then they have no point in common.

    HINT Assume that they have a point in common and see that this assumption leads to a contradiction.

98. *Unequal Slopes*   Show that the lines $y = m_1 x + b_1$ and $y = m_2 x + b_2$ intersect at a point with x-coordinate $(b_2 - b_1)/(m_1 - m_2)$ provided $m_1 \neq m_2$. Explain how this exercise and the previous exercise prove the theorem that two nonvertical lines are parallel if and only if they have equal slopes.

99. *Summing Angles*   Consider the angles shown in the accompanying figure. Show that the degree measure of angle A plus the degree measure of angle B is equal to the degree measure of angle C.

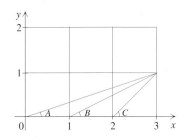

■ **Figure for Exercise 99**

100. *Radius of a Circle*   Find the radius of the circle shown in the figure.

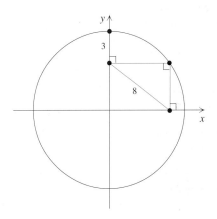

■ **Figure for Exercise 100**

Thinking Outside the Box XV

Army of Ants   An army of ants is marching across the kitchen floor. If they form columns with 10 ants in each column, then there are 6 ants left over. If they form columns with 7, 11, or 13 ants in each column, then there are 2 ants left over. What is the smallest number of ants that could be in this army?

1.4 Pop Quiz

1. Find the slope of the line through $(-4, 9)$ and $(5, 6)$.

2. Find the equation of the line through $(3, 4)$ and $(6, 8)$.

3. What is the slope of the line $2x - 7y = 1$?

4. Find the equation of the line with y-intercept $(0, 7)$ that is parallel to $y = 3x - 1$.

5. Find the equation of the line with y-intercept $(0, 8)$ that is perpendicular to $y = \frac{1}{2}x + 4$.

Linking Concepts

The Negative Income Tax Model

One idea for income tax reform is that of a negative income tax. Under this plan families with an income below a certain level receive a payment from the government in addition to their income while families above a certain level pay taxes. In one proposal a family with an earned income of $6000 would receive a $4000 payment, giving the family a disposable income of $10,000. A family with an earned income of $24,000 would pay $2000 in taxes, giving the family a disposable income of $22,000. A family's disposable income D is a linear function of the family's earned income E.

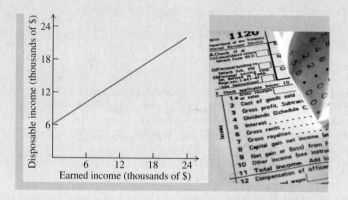

a) Use the given data to write a formula for D as a linear function of E.

b) What is the disposable income for a family with an earned income of $60,000?

c) Find the break-even point, the income at which a family pays no taxes and receives no payment.

d) What percentage of their earned income is paid in taxes for a family with an earned income of $25,000? $100,000? $2,000,000?

e) What is the maximum percentage of earned income that anyone will pay in taxes?

f) Compare your answer to part (e) with current tax laws. Does this negative income tax favor the rich, poor, or middle class?

1.5

Scatter Diagrams and Curve Fitting

The Cartesian coordinate system is often used to illustrate real data and relationships between variables. In this section we will graph real data and see how to fit lines to that data.

Scatter Diagrams and Types of Relationships

In statistics we often gather paired data such as height and weight of an individual. We seek a relationship between the two variables. We can graph pairs of data as ordered pairs in the Cartesian coordinate system as shown in Fig. 1.40. The graph is called a **scatter diagram** because real data are more likely to be scattered about rather than perfectly lined up. If a pattern appears in the scatter diagram, then there is a relationship between the variables. If the points in the scatter diagram look like they are scattered about a line, then the relationship is **linear.** Otherwise the relationship is **nonlinear.**

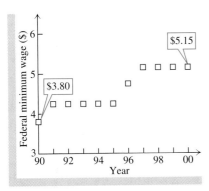

▪ **Figure 1.40**

Example **1** Classifying scatter diagrams

For each given scatter diagram determine whether there is a relationship between the variables. If there is a relationship then determine whether it is linear or nonlinear.

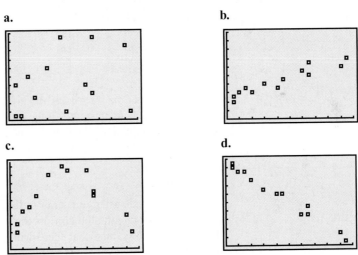

Solution

a. The points are just scattered about and it does not appear that there is a relationship between the variables.

b. It appears that there is a linear relationship between the variables.

c. It appears that there is a pattern to the data and that there is a nonlinear relationship between the variables.

d. It appears that there is a linear relationship between the variables.

Try This. Determine whether the variables depicted in the following scatter diagrams have a linear or nonlinear relationship.

a.

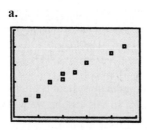

b.

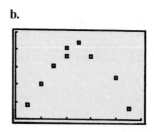

Using a scatter diagram to determine whether there is a relationship between variables is not precise. We will see a more precise method in Example 3.

Finding a Line of Best Fit

If there appears to be a linear relationship in a scatter diagram, we can draw in a line and use it to make predictions.

Example **2** **Fitting a line to data**

The following table shows the year and the number of persons in the labor force in the United States (Bureau of Labor Statistics, www.bls.gov).

Year	Labor Force (millions)
1960	70
1970	82
1980	106
1990	125
2000	141

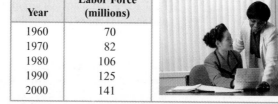

a. Make a scatter diagram using the given data.
b. Draw a line that appears to fit the data.
c. Use your graph to estimate the number of persons in the labor force in 1985.
d. Use your graph to predict the number of persons in the labor force in 2010.

■ Figure 1.41

Solution

Make a scatter diagram and draw a line that appears to fit the data as shown in Fig. 1.41. From the graph it appears that there were about 115 million persons in the labor force in 1985 and that there will be about 160 million persons in 2010. Of course, answers will vary depending on the line that is drawn.

Try This. Predict the cost in 2010 by using a scatter diagram for the given data.

Year	1980	1985	1990	2000	2005
Cost ($)	12	21	33	49	61

■

Making a prediction within the range of the data (as in Example 2(c)) is called **interpolating** and making a prediction outside the range of the data (as in Example 2(d)) is called **extrapolating.** Remember that a prediction made by extrapolating or interpolating the data is a guess and it may be incorrect.

We could have drawn many different lines to fit the data in Example 2. However, there is one line that is preferred over all others. It is called the **line of best fit** or the **regression line.** The line that best fits the data minimizes the total "distance" between the points in the scatter diagram and the line. More details on the line of best fit can be found in any introductory statistics text and we will not give them here. However, it is relatively easy to find an equation for the regression line because a graphing calculator performs all of the computations for you. The calculator even indicates how well the equation fits the data by giving a number r called the **correlation coefficient.** The correlation coefficient is between -1 and 1. If $r = 1$ (a **positive correlation**) the data are perfectly in line and increasing values of one variable correspond to increasing values of the other. If $r = -1$ (a **negative correlation**) the data are also perfectly in line, but increasing values of one variable correspond to decreasing values of the other. If $r = 0$ (or r is close to 0) the data are scattered about with no linear correlation. The closer r is to 1 or -1, the more the scatter diagram will look linear. See Fig. 1.42.

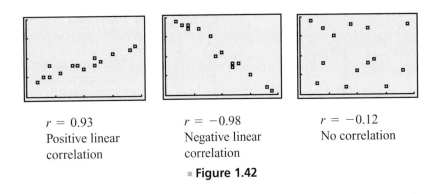

$r = 0.93$
Positive linear
correlation

$r = -0.98$
Negative linear
correlation

$r = -0.12$
No correlation

■ **Figure 1.42**

Example 3   Finding the line of best fit

The table shows the temperature at the start of the race and the percentage of runners injured in eight runnings of the Boston Marathon.

a. Use a graphing calculator to make a scatter diagram and find the equation of the regression line for the data.
b. Use the regression equation to estimate the percentage of runners that would be injured if the marathon was run on a day when the temperature was 82°F.
c. Use the regression equation to estimate the temperature at the start of the race for a race in which 9% of the runners were injured.

Temp (°F)	% Injured
46	4.7
72	12.3
59	6.5
56	5.9
54	6.6
70	10.3
68	8.4
48	4.0

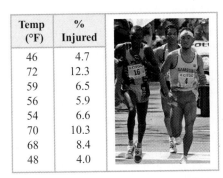

Solution

a. First enter the data into the calculator using the STAT EDIT feature as shown in Fig. 1.43(a) on the next page. Enter the temperature in the first list or x-list and the percent injured in the second list or y-list. Consult your calculator manual if you have trouble or your calculator does not accept data in this manner. Use the STAT CALC feature and the choice LinReg, to find the equation as shown in Fig. 1.43(b). The calculator gives the equation in the form $y = ax + b$ where $a \approx 0.27$ and $b \approx -8.46$. So the equation is $y = 0.27x - 8.46$. Since $r \approx 0.94$, the equation is a good model for the data. The scatter diagram and the regression line are shown in Fig. 1.43(c).

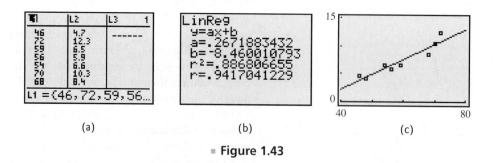

(a)   (b)   (c)

■ **Figure 1.43**

b. If $x = 82$, then $y = 0.27(82) - 8.46 = 13.68$. When the temperature is 82°F about 13.7% of the runners would be injured.

c. If 9% of the runners are injured, then $9 = 0.27x - 8.46$. Solve for x to get $x \approx 64.7$. So the temperature should be about 65°F if 9% are injured.

Try This. Predict the cost in 2010 to the nearest cent by using linear regression and the given data. Use the equation given by the calculator to make the prediction.

Year	1980	1985	1990	2000	2005
Cost ($)	12	21	33	49	61

■

For Thought

True or False? Explain.

1. A scatter diagram is a graph.

2. If data are roughly in line in a scatter diagram, then there is a linear relationship between the variables.

3. If there is no pattern in a scatter diagram, then there is a nonlinear relationship between the variables.

4. The line of best fit is the regression line.

5. If $r = 0.999$, then there is a positive correlation between the variables.

6. If $r = -0.998$, then there is a negative correlation between the variables.

7. If $r = 0.002$, then there is a positive correlation between the variables.

8. If $r = -0.001$, then there is a negative correlation between the variables.

9. If we make a prediction outside the range of the data, then we are interpolating.

10. If we make a prediction within the range of the data, then we are extrapolating.

1.5 Exercises

For each given scatter diagram determine whether there is a linear relationship, a nonlinear relationship, or no relationship between the variables.

1.

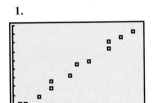

2.

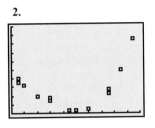

3.

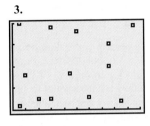

4.

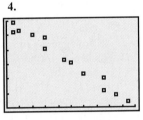

5.

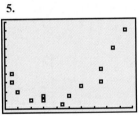

6.

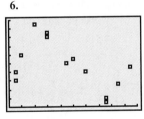

7.

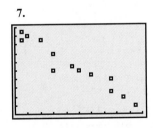

8.

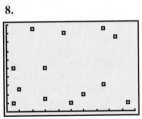

Draw a scatter diagram for each given data set. Use graph paper. From your graph determine whether there is a linear relationship between the variables.

9.

Height (in.)	Weight (lb)
24	52
30	62
32	66
18	40
22	50
36	68
40	78
28	60
22	42

10.

Age (yr)	Income (thousands of $)
33	24
30	23
32	26
30	22
38	36
35	34
32	30
40	44
39	42

11.

ACT Score	Grade Point Average
18	2.2
26	3.4
20	2.8
17	2.0
19	2.6
23	3.2
25	3.6
30	4.0

12.

Distance (mi)	Time (min)
8	34
5	20
6	26
16	70
12	50
6	32
20	90
18	70

13.

Height (in.)	Grade Point Average
61	4.0
59	1.9
66	3.8
72	1.6
62	2.5
74	3.8
60	1.2
68	2.6
65	2.0

14.

Price (thousands of $)	Miles per gal
16	40
18	20
22	16
24	33
29	8
32	30
35	40
38	12
40	66

Draw a scatter diagram using the given ordered pairs and a line that you think fits the data. Use graph paper. Complete the missing entries in the table by reading them from your graph. Answers may vary.

15.

Weight (thousands of lb)	Stopping Distance (ft)
2.0	96
2.4	110
2.7	130
2.8	130
2.8	142
?	160
3.7	180
3.8	180
4.0	?

16.

Prime Rate (%)	Unemployment Rate (%)
8.2	6.1
5.1	3.0
?	4.6
9.7	7.7
5.5	3.2
7.3	5.2
5.5	3.7
7.8	5.6
9.9	?

17.

Pressure (lb/in.^{2})	Volume (in.^{3})
122	40
110	50
125	38
132	?
141	30
138	32
117	40
144	26
?	25

18.

Rate (ft/sec)	Time (sec)
10.0	6
10.1	6
10.2	5.7
10.3	?
10.4	5.6
10.5	5.4
10.6	5.5
10.7	5.4
10.8	5.2
?	5.0

Solve each problem.

19. *Toyota Production*   The accompanying table shows the number of vehicles produced by the Toyota Motor Corporation for the years 2000 through 2004 (www.toyota.com).

Year	Vehicles (thousands)
2000	5132
2001	5297
2002	5306
2003	5850
2004	6513

a. Use linear regression on your graphing calculator to find the regression equation. Let $x = 0$ correspond to 2000.

b. Use the regression equation to predict the number of vehicles produced in 2008.

c. According to the equation, in what year were there no vehicles produced by Toyota Motor Corporation?

20. *Women in the Labor Force*   The accompanying table shows the percentage of the labor force population aged 16 and over that is made up of women (www.infoplease.com).

Year	Women in Labor Force (%)
1993	45.5
1994	46.0
1995	46.1
1996	46.2
1997	46.2
1998	46.3
2002	46.6

a. Use linear regression on your graphing calculator to find the regression equation. Let $x = 0$ correspond to 1990.

b. Use the equation to predict the year in which women will make up 50% of the labor force aged 16 and over.

c. According to the equation, what percent of the labor force was female in 1980?

21. *Money Supply*   The M1 money supply is a measure of money in its most liquid forms, including currency and checking accounts. The accompanying table shows the M1 in billions of dollars in the United States for the years 2000 through 2005 (www.economagic.com).

Year	MI Money Supply ($ billions)
2000	1121
2001	1096
2002	1185
2003	1221
2004	1290
2005	1354

a. Use linear regression on your graphing calculator to find the regression equation. Let $x = 0$ correspond to 2000.

b. Use the regression equation to predict the year in which the money supply will reach $2 trillion.

c. According to the equation, what was the money supply in 1995?

22. *Cost of Oil*   The accompanying table shows the price of West Texas Intermediate Crude at the beginning of each year for 2000 through 2005 (www.economagic.com).

Year	Price per Barrel (dollars)
2000	27.18
2001	29.58
2002	19.67
2003	32.94
2004	34.27
2005	46.84

a. Use linear regression on your graphing calculator to find the regression equation. Let $x = 0$ correspond to 2000.

b. Use the regression equation to predict the year in which the price will reach $75 per barrel.

c. According to the equation, what was the price in 1995?

23. *Injuries in the Boston Marathon* A study of the 4386 male runners in the Boston Marathon showed a high correlation between age and the percentage of injured runners for men aged 59 and below (The Boston Globe, www.boston.com/marathon). The bar graph shows the percentage of injured men age 59 and below grouped into five categories. There appears to be a relationship between the age group A and the percentage p of those injured. Use the linear regression feature of a graphing calculator to find an equation that expresses p in terms of A. Find the percentage of runners that are predicted to be injured in age group 4 according to the equation and compare the answer to the actual percentage injured.

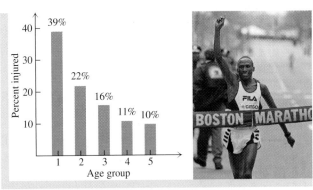

1: under 20, 2: 20–29, 3: 30–39, 4: 40–49, 5: 50–59

■ **Figure for Exercise 23**

24. *Injuries in the Boston Marathon* Use the equation of Exercise 23 to predict the percentage of runners injured in age group 6 (60-plus). Actually, 21% of the runners in the 60-plus age group were injured. Can you offer an explanation for the difference between the predicted and the actual figures?

Thinking Outside the Box XVI

Counting Coworkers Chris and Pat work at a Tokyo Telemarketing. One day Chris said to Pat, "19/40 of my coworkers are female." Pat replied, "That's strange, 12/25 of my coworkers are female." If both are correct, then how many workers are there at Tokyo Telemarketing and what are the genders of Chris and Pat?

1.5 Pop Quiz

1. Use your calculator to find the regression equation for the data given in the table:

Year	01	02	03	04	05
Cost ($)	10	12	13	16	17

2. Use the regression equation to estimate the cost in 2010.

3. Use the regression equation to estimate the cost in 1998.

1.6

Quadratic Equations

One of our main goals is to solve polynomial equations. In Section 1.1 we learned to solve equations of the form $ax + b = 0$, the linear equations. Linear equations are first-degree polynomial equations. In this section we will solve second-degree polynomial equations, the quadratic equations.

Definition

Quadratic equations have second-degree terms that linear equations do not have.

Definition:
Quadratic Equation

> A **quadratic equation** is an equation of the form
>
> $$ax^2 + bx + c = 0,$$
>
> where a, b, and c are real numbers with $a \neq 0$.

The condition that $a \neq 0$ in the definition ensures that the equation actually does have an x^2-term. There are several methods for solving quadratic equations. Which method is most appropriate depends on the type of quadratic equation that we are solving. We first consider a method for solving quadratic equations in which $b = 0$ and later consider methods for solving equations in which $b \neq 0$.

Solving $ax^2 + c = 0$

If $b = 0$ in the general form of the quadratic equation, then the equation is of the form $ax^2 + c = 0$. Equations of this form are solved by using the square root property.

The Square Root Property

> For any real number k, the equation $x^2 = k$ is equivalent to $x = \pm \sqrt{k}$.

If $k > 0$, then $x^2 = k$ has two real solutions. If $k < 0$, then $\sqrt{k}$ is an imaginary number and $x^2 = k$ has two imaginary solutions, but no real solutions. (Imaginary numbers were discussed in Section P.4.) If $k = 0$, then 0 is the only solution to $x^2 = k$.

Example **1** Using the square root property

Find all complex solutions to each equation.

a. $x^2 - 9 = 0$ **b.** $(2x - 1)^2 = 0$ **c.** $(x - 3)^2 + 8 = 0$

Solution

a. Before using the square root property, isolate x^2.

$$x^2 - 9 = 0$$
$$x^2 = 9$$
$$x = \pm \sqrt{9} = \pm 3 \quad \text{Square root property}$$

Check: $3^2 - 9 = 0$ and $(-3)^2 - 9 = 0$. The solution set is $\{-3, 3\}$.

b. $(2x - 1)^2 = 0$

$$2x - 1 = \pm \sqrt{0} \quad \text{Square root property}$$
$$2x - 1 = 0$$
$$2x = 1$$
$$x = \frac{1}{2}$$

The solution set is $\left\{ \frac{1}{2} \right\}$.

c. $(x - 3)^2 + 8 = 0$

$\qquad (x - 3)^2 = -8$

$\qquad x - 3 = \pm \sqrt{-8}$ Square root property

$\qquad x = 3 \pm 2i\sqrt{2}$ $\sqrt{-8} = i\sqrt{8} = i\sqrt{4}\sqrt{2} = 2i\sqrt{2}$

The solution set is $\left\{ 3 - 2i\sqrt{2}, 3 + 2i\sqrt{2} \right\}$. Note that there are no real solutions to this equation, because the square of every real number is nonnegative.

Try This. Solve $(x - 3)^2 = 16$ by the square root property. ■

In Section 1.3 we learned that the solution to an equation in one variable corresponds to an x-intercept on a graph in two variables. Note that the solutions to $x^2 - 9 = 0$ and $(2x - 1)^2 = 0$ in Example 1 correspond to the x-intercepts on the graphs of $y = x^2 - 9$ and $y = (2x - 1)^2$ in Figs. 1.44(a) and 1.44(b). Because there are no real solutions to $(x - 3)^2 + 8 = 0$ in Example 1(c), the graph of $y = (x - 3)^2 + 8$ in Fig. 1.44(c) has no x-intercepts.

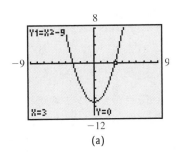

(a)

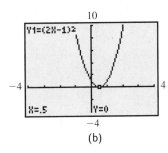

(b)

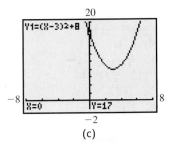

(c)

■ **Figure 1.44**

We will study graphs and their properties extensively in the coming chapters. In this chapter we will point out, using calculator graphs, how graphs in two variables can help us understand the solutions to equations and inequalities in one variable. So even if you are not using a graphing calculator, you should pay close attention to the calculator graphs shown in this chapter.

The key idea in solving $ax^2 + c = 0$ is to solve for x^2 and then apply the square root property. We now turn our attention to equations of the form $ax^2 + bx + c = 0$ in which the trinomial can be factored.

Solving Quadratic Equations by Factoring

Many second-degree polynomials can be factored as a product of first-degree binomials. When one side of an equation is a product and the other side is 0, we can write an equivalent statement by setting each factor equal to zero. This idea is called the **zero factor property.**

The Zero Factor Property

If A and B are algebraic expressions, then the equation $AB = 0$ is equivalent to the compound statement $A = 0$ or $B = 0$.

Example **2** Quadratic equations solved by factoring

Solve each equation by factoring.

a. $x^2 - x - 12 = 0$

b. $(x + 3)(x - 4) = 8$

Solution

a. $x^2 - x - 12 = 0$

$(x - 4)(x + 3) = 0$ Factor the left-hand side.

$x - 4 = 0$ or $x + 3 = 0$ Zero factor property

$x = 4$ or $x = -3$

Check: $(-3)^2 - (-3) - 12 = 0$ and $4^2 - 4 - 12 = 0$. The solution set is $\{-3, 4\}$.

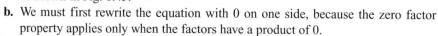 The graph of $y = x^2 - x - 12$ intersects the x-axis at $(-3, 0)$ and $(4, 0)$ as shown in Fig. 1.45. □

b. We must first rewrite the equation with 0 on one side, because the zero factor property applies only when the factors have a product of 0.

$(x + 3)(x - 4) = 8$

$x^2 - x - 12 = 8$ Multiply on the left-hand side.

$x^2 - x - 20 = 0$ Get 0 on the right-hand side.

$(x - 5)(x + 4) = 0$ Factor.

$x - 5 = 0$ or $x + 4 = 0$ Zero factor property

$x = 5$ or $x = -4$

Check in the original equation. The solution set is $\{-4, 5\}$.

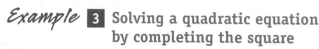 The graph of $y = x^2 - x - 20$ intersects the x-axis at $(-4, 0)$ and $(5, 0)$ as shown in Fig. 1.46.

Try This. Solve $x^2 - 7x - 18 = 0$ by factoring. ■

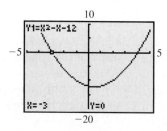

■ **Figure 1.45**

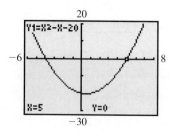

■ **Figure 1.46**

Completing the Square

Solving quadratic equations by factoring is limited to equations that we can factor. However, we can solve any quadratic equation by completing the square and then using the square root property. We used completing the square in Section 1.3 to convert an equation of a circle into standard form.

Since it is simplest to complete the square when the leading coefficient of the polynomial is 1, we divide by the leading coefficient before completing the square. This is illustrated in part (b) of the next example.

Example **3** Solving a quadratic equation
by completing the square

Solve each equation by completing the square.

a. $x^2 + 6x + 7 = 0$ **b.** $2x^2 - 3x - 4 = 0$

Solution

a. Since $x^2 + 6x + 7$ is not a perfect square trinomial, we must find a perfect square trinomial that has $x^2 + 6x$ as its first two terms. Since one-half of 6 is 3 and $3^2 = 9$, our goal is to get $x^2 + 6x + 9$ on the left-hand side:

$$x^2 + 6x + 7 = 0$$

$$x^2 + 6x \qquad = -7 \qquad\qquad \text{Subtract 7 from each side.}$$

$$x^2 + 6x + 9 = -7 + 9 \qquad \text{Add 9 to each side.}$$

$$(x + 3)^2 = 2 \qquad\qquad \text{Factor the left-hand side.}$$

$$x + 3 = \pm\sqrt{2} \qquad\qquad \text{Square root property}$$

$$x = -3 \pm \sqrt{2}$$

Check in the original equation. The solution set is $\{-3 - \sqrt{2}, -3 + \sqrt{2}\}$.

b. $2x^2 - 3x - 4 = 0$

$$x^2 - \frac{3}{2}x - 2 = 0 \qquad\qquad\qquad \text{Divide each side by 2 to get } a = 1.$$

$$x^2 - \frac{3}{2}x \qquad = 2 \qquad\qquad\qquad \text{Add 2 to each side.}$$

$$x^2 - \frac{3}{2}x + \frac{9}{16} = 2 + \frac{9}{16} \qquad\qquad \frac{1}{2} \cdot \frac{3}{2} = \frac{3}{4} \text{ and } \left(\frac{3}{4}\right)^2 = \frac{9}{16}.$$

$$\left(x - \frac{3}{4}\right)^2 = \frac{41}{16} \qquad\qquad\qquad \text{Factor the left-hand side.}$$

$$x - \frac{3}{4} = \pm\frac{\sqrt{41}}{4} \qquad\qquad\qquad \text{Square root property}$$

$$x = \frac{3}{4} \pm \frac{\sqrt{41}}{4} = \frac{3 \pm \sqrt{41}}{4}$$

The solution set is $\left\{\dfrac{3 - \sqrt{41}}{4}, \dfrac{3 + \sqrt{41}}{4}\right\}$.

To check, enter $y_1 = 2x^2 - 3x - 4$ as in Fig. 1.47(a). On the home screen in Fig. 1.47(b), enter y_1 followed in parentheses by the x-value at which you want to calculate y_1.

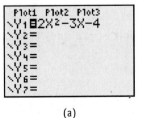

(a)　　　　　　　　(b)

■ **Figure 1.47**

Try This. Solve $2x^2 - 4x - 1 = 0$ by completing the square.　　　　　■

| S | T | R | A | T | E | G | Y | Completing the Square |

The goal is to get a perfect square trinomial on one side of the equation.
1. If the coefficient of x is not 1, divide by it.
2. To complete $x^2 + bx$, add $(b/2)^2$ to both sides of the equation.
3. Factor the perfect square trinomial.
4. Apply the square root property.

The Quadratic Formula

The method of completing the square can be applied to any quadratic equation

$$ax^2 + bx + c = 0, \qquad \text{where } a \neq 0.$$

Assume for now that $a > 0$, and divide each side by a.

$$x^2 + \frac{b}{a}x + \frac{c}{a} = 0$$

$$x^2 + \frac{b}{a}x = -\frac{c}{a} \qquad \text{Subtract } \frac{c}{a} \text{ from each side.}$$

$$x^2 + \frac{b}{a}x + \frac{b^2}{4a^2} = -\frac{c}{a} + \frac{b^2}{4a^2} \qquad \frac{1}{2} \cdot \frac{b}{a} = \frac{b}{2a} \text{ and } \left(\frac{b}{2a}\right)^2 = \frac{b^2}{4a^2}.$$

Now factor the perfect square trinomial on the left-hand side. On the right-hand side get a common denominator and add.

$$\left(x + \frac{b}{2a}\right)^2 = \frac{b^2 - 4ac}{4a^2} \qquad \frac{c}{a} \cdot \frac{4a}{4a} = \frac{4ac}{4a^2}$$

$$x + \frac{b}{2a} = \pm\sqrt{\frac{b^2 - 4ac}{4a^2}} \qquad \text{Square root property, assuming } b^2 - 4ac \geq 0$$

$$x = -\frac{b}{2a} \pm \frac{\sqrt{b^2 - 4ac}}{2a} \qquad \text{Because } a > 0, \sqrt{4a^2} = 2a.$$

$$x = \frac{-b \pm \sqrt{b^2 - 4ac}}{2a}$$

We assumed that $a > 0$ so that $\sqrt{4a^2} = 2a$ would be correct. If a is negative, then $\sqrt{4a^2} = -2a$, and we get

$$x = -\frac{b}{2a} \pm \frac{\sqrt{b^2 - 4ac}}{-2a}.$$

However, the negative sign in $-2a$ can be deleted because of the $\pm$ symbol preceding it. For example, $5 \pm (-3)$ gives the same values as 5 ± 3. After deleting the negative sign on $-2a$, we get the same formula for the solution. It is called the **quadratic formula.** Its importance lies in its wide applicability. Any quadratic equation can be solved by using this formula.

The Quadratic Formula

The solutions to $ax^2 + bx + c = 0$, with $a \neq 0$, are given by the formula

$$x = \frac{-b \pm \sqrt{b^2 - 4ac}}{2a}.$$

Example 4 Using the quadratic formula

Use the quadratic formula to find all real or imaginary solutions to each equation.

a. $x^2 + 8x + 6 = 0$ **b.** $x^2 - 6x + 11 = 0$ **c.** $4x^2 + 9 = 12x$

Solution

a. For $x^2 + 8x + 6 = 0$ we use $a = 1$, $b = 8$, and $c = 6$ in the formula:

$$x = \frac{-8 \pm \sqrt{8^2 - 4(1)(6)}}{2(1)} = \frac{-8 \pm \sqrt{40}}{2} = \frac{-8 \pm 2\sqrt{10}}{2}$$

$$= \frac{2(-4 \pm \sqrt{10})}{2} = -4 \pm \sqrt{10}$$

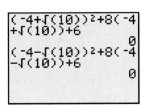

■ **Figure 1.48**

The solution set is $\{-4 - \sqrt{10}, -4 + \sqrt{10}\}$.

〜〜 Check by using a calculator as in Fig. 1.48. □

b. Use $a = 1$, $b = -6$, and $c = 11$ in the quadratic formula:

$$x = \frac{-(-6) \pm \sqrt{(-6)^2 - 4(1)(11)}}{2(1)}$$

$$= \frac{6 \pm \sqrt{-8}}{2} = \frac{6 \pm 2i\sqrt{2}}{2} = 3 \pm i\sqrt{2}$$

Check by evaluating $x^2 - 6x + 11$ with $x = 3 + i\sqrt{2}$ as follows:

$$(3 + i\sqrt{2})^2 - 6(3 + i\sqrt{2}) + 11 = 9 + 6i\sqrt{2} + (i\sqrt{2})^2 - 18 - 6i\sqrt{2} + 11$$

$$= 9 + 6i\sqrt{2} - 2 - 18 - 6i\sqrt{2} + 11$$

$$= 0$$

■ **Figure 1.49**

The reader should check $3 - i\sqrt{2}$. The imaginary solutions are $3 - i\sqrt{2}$ and $3 + i\sqrt{2}$.

〜〜 The solutions can be checked with a calculator as in Fig. 1.49. □

c. To use the quadratic formula, the equation must be in the form $ax^2 + bx + c = 0$. So rewrite the equation as $4x^2 - 12x + 9 = 0$. Now use $a = 4$, $b = -12$, and $c = 9$ in the formula:

$$x = \frac{-(-12) \pm \sqrt{(-12)^2 - 4(4)(9)}}{2(4)} = \frac{12 \pm \sqrt{0}}{8} = \frac{3}{2}$$

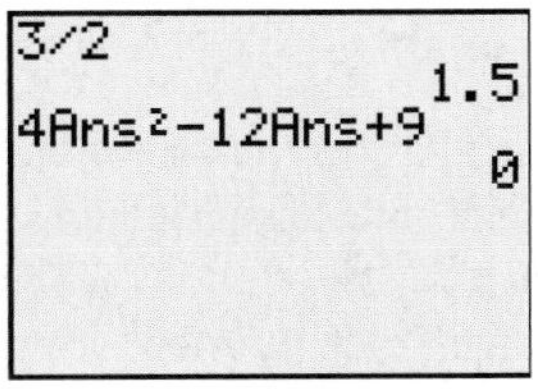

■ **Figure 1.50**

You should check that $4\left(\frac{3}{2}\right)^2 - 12\left(\frac{3}{2}\right) + 9 = 0$. The solution set is $\left\{\frac{3}{2}\right\}$.

〜〜 You can check with a calculator as in Fig. 1.50.

Try This. Solve $2x^2 - 3x - 2 = 0$ by the quadratic formula.   ■

To decide which of the four methods to use for solving a given quadratic equation, use the following strategy.

S T R A T E G Y   **Solving $ax^2 + bx + c = 0$**

1. If $b = 0$, solve $ax^2 + c = 0$ for x^2 and apply the square root property.
2. If $ax^2 + bx + c$ can be easily factored, then solve by factoring.
3. The quadratic formula or completing the square may be used on any quadratic equation, but the quadratic formula is usually easier to use.

The Discriminant

The expression $b^2 - 4ac$ in the quadratic formula is called the **discriminant,** because its value determines the number and type of solutions to a quadratic equation. If $b^2 - 4ac > 0$, then $\sqrt{b^2 - 4ac}$ is real and the equation has two real solutions. If $b^2 - 4ac < 0$, then $\sqrt{b^2 - 4ac}$ is not a real number and there are no real solutions. There are two imaginary solutions. If $b^2 - 4ac = 0$, then $-b/(2a)$ is the only solution. See Table 1.2.

■ **Table 1.2**   Number of real solutions to a quadratic equation

Value of $b^2 - 4ac$	Number of Real Solutions
Positive	2
Zero	1
Negative	0

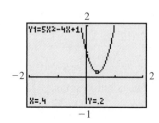

■ **Figure 1.51**

■ **Figure 1.52**

Example ⑤ Using the discriminant

For each equation, state the value of the discriminant and the number of real solutions.

a. $x^2 + 8x + 6 = 0$ **b.** $5x^2 - 4x + 1 = 0$ **c.** $4x^2 + 12x + 9 = 0$

Solution

a. Find the value of $b^2 - 4ac$ using $a = 1$, $b = 8$, and $c = 6$:

$$b^2 - 4ac = 8^2 - 4(1)(6) = 40$$

The value of the discriminant is 40 and the equation has two real solutions.
    The graph of $y = x^2 + 8x + 6$ has two x-intercepts as shown in Fig. 1.51.   □

b. Find the value of the discriminant for the equation $5x^2 - 4x + 1 = 0$:

$$b^2 - 4ac = (-4)^2 - 4(5)(1) = -4$$

Because the discriminant is negative, the equation has no real solutions.
    The graph of $y = 5x^2 - 4x + 1$ has no x-intercepts as shown in Fig. 1.52.   □

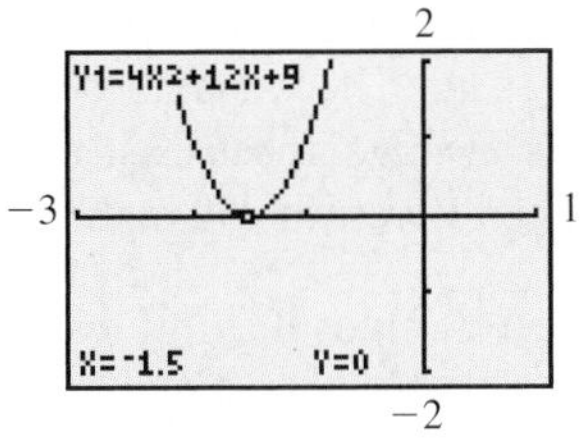

■ **Figure 1.53**

c. For $4x^2 + 12x + 9 = 0$, we have $b^2 - 4ac = 12^2 - 4(4)(9) = 0$. So the equation has one real solution.

    ⊞ The graph of $y = 4x^2 + 12x + 9$ has one x-intercept as shown in Fig. 1.53.

Try This. State the value of the discriminant and the number of real solutions to $5x^2 - 7x + 9 = 0$. ∎

The number of solutions to polynomial equations is a topic of considerable interest in algebra. In Section 3.2 we will learn that the number of solutions to any polynomial equation of degree n is less than or equal to n (the fundamental theorem of algebra). Notice how the number of solutions to linear and quadratic equations agree with these results.

Using Quadratic Models in Applications

The problems that we solve in this section are very similar to those in Section 1.2. However, in this section the mathematical model of the situation results in a quadratic equation.

Example **6** **A problem solved with a quadratic equation**

It took Susan 30 minutes longer to drive 275 miles on I-70 west of Green River, Utah, than it took her to drive 300 miles east of Green River. Because of a sand storm and a full load of cantaloupes, she averaged 10 mph less while traveling west of Green River. What was her average speed for each part of the trip?

Solution

Let x represent Susan's average speed in mph east of Green River and $x - 10$ represent her average speed in mph west of Green River. We can organize all of the given information as in the following table. Since $D = RT$, the time is determined by $T = D/R$.

	Distance	Rate	Time
East	300 mi	x mi/hr	$\dfrac{300}{x}$ hr
West	275 mi	$x - 10$ mi/hr	$\dfrac{275}{x - 10}$ hr

The following equation expresses the fact that her time west of Green River was $\frac{1}{2}$ hour greater than her time east of Green River.

$$\frac{275}{x - 10} = \frac{300}{x} + \frac{1}{2}$$

$$2x(x - 10) \cdot \frac{275}{x - 10} = 2x(x - 10)\left(\frac{300}{x} + \frac{1}{2}\right)$$

$$550x = 600(x - 10) + x(x - 10)$$

$$-x^2 - 40x + 6000 = 0$$

$$x^2 + 40x - 6000 = 0$$

$$(x + 100)(x - 60) = 0 \qquad \text{Factor.}$$

$$x = -100 \quad \text{or} \quad x = 60$$

The solution $x = -100$ is a solution to the equation, but not a solution to the problem. The other solution, $x = 60$, means that $x - 10 = 50$. Check that these two average speeds are a solution to the problem. Susan's average speed east of Green River was 60 mph, and her average speed west of Green River was 50 mph.

Try This. It took Josh 30 minutes longer to drive 100 miles than it took Bree to drive 90 miles. If Josh averaged 5 mph less than Bree, then what was the average speed of each driver?   ■

In the next example we use the formula for the height of a projectile under the influence of gravity.

Example **7** Applying the quadratic formula

In tennis a lob can be used to buy time to get into position. The approximate height S in feet for a tennis ball that is hit straight upward at v_0 ft/sec from a height of s_0 ft is modeled by

$$S = -16t^2 + v_0t + s_0,$$

where t is time in seconds.

a. Use Fig. 1.54 to estimate the time that it takes a ball to return to the court when it is hit straight upward with velocity 60 ft/sec from a height of 5 ft.
b. Find the time for part (a) by using the quadratic formula.

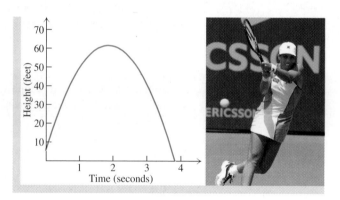

■ **Figure 1.54**

Solution

a. At the t-intercept in Fig. 1.54, $S = 0$. Since the t-intercept is approximately $(3.7, 0)$, the ball is in the air for about 3.7 sec.
b. We are looking for the value of t for which the height S is zero. Use $v_0 = 60$, $s_0 = 5$, and $S = 0$ in the given formula:

$$0 = -16t^2 + 60t + 5$$

Use the quadratic formula to solve the equation:

$$t = \frac{-60 \pm \sqrt{60^2 - 4(-16)(5)}}{2(-16)} = \frac{-60 \pm \sqrt{3920}}{-32} = \frac{-60 \pm 28\sqrt{5}}{-32}$$

$$= \frac{15 \pm 7\sqrt{5}}{8}$$

```
(15+7√(5))/8
          3.83155948
-16Ans²+60Ans+5
              -1ᴇ-11
```

■ **Figure 1.55**

Now $(15 - 7\sqrt{5})/8 \approx -0.082$ and $(15 + 7\sqrt{5})/8 \approx 3.83$. Since the time at which the ball returns to the earth must be positive, it will take exactly $(15 + 7\sqrt{5})/8$ seconds or approximately 3.83 seconds for the ball to hit the earth.

⊞   Check as shown in Fig. 1.55. Due to round-off errors the calculator answer is not zero, but really close to zero ($-1 \times 10^{-11} = -0.00000000001$).

Try This. A ball is tossed straight upward. Its height h in feet at time t in seconds is given by $h = -16t^2 + 40t + 6$. For how long is the ball in the air?   ■

Quadratic equations often arise from applications that involve the Pythagorean theorem from geometry: *A triangle is a right triangle if and only if the sum of the squares of the legs is equal to the square of the hypotenuse.*

Example ⬛**8** **Using the Pythagorean theorem**

In the house shown in Fig. 1.56 the ridge of the roof at point B is 6 feet above point C. If the distance from A to C is 18 feet, then what is the length of a rafter from A to B?

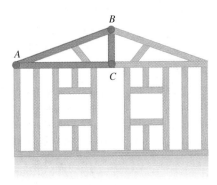

■ **Figure 1.56**

Solution

Let x be the distance from A to B. Use the Pythagorean theorem to write the following equation.

$$x^2 = 18^2 + 6^2$$
$$x^2 = 360$$
$$x = \pm\sqrt{360} = \pm 6\sqrt{10}$$

Since x must be positive in this problem, $x = 6\sqrt{10}$ feet or $x \approx 18.97$ feet.

Try This. One leg of a right triangle is 2 feet longer than the other leg. If the hypotenuse is 6 feet, then what are the lengths of the legs?   ■

Quadratic Regression

In Section 1.5 we used a graphing calculator to find the line of best fit. In the next example we will use the quadratic regression feature of a calculator to find an equation of the form $y = ax^2 + bx + c$ that fits a set of data points.

Example **9** Quadratic regression

The accompanying table shows the number of new housing starts each year for 2000 through 2005 (www.economagic.com).

Year	Housing Starts (thousands)
2000	1559
2001	1636
2002	1717
2003	1850
2004	1817
2005	1744

a. Use quadratic regression on your graphing calculator to find the regression equation.
b. Draw the scatter diagram and the regression curve on the calculator.
c. Use the equation to determine the number of new housing starts in 2006.

Solution

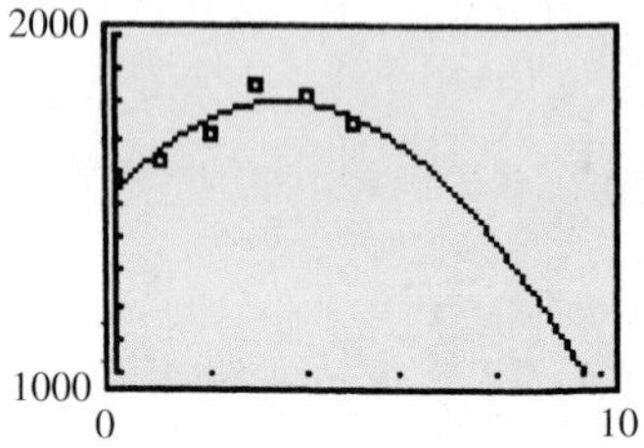

a. Use 0 through 5 for the year to get a simpler equation. Enter the data and choose quadratic regression from the STAT CALC menu. The equation

$$y = -21.5x^2 + 153.4x + 1534.4$$

expresses the housing starts y in terms of the year x, where x is the number of years since 2000.

b. The quadratic regression curve and the scatter diagram are shown in Fig. 1.57.

c. Let $x = 6$ in the equation given by the calculator or use the trace feature of the calculator to find $y \approx 1680$. So the equation predicts 1,680,000 housing starts for 2006. Rounding the coefficients in the equation to the nearest tenth yields 1,681,000.

■ **Figure 1.57**

Try This. Use quadratic regression and the given data to predict the cost in 2010.

Year	1980	1985	1990	2000	2005
Cost ($)	11.9	8.15	7.04	12.3	18.40

■

For Thought

True or False? Explain.

1. The equation $(x - 3)^2 = 4$ is equivalent to $x - 3 = 2$.

2. Every quadratic equation can be solved by factoring.

3. The trinomial $x^2 + \frac{2}{3}x + \frac{4}{9}$ is a perfect square trinomial.

4. The equation $(x - 3)(2x + 5) = 0$ is equivalent to $x = 3$ or $x = \frac{5}{2}$.

5. All quadratic equations have two distinct solutions.

6. If $x^2 - 3x + 1 = 0$, then $x = \frac{3 \pm \sqrt{5}}{2}$.

7. If $b = 0$, then $ax^2 + bx + c = 0$ cannot be solved by the quadratic formula.

8. All quadratic equations have at least one real solution.

9. For $4x^2 + 12x + 9 = 0$ the discriminant is 0.

10. A quadratic equation with real coefficients might have only one real solution.

1.6 Exercises

Use the square root property to find all real or imaginary solutions to each equation.

1. $x^2 - 5 = 0$

2. $x^2 - 8 = 0$

3. $3x^2 + 2 = 0$

4. $2x^2 + 16 = 0$

5. $(x - 3)^2 = 9$

6. $(x + 1)^2 = \dfrac{9}{4}$

7. $(3x - 1)^2 = 0$

8. $(5x + 2)^2 = 0$

9. $\left(x - \dfrac{1}{2}\right)^2 = \dfrac{25}{4}$

10. $(3x - 1)^2 = \dfrac{1}{4}$

11. $(x + 2)^2 = -4$

12. $(x - 3)^2 = -20$

13. $\left(x - \dfrac{2}{3}\right)^2 = \dfrac{4}{9}$

14. $\left(x + \dfrac{3}{2}\right)^2 = \dfrac{1}{2}$

Solve each equation by factoring.

15. $x^2 - x - 20 = 0$

16. $x^2 + 2x - 8 = 0$

17. $a^2 + 3a = -2$

18. $b^2 - 4b = 12$

19. $2x^2 - 5x - 3 = 0$

20. $2x^2 - 5x + 2 = 0$

21. $6x^2 - 7x + 2 = 0$

22. $12x^2 - 17x + 6 = 0$

23. $(y - 3)(y + 4) = 30$

24. $(w - 1)(w - 2) = 6$

Find the perfect square trinomial whose first two terms are given.

25. $x^2 - 12x$

26. $y^2 + 20y$

27. $r^2 + 3r$

28. $t^2 - 7t$

29. $w^2 + \dfrac{1}{2}w$

30. $p^2 - \dfrac{2}{3}p$

Find the real or imaginary solutions by completing the square. See the strategy for completing the square on page 150.

31. $x^2 + 6x + 1 = 0$

32. $x^2 - 10x + 5 = 0$

33. $n^2 - 2n - 1 = 0$

34. $m^2 - 12m + 33 = 0$

35. $h^2 + 3h - 1 = 0$

36. $t^2 - 5t + 2 = 0$

37. $2x^2 + 5x = 12$

38. $3x^2 + x = 2$

39. $3x^2 + 2x + 1 = 0$

40. $5x^2 + 4x + 3 = 0$

Find the real or imaginary solutions to each equation by using the quadratic formula.

41. $x^2 + 3x - 4 = 0$

42. $x^2 + 8x + 12 = 0$

43. $2x^2 - 5x - 3 = 0$

44. $2x^2 + 3x - 2 = 0$

45. $9x^2 + 6x + 1 = 0$

46. $16x^2 - 24x + 9 = 0$

47. $2x^2 - 3 = 0$

48. $-2x^2 + 5 = 0$

49. $x^2 + 5 = 4x$

50. $x^2 = 6x - 13$

51. $x^2 - 2x + 4 = 0$

52. $x^2 - 4x + 9 = 0$

53. $-2x^2 + 2x = 5$

54. $12x - 5 = 9x^2$

55. $4x^2 - 8x + 7 = 0$

56. $9x^2 - 6x + 4 = 0$

Use a calculator and the quadratic formula to find all real solutions to each equation. Round answers to two decimal places.

57. $3.2x^2 + 7.6x - 9 = 0$

58. $1.5x^2 - 6.3x - 10.1 = 0$

59. $3.25x^2 - 4.6x + 20 = 42$

60. $4.76x^2 + 6.12x = 55.3$

For each equation, state the value of the discriminant and the number of real solutions.

61. $9x^2 - 30x + 25 = 0$

62. $4x^2 + 28x + 49 = 0$

63. $5x^2 - 6x + 2 = 0$

64. $3x^2 + 5x + 5 = 0$

65. $7x^2 + 12x - 1 = 0$

66. $3x^2 - 7x + 3 = 0$

Find the solutions to each equation by reading the accompanying graph.

67. $6x^2 + x - 2 = 0$

68. $-2x^2 - 2x + 12 = 0$

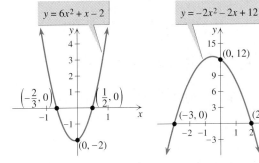

69. $-0.5x^2 + x + 7.5 = 0$ **70.** $0.5x^2 - 2.5x + 2 = 0$

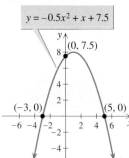

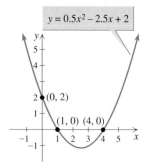

For each equation of the form $ax^2 + bx + c = 0$, determine the number of real solutions by examining the graph of $y = ax^2 + bx + c$.

71. $1.44x^2 - 8.4x + 12.25 = 0$

72. $4.41x^2 - 5.46x + 1.69 = 0$

73. $x^2 + 3x + 15 = 0$ **74.** $-2x^2 + 5x - 40 = 0$

75. $x^2 + 3x - 160 = 0$ **76.** $-x^2 + 5x - 6.1 = 0$

Use the method of your choice to find all real solutions to each equation. See the strategy for solving $ax^2 + bx + c = 0$ on page 152.

77. $x^2 = \dfrac{4}{3}x + \dfrac{5}{9}$ **78.** $x^2 = \dfrac{2}{7}x + \dfrac{2}{49}$

79. $x^2 - \sqrt{2} = 0$ **80.** $\sqrt{2}x^2 - 1 = 0$

81. $12x^2 + x\sqrt{6} - 1 = 0$ **82.** $-10x^2 - x\sqrt{5} + 1 = 0$

83. $x(x + 6) = 72$ **84.** $x = \dfrac{96}{x + 4}$

85. $x = 1 + \dfrac{1}{x}$ **86.** $x = \dfrac{1}{x}$

87. $\dfrac{28}{x} - \dfrac{7}{x^2} = 7$ **88.** $\dfrac{20}{x} - \dfrac{46}{x^2} = 2$

89. $\dfrac{x - 12}{3 - x} = \dfrac{x + 4}{x + 7}$ **90.** $\dfrac{x - 9}{x - 2} = -\dfrac{x + 3}{x + 1}$

91. $\dfrac{x - 8}{x + 2} = -\dfrac{1 - 2x}{x + 3}$ **92.** $\dfrac{x - 1}{x} = \dfrac{3x}{x + 1}$

93. $\dfrac{2x + 3}{2x + 1} = \dfrac{8}{2x + 3}$ **94.** $\dfrac{2x + 3}{6x + 5} = \dfrac{2}{2x + 3}$

Use the methods for solving quadratic equations to solve each formula for the indicated variable.

95. $A = \pi r^2$ for r **96.** $S = 2\pi rh + 2\pi r^2$ for r

97. $x^2 + 2kx + 3 = 0$ for x **98.** $hy^2 - ky = p$ for y

99. $2y^2 + 4xy = x^2$ for y **100.** $\dfrac{\dfrac{1}{x + h} - \dfrac{1}{x}}{h} = 1$ for x

Find an exact solution to each problem. If the solution is irrational, then find an approximate solution also.

101. *Demand Equation* The demand equation for a certain product is $P = 40 - 0.001x$, where x is the number of units sold per week and P is the price in dollars at which each one is sold. The weekly revenue R is given by $R = xP$. What number of units sold produces a weekly revenue of $175,000?

102. *Average Cost* The total cost in dollars of producing x items is given by $C = 0.02x^3 + 5x$. For what number of items is the average cost per item equal to $5.50?
    HINT Average cost is total cost divided by the number of items.

103. *Height of a Ball* A juggler tosses a ball into the air with a velocity of 40 ft/sec from a height of 4 ft. Use $S = -16t^2 + v_0t + s_0$ to find how long it takes for the ball to return to the height of 4 ft.

104. *Height of a Sky Diver* A sky diver steps out of an airplane at 5000 ft. Use the formula $S = -16t^2 + v_0t + s_0$ to find how long it takes the sky diver to reach 4000 ft.

105. *Diagonal of a Football Field* A football field is 100 yd long from goal line to goal line and 160 ft wide. If a player ran diagonally across the field from one goal line to the other, then how far did he run?

106. *Dimensions of a Flag* If the perimeter of a rectangular flag is 34 in. and the diagonal is 13 in., then what are the length and width?

107. *Long Shot* To avoid hitting the ball out, a tennis player in one corner of the 312 yd² court hits the ball to the farthest corner of the opponent's court as shown in the diagram. If the length L of the tennis court is 2 yd longer than twice the width W, then how far did the player hit the ball?

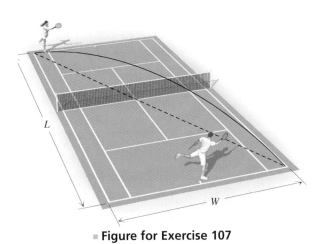

■ **Figure for Exercise 107**

$A = 822 \text{ ft}^2$
$S = 18.8$

$2^{-12}d^2S^3 - A^3 = 0$

■ **Figure for Exercise 109**

108. *Open-Top Box* Imogene wants to make an open-top box for packing baked goods by cutting equal squares from each corner of an 11 in. by 14 in. piece of cardboard as shown in the diagram. She figures that for versatility the area of the bottom must be 80 in.^{2}. What size square should she cut from each corner?

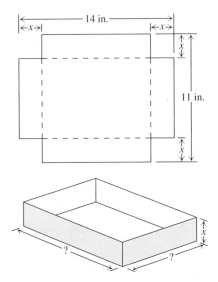

■ **Figure for Exercise 108**

109. *Finding the Displacement* The sail area-displacement ratio S measures the amount of power available to drive a sailboat in moderate to heavy winds (Ted Brewer Yacht Design, www.tedbrewer.com). For the Sabre 402 shown in the figure, the sail area A is 822 ft^2 and $S = 18.8$. The displacement d (in pounds) satisfies $2^{-12}d^2S^3 - A^3 = 0$. Find d.

110. *Charleston Earthquake* The Charleston, South Carolina, earthquake of 1886 registered 7.6 on the Richter scale and was felt over an area of 1.5 million square miles (U.S. Geological Survey, www.usgs.gov). If the area in which it was felt was circular and centered at Charleston, then how far away was it felt?

111. *Radius of a Pipe* A large pipe is placed next to a wall and a 1-ft-high block is placed 5 ft from the wall to keep the pipe in place as shown in the figure. What is the radius of the pipe?

  HINT Draw in the radius at several locations.

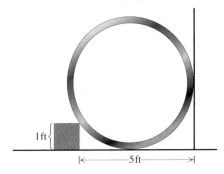

1 ft

⊢——5 ft——⊣

■ **Figure for Exercise 111**

112. *Another Pipe* A small pipe is placed against a wall as in the previous exercise, but no block is used to keep it in place. There is a point on the edge of this pipe that is both 5 in. from the ground and 10 in. from the wall. Find two possibilities for the radius of the pipe.

113. *Speed of a Tortoise* When a tortoise crosses a highway, his speed is 2 ft/hr faster than normal. If he can cross a 24-ft lane in 24 min less time than he can travel that same distance off the highway, then what is his normal speed?

114. *Speed of an Electric Car* An experimental electric-solar car completed a 1000-mi race in 35 hr. For the 600 mi traveled during daylight the car averaged 20 mph more than it did for the 400 mi traveled at night. What was the average speed of the car during the daytime?

115. *Initial Velocity of a Basketball Player*   Vince Carter is famous for his high leaps and "hang time," especially while slam-dunking. If Carter leaped for a dunk and reached a peak height at the basket of 1.07 m, what was his upward velocity in meters per second at the moment his feet left the floor? The formula $v_1^2 = v_0^2 + 2gS$ gives the relationship between final velocity v_1, initial velocity v_0, acceleration of gravity g, and his height S. Use $g = -9.8$ m/sec^2 and the fact that his final velocity was zero at his peak height. Use $S = \frac{1}{2}gt^2 + v_0 t$ to find the amount of time he was in the air.

▪ **Figure for Exercise 115**

116. *Hazards of Altitude*   As shown in the table (www.nasa.gov), atmospheric pressure a (atm) decreases as height above sea level h (in feet) increases. The equation

$$a = 3.89 \times 10^{-10}h^2 - 3.48 \times 10^{-5}h + 1$$

can be used to model this relationship.
a. Mountain climbers begin to deteriorate at 18,000 ft. Find a at that height.

b. The atmospheric pressure at the highest human settlements is 0.52 atm. Find the height for $a = 0.52$.

c. Use the quadratic regression feature of your graphing calculator and the data given in the table to find an equation expressing atmospheric pressure a in terms of h.

▪ **Table for Exercise 116**

Altitude (ft)	Atmospheric Pressure (atm)	
28,000	0.33	
20,000	0.46	
18,000		
	0.52	
10,000	0.69	
0	1	

 117. *Teen Birth Rate*   The accompanying table gives the number of births per 1000 females ages 15–19 (www.infoplease.com).
a. Use your graphing calculator to find a quadratic regression equation that expresses the birth rate in terms of the year.

b. In what year will the birth rate reach zero according to the quadratic equation of part (a)?

Year	Births per 1000 Females (ages 15–19)
1980	53.0
1985	51.0
1990	59.9
1995	56.8
2000	47.7
2005	41.4

 118. *Teen Birth Rate*   Use the data in the previous exercise.
a. Find the equation of the regression line that expresses the birth rate in terms of the year.

b. In what year will the birth rate reach zero according to the regression line?

119. *Computer Design*   Using a computer design package, Tina can write and design a direct-mail package in two days less time than it takes to create the same package using traditional design methods. If Tina uses the computer and her assistant Curt uses traditional methods, and together they complete the job in 3.5 days, then how long would it have taken Curt to do the job alone using traditional methods?
HINT Curt's rate is $1/x$ job/day and Tina's rate is $1/(x - 2)$ job/day.

120. *Making a Dress*   Rafael designed a sequined dress to be worn at the Academy Awards. His top seamstress, Maria, could sew on all the sequins in 10 hr less time than his next-best seamstress, Stephanie. To save time, he gave the job to both women and got all of the sequins attached in 17 hr. How long would it have taken Stephanie working alone?

121. *Percentage of White Meat*   The Kansas Fried Chicken store sells a Party Size bucket that weighs 10 lb more than the Big Family Size bucket. The Party size bucket contains 8 lb of white meat, while the Big Family Size bucket contains 3 lb of white meat. If the percentage of white meat in the Party Size is 10 percentage points greater than the percentage of white meat in the Big Family Size, then how much does the Party Size bucket weigh?

122. *Mixing Antifreeze in a Radiator*   Steve's car had a large radiator that contained an unknown amount of pure water. He added two quarts of antifreeze to the radiator. After testing, he decided that the percentage of antifreeze in the radiator was not large enough. Not knowing how to solve mixture problems, Steve decided to add one quart of water and another quart of antifreeze to the radiator to see what he would get. After testing he found that the last addition increased the percentage of antifreeze by three percentage points. How much water did the radiator contain originally?

Thinking Outside the Box XVII

As the Crow Flies   In Perfect City the avenues run east and west, the streets run north and south, and all of the blocks are square. A crow flies from the corner of 1st Ave and 1st Street to the corner of mth Ave and nth Street, "as the crow flies." Assume that m and n are positive integers greater than 1 and the streets and avenues are simply lines on a map. If the crow flies over an intersection, then he flies over only two of the blocks that meet at the intersection.

a. If $m - 1$ and $n - 1$ are relatively prime (no common factors), then how many city blocks does the crow fly over?

b. If d is the greatest common factor for m and n, then how many city blocks does the crow fly over?

1.6 Pop Quiz

1. Solve $x^2 = 2$.

2. Solve $x^2 - 2x = 48$ by factoring.

3. Solve $x^2 - 4x = 1$ by completing the square.

4. Solve $2x^2 - 4x = 3$ by the quadratic formula.

5. How many real solutions are there to $5x^2 - 9x + 5 = 0$?

Linking Concepts

For Individual or Group Explorations

Baseball Statistics

Baseball fans keep up with their favorite teams through charts where the teams are ranked according to the percentage of games won. The chart usually has a column indicating the games behind (GB) for each team. If the win-loss record of the number one team is (A, B), then the games behind of another team whose win-loss record is (a, b) is calculated by the formula

$$GB = \frac{(A - a) + (b - B)}{2}.$$

a)   Find *GB* for Atlanta and Philadelphia in the following table.

Team	Won	Lost	Pct.	GB
New York	38	24	.613	—
Atlanta	35	29	.547	?
Philadelphia	34	31	.523	?

b)   In the following table Pittsburgh has a higher percentage of wins than Chicago, and so Pittsburgh is in first place. Find *GB* for Chicago.

Team	Won	Lost	Pct.	GB
Pittsburgh	18	13	.581	—
Chicago	22	16	.579	?

(continued on next page)

c) Is Chicago actually behind Pittsburgh in terms of the statistic *GB*?

d) Another measure of how far a team is from first place, called the deficit *D*, is the number of games that the two teams would have to play against each other to get equal percentages of wins, with the higher-ranked team losing all of the games. Find the deficits for Atlanta, Philadelphia, and Chicago.

e) Is it possible for *GB* or *D* to be negative? Does it make any sense if they are?

f) Compare the values of *D* and *GB* for each of the three teams. Which is a better measure of how far a team is from first place?

1.7

Linear and Absolute Value Inequalities

An equation states that two algebraic expressions are equal, while an **inequality** or **simple inequality** is a statement that two algebraic expressions are not equal in a particular way. Inequalities are stated using less than ($<$), less than or equal to ($\leq$), greater than ($>$), or greater than or equal to ($\geq$). In this section we study some basic inequalities. We will see more inequalities in Chapter 3.

Interval Notation

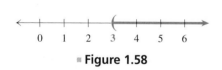

■ **Figure 1.58**

The solution set to an inequality is the set of all real numbers for which the inequality is true. The solution set to the inequality $x > 3$ is written $\{x \mid x > 3\}$ and consists of all real numbers to the right of 3 on the number line. This set is also called the **interval** of numbers greater than 3, and it is written in **interval notation** as $(3, \infty)$. The graph of the interval $(3, \infty)$ is shown in Fig. 1.58. A parenthesis is used next to the 3 to indicate that 3 is not in the interval or the solution set. The infinity symbol (∞) is not used as a number, but only to indicate that there is no bound on the numbers greater than 3.

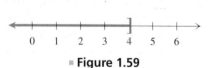

■ **Figure 1.59**

The solution set to $x \leq 4$ is written in set notation as $\{x \mid x \leq 4\}$ and in interval notation as $(-\infty, 4]$. The symbol $-\infty$ means that all numbers to the left of 4 on the number line are in the set and the bracket means that 4 is in the set. The graph of the interval $(-\infty, 4]$ is shown in Fig. 1.59. Intervals that use the infinity symbol are **unbounded** intervals. The following summary lists the different types of unbounded intervals used in interval notation and the graphs of those intervals on a number line. An unbounded interval with an endpoint is **open** if the endpoint is not included in the interval and **closed** if the endpoint is included.

SUMMARY   **Interval Notation for Unbounded Intervals**

Set	Interval notation	Type	Graph
$\{x \mid x > a\}$	(a, ∞)	Open	
$\{x \mid x < a\}$	$(-\infty, a)$	Open	
$\{x \mid x \geq a\}$	$[a, \infty)$	Closed	
$\{x \mid x \leq a\}$	$(-\infty, a]$	Closed	
Real numbers	$(-\infty, \infty)$	Open	

We use a parenthesis when an endpoint of an interval is not included in the solution set and a bracket when an endpoint is included. A bracket is never used next to ∞ because infinity is not a number. On the graphs above, the number lines are shaded, showing that the solutions include all real numbers in the given interval.

Example 1 Interval notation

Write an inequality whose solution set is the given interval.

a. $(-\infty, -9)$ **b.** $[0, \infty)$

Solution

a. The interval $(-\infty, -9)$ represents all real numbers less than -9. It is the solution set to $x < -9$.
b. The interval $[0, \infty)$ represents all real numbers greater than or equal to 0. It is the solution set to $x \geq 0$.

Try This. Write an inequality whose solution set is $(-\infty, 5]$.   ■

Linear Inequalities

Replacing the equal sign in the general linear equation $ax + b = 0$ by any of the symbols $<, \leq, >$, or $\geq$ gives a **linear inequality.** Two inequalities are **equivalent** if they have the same solution set. We solve linear inequalities like we solve linear equations by performing operations on each side to get equivalent inequalities. However, the rules for inequalities are slightly different from the rules for equations.

Adding any real number to both sides of an inequality results in an equivalent inequality. For example, adding 3 to both sides of $-4 < 5$ yields $-1 < 8$, which is true. Adding or subtracting the same number simply moves the original numbers to the right or left along the number line and does not change their order.

The order of two numbers will also be unchanged when they are multiplied or divided by the same positive real number. For example, $10 < 20$, and after dividing both numbers by 10 we have $1 < 2$. Multiplying or dividing two numbers by a negative number will change the order. For example, $4 > 2$, but after multiplying both numbers by -1 we have $-4 < -2$. Likewise, $-10 < 20$, but after dividing both numbers by -10 we have $1 > -2$. *When an inequality is multiplied or divided by a negative number, the direction of the inequality symbol is reversed.* These ideas are stated symbolically in the following box for $<$, but they also hold for $>$, $\leq$, and $\geq$.

Properties of Inequality

If A and B are algebraic expressions and C is a nonzero real number, then the inequality $A < B$ is equivalent to

1. $A \pm C < B \pm C$,
2. $CA < CB$ (for C positive), $CA > CB$ (for C negative),
3. $\dfrac{A}{C} < \dfrac{B}{C}$ (for C positive), $\dfrac{A}{C} > \dfrac{B}{C}$ (for C negative).

Example 2 Solving a linear inequality

Solve $-3x - 9 < 0$. Write the solution set in interval notation and graph it.

Solution

Isolate the variable as is done in solving equations.

$$-3x - 9 < 0$$

$$-3x - 9 + 9 < 0 + 9 \qquad \text{Add 9 to each side.}$$

$$-3x < 9$$

$$x > -3 \qquad \text{Divide each side by } -3, \text{ reversing the inequality.}$$

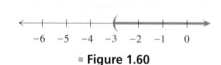

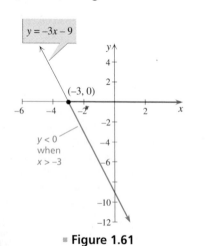

■ **Figure 1.60**

■ **Figure 1.61**

The solution set is the interval $(-3, \infty)$ and its graph is shown in Fig. 1.60. Checking the solution to an inequality is generally not as simple as checking an equation, because usually there are infinitely many solutions. We can do a "partial check" by checking one number in $(-3, \infty)$ and one number not in $(-3, \infty)$. For example, $0 > -3$ and $-3(0) - 9 < 0$ is correct, while $-6 < -3$ and $-3(-6) - 9 < 0$ is incorrect.

Try This. Solve $2 - 5x \leq 7$. Write the solution set in interval notation and graph it. ■

We can read the solution to an inequality in one variable from a graph in two variables in the same manner that we read the solution to an equation in one variable. Figure 1.61 shows the graph of $y = -3x - 9$. From this figure we see that the y-coordinates on this line are negative when the x-coordinates are greater than -3. In other words, $-3x - 9 < 0$ when $x > -3$.

We can also perform operations on each side of an inequality using a variable expression. Addition or subtraction with variable expressions will give equivalent inequalities. However, we must always watch for undefined expressions. *Multiplication and division with a variable expression is usually avoided because we do not know whether the expression is positive or negative.*

Example **3**   Solving a linear inequality

Solve $\frac{1}{2}x - 3 \geq \frac{1}{4}x + 2$ and graph the solution set.

Solution

Multiply each side by the LCD to eliminate the fractions.

$$\frac{1}{2}x - 3 \geq \frac{1}{4}x + 2$$

$$4\left(\frac{1}{2}x - 3\right) \geq 4\left(\frac{1}{4}x + 2\right) \quad \text{Multiply each side by 4.}$$

$$2x - 12 \geq x + 8$$

$$x - 12 \geq 8$$

$$x \geq 20$$

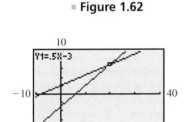

■ **Figure 1.62**

The solution set is the interval $[20, \infty)$. See Fig. 1.62 for its graph.

The algebraic solution to $\frac{1}{2}x - 3 \geq \frac{1}{4}x + 2$ proves that the graph of $y = \frac{1}{2}x - 3$ is at or above the graph of $y = \frac{1}{4}x + 2$ when $x \geq 20$, as it appears to be in Fig. 1.63. Conversely, the graphs in Fig. 1.63 support the conclusion that $x \geq 20$ causes the inequality to be true.

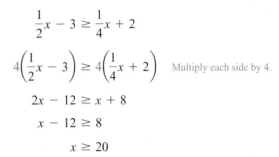

■ **Figure 1.63**

Try This.   Solve $\frac{1}{2}x + \frac{1}{3} \leq \frac{1}{3}x + 1$ and graph the solution set.   ■

Compound Inequalities

A **compound inequality** is a sentence containing two simple inequalities connected with "and" or "or." The solution to a compound inequality can be an interval of real numbers that does not involve infinity, a **bounded** interval of real numbers. For example, the solution set to the compound inequality $x \geq 2$ and $x \leq 5$ is the set of real numbers between 2 and 5, inclusive. This inequality is also written as $2 \leq x \leq 5$. Its solution set is $\{x \mid 2 \leq x \leq 5\}$, which is written in interval notation as $[2, 5]$. Because $[2, 5]$ contains both of its endpoints, the interval is **closed.** The following summary lists the different types of bounded intervals used in interval notation and the graphs of those intervals on a number line.

S U M M A R Y   **Interval Notation for Bounded Intervals**

Set	Interval notation	Type	Graph
$\{x \mid a < x < b\}$	(a, b)	Open	
$\{x \mid a \leq x \leq b\}$	$[a, b]$	Closed	
$\{x \mid a \leq x < b\}$	$[a, b)$	Half open or half closed	
$\{x \mid a < x \leq b\}$	$(a, b]$	Half open or half closed	

The notation $a < x < b$ is used only when x is between a and b, and a is less than b. We do *not* write inequalities such as $5 < x < 3$, $4 > x < 9$, or $2 < x > 8$.

The **intersection** of sets A and B is the set $A \cap B$ (read "A intersect B"), where $x \in A \cap B$ if and only if $x \in A$ and $x \in B$. (The symbol $\in$ means "belongs to.") The **union** of sets A and B is the set $A \cup B$ (read "A union B"), where $x \in A \cup B$ if and only if $x \in A$ or $x \in B$. In solving compound inequalities it is often necessary to find intersections and unions of intervals.

Example 4 Intersections and unions of intervals

Let $A = (1, 5)$, $B = [3, 7)$, and $C = (6, \infty)$. Write $A \cup B$, $A \cap B$, $A \cap C$, and $A \cup C$ in interval notation.

Solution

Figure 1.64 shows the intervals A, B, and C on the number line. Since $A \cup B$ consists of points that are either in A or in B, $A \cup B = (1, 7)$. Since $A \cap B$ consists of points that are in both A and B, $A \cap B = [3, 5)$. Since A and C have no points in common, $A \cap C = \emptyset$ and $A \cup C = (1, 5) \cup (6, \infty)$.

Try This. Find $A \cup B$ and $A \cap B$ if $A = (1, 6)$ and $B = [4, 9)$.   ▪

The solution set to a compound inequality using the connector "or" is the union of the two solution sets, and the solution set to a compound inequality using "and" is the intersection of the two solution sets.

Example 5 Solving compound inequalities

Solve each compound inequality. Write the solution set using interval notation and graph it.

a. $2x - 3 > 5$ and $4 - x \le 3$ **b.** $4 - 3x < -2$ or $3(x - 2) \le -6$
c. $-4 \le 3x - 1 < 5$

Solution

a. $2x - 3 > 5$ and $4 - x \le 3$

$\qquad 2x > 8$ and $\quad -x \le -1$

$\qquad\ \ x > 4$ and $\qquad x \ge 1$

The intersection of the intervals $(4, \infty)$ and $[1, \infty)$ is $(4, \infty)$, because only numbers larger than 4 belong to both intervals. The solution set to the compound inequality is the open interval $(4, \infty)$ Its graph is shown in Fig. 1.65.

b. $4 - 3x < -2$ or $3(x - 2) \le -6$

$\qquad -3x < -6$ or $\quad x - 2 \le -2$

$\qquad\ \ x > 2$ or $\qquad x \le 0$

The solution set is $(-\infty, 0] \cup (2, \infty)$, and its graph is shown in Fig. 1.66.

c. We could write $-4 \le 3x - 1 < 5$ as the compound inequality $-4 \le 3x - 1$ and $3x - 1 < 5$, and then solve each simple inequality. Since each is solved

Figures (left column)

A
```
←+─(──────)─+─+─+─→
  0 1 2 3 4 5 6 7 8
```

B
```
←+─+─+─[──────)─+→
  0 1 2 3 4 5 6 7 8
```

C
```
←+─+─+─+─+─+─(───→
  0 1 2 3 4 5 6 7 8
```

▪ **Figure 1.64**

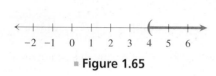

▪ **Figure 1.65**

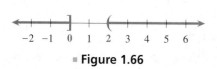

▪ **Figure 1.66**

using the same sequence of steps, we can solve the original inequality without separating it:

$$-4 \le 3x - 1 < 5$$

$$-4 + 1 \le 3x - 1 + 1 < 5 + 1 \quad \text{Add 1 to each part of the inequality.}$$

$$-3 \le 3x < 6$$

$$\frac{-3}{3} \le \frac{3x}{3} < \frac{6}{3} \quad \text{Divide each part by 3.}$$

$$-1 \le x < 2$$

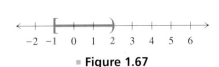

■ **Figure 1.67**

The solution set is the half-open interval $[-1, 2)$, graphed in Fig. 1.67.

Try This. Solve $2x > -4$ and $4 - x \ge 0$ and graph the solution set.   ■

It is possible that all real numbers satisfy a compound inequality or no real numbers satisfy a compound inequality.

Example 6 Solving compound inequalities

Solve each compound inequality.

a. $3x - 9 \le 9$ or $4 - x \le 3$

b. $-\frac{2}{3}x < 4$ and $\frac{3}{4}x < -6$

Solution

a. Solve each simple inequality and find the union of their solution sets:

$$3x - 9 \le 9 \quad \text{or} \quad 4 - x \le 3$$

$$3x \le 18 \quad \text{or} \quad -x \le -1$$

$$x \le 6 \quad \text{or} \quad x \ge 1$$

The union of $(-\infty, 6]$ and $[1, \infty)$ is the set of all real numbers, $(-\infty, \infty)$.

b. Solve each simple inequality and find the intersection of their solution sets:

$$-\frac{2}{3}x < 4 \qquad \text{and} \qquad \frac{3}{4}x < -6$$

$$\left(-\frac{3}{2}\right)\left(-\frac{2}{3}x\right) > \left(-\frac{3}{2}\right)4 \quad \text{and} \quad \left(\frac{4}{3}\right)\left(\frac{3}{4}x\right) < \left(\frac{4}{3}\right)(-6)$$

$$x > -6 \qquad \text{and} \qquad x < -8$$

Since $(-6, \infty) \cap (-\infty, -8) = \varnothing$, there is no solution to the compound inequality.

Try This. Solve $3x + 2 > -1$ and $5 < -3 - 4x$.   ■

Absolute Value Inequalities

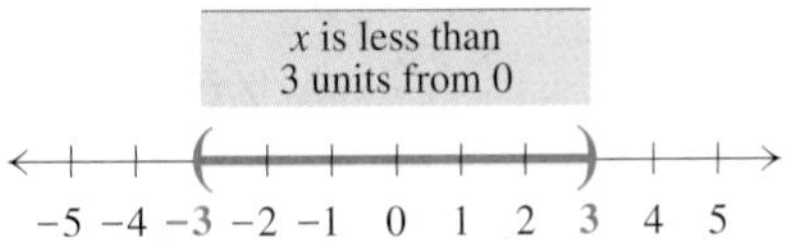

Recall that the absolute value of a number is the number's distance from 0 on the number line. The inequality $|x| < 3$ means that x is less than three units from 0. See Fig. 1.68. The real numbers that are less than three units from 0 are precisely the

> *x* is less than
> 3 units from 0

■ **Figure 1.68**

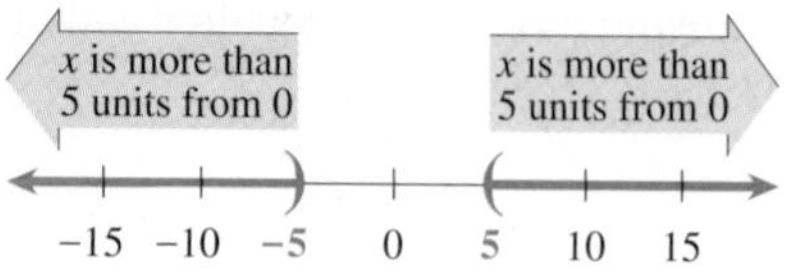

x is more than 5 units from 0   *x is more than 5 units from 0*

⟵——┼——┼——┼——)——┼——(——┼——┼——┼——⟶
-15 -10 -5 0 5 10 15

■ **Figure 1.69**

numbers that satisfy $-3 < x < 3$. So the solution set to $|x| < 3$ is the open interval $(-3, 3)$.

The inequality $|x| > 5$ means that x is more than five units from 0 on the number line, which is equivalent to the compound inequality $x > 5$ or $x < -5$. See Fig. 1.69. So the solution to $|x| > 5$ is the union of two intervals, $(-\infty, -5) \cup (5, \infty)$. These ideas about absolute value inequalities are summarized as follows.

S U M M A R Y   **Basic Absolute Value Inequalities (for $k > 0$)**

Absolute value inequality	Equivalent statement	Solution set in interval notation	Graph of solution set
$|x| > k$	$x < -k$ or $x > k$	$(-\infty, -k) \cup (k, \infty)$	 ⟵—)——————(—⟶ $-k$ k
$|x| \geq k$	$x \leq -k$ or $x \geq k$	$(-\infty, -k] \cup [k, \infty)$	⟵—]——————[—⟶ $-k$ k
$|x| < k$	$-k < x < k$	$(-k, k)$	 ⟵——(——————)——⟶ $-k$ k
$|x| \leq k$	$-k \leq x \leq k$	$[-k, k]$	⟵——]——————[——⟶ $-k$ k

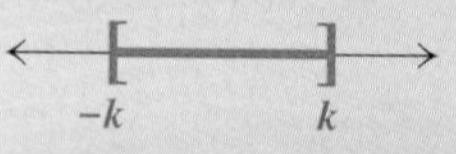

In the next example we use the rules for basic absolute value inequalities to solve more complicated absolute value inequalities.

Example **7** **Absolute value inequalities**

Solve each absolute value inequality and graph the solution set.

a. $|3x + 2| < 7$ **b.** $-2|4 - x| \leq -4$ **c.** $|7x - 9| \geq -3$

Solution

a.
$$|3x + 2| < 7$$
$$-7 < 3x + 2 < 7 \qquad \text{Write the equivalent compound inequality.}$$
$$-7 - 2 < 3x + 2 - 2 < 7 - 2 \qquad \text{Subtract 2 from each part.}$$
$$-9 < 3x < 5$$
$$-3 < x < \frac{5}{3} \qquad \text{Divide each part by 3.}$$

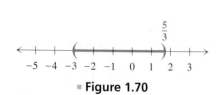

$\frac{5}{3}$

⟵—┼——┼——(——┼——┼——┼——┼——)—┼——┼——⟶
-5 -4 -3 -2 -1 0 1 2 3

■ **Figure 1.70**

The solution set is the open interval $\left(-3, \frac{5}{3}\right)$. The graph is shown in Fig. 1.70.

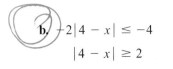

b. $-2|4 - x| \leq -4$ Divide each side by -2, reversing the inequality.

$$|4 - x| \geq 2$$

$4 - x \leq -2$ or $4 - x \geq 2$ Write the equivalent compound inequality.

$-x \leq -6$ or $-x \geq -2$

$x \geq 6$ or $x \leq 2$ Multiply each side by -1.

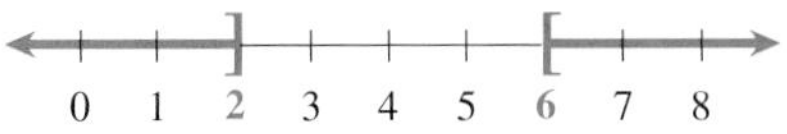

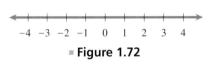

■ **Figure 1.71**

The solution set is $(-\infty, 2] \cup [6, \infty)$. Its graph is shown in Fig. 1.71. Check 8, 4, and 0 in the original inequality. If $x = 8$, we get $-2|4 - 8| \leq -4$, which is true. If $x = 4$, we get $-2|4 - 4| \leq -4$, which is false. If $x = 0$, we get $-2|4 - 0| \leq -4$, which is true.

c. The expression $|7x - 9|$ has a nonnegative value for every real number x. So the inequality $|7x - 9| \geq -3$ is satisfied by every real number. The solution set is $(-\infty, \infty)$, and its graph is shown in Fig. 1.72.

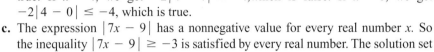

■ **Figure 1.72**

Try This. Solve $|x - 6| - 3 \leq -2$ and graph the solution set.   ■

Modeling with Inequalities

Inequalities occur in applications just as equations do. In fact, in real life, equality in anything is usually the exception. The solution to a problem involving an inequality is generally an interval of real numbers. In this case we often ask for the range of values that solve the problem.

Example **8** An application involving inequality

Remington scored 74 on his midterm exam in history. If he is to get a B, the average of his midterm and final exam must be between 80 and 89 inclusive. In what range must his final exam score lie for him to get a B in the course?

Solution

Let x represent Remington's final exam score. The average of his midterm and final must satisfy the following inequality.

$$80 \leq \frac{74 + x}{2} \leq 89$$

$$160 \leq 74 + x \leq 178$$

$$86 \leq x \leq 104$$

His final exam score must lie in the interval [86, 104] if he is to get a B.

Try This. Kelly's commissions for her first two sales of the Whirlwind Vacuum Cleaner were \$80 and \$90. In what range must her third commission lie for her average to be between \$100 and \$110?   ■

When discussing the error made in a measurement, we may refer to the *absolute error* or the *relative error*. For example, if L is the actual length of an object and x is the length determined by a measurement, then the absolute error is $|x - L|$ and the relative error is $|x - L|/L$.

Example **9** **Application of absolute value inequality**

A technician is testing a scale with a 50-lb block of steel. The scale passes this test if the relative error when weighing this block is less than 0.1%. If x is the reading on the scale, then for what values of x does the scale pass this test?

Solution

If the relative error must be less than 0.1%, then x must satisfy the following inequality:

$$\frac{|x - 50|}{50} < 0.001$$

Solve the inequality for x:

$$|x - 50| < 0.05$$
$$-0.05 < x - 50 < 0.05$$
$$49.95 < x < 50.05$$

So the scale passes the test if it shows a weight in the interval (49.95, 50.05).

Try This. A gas pump is certified as accurate if the relative error when dispensing 10 gallons of gas is less than 1%. If x is the actual amount of gas dispensed, then for what values of x is the pump certified as accurate?   ■

For Thought

True or False? Explain.

1. The inequality $-3 < x + 6$ is equivalent to $x + 6 > -3$.

2. The inequality $-2x < -6$ is equivalent to $\frac{-2x}{-2} < \frac{-6}{-2}$.

3. The smallest real number that satisfies $x > 12$ is 13.

4. The number -6 satisfies $|x - 6| > -1$.

5. $(-\infty, -3) \cap (-\infty, -2) = (-\infty, -2)$

6. $(5, \infty) \cap (-\infty, -3) = (-3, 5)$

7. All negative numbers satisfy $|x - 2| < 0$.

8. The compound inequality $x < -3$ or $x > 3$ is equivalent to $|x| < -3$.

9. The inequality $|x| + 2 < 5$ is equivalent to $-5 < x + 2 < 5$.

10. The fact that the difference between your age, y, and my age, m, is at most 5 years is written $|y - m| \le 5$.

1.7 Exercises

For each interval write an inequality whose solution set is the interval, and for each inequality, write the solution set in interval notation. See the summary of interval notation for unbounded intervals on page 163.

1. $(-\infty, 12)$ **2.** $(-\infty, -3]$ **3.** $[-7, \infty)$

4. $(1.2, \infty)$ **5.** $x \ge -8$ **6.** $x < 54$

7. $x < \pi/2$ **8.** $x \ge \sqrt{3}$

Solve each inequality. Write the solution set using interval notation and graph it.

9. $3x - 6 > 9$ **10.** $2x + 1 < 6$

11. $7 - 5x \le -3$ **12.** $-1 - 4x \ge 7$

13. $\frac{1}{2}x - 4 < \frac{1}{3}x + 5$ **14.** $\frac{1}{2} - x > \frac{x}{3} + \frac{1}{4}$

15. $\dfrac{7 - 3x}{2} \geq -3$

16. $\dfrac{5 - x}{3} \leq -2$

17. $\dfrac{2x - 3}{-5} \geq 0$

18. $\dfrac{5 - 3x}{-7} \leq 0$

19. $-2(3x - 2) \geq 4 - x$

20. $-5x \leq 3(x - 9)$

Solve each inequality by reading the accompanying graph.

21. $1.8x + 6.3 < 0$

22. $1.2x - 3 \geq 0$

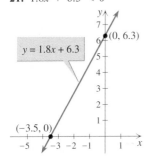

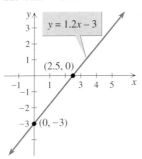

23. $6.3 - 4.5x \geq 0$

24. $-5.1 - 1.7x < 0$

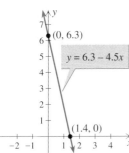

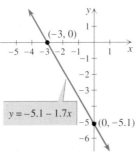

Solve the inequalities in Exercises 25–28 by reading the following graphs.

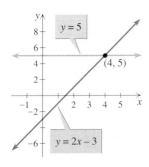

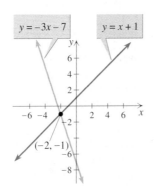

25. $2x - 3 > 5$

26. $5 \geq 2x - 3$

27. $x + 1 \leq -3x - 7$

28. $x + 1 > -3x - 7$

Write as a single interval. See the summary of interval notation for bounded intervals on page 165.

29. $(-3, \infty) \cup (5, \infty)$

30. $(-\infty, 0) \cup (-\infty, 6)$

31. $(3, 5) \cup (-3, \infty)$

32. $(4, 7) \cap (3, \infty)$

33. $(-\infty, -2) \cap (-5, \infty)$

34. $(-3, \infty) \cap (2, \infty)$

35. $(-\infty, -5) \cap (-2, \infty)$

36. $(-\infty, -3) \cup (-7, \infty)$

37. $(-\infty, 4) \cup [4, 5]$

38. $[3, 5] \cup [5, 7]$

Solve each compound inequality. Write the solution set using interval notation and graph it.

39. $5 > 8 - x$ and $1 + 0.5x < 4$

40. $5 - x < 4$ and $0.2x - 5 < 1$

41. $\dfrac{2x - 5}{-2} < 2$ and $\dfrac{2x + 1}{3} > 0$

42. $\dfrac{4 - x}{2} > 1$ and $\dfrac{2x - 7}{-3} < 1$

43. $1 - x < 7 + x$ or $4x + 3 > x$

44. $5 + x > 3 - x$ or $2x - 3 > x$

45. $\dfrac{1}{2}(x + 1) > 3$ or $0 < 7 - x$

46. $\dfrac{1}{2}(x + 6) > 3$ or $4(x - 1) < 3x - 4$

47. $1 - \dfrac{3}{2}x < 4$ and $\dfrac{1}{4}x - 2 \leq -3$

48. $\dfrac{3}{5}x - 1 > 2$ and $5 - \dfrac{2}{5}x \geq 3$

49. $1 < 3x - 5 < 7$

50. $-3 \leq 4x + 9 \leq 17$

51. $-2 \leq 4 - 6x < 22$

52. $-13 < 5 - 9x \leq 41$

Solve each absolute value inequality. Write the solution set using interval notation and graph it. See the summary of basic absolute value inequalities on page 168.

53. $|3x - 1| < 2$

54. $|4x - 3| \leq 5$

55. $|5 - 4x| \leq 1$

56. $|6 - x| < 6$

57. $|x - 1| \geq 1$

58. $|x + 2| > 5$

59. $|5 - x| > 3$

60. $|3 - 2x| \geq 5$

61. $5 \geq |4 - x|$

62. $3 < |2x - 1|$

63. $|5 - 4x| < 0$

64. $|3x - 7| \geq -5$

65. $|4 - 5x| < -1$

66. $|2 - 9x| \geq 0$

67. $3|x - 2| + 6 > 9$

68. $3|x - 1| + 2 < 8$

69. $\left|\dfrac{x - 3}{2}\right| > 1$

70. $\left|\dfrac{9 - 4x}{2}\right| < 3$

Write an inequality of the form $|x - a| < k$ or of the form $|x - a| > k$ so that the inequality has the given solution set.
HINT $|x - a| < k$ means that x is less than k units from a and $|x - a| > k$ means that x is more than k units from a on the number line.

71. $(-5, 5)$

72. $(-2, 2)$

73. $(-\infty, -3) \cup (3, \infty)$

74. $(-\infty, -1) \cup (1, \infty)$

75. $(4, 8)$

76. $(-3, 9)$

77. $(-\infty, 3) \cup (5, \infty)$

78. $(-\infty, -1) \cup (5, \infty)$

For each graph write an absolute value inequality that has the given solution set.

79.

$-15\ -12\ -9\ -6\ -3\ \ 0\ \ 3\ \ 6\ \ 9\ \ 12\ 15$

80.

$-10\ -8\ -6\ -4\ -2\ \ 0\ \ 2\ \ 4\ \ 6\ \ 8\ \ 10$

81.

$2\ \ 3\ \ 4\ \ 5\ \ 6\ \ 7\ \ 8\ \ 9\ \ 10\ \ 11\ \ 12$

82.

$-6\ -4\ -2\ \ 0\ \ 2\ \ 4\ \ 6\ \ 8\ \ 10$

83.

$1\ \ 2\ \ 3\ \ 4\ \ 5\ \ 6\ \ 7\ \ 8\ \ 9$

84.

$-4\ \ \ -3\ \ \ -2\ \ \ -1\ \ \ 0$

Recall that $\sqrt{w}$ is a real number only if $w \geq 0$ and $1/w$ is a real number only if $w \neq 0$. For what values of x is each of the following expressions a real number?

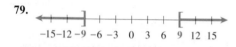

85. $\sqrt{x - 2}$

86. $\sqrt{3x - 1}$

87. $\dfrac{1}{\sqrt{2 - x}}$

88. $\dfrac{5}{\sqrt{3 - 2x}}$

89. $\sqrt{|x| - 3}$

90. $\sqrt{5 - |x|}$

Solve each problem.

91. *Price Range for a Car*   Yolanda is shopping for a used car in a city where the sales tax is 10% and the title and license fee is $300. If the maximum that she can spend is $8000, then she should look at cars in what price range?

92. *Price of a Burger*   The price of Elaine's favorite Big Salad at the corner restaurant is 10 cents more than the price of Jerry's hamburger. After treating a group of friends to lunch, Jerry is certain that for 10 hamburgers and 5 salads he spent more than $9.14, but not more than $13.19, including tax at 8% and a 50 cent tip. In what price range is a hamburger?

93. *Final Exam Score*   Lucky scored 65 on his Psychology 101 midterm. If the average of his midterm and final must be between 79 and 90 for a B, then for what range of scores on the final exam would Lucky get a B?
HINT Write a compound inequality with his average between 79 and 90.

94. *Bringing Up Your Average*   Felix scored 52 and 64 on his first two tests in Sociology 212. What must he get on the third test to get an average for the three tests above 70?

95. *Weighted Average with Whole Numbers*   Ingrid scored 65 on her calculus midterm. If her final exam counts twice as much as her midterm exam, then for what range of scores on her final would she get an average between 79 and 90?
HINT For this weighted average multiply the final exam score by 2 and the midterm score by 1.

96. *Weighted Average with Fractions*   Elizabeth scored 62, 76, and 80 on three equally weighted tests in French. If the final exam score counts for two-thirds of the grade and the tests count for one-third, then what range of scores on the final exam would give her a final average over 70?
HINT For this weighted average multiply the final exam score by 2/3 and the average of the test scores by 1/3.

97. *Maximum Girth*   United Parcel Service (UPS) defines girth of a box as the sum of the length, twice the width, and twice the height. The maximum length that can be shipped with UPS is 108 in. and the maximum girth is 130 in. If a box has a length of 40 in. and a width of 30 in. then in what range must the height fall?

98. *Raising a Batting Average*   At one point during the 2001 season, Javy Lopez had 97 hits in 387 times at bat for an average of 0.251 (www.espn.com).
   a. How many more times would he have to bat to get his average over 0.300, assuming he got a hit every time?

   b. How many more times would he have to bat to get his average over 0.300, assuming he got a hit 50% of the time?

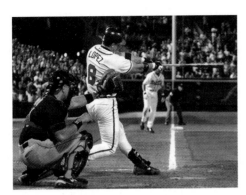

■ **Figure for Exercise 98**

99. *Bicycle Gear Ratio*   The gear ratio r for a bicycle is defined by the following formula

$$r = \frac{Nw}{n},$$

where N is the number of teeth on the chainring (by the pedal), n is the number of teeth on the cog (by the wheel), and w is the wheel diameter in inches. The following chart gives uses for the various gear ratios.

Ratio	Use	
$r > 90$	down hill	
$70 < r \leq 90$	level	
$50 < r \leq 70$	mild hill	
$35 < r \leq 50$	long hill	

A bicycle with a 27-in. diameter wheel has 50 teeth on the chainring and 5 cogs with 14, 17, 20, 24, and 29 teeth. Find the gear ratio with each of the five cogs. Does this bicycle have a gear ratio for each of the four types of pedaling described in the table?

100. *Selecting the Cogs*   Use the formula from the previous exercise to answer the following.
 a. If a single-speed 27-in. bicycle has 17 teeth on the cog, then for what numbers of teeth on the chainring will the gear ratio be between 60 and 80?

 b. If a 26-in. bicycle has 40 teeth on the chainring, then for what numbers of teeth on the cog will the gear ratio be between 60 and 75?

101. *Expensive Models*   The prices of the 2005 BMW 745 Li and 760 Li differ by more than $25,000. The list price of the 745 Li is $74,595. Assuming that you do not know which model is more expensive, write an absolute value inequality that describes the price of the 760 Li. What are the possibilities for the price of the 760 Li?

102. *Difference in Prices*   There is less than $800 difference between the price of a $14,200 Ford and a comparable Chevrolet. Express this as an absolute value inequality. What are the possibilities for the price of the Chevrolet?

103. *Controlling Temperature*   Michelle is trying to keep the water temperature in her chemistry experiment at 35°C. For the experiment to work, the relative error for the actual temperature must be less than 1%. Write an absolute value inequality for the actual temperature. Find the interval in which the actual temperature must lie.

104. *Laying Out a Track*   Melvin is attempting to lay out a track for a 100-m race. According to the rules of competition, the relative error of the actual length must be less than 0.5%. Write an absolute value inequality for the actual length. Find the interval in which the actual length must fall.

105. *Acceptable Bearings*   A spherical bearing is to have a circumference of 7.2 cm with an error of no more than 0.1 cm. Use an absolute value inequality to find the acceptable range of values for the diameter of the bearing.

106. *Acceptable Targets*   A manufacturer makes circular targets that have an area of 15 ft^2. According to competition rules, the area can be in error by no more than 0.5 ft^2. Use an absolute value inequality to find the acceptable range of values for the radius.

Area: 15 ± 0.5 ft^2

■ **Figure for Exercise 106**

107. *Per Capita Income*   The 2003 per capita income for the United States was $31,632 (U.S. Census Bureau, www.census.gov). The following table shows per capita income for 2003 for selected states.

State	Per capita income
Alabama	26,338
Maine	28,831
Florida	30,446
Hawaii	30,913
Alaska	33,568
California	33,749
Colorado	34,283
Maryland	37,331
New Jersey	40,427
Connecticut	43,173

a. If a is the per capita income for a state, then for which states is $|a - 31{,}632| < 2000$?

b. For which states is $|a - 31{,}632| > 3000$?

108. *Making a Profit*   A strawberry farmer paid $4200 for planting and fertilizing her strawberry crop. She must also pay $2.40 per flat (12 pints) for picking and packing the berries and $300 rent for space in a farmers' market where she sells the berries for $11 per flat. For what number of flats will her revenue exceed her costs?

Thinking Outside the Box XVIII

One in a Million   If you write the integers from 1 through 1,000,000 inclusive, then how many ones will you write?

1.7 Pop Quiz

Solve each inequality. Write the solution set using interval notation.

1. $x - \sqrt{2} \geq 0$

2. $6 - 2x < 0$

3. $x > 4$ or $x \geq -1$

4. $x - 1 > 5$ and $2x < 18$

5. $|x| > 6$

6. $|x - 1| \leq 2$

Linking Concepts

For Individual or Group Explorations

Modeling the Cost of Copying

A company can rent a copy machine for five years from American Business Machines for $105 per month plus $0.08 per copy. The same copy machine can be purchased for $6500 with a per copy cost of $0.04 plus $25 per month for a maintenance contract. After five years the copier is worn out and worthless. To help the company make its choice, ABM sent the accompanying graph.

a)   Write a formula for the total cost of renting and using the copier for five years in terms of the number of copies made during five years.

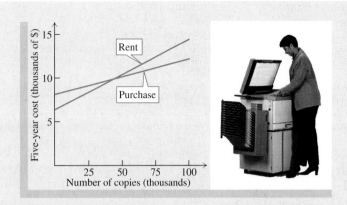

b) Write a formula for the total cost of buying and using the copier for five years in terms of the number of copies made during five years.

c) For what number of copies does the total cost of renting exceed $10,000?

d) For what number of copies does the total cost of buying exceed $10,000?

e) Use an absolute value inequality and solve it to find the number of copies for which the two plans differ by less than $1000.

f) For what number of copies is the cost of renting equal to the cost of buying?

g) If the company estimates that it will make between 40,000 and 50,000 copies in five years, then which plan is better?

▪ ▪ ▪ Highlights

1.1 Equations in One Variable

Linear Equation	$ax + b = 0$ where a and b are real numbers with $a \neq 0$.	$5x - 10 = 0$
Properties of Equality	An equivalent equation is obtained by adding, subtracting, multiplying, or dividing each side of an equation by the same nonzero real number.	$5x = 10$ $x = 2$
Solution Set	The set of all real numbers that satisfy an equation	Solution set to $5x - 10 = 0$ is $\{2\}$.
Identity	An equation satisfied by all real numbers for which both sides are defined	$\frac{y}{y} = 1$, Solution set: $(-\infty, 0) \cup (0, \infty)$
Conditional Equation	An equation that has at least one solution but is not an indentity	$2z - 1 = 0$ Solution set: $\{1/2\}$

Inconsistent Equation	An equation that has no solutions	$w = w + 1$, Solution set: $\varnothing$
Absolute Value Equations	$\|x\| = k$ for $k > 0 \Leftrightarrow x = k$ or $x = -k$ $\|x\| = k$ for $k < 0$ is inconsistent $\|x\| = 0 \Leftrightarrow x = 0$	$\|t\| = 1 \Leftrightarrow t = 1$ or $t = -1$ $\|m\| = -2$, solution set: $\varnothing$ $\|q - 4\| = 0 \Leftrightarrow q = 4$

1.2 Constructing Models to Solve Problems

Formula	An equation in two or more variables that is usually of an applied nature	$A = \pi r^2, P = 2L + 2W$ $E = mc^2, F = \frac{9}{5}C + 32$
Solving for a Variable	The properties of equality are used to solve a formula for one variable in terms of the others.	$P = 2L + 2W$ $2L = P - 2W$ $L = (P - 2W)/2$

1.3 Equations and Graphs in Two Variables

Distance Formula	Distance between (x_1, y_1) and (x_2, y_2) is $\sqrt{(x_2 - x_1)^2 + (y_2 - y_1)^2}$.	For $(1, 2)$ and $(4, -2)$, $\sqrt{(4 - 1)^2 + (-2 - 2)^2} = 5$.
Midpoint Formula	The midpoint of the line segment with endpoints (x_1, y_1) and (x_2, y_2) is $\left(\frac{x_1 + x_2}{2}, \frac{y_1 + y_2}{2}\right)$.	For $(0, -4)$ and $(6, 2)$, $\left(\frac{0 + 6}{2}, \frac{-4 + 2}{2}\right) = (3, -1)$.
Equation of a Circle	The graph of $(x - h)^2 + (y - k)^2 = r^2 (r > 0)$ is a circle with center (h, k) and radius r.	Circle: $(x - 1)^2 + (y + 2)^2 = 9$ Center $(1, -2)$, radius 3
Linear Equation: Standard Form	$Ax + By = C$ where A and B are not both zero, $x = h$ is a vertical line, $y = k$ is a horizontal line	$2x + 3y = 6$ is a line. $x = 5$ is a vertical line. $y = 7$ is a horizontal line.

1.4 Linear Equations in Two Variables

Slope Formula	The slope of a line through (x_1, y_1) and (x_2, y_2) is $(y_2 - y_1)/(x_2 - x_1)$ provided $x_1 \neq x_2$.	$(1, 2), (3, -6)$ slope $\frac{-6 - 2}{3 - 1} = -4$
Slope-Intercept Form	$y = mx + b$, slope m, y-intercept $(0, b)$ y is a linear function of x.	$y = 2x + 5$, slope 2, y-intercept $(0, 5)$
Point-Slope Form	The line through (x_1, y_1) with slope m is $y - y_1 = m(x - x_1)$.	Point $(-3, 2)$, $m = 5$ $y - 2 = 5(x - (-3))$
Parallel Lines	Two nonvertical lines are parallel if and only if their slopes are equal.	$y = 7x - 1$ and $y = 7x + 4$ are parallel.
Perpendicular Lines	Two lines with slopes m_1 and m_2 are perpendicular if and only if $m_1 m_2 = -1$.	$y = \frac{1}{2}x + 4$ and $y = -2x - 3$ are perpendicular.

1.5 Scatter Diagrams and Curve Fitting

Scatter Diagram	A graph of a set of ordered pairs
Linear Relationship	If the points in a scatter diagram appear to be scattered about a line, then there is a linear relationship between the two variables.

1.6 Quadratic Equations

Quadratic Equation	$ax^2 + bx + c = 0$ where a, b, and c are real and $a \neq 0$	$x^2 + 2x - 3 = 0$
Methods for Solving Quadratic Equations	Factoring: factor the quadratic polynomial and set the factors equal to zero.	$(w + 3)(w - 1) = 0 \Rightarrow$ $w + 3 = 0$ or $w - 1 = 0$
	Square root property: $x^2 = k$ is equivalent to $x = \pm\sqrt{k}$	$m^2 = 5 \Rightarrow m = \pm\sqrt{5}$
	Completing the square: complete the square and then apply the square root property	$(w + 1)^2 = 4 \Rightarrow$ $w + 1 = \pm 2$
	Quadratic formula: solutions to $ax^2 + bx + c = 0$ are $x = \dfrac{-b \pm \sqrt{b^2 - 4ac}}{2a}$	$t^2 + 2t - 3 = 0 \Rightarrow$ $t = \dfrac{-2 \pm \sqrt{2^2 - 4(1)(-3)}}{2(1)}$

1.7 Linear and Absolute Value Inequalities

Linear Inequalities	The inequality symbol is reversed if the inequality is multiplied or divided by a negative number.	$4 - 2x > 10$ $-2x > 6$ $x < -3$
Absolute Value Inequalities	$\lvert x \rvert > k \ (k > 0) \Leftrightarrow x > k$ or $x < -k$ $\lvert x \rvert < k \ (k > 0) \Leftrightarrow -k < x < k$ $\lvert x \rvert \leq 0 \Leftrightarrow x = 0$ $\lvert x \rvert \geq 0 \Leftrightarrow x$ is any real number	$\lvert y \rvert > 1 \Leftrightarrow y > 1$ or $y < -1$ $\lvert z \rvert < 2 \Leftrightarrow -2 < z < 2$ $\lvert 2b - 5 \rvert \leq 0 \Leftrightarrow 2b - 5 = 0$ All real numbers satisfy $\lvert 3s - 7 \rvert \geq 0$.

■ ■ ■ Chapter 1 Review Exercises

Find all real solutions to each equation.

1. $3x - 2 = 0$

2. $3x - 5 = 5(x + 7)$

3. $\dfrac{1}{2}y - \dfrac{1}{3} = \dfrac{1}{4}y + \dfrac{1}{5}$

4. $\dfrac{1}{2} - \dfrac{w}{5} = \dfrac{w}{4} - \dfrac{1}{8}$

5. $\dfrac{2}{x} = \dfrac{3}{x - 1}$

6. $\dfrac{5}{x + 1} = \dfrac{2}{x - 3}$

7. $\dfrac{x + 1}{x - 1} = \dfrac{x + 2}{x - 3}$

8. $\dfrac{x + 3}{x - 8} = \dfrac{x + 7}{x - 4}$

For each pair of points, find the distance between them and the midpoint of the line segment joining them.

9. $(-3, 5), (2, -6)$

10. $(-1, 1), (-2, -3)$

11. $(1/2, 1/3), (1/4, 1)$

12. $(0.5, 0.2), (-1.2, 2.1)$

Sketch the graph of each equation. For the circles, state the center and the radius. For the lines state the intercepts.

13. $x^2 + y^2 = 25$

14. $(x - 2)^2 + y^2 = 1$

15. $x^2 + 4x + y^2 = 0$

16. $x^2 - 6x = 2y - y^2 - 1$

17. $x + y = 25$

18. $2x - y = 40$

19. $y = 3x - 4$

20. $y = -\dfrac{1}{2}x + 4$

21. $x = 5$

22. $y = 6$

Solve each problem.

23. Write in standard form the equation of the circle that has center $(-3, 5)$ and radius $\sqrt{3}$.

24. Find the center and radius for the circle $x^2 + y^2 = x - 2y + 1$.

25. Find the x- and y-intercepts for the graph of $3x - 4y = 12$.

26. What is the y-intercept for the graph of $y = 5$?

27. Find the slope of the line that goes through $(3, -6)$ and $(-1, 2)$.

28. Find the slope of the line $3x - 4y = 9$.

29. Find the equation (in slope-intercept form) for the line through $(-2, 3)$ and $(5, -1)$.

30. Find the equation (in standard form using only integers) for the line through $(-1, -3)$ and $(2, -1)$.

31. Find the equation (in standard form using only integers) for the line through $(2, -4)$ that is perpendicular to $3x + y = -5$.

32. Find the equation (in slope-intercept form) for the line through $(2, -5)$ that is parallel to $2x - 3y = 5$.

Solve each equation for y.

33. $2x - 3y = 6$

34. $x(y - 2) = 1$

35. $xy = 1 + 3y$

36. $x^2 y = 1 + 9y$

37. $ax + by = c$

38. $\dfrac{1}{y} = \dfrac{1}{x} + \dfrac{1}{2}$

Evaluate the discriminant for each equation, and use it to determine the number of real solutions to the equation.

39. $x^2 + 2 = 4x$

40. $y^2 + 2 = 3y$

41. $4w^2 = 20w - 25$

42. $2x^2 - 3x + 10 = 0$

Find all real or imaginary solutions to each equation. Use the method of your choice on each problem.

43. $x^2 - 5 = 0$

44. $3x^2 - 54 = 0$

45. $x^2 + 8 = 0$

46. $x^2 + 27 = 0$

47. $2x^2 + 1 = 0$

48. $3x^2 + 2 = 0$

49. $(x - 2)^2 = 17$

50. $(2x - 1)^2 = 9$

51. $x^2 - x - 12 = 0$

52. $2x^2 - 11x + 5 = 0$

53. $b^2 + 10 = 6b$

54. $4t^2 + 17 = 16t$

55. $s^2 - 4s + 1 = 0$

56. $3z^2 - 2z - 1 = 0$

57. $4x^2 - 4x - 5 = 0$

58. $9x^2 - 30x + 23 = 0$

59. $x^2 - 2x + 2 = 0$

60. $x^2 - 4x + 5 = 0$

61. $\dfrac{1}{x} + \dfrac{1}{x - 1} = \dfrac{3}{2}$

62. $\dfrac{2}{x - 2} - \dfrac{3}{x + 2} = \dfrac{1}{2}$

Solve each equation.

63. $|3q - 4| = 2$

64. $|2v - 1| = 3$

65. $|2h - 3| = 0$

66. $4|x - 3| = 0$

67. $|5 - x| = -1$

68. $|3y - 1| = -2$

Solve each inequality. State the solution set using interval notation and graph the solution set.

69. $4x - 1 > 3x + 2$

70. $6(x - 3) < 5(x + 4)$

71. $5 - 2x > -3$

72. $7 - x > -6$

73. $\dfrac{1}{2}x - \dfrac{1}{3} > x + 2$

74. $0.06x + 1000 > x + 60$

75. $-2 < \dfrac{x - 3}{2} \le 5$

76. $-1 \le \dfrac{3 - 2x}{4} < 3$

77. $3 - 4x < 1$ and $5 + 3x < 8$

78. $-3x < 6$ and $2x + 1 > -1$

79. $-2x < 8$ or $3x > -3$

80. $1 - x < 6$ or $-5 + x < 1$

81. $|x - 3| > 2$

82. $|4 - x| \le 3$

83. $|2x - 7| \le 0$

84. $|6 - 5x| < 0$

85. $|7 - 3x| > -4$

86. $|4 - 3x| \ge 1$

Solve each equation or inequality by reading the accompanying graph.

87. $x - 0.5(30 - x) = 0$

88. $x^2 + 4x - 780 = 0$

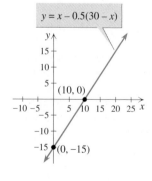

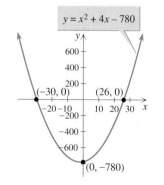

89. $0.6x - 4.8 < 0$

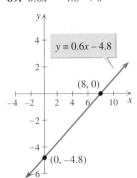

90. $36 - 1.2x \geq 0$

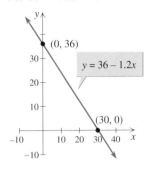

Solve each problem. Use an equation or an inequality as appropriate. For problems in which the answer involves an irrational number, find the exact answer and an approximate answer.

91. *Folding Sheet Metal*   A square is to be cut from each corner of an 8-in. by 11-in. rectangular piece of copper and the sides are to be folded up to form a box as shown in the figure. If the area of the bottom is to be 50 in.^{2}, what is the length of the side of the square to be cut from the corner?

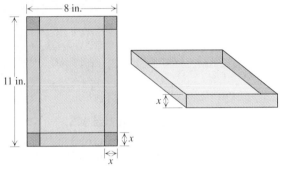

■ **Figure for Exercise 91**

92. *Peeling Apples*   Bart can peel a batch of 3000 lb of apples in 12 hr with the old machine, while Mona can do the same job in 8 hr using the new machine. If Bart starts peeling at 8 A.M. and Mona joins him at 9 A.M., then at what time will they finish the batch working together?

93. *Meeting Between Two Cities*   Lisa, an architect from Huntsville, has arranged a business luncheon with her client, Taro Numato from Norwood. They plan to meet at a restaurant off the 300-mi highway connecting their two cities. If they leave their offices simultaneously and arrive at the restaurant simultaneously, and Lisa averages 50 mph while Taro averages 60 mph, then how far from Huntsville is the restaurant?

94. *Driving Speed*   Lisa and Taro agree to meet at the construction site located 100 mi from Norwood on the 300-mi highway connecting Huntsville and Norwood. Lisa leaves Huntsville at noon, while Taro departs from Norwood 1 hr later. If they arrive at the construction site simultaneously and Lisa's driving speed averaged 10 mph faster than Taro's, then how fast did she drive?

95. *Fish Population*   A channel was dug to connect Homer and Mirror Lakes. Before the channel was dug, a biologist estimated that 20% of the fish in Homer Lake and 30% of the fish in Mirror Lake were bass. After the lakes were joined, the biologist made another estimate of the bass population in order to set fishing quotas. If she decided that 28% of the total fish population of 8000 were bass, then how many fish were in Homer Lake originally?

96. *Support for Gambling*   Eighteen pro-gambling representatives in the state house of representatives bring up a gambling bill every year. After redistricting, four new representatives are added to the house, causing the percentage of pro-gambling representatives to increase by 5 percentage points. If all four of the new representatives are pro-gambling and they still do not constitute a majority of the house, then how many representatives are there in the house after redistricting?

97. *Hiking Distance*   The distance between Marjorie and the dude ranch was $4\sqrt{34}$ mi straight across a rattlesnake-infested canyon. Instead of crossing the canyon, she hiked due north for a long time and then hiked due east for the shorter leg of the journey to the ranch. If she averaged 4 mph and it took her 8 hr to get to the ranch, then how far did she hike in a northerly direction?

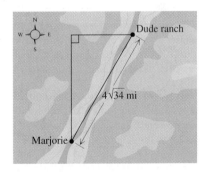

■ **Figure for Exercise 97**

98. *Golden Rectangle*   The *golden rectangle* of the ancient Greeks was thought to be the rectangle with the shape that was most pleasing to the eye. The golden rectangle was defined as a rectangle that would retain its original shape after removal of a *W* by *W* square from one end, as shown in the figure. So the length and width of the original rectangle must satisfy

$$\frac{L}{W} = \frac{W}{L - W}.$$

If the length of a golden rectangle is 20 m, then what is its width? If the width of a golden rectangle is 8 m, then what is its length?

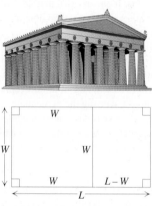

■ **Figure for Exercise 98**

99. *Price Range for a Haircut*   A haircut at Joe's Barber Shop costs $50 less than a haircut at Renee's French Salon. In fact, you can get five haircuts at Joe's for less than the cost of one haircut at Renee's. What is the price range for a haircut at Joe's?

100. *Selling Price of a Home*   Elena must get at least $120,000 for her house in order to break even with the original purchase price. If the real estate agent gets 6% of the selling price, then what should the selling price be?

101. *Dimensions of a Picture Frame*   The length of a picture frame must be 2 in. longer than the width. If Reginald can use between 32 and 50 in. of frame molding for the frame, then what are the possibilities for the width?

102. *Saving Gasoline*   If Americans drive 10^{12} mi per year and the average gas mileage is raised from 27.5 mpg to 29.5 mpg, then how many gallons of gasoline are saved?

103. *Increasing Gas Mileage*   If Americans continue to drive 10^{12} mi per year and the average gas mileage is raised from 29.5 to 31.5 mpg, then how many gallons of gasoline are saved? If the average mpg is presently 29.5, then what should it be increased to in order to achieve the same savings in gallons as the increase from 27.5 to 29.5 mpg?

104. *Thickness of Concrete*   Alfred used 40 yd³ of concrete to pour a section of interstate highway that was 12 ft wide and 54 ft long. How many inches thick was the concrete?

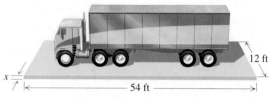

■ **Figure for Exercise 104**

105. *U.S. Computer Usage*   The number of computers in use in the United States went from 21.5 million in 1985 to 240 million in 2005 as shown in the accompanying graph (*Computer Industry Almanac*, www.c-i-a.com). Find the equation of the line through the two given points and use the equation to estimate the number of computers in use in 1993.

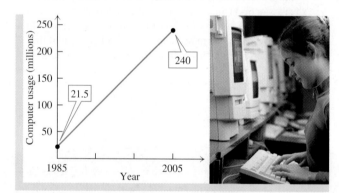

■ **Figure for Exercise 105**

106. *Worldwide Computer Usage*   The number of computers in use worldwide went from 38.1 million in 1985 to 920 million in 2005. (*Computer Industry Almanac*, www.c-i-a.com). Find the equation of the line through the two given points and use the equation to estimate the number of computers that will be in use in 2010.

107. *Percent of Body Fat*   The average 20-year-old woman has 23% body fat and the average 50-year-old woman has 47% body fat (Infoplease, www.infoplease.com). Assuming that the percentage of body fat is a linear function of age, find the function. Use the function to determine the percentage of body fat in the average 65-year-old woman.

108. *Olympic Gold*   In 2000 Konstantinos Kinteris won the 200-meter race with a time of 20.09 seconds (www.infoplease.com). In 2004 Shawn Crawford won with a time 19.79 seconds. Assuming that the winning time is decreasing linearly as shown in the accompanying figure, express the winning time as a linear function of the year. Use the function to predict the winning time in the 2008 Olympics.

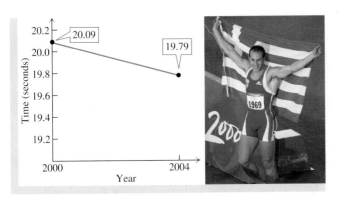

■ **Figure for Exercise 108**

109. *Cost of Health Care per Capita* The accompanying table shows the cost of health care per capita in the United States for the years 1994 through 2004 (U.S. Census Bureau, www.census.gov).
   a. Use linear regression with a graphing calculator to find a linear equation that expresses the cost per capita C as a linear function of the year y.

   b. Use the equation found in part (a) to predict the cost of health care per capita for 2010.

■ **Table for Exercise 109**

Year	Cost per capita ($)
1994	3534
1996	3847
1998	4179
2000	4670
2002	5440
2004	6167

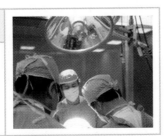

110. *National Health Spending* The accompanying table gives the total amount spent on health care for the years 2000 through 2004 (U.S. Census Bureau, www.census.gov).

Year	Amount ($ billions)
2000	1309
2001	1421
2002	1553
2003	1674
2004	1794

   a. Use linear regression on your graphing calculator to find a linear equation that expresses the amount A as a linear function of the year y.

   b. Use the equation to predict the amount spent nationally in 2010.

Thinking Outside the Box XIX & XX

Unlucky Number What is the smallest positive integer that is 13 times the sum of its digits?

Run for Your Life A hiker is one-fourth of the way through a narrow train tunnel when he looks over his shoulder to see a train approaching the tunnel at 30 mph. He quickly figures that if he runs in either direction at his top speed he can just make it out of the tunnel. What is his top running speed?

■ ■ ■ Chapter 1 Test

Find all real or imaginary solutions to each equation.

1. $2x + 1 = x - 6$

2. $\frac{1}{2}x - \frac{1}{6} = \frac{1}{3}x$

3. $3x^2 - 2 = 0$

4. $x^2 + 1 = 6x$

5. $x^2 + 14 = 9x$

6. $\frac{x - 1}{x + 3} = \frac{x + 2}{x - 6}$

7. $x^2 = 2x - 5$

8. $x^2 + 1 = 0$

Sketch the graph of each equation in the xy-coordinate system.

9. $3x - 4y = 120$

10. $x^2 + y^2 = 400$

11. $x^2 + y^2 + 4y = 0$

12. $y = -\frac{2}{3}x + 4$

13. $y = 4$

14. $x = -2$

Solve each problem.

15. Find the slope of the line $3x - 5y = 8$.

16. Find the slope of the line through $(-3, 6)$ and $(5, -4)$.

17. Find the slope-intercept form of the equation of the line that goes through $(1, -2)$ and is perpendicular to the line $2x - 3y = 6$.

18. Find the equation of the line in slope-intercept form that goes through $(3, -4)$ and is parallel to the line through $(0, 2)$ and $(3, -1)$.

19. Find the exact distance between $(-3, 1)$ and $(2, 4)$.

20. Find the midpoint of the line segment with endpoints $(-1, 1)$ and $(1, 0)$.

21. Find the value of the discriminant for $x^2 - 5x + 9 = 0$. How many real solutions are there for this equation?

22. Solve $5 - 2y = 4 + 3xy$ for y.

Solve each inequality in one variable. State the solution set using interval notation and graph it on the number line.

23. $3 - 2x > 7$

24. $\dfrac{x}{2} > 3$ and $5 < x$

25. $|2x - 1| \le 3$

26. $2|x - 3| + 1 > 5$

Solve each problem.

27. Terry had a square patio. After expanding the length by 20 ft and the width by 10 ft the area was 999 ft². What was the original area?

28. How many gallons of 20% alcohol solution must be mixed with 10 gal of a 50% alcohol solution to obtain a 30% alcohol solution?

29. The median price of a new home in Springfield in 1994 was $88,000 and in 2004 it was $122,000. Assuming that the price is a linear function of the year, write a formula for that function. Use the formula to predict the median price in the year 2015.

30. *Number of Farms* The accompanying table gives the number of farms in the United States for the years 1999 through 2003 (U.S. Department of Agriculture, www.usda.gov).

Year	Number of Farms (thousands)	
1999	2192	
2000	2172	
2001	2156	
2002	2158	
2003	2127	

a. Find equations that express the number of farms in terms of the year by using linear regression and quadratic regression. Let $x = 0$ correspond to 1999.

b. Predict the number of farms in the United States in 2009 using linear regression and quadratic regression. By how much do these predictions differ?

Tying it all Together

Chapters P–1

Perform the indicated operations.

1. $3x + 4x$

2. $5x \cdot 6x$

3. $\dfrac{1}{x} + \dfrac{1}{2x}$

4. $(x + 3)^2$

5. $(2x - 1)(3x + 2)$

6. $\dfrac{(x + h)^2 - x^2}{h}$

7. $\dfrac{1}{x - 1} + \dfrac{1}{x + 1}$

8. $\left(x + \dfrac{3}{2}\right)^2$

Solve each equation.

9. $3x + 4x = 7x$

10. $5x \cdot 6x = 11x$

11. $\dfrac{1}{x} + \dfrac{1}{2x} = \dfrac{3}{2x}$

12. $(x + 3)^2 = x^2 + 9$

13. $6x^2 + x - 2 = 0$

14. $\dfrac{1}{x - 1} + \dfrac{1}{x + 1} = \dfrac{5}{8}$

15. $3x + 4x = 7x^2$

16. $3x + 4x = 7$

Given that $P(x) = x^3 - 3x^2 + 2$ and $T(x) = -x^2 + 3x - 4$, find the value of each polynomial for the indicated values of x.

17. $P(1)$

18. $P(-1)$

19. $P(-1/2)$

20. $P(-1/3)$

21. $T(-1)$

22. $T(2)$

23. $T(0)$

24. $T(0.5)$

2 Functions and Graphs

THE rainforests cover less than an eighth of a percent of the earth's surface, yet they are home to over half of our animal and plant species. Rainforests are also a source of foods such as chocolate, vanilla, pineapples, and cinnamon, and about twenty percent of all medicines.

However, the rainforests are disappearing at an alarming rate and along with them a thousand species of plants and animals per year. Destroying the rainforests will irrevocably change our planet's future.

Depletion of the rainforests is only one of many ecological issues that include the ozone layer, global warming, and hazardous waste. In assessing these problems, scientists often look for relationships between variables.

WHAT YOU WILL LEARN In this chapter you will study relationships between variables using graphic, numeric, and algebraic points of view.

2.1

Functions

If one variable determines the value of another, then the second variable is a function of the first. We introduced this terminology in Section 1.4. In this section we will study this idea in depth.

The Function Concept

If you spend $10 on gasoline, then the price per gallon determines the number of gallons that you get. There is a rule: the number of gallons is $10 divided by the price per gallon. The number of hours that you sleep before a test might be related to your grade on the test, but does not determine your grade. There is no rule that will determine your grade from the number of hours of sleep. If the value of a variable y is determined by the value of another variable x, then y **is a function of** x. The phrase "is a function of" means "is determined by." If there is more than one value for y corresponding to a particular x-value, then y is not determined by x and y is not a function of x.

Example **1** Using the phrase "is a function of"

Decide whether a is a function of b, b is a function of a, or neither.

a. Let a represent a positive integer smaller than 100 and b represent the number of divisors of a.
b. Let a represent the age of a United States citizen and b represent the number of days since his/her birth.
c. Let a represent the age of a United States citizen and b represent his/her annual income.

Solution

a. We can determine the number of divisors of any positive integer smaller than 100. So b is a function of a. We cannot determine the integer knowing the number of its divisors, because different integers have the same number of divisors. So a is not a function of b.
b. The number of days since a person's birth certainly determines the age of the person in the usual way. So a is a function of b. However, you cannot determine the number of days since a person's birth from their age. You need more information. For example, the number of days since birth for two 1-year-olds could be 370 and 380 days. So b is not a function of a.
c. We cannot determine the income from the age or the age from the income. We would need more information. Even though age and income are related, the relationship is not strong enough to say that either one is a function of the other.

Try This. Let p be the price of a grocery item and t be the amount of sales tax at 5% on that item. Determine whether p is a function of t, t is a function of p, or neither.

∎

To make the function concept clearer, we now define the noun "function." A *function* is often defined as a rule that assigns each element in one set to a unique element in a second set. Of course understanding this definition depends on knowing the meanings of the words *rule, assigns,* and *unique.* Using the language of ordered pairs we can make the following precise definition in which there are no undefined words.

Definition: Function

> A **function** is a set of ordered pairs in which no two ordered pairs have the same first coordinate and different second coordinates.

If we start with two related variables, we can identify one as the first variable and the other as the second variable and consider the set of ordered pairs containing their corresponding values. If the set of ordered pairs satisfies the function definition, then we say that the second variable is a function of the first. The variable corresponding to the first coordinate is the **independent variable** and the variable corresponding to the second coordinate is the **dependent variable.**

Identifying Functions

A **relation** is a set of ordered pairs. A relation can be indicated by a verbal description, a graph, a formula or equation, a table, or a list of ordered pairs. Not every relation is a function. A function is a special relation.

When a relation is given by a graph, we can visually check whether there are two ordered pairs with the same first coordinate and different second coordinates. For example, the circle shown in Fig. 2.1 is not the graph of a function, because there are two points on the circle with the same first coordinate. These points lie on a vertical line. In general, if there is a vertical line that crosses a graph more than once, the graph is not the graph of a function. This criterion is known as the **vertical line test.**

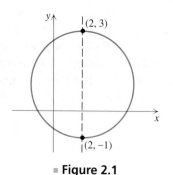

■ **Figure 2.1**

Theorem: The Vertical Line Test

> A graph is the graph of a function if and only if there is no vertical line that crosses the graph more than once.

Every nonvertical line is the graph of a function, because every vertical line crosses a nonvertical line exactly once. Note that the vertical line test makes sense only because we always put the independent variable on the horizontal axis.

Example **2** **Identifying a function from a graph**

Determine which of the graphs shown in Fig. 2.2 are graphs of functions.

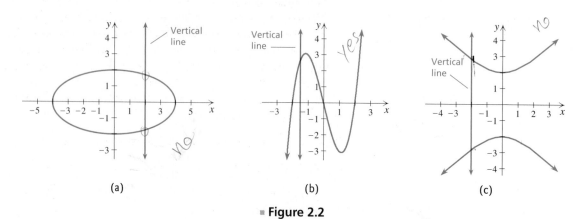

■ **Figure 2.2**

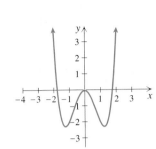

■ **Figure 2.3**

Solution

For each graph we want to decide whether y is a function of x. That is, can y be uniquely determined from x? In parts (a) and (c) we can draw a vertical line that crosses the graph more than once. So in each case there is an x-coordinate that corresponds to two different y-coordinates. So in parts (a) and (c) we cannot always determine a unique value of y from a given x-coordinate. So y is not a function of x in parts (a) and (c). The graph in Fig. 2.2(b) is the graph of a function because every vertical line appears to cross the graph at most once. So in this case we can determine y for a given x-coordinate and y is a function of x.

Try This. Use the graph in Fig. 2.3 to determine whether y is a function of x. ■

A calculator is a virtual function machine. Built-in functions on a calculator are marked with symbols such as $\sqrt{x}$, x^2, $x!$, 10^x, e^x, $\ln(x)$, $\sin(x)$, $\cos(x)$, etc. When you provide an x-coordinate and use one of these symbols, the calculator finds the appropriate y-coordinate. The ordered pairs certainly satisfy the definition of function because the calculator will not produce two different second coordinates corresponding to one first coordinate.

In the next example we determine whether a relation is a function when the relation is given as a list of ordered pairs or as a table.

Example **3** **Identifying a function from a list or table**

Determine whether each relation is a function.

a. {(1, 3), (2, 3), (4, 9)}   **b.** {(9, −3), (9, 3), (4, 2), (0, 0)}

c.

Quantity	Price each
1–5	$9.40
5–10	$8.75

Solution

a. This set of ordered pairs is a function because no two ordered pairs have the same first coordinate and different second coordinates. The second coordinate is a function of the first coordinate.

b. This set of ordered pairs is not a function because both (9, 3) and (9, −3) are in the set and they have the same first coordinate and different second coordinates. The second coordinate is not a function of the first coordinate.

c. The quantity 5 corresponds to a price of $9.40 and also to a price of $8.75. Assuming that quantity is the first coordinate, the ordered pairs (5, $9.40) and (5, $8.75) both belong to this relation. If you are purchasing items whose price was determined from this table, you would certainly say that something is wrong, the table has a mistake in it, or the price you should pay is not clear. The price is not a function of the quantity purchased.

Try This. Determine whether each relation is a function.

a. {(4, 5), (5, 5), (5, 7)}   **b.**

Time (minutes)	Cost ($)
0–30	40
31–60	75

■

We have seen and used many functions as formulas. For example, the formula $c = \pi d$ defines a set of ordered pairs in which the first coordinate is the diameter of a circle and the second coordinate is the circumference. Since each diameter corresponds to a unique circumference, the circumference is a function of the diameter. If the set of ordered pairs satisfying an equation is a function, then we say that the equation is a function or the equation defines a function. Other well-known formulas such as $C = \frac{5}{9}(F - 32)$, $A = \pi r^2$, and $V = \frac{4}{3}\pi r^3$ are also functions, but, as we will see in the next example, not every equation defines a function. The variables in the next example and all others in this text represent real numbers unless indicated otherwise.

Example 4 Identifying a function from an equation

Determine whether each equation defines y as a function of x.

a. $|y| = x$ **b.** $y = x^2 - 3x + 2$ **c.** $x^2 + y^2 = 1$ **d.** $3x - 4y = 8$

Solution

a. We must determine whether there are any values of x for which there is more than one y-value satisfying the equation. We arbitrarily select a number for x and see. If we select $x = 2$, then the equation is $|y| = 2$, which is satisfied if $y = \pm 2$. So both $(2, 2)$ and $(2, -2)$ satisfy $|y| = x$. So this equation does *not* define y as a function of x.

b. If we select any number for x, then y is calculated by the equation $y = x^2 - 3x + 2$. Since there is only one result when $x^2 - 3x + 2$ is calculated, there is only one y for any given x. So this equation *does* define y as a function of x.

c. Is it possible to pick a number for x for which there is more than one y-value? If we select $x = 0$, then the equation is $0^2 + y^2 = 1$ or $y = \pm 1$. So $(0, 1)$ and $(0, -1)$ both satisfy $x^2 + y^2 = 1$ and this equation does *not* define y as a function of x. Note that $x^2 + y^2 = 1$ is equivalent to $y = \pm\sqrt{1 - x^2}$, which indicates that there are many values for x that would produce two different y-coordinates.

d. The equation $3x - 4y = 8$ is equivalent to $y = \frac{3}{4}x - 2$. Since there is only one result when $\frac{3}{4}x - 2$ is calculated, there is only one y corresponding to any given x. So $3x - 4y = 8$ *does* define y as a function of x.

Try This. Determine whether $x^3 + y^2 = 0$ defines y as a function of x. ■

Domain and Range

A relation is a set of ordered pairs. The **domain** of a relation is the set of all first coordinates of the ordered pairs. The **range** of a relation is the set of all second coordinates of the ordered pairs. The relation

$$\{(19, 2.4), (27, 3.0), (19, 3.6), (22, 2.4), (36, 3.8)\}$$

shows the ages and grade point averages of five randomly selected students. The domain of this relation is the set of ages, $\{19, 22, 27, 36\}$. The range is the set of grade point averages, $\{2.4, 3.0, 3.6, 3.8\}$. This relation matches elements of the domain (ages) with elements of the range (grade point averages), as shown in Fig. 2.4. Is this relation a function?

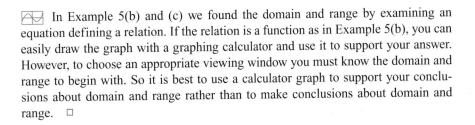

■ **Figure 2.4**

For some relations, all of the ordered pairs are listed, but for others, only an equation is given for determining the ordered pairs. When the domain of the relation is not stated, it is understood that the domain consists of only values of the independent variable that can be used in the expression defining the relation. When we use x and y for the variables, we always assume that x is the independent variable and y is the dependent variable.

Example **5** Determining domain and range

State the domain and range of each relation and whether the relation is a function.

a. $\{(-1, 1), (3, 9), (3, -9)\}$ **b.** $y = \sqrt{2x - 1}$ **c.** $x = |y|$

Solution

a. The domain is the set of first coordinates $\{-1, 3\}$, and the range is the set of second coordinates $\{1, 9, -9\}$. Note that an element of a set is not listed more than once. Since $(3, 9)$ and $(3, -9)$ are in the relation, the relation is not a function.

b. Since y is determined uniquely from x by the formula $y = \sqrt{2x - 1}$, y is a function of x. We are discussing only real numbers. So $\sqrt{2x - 1}$ is a real number provided that $2x - 1 \geq 0$, or $x \geq 1/2$. So the domain of the function is the interval $[1/2, \infty)$. If $2x - 1 \geq 0$ and $y = \sqrt{2x - 1}$, we have $y \geq 0$. So the range of the function is the interval $[0, \infty)$.

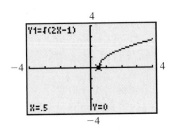

The graph of $y = \sqrt{2x - 1}$ shown in Fig. 2.5 supports these answers because the points that are plotted appear to have $x \geq 1/2$ and $y \geq 0$.   □

c. The expression $|y|$ is defined for any real number y. So the range is the interval of all real numbers, $(-\infty, \infty)$. Since $|y|$ is nonnegative, the values of x must be nonnegative. So the domain is $[0, \infty)$. Since ordered pairs such as $(2, 2)$ and $(2, -2)$ satisfy $x = |y|$, this equation does not give y as a function of x.

Try This. Determine whether $y = \sqrt{x + 3}$ is a function and find its domain and range.   ■

■ **Figure 2.5**

In Example 5(b) and (c) we found the domain and range by examining an equation defining a relation. If the relation is a function as in Example 5(b), you can easily draw the graph with a graphing calculator and use it to support your answer. However, to choose an appropriate viewing window you must know the domain and range to begin with. So it is best to use a calculator graph to support your conclusions about domain and range rather than to make conclusions about domain and range.   □

Function Notation

A function defined by a set of ordered pairs can be named with a letter. For example,

$$f = \{(2, 5), (3, 8)\}.$$

Since the function f pairs 2 with 5 we write $f(2) = 5$, which is read as "the value of f at 2 is 5" or simply "f of 2 is 5." We also have $f(3) = 8$.

A function defined by an equation can also be named with a letter. For example, the function $y = x^2$ could be named by a new letter, say g. We can then use $g(x)$, read "g of x" as a symbol for the second coordinate when the first coordinate is x. Since y and $g(x)$ are both symbols for the second coordinate we can write $y = g(x)$ and $g(x) = x^2$. Since $3^2 = 9$, the function g pairs 3 with 9 and we write $g(3) = 9$. This notation is called **function notation.**

Example **6** **Using function notation**

Let $h = \{(1, 4), (6, 0), (7, 9)\}$ and $f(x) = \sqrt{x - 3}$. Find each of the following.

a. $h(7)$ **b.** w, if $h(w) = 0$ **c.** $f(7)$ **d.** x, if $f(x) = 5$

Solution

a. The expression $h(7)$ is the second coordinate when the first coordinate is 7 in the function named h. So $h(7) = 9$.

b. We are looking for a number w for which $h(w) = 0$. That is, the second coordinate is 0 for some unknown first coordinate w. By examining the function h we see that $w = 6$.

c. To find $f(7)$ replace x by 7 in $f(x) = \sqrt{x - 3}$:

$$f(7) = \sqrt{7 - 3} = \sqrt{4} = 2$$

d. To find x for which $f(x) = 5$ we replace $f(x)$ by 5 in $f(x) = \sqrt{x - 3}$:

$$5 = \sqrt{x - 3}$$
$$25 = x - 3$$
$$28 = x$$

Try This. Let $f(x) = x - 3$. **a.** Find $f(4)$. **b.** Find x if $f(x) = 9$. ■

Function notation such as $f(x) = 3x + 1$ provides a rule for finding the second coordinate: Multiply the first coordinate (whatever it is) by 3 and then add 1. The x in this notation is called a **dummy variable** because the letter used is unimportant. We could write $f(t) = 3t + 1$,

$$f(\text{first coordinate}) = 3(\text{first coordinate}) + 1,$$

or even $f(\ ) = 3(\ ) + 1$ to convey the same idea. Whatever appears in the parentheses following f must be used in place of x on the other side of the equation.

Example **7** **Using function notation with variables**

Given that $f(x) = x^2 - 2$ and $g(x) = 2x - 3$, find and simplify each of the following expressions.

a. $f(a)$ **b.** $f(a + 1)$ **c.** $g(x - 2)$ **d.** $g(x + h)$

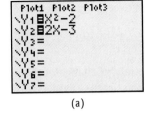

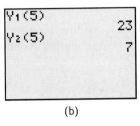

(a)

(b)

■ **Figure 2.6**

Solution

a. Replace x by a in $f(x) = x^2 - 2$ to get $f(a) = a^2 - 2$.

b. $f(a + 1) = (a + 1)^2 - 2$ Replace x by $a + 1$ in $f(x) = x^2 - 2$.

$\qquad\qquad\ = a^2 + 2a + 1 - 2$

$\qquad\qquad\ = a^2 + 2a - 1$

c. $g(x - 2) = 2(x - 2) - 3$ Replace x by $x - 2$ in $g(x) = 2x - 3$.

$\qquad\qquad\quad = 2x - 7$

d. $g(x + h) = 2(x + h) - 3$

$\qquad\qquad\quad = 2x + 2h - 3$

Try This. Let $f(x) = x^2 - 4$. Find and simplify $f(x + 2)$.   ■

A graphing calculator uses subscripts to indicate different functions. For example, if $y_1 = x^2 - 2$ and $y_2 = 2x - 3$, then $y_1(5) = 23$ and $y_2(5) = 7$ as shown in Fig. 2.6(a) and (b).   □

    If a function describes some real application, then a letter that fits the situation is usually used. For example, if watermelons are \$3 each, then the cost of x watermelons is given by the function $C(x) = 3x$. The cost of five watermelons is $C(5) = 3 \cdot 5 = \$15$. In trigonometry the abbreviations sin, cos, and tan are used rather than a single letter to name the trigonometric functions. The dependent variables are written as $\sin(x)$, $\cos(x)$, and $\tan(x)$.

The Average Rate of Change of a Function

In Section 1.4 we defined the slope of the line through (x_1, y_1) and (x_2, y_2) as $\frac{y_2 - y_1}{x_2 - x_1}$. We now extend that idea to any function (linear or not).

Definition: Average Rate of Change from x_1 to x_2

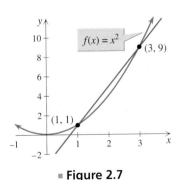

■ **Figure 2.7**

> If (x_1, y_1) and (x_2, y_2) are two ordered pairs of a function, we define the **average rate of change** of the function as x varies from x_1 to x_2 as
>
> $$\frac{y_2 - y_1}{x_2 - x_1}.$$

Note that the x-values can be specified with interval notation. For example, the average rate of change $f(x) = x^2$ on $[1, 3]$ is found as follows:

$$\frac{f(3) - f(1)}{3 - 1} = \frac{9 - 1}{3 - 1} = 4$$

The average rate of change is simply the slope of the line that passes through two points on the graph of the function as shown in Fig. 2.7.

    It is not necessary to have a formula for a function to find an average rate of change, as is shown in the next example.

Example **8** Finding the average rate of change

The population of California was 29.8 million in 1990 and 33.9 million in 2000 (U.S. Census Bureau, www.census.gov). What was the average rate of change of the population over that time interval?

Solution

The population is a function of the year. The average rate of change of the population is the change in population divided by the change in time:

$$\frac{33.9 - 29.8}{2000 - 1990} = \frac{4.1}{10} = 0.41$$

The average rate of change of the population was 0.41 million people/year or 410,000 people/year. Note that 410,000 people/year is an average and that the population did not actually increase by 410,000 every year.

Try This. A BMW was purchased for $28,645 in 2000 and sold for $13,837 in 2006. What was the average rate of change of the car's value for that time interval?

■

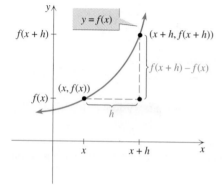

■ **Figure 2.8**

The expression $\dfrac{f(x + h) - f(x)}{h}$ is called the **difference quotient.** The difference quotient is the average rate of change between two points that are labeled as shown in Fig. 2.8. In calculus it is often necessary to find and simplify the difference quotient for a function.

Example **9** Finding a difference quotient

Find and simplify the difference quotient for each of the following functions.

a. $j(x) = 3x + 2$ **b.** $f(x) = x^2 - 2x$ **c.** $g(x) = \sqrt{x}$ **d.** $y = \dfrac{5}{x}$

Solution

a.
$$\frac{j(x + h) - j(x)}{h} = \frac{3(x + h) + 2 - (3x + 2)}{h}$$

$$= \frac{3x + 3h + 2 - 3x - 2}{h}$$

$$= \frac{3h}{h} = 3$$

b.
$$\frac{f(x + h) - f(x)}{h} = \frac{[(x + h)^2 - 2(x + h)] - (x^2 - 2x)}{h}$$

$$= \frac{x^2 + 2xh + h^2 - 2x - 2h - x^2 + 2x}{h}$$

$$= \frac{2xh + h^2 - 2h}{h}$$

$$= 2x + h - 2$$

c. $\dfrac{g(x + h) - g(x)}{h} = \dfrac{\sqrt{x + h} - \sqrt{x}}{h}$

$= \dfrac{\left(\sqrt{x + h} - \sqrt{x}\right)\left(\sqrt{x + h} + \sqrt{x}\right)}{h\left(\sqrt{x + h} + \sqrt{x}\right)}$ Rationalize the numerator.

$= \dfrac{x + h - x}{h(\sqrt{x + h} + \sqrt{x})}$

$= \dfrac{1}{\sqrt{x + h} + \sqrt{x}}$

d. Use the function notation $f(x) = \frac{5}{x}$ for the function $y = \frac{5}{x}$:

$\dfrac{f(x + h) - f(x)}{h} = \dfrac{\dfrac{5}{x + h} - \dfrac{5}{x}}{h} = \dfrac{\left(\dfrac{5}{x + h} - \dfrac{5}{x}\right)x(x + h)}{h \cdot x(x + h)}$

$= \dfrac{5x - 5(x + h)}{hx(x + h)} = \dfrac{-5h}{hx(x + h)} = \dfrac{-5}{x(x + h)}$

Try This. Find and simplify the difference quotient for $f(x) = x^2 - x$. ■

Note that in Examples 9(b), 9(c), and 9(d) the average rate of change of the function depends on the values of x and h, while in Example 9(a) the average rate of change of the function is constant. In Example 9(c) the expression does not look much different after rationalizing the numerator than it did before. However, we did remove h as a factor of the denominator, and in calculus it is often necessary to perform this step.

Constructing Functions

In the next example we find a formula for, or **construct,** a function relating two variables in a geometric figure.

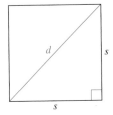

■ **Figure 2.9**

Example **10** Constructing a function

Given that a square has diagonal of length d and side of length s, write the area A as a function of the length of the diagonal.

Solution

The area of any square is given by $A = s^2$. The diagonal is the hypotenuse of a right triangle as shown in Fig. 2.9. By the Pythagorean theorem, $d^2 = s^2 + s^2, d^2 = 2s^2$, or $s^2 = d^2/2$. Since $A = s^2$ and $s^2 = d^2/2$, we get the formula

$$A = \dfrac{d^2}{2}$$

expressing the area of the square as a function of the length of the diagonal.

Try This. A square has perimeter P and sides of length s. Write the side as a function of the perimeter. ▪

For Thought

True or False? Explain.

1. Any set of ordered pairs is a function.

2. If $f = \{(1, 1), (2, 4), (3, 9)\}$, then $f(5) = 25$.

3. The domain of $f(x) = 1/x$ is $(-\infty, 0) \cup (0, \infty)$.

4. Each student's exam grade is a function of the student's IQ.

5. If $f(x) = x^2$, then $f(x + h) = x^2 + h$.

6. The domain of $g(x) = |x - 3|$ is $[3, \infty)$. *false*

7. The range of $y = 8 - x^2$ is $(-\infty, 8]$.

8. The equation $x = y^2$ does not define y as a function of x.

9. If $f(t) = \dfrac{t - 2}{t + 2}$, then $f(0) = -1$.

10. The set $\left\{\left(\frac{3}{8}, 8\right), \left(\frac{4}{7}, 7\right), \left(0.16, 6\right), \left(\frac{3}{8}, 5\right)\right\}$ is a function.

2.1 Exercises

For each pair of variables determine whether a is a function of b, b is a function of a, or neither.

1. a is the radius of any U.S. coin and b is its circumference.

2. a is the length of any rectangle with a width of 5 in. and b is its perimeter.

3. a is the length of any piece of U.S. paper currency and b is its denomination.

4. a is the diameter of any U.S. coin and b is its value.

5. a is the universal product code for an item at Wal-Mart and b is its price.

6. a is the final exam score for a student in your class and b is his/her semester grade.

7. a is the time spent studying for the final exam for a student in your class and b is the student's final exam score.

8. a is the age of an adult male and b is his shoe size.

9. a is the height of a car in inches and b is its height in centimeters.

10. a is the cost for mailing a first-class letter and b is its weight.

Use the vertical line test on each graph in Exercises 11–16 to determine whether y is a function of x.

11.

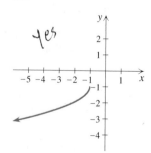

12.

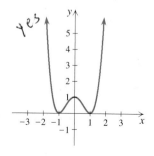

13.

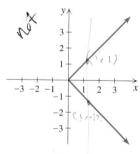

14.

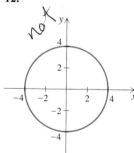

15. *yes*

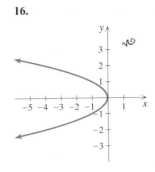

16. *no*

Determine whether each relation is a function.

17. $\{(-1, -1), (2, 2), (3, 3)\}$ *yes*

18. $\{(0.5, 7), (0, 7), (1, 7), (9, 7)\}$ *yes*

19. $\{(25, 5), (25, -5), (0, 0)\}$ *not*

20. $\{(1, \pi), (30, \pi/2), (60, \pi/4)\}$ *yes*

21.

x	y
3	6
4	9
3	12

no

22.

x	y
1	6.98
5	5.98
9	6.98

yes

23.

x	y
-1	1
1	1
-5	1
5	1

yes

24.

x	y
1	1
2	4
3	9
4	16

yes

Determine whether each equation defines y as a function of x.

25. $y = 3x - 8$

26. $y = x^2 - 3x + 7$

27. $x = 3y - 9$

28. $x = y^3$

29. $x^2 = y^2$

30. $y^2 - x^2 = 9$

31. $x = \sqrt{y}$

32. $x = \sqrt[3]{y}$

33. $y + 2 = |x|$

34. $y - 1 = x^2$

35. $x = |2y|$

36. $x = y^2 + 1$

Determine the domain and range of each relation.

37. $\{(-3, 1), (4, 2), (-3, 6), (5, 6)\}$

38. $\{(1, 2), (2, 4), (3, 8), (4, 16)\}$

39. $\{(x, y) | y = 4\}$

40. $\{(x, y) | x = 5\}$

41. $y = |x| + 5$

42. $y = x^2 + 8$

43. $x + 3 = |y|$

44. $x + 2 = \sqrt{y}$

45. $y = \sqrt{x - 4}$

46. $y = \sqrt{5 - x}$

47. $x = -y^2$

48. $x = -|y|$

Let $f = \{(2, 6), (3, 8), (4, 5)\}$ and $g(x) = 3x + 5$. Find the following.

49. $f(2)$

50. $f(4)$

51. $g(2)$

52. $g(4)$

53. x, if $f(x) = 8$

54. x, if $f(x) = 6$

55. x, if $g(x) = 26$

56. x, if $g(x) = -4$

57. $f(4) + g(4)$

58. $f(3) - g(3)$

Let $f(x) = 3x^2 - x$ and $g(x) = 4x - 2$. Find the following.

59. $f(a)$

60. $f(w)$

61. $g(a + 2)$

62. $g(a - 5)$

63. $f(x + 1)$

64. $f(x - 3)$

65. $g(x + h)$

66. $f(x + h)$

67. $f(x + h) - f(x)$

68. $g(x + h) - g(x)$

The following problems involve average rate of change.

69. *Depreciation of a Mustang* If a new Mustang is valued at $16,000 and five years later it is valued at $4000, then what is the average rate of change of its value during those five years?

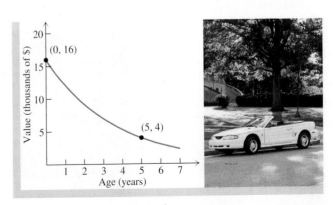

■ **Figure for Exercise 69**

70. *Cost of Gravel* Wilson's Sand and Gravel will deliver 12 yd³ of gravel for $240, 30 yd³ for $528, and 60 yd³ for $948. What is the average rate of change of the cost as the number of cubic yards varies from 12 to 30? What is the average rate of change as the number of cubic yards varies from 30 to 60?

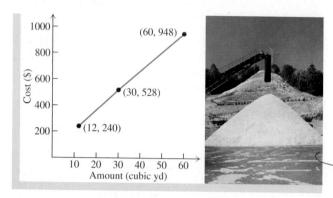

■ **Figure for Exercise 70**

71. *Dropping a Watermelon* If a comedian drops a watermelon from a height of 64 ft, then its height (in feet) above the ground is given by the function $h(t) = -16t^2 + 64$ where t is time (in seconds). To get an idea of how fast the watermelon is traveling when it hits the ground find the average rate of change of the height on each of the time intervals $[0, 2]$, $[1, 2]$, $[1.9, 2]$, $[1.99, 2]$, and $[1.999, 2]$.

72. *Bungee Jumping* Billy Joe McCallister jumped off the Tallahatchie Bridge, 70 ft above the water, with a bungee cord tied to his legs. If he was 6 ft above the water 2 sec after jumping, then what was the average rate of change of his altitude as the time varied from 0 to 2 sec?

73. *Deforestation* In 1988 the world's tropical moist forest covered approximately 1056 million hectares (1 hectare = 10,000 m²). In 2004 the world's tropical moist forest covered approximately 896 million hectares. What was the average rate of change of the area of tropical moist forest over those 16 years?

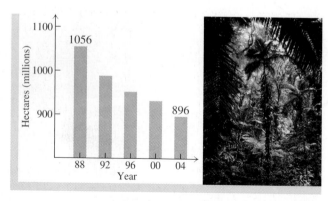

■ **Figure for Exercises 73 and 74**

74. *Elimination* If the deforestation described in the previous exercise continues at the same rate, then in which year will the tropical moist forest be totally eliminated?

Find the difference quotient $\dfrac{f(x + h) - f(x)}{h}$ for each function and simplify it.

75. $f(x) = 4x$

76. $f(x) = \dfrac{1}{2}x$

77. $f(x) = 3x + 5$

78. $f(x) = -2x + 3$

79. $y = x^2 + x$

80. $y = x^2 - 2x$

81. $y = -x^2 + x - 2$

82. $y = x^2 - x + 3$

83. $g(x) = 3\sqrt{x}$

84. $g(x) = -2\sqrt{x}$

85. $f(x) = \sqrt{x + 2}$

86. $f(x) = \sqrt{\dfrac{x}{2}}$

87. $g(x) = \dfrac{1}{x}$

88. $g(x) = \dfrac{3}{x}$

89. $g(x) = \dfrac{3}{x + 2}$

90. $g(x) = 3 + \dfrac{2}{x - 1}$

Solve each problem.

91. *Constructing Functions* Consider a square with side of length s, diagonal of length d, perimeter P, and area A.
a. Write A as a function of s. **b.** Write s as a function of A.

c. Write s as a function of d. **d.** Write d as a function of s.

e. Write P as a function of s. **f.** Write s as a function of P.

g. Write A as a function of P. **h.** Write d as a function of A.

92. *Constructing Functions* Consider a circle with area A, circumference C, radius r, and diameter d.
a. Write A as a function of r. **b.** Write r as a function of A.

c. Write C as a function of r. **d.** Write d as a function of r.

e. Write d as a function of C. **f.** Write A as a function of d.

g. Write d as a function of A.

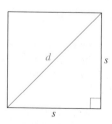 

■ **Figure for Exercise 91** ■ **Figure for Exercise 92**

93. *Cost of Window Cleaning* A window cleaner charges $50 per visit plus $35 per hour. Express the total charge as a function of the number of hours worked, *n*.

94. *Below Sea Level* The accompanying table shows the depth below sea level *d* and the atmospheric pressure *A* (www.sportsfigures.espn.com). The equation $A(d) = 0.03d + 1$ expresses *A* as a function of *d*.
   a. Find the atmospheric pressure for a depth of 100 ft, where nitrogen narcosis begins.

   b. Find the depth at which the pressure is 4.9 atm, the maximum depth for intermediate divers.

■ **Table for Exercise 94**

Depth (ft)	Atmospheric Pressure (atm)	
21	1.63	
60	2.8	
100		
	4.9	
200	7.0	
250	8.5	

95. *Computer Spending* The amount spent online for computers in the year 2000 + *n* can be modeled by the function $C(n) = 0.95n + 5.8$ where *n* is a whole number and $C(n)$ is billions of dollars.
   a. What does $C(4)$ represent and what is it?

   b. Find the year in which online spending for computers will reach $15 billion?

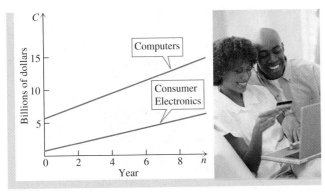

■ **Figure for Exercises 95 and 96**

96. *Electronics Spending* The amount spent online for consumer electronics (excluding computers) in the year 2000 + *n* can be modeled by the function $E(n) = 0.5n + 1$ where *n* is a whole number and $E(n)$ is billions of dollars.

   a. What does $E(4) + C(4)$ represent and what is it?

   b. Find the year in which the total spending for computers and electronics will reach $20 billion?

   c. In which category is spending growing faster? See the accompanying figure.

97. *Pile of Pipes* Six pipes, each with radius *a*, are stacked as shown in the accompanying figure. Construct a function that gives the height *h* of the pile in terms of *a*.
   HINT Connect the centers of three circles to form an equilateral triangle.

■ **Figure for Exercise 97**

98. *Angle Bisectors* The angle bisectors of any triangle meet at a single point. Let *a* be the length of the hypotenuse of a 30-60-90 triangle and *d* be the distance from the vertex of the right angle to the point where the angle bisectors meet. Write *d* as function of *a*.
   HINT Draw a diagram and label as many sides and angles as you can.

99. *Concert Revenue* The revenue in dollars from the sale of concert tickets at *x* dollars each is given by the function

$$R(x) = 20,000x - 500x^2.$$

Find the difference quotient when $x = 18$ and $h = 0.1$ and when $x = 22$ and $h = 0.1$. Interpret your answers.

100. *Surface Area* The amount of tin *A* (in square inches) needed to make a tin can with radius *r* inches and volume 36 in.³ can be found by the function

$$A(r) = \frac{72}{r} + 2\pi r^2.$$

Find the difference quotient when $r = 1.4$ and $h = 0.1$ and when $r = 2$ and $h = 0.1$. Interpret your answers.

For Writing/Discussion

101. *Find a Function*   Give an example of a function and an example of a relation that is not a function, from situations that you have encountered outside of this textbook.

102. *Cooperative Learning*   Work in a small group to consider the equation $y^n = x^m$ for any integers n and m. For which integers n and m does the equation define y as a function of x?

Thinking Outside the Box XXI

Lucky Lucy   Lucy's teacher asked her to evaluate $(20 + 25)^2$. As she was trying to figure out what to do she mumbled, "twenty twenty-five." Her teacher said, "Good, 2025 is correct." Find another pair of two digit whole numbers for which the square of their sum can be found by Lucy's method.

2.1 Pop Quiz

1. Is the radius of a circle a function of its area?

2. Is $\{(2, 4), (1, 8), (2, -4)\}$ a function?

3. Does $x^2 + y^2 = 1$ define y as a function of x?

4. What is the domain of $y = \sqrt{x - 1}$?

5. What is the range of $y = x^2 + 2$?

6. What is $f(2)$ if $f = \{(1, 8), (2, 9)\}$?

7. What is a if $f(a) = 1$ and $f(x) = 2x$?

8. If the cost was $20 in 1998 and $40 in 2008, then what is the average rate of change of the cost for that time period?

9. Find and simplify the difference quotient for $f(x) = x^2 + 3$.

Linking Concepts

For Individual or Group Explorations

Modeling Debt and Population Growth

The following table gives the U.S. federal debt in billions of dollars as a function of the year and the population in millions of people as a function of the year (U.S. Treasury Department, www.treas.gov). Do parts (a) through (e) for each function.

Year	Debt	Pop
1940	51	131.7
1950	257	150.7
1960	291	179.3
1970	381	203.3
1980	909	226.5
1990	3207	248.7
2000	5666	274.8

a)   Draw an accurate graph of the function.

b)   Find the average rate of change of the function over each ten-year period.

c)   Take the average rate of change for each ten-year period (starting with 1950–1960) and subtract from it the average rate of change for the previous ten-year period.

d) Are the average rates of change for the function positive or negative?

e) Are the answers to part (c) mostly positive or mostly negative?

f) Judging from the graphs and the average rates of change, which is growing out of control, the federal debt or the population?

g) Explain the relationship between your answer to part (f) and your answer to part (e).

2.2

Graphs of Relations and Functions

When we graph the set of ordered pairs that satisfy an equation, we are combining algebra with geometry. We saw in Section 1.3 that the graph of any equation of the form $(x - h)^2 + (y - k)^2 = r^2$ is a circle and that the graph of any equation of the form $Ax + By = C$ is a straight line. In this section we will see that graphs of equations have many different geometric shapes.

Graphing Equations

The circle and the line provide nice examples of how algebra and geometry are interrelated. When you see an equation that you recognize as the equation of a circle or a line, sketching a graph is easy to do. Other equations have graphs that are not such familiar shapes. Until we learn to recognize the kinds of graphs that other equations have, we graph other equations by calculating enough ordered pairs to determine the shape of the graph. When you graph equations, try to anticipate what the graph will look like, and after the graph is drawn, pause to reflect on the shape of the graph and the type of equation that produced it.

◪◪ Of course a graphing calculator can speed up this process. Remember that a graphing calculator shows only finitely many points and a graph usually consists of infinitely many points. After looking at the display of a graphing calculator, you must still decide what the entire graph looks like.   □

Example 1 The square function

Graph the equation $y = x^2$ and state the domain and range. Determine whether the relation is a function.

Solution

Make a table of ordered pairs that satisfy $y = x^2$:

x is any real number

x	0	1	-1	2	-2
$y = x^2$	0	1	1	4	4

$y \geq 0$

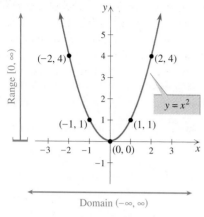

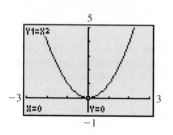

■ **Figure 2.10**

■ **Figure 2.11**

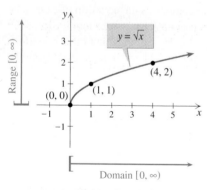

■ **Figure 2.12**

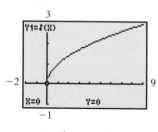

■ **Figure 2.13**

These ordered pairs indicate a graph in the shape shown in Fig. 2.10. The domain is $(-\infty, \infty)$ because any real number can be used for x in $y = x^2$. Since all y-coordinates are nonnegative, the range is $[0, \infty)$. Because no vertical line crosses this curve more than once, $y = x^2$ is a function.

The calculator graph shown in Fig. 2.11 supports these conclusions.

Try This. Determine whether $y = \frac{1}{2}x^2$ is a function, graph it, and state the domain and range. ■

The graph of $y = x^2$ is called a **parabola**. We study parabolas in detail in Chapter 3. The graph of the **square-root function** $y = \sqrt{x}$ is half of a parabola.

Example **2** The square-root function

Graph $y = \sqrt{x}$ and state the domain and range of the relation. Determine whether the relation is a function.

Solution

Make a table listing ordered pairs that satisfy $y = \sqrt{x}$:

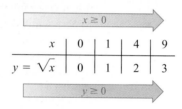

	$x \geq 0$			
x	0	1	4	9
$y = \sqrt{x}$	0	1	2	3
	$y \geq 0$			

Plotting these ordered pairs suggests the graph shown in Fig. 2.12. The domain of the relation is $[0, \infty)$ and the range is $[0, \infty)$. Because no vertical line can cross this graph more than once, $y = \sqrt{x}$ is a function.

The calculator graph in Fig. 2.13 supports these conclusions.

Try This. Determine whether $y = \sqrt{1 - x}$ is a function, graph it, and state the domain and range. ■

In the next example we graph $x = y^2$ and see that its graph is also a parabola.

Example **3** A parabola opening to the right

Graph $x = y^2$ and state the domain and range of the relation. Determine whether the relation is a function.

Solution

Make a table listing ordered pairs that satisfy $x = y^2$. In this case choose y and calculate x:

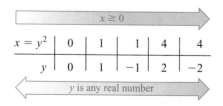

	$x \geq 0$				
$x = y^2$	0	1	1	4	4
y	0	1	-1	2	-2
	y is any real number				

Note that these ordered pairs are the same ones that satisfy $y = x^2$ except that the coordinates are reversed. For this reason the graph of $x = y^2$ in Fig. 2.14 has the same shape as the parabola in Fig. 2.10 and it is also a parabola. The domain of $x = y^2$ is $[0, \infty)$ and the range is $(-\infty, \infty)$. Because we can draw a vertical line that crosses this parabola twice, $x = y^2$ does not define y as a function of x. (Of course, $x = y^2$ does express x as a function of y.)

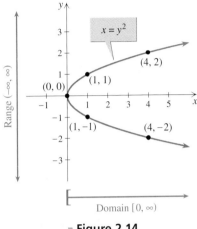

■ **Figure 2.14**

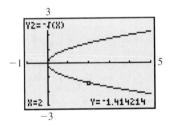

■ **Figure 2.15**

Try This. Determine whether $x = -y^2$ is a function, graph it, and state the domain and range.   ■

Because $x = y^2$ is equivalent to $y = \pm\sqrt{x}$, the top half of the graph of $x = y^2$ is $y = \sqrt{x}$ and the bottom half is $y = -\sqrt{x}$.

To support these conclusions with a graphing calculator, graph $y_1 = \sqrt{x}$ and $y_2 = -\sqrt{x}$ as shown in Fig. 2.15. □

In the next example we graph the **cube-root function** $y = \sqrt[3]{x}$.

Example **4** The cube-root function

Graph the equation $y = \sqrt[3]{x}$ and state the domain and range. Determine whether the relation is a function.

Solution

Make a table listing ordered pairs that satisfy $y = \sqrt[3]{x}$:

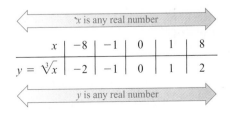

	x is any real number				
x	-8	-1	0	1	8
$y = \sqrt[3]{x}$	-2	-1	0	1	2
	y is any real number				

These ordered pairs indicate a graph in the shape shown in Fig. 2.16. The domain is $(-\infty, \infty)$ and the range is $(-\infty, \infty)$. Because no vertical line crosses this graph more than once, $y = \sqrt[3]{x}$ is a function.

The calculator graph in Fig. 2.17 supports these conclusions.

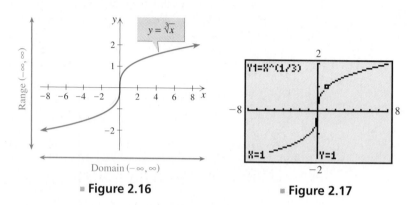

■ **Figure 2.16**   ■ **Figure 2.17**

Try This. Determine whether $y = -\sqrt[3]{x}$ is a function, graph it, and state the domain and range.   ■

Semicircles

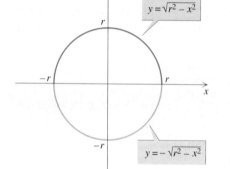

■ **Figure 2.18**

The graph of $x^2 + y^2 = r^2$ $(r > 0)$ is a circle centered at the origin of radius r. The circle does not pass the vertical line test, and a circle is not the graph of a function. We can find an equivalent equation by solving for y:

$$x^2 + y^2 = r^2$$
$$y^2 = r^2 - x^2$$
$$y = \pm\sqrt{r^2 - x^2}$$

The equation $y = \sqrt{r^2 - x^2}$ does define y as a function of x. Because y is nonnegative in this equation, the graph is the top semicircle in Fig. 2.18. The top semicircle passes the vertical line test. Likewise, the equation $y = -\sqrt{r^2 - x^2}$ defines y as a function of x, and its graph is the bottom semicircle in Fig. 2.18.

Example **5** Graphing a semicircle

Sketch the graph of each function and state the domain and range of the function.

a. $y = -\sqrt{4 - x^2}$ **b.** $y = \sqrt{9 - x^2}$

Solution

a. Rewrite the equation in the standard form for a circle:

$$y = -\sqrt{4 - x^2}$$
$$y^2 = 4 - x^2 \qquad \text{Square each side.}$$
$$x^2 + y^2 = 4 \qquad \text{Standard form for the equation of a circle.}$$

The graph of $x^2 + y^2 = 4$ is a circle of radius 2 centered at (0, 0). Since y must be negative in $y = -\sqrt{4 - x^2}$, the graph of $y = -\sqrt{4 - x^2}$ is the semicircle

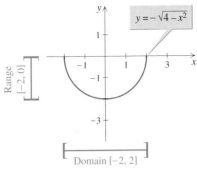

■ **Figure 2.19**

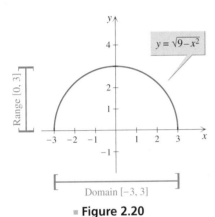

■ **Figure 2.20**

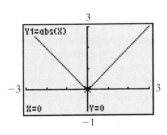

■ **Figure 2.21**

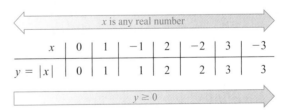

■ **Figure 2.22**

shown in Fig. 2.19. We can see from the graph that the domain is $[-2, 2]$ and the range is $[-2, 0]$.

b. Rewrite the equation in the standard form for a circle:

$$y = \sqrt{9 - x^2}$$

$$y^2 = 9 - x^2 \qquad \text{Square each side.}$$

$$x^2 + y^2 = 9$$

The graph of $x^2 + y^2 = 9$ is a circle with center $(0, 0)$ and radius 3. But this equation is not equivalent to the original. The value of y in $y = \sqrt{9 - x^2}$ is nonnegative. So the graph of the original equation is the semicircle shown in Fig. 2.20. We can read the domain $[-3, 3]$ and the range $[0, 3]$ from the graph.

Try This. Graph $y = -\sqrt{9 - x^2}$ and state the domain and range. ■

Piecewise Functions

For some functions, different formulas are used in different regions of the domain. Such functions are called **piecewise functions.** The simplest example of such a function is the **absolute value function** $f(x) = |x|$, which can be written as

$$f(x) = \begin{cases} x & \text{for } x \geq 0 \\ -x & \text{for } x < 0. \end{cases}$$

For $x \geq 0$ the equation $f(x) = x$ is used to obtain the second coordinate, and for $x < 0$ the equation $f(x) = -x$ is used. The graph of the absolute value function is shown in the next example.

Example **6** **The absolute value function**

Graph the equation $y = |x|$ and state the domain and range. Determine whether the relation is a function.

Solution

Make a table of ordered pairs that satisfy $y = |x|$:

				x is any real number			
x	0	1	-1	2	-2	3	-3
$y = |x|$	0	1	1	2	2	3	3
				$y \geq 0$			

These ordered pairs suggest the V-shaped graph of Fig. 2.21. Because no vertical line can cross this graph more than once, $y = |x|$ is a function. The domain is $(-\infty, \infty)$ and the range is $[0, \infty)$.

To support these conclusions with a graphing calculator, graph $y_1 = \text{abs}(x)$ as shown in Fig. 2.22.

Try This. Determine whether $y = |x| + 2$ is a function, graph it, and state the domain and range. ■

In the next example we graph two more piecewise functions.

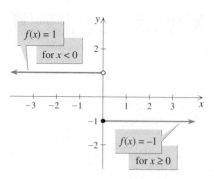

$f(x) = 1$
for $x < 0$

$f(x) = -1$
for $x \geq 0$

■ Figure 2.23

Example **7** Graphing a piecewise function

Sketch the graph of each function and state the domain and range.

a. $f(x) = \begin{cases} 1 & \text{for } x < 0 \\ -1 & \text{for } x \geq 0 \end{cases}$ **b.** $f(x) = \begin{cases} x^2 - 4 & \text{for } -2 \leq x \leq 2 \\ x - 2 & \text{for } x > 2 \end{cases}$

Solution

a. For $x < 0$ the graph is the horizontal line $y = 1$. For $x \geq 0$ the graph is the horizontal line $y = -1$. Note that $(0, -1)$ is on the graph shown in Fig. 2.23 but $(0, 1)$ is not, because when $x = 0$ we have $y = -1$. The domain is the interval $(-\infty, \infty)$ and the range consists of only two numbers, -1 and 1. The range is not an interval. It is written in set notation as $\{-1, 1\}$.

b. Make a table of ordered pairs using $y = x^2 - 4$ for x between -2 and 2 and $y = x - 2$ for $x > 2$.

x	-2	-1	0	1	2
$y = x^2 - 4$	0	-3	-4	-3	0

x	2.1	3	4	5
$y = x - 2$	0.1	1	2	3

For x in the interval $[-2, 2]$ the graph is a portion of a parabola as shown in Fig. 2.24. For $x > 2$, the graph is a portion of a straight line through $(3,1)$, $(4, 2)$, and $(5, 3)$. The domain is $[-2, \infty)$, and the range is $[-4, \infty)$.

Figure 2.25(a) shows how to use the inequality symbols from the TEST menu to enter a piecewise function on a calculator. The inequality $x \geq -2$ is not treated as a normal inequality, but instead the calculator gives it a value of 1 when it is satisfied and 0 when it is not satisfied. So, dividing $x^2 - 4$ by the inequalities $x \geq -2$ and $x \leq 2$ will cause $x^2 - 4$ to be graphed only when they are both satisfied and not graphed when they are not both satisfied. The calculator graph in Fig. 2.25(b) is consistent with the conclusions that we have made.

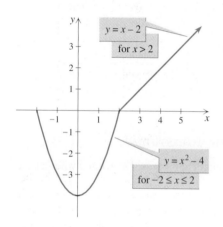

$y = x - 2$
for $x > 2$

$y = x^2 - 4$
for $-2 \leq x \leq 2$

■ Figure 2.24

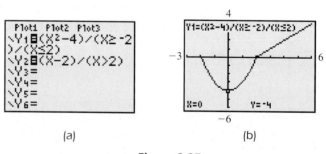

(a)   (b)

■ Figure 2.25

Try This. Graph $y = \begin{cases} x & \text{for } x \geq 0 \\ -2x & \text{for } x < 0 \end{cases}$ and state the domain and range.   ■

Piecewise functions are often found in shipping charges. For example, if the weight in pounds of an order is in the interval $(0, 1]$, the shipping and handling charge is \$3. If the weight is in the interval $(1, 2]$, the shipping and handling charge is \$4, and so on. The next example is a function that is similar to a shipping and handling charge. This function is referred to as the **greatest integer function** and is written $f(x) = [\![x]\!]$ or $f(x) = \text{int}(x)$. The symbol $[\![x]\!]$ is defined to be the largest integer that is less than or equal to x. For example, $[\![5.01]\!] = 5$, because the greatest integer less than or equal to 5.01 is 5. Likewise, $[\![3.2]\!] = 3$, $[\![-2.2]\!] = -3$, and $[\![7]\!] = 7$.

Example 8 Graphing the greatest integer function

Sketch the graph of $f(x) = [\![x]\!]$ and state the domain and range.

Solution

For any x in the interval $[0, 1)$ the greatest integer less than or equal to x is 0. For any x in $[1, 2)$ the greatest integer less than or equal to x is 1. For any x in $[-1, 0)$ the greatest integer less than or equal to x is -1. The definition of $[\![x]\!]$ causes the function to be constant between the integers and to "jump" at each integer. The graph of $f(x) = [\![x]\!]$ is shown in Fig. 2.26. The domain is $(-\infty, \infty)$, and the range is the set of integers.

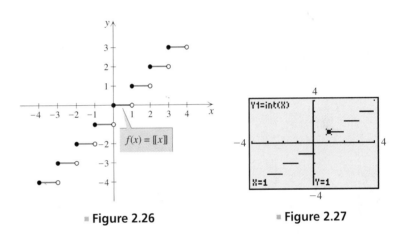

■ Figure 2.26   ■ Figure 2.27

The calculator graph of $y_1 = \text{int}(x)$ looks best in dot mode as in Fig. 2.27, because in connected mode the calculator connects the disjoint pieces of the graph. The calculator graph in Fig. 2.27 supports our conclusion that the graph of this function looks like the one drawn in Fig. 2.26. Note that the calculator is incapable of showing whether the endpoints of the line segments are included.

Try This. Graph $y = -[\![x]\!]$ and state the domain and range.   ■

In the next example we vary the form of the greatest integer function, but the graph is still similar to the graph in Fig. 2.26.

Example 9 A variation of the greatest integer function

Sketch the graph of $f(x) = [\![x - 2]\!]$ for $0 \le x \le 5$.

Solution

If $x = 0$, $f(0) = [\![-2]\!] = -2$. If $x = 0.5$, $f(0.5) = [\![-1.5]\!] = -2$. In fact, $f(x) = -2$ for any x in the interval $[0, 1)$. Similarly, $f(x) = -1$ for any x in the interval $[1, 2)$. This pattern continues with $f(x) = 2$ for any x in the interval $[4, 5)$, and $f(x) = 3$ for $x = 5$. The graph of $f(x) = [\![x - 2]\!]$ is shown in Fig. 2.28.

Try This. Graph $y = [\![x + 2]\!]$ and state the domain and range.   ■

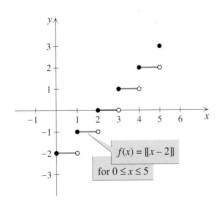

■ Figure 2.28

Increasing, Decreasing, and Constant

Imagine a point moving from left to right along the graph of a function. If the y-coordinate of the point is getting larger, getting smaller, or staying the same, then the function is said to be **increasing, decreasing,** or **constant,** respectively.

Example 10 Increasing, decreasing, or constant functions

Determine whether each function is increasing, decreasing, or constant by examining its graph.

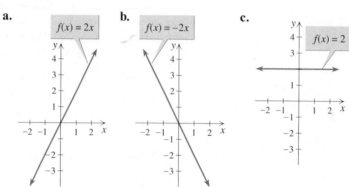

a. $f(x) = 2x$ **b.** $f(x) = -2x$ **c.** $f(x) = 2$

Solution

If a point is moving left to right along the graph of $f(x) = 2x$, then its y-coordinate is increasing. So $f(x) = 2x$ is an increasing function. If a point is moving left to right along the graph of $f(x) = -2x$, then its y-coordinate is decreasing. So $f(x) = -2x$ is a decreasing function. The function $f(x) = 2$ has a constant y-coordinate and it is a constant function.

Try This. Determine whether $f(x) = -3x$ is increasing, decreasing, or constant by examining its graph. ■

If the y-coordinates are getting larger, getting smaller, or staying the same when x is in an open interval, then the function is said to be increasing, decreasing, or constant, respectively, on that interval.

Example 11 Increasing, decreasing, or constant on an interval

Sketch the graph of each function and identify any open intervals on which the function is increasing, decreasing, or constant.

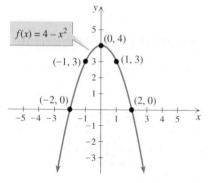

■ **Figure 2.29**

a. $f(x) = 4 - x^2$ **b.** $g(x) = \begin{cases} 0 & \text{for} & x \le 0 \\ \sqrt{x} & \text{for} & 0 < x < 4 \\ 2 & \text{for} & x \ge 4 \end{cases}$

Solution

a. The graph of $f(x) = 4 - x^2$ includes the points $(-2, 0)$, $(-1, 3)$, $(0, 4)$, $(1, 3)$, and $(2, 0)$. The graph is shown in Fig. 2.29. The function is increasing on the interval $(-\infty, 0)$ and decreasing on $(0, \infty)$.

b. The graph of g is shown in Fig. 2.30. The function g is constant on the intervals $(-\infty, 0)$ and $(4, \infty)$, and increasing on $(0, 4)$.

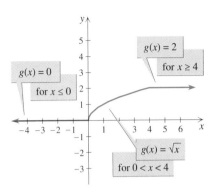

■ **Figure 2.30**

Try This. Graph $f(x) = \begin{cases} x & \text{for} \quad x \geq 0 \\ -2x & \text{for} \quad x < 0 \end{cases}$ and identify any open intervals on which the function is increasing, decreasing, or constant. ■

For Thought

True or False? Explain

1. The range of $y = -x^2$ is $(-\infty, 0]$.

2. The function $y = -\sqrt{x}$ is increasing on $(0, \infty)$.

3. The function $f(x) = \sqrt[3]{x}$ is increasing on $(-\infty, \infty)$.

4. If $f(x) = [\![x + 3]\!]$, then $f(-4.5) = -2$.

5. The range of the function $f(x) = \dfrac{|x|}{x}$ is the interval $(-1, 1)$.

6. The range of $f(x) = [\![x - 1]\!]$ is the set of integers.

7. The only ordered pair that satisfies $(x - 5)^2 + (y + 6)^2 = 0$ is $(5, -6)$.

8. The domain of the function $y = \sqrt{4 - x^2}$ is the interval $[-2, 2]$.

9. The range of the function $y = \sqrt{16 - x^2}$ is the interval $[0, \infty)$.

10. The function $y = \sqrt{4 - x^2}$ is increasing on $(-2, 0)$ and decreasing on $(0, 2)$.

2.2 Exercises

Make a table listing ordered pairs that satisfy each equation. Then graph the equation. Determine the domain and range, and whether y is a function of x.

1. $y = 2x$

2. $x = 2y$

3. $x - y = 0$

4. $x - y = 2$

5. $y = 5$

6. $x = 3$

7. $y = 2x^2$

8. $y = x^2 - 1$

9. $y = 1 - x^2$

10. $y = -1 - x^2$

11. $y = 1 + \sqrt{x}$

12. $y = 2 - \sqrt{x}$

13. $x = y^2 + 1$

14. $x = 1 - y^2$

15. $x = \sqrt{y}$

16. $x - 1 = \sqrt{y}$

17. $y = \sqrt[3]{x} + 1$

18. $y = \sqrt[3]{x} - 2$

19. $x = \sqrt[3]{y}$

20. $x = \sqrt[3]{y} - 1$

21. $y^2 = 1 - x^2$

22. $x^2 + y^2 = 4$

23. $y = \sqrt{1 - x^2}$

24. $y = -\sqrt{25 - x^2}$

25. $y = x^3$

26. $y = -x^3$

27. $y = 2|x|$

28. $y = |x - 1|$

29. $y = -|x|$

30. $y = -|x + 1|$

31. $x = |y|$

32. $x = |y| + 1$

Make a table listing ordered pairs for each function. Then sketch the graph and state the domain and range.

33. $f(x) = \begin{cases} 2 & \text{for } x < -1 \\ -2 & \text{for } x \geq -1 \end{cases}$

34. $f(x) = \begin{cases} 3 & \text{for } x < 2 \\ 1 & \text{for } x \geq 2 \end{cases}$

35. $f(x) = \begin{cases} x + 1 & \text{for } x > 1 \\ x - 3 & \text{for } x \leq 1 \end{cases}$

36. $f(x) = \begin{cases} 5 - x & \text{for } x \leq 2 \\ x + 1 & \text{for } x > 2 \end{cases}$

37. $f(x) = \begin{cases} \sqrt{x + 2} & \text{for } -2 \leq x \leq 2 \\ 4 - x & \text{for } x > 2 \end{cases}$

38. $f(x) = \begin{cases} \sqrt{x} & \text{for } x \geq 1 \\ -x & \text{for } x < 1 \end{cases}$

39. $f(x) = \begin{cases} \sqrt{-x} & \text{for } x < 0 \\ \sqrt{x} & \text{for } x \geq 0 \end{cases}$

40. $f(x) = \begin{cases} 3 & \text{for } x < 0 \\ 3 + \sqrt{x} & \text{for } x \geq 0 \end{cases}$

41. $f(x) = \begin{cases} x^2 & \text{for } x < -1 \\ -x & \text{for } x \geq -1 \end{cases}$

42. $f(x) = \begin{cases} 4 - x^2 & \text{for } -2 \leq x \leq 2 \\ x - 2 & \text{for } x > 2 \end{cases}$

43. $f(x) = [\![x + 1]\!]$

44. $f(x) = 2[\![x]\!]$

45. $f(x) = [\![x]\!] + 2 \quad \text{for } 0 \leq x < 4$

46. $f(x) = [\![x - 3]\!] \quad \text{for } 0 < x \leq 5$

From the graph of each function in Exercises 47–54, state the domain, the range, and the intervals on which the function is increasing, decreasing, or constant.

47. a.   **b.**

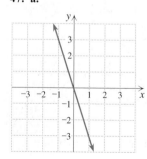
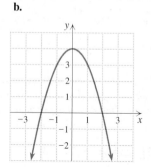

48. a.   **b.**

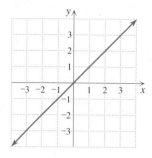
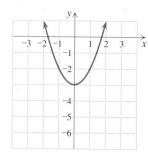

49. a.   **b.**

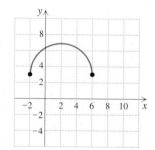
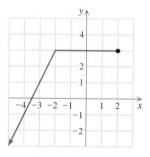

50. a.   **b.**

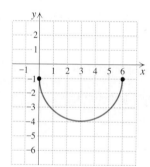
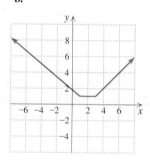

51. a.   **b.**

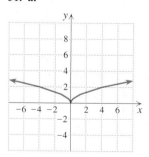
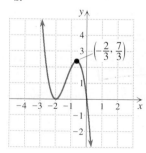

52. a.
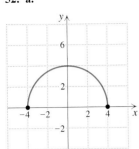

b.
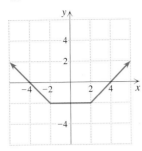

53. a.
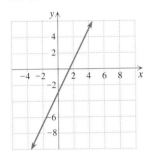

b.
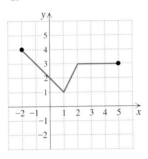

54. a.
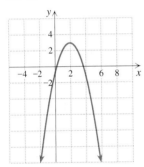

b.
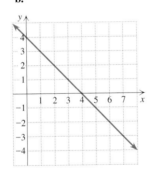

Make a table listing ordered pairs for each function. Then sketch the graph and state the domain and range. Identify any intervals on which f is increasing, decreasing, or constant.

55. $f(x) = 2x + 1$

56. $f(x) = -3x$

57. $f(x) = |x - 1|$

58. $f(x) = |x| + 1$

59. $f(x) = \dfrac{|x|}{x}$

60. $f(x) = \dfrac{2x}{|x|}$

61. $f(x) = \sqrt{9 - x^2}$

62. $f(x) = -\sqrt{1 - x^2}$

63. $f(x) = \begin{cases} x + 1 & \text{for } x \geq 3 \\ x + 2 & \text{for } x < 3 \end{cases}$

64. $f(x) = \begin{cases} \sqrt{-x} & \text{for } x < 0 \\ -\sqrt{x} & \text{for } x \geq 0 \end{cases}$

65. $f(x) = \begin{cases} x + 3 & \text{for } x \leq -2 \\ \sqrt{4 - x^2} & \text{for } -2 < x < 2 \\ -x + 3 & \text{for } x \geq 2 \end{cases}$

66. $f(x) = \begin{cases} 8 + 2x & \text{for } x \leq -2 \\ x^2 & \text{for } -2 < x < 2 \\ 8 - 2x & \text{for } x \geq 2 \end{cases}$

Write a piecewise function for each given graph.

67.
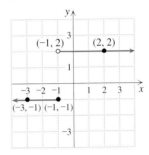

68.
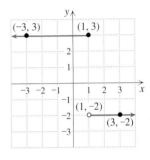

69.
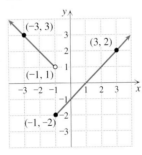

70.
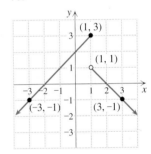

71.
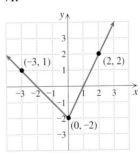

72.
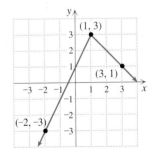

Use the minimum and maximum features of a graphing calculator to find the intervals on which each function is increasing or decreasing. Round approximate answers to two decimal places.

73. $y = 3x^2 - 5x - 4$

74. $y = -6x^2 + 2x - 9$

75. $y = x^3 - 3x$

76. $y = x^4 - 11x^2 + 18$

77. $y = 2x^4 - 12x^2 + 25$

78. $y = x^5 - 13x^3 + 36x$

79. $y = \left| 20 - |x - 50| \right|$

80. $y = x + |x + 30| - |x + 50|$

Select the graph (a, b, c, or d) that best depicts each of the following scenarios.

81. Profits for Cajun Drilling Supplies soared during the seventies, but went flat when the bottom fell out of the oil industry in the eighties. Cajun saw a period of moderate growth in the nineties.

82. To attain a minimum therapeutic level of Flexeril in her blood, Millie takes three Flexeril tablets, one every four hours. Because the half-life of Flexeril is four hours, one-half of the Flexeril in her blood is eliminated after four hours.

83. The bears dominated the market during the first quarter with massive sell-offs. The second quarter was an erratic period in which investors could not make up their minds. The bulls returned during the third quarter sending the market to record highs.

84. Medicare spending soared during the eighties. Congress managed to slow the rate of growth of Medicare during the nineties and actually managed to decrease Medicare spending in the first decade of the twenty-first century.

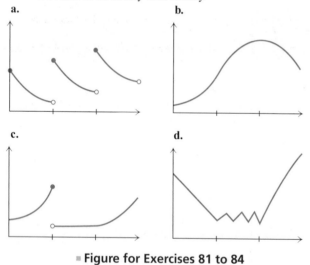

a.

b.

c.

d.

▪ **Figure for Exercises 81 to 84**

Draw a graph that pictures each situation. Explain any choices that you make. Identify the independent and dependent variables. Determine the intervals on which the dependent variable is increasing, decreasing, or constant. Answers may vary.

85. Captain Janeway left the holodeck at 7:45 to meet Tuvok, her chief of security, on the Bridge. After walking for 3 minutes, she realized she had forgotten her tricorder and returned to get

it. She picked up the tricorder and resumed her walk, arriving on the Bridge at 8:00. After 15 minutes the discussion was over, and Janeway returned to the holodeck. Graph Janeway's distance from the holodeck as a function of time.

86. Starting from the pit, Helen made three laps around a circular race track at 40 seconds per lap. She then made a 30-second pit stop and two and a half laps before running off the track and getting stuck in the mud for the remainder of the five-minute race. Graph Helen's distance from the pit as a function of time.

87. Winona deposited $30 per week into her cookie jar. After 2 years she spent half of her savings on a stereo. After spending 6 months in the outback she proceeded to spend $15 per week on CDs until all of the money in the jar was gone. Graph the amount in her cookie jar as a function of time.

88. Michael started buying Navajo crafts with $8000 in his checking account. He spent $200 per day for 10 days on pottery, then $400 per day for the next 10 days on turquoise and silver jewelry. For the next 20 days, he spent $50 per day on supplies while he set up his retail shop. He rested for 5 days then took in $800 per day for the next 20 days from the resale of his collection. Graph the amount in his checking account as a function of time.

Solve each problem.

89. *Motor Vehicle Ownership*   World motor vehicle ownership in developed countries can be modeled by the function

$$M(t) = \begin{cases} 17.5t + 250 & 0 \le t \le 20 \\ 10t + 400 & 20 < t \le 40 \end{cases}$$

where t is the number of years since 1970 and $M(t)$ is the number of motor vehicles in millions in the year $1970 + t$ (World Resources Institute, www.wri.org). See the accompanying figure. How many vehicles were there in developed countries in 1988? How many will there be in 2010? What was the average rate of change of motor vehicle ownership from 1984 to 1994?

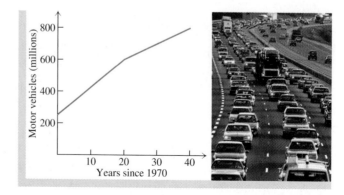

▪ **Figure for Exercise 89**

90. *Motor Vehicle Ownership*   World motor vehicle ownership in developing countries and Eastern Europe can be modeled by the function

$$M(t) = 6.25t + 50$$

where t is the number of years since 1970 and $M(t)$ is in millions of vehicles (World Resources Institute, www.wri.org). Graph this function. What is the expected average rate of change of motor vehicle ownership from 1990 through 2010? Is motor vehicle ownership expected to grow faster in developed or developing countries in the period 1990 through 2010? (See the previous exercise.)

91. *Water Bill*   The monthly water bill in Hammond is a function of the number of gallons used. The cost is $10.30 for 10,000 gal or less. Over 10,000 gal, the cost is $10.30 plus $1 for each 1000 gal over 10,000 with any fraction of 1000 gal charged at a fraction of $1. On what interval is the cost constant? On what interval is the cost increasing?

92. *Gas Mileage*   The number of miles per gallon obtained with a new Firebird is a function of the speed at which it is driven. Is this function increasing or decreasing on its domain? Explain.

93. *Filing a Tax Return*   An accountant determines the charge for filing a tax return by using the function $C = 50 + 40[\![t]\!]$ for $t > 0$, where C is in dollars and t is in hours. Sketch the graph of this function. For what values of t is the charge over $235?

94. *Shipping Machinery*   The cost in dollars of shipping a machine is given by the function $C = 200 + 37[\![w/100]\!]$ for $w > 0$, where w is the weight of the machine in pounds. For which values of w is the cost less than $862?

95. *Delivering Concrete*   A concrete company charges $150 for delivering less than 3 yd^3 of concrete. For 3 yd^3 and more, the charge is $50/$yd^3$ with a fraction of a yard charged as a fraction of $50. Use function notation to write the charge as a function of the number x of cubic yards delivered, where $0 < x \le 10$, and graph this function.

96. *Parking Charges*   A garage charges $4/hr up to 3 hr, with any fraction of an hour charged as a whole hour. Any time over 3 hr is charged at the all-day rate of $15. Use function notation to write the charge as a function of the number of hours x, where $0 < x \le 8$, and graph this function.

For Writing/Discussion

97. *Steps*   Find an example of a function in real life whose graph has "steps" like that of the greatest integer function. Graph your function and find a formula for it if possible.

98. *Cooperative Learning*   Select two numbers a and b. Then define a piecewise function using different formulas on the intervals $(-\infty, a]$, (a, b), and $[b, \infty)$ so that the graph does not "jump" at a or b. Give your function to a classmate to graph and check.

Thinking Outside the Box XXII

Best Fitting Pipe   A work crew is digging a pipeline through a frozen wilderness in Alaska. The cross section of the trench is in the shape of the parabola $y = x^2$. The pipe has a circular cross section. If the pipe is too large, then the pipe will not lay on the bottom of the trench.

a. What is the radius of the largest pipe that will lay on the bottom of the trench?

b. If the radius of the pipe is 3 and the trench is in the shape of $y = ax^2$, then what is the largest value of a for which the pipe will lay in the bottom of the trench?

2.2 Pop Quiz

1. Find the domain and range for $y = 1 - \sqrt{x}$.

2. Find the domain and range for $y = \sqrt{9 - x^2}$.

3. Find the range for

$$f(x) = \begin{cases} 2x & \text{for} \quad x \ge 1 \\ 3 - x & \text{for} \quad x < 1 \end{cases}.$$

4. On what interval is $y = x^2$ increasing?

5. On what interval is $y = |x - 3|$ decreasing?

Linking Concepts

For Individual or Group Explorations

Social Security and Life Expectancy

The annual Social Security benefit of a retiree who earned a lifetime average annual salary of $25,000 depends on the age at the time of retirement. The following function gives the annual benefit in dollars for persons retiring at ages 62 through 70 in the year 2005 or later based on current legislation.

$$B = \begin{cases} 500a - 24{,}000 & 62 \le a < 64 \\ \dfrac{2000a - 104{,}000}{3} & 64 \le a < 67 \\ 800a - 43{,}600 & 67 \le a \le 70 \end{cases}$$

a) Graph the benefit function.

b) What is the annual benefit for a person who retires at age 64?

c) At what age does a person receive an annual benefit of $11,600?

d) What is the average rate of change of the benefit for the ages 62 through 64? Ages 64 through 67? Ages 67 through 70?

e) Do the answers to part (d) appear in the annual benefit formula?

The life expectancy L for a U.S. white male with present age a can be modeled by the formula

$$L = 67.0166(1.00308)^a.$$

f) If Bob is a white male retiring in 2005 at age 62, then what total amount can he be expected to draw from Social Security before he dies?

g) If Bill is a white male retiring in 2005 at age 70, then what total amount can he be expected to draw from Social Security before he dies?

h) Why does Bill draw more than Bob?

2.3

Families of Functions, Transformations, and Symmetry

If a, h, and k are real numbers with $a \ne 0$, then the graph of $y = af(x - h) + k$ is a **transformation** of the graph of $y = f(x)$. All of the transformations of a function form a **family of functions.** For example, any function of the form $y = a\sqrt{x - h} + k$ is in the square-root family because it is a transformation of $y = \sqrt{x}$. If a transformation changes the shape of a graph then it is a **nonrigid** transformation. Otherwise, it is **rigid.** We will study two types of rigid transformations—*translating* and *reflecting,* and two types of nonrigid transformations—*stretching* and *shrinking.* We start with translating.

Translation

The idea of translation is to move a graph either vertically or horizontally without changing its shape.

Definition: Translation Upward or Downward

> If $k > 0$, then the graph of $y = f(x) + k$ is a **translation of k units upward** of the graph of $y = f(x)$. If $k < 0$, then the graph of $y = f(x) + k$ is a **translation of $|k|$ units downward** of the graph of $y = f(x)$.

Example ☐1 **Translations upward or downward**

Graph the three given functions on the same coordinate plane.

a. $f(x) = \sqrt{x}, g(x) = \sqrt{x} + 3, h(x) = \sqrt{x} - 5$
b. $f(x) = x^2, g(x) = x^2 + 2, h(x) = x^2 - 3$

Solution

a. First sketch $f(x) = \sqrt{x}$ through $(0, 0)$, $(1, 1)$, and $(4, 2)$ as shown in Fig. 2.31. Since $g(x) = \sqrt{x} + 3$ we can add 3 to the y-coordinate of each point to get $(0, 3)$, $(1, 4)$, and $(4, 5)$. Sketch g through these points. Every point on f can be moved up 3 units to obtain a corresponding point on g. We now subtract 5 from the y-coordinates on f to obtain points on h. So h goes through $(0, -5)$, $(1, -4)$, and $(4, -3)$.

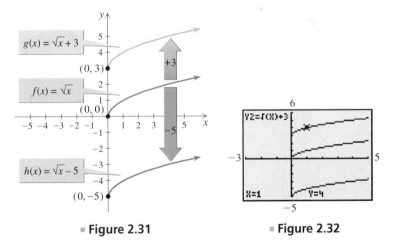

■ **Figure 2.31**   ■ **Figure 2.32**

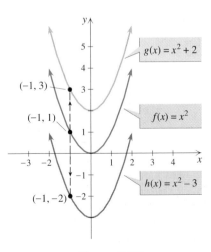

■ **Figure 2.33**

The relationship between f, g, and h can be seen with a graphing calculator in Fig. 2.32. You should experiment with your graphing calculator to see how a change in the formula changes the graph.   ☐

b. First sketch the familiar graph of $f(x) = x^2$ through $(\pm 2, 4)$, $(\pm 1, 1)$, and $(0, 0)$ as shown in Fig. 2.33. Since $g(x) = f(x) + 2$, the graph of g can be obtained by translating the graph of f upward two units. Since $h(x) = f(x) - 3$, the graph of h can be obtained by translating the graph of f downward three units. For example, the point $(-1, 1)$ on the graph of f moves up to $(-1, 3)$ on the graph of g and down to $(-1, -2)$ on the graph of h as shown in Fig. 2.33.

Try This. Graph $f(x) = \sqrt{x}$, $g(x) = \sqrt{x} + 1$, and $h(x) = \sqrt{x} - 2$ on the same coordinate plane. ■

Any function of the form $y = a(x - h)^2 + k$ is in the **square family** because it is a transformation of $y = x^2$. The graph of any function in the square family is called a **parabola.**

The graph of $y = f(x) + k$ for $k > 0$ is an upward translation of $y = f(x)$ because the last operation performed is addition of k. To make a graph move horizontally, the first operation performed must be addition or subtraction.

Definition: Translation to the Right or Left

> If $h > 0$, then the graph of $y = f(x - h)$ is a **translation of h units to the right** of the graph of $y = f(x)$. If $h < 0$, then the graph of $y = f(x - h)$ is a **translation of $|h|$ units to the left** of the graph of $y = f(x)$.

Example **2** **Translations to the right or left**

Graph $f(x) = \sqrt{x}$, $g(x) = \sqrt{x - 3}$, and $h(x) = \sqrt{x + 5}$ on the same coordinate plane.

Solution

First sketch $f(x) = \sqrt{x}$ through $(0, 0)$, $(1, 1)$, and $(4, 2)$ as shown in Fig. 2.34. Since the first operation of g is to subtract 3, we get the corresponding points by adding 3 to each x-coordinate. So g goes through $(3, 0)$, $(4, 1)$, and $(7, 2)$. Since the first operation of h is to add 5, we get corresponding points by subtracting 5 from the x-coordinates. So h goes through $(-5, 0)$, $(-4, 1)$, and $(-1, 2)$.

The calculator graphs of f, g, and h are shown in Fig. 2.35. Be sure to note the difference between $y = \sqrt{}(x) - 3$ and $y = \sqrt{}(x - 3)$ on a calculator.

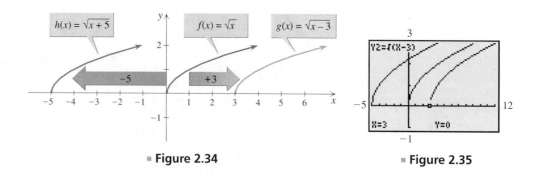

■ **Figure 2.34**   ■ **Figure 2.35**

Try This. Graph $f(x) = \sqrt{x}$, $g(x) = \sqrt{x - 2}$, and $h(x) = \sqrt{x + 1}$ on the same coordinate plane. ■

Notice that $y = \sqrt{x - 3}$ lies 3 units to the *right* and $y = \sqrt{x + 5}$ lies 5 units to the *left* of $y = \sqrt{x}$. The next example shows two more horizontal translations.

Example ③ Horizontal translations

Sketch the graph of each function.

a. $f(x) = |x - 1|$ **b.** $f(x) = (x + 3)^2$

Solution

a. The function $f(x) = |x - 1|$ is in the absolute value family and its graph is a translation one unit to the right of $g(x) = |x|$. Calculate a few ordered pairs to get an accurate graph. The points $(0, 1)$, $(1, 0)$, and $(2, 1)$ are on the graph of $f(x) = |x - 1|$ shown in Fig. 2.36.

b. The function $f(x) = (x + 3)^2$ is in the square family and its graph is a translation three units to the left of the graph of $g(x) = x^2$. Calculate a few ordered pairs to get an accurate graph. The points $(-3, 0)$, $(-2, 1)$, and $(-4, 1)$ are on the graph shown in Fig. 2.37.

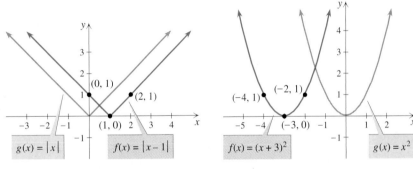

■ **Figure 2.36**   ■ **Figure 2.37**

Try This. Graph $f(x) = (x - 2)^2$.   ■

Reflection

The idea in *reflection* is to get a graph that is a mirror image of another graph, where the mirror is placed on the *x*-axis. To find the mirror image of a point, we simply change the sign of its *y*-coordinate.

Definition: Reflection

> The graph of $y = -f(x)$ is a **reflection** in the *x*-axis of the graph of $y = f(x)$.

Example ④ Graphing using reflection

Graph each pair of functions on the same coordinate plane.

a. $f(x) = x^2, g(x) = -x^2$ **b.** $f(x) = x^3, g(x) = -x^3$
c. $f(x) = |x|, g(x) = -|x|$

Solution

a. The graph of $f(x) = x^2$ goes through $(0, 0)$, $(\pm 1, 1)$, and $(\pm 2, 4)$. The graph of $g(x) = -x^2$ goes through $(0, 0)$, $(\pm 1, -1)$, and $(\pm 2, -4)$ as shown in Fig. 2.38.

b. Make a table of ordered pairs for f as follows:

x	-2	-1	0	1	2
$f(x) = x^3$	-8	-1	0	1	8

Sketch the graph of f through these ordered pairs as shown in Fig. 2.39. Since $g(x) = -f(x)$, the graph of g can be obtained by reflecting the graph of f in the x-axis. Each point on the graph of f corresponds to a point on the graph of g with the opposite y-coordinate. For example, $(2, 8)$ on the graph of f corresponds to $(2, -8)$ on the graph of g. Both graphs are shown in Fig. 2.39.

c. The graph of f is the familiar V-shaped graph of the absolute value function as shown in Fig. 2.40. Since $g(x) = -f(x)$, the graph of g can be obtained by reflecting the graph of f in the x-axis. Each point on the graph of f corresponds to a point on the graph of g with the opposite y-coordinate. For example, $(2, 2)$ on f corresponds to $(2, -2)$ on g. Both graphs are shown in Fig. 2.40.

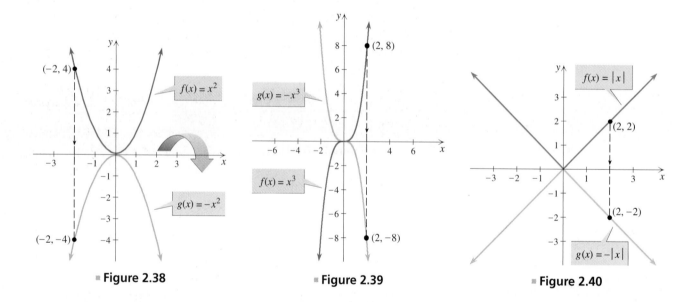

■ **Figure 2.38**   ■ **Figure 2.39**   ■ **Figure 2.40**

Try This. Graph $f(x) = \sqrt{x}$ and $g(x) = -\sqrt{x}$ on the same coordinate plane.

■

Note that if $y = x^2$ is reflected in the x-axis and then translated one unit upward, the equation for the graph in the final position is $y = -x^2 + 1$. If $y = x^2$ is translated one unit upward and then reflected in the x-axis, the equation for the graph in the final position is $y = -(x^2 + 1)$ or $y = -x^2 - 1$. The order in which the transformations are done can make the final functions different.

Stretching and Shrinking

To *stretch* a graph we multiply the *y*-coordinates by a number larger than 1. To *shrink* a graph we multiply the *y*-coordinates by a number between 0 and 1.

Definitions:
Stretching and Shrinking

> The graph of $y = af(x)$ is obtained from the graph of $y = f(x)$ by
>
> 1. **stretching** the graph of $y = f(x)$ by a when $a > 1$, or
> 2. **shrinking** the graph of $y = f(x)$ by a when $0 < a < 1$.

Example **5** **Graphing using stretching and shrinking**

In each case graph the three functions on the same coordinate plane.

a. $f(x) = \sqrt{x}, g(x) = 2\sqrt{x}, h(x) = \dfrac{1}{2}\sqrt{x}$

b. $f(x) = x^2, g(x) = 2x^2, h(x) = \dfrac{1}{2}x^2$

Solution

a. The graph of $f(x) = \sqrt{x}$ goes through (0, 0), (1, 1), and (4, 2) as shown in Fig. 2.41. The graph of g is obtained by stretching the graph of f by a factor of 2. So g goes through (0, 0), (1, 2), and (4, 4). The graph of h is obtained by shrinking the graph of f by a factor of $\frac{1}{2}$. So h goes through (0, 0), $\left(1, \frac{1}{2}\right)$, and (4, 1).

The functions f, g, and h are shown on a graphing calculator in Fig. 2.42. Note how the viewing window affects the shape of the graph. They do not appear as separated on the calculator as they do in Fig. 2.41. □

b. The graph of $f(x) = x^2$ is the familiar parabola shown in Fig. 2.43. We stretch it by a factor of 2 to get the graph of g and shrink it by a factor of $\frac{1}{2}$ to get the graph of h.

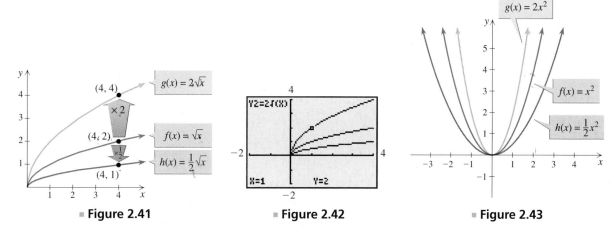

■ **Figure 2.41**   ■ **Figure 2.42**   ■ **Figure 2.43**

Try This. Graph $f(x) = |x|, g(x) = 3|x|,$ and $h(x) = \dfrac{1}{3}|x|$ on the same coordinate plane.   ■

A function involving more than one transformation may be graphed using the following procedure.

PROCEDURE **Multiple Transformations**

Graph a function involving more than one transformation in the following order:

1. Horizontal translation
2. Stretching or shrinking
3. Reflecting
4. Vertical translation

The function in the next example involves all four of the above transformations.

Example 6 Graphing using several transformations

Graph the function $y = 4 - 2\sqrt{x + 1}$.

Solution

First recognize that this function is in the square root family. So its graph is a transformation of the graph of $y = \sqrt{x}$. The graph of $y = \sqrt{x + 1}$ is a *horizontal* translation one unit to the left of the graph of $y = \sqrt{x}$. The graph of $y = 2\sqrt{x + 1}$ is obtained from $y = \sqrt{x + 1}$ by *stretching* it by a factor of 2. *Reflect* $y = 2\sqrt{x + 1}$ in the x-axis to obtain the graph of $y = -2\sqrt{x + 1}$. Finally, the graph of $y = 4 - 2\sqrt{x + 1}$ is a *vertical* translation of $y = -2\sqrt{x + 1}$, four units upward. All of these graphs are shown in Fig. 2.44.

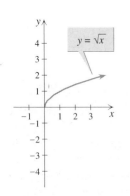 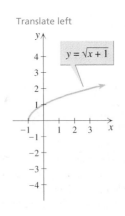 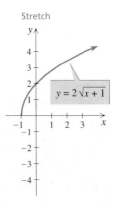 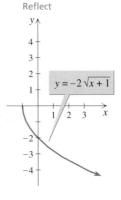 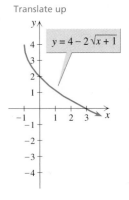

■ **Figure 2.44**

Try This. Graph $y = 4 - 2|x + 1|$. ■

The Linear Family of Functions

The function $f(x) = x$ is called the **identity function** because the coordinates in each ordered pair are identical. Its graph is a line through $(0, 0)$ with slope 1. A member of the **linear family**, a **linear function**, is a transformation of the identity function: $f(x) = a(x - h) + k$ where $a \neq 0$. Since a, h, and k are real numbers, we can rewrite this form as a multiple of x plus a constant. So a linear function has the form $f(x) = mx + b$, with $m \neq 0$ (the slope-intercept form). If $m = 0$, then the function has the form $f(x) = b$ and it is a **constant function.**

Example 7 **Graphing linear functions using transformations**

Sketch the graphs of $y = x$, $y = 2x$, $y = -2x$, and $y = -2x - 3$.

Solution

The graph of $y = x$ is a line through $(0, 0)$, $(1, 1)$, and $(2, 2)$. Stretch the graph of $y = x$ by a factor of 2 to get the graph of $y = 2x$. Reflect in the x-axis to get the graph of $y = -2x$. Translate downward three units to get the graph of $y = -2x - 3$. See Fig. 2.45.

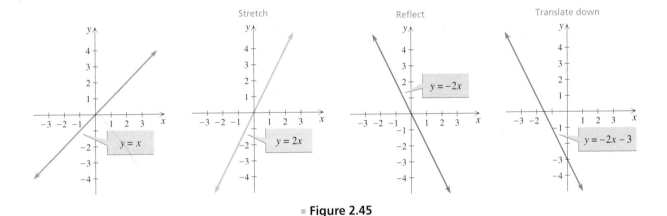

■ **Figure 2.45**

Try This. Graph $y = -2x + 5$.   ■

Symmetry

The graph of $g(x) = -x^2$ is a reflection in the x-axis of the graph of $f(x) = x^2$. If the paper were folded along the x-axis, the graphs would coincide. See Fig. 2.46 on the next page. The symmetry that we call reflection occurs between two functions, but the graph of $f(x) = x^2$ has a symmetry within itself. Points such as $(2, 4)$ and $(-2, 4)$ are on the graph and are equidistant from the y-axis. Folding the paper along the y-axis brings all such pairs of points together. See Fig. 2.47. The reason for this symmetry about the y-axis is the fact that $f(-x) = f(x)$ for every value of x. We get the same y-coordinate whether we evaluate the function at a number or at its opposite.

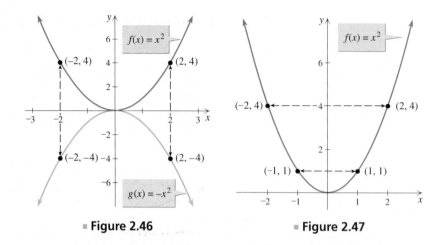

■ **Figure 2.46**   ■ **Figure 2.47**

Definition: Symmetric about the *y*-Axis

If $f(-x) = f(x)$ for every value of x in the domain of the function f, then f is called an **even function** and its graph is **symmetric about the *y*-axis.**

Consider the graph of $f(x) = x^3$ shown in Fig. 2.48. On the graph of $f(x) = x^3$ we find pairs of points such as $(2, 8)$ and $(-2, -8)$. The odd exponent in x^3 causes the second coordinate to be negative when the sign of the first coordinate is changed. These points are equidistant from the origin and on opposite sides of the origin. So the symmetry of this graph is about the origin. In this case $f(x)$ and $f(-x)$ are not equal, but $f(-x) = -f(x)$.

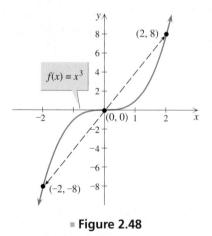

■ **Figure 2.48**

Definition: Symmetric about the Origin

If $f(-x) = -f(x)$ for every value of x in the domain of the function f, then f is called an **odd function** and its graph is **symmetric about the origin.**

A graph might look like it is symmetric about the *y*-axis or the origin, but the only way to be sure is to use the definitions of these terms as shown in the following example. Note that an odd power of a negative number is negative and an even power of a negative number is positive. So for any real number x we have $(-x)^n = -x^n$ if n is odd and $(-x)^n = x^n$ if n is even.

Example **8** Determining symmetry in a graph

Discuss the symmetry of the graph of each function.

a. $f(x) = 5x^3 - x$ **b.** $f(x) = |x| + 3$ **c.** $f(x) = x^2 - 3x + 6$

Solution

a. Replace x by $-x$ in the formula for $f(x)$ and simplify:

$$f(-x) = 5(-x)^3 - (-x) = -5x^3 + x$$

Is $f(-x)$ equal to $f(x)$ or the opposite of $f(x)$? Since $-f(x) = -5x^3 + x$, we have $f(-x) = -f(x)$. So f is an odd function and the graph is symmetric about the origin.

b. Since $|-x| = |x|$ for any x, we have $f(-x) = |-x| + 3 = |x| + 3$. Because $f(-x) = f(x)$, the function is even and the graph is symmetric about the y-axis.

c. In this case, $f(-x) = (-x)^2 - 3(-x) + 6 = x^2 + 3x + 6$. So $f(-x) \neq f(x)$, and $f(-x) \neq -f(x)$. This function is neither odd nor even and its graph has neither type of symmetry.

Try This. Discuss the symmetry of the graph of $f(x) = -2x^2 + 5$. ■

Do you see why functions symmetric about the y-axis are called *even* and functions symmetric about the origin are called *odd?* In general, a function defined by a polynomial with even exponents only, such as $f(x) = x^2$ or $f(x) = x^6 - 5x^4 + 2x^2 + 3$, is symmetric about the y-axis. (The constant term 3 has even degree because $3 = 3x^0$.) A function with only odd exponents such as $f(x) = x^3$ or $f(x) = x^5 - 6x^3 + 4x$ is symmetric about the origin. A function containing both even and odd powered terms such as $f(x) = x^2 + 3x$ has neither symmetry. For other types of functions (such as absolute value) you must examine the function more carefully to determine symmetry.

Reading Graphs to Solve Inequalities

In Chapter 1 we learned that the solution to an inequality in one variable could be read from a graph of an equation in two variables. Now that we have more experience with graphing, we will solve more inequalities by reading graphs.

Example **9** Using a graph to solve an inequality

Solve the inequality $(x - 1)^2 - 2 < 0$ by graphing.

Solution

The graph of $y = (x - 1)^2 - 2$ is obtained by translating the graph of $y = x^2$ one unit to the right and two units downward. See Fig. 2.49. To find the x-intercepts we solve $(x - 1)^2 - 2 = 0$:

$$(x - 1)^2 = 2$$
$$x - 1 = \pm\sqrt{2}$$
$$x = 1 \pm \sqrt{2}$$

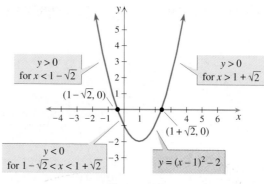

■ **Figure 2.49**

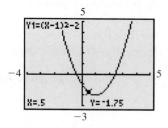

■ **Figure 2.50**

The x-intercepts are $\left(1 - \sqrt{2}, 0\right)$ and $\left(1 + \sqrt{2}, 0\right)$. If the y-coordinate of a point on the graph is negative, then the x-coordinate satisfies $(x - 1)^2 - 2 < 0$. So the solution set to $(x - 1)^2 - 2 < 0$ is the open interval $\left(1 - \sqrt{2}, 1 + \sqrt{2}\right)$.

Although a graphing calculator will not find the exact solution to this inequality, you can use TRACE to support the answer and see that y is negative between the x-intercepts. See Fig. 2.50.

Try This. Solve $2 - |x - 1| \le 0$ by graphing.   ■

Note that the solution set to $(x - 1)^2 - 2 \ge 0$ can also be obtained from the graph in Fig. 2.49. If the y-coordinate of a point on the graph is positive or zero, then the x-coordinate satisfies $(x - 1)^2 - 2 \ge 0$. So the solution set to $(x - 1)^2 - 2 \ge 0$ is $\left(-\infty, 1 - \sqrt{2}\right] \cup \left[1 + \sqrt{2}, \infty\right)$.

Function Gallery: **Some Basic Functions and Their Properties**

Constant Function

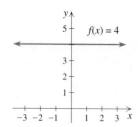

Domain $(-\infty, \infty)$
Range $\{4\}$
Constant on $(-\infty, \infty)$
Symmetric about y-axis

Identity Function

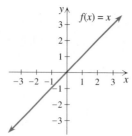

Domain $(-\infty, \infty)$
Range $(-\infty, \infty)$
Increasing on $(-\infty, \infty)$
Symmetric about origin

Linear Function

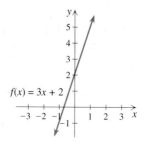

Domain $(-\infty, \infty)$
Range $(-\infty, \infty)$
Increasing on $(-\infty, \infty)$

Absolute-Value Function

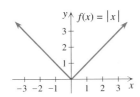

$f(x) = |x|$

Domain $(-\infty, \infty)$
Range $[0, \infty)$
Increasing on $[0, \infty)$
Decreasing on $(-\infty, 0)$
Symmetric about y-axis

Square Function

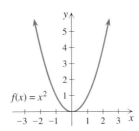

$f(x) = x^2$

Domain $(-\infty, \infty)$
Range $[0, \infty)$
Increasing on $(0, \infty)$
Decreasing on $(-\infty, 0)$
Symmetric about y-axis

Square-Root Function

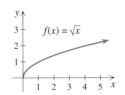

$f(x) = \sqrt{x}$

Domain $[0, \infty)$
Range $[0, \infty)$
Increasing on $(0, \infty)$

Cube Function

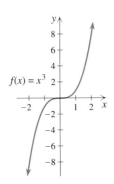

$f(x) = x^3$

Domain $(-\infty, \infty)$
Range $(-\infty, \infty)$
Increasing on $(-\infty, \infty)$
Symmetric about origin

Cube-Root Function

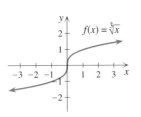

$f(x) = \sqrt[3]{x}$

Domain $(-\infty, \infty)$
Range $(-\infty, \infty)$
Increasing on $(-\infty, \infty)$
Symmetric about origin

Greatest Integer Function

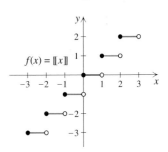

$f(x) = [\![x]\!]$

Domain $(-\infty, \infty)$
Range $\{n \mid n$ is an integer$\}$
Constant on $[n, n + 1)$
for every integer n

For Thought

True or False? Explain.

1. The graph of $f(x) = (-x)^4$ is a reflection in the x-axis of the graph of $g(x) = x^4$.

2. The graph of $f(x) = x - 4$ lies four units to the right of the graph of $f(x) = x$.

3. The graph of $y = |x + 2| + 3$ is a translation two units to the right and three units upward of the graph of $y = |x|$.

4. The graph of $f(x) = -3$ is a reflection in the x-axis of the graph of $g(x) = 3$.

5. The functions $y = x^2 + 4x + 1$ and $y = (x + 2)^2 - 3$ have the same graph.

6. The graph of $y = -(x - 3)^2 - 4$ can be obtained by moving $y = x^2$ three units to the right and down four units, and then reflecting in the x-axis.

7. If $f(x) = -x^3 + 2x^2 - 3x + 5$, then $f(-x) = x^3 + 2x^2 + 3x + 5$.

8. The graphs of $f(x) = -\sqrt{x}$ and $g(x) = \sqrt{-x}$ are identical.

9. If $f(x) = x^3 - x$, then $f(-x) = -f(x)$.

10. The solution set to $|x| - 1 \le 0$ is $[-1, 1]$.

2.3 Exercises

Sketch the graphs of each pair of functions on the same coordinate plane.

1. $f(x) = |x|, g(x) = |x| - 4$

2. $f(x) = \sqrt{x}, g(x) = \sqrt{x} + 3$

3. $f(x) = x, g(x) = x + 3$ **4.** $f(x) = x^2, g(x) = x^2 - 5$

5. $y = x^2, y = (x - 3)^2$ **6.** $y = |x|, y = |x + 2|$

7. $y = \sqrt{x}, y = \sqrt{x + 9}$ **8.** $y = x^2, y = (x - 1)^2$

9. $f(x) = \sqrt{x}, g(x) = -\sqrt{x}$ **10.** $f(x) = x, g(x) = -x$

11. $y = \sqrt{x}, y = 3\sqrt{x}$

12. $y = \sqrt{1 - x^2}, y = 4\sqrt{1 - x^2}$

13. $y = x^2, y = \dfrac{1}{4}x^2$ **14.** $y = |x|, y = \dfrac{1}{3}|x|$

15. $y = \sqrt{4 - x^2}, y = -\sqrt{4 - x^2}$

16. $f(x) = x^2 + 1, g(x) = -(x^2 + 1)$

Match each function in Exercises 17–24 with its graph (a)–(h). See the procedure for multiple transformations on page 218.

17. $y = x^2$

18. $y = (x - 4)^2 + 2$

19. $y = (x + 4)^2 - 2$

20. $y = -2(x - 2)^2$

21. $y = -2(x + 2)^2$

22. $y = -\dfrac{1}{2}x^2 - 4$

23. $y = \dfrac{1}{2}(x + 4)^2 + 2$

24. $y = -2(x - 4)^2 - 2$

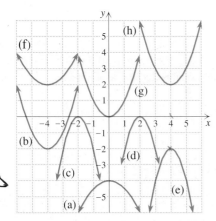

■ **Figure for Exercises 17 to 24**

Write the equation of each graph after the indicated transformation(s).

25. The graph of $y = \sqrt{x}$ is translated two units upward.

26. The graph of $y = \sqrt{x}$ is translated three units downward.

27. The graph of $y = x^2$ is translated five units to the right.

28. The graph of $y = x^2$ is translated seven units to the left.

29. The graph of $y = x^2$ is translated ten units to the right and four units upward.

30. The graph of $y = \sqrt{x}$ is translated five units to the left and twelve units downward.

31. The graph of $y = \sqrt{x}$ is stretched by a factor of 3, translated five units upward, then reflected in the x-axis.

32. The graph of $y = x^2$ is translated thirteen units to the right and six units downward, then reflected in the x-axis.

33. The graph of $y = |x|$ is reflected in the x-axis, stretched by a factor of 3, then translated seven units to the right and nine units upward.

34. The graph of $y = x$ is stretched by a factor of 2, reflected in the x-axis, then translated eight units downward and six units to the left.

Use transformations to graph each function and state the domain and range.

35. $y = (x - 1)^2 + 2$ **36.** $y = (x + 5)^2 - 4$

37. $y = |x - 1| + 3$ **38.** $y = |x + 3| - 4$

39. $y = 3x - 40$ **40.** $y = -4x + 200$

41. $y = \dfrac{1}{2}x - 20$ **42.** $y = -\dfrac{1}{2}x + 40$

43. $y = -\dfrac{1}{2}|x| + 40$ **44.** $y = 3|x| - 200$

45. $y = -\dfrac{1}{2}|x + 4|$ **46.** $y = 3|x - 2|$

47. $y = -\sqrt{x - 3} + 1$ **48.** $y = -\sqrt{x + 2} - 4$

49. $y = -2\sqrt{x + 3} + 2$ **50.** $y = -\dfrac{1}{2}\sqrt{x + 2} + 4$

Determine algebraically whether the function is even, odd, or neither. Discuss the symmetry of each function.

51. $f(x) = x^4$

52. $f(x) = x^4 - 2x^2$

53. $f(x) = x^4 - x^3$

54. $f(x) = x^3 - x$

55. $f(x) = (x + 3)^2$

56. $f(x) = (x - 1)^2$

57. $f(x) = |x - 2|$

58. $f(x) = |x| - 9$

59. $f(x) = x$

60. $f(x) = -x$

61. $f(x) = 3x + 2$

62. $f(x) = x - 3$

63. $f(x) = x^3 - 5x + 1$

64. $f(x) = x^6 - x^4 + x^2$

65. $f(x) = 1 + \dfrac{1}{x^2}$

66. $f(x) = (x^2 - 2)^3$

67. $f(x) = \sqrt{x}$

68. $f(x) = \sqrt{9 - x^2}$

69. $f(x) = |x^2 - 3|$

70. $f(x) = \sqrt{x^2 + 3}$

Match each function with its graph (a)–(h).

71. $y = 2 + \sqrt{x}$ e

72. $y = \sqrt{2 + x}$ g

73. $y = \sqrt{x^2}$ g

74. $y = \sqrt{\dfrac{x}{2}}$ h

75. $y = \dfrac{1}{2}\sqrt{x}$

76. $y = 2 - \sqrt{x - 2}$ d

77. $y = -2\sqrt{x}$ c

78. $y = -\sqrt{-x}$ f

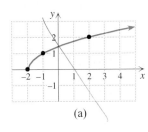

(a)

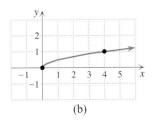

(b)

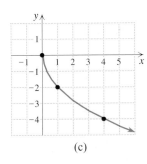

(c)

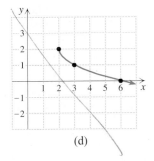

(d)

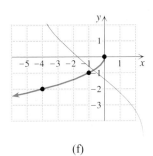

(e)

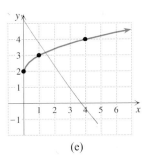

(f)

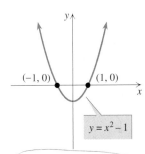

(g)

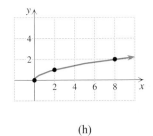

(h)

▪ **Figure for Exercises 71 to 78**

Solve each inequality by reading the corresponding graph.

79. $x^2 - 1 \geq 0$

80. $2x^2 - 3 < 0$

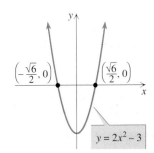

81. $|x - 2| - 3 > 0$

82. $2 - |x + 1| \geq 0$

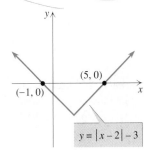

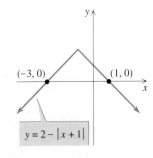

Solve each inequality by graphing an appropriate function. State the solution set using interval notation.

83. $(x - 1)^2 - 9 < 0$

84. $\left(x - \dfrac{1}{2}\right)^2 - \dfrac{9}{4} \geq 0$

85. $5 - \sqrt{x} \geq 0$

86. $\sqrt{x + 3} - 2 \geq 0$

87. $(x - 2)^2 > 3$ **88.** $(x - 1)^2 < 4$

89. $\sqrt{25 - x^2} > 0$ **90.** $\sqrt{4 - x^2} \geq 0$

Use a graphing calculator to find an approximate solution to each inequality by reading the graph of an appropriate function. Round to two decimal places.

91. $\sqrt{3}x^2 + \pi x - 9 < 0$ **92.** $x^3 - 5x^2 + 6x - 1 > 0$

Graph each of the following functions by transforming the given graph of $y = f(x)$.

93. a. $y = 2f(x)$

   b. $y = -f(x)$

   c. $y = f(x + 1)$

   d. $y = f(x - 3)$

   e. $y = -3f(x)$

   f. $y = f(x + 2) - 1$

   g. $y = f(x - 1) + 3$

   h. $y = 3f(x - 2) + 1$

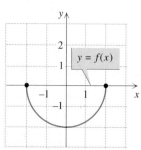

■ **Figure for Exercise 93**

94. a. $y = -f(x)$

   b. $y = 2f(x)$

   c. $y = -3f(x)$

   d. $y = f(x + 2)$

   e. $y = f(x - 1)$

   f. $y = f(x - 2) + 1$

   g. $y = -2f(x + 4)$

   h. $y = 2f(x - 3) + 1$

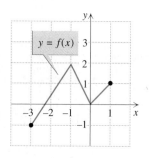

■ **Figure for Exercise 94**

Solve each problem.

95. *Across-the-Board Raise*   Each teacher at C. F. Gauss Elementary School is given an across-the-board raise of $2000. Write a function that *transforms* each old salary x into a new salary $N(x)$.

96. *Cost-of-Living Raise*   Each registered nurse at Blue Hills Memorial Hospital is first given a 5% cost-of-living raise and then a $3000 merit raise. Write a function that *transforms* each old salary x into a new salary $N(x)$. Does the order in which these raises are given make any difference? Explain.

97. *Unemployment Versus Inflation*   The Phillips curve shows the relationship between the unemployment rate x and the

inflation rate y. If the equation of the curve is $y = 1 - \sqrt{x}$ for a certain Third World country, then for what values of x is the inflation rate less than 50%?

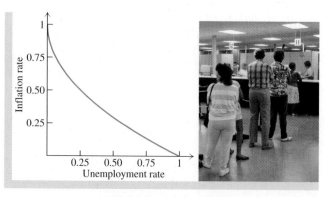

■ **Figure for Exercise 97**

98. *Production Function*   The production function shows the relationship between inputs and outputs. A manufacturer of custom windows produces y windows per week using x hours of labor per week, where $y = 1.75\sqrt{x}$. How many hours of labor are required to keep production at or above 28 windows per week?

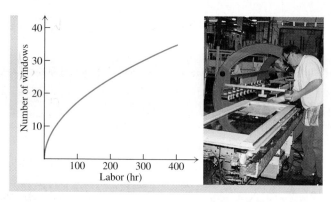

■ **Figure for Exercise 98**

For Writing/Discussion

99. Graph each pair of functions (without simplifying the second function) on the same screen of a graphing calculator and explain what each exercise illustrates.

   a. $y = x^4 - x^2$, $y = (-x)^4 - (-x)^2$

   b. $y = x^3 - x$, $y = (-x)^3 - (-x)$

   c. $y = x^4 - x^2$, $y = (x + 1)^4 - (x + 1)^2$

   d. $y = x^3 - x$, $y = (x - 2)^3 - (x - 2) + 3$

100. Graph $y = x^3 + 6x^2 + 12x + 8$ on a graphing calculator. Explain how this graph could be obtained as a transformation of a simpler function.

Thinking Outside the Box XXIII

Lucky Lucy   Ms. Willis asked Lucy to come to the board to find the mean of a pair of one-digit positive integers. Lucy slowly wrote the numbers on the board. While trying to think of what to do next, she rested the chalk between the numbers to make a mark that looked like a decimal point to Ms. Willis. Ms. Willis said "correct" and asked her to find the mean for a pair of two-digit positive integers. Being a quick learner, Lucy again wrote the numbers on the board, rested the chalk between the numbers, and again Ms. Willis said "correct." Lucy had to demonstrate her ability to find the mean for a pair of three-digit and a pair of four-digit positive integers before Ms. Willis was satisfied that she understood the concept. What four pairs of integers did Ms. Willis give to Lucy? Explain why Lucy's method will not work for any other pairs of one-, two-, three-, or four-digit positive integers.

2.3 Pop Quiz

1. What is the equation of the curve $y = \sqrt{x}$ after it is translated 8 units upward?

2. What is the equation of the curve $y = x^2$ after it is translated 9 units to the right?

3. What is the equation of the curve $y = x^3$ after it is reflected in the x-axis?

4. Find the domain and range for $y = -2\sqrt{x - 1} + 5$.

5. If the curve $y = x^2$ is translated 6 units to the right, stretched by a factor of 3, reflected in the x-axis, and translated 4 units upward, then what is the equation of the curve in its final position?

6. Is $y = \sqrt{4 - x^2}$ even, odd, or neither?

Linking Concepts

For Individual or Group Explorations

Designing a Racing Boat

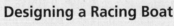 *According to International America's Cup Class rules, the basic dimensions of any yacht competing for the silver trophy must satisfy the inequality*

$$16.96 + 9.8D^{1/3} - L - 1.25S^{1/2} \geq 0,$$

where L is the length in meters, S is the sail area in square meters, and D is the displacement in cubic meters (Americas Cup, www.americascup.org). Use a graphing utility to get approximate answers to the following questions.

a) A team of British designers wants its new boat to have a length of 20.85 m and a displacement of 17.67 m³. Write the inequality that must be satisfied by the sail area S. Use the graphing technique described in this section to find the interval in which S must lie.

b) A team of Australian designers wants its new boat to have a sail area of 312.54 m² and length of 21.45 m. Write the inequality that must be satisfied by the displacement D. Use the graphing technique to find the interval in which D must lie.

c) A team of American designers wants its new boat to have a sail area of 310.28 m² and a displacement of 17.26 m³. Write the inequality that must be satisfied by the length L. Use the graphing technique to find the interval in which L must lie.

d) When two of the three variables are fixed, the third variable either has a maximum or minimum value. Explain how you can determine whether it is a maximum or minimum by looking at the original inequality.

2.4

Operations with Functions

In Sections 2.2 and 2.3 we studied the graphs of functions to see the relationships between types of functions and their graphs. In this section we will study various ways in which two or more functions can be combined to make new functions. The emphasis here will be on formulas that define functions.

Basic Operations with Functions

A college student is hired to deliver new telephone books and collect the old ones for recycling. She is paid $6 per hour plus $0.30 for each old phone book she collects. Her salary for a 40-hour week is a function of the number of phone books collected. If x represents the number of phone books collected in one week, then the function $S(x) = 0.30x + 240$ gives her salary in dollars. However, she must use her own car for this job. She figures that her car expenses average $0.20 per phone book collected plus a fixed cost of $20 per week for insurance. We can write her expenses as a function of the number of phone books collected, $E(x) = 0.20x + 20$. Her profit for one week is her salary minus her expenses:

$$P(x) = S(x) - E(x)$$

$$= 0.30x + 240 - (0.20x + 20)$$

$$= 0.10x + 220$$

By subtracting, we get $P(x) = 0.10x + 220$. Her weekly profit is written as a function of the number of phone books collected. In this example we obtained a new function by subtracting two functions. In general, there are four basic arithmetic operations defined for functions.

Definition: Sum, Difference, Product, and Quotient Functions

For two functions f and g, the **sum, difference, product,** and **quotient functions,** functions $f + g, f - g, f \cdot g$, and f/g, respectively, are defined as follows:

$$(f + g)(x) = f(x) + g(x)$$

$$(f - g)(x) = f(x) - g(x)$$

$$(f \cdot g)(x) = f(x) \cdot g(x)$$

$$(f/g)(x) = f(x)/g(x) \qquad \text{provided that } g(x) \neq 0.$$

Example **1** **Evaluating functions**

Let $f(x) = 3\sqrt{x} - 2$ and $g(x) = x^2 + 5$. Find and simplify each expression.

a. $(f + g)(4)$ **b.** $(f - g)(x)$ **c.** $(f \cdot g)(0)$ **d.** $\left(\dfrac{f}{g}\right)(9)$

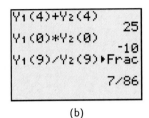

(a)

```
Y1(4)+Y2(4)
              25
Y1(0)*Y2(0)
             -10
Y1(9)/Y2(9)▶Frac
            7/86
```

(b)

■ **Figure 2.51**

Solution

a. $(f + g)(4) = f(4) + g(4) = 3\sqrt{4} - 2 + 4^2 + 5 = 25$

b. $(f - g)(x) = f(x) - g(x) = 3\sqrt{x} - 2 - (x^2 + 5) = 3\sqrt{x} - x^2 - 7$

c. $(f \cdot g)(0) = f(0) \cdot g(0) = (3\sqrt{0} - 2)(0^2 + 5) = (-2)(5) = -10$

d. $\left(\dfrac{f}{g}\right)(9) = \dfrac{f(9)}{g(9)} = \dfrac{3\sqrt{9} - 2}{9^2 + 5} = \dfrac{7}{86}$

Parts (a), (c), and (d) can be checked with a graphing calculator as shown in Figs. 2.51(a) and (b).

Try This. Let $h(x) = x^2$ and $j(x) = 3x$. Find $(h + j)(5)$, $(h \cdot j)(2)$, and $(h/j)(a)$. ■

Think of $f + g$, $f - g$, $f \cdot g$, and f/g as generic names for the sum, difference, product, and quotient of the functions f and g. If any of these functions has a particular meaning, as in the phone book example, we can use a new letter to identify it. The domain of $f + g$, $f - g$, $f \cdot g$, or f/g is the intersection of the domain of f with the domain of g. Of course, we exclude from the domain of f/g any number for which $g(x) = 0$.

review

Example **2** **The sum, product, and quotient functions**

Let $f = \{(1, 3), (2, 8), (3, 6), (5, 9)\}$ and $g = \{(1, 6), (2, 11), (3, 0), (4, 1)\}$. Find $f + g$, $f \cdot g$, and f/g. State the domain of each function.

Solution

The domain of $f + g$ and $f \cdot g$ is $\{1, 2, 3\}$ because that is the intersection of the domains of f and g. The ordered pair $(1, 9)$ belongs to $f + g$ because

$$(f + g)(1) = f(1) + g(1) = 3 + 6 = 9.$$

The ordered pair $(2, 19)$ belongs to $f + g$ because $(f + g)(2) = 19$. The pair $(3, 6)$ also belongs to $f + g$. So

$$f + g = \{(1, 9), (2, 19), (3, 6)\}.$$

Since $(f \cdot g)(1) = f(1) \cdot g(1) = 3 \cdot 6 = 18$, the pair $(1, 18)$ belongs to $f \cdot g$. Likewise, $(2, 88)$ and $(3, 0)$ also belong to $f \cdot g$. So

$$f \cdot g = \{(1, 18), (2, 88), (3, 0)\}.$$

The domain of f/g is $\{1, 2\}$ because $g(3) = 0$. So

$$\frac{f}{g} = \left\{\left(1, \frac{1}{2}\right), \left(2, \frac{8}{11}\right)\right\}.$$

Try This. Let $h = \{(2, 10), (4, 0), (6, 8)\}$ and $j = \{(2, 5), (6, 0)\}$. Find $h + j$ and h/j and state the domain of each function. ■

In Example 2 the functions are given as sets of ordered pairs, and the results of performing operations with these functions are sets of ordered pairs. In the next example the sets of ordered pairs are defined by means of equations, so the result of performing operations with these functions will be new equations that determine the ordered pairs of the function.

Example **3** **The sum, quotient, product, and difference functions**

Let $f(x) = \sqrt{x}$, $g(x) = 3x + 1$, and $h(x) = x - 1$. Find each function and state its domain.

a. $f + g$ **b.** $\dfrac{g}{f}$ **c.** $g \cdot h$ **d.** $g - h$

Solution

a. Since the domain of f is $[0, \infty)$ and the domain of g is $(-\infty, \infty)$, the domain of $f + g$ is $[0, \infty)$. Since $(f + g)(x) = f(x) + g(x) = \sqrt{x} + 3x + 1$, the equation defining the function $f + g$ is

$$(f + g)(x) = \sqrt{x} + 3x + 1.$$

b. The number 0 is not in the domain of g/f because $f(0) = 0$. So the domain of g/f is $(0, \infty)$. The equation defining g/f is

$$\left(\frac{g}{f}\right)(x) = \frac{3x + 1}{\sqrt{x}}.$$

c. The domain of both g and h is $(-\infty, \infty)$. So the domain of $g \cdot h$ is $(-\infty, \infty)$. Since $(3x + 1)(x - 1) = 3x^2 - 2x - 1$, the equation defining the function $g \cdot h$ is

$$(g \cdot h)(x) = 3x^2 - 2x - 1.$$

d. The domain of both g and h is $(-\infty, \infty)$. So the domain of $g - h$ is $(-\infty, \infty)$. Since $(3x + 1) - (x - 1) = 2x + 2$, the equation defining $g - h$ is

$$(g - h)(x) = 2x + 2.$$

Try This. Let $h(x) = \sqrt{x}$ and $j(x) = x$. Find $h + j$ and h/j and state the domain of each function.   ■

Composition of Functions

It is often the case that the output of one function is the input for another function. For example, the number of hamburgers purchased at $1.49 each determines the subtotal. The subtotal is then used to determine the total (including sales tax). So the number of hamburgers actually determines the total and that function is called the *composition* of the other two functions. The composition of functions is defined using function notation as follows.

Definition: Composition of Functions

If f and g are two functions, the **composition** of f and g, written $f \circ g$, is defined by the equation

$$(f \circ g)(x) = f(g(x)),$$

provided that $g(x)$ is in the domain of f. The composition of g and f, written $g \circ f$, is defined by

$$(g \circ f)(x) = g(f(x)),$$

provided that $f(x)$ is in the domain of g.

Note that $f \circ g$ is not the same function as $f \cdot g$, the product of f and g.

For the composition $f \circ g$ to be defined at x, $g(x)$ must be in the domain of f. So the domain of $f \circ g$ is the set of all values of x in the domain of g for which $g(x)$ is in the domain of f. The diagram shown in Fig. 2.52 will help you to understand the composition of functions.

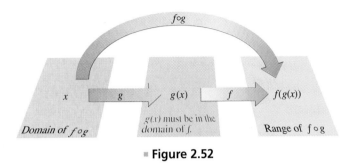

■ **Figure 2.52**

Example 4 Composition of functions defined by sets

Let $g = \{(1, 4), (2, 5), (3, 6)\}$ and $f = \{(3, 8), (4, 9), (5, 10)\}$. Find $f \circ g$.

Solution

Since $g(1) = 4$ and $f(4) = 9$, $(f \circ g)(1) = 9$. So the ordered pair $(1, 9)$ is in $f \circ g$. Since $g(2) = 5$ and $f(5) = 10$, $(f \circ g)(2) = 10$. So $(2, 10)$ is in $f \circ g$. Now $g(3) = 6$, but 6 is not in the domain of f. So there are only two ordered pairs in $f \circ g$:

$$f \circ g = \{(1, 9), (2, 10)\}$$

Try This. Let $h = \{(2, 0), (4, 0), (6, 8)\}$ and $j = \{(0, 7), (6, 0)\}$. Find $j \circ h$ and state the domain. ■

$j \circ h = \{(2, 7) \ (4, 7)\} \langle \qquad Dom. \ \{2, 4\}$

In the next example we find specific values of compositions that are defined by equations.

Example 5 Evaluating compositions defined by equations

Let $f(x) = \sqrt{x}$, $g(x) = 2x - 1$, and $h(x) = x^2$. Find the value of each expression.

a. $(f \circ g)(5)$ **b.** $(g \circ f)(5)$ **c.** $(h \circ g \circ f)(9)$

Solution

a. $(f \circ g)(5) = f(g(5))$ Definition of composition

$\qquad\qquad = f(9)$ $g(5) = 2 \cdot 5 - 1 = 9$

$\qquad\qquad = \sqrt{9}$

$\qquad\qquad = 3$

b. $(g \circ f)(5) = g(f(5)) = g(\sqrt{5}) = 2\sqrt{5} - 1$

c. $(h \circ g \circ f)(9) = h(g(f(9))) = h(g(3)) = h(5) = 5^2 = 25$

You can check these answers with a graphing calculator as shown in Figs. 2.53(a) and (b) on the next page. Enter the functions using the Y = key, then go back to the home screen to evaluate. The symbols Y_1, Y_2, and Y_3 are found in the variables menu (VARS) on a TI-83.

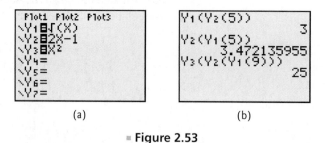

(a)   (b)

■ **Figure 2.53**

Try This. Let $h(x) = \sqrt{x - 1}$ and $j(x) = 2x$. Find $(h \circ j)(5)$ and $(j \circ h)(5)$.

■

In Example 4, the domain of g is $\{1, 2, 3\}$ while the domain of $f \circ g$ is $\{1, 2\}$. To find the domain of $f \circ g$, we remove from the domain of g any number x such that $g(x)$ is not in the domain of f. In the next example we construct compositions of functions defined by equations and determine their domains.

Example **6** **Composition of functions defined by equations**

Let $f(x) = \sqrt{x}$, $g(x) = 2x - 1$, and $h(x) = x^2$. Find each composition function and state its domain.

a. $f \circ g$ **b.** $g \circ f$ **c.** $h \circ g$

Solution

a. In the composition $f \circ g$, $g(x)$ must be in the domain of f. The domain of f is $[0, \infty)$. If $g(x)$ is in $[0, \infty)$, then $2x - 1 \geq 0$, or $x \geq \frac{1}{2}$. So the domain of $f \circ g$ is $\left[\frac{1}{2}, \infty\right)$. Since

$$(f \circ g)(x) = f(g(x)) = f(2x - 1) = \sqrt{2x - 1},$$

the function $f \circ g$ is defined by the equation $(f \circ g)(x) = \sqrt{2x - 1}$, for $x \geq \frac{1}{2}$.

b. Since the domain of g is $(-\infty, \infty)$, $f(x)$ is certainly in the domain of g. So the domain of $g \circ f$ is the same as the domain of f, $[0, \infty)$. Since

$$(g \circ f)(x) = g(f(x)) = g(\sqrt{x}) = 2\sqrt{x} - 1,$$

the function $g \circ f$ is defined by the equation $(g \circ f)(x) = 2\sqrt{x} - 1$. Note that $g \circ f$ is generally not equal to $f \circ g$, but in Section 2.5 we will study special types of functions for which they are equal.

c. Since the domain of h is $(-\infty, \infty)$, $g(x)$ is certainly in the domain of h. So the domain of $h \circ g$ is the same as the domain of g, $(-\infty, \infty)$. Since

$$(h \circ g)(x) = h(g(x)) = h(2x - 1) = (2x - 1)^2 = 4x^2 - 4x + 1,$$

the function $h \circ g$ is defined by $(h \circ g)(x) = 4x^2 - 4x + 1$.

Try This. Let $h(x) = \sqrt{x + 3}$ and $j(x) = 3x$. Find $h \circ j$ and state its domain.

■

Some complicated functions can be thought of as a composition of simpler functions. For example, if $H(x) = (x + 3)^2$, we start with x, add 3 to x, then square the result. These two operations can be accomplished by composition, using $f(x) = x + 3$ followed by $g(x) = x^2$:

$$(g \circ f)(x) = g(f(x)) = g(x + 3) = (x + 3)^2$$

So the function H is the same as the composition of g and f, $H = g \circ f$. Notice that $(f \circ g)(x) = x^2 + 3$ and it is not the same as $H(x)$.

Example 7 Writing a function as a composition

Let $f(x) = \sqrt{x}$, $g(x) = x - 3$, and $h(x) = 2x$. Write each given function as a composition of appropriate functions chosen from f, g, and h.

a. $F(x) = \sqrt{x - 3}$ **b.** $G(x) = x - 6$ **c.** $H(x) = 2\sqrt{x} - 3$

Solution

a. The function F consists of subtracting 3 from x and finding the square root of that result. These two operations can be accomplished by composition, using g followed by f. So $F = f \circ g$. Check this answer as follows:

$$(f \circ g)(x) = f(g(x)) = f(x - 3) = \sqrt{x - 3} = F(x)$$

b. Subtracting 6 from x can be accomplished by subtracting 3 from x and then subtracting 3 from that result, so $G = g \circ g$. Check as follows:

$$(g \circ g)(x) = g(g(x)) = g(x - 3) = (x - 3) - 3 = x - 6 = G(x)$$

c. For the function H, find the square root of x, then multiply by 2, and finally subtract 3. These three operations can be accomplished by composition, using f, then h, and then g. So $H = g \circ h \circ f$. Check as follows:

$$(g \circ h \circ f)(x) = g(h(f(x))) = g\big(h(\sqrt{x})\big) = g\big(2\sqrt{x}\big) = 2\sqrt{x} - 3 = H(x)$$

Try This. Write $K(x) = 2\sqrt{x - 3}$ as a composition of f, g, and h.   ■

Applications

In applied situations, functions are often defined with formulas rather than function notation. In this case, composition can be simply a matter of substitution.

Example 8 Composition with formulas

The radius of a circle is a function of the diameter ($r = d/2$) and the area is a function of the radius ($A = \pi r^2$). Construct a formula that expresses the area as a function of the diameter.

Solution

The formula for A as a function of d is obtained by substituting $d/2$ for r:

$$A = \pi r^2 = \pi \left(\frac{d}{2}\right)^2 = \pi \frac{d^2}{4}$$

The function $A = \pi d^2/4$ is the composition of $r = d/2$ and $A = \pi r^2$.

Try This. The diameter of a circle is a function of the radius ($d = 2r$) and the radius is a function of the circumference ($r = C/(2\pi)$). Construct a formula that expresses d as a function of C. ■

In the next example we find a composition using function notation for the salary of the phone book collector mentioned at the beginning of this section.

Example 9 Composition with function notation

A student's salary (in dollars) for collecting x phone books is given by $S(x) = 0.30x + 240$. The amount of withholding (for taxes) is given by $W(x) = 0.20x$, where x is the salary. Express the withholding as a function of the number of phone books collected.

Solution

Note that x represents the number of phone books in $S(x) = 0.30x + 240$ and x represents salary in $W(x) = 0.20x$. So we can replace the salary x in $W(x)$ with $S(x)$ or $0.30x + 240$, which also represents salary:

$$W(S(x)) = W(0.30x + 240) = 0.20(0.30x + 240) = 0.06x + 48$$

Use a new letter to name this function, say T. Then $T(x) = 0.06x + 48$ gives the amount of withholding (for taxes) as a function of x, where x is the number of phone books.

Try This. The revenue for a sale of x books at \$80 each is given by $R(x) = 80x$. The commission for a revenue of x dollars is given by $C(x) = 0.10x$. Express the commission as a function of the number of books sold. ■

For Thought

True or False? Explain.

1. If $f = \{(2, 4)\}$ and $g = \{(1, 5)\}$, then $f + g = \{(3, 9)\}$.

2. If $f = \{(1, 6), (9, 5)\}$ and $g = \{(1, 3), (9, 0)\}$, then $f/g = \{(1, 2)\}$.

3. If $f = \{(1, 6), (9, 5)\}$ and $g = \{(1, 3), (9, 0)\}$, then $f \cdot g = \{(1, 18), (9, 0)\}$.

4. If $f(x) = x + 2$ and $g(x) = x - 3$, then $(f \cdot g)(5) = 14$.

5. If $s = P/4$ and $A = s^2$, then A is a function of P.

6. If $f(3) = 19$ and $g(19) = 99$, then $(g \circ f)(3) = 99$.

7. If $f(x) = \sqrt{x}$ and $g(x) = x - 2$, then $(f \circ g)(x) = \sqrt{x} - 2$.

8. If $f(x) = 5x$ and $g(x) = x/5$, then $(f \circ g)(x) = (g \circ f)(x) = x$.

9. If $F(x) = (x - 9)^2$, $g(x) = x^2$, and $h(x) = x - 9$, then $F = h \circ g$.

10. If $f(x) = \sqrt{x}$ and $g(x) = x - 2$, then the domain of $f \circ g$ is $[2, \infty)$.

2.4 Exercises

Let $f(x) = x - 3$ and $g(x) = x^2 - x$. Find and simplify each expression.

1. $(f + g)(2)$ **2.** $(g + f)(3)$ **3.** $(f - g)(-2)$

4. $(g - f)(-6)$ **5.** $(f \cdot g)(-1)$ **6.** $(g \cdot f)(0)$

7. $(f/g)(4)$ **8.** $(g/f)(4)$ **9.** $(f + g)(a)$

10. $(f - g)(b)$ **11.** $(f \cdot g)(a)$ **12.** $(f/g)(b)$

Let $f = \{(-3, 1), (0, 4), (2, 0)\}$, $g = \{(-3, 2), (1, 2), (2, 6), (4, 0)\}$, and $h = \{(2, 4), (1, 0)\}$. Find each function and state the domain of each function.

13. $f + g$ **14.** $f + h$ **15.** $f - g$ **16.** $f - h$

17. $f \cdot g$ **18.** $f \cdot h$ **19.** g/f **20.** f/g

Let $f(x) = \sqrt{x}$, $g(x) = x - 4$, and $h(x) = \frac{1}{x - 2}$. Find an equation defining each function and state the domain of the function.

21. $f + g$ **22.** $f + h$ **23.** $f - h$ **24.** $h - g$

25. $g \cdot h$ **26.** $f \cdot h$ **27.** g/f **28.** f/g

Let $f = \{(-3, 1), (0, 4), (2, 0)\}$, $g = \{(-3, 2), (1, 2), (2, 6), (4, 0)\}$, and $h = \{(2, 4), (1, 0)\}$. Find each function.

29. $f \circ g$ **30.** $g \circ f$ **31.** $f \circ h$

32. $h \circ f$ **33.** $h \circ g$ **34.** $g \circ h$

Let $f(x) = 3x - 1$, $g(x) = x^2 + 1$, and $h(x) = \frac{x + 1}{3}$. Evaluate each expression. Round approximate answers to three decimal places.

35. $f(g(-1))$ **36.** $g(f(-1))$ **37.** $(f \circ h)(5)$

38. $(h \circ f)(-7)$ **39.** $(f \circ g)(4.39)$ **40.** $(g \circ h)(-9.87)$

41. $(g \circ h \circ f)(2)$ **42.** $(h \circ f \circ g)(3)$ **43.** $(f \circ g \circ h)(2)$

44. $(h \circ g \circ f)(0)$ **45.** $(f \circ h)(a)$ **46.** $(h \circ f)(w)$

47. $(f \circ g)(t)$ **48.** $(g \circ f)(m)$

Let $f(x) = x - 2$, $g(x) = \sqrt{x}$, and $h(x) = \frac{1}{x}$. Find an equation defining each function and state the domain of the function.

49. $f \circ g$ **50.** $g \circ f$ **51.** $f \circ h$

52. $h \circ f$ **53.** $h \circ g$ **54.** $g \circ h$

55. $f \circ f$ **56.** $g \circ g$ **57.** $h \circ g \circ f$

58. $f \circ g \circ h$ **59.** $h \circ f \circ g$ **60.** $g \circ h \circ f$

Let $f(x) = |x|$, $g(x) = x - 7$, and $h(x) = x^2$. Write each of the following functions as a composition of functions chosen from f, g, and h.

61. $F(x) = x^2 - 7$ **62.** $G(x) = |x| - 7$

63. $H(x) = (x - 7)^2$ **64.** $M(x) = |x - 7|$

65. $N(x) = (|x| - 7)^2$ **66.** $R(x) = |x^2 - 7|$

67. $P(x) = |x - 7| - 7$ **68.** $Q(x) = (x^2 - 7)^2$

69. $S(x) = x - 14$ **70.** $T(x) = x^4$

For each given function $f(x)$, find two functions $g(x)$ and $h(x)$ such that $f = h \circ g$. Answers may vary.

71. $f(x) = x^3 - 2$ **72.** $f(x) = (x - 2)^3$

73. $f(x) = \sqrt{x + 5}$ **74.** $f(x) = \sqrt{x} + 5$

75. $f(x) = \sqrt{3x - 1}$ **76.** $f(x) = 3\sqrt{x} - 1$

77. $f(x) = 4|x| + 5$ **78.** $f(x) = |4x + 5|$

Use the two given functions to write y as a function of x.

79. $y = 2a - 3, a = 3x + 1$

80. $y = -4d - 1, d = -3x - 2$

81. $y = w^2 - 2, w = x + 3$

82. $y = 3t^2 - 3, t = x - 1$

83. $y = 3m - 1, m = \dfrac{x + 1}{3}$ **84.** $y = 2z + 5, z = \dfrac{1}{2}x - \dfrac{5}{2}$

Find each function from the given verbal description of the function.

85. If m is n minus 4, and y is the square of m, then write y as a function of n.

86. If u is the sum of t and 9, and v is u divided by 3, then write v as a function of t.

87. If w is equal to the sum of x and 16, z is the square root of w, and y is z divided by 8, then write y as a function of x.

88. If a is the cube of b, c is the sum of a and 25, and d is the square root of c, then write d as a function of b.

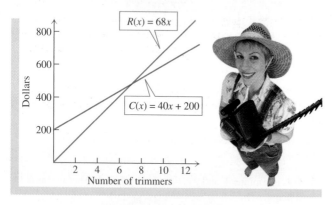

Graph each function on a graphing calculator without simplifying the given expression. Examine the graph and write a function for the graph. Then simplify the original function and see whether it is the same as your function. Is the domain of the original function equal to the domain of the function after it is simplified?

89. $y = ((x - 1)/(x + 1) + 1)/((x - 1)/(x + 1) - 1)$

90. $y = ((3x + 1)/(x - 1) + 1)/((3x + 1)/(x - 1) - 3)$

Define $y_1 = \sqrt{x + 1}$ and $y_2 = 3x - 4$ on your graphing calculator. For each function y_3, defined in terms of y_1 and y_2, determine the domain and range of y_3 from its graph on your calculator and explain what each graph illustrates.

91. $y_3 = y_1 + y_2$ **92.** $y_3 = 3y_1 - 4$

93. $y_3 = \sqrt{y_2 + 1}$ **94.** $y_3 = \sqrt{y_1 + 1}$

Define $y_1 = \sqrt[3]{x}$, $y_2 = \sqrt{x}$, and $y_3 = x + 4$. For each function y_4, determine the domain and range of y_4 from its graph on your calculator and explain what each graph illustrates.

95. $y_4 = y_1 + y_2 + y_3$ **96.** $y_4 = \sqrt{y_1 + 4}$

Solve each problem.

97. *Profitable Business* Charles buys factory reconditioned hedge trimmers for $40 each and sells them on Ebay for $68 each. He has a fixed cost of $200 per month. If x is the number of hedge trimmers he sells per month, then his revenue and cost (in dollars) are given by $R(x) = 68x$ and $C(x) = 40x + 200$. Find a formula for the function $P(x) = R(x) - C(x)$. For what values of x is his profit positive?

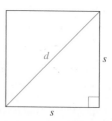

■ **Figure for Exercise 97**

98. *Profit* The revenue in dollars that a company receives for installing x alarm systems per month is given by $R(x) = 3000x - 20x^2$, while the cost in dollars is given by $C(x) = 600x + 4000$. The function $P(x) = R(x) - C(x)$ gives the profit for installing x alarm systems per month. Find $P(x)$ and simplify it.

99. Write the area A of a square with a side of length s as a function of its diagonal d.

     HINT Write down the formulas involving the area, perimeter, and diagonal of square.

100. Write the perimeter of a square P as a function of the area A.

■ **Figure for Exercises 99 and 100**

101. *Deforestation in Nigeria* In Nigeria deforestation occurs at the rate of about 5.2% per year. If x is the total forest area of Nigeria at the start of 2000, then $f(x) = 0.948x$ is the amount of forest land in Nigeria at the start of 2001. Find and interpret $(f \circ f)(x)$. Find and interpret $(f \circ f \circ f)(x)$.

102. *Doubling Your Money* On the average, money invested in bonds doubles every 12 years. So $f(x) = 2x$ gives the value of x dollars invested in bonds after 12 years. Find and interpret $(f \circ f)(x)$. Find and interpret $(f \circ f \circ f)(x)$.

103. *Hamburgers* If hamburgers are $1.20 each, then $C(x) = 1.20x$ gives the pre-tax cost in dollars for x hamburgers. If sales tax is 5%, then $T(x) = 1.05x$ gives the total cost when the pre-tax cost is x dollars. Write the total cost as a function of the number of hamburgers.

104. *Laying Sod* Southern Sod will deliver and install 20 pallets of St. Augustine sod for $2200 or 30 pallets for $3200 not including tax.
     a. Write the cost (not including tax) as a linear function of x, where x is the number of pallets.
     b. Write a function that gives the total cost (including tax at 9%) as a function of x, where x is the cost (not including tax).
     c. Find the function that gives the total cost as a function of x, where x is the number of pallets.

105. *Displacement-Length Ratio* The displacement-length ratio D indicates whether a sailboat is relatively heavy or relatively light:

$$D = (d \div 2240) \div x$$

where d is the displacement in pounds and

$$x = (L \div 100)^3.$$

where L is the length at the waterline in feet (*Sailing*, www.sailing.com). Assuming that the displacement is 26,000 pounds, write D as a function of L and simplify it.

106. *Sail Area-Displacement Ratio*   The sail area-displacement ratio S measures the sail power available to drive a sailboat:

$$S = A \div y$$

where A is the sail area in square feet and

$$y = (d \div 64)^{2/3}$$

where d is the displacement in pounds. Assuming that the sail area is 6500 square feet, write S as a function of d and simplify it.

107. *Area of a Window*   A window is in the shape of a square with a side of length s, with a semicircle of diameter s adjoining the top of the square. Write the total area of the window W as a function of s.

  HINT Start with the formulas for the area of a square and the area of a circle.

■ **Figure for Exercises 107 and 108**

108. *Area of a Window*   Using the window of Exercise 107, write the area of the square A as a function of the area of the semicircle S.

109. *Packing a Square Piece of Glass*   A glass prism with a square cross section is to be shipped in a cylindrical cardboard tube that has an inside diameter of d inches. Given that s is the length of a side of the square and the glass fits snugly into the tube, write s as a function of d.

  HINT Use the Pythagorean theorem and the formula for the area of a circle.

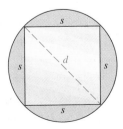

■ **Figure for Exercise 109**

110. *Packing a Triangular Piece of Glass*   A glass prism is to be shipped in a cylindrical cardboard tube that has an inside diameter of d inches. Given that the cross section of the prism is an equilateral triangle with side of length p and the prism fits snugly into the tube, write p as a function of d.

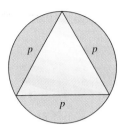

■ **Figure for Exercise 110**

For Writing/Discussion

111. Sears often has Super Saturday sales during which everything is 10% off, even items already on sale. If a coat is on sale at 25%, then is it 35% off on Super Saturday? What operation with functions is at work here? Explain your answers.

112. We can combine two functions to obtain a new function by addition, subtraction, multiplication, division, and composition. Are any of these operations commutative? Associative? Explain your answers and give examples.

Thinking Outside the Box XXIV & XXV

Thirty-Percent Reduction   If each of the numbers in the expression $ab^2c^3d^4$ is decreased by 30%, then the value of the expression is decreased by what percent? Give the answer to the nearest tenth of a percent.

Whole Number Expression   What is the largest whole number N that cannot be expressed as $N = 3x + 11y$ where x and y are whole numbers?

2.4 Pop Quiz

1. Write the area of a circle as a function of its diameter.

Let $f(x) = x^2$ and $g(x) = x - 2$. Find and simplify.

2. $(f + g)(3)$ 3. $(f \cdot g)(4)$ 4. $(f \circ g)(5)$

Let $m = \{(1, 3), (4, 8)\}$ and $n = \{(3, 5), (4, 9)\}$. Find each function.

5. $m + n$ 6. $n \circ m$

Let $h(x) = x^2$ and $j(x) = \sqrt{x + 2}$. Find the domain of each function.

7. $h + j$ 8. $h \circ j$ 9. $j \circ h$

Linking Concepts

For Individual or Group Explorations

Modeling Glue Coverage

Frank is designing a notched trowel for spreading glue. The notches on the trowel are in the shape of semicircles. The diameter of each notch and the space between consecutive notches are d inches as shown in the figure. Suppose that the trowel is used to make parallel beads of glue on one square foot of floor.

a) Write the number of parallel beads made as a function of d.

b) The cross section of each parallel bead is a semicircle with diameter d. Write the area of a cross section as a function of d.

c) Write the volume of one of the parallel beads as a function of d.

d) Write the volume of glue on one square foot of floor as a function of d.

e) Write the number of square feet that one gallon of glue will cover as a function of d. (Use 1 ft³ = 7.5 gal.)

f) Suppose that the trowel has square notches where the sides of the squares are d inches in length and the distance between consecutive notches is d inches. Write the number of square feet that one gallon of glue will cover as a function of d.

2.5

Inverse Functions

It is possible for one function to undo what another function does. For example, squaring undoes the operation of taking a square root. The composition of two such functions is the identity function. In this section we explore this idea in detail.

One-to-One and Invertible Functions

Consider a medium pizza that costs $5 plus $2 per topping. Table 2.1 shows the ordered pairs of the function that determines the cost. Note that for every number of toppings there is a unique cost and for every cost there is a unique number of toppings. There is a **one-to-one correspondence** between the domain and range of this function and the function is a *one-to-one function*. For a function that is not one-to-one, consider a Wendy's menu. Every item corresponds to a unique price, but the price $0.99 corresponds to many different items.

■ **Table 2.1**

Toppings x	Cost y	
0	$ 5	
1	7	
2	9	
3	11	
4	13	

**Definition:
One-to-One Function**

> If a function has no two ordered pairs with different first coordinates and the same second coordinate, then the function is called **one-to-one.**

Because the function in Table 2.1 is one-to-one, we can determine the number of toppings when given the cost as shown in Table 2.2. Of course we could just read Table 2.1 backwards, but we make a new table to emphasize that there is a new function under discussion. Table 2.2 is the *inverse function* of the function in Table 2.1.

■ **Table 2.2**

Cost x	Toppings y	
$ 5	0	
7	1	
9	2	
11	3	
13	4	

A function is a set of ordered pairs in which no two ordered pairs have the same first coordinates and different second coordinates. If we interchange the x- and y-coordinates in each ordered pair of a function, as in Tables 2.1 and 2.2, the resulting set of ordered pairs might or might not be a function. If the original function is one-to-one, then the set obtained by interchanging the coordinates in each ordered pair is a function, the inverse function. If a function is one-to-one, then it has an inverse function or it is **invertible.**

**Definition:
Inverse Function**

> The **inverse** of a one-to-one function f is the function f^{-1} (read "f inverse"), where the ordered pairs of f^{-1} are obtained by interchanging the coordinates in each ordered pair of f.

In this notation, the number -1 in f^{-1} does not represent a negative exponent. It is merely a symbol for denoting the inverse function.

Example 1 Finding an inverse function

For each function, determine whether it is invertible. If it is invertible, then find the inverse.

a. $f = \{(-2, 3), (4, 5), (2, 3)\}$ **b.** $g = \{(3, 1), (5, 2), (7, 4), (9, 8)\}$

Solution

a. This function f is *not* one-to-one because of the ordered pairs $(-2, 3)$ and $(2, 3)$. So f is not invertible.
b. The function g is one-to-one, and so g is invertible. The inverse of g is the function $g^{-1} = \{(1, 3), (2, 5), (4, 7), (8, 9)\}$.

Try This. Determine whether $h = \{(2, 1), (3, 4), (4, 0)\}$ is invertible. If it is invertible, then find the inverse function.   ■

Example 2 Using inverse function notation

Let $f = \{(1, 3), (2, 4), (5, 7)\}$. Find f^{-1}, $f^{-1}(3)$, and $(f^{-1} \circ f)(1)$.

Solution

Interchange the x- and y-coordinates of each ordered pair of f to find f^{-1}:

$$f^{-1} = \{(3, 1), (4, 2), (7, 5)\}$$

To find the value of $f^{-1}(3)$, notice that $f^{-1}(3)$ is the second coordinate when the first coordinate is 3 in the function f^{-1}. So $f^{-1}(3) = 1$. To find the composition, use the definition of composition of functions:

$$(f^{-1} \circ f)(1) = f^{-1}(f(1)) = f^{-1}(3) = 1$$

Try This. Let $h = \{(2, 1), (3, 4), (4, 0)\}$. Find $h(3)$, $h^{-1}(4)$, and $(h \circ h^{-1})(1)$.   ■

Since the coordinates in the ordered pairs are interchanged, the domain of f^{-1} is the range of f, and the range of f^{-1} is the domain of f. If f^{-1} is the inverse function of f, then certainly f is the inverse of f^{-1}. The functions f and f^{-1} are inverses of each other.

Determining Whether a Function Is One-to-One

If a function is given as a short list of ordered pairs, then it is easy to determine whether the function is one-to-one. A graph of a function can also be used to determine whether a function is one-to-one using the **horizontal line test.**

Horizontal Line Test

If each horizontal line crosses the graph of a function at no more than one point, then the function is one-to-one.

The graph of a one-to-one function never has the same *y*-coordinate for two different *x*-coordinates on the graph. So if it is possible to draw a horizontal line that crosses the graph of a function two or more times, then the function is not one-to-one.

Example **3** **The horizontal line test**

Use the horizontal line test to determine whether the functions shown in Fig. 2.54 are one-to-one.

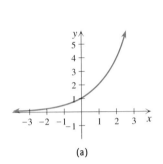

(a)

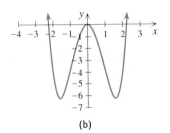

(b)

■ **Figure 2.55**

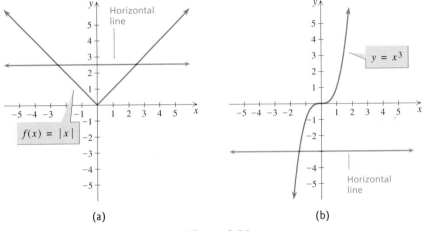

(a)   (b)

■ **Figure 2.54**

Solution

The function $f(x) = |x|$ is not one-to-one because it is possible to draw a horizontal line that crosses the graph twice as shown in Fig. 2.54(a). The function $y = x^3$ in Fig. 2.54(b) is one-to-one because it appears to be impossible to draw a horizontal line that crosses the graph more than once.

Try This. Use the horizontal line test to determine whether the functions shown in Fig. 2.55 are one-to-one.   ■

The horizontal line test explains the visual difference between the graph of a one-to-one function and the graph of a function that is not one-to-one. Because no graph of a function is perfectly accurate, conclusions made from a graph alone may not be correct. For example, from the graph of $y = x^3 - 0.01x$ shown in Fig. 2.56(a) we would conclude that the function is one-to-one and invertible. However, another view of the same function in Fig. 2.56(b) shows that the function is not one-to-one.

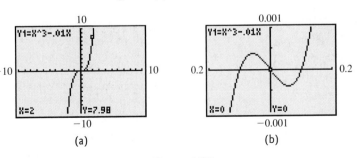

(a)   (b)

■ **Figure 2.56**

Using the equation for a function we can often determine conclusively whether the function is one-to-one. A function f is not one-to-one if it is possible to find two different numbers x_1 and x_2 such that $f(x_1) = f(x_2)$. For example, for the function $f(x) = x^2$, it is possible to find two different numbers, 2 and -2, such that $f(2) = f(-2)$. So $f(x) = x^2$ is not one-to-one. To prove that a function f is one-to-one, we must show that $f(x_1) = f(x_2)$ implies that $x_1 = x_2$.

Example **4** **Using the definition of one-to-one**

Determine whether each function is one-to-one.

a. $f(x) = \dfrac{2x + 1}{x - 3}$ **b.** $g(x) = |x|$

Solution

a. If $f(x_1) = f(x_2)$, then we have the following equation:

$$\frac{2x_1 + 1}{x_1 - 3} = \frac{2x_2 + 1}{x_2 - 3}$$

$$(2x_1 + 1)(x_2 - 3) = (2x_2 + 1)(x_1 - 3) \qquad \text{Multiply by the LCD.}$$

$$2x_1x_2 + x_2 - 6x_1 - 3 = 2x_2x_1 + x_1 - 6x_2 - 3$$

$$7x_2 = 7x_1$$

$$x_2 = x_1$$

Since $f(x_1) = f(x_2)$ implies that $x_1 = x_2$, $f(x)$ is a one-to-one function.

b. If $g(x_1) = g(x_2)$, then $|x_1| = |x_2|$. But this does not imply that $x_1 = x_2$, because two different numbers can have the same absolute value. For example $|3| = |-3|$ but $3 \neq -3$. So g is not one-to-one.

Try This. Determine whether $h(x) = 5x^2$ is one-to-one.   ■

Inverse Functions Using Function Notation

The function $f(x) = 2x + 5$ gives the cost of a pizza where \$5 is the basic cost and x is the number of toppings at \$2 each. If we assume that x could be any real number, this function defines the set

$$f = \{(x, y) \mid y = 2x + 5\}.$$

Since the coordinates of each ordered pair are interchanged for f^{-1}, the ordered pairs of f^{-1} must satisfy $x = 2y + 5$:

$$f^{-1} = \{(x, y) \mid x = 2y + 5\}.$$

Since $x = 2y + 5$ is equivalent to $y = \dfrac{x - 5}{2}$,

$$f^{-1} = \left\{(x, y) \mid y = \frac{x - 5}{2}\right\}.$$

The function f^{-1} is described in function notation as $f^{-1}(x) = \frac{x-5}{2}$. The function f^{-1} gives the number of toppings as a function of cost.

A one-to-one function f pairs members of the domain of f with members of the range of f, and its inverse function f^{-1} exactly reverses those pairings as shown in Fig. 2.57. The above example suggests the following steps for finding an inverse function using function notation.

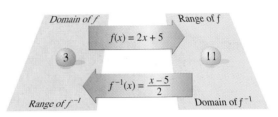

Domain of f | Range of f

$f(x) = 2x + 5$

3 | 11

$f^{-1}(x) = \frac{x-5}{2}$

Range of f^{-1} | Domain of f^{-1}

■ **Figure 2.57**

PROCEDURE   Finding $f^{-1}(x)$ by the Switch-and-Solve Method

To find the inverse of a one-to-one function given in function notation:

1. Replace $f(x)$ by y.

2. Interchange x and y.

3. Solve the equation for y.

4. Replace y by $f^{-1}(x)$.

5. Check that the domain of f is the range of f^{-1} and the range of f is the domain of f^{-1}.

Example 5 **The switch-and-solve method**

Find the inverse of each function.

a. $f(x) = 4x - 1$ **b.** $f(x) = \dfrac{2x+1}{x-3}$

Solution

a. The graph of $f(x) = 4x - 1$ is a line with slope 4. By the horizontal line test the function is one-to-one and invertible. Replace $f(x)$ with y to get $y = 4x - 1$. Next, interchange x and y to get $x = 4y - 1$. Now solve for y:

$$x = 4y - 1$$

$$x + 1 = 4y$$

$$\frac{x+1}{4} = y$$

Replace y by $f^{-1}(x)$ to get $f^{-1}(x) = \frac{x+1}{4}$. The domain of f is $(-\infty, \infty)$ and that is the range of f^{-1}. The range of f is $(-\infty, \infty)$ and that is the domain of f^{-1}.

b. In Example 4(a) we showed that f is a one-to-one function. So we can find f^{-1} by interchanging x and y and solving for y:

$$y = \frac{2x + 1}{x - 3} \quad \text{Replace } f(x) \text{ by } y.$$

$$x = \frac{2y + 1}{y - 3} \quad \text{Interchange } x \text{ and } y.$$

$$x(y - 3) = 2y + 1 \quad \text{Solve for } y.$$

$$xy - 3x = 2y + 1$$

$$xy - 2y = 3x + 1$$

$$y(x - 2) = 3x + 1$$

$$y = \frac{3x + 1}{x - 2}$$

Replace y by $f^{-1}(x)$ to get $f^{-1}(x) = \frac{3x + 1}{x - 2}$. The domain of f is all real numbers except 3. The range of f^{-1} is the set of all real numbers except 3. We exclude 3 because $\frac{3x + 1}{x - 2} = 3$ has no solution. Check that the range of f is equal to the domain of f^{-1}.

Try This. Use switch-and-solve to find the inverse of $h(x) = x^3 - 5$. ■

If a function f is defined by a short list of ordered pairs, then we simply write all of the pairs in reverse to find f^{-1}. If two functions are defined by formulas, it may not be obvious whether the ordered pairs of one are the reverse of the ordered pairs of the other. However, we can use the compositions of the functions to make the determination as stated in the following theorem.

Theorem: Verifying Whether f and g Are Inverses

The functions f and g are inverses of each other if and only if

1. $g(f(x)) = x$ for every x in the domain of f and
2. $f(g(x)) = x$ for every x in the domain of g.

Example 6 **Using composition to verify inverse functions**

Determine whether the functions $f(x) = x^3 - 1$ and $g(x) = \sqrt[3]{x + 1}$ are inverse functions.

Solution

Find $f(g(x))$ and $g(f(x))$:

$$f(g(x)) = f\left(\sqrt[3]{x + 1}\right) = \left(\sqrt[3]{x + 1}\right)^3 - 1 = x + 1 - 1 = x$$

$$g(f(x)) = g(x^3 - 1) = \sqrt[3]{x^3 - 1 + 1} = \sqrt[3]{x^3} = x$$

Since $f(g(x)) = x$ is true for any real number (the domain of g) and $g(f(x)) = x$ is true for any real number (the domain of f), the functions f and g are inverses of each other.

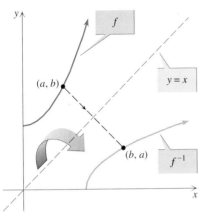

■ **Figure 2.58**

Try This. Determine whether $f(x) = 2x + 1$ and $g(x) = \dfrac{x-1}{2}$ are inverse functions. ■

Only one-to-one functions are invertible. However, sometimes it is possible to restrict the domain of a function that is not one-to-one so that it is one-to-one on the restricted domain. For example, $f(x) = x^2$ for x in $(-\infty, \infty)$ is not one-to-one and not invertible. But $f(x) = x^2$ for $x \geq 0$ is one-to-one and invertible. The inverse of $f(x) = x^2$ for $x \geq 0$ is the square-root function $f^{-1}(x) = \sqrt{x}$.

Graphs of f and f^{-1}

If a point (a, b) is on the graph of an invertible function f, then (b, a) is on the graph of f^{-1}. See Fig. 2.58. Since the points (a, b) and (b, a) are symmetric with respect to the line $y = x$, the graph of f^{-1} is a reflection of f with respect to the line $y = x$. This reflection property of inverse functions gives us a way to visualize the graphs of inverse functions.

Example **7** Graphing a function and its inverse

Find the inverse of the function $f(x) = \sqrt{x - 1}$ and graph both f and f^{-1} on the same coordinate axes.

Solution

If $\sqrt{x_1 - 1} = \sqrt{x_2 - 1}$, then $x_1 = x_2$. So f is one-to-one. Since the range of $f(x) = \sqrt{x - 1}$ is $[0, \infty)$, f^{-1} must be a function with domain $[0, \infty)$. To find a formula for the inverse, interchange x and y in $y = \sqrt{x - 1}$ to get $x = \sqrt{y - 1}$. Solving this equation for y gives $y = x^2 + 1$. So $f^{-1}(x) = x^2 + 1$ for $x \geq 0$. The graph of f is a translation one unit to the right of the graph of $y = \sqrt{x}$, and the graph of f^{-1} is a translation one unit upward of the graph of $y = x^2$ for $x \geq 0$. Both graphs are shown in Fig. 2.59 along with the line $y = x$. Notice that the graph of f^{-1} is a reflection of the graph of f, the domain of f is the range of f^{-1}, and the range of f is the domain of f^{-1}.

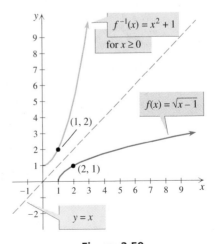

■ **Figure 2.59**

Try This. Find the inverse of $f(x) = \sqrt{x + 2}$ and graph f and f^{-1}. ■

Finding Inverse Functions Mentally

It is no surprise that the inverse of $f(x) = x^2$ for $x \geq 0$ is the function $f^{-1}(x) = \sqrt{x}$. For nonnegative numbers, taking the square root undoes what squaring does. It is also no surprise that the inverse of $f(x) = 3x$ is $f^{-1}(x) = x/3$ or that the inverse of $f(x) = x + 9$ is $f^{-1}(x) = x - 9$. If an invertible function involves a single operation, it is usually easy to write the inverse function because for most operations there is an inverse operation. See Table 2.3. If an invertible function involves more than one operation, we find the inverse function by applying the inverse operations in the opposite order from the order in which they appear in the original function.

■ **Table 2.3**

Function	Inverse
$f(x) = 2x$	$f^{-1}(x) = x/2$
$f(x) = x - 5$	$f^{-1}(x) = x + 5$
$f(x) = \sqrt{x}$	$f^{-1}(x) = x^2 \ (x \geq 0)$
$f(x) = x^3$	$f^{-1}(x) = \sqrt[3]{x}$
$f(x) = 1/x$	$f^{-1}(x) = 1/x$
$f(x) = -x$	$f^{-1}(x) = -x$

Example **8** **Finding inverses mentally**

Find the inverse of each function mentally.

a. $f(x) = 2x + 1$ **b.** $g(x) = \dfrac{x^3 + 5}{2}$

Solution

a. The function $f(x) = 2x + 1$ is a composition of multiplying x by 2 and then adding 1. So the inverse function is a composition of subtracting 1 and then dividing by 2: $f^{-1}(x) = (x - 1)/2$. See Fig. 2.60.

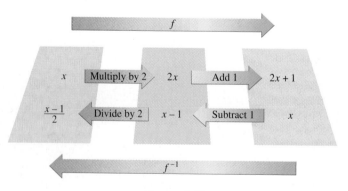

■ **Figure 2.60**

b. The function $g(x) = (x^3 + 5)/2$ is a composition of cubing x, adding 5, and dividing by 2. The inverse is a composition of multiplying by 2, subtracting 5, and then taking the cube root: $g^{-1}(x) = \sqrt[3]{2x - 5}$. See Fig. 2.61.

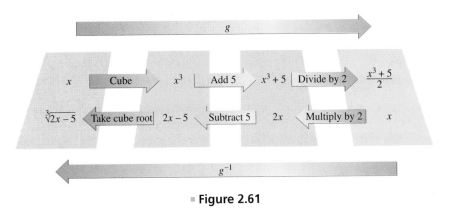

■ **Figure 2.61**

Try This.   Find the inverse of $f(x) = \frac{2}{3}x + 6$ mentally.   ■

Function Gallery: Some Inverse Functions

Linear

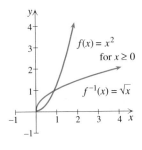

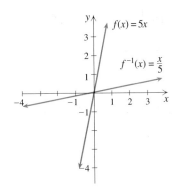

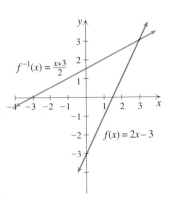

Powers and Roots

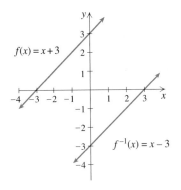

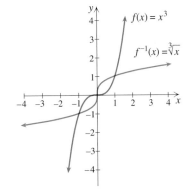

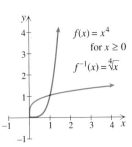

For Thought

True or False? Explain.

1. The inverse of the function $\{(2, 3), (5, 5)\}$ is $\{(5, 2), (5, 3)\}$.

2. The function $f(x) = -2$ is an invertible function.

3. If $g(x) = x^2$, then $g^{-1}(x) = \sqrt{x}$.

4. The only functions that are invertible are the one-to-one functions.

5. Every function has an inverse function.

6. The function $f(x) = x^4$ is invertible.

7. If $f(x) = 3\sqrt{x - 2}$, then $f^{-1}(x) = \dfrac{(x + 2)^2}{3}$ for $x \geq 0$.

8. If $f(x) = |x - 3|$, then $f^{-1}(x) = |x| + 3$.

9. According to the horizontal line test, $y = |x|$ is one-to-one.

10. The function $g(x) = -x$ is the inverse of the function $f(x) = -x$.

2.5 Exercises

Determine whether each function is invertible and explain your answer.

1. The function that pairs the universal product code of an item at Sears with a price.

2. The function that pairs the number of days since your birth with your age in years.

3. The function that pairs the length of a VCR tape in feet with the playing time in minutes.

4. The function that pairs the speed of your car in miles per hour with the speed in kilometers per hour.

5. The function that pairs the number of days of a hotel stay with the total cost for the stay.

6. The function that pairs the number of days that a deposit of $100 earns interest at 6% compounded daily with the amount of interest.

Determine whether each function is invertible. If it is invertible, find the inverse.

7. $\{(9, 3), (2, 2)\}$

8. $\{(4, 5), (5, 6)\}$

9. $\{(-1, 0), (1, 0), (5, 0)\}$

10. $\{(1, 2), (5, 2), (6, 7)\}$

11. $\{(3, 3), (2, 2), (4, 4), (7, 7)\}$

12. $\{(1, 1), (2, 4), (4, 16), (7, 49)\}$

13. $\{(1, 1), (2, 2), (4.5, 2), (5, 5)\}$

14. $\{(0, 2), (2, 0), (1, 0), (4, 6)\}$

For each function f, find f^{-1}, $f^{-1}(5)$, and $(f^{-1} \circ f)(2)$.

15. $f = \{(2, 1), (3, 5)\}$

16. $f = \{(-1, 5), (0, 0), (2, 6)\}$

17. $f = \{(-3, -3), (0, 5), (2, -7)\}$

18. $f = \{(3.2, 5), (2, 1.99)\}$

Use the horizontal line test to determine whether each function is one-to-one.

19. $f(x) = x^2 - 3x$

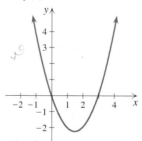

20. $g(x) = |x - 2| + 1$

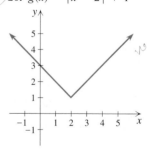

21. $y = \sqrt[3]{x} + 2$

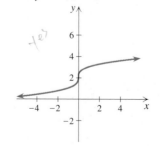

22. $y = (x - 2)^3$

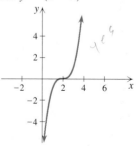

23. $y = x^3 - x$

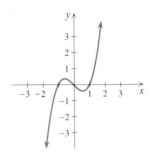

24. $y = \dfrac{1}{x}$

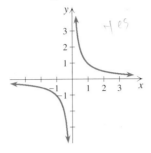

Determine whether each function is one-to-one.

25. $f(x) = 2x - 3$

26. $h(x) = 4x - 9$

27. $q(x) = \dfrac{1 - x}{x - 5}$

28. $g(x) = \dfrac{x + 2}{x - 3}$

29. $p(x) = |x + 1|$

30. $r(x) = 2|x - 1|$

31. $w(x) = x^2 + 3$

32. $v(x) = 2x^2 - 1$

33. $k(x) = \sqrt[3]{x} + 9$

34. $t(x) = \sqrt{x + 3}$

 Determine whether each function is invertible by inspecting its graph on a graphing calculator.

35. $f(x) = (x + 0.01)(x + 0.02)(x + 0.03)$

36. $f(x) = x^3 - 0.6x^2 + 0.11x - 0.006$

37. $f(x) = |x - 2| - |5 - x|$

38. $f(x) = \sqrt[3]{0.1x + 3} + \sqrt[3]{-0.1x}$

Find the inverse of each function using the procedure for the switch-and-solve method on page 243.

39. $f(x) = 3x - 7$

40. $f(x) = -2x + 5$

41. $f(x) = 2 + \sqrt{x - 3}$

42. $f(x) = \sqrt{3x - 1}$

43. $f(x) = -x - 9$

44. $f(x) = -x + 3$

45. $f(x) = \dfrac{x + 3}{x - 5}$

46. $f(x) = \dfrac{2x - 1}{x - 6}$

47. $f(x) = -\dfrac{1}{x}$

48. $f(x) = x$

49. $f(x) = \sqrt[3]{x - 9} + 5$

50. $f(x) = \sqrt[3]{\dfrac{x}{2}} + 5$

51. $f(x) = (x - 2)^2$ for $x \geq 2$

52. $f(x) = x^2$ for $x \leq 0$

In each case find $f(g(x))$ and $g(f(x))$. Then determine whether g and f are inverse functions.

53. $f(x) = 4x + 4,\ g(x) = 0.25x - 1$

54. $f(x) = 20 - 5x,\ g(x) = -0.2x + 4$

55. $f(x) = x^2 + 1,\ g(x) = \sqrt{x - 1}$

56. $f(x) = \sqrt[4]{x},\ g(x) = x^4$

57. $f(x) = \dfrac{1}{x} + 3,\ g(x) = \dfrac{1}{x - 3}$

58. $f(x) = 4 - \dfrac{1}{x},\ g(x) = \dfrac{1}{4 - x}$

59. $f(x) = \sqrt[3]{\dfrac{x - 2}{5}},\ g(x) = 5x^3 + 2$

60. $f(x) = x^3 - 27,\ g(x) = \sqrt[3]{x} + 3$

 For each exercise, graph the three functions on the same screen of a graphing calculator. Enter these functions as shown without simplifying any expressions. Explain what these exercises illustrate.

61. $y_1 = \sqrt[3]{x} - 1,\ y_2 = (x + 1)^3,\ y_3 = \left(\sqrt[3]{x} - 1 + 1\right)^3$

62. $y_1 = (2x - 1)^{1/3},\ y_2 = (x^3 + 1)/2,$
$\quad y_3 = (2((x^3 + 1)/2) - 1)^{1/3}$

Determine whether each pair of functions f and g are inverses of each other.

63.

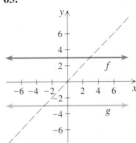

64.

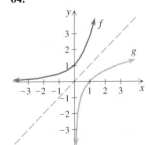

65.

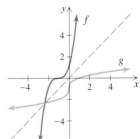

66.

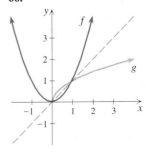

For each function f, sketch the graph of f^{-1}.

67.

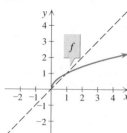

68.

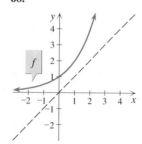

69.

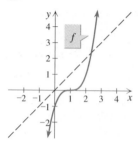

70.

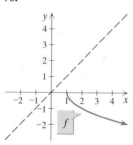

Find the inverse of each function and graph both f and f^{-1} on the same coordinate plane.

71. $f(x) = 3x + 2$

72. $f(x) = -x - 8$

73. $f(x) = x^2 - 4$ for $x \geq 0$

74. $f(x) = 1 - x^2$ for $x \geq 0$

75. $f(x) = x^3$

76. $f(x) = -x^3$

77. $f(x) = \sqrt{x} - 3$

78. $f(x) = \sqrt{x - 3}$

Find the inverse of each function mentally.

79. a. $f(x) = 5x$

b. $f(x) = x - 88$

c. $f(x) = 3x - 7$

d. $f(x) = 4 - 3x$

e. $f(x) = \dfrac{1}{2}x - 9$

f. $f(x) = -x$

g. $f(x) = \sqrt[3]{x} - 9$

h. $f(x) = 3x^3 - 7$

80. a. $f(x) = \dfrac{x}{2}$

b. $f(x) = x + 99$

c. $f(x) = 5x + 1$

d. $f(x) = 5 - 2x$

e. $f(x) = \dfrac{x}{3} + 6$

f. $f(x) = \dfrac{1}{x}$

g. $f(x) = \sqrt[3]{x} - 9$

h. $f(x) = -x^3 + 4$

Solve each problem.

81. *Price of a Car* The tax on a new car is 8% of the purchase price P. Express the total cost C as a function of the purchase price P. Express the purchase price P as a function of the total cost C.

82. *Volume of a Cube* Express the volume of a cube $V(x)$ as a function of the length of a side x. Express the length of a side of a cube $S(x)$ as a function of the volume x.

83. *Rowers and Speed* The world record times in the 2000-m race are a function of the number of rowers as shown in the accompanying figure (www.cbs.sportsline.com). If r is the number of rowers and t is the time in minutes, then the formula $t = -0.39r + 7.89$ models this relationship. Is this function invertible? Find the inverse. If the time for a 2000-m race was 5.55 min, then how many rowers were probably in the boat?

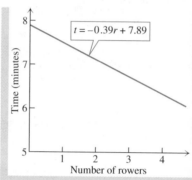

■ **Figure for Exercise 83**

84. *Temperature* The function $C = \dfrac{5}{9}(F - 32)$ expresses the Celsius temperature as a function of the Fahrenheit temperature. Find the inverse function. What is it used for?

85. *Landing Speed* The function $V = \sqrt{1.496w}$ expresses the landing speed V (in feet per second) as a function of the gross weight (in pounds) of the Piper Cheyenne aircraft. Find the inverse function. Use the inverse function to find the gross weight for a Piper Cheyenne for which the proper landing speed is 115 ft/sec.

86. *Poiseuille's Law* Under certain conditions, the velocity V of blood in a vessel at distance r from the center of the vessel is given by $V = 500(5.625 \times 10^{-5} - r^2)$ where $0 \leq r \leq 7.5 \times 10^{-3}$. Write r as a function of V.

87. *Depreciation Rate* The depreciation rate r for a \$50,000 new car is given by the function $r = 1 - \left(\dfrac{V}{50,000}\right)^{1/5}$, where V is the value of the car when it is five years old.
a. What is the depreciation rate for a \$50,000 BMW that is worth \$28,000 after 5 years?

b. Write V as a function of r.

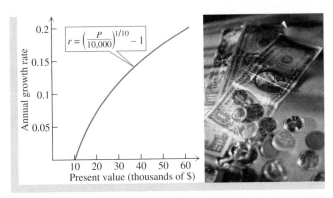

■ **Figure for Exercise 87**

88. *Annual Growth Rate*   One measurement of the quality of a mutual fund is its average annual growth rate over the last 10 years. The function $r = \left(\frac{P}{10,000}\right)^{1/10} - 1$ expresses the average annual growth rate r as a function of the present value P of an investment of \$10,000 made 10 years ago. An investment of \$10,000 in Fidelity's Contrafund in 1995 was worth \$36,555 in 2005 (Fidelity Investments, www.fidelity.com). What was the annual growth rate for that period? Write P as a function of r.

■ **Figure for Exercise 88**

For Writing/Discussion

89. If $f(x) = 2x + 1$ and $g(x) = 3x - 5$, verify the equation $(f \circ g)^{-1} = g^{-1} \circ f^{-1}$.

HINT Find formulas for $f \circ g$, $(f \circ g)^{-1}$, f^{-1}, g^{-1}, and $g^{-1} \circ f^{-1}$.

90. Explain why $(f \circ g)^{-1} = g^{-1} \circ f^{-1}$ for any invertible functions f and g. Discuss any restrictions on the domains and ranges of f and g for this equation to be correct.

91. Use a geometric argument to prove that if $a \neq b$, then (a, b) and (b, a) lie on a line perpendicular to the line $y = x$ and are equidistant from $y = x$.

92. Why is it difficult to find the inverse of a function such as $f(x) = (x - 3)/(x + 2)$ mentally?

93. Find an expression equivalent to $(x - 3)/(x + 2)$ in which x appears only once. Explain.

94. Use the result of Exercise 93 to find the inverse of $f(x) = (x - 3)/(x + 2)$ mentally. Explain.

Thinking Outside the Box XXVI

Costly Computers   A school district purchased x computers at y dollars each for a total of \$640,000. Both x and y are whole numbers and neither is a multiple of 10. Find the absolute value of the difference between x and y.

2.5 Pop Quiz

1. Is the function $\{(1, 3), (4, 5), (2, 3)\}$ invertible?

2. If $f = \{(5, 4), (3, 6), (2, 5)\}$, then what is $f^{-1}(5)$?

3. If $f(x) = 2x$, then what is $f^{-1}(8)$?

4. Is $f(x) = x^4$ a one-to-one function?

5. If $f(x) = 2x - 1$, then what is $f^{-1}(x)$?

6. If $g(x) = \sqrt[3]{x + 1} - 4$, then what is $g^{-1}(x)$?

7. Find $(h \circ j)(x)$ if $h(x) = x^3 - 5$ and $j(x) = \sqrt[3]{x + 5}$.

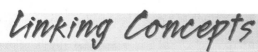

Linking Concepts

For Individual or
Group Explorations

Pay Day Loans

Star Financial Corporation offers short-term loans as shown in the accompanying advertisement. If you borrow P dollars for n days at the annual percentage rate r, then you pay back A dollars where

$$A = P\left(1 + \frac{r}{365}\right)^n.$$

a) Solve the formula for r in terms of A, P, and n.

b) Find the annual percentage rate (APR) for each of the loans described in the advertisement.

c) If you borrowed $200, then how much would you have to pay back in one year using the APR that you determined in part (b)?

d) Find the current annual percentage rates for a car loan, house mortgage, and a credit card.

e) Why do you think Star Financial Corporation charges such high rates?

2.6

Constructing Functions with Variation

The area of a circle is a function of the radius, $A = \pi r^2$. As the values of r change or vary, the values of A vary also. Certain relationships are traditionally described by indicating how one variable changes according to the values of another variable. In this section we will construct functions for those relationships.

Direct Variation

If we save 3 cubic yards of landfill space for every ton of paper that we recycle, then the volume V of space saved is a linear function of the number of tons recycled n, $V = 3n$. If n is 20 tons, then V is 60 cubic yards. If n is doubled to 40 tons, then V is also doubled to 120 cubic yards. If n is tripled, V is also tripled. This relationship is called **direct variation.**

Definition: Direct Variation

> The statement y **varies directly as** x or y **is directly proportional to** x means that
>
> $$y = kx$$
>
> for some fixed nonzero real number k. The constant k is called the **variation constant** or **proportionality constant.**

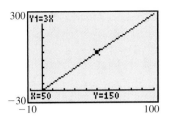

■ **Figure 2.62**

Compare the landfill savings to a pizza that costs $5 plus $2 per topping. The cost is a linear function of the number of toppings n, $C = 2n + 5$. If $n = 2$, then $C = \$9$ and if $n = 4$, then $C = \$13$. The cost increases as n increases, but doubling n does not double C. The savings in landfill space is directly proportional to the number of tons of paper that is recycled, but the cost of the pizza is not directly proportional to the number of toppings. The graph of $V = 3n$ in Fig. 2.62 is a straight line through the origin. In general, if y varies directly as x, then y is a linear function of x whose graph is a straight line through the origin. We are merely introducing some new terms to describe an old idea.

Example ■1 Direct variation

The cost, C, of a house in Wedgewood Estates is directly proportional to the size of the house, s. If a 2850-square-foot house costs $182,400, then what is the cost of a 3640-square-foot house?

Solution

Because C is directly proportional to s, there is a constant k such that

$$C = ks.$$

We can find the constant by using $C = \$182,400$ when $s = 2850$:

$$182,400 = k(2850)$$

$$64 = k$$

The cost for a house is $64 per square foot. To find the cost of a 3640-square-foot house, use the formula $C = 64s$:

$$C = 64(3640)$$

$$= 232,960$$

The 3640-square-foot house costs $232,960.

Try This. The cost of a smoothie is directly proportional to its size. If a 12-ounce smoothie is $3.60, then what is the cost of a 16-ounce smoothie? ■

Inverse Variation

The time that it takes to make the 3470-mile trip by air from New York to London is a function of the rate of the airplane, $T = 3470/R$. A Boeing 747 averaging 675 mph can make the trip in about 5 hours. The Concorde, traveling twice as fast at 1350 mph, made it in half the time, about 2.5 hours. If you travel three times as fast as the 747 you could make the trip in one-third the time. This relationship is called **inverse variation.**

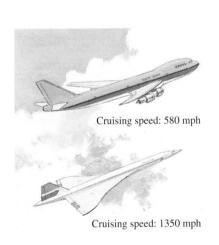

Cruising speed: 580 mph

Cruising speed: 1350 mph

Definition: Inverse Variation

The statement y **varies inversely as** x or y **is inversely proportional to** x means that

$$y = \frac{k}{x}$$

for a fixed nonzero real number k.

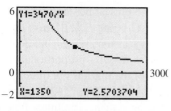

■ **Figure 2.63**

The graph of $T = 3470/R$ is shown in Fig. 2.63. As the speed increases the time decreases, but does not reach zero for any speed.

Example **2** **Inverse variation**

The time it takes to remove all of the campaign signs in Bakersfield after an election varies inversely with the number of volunteers working on the job. If 12 volunteers could complete the job in 7 days, then how many days would it take for 15 volunteers?

Solution

Since the time T varies inversely as the number of volunteers n, we have

$$T = \frac{k}{n}.$$

Since $T = 7$ when $n = 12$, we can find k:

$$7 = \frac{k}{12}$$

$$84 = k$$

So the formula is $T = 84/n$. Now if $n = 15$, we get

$$T = \frac{84}{15} = 5.6.$$

It would take 15 volunteers 5.6 days to remove all of the political signs.

Try This. The time required to rake the grounds at Rockwood Manor varies inversely with the number of rakers. If 4 rakers can complete the job in 12 hours, then how long would it take for 6 rakers to complete the job?   ■

Joint Variation

We have been studying functions of one variable, but situations often arise in which one variable depends on the values of two or more variables. For example, the area of a rectangular room depends on the length *and* the width, $A = LW$. If carpeting costs $25 per square yard, then the total cost is a function of the length and width, $C = 25LW$. This example illustrates **joint variation.**

Definition: Joint Variation

> The statement *y* **varies jointly as** *x* **and** *z* or *y* **is jointly proportional to** *x* **and** *z* means that
>
> $$y = kxz$$
>
> for a fixed nonzero real number k.

Example **3** **Joint variation**

The cost of constructing a 9-foot by 12-foot patio is $734.40. If the cost varies jointly as the length and width, then what does an 8-foot by 14-foot patio cost?

Solution

Since the cost varies jointly as L and W, we have $C = kLW$ for some constant k. Use $W = 9$, $L = 12$, and $C = \$734.40$ to find k:

$$734.40 = k(12)(9)$$

$$6.80 = k$$

So the formula is $C = 6.80LW$. Use $W = 8$ and $L = 14$ to find C:

$$C = 6.80(14)(8)$$

$$= 761.60$$

The cost of constructing an 8-foot by 14-foot patio is $761.60.

Try This. The cost of a fence that is 200 feet long and 5 feet high is $3,000. If the cost varies jointly with the length and height, then what is the cost of a fence that is 250 feet long and 6 feet high?   ■

Combined Variation

Some relationships are combinations of direct and inverse variation. For example, the formula $V = kT/P$ expresses the volume of a gas in terms of its temperature T and its pressure P. The volume varies directly as the temperature and inversely as the pressure. The next example gives more illustrations of combining direct, inverse, and joint variation.

Example **4** **Constructing functions for combined variation**

Write each sentence as a function involving a constant of variation k.

a. y varies directly as x and inversely as z.
b. y varies jointly as x and the square root of z.
c. y varies jointly as x and the square of z and inversely as the cube of w.

Solution

a. $y = \dfrac{kx}{z}$ **b.** $y = kx\sqrt{z}$ **c.** $y = \dfrac{kxz^2}{w^3}$

Try This. Suppose that M varies directly as w and the cube of z. Write a formula that describes this variation with k as the constant.   ■

The language of variation evolved as a means of describing functions that involve only multiplication and/or division. The terms "varies directly," "directly proportional," and "jointly proportional" indicate multiplication. The term "inverse" indicates division. The term "variation" is *never* used in this text to refer to a formula involving addition or subtraction. Note that the formula for the cost of a pizza $C = 2n + 5$ is *not* the kind of formula that is described in terms of variation.

Example **5** **Solving a combined variation problem**

The recommended maximum load for the ceiling joist with rectangular cross section shown in Fig. 2.64 varies directly as the product of the width and the square of the depth of the cross section, and inversely as the length of the joist. If the recommended

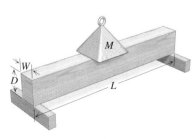
■ **Figure 2.64**

maximum load is 800 pounds for a 2-inch by 6-inch joist that is 12 feet long, then what is the recommended maximum load for a 2-inch by 12-inch joist that is 16 feet long?

Solution

The maximum load M varies directly as the product of the width W and the square of the depth D, and inversely as the length L:

$$M = \frac{kWD^2}{L}$$

Because the depth is squared, the maximum load is always obtained by positioning the joist so that the larger dimension is the depth. Use $W = 2, D = 6, L = 12$, and $M = 800$ to find k:

$$800 = \frac{k \cdot 2 \cdot 6^2}{12}$$

$$800 = 6k$$

$$\frac{400}{3} = k$$

Now use $W = 2, D = 12, L = 16$, and $k = 400/3$ in $M = \frac{kWD^2}{L}$ to find M:

$$M = \frac{400 \cdot 2 \cdot 12^2}{3 \cdot 16} = 2400$$

The maximum load on the joist is 2400 pounds. Notice that feet and inches were used in the formula without converting one to the other. If we solve the problem using only feet or only inches, then the value of k is different but we get the same value for M.

Try This. W varies directly with m and inversely as the square of t. If $W = 480$ when $m = 5$ and $t = 4$, then what is W when $m = 3$ and $t = 6$? ▪

For Thought

True or False? Explain.

1. If the cost of an 800 number is $29.95 per month plus 28 cents per minute, then the monthly cost varies directly as the number of minutes.

2. If bananas are $0.39 per pound, then your cost for a bunch of bananas varies directly with the number of bananas purchased.

3. The area of a circle varies directly as the square of the radius.

4. The area of a triangle varies jointly as the length of the base and the height.

5. The number π is a constant of proportionality.

6. If y is inversely proportional to x, then for $x = 0$, we get $y = 0$.

7. If the cost of potatoes varies directly with the number of pounds purchased, then the proportionality constant is the price per pound.

8. The height of a student in centimeters is directly proportional to the height of the same student in inches.

9. The surface area of a cube varies directly as the square of the length of an edge.

10. The surface area of a rectangular box varies jointly as its length, width, and height.

2.6   Exercises

Construct a function that expresses the relationship in each statement. Use k as the constant of variation.

1. Your grade on the next test, G, varies directly with the number of hours, n, that you study for it.

2. The amount of sales tax on a new car is directly proportional to the purchase price of the car.

3. The volume of a gas in a cylinder, V, is inversely proportional to the pressure on the gas, P.

4. For two lines that are perpendicular and have slopes, the slope of one is inversely proportional to the slope of the other.

5. The cost of constructing a silo varies jointly as the height and the radius.

6. The volume of grain that a silo can hold varies jointly as the height and the square of the radius.

7. Y varies directly as x and inversely as the square root of z.

8. W varies directly with r and t and inversely with v.

For each formula that illustrates a type of variation, write the formula in words using the vocabulary of variation. The letters k and π represent constants, and all other letters represent variables.

9. $A = \pi r^2$

10. $C = \pi D$

11. $y = \dfrac{1}{x}$

12. $m_1 = \dfrac{-1}{m_2}$

13. $T = 3W + 5$

14. $y = k\sqrt{x - 3}$

15. $a = kzw$

16. $V = LWH$

17. $H = \dfrac{k\sqrt{t}}{s}$

18. $B = \dfrac{y^2}{2\sqrt{x}}$

19. $D = \dfrac{2LJ}{W}$

20. $E = mc^2$

Find the constant of variation and construct the function that is expressed in each statement.

21. y varies directly as x, and $y = 5$ when $x = 9$.

22. h is directly proportional to z, and $h = 210$ when $z = 200$.

23. T is inversely proportional to y, and $T = -30$ when $y = 5$.

24. H is inversely proportional to n, and $H = 9$ when $n = -6$.

25. m varies directly as the square of t, and $m = 54$ when $t = 3\sqrt{2}$.

26. p varies directly as the cube root of w, and $p = \sqrt[3]{2}/2$ when $w = 4$.

27. y varies directly as x and inversely as the square root of z, and $y = 2.192$ when $x = 2.4$ and $z = 2.25$.

28. n is jointly proportional to x and the square root of b, and $n = -18.954$ when $x = -1.35$ and $b = 15.21$.

Solve each variation problem.

29. If y varies directly as x, and $y = 9$ when $x = 2$, what is y when $x = -3$?

30. If y is directly proportional to z, and $y = 6$ when $z = \sqrt{12}$, what is y when $z = \sqrt{75}$?

31. If P is inversely proportional to w, and $P = 2/3$ when $w = 1/4$, what is P when $w = 1/6$?

32. If H varies inversely as q, and $H = 0.03$ when $q = 0.01$, what is H when $q = 0.05$?

33. If A varies jointly as L and W, and $A = 30$ when $L = 3$ and $W = 5\sqrt{2}$, what is A when $L = 2\sqrt{3}$ and $W = 1/2$?

34. If J is jointly proportional to G and V, and $J = \sqrt{3}$ when $G = \sqrt{2}$ and $V = \sqrt{8}$, what is J when $G = \sqrt{6}$ and $V = 8$?

35. If y is directly proportional to u and inversely proportional to the square of v, and $y = 7$ when $u = 9$ and $v = 6$, what is y when $u = 4$ and $v = 8$?

36. If q is directly proportional to the square root of h and inversely proportional to the cube of j, and $q = 18$ when $h = 9$ and $j = 2$, what is q when $h = 16$ and $j = 1/2$?

Determine whether the first variable varies directly or inversely with the other variable. Construct a function for the variation using the appropriate variation constant.

37. The length of a car in inches, the length of the same car in feet

38. The time in seconds to pop a bag of microwave popcorn, the time in minutes to pop the same bag of popcorn

39. The price for each person sharing a $20 pizza, the number of hungry people

40. The number of identical steel rods with a total weight of 40,000 lb, the weight of one of those rods

41. The speed of a car in miles per hour, the speed of the same car at the same time in kilometers per hour

42. The weight of a sumo wrestler in pounds, the weight of the same sumo wrestler in kilograms. (Chad Rowan of the United States, at 455 lb or 206 kg, was the first-ever foreign grand champion in Tokyo's annual tournament.)

■ **Figure for Exercise 42**

43. The temperature in degrees Celsius, the temperature at the same time and place in degrees Fahrenheit

44. The perimeter of a rectangle with a length of 4 ft, the width of the same rectangle

45. The area of a rectangle with a length of 30 in., the width of the same rectangle

46. The area of a triangle with a height of 10 cm, the base of the same triangle

47. The number of gallons of gasoline that you can buy for $5, the price per gallon of that same gasoline

48. The length of Yolanda's 40 ft^2 closet, the width of that closet

Solve each problem.

49. *First-Class Diver* The pressure exerted by water at a point below the surface varies directly with the depth. The pressure is 4.34 lb/in.^{2} at a depth of 10 ft. What pressure does the sperm whale (the deepest diver among the air-breathing mammals) experience when it dives 6000 ft below the surface?

50. *Second Place* The elephant seal (second only to the sperm whale) experiences a pressure of 2170 lb/in.^{2} when it makes its deepest dives. At what depth does the elephant seal experience this pressure? See the previous exercise.

51. *Processing Oysters* The time required to process a shipment of oysters varies directly with the number of pounds in the shipment and inversely with the number of workers assigned. If 3000 lb can be processed by 6 workers in 8 hr, then how long would it take 5 workers to process 4000 lb?

52. *View from an Airplane* The view V from the air is directly proportional to the square root of the altitude A. If the view from horizon to horizon at an altitude of 16,000 ft is approximately 154 mi, then what is the view from 36,000 ft?

53. *Simple Interest* The amount of simple interest on a deposit varies jointly with the principal and the time in days. If $20.80 is earned on a deposit of $4000 for 16 days, how much interest would be earned on a deposit of $6500 for 24 days?

54. *Free Fall at Six Flags* Visitors at Six Flags in Atlanta get a brief thrill in the park's "free fall" ride. Passengers seated in a small cab drop from a 10-story tower down a track that eventually curves level with the ground. The distance the cab falls varies directly with the square of the time it is falling. If the cab falls 16 ft in the first second, then how far has it fallen after 2 sec?

55. *Cost of Plastic Sewer Pipe* The cost of a plastic sewer pipe varies jointly as its diameter and its length. If a 20-ft pipe with a diameter of 6 in. costs $18.60, then what is the cost of a 16-ft pipe with a diameter of 8 in.?

56. *Cost of Copper Tubing* A plumber purchasing materials for a renovation observed that the cost of copper tubing varies jointly as the length and the diameter of the tubing. If 20 ft of $\frac{1}{2}$-in.-diameter tubing costs $36.60, then what is the cost of 100 ft of $\frac{3}{4}$-in.-diameter tubing?

57. *Weight of a Can* The weight of a can of baked beans varies jointly with the height and the square of the diameter. If a 4-in.-high can with a 3-in. radius weighs 14.5 oz, then what is the weight of a 5-in.-high can with a diameter of 6 in.?

58. *Nitrogen Gas Shock Absorber* The volume of gas in a nitrogen gas shock absorber varies directly with the temperature of the gas and inversely with the amount of weight on the piston. If the volume is 10 in.^{3} at 80° F with a weight of 600 lb, then what is the volume at 90° F with a weight of 800 lb?

59. *Velocity of Underground Water* Darcy's law states that the velocity V of underground water through sandstone varies directly as the head h and inversely as the length l of the flow. The head is the vertical distance between the point of intake into the rock and the point of discharge such as a spring, and the length is the length of the flow from intake to discharge. In a certain sandstone a velocity of 10 ft per year has been recorded with a head of 50 ft and length of 200 ft. What would we expect the velocity to be if the head is 60 ft and the length is 300 ft?

60. *Cross-Sectional Area of a Well* The rate of discharge of a well, V, varies jointly as the hydraulic gradient, i, and the cross-sectional area of the well wall, A. Suppose that a well with a cross-sectional area of 10 ft^2 discharges 3 gal of water per minute in an area where the hydraulic gradient is 0.3. If we dig another well nearby where the hydraulic gradient is 0.4, and we want a discharge of 5 gal/min, then what should be the cross-sectional area for the well?

61. *Dollars Per Death*   The accompanying table shows the annual amount of money that the National Institutes of Health spends on disease research and the annual number of deaths for three diseases (www.nih.gov). Is the annual spending directly proportional to the annual number of deaths? Should it be?

■ **Table for Exercise 61**

Disease	Research spending	Annual deaths
AIDS	$1.34 billion	42,506
Diabetes	$295 million	59,085
Heart disease	$958 million	738,781

62. *Bicycle Gear Ratio*   The gear ratio r for a bicycle varies jointly with the number of teeth on the chainring n, by the pedals, and the diameter of the wheel w (in inches), and inversely with the number of teeth on the cog c, by the wheel (www. harriscyclery.com). Find the missing entries in the accompanying table.

■ **Table for Exercise 62**

n	w	c	r	
50	27	25	54	
40	26	13		
45	27		67.5	

63. *Grade on an Algebra Test*   Calvin believes that his grade on a college algebra test varies directly with the number of hours spent studying during the week prior to the test and inversely with the number of hours spent at the Beach Club playing volleyball during the week prior to the test. If he scored 76 on a test when he studied 12 hr and played 10 hr during the week prior to the test, then what score should he expect if he studies 9 hr and plays 15 hr?

64. *Carpeting a Room*   The cost of carpeting a room varies jointly as the length and width. Julie advertises a price of $263.40 to carpet a 9-ft by 12-ft room with Dupont Stainmaster. Julie gave a customer a price of $482.90 for carpeting a room with a width of 4 yd with that same carpeting, but she has forgotten the length of the room. What is the length of the room?

65. *Pole-Vault Principle*   The height a pole-vaulter attains is directly proportional to the square of his velocity on the runway. Given that a speed of 32 ft/sec will loft a vaulter to 16 ft, find the speed necessary for world record holder Sergie Bubka to reach a record height of 20 ft 2.5 in.

For Writing/Discussion

66. *Cooperative Learning*   Working in groups, drop a ball from various heights and measure the distance that the ball rebounds after hitting the floor. Enter your pairs of data into a graphing calculator and calculate the regression line. What is the value of the correlation coefficient, r? Do you think that the rebound distance is directly proportional to the drop distance? Compare your results with other groups.

67. *Moving Melons*   To move more watermelons, a retailer decided to make the price of each watermelon purchased by a single customer inversely proportional to the number of watermelons purchased by the customer. Is this a good idea? Explain.

68. *Common Units*   In general we do not perform computations involving inches and feet until they are converted to a common unit of measure. Explain how it is possible to use inches and feet in Example 5 without converting to a common unit of measure.

Thinking Outside the Box XXVII

Good Timing   Sharon leaves at the same time every day and walks to her 8 o'clock class. On Monday she averaged 4 miles per hour and was one minute late. On Wednesday she averaged 5 miles per hour and was one minute early. Find the exact speed that she should walk on Friday to get to class on time.

2.6 Pop Quiz

1. Find the constant of variation if y varies directly as x and $y = 4$ when $x = 20$.

2. If a varies inversely as b and the constant is 10, then what is a when $b = 2$?

3. The cost of a round rug varies directly as the square of its radius. If a rug with a 3-foot radius costs $108, then what is the cost of a rug with a 4-foot radius?

4. The cost of a rectangular tapestry varies jointly as its length and width. If a 6 × 2 foot tapestry costs $180, then what is a the cost of a 5 × 4 foot tapestry?

Linking Concepts

For Individual or
Group Explorations

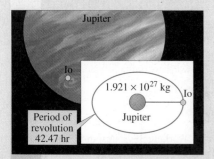

Determining the Mass of a Planet

A result of Kepler's harmonic law is that the mass of a planet with a satellite is directly proportional to the cube of the mean distance from the satellite to the planet, and inversely proportional to the square of the period of revolution. Early astronomers estimated the mass of the earth to be 5.976×10^{24} kg and observed that the moon orbited the earth with a period of 27.322 days at a mean distance of 384.4×10^3 km.

a) Write a formula for the mass of a planet according to Kepler.

b) Find the proportionality constant using the observations and estimates for the earth and the moon.

c) Extend your knowledge across the solar system to find the mass of Mars based on observations that Phobos orbits Mars in 7.65 hr at a mean distance of 9330 km.

d) What is the approximate ratio of the mass of the earth to the mass of Mars?

e) If the mass of Jupiter is 1.921×10^{27} kg and the period of revolution of Io around Jupiter is 42.47 hr, then what is the mean distance between Io and Jupiter?

■ ■ ■ Highlights

2.1 Functions

Relation	Any set of ordered pairs	$\{(1, 5), (1, 3), (3, 5)\}$
Function	A relation in which no two ordered pairs have the same first coordinate and different second coordinates	$\{(1, 5), (2, 3), (3, 5)\}$
Vertical Line Test	If no vertical line crosses a graph more than once then the graph is a function.	Function Not function
Average Rate of Change	The slope of the line through $(a, f(a))$ and $(b, f(b))$: $\dfrac{f(b) - f(a)}{b - a}$	Average rate of change of $f(x) = x^2$ on $[2, 9]$ is $\dfrac{9^2 - 2^2}{9 - 2}$.
Difference Quotient	Average rate of change of f on $[x, x + h]$: $\dfrac{f(x + h) - f(x)}{h}$	$f(x) = x^2$, difference quotient $= \dfrac{(x + h)^2 - x^2}{h} = 2x + h$

2.2 Graphs of Relations and Functions

Graph of a Relation
An illustration of all ordered pairs of a relation

Graph of $y = x + 1$ shows all ordered pairs in this function.

Circle
A circle is not the graph of a function

Since $(0, \pm2)$ satisfies $x^2 + y^2 = 4$, it is not a function.

Increasing, Decreasing, or Constant
When going from left to right, a function is increasing if its graph is rising, decreasing if its graph is falling, constant if it is staying the same.

$y = |x|$ is increasing on $(0, \infty)$ and decreasing on $(-\infty, 0)$
$y = 5$ is constant on $(-\infty, \infty)$

2.3 Families of Functions, Transformations, and Symmetry

Transformations of $y = f(x)$
Horizontal: $y = f(x - h)$
Vertical: $y = f(x) + k$
Stretching: $y = af(x)$ for $a > 1$
Shrinking: $y = af(x)$ for $0 < a < 1$
Reflection: $y = -f(x)$

$y = (x - 4)^2$
$y = x^2 + 9$
$y = 3x^2$
$y = 0.5x^2$
$y = -x^2$

Family of Functions
All functions of the form $f(x) = af(x - h) + k$ $(a \neq 0)$ for a given function $y = f(x)$.

The square root family:
$y = a\sqrt{x - h} + k$

Even Function
Graph is symmetric with respect to y-axis.
$f(-x) = f(x)$

$f(x) = x^2, g(x) = |x|$

Odd Function
Graph is symmetric about the origin.
$f(-x) = -f(x)$

$f(x) = x, g(x) = x^3$

Inequalities
$f(x) > 0$ is satisfied on all intervals where the graph of $y = f(x)$ is above the x-axis
$f(x) < 0$ is satisfied on all intervals where the graph of $y = f(x)$ is below the x-axis

$4 - x^2 > 0$ on $(-2, 2)$

$4 - x^2 < 0$ on $(-\infty, -2) \cup (2, \infty)$

2.4 Operations with Functions

Sum
$(f + g)(x) = f(x) + g(x)$

$f(x) = x^2 - 4, g(x) = x + 2$
$(f + g)(x) = x^2 + x - 2$

Difference
$(f - g)(x) = f(x) - g(x)$

$(f - g)(x) = x^2 - x - 6$

Product
$(f \cdot g)(x) = f(x) \cdot g(x)$

$(f \cdot g)(x) = x^3 + 2x^2 - 4x - 8$

Quotient
$(f/g)(x) = f(x)/g(x)$

$(f/g)(x) = x - 2$

Composition
$(f \circ g)(x) = f(g(x))$

$(f \circ g)(x) = (x + 2)^2 - 4$
$(g \circ f)(x) = x^2 - 2$

2.5 Inverse Functions

One-to-One Function
A function that has no two ordered pairs with different first coordinates and the same second coordinate.

$\{(1, 2), (3, 5), (6, 9)\}$
$g(x) = x + 3$
$f(x) = x^2$ is not one-to-one.

Inverse Function
A one-to-one function has an inverse. The inverse function has the same ordered pairs, with the coordinates reversed.

$f = \{(1, 2), (3, 5), (6, 9)\}$
$f^{-1} = \{(2, 1), (5, 3), (9, 6)\}$
$g(x) = x + 3, g^{-1}(x) = x - 3$

Horizontal Line Test	If there is a horizontal line that crosses the graph of f more than once, then f is not invertible.	$y = 4$ crosses $f(x) = x^2$ twice, so f is not invertible.
Graph of f^{-1}	Reflect the graph of f about the line $y = x$ to get the graph of f^{-1}.	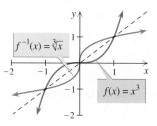

2.6 Constructing Functions with Variation

Direct Variation	$y = kx$ for a constant $k \neq 0$.	$D = 40T$
Inverse Variation	$y = k/x$ for a constant $k \neq 0$.	$T = 200/R$
Joint Variation	$y = kxz$ for a constant $k \neq 0$.	$C = 25LW$

▪▪▪ Chapter 2 Review Exercises

Graph each set of ordered pairs. State the domain and range of each relation. Determine whether each set is a function.

1. $\{(0, 0), (1, 1), (-2, -2)\}$

2. $\{(0, -3), (1, -1), (2, 1), (2, 3)\}$

3. $\{(x, y) \mid y = 3 - x\}$ **4.** $\{(x, y) \mid 2x + y = 5\}$

5. $\{(x, y) \mid x = 2\}$ **6.** $\{(x, y) \mid y = 3\}$

7. $\{(x, y) \mid x^2 + y^2 = 0.01\}$

8. $\{(x, y) \mid x^2 + y^2 = 2x - 4y\}$

9. $\{(x, y) \mid x = y^2 + 1\}$ **10.** $\{(x, y) \mid y = |x - 2|\}$

11. $\{(x, y) \mid y = \sqrt{x} - 3\}$ **12.** $\{(x, y) \mid x = \sqrt{y}\}$

Let $f(x) = x^2 + 3$ and $g(x) = 2x - 7$. Find and simplify each expression.

13. $f(-3)$ **14.** $g(3)$

15. $g(12)$ **16.** $f(-1)$

17. x, if $f(x) = 19$ **18.** x, if $g(x) = 9$

19. $(g \circ f)(-3)$ **20.** $(f \circ g)(3)$

21. $(f + g)(2)$ **22.** $(f - g)(-2)$

23. $(f \cdot g)(-1)$ **24.** $(f/g)(4)$

25. $f(g(2))$ **26.** $g(f(-2))$

27. $(f \circ g)(x)$ **28.** $(g \circ f)(x)$

29. $(f \circ f)(x)$ **30.** $(g \circ g)(x)$

31. $f(a + 1)$ **32.** $g(a + 2)$

33. $\dfrac{f(3 + h) - f(3)}{h}$ **34.** $\dfrac{g(5 + h) - g(5)}{h}$

35. $\dfrac{f(x + h) - f(x)}{h}$ **36.** $\dfrac{g(x + h) - g(x)}{h}$

37. $g\left(\dfrac{x + 7}{2}\right)$ **38.** $f\left(\sqrt{x - 3}\right)$

39. $g^{-1}(x)$ **40.** $g^{-1}(-3)$

Use transformations to graph each pair of functions on the same coordinate plane.

41. $f(x) = \sqrt{x}, g(x) = 2\sqrt{x + 3}$

42. $f(x) = \sqrt{x}, g(x) = -2\sqrt{x} + 3$

43. $f(x) = |x|, g(x) = -2|x + 2| + 4$

44. $f(x) = |x|, g(x) = \dfrac{1}{2}|x - 1| - 3$

45. $f(x) = x^2, g(x) = \dfrac{1}{2}(x - 2)^2 + 1$

46. $f(x) = x^2, g(x) = -2x^2 + 4$

For each exercise, graph the function by transforming the given graph of $y = f(x).$

47. $y = 2f(x - 2) + 1$

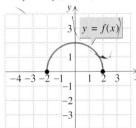

48. $y = 2f(x + 3) - 1$

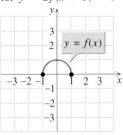

49. $y = -f(x + 1) - 3$

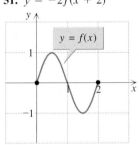

50. $y = -f(x - 1) + 2$

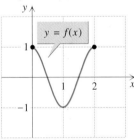

51. $y = -2f(x + 2)$

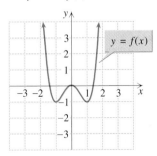

52. $y = -3f(x) + 1$

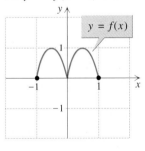

53. $y = -2f(x) + 3$

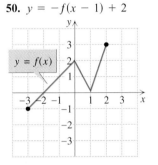

54. $y = 4f(x - 1) + 3$

Let $f(x) = \sqrt[3]{x}$, $g(x) = x - 4$, $h(x) = x/3$, *and* $j(x) = x^2$. *Write each function as a composition of the appropriate functions chosen from f, g, h, and j.*

55. $F(x) = \sqrt[3]{x - 4}$

56. $G(x) = -4 + \sqrt[3]{x}$

57. $H(x) = \sqrt[3]{\dfrac{x^2 - 4}{3}}$

58. $M(x) = \left(\dfrac{x - 4}{3}\right)^2$

59. $N(x) = \dfrac{1}{3}x^{2/3}$

60. $P(x) = x^2 - 16$

61. $R(x) = \dfrac{x^2}{3} - 4$

62. $Q(x) = x^2 - 8x + 16$

Find the difference quotient for each function and simplify it.

63. $f(x) = -5x + 9$

64. $f(x) = \sqrt{x - 7}$

65. $f(x) = \dfrac{1}{2x}$

66. $f(x) = -5x^2 + x$

Sketch the graph of each function and state its domain and range. Determine the intervals on which the function is increasing, decreasing, or constant.

67. $f(x) = \sqrt{100 - x^2}$

68. $f(x) = -\sqrt{7 - x^2}$

69. $f(x) = \begin{cases} -x^2 & \text{for } x \le 0 \\ x^2 & \text{for } x > 0 \end{cases}$

70. $f(x) = \begin{cases} x^2 & \text{for } x \le 0 \\ x & \text{for } 0 < x \le 4 \end{cases}$

71. $f(x) = \begin{cases} -x - 4 & \text{for } x \le -2 \\ -|x| & \text{for } -2 < x < 2 \\ x - 4 & \text{for } x \ge 2 \end{cases}$

72. $f(x) = \begin{cases} (x + 1)^2 - 1 & \text{for } x \le -1 \\ -1 & \text{for } -1 < x < 1 \\ (x - 1)^2 - 1 & \text{for } x \ge 1 \end{cases}$

Each of the following graphs is from the absolute-value family. Construct a function for each graph and state the domain and range of the function.

73.

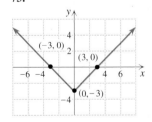

74.

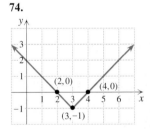

75.

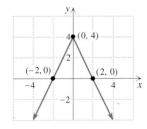

76.

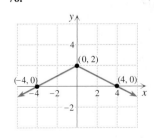

77.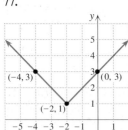

78.

Determine whether the graph of each function is symmetric with respect to the y-axis or the origin.

79. $f(x) = x^4 - 9x^2$

80. $f(x) = |x| - 99$

81. $f(x) = -x^3 - 5x$

82. $f(x) = \dfrac{15}{x}$

83. $f(x) = -x + 1$

84. $f(x) = |x - 1|$

85. $f(x) = \sqrt{x^2}$

86. $f(x) = \sqrt{16 - x^2}$

Use the graph of an appropriate function to find the solution set to each inequality.

87. $|x - 3| \geq 1$

88. $|x + 2| + 1 < 0$

89. $-2x^2 + 4 > 0$

90. $-\dfrac{1}{2}x^2 + 2 \leq 0$

91. $-\sqrt{x + 1} - 2 > 0$

92. $\sqrt{x} - 3 < 0$

Graph each pair of functions on the same coordinate plane. What is the relationship between the functions in each pair?

93. $f(x) = \sqrt{x + 3},\ g(x) = x^2 - 3$ for $x \geq 0$

94. $f(x) = (x - 2)^3,\ g(x) = \sqrt[3]{x} + 2$

95. $f(x) = 2x - 4,\ g(x) = \dfrac{1}{2}x + 2$

96. $f(x) = -\dfrac{1}{2}x + 4,\ g(x) = -2x + 8$

Determine whether each function is invertible. If the function is invertible, then find the inverse function, and state the domain and range of the inverse function.

97. $\{(\pi, 0), (\pi/2, 1), (2\pi/3, 0)\}$

98. $\{(-2, 1/3), (-3, 1/4), (-4, 1/5)\}$

99. $f(x) = 3x - 21$

100. $f(x) = 3|x|$

101. $y = \sqrt{9 - x^2}$

102. $y = 7 - x$

103. $f(x) = \sqrt{x - 9}$

104. $f(x) = \sqrt{x} - 9$

105. $f(x) = \dfrac{x - 7}{x + 5}$

106. $f(x) = \dfrac{2x - 3}{5 - x}$

107. $f(x) = x^2 + 1$ for $x \leq 0$

108. $f(x) = (x + 3)^4$ for $x \geq 0$

Solve each problem.

109. *Turning a Profit*   Mary Beth buys roses for $1.20 each and sells them for $2 each at an outdoor market where she rents space for $40 per day. She buys and sells x roses per day. Construct her daily cost, revenue, and profit functions. How many roses must she sell to make a profit?

110. *Tin Can*   The surface area S and volume V for a can with a top and bottom are given by

$$S = 2\pi r^2 + 2\pi rh \quad \text{and} \quad V = \pi r^2 h,$$

where r is the radius and h is the height. Suppose that a can has a volume of 1 ft³.
a. Write its height as a function of its radius.

b. Write its radius as a function of its height.

c. Write its surface area as a function of its radius.

111. *Dropping the Ball*   A ball is dropped from a height of 64 feet. Its height h (in feet) is a function of time t (in seconds), where $h = -16t^2 + 64$ for t in the interval [0, 2]. Find the inverse of this function and state the domain of the inverse function.

112. *Sales Tax Function*   If S is the subtotal in dollars of your groceries before a 5% sales tax, then the function $T = 1.05S$ gives the total cost in dollars including tax. Write S as a function of T.

113. Write the diameter of a circle, d, as a function of its area, A.

114. *Circle Inscribed in a Square*   A cylindrical pipe with an outer radius r must fit snugly through a square hole in a wall. Write the area of the square as a function of the radius of the pipe.

115. *Load on a Spring*   When a load of 5 lb is placed on a spring, its length is 6 in., and when a load of 9 lb is placed on the spring, its length is 8 in. What is the average rate of change of the length of the spring as the load varies from 5 lb to 9 lb?

116. *Changing Speed of a Dragster*   Suppose that 2 sec after starting, a dragster is traveling 40 mph, and 5 sec after starting, the dragster is traveling 130 mph. What is the average rate of change of the speed of the dragster over the time interval from 2 sec to 5 sec? What are the units for this measurement?

117. Given that D is directly proportional to W and $D = 9$ when $W = 25$, find D when $W = 100$.

118. Given that t varies directly as u and inversely as v, and $t = 6$ when $u = 8$ and $v = 2$, find t when $u = 19$ and $v = 3$.

119. *Dinosaur Speed* R. McNeill Alexander, a British paleontologist, uses observations of living animals to estimate the speed of dinosaurs. Alexander believes that two animals of different sizes but geometrically similar shapes will move in a similar fashion. For animals of similar shapes, their velocity is directly proportional to the square root of their hip height. Alexander has compared a white rhinoceros, with a hip height of 1.5 m, and a member of the genus Triceratops, with a hip height of 2.8 m. If a white rhinoceros can move at 45 km/hr, then what is the estimated velocity of the dinosaur?

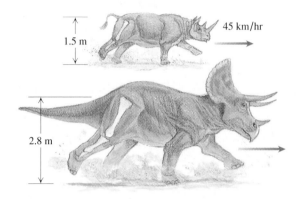

45 km/hr

1.5 m

2.8 m

■ **Figure for Exercise 119**

120. *Arranging Plants in a Row* Timothy has a shipment of strawberry plants to plant in his field, which is prepared with rows of equal length. He knows that the number of rows that he will need varies inversely with the number of plants per row. If he plants 240 plants per row, then he needs 21 rows. How many rows does he need if he plants 224 plants per row?

121. *Handling Charge for a Globe* The shipping and handling charge for a globe from Mercator's Map Company varies directly as the square of the diameter. If the charge is $4.32 for a 6-in.-diameter globe, then what is the charge for a 16-in.-diameter globe?

122. *Newton's Law of Gravity* Newton's law of gravity states that the force of gravity acting between two objects varies directly as the product of the masses of the two objects and inversely as the square of the distance between their centers. Write an equation for Newton's law of gravity.

For Writing/Discussion

123. Explain what the vertical and horizontal line tests are and how to apply them.

124. Is a vertical line the graph of a linear function? Explain.

Thinking Outside the Box XXVIII & XXIX

Area of a Polygon Find the exact area of the polygon bounded by the x-axis, the y-axis, the line $2x + 3y = 7$, and the line $3x - 4y = 18$.

Whole Lotta Shakin At the start of an economic conference between the eastern delegation and the western delegation, each delegate shook hands with every other member of his own delegation for a total of 466 handshakes. Next, each delegate shook hands with every person in the other delegation for 480 more handshakes. What was the total number of delegates at the conference?

■ ■ ■ ■ Chapter 2 Test

Determine whether each equation defines y as a function of x.

1. $x^2 + y^2 = 25$

2. $3x - 5y = 20$

3. $x = |y + 2|$

4. $y = x^3 - 3x^2 + 2x - 1$

State the domain and range of each relation.

5. $\{(2, -3), (2, -4), (5, 7)\}$

6. $y = \sqrt{x - 9}$

7. $x = |y - 1|$

Sketch the graph of each function.

8. $3x - 4y = 12$

9. $y = 2x - 3$

10. $y = \sqrt{25 - x^2}$

11. $y = -(x - 2)^2 + 5$

12. $y = 2|x| - 4$

13. $y = \sqrt{x + 3} - 5$

14. $f(x) = \begin{cases} x & \text{for } x < 0 \\ 2 & \text{for } x \geq 0 \end{cases}$

Let $f(x) = \sqrt{x + 2}$ and $g(x) = 3x - 1$. Find and simplify each of the following expressions.

15. $f(7)$

16. $(f \circ g)(2)$

17. $(f \circ g)(x)$

18. $g^{-1}(x)$

19. $(f + g)(14)$

20. $\dfrac{g(x + h) - g(x)}{h}$

Solve each problem.

21. State the intervals on which $f(x) = (x - 3)^2$ is increasing and those on which $f(x)$ is decreasing.

22. Discuss the symmetry of the graph of the function $f(x) = x^4 - 3x^2 + 9$.

23. State the solution set to the inequality $|x - 1| - 3 < 0$ using interval notation.

24. Find the inverse of the function $g(x) = 3 + \sqrt[3]{x - 2}$.

25. Jang's Postal Service charges $35 for addressing 200 envelopes, and $60 for addressing 400 envelopes. What is the average rate of change of the cost as the number of envelopes goes from 200 to 400?

26. The intensity of illumination I from a light source varies inversely as the square of the distance d from the source. If a flash on a camera has an intensity of 300 candlepower at a distance of 2 m, then what is the intensity of the flash at a distance of 10 m?

27. Construct a function expressing the volume V of a cube as a function of the length of the diagonal d of a side of the cube.

Tying it all Together

Chapters 1–2

Find all real solutions to each equation.

1. $2x - 3 = 0$

2. $-2x + 6 = 0$

3. $|x| - 100 = 0$

4. $1 - 2|x + 90| = 0$

5. $3 - 2\sqrt{x + 30} = 0$

6. $\sqrt{x - 3} + 15 = 0$

7. $-2(x - 2)^2 + 1 = 0$

8. $4(x + 2)^2 - 1 = 0$

9. $\sqrt{9 - x^2} - 2 = 0$

10. $\sqrt{49 - x^2} + 3 = 0$

Graph each function. Identify the domain, range, and x-intercepts.

11. $y = 2x - 3$

12. $y = -2x + 6$

13. $y = |x| - 100$

14. $y = 1 - 2|x + 90|$

15. $y = 3 - 2\sqrt{x + 30}$

16. $y = \sqrt{x - 3} + 15$

17. $y = -2(x - 2)^2 + 1$

18. $y = 4(x + 2)^2 - 1$

19. $y = \sqrt{9 - x^2} - 2$

20. $y = \sqrt{49 - x^2} + 3$

Solve each inequality. State the solution set using interval notation.

21. $2x - 3 > 0$

22. $-2x + 6 \leq 0$

23. $|x| - 100 \geq 0$

24. $1 - 2|x + 90| > 0$

25. $3 - 2\sqrt{x + 30} \leq 0$

26. $\sqrt{x - 3} + 15 > 0$

27. $-2(x - 2)^2 + 1 < 0$

28. $4(x + 2)^2 - 1 \geq 0$

29. $\sqrt{9 - x^2} - 2 \geq 0$

30. $\sqrt{49 - x^2} + 3 \leq 0$

Let $f(x) = 3(x + 1)^3 - 24$ for the following exercises.

31. Find $f(2)$ without using a calculator.

32. Graph the function f.

33. Express f as a composition using the functions $F(x) = 3x$, $G(x) = x + 1$, $H(x) = x^3$, and $K(x) = x - 24$.

34. Solve the equation $3(x + 1)^3 - 24 = 0$.

35. Solve $3(x + 1)^3 - 24 \geq 0$.

36. Solve $y = 3(x + 1)^3 - 24$ for x.

37. Find the inverse of the function f.

38. Graph the function f^{-1}.

39. Solve the inequality $f^{-1}(x) > 0$.

40. Express f^{-1} as a composition using the functions $F^{-1}(x) = x/3$, $G^{-1}(x) = x - 1$, $H^{-1}(x) = \sqrt[3]{x}$, and $K^{-1}(x) = x + 24$.

3 Polynomial and Rational Functions

OTHERS had died trying, but in May of 1927, Charles Lindbergh claimed the $25,000 prize for flying across the Atlantic Ocean. He was the first aviator to make the 3610-mile nonstop solo flight from New York to Paris.

Lindbergh did not just get in a plane and start flying. He spent years planning, designing, and practicing for his attempt to cross the Atlantic. Lindbergh collected and analyzed a vast amount of data to determine relationships between wind velocity, air speed, fuel consumption, and time elapsed. As he burned fuel, the plane got lighter and he flew slower.

WHAT YOU WILL LEARN In this chapter we explore polynomial and rational functions and their graphs. We'll learn how to use these basic functions of algebra to model many real-life situations, and we'll see how Lindbergh used them in planning his historic flight.

3.1

Quadratic Functions and Inequalities

A **polynomial function** is defined by a polynomial. A **quadratic function** is defined by a quadratic or second-degree polynomial. So a quadratic function has the form $f(x) = ax^2 + bx + c$ where $a \neq 0$. We studied functions of the form $f(x) = a(x - h)^2 + k$ in Section 2.3 (though we did not call them quadratic functions there). Once we see that these forms are equivalent, we can use what we learned in Section 2.3 in our study of quadratic functions here.

Two Forms for Quadratic Functions

It is easy to convert $f(x) = a(x - h)^2 + k$ to the form $f(x) = ax^2 + bx + c$ by squaring the binomial. To convert in the other direction we use completing the square from Section 1.3. Recall that the perfect square trinomial whose first two terms are $x^2 + bx$ is $x^2 + bx + \left(\frac{b}{2}\right)^2$. So to get the last term you *take one-half of the coefficient of the middle term and square it.*

When completing the square in Section 1.3, we added the last term of the perfect square trinomial to both sides of the equation. Here we want to keep $f(x)$ on the left side. To complete the square and change only the right side, we add and subtract on the right side as shown in Example 1(a). Another complication here is that the coefficient of x^2 is not always 1. To overcome this problem, you must factor before completing the square as shown in Example 1(b).

Example **1** Completing the square for a
معادلة من الدرجة الثانية ← quadratic function

Rewrite each function in the form $f(x) = a(x - h)^2 + k$.

a. $f(x) = x^2 + 6x$ **b.** $f(x) = 2x^2 - 20x + 3$

Solution

a. One-half of 6 is 3 and 3^2 is 9. Adding and subtracting 9 on the right side of $f(x) = x^2 + 6x$ does not change the function.

$$f(x) = x^2 + 6x$$
$$= x^2 + 6x + 9 - 9 \qquad 9 = \left(\frac{1}{2} \cdot 6\right)^2$$
$$= (x + 3)^2 - 9 \qquad \text{Factor.}$$

b. First factor 2 out of the first two terms, because the leading coefficient in $(x - h)^2$ is 1.

$$f(x) = 2x^2 - 20x + 3$$
$$= 2(x^2 - 10x) + 3$$
$$= 2(x^2 - 10x + 25 - 25) + 3 \qquad 25 = \left(\frac{1}{2} \cdot 10\right)^2$$
$$= 2(x^2 - 10x + 25) - 50 + 3 \qquad \text{Remove } -25 \text{ from the parentheses.}$$
$$= 2(x - 5)^2 - 47$$

Because of the 2 preceding the parentheses, the second 25 was doubled when it was removed from the parentheses.

Try This. Rewrite $f(x) = 2x^2 - 8x + 9$ in the form $f(x) = a(x - h)^2 + k$. ■

In the next example we complete the square and graph a quadratic function.

Example 2 Graphing a quadratic function

Rewrite $f(x) = -2x^2 - 4x + 3$ in the form $f(x) = a(x - h)^2 + k$ and sketch its graph.

Solution

Start by completing the square:

$$f(x) = -2(x^2 + 2x) + 3 \qquad \text{Factor out } -2 \text{ from the first two terms.}$$

$$= -2(x^2 + 2x + 1 - 1) + 3 \qquad \text{Complete the square for } x^2 + 2x.$$

$$= -2(x^2 + 2x + 1) + 2 + 3 \qquad \text{Remove } -1 \text{ from the parentheses.}$$

$$= -2(x + 1)^2 + 5$$

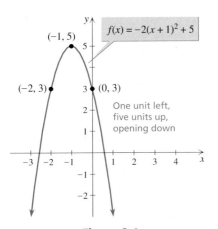

■ **Figure 3.1**

The function is now in the form $f(x) = a(x - h)^2 + k$. The number 1 indicates that the graph of $f(x) = x^2$ is translated one unit to the left. Since $a = -2$, the graph is stretched by a factor of 2 and reflected below the x-axis. Finally, the graph is translated five units upward. The graph shown in Fig. 3.1 includes the points $(0, 3)$, $(-1, 5)$, and $(-2, 3)$.

Try This. Rewrite $f(x) = -2x^2 - 4x + 1$ in the form $f(x) = a(x - h)^2 + k$ and sketch its graph. ■

We can use completing the square as in Example 2 to prove the following theorem.

Theorem:
Quadratic Functions

> The graph of any quadratic function is a transformation of the graph of $f(x) = x^2$.

■ **PROOF** Since $f(x) = a(x - h)^2 + k$ is a transformation of $f(x) = x^2$, we will show that $f(x) = ax^2 + bx + c$ can be written in that form:

$$f(x) = ax^2 + bx + c$$

$$= a\left(x^2 + \frac{b}{a}x\right) + c \qquad \text{Factor } a \text{ out of the first two terms.}$$

To complete the square for $x^2 + \frac{b}{a}x$, add and subtract $\frac{b^2}{4a^2}$ inside the parentheses:

$$f(x) = a\left(x^2 + \frac{b}{a}x + \frac{b^2}{4a^2} - \frac{b^2}{4a^2}\right) + c$$

$$= a\left(x^2 + \frac{b}{a}x + \frac{b^2}{4a^2}\right) - \frac{b^2}{4a} + c \qquad \text{Remove } -\frac{b^2}{4a^2} \text{ from the parentheses.}$$

$$= a\left(x + \frac{b}{2a}\right)^2 + \frac{4ac - b^2}{4a} \qquad \text{Factor and get a common denominator.}$$

$$= a(x - h)^2 + k. \qquad \text{Let } h = -\frac{b}{2a} \text{ and } k = \frac{4ac - b^2}{4a}.$$

So the graph of any quadratic function is a transformation of $f(x) = x^2$. ■

In Section 2.3 we stated that any transformation of $f(x) = x^2$ is called a parabola. Therefore, the graph of any quadratic function is a parabola.

Opening, Vertex, and Axis of Symmetry

If $a > 0$, the graph of $f(x) = a(x - h)^2 + k$ **opens upward;** if $a < 0$, the graph **opens downward** as shown in Fig. 3.2. Notice that h determines the amount of horizontal translation and k determines the amount of vertical translation of the graph of $f(x) = x^2$. Because of the translations, the point $(0, 0)$ on the graph of $f(x) = x^2$ moves to the point (h, k) on the graph of $f(x) = a(x - h)^2 + k$. Since $(0, 0)$ is the lowest point on the graph of $f(x) = x^2$, the lowest point on any parabola that opens upward is (h, k). Since $(0, 0)$ is the highest point on the graph of $f(x) = -x^2$, the highest point on any parabola that opens downward is (h, k). The point (h, k) is called the **vertex** of the parabola.

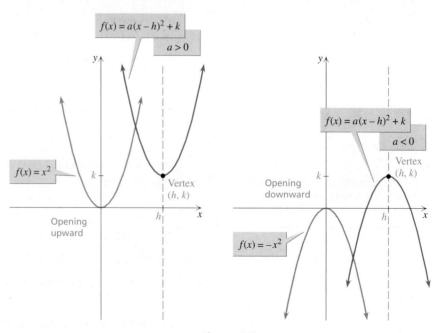

■ **Figure 3.2**

You can use a graphing calculator to experiment with different values for a, h, and k. Two possibilities are shown in Fig. 3.3(a) and (b). The graphing calculator allows you to see results quickly and to use values that you would not use otherwise. □

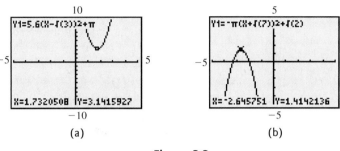

(a)   (b)

■ **Figure 3.3**

We can also find the vertex of the parabola when the function is written in the form $f(x) = ax^2 + bx + c$. The x-coordinate of the vertex is $-b/(2a)$ because we used $h = -b/(2a)$ when we obtained $y = a(x - h)^2 + k$ from $y = ax^2 + bx + c$. The y-coordinate is found by substitution.

S U M M A R Y   Vertex of a Parabola

1. For a quadratic function in the form $f(x) = a(x - h)^2$ the vertex of the parabola is (h, k).
2. For the form $f(x) = ax^2 + bx + c$, the x-coordinate of the vertex is $\frac{-b}{2a}$. The y-coordinate is $f\left(\frac{-b}{2a}\right)$.

Example **3** **Finding the vertex**

Find the vertex of the graph of $f(x) = -2x^2 - 4x + 3$.

Solution

Use $x = -b/(2a)$ to find the x-coordinate of the vertex:

$$x = \frac{-b}{2a} = \frac{-(-4)}{2(-2)} = -1$$

Now find $f(-1)$:

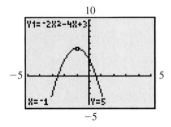

■ **Figure 3.4**

$$f(-1) = -2(-1)^2 - 4(-1) + 3 = 5$$

The vertex is $(-1, 5)$. In Example 2, $f(x) = -2x^2 - 4x + 3$ was rewritten as $f(x) = -2(x + 1)^2 + 5$. In this form we see immediately that the vertex is $(-1, 5)$. The graph in Fig. 3.4 supports these conclusions.

Try This. Find the vertex of the graph of $f(x) = 3x^2 - 12x + 5$.   ■

The domain of every quadratic function $f(x) = a(x - h)^2 + k$ is the set of real numbers, $(-\infty, \infty)$. The range of a quadratic function is determined from the second

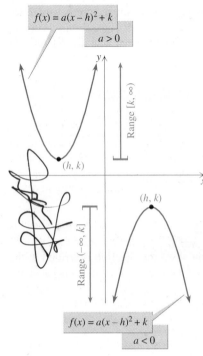

■ **Figure 3.5**

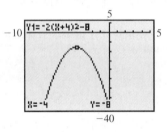

■ **Figure 3.6**

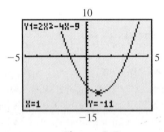

■ **Figure 3.7**

coordinate of the vertex. If $a > 0$, the range is $[k, \infty)$ and k is called the **minimum value of the function.** The function is decreasing on $(-\infty, h)$ and increasing on (h, ∞). See Fig. 3.5. If $a < 0$, the range is $(-\infty, k]$ and k is called the **maximum value of the function.** The function is increasing on $(-\infty, h)$ and decreasing on (h, ∞).

The graph of $f(x) = x^2$ is symmetric about the y-axis, which runs vertically through the vertex of the parabola. Since this symmetry is preserved in transformations, the graph of any quadratic function is symmetric about the vertical line through its vertex. The vertical line $x = -b/(2a)$ is called the **axis of symmetry** for the graph of $f(x) = ax^2 + bx + c$. Identifying these characteristics of a parabola before drawing a graph makes graphing easier and more accurate.

Example ④ Identifying the characteristics of a parabola

For each parabola, determine whether the parabola opens upward or downward, and find the vertex, axis of symmetry, and range of the function. Find the maximum or minimum value of the function and the intervals on which the function is increasing or decreasing.

a. $y = -2(x + 4)^2 - 8$ **b.** $y = 2x^2 - 4x - 9$

Solution

a. Since $a = -2$, the parabola opens downward. In $y = a(x - h)^2 + k$, the vertex is (h, k). So the vertex is $(-4, -8)$. The axis of symmetry is the vertical line through the vertex, $x = -4$. Since the parabola opens downward from $(-4, -8)$, the range of the function is $(-\infty, -8]$. The maximum value of the function is -8, and the function is increasing on $(-\infty, -4)$ and decreasing on $(-4, \infty)$.

The graph in Fig. 3.6 supports these results.   □

b. Since $a = 2$, the parabola opens upward. In the form $y = ax^2 + bx + c$, the x-coordinate of the vertex is $x = -b/(2a)$. In this case,

$$x = \frac{-b}{2a} = \frac{-(-4)}{2(2)} = 1.$$

Use $x = 1$ to find $y = 2(1)^2 - 4(1) - 9 = -11$. So the vertex is $(1, -11)$, and the axis of symmetry is the vertical line $x = 1$. Since the parabola opens upward, the vertex is the lowest point and the range is $[-11, \infty)$. The function is decreasing on $(-\infty, 1)$ and increasing on $(1, \infty)$. The minimum value of the function is -11.

The graph in Fig. 3.7 supports these results.

Try This. Identify all of the characteristics of the parabola $y = 3(x - 5)^2 - 4$. ■

Intercepts

The x-intercepts and the y-intercept are important points on the graph of a parabola. The x-intercepts are used in solving quadratic inequalities and the y-intercept is the starting point on the graph of a function whose domain is the nonnegative real numbers. The y-intercept is easily found by letting $x = 0$. The x-intercepts are found by letting $y = 0$ and solving the resulting quadratic equation.

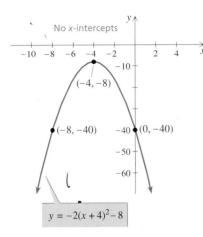

No x-intercepts

$y = -2(x + 4)^2 - 8$

■ **Figure 3.8**

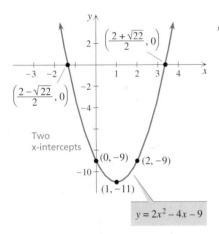

$\left(\frac{2 + \sqrt{22}}{2}, 0\right)$

$\left(\frac{2 - \sqrt{22}}{2}, 0\right)$

Two x-intercepts

$(0, -9)$ $(2, -9)$

$(1, -11)$

$y = 2x^2 - 4x - 9$

■ **Figure 3.9**

$x - 3$

■ **Figure 3.10**

Example **5** **Finding the intercepts**

Find the y-intercept and the x-intercepts for each parabola and sketch the graph of each parabola.

a. $y = -2(x + 4)^2 - 8$ **b.** $y = 2x^2 - 4x - 9$

Solution

a. If $x = 0$, $y = -2(0 + 4)^2 - 8 = -40$. The y-intercept is $(0, -40)$. Because the graph is symmetric about the line $x = -4$, the point $(-8, -40)$ is also on the graph. Since the parabola opens downward from $(-4, -8)$, which is below the x-axis, there are no x-intercepts. If we try to solve $-2(x + 4)^2 - 8 = 0$ to find the x-intercepts, we get $(x + 4)^2 = -4$, which has no real solution. The graph is shown in Fig. 3.8.

b. If $x = 0$, $y = 2(0)^2 - 4(0) - 9 = -9$. The y-intercept is $(0, -9)$. From Example 4(b), the vertex is $(1, -11)$. Because the graph is symmetric about the line $x = 1$, the point $(2, -9)$ is also on the graph. The x-intercepts are found by solving $2x^2 - 4x - 9 = 0$:

$$x = \frac{4 \pm \sqrt{(-4)^2 - 4(2)(-9)}}{2(2)} = \frac{4 \pm \sqrt{88}}{4} = \frac{2 \pm \sqrt{22}}{2}$$

The x-intercepts are $\left(\frac{2 + \sqrt{22}}{2}, 0\right)$ and $\left(\frac{2 - \sqrt{22}}{2}, 0\right)$. See Fig. 3.9.

Try This. Find all intercepts and sketch the graph of $y = 3(x - 5)^2 - 4$. ■

Note that if $y = ax^2 + bx + c$ has x-intercepts, they can always be found by using the quadratic formula. The x-coordinates of the x-intercepts are

$$x = \frac{-b \pm \sqrt{b^2 - 4ac}}{2a} = \frac{-b}{2a} \pm \frac{\sqrt{b^2 - 4ac}}{2a}.$$

Note how the axis of symmetry, $x = -b/(2a)$, appears in this formula. The x-intercepts are on opposite sides of the graph and are equidistant from the axis of symmetry.

Quadratic Inequalities

A **sign graph** is a number line that shows where the value of an expression is positive, negative, or zero. For example, the expression $x - 3$ has a positive value if $x > 3$ and a negative value if $x < 3$. If $x = 3$, then the value of $x - 3$ is zero. This information is shown on the sign graph in Fig. 3.10. We can use sign graphs and the rules for multiplying signed numbers to solve **quadratic inequalities**—inequalities that involve quadratic polynomials.

Example **6** **Solving a quadratic inequality using a sign graph of the factors**

Solve each inequality. Write the solution set in interval notation and graph it.

a. $x^2 - x > 6$ **b.** $x^2 - x \leq 6$

Solution

a. Write the inequality as $x^2 - x - 6 > 0$ and factor to get

$$(x - 3)(x + 2) > 0.$$

Note that $x + 2 > 0$ if $x > -2$ and $x + 2 < 0$ if $x < -2$. If $x = -2$, then $x + 2 = 0$. Combine this information with the sign graph for $x - 3$ in Fig. 3.10 to get Fig. 3.11, which shows the signs of both factors.

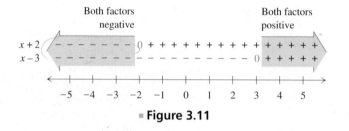

▪ **Figure 3.11**

The product $(x - 3)(x + 2)$ is positive only where both factors are positive or both factors are negative. So the solution set is $(-\infty, -2) \cup (3, \infty)$ and its graph is in Fig. 3.12.

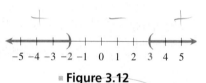

▪ **Figure 3.12**

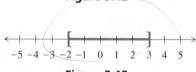

▪ **Figure 3.13**

b. The inequality $x^2 - x \le 6$ is equivalent to $(x - 3)(x + 2) \le 0$. The product $(x - 3)(x + 2)$ is negative when x is between -2 and 3, because that is when the factors have opposite signs on the sign graph in Fig. 3.11. Because $(x - 3)(x + 2) = 0$ if $x = 3$ or $x = -2$, these points are included in the solution set $[-2, 3]$. The graph is shown in Fig. 3.13.

The graph of $y = x^2 - x - 6$ in Fig. 3.14 is above the x-axis (y-coordinate greater than 0) when x is in $(-\infty, -2) \cup (3, \infty)$ and on or below the x-axis (y-coordinate less than or equal to 0) when x is in $[-2, 3]$.

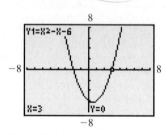

▪ **Figure 3.14**

Try This. Solve $x^2 - 2x > 8$ and graph the solution set. ■

Making a sign graph of the factors as in Example 6 shows how the signs of the linear factors determine the solution to a quadratic inequality. Of course, that method works only if you can factor the quadratic polynomial. The **test-point method** works on all quadratic polynomials. For this method, we find the roots of the quadratic polynomial (using the quadratic formula if necessary) and then make a sign graph for the quadratic polynomial itself, not the factors. The signs of the quadratic polynomial are determined by testing numbers in the intervals determined by the roots.

Example **7** Solving a quadratic inequality using the test-point method

Solve $2x^2 - 4x - 9 < 0$. Write the solution set in interval notation.

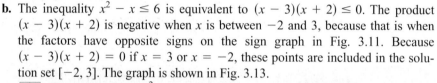

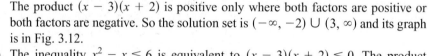

▪ **Figure 3.15**

Solution

The roots to $2x^2 - 4x - 9 = 0$ were found to be $\dfrac{2 \pm \sqrt{22}}{2}$ in Example 5(b). Since $\dfrac{2 - \sqrt{22}}{2} \approx -1.3$ and $\dfrac{2 + \sqrt{22}}{2} \approx 3.3$, we make the number line as in Fig. 3.15. Select a convenient test point in each of the three intervals determined by the roots.

Our selections -2, 1, and 5 are shown in red on the number line. Now evaluate $2x^2 - 4x - 9$ for each test point:

$$2(-2)^2 - 4(-2) - 9 = 7 \qquad \text{Positive}$$
$$2(1)^2 - 4(1) - 9 = -11 \qquad \text{Negative}$$
$$2(5)^2 - 4(5) - 9 = 21 \qquad \text{Positive}$$

The signs of these results are shown on the number line in Fig. 3.15. Since $2x^2 - 4x - 9 < 0$ between the roots, the solution set is

$$\left(\frac{2 - \sqrt{22}}{2}, \frac{2 + \sqrt{22}}{2} \right).$$

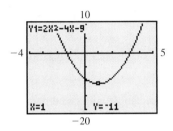

■ **Figure 3.16**

The graph of $y = 2x^2 - 4x - 9$ in Fig. 3.16 is below the x-axis between the x-intercepts and supports our conclusion. Note that every point that appears on the graph can be viewed as a test point.

Try This. Solve $x^2 - 2x - 4 < 0$ using the test-point method.   ■

If there are no real solutions to the quadratic equation that corresponds to a quadratic inequality, then the solution set to the inequality is either the empty set or all real numbers. For example, $x^2 + 4 = 0$ has no real solutions. Use any number as a test point to find that the solution set to $x^2 + 4 > 0$ is $(-\infty, \infty)$ and the solution set to $x^2 + 4 < 0$ is the empty set $\varnothing$.

Applications of Maximum and Minimum

In applications we often seek to minimize cost, maximize profit, or maximize area. If one variable is a quadratic function of another, then the maximum or minimum value of the dependent variable occurs at the vertex of the parabola.

Example **8** **Maximizing area of a rectangle**

If 100 m of fencing will be used to fence a rectangular region, then what dimensions for the rectangle will maximize the area of the region?

Solution

■ **Figure 3.17**

Since the 100 m of fencing forms the perimeter of a rectangle as shown in Fig. 3.17, we have $2L + 2W = 100$, where W is its width and L is its length. Dividing by 2 we get $L + W = 50$ or $W = 50 - L$. Since $A = LW$ for a rectangle, by substituting we get

$$A = LW = L(50 - L) = -L^2 + 50L.$$

So the area is a quadratic function of the length. The graph of this function is the parabola in Fig. 3.18. Since the parabola opens downward, the maximum value of A occurs when

$$L = \frac{-b}{2a} = \frac{-50}{2(-1)} = 25.$$

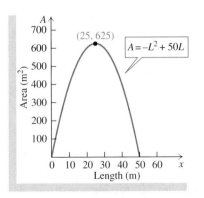

■ **Figure 3.18**

If $L = 25$ then $W = 25$ also, since $W = 50 - L$. So the length should be 25 meters and the width 25 meters to get the maximum area. The rectangle that gives the maximum area is actually a square with an area of 625 m^2.

Try This. If a rectangle has a perimeter of 50 meters, then what dimensions will maximize its area?   ■

For Thought

True or False? Explain.

1. The domain and range of a quadratic function are $(-\infty, \infty)$.

2. The vertex of the graph of $y = 2(x - 3)^2 - 1$ is $(3, 1)$.

3. The graph of $y = -3(x + 2)^2 - 9$ has no x-intercepts.

4. The maximum value of y in the function $y = -4(x - 1)^2 + 9$ is 9.

5. For $y = 3x^2 - 6x + 7$, the value of y is at its minimum when $x = 1$.

6. The graph of $f(x) = 9x^2 + 12x + 4$ has one x-intercept and one y-intercept.

7. The graph of every quadratic function has exactly one y-intercept.

8. The inequality $\pi\left(x - \sqrt{3}\right)^2 + \pi/2 \le 0$ has no solution.

9. The maximum area of a rectangle with fixed perimeter p is $p^2/16$.

10. The function $f(x) = (x - 3)^2$ is increasing on the interval $[-3, \infty)$.

3.1 Exercises

Write each quadratic function in the form $y = a(x - h)^2 + k$ and sketch its graph.

1. $y = x^2 + 4x$

2. $y = x^2 - 6x$

3. $y = x^2 - 3x$

4. $y = x^2 + 5x$

5. $y = 2x^2 - 12x + 22$

6. $y = 3x^2 - 12x + 1$

7. $y = -3x^2 + 6x - 3$

8. $y = -2x^2 - 4x + 8$

9. $y = x^2 + 3x + \dfrac{5}{2}$

10. $y = x^2 - x + 1$

11. $y = -2x^2 + 3x - 1$

12. $y = 3x^2 + 4x + 2$

Find the vertex of the graph of each quadratic function. See the summary on finding the vertex on page 271.

13. $f(x) = 3x^2 - 12x + 1$

14. $f(x) = -2x^2 - 8x + 9$

15. $f(x) = -3(x - 4)^2 + 1$

16. $f(x) = \dfrac{1}{2}(x + 6)^2 - \dfrac{1}{4}$

17. $y = -\dfrac{1}{2}x^2 - \dfrac{1}{3}x$

18. $y = \dfrac{1}{4}x^2 + \dfrac{1}{2}x - 1$

From the graph of each parabola, determine whether the parabola opens upward or downward, and find the vertex, axis of symmetry, and range of the function. Find the maximum or minimum value of the function and the intervals on which the function is increasing or decreasing.

19.

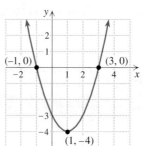

20.

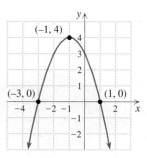

Find the range of each quadratic function and the maximum or minimum value of the function. Identify the intervals on which each function is increasing or decreasing.

21. $f(x) = 3 - x^2$

22. $f(x) = 5 - x^2$

23. $y = (x - 1)^2 - 1$

24. $y = (x + 3)^2 + 4$

25. $y = x^2 + 8x - 2$

26. $y = x^2 - 2x - 3$

27. $y = \dfrac{1}{2}(x - 3)^2 + 4$

28. $y = -\dfrac{1}{3}(x + 6)^2 + 37$

29. $f(x) = -2x^2 + 6x + 9$

30. $f(x) = -3x^2 - 9x + 4$

31. $y = -\dfrac{3}{4}\left(x - \dfrac{1}{2}\right)^2 + 9$

32. $y = \dfrac{3}{2}\left(x - \dfrac{1}{3}\right)^2 - 6$

Identify the vertex, axis of symmetry, y-intercept, x-intercepts, and opening of each parabola, then sketch the graph.

33. $y = x^2 - 3$

34. $y = 8 - x^2$

35. $y = x^2 - x$

36. $y = 2x - x^2$

37. $f(x) = x^2 + 6x + 9$

38. $f(x) = x^2 - 6x$

39. $f(x) = (x - 3)^2 - 4$

40. $f(x) = (x + 1)^2 - 9$

41. $y = -3(x - 2)^2 + 12$

42. $y = -2(x + 3)^2 + 8$

43. $y = -2x^2 + 4x + 1$

44. $y = -x^2 + 2x - 6$

Solve each inequality by making a sign graph of the factors. State the solution set in interval notation and graph it.

45. $2x^2 - x - 3 < 0$

46. $3x^2 - 4x - 4 \leq 0$

47. $2x + 15 < x^2$

48. $5x - x^2 < 4$

49. $w^2 - 4w - 12 \geq 0$

50. $y^2 + 8y + 15 \leq 0$

51. $t^2 \leq 16$

52. $36 \leq h^2$

53. $a^2 + 6a + 9 \leq 0$

54. $c^2 + 4 \leq 4c$

55. $4z^2 - 12z + 9 > 0$

56. $9s^2 + 6s + 1 \geq 0$

Solve each inequality by using the test-point method. State the solution set in interval notation and graph it.

57. $x^2 - 4x + 2 < 0$

58. $x^2 - 4x + 1 \leq 0$

59. $x^2 - 9 > 1$

60. $6 < x^2 - 1$

61. $y^2 + 18 > 10y$

62. $y^2 + 3 \geq 6y$

63. $p^2 + 9 > 0$

64. $-5 - s^2 < 0$

65. $a^2 + 20 \leq 8a$

66. $6t \leq t^2 + 25$

67. $-2w^2 + 5w < 6$

68. $-3z^2 - 5 > 2z$

Identify the solution set to each quadratic inequality by inspecting the graphs of $y = x^2 - 2x - 3$ and $y = -x^2 - 2x + 3$ as shown.

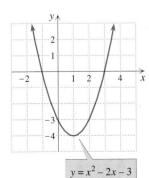

$y = x^2 - 2x - 3$

$y = -x^2 - 2x + 3$

69. $x^2 - 2x - 3 \geq 0$

70. $x^2 - 2x - 3 < 0$

71. $-x^2 - 2x + 3 > 0$

72. $-x^2 - 2x + 3 \leq 0$

73. $x^2 + 2x \leq 3$

74. $x^2 \leq 2x + 3$

The next two exercises incorporate many concepts of quadratics.

75. Let $f(x) = x^2 - 3x - 10$.
   a. Solve $f(x) = 0$.   **b.** Solve $f(x) = -10$.

   c. Solve $f(x) > 0$.   **d.** Solve $f(x) \leq 0$.

   e. Write f in the form $f(x) = a(x - h)^2 + k$ and describe the graph of f as a transformation of the graph of $y = x^2$.

   f. Graph f and state the domain, range, and the maximum or minimum y-coordinate on the graph.

   g. What is the relationship between the graph of f and the answers to parts (c) and (d)?

   h. Find the intercepts, axis of symmetry, vertex, opening, and intervals on which f is increasing or decreasing.

76. Repeat parts (a) through (h) from the previous exercise for $f(x) = -x^2 + 2x + 1$.

Solve each problem.

77. *Maximum Height of a Football* If a football is kicked straight up with an initial velocity of 128 ft/sec from a height of 5 ft, then its height above the earth is a function of time given by $h(t) = -16t^2 + 128t + 5$. What is the maximum height reached by this ball?

   HINT Find the vertex of the graph of the quadratic function.

78. *Maximum Height of a Ball* If a juggler can toss a ball into the air at a velocity of 64 ft/sec from a height of 6 ft, then what is the maximum height reached by the ball?

79. *Shooting an Arrow* If an archer shoots an arrow straight upward with an initial velocity of 160 ft/sec from a height of 8 ft, then its height above the ground in feet at time t in seconds is given by the function

$$h(t) = -16t^2 + 160t + 8.$$

   a. What is the maximum height reached by the arrow?

   b. How long does it take for the arrow to reach the ground?

80. *Rocket Propelled Grenade* If a soldier in basic training fires a rocket propelled grenade (RPG) straight up from ground level with an initial velocity of 256 ft/sec, then its height above the ground in feet at time t in seconds is given by the function

$$h(t) = -16t^2 + 256t.$$

   a. What is the maximum height reached by the RPG?

   b. How long does it take for the RPG to reach the ground?

81. *Lindbergh's Air Speed*   Flying too fast or slow wastes fuel. For the Spirit of St. Louis, miles per pound of fuel M was a function of air speed A in miles per hour, modeled by the formula

$$M = -0.000653A^2 + 0.127A - 5.01.$$

a. Use the accompanying graph to estimate the most economical air speed.

b. Use the formula to find the value of A that maximizes M.

c. How many gallons of fuel did he burn per hour if he flew at 97 mph and got 1.2 mi/lb of fuel, which weighed 6.12 lb/gal?

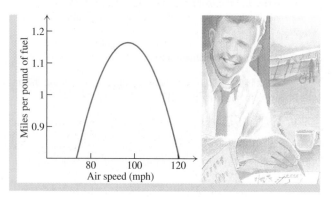

■ **Figure for Exercise 81**

82. *Average Farm Size*   The average size in acres of a farm in the U.S. in the year 1994 $+ x$ can be modeled by the quadratic function

$$A(x) = 0.37x^2 - 3.06x + 440.3.$$

a. Find the year in which the average size of a farm reached its maximum.

b. Find the year in which the average size will be 470 acres.

c. For what years shown in the accompanying figure is the average size decreasing? Increasing?

d. When was the average size less than 435 acres?

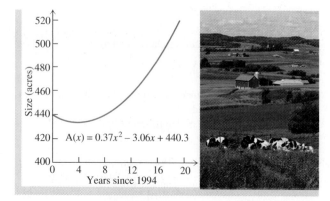

■ **Figure for Exercise 82**

83. *Maximum Area*   Shondra wants to enclose a rectangular garden with 200 yards of fencing. What dimensions for the garden will maximize its area.

> **HINT** Write a quadratic function for the area and find the vertex.

84. *Mirror Mirror*   Chantel wants to make a rectangular frame for a mirror using 10 feet of frame molding. What dimensions will maximize the area of the mirror assuming that there is no waste?

85. *Twin Kennels*   Martin plans to construct a rectangular kennel for two dogs using 120 feet of chain-link fencing. He plans to fence all four sides and down the middle to keep the dogs separate. What overall dimensions will maximize the total area fenced?

> **HINT** Write a quadratic function for the area and find the vertex.

86. *Cross Fenced*   Kim wants to construct rectangular pens for four animals with 400 feet of fencing. To get four separate pens she will fence a large rectangle and then fence through the middle of the rectangle parallel to the length and parallel to the width. What overall dimensions will maximize the total area of the pens?

87. *Big Barn*   Mike wants to enclose a rectangular area for his rabbits alongside his large barn using 30 feet of fencing. What dimensions will maximize the area fenced if the barn is used for one side of the rectangle?

88. *Maximum Area*   Kevin wants to enclose a rectangular garden using 14 eight-ft railroad ties, which he cannot cut. What are the dimensions of the rectangle that maximize the area enclosed?

89. *Cross Section of a Gutter*   Seth has a piece of aluminum that is 10 in. wide and 12 ft long. He plans to form a gutter with a rectangular cross section and an open top by folding up the sides as shown in the figure. What dimensions of the gutter would maximize the amount of water that it can hold?

> **HINT** Write a quadratic function for the cross-sectional area and find the vertex.

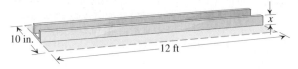

■ **Figure for Exercise 89**

90. *Maximum Volume of a Cage*   Sharon has a 12-ft board that is 12 in. wide. She wants to cut it into five pieces to make a cage for two pigeons, as shown in the figure. The front and back will be covered with chicken wire. What should be the dimensions of the cage to maximize the volume and use all of the 12-ft board?

▪ **Figure for Exercise 90**

91. *Maximizing Revenue* Mona Kalini gives a walking tour of Honolulu to one person for $49. To increase her business, she advertised at the National Orthodontist Convention that she would lower the price by $1 per person for each additional person, up to 49 people.

 a. Write the price per person p as a function of the number of people n.

 HINT See Exercise 91 of Section 1.4.

 b. Write her revenue as a function of the number of people on the tour.

 HINT Her revenue is the product of n and p.

 c. What is the maximum revenue for her tour?

92. *Concert Tickets* At $10 per ticket, Willie Williams and the Wranglers will fill all 8000 seats in the Assembly Center. The manager knows that for every $1 increase in the price, 500 tickets will go unsold.

 a. Write the number of tickets sold n as a function of ticket price p.

 HINT See Exercise 92 of Section 1.4.

 b. Write the total revenue as a function of the ticket price.

 HINT Revenue is the product of n and p.

 c. What ticket price will maximize the revenue?

93. *Variance of the Number of Smokers* If p is the probability that a randomly selected person in Chicago is a smoker, then $1 - p$ is the probability that the person is not a smoker. The variance of the number of smokers in a random sample of 50 Chicagoans is $50p(1 - p)$. What value of p maximizes the variance?

94. *Rate of Flu Infection* In a town of 5000 people the daily rate of infection with a flu virus varies directly with the product of the number of people who have been infected and the number of people not infected. When 1000 people have been infected, the flu is spreading at a rate of 40 new cases per day. For what number of people infected is the daily rate of infection at its maximum?

95. *Altitude and Pressure* At an altitude of h feet the atmospheric pressure a (atm) is given by

$$a = 3.89 \times 10^{-10}h^2 - 3.48 \times 10^{-5}h + 1$$

 (Sportscience, www.sportsci.org).

 a. Graph this quadratic function.

 b. In 1953, Edmund Hillary and Tenzing Norgay were the first persons to climb Mount Everest (see accompanying figure). Was the atmospheric pressure increasing or decreasing as they went to the 29,029-ft summit?

 c. Where is the function decreasing? Increasing?

 d. Does your answer to part (c) make sense?

 e. For what altitudes do you think the function is valid?

▪ **Figure for Exercise 95**

96. *Increasing Revenue* A company's weekly revenue in dollars is given by $R(x) = 2000x - 2x^2$, where x is the number of items produced during a week.

 a. For what x is $R(x) > 0$?

 b. On what interval is $R(x)$ increasing? Decreasing?

 c. For what number of items is the revenue maximized?

 d. What is the maximum revenue?

 e. Find the marginal revenue function $MR(x)$, where $MR(x)$ is defined by

$$MR(x) = R(x + 1) - R(x).$$

 f. On what interval is $MR(x)$ positive? Negative?

Use a calculator or a computer for the following regression problems.

97. *Quadratic Versus Linear* The average retail price in the spring of 2005 for a used Corvette depended on the age of the car, as shown in the following table (Edmund's, www.edmunds.com).

Age (yrs)	Price ($)
1	37,520
2	35,292
3	31,935
4	29,000
5	26,359
6	23,786
7	21,849
8	18,248
9	16,825

a. Use both linear regression and quadratic regression on a graphing calculator to express the price as a function of the age of the car.

b. Plot the data, the linear function, and the quadratic function on your calculator. Judging from the graph, which function appears to fit the data better?

c. Predict the price of an eleven-year-old car using both the linear function and the quadratic function. Which prediction seems more reasonable and why?

98. *Homicides in the United States*   The following table gives the number of homicides per 100,000 population in the United States (Federal Bureau of Investigation, www.fbi.gov).
 a. Use quadratic regression on a graphing calculator to express the number of homicides as a function of x, where x is the number of years after 1992.

b. Plot the data and the quadratic function on your calculator. Judging from the graph, does the quadratic function appear to be a good model for the data?

c. Use the quadratic function to find the year in which the number of homicides will be a minimum.

d. Use your calculator to find the year in which there will be 10 homicides per 100,000 population.

Year	Homicides	Year	Homicides
1992	9.3	1998	6.3
1993	9.5	1999	5.7
1994	9.0	2000	5.5
1995	8.2	2001	5.6
1996	7.4	2002	5.6
1997	6.8	2003	5.5

Thinking Outside the Box XXX

Overlapping Region   A right triangle with sides of length 3, 4, and 5 is drawn so that the endpoints of the side of length 5 are $(0, 0)$ and $(5, 0)$. A square with sides of length 1 is drawn so that its center is the vertex of the right angle and its sides are parallel to the x- and y-axes. What is the area of the region where the square and the triangle overlap?

3.1 Pop Quiz

1. Write $y = 2x^2 + 16x - 1$ in the form $y = a(x - h)^2 + k$.

2. Find the vertex of the graph of $y = 3(x + 4)^2 + 8$.

3. Find the range of $f(x) = -x^2 - 4x + 9$.

4. Find the minimum y-value for $y = x^2 - 3x$.

5. Find the x-intercepts and axis of symmetry for $y = x^2 - 2x - 8$.

6. Solve the inequality $x^2 + 4x < 0$.

Linking Concepts

For Individual or Group Explorations

Designing a Race Track

An architect is designing a race track that will be 60 feet wide and have semicircular ends, as shown in the accompanying figure. The length of the track is to be one mile, measured on the inside edge of the track.

a) Find the exact dimensions for x and y in the figure that will maximize the area of the rectangular center section.

b) Round the answers to part (a) to the nearest tenth of a foot and make an accurate drawing of the track.

c) Find the exact dimensions for x and y that will maximize the total area enclosed by the track.

d) Repeat parts (a), (b), and (c) assuming that the one-mile length is measured on the outside edge of the track.

3.2

Zeros of Polynomial Functions

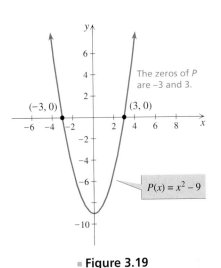

The zeros of P are -3 and 3.

$(-3, 0)$ $(3, 0)$

$P(x) = x^2 - 9$

■ **Figure 3.19**

We have studied linear and quadratic functions extensively. In this section we will study general polynomial functions.

The Remainder Theorem

If $y = P(x)$ is a polynomial function, then a value of x that satisfies $P(x) = 0$ is called a **zero** of the polynomial function or a zero of the polynomial. For example, 3 and -3 are zeros of the function $P(x) = x^2 - 9$, because $P(3) = 0$ and $P(-3) = 0$. Note that the zeros of $P(x) = x^2 - 9$ are the same as the solutions to the equation $x^2 - 9 = 0$. The real zeros of a polynomial function appear on the graph of the function as the x-coordinates of the x-intercepts. The x-intercepts of the graph of $P(x) = x^2 - 9$ shown in Fig. 3.19 are $(-3, 0)$ and $(3, 0)$.

For polynomial functions of degree 2 or less, the zeros can be found by solving quadratic or linear equations. Our goal in this section is to find all of the zeros of a polynomial function when possible. For polynomials of degree higher than 2, the difficulty of this task ranges from easy to impossible, but we have some theorems to assist us. The remainder theorem relates evaluating a polynomial to division of polynomials.

The Remainder Theorem	If R is the remainder when a polynomial $P(x)$ is divided by $x - c$, then $R = P(c)$.

■ **PROOF**   Let $Q(x)$ be the quotient and R be the remainder when $P(x)$ is divided by $x - c$. Since the dividend is equal to the divisor times the quotient plus the remainder, we have

$$P(x) = (x - c)Q(x) + R.$$

This statement is true for any value of x, and so it is also true for $x = c$:

$$P(c) = (c - c)Q(c) + R$$
$$= 0 \cdot Q(c) + R$$
$$= R$$

So $P(c)$ is equal to the remainder when $P(x)$ is divided by $x - c$. ■

To illustrate the remainder theorem, we will now use long division to evaluate a polynomial. Long division was discussed in Section P.5.

Example ▮ Using the remainder theorem to evaluate a polynomial

Use the remainder theorem to find $P(3)$ if $P(x) = 2x^3 - 5x^2 + 4x - 6$.

Solution

By the remainder theorem $P(3)$ is the remainder when $P(x)$ is divided by $x - 3$:

$$
\require{enclose}
\begin{array}{r}
2x^2 + x + 7 \\
x - 3 \enclose{longdiv}{2x^3 - 5x^2 + 4x - 6} \\
\underline{2x^3 - 6x^2 } \\
x^2 + 4x \\
\underline{x^2 - 3x } \\
7x - 6 \\
\underline{7x - 21} \\
15
\end{array}
$$

The remainder is 15 and therefore $P(3) = 15$. We can check by finding $P(3)$ in the usual manner:

$$P(3) = 2 \cdot 3^3 - 5 \cdot 3^2 + 4 \cdot 3 - 6 = 54 - 45 + 12 - 6 = 15$$

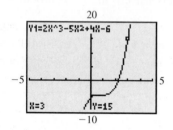 You can check this with a graphing calculator as shown in Fig. 3.20.

Try This. Use the remainder theorem to find $P(1)$ if $P(x) = x^2 - 7x + 9$. ■

■ Figure 3.20

Synthetic Division

In Example 1 we found $P(3) = 15$ in two different ways. Certainly, evaluating $P(x)$ for $x = 3$ in the usual manner is faster than dividing $P(x)$ by $x - 3$ using the ordinary method of dividing polynomials. However, there is a faster method, called **synthetic division,** for dividing by $x - 3$. Compare the two methods side by side, both showing $2x^3 - 5x^2 + 4x - 6$ divided by $x - 3$:

Ordinary Division of Polynomials

$$
\require{enclose}
\begin{array}{r}
2x^2 + x + 7 \leftarrow \text{Quotient} \\
x - 3 \enclose{longdiv}{2x^3 - 5x^2 + 4x - 6} \\
\underline{2x^3 - 6x^2 } \\
x^2 + 4x \\
\underline{x^2 - 3x } \\
7x - 6 \\
\underline{7x - 21} \\
15 \leftarrow \text{Remainder}
\end{array}
$$

Synthetic Division

$$
\begin{array}{r|rrrr}
3 & 2 & -5 & 4 & -6 \\
  &   & 6 & 3 & 21 \\
\hline
  & \underbrace{2 \quad\quad 1 \quad\quad 7}_{\text{Quotient}} & & & 15 \leftarrow \text{Remainder}
\end{array}
$$

Synthetic division certainly looks easier than ordinary division, and in general it is faster than evaluating the polynomial by substitution. Synthetic division is used as a quick means of dividing a polynomial by a binomial of the form $x - c$.

In synthetic division we write just the necessary parts of the ordinary division. Instead of writing $2x^3 - 5x^2 + 4x - 6$, write the coefficients 2, −5, 4, and −6. For $x - 3$, write only the 3. The bottom row in synthetic division gives the coefficients of the quotient and the remainder.

To actually perform the synthetic division, start with the following arrangement of coefficients:

$$3 \,\big|\, \begin{array}{cccc} 2 & -5 & 4 & -6 \end{array}$$

Bring down the first coefficient, 2. Multiply 2 by 3 and write the answer beneath −5. Then add:

$$3 \,\big|\, \begin{array}{cccc} 2 & -5 & 4 & -6 \\ & \downarrow\ 6 \\ \hline \text{Multiply} \to 2 & 1 \end{array}$$

Add

Using 3 rather than −3 when dividing by $x - 3$ allows us to multiply and add rather than multiply and subtract as in ordinary division. Now multiply 1 by 3 and write the answer beneath 4. Then add. Repeat the multiply-and-add step for the remaining column:

$$3 \,\big|\, \begin{array}{cccc} 2 & -5 & 4 & -6 \\ & 6 & 3 & 21 \\ \hline 2 & 1 & 7 & 15 \end{array}$$

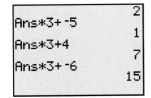

■ **Figure 3.21**

To perform this arithmetic on a graphing calculator, start with the leading coefficient 2 as the answer. Then repeatedly multiply the answer by 3 and add the next coefficient as shown in Fig. 3.21. □

The quotient is $2x^2 + x + 7$, and the remainder is 15. Since the divisor in synthetic division is of the form $x - c$, the degree of the quotient is always one less than the degree of the dividend.

Example **2** Synthetic division

Use synthetic division to find the quotient and remainder when $x^4 - 14x^2 + 5x - 9$ is divided by $x + 4$.

Solution

Since $x + 4 = x - (-4)$, we use −4 in the synthetic division. Use 1, 0, −14, 5, and −9 as the coefficients of the polynomial. We use 0 for the coefficient of the missing x^3-term, as we would in ordinary division of polynomials.

$$-4 \,\big|\, \begin{array}{ccccc} 1 & 0 & -14 & 5 & -9 \\ & -4 & 16 & -8 & 12 \\ \hline \text{Multiply} \to 1 & -4 & 2 & -3 & 3 \end{array}$$

Add

The quotient is $x^3 - 4x^2 + 2x - 3$ and the remainder is 3.

Try This. Use synthetic division to find the quotient and remainder when $x^3 - 7x + 5$ is divided by $x + 2$. ■

To find the value of a polynomial $P(x)$ for $x = c$, we can divide $P(x)$ by $x - c$ or we can substitute c for x and compute. Using long division to divide by $x - c$ is not an efficient way to evaluate a polynomial. However, if we use synthetic division to divide by $x - c$ we can actually evaluate some polynomials using fewer arithmetic operations than we use in the substitution method. Note that in part (a) of the next example, synthetic division takes more steps than substitution, but in part (b) synthetic division takes fewer steps. To gain a better understanding of why synthetic division gives the value of a polynomial, see Linking Concepts at the end of this section.

Example ③ **Using synthetic division to evaluate a polynomial**

Let $f(x) = x^3$ and $g(x) = x^3 - 3x^2 + 5x - 12$. Use synthetic division to find the following function values.

a. $f(-2)$ **b.** $g(4)$

Solution

a. To find $f(-2)$, divide the polynomial x^3 by $x - (-2)$ or $x + 2$ using synthetic division. Write x^3 as $x^3 + 0x^2 + 0x + 0$, and use 1, 0, 0, and 0 as the coefficients. We use a zero for each power of x below x^3.

$$
\begin{array}{r|rrrr}
-2 & 1 & 0 & 0 & 0 \\
   &   & -2 & 4 & -8 \\
\hline
   & 1 & -2 & 4 & -8 \\
\end{array}
$$

The remainder is -8, so $f(-2) = -8$. To check, find $f(-2) = (-2)^3 = -8$.

b. To find $g(4)$, use synthetic division to divide $x^3 - 3x^2 + 5x - 12$ by $x - 4$:

$$
\begin{array}{r|rrrr}
4 & 1 & -3 & 5 & -12 \\
  &   & 4 & 4 & 36 \\
\hline
  & 1 & 1 & 9 & 24 \\
\end{array}
$$

The remainder is 24, so $g(4) = 24$. Check this answer by finding $g(4) = 4^3 - 3(4^2) + 5(4) - 12 = 24$.

Try This. Use synthetic division to find $P(3)$ if $P(x) = x^3 + x^2 - 9$.   ■

The Factor Theorem

Consider the polynomial function $P(x) = x^2 - x - 6$. We can find the zeros of the function by solving $x^2 - x - 6 = 0$ by factoring:

$$(x - 3)(x + 2) = 0$$

$$x - 3 = 0 \quad \text{or} \quad x + 2 = 0$$

$$x = 3 \quad \text{or} \qquad x = -2$$

Both 3 and -2 are zeros of the function $P(x) = x^2 - x - 6$. Note how each factor of the polynomial corresponds to a zero of the function. This example suggests the following theorem.

The Factor Theorem

> The number c is a zero of the polynomial function $y = P(x)$ if and only if $x - c$ is a factor of the polynomial $P(x)$.

■ **PROOF**   If c is a zero of the polynomial function $y = P(x)$, then $P(c) = 0$. If $P(x)$ is divided by $x - c$, we get a quotient $Q(x)$ and a remainder R such that

$$P(x) = (x - c)Q(x) + R.$$

By the remainder theorem, $R = P(c)$. Since $P(c) = 0$, we have $R = 0$ and $P(x) = (x - c)Q(x)$, which proves that $x - c$ is a factor of $P(x)$.

In Exercise 87 you will be asked to prove that if $x - c$ is a factor of $P(x)$, then c is a zero of the polynomial function. These two arguments together establish the truth of the factor theorem.   ■

Synthetic division can be used in conjunction with the factor theorem. If the remainder of dividing $P(x)$ by $x - c$ is 0, then $P(c) = 0$ and c is a zero of the polynomial function. By the factor theorem, $x - c$ is a factor of $P(x)$.

Example **4**   **Using the factor theorem to factor a polynomial**

Determine whether $x + 4$ is a factor of the polynomial $P(x) = x^3 - 13x + 12$. If it is a factor, then factor $P(x)$ completely.

Solution

By the factor theorem, $x + 4$ is a factor of $P(x)$ if and only if $P(-4) = 0$. We can find $P(-4)$ using synthetic division:

$$
\begin{array}{r|rrrr}
-4 & 1 & 0 & -13 & 12 \\
   &   & -4 & 16 & -12 \\
\hline
   & 1 & -4 & 3 & 0
\end{array}
$$

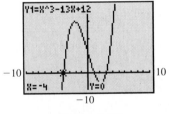

■ **Figure 3.22**

Since $P(-4)$ is equal to the remainder, $P(-4) = 0$ and -4 is a zero of $P(x)$. By the factor theorem, $x + 4$ is a factor of $P(x)$. Since the other factor is the quotient from the synthetic division, $P(x) = (x + 4)(x^2 - 4x + 3)$. Factor the quadratic polynomial to get $P(x) = (x + 4)(x - 1)(x - 3)$.

You can check this result by examining the calculator graph of $y = x^3 - 13x + 12$ shown in Fig. 3.22. The graph appears to cross the x-axis at $-4, 1$, and 3, supporting the conclusion that $P(x) = (x + 4)(x - 1)(x - 3)$.

Try This.   Is $x - 1$ a factor of $P(x) = x^3 - 3x + 2$? If it is a factor, then factor $P(x)$ completely.   ■

The Fundamental Theorem of Algebra

Whether a number is a zero of a polynomial function can be determined by synthetic division. But does every polynomial function have a zero? This question was answered in the affirmative by Carl F. Gauss when he proved the fundamental theorem of algebra in his doctoral thesis in 1799 at the age of 22.

The Fundamental Theorem of Algebra

> If $y = P(x)$ is a polynomial function of positive degree, then $y = P(x)$ has at least one zero in the set of complex numbers.

Gauss also proved the n-root theorem of Section 3.3 that says that the number of zeros of a polynomial (or polynomial function) of degree n is at most n. For example, a fifth-degree polynomial function has at least one zero and at most five.

Note that the zeros guaranteed by Gauss are in the set of complex numbers. So the zero might be real or imaginary. The theorem applies only to polynomial functions of degree 1 or more, because a polynomial function of zero degree such as $P(x) = 7$ has no zeros. A polynomial function of degree 1, $f(x) = ax + b$, has exactly one zero which is found by solving $ax + b = 0$. A polynomial function of degree 2, $f(x) = ax^2 + bx + c$, has one or two zeros that can be found by the quadratic formula. For higher degree polynomials the situation is not as simple. The fundamental theorem tells us that a polynomial function has at least one zero but not how to find it. For this purpose we have some other theorems.

The Rational Zero Theorem

Zeros or roots that are rational numbers, the **rational zeros,** are generally the easiest to find. The polynomial function $f(x) = 6x^2 - x - 35$ has two rational zeros that can be found as follows:

$$6x^2 - x - 35 = 0$$

$$(2x - 5)(3x + 7) = 0 \quad \text{Factor.}$$

$$x = \frac{5}{2} \quad \text{or} \quad x = -\frac{7}{3}$$

Note that in $5/2$, 5 is a factor of -35 (the constant term) and 2 is a factor of 6 (the leading coefficient). For the zero $-7/3$, -7 is a factor of -35 and 3 is a factor of 6. Of course, these observations are not surprising, because we used these facts to factor the quadratic polynomial in the first place. Note that there are a lot of other factors of -35 and 6 for which the ratio is *not* a zero of this function. This example illustrates the rational zero theorem, which is also called the rational root theorem.

The Rational Zero Theorem

> If $f(x) = a_nx^n + a_{n-1}x^{n-1} + a_{n-2}x^{n-2} + \cdots + a_1x + a_0$ is a polynomial function with integral coefficients ($a_n \neq 0$ and $a_0 \neq 0$) and p/q (in lowest terms) is a rational zero of $f(x)$, then p is a factor of the constant term a_0 and q is a factor of the leading coefficient a_n.

■ **PROOF** If p/q is a zero of $f(x)$, then $f(p/q) = 0$:

$$a_n\left(\frac{p}{q}\right)^n + a_{n-1}\left(\frac{p}{q}\right)^{n-1} + a_{n-2}\left(\frac{p}{q}\right)^{n-2} + \cdots + a_1\frac{p}{q} + a_0 = 0$$

Subtract a_0 from each side of this equation.

$$a_n\left(\frac{p}{q}\right)^n + a_{n-1}\left(\frac{p}{q}\right)^{n-1} + a_{n-2}\left(\frac{p}{q}\right)^{n-2} + \cdots + a_1\frac{p}{q} = -a_0$$

Since p/q is in lowest terms, p is not a factor of q. So p appears at least once in each term on the left-hand side of the equation, and p is a factor of the left-hand side. Since the left-hand side is equal to the integer $-a_0$, p is also a factor of a_0.

To prove that q is a factor of a_n, multiply each side of the original equation by $(q/p)^n$, then subtract a_n from each side:

$$a_{n-1}\left(\frac{q}{p}\right) + a_{n-2}\left(\frac{q}{p}\right)^2 + a_{n-3}\left(\frac{q}{p}\right)^3 + \cdots + a_0\left(\frac{q}{p}\right)^n = -a_n$$

Now q is a factor of a_n by the same argument that we used above. ▪

The rational zero theorem does not identify exactly which rational numbers are zeros of a function; it only gives *possibilities* for the rational zeros.

Example 5 Using the rational zero theorem

Find all possible rational zeros for each polynomial function.

a. $f(x) = 2x^3 - 3x^2 - 11x + 6$ **b.** $g(x) = 3x^3 - 8x^2 - 8x + 8$

Solution

a. If the rational number p/q is a zero of $f(x)$, then p is a factor of 6 and q is a factor of 2. The positive factors of 6 are 1, 2, 3, and 6. The positive factors of 2 are 1 and 2. Take each factor of 6 and divide by 1 to get $1/1$, $2/1$, $3/1$, and $6/1$. Take each factor of 6 and divide by 2 to get $1/2$, $2/2$, $3/2$, and $6/2$. Simplify the ratios, eliminate duplications, and put in the negative factors to get

$$\pm 1, \quad \pm 2, \quad \pm 3, \quad \pm 6, \quad \pm\frac{1}{2}, \quad \text{and} \quad \pm\frac{3}{2}$$

as the possible rational zeros to the function $f(x)$.

b. If the rational number p/q is a zero of $g(x)$, then p is a factor of 8 and q is a factor of 3. The factors of 8 are 1, 2, 4, and 8. The factors of 3 are 1 and 3. If we take all possible ratios of a factor of 8 over a factor of 3, we get

$$\pm 1, \quad \pm 2, \quad \pm 4, \quad \pm 8, \quad \pm\frac{1}{3}, \quad \pm\frac{2}{3}, \quad \pm\frac{4}{3}, \quad \text{and} \quad \pm\frac{8}{3}$$

as the possible rational zeros of the function $g(x)$.

Try This. Find all possible rational zeros for $h(x) = 2x^3 + x^2 - 4x - 3$. ▪

Our goal is to find all of the zeros to a polynomial function. The zeros to a polynomial function might be rational, irrational, or imaginary. We can determine the rational zeros by simply evaluating the polynomial function for every number in the list of possible rational zeros. If the list is long, looking at a graph of the function can speed up the process. We will use synthetic division to evaluate the polynomial, because synthetic division gives the quotient polynomial as well as the value of the polynomial.

Example 6 Finding all zeros of a polynomial function

Find all of the real and imaginary zeros for each polynomial function of Example 5.

a. $f(x) = 2x^3 - 3x^2 - 11x + 6$ **b.** $g(x) = 3x^3 - 8x^2 - 8x + 8$

Solution

a. The possible rational zeros of $f(x)$ are listed in Example 5(a). Use synthetic division to check each possible zero to see whether it is actually a zero. Try 1 first.

$$
\begin{array}{r|rrrr}
1 & 2 & -3 & -11 & 6 \\
  &   & 2 & -1 & -12 \\
\hline
  & 2 & -1 & -12 & -6
\end{array}
$$

Since the remainder is -6, 1 is not a zero of the function. Keep on trying numbers from the list of possible rational zeros. To save space, we will not show any more failures. So try $1/2$ next.

$$
\begin{array}{r|rrrr}
\dfrac{1}{2} & 2 & -3 & -11 & 6 \\
  &   & 1 & -1 & -6 \\
\hline
  & 2 & -2 & -12 & 0
\end{array}
$$

Since the remainder in the synthetic division is 0, $1/2$ is a zero of $f(x)$. By the factor theorem, $x - 1/2$ is a factor of the polynomial. The quotient is the other factor.

$$2x^3 - 3x^2 - 11x + 6 = 0$$

$$\left(x - \frac{1}{2}\right)(2x^2 - 2x - 12) = 0 \quad \text{Factor.}$$

$$(2x - 1)(x^2 - x - 6) = 0 \quad \begin{array}{l}\text{Factor 2 out of the second "factor" and}\\ \text{distribute it into the first factor.}\end{array}$$

$$(2x - 1)(x - 3)(x + 2) = 0 \quad \text{Factor completely.}$$

$$2x - 1 = 0 \quad \text{or} \quad x - 3 = 0 \quad \text{or} \quad x + 2 = 0$$

$$x = \frac{1}{2} \quad \text{or} \qquad x = 3 \quad \text{or} \qquad x = -2$$

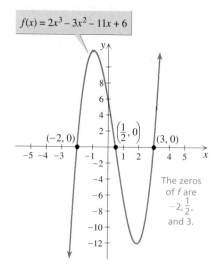

$f(x) = 2x^3 - 3x^2 - 11x + 6$

$(-2, 0)$ $\left(\frac{1}{2}, 0\right)$ $(3, 0)$

The zeros of f are $-2, \frac{1}{2},$ and 3.

■ **Figure 3.23**

The zeros of the function f are $1/2$, 3, and -2. Note that each zero of f corresponds to an x-intercept on the graph of f shown in Fig. 3.23. Because this polynomial had three rational zeros, we could have found them all by using synthetic division or by examining the calculator graph. However, it is good to factor the polynomial to see the correspondence between the three zeros, the three factors, and the three x-intercepts.

You could speed up this process of finding a zero with the calculator graph shown in Fig. 3.24. It is not too hard to discover that $1/2$ is a zero by looking at the graph and the list of possible rational zeros that we listed in Example 5. If you use a graph to find that $1/2$ is a zero, you still need to do synthetic division to factor the polynomial.

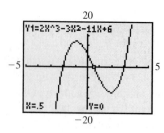

■ **Figure 3.24**

b. The possible rational zeros of $g(x)$ are listed in Example 5(b). First check $2/3$ to see whether it produces a remainder of 0.

$$
\begin{array}{r|rrrr}
\dfrac{2}{3} & 3 & -8 & -8 & 8 \\
  &   & 2 & -4 & -8 \\
\hline
  & 3 & -6 & -12 & 0
\end{array}
$$

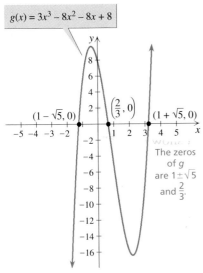

$g(x) = 3x^3 - 8x^2 - 8x + 8$

$(1 - \sqrt{5}, 0)$ $\left(\frac{2}{3}, 0\right)$ $(1 + \sqrt{5}, 0)$

What?
The zeros
of g
are $1 \pm \sqrt{5}$
and $\frac{2}{3}$.

▪ **Figure 3.25**

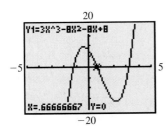

Y1=3X^3-8X2-8X+8

X=.66666667 Y=0

▪ **Figure 3.26**

Since the remainder in the synthetic division is 0, 2/3 is a zero of $g(x)$. By the factor theorem, $x - 2/3$ is a factor of the polynomial. The quotient is the other factor.

$$3x^3 - 8x^2 - 8x + 8 = 0$$

$$\left(x - \frac{2}{3}\right)(3x^2 - 6x - 12) = 0$$

$$(3x - 2)(x^2 - 2x - 4) = 0$$

$$3x - 2 = 0 \quad \text{or} \quad x^2 - 2x - 4 = 0$$

$$x = \frac{2}{3} \quad \text{or} \quad x = \frac{2 \pm \sqrt{20}}{2}$$

$$x = \frac{2}{3} \quad \text{or} \quad x = 1 \pm \sqrt{5}$$

There are one rational and two irrational roots to the equation. So the zeros of the function g are 2/3, $1 + \sqrt{5}$, and $1 - \sqrt{5}$. Each zero corresponds to an x-intercept on the graph of g shown in Fig. 3.25.

You could graph the function with a calculator as shown in Fig. 3.26. Keeping in mind the list of possible rational zeros, it is not hard to discover that 2/3 is a zero.

Try This. Find all zeros for $h(x) = 2x^3 + x^2 - 4x - 3$. ▪

Note that in Example 6(a) all of the zeros were rational. All three could have been found by continuing to check the possible rational zeros using synthetic division. In Example 6(b) we would be wasting time if we continued to check the possible rational zeros, because there is only one. When we get to a quadratic polynomial, it is best to either factor the quadratic polynomial or use the quadratic formula to find the remaining zeros.

For Thought

True or False? Explain.

1. The function $f(x) = 1/x$ has at least one zero.

2. If $P(x) = x^4 - 6x^2 - 8$ is divided by $x^2 - 2$, then the remainder is $P(\sqrt{2})$.

3. If $1 - 2i$ and $1 + 2i$ are zeros of $P(x) = x^3 - 5x^2 + 11x - 15$, then $x - 1 - 2i$ and $x - 1 + 2i$ are factors of $P(x)$.

4. If we divide $x^5 - 1$ by $x - 2$, then the remainder is 31.

5. Every polynomial function has at least one zero.

6. If $P(x) = x^3 - x^2 + 4x - 5$ and b is the remainder from division of $P(x)$ by $x - c$, then $b^3 - b^2 + 4b - 5 = c$.

7. If $P(x) = x^3 - 5x^2 + 4x - 15$, then $P(4) = 0$.

8. The equation $\pi^2 x^4 - \frac{1}{\sqrt{2}} x^3 + \frac{1}{\sqrt{7} + \pi} = 0$ has at least one complex solution.

9. The binomial $x - 1$ is a factor of $x^5 + x^4 - x^3 - x^2 - x + 1$.

10. The binomial $x + 3$ is a factor of $3x^4 - 5x^3 + 7x^2 - 9x - 2$.

3.2 Exercises

Use ordinary division of polynomials to find the quotient and remainder when the first polynomial is divided by the second.

1. $x^2 - 5x + 7, x - 2$

2. $x^2 - 3x + 9, x - 4$

3. $-2x^3 + 4x - 9, x + 3$

4. $-4w^3 + 5w^2 - 7, w - 3$

5. $s^4 - 3s^2 + 6, s^2 - 5$

6. $h^4 + 3h^3 + h - 5, h^2 - 3$

Use synthetic division to find the quotient and remainder when the first polynomial is divided by the second.

7. $x^2 + 4x + 1, x - 2$

8. $2x^2 - 3x + 6, x - 5$

9. $-x^3 + x^2 - 4x + 9, x + 3$

10. $-3x^3 + 5x^2 - 6x + 1, x + 1$

11. $4x^3 - 5x + 2, x - \dfrac{1}{2}$

12. $-6x^3 + 25x^2 - 9, x - \dfrac{3}{2}$

13. $2a^3 - 3a^2 + 4a + 3, a + \dfrac{1}{2}$

14. $-3b^3 - b^2 - 3b - 1, b + \dfrac{1}{3}$

15. $x^4 - 3, x - 1$

16. $x^4 - 16, x - 2$

17. $x^5 - 6x^3 + 4x - 5, x - 2$

18. $2x^5 - 5x^4 - 5x + 7, x - 3$

Let $f(x) = x^5 - 1, g(x) = x^3 - 4x^2 + 8$, and $h(x) = 2x^4 + x^3 - x^2 + 3x + 3$. Find the following function values by using synthetic division. Check by using substitution.

19. $f(1)$ **20.** $f(-1)$ **21.** $f(-2)$ **22.** $f(3)$

23. $g(1)$ **24.** $g(-1)$ **25.** $g\left(-\dfrac{1}{2}\right)$ **26.** $g\left(\dfrac{1}{2}\right)$

27. $h(-1)$ **28.** $h(2)$ **29.** $h(1)$ **30.** $h(-3)$

Determine whether the given binomial is a factor of the polynomial following it. If it is a factor, then factor the polynomial completely.

31. $x + 3, x^3 + 4x^2 + x - 6$

32. $x + 5, x^3 + 8x^2 + 11x - 20$

33. $x - 4, x^3 + 4x^2 - 17x - 60$

34. $x - 2, x^3 - 12x^2 + 44x - 48$

Determine whether each given number is a zero of the polynomial function following the number.

35. $3, f(x) = 2x^3 - 5x^2 - 4x + 3$

36. $-2, g(x) = 3x^3 - 6x^2 - 3x - 19$

37. $-2, g(d) = d^3 + 2d^2 + 3d + 1$

38. $-1, w(x) = 3x^3 + 2x^2 - 2x - 1$

39. $-1, P(x) = x^4 + 2x^3 + 4x^2 + 6x + 3$

40. $3, G(r) = r^4 + 4r^3 + 5r^2 + 3r + 17$

41. $\dfrac{1}{2}, H(x) = x^3 + 3x^2 - 5x + 7$

42. $-\dfrac{1}{2}, T(x) = 2x^3 + 3x^2 - 3x - 2$

Use the rational zero theorem to find all possible rational zeros for each polynomial function.

43. $f(x) = x^3 - 9x^2 + 26x - 24$

44. $g(x) = x^3 - 2x^2 - 5x + 6$

45. $h(x) = x^3 - x^2 - 7x + 15$

46. $m(x) = x^3 + 4x^2 + 4x + 3$

47. $P(x) = 8x^3 - 36x^2 + 46x - 15$

48. $T(x) = 18x^3 - 9x^2 - 5x + 2$

49. $M(x) = 18x^3 - 21x^2 + 10x - 2$

50. $N(x) = 4x^3 - 10x^2 + 4x + 5$

Find all of the real and imaginary zeros for each polynomial function.

51. $f(x) = x^3 - 9x^2 + 26x - 24$

52. $g(x) = x^3 - 2x^2 - 5x + 6$

53. $h(x) = x^3 - x^2 - 7x + 15$

54. $m(x) = x^3 + 4x^2 + 4x + 3$

55. $P(a) = 8a^3 - 36a^2 + 46a - 15$

56. $T(b) = 18b^3 - 9b^2 - 5b + 2$

57. $M(t) = 18t^3 - 21t^2 + 10t - 2$

58. $N(t) = 4t^3 - 10t^2 + 4t + 5$

59. $S(w) = w^4 + w^3 - w^2 + w - 2$

60. $W(v) = 2v^4 + 5v^3 + 3v^2 + 15v - 9$

61. $V(x) = x^4 + 2x^3 - x^2 - 4x - 2$

62. $U(x) = x^4 - 4x^3 + x^2 + 12x - 12$

63. $f(x) = 24x^3 - 26x^2 + 9x - 1$

64. $f(x) = 30x^3 - 47x^2 - x + 6$

65. $y = 16x^3 - 33x^2 + 82x - 5$

66. $y = 15x^3 - 37x^2 + 44x - 14$

67. $f(x) = 21x^4 - 31x^3 - 21x^2 - 31x - 42$

68. $f(x) = 119x^4 - 5x^3 + 214x^2 - 10x - 48$

69. $f(x) = (x^2 + 9)(x^3 + 6x^2 + 3x - 10)$

70. $f(x) = (x^2 - 5)(x^3 - 5x^2 - 12x + 36)$

71. $f(x) = (x^2 - 4x + 1)(x^3 - 9x^2 + 23x - 15)$

72. $f(x) = (x^2 - 4x + 13)(x^3 - 4x^2 - 17x + 60)$

*Use division to write each rational expression in the form quotient +
remainder/divisor. Use synthetic division when possible.*

73. $\dfrac{2x + 1}{x - 2}$

74. $\dfrac{x - 1}{x + 3}$

75. $\dfrac{a^2 - 3a + 5}{a - 3}$

76. $\dfrac{2b^2 - 3b + 1}{b + 2}$

77. $\dfrac{c^2 - 3c - 4}{c^2 - 4}$

78. $\dfrac{2h^2 + h - 2}{h^2 - 1}$

79. $\dfrac{4t - 5}{2t + 1}$

80. $\dfrac{6y - 1}{3y - 1}$

Solve each problem.

81. *Drug Testing* The concentration of a drug (in parts per
million) in a patient's bloodstream t hours after administration
of the drug is given by the function

$$P(t) = -t^4 + 12t^3 - 58t^2 + 132t.$$

a. Use the formula to determine when the drug will be totally
eliminated from the bloodstream.

b. Use the graph to estimate the maximum concentration of
the drug.

c. Use the graph to estimate the time at which the maximum
concentration occurs.

d. Use the graph to estimate the amount of time for which the
concentration is above 80 ppm.

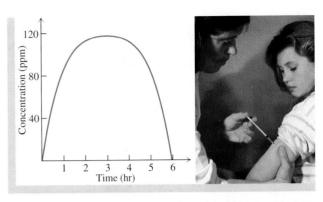

■ **Figure for Exercise 81**

82. *Open-Top Box* Joan intends to make an 18-in.³ open-top box
out of a 6 in. by 7 in. piece of copper by cutting equal squares
(x in. by x in.) from the corners and folding up the sides. Write
the difference between the intended volume and the actual volume as a function of x. For what value of x is there no difference between the intended volume and the actual volume?

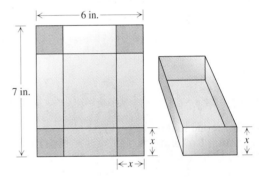

■ **Figure for Exercise 82**

83. *Cartridge Box* The height of a box containing an HP Laser
Jet III printer cartridge is 4 in. more than the width and the
length is 9 in. more than the width. If the volume of the box is
630 in.³, then what are the dimensions of the box?

84. *Computer Case* The width of the case for a 733 megahertz
Pentium computer is 4 in. more than twice the height and the
depth is 1 in. more than the width. If the volume of the case is
1632 in.³, then what are the dimensions of the case?

For Writing/Discussion

85. For what value of c is the equation

$\dfrac{x^2 - 2x + 7}{x - 3} = x + 1 + \dfrac{c}{x - 3}$ an identity?

86. *Synthetic Division* Explain how synthetic division can be used
to find the quotient and remainder when $3x^3 + 4x^2 + 2x - 4$
is divided by $3x - 2$.

87. Prove that if $x - c$ is a factor of $P(x)$, then c is a zero of the
polynomial function.

Thinking Outside the Box XXXI

Moving a Refrigerator   A box containing a refrigerator is 3 ft wide, 3 ft deep, and 6 ft high. To move it, Wally lays it on its side, then on its top, then on its other side, and finally stands it upright as shown in the figure. Exactly how far has point A traveled in going from its initial location to its final location?

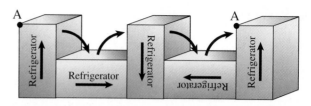

■ **Figure for Thinking Outside the Box XXXI**

3.2 Pop Quiz

1. Use ordinary division to find the quotient and remainder when $x^3 - 5x + 7$ is divided by $x + 4$.

2. Use synthetic division to find the quotient and remainder when $x^2 - 3x + 9$ is divided by $x - 5$.

3. Use synthetic division to find $f(3)$ if $f(x) = x^3 - 2x^2 + 4x - 1$.

4. List the possible rational zeros for $f(x) = 2x^3 - 3x + 8$.

5. Find all real and imaginary zeros for $f(x) = 2x^3 - x^2 + 18x - 9$.

Linking Concepts

For Individual or Group Explorations

Horner's Method

A fourth-degree polynomial in x such as $3x^4 + 5x^3 + 4x^2 + 3x + 1$ contains all of the powers of x from the first through the fourth. However, any polynomial can be written without powers of x. Evaluating a polynomial without powers of x (Horner's method) is somewhat easier than evaluating a polynomial with powers.

a)   Show that $\{[(3x + 5)x + 4]x + 3\}x + 1 = 3x^4 + 5x^3 + 4x^2 + 3x + 1$ is an identity.

b)   Rewrite the polynomial $P(x) = 6x^5 - 3x^4 + 9x^3 + 6x^2 - 8x + 12$ without powers of x as in part (a).

c)   Find $P(2)$ without a calculator using both forms of the polynomial.

d)   For which form did you perform fewer arithmetic operations?

e)   Explain in detail how to rewrite any polynomial without powers of x.

f)   Explain how this new form relates to synthetic division and the remainder theorem.

The Theory of Equations

One of the main goals in algebra is to keep expanding our knowledge of solving equations. The solutions (roots) of a polynomial equation $P(x) = 0$ are precisely the zeros of a polynomial function $y = P(x)$. Therefore the theorems of Section 3.2 concerning zeros of polynomial functions apply also to the roots of polynomial equations. In this section we study several additional theorems that are useful in solving polynomial equations.

The Number of Roots of a Polynomial Equation

When a polynomial equation is solved by factoring, a factor may occur more than once. For example, $x^2 - 10x + 25 = 0$ is equivalent to $(x - 5)^2 = 0$. Since the factor $x - 5$ occurs twice, we say that 5 is a root of the equation with *multiplicity* 2.

Definition: Multiplicity

> If the factor $x - c$ occurs k times in the complete factorization of the polynomial $P(x)$, then c is called a root of $P(x) = 0$ with **multiplicity** k.

If a quadratic equation has a single root, as in $x^2 - 10x + 25 = 0$, then that root has multiplicity 2. If a root with multiplicity 2 is counted as two roots, then every quadratic equation has two roots in the set of complex numbers. This situation is generalized in the following theorem, where the phrase "when multiplicity is considered" means that a root with multiplicity k is counted as k individual roots.

n-Root Theorem

> If $P(x) = 0$ is a polynomial equation with real or complex coefficients and positive degree n, then, when multiplicity is considered, $P(x) = 0$ has n roots.

■ **PROOF**   By the fundamental theorem of algebra, the polynomial equation $P(x) = 0$ with degree n has at least one complex root c_1. By the factor theorem, $P(x) = 0$ is equivalent to

$$(x - c_1)Q_1(x) = 0,$$

where $Q_1(x)$ is a polynomial with degree $n - 1$ (the quotient when $P(x)$ is divided by $x - c_1$). Again, by the fundamental theorem of algebra, there is at least one complex root c_2 of $Q_1(x) = 0$. By the factor theorem, $P(x) = 0$ can be written as

$$(x - c_1)(x - c_2)Q_2(x) = 0,$$

where $Q_2(x)$ is a polynomial with degree $n - 2$. Reasoning in this manner n times, we get a quotient polynomial that has 0 degree, n factors for $P(x)$, and n complex roots, not necessarily all different.   ■

Example **1**   Finding all roots of a polynomial equation

State the degree of each polynomial equation. Find all real and imaginary roots of each equation, stating multiplicity when it is greater than one.

a. $6x^5 + 24x^3 = 0$ **b.** $(x - 3)^2(x + 14)^5 = 0$

Solution

a. This fifth-degree equation can be solved by factoring:

$$6x^3(x^2 + 4) = 0$$

$$6x^3 = 0 \quad \text{or} \quad x^2 + 4 = 0$$

$$x^3 = 0 \quad \text{or} \quad x^2 = -4$$

$$x = 0 \quad \text{or} \quad x = \pm 2i$$

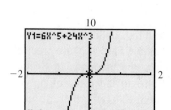

The roots are $\pm 2i$ and 0. Since there are two imaginary roots and 0 is a root with multiplicity 3, there are five roots when multiplicity is considered.

 Because 0 is the only real root, the graph of $y = 6x^5 + 24x^3$ has only one x-intercept at $(0, 0)$ as shown in Fig. 3.27. □

▪ **Figure 3.27**

b. The highest power of x in $(x - 3)^2$ is 2, and in $(x + 14)^5$ is 5. By the product rule for exponents, the highest power of x in this equation is 7. The only roots of this seventh-degree equation are 3 and -14. The root 3 has multiplicity 2, and -14 has multiplicity 5. So there are seven roots when multiplicity is considered.

 Because the equation has two real solutions, the graph of $y = (x - 3)^2(x + 14)^5$ has two x-intercepts at $(3, 0)$ and $(-14, 0)$ as shown in Fig. 3.28.

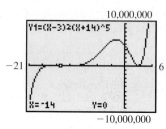

▪ **Figure 3.28**

Try This. Find all roots to $x^3(x + 2)^2(2x - 5) = 0$, including multiplicities.   ▪

Note that graphing polynomial functions and solving polynomial equations go hand in hand. The solutions to the equation can help us find an appropriate viewing window for the graph as they did in Example 1, and the graph can help us find solutions to the equation.

The Conjugate Pairs Theorem

For second-degree polynomial equations, the imaginary roots occur in pairs. For example, the roots of $x^2 - 2x + 5 = 0$ are

$$x = \frac{2 \pm \sqrt{(-2)^2 - 4(1)(5)}}{2} = 1 \pm 2i.$$

The roots $1 - 2i$ and $1 + 2i$ are complex conjugates. The $\pm$ symbol in the quadratic formula causes the complex solutions of a quadratic equation with real coefficients to occur in conjugate pairs. The conjugate pairs theorem indicates that this situation occurs also for polynomial equations of higher degree.

Conjugate Pairs Theorem

> If $P(x) = 0$ is a polynomial equation with real coefficients and the complex number $a + bi$ $(b \neq 0)$ is a root, then $a - bi$ is also a root.

The proof for this theorem is left for the exercises.

 **Using the conjugate pairs theorem**

Find a polynomial equation with real coefficients that has 2 and $1 - i$ as roots.

Solution

If the polynomial has real coefficients, then its imaginary roots occur in conjugate pairs. So a polynomial with these two roots must actually have at least three roots:

2, $1 - i$, and $1 + i$. Since each root of the equation corresponds to a factor of the polynomial, we can write the following equation.

$$(x - 2)[x - (1 - i)][x - (1 + i)] = 0$$

$$(x - 2)[(x - 1) + i][(x - 1) - i] = 0 \quad \text{Regroup.}$$

$$(x - 2)[(x - 1)^2 - i^2] = 0 \quad (a + b)(a - b) = a^2 - b^2$$

$$(x - 2)[x^2 - 2x + 1 + 1] = 0 \quad i^2 = -1$$

$$(x - 2)(x^2 - 2x + 2) = 0$$

$$x^3 - 4x^2 + 6x - 4 = 0$$

This equation has the required roots and the smallest degree. Any multiple of this equation would also have the required roots but would not be as simple.

Try This. Find a polynomial equation with real coefficients that has 3 and $-i$ as roots.   ■

Descartes's Rule of Signs

None of the theorems in this chapter tells us how to find all of the n roots to a polynomial equation of degree n. However, the theorems and rules presented here add to our knowledge of polynomial equations and help us to predict the type and number of solutions to expect for a particular equation. Descartes's rule of signs is a method for determining the number of positive, negative, and imaginary solutions. For this rule, a solution with multiplicity k is counted as k solutions.

When a polynomial is written in descending order, a **variation of sign** occurs when the signs of consecutive terms change. For example, if

$$P(x) = 3x^5 - 7x^4 - 8x^3 - x^2 + 3x - 9,$$

there are sign changes in going from the first to the second term, from the fourth to the fifth term, and from the fifth to the sixth term. So there are three variations of sign for $P(x)$. This information determines the number of positive real solutions to $P(x) = 0$. Descartes's rule requires that we look at $P(-x)$ and also count the variations of sign after it is simplified:

$$P(-x) = 3(-x)^5 - 7(-x)^4 - 8(-x)^3 - (-x)^2 + 3(-x) - 9$$

$$= -3x^5 - 7x^4 + 8x^3 - x^2 - 3x - 9$$

In $P(-x)$ the signs of the terms change from the second to the third term and again from the third to the fourth term. So there are two variations of sign for $P(-x)$. This information determines the number of negative real solutions to $P(x) = 0$.

Descartes's Rule of Signs

Suppose $P(x) = 0$ is a polynomial equation with real coefficients and with terms written in descending order.

- ■ The number of positive real roots of the equation is either equal to the number of variations of sign of $P(x)$ or less than that by an even number.
- ■ The number of negative real roots of the equation is either equal to the number of variations of sign of $P(-x)$ or less than that by an even number.

The proof of Descartes's rule of signs is beyond the scope of this text, but we can apply the rule to polynomial equations. Descartes's rule of signs is especially helpful when the number of variations of sign is 0 or 1.

Example 3 Using Descartes's rule of signs

Discuss the possibilities for the roots to $2x^3 - 5x^2 - 6x + 4 = 0$.

Solution

The number of variations of sign in

$$P(x) = 2x^3 - 5x^2 - 6x + 4$$

is 2. By Descartes's rule, the number of positive real roots is either 2 or 0. Since

$$P(-x) = 2(-x)^3 - 5(-x)^2 - 6(-x) + 4$$
$$= -2x^3 - 5x^2 + 6x + 4,$$

there is one variation of sign in $P(-x)$. So there is exactly one negative real root.

The equation must have three roots, because it is a third-degree polynomial equation. Since there must be three roots and one is negative, the other two roots must be either both imaginary numbers or both positive real numbers. Table 3.1 summarizes these two possibilities.

■ **Table 3.1**   Number of roots

Positive	Negative	Imaginary
2	1	0
0	1	2

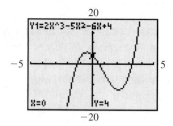

■ **Figure 3.29**

The graph of $y = 2x^3 - 5x^2 - 6x + 4$ shown in Fig. 3.29 crosses the positive x-axis twice and the negative x-axis once. So the first case in Table 3.1 is actually correct.

Try This. Discuss the possibilities for the roots to $x^3 - 5x^2 + 4x + 3 = 0$. ■

Example 4 Using Descartes's rule of signs

Discuss the possibilities for the roots to $3x^4 - 5x^3 - x^2 - 8x + 4 = 0$.

Solution

There are two variations of sign in the polynomial

$$P(x) = 3x^4 - 5x^3 - x^2 - 8x + 4.$$

According to Descartes's rule, there are either two or zero positive real roots to the equation. Since

$$P(-x) = 3(-x)^4 - 5(-x)^3 - (-x)^2 - 8(-x) + 4$$
$$= 3x^4 + 5x^3 - x^2 + 8x + 4,$$

there are two variations of sign in $P(-x)$. So the number of negative real roots is either two or zero. Since the degree of the polynomial is 4, there must be four roots.

Each line of Table 3.2 gives a possible distribution of the type of those four roots. Note that the number of imaginary roots is even in each case, as we would expect from the conjugate pairs theorem.

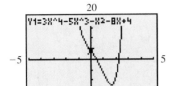

■ **Figure 3.30**

■ **Table 3.2** Number of roots

Positive	Negative	Imaginary
2	2	0
2	0	2
0	2	2
0	0	4

The calculator graph of $y = 3x^4 - 5x^3 - x^2 - 8x + 4$ shown in Fig. 3.30 shows two positive intercepts and no negative intercepts. However, we might not have the appropriate viewing window. The negative intercepts might be less than -5. In this case the graph did not allow us to conclude which line in Table 3.2 is correct.

Try This. Discuss the possibilities for the roots to $x^4 - 6x^2 + 10 = 0$. ■

Bounds on the Roots

If a polynomial equation has no roots greater than c, then c is called an **upper bound** for the roots. If there are no roots less than c, then c is called a **lower bound** for the roots. The next theorem is used to determine upper and lower bounds for the roots of a polynomial equation. We will not prove this theorem.

Theorem on Bounds

> Suppose that $P(x)$ is a polynomial with real coefficients and a positive leading coefficient, and synthetic division with $x - c$ is performed.
>
> ■ If $c > 0$ and all terms in the bottom row are nonnegative, then c is an upper bound for the roots of $P(x) = 0$.
> ■ If $c < 0$ and the terms in the bottom row alternate in sign, then c is a lower bound for the roots of $P(x) = 0$.

If 0 appears in the bottom row of the synthetic division, then it may be assigned either a positive or negative sign in determining whether the signs alternate. For example, the numbers 3, 0, 5, and -6 would be alternating in sign if we assign a negative sign to 0. The numbers $-7, 5, -8, 0$, and -2 would be alternating in sign if we assign a positive sign to 0.

Example **5** Finding bounds for the roots

Use the theorem on bounds to establish the best integral bounds for the roots of $2x^3 - 5x^2 - 6x + 4 = 0$.

Solution

Try synthetic division with the integers 1, 2, 3, and so on. The first integer for which all terms on the bottom row are nonnegative is the best upper bound for the roots

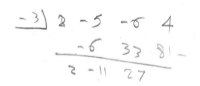

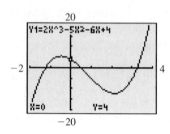

■ **Figure 3.31**

according to the theorem on bounds. Note that we are not trying fractions because we are looking for integral bounds for the roots and not the actual roots.

$$
\begin{array}{r|rrrr}
1 & 2 & -5 & -6 & 4 \\
  &   & 2 & -3 & -9 \\
\hline
  & 2 & -3 & -9 & -5
\end{array}
\qquad
\begin{array}{r|rrrr}
2 & 2 & -5 & -6 & 4 \\
  &   & 4 & -2 & -16 \\
\hline
  & 2 & -1 & -8 & -12
\end{array}
$$

$$
\begin{array}{r|rrrr}
3 & 2 & -5 & -6 & 4 \\
  &   & 6 & 3 & -9 \\
\hline
  & 2 & 1 & -3 & -5
\end{array}
\qquad
\begin{array}{r|rrrr}
4 & 2 & -5 & -6 & 4 \\
  &   & 8 & 12 & 24 \\
\hline
  & 2 & 3 & 6 & 28
\end{array}
$$

By the theorem on bounds, no number greater than 4 can be a root to the equation. Now try synthetic division with the integers -1, -2, -3, and so on. The first negative integer for which the terms on the bottom row alternate in sign is the best lower bound for the roots.

$$
\begin{array}{r|rrrr}
-1 & 2 & -5 & -6 & 4 \\
   &   & -2 & 7 & -1 \\
\hline
   & 2 & -7 & 1 & 3
\end{array}
\qquad
\begin{array}{r|rrrr}
-2 & 2 & -5 & -6 & 4 \\
   &   & -4 & 18 & -24 \\
\hline
   & 2 & -9 & 12 & -20
\end{array}
$$

By the theorem on bounds, no number less than -2 can be a root to the equation. So all of the real roots to this equation are between -2 and 4.

The graph of $y = 2x^3 - 5x^2 - 6x + 4$ in Fig. 3.31 shows three x-intercepts between -2 and 4, which supports our conclusion about the bounds for the roots. Note that the bounds for the roots help you determine a good window.

Try This. Use the theorem on bounds to establish the best integral bounds for the roots to $x^3 + 3x^2 - 5x - 15 = 0$. ■

In the next example we use all of the available information about roots.

Example 6 Using all of the theorems about roots

Find all of the solutions to $2x^3 - 5x^2 - 6x + 4 = 0$.

Solution

In Example 3 we used Descartes's rule of signs on this equation to determine that it has either two positive roots and one negative root or one negative root and two imaginary roots. In Example 5 we used the theorem on bounds to determine that all of the real roots to this equation are between -2 and 4. From the rational zero theorem, the possible rational roots are ± 1, ± 2, ± 4, and $\pm 1/2$. Since there must be one negative root and it must be greater than -2, the only possible numbers from the list are -1 and $-1/2$. So start by checking $-1/2$ and -1 with synthetic division.

$$
\begin{array}{r|rrrr}
-\dfrac{1}{2} & 2 & -5 & -6 & 4 \\[2mm]
   &   & -1 & 3 & \dfrac{3}{2} \\[2mm]
\hline
   & 2 & -6 & -3 & \dfrac{11}{2}
\end{array}
\qquad
\begin{array}{r|rrrr}
-1 & 2 & -5 & -6 & 4 \\
   &   & -2 & 7 & -1 \\
\hline
   & 2 & -7 & 1 & 3
\end{array}
$$

Since neither -1 nor $-1/2$ is a root, the negative root must be irrational. The only rational possibilities for the two positive roots smaller than 4 are $1/2$, 1, and 2.

$$\frac{1}{2} \begin{array}{|rrrr} 2 & -5 & -6 & 4 \\ & 1 & -2 & -4 \\ \hline 2 & -4 & -8 & 0 \end{array}$$

Since $1/2$ is a root of the equation, $x - 1/2$ is a factor of the polynomial. The last line in the synthetic division indicates that the other factor is $2x^2 - 4x - 8$.

$$\left(x - \frac{1}{2}\right)(2x^2 - 4x - 8) = 0$$

$$(2x - 1)(x^2 - 2x - 4) = 0$$

$$2x - 1 = 0 \quad \text{or} \quad x^2 - 2x - 4 = 0$$

$$x = \frac{1}{2} \quad \text{or} \quad x = \frac{2 \pm \sqrt{4 - 4(1)(-4)}}{2} = 1 \pm \sqrt{5}$$

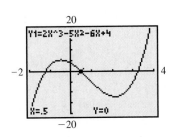

Figure 3.32

There are two positive roots, $1/2$ and $1 + \sqrt{5}$. The negative root is $1 - \sqrt{5}$. Note that the roots guaranteed by Descartes's rule of signs are real numbers but not necessarily rational numbers.

The graph of $y = 2x^3 - 5x^2 - 6x + 4$ in Fig. 3.32 supports these conclusions, because its x-intercepts appear to be $\left(1 - \sqrt{5}, 0\right)$, $(1/2, 0)$, and $\left(1 + \sqrt{5}, 0\right)$.

Try This. Find all solutions to $2x^3 - 5x^2 - 8x + 5 = 0$.   ■

For Thought

True or False? Explain.

1. The number 1 is a root of $x^3 - 1 = 0$ with multiplicity 3.

2. The equation $x^3 = 125$ has three complex number solutions.

3. For $(x + 1)^3(x^2 - 2x + 1) = 0$, -1 is a root with multiplicity 3.

4. For $(x - 5)^3(x^2 - 3x - 10) = 0$, 5 is a root with multiplicity 3.

5. If $4 - 5i$ is a solution to a polynomial equation with real coefficients, then $5i - 4$ is also a solution to the equation.

6. If $P(x) = 0$ is a polynomial equation with real coefficients and i, $2 - 3i$, and $5 + 7i$ are roots, then the degree of $P(x)$ is at least 6.

7. Both $-3 - i\sqrt{5}$ and $3 - i\sqrt{5}$ are solutions to $5x^3 - 9x^2 + 17x - 23 = 0$.

8. Both $3/2$ and 2 are solutions to $2x^5 - 4x^3 - 6x^2 - 3x - 6 = 0$.

9. The equation $x^3 - 5x^2 + 6x - 1 = 0$ has no negative roots.

10. The equation $5x^3 - 171 = 0$ has two imaginary solutions.

3.3 Exercises

State the degree of each polynomial equation. Find all of the real and imaginary roots of each equation, stating multiplicity when it is greater than one.

1. $x^2 - 10x + 25 = 0$

2. $x^2 - 18x + 81 = 0$

3. $x^5 - 9x^3 = 0$

4. $x^6 + x^4 = 0$

5. $x^4 - 2x^3 + x^2 = 0$

6. $x^5 - 6x^4 + 9x^3 = 0$

7. $(2x - 3)^2(3x + 4)^2 = 0$

8. $(2x^2 + x)^2(3x - 1)^4 = 0$

9. $x^3 - 4x^2 - 6x = 0$

10. $-x^3 + 8x^2 - 14x = 0$

Find each product.

11. $(x - 3i)(x + 3i)$

12. $(x + 6i)(x - 6i)$

13. $\left[x - \left(1 + \sqrt{2}\right)\right]\left[x - \left(1 - \sqrt{2}\right)\right]$

14. $\left[x - \left(3 - \sqrt{5}\right)\right]\left[x - \left(3 + \sqrt{5}\right)\right]$

15. $[x - (3 + 2i)][x - (3 - 2i)]$

16. $[x - (3 - i)][x - (3 + i)]$

17. $(x - 2)[x - (3 + 4i)][x - (3 - 4i)]$

18. $(x + 1)(x - (1 - i))(x - (1 + i))$

Find a polynomial equation with real coefficients that has the given roots.

19. $-3, 5$

20. $6, -1$

21. $-4i, 4i$

22. $-9i, 9i$

23. $3 - i$

24. $4 + i$

25. $-2, i$

26. $4, -i$

27. $0, i\sqrt{3}$

28. $-2, i\sqrt{2}$

29. $3, 1 - i$

30. $5, 4 - 3i$

31. $1, 2, 3$

32. $-1, 2, -3$

33. $1, 2 - 3i$

34. $-1, 4 - 2i$

35. $\dfrac{1}{2}, \dfrac{1}{3}, \dfrac{1}{4}$

36. $-\dfrac{1}{2}, -\dfrac{1}{3}, 1$

37. $i, 1 + i$

38. $3i, 3 - i$

Use Descartes's rule of signs to discuss the possibilities for the roots of each equation. Do not solve the equation.

39. $x^3 + 5x^2 + 7x + 1 = 0$

40. $2x^3 - 3x^2 + 5x - 6 = 0$

41. $-x^3 - x^2 + 7x + 6 = 0$

42. $-x^4 - 5x^2 - x + 7 = 0$

43. $y^4 + 5y^2 + 7 = 0$

44. $-3y^4 - 6y^2 + 7 = 0$

45. $t^4 - 3t^3 + 2t^2 - 5t + 7 = 0$

46. $-5r^4 + 4r^3 + 7r - 16 = 0$

47. $x^5 + x^3 + 5x = 0$

48. $x^4 - x^2 + 1 = 0$

Use the theorem on bounds to establish the best integral bounds for the roots of each equation.

49. $2x^3 - 5x^2 + 6 = 0$

50. $2x^3 - x^2 - 5x + 3 = 0$

51. $4x^3 + 8x^2 - 11x - 15 = 0$

52. $6x^3 + 5x^2 - 36x - 35 = 0$

53. $w^4 - 5w^3 + 3w^2 + 2w - 1 = 0$

54. $3z^4 - 7z^2 + 5 = 0$

55. $-2x^3 + 5x^2 - 3x + 9 = 0$

56. $-x^3 + 8x - 12 = 0$

Use the rational zero theorem, Descartes's rule of signs, and the theorem on bounds as aids in finding all real and imaginary roots to each equation.

57. $x^3 - 4x^2 - 7x + 10 = 0$

58. $x^3 + 9x^2 + 26x + 24 = 0$

59. $x^3 - 10x - 3 = 0$

60. $2x^3 - 7x^2 - 16 = 0$

61. $x^4 + 2x^3 - 7x^2 + 2x - 8 = 0$

62. $x^4 - 4x^3 + 7x^2 - 16x + 12 = 0$

63. $6x^3 + 25x^2 - 24x + 5 = 0$

64. $6x^3 - 11x^2 - 46x - 24 = 0$

65. $x^4 + 2x^3 - 3x^2 - 4x + 4 = 0$

66. $x^5 + 3x^3 + 2x = 0$

67. $x^4 - 6x^3 + 12x^2 - 8x = 0$

68. $x^4 + 9x^3 + 27x^2 + 27x = 0$

69. $x^6 - x^5 - x^4 + x^3 - 12x^2 + 12x = 0$

70. $2x^7 - 2x^6 + 7x^5 - 7x^4 - 4x^3 + 4x^2 = 0$

71. $8x^5 + 2x^4 - 33x^3 + 4x^2 + 25x - 6 = 0$

72. $6x^5 + x^4 - 28x^3 - 3x^2 + 16x - 4 = 0$

For each of the following functions use synthetic division and the theorem on bounds to find integers a and b, such that the interval (a, b) contains all real zeros of the function. This method does not necessarily give the shortest interval containing all real zeros. By inspecting the graph of each function, find the shortest interval (c, d) that contains all real zeros of the function with c and d integers. The second interval should be a subinterval of the first.

73. $y = 2x^3 - 3x^2 - 50x + 18$ **74.** $f(x) = x^3 - 33x - 58$

75. $f(x) = x^4 - 26x^2 + 153$

76. $y = x^4 + x^3 - 16x^2 - 10x + 60$

77. $y = 4x^3 - 90x^2 - 2x + 45$

78. $f(x) = x^4 - 12x^3 + 27x^2 - 6x - 8$

Solve each problem.

79. *Growth Rate for Bacteria*   A car's speedometer indicates velocity at every instant in time. The instantaneous growth rate of a population is the rate at which it is growing at every instant in time. The instantaneous growth rate r of a colony of bacteria t hours after the start of an experiment is given by the function

$$r = 0.01t^3 - 0.08t^2 + 0.11t + 0.20$$

for $0 \le t \le 7$. Find the times for which the instantaneous growth rate is zero.

80. *Retail Store Profit*   The manager of a retail store has figured that her monthly profit P (in thousands of dollars) is determined by her monthly advertising expense x (in tens of thousands of dollars) according to the formula

$$P = x^3 - 20x^2 + 100x \qquad \text{for} \qquad 0 \le x \le 4.$$

For what value of x does she get $147,000 in profit?

81. *Designing Fireworks*   Marshall is designing a rocket for the Red Rocket Fireworks Company. The rocket will consist of a cardboard circular cylinder with a height that is four times as large as the radius. On top of the cylinder will be a cone with a height of 2 in. and a radius equal to the radius of the base as shown in the figure. If he wants to fill the cone and the cylinder with a total of 114π in.^{3} of powder, then what should be the radius of the cylinder?

HINT Use the formulas for the volume of a cone and a cylinder.

■ **Figure for Exercise 81**

82. *Heating and Air*   An observatory is built in the shape of a right circular cylinder with a hemispherical roof as shown in the figure. The heating and air contractor has figured the volume of the structure as 3168π ft^3. If the height of the cylindrical walls is 2 ft more than the radius of the building, then what is the radius of the building?

HINT Use the formulas for the volume of a sphere and a cylinder.

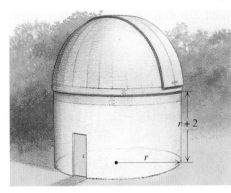

■ **Figure for Exercise 82**

For Writing/Discussion

83. *Conjugate of a Sum*   Show that the conjugate of the sum of two complex numbers is equal to the sum of their conjugates.

84. *Conjugate of a Product*   Show that the conjugate of the product of two complex numbers is equal to the product of their conjugates.

85. *Conjugate of a Real Number*   Show that $\bar{a} = a$ for any real number a, where $\bar{a}$ is the conjugate of a.

86. *Conjugate Pairs*   Assume that $a + bi$ is a root of $a_n x^n + a_{n-1} x^{n-1} + \cdots + a_1 x + a_0 = 0$ and substitute $a + bi$ for x. Take the conjugate of each side of the resulting equation and use the results of Exercises 83–85 to simplify it. Your final equation should show that $a - bi$ is a root of $a_n x^n + a_{n-1} x^{n-1} + \cdots + a_1 x + a_0 = 0$. This proves the conjugate pairs theorem.

87. *Finding Polynomials*   Find a third-degree polynomial function such that $f(0) = 3$ and whose zeros are 1, 2, and 3. Explain how you found it.

88. *Finding Polynomials*   Is there a third-degree polynomial function such that $f(0) = 6$ and $f(-1) = 12$ and whose zeros are 1, 2, and 3? Explain.

Thinking Outside the Box XXXII

Packing Billiard Balls There are several ways to tightly pack nine billiard balls each with radius 1 into a rectangular box. Find the volume of the box in each of the following cases and determine which box has the least volume.

a. Four balls are placed so that they just fit into the bottom of the box, then another layer of four, then one ball in the middle tangent to all four in the second layer, as shown in this side view.

b. Four balls are placed so that they just fit into the bottom of the box as in (a), then one is placed in the middle on top of the first four. Finally, four more are placed so that they just fit at the top of the box.

c. The box is packed with layers of four, one, and four as in (b), but the box is required to be cubic. In this case, the four balls in the bottom layer will not touch each other and the four balls in the top layer will not touch each other. The ball in the middle will be tangent to all of the other eight balls.

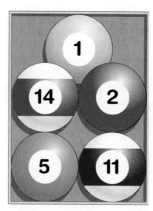

■ **Figure for Thinking Outside the Box XXXII**

3.3 Pop Quiz

1. Find all real and imaginary roots to $x^5 - x^3 = 0$, including multiplicities.

2. Find a polynomial equation with real coefficients that has the roots $-4i$ and 5.

3. By Descartes's rule of signs, how many positive roots can $x^4 + x^3 - 3x^2 + 5x + 9 = 0$ have?

4. By Descartes's rule of signs, how many negative roots can $3x^3 + 5x^2 - x + 9 = 0$ have?

5. Find all real and imaginary roots to $x^3 - 3x^2 - 6x + 8 = 0$.

Linking Concepts

For Individual or Group Explorations

Designing a Crystal Ball

The wizard Gandalf is creating a massive crystal ball. To achieve maximum power, the orb must be a perfect solid sphere of clear crystal mounted on a square solid silver base, as shown in the figure.

a) Write a formula for the total volume of material used in terms of the radius of the sphere, the thickness of the base, and the length of the side of the base.

b) The wizard has determined that the diameter of the sphere must equal the length of the side of the square base, and the thickness of the base must be π in.

Find the exact radius of the sphere if the total amount of material used in the sphere and base must be 1296π in.^{3}.

c) If it turns out that the sphere in part (b) is not powerful enough, Gandalf plans to make a sphere and base of solid dilitheum crystal. For this project the diameter of the sphere must be 2 in. less than the length of the side of the base and the thickness of the base must be 2 in. Find the approximate radius of the sphere if the total volume of material used in the sphere and base must be 5000 in.^{3}.

3.4

Miscellaneous Equations

In Section 3.3 we learned that an nth-degree polynomial equation has n roots. However, it is not always obvious how to find them. In this section we will solve polynomial equations using some new techniques and we will solve several other types of equations. Unfortunately, we cannot generally predict the number of roots to nonpolynomial equations.

Factoring Higher-Degree Equations

We usually use factoring to solve quadratic equations. However, since we can also factor many higher-degree polynomials, we can solve many higher-degree equations by factoring. Factoring is often the fastest method for solving an equation.

Example 1 Solving an equation by factoring

Solve $x^3 + 3x^2 + x + 3 = 0$.

Solution

Factor the polynomial on the left-hand side by grouping.

$$x^2(x + 3) + 1(x + 3) = 0 \quad \text{Factor by grouping.}$$
$$(x^2 + 1)(x + 3) = 0 \quad \text{Factor out } x + 3.$$
$$x^2 + 1 = 0 \quad \text{or} \quad x + 3 = 0 \quad \text{Zero factor property}$$
$$x^2 = -1 \quad \text{or} \quad x = -3$$
$$x = \pm i \quad \text{or} \quad x = -3$$

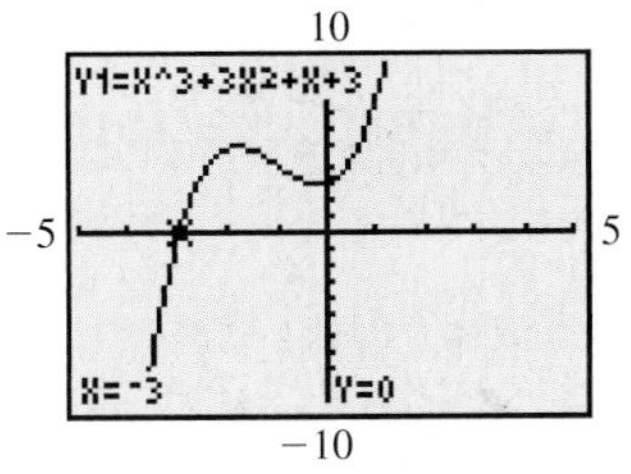

The solution set is $\{-3, -i, i\}$.

The graph in Fig. 3.33 supports these solutions. Because there is only one real solution, the graph crosses the x-axis only once.

Y1=X^3+3X²+X+3

X=-3   Y=0

■ **Figure 3.33**

Try This. Solve $x^3 - 2x^2 + 5x - 10 = 0$ by factoring. ■

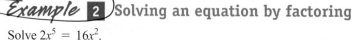

Example **2** Solving an equation by factoring

Solve $2x^5 = 16x^2$.

Solution

Write the equation with 0 on the right-hand side, then factor completely.

$$2x^5 - 16x^2 = 0$$

$$2x^2(x^3 - 8) = 0 \qquad \text{Factor out the greatest common factor.}$$

$$2x^2(x - 2)(x^2 + 2x + 4) = 0 \qquad \text{Factor the difference of two cubes.}$$

$$2x^2 = 0 \quad \text{or} \quad x - 2 = 0 \quad \text{or} \quad x^2 + 2x + 4 = 0$$

$$x = 0 \quad \text{or} \qquad x = 2 \quad \text{or} \quad x = \frac{-2 \pm \sqrt{-12}}{2} = -1 \pm i\sqrt{3}$$

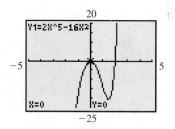

■ **Figure 3.34**

The solution set is $\{0, 2, -1 \pm i\sqrt{3}\}$. Since 0 is a root with multiplicity 2, there are five roots, counting multiplicity, to this fifth-degree equation.

The graph in Fig. 3.34 supports this solution.

Try This. Solve $x^5 = 27x^2$.   ■

Note that in Example 2, if we had divided each side by x^2 as our first step, we would have lost the solution $x = 0$. *We do not usually divide each side of an equation by a variable expression.* Instead, bring all expressions to the same side and factor out the common factors.

Equations Involving Square Roots

Recall that $\sqrt{x}$ represents the nonnegative square root of x. To solve $\sqrt{x} = 3$ we can use the definition of square root. Since the nonnegative square root of 9 is 3, the solution to $\sqrt{x} = 3$ is 9. To solve $\sqrt{x} = -3$ we again use the definition of square root. Because $\sqrt{x}$ is nonnegative while -3 is negative, this equation has no solution.

More complicated equations involving square roots are usually solved by squaring both sides. However, squaring both sides does not always lead to an equivalent equation. If we square both sides of $\sqrt{x} = 3$, we get $x = 9$, which is equivalent to $\sqrt{x} = 3$. But if we square both sides of $\sqrt{x} = -3$, we also get $x = 9$, which is not equivalent to $\sqrt{x} = -3$. Because 9 appeared in the attempt to solve $\sqrt{x} = -3$, but does not satisfy the equation, it is called an *extraneous root*. This same situation can occur with an equation involving a fourth root or any other even root. So if you raise each side of an equation to an even power, you must check for extraneous roots.

Example **3** Squaring each side to solve an equation

Solve $\sqrt{x} + 2 = x$.

Solution

Isolate the radical before squaring each side.

$$\sqrt{x} = x - 2$$

$$(\sqrt{x})^2 = (x - 2)^2 \qquad \text{Square each side.}$$

$$x = x^2 - 4x + 4 \qquad \text{Use the special product } (a - b)^2 = a^2 - 2ab + b^2.$$

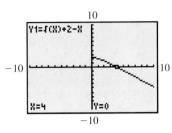

■ **Figure 3.35**

$$0 = x^2 - 5x + 4 \qquad \text{Write in the form } ax^2 + bx + c = 0.$$
$$0 = (x - 4)(x - 1) \qquad \text{Factor the quadratic polynomial.}$$
$$x - 4 = 0 \quad \text{or} \quad x - 1 = 0 \qquad \text{Zero factor property}$$
$$x = 4 \quad \text{or} \qquad x = 1$$

Checking $x = 4$, we get $\sqrt{4} + 2 = 4$, which is correct. Checking $x = 1$, we get $\sqrt{1} + 2 = 1$, which is incorrect. So 1 is an extraneous root and the solution set is $\{4\}$.

The graph in Fig. 3.35 supports this solution.

Try This. Solve $\sqrt{x} + 12 = x$.   ■

The next example involves two radicals. In this example we will isolate the more complicated radical before squaring each side. But not all radicals are eliminated upon squaring each side. So we isolate the remaining radical and square each side again.

Example 4 Squaring each side twice

Solve $\sqrt{2x + 1} - \sqrt{x} = 1$.

Solution

First we write the equation so that the more complicated radical is isolated. Then we square each side. On the left side, when we square $\sqrt{2x + 1}$, we get $2x + 1$. On the right side, when we square $1 + \sqrt{x}$, we use the special product rule $(a + b)^2 = a^2 + 2ab + b^2$.

$$\sqrt{2x + 1} = 1 + \sqrt{x}$$
$$\left(\sqrt{2x + 1}\right)^2 = \left(1 + \sqrt{x}\right)^2 \qquad \text{Square each side.}$$
$$2x + 1 = 1 + 2\sqrt{x} + x$$
$$x = 2\sqrt{x} \qquad \text{All radicals are not eliminated by the first squaring.}$$
$$x^2 = \left(2\sqrt{x}\right)^2 \qquad \text{Square each side a second time.}$$
$$x^2 = 4x$$
$$x^2 - 4x = 0$$
$$x(x - 4) = 0$$
$$x = 0 \quad \text{or} \quad x - 4 = 0$$
$$x = 0 \quad \text{or} \qquad x = 4$$

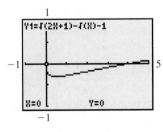

■ **Figure 3.36**

Both 0 and 4 satisfy the original equation. So the solution set is $\{0, 4\}$.

The graph in Fig. 3.36 supports this solution.

Try This. Solve $\sqrt{3x - 2} - \sqrt{x} = 2$.   ■

Equations with Rational Exponents

To solve equations of the form $x^{m/n} = k$ in which m and n are positive integers and m/n is in lowest terms, we adapt the methods of Examples 3 and 4 of raising each side to a power. Cubing each side of $x^{2/3} = 4$, yields $(x^{2/3})^3 = 4^3$ or $x^2 = 64$. By the square root property, $x = \pm 8$. We can shorten this solution by raising each side of the equation to the power 3/2 (the reciprocal of 2/3) and inserting the $\pm$ symbol to obtain the two square roots.

$$x^{2/3} = 4$$

$$(x^{2/3})^{3/2} = \pm 4^{3/2} \qquad \text{Raise each side to the power 3/2 and insert } \pm.$$

$$x = \pm 8$$

The equation $x^{2/3} = 4$ has two solutions because the numerator of the exponent 2/3 is an even number. An equation such as $x^{-3/2} = 1/8$ has only one solution because the numerator of the exponent $-3/2$ is odd. To solve $x^{-3/2} = 1/8$, raise each side to the power $-2/3$ (the reciprocal of $-3/2$).

$$x^{-3/2} = \frac{1}{8}$$

$$(x^{-3/2})^{-2/3} = \left(\frac{1}{8}\right)^{-2/3} \qquad \text{Raise each side to the power } -2/3.$$

$$x = 4$$

In the next example we solve two more equations of this type by raising each side to a fractional power. Note that we use the $\pm$ symbol only when the numerator of the original exponent is even.

Example **5** **Equations with rational exponents**

Solve each equation.

a. $x^{4/3} = 625$ **b.** $(y - 2)^{-5/2} = 32$

Solution

a. Raise each side of the equation to the power 3/4. Use the $\pm$ symbol because the numerator of 4/3 is even.

$$x^{4/3} = 625$$

$$(x^{4/3})^{3/4} = \pm 625^{3/4}$$

$$x = \pm 125$$

Check in the original equation. The solution set is $\{-125, 125\}$.
The graph in Fig. 3.37 supports this solution. □

b. Raise each side to the power $-2/5$. Because the numerator in $-5/2$ is an odd number, there is only one solution.

$$(y - 2)^{-5/2} = 32$$

$$((y - 2)^{-5/2})^{-2/5} = 32^{-2/5} \qquad \text{Raise each side to the power } -2/5.$$

$$y - 2 = \frac{1}{4}$$

$$y = 2 + \frac{1}{4} = \frac{9}{4}$$

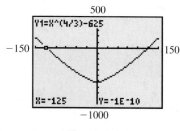

■ **Figure 3.37**

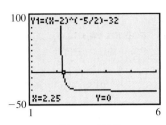

■ **Figure 3.38**

Check 9/4 in the original equation. The solution set is $\left\{\frac{9}{4}\right\}$.

The graph in Fig. 3.38 supports this solution.

Try This. Solve $x^{-4/5} = 16$. ■

Equations of Quadratic Type

In some cases, an equation can be converted to a quadratic equation by substituting a single variable for a more complicated expression. Such equations are called **equations of quadratic type.** An equation of quadratic type has the form $au^2 + bu + c = 0$, where $a \neq 0$ and u is an algebraic expression.

In the next example, the expression x^2 in a fourth-degree equation is replaced by u, yielding a quadratic equation. After the quadratic equation is solved, u is replaced by x^2 so that we find values for x that satisfy the original fourth-degree equation.

Example **6** Solving a fourth-degree polynomial equation

Solve $x^4 - 14x^2 + 45 = 0$.

Solution

We let $u = x^2$ so that $u^2 = (x^2)^2 = x^4$.

$$(x^2)^2 - 14x^2 + 45 = 0$$
$$u^2 - 14u + 45 = 0 \quad \text{Replace } x^2 \text{ by } u.$$
$$(u - 9)(u - 5) = 0$$
$$u - 9 = 0 \quad \text{or} \quad u - 5 = 0$$
$$u = 9 \quad \text{or} \quad u = 5$$
$$x^2 = 9 \quad \text{or} \quad x^2 = 5 \quad \text{Replace } u \text{ by } x^2.$$
$$x = \pm 3 \quad \text{or} \quad x = \pm\sqrt{5}$$

Check in the original equation. The solution set is $\left\{-3, -\sqrt{5}, \sqrt{5}, 3\right\}$.

The graph in Fig. 3.39 supports this solution.

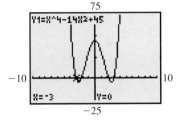

■ **Figure 3.39**

Try This. Solve $x^4 - 9x^2 + 20 = 0$. ■

Note that the equation of Example 6 could be solved by factoring without doing substitution, because $x^4 - 14x^2 + 45 = (x^2 - 9)(x^2 - 5)$. Since the next example involves a more complicated algebraic expression, we use substitution to simplify it, although it too could be solved by factoring, without substitution.

Example **7** Another equation of quadratic type

Solve $(x^2 - x)^2 - 18(x^2 - x) + 72 = 0$.

Solution

If we let $u = x^2 - x$, then the equation becomes a quadratic equation.

$$(x^2 - x)^2 - 18(x^2 - x) + 72 = 0$$

$$u^2 - 18u + 72 = 0 \qquad \text{Replace } x^2 - x \text{ by } u.$$

$$(u - 6)(u - 12) = 0$$

$$u - 6 = 0 \qquad \text{or} \qquad u - 12 = 0$$

$$u = 6 \qquad \text{or} \qquad u = 12$$

$$x^2 - x = 6 \qquad \text{or} \qquad x^2 - x = 12 \qquad \text{Replace } u \text{ by } x^2 - x.$$

$$x^2 - x - 6 = 0 \qquad \text{or} \qquad x^2 - x - 12 = 0$$

$$(x - 3)(x + 2) = 0 \qquad \text{or} \qquad (x - 4)(x + 3) = 0$$

$$x = 3 \quad \text{or} \quad x = -2 \quad \text{or} \quad x = 4 \quad \text{or} \quad x = -3$$

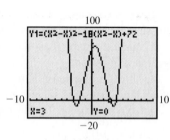

▪ **Figure 3.40**

Check in the original equation. The solution set is $\{-3, -2, 3, 4\}$. The graph in Fig. 3.40 supports this solution.

Try This. Solve $(x^2 + x)^2 - 8(x^2 + x) + 12 = 0$.   ▪

The next equations of quadratic type have rational exponents.

Example **8**   **Quadratic type and rational exponents**

Find all real solutions to each equation.

a. $x^{2/3} - 9x^{1/3} + 8 = 0$ **b.** $(11x^2 - 18)^{1/4} = x$

Solution

a. If we let $u = x^{1/3}$, then $u^2 = (x^{1/3})^2 = x^{2/3}$.

$$u^2 - 9u + 8 = 0 \qquad \text{Replace } x^{2/3} \text{ by } u^2 \text{ and } x^{1/3} \text{ by } u.$$

$$(u - 8)(u - 1) = 0$$

$$u = 8 \qquad \text{or} \qquad u = 1$$

$$x^{1/3} = 8 \qquad \text{or} \qquad x^{1/3} = 1 \qquad \text{Replace } u \text{ by } x^{1/3}.$$

$$(x^{1/3})^3 = 8^3 \qquad \text{or} \qquad (x^{1/3})^3 = 1^3$$

$$x = 512 \qquad \text{or} \qquad x = 1$$

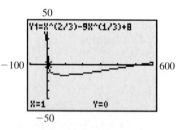

▪ **Figure 3.41**

Check in the original equation. The solution set is $\{1, 512\}$. The graph in Fig. 3.41 supports this solution.   □

b. $(11x^2 - 18)^{1/4} = x$

$$((11x^2 - 18)^{1/4})^4 = x^4 \qquad \text{Raise each side to the power 4.}$$

$$11x^2 - 18 = x^4$$

$$x^4 - 11x^2 + 18 = 0$$

$$(x^2 - 9)(x^2 - 2) = 0$$

$$x^2 = 9 \qquad \text{or} \quad x^2 = 2$$

$$x = \pm 3 \quad \text{or} \quad x = \pm\sqrt{2}$$

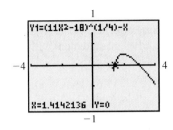

■ **Figure 3.42**

Since the exponent $1/4$ means principal fourth root, the right-hand side of the equation cannot be negative. So -3 and $-\sqrt{2}$ are extraneous roots. Since 3 and $\sqrt{2}$ satisfy the original equation, the solution set is $\left\{\sqrt{2}, 3\right\}$.

The graph in Fig. 3.42 supports this solution.

Try This. Solve $x^{2/3} - x^{1/3} - 6 = 0$. ■

Equations Involving Absolute Value

We solved basic absolute value equations in Section 1.1. In the next two examples we solve some more complicated absolute value equations.

Example **9** **An equation involving absolute value**

Solve $|x^2 - 2x - 16| = 8$.

Solution

First write an equivalent statement without using absolute value symbols.

$$x^2 - 2x - 16 = 8 \quad \text{or} \quad x^2 - 2x - 16 = -8$$
$$x^2 - 2x - 24 = 0 \quad \text{or} \quad x^2 - 2x - 8 = 0$$
$$(x - 6)(x + 4) = 0 \quad \text{or} \quad (x - 4)(x + 2) = 0$$
$$x = 6 \quad \text{or} \quad x = -4 \quad \text{or} \quad x = 4 \quad \text{or} \quad x = -2$$

The solution set is $\{-4, -2, 4, 6\}$.

The graph in Fig. 3.43 supports this solution.

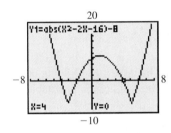

■ **Figure 3.43**

Try This. Solve $|x^2 - x - 4| = 2$. ■

In the next example we have an equation in which an absolute value expression is equal to an expression that could be positive or negative and an equation with two absolute value expressions.

Example **10** **More equations involving absolute value**

Solve each equation.

a. $|x^2 - 6| = 5x$ **b.** $|a - 1| = |2a - 3|$

Solution

a. Since $|x^2 - 6|$ is nonnegative for any value of x, $5x$ must be nonnegative. Write the equivalent statement assuming that $5x$ is nonnegative:

$$x^2 - 6 = 5x \quad \text{or} \quad x^2 - 6 = -5x$$
$$x^2 - 5x - 6 = 0 \quad \text{or} \quad x^2 + 5x - 6 = 0$$
$$(x - 6)(x + 1) = 0 \quad \text{or} \quad (x + 6)(x - 1) = 0$$
$$x = 6 \quad \text{or} \quad x = -1 \quad \text{or} \quad x = -6 \quad \text{or} \quad x = 1$$

The expression $|x^2 - 6|$ is nonnegative for any real number x. But $5x$ is negative if $x = -1$ or if $x = -6$. So -1 and -6 are extraneous roots. They do not satisfy the original equation. The solution set is $\{1, 6\}$.

The graph in Fig. 3.44 supports this solution. □

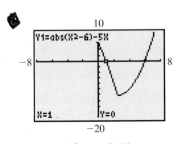

■ **Figure 3.44**

b. The equation $|a - 1| = |2a - 3|$ indicates that $a - 1$ and $2a - 3$ have the same absolute value. If two quantities have the same absolute value, they are either equal or opposites. Use this fact to write an equivalent statement without absolute value signs.

$$a - 1 = 2a - 3 \quad \text{or} \quad a - 1 = -(2a - 3)$$
$$a + 2 = 2a \qquad \text{or} \quad a - 1 = -2a + 3$$
$$2 = a \qquad\qquad \text{or} \qquad a = \frac{4}{3}$$

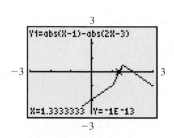

Figure 3.45

Check that both 2 and $\frac{4}{3}$ satisfy the original absolute value equation. The solution set is $\left\{\frac{4}{3}, 2\right\}$.

The graph in Fig. 3.45 supports this solution.

Try This. Solve $|x| = |x - 1|$. ■

Applications

The **break-even point** for a business is the point at which the cost of doing business is equal to the revenue generated by the business. The business is profitable when the revenue is greater than the cost.

Example **11** **Break-even point for a bus tour**

A tour operator uses the equation $C = 3x + \sqrt{50x + 9000}$ to find his cost in dollars for taking x people on a tour of San Francisco.

a. For what value of x is the cost $160?
b. If he charges $10 per person for the tour, then what is his break-even point?

Solution

a. Replace C by $160 and solve the equation.

$$160 = 3x + \sqrt{50x + 9000}$$
$$160 - 3x = \sqrt{50x + 9000} \quad \text{Isolate the radical.}$$
$$25{,}600 - 960x + 9x^2 = 50x + 9000 \quad \text{Square each side.}$$
$$9x^2 - 1010x + 16{,}600 = 0$$
$$x = \frac{-(-1010) \pm \sqrt{1010^2 - 4(9)(16{,}600)}}{2(9)}$$
$$x = 20 \quad \text{or} \quad x \approx 92.2$$

Check that 20 satisfies the original equation but 92.2 does not. The cost is $160 when 20 people take the tour.

b. At $10 per person, the revenue in dollars is given by $R = 10x$. When the revenue is equal to the cost, as shown in Fig. 3.46, the operator breaks even.

$$10x = 3x + \sqrt{50x + 9000}$$

$$7x = \sqrt{50x + 9000}$$

$$49x^2 = 50x + 9000 \quad \text{Square each side.}$$

$$49x^2 - 50x - 9000 = 0$$

$$x = \frac{50 \pm \sqrt{50^2 - 4(49)(-9000)}}{2(49)}$$

$$x \approx -13.052 \quad \text{or} \quad x \approx 14.072$$

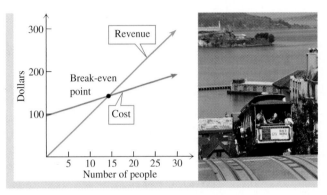

■ **Figure 3.46**

If 14.072 people took the tour, the operator would break even. Since the break-even point is not a whole number, the operator actually needs 15 people to make a profit.

Try This. The function $D = \sqrt{P} - \sqrt{P - 400}$ gives the monthly demand D for custom-made suits in terms of the price P in dollars. For what price is the demand 10 suits?   ■

For Thought

True or False? Explain.

1. Squaring each side of $\sqrt{x - 1} + \sqrt{x} = 6$ yields $x - 1 + x = 36$.

2. The equations $(2x - 1)^2 = 9$ and $2x - 1 = 3$ are equivalent.

3. The equations $x^{2/3} = 9$ and $x = 27$ have the same solution set.

4. To solve $2x^{1/4} - x^{1/2} + 3 = 0$, we let $u = x^{1/2}$ and $u^2 = x^{1/4}$.

5. If $(x - 1)^{-2/3} = 4$, then $x = 1 \pm 4^{-3/2}$.

6. No negative number satisfies $x^{-2/5} = 4$.

7. The solution set to $|2x + 10| = 3x$ is $\{-2, 10\}$.

8. No negative number satisfies $|x^2 - 3x + 2| = 7x$.

9. The equation $|2x + 1| = |x|$ is equivalent to $2x + 1 = x$ or $2x + 1 = -x$.

10. The equation $x^9 - 5x^3 + 6 = 0$ is an equation of quadratic type.

3.4 Exercises

Find all real and imaginary solutions to each equation. Check your answers.

1. $x^3 + 3x^2 - 4x - 12 = 0$

2. $x^3 - x^2 - 5x + 5 = 0$

3. $2x^3 + 1000x^2 - x - 500 = 0$

4. $3x^3 - 1200x^2 - 2x + 800 = 0$

5. $a^3 + 5a = 15a^2$ **6.** $b^3 + 20b = 9b^2$

7. $3y^4 - 12y^2 = 0$ **8.** $5m^4 - 10m^3 + 5m^2 = 0$

9. $a^4 - 16 = 0$ **10.** $w^4 + 8w = 0$

Find all real solutions to each equation. Check your answers.

11. $\sqrt{x + 1} = x - 5$ **12.** $\sqrt{x - 1} = x - 7$

13. $\sqrt{x - 2} = x - 22$ **14.** $3 + \sqrt{x} = 1 + x$

15. $w = \dfrac{\sqrt{1 - 3w}}{2}$ **16.** $t = \dfrac{\sqrt{2 - 3t}}{3}$

17. $\dfrac{1}{z} = \dfrac{3}{\sqrt{4z + 1}}$ **18.** $\dfrac{1}{p} - \dfrac{2}{\sqrt{9p + 1}} = 0$

19. $\sqrt{x^2 - 2x - 15} = 3$ **20.** $\sqrt{3x^2 + 5x - 3} = x$

21. $\sqrt{x + 40} - \sqrt{x} = 4$ **22.** $\sqrt{x} + \sqrt{x - 36} = 2$

23. $\sqrt{n + 4} + \sqrt{n - 1} = 5$

24. $\sqrt{y + 10} - \sqrt{y - 2} = 2$

25. $\sqrt{2x + 5} + \sqrt{x + 6} = 9$

26. $\sqrt{3x - 2} - \sqrt{x - 2} = 2$

Find all real solutions to each equation. Check your answers.

27. $x^{2/3} = 2$ **28.** $x^{2/3} = \dfrac{1}{2}$ **29.** $w^{-4/3} = 16$

30. $w^{-3/2} = 27$ **31.** $t^{-1/2} = 7$ **32.** $t^{-1/2} = \dfrac{1}{2}$

33. $(s - 1)^{-1/2} = 2$ **34.** $(s - 2)^{-1/2} = \dfrac{1}{3}$

Find all real and imaginary solutions to each equation. Check your answers.

35. $x^4 - 12x^2 + 27 = 0$ **36.** $x^4 + 10 = 7x^2$

37. $\left(\dfrac{2c - 3}{5}\right)^2 + 2\left(\dfrac{2c - 3}{5}\right) = 8$

38. $\left(\dfrac{b - 5}{6}\right)^2 - \left(\dfrac{b - 5}{6}\right) - 6 = 0$

39. $\dfrac{1}{(5x - 1)^2} + \dfrac{1}{5x - 1} - 12 = 0$

40. $\dfrac{1}{(x - 3)^2} + \dfrac{2}{x - 3} - 24 = 0$

41. $(v^2 - 4v)^2 - 17(v^2 - 4v) + 60 = 0$

42. $(u^2 + 2u)^2 - 2(u^2 + 2u) - 3 = 0$

43. $x - 4\sqrt{x} + 3 = 0$ **44.** $2x + 3\sqrt{x} - 20 = 0$

45. $q - 7q^{1/2} + 12 = 0$ **46.** $h + 1 = 2h^{1/2}$

47. $x^{2/3} + 10 = 7x^{1/3}$ **48.** $x^{1/2} - 3x^{1/4} + 2 = 0$

Solve each absolute value equation.

49. $|w^2 - 4| = 3$ **50.** $|a^2 - 1| = 1$

51. $|v^2 - 3v| = 5v$ **52.** $|z^2 - 12| = z$

53. $|x^2 - x - 6| = 6$ **54.** $|2x^2 - x - 2| = 1$

55. $|x + 5| = |2x + 1|$ **56.** $|3x - 4| = |x|$

Solve each equation. Find imaginary solutions when possible.

57. $\sqrt{16x + 1} - \sqrt{6x + 13} = -1$

58. $\sqrt{16x + 1} - \sqrt{6x + 13} = 1$

59. $v^6 - 64 = 0$ **60.** $t^4 - 1 = 0$

61. $(7x^2 - 12)^{1/4} = x$ **62.** $(10x^2 - 1)^{1/4} = 2x$

63. $\sqrt[3]{2 + x - 2x^2} = x$

64. $\sqrt{48 + \sqrt{x}} - 4 = \sqrt[4]{x}$

65. $\left(\dfrac{x - 2}{3}\right)^2 - 2\left(\dfrac{x - 2}{3}\right) + 10 = 0$

66. $\dfrac{1}{(x + 1)^2} - \dfrac{2}{x + 1} + 2 = 0$

67. $(3u - 1)^{2/5} = 2$ **68.** $(2u + 1)^{2/3} = 3$

69. $x^2 - 11\sqrt{x^2 + 1} + 31 = 0$

70. $2x^2 - 3\sqrt{2x^2 - 3} - 1 = 0$

71. $|x^2 - 2x| = |3x - 6|$

72. $|x^2 + 5x| = |3 - x^2|$

73. $(3m + 1)^{-3/5} = -\dfrac{1}{8}$ **74.** $(1 - 2m)^{-5/3} = -\dfrac{1}{32}$

75. $|x^2 - 4| = x - 2$ **76.** $|x^2 + 7x| = x^2 - 4$

Solve each problem.

77. *Maximum Sail Area*   According to the International America's Cup Rules, the maximum sail area S for a boat with length L (in meters) and displacement D (in cubic meters) is determined by the equation

$$L + 1.25S^{1/2} - 9.8D^{1/3} = 16.296$$

(America's Cup, www.americascup.org). Find S for a boat with length 21.24 m and displacement 18.34 m³.

$$L + 1.25S^{1/2} - 9.8D^{1/3} = 16.296$$

■ **Figure for Exercises 77 and 78**

78. *Minimum Displacement for a Yacht*   The minimum displacement D for a boat with length 21.52 m and a sail area of 310.64 m² is determined by the equation $L + 1.25S^{1/2} - 9.8D^{1/3} = 16.296$. Find this boat's minimum displacement.

79. *Cost of Baking Bread*   The daily cost for baking x loaves of bread at Juanita's Bakery is given in dollars by $C = 0.5x + \sqrt{8x + 5000}$. Find the number of loaves for which the cost is $83.50.

80. *Break-Even Analysis*   If the bread in Exercise 79 sells for $1.10 per loaf, then what is the minimum number of loaves that Juanita must bake and sell to make a profit?

81. *Square Roots*   Find two numbers that differ by 6 and whose square roots differ by 1.

82. *Right Triangle*   One leg of a right triangle is 1 cm longer than the other leg. What is the length of the short leg if the total length of the hypotenuse and the short leg is 10 cm?

83. *Perimeter of a Right Triangle*   A sign in the shape of a right triangle has one leg that is 7 in. longer than the other leg. What is the length of the shorter leg if the perimeter is 30 in.?

84. *Right Triangle Inscribed in a Semicircle*   One leg of a right triangle is 1 ft longer than the other leg. If the triangle is inscribed in a circle and the hypotenuse of the triangle is the diameter of the circle, then what is the length of the radius for which the length of the radius is equal to the length of the shortest side?

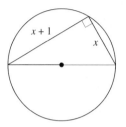

■ **Figure for Exercise 84**

85. *Area of a Foundation*   The original plans for Jennifer's house called for a square foundation. After increasing one side by 30 ft and decreasing the other by 10 ft, the area of the rectangular foundation was 2100 ft². What was the area of the original square foundation?

86. *Shipping Carton*   Heloise designed a cubic box for shipping paper. The height of the box was acceptable, but the paper would not fit into the box. After increasing the length by 0.5 in. and the width by 6 in. the area of the bottom was 119 in.² and the paper would fit. What was the original volume of the cubic box?

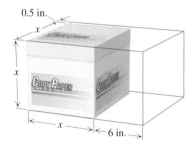

■ **Figure for Exercise 86**

87. *Sail Area-Displacement Ratio*   The sail area-displacement ratio S is defined as

$$S = A\left(\dfrac{d}{64}\right)^{-2/3},$$

where A is the sail area in square feet and d is the displacement in pounds. The Oceanis 381 is a 39 ft sailboat with a sail area-displacement ratio of 14.26 and a sail area of 598.9 ft². Find the displacement for the Oceanis 381.

88. *Capsize Screening Value* The capsize screening value C is defined as

$$C = b\left(\frac{d}{64}\right)^{-1/3},$$

where b is the beam (or width) in feet and d is the displacement in pounds. The Bahia 46 is a 46 ft catamaran with a capsize screening value of 3.91 and a beam of 26.1 ft. Find the displacement for the Bahia 46.

89. *Insulated Carton* Nina is designing a box for shipping frozen shrimp. The box is to have a square base and a height that is 2 in. greater than the width of the base. The box will be surrounded with a 1-in. thick layer of styrofoam. If the volume of the inside of the box must be equal to the volume of the styrofoam used, then what volume of shrimp can be shipped in the box?

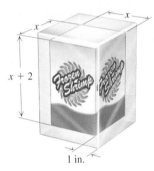

■ **Figure for Exercise 89**

90. *Volume of a Cubic Box* If the width of the base of a cubic container is increased by 3 m and the length of the base decreased by 1 m, then the volume of the new container is 6 m³. What is the height of the cubic container?

91. *Hiking Time* William, Nancy, and Edgar met at the lodge at 8 A.M., and William began hiking west at 4 mph. At 10 A.M., Nancy began hiking north at 5 mph and Edgar went east on a three-wheeler at 12 mph. At what time was the distance between Nancy and Edgar 14 mi greater than the distance between Nancy and William?

92. *Accuracy of Transducers* Setra Systems Inc., of Acton, MA, calculates the accuracy A of its pressure transducers using the formula

$$A = \sqrt{(NL)^2 + (HY)^2 + (NR)^2}$$

where NL represents nonlinearity, HY represents hysteresis, and NR represents nonrepeatability. If the nonlinearity is 0.1%, the hysteresis is 0.05%, and the nonrepeatability is 0.02%, then what is the accuracy? Solve the formula for hysteresis.

93. *Geometric Mean* The geometric mean of the numbers $x_1, x_2, \ldots, x_n$ is defined as

$$GM = \sqrt[n]{x_1 \cdot x_2 \cdot \cdots \cdot x_n}.$$

The accompanying graph shows the quarterly net income for Wal-Mart Stores for the first three quarters of 2004 (www.walmartstores.com).

a. Find the geometric mean of the net incomes for these three quarters.

b. What would the net income have to be for the next quarter so that the geometric mean for the four quarters would be $2.6 billion?

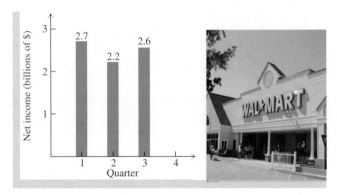

■ **Figure for Exercise 93**

94. *Time Swimming and Running* Lauren is competing in her town's cross-country competition. Early in the event, she must race from point A on the Greenbriar River to point B, which is 5 mi downstream and on the opposite bank. The Greenbriar is 1 mi wide. In planning her strategy, Lauren knows she can use any combination of running and swimming. She can run 10 mph and swim 8 mph. How long would it take if she ran 5 mi downstream and then swam across? Find the time it would take if she swam diagonally from A to B. Find x so that she could run x miles along the bank, swim diagonally to B, and complete the race in 36 min. (Ignore the current in the river.)

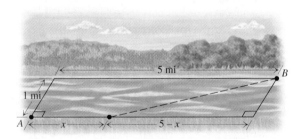

■ **Figure for Exercise 94**

95. *Boston Molasses Disaster* In the city of Boston, during the afternoon of January 15, 1919, a cylindrical metal tank containing 25,850,000 kg of molasses ruptured. The sticky liquid poured into the streets in a 9-m-deep stream that knocked down buildings and killed pedestrians and horses (www.discovery.com). If the diameter of the tank was equal to its height and the weight of molasses is 1600 kg/m³, then what was the height of the tank in meters?

■ **Figure for Exercise 95**

HINT The volume is the area of the triangular end times the length of the tent.

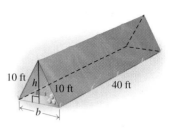

■ **Figure for Exercise 96**

96. *Storing Supplies* An army sergeant wants to use a 20-ft by 40-ft piece of canvas to make a two-sided tent for holding supplies as shown in the figure. Write the volume of the tent as a function of b. For what value of b is the volume 1600 ft³? Use a graphing calculator to find the values for b and h that will maximize the volume of the tent.

Thinking Outside the Box XXXIII

Painting Problem A painter has seven 3-ft by 5-ft rectangular drop cloths. If he lays each drop cloth on the carpet as a 3-ft by 5-ft rectangle, without folding, cutting, or tearing them, then what is the maximum area that he can cover with these drop cloths in an 8-ft by 13-ft room?

3.4 Pop Quiz

Solve each equation. Find imaginary solutions when possible.

1. $x^3 + x^2 + x + 1 = 0$

2. $\sqrt{x + 4} = x - 2$

3. $x^{-2/3} = 4$

4. $x^4 - 3x^2 = 4$

5. $|x + 3| = |2x - 5|$

Linking Concepts

For Individual or Group Explorations

Minimizing Construction Cost

A homeowner needs to run a new water pipe from his house to a water terminal as shown in the accompanying diagram. The terminal is 30 ft down the 10-ft-wide driveway and on the other side. A contractor charges $3/ft alongside the driveway and $4/ft for underneath the driveway.

a) What will it cost if the contractor runs the pipe entirely under the driveway along the diagonal of the 30-ft by 10-ft rectangle?

b) What will it cost if the contractor runs the pipe 30 ft alongside the driveway and then 10 ft straight across?

c) The contractor claims that he can do the job for $120 by going alongside the driveway for some distance and then going under the drive diagonally to the terminal. Find x, the distance alongside the driveway.

d) Write the cost as a function of x and sketch the graph of the function.

e) Use the minimum feature of a graphing calculator to find the approximate value for x that will minimize the cost.

f) What is the minimum cost (to the nearest cent) for which the job can be done?

3.5

Graphs of Polynomial Functions

In Chapter 1 we learned that the graph of a polynomial function of degree 0 or 1 is a straight line. In Section 3.1 we learned that the graph of a second-degree polynomial function is a parabola. In this section we will concentrate on graphs of polynomial functions of degree greater than 2.

Drawing Good Graphs

A graph of an equation is a picture of all of the ordered pairs that satisfy the equation. However, it is impossible to draw a perfect picture of any set of ordered pairs. We usually find a few important features of the graph and make sure that our picture brings out those features. For example, you can make a good graph of a linear function by drawing a line through the intercepts using a ruler and a sharp pencil. A good parabola should look smooth and symmetric and pass through the vertex and intercepts.

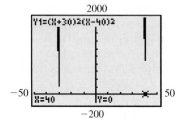

■ **Figure 3.47**

A graphing calculator is a tremendous aid in graphing because it can quickly plot many points. However, the calculator does not know if it has drawn a graph that shows the important features. For example, the graph of $y = (x + 30)^2(x - 40)^2$ has x-intercepts at $(-30, 0)$ and $(40, 0)$, but they do not appear on the graph in Fig. 3.47. □

To draw good graphs of polynomial functions we must understand symmetry, the behavior of the graph at its x-intercepts, and how the leading coefficient affects the shape of the graph.

Symmetry

Symmetry is a very special property of graphs of some functions but not others. Recognizing that the graph of a function has some symmetry usually cuts in half the work required to obtain the graph and also helps cut down on errors in graphing. So far we have discussed the following types of symmetry.

S U M M A R Y	**Types of Symmetry**

1. The graph of a function $f(x)$ is *symmetric about the y-axis* and f is an even function if $f(-x) = f(x)$ for any value of x in the domain of the function. (Section 2.3)

2. The graph of a function $f(x)$ is *symmetric about the origin* and f is an odd function if $f(-x) = -f(x)$ for any value of x in the domain of the function. (Section 2.3)

3. The graph of a quadratic function $f(x) = ax^2 + bx + c$ is *symmetric about its axis of symmetry, $x = -b/(2a)$.* (Section 3.1)

The graphs of $f(x) = x^2$ and $f(x) = x^3$ shown in Figs. 3.48 and 3.49 are nice examples of symmetry about the y-axis and symmetry about the origin, respectively. The axis of symmetry of $f(x) = x^2$ is the y-axis. The symmetry of the other quadratic functions comes from the fact that the graph of every quadratic function is a transformation of the graph of $f(x) = x^2$.

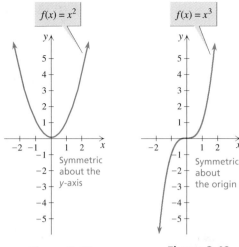

■ **Figure 3.48**   ■ **Figure 3.49**

Example 1 Determining the symmetry of a graph

Discuss the symmetry of the graph of each polynomial function.

a. $f(x) = 5x^3 - x$ **b.** $g(x) = 2x^4 - 3x^2$ **c.** $h(x) = x^2 - 3x + 6$
d. $j(x) = x^4 - x^3$

Solution

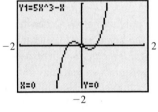

■ **Figure 3.50**

a. Replace x by $-x$ in $f(x) = 5x^3 - x$ and simplify:

$$f(-x) = 5(-x)^3 - (-x)$$
$$= -5x^3 + x$$

Since $f(-x)$ is the opposite of $f(x)$, the graph is symmetric about the origin. Fig. 3.50 supports this conclusion. □

b. Replace x by $-x$ in $g(x) = 2x^4 - 3x^2$ and simplify:

$$g(-x) = 2(-x)^4 - 3(-x)^2$$
$$= 2x^4 - 3x^2$$

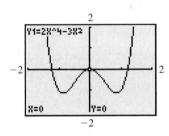

■ **Figure 3.51**

Since $g(-x) = g(x)$, the graph is symmetric about the y-axis. Fig. 3.51 supports this conclusion. □

c. Because h is a quadratic function, its graph is symmetric about the line $x = -b/(2a)$, which in this case is the line $x = 3/2$.

d. In this case

$$j(-x) = (-x)^4 - (-x)^3$$
$$= x^4 + x^3.$$

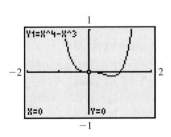

■ **Figure 3.52**

So $j(-x) \neq j(x)$ and $j(-x) \neq -j(x)$. The graph of j is not symmetric about the y-axis and is not symmetric about the origin. Fig. 3.52 supports this conclusion.

Try This. Discuss the symmetry of $f(x) = -x^3 - 4x$.   ■

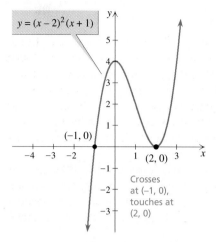

$y = (x - 2)^2(x + 1)$

$(-1, 0)$

$(2, 0)$

Crosses
at $(-1, 0)$,
touches at
$(2, 0)$

■ **Figure 3.53**

Behavior at the x-Intercepts

The x-intercepts are key points for the graph of a polynomial function, as they are for any function. Consider the graph of $y = (x - 2)^2(x + 1)$ in Fig. 3.53. Near 2 the values of y are positive, as shown in Fig. 3.54, and the graph does not cross the x-axis. Near -1 the values of y are negative for $x < -1$ and positive for $x > -1$, as shown in Fig. 3.55, and the graph crosses the x-axis. The reason for this behavior is the exponents in $(x - 2)^2$ and $(x + 1)^1$. Because $(x - 2)^2$ has an even exponent, $(x - 2)^2$ cannot be negative. Because $(x + 1)^1$ has an odd exponent, $(x + 1)^1$ changes sign at -1 but does not change sign at 2. So the product $(x - 2)^2(x + 1)$ does not change sign at 2, but does change sign at -1.

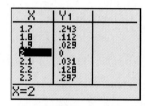

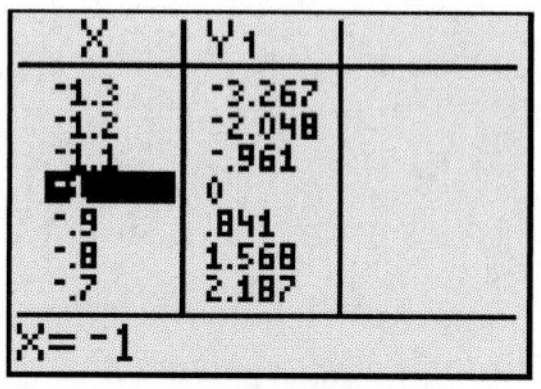

■ **Figure 3.54**   ■ **Figure 3.55**

Every x-intercept corresponds to a factor of the polynomial. Whether that factor occurs an odd or even number of times determines the behavior of the graph at that intercept. *The graph crosses the x-axis at an x-intercept if the factor corresponding to that intercept is raised to an odd power, and the graph touches but does not cross if the factor is raised to an even power.* As another example, consider the graphs of $f(x) = x^2$ and $f(x) = x^3$ shown in Figs. 3.48 and 3.49. Each has only one x-intercept. Note that $f(x) = x^2$ does not cross the x-axis at $(0, 0)$, but $f(x) = x^3$ does cross the x-axis at $(0, 0)$.

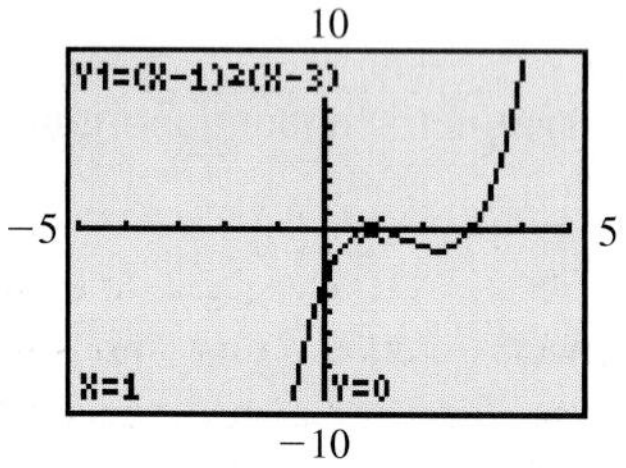

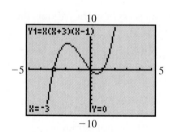

■ **Figure 3.56**

Example **2** **Crossing at the x-intercepts**

Find the x-intercepts and determine whether the graph of the function crosses the x-axis at each x-intercept.

a. $f(x) = (x - 1)^2(x - 3)$ **b.** $f(x) = x^3 + 2x^2 - 3x$

Solution

a. The x-intercepts are found by solving $(x - 1)^2(x - 3) = 0$. The x-intercepts are $(1, 0)$ and $(3, 0)$. The graph does not cross the x-axis at $(1, 0)$ because the factor $x - 1$ occurs to an even power. The graph crosses the x-axis at $(3, 0)$ because $x - 3$ occurs to an odd power.

  The graph in Fig. 3.56 supports these conclusions.   □

b. The x-intercepts are found by solving $x^3 + 2x^2 - 3x = 0$. By factoring, we get $x(x + 3)(x - 1) = 0$. The x-intercepts are $(0, 0)$, $(-3, 0)$, and $(1, 0)$. Since each factor occurs an odd number of times (once), the graph crosses the x-axis at each of the x-intercepts.

  The graph in Fig. 3.57 supports these conclusions.

■ **Figure 3.57**

Try This. Determine whether the graph of $f(x) = (x - 1)^3(x + 5)^2$ crosses the x-axis at each of its x-intercepts. ■

The Leading Coefficient Test

We now consider the behavior of a polynomial function as the x-coordinate goes to or approaches infinity or negative infinity. In symbols, $x \to \infty$ or $x \to -\infty$. Since we seek only an intuitive understanding of the ideas presented here, we will not give precise definitions of these terms. Precise definitions are given in a calculus course.

To say that $x \to \infty$ means that x gets larger and larger without bound. For our purposes we can think of x assuming the values 1, 2, 3, and so on, without end. Similarly, $x \to -\infty$ means that x gets smaller and smaller without bound. Think of x assuming the values -1, -2, -3, and so on, without end.

As x approaches ∞ the y-coordinates of any polynomial function approach ∞ or $-\infty$. Likewise, when $x \to -\infty$ the y-coordinates approach ∞ or $-\infty$. The direction that y goes is determined by the degree of the polynomial and the sign of the leading coefficient. The four possible types of behavior are illustrated in the next example.

Example **3** Behavior as $x \to \infty$ or $x \to -\infty$

Determine the behavior of the graph of each function as $x \to \infty$ or $x \to -\infty$.

a. $y = x^3 - x$ **b.** $y = -x^3 + 1$

c. $y = x^4 - 4x^2$ **d.** $y = -x^4 + 4x^2 + x$

Solution

a. Considering the following table. If you have a graphing calculator, make a table like this and scroll through it.

x	-30	-20	-10	0	10	20	30
$y = x^3 - x$	$-26{,}970$	-7980	-990	0	990	7980	26{,}970

$\longleftarrow$ $y \to -\infty$ $y \to \infty$ $\longrightarrow$

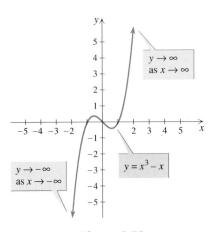

■ **Figure 3.58**

The graph of $y = x^3 - x$ is shown in Fig. 3.58. As x gets larger and larger $(x \to \infty)$, y increases without bound $(y \to \infty)$. As x gets smaller and smaller $(x \to -\infty)$, y decreases without bound $(y \to -\infty)$. Notice that the degree of the polynomial is odd and the sign of the leading coefficient is positive.

b. Consider the following table. If you have a graphing calculator, make a table like this and scroll through it.

x	-30	-20	-10	0	10	20	30
$y = -x^3 + 1$	27{,}001	8001	1001	1	-999	-7999	$-26{,}999$

$\longleftarrow$ $y \to \infty$ $y \to -\infty$ $\longrightarrow$

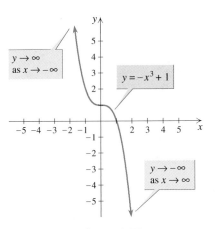

■ **Figure 3.59**

The graph of $y = -x^3 + 1$ is shown in Fig. 3.59. As x gets larger and larger $(x \to \infty)$, y decreases without bound $(y \to -\infty)$. As x gets smaller and smaller $(x \to -\infty)$, y increases without bound $(y \to \infty)$. Notice that the degree of this polynomial is odd and the sign of the leading coefficient is negative.

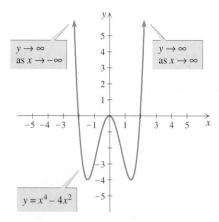

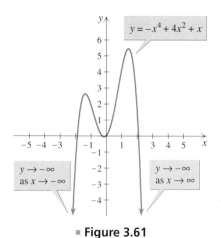

■ **Figure 3.60**

■ **Figure 3.61**

c. Consider the following table. If you have a graphing calculator, make a table like this and scroll through it.

x	0	± 10	± 20	± 30	± 40
$y = x^4 - 4x^2$	0	9600	158,400	806,400	2,553,600

$y \to \infty$

The graph of $y = x^4 - 4x^2$ is shown in Fig. 3.60. As $x \to \infty$ or $x \to -\infty$, y increases without bound ($y \to \infty$). Notice that the degree of this polynomial is even and the sign of the leading coefficient is positive.

d. Consider the following table. If you have a graphing calculator, make a table like this and scroll through it.

x	-20	-10	0	10	20
$y = -x^4 + 4x^2 + x$	$-158,420$	-9610	0	-9590	$-158,380$

$y \to -\infty$ $y \to -\infty$

The graph of $y = -x^4 + 4x^2 + x$ is shown in Fig. 3.61. As $x \to \infty$ or $x \to -\infty$, y decreases without bound ($y \to -\infty$). Notice that the degree of this polynomial is even and the sign of the leading coefficient is negative.

Try This. Discuss the behavior of the graph of $f(x) = -x^3 - 4x$.   ■

As $x \to \infty$ or $x \to -\infty$ all first-degree polynomial functions have "end behavior" like the lines $y = x$ or $y = -x$, all second-degree polynomial functions behave like the parabolas $y = x^2$ or $y = -x^2$, and all third-degree polynomial functions behave like $y = x^3$ or $y = -x^3$, and so on. The end behavior is determined by the degree and the sign of the first term. The smaller-degree terms in the polynomial determine the number of "hills" and "valleys" between the ends of the curve. With degree n there are at most $n - 1$ hills and valleys. The end behavior of polynomial functions is summarized in the **leading coefficient test.**

Leading Coefficient Test

If $f(x) = a_n x^n + a_{n-1} x^{n-1} + \cdots + a_1 x + a_0$, the behavior of the graph of f to the left and right is determined as follows:

For n odd and $a_n > 0$, $y \to \infty$ as $x \to \infty$, and $y \to -\infty$ as $x \to -\infty$.

For n odd and $a_n < 0$, $y \to -\infty$ as $x \to \infty$, and $y \to \infty$ as $x \to -\infty$.

For n even and $a_n > 0$, $y \to \infty$ as $x \to \infty$ or $x \to -\infty$.

For n even and $a_n < 0$, $y \to -\infty$ as $x \to \infty$ or $x \to -\infty$.

The leading coefficient test is shown visually in Fig. 3.62, which shows only the "ends" of the graphs of the polynomial functions.

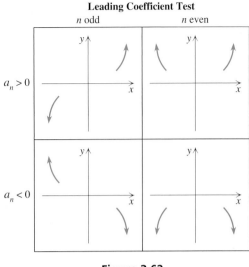

Leading Coefficient Test

■ Figure 3.62

Sketching Graphs of Polynomial Functions

A good graph of a polynomial function should include the features that we have been discussing. The following strategy will help you graph polynomial functions.

STRATEGY   **Graphing a Polynomial Function**

1. Check for symmetry.
2. Find all real zeros of the polynomial function.
3. Determine the behavior at the corresponding *x*-intercepts.
4. Determine the behavior as $x \to \infty$ and as $x \to -\infty$.
5. Calculate several ordered pairs including the *y*-intercept to verify your suspicions about the shape of the graph.
6. Draw a smooth curve through the points to make the graph.

Example **4** **Graphing polynomial functions**

Sketch the graph of each polynomial function.

a. $f(x) = x^3 - 5x^2 + 7x - 3$ **b.** $f(x) = x^4 - 200x^2 + 10,000$

Solution

a. First find $f(-x)$ to determine symmetry and the number of negative roots.

$$f(-x) = (-x)^3 - 5(-x)^2 + 7(-x) - 3$$
$$= -x^3 - 5x^2 - 7x - 3$$

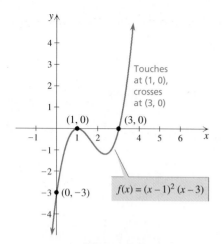

▪ **Figure 3.63**

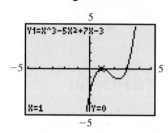

▪ **Figure 3.64**

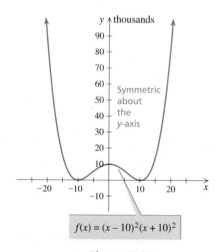

▪ **Figure 3.65**

From $f(-x)$, we see that the graph has neither type of symmetry. Because $f(-x)$ has no sign changes, $x^3 - 5x^2 + 7x - 3 = 0$ has no negative roots by Descartes's rule of signs. The only possible rational roots are 1 and 3.

$$
\begin{array}{r|rrrr}
7=1 & 1 & -5 & 7 & -3 \\
(x-1) & & 1 & -4 & 3 \\
\hline
& 1 & -4 & 3 & 0
\end{array}
$$

From the synthetic division we know that 1 is a root and we can factor $f(x)$:

$$f(x) = (x - 1)(x^2 - 4x + 3)$$
$$= (x - 1)^2(x - 3) \qquad \text{Factor completely.}$$

The x-intercepts are $(1, 0)$ and $(3, 0)$. The graph of f does not cross the x-axis at $(1, 0)$ because $x - 1$ occurs to an even power, while the graph crosses at $(3, 0)$ because $x - 3$ occurs to an odd power. The y-intercept is $(0, -3)$. Since the leading coefficient is positive and the degree is odd, $y \to \infty$ as $x \to \infty$ and $y \to -\infty$ as $x \to -\infty$. Calculate two more ordered pairs for accuracy, say $(2, -1)$ and $(4, 9)$. Draw a smooth curve as in Fig. 3.63.

⊿ The calculator graph shown in Fig. 3.64 supports these conclusions. ☐

b. First find $f(-x)$:

$$f(-x) = (-x)^4 - 200(-x)^2 + 10{,}000$$
$$= x^4 - 200x^2 + 10{,}000$$

Since $f(x) = f(-x)$, the graph is symmetric about the y-axis. We can factor the polynomial as follows.

$$f(x) = x^4 - 200x^2 + 10{,}000$$
$$= (x^2 - 100)(x^2 - 100)$$
$$= (x - 10)(x + 10)(x - 10)(x + 10)$$
$$= (x - 10)^2(x + 10)^2$$

The x-intercepts are $(10, 0)$ and $(-10, 0)$. Since each factor for these intercepts has an even power, the graph does not cross the x-axis at the intercepts. The y-intercept is $(0, 10{,}000)$. Since the leading coefficient is positive and the degree is even, $y \to \infty$ as $x \to \infty$ or as $x \to -\infty$. The graph also goes through $(-20, 90{,}000)$ and $(20, 90{,}000)$. Draw a smooth curve through these points and the intercepts as shown in Fig. 3.65.

⊿ The calculator graph shown in Fig. 3.66 supports these conclusions.

Try This. Sketch the graph of $f(x) = x^4 - 4x^2$. ▪

Polynomial Inequalities

We can solve polynomial inequalities by the same methods that we used on quadratic inequalities in Section 3.1. The next example illustrates the test-point method.

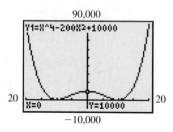

▪ **Figure 3.66**

This method depends on the fact that a polynomial function can change sign only at a zero of the function.

Example 5 Solving a polynomial inequality with test points

Solve $x^4 + x^3 - 15x^2 - 3x + 36 < 0$ using the test-point method.

Solution

Use synthetic division to see that 3 and -4 are zeros of the function $f(x) = x^4 + x^3 - 15x^2 - 3x + 36$:

$$
\begin{array}{r|rrrrr}
3 & 1 & 1 & -15 & -3 & 36 \\
  &   & 3 & 12 & -9 & -36 \\
\hline
-4 & 1 & 4 & -3 & -12 & 0 \\
   &   & -4 & 0 & 12 & \\
\hline
   & 1 & 0 & -3 & 0 &
\end{array}
$$

Since $x^4 + x^3 - 15x^2 - 3x + 36 = (x - 3)(x + 4)(x^2 - 3)$, the other two zeros are $\pm\sqrt{3}$. The four zeros determine five intervals on the number line in Fig. 3.67. Select an arbitrary test point in each of these intervals. The selected points $-5, -3, 0, 2$, and 4 are shown in red in the figure:

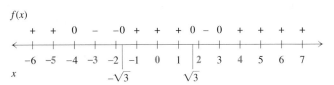

■ **Figure 3.67**

Now evaluate $f(x) = x^4 + x^3 - 15x^2 - 3x + 36$ at each test point:

$$f(-5) = 176, \quad f(-3) = -36, \quad f(0) = 36, \quad f(2) = -6, \quad f(4) = 104$$

These values indicate that the signs of the function are $+, -, +, -,$ and $+$ on the intervals shown in Fig. 3.67. The values of x that satisfy the original inequality are the values of x for which $f(x)$ is negative. So the solution set is $\left(-4, -\sqrt{3}\right) \cup \left(\sqrt{3}, 3\right)$.

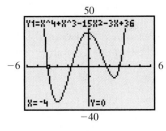

■ **Figure 3.68**

The calculator graph of $y = x^4 + x^3 - 15x^2 - 3x + 36$ in Fig. 3.68 confirms that y is negative for x in $\left(-4, -\sqrt{3}\right) \cup \left(\sqrt{3}, 3\right)$. Since the multiplicity of each zero of the function is one, the graph crosses the x-axis at each intercept and the y-coordinates change sign at each intercept.

Try This. Solve $x^4 - 4x^2 < 0$ using test points.   ■

Function Gallery: **Polynomial Functions**

Linear: $f(x) = mx + b$

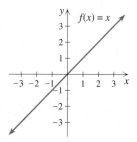

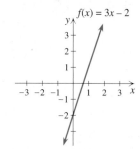

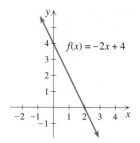

Slope 1, y-intercept (0, 0)   Slope 3, y-intercept (0, −2)   Slope −2, y-intercept (0, 4)

Quadratic: $f(x) = ax^2 + bx + c$ or $f(x) = a(x − h)^2 + k$

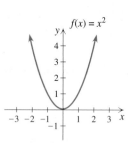

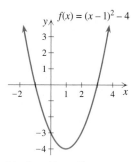

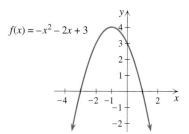

Vertex (0, 0)   Vertex (1, −4)   Vertex (−1, 4)
Range [0, ∞)   Range [−4, ∞)   Range (−∞, 4]

Cubic: $f(x) = ax^3 + bx^2 + cx + d$

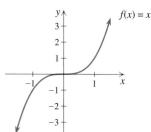

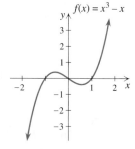

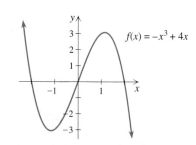

Quartic or Fourth-Degree: $f(x) = ax^4 + bx^3 + cx^2 + dx + e$

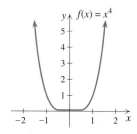

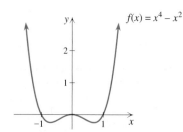

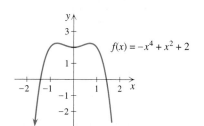

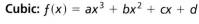

For Thought

True or False? Explain.

1. If P is a function for which $P(2) = 8$ and $P(-2) = -8$, then the graph of P is symmetric about the origin.

2. If $y = -3x^3 + 4x^2 - 6x + 9$, then $y \to -\infty$ as $x \to \infty$.

3. If the graph of $y = P(x)$ is symmetric about the origin and $P(8) = 4$, then $-P(-8) = 4$.

4. If $f(x) = x^3 - 3x$, then $f(x) = f(-x)$ for any value of x.

5. If $f(x) = x^4 - x^3 + x^2 - 6x + 7$, then
$f(-x) = x^4 + x^3 + x^2 + 6x + 7$.

6. The graph of $f(x) = x^2 - 6x + 9$ has only one x-intercept.

7. The x-intercepts for $P(x) = (x - 1)^2(x + 1)$ are $(0, 1)$ and $(0, -1)$.

8. The y-intercept for $P(x) = 4(x - 3)^2 + 2$ is $(0, 2)$.

9. The graph of $f(x) = x^2(x + 8)^2$ has no points in quadrants III and IV.

10. The graph of $f(x) = x^3 - 1$ has three x-intercepts.

3.5   Exercises

Discuss the symmetry of the graph of each polynomial function. See the summary of the types of symmetry on page 316.

1. $f(x) = x^6$

2. $f(x) = x^5 - x$

3. $f(x) = x^2 - 3x + 5$

4. $f(x) = 5x^2 + 10x + 1$

5. $f(x) = 3x^6 - 5x^2 + 3x$

6. $f(x) = x^6 - x^4 + x^2 - 8$

7. $f(x) = 4x^3 - x$

8. $f(x) = 7x^3 + x^2$

9. $f(x) = (x - 5)^2$

10. $f(x) = (x^2 - 1)^2$

11. $f(x) = -x$

12. $f(x) = 3x$

Find the x-intercepts and discuss the behavior of the graph of each polynomial function at its x-intercepts.

13. $f(x) = (x - 4)^2$

14. $f(x) = (x - 1)^2(x + 3)^2$

15. $f(x) = (2x - 1)^3$

16. $f(x) = x^6$

17. $f(x) = 4x - 1$

18. $f(x) = x^2 - 5x - 6$

19. $f(x) = x^2 - 3x + 10$

20. $f(x) = x^4 - 16$

21. $f(x) = x^3 - 3x^2$

22. $f(x) = x^3 - x^2 - x + 1$

23. $f(x) = 2x^3 - 5x^2 + 4x - 1$ **24.** $f(x) = x^3 - 3x^2 + 4$

25. $f(x) = -2x^3 - 8x^2 + 6x + 36$

26. $f(x) = -x^3 + 7x - 6$

For each function use the leading coefficient test to determine whether $y \to \infty$ or $y \to -\infty$ as $x \to \infty$.

27. $y = 2x^3 - x^2 + 9$

28. $y = -3x + 7$

29. $y = -3x^4 + 5$

30. $y = 6x^4 - 5x^2 - 1$

31. $y = x - 3x^3$

32. $y = 5x - 7x^4$

For each function use the leading coefficient test to determine whether $y \to \infty$ or $y \to -\infty$ as $x \to -\infty$.

33. $y = -2x^5 - 3x^2$

34. $y = x^3 + 8x + \sqrt{2}$

35. $y = 3x^6 - 999x^3$

36. $y = -12x^4 - 5x$

For each graph discuss its symmetry, indicate whether the graph crosses the x-axis at each x-intercept, and determine whether $y \to \infty$ or $y \to -\infty$ as $x \to \infty$ and $x \to -\infty$.

37.

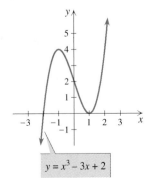

$y = x^3 - 3x + 2$

38.

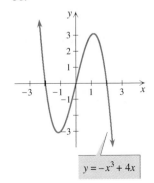

$y = -x^3 + 4x$

39.   **40.**

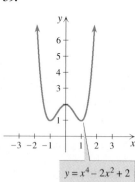

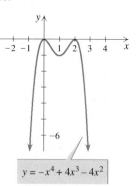

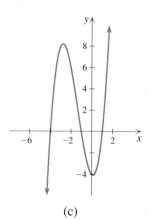

(c)

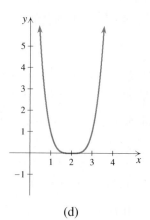

(d)

For each given function make a rough sketch of the graph that shows the behavior at the x-intercepts and the behavior as x approaches ∞ and −∞.

41. $f(x) = (x - 1)^2(x + 3)$ **42.** $f(x) = (x + 2)^2(x - 5)^2$

43. $f(x) = -2(2x - 1)^2(x + 1)^3$

44. $f(x) = -3(3x - 4)^2(2x + 1)^4$

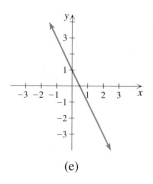

(e)

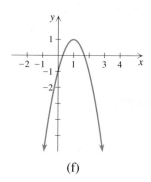

(f)

Match each polynomial function with its graph (a)–(h).

45. $f(x) = -2x + 1$ **46.** $f(x) = -2x^2 + 1$

47. $f(x) = -2x^3 + 1$ **48.** $f(x) = -2x^2 + 4x - 1$

49. $f(x) = -2x^4 + 6$ **50.** $f(x) = -2x^4 + 6x^2$

51. $f(x) = x^3 + 4x^2 - x - 4$ **52.** $f(x) = (x - 2)^4$

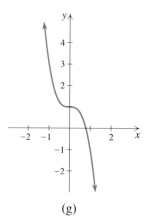

(g)

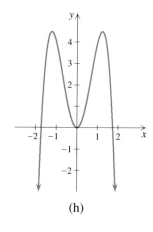

(h)

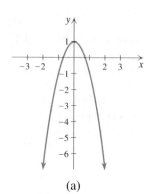

(a)

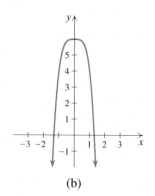

(b)

Sketch the graph of each function. See the strategy for graphing polynomial functions on page 321.

53. $f(x) = x - 30$ **54.** $f(x) = 40 - x$

55. $f(x) = (x - 30)^2$ **56.** $f(x) = (40 - x)^2$

57. $f(x) = x^3 - 40x^2$ **58.** $f(x) = x^3 - 900x$

59. $f(x) = (x - 20)^2(x + 20)^2$

60. $f(x) = (x - 20)^2(x + 12)$

61. $f(x) = -x^3 - x^2 + 5x - 3$ **62.** $f(x) = -x^4 + 6x^3 - 9x^2$

63. $f(x) = x^3 - 10x^2 - 600x$

64. $f(x) = -x^4 + 24x^3 - 144x^2$

65. $f(x) = x^3 + 18x^2 - 37x + 60$

66. $f(x) = x^3 - 7x^2 - 25x - 50$

67. $f(x) = -x^4 + 196x^2$ **68.** $f(x) = -x^4 + x^2 + 12$

69. $f(x) = x^3 + 3x^2 + 3x + 1$ **70.** $f(x) = -x^3 + 3x + 2$

71. $f(x) = (x - 3)^2(x + 5)^2(x + 7)$

72. $f(x) = x(x + 6)^2(x^2 - x - 12)$

Solve each polynomial inequality using the test-point method.

73. $x^3 - 3x > 0$ **74.** $-x^3 + 3x + 2 < 0$

75. $2x^2 - x^4 \leq 0$ **76.** $-x^4 + x^2 + 12 \geq 0$

77. $x^3 + 4x^2 - x - 4 > 0$ **78.** $x^3 + 2x^2 - 2x - 4 < 0$

79. $x^3 - 4x^2 - 20x + 48 \geq 0$ **80.** $x^3 + 7x^2 - 36 \leq 0$

81. $x^3 - x^2 + x - 1 < 0$ **82.** $x^3 + x^2 + 2x - 4 > 0$

83. $x^4 - 19x^2 + 90 \leq 0$

84. $x^4 - 5x^3 + 3x^2 + 15x - 18 \geq 0$

Determine which of the given functions is shown in the accompanying graph.

85. a. $f(x) = (x - 3)(x + 2)$ **b.** $f(x) = (x + 3)(x - 2)$

　　c. $f(x) = \frac{1}{3}(x - 3)(x + 2)$ **d.** $f(x) = \frac{1}{3}(x + 3)(x - 2)$

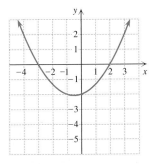

▪ **Figure for Exercise 85**

86. a. $f(x) = (x + 3)(x + 1)(x - 1)$

　　b. $f(x) = (x - 3)(x^2 - 1)$

c. $f(x) = \frac{2}{3}(x + 3)(x + 1)(x - 1)$

d. $f(x) = -\frac{2}{3}(x + 3)(x^2 - 1)$

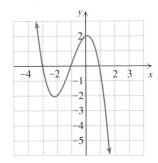

▪ **Figure for Exercise 86**

Find the equation and sketch the graph for each function.

87. A quadratic function with x-intercepts $(-5, 0)$ and $(4, 0)$ and y-intercept $(0, 3)$

88. A quadratic function with x-intercepts $(3, 0)$ and $(-6, 0)$ and y-intercept $(0, -5)$

89. A quadratic function that passes through $(1, 2)$ and has x-intercepts $(-1, 0)$ and $(3, 0)$

90. A quadratic function that passes through $(-1, -3)$ and has x-intercepts $(2, 0)$ and $(4, 0)$

91. A cubic function (a third-degree polynomial function) with x-intercepts $(2, 0)$, $(-3, 0)$, and $(4, 0)$ and y-intercept $(0, 6)$

92. A cubic function with x-intercepts $(1, 0)$, $(-1/2, 0)$, and $(1/4, 0)$ and y-intercept $(0, 2)$

93. A quartic function (a fourth-degree polynomial function) that passes through the point $(1, 3)$ and has x-intercepts $(\pm 2, 0)$ and $(\pm 4, 0)$

94. A quartic function that passes through the point $(1, 5)$ and has x-intercepts $(1/2, 0)$, $(1/3, 0)$, $(1/4, 0)$, and $(-2/3, 0)$

📈 *Each of the following third-degree polynomial functions is increasing on $(-\infty, a)$, decreasing on (a, b), and increasing on (b, ∞), where $a < b$. There is no maximum or minimum value for this type of function, but we say that the function has a **local maximum** value of $f(a)$ at $x = a$ and a **local minimum** value of $f(b)$ at $x = b$. Use a calculator graph to estimate the local maximum and local minimum values of the function to the nearest hundredth. (Using calculus, exact answers could be found for these questions.)*

95. $f(x) = x^3 + x^2 - 2x + 1$

96. $y = x^3 - x^2 - 20x - 5$

97. $y = x^3 - 9x^2 - 12x + 20$

98. $f(x) = x^3 - x^2 - 4x + 6$

99. $f(x) = 2x^3 - 3x^2 - 4x + 20$

100. $y = x^3 - 7x^2 + 12x - 14$

Solve each problem.

101. *Maximum Profit*   A company's weekly profit (in thousands of dollars) is given by the function $P(x) = x^3 - 3x^2 + 2x + 3$, where x is the amount (in thousands of dollars) spent per week on advertising. Use a graphing calculator to estimate the local maximum and the local minimum value for the profit. What amount must be spent on advertising to get the profit higher than the local maximum profit?

102. *Maximum Volume*   An open-top box is to be made from a 6 in. by 7 in. piece of copper by cutting equal squares (x in. by x in.) from each corner and folding up the sides. Write the volume of the box as a function of x. Use a graphing calculator to find the maximum possible volume to the nearest hundredth of a cubic inch.

103. *Economic Forecast*   The total annual profit (in thousands of dollars) for a department store chain is determined by $P = 90x - 6x^2 + 0.1x^3$, where x is the number of stores in the chain. The company now has 10 stores in operation and plans to pursue an aggressive expansion program. What happens to the total annual profit as the company opens more and more stores? For what values of x is profit increasing?

104. *Contaminated Chicken*   The number of bacteria of a certain type found on a chicken t minutes after processing in a contaminated plant is given by the function

$$N = 30t + 25t^2 + 44t^3 - 0.01t^4.$$

Find N for $t = 30$, 300, and 3000. What happens to N as t keeps increasing?

105. *Packing Cheese*   Workers at the Green Bay Cheese Factory are trying to cover a block of cheese with an 8 in. by 12 in. piece of foil paper as shown in the figure. The ratio of the length and width of the block must be 4 to 3 to accommodate the label. Find a polynomial function that gives the volume of the block of cheese covered in this manner as a function of the thickness x. Use a graphing calculator to find the dimensions of the block that will maximize the volume.

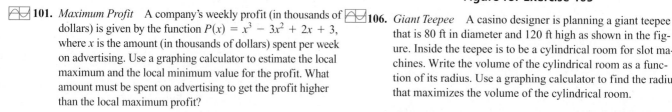

■ **Figure for Exercise 105**

106. *Giant Teepee*   A casino designer is planning a giant teepee that is 80 ft in diameter and 120 ft high as shown in the figure. Inside the teepee is to be a cylindrical room for slot machines. Write the volume of the cylindrical room as a function of its radius. Use a graphing calculator to find the radius that maximizes the volume of the cylindrical room.

> **HINT** If h is the height of the room and r the radius, then the ratio of h to $40 - r$ is 3 to 1.

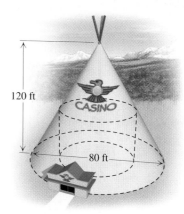

120 ft

80 ft

■ **Figure for Exercise 106**

107. *Paint Coverage*   A one pint can of spray paint will cover 50 ft². Of course a small amount of paint coats the inside of the cylindrical can and it is not used. What would happen if the inside of the can had a surface area of 50 ft²? Find the dimensions for the two one-pint cans that have a surface area of 50 ft².

> **HINT** For a cylinder, $V = \pi r^2 h$ and $S = 2\pi r^2 + 2\pi rh$. Use $1 \text{ ft}^3 \approx 7.5$ gallons.

■ **Figure for Exercise 107**

108. *Booming Business*   When Computer Recyclers opened its doors, business started booming. After a few months, there was a temporary slowdown in sales, after which sales took off again. We can model sales for this business with the function

$$N = 8t^3 - 133t^2 + 653t,$$

where N is the number of computers sold in month t ($t = 0$ corresponds to the opening of the business).
   a. Use a graphing calculator to estimate the month in which the temporary slowdown was the worst.

   b. What percentage drop in sales occurred at the bottom of the slowdown compared to the previous high point in sales?

Thinking Outside the Box XXXIV

Leaning Ladder   A 7-ft ladder is leaning against a vertical wall. There is a point near the bottom of the ladder that is 1 ft from the ground and 1 ft from the wall. Find the exact or approximate distance from the top of the ladder to the ground.

1 ft

1 ft

■ **Figure for Thinking Outside the Box XXXIV**

3.5 Pop Quiz

1. Discuss the symmetry of the graph of $y = x^4 - 3x^2$.

2. Discuss the symmetry of the graph of $y = x^3 - 3x$.

3. Does $f(x) = (x - 4)^3$ cross the x-axis at $(4, 0)$?

4. If $y = x^4 - 3x^3$, does y go to ∞ or $-\infty$ as $x \to \infty$?

5. If $y = -2x^4 + 5x^2$, does y go to ∞ or $-\infty$ as $x \to -\infty$?

6. Solve $(x - 3)^2(x + 1)^3 > 0$.

Linking Concepts

For Individual or Group Explorations

x

10 in.

14 in.

Maximizing Volume

A sheet metal worker is planning to make an open-top box by cutting equal squares (x-in. by x-in.) from the corners of a 10-in. by 14-in. piece of copper. A second box is to be made in the same manner from an 8-in. by 10-in. piece of aluminum, but its height is to be one-half that of the first box.

a) Find polynomial functions for the volume of each box.

b) Find the values of x for which the copper box is 72 in.^{3} larger than the aluminum box.

c) Write the difference between the two volumes d as a function of x and graph it.

d) Find d for $x = 1.5$ in. and $x = 8$ in.

e) For what value of x is the difference between the two volumes the largest?

f) For what value of x is the total volume the largest?

3.6

Rational Functions and Inequalities

In this section we will use our knowledge of polynomial functions to study functions that are ratios of polynomial functions.

Rational Functions and Their Domains

Functions such as

$$y = \frac{1}{x}, \qquad f(x) = \frac{x-3}{x-1}, \qquad \text{and} \qquad g(x) = \frac{2x-3}{x^2-4}$$

are rational functions.

Definition:
Rational Function

If $P(x)$ and $Q(x)$ are polynomials, then a function of the form

$$f(x) = \frac{P(x)}{Q(x)}$$

is called a **rational function,** provided that $Q(x)$ is not the zero polynomial.

To simplify discussions of rational functions we will assume that $f(x)$ is in lowest terms ($P(x)$ and $Q(x)$ have no common factors) unless it is stated otherwise.

The domain of a polynomial function is the set of all real numbers, while the domain of a rational function is restricted to real numbers that do not cause the denominator to have a value of 0. The domain of $y = 1/x$ is the set of all real numbers except 0.

Example **1** **The domain of a rational function**

Find the domain of each rational function.

a. $f(x) = \dfrac{x-3}{x-1}$ **b.** $g(x) = \dfrac{2x-3}{x^2-4}$

Solution

a. Since $x - 1 = 0$ only for $x = 1$, the domain of f is the set of all real numbers except 1. The domain is written in interval notation as $(-\infty, 1) \cup (1, \infty)$.
b. Since $x^2 - 4 = 0$ for $x = \pm 2$, any real number except 2 and -2 can be used for x. So the domain of g is $(-\infty, -2) \cup (-2, 2) \cup (2, \infty)$.

Try This. Find the domain of $f(x) = \dfrac{x-1}{x^2-9}$. ∎

Horizontal and Vertical Asymptotes

The graph of a rational function such as $f(x) = 1/x$ does not look like the graph of a polynomial function. The domain of $f(x) = 1/x$ is the set of all real numbers

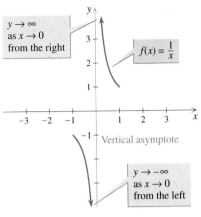

■ **Figure 3.69**

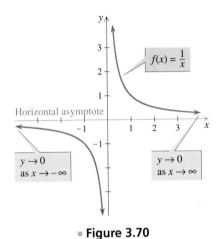

■ **Figure 3.70**

except 0. However, 0 is an important number for the graph of this function because of the behavior of the graph when x is close to 0. The following table shows ordered pairs in which x is close to 0.

	$x \to 0$ from left				$x \to 0$ from right		
x	-0.1	-0.01	-0.001	0	0.001	0.01	0.1
$y = 1/x$	-10	-100	-1000		1000	100	10
		$y \to -\infty$			$y \to \infty$		

Notice that the closer x is to 0, the farther y is from 0. In symbols, $y \to \infty$ as $x \to 0$ from the right and $y \to -\infty$ as $x \to 0$ from the left. Plotting these ordered pairs suggests a curve that gets closer and closer to the vertical line $x = 0$ (the y-axis) but never touches it as shown in Fig. 3.69. The y-axis is called a *vertical asymptote* for this curve.

The following table shows ordered pairs in which x is far from 0.

	$x \to -\infty$				$x \to \infty$		
x	-1000	-100	-10	0	10	100	1000
$y = 1/x$	-0.001	-0.01	-0.1		0.1	0.01	0.001
		$y \to 0$ from below			$y \to 0$ from above		

Notice that the farther x is from 0, the closer y is to 0. These ordered pairs suggest a curve that lies just above the positive x-axis and just below the negative x-axis. The x-axis is called a *horizontal asymptote* for the graph of f. The complete graph of $f(x) = 1/x$ is shown in Fig. 3.70. Since $f(-x) = 1/(-x) = -f(x)$, the graph is symmetric about the origin.

For any rational function expressed in lowest terms, a horizontal asymptote is determined by the value approached by the rational expression as $|x| \to \infty$ ($x \to \infty$ or $x \to -\infty$). A vertical asymptote occurs for every number that causes the denominator of the function to have a value of 0, provided the rational function is in lowest terms. As we will learn shortly, not every rational function has a vertical and a horizontal asymptote. We can give a formal definition of asymptotes as described below.

Definition: Vertical and Horizontal Asymptotes

Let $f(x) = P(x)/Q(x)$ be a rational function written in lowest terms.

If $|f(x)| \to \infty$ as $x \to a$, then the vertical line $x = a$ is a **vertical asymptote.**
The line $y = a$ is a **horizontal asymptote** if $f(x) \to a$ as $x \to \infty$ or $f(x) \to a$ as $x \to -\infty$.

To find a horizontal asymptote we need to approximate the value of a rational expression when x is arbitrarily large. If x is large, then expressions such as

$$\frac{500}{x}, \qquad -\frac{14}{x}, \qquad \frac{3}{x^2}, \qquad \frac{6}{x^3}, \qquad \text{and} \qquad \frac{4}{x - 5},$$

which consist of a fixed number over a polynomial, are approximately zero. To approximate a ratio of two polynomials that both involve x, such as $\frac{x-2}{2x+3}$, we rewrite the expression by dividing by the highest power of x:

$$\frac{x-2}{2x+3} = \frac{\dfrac{x}{x} - \dfrac{2}{x}}{\dfrac{2x}{x} + \dfrac{3}{x}} = \frac{1 - \dfrac{2}{x}}{2 + \dfrac{3}{x}}$$

Since $2/x$ and $3/x$ are approximately zero when x is large, the approximate value of this expression is $1/2$. We use this idea in the next example.

Example 2 Identifying horizontal and vertical asymptotes

Find the horizontal and vertical asymptotes for each rational function.

a. $f(x) = \dfrac{3}{x^2 - 1}$ **b.** $g(x) = \dfrac{x}{x^2 - 4}$ **c.** $h(x) = \dfrac{2x+1}{x+3}$

Solution

a. The denominator $x^2 - 1$ has a value of 0 if $x = \pm 1$. So the lines $x = 1$ and $x = -1$ are vertical asymptotes. As $x \to \infty$ or $x \to -\infty$, $x^2 - 1$ gets larger, making $3/(x^2 - 1)$ the ratio of 3 and a large number. Thus $3/(x^2 - 1) \to 0$ and the x-axis is a horizontal asymptote.

The calculator table shown in Fig. 3.71 supports the conclusion that the x-axis is a horizontal asymptote. □

b. The denominator $x^2 - 4$ has a value of 0 if $x = \pm 2$. So the lines $x = 2$ and $x = -2$ are vertical asymptotes. As $x \to \infty$, $x/(x^2 - 4)$ is a ratio of two large numbers. The approximate value of this ratio is not clear. However, if we divide the numerator and denominator by x^2, the highest power of x, we can get a clearer picture of the value of this ratio:

$$g(x) = \frac{x}{x^2 - 4} = \frac{\dfrac{x}{x^2}}{\dfrac{x^2}{x^2} - \dfrac{4}{x^2}} = \frac{\dfrac{1}{x}}{1 - \dfrac{4}{x^2}}$$

As $|x| \to \infty$, the values of $1/x$ and $4/x^2$ go to 0. So

$$g(x) \to \frac{0}{1-0} = 0.$$

So the x-axis is a horizontal asymptote.

The calculator table shown in Fig. 3.72 supports the conclusion that the x-axis is a horizontal asymptote. □

c. The denominator $x + 3$ has a value of 0 if $x = -3$. So the line $x = -3$ is a vertical asymptote. To find any horizontal asymptotes, rewrite the rational expression by dividing the numerator and denominator by the highest power of x:

$$h(x) = \frac{2x+1}{x+3} = \frac{\dfrac{2x}{x} + \dfrac{1}{x}}{\dfrac{x}{x} + \dfrac{3}{x}} = \frac{2 + \dfrac{1}{x}}{1 + \dfrac{3}{x}}$$

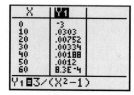

X	Y1
0	-3
10	.0303
20	.00752
30	.00334
40	.00188
50	.0012
60	8.3E-4

Y1◻3/(X²-1)

▪ **Figure 3.71**

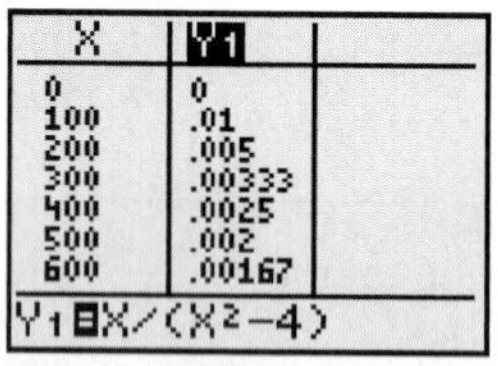

X	Y1
0	0
100	.01
200	.005
300	.00333
400	.0025
500	.002
600	.00167

Y1◻X/(X²-4)

▪ **Figure 3.72**

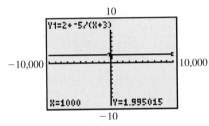

■ **Figure 3.73**

As $x \to \infty$ or $x \to -\infty$ the values of $1/x$ and $3/x$ approach 0. So

$$h(x) \to \frac{2 + 0}{1 + 0} = 2.$$

So the line $y = 2$ is a horizontal asymptote.

The calculator table shown in Fig. 3.73 supports the conclusion that $y = 2$ is a horizontal asymptote.

Try This.   Find all asymptotes for $f(x) = \dfrac{x - 1}{x^2 - 9}$.   ■

If the degree of the numerator of a rational function is less than the degree of the denominator, as in Examples 2(a) and (b), then the x-axis is a horizontal asymptote for the graph of the function. If the degree of the numerator is equal to the degree of the denominator, as in Example 2(c), then the x-axis is not a horizontal asymptote. We can see this clearly if we use division to rewrite the expression as quotient + remainder/divisor. For the function of Example 2(c), we get

$$h(x) = \frac{2x + 1}{x + 3} = 2 + \frac{-5}{x + 3}.$$

The graph of h is a translation two units upward of the graph of $y = -5/(x + 3)$, which has the x-axis as a horizontal asymptote. So $y = 2$ is a horizontal asymptote for h. Note that 2 is simply the ratio of the leading coefficients.

Since the graph of h is very close to $y = 2$, the view of h in Fig. 3.74 looks like $y = 2$. Note how this choice of viewing window causes some of the important features of the graph to disappear.   □

Oblique Asymptotes

Each rational function of Example 2 had one horizontal asymptote and had a vertical asymptote for each zero of the polynomial in the denominator. The horizontal asymptote $y = 0$ occurs because the y-coordinate gets closer and closer to 0 as $x \to \infty$ or $x \to -\infty$. Some rational functions have a nonhorizontal line for an asymptote. An asymptote that is neither horizontal nor vertical is called an **oblique asymptote** or **slant asymptote**.

■ **Figure 3.74**

Example **3**   **A rational function with an oblique asymptote**

Determine all of the asymptotes for $g(x) = \dfrac{2x^2 + 3x - 5}{x + 2}$.

Solution

If $x + 2 = 0$, then $x = -2$. So the line $x = -2$ is a vertical asymptote. Because the degree of the numerator is larger than the degree of the denominator, we use long division (or in this case, synthetic division) to rewrite the function as quotient + remainder/divisor (dividing the numerator and denominator by x^2 will not work in this case):

$$g(x) = \frac{2x^2 + 3x - 5}{x + 2} = 2x - 1 + \frac{-3}{x + 2}$$

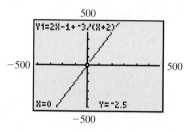

If $|x| \to \infty$, then $-3/(x + 2) \to 0$. So the value of $g(x)$ approaches $2x - 1$ as $|x| \to \infty$. The line $y = 2x - 1$ is an oblique asymptote for the graph of g. You may look ahead to Fig. 3.83 to see the graph of this function with its oblique asymptote. Note that the wide view of g in Fig. 3.75 looks like the line $y = 2x - 1$.

Try This. Find all asymptotes for $f(x) = \dfrac{3x^2 - 4}{x - 1}$.   ■

If the degree of $P(x)$ is 1 greater than the degree of $Q(x)$ and the degree of $Q(x)$ is at least 1, then the rational function has an oblique asymptote. In this case use division to rewrite the function as quotient + remainder/divisor. The graph of the equation formed by setting y equal to the quotient is an oblique asymptote. We conclude this discussion of asymptotes with a summary.

S U M M A R Y   Finding Asymptotes for a Rational Function

Let $f(x) = P(x)/Q(x)$ be a rational function in lowest terms with the degree of $Q(x)$ at least 1.

1. The graph of f has a vertical asymptote corresponding to each root of $Q(x) = 0$.
2. If the degree of $P(x)$ is less than the degree of $Q(x)$, then the x-axis is a horizontal asymptote.
3. If the degree of $P(x)$ equals the degree of $Q(x)$, then the horizontal asymptote is determined by the ratio of the leading coefficients.
4. If the degree of $P(x)$ is greater than the degree of $Q(x)$, then use division to rewrite the function as quotient + remainder/divisor. The graph of the equation formed by setting y equal to the quotient is an asymptote. This asymptote is an oblique or slant asymptote if the degree of $P(x)$ is 1 larger than the degree of $Q(x)$.

Sketching Graphs of Rational Functions

We now use asymptotes and symmetry to help us sketch the graphs of the rational functions discussed in Examples 2 and 3. Use the following steps to graph a rational function.

P R O C E D U R E   Graphing a Rational Function

To graph a rational function in lowest terms:

1. Determine the asymptotes and draw them as dashed lines.
2. Check for symmetry.
3. Find any intercepts.
4. Plot several selected points to determine how the graph approaches the asymptotes.
5. Draw curves through the selected points, approaching the asymptotes.

Example **4** **Functions with horizontal and vertical asymptotes**

Sketch the graph of each rational function.

a. $f(x) = \dfrac{3}{x^2 - 1}$ **b.** $g(x) = \dfrac{x}{x^2 - 4}$ **c.** $h(x) = \dfrac{2x + 1}{x + 3}$

Solution

a. From Example 2(a), $x = 1$ and $x = -1$ are vertical asymptotes, and the x-axis is a horizontal asymptote. Draw the vertical asymptotes using dashed lines. Since all of the powers of x are even, $f(-x) = f(x)$ and the graph is symmetric about the y-axis. The y-intercept is $(0, -3)$. There are no x-intercepts because $f(x) = 0$ has no solution. Evaluate the function at $x = 0.9$ and $x = 1.1$ to see how the curve approaches the asymptote at $x = 1$. Evaluate at $x = 2$ and $x = 3$ to see whether the curve approaches the horizontal asymptote from above or below. We get $(0.9, -15.789)$, $(1.1, 14.286)$, $(2, 1)$, and $(3, 3/8)$. From these points you can see that the curve is going downward toward $x = 1$ from the left and upward toward $x = 1$ from the right. It is approaching its horizontal asymptote from above. Now use the symmetry with respect to the y-axis to draw the curve approaching its asymptotes as shown in Fig. 3.76.

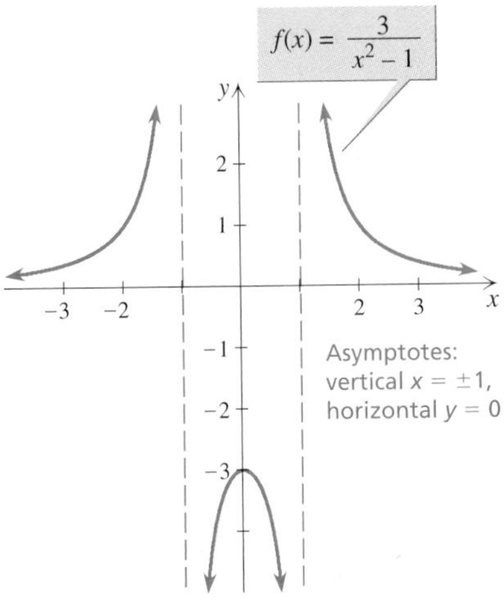

The calculator graph in dot mode in Fig. 3.77 supports these conclusions. □

b. Draw the vertical asymptotes $x = 2$ and $x = -2$ from Example 2(b) as dashed lines. The x-axis is a horizontal asymptote. Because $f(-x) = -f(x)$, the graph is symmetric about the origin. The x-intercept is $(0, 0)$. Evaluate the function near the vertical asymptote $x = 2$, and for larger values of x to see how the curve approaches the horizontal asymptote. We get $(1.9, -4.872)$, $(2.1, 5.122)$, $(3, 3/5)$, and $(4, 1/3)$. From these points you can see that the curve is going downward toward $x = 2$ from the left and upward toward $x = 2$ from the right. It is approaching its horizontal asymptote from above. Now use the symmetry with respect to the origin to draw the curve approaching its asymptotes as shown in Fig. 3.78.

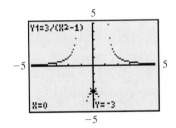

$f(x) = \dfrac{3}{x^2 - 1}$

Asymptotes:
vertical $x = \pm 1$,
horizontal $y = 0$

■ **Figure 3.76**

Y1=3/(X2-1)

X=0   Y=-3

■ **Figure 3.77**

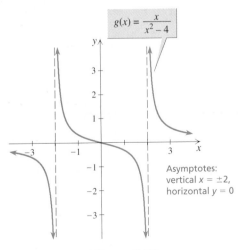

$g(x) = \dfrac{x}{x^2 - 4}$

Asymptotes:
vertical $x = \pm 2$,
horizontal $y = 0$

■ **Figure 3.78**

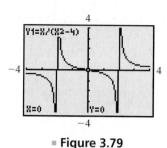

■ **Figure 3.79**

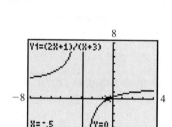

■ **Figure 3.81**

The calculator graph in connected mode in Fig. 3.79 confirms these conclusions. Note how the calculator appears to draw the vertical asymptotes as it connects the points that are close to but on opposite sides of the asymptotes. □

c. Draw the vertical asymptote $x = -3$ and the horizontal asymptote $y = 2$ from Example 2(c) as dashed lines. The x-intercept is $(-1/2, 0)$ and the y-intercept is $(0, 1/3)$. The points $(-2, -3)$, $(7, 1.5)$, $(-4, 7)$, and $(-13, 2.5)$ are also on the graph. From these points we can conclude that the curve goes upward toward $x = -3$ from the left and downward toward $x = -3$ from the right. It approaches $y = 2$ from above as x goes to $-\infty$ and from below as x goes to ∞. Draw the graph approaching its asymptotes as shown in Fig. 3.80.

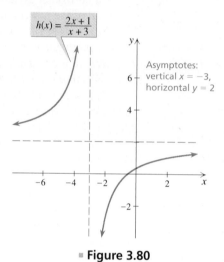

$$h(x) = \frac{2x+1}{x+3}$$

Asymptotes:
vertical $x = -3$,
horizontal $y = 2$

■ **Figure 3.80**

The calculator graph in Fig. 3.81 supports these conclusions.

Try This. Sketch the graph of $f(x) = \dfrac{x-1}{x^2-9}$. ■

The graph of a rational function cannot cross a vertical asymptote, but it can cross a nonvertical asymptote. The graph of a rational function gets closer and closer to a nonvertical asymptote as $|x| \to \infty$, but it can also cross a nonvertical asymptote as illustrated in the next example.

Example 5 A rational function that crosses an asymptote

Sketch the graph of $f(x) = \dfrac{4x}{(x+1)^2}$.

Solution

The graph of f has a vertical asymptote at $x = -1$, and since the degree of the numerator is less than the degree of the denominator, the x-axis is a horizontal asymptote. If $x > 0$ then $f(x) > 0$, and if $x < 0$ then $f(x) < 0$. So the graph approaches the horizontal axis from above for $x > 0$, and from below for $x < 0$. However, the x-intercept is $(0, 0)$. So the graph crosses its horizontal asymptote at $(0, 0)$. The graph shown in Fig. 3.82 goes through $(1, 1)$, $(2, 8/9)$, $(-2, -8)$, and $(-3, -3)$.

Try This. Sketch the graph of $f(x) = \dfrac{x}{(x-2)^2}$. ■

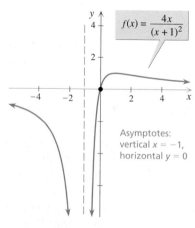

$$f(x) = \frac{4x}{(x+1)^2}$$

Asymptotes:
vertical $x = -1$,
horizontal $y = 0$

■ **Figure 3.82**

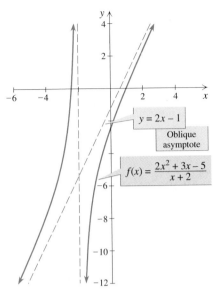

■ **Figure 3.83**

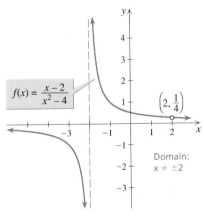

■ **Figure 3.84**

Although the graphing calculator is very helpful for graphing rational functions, the calculator has two serious limitations. The graph on a calculator crosses its vertical asymptotes in connected mode and it usually appears to touch its horizontal asymptotes.

Example **6** **Graphing a function with an oblique asymptote**

Sketch the graph of $k(x) = \dfrac{2x^2 + 3x - 5}{x + 2}$.

Solution

Draw the vertical asymptote $x = -2$ and the oblique asymptote $y = 2x - 1$ determined in Example 3 as dashed lines. The x-intercepts, $(1, 0)$ and $(-2.5, 0)$, are found by solving $2x^2 + 3x - 5 = 0$. The y-intercept is $(0, -2.5)$. The points $(-1, -6)$, $(4, 6.5)$, and $(-3, -4)$ are also on the graph. From these points we can conclude that the curve goes upward toward $x = -2$ from the left and downward toward $x = -2$ from the right. It approaches $y = 2x - 1$ from above as x goes to $-\infty$ and from below as x goes to ∞. Draw the graph approaching its asymptotes as shown in Fig. 3.83.

The calculator graph in Fig. 3.84 supports these conclusions.

Try This. Sketch the graph of $f(x) = \dfrac{3x^2 - 4}{x - 1}$. ■

All rational functions so far have been given in lowest terms. In the next example the numerator and denominator have a common factor. In this case the graph does not have as many vertical asymptotes as you might expect. The graph is almost identical to the graph of the rational function obtained by reducing the expression to lowest terms.

Example **7** **A graph with a hole in it**

Sketch the graph of $f(x) = \dfrac{x - 2}{x^2 - 4}$.

Solution

Since $x^2 - 4 = (x - 2)(x + 2)$, the domain of f is the set of all real numbers except 2 and -2. Since the rational expression can be reduced, the function f could also be defined as

$$f(x) = \frac{1}{x + 2} \qquad \text{for } x \neq 2 \text{ and } x \neq -2.$$

Note that the domain is determined before the function is simplified. The graph of $y = 1/(x + 2)$ has only one vertical asymptote $x = -2$. In fact, the graph of $y = 1/(x + 2)$ is a translation two units to the left of $y = 1/x$. Since $f(2)$ is undefined, the point $(2, 1/4)$ that would normally be on the graph of $y = 1/(x + 2)$ is omitted. The missing point is indicated on the graph in Fig. 3.85 as a small open circle. Note that the graph made by a graphing calculator will usually not show the missing point.

Try This. Sketch the graph of $f(x) = \dfrac{x - 1}{x^2 - 1}$. ■

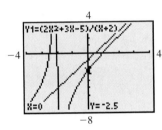

■ **Figure 3.85**

Rational Inequalities

An inequality that involves a rational expression, such as $\frac{x+3}{x-2} \le 2$, is called a **rational inequality.** Our first thought for solving this inequality might be to clear the denominator by multiplying each side by $x - 2$. But, when we multiply each side of an inequality by a real number, we must know whether the real number is positive or negative. Whether $x - 2$ is positive or negative depends on x. So multiplying by $x - 2$ is not a good idea. *We usually do not multiply a rational inequality by an expression that involves a variable.* However, we can solve rational inequalities using sign graphs as we did with quadratic inequalities in Section 3.1.

Example **8** **Solving a rational inequality with a sign graph of the factors**

Solve $\frac{x+3}{x-2} \le 2$. State the solution set using interval notation.

Solution

As with quadratic inequalities, we must have 0 on one side. Note that we do not multiply each side $x - 2$.

$$\frac{x+3}{x-2} \le 2$$

$$\frac{x+3}{x-2} - 2 \le 0$$

$$\frac{x+3}{x-2} - \frac{2(x-2)}{x-2} \le 0 \qquad \text{Get a common denominator.}$$

$$\frac{x+3-2(x-2)}{x-2} \le 0 \qquad \text{Subtract.}$$

$$\frac{7-x}{x-2} \le 0 \qquad \text{Combine like terms.}$$

The sign graph in Fig. 3.86 shows that $7 - x > 0$ if $x < 7$, and $7 - x < 0$ if $x > 7$. If $x > 2$, then $x - 2 > 0$, and if $x < 2$, then $x - 2 < 0$.

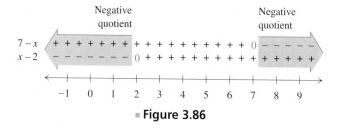

▪ **Figure 3.86**

The inequality $\frac{7-x}{x-2} \le 0$ is satisfied if the quotient of $7 - x$ and $x - 2$ is negative or 0. Fig. 3.86 indicates that $7 - x$ and $x - 2$ have opposite signs and therefore a negative quotient for $x > 7$ or $x < 2$. If $x = 7$ then $7 - x = 0$ and the quotient is 0. If $x = 2$ then $x - 2 = 0$ and the quotient is undefined. So 7 is in the solution set but 2 is not. The solution set is $(-\infty, 2) \cup [7, \infty)$.

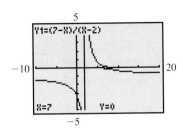

■ **Figure 3.87**

$\dfrac{x + 3}{x^2 - 1}$

■ **Figure 3.88**

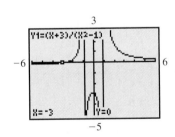

■ **Figure 3.89**

The graph of $y = (7 - x)/(x - 2)$ in Fig. 3.87 supports our conclusion that the inequality is satisfied if $x < 2$ or if $x \geq 7$.

Try This. Solve $\dfrac{1}{x - 1} > 1$ with a sign graph. ■

Making a sign graph of the factors of the numerator and denominator as in Example 8 shows how the signs of the linear factors determine the solution to a rational inequality. Of course that method works only if you can factor the numerator and denominator. The test-point method works on any rational inequality for which we can find all zeros of the numerator and denominator. The test-point method depends on the fact that the graph of a rational function can go from one side of the x-axis to the other only at a vertical asymptote (where the function is undefined) or at an x-intercept (where the value of the function is 0).

Example **9** **Solving a rational inequality with the test-point method**

Solve $\dfrac{x + 3}{x^2 - 1} \geq 0$. State the solution set using interval notation.

Solution

Since the denominator can be factored, the inequality is equivalent to

$$\frac{x + 3}{(x - 1)(x + 1)} \geq 0.$$

The rational expression is undefined if $x = \pm 1$ and has value 0 if $x = -3$. Above each of these numbers on a number line put a 0 or a "U" (for undefined) as shown in Fig. 3.88. Now we select $-4, -2, 0$, and 3 as test points (shown in red in Fig. 3.88). Let $R(x) = \dfrac{x + 3}{x^2 - 1}$ and evaluate $R(x)$ at each test point to determine the sign of $R(x)$ in the interval of the test point:

$$R(-4) = -\frac{1}{15}, \qquad R(-2) = \frac{1}{3}, \qquad R(0) = -3, \qquad R(3) = \frac{3}{4}$$

So $-, +, -$, and $+$ are the signs of the rational expression in the four intervals in Fig. 3.88. The inequality is satisfied whenever the rational expression has a positive or 0 value. Since $R(-1)$ and $R(1)$ are undefined, -1 and 1 are not in the solution set. The solution set is $[-3, -1) \cup (1, \infty)$.

The graph of $y = (x + 3)/(x^2 - 1)$ in Fig. 3.89 supports the conclusion that the inequality is satisfied if $-3 \leq x < -1$ or $x > 1$.

Try This. Solve $\dfrac{x + 3}{x - 1} > 0$ with test points. ■

Applications

Rational functions can occur in many applied situations. A horizontal asymptote might indicate that the average cost of producing a product approaches a fixed value in the long run. A vertical asymptote might show that the cost of a project goes up astronomically as a certain barrier is approached.

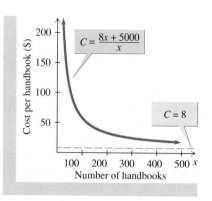

■ **Figure 3.90**

Example **10** Average cost of a handbook

Eco Publishing spent $5000 to produce an environmental handbook and $8 each for printing. Write a function that gives the average cost to the company per printed handbook. Graph the function for $0 < x \le 500$. What happens to the average cost if the book becomes very, very popular?

Solution

The total cost of producing and printing x handbooks is $8x + 5000$. To find the average cost per book, divide the total cost by the number of books:

$$C = \frac{8x + 5000}{x}$$

This rational function has a vertical asymptote at $x = 0$ and a horizontal asymptote $C = 8$. The graph is shown in Fig. 3.90. As x gets larger and larger, the $5000 production cost is spread out over more and more books, and the average cost per book approaches $8, as shown by the horizontal asymptote.

Try This. A truck rents for $100 per day plus $0.80 per mile that it is driven. Write a function that gives the average cost per mile for a one-day rental. What happens to the average cost per mile as the number of miles gets very large? ■

Function Gallery: **Some Basic Rational Functions**

Horizontal Asymptote *x*-axis and Vertical Asymptote *y*-axis

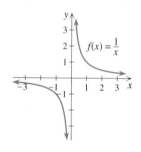

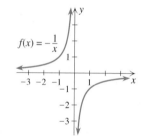

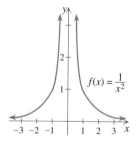

Various Asymptotes

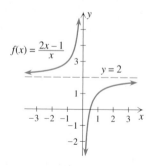

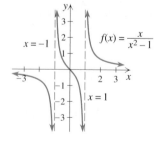

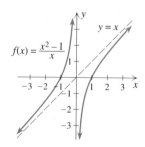

For Thought

True or False? Explain.

1. The function $f(x) = \dfrac{1}{\sqrt{x} - 3}$ is a rational function.

2. The domain of $f(x) = \dfrac{x + 2}{x - 2}$ is

$(-\infty, -2) \cup (-2, 2) \cup (2, \infty)$.

3. The number of vertical asymptotes for a rational function equals the degree of the denominator.

4. The graph of $f(x) = \dfrac{1}{x^4 - 4x^2}$ has four vertical asymptotes.

5. The x-axis is a horizontal asymptote for

$f(x) = \dfrac{x^2 - 2x + 7}{x^3 - 4x}$.

6. The x-axis is a horizontal asymptote for the graph of

$f(x) = \dfrac{5x - 1}{x + 2}$.

7. The line $y = x - 3$ is an asymptote for the graph of

$f(x) = x - 3 + \dfrac{1}{x - 1}$.

8. The graph of a rational function cannot intersect its asymptote.

9. The graph of $f(x) = \dfrac{4}{x^2 - 16}$ is symmetric about the y-axis.

10. The graph of $f(x) = \dfrac{2x + 6}{x^2 - 9}$ has only one vertical asymptote.

3.6 Exercises

Find the domain of each rational function.

1. $f(x) = \dfrac{4}{x + 2}$

2. $f(x) = \dfrac{-1}{x - 2}$

3. $f(x) = \dfrac{-x}{x^2 - 4}$

4. $f(x) = \dfrac{2}{x^2 - x - 2}$

5. $f(x) = \dfrac{2x + 3}{x - 3}$

6. $f(x) = \dfrac{4 - x}{x + 2}$

7. $f(x) = \dfrac{x^2 - 2x + 4}{x}$

8. $f(x) = \dfrac{x^3 + 2}{x^2}$

9. $f(x) = \dfrac{3x^2 - 1}{x^3 - x}$

10. $f(x) = \dfrac{x^2 + 1}{8x^3 - 2x}$

11. $f(x) = \dfrac{-x^2 + x}{x^2 + 5x + 6}$

12. $f(x) = \dfrac{-x^2 + 2x - 3}{x^2 + x - 12}$

Determine the domain and the equations of the asymptotes for the graph of each rational function.

13.

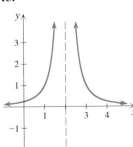

14.

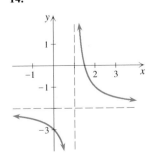

15.

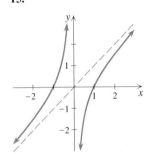

16.

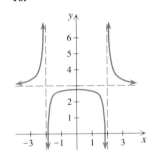

Determine the equations of all asymptotes for the graph of each function. See the summary for finding asymptotes for a rational function on page 334.

17. $f(x) = \dfrac{5}{x - 2}$

18. $f(x) = \dfrac{-1}{x + 12}$

19. $f(x) = \dfrac{-x}{x^2 - 9}$

20. $f(x) = \dfrac{-2}{x^2 - 5x + 6}$

21. $f(x) = \dfrac{2x + 4}{x - 1}$

22. $f(x) = \dfrac{5 - x}{x + 5}$

23. $f(x) = \dfrac{x^2 - 2x + 1}{x}$

24. $f(x) = \dfrac{x^3 - 8}{x^2}$

25. $f(x) = \dfrac{3x^2 + 4}{x + 1}$

26. $f(x) = \dfrac{x^2}{x - 9}$

27. $f(x) = \dfrac{-x^2 + 4x}{x + 2}$

28. $f(x) = \dfrac{-x^2 + 3x - 7}{x - 3}$

Find all asymptotes, x-intercepts, and y-intercepts for the graph of each rational function and sketch the graph of the function. See the procedure for graphing a rational function on page 334.

29. $f(x) = \dfrac{-1}{x}$ **30.** $f(x) = \dfrac{1}{x^2}$

31. $f(x) = \dfrac{1}{x - 2}$ **32.** $f(x) = \dfrac{-1}{x + 1}$

33. $f(x) = \dfrac{1}{x^2 - 4}$ **34.** $f(x) = \dfrac{1}{x^2 - 2x + 1}$

35. $f(x) = \dfrac{-1}{(x + 1)^2}$ **36.** $f(x) = \dfrac{-2}{x^2 - 9}$

37. $f(x) = \dfrac{2x + 1}{x - 1}$ **38.** $f(x) = \dfrac{3x - 1}{x + 1}$

39. $f(x) = \dfrac{x - 3}{x + 2}$ **40.** $f(x) = \dfrac{2 - x}{x + 2}$

41. $f(x) = \dfrac{x}{x^2 - 1}$ **42.** $f(x) = \dfrac{-x}{x^2 - 9}$

43. $f(x) = \dfrac{4x}{x^2 - 2x + 1}$ **44.** $f(x) = \dfrac{-2x}{x^2 + 6x + 9}$

45. $f(x) = \dfrac{8 - x^2}{x^2 - 9}$ **46.** $f(x) = \dfrac{2x^2 + x - 8}{x^2 - 4}$

47. $f(x) = \dfrac{2x^2 + 8x + 2}{x^2 + 2x + 1}$ **48.** $f(x) = \dfrac{-x^2 + 7x - 9}{x^2 - 6x + 9}$

Find the oblique asymptote and sketch the graph of each rational function.

49. $f(x) = \dfrac{x^2 + 1}{x}$ **50.** $f(x) = \dfrac{x^2 - 1}{x}$

51. $f(x) = \dfrac{x^3 - 1}{x^2}$ **52.** $f(x) = \dfrac{x^3 + 1}{x^2}$

53. $f(x) = \dfrac{x^2}{x + 1}$ **54.** $f(x) = \dfrac{x^2}{x - 1}$

55. $f(x) = \dfrac{2x^2 - x}{x - 1}$ **56.** $f(x) = \dfrac{-x^2 + x + 1}{x + 1}$

Match each rational function with its graph (a)–(h), without using a graphing calculator.

57. $f(x) = -\dfrac{3}{x}$ **58.** $f(x) = \dfrac{1}{3 - x}$

59. $f(x) = \dfrac{x}{x - 3}$ **60.** $f(x) = \dfrac{x - 3}{x}$

61. $f(x) = \dfrac{1}{x^2 - 3x}$ **62.** $f(x) = \dfrac{x^2}{x^2 - 9}$

63. $f(x) = \dfrac{x^2 - 3}{x}$ **64.** $f(x) = \dfrac{-x^3 + 1}{x^2}$

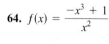

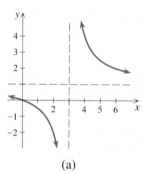

(a)

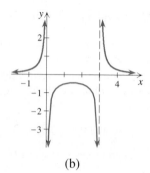

(b)

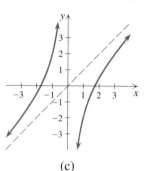

(c)

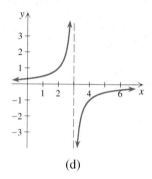

(d)

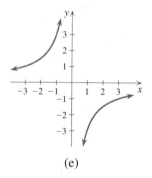

(e)

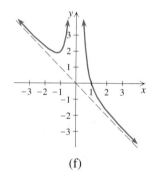

(f)

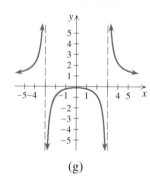

(g)

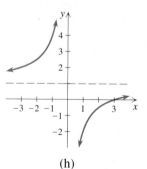

(h)

Sketch the graph of each rational function. Note that the functions are not in lowest terms. Find the domain first.

65. $f(x) = \dfrac{x + 1}{x^2 - 1}$

66. $f(x) = \dfrac{x}{x^2 + 2x}$

67. $f(x) = \dfrac{x^2 - 1}{x - 1}$

68. $f(x) = \dfrac{x^2 - 5x + 6}{x - 2}$

Sketch the graph of each rational function.

69. $f(x) = \dfrac{2}{x^2 + 1}$

70. $f(x) = \dfrac{x}{x^2 + 1}$

71. $f(x) = \dfrac{x - 1}{x^3 - 9x}$

72. $f(x) = \dfrac{x^2 + 1}{x^3 - 4x}$

73. $f(x) = \dfrac{x + 1}{x^2}$

74. $f(x) = \dfrac{x - 1}{x^2}$

Solve with a sign graph of the factors. State the solution set using interval notation.

75. $\dfrac{x - 4}{x + 2} \le 0$

76. $\dfrac{x + 3}{x + 5} \ge 0$

77. $\dfrac{q - 2}{q + 3} < 2$

78. $\dfrac{p + 1}{2p - 1} \ge 1$

79. $\dfrac{w^2 - w - 6}{w - 6} \ge 0$

80. $\dfrac{z - 5}{z^2 + 2z - 8} \le 0$

81. $\dfrac{1}{x + 2} > \dfrac{1}{x - 3}$

82. $\dfrac{1}{x + 1} > \dfrac{2}{x - 1}$

83. $x < \dfrac{3x - 8}{5 - x}$

84. $\dfrac{2}{x + 3} \ge \dfrac{1}{x - 1}$

Solve with the test-point method. State the solution set using interval notation.

85. $\dfrac{(x - 3)(x + 1)}{x - 5} \ge 0$

86. $\dfrac{(x + 2)(x - 1)}{(x + 4)^2} \le 0$

87. $\dfrac{x^2 - 7}{2 - x^2} \le 0$

88. $\dfrac{x^2 + 1}{5 - x^2} \ge 0$

89. $\dfrac{x^2 + 2x + 1}{x^2 - 2x - 15} \ge 0$

90. $\dfrac{x^2 - 2x - 8}{x^2 + 10x + 25} \le 0$

91. $\dfrac{1}{w} > \dfrac{1}{w^2}$

92. $\dfrac{1}{w} > w^2$

93. $w > \dfrac{w - 5}{w - 3}$

94. $w < \dfrac{w - 2}{w + 1}$

Find the equation and sketch the graph of each function.

95. A rational function that passes through $(3, 1)$, has the x-axis as a horizontal asymptote, and has the line $x = 1$ as its only vertical asymptote.

96. A rational function that passes through $(0, 4)$, has the x-axis as a horizontal asymptote, and has the line $x = 3$ as its only vertical asymptote.

97. A rational function that passes through $(0, 5)$, has the line $y = 1$ as a horizontal asymptote, and has the line $x = 2$ as its only vertical asymptote.

98. A rational function that passes through $(-1, 2)$, has the line $y = -2$ as a horizontal asymptote, and has the line $x = -3$ as its only vertical asymptote.

99. A rational function that passes through $(0, 0)$ and $(4, 8/7)$, has the x-axis as a horizontal asymptote, and has two vertical asymptotes $x = 3$ and $x = -3$.

100. A rational function that passes through $(0, 1)$ and $(2, 1)$, has the x-axis as a horizontal asymptote, and has two vertical asymptotes $x = 3$ and $x = -3$.

101. A rational function that passes through $(0, 3)$, has $y = 2x + 1$ as an oblique asymptote, and has $x = 1$ as its only vertical asymptote.

102. A rational function that passes through $(3, 3)$, has $y = x - 2$ as an oblique asymptote, and has $x = -5$ as its only vertical asymptote.

Solve each problem.

103. *Admission to the Zoo*   Winona paid $100 for a lifetime membership to Friends of the Zoo, so that she could gain admittance to the zoo for only $1 per visit. Write Winona's average cost per visit C as a function of the number of visits when she has visited x times. What is her average cost per visit when she has visited the zoo 100 times? Graph the function for $x > 0$. What happens to her average cost per visit if she starts when she is young and visits the zoo every day?

104. *Renting a Car*   The cost of renting a car for one day is $19 plus 30 cents per mile. Write the average cost per mile C as a function of the number of miles driven in one day x. Graph the function for $x > 0$. What happens to C as the number of miles gets very large?

105. *Average Speed of an Auto Trip*   A 200-mi trip by an electric car must be completed in 4 hr. Let x be the number of hours it takes to travel the first half of the distance and write the average speed for the second half of the trip as a function of x. Graph this function for $0 < x < 4$. What is the significance of the vertical asymptote?

    HINT Average speed is distance divided by time.

106. *Billboard Advertising* An Atlanta marketing agency figures that the monthly cost of a billboard advertising campaign depends on the fraction of the market p that the client wishes to reach. For $0 \le p < 1$ the cost in dollars is determined by the formula $C = (4p - 1200)/(p - 1)$. What is the monthly cost for a campaign intended to reach 95% of the market? Graph this function for $0 \le p < 1$. What happens to the cost for a client who wants to reach 100% of the market?

107. *Carbon Monoxide Poisoning* The accompanying graph shows the relationship between carbon monoxide levels and time for which permanent brain damage will occur (*American Sensors, Carbon Monoxide Detector Owner's Manual*).
  a. At 1800 PPM, how long will it take to get permanent brain damage?

  b. What level of carbon monoxide takes 240 minutes to produce permanent brain damage?

  c. What are the horizontal and vertical asymptotes for the curve? Explain what these asymptotes mean in simple terms.

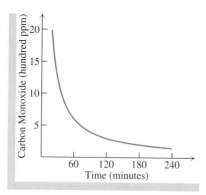

■ **Figure for Exercise 107**

108. *Balancing the Costs* A furniture maker buys foam rubber x times per year. The delivery charge is $400 per purchase regardless of the amount purchased. The annual cost of storage is figured as $10,000/x$, because the more frequent the purchase, the less it costs for storage. So the annual cost of delivery and storage is given by

$$C = 400x + \frac{10,000}{x}.$$

  a. Graph the function with a graphing calculator.

  b. Find the number of purchases per year that minimizes the annual cost of delivery and storage.

109. *Making a Gas Tank* An engineer is designing a cylindrical metal tank that is to hold 500 ft³ of gasoline.
  a. Write the height h as a function of the radius r.
    HINT The volume is given by $V = \pi r^2 h$.

  b. Use the result of part (a) to write the surface area S as a function of r and graph it.
    HINT The surface area is given by $S = 2\pi r^2 + 2\pi rh$.

  c. Use the minimum feature of a graphing calculator to find the radius to the nearest tenth of a foot that minimizes the surface area. Ignore the thickness of the metal.

  d. If the tank costs $8 per square foot to construct, then what is the minimum cost for making the tank?

110. *Making a Glass Tank* An architect for the Aquarium of the Americas is designing a cylindrical fish tank to hold 1000 ft³ of water. The bottom and side of the tank will be 3 in.-thick glass and the tank will have no top.

  a. Write the inside surface area of the tank as a function of the inside radius and graph the function.

  b. Use the minimum feature of a graphing calculator to find the inside radius that minimizes that inside surface area.

  c. If Owens Corning will build the tank for $250 per cubic foot of glass used, then what will be the cost for the minimal tank?

For Writing/Discussion

111. *Approximate Value* Given an arbitrary rational function $R(x)$, explain how you can find the approximate value of $R(x)$ when the absolute value of x is very large.

112. *Cooperative Learning* Each student in your small group should write the equations of the horizontal, vertical, and/or oblique asymptotes of an unknown rational function. Then work as a group to find a rational function that has the specified asymptotes. Is there a rational function with any specified asymptotes?

Thinking Outside the Box XXXV

Filling a Triangle Fiber-optic cables just fit inside a triangular pipe as shown in the figure. The cables have circular cross sections and the cross section of the pipe is an equilateral triangle with sides of length 1. Suppose that there are n cables of the same size in the bottom row, $n - 1$ of that size in the next row, and so on.
a. Write the total cross-sectional area of the cables as a function of n.

b. As n approaches infinity, will the triangular pipe get totally filled with cables? Explain.

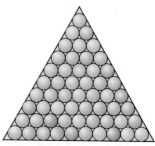

■ **Figure for Thinking Outside the Box XXXV**

3.6 Pop Quiz

1. What is the domain of $f(x) = \frac{x - 1}{x + 4}$?

2. Find the equations of all asymptotes for the graph of

   $y = \frac{x - 5}{x + 2}$.

3. Find the x-intercept and y-intercept for $f(x) = \frac{x^2 - 9}{x^2 - 1}$.

4. What is the horizontal asymptote for $y = \frac{x - 8}{x + 3}$?

5. What is the oblique asymptote for $f(x) = \frac{x^2 - 2x}{x - 3}$?

6. Solve $\frac{x - 1}{x + 4} \geq 0$.

Linking Concepts

For Individual or Group Explorations

Modeling Capital and Operating Cost

To decide when to replace company cars, an accountant looks at two cost components: capital cost and operating cost. The capital cost is the difference between the original cost and the salvage value, and the operating cost is the cost of operating the vehicle.

a) For a new Ford Taurus, the capital cost is $3000 plus $0.12 for each mile that the car is driven. Write the capital cost C as a function of the number of miles the car is driven x.

b) If a Ford Taurus is purchased for $19,846 and driven 80,000 miles, then what is its salvage value?

c) The operating cost P for the Taurus starts at $0.15 per mile when the car is new and increases linearly to $0.25 per mile when the car reaches 100,000 miles. Write P as a function of x.

d) What is the operating cost at 70,000 miles?

e) The total cost per mile is given by $T = \frac{C}{x} + P$. Identify all asymptotes and graph T as a function of x.

f) Explain each asymptote in terms of this application.

g) Find T for x = 20,000, 30,000, and 90,000 miles.

h) The senior vice president has told the accountant that the goal is to keep the total operating cost less than $0.38 per mile. Is this goal practical?

▪▪▪ Highlights

3.1 Quadratic Functions and Inequalities

Quadratic Function
$y = ax^2 + bx + c$ where $a \neq 0$
Graph is a parabola opening upward for $a > 0$
and downward for $a < 0$.

$y = 2x^2 + 8x - 1$
Opens upward

Vertex
$x = -b/(2a)$ is the x-coordinate of the vertex.
Use $y = ax^2 + bx + c$ to find the y-coordinate.

$x = -8/(2 \cdot 2) = -2$
Vertex: $(-2, -9)$

Maximum Minimum
The y-coordinate of the vertex is the max value
of the function if $a < 0$ and the min if $a > 0$.

$y = 2x^2 + 8x - 1$
Min y-value is -9.

Two Forms
By completing the square, $y = ax^2 + bx + c$
can be written as $y = a(x - h)^2 + k$, which
has vertex (h, k).

$y = 2x^2 + 8x - 1$
$y = 2(x + 2)^2 - 9$

Inequalities
To solve $ax^2 + bx + c > 0$ find all roots to
$ax^2 + bx + c = 0$, then test a point in each
interval determined by the roots.

$(x - 2)(x + 3) > 0$
Roots are -3 and 2.
Solution set:
$(-\infty, -3) \cup (2, \infty)$

3.2 Zeros of Polynomial Functions

Remainder Theorem
The remainder when $P(x)$ is divided by
$x - c$ is $P(c)$.

$P(x) = x^2 + 3x - 4$
$P(x)$ divided by $x + 1$ has
remainder $P(-1)$ or -6.

Synthetic Division
An abbreviated version of long division, used
only for dividing a polynomial by $x - c$.

$$-1 \;\big|\; \begin{array}{rrr} 1 & 3 & -4 \\ & -1 & -2 \\ \hline 1 & 2 & -6 \end{array}$$

Dividend: $P(x)$, divisor $x + 1$
quotient $x + 2$, remainder -6

Factor Theorem
c is a zero of $y = P(x)$ if and only if $x - c$
is a factor of $P(x)$.

Since $x - 2$ is a factor of
$P(x) = x^2 + x - 6$, $P(2) = 0$.

Fundamental Theorem of Algebra
If $y = P(x)$ is a polynomial function of positive
degree, then $y = P(x)$ has at least one zero
in the set of complex numbers.

$f(x) = x^7 - x^5 + 3x^2 - 9$
has at least one complex zero
and in fact has seven of them.

Rational Zero Theorem
If p/q is a rational zero in lowest terms for
$y = P(x)$ with integral coefficients, then p is a
factor of the constant term and q is a factor of
the leading coefficient.

$P(x) = 6x^2 + x - 15$
$P(3/2) = 0$
3 is a factor of -15,
2 is a factor of 6

3.3 The Theory of Equations

n-Root Theorem
If $P(x)$ has positive degree and complex
coefficients, then $P(x) = 0$ has n roots
counting multiplicity.

$(x - 2)^3(x^4 - 9) = 0$
has seven complex solutions
counting multiplicity.

Conjugate Pairs Theorem
If $P(x)$ has real coefficients, then the imaginary
roots of $P(x) = 0$ occur in conjugate pairs.

$x^2 - 4x + 5 = 0$
$x = 2 \pm i$

Descartes's Rule of Signs	The changes in sign of $P(x)$ determine the number of positive roots and the changes in sign of $P(-x)$ determine the number of negative roots to $P(x) = 0$.	$P(x) = x^5 - x^2$ has 1 positive root, no negative roots
Theorem on Bounds	Use synthetic division to obtain an upper bound and lower bound for the roots of $P(x) = 0$.	See Example 5 in Section 3.3

3.4 Miscellaneous Equations

Higher Degree	Solve by factoring.	$x^4 - x^2 - 6 = 0$ $(x^2 - 3)(x^2 + 2) = 0$
Squaring Each Side	Possibly extraneous roots	$\sqrt{x - 3} = -2, x - 3 = 4$ $x = 7$ extraneous root
Rational Exponents	Raise each side to a whole number power and then apply the even or odd root property.	$x^{2/3} = 4, x^2 = 64, x = \pm 8$
Quadratic Type	Make a substitution to get a quadratic equation.	$x^{1/3} + x^{1/6} - 12 = 0$ $a^2 + a - 12 = 0$ if $a = x^{1/6}$
Absolute Value	Write equivalent equations without absolute value.	$|x^2 - 4| = 2$ $x^2 - 4 = 2$ or $x^2 - 4 = -2$

3.5 Graphs of Polynomial Functions

Axis of Symmetry	The parabola $y = ax^2 + bx + c$ is symmetric about the line $x = -b/(2a)$.	Axis of symmetry for $y = x^2 - 4x$ is $x = 2$.
Behavior at the x-Intercepts	A polynomial function crosses the x-axis at $(c, 0)$ if $x - c$ occurs with an odd power or touches the x-axis if $x - c$ occurs with an even power.	$f(x) = (x - 3)^2(x + 1)^3$ crosses at $(-1, 0)$, does not cross at $(3, 0)$
End Behavior	The leading coefficient and the degree of the polynomial determine the behavior as $x \to -\infty$ or $x \to \infty$.	$y = x^3$ $y \to \infty$ as $x \to \infty$ $y \to -\infty$ as $x \to -\infty$
Polynomial Inequality	Locate all zeros and then test a point in each interval determined by the zeros.	$x^3 - 4x > 0$ when x is in $(-2, 0) \cup (2, \infty)$

3.6 Rational Functions and Inequalities

Vertical Asymptote	A rational function in lowest terms has a vertical asymptote wherever the denominator is zero.	$y = x/(x^2 - 4)$ has vertical asymptotes $x = -2$ and $x = 2$.
Horizontal Asymptote	If degree of denominator exceeds degree of numerator, then the x-axis is the horizontal asymptote. If degree of numerator equals degree of denominator, then the ratio of leading coefficients is the horizontal asymptote.	$y = x/(x^2 - 4)$ Horizontal asymptote: $y = 0$ $y = (2x - 1)/(3x - 2)$ Horizontal asymptote: $y = 2/3$
Oblique or Slant Asymptote	If degree of numerator exceeds degree of denominator by 1, then use division to determine the slant asymptote.	$y = \dfrac{x^2}{x - 1} = x + 1 + \dfrac{1}{x - 1}$ Slant asymptote: $y = x + 1$
Rational Inequality	Test a point in each interval determined by the zeros and the horizontal asymptotes.	$\dfrac{x - 2}{x + 3} \le 0$ when x is in the interval $(-3, 2]$.

■ ■ ■ Chapter 3 Review Exercises

Solve each problem.

1. Write the function $f(x) = 3x^2 - 2x + 1$ in the form $f(x) = a(x - h)^2 + k$.

2. Write the function $f(x) = -4\left(x - \frac{1}{3}\right)^2 - \frac{1}{2}$ in the form $f(x) = ax^2 + bx + c$.

3. Find the vertex, axis of symmetry, x-intercepts, and y-intercept for the parabola $y = 2x^2 - 4x - 1$.

4. Find the maximum value of the function $y = -x^2 + 3x - 5$.

5. Write the equation of a parabola that has x-intercepts $(-1, 0)$ and $(3, 0)$ and y-intercept $(0, 6)$.

6. Write the equation of the parabola that has vertex $(1, 2)$ and y-intercept $(0, 5)$.

Find all the real and imaginary zeros for each polynomial function.

7. $f(x) = 3x - 1$

8. $g(x) = 7$

9. $h(x) = x^2 - 8$

10. $m(x) = x^3 - 8$

11. $n(x) = 8x^3 - 1$

12. $C(x) = 3x^2 - 2$

13. $P(t) = t^4 - 100$

14. $S(t) = 25t^4 - 1$

15. $R(s) = 8s^3 - 4s^2 - 2s + 1$

16. $W(s) = s^3 + s^2 + s + 1$

17. $f(x) = x^3 + 2x^2 - 6x$

18. $f(x) = 2x^3 - 4x^2 + 3x$

For each polynomial, find the indicated value in two different ways.

19. $P(x) = 4x^3 - 3x^2 + x - 1, P(3)$

20. $P(x) = 2x^3 + 5x^2 - 3x - 2, P(-2)$

21. $P(x) = -8x^5 + 2x^3 - 6x + 2, P\left(-\frac{1}{2}\right)$

22. $P(x) = -4x^4 + 3x^2 + 1, P\left(\frac{1}{2}\right)$

Use the rational zero theorem to list all possible rational zeros for each polynomial function.

23. $f(x) = -3x^3 + 6x^2 + 5x - 2$

24. $f(x) = 2x^4 + 9x^2 - 8x - 3$

25. $f(x) = 6x^4 - x^2 - 9x + 3$

26. $f(x) = 4x^3 - 5x^2 - 13x - 8$

Find a polynomial equation with integral coefficients (and lowest degree) that has the given roots.

27. $-\frac{1}{2}, 3$

28. $\frac{1}{2}, -5$

29. $3 - 2i$

30. $4 + 2i$

31. $2, 1 - 2i$

32. $-3, 3 - 4i$

33. $2 - \sqrt{3}$

34. $1 + \sqrt{2}$

Use Descartes's rule of signs to discuss the possibilities for the roots to each equation. Do not solve the equation.

35. $x^8 + x^6 + 2x^2 = 0$

36. $-x^3 - x - 3 = 0$

37. $4x^3 - 3x^2 + 2x - 9 = 0$

38. $5x^5 + x^3 + 5x = 0$

39. $x^3 + 2x^2 + 2x + 1 = 0$

40. $-x^4 - x^3 + 3x^2 + 5x - 8 = 0$

Establish the best integral bounds for the roots of each equation according to the theorem on bounds.

41. $6x^2 + 5x - 50 = 0$

42. $4x^2 - 12x - 27 = 0$

43. $2x^3 - 15x^2 + 31x - 12 = 0$

44. $x^3 + 6x^2 + 11x + 7 = 0$

45. $12x^3 - 4x^2 - 3x + 1 = 0$

46. $4x^3 - 4x^2 - 9x + 10 = 0$

Find all real and imaginary solutions to each equation, stating multiplicity when it is greater than one.

47. $x^3 - 6x^2 + 11x - 6 = 0$

48. $x^3 + 7x^2 + 16x + 12 = 0$

49. $6x^4 - 5x^3 + 7x^2 - 5x + 1 = 0$

50. $6x^4 + 5x^3 + 25x^2 + 20x + 4 = 0$

51. $x^3 - 9x^2 + 28x - 30 = 0$

52. $x^3 - 4x^2 + 6x - 4 = 0$

53. $x^3 - 4x^2 + 7x - 6 = 0$

54. $2x^3 - 5x^2 + 10x - 4 = 0$

55. $2x^4 - 5x^3 - 2x^2 + 2x = 0$

56. $2x^5 - 15x^4 + 26x^3 - 12x^2 = 0$

Find all real solutions to each equation.

57. $|2v - 1| = 3v$

58. $|2h - 3| = |h|$

59. $x^4 + 7x^2 = 18$

60. $2x^{-2} + 5x^{-1} = 12$

61. $\sqrt{x + 6} - \sqrt{x - 5} = 1$

62. $\sqrt{2x - 1} = \sqrt{x - 1} + 1$

63. $\sqrt{y} + \sqrt[4]{y} = 6$

64. $\sqrt[3]{x^2} + \sqrt[3]{x} = 2$

65. $x^4 - 3x^2 - 4 = 0$

66. $(y - 1)^2 - (y - 1) = 2$

67. $(x - 1)^{2/3} = 4$

68. $(2x - 3)^{-1/2} = \dfrac{1}{2}$

69. $(x + 3)^{-3/4} = -8$

70. $\left(\dfrac{1}{x - 3}\right)^{-1/4} = \dfrac{1}{2}$

71. $\sqrt[3]{3x - 7} = \sqrt[3]{4 - x}$

72. $\sqrt[3]{x + 1} = \sqrt[6]{4x + 9}$

Discuss the symmetry of the graph of each function.

73. $f(x) = 2x^2 - 3x + 9$

74. $f(x) = -3x^2 + 12x - 1$

75. $f(x) = -3x^4 - 2$

76. $f(x) = \dfrac{-x^3}{x^2 - 1}$

77. $f(x) = \dfrac{x}{x^2 + 1}$

78. $f(x) = 2x^4 + 3x^2 + 1$

Find the domain of each rational function.

79. $f(x) = \dfrac{x^2 - 4}{2x + 5}$

80. $f(x) = \dfrac{4x + 1}{x^2 - x - 6}$

81. $f(x) = \dfrac{1}{x^2 + 1}$

82. $f(x) = \dfrac{x - 9}{x^2 - 1}$

Find the x-intercepts, y-intercept, and asymptotes for the graph of each function and sketch the graph.

83. $f(x) = x^2 - x - 2$

84. $f(x) = -2(x - 1)^2 + 6$

85. $f(x) = x^3 - 3x - 2$

86. $f(x) = x^3 - 3x^2 + 4$

87. $f(x) = \dfrac{1}{2}x^3 - \dfrac{1}{2}x^2 - 2x + 2$

88. $f(x) = \dfrac{1}{2}x^3 - 3x^2 + 4x$

89. $f(x) = \dfrac{1}{4}x^4 - 2x^2 + 4$

90. $f(x) = \dfrac{1}{2}x^4 + 2x^3 + 2x^2$

91. $f(x) = \dfrac{2}{x + 3}$

92. $f(x) = \dfrac{1}{2 - x}$

93. $f(x) = \dfrac{2x}{x^2 - 4}$

94. $f(x) = \dfrac{2x^2}{x^2 - 4}$

95. $f(x) = \dfrac{x^2 - 2x + 1}{x - 2}$

96. $f(x) = \dfrac{-x^2 + x + 2}{x - 1}$

97. $f(x) = \dfrac{2x - 1}{2 - x}$

98. $f(x) = \dfrac{1 - x}{x + 1}$

99. $f(x) = \dfrac{x^2 - 4}{x - 2}$

100. $f(x) = \dfrac{x^3 + x}{x}$

Solve each inequality. State the solution set using interval notation.

101. $8x^2 + 1 < 6x$

102. $x^2 + 2x < 63$

103. $(3 - x)(x + 5) \geq 0$

104. $-x^2 - 2x + 15 < 0$

105. $4x^3 - 400x^2 - x + 100 \geq 0$

106. $x^3 - 49x^2 - 52x + 100 < 0$

107. $\dfrac{x + 10}{x + 2} < 5$

108. $\dfrac{x - 6}{2x + 1} \geq 1$

109. $\dfrac{12 - 7x}{x^2} > -1$

110. $x - \dfrac{2}{x} \leq -1$

111. $\dfrac{x^2 - 3x + 2}{x^2 - 7x + 12} \geq 0$

112. $\dfrac{x^2 + 4x + 3}{x^2 - 2x - 15} \leq 0$

Solve each problem.

113. Find the quotient and remainder when $x^3 - 6x^2 + 9x - 15$ is divided by $x - 3$.

114. Find the quotient and remainder when $3x^3 + 4x^2 + 2x - 4$ is divided by $3x - 2$.

115. *Altitude of a Rocket*   If the altitude in feet of a model rocket is given by the equation $S = -16t^2 + 156t$, where t is the time in seconds after ignition, then what is the maximum height attained by the rocket?

116. *Antique Saw*   Willard is making a reproduction of an antique saw. The handle consists of two pieces of wood joined at a right angle, with the blade being the hypotenuse of the right triangle as shown in the figure. If the total length of the handle is to be 36 in., then what length for each piece would minimize the square of the length of the blade?

■ **Figure for Exercise 116**

117. *Bonus Room*   A homeowner wants to put a room in the attic of her house. The house is 48 ft wide and the roof has a 7–12 pitch. (The roof rises 7 ft in a horizontal distance of 12 ft.) Find the dimensions of the room that will maximize the area of the cross section shown in the figure.

■ **Figure for Exercise 117**

118. *Maximizing Area*   An isosceles triangle has one vertex at the origin and a horizontal base below the *x*-axis with its endpoints on the curve $y = x^2 - 16$. See the figure. Let (a, b) be the vertex in the fourth quadrant and write the area of the triangle as a function of *a*. Use a graphing calculator to find the point (a, b) for which the triangle has the maximum possible area.

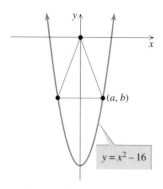

■ **Figure for Exercise 118**

119. *Limiting Velocity*   As a skydiver falls, his velocity keeps increasing. However, because of air resistance, the rate at which the velocity increases keeps decreasing and there is a limit velocity that the skydiver cannot exceed. To see this behavior, graph

$$V = \frac{1000t}{5t + 8},$$

where *V* is the velocity in feet per second and *t* is the time in seconds.

a. What is the velocity at time $t = 10$ sec?

b. What is the horizontal asymptote for this graph?

c. What is the limiting velocity that cannot be exceeded?

■ **Figure for Exercise 119**

Thinking Outside the Box XXXVI & XXXVII

Polynomial Equation   Find a polynomial equation with integral coefficients for which $\sqrt{3} + \sqrt{5}$ is a root.

Bad Arithmetic   Each letter in the following addition problem represents a different digit. Find the digits that will make the addition problem correct. There is more than one possibility.

```
  SEVEN
 +EIGHT
 ------
 TWELVE
```

∎∎∎ **Chapter 3 Test**

Solve each problem.

1. Write $y = 3x^2 - 12x + 1$ in the form $y = a(x - h)^2 + k$.

2. Identify the vertex, axis of symmetry, y-intercept, x-intercepts, and range for $y = 3x^2 - 12x + 1$.

3. What is the minimum value of y in the function $y = 3x^2 - 12x + 1$?

4. Use synthetic division to find the quotient and remainder when $2x^3 - 4x + 5$ is divided by $x + 3$.

5. What is the remainder when
$$x^{98} - 19x^{73} + 17x^{44} - 12x^{23} + 2x^9 - 3$$
is divided by $x - 1$?

6. List the possible rational zeros for the function $f(x) = 3x^3 - 4x^2 + 5x - 6$ according to the rational zero theorem.

7. Find a polynomial equation with real coefficients that has the roots -3 and $4i$.

8. Use Descartes's rule of signs to discuss the possibilities for the roots of $x^3 - 3x^2 + 5x + 7 = 0$.

9. The altitude in feet of a toy rocket t seconds after launch is given by the function $S(t) = -16t^2 + 128t$. Find the maximum altitude reached by the rocket.

Find all real and imaginary zeros for each polynomial function.

10. $f(x) = x^2 - 9$

11. $f(x) = x^4 - 16$

12. $f(x) = x^3 - 4x^2 - x + 10$

Find all real and imaginary roots of each equation. State the multiplicity of a root when it is greater than one.

13. $x^4 + 2x^2 + 1 = 0$

14. $(x^3 - 2x^2)(2x + 3)^3 = 0$

15. $2x^3 - 9x^2 + 14x - 5 = 0$

Sketch the graph of each function.

16. $y = 2(x - 3)^2 + 1$

17. $y = (x - 2)^2(x + 1)$

18. $y = x^3 - 4x$

Find all asymptotes and sketch the graph of each function.

19. $y = \dfrac{1}{x - 2}$

20. $y = \dfrac{2x - 3}{x - 2}$

21. $f(x) = \dfrac{x^2 + 1}{x}$

22. $y = \dfrac{4}{x^2 - 4}$

23. $y = \dfrac{x^2 - 2x + 1}{x - 1}$

Solve each inequality.

24. $x^2 - 2x < 8$

25. $\dfrac{x + 2}{x - 3} > -1$

26. $\dfrac{x + 3}{(4 - x)(x + 1)} \geq 0$

27. $x^3 - 7x > 0$

Find all real solutions to each equation.

28. $(x - 3)^{-2/3} = \dfrac{1}{3}$

29. $\sqrt{x} - \sqrt{x - 7} = 1$

Tying it all Together

Chapters 1–3

Find all real and imaginary solutions to each equation.

1. $2x - 3 = 0$

2. $2x^2 - 3x = 0$

3. $x^3 + 3x^2 + 3x + 1 = 0$

4. $\dfrac{x - 1}{x - 2} = 0$

5. $x^{2/3} - 9 = 0$

Graph each function. State the domain, range, and x-intercepts. Determine the intervals on which the function is increasing or decreasing.

6. $f(x) = 2x - 3$

7. $f(x) = 2x^2 - 3x$

8. $f(x) = x^3 + 3x^2 + 3x + 1$

9. $f(x) = \dfrac{x - 1}{x - 2}$

10. $f(x) = x^{2/3} - 9$

Solve each inequality.

11. $2x - 3 < 0$

12. $2x^2 - 3x < 0$

13. $x^3 + 3x^2 + 3x + 1 > 0$

14. $\dfrac{x - 1}{x - 2} > 0$

15. $x^{2/3} - 9 < 0$

Find the inverse for each function that has an inverse.

16. $f(x) = 2x - 3$

17. $f(x) = 2x^2 - 3x$

18. $f(x) = x^3 + 3x^2 + 3x + 1$

19. $f(x) = \dfrac{x - 1}{x - 2}$

20. $f(x) = x^{2/3} - 9$

4 Exponential and Logarithmic Functions

I N the summer of 1964, paleontologist John Ostrom was exploring the Montana badlands when he came across the fossils of a new dinosaur. He named it *Deinonychus* or "terrible claw" because of the 6-inch sickle-like claw on a toe of each hind foot.

Paleontologists use data gathered from fossils to determine when a dinosaur lived and to determine dinosaur behavior. When alive, 107 million years ago, Deinonychus was about 9 feet long, weighed about 170 pounds, and used its claws to slash its prey.

WHAT YOU WILL LEARN   In this chapter we learn how scientists use exponential functions to date archeological finds that are millions of years old. Through applications ranging from population growth to growth of financial investments, we'll see how exponential functions can help us predict the future and discover the past.

4.1 Exponential Functions and Their Applications

4.2 Logarithmic Functions and Their Applications

4.3 Rules of Logarithms

4.4 More Equations and Applications

Exponential Functions and Their Applications

The functions that involve some combination of basic arithmetic operations, powers, or roots are called **algebraic functions.** Most of the functions studied so far are algebraic functions. In this chapter we turn to the exponential and logarithmic functions. These functions are used to describe phenomena ranging from growth of investments to the decay of radioactive materials, which cannot be described with algebraic functions. Since the exponential and logarithmic functions transcend what can be described with algebraic functions, they are called **transcendental functions.**

The Definition

In algebraic functions such as

$$j(x) = x^2, \qquad p(x) = x^5, \qquad \text{and} \qquad m(x) = x^{1/3}$$

the base is a variable and the exponent is constant. For the exponential functions the base is constant and the exponent is a variable. The functions

$$f(x) = 2^x, \qquad g(x) = 5^x, \qquad \text{and} \qquad h(x) = \left(\frac{1}{3}\right)^x$$

are exponential functions.

Definition:
Exponential Function

> An **exponential function** with **base a** is a function of the form
> $$f(x) = a^x$$
> where a and x are real numbers such that $a > 0$ and $a \neq 1$.

We rule out the base $a = 1$ in the definition because $f(x) = 1^x$ is the constant function $f(x) = 1$. Negative numbers and 0 are not used as bases because powers such as $(-4)^{1/2}$ and 0^{-2} are not real numbers or not defined. Other functions with variable exponents, such as $f(x) = 3^{2x+1}$ or $f(x) = 10 \cdot 7^{-x^2}$, may also be called exponential functions.

To evaluate an exponential function we use our knowledge of exponents. A review of exponents can be found in Section P.3.

Example **1** **Evaluating exponential functions**

Let $f(x) = 4^x, g(x) = 5^{2-x}$, and $h(x) = \left(\frac{1}{3}\right)^x$. Find the following values. Check your answers with a calculator.

a. $f(3/2)$ **b.** $g(3)$ **c.** $h(-2)$

Solution

a. $f(3/2) = 4^{3/2} = \left(\sqrt{4}\right)^3 = 2^3 = 8$ Take the square root, then cube.

b. $g(3) = 5^{2-3} = 5^{-1} = \dfrac{1}{5}$ Exponent -1 means reciprocal.

c. $h(-2) = \left(\dfrac{1}{3}\right)^{-2} = 3^2 = 9$ Find the reciprocal, then square.

⊞ Evaluating these functions with a calculator, as in Fig. 4.1, yields the same results.

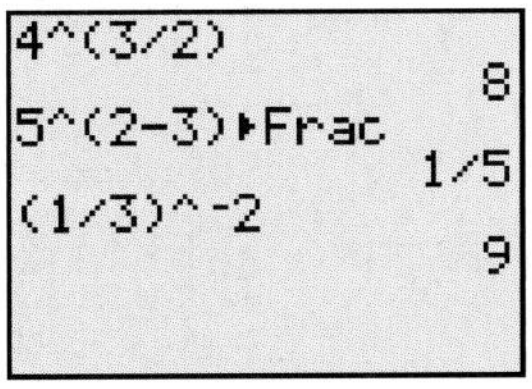

```
4^(3/2)
                    8
5^(2-3)▸Frac
                  1/5
(1/3)^-2
                    9
```

■ **Figure 4.1**

Try This. Find $f(-2)$ and $f(1/2)$ if $f(x) = 9^x$.   ■

Domain of Exponential Functions

The domain of the exponential function $f(x) = 2^x$ is the set of all real numbers. To understand how this function can be evaluated for any *real* number, first note that the function is evaluated for any *rational* number using powers and roots. For example,

$$f(1.7) = f(17/10) = 2^{17/10} = \sqrt[10]{2^{17}} \approx 3.249009585.$$

Until now, we have not considered *irrational* exponents.

Consider the expression $2^{\sqrt{3}}$ and recall that the irrational number $\sqrt{3}$ is an infinite nonterminating nonrepeating decimal number:

$$\sqrt{3} = 1.7320508075 \ldots$$

If we use rational approximations to $\sqrt{3}$ as exponents, we see a pattern:

$$2^{1.7} = 3.249009585 \ldots$$

$$2^{1.73} = 3.317278183 \ldots$$

$$2^{1.732} = 3.321880096 \ldots$$

$$2^{1.73205} = 3.321995226 \ldots$$

$$2^{1.7320508} = 3.321997068 \ldots$$

As the exponents get closer and closer to $\sqrt{3}$ we get results that are approaching some number. We define $2^{\sqrt{3}}$ to be that number. Of course, it is impossible to write the exact value of $\sqrt{3}$ or $2^{\sqrt{3}}$ as a decimal, but you can use a calculator to get $2^{\sqrt{3}} \approx 3.321997085$. Since any irrational number can be approximated by rational numbers in this same manner, 2^x is defined similarly for any irrational number. This idea extends to any exponential function.

Domain of an Exponential Function

> The domain of $f(x) = a^x$ for $a > 0$ and $a \neq 1$ is the set of all real numbers.

Graphing Exponential Functions

Even though the domain of an exponential function is the set of real numbers, for ease of computation, we generally choose only rational numbers for x to find ordered pairs on the graph of the function.

Example **2** **Graphing an exponential function (a > 1)**

Sketch the graph of each exponential function and state the domain and range.

a. $f(x) = 2^x$ **b.** $g(x) = 10^x$

Solution

a. Find some ordered pairs satisfying $f(x) = 2^x$ as follows:

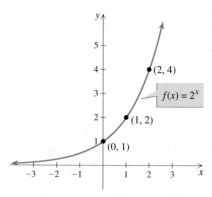

■ **Figure 4.2**

x	-2	-1	0	1	2	3
$y = 2^x$	1/4	1/2	1	2	4	8

y-values increasing ⟶

Plot these points. As x gets larger, so does 2^x. As x approaches $-\infty$, 2^x approaches but never reaches 0. So the x-axis is a horizontal asymptote for the curve. It can be shown that the graph of $f(x) = 2^x$ increases in a continuous manner, with no "jumps" or "breaks" in the graph. This fact allows us to draw a smooth curve through the points as shown in Fig. 4.2. The domain of $f(x) = 2^x$ is $(-\infty, \infty)$. The range is $(0, \infty)$.

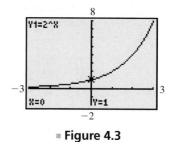

■ **Figure 4.3**

The calculator graph shown in Fig. 4.3 supports these conclusions. Because of the limited resolution of the calculator screen, the calculator graph appears to touch its horizontal asymptote in this window. You can have a mental picture of a curve that gets closer and closer but never touches its asymptote, but it is impossible to accurately draw such a picture. □

b. Find some ordered pairs satisfying $g(x) = 10^x$ as follows:

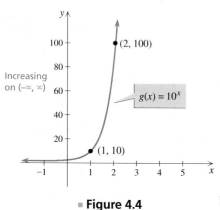

■ **Figure 4.4**

x	-2	-1	0	1	2	3
$y = 10^x$	0.01	0.1	1	10	100	1000

y-values increasing ⟶

These points indicate that the graph of $g(x) = 10^x$ increases in the same manner as the graph of $f(x) = 2^x$. The domain of $g(x) = 10^x$ is $(-\infty, \infty)$. The range is $(0, \infty)$. Plot some points and sketch the curve shown in Fig. 4.4. Make sure that your hand-drawn curve does not touch its asymptote, the x-axis.

The calculator graph in Fig. 4.5 supports these conclusions.

■ **Figure 4.5**

Try This. Graph $f(x) = 9^x$ and state the domain and range.   ■

Note the similarities between the graphs shown in Figs. 4.2 and 4.4. Both pass through the point (0, 1), both functions are increasing, and both have the x-axis as a horizontal asymptote. Since a horizontal line can cross these graphs only once, the functions $f(x) = 2^x$ and $g(x) = 10^x$ are one-to-one by the horizontal line test. All functions of the form $f(x) = a^x$ for $a > 1$ have graphs similar to those in Figs. 4.2 and 4.4.

The phrase "growing exponentially" is often used to describe a population or other quantity whose growth can be modeled with an increasing exponential function. For example, the earth's population is said to be growing exponentially because of the shape of the graph shown in Fig. 4.6.

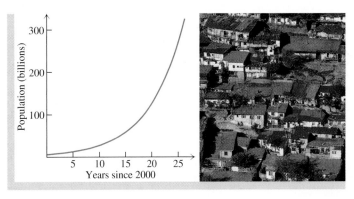

▪ **Figure 4.6**

If $a > 1$, then the graph of $f(x) = a^x$ is increasing. In the next example we graph exponential functions in which $0 < a < 1$ and we will see that these functions are decreasing.

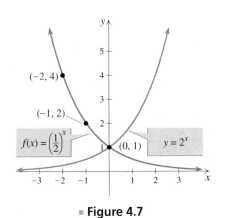

▪ **Figure 4.7**

▪ **Figure 4.8**

Example 3 Graphing an exponential function $(0 < a < 1)$

Sketch the graph of each function and state the domain and range of the function.

a. $f(x) = (1/2)^x$ **b.** $g(x) = 3^{-x}$

Solution

a. Find some ordered pairs satisfying $f(x) = (1/2)^x$ as follows:

x	-2	-1	0	1	2	3
$y = \left(\dfrac{1}{2}\right)^x$	4	2	1	1/2	1/4	1/8

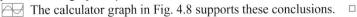
y-values decreasing

Plot these points and draw a smooth curve through them as shown in Fig. 4.7. From the graph in Fig. 4.7 we see that the domain is $(-\infty, \infty)$ and the range is $(0, \infty)$. Because $(1/2)^x = 2^{-x}$ the graph of $f(x) = (1/2)^x$ is a reflection in the y-axis of the graph of $y = 2^x$.

◄◄ The calculator graph in Fig. 4.8 supports these conclusions. □

b. Since $3^{-x} = 1/3^x = (1/3)^x$, this function is of the form $y = a^x$ for $0 < a < 1$. Find some ordered pairs satisfying $g(x) = 3^{-x}$ or $g(x) = (1/3)^x$ as follows:

x	-2	-1	0	1	2	3
$y = 3^{-x}$	9	3	1	$1/3$	$1/9$	$1/27$

y-values decreasing ⟶

Plot these points and draw a smooth curve through them as shown in Fig. 4.9. The domain is $(-\infty, \infty)$ and the range is $(0, \infty)$. The x-axis is again a horizontal asymptote.

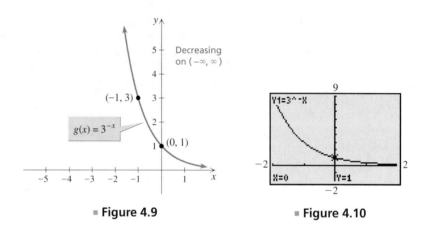

▪ **Figure 4.9**   ▪ **Figure 4.10**

The calculator graph in Fig. 4.10 supports these conclusions.

Try This. Graph $f(x) = 9^{-x}$ and state the domain and range.   ▪

Again note the similarities between the graphs in Figs. 4.7 and 4.9. Both pass through $(0, 1)$, both functions are decreasing, and both have the x-axis as a horizontal asymptote. By the horizontal line test, these functions are also one-to-one. The base determines whether the exponential function is increasing or decreasing. In general, we have the following properties of exponential functions.

Properties of Exponential Functions

The exponential function $f(x) = a^x$ has the following properties:

1. The function f is increasing for $a > 1$ and decreasing for $0 < a < 1$.
2. The y-intercept of the graph of f is $(0, 1)$.
3. The graph has the x-axis as a horizontal asymptote.
4. The domain of f is $(-\infty, \infty)$, and the range of f is $(0, \infty)$.
5. The function f is one-to-one.

The Exponential Family of Functions

Any function of the form $f(x) = a \cdot b^{x-h} + k$ is a member of the **exponential family** of functions. If $h > 0$, then the graph of f is h units to the right of $y = b^x$, and if $h < 0$, the graph of f is $|h|$ units to the left of the graph of $y = b^x$. The graph of $y = b^x$ is translated upward if $k > 0$ or downward if $k < 0$. The graph of $y = b^x$ is stretched if $a > 1$ and shrunk if $0 < a < 1$. A negative a produces a reflection in the x-axis. These ideas were discussed in Section 2.3.

Example **4** **Graphing members of the exponential family**

Sketch the graph of each function and state its domain and range.

a. $y = 2^{x-3}$ **b.** $f(x) = -4 + 3^{x+2}$

Solution

a. The graph of $y = 2^{x-3}$ is a translation three units to the right of the graph of $y = 2^x$. For accuracy, find a few ordered pairs that satisfy $y = 2^{x-3}$:

x	3	4	5
$y = 2^{x-3}$	1	2	4

Draw the graph of $y = 2^{x-3}$ through these points as shown in Fig. 4.11. The domain of $y = 2^{x-3}$ is $(-\infty, \infty)$, and the range is $(0, \infty)$.

The calculator graph in Fig. 4.12 supports these conclusions.   □

b. The graph of $f(x) = -4 + 3^{x+2}$ is obtained by translating $y = 3^x$ to the left two units and downward four units. For accuracy, find a few points on the graph of $f(x) = -4 + 3^{x+2}$:

x	-3	-2	-1	0
$y = -4 + 3^{x+2}$	$-11/3$	-3	-1	5

Sketch a smooth curve through these points as shown in Fig. 4.13. The horizontal asymptote is the line $y = -4$. The domain of f is $(-\infty, \infty)$, and the range is $(-4, \infty)$.

The calculator graph of $y = -4 + 3^{x+2}$ and the asymptote $y = -4$ in Fig. 4.14 supports these conclusions.

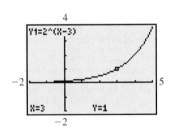

■ **Figure 4.11**

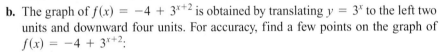

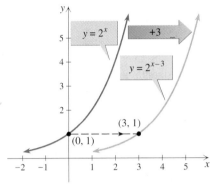

■ **Figure 4.12**

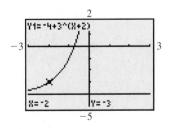

■ **Figure 4.14**

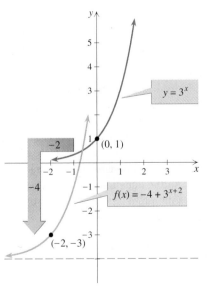

■ **Figure 4.13**

Try This. Graph $f(x) = 2^{x-2} + 1$ and state the domain and range.   ■

asymptote

If you recognize that a function is a transformation of a simpler function, then you know what its graph looks like. This information along with a few ordered pairs will help you to make accurate graphs.

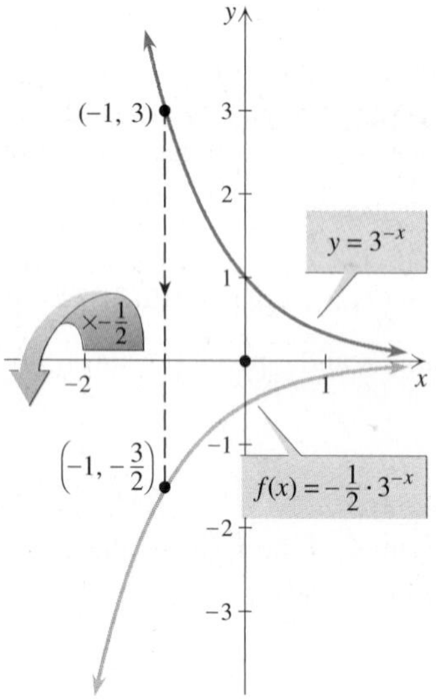

Example **5** Graphing an exponential function with a reflection

Sketch the graph of $f(x) = -\frac{1}{2} \cdot 3^{-x}$ and state the domain and range.

Solution

The graph of $y = 3^{-x}$ is shown in Fig. 4.9. Multiplying by $-1/2$ shrinks the graph and reflects it below the x-axis. The x-axis is the horizontal asymptote for the graph of f. Find a few ordered pairs as follows:

x	-2	-1	0	1	2
$y = -\dfrac{1}{2} \cdot 3^{-x}$	$-9/2$	$-3/2$	$-1/2$	$-1/6$	$-1/18$

y-values increasing →

Sketch the curve through these points as in Fig. 4.15. The domain is $(-\infty, \infty)$ and the range is $(-\infty, 0)$.

Try This. Graph $f(x) = -4^{-x}$ and state the domain and range. ■

The form $y = ab^x$ is often used when we must write a function from some given information. For example, if an initial order of 30 pencils is doubled every day, then the number ordered after x days is given by $y = 30 \cdot 2^x$. If the order is increased by 50% every day then $y = 30 \cdot 1.5^x$.

Exponential Equations

From the graphs of exponential functions, we observed that they are one-to-one. For an exponential function, one-to-one means that if two exponential expressions with the same base are equal, then the exponents are equal. For example, if $2^m = 2^n$, then $m = n$.

One-to-One Property of Exponential Functions

For $a > 0$ and $a \neq 1$,

$$\text{if} \quad a^{x_1} = a^{x_2}, \quad \text{then} \quad x_1 = x_2.$$

The one-to-one property is used in solving simple exponential equations. For example, to solve $2^x = 8$, we recall that $8 = 2^3$. So the equation becomes $2^x = 2^3$. By the one-to-one property, $x = 3$ is the only solution. The one-to-one property applies only to equations in which each side is a power of the same base.

Example **6** Solving exponential equations

Solve each exponential equation.

a. $4^x = \dfrac{1}{4}$ **b.** $\left(\dfrac{1}{10}\right)^x = 100$

$(-1, 3)$

$y = 3^{-x}$

$\times -\frac{1}{2}$

$\left(-1, -\frac{3}{2}\right)$

$f(x) = -\frac{1}{2} \cdot 3^{-x}$

■ **Figure 4.15**

Solution

a. Since $1/4 = 4^{-1}$, we can write the right-hand side as a power of 4 and use the one-to-one property:

$$4^x = \frac{1}{4} = 4^{-1} \qquad \text{Both sides are powers of 4.}$$

$$x = -1 \qquad \text{One-to-one property}$$

b. Since $(1/10)^x = 10^{-x}$ and $100 = 10^2$, we can write each side as a power of 10:

$$\left(\frac{1}{10}\right)^x = 100 \qquad \text{Original equation}$$

$$10^{-x} = 10^2 \qquad \text{Both sides are powers of 10.}$$

$$-x = 2 \qquad \text{One-to-one property}$$

$$x = -2$$

Try This. Solve $2^{-x} = \frac{1}{8}$. $X = 3$ ▪

The type of equation that we solved in Example 6 arises naturally when we try to find the first coordinate of an ordered pair of an exponential function when given the second coordinate, as shown in the next example.

Example **7** **Finding the first coordinate given the second**

Let $f(x) = 5^{2-x}$. Find x such that $f(x) = 125$.

Solution

To find x such that $f(x) = 125$, we must solve $5^{2-x} = 125$:

$$5^{2-x} = 5^3 \qquad \text{Since } 125 = 5^3$$

$$2 - x = 3 \qquad \text{One-to-one property}$$

$$x = -1$$

Try This. Let $f(x) = 3^{4-x}$. Find x such that $f(x) = \frac{1}{9}$. $x = 2$ ▪

The Compound Interest Model

Exponential functions are used to model phenomena such as population growth, radioactive decay, and compound interest. Here we will show how these functions are used to determine the amount of an investment earning compound interest.

If P dollars are deposited in an account with a simple annual interest rate r for t years, then the amount A in the account at the end of t years is found by using the formula $A = P + Prt$ or $A = P(1 + rt)$. If $1000 is deposited at 6% annual rate for one quarter of a year, then at the end of three months the amount is

$$A = 1000\left(1 + 0.06 \cdot \frac{1}{4}\right) = 1000(1.015) = \$1015.$$

If the account begins the next quarter with $1015, then at the end of the second quarter we again multiply by 1.015 to get the amount

$$A = 1000(1.015)^2 \approx \$1030.23.$$

This process is referred to as compound interest because interest is put back into the account and the interest also earns interest. At the end of 20 years, or 80 quarters, the amount is

$$A = 1000(1.015)^{80} \approx \$3290.66.$$

Compound interest can be thought of as simple interest computed over and over. The general **compound interest formula** follows.

Compound Interest Formula

> If a principal P is invested for t years at an annual rate r compounded n times per year, then the amount A, or ending balance, is given by
>
> $$A = P\left(1 + \frac{r}{n}\right)^{nt}.$$

The principal P is also called **present value** and the amount A is called **future value**.

Example 8 Using the compound interest formula

Find the amount or future value when a principal of $20,000 is invested at 6% compounded daily for three years.

Solution

Use $P = \$20,000$, $r = 0.06$, $n = 365$, and $t = 3$ in the compound interest formula:

$$A = \$20,000\left(1 + \frac{0.06}{365}\right)^{365 \cdot 3} \approx \$23,943.99$$

Figure 4.16 shows this expression on a graphing calculator.

Try This. Find the amount when $9000 is invested at 5.4% compounded monthly for 6 years. ■

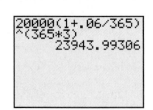

```
20000(1+.06/365)
^(365*3)
        23943.99306
```

■ **Figure 4.16**

When interest is compounded daily, financial institutions usually use the exact number of days. To keep our discussion of interest simple, we assume that all years have 365 days and ignore leap years. When discussing months we assume that all months have 30 days. There may be other rules involved in computing interest that are specific to the institution doing the computing. One popular method is to compound quarterly and only give interest on money that is on deposit for full quarters. With this rule you could have money on deposit for nearly six months (say January 5 to June 28) and receive no interest.

Continuous Compounding and the Number *e*

The more often that interest is figured during the year, the more interest an investment will earn. The first five lines of Table 4.1 show the future value of $10,000 invested at 12% for one year for more and more frequent compounding. The last line shows the limiting amount $11,274.97, which we cannot exceed no matter how often we compound the interest on this investment for one year.

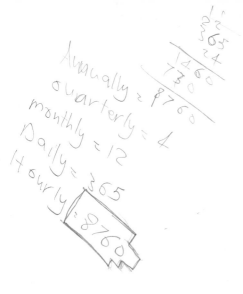

■ **Table 4.1**

Compounding	Future value in one year
Annually	$\$10,000\left(1 + \dfrac{0.12}{1}\right)^{1} = \$11,200$
Quarterly	$\$10,000\left(1 + \dfrac{0.12}{4}\right)^{4} \approx \$11,255.09$
Monthly	$\$10,000\left(1 + \dfrac{0.12}{12}\right)^{12} \approx \$11,268.25$
Daily	$\$10,000\left(1 + \dfrac{0.12}{365}\right)^{365} \approx \$11,274.75$
Hourly	$\$10,000\left(1 + \dfrac{0.12}{8760}\right)^{8760} \approx \$11,274.96$
Continuously	$\$10,000e^{0.12(1)} \approx \$11,274.97$

To better understand the last line of Table 4.1, we need to use a fact from calculus. The expression $[1 + (r/n)]^{nt}$ used in calculating the first five values in the table can be shown to approach e^{rt} as $n \to \infty$. The number e (like π) is an irrational number that occurs in many areas of mathematics. The factor e^{rt} is used in the last line of the table to find the limit to the values of the investment obtained by more and more frequent compounding. Using e^{rt} to find the future value is called **continuous compounding.**

You can find approximate values for powers of e using a calculator with an e^{x} key. To find an approximate value for e itself, find e^{1} on a calculator:

$$e \approx 2.718281828459$$

```
10000e^(.12*1)
      11274.96852
e^(1)
      2.718281828
```

■ **Figure 4.17**

Figure 4.17 shows the graphing calculator computation of the future value of $\$10,000$ at 12% compounded continuously for one year and the value of e^{1}.   □

Continuous Compounding Formula

If a principal P is invested for t years at an annual rate r compounded continuously, then the amount A, or ending balance, is given by

$$A = P \cdot e^{rt}.$$

Example **9** Interest compounded continuously

Find the amount when a principal of $\$5600$ is invested at $6\frac{1}{4}\%$ annual rate compounded continuously for 5 years and 9 months.

Solution

Convert 5 years and 9 months to 5.75 years. Use $r = 0.0625$, $t = 5.75$, and $P = \$5600$ in the continuous compounding formula:

$$A = 5600 \cdot e^{(0.0625)(5.75)} \approx \$8021.63$$

Try This.   Find the amount when $\$8000$ is invested at 6.3% compounded continuously for 7 years and 3 months.   ■

The function $f(x) = e^{x}$ is called the **base-e exponential function.** Variations of this function are used to model many types of growth and decay. The graph of

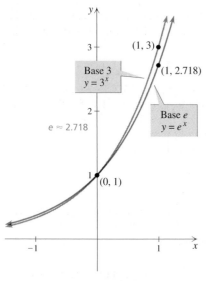

■ **Figure 4.18**

$y = e^x$ looks like the graph of $y = 3^x$, because the value of e is close to 3. Use your calculator to check that the graph of $y = e^x$ shown in Fig. 4.18 goes approximately through the points $(-1, 0.368)$, $(0, 1)$, $(1, 2.718)$, and $(2, 7.389)$.

The Radioactive Decay Model

The mathematical model of radioactive decay is based on the formula

$$A = A_0 e^{rt},$$

which gives the amount A of a radioactive substance remaining after t years, where A_0 is the initial amount present and r is the annual rate of decay. The only difference between this formula and the continuous compounding formula is that when decay is involved, the rate r is negative.

Example 10 Radioactive decay of carbon-14

Finely chipped spear points shaped by nomadic hunters in the Ice Age were found at the scene of a large kill in Lubbock, Texas. From the amount of decay of carbon-14 in the bone fragments of the giant bison, scientists determined that the kill took place about 9833 years ago (*National Geographic*, December 1955). Carbon-14 is a radioactive substance that is absorbed by an organism while it is alive but begins to decay upon the death of the organism. The number of grams of carbon-14 remaining in a fragment of charred bone from the giant bison after t years is given by the formula $A = 3.6e^{rt}$, where $r = -1.21 \times 10^{-4}$. Find the amount present initially and after 9833 years.

Solution

To find the initial amount, let $t = 0$ and $r = -1.21 \times 10^{-4}$ in the formula:

$$A = 3.6e^{(-1.21 \times 10^{-4})(0)} = 3.6 \text{ grams}$$

To find the amount present after 9833 years, let $t = 9833$ and $r = -1.21 \times 10^{-4}$:

$$A = 3.6e^{(-1.21 \times 10^{-4})(9833)}$$

$$\approx 1.1 \text{ grams}$$

The initial amount of carbon-14 was 3.6 grams, and after 9833 years, approximately 1.1 grams of carbon-14 remained. See Fig. 4.19.

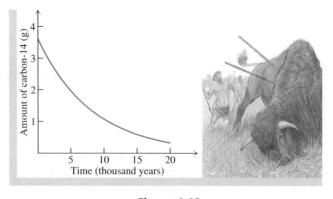

■ **Figure 4.19**

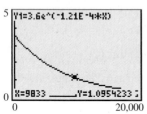

■ **Figure 4.20**

The calculator graph in Fig. 4.20 can be used to find the amount of radioactive substance present after 9833 years.

Try This. If the amount (in grams) of carbon-14 present after t years is given by $A = 2.9e^{rt}$, where $r = -1.21 \times 10^{-4}$, then what is the initial amount and what amount is present after 6500 years? ■

The formula for radioactive decay is used to determine the age of ancient objects such as bones, campfire charcoal, and meteorites. To date objects, we must solve the radioactive decay formula for t, which will be done in Section 4.4 after we have studied logarithms.

For Thought

True or False? Explain.

1. The function $f(x) = (-2)^x$ is an exponential function.

2. The function $f(x) = 2^x$ is invertible.

3. If $2^x = \frac{1}{8}$, then $x = -3$.

4. If $f(x) = 3^x$, then $f(0.5) = \sqrt{3}$.

5. If $f(x) = e^x$, then $f(0) = 1$.

6. If $f(x) = e^x$ and $f(t) = e^2$, then $t = 2$.

7. The x-axis is a horizontal asymptote for the graph of $y = e^x$.

8. The function $f(x) = (0.5)^x$ is increasing.

9. The functions $f(x) = 4^{x-1}$ and $g(x) = (0.25)^{1-x}$ have the same graph.

10. $2^{1.73} = \sqrt[100]{2^{173}}$

4.1 Exercises

Evaluate each exponential expression without using a calculator.

1. 3^3 **2.** 2^5 **3.** -2^0 **4.** -4^0

5. 2^{-3} **6.** 3^{-2} **7.** $\left(\frac{1}{2}\right)^{-4}$ **8.** $\left(\frac{1}{3}\right)^{-2}$

9. $8^{2/3}$ **10.** $9^{3/2}$ **11.** $-9^{-3/2}$ **12.** $-4^{-3/2}$

Let $f(x) = 3^x$, $g(x) = 2^{1-x}$, and $h(x) = (1/4)^x$. Find the following values.

13. $f(2)$ **14.** $f(4)$ **15.** $f(-2)$

16. $f(-3)$ **17.** $g(2)$ **18.** $g(1)$

19. $g(-2)$ **20.** $g(-3)$ **21.** $h(-1)$

22. $h(-2)$ **23.** $h(-1/2)$ **24.** $h(3/2)$

Sketch the graph of each function by finding at least three ordered pairs on the graph. State the domain, the range, and whether the function is increasing or decreasing.

25. $f(x) = 5^x$ **26.** $f(x) = 4^x$ **27.** $y = 10^{-x}$

28. $y = e^{-x}$ **29.** $f(x) = (1/4)^x$ **30.** $f(x) = (0.2)^x$

Use transformations to help you graph each function. Identify the domain, range, and horizontal asymptote. Determine whether the function is increasing or decreasing.

31. $f(x) = 2^x - 3$ **32.** $f(x) = 3^{-x} + 1$

33. $f(x) = 2^{x+3} - 5$ **34.** $f(x) = 3^{1-x} - 4$

35. $y = -2^{-x}$ **36.** $y = -10^{-x}$

37. $y = 1 - 2^x$ **38.** $y = -1 - 2^{-x}$

39. $f(x) = 0.5 \cdot 3^{x-2}$ **40.** $f(x) = -0.1 \cdot 5^{x+4}$

41. $y = 500(0.5)^x$ **42.** $y = 100 \cdot 2^x$

Write the equation of each graph in its final position.

43. The graph of $y = 2^x$ is translated five units to the right and then two units downward.

44. The graph of $y = e^x$ is translated three units to the left and then one unit upward.

45. The graph of $y = (1/4)^x$ is translated one unit to the right, reflected in the x-axis, and then translated two units downward.

46. The graph of $y = 10^x$ is translated three units upward, two units to the left, and then reflected in the x-axis.

Solve each equation.

47. $2^x = 64$ **48.** $5^x = 1$ **49.** $10^x = 0.1$

50. $10^{2x} = 1000$ **51.** $-3^x = -27$ **52.** $-2^x = -\dfrac{1}{2}$

53. $3^{-x} = 9$ **54.** $2^x = \dfrac{1}{8}$ **55.** $8^x = 2$ **56.** $9^x = 3$

57. $e^x = \dfrac{1}{e^2}$ **58.** $e^{-x} = \dfrac{1}{e}$ **59.** $\left(\dfrac{1}{2}\right)^x = 8$

60. $\left(\dfrac{2}{3}\right)^x = \dfrac{9}{4}$ **61.** $10^{x-1} = 0.01$ **62.** $10^{|x|} = 1000$

Let $f(x) = 2^x$, $g(x) = (1/3)^x$, $h(x) = 10^x$, and $m(x) = e^x$. Find the value of x in each equation.

63. $f(x) = 4$ **64.** $f(x) = 32$ **65.** $f(x) = \dfrac{1}{2}$

66. $f(x) = 1$ **67.** $g(x) = 1$ **68.** $g(x) = 9$

69. $g(x) = 27$ **70.** $g(x) = \dfrac{1}{9}$ **71.** $h(x) = 1000$

72. $h(x) = 10^5$ **73.** $h(x) = 0.1$ **74.** $h(x) = 0.0001$

75. $m(x) = e$ **76.** $m(x) = e^3$

77. $m(x) = \dfrac{1}{e}$ **78.** $m(x) = 1$

Fill in the missing coordinate in each ordered pair so that the pair is a solution to the given equation.

79. $y = 3^x$ $(2, \ ), (\ , 3), (-1, \ ), (\ , 1/9)$

80. $y = 10^x$ $(3, \ ), (\ , 1), (-2, \ ), (\ , 0.01)$

81. $f(x) = 5^{-x}$ $(0, \ ), (\ , 25), (-1, \ ), (\ , 1/5)$

82. $f(x) = e^{-x}$ $(1, \ ), (\ , e), (0, \ ), (\ , e^2)$

83. $f(x) = -2^x$ $(4, \ ), (\ , -1/4), (-1, \ ), (\ , -32)$

84. $f(x) = -(1/4)^{x-1}$ $(3, \ ), (\ , -4), (-1, \ ), (\ , -1/16)$

Solve each problem. When needed, use 365 days per year and 30 days per month.

85. *Periodic Compounding*   A deposit of $5000 earns 8% annual interest. Find the amount in the account at the end of 6 years and the amount of interest earned during the 6 years if the interest is compounded
 a. annually

 b. quarterly

 c. monthly

 d. daily.

86. *Periodic Compounding*   Melinda invests her $80,000 winnings from Publishers Clearing House at a 9% annual percentage rate. Find the amount of the investment at the end of 20 years and the amount of interest earned during the 20 years if the interest is compounded
 a. annually

 b. quarterly

 c. monthly

 d. daily.

87. *Compounded Continuously*   The Lakewood Savings Bank pays 8% annual interest compounded continuously. How much will a deposit of $5000 amount to for each time period?
 HINT Convert months to days.
 a. 6 years

 b. 8 years 3 months

 c. 5 years 4 months 22 days

 d. 20 years 321 days

88. *Compounding Continuously*   The Commercial Federal Credit Union pays $6\frac{3}{4}$% annual interest compounded continuously. How much will a deposit of $9000 amount to for each time period?
 HINT Convert months to days.
 a. 13 years

 b. 12 years 8 months

 c. 10 years 6 months 14 days

 d. 40 years 66 days

89. *Present Value Compounding Daily*   A credit union pays 6.5% annual interest compounded daily. What deposit today (present value) would amount to $3000 in 5 years and 4 months?

   HINT Solve the compound interest formula for *P*.

90. *Higher Yields*   Federal Savings and Loan offers $6\frac{3}{4}$% annual interest compounded daily on certificates of deposit. What dollar amount deposited now would amount to $40,000 in 3 years and 2 months?

91. *Present Value Compounding Continuously*   Peoples Bank offers 5.42% compounded continuously on CDs. What amount invested now would grow to $20,000 in 30 years.

   HINT Solve the continuous compounding formula for *P*.

92. *Saving for Retirement*   An investor wants to have a retirement nest egg of $100,000 and estimates that her investment now will grow at 3% compounded continuously for 40 years. What amount should she invest now to achieve this goal?

93. *Working by the Hour*   One million dollars is deposited in an account paying 6% compounded continuously.
   a. What amount of interest will it earn in its first hour on deposit?
      HINT One hour is 1/8760 of a year.

   b. What amount of interest will it earn during its 500th hour on deposit?
      HINT Subtract the amount for $t = 499$ from the amount for $t = 500$.

94. *Big Debt*   The national debt is approximately $8 trillion. If Uncle Sam paid 5% annual interest compounded continuously on that debt, then what amount of interest would he pay for one day? How much could he save if he paid 5% annual interest compounded daily?

95. *Radioactive Decay*   The number of grams of a certain radioactive substance present at time *t* is given by the formula $A = 200e^{-0.001t}$, where *t* is the number of years. How many grams are present at time $t = 0$? How many grams are present at time $t = 500$?

96. *Population Growth*   The function $P = 2.4e^{0.03t}$ models the size of the population of a small country, where *P* is in millions of people in the year $2000 + t$.
   a. What was the population in 2000?

   b. Use the formula to estimate the population to the nearest tenth of a million in 2020.

97. *Cell Phone Subscribers*   The number of cell phone subscribers (in millions) in the U.S. is given in the following table (www.census.gov).

Year	Cell Phone Subscribers (millions)
1990	5.3
1995	22.1
2000	109.5
2001	128.4
2002	130.8
2003	158.7

   a. Use the exponential regression feature of a graphing calculator to find an equation of the form $y = ab^x$ that fits the data, where *x* is years since 1990.

   b. Judging from the graph of the data and the curve, does the exponential model look like a good model?

   c. Use the equation to estimate the number of cell phone subscribers in 2006.

98. *Newspaper Circulation*   The number of daily newspapers sold (in millions) in the U.S. is given in the following table (www.census.gov).

Year	Daily Paper Circulation (millions)
1990	62.3
1995	58.1
2000	55.8
2001	55.6
2002	55.2
2003	55.2

   a. Use the exponential regression feature of a graphing calculator to find an equation of the form $y = ab^x$ that fits the data, where *x* is years since 1990.

   b. Judging from the graph of the data and the curve, does the exponential model look like a good model?

   c. Use the equation to estimate the daily newspaper circulation in 2006.

99. *Position of a Football*   The football is on the 10-yard line. Several penalties in a row are given, and each penalty moves the ball half the distance to the closer goal line. Write a formula that gives the ball's position *P* after *n*th such penalty.
   HINT First write some ordered pairs starting with (1, 5).

100. *Cost of a Parking Ticket*   The cost of a parking ticket on campus is $15 for the first offense. Given that the cost doubles for each additional offense, write a formula for the cost *C* as a function of the number of tickets *n*.
   HINT First write some ordered pairs starting with (1, 15).

101. *Challenger Disaster*   Using data on O-ring damage from 24 previous space shuttle launches, Professor Edward R. Tufte of Yale University concluded that the number of O-rings damaged per launch is an exponential function of the temperature at the time of the launch. If NASA had used a model such as $n = 644e^{-0.15t}$, where t is the Fahrenheit temperature at the time of launch and n is the number of O-rings damaged, then the tragic end to the flight of the space shuttle Challenger might have been avoided. Using this model, find the number of O-rings that would be expected to fail at 31°F, the temperature at the time of the Challenger launch on January 28, 1986.

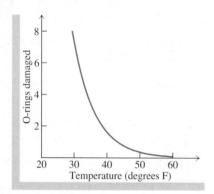

▪ **Figure for Exercise 101**

102. *Manufacturing Cost*   The cost in dollars for manufacturing x units of a certain drug is given by the function $C(x) = xe^{0.001x}$. Find the cost for manufacturing 500 units. Find the function $AC(x)$ that gives the average cost per unit for manufacturing x units. What happens to the average cost per unit as x gets larger and larger?

For Writing/Discussion

103. *Cooperative Learning*   Work in a small group to write a summary (including drawings) of the types of graphs that can be obtained for exponential functions of the form $y = a^x$ for $a > 0$ and $a \neq 1$.

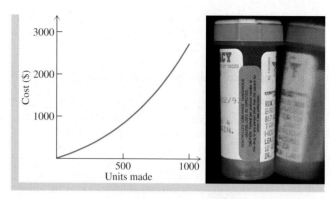

▪ **Figure for Exercise 102**

104. *Group Toss*   Have all students in your class stand and toss a coin. Those that get heads sit down. Those that are left standing toss again and all who obtain heads must sit down. Repeat until no one is left standing. For each toss record the number of coins that are tossed. For example, (1, 30), (2, 14), and so on. Do not include a pair with zero coins tossed. Enter the data into a graphing calculator and use exponential regression to find an equation of the form $y = a \cdot b^x$ that fits the data. What percent of those standing did you expect would sit down after each toss? Is the value of b close to this number?

Thinking Outside the Box XXXVIII & XXXIX

Swim Meet   Two swimmers start out from opposite sides of a pool. Each swims the length of the pool and back at a constant rate. They pass each other for the first time 40 feet from one side of the pool and for the second time 45 feet from the other side of the pool. What is the length of the pool?

One-seventh to one   Six matches are arranged to form the fraction one-seventh as shown here:

$$\frac{\text{I}}{\text{VII}}$$

Move one match, but not the fraction bar, to make a fraction that is equal to one.

4.1 Pop Quiz

1. What is $f(4)$ if $f(x) = -2^x$?

2. Is $f(x) = 3^{-x}$ increasing or decreasing?

3. Find the domain and range for $y = e^{x-1} + 2$.

4. What is the horizontal asymptote for $y = 2^x - 1$?

5. Solve $(1/4)^x = 64$.

6. What is a, if $f(a) = 8$ and $f(x) = 2^{-x}$?

7. If $1000 earns 4% annual interest compounded quarterly, then what is the amount after 20 years?

8. Find the amount in the last problem if the interest is compounded continuously.

Linking Concepts

For Individual or
Group Explorations

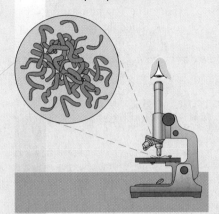

Comparing Exponential and Linear Models

The function $f(t) = 300e^{0.5t}$ gives the number of bacteria present in a culture t hours after the start of an experiment in which the bacteria are growing continuously at a rate of 50% per hour. Let $A[a, b]$ represent the average rate of change of f on the time interval $[a, b]$.

a) Fill in the following table.

Interval [a, b]	f(a)	A[a, b]	A[a, b]/f(a)
[3.00, 3.05]			
[7.50, 7.51]			
[8.623, 8.624]			

b) What are the units for the quantity $A[a, b]$?

c) What can you conjecture about the ratio $A[a, b]/f(a)$?

d) Test your conjecture on a few more intervals of various lengths. Explain how the length of the interval affects the ratio.

e) State your conjecture in terms of variation (Section 2.6). What is the constant of proportionality?

f) Suppose the bacteria were growing in a linear manner, say $f(t) = 800t + 300$. Make a table like the given table and make a conjecture about the ratio $A[a, b]/f(a)$.

g) Explain how your conclusions about average rate of change can be used to justify an exponential model as better than a linear model for modeling growth of a bacteria (or human) population.

4.2

Logarithmic Functions and Their Applications

Since exponential functions are one-to-one functions (Section 4.1), they are invertible. In this section we will study the inverses of the exponential functions.

The Definition

The inverses of the exponential functions are called **logarithmic functions.** Since f^{-1} is a general name for an inverse function, we adopt a more descriptive notation for these inverses. If $f(x) = a^x$, then instead of $f^{-1}(x)$, we write $\log_a(x)$ for the inverse of the base-*a* exponential function. We read $\log_a(x)$ as "log of *x* with base *a*," and we call the expression $\log_a(x)$ a **logarithm.**

The meaning of $\log_a(x)$ will be clearer if we consider the exponential function

$$f(x) = 2^x$$

as an example. Since $f(3) = 2^3 = 8$, the base-2 exponential function pairs the exponent 3 with the value of the exponential expression 8. Since the function $\log_2(x)$ reverses that pairing, we have $\log_2(8) = 3$. So $\log_2(8)$ is the exponent that is used on the base 2 to obtain 8. *In general, $\log_a(x)$ is the exponent that is used on the base a to obtain the value x.*

Definition:
Logarithmic Function

> For $a > 0$ and $a \neq 1$, the **logarithmic function with base *a*** is denoted $f(x) = \log_a(x)$, where
>
> $$y = \log_a(x) \qquad \text{if and only if} \qquad a^y = x.$$

Example **1** **Evaluating logarithmic functions**

Find the indicated values of the logarithmic functions.

a. $\log_3(9)$ **b.** $\log_2\left(\dfrac{1}{4}\right)$ **c.** $\log_{1/2}(8)$

Solution

a. By the definition of logarithm, $\log_3(9)$ is the exponent that is used on the base 3 to obtain 9. Since $3^2 = 9$, we have $\log_3(9) = 2$.

b. Since $\log_2(1/4)$ is the exponent that is used on the base 2 to obtain $1/4$, we try various powers of 2 until we find $2^{-2} = 1/4$. So $\log_2(1/4) = -2$.

 You can use the graph of $y = 2^x$ as in Fig. 4.21 to check that $2^{-2} = 1/4$. □

c. Since $\log_{1/2}(8)$ is the exponent that is used on a base of $1/2$ to obtain 8, we try various powers of $1/2$ until we find $(1/2)^{-3} = 8$. So $\log_{1/2}(8) = -3$.

Try This. Find $\log_4(1/16)$. ⌐⊃ ■

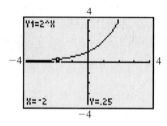

■ **Figure 4.21**

Since the exponential function $f(x) = a^x$ has domain $(-\infty, \infty)$ and range $(0, \infty)$, the logarithmic function $f(x) = \log_a(x)$ has domain $(0, \infty)$ and range $(-\infty, \infty)$. So there are no logarithms of negative numbers or zero. Expressions such as $\log_2(-4)$ and $\log_3(0)$ are undefined. Note that $\log_a(1) = 0$ for any base *a*, because $a^0 = 1$ for any base *a*.

There are two bases that are used more frequently than the others; they are 10 and *e*. The notation $\log_{10}(x)$ is abbreviated $\log(x)$, and $\log_e(x)$ is abbreviated $\ln(x)$. Most scientific calculators have function keys for the exponential functions 10^x and e^x and their inverses $\log(x)$ and $\ln(x)$, which are called **common logarithms** and **natural logarithms,** respectively. Natural logarithms are also called **Napierian logarithms** after John Napier (1550–1617).

 Note that $\log(76)$ is approximately 1.8808 and $10^{1.8808}$ is approximately 76 as shown in Fig. 4.22. If you use more digits for $\log(76)$ as the power of 10, then the calculator gets closer to 76. □

■ **Figure 4.22**

You can use a calculator to find common or natural logarithms, but you should know how to find the values of logarithms such as those in Examples 1 and 2 without using a calculator.

Example **2** **Evaluating logarithmic functions**

Find the indicated values of the logarithmic functions without a calculator. Use a calculator to check.

a. $\log(1000)$ **b.** $\ln(1)$ **c.** $\ln(-6)$

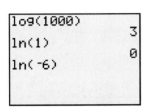

■ **Figure 4.23**

Solution

a. To find $\log(1000)$, we must find the exponent that is used on the base 10 to obtain 1000. Since $10^3 = 1000$, we have $\log(1000) = 3$.
b. Since $e^0 = 1$, $\ln(1) = 0$.
c. The expression $\ln(-6)$ is undefined because -6 is not in the domain of the natural logarithm function. There is no power of e that results in -6.

The calculator results for parts (a) and (b) are shown in Fig. 4.23. If you ask a calculator for $\ln(-6)$, it will give you an error message.

Try This. Find $\log(100)$. ■

Graphs of Logarithmic Functions

The functions of $y = a^x$ and $y = \log_a(x)$ for $a > 0$ and $a \neq 1$ are inverse functions. So the graph of $y = \log_a(x)$ is a reflection about the line $y = x$ of the graph of $y = a^x$. The graph of $y = a^x$ has the x-axis as its horizontal asymptote, while the graph of $y = \log_a(x)$ has the y-axis as its vertical asymptote.

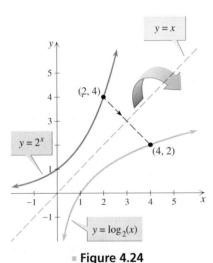

■ **Figure 4.24**

Example **3** **Graph of a base-a logarithmic function with $a > 1$**

Sketch the graphs of $y = 2^x$ and $y = \log_2(x)$ on the same coordinate system. State the domain and range of each function.

Solution

Since these two functions are inverses of each other, the graph $y = \log_2(x)$ is a reflection of the graph of $y = 2^x$ about the line $y = x$. Make a table of ordered pairs for each function:

x	-1	0	1	2
$y = 2^x$	$1/2$	1	2	4

x	$1/2$	1	2	4
$y = \log_2(x)$	-1	0	1	2

Sketch $y = 2^x$ as in Section 4.1. Then sketch $y = \log_2(x)$ through the points given in the table, keeping in mind that it is a reflection of $y = 2^x$. Both curves are shown in Fig. 4.24 along with the line $y = x$. The domain of $y = 2^x$ is $(-\infty, \infty)$, and its range is $(0, \infty)$. The domain of $y = \log_2(x)$ is $(0, \infty)$, and its range is $(-\infty, \infty)$.

Try This. Graph $f(x) = \log_6(x)$ and state the domain and range. ■

Note that the function $y = \log_2(x)$ graphed in Fig. 4.24 is one-to-one by the horizontal line test and it is an increasing function. The graphs of the common logarithm function $y = \log(x)$ and the natural logarithm function $y = \ln(x)$ are shown in Fig. 4.25 and they are also increasing functions. The function $y = \log_a(x)$ is

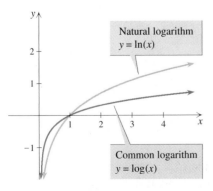

■ **Figure 4.25**

increasing if $a > 1$. By contrast, if $0 < a < 1$, the function $y = \log_a(x)$ is decreasing, as illustrated in the next example.

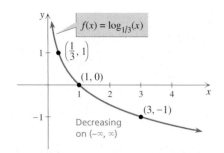

Example **4** Graph of a base-*a* logarithmic function with $0 < a < 1$

Sketch the graph of $f(x) = \log_{1/3}(x)$ and state its domain and range.

Solution

Make a table of ordered pairs for the function:

x	1/9	1/3	1	3	9
$y = \log_{1/3}(x)$	2	1	0	−1	−2

▪ **Figure 4.26**

Sketch the curve through these points as shown in Fig. 4.26. The y-axis is a vertical asymptote for the curve. The domain of $f(x) = \log_{1/3}(x)$ is $(0, \infty)$, and its range is $(-\infty, \infty)$.

Try This. Graph $f(x) = \log_{1/6}(x)$ and state the domain and range. ▪

The logarithmic functions have properties corresponding to the properties of the exponential functions stated in Section 4.1.

Properties of Logarithmic Functions

The logarithmic function $f(x) = \log_a(x)$ has the following properties:

1. The function f is increasing for $a > 1$ and decreasing for $0 < a < 1$.
2. The x-intercept of the graph of f is $(1, 0)$.
3. The graph has the y-axis as a vertical asymptote.
4. The domain of f is $(0, \infty)$, and the range of f is $(-\infty, \infty)$.
5. The function f is one-to-one.

The Logarithmic Family of Functions

Any function of the form $f(x) = a \cdot \log_b(x - h) + k$ is a member of the **logarithmic family** of functions. If $h > 0$, then the graph of f is h units to the right of $y = \log_b(x)$, and if $h < 0$, the graph of f is $|h|$ units to the left of the graph of $y = \log_b(x)$. The graph of $y = \log_b(x)$ is translated upward if $k > 0$ or downward if $k < 0$. The graph of $y = \log_b(x)$ is stretched if $a > 1$ and shrunk if $0 < a < 1$. A negative a produces a reflection in the x-axis.

Example **5** Graphing members of the logarithmic family

Sketch the graph of each function and state its domain and range.

a. $y = \log_2(x - 1)$ **b.** $f(x) = -\dfrac{1}{2}\log_2(x + 3)$

Solution

a. The graph of $y = \log_2(x - 1)$ is obtained by translating the graph of $y = \log_2(x)$ to the right one unit. Since the domain of $y = \log_2(x)$ is $(0, \infty)$, the domain of $y = \log_2(x - 1)$ is $(1, \infty)$. The line $x = 1$ is the vertical asymptote. Calculate a few ordered pairs to get an accurate graph.

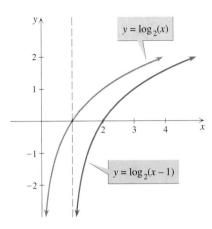

■ **Figure 4.27**

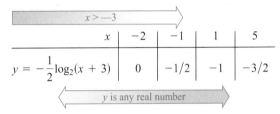

Sketch the curve as shown in Fig. 4.27. The range is $(-\infty, \infty)$.

b. The graph of f is obtained by translating the graph of $y = \log_2(x)$ to the left three units. The vertical asymptote for f is the line $x = -3$. Multiplication by $-1/2$ shrinks the graph and reflects it in the x-axis. Calculate a few ordered pairs for accuracy.

	$x > -3$			
x	-2	-1	1	5
$y = -\dfrac{1}{2}\log_2(x + 3)$	0	$-1/2$	-1	$-3/2$
	y is any real number			

Sketch the curve through these points as shown in Fig. 4.28. The domain of f is $(-3, \infty)$, and the range is $(-\infty, \infty)$.

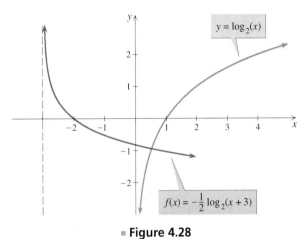

■ **Figure 4.28**

Try This. Graph $f(x) = -\log_2(x + 1)$ and state the domain and range. ■

Note that logarithmic functions involving common or natural logarithms can be graphed with a graphing calculator because the calculator has keys for log and ln. To graph logarithmic functions involving other bases, we need the base-change formula that is discussed in Section 4.3. ☐

Logarithmic and Exponential Equations

Some equations involving logarithms can be solved by writing an equivalent exponential equation, and some equations involving exponents can be solved by writing an equivalent logarithmic equation. Rewriting logarithmic and exponential equations is possible because of the definition of logarithms:

$$y = \log_a(x) \qquad \text{if and only if} \qquad a^y = x$$

Example **6** **Rewriting logarithmic and exponential equations**

Write each equation involving logarithms as an equivalent exponential equation, and write each equation involving exponents as an equivalent logarithmic equation.

a. $\log_5(625) = 4$ **b.** $\log_3(n) = 5$ **c.** $3^x = 50$ **d.** $e^{x-1} = 9$

Solution

a. $\log_5(625) = 4$ is equivalent to $5^4 = 625$.
b. $\log_3(n) = 5$ is equivalent to $3^5 = n$.
c. $3^x = 50$ is equivalent to $x = \log_3(50)$.
d. $e^{x-1} = 9$ is equivalent to $x - 1 = \ln(9)$.

Try This. Write $\log_4(w) = 7$ as an exponential equation. ■

In the next example we use the definition of logarithms to find the inverse of an exponential function.

Example **7** **The inverse of an exponential function**

Find f^{-1} for $f(x) = \frac{1}{2} \cdot 6^{x-3}$.

Solution

Using the switch-and-solve method, we switch x and y then solve for y:

$$x = \frac{1}{2} \cdot 6^{y-3} \qquad \text{Switch } x \text{ and } y \text{ in } y = \frac{1}{2} \cdot 6^{x-3}.$$

$$2x = 6^{y-3} \qquad \text{Multiply each side by 2.}$$

$$y - 3 = \log_6(2x) \qquad \text{Definition of logarithm}$$

$$y = \log_6(2x) + 3 \qquad \text{Add 3 to each side.}$$

So $f^{-1}(x) = \log_6(2x) + 3$.

Try This. Find f^{-1} for $f(x) = 3 \cdot e^{5x-1}$. ■

The one-to-one property of exponential functions was used to solve exponential equations in Section 4.1. Likewise, the one-to-one property of logarithmic functions is used in solving logarithmic equations. The one-to-one property says that *if two quantities have the same base-a logarithm, then the quantities are equal.* For example, if $\log_2(m) = \log_2(n)$, then $m = n$.

One-to-One Property of Logarithms

For $a > 0$ and $a \neq 1$,

if $\log_a(x_1) = \log_a(x_2)$, then $x_1 = x_2$.

The one-to-one properties and the definition of logarithm are at present the only new tools that we have for solving equations. Later in this chapter we will develop more properties of logarithms and solve more complicated equations.

Example **8** Solving equations involving logarithms

Solve each equation.

a. $\log_3(x) = -2$ **b.** $\log_x(5) = 2$ **c.** $5^x = 9$ **d.** $\ln(x^2) = \ln(3x)$

Solution

a. Use the definition of logarithm to write the equivalent exponential equation.

$$\log_3(x) = -2 \qquad \text{Original equation}$$

$$x = 3^{-2} \qquad \text{Definition of logarithm}$$

$$= \frac{1}{9}$$

Since $\log_3(1/9) = -2$ is correct, the solution to the equation is $1/9$.

b. $\log_x(5) = 2$ Original equation

$$x^2 = 5 \qquad \text{Definition of logarithm}$$

$$x = \pm\sqrt{5}$$

Since the base of a logarithm is always nonnegative, the only solution is $\sqrt{5}$.

c. $5^x = 9$ Original equation

$$x = \log_5(9) \qquad \text{Definition of logarithm}$$

The exact solution to $5^x = 9$ is the irrational number $\log_5(9)$. In the next section we will learn how to find a rational approximation for $\log_5(9)$.

d. $\ln(x^2) = \ln(3x)$ Original equation

$$x^2 = 3x \qquad \text{One-to-one property of logarithms}$$

$$x^2 - 3x = 0$$

$$x(x - 3) = 0$$

$$x = 0 \quad \text{or} \quad x = 3$$

Checking $x = 0$ in the original equation, we get the undefined expression $\ln(0)$. So the only solution to the equation is 3.

Try This. Solve $\log_5(2x) = -3$. $1/_2 50$ ■

Example **9** Equations involving common
and natural logarithms

Use a calculator to find the value of x rounded to four decimal places.

a. $10^x = 50$ **b.** $2e^{0.5x} = 6$

Solution

a. $10^x = 50$ Original equation

$$x = \log(50) \qquad \text{Definition of logarithm}$$

$$x \approx 1.6990$$

```
log(50)
        1.698970004
10^1.6990
        50.0034535
10^(log(50))
                 50
```

▪ **Figure 4.29**

Figure 4.29 shows how to find the approximate value of $\log(50)$ and how to check the approximate answer and exact answer. ☐

b. $2e^{0.5x} = 6$ Original equation

$e^{0.5x} = 3$ Divide by 2 to get the form $a^x = y$.

$0.5x = \ln(3)$ Definition of logarithm

$x = \dfrac{\ln(3)}{0.5}$

$x \approx 2.1972$

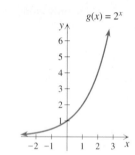

■ **Figure 4.30**

Figure 4.30 shows how to find the approximate value of x and how to check it in the original equation.

Try This. Solve $10^{3x} = 70$.   ■

Function Gallery: Exponential and Logarithmic Functions

Exponential: $f(x) = a^x$, domain $(-\infty, \infty)$, range $(0, \infty)$

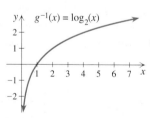

$g(x) = 2^x$

Increasing on $(-\infty, \infty)$
y-intercept $(0, 1)$

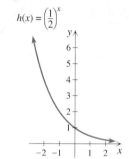

$h(x) = \left(\dfrac{1}{2}\right)^x$

Decreasing on $(-\infty, \infty)$
y-intercept $(0, 1)$

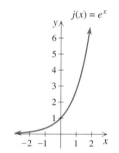

$j(x) = e^x$

Increasing on $(-\infty, \infty)$
y-intercept $(0, 1)$

Logarithmic: $f^{-1}(x) = \log_a(x)$, domain $(0, \infty)$, range $(-\infty, \infty)$

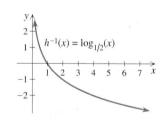

$g^{-1}(x) = \log_2(x)$

Increasing on $(0, \infty)$
x-intercept $(1, 0)$

$h^{-1}(x) = \log_{1/2}(x)$

Decreasing on $(0, \infty)$
x-intercept $(1, 0)$

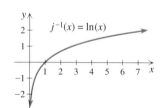

$j^{-1}(x) = \ln(x)$

Increasing on $(0, \infty)$
x-intercept $(1, 0)$

Applications

We saw in Section 4.1 that if a principal of P dollars earns interest at an annual rate r compounded continuously, then the amount after t years is given by

$$A = Pe^{rt}.$$

If either the rate or the time is the only unknown in this formula, then the definition of logarithms can be used to solve for the rate or the time.

Example **10** **Finding the time in a continuous compounding problem**

If $8000 is invested at 9% compounded continuously, then how long will it take for the investment to grow to $20,000?

Solution

Use the formula $A = Pe^{rt}$ with $A = \$20,000$, $P = \$8000$, and $r = 0.09$:

$$20{,}000 = 8000e^{0.09t}$$

$$2.5 = e^{0.09t} \qquad \text{Divide by 8000 to get the form } y = a^x.$$

$$0.09t = \ln(2.5) \qquad \begin{array}{l}\text{Definition of logarithm: } y = a^x \text{ if}\\ \text{and only if } x = \log_a(y).\end{array}$$

$$t = \frac{\ln(2.5)}{0.09}$$

$$\approx 10.181 \text{ years}$$

We can multiply 365 by 0.181 to get approximately 66 days. So the investment grows to $20,000 in approximately 10 years and 66 days.

You can use a graphing calculator to check as shown in Fig. 4.31.

```
ln(2.5)/.09
            10.18100813
8000e^(.09*Ans)
                  20000
```

▪ **Figure 4.31**

Try This. How long does it take for $4 to grow to $12 at 5% compounded continuously?   ▪

The formula $A = Pe^{rt}$ was introduced to model the continuous growth of money. However, this type of formula is used in a wide variety of applications. In the following exercises you will find problems involving population growth, declining forests, and global warming.

For Thought

True or False? Explain.

1. The first coordinate of an ordered pair in an exponential function is a logarithm.

2. $\log_{100}(10) = 2$

3. If $f(x) = \log_3(x)$, then $f^{-1}(x) = 3^x$.

4. $10^{\log(1000)} = 1000$

5. The domain of $f(x) = \ln(x)$ is $(-\infty, \infty)$.

6. $\ln(e^{2.451}) = 2.451$

7. For any positive real number x, $e^{\ln(x)} = x$.

8. For any base a, where $a > 0$ and $a \neq 1$, $\log_a(0) = 1$.

9. $\log(10^3) + \log(10^5) = \log(10^8)$

10. $\log_2(32) - \log_2(8) = \log_2(4)$

4.2 Exercises

Determine the number that can be used in place of the question mark to make the equation true.

1. $2^? = 64$ **2.** $2^? = 16$ **3.** $3^? = \dfrac{1}{81}$

4. $3^? = 1$ **5.** $16^? = 2$ **6.** $16^? = 16$

7. $\left(\dfrac{1}{5}\right)^? = 125$ **8.** $\left(\dfrac{1}{5}\right)^? = \dfrac{1}{125}$

Find the indicated value of the logarithmic functions.

9. $\log_2(64)$ **10.** $\log_2(16)$ **11.** $\log_3(1/81)$

12. $\log_3(1)$ **13.** $\log_{16}(2)$ **14.** $\log_{16}(16)$

15. $\log_{1/5}(125)$ **16.** $\log_{1/5}(1/125)$ **17.** $\log(0.1)$

18. $\log(10^6)$ **19.** $\log(1)$ **20.** $\log(10)$

21. $\ln(e)$ **22.** $\ln(0)$ **23.** $\ln(e^{-5})$ **24.** $\ln(e^9)$

Sketch the graph of each function, and state the domain and range of each function.

25. $y = \log_3(x)$ **26.** $y = \log_4(x)$

27. $f(x) = \log_5(x)$ **28.** $g(x) = \log_8(x)$

29. $y = \log_{1/2}(x)$ **30.** $y = \log_{1/4}(x)$

31. $h(x) = \log_{1/5}(x)$ **32.** $s(x) = \log_{1/10}(x)$

33. $f(x) = \ln(x - 1)$ **34.** $f(x) = \log_3(x + 2)$

35. $f(x) = -3 + \log(x + 2)$ **36.** $f(x) = 4 - \log(x + 6)$

37. $f(x) = -\dfrac{1}{2}\log(x - 1)$ **38.** $f(x) = -2 \cdot \log_2(x + 2)$

Write the equation of each graph in its final position.

39. The graph of $y = \ln(x)$ is translated three units to the right and then four units downward.

40. The graph of $y = \log(x)$ is translated five units to the left and then seven units upward.

41. The graph of $y = \log_2(x)$ is translated five units to the right, reflected in the *x*-axis, and then translated one unit downward.

42. The graph of $y = \log_3(x)$ is translated four units upward, six units to the left, and then reflected in the *x*-axis.

Write each equation as an equivalent exponential equation.

43. $\log_2(32) = 5$ **44.** $\log_3(81) = 4$

45. $\log_5(x) = y$ **46.** $\log_4(a) = b$

47. $\log(1000) = z$ **48.** $\ln(y) = 3$

49. $\ln(5) = x$ **50.** $\log(y) = 2$

51. $\log_a(x) = m$ **52.** $\log_b(q) = t$

Write each equation as an equivalent logarithmic equation.

53. $5^3 = 125$ **54.** $2^7 = 128$ **55.** $e^3 = y$

56. $10^5 = w$ **57.** $y = 10^m$ **58.** $p = e^x$

59. $y = a^z$ **60.** $w = b^k$ **61.** $a^{x-1} = n$

62. $w^{x+2} = r$

For each function, find f^{-1}.

63. $f(x) = 2^x$ **64.** $f(x) = 5^x$

65. $f(x) = \log_7(x)$ **66.** $f(x) = \log(x)$

67. $f(x) = \ln(x - 1)$ **68.** $f(x) = \log(x + 4)$

69. $f(x) = 3^{x+2}$ **70.** $f(x) = 6^{x-1}$

71. $f(x) = \dfrac{1}{2} \cdot 10^{x-1} + 5$ **72.** $f(x) = 2^{3x+1} - 6$

Solve each equation. Find the exact solutions.

73. $\log_2(x) = 8$ **74.** $\log_5(x) = 3$

75. $\log_3(x) = \dfrac{1}{2}$ **76.** $\log_4(x) = \dfrac{1}{3}$

77. $\log_x(16) = 2$ **78.** $\log_x(16) = 4$

79. $3^x = 77$ **80.** $\dfrac{1}{2^x} = 5$

81. $\ln(x - 3) = \ln(2x - 9)$ **82.** $\log_2(4x) = \log_2(x + 6)$

83. $\log_x(18) = 2$ **84.** $\log_x(9) = \dfrac{1}{2}$

85. $3^{x+1} = 7$ **86.** $5^{3-x} = 12$

87. $\log(x) = \log(6 - x^2)$ **88.** $\log_3(2x) = \log_3(24 - x^2)$

89. $\log_x\left(\dfrac{1}{9}\right) = -\dfrac{2}{3}$ **90.** $\log_x\left(\dfrac{1}{16}\right) = \dfrac{4}{3}$

91. $4^{2x-1} = \dfrac{1}{2}$ **92.** $e^{3x-4} = 1$

93. $\log_{32}(64) = x$ **94.** $\ln\left(\dfrac{1}{\sqrt{e}}\right) = x$

Find the approximate solution to each equation. Round to four decimal places.

95. $10^x = 25$ **96.** $e^x = 2$

97. $e^{2x} = 3$ **98.** $10^{3x} = 5$

99. $5e^x = 4$ **100.** $10^x - 3 = 5$

101. $\dfrac{1}{10^x} = 2$ **102.** $\dfrac{1}{e^{x-1}} = 5$

Solve each problem.

103. *Finding Time* Find the amount of time to the nearest tenth of a year that it would take for $10 to grow to $20 at each of the following annual rates compounded continuously.
   a. 2%   **b.** 4%

   c. 8%   **d.** 16%

104. *Finding Time* Find the amount of time, to the nearest tenth of a year that it would take for $10 to grow to $40 at each of the following annual rates compounded continuously.
   a. 1%   **b.** 2%

   c. 8.327%   **d.** $7\dfrac{2}{3}$%

105. *Finding Rate* Find the annual percentage rate compounded continuously to the nearest tenth of a percent for which $10 would grow to $30 for each of the following time periods.
   a. 5 years   **b.** 10 years

   c. 20 years   **d.** 40 years

106. *Finding Rate* Find the annual percentage rate to the nearest tenth of a percent for which $10 would grow to $50 for each of the following time periods.
   a. 4 years   **b.** 8 years

   c. 16 years   **d.** 32 years

107. *Becoming a Millionaire* Find the amount of time to the nearest day that it would take for a deposit of $1000 to grow to $1 million at 14% compounded continuously.

108. *Doubling Your Money* How long does it take for a deposit of $1000 to double at 8% compounded continuously?

109. *Finding the Rate* Solve the equation $A = Pe^{rt}$ for r, then find the rate at which a deposit of $1000 would double in 3 years compounded continuously.

110. *Finding the Rate* At what interest rate would a deposit of $30,000 grow to $2,540,689 in 40 years with continuous compounding?

111. *Rule of 70*
   a. Find the time that it takes for an investment to double at 10% compounded continuously.

   b. The time that it takes for an investment to double at r% is approximately 70 divided by r (the rule of 70). So at 10%, an investment will double in about 7 years. Explain why this rule works.

112. *Using the Rule of 70* Find approximate answers to these questions without using a calculator. See Exercise 111.
   a. Connie deposits $1000 in a bank at an annual interest rate of 2%. How long does it take for her money to double?

   b. Celeste invests $1000 in the stock market and her money grows at an annual rate of 10% (the historical average rate of return for the stock market). How long does it take for her money to double?

   c. What is the ratio of the value of Celeste's investment after 35 years to the value of Connie's investment after 35 years?

113. *Miracle in Boston* To illustrate the "miracle" of compound interest, Ben Franklin bequeathed $4000 to the city of Boston in 1790. The fund grew to $4.5 million in 200 years. Find the annual rate compounded continuously that would cause this "miracle" to happen.

114. *Miracle in Philadelphia* Ben Franklin's gift of $4000 to the city of Philadelphia in 1790 was not managed as well as his gift to Boston. The Philadelphia fund grew to only $2 million in 200 years. Find the annual rate compounded continuously that would yield this total value.

115. *Deforestation in Nigeria* In Nigeria, deforestation occurs at the rate of about 5.2% per year. Assuming that the amount of forest remaining is determined by the function

$$F = F_0 e^{-0.052t},$$

where F_0 is the present acreage of forest land and t is the time in years from the present. In how many years will there be only 60% of the present acreage remaining?
   HINT Find the amount of time it takes for F_0 to become $0.60F_0$.

116. *Deforestation in El Salvador* It is estimated that at the present rate of deforestation in El Salvador, in 20 years only 53% of the present forest will be remaining. Use the exponential model $F = F_0 e^{rt}$ to determine the annual rate of deforestation in El Salvador.

117. *World Population* The population of the world doubled from 1950 to 1987, going from 2.5 billion to 5 billion people. Using the exponential model,

$$P = P_0 e^{rt},$$

find the annual growth rate r for that period. Although the annual growth rate has declined slightly to 1.63% annually, the population of the world is still growing at a tremendous rate. Using the initial population of 5 billion in 1987 and an annual rate of 1.63%, estimate the world population in the year 2010.

118. *Black Death* Because of the Black Death, or plague, the only substantial period in recorded history when the earth's population was not increasing was from 1348 to 1400. During that period the world population decreased by about 100 million people. Use the exponential model $P = P_0 e^{rt}$ and the data from the accompanying table to find the annual growth rate for the period 1400 to 2000. If the 100 million people had not been lost, then how many people would they have grown to in 600 years using the growth rate that you just found?

Year	World Population
1348	0.47×10^9
1400	0.37×10^9
1900	1.60×10^9
2000	6.07×10^9

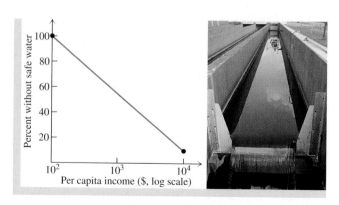

■ **Figure for Exercise 118**

119. *Traffic Jam* The amount of U.S. long-distance traffic for data transmission is expected to grow exponentially in the coming years. See the accompanying figure. The expected growth can be modeled by the function

$$d = 0.1 e^{0.46t},$$

where t is the number of years since 1994 and d is measured in billions of gigabits per year.

a. What will be the amount of long-distance data transmission in 2009?

b. In what year will long-distance data transmission reach 14 billion gigabits per year?

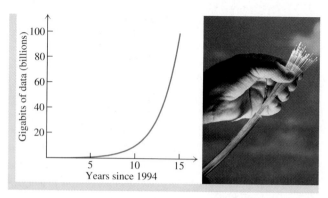

■ **Figure for Exercise 119**

120. *Global Warming* The increasing global temperature can be modeled by the function

$$I = 0.1 e^{0.02t},$$

where I is the increase in global temperature in degrees Celsius since 1900, and t is the number of years since 1900 (NASA, www.science.nasa.gov).

a. How much warmer will it be in 2010 than it was in 1950?

b. In what year will the global temperature be 4° greater than the global temperature in 2000?

121. *Safe Water* The accompanying graph shows the percentage of population without safe water p as a function of per capita income I (World Resources Institute, www.wri.org). Because the horizontal axis has a *log scale* (each mark is 10 times the previous mark), the relationship looks linear but it is not.

a. Find a formula for this function.

 HINT First find the equation of the line through (2, 100) and (4, 10) in the form $p = mx + b$, then replace x with $\log(I)$.

b. What percent of the population would be expected to be without safe drinking water in a city with a per capita income of $100,000?

■ **Figure for Exercise 121**

122. *Municipal Waste*   The accompanying graph shows the amount of municipal waste per capita w as a function of per capita income I (World Resources Institute, www.wri.org).

   a. Use the technique of the previous exercise to find a formula for w as a function of I.

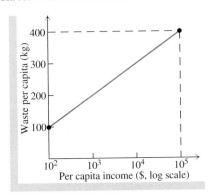

■ **Figure for Exercise 122**

   b. How much municipal waste per capita would you expect in a city where the per capita income is $200,000?

Use the following definition for Exercises 123–126.

   In chemistry, the pH of a solution is defined to be

$$\text{pH} = -\log[H^+],$$

where H^+ is the hydrogen ion concentration of the solution in moles per liter. Distilled water has a pH of approximately 7. A substance with a pH under 7 is called an acid and one with a pH over 7 is called a base.

123. *Acidity of Tomato Juice*   Tomato juice has a hydrogen ion concentration of $10^{-4.1}$ moles per liter. Find the pH of tomato juice.

124. *Acidity in Your Stomach*   The gastric juices in your stomach have a hydrogen ion concentration of 10^{-1} moles per liter. Find the pH of your gastric juices.

125. *Acidity of Orange Juice*   The hydrogen ion concentration of orange juice is $10^{-3.7}$ moles per liter. Find the pH of orange juice.

126. *Acidosis*   A healthy body maintains the hydrogen ion concentration of human blood at $10^{-7.4}$ moles per liter. What is the pH of normal healthy blood? The condition of low blood pH is called acidosis. Its symptoms are sickly sweet breath, headache, and nausea.

For Writing/Discussion

127. *Different Models*   Calculators that perform exponential regression often use $y = a \cdot b^x$ as the exponential growth model instead of $y = a \cdot e^{cx}$. For what value of c is $a \cdot b^x = a \cdot e^{cx}$? If a calculator gives $y = 500(1.036)^x$ for a growth model, then what is the continuous growth rate to the nearest hundredth of a percent?

128. *Making Conjectures*   Consider the function $y = \log(10^n \cdot x)$ where n is an integer. Use a graphing calculator to graph this function for several choices of n. Make a conjecture about the relationship between the graph of $y = \log(10^n \cdot x)$ and the graph of $y = \log(x)$. Save your conjecture and attempt to prove it after you have studied the properties of logarithms, which are coming in Section 4.3. Repeat this exercise with $y = \log(x^n)$ where n is an integer.

129. *Increasing or Decreasing*   Which exponential and logarithmic functions are increasing? Decreasing? Is the inverse of an increasing function increasing or decreasing? Is the inverse of a decreasing function increasing or decreasing? Explain.

130. *Cooperative Learning*   Work in a small group to write a summary (including drawings) of the types of graphs that can be obtained for logarithmic functions of the form $y = \log_a(x)$ for $a > 0$ and $a \neq 1$.

Thinking Outside the Box XL

Seven-Eleven   A convenience store sells a gallon of milk for $7 and a loaf of bread for $11. You are allowed to buy any combination of milk and bread, including only milk or only bread. Your total bill is always a whole number of dollars, but there are many whole numbers that cannot be the total. For example, the total cannot be $15. What is the largest whole number of dollars that cannot be the total?

4.2 Pop Quiz

1. What is x if $2^x = 32$?   **2.** Find $\log_2(32)$.

3. Is $y = \log_3(x)$ increasing or decreasing?

4. Find all asymptotes for $f(x) = \ln(x - 1)$.

5. Write $3^a = b$ as a logarithmic equation.

6. Find f^{-1} if $f(x) = \log(x + 3)$.

7. Solve $\log_5(x) = 3$.

8. Solve $\log_x(36) = 2$.

9. How long (to the nearest day) does it take for $2 to grow to $4 at 5% compounded continuously?

Linking Concepts

Modeling the U.S. Population

To effectively plan for the future one must attempt to predict the future. Government agencies use data about the past to construct a model and predict the future. The following table gives the population of the United States in millions every 10 years since 1900 (Census Bureau, www.census.gov).

Year	Pop.	Year	Pop.
1900	76	1960	180
1910	92	1970	204
1920	106	1980	227
1930	123	1990	249
1940	132	2000	279
1950	152		

a) Draw a bar graph of the data in the accompanying table. Use a computer graphics program if one is available.

b) Enter the data into your calculator and use exponential regression to find an exponential model of the form $y = a \cdot b^x$, where $x = 0$ corresponds to 1900.

c) Make a table (like the given table) that shows the predicted population according to the exponential model rather than the actual population.

d) Plot the points from part (c) on your bar graph and sketch an exponential curve through the points.

e) Use the given population data with linear regression to find a linear model of the form $y = ax + b$ and graph the line on your bar graph.

f) Predict the population in the year 2010 using the exponential model and the linear model.

g) Judging from your exponential curve and the line on your bar graph, in which prediction do you have the most confidence?

h) Use only the fact that the population grew from 76 million at time $t = 0$ (the year 1900) to 279 million at time $t = 100$ (the year 2000) to find a formula of the form $P(t) = P_0 \cdot e^{rt}$ for the population at any time t. Do not use regression.

i) Use the formula from part (h) to predict the population in the year 2010. Compare this prediction to the prediction that you found using exponential regression.

4.3

Rules of Logarithms

The rules of logarithms are closely related to the rules of exponents, because logarithms are exponents. In this section we use the rules of exponents to develop some rules of logarithms. With these rules of logarithms we will be able to solve more equations involving exponents and logarithms. The rules of exponents were discussed in Sections P.2 and P.3.

The Logarithm of a Product

By the product rule for exponents, we add exponents when multiplying exponential expressions having the same base. To find a corresponding rule for logarithms, let's examine the equation $2^3 \cdot 2^2 = 2^5$. Notice that the exponents 3, 2, and 5 are the base-2 logarithms of 8, 4, and 32, respectively.

$$\log_2(8) \qquad \log_2(4) \qquad \log_2(32)$$
$$2^3 \cdot 2^2 = 2^5$$

When we add the exponents 3 and 2 to get 5, we are adding logarithms and getting another logarithm as the result. So the base-2 logarithm of 32 (the product of 8 and 4) is the sum of the base-2 logarithms of 8 and 4:

$$\log_2(8 \cdot 4) = \log_2(8) + \log_2(4)$$

This example suggests the **product rule for logarithms.**

Product Rule for Logarithms

For $M > 0$ and $N > 0$,

$$\log_a(MN) = \log_a(M) + \log_a(N).$$

■ **PROOF**   If $M = a^x$ and $N = a^y$, then

$$MN = a^x a^y = a^{x+y}.$$

By the definition of logarithm, $MN = a^{x+y}$ is equivalent to

$$\log_a(MN) = x + y.$$

Since $M = a^x$, we have $x = \log_a(M)$, and since $N = a^y$, we have $y = \log_a(N)$. So

$$\log_a(MN) = \log_a(M) + \log_a(N). \qquad ■$$

The product rule for logarithms says that *the logarithm of a product of two numbers is equal to the sum of their logarithms*, provided that all of the logarithms are defined and all have the same base. There is no rule about the logarithm of a sum, and the logarithm of a sum is generally *not* equal to the sum of the logarithms. For example, $\log_2(8 + 8) \neq \log_2(8) + \log_2(8)$ because $\log_2(8 + 8) = 4$ while $\log_2(8) + \log_2(8) = 6$.

You can use a calculator to illustrate the product rule, as in Fig. 4.32.   □

```
log(36)
        1.556302501
log(9)+log(4)
        1.556302501
```

■ **Figure 4.32**

Example **1** Using the product rule for logarithms

Write each expression as a single logarithm. All variables represent positive real numbers.

a. $\log_3(x) + \log_3(6)$ **b.** $\ln(3) + \ln(x^2) + \ln(y)$

Solution

a. By the product rule for logarithms, $\log_3(x) + \log_3(6) = \log_3(6x)$.
b. By the product rule for logarithms, $\ln(3) + \ln(x^2) + \ln(y) = \ln(3x^2y)$.

Try This. Write $\log(x) + \log(y)$ as a single logarithm.   ■

The Logarithm of a Quotient

By the quotient rule for exponents, we subtract the exponents when dividing exponential expressions having the same base. To find a corresponding rule for logarithms, examine the equation $2^5/2^2 = 2^3$. Notice that the exponents 5, 2, and 3 are the base-2 logarithms of 32, 4, and 8, respectively:

$$\frac{2^5}{2^2} = 2^3$$

When we subtract the exponents 5 and 2 to get 3, we are subtracting logarithms and getting another logarithm as the result. So the base-2 logarithm of 8 (the quotient of 32 and 4) is the difference of the base-2 logarithms of 32 and 4:

$$\log_2\left(\frac{32}{4}\right) = \log_2(32) - \log_2(4)$$

This example suggests the **quotient rule for logarithms.**

Quotient Rule for Logarithms

For $M > 0$ and $N > 0$,

$$\log_a\left(\frac{M}{N}\right) = \log_a(M) - \log_a(N).$$

The quotient rule for logarithms says that *the logarithm of a quotient of two numbers is equal to the difference of their logarithms*, provided that all logarithms are defined and all have the same base. (The proof of the quotient rule is similar to that of the product rule and so it is left as an exercise.) Note that the quotient rule does not apply to division of logarithms. For example,

$$\frac{\log_2(32)}{\log_2(4)} \neq \log_2(32) - \log_2(4),$$

because $\log_2(32)/\log_2(4) = 5/2$, while $\log_2(32) - \log_2(4) = 3$.

You can use a calculator to illustrate the quotient rule, as shown in Fig. 4.33.   □

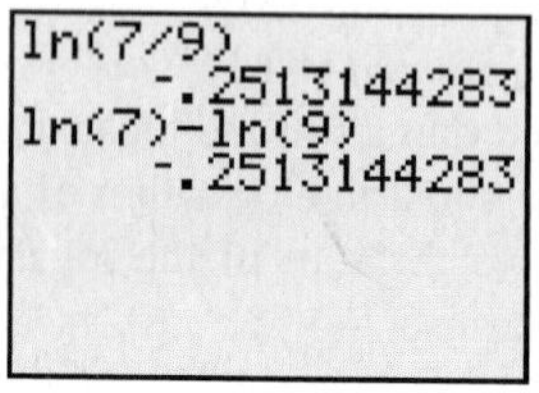

```
ln(7/9)
        -.2513144283
ln(7)-ln(9)
        -.2513144283
```

■ **Figure 4.33**

Example **2** **Using the quotient rule for logarithms**

Write each expression as a single logarithm. All variables represent positive real numbers.

a. $\log_3(24) - \log_3(4)$ **b.** $\ln(x^6) - \ln(x^2)$

Solution

a. By the quotient rule, $\log_3(24) - \log_3(4) = \log_3(24/4) = \log_3(6)$.

b. $\ln(x^6) - \ln(x^2) = \ln\left(\dfrac{x^6}{x^2}\right)$ By the quotient rule for logarithms

$\qquad\qquad\qquad = \ln(x^4)$ By the quotient rule for exponents

Try This.   Write $\log(5x) - \log(5)$ as a single logarithm.   ▪

The Logarithm of a Power

By the power rule for exponents, we multiply the exponents when finding a power of an exponential expression. For example, $(2^3)^2 = 2^6$. Notice that the exponents 3 and 6 are the base-2 logarithms of 8 and 64, respectively.

$$\overset{\log_2(8)}{\underset{\downarrow}{}} \quad \overset{\log_2(64)}{\underset{\downarrow}{}}$$
$$(2^3)^2 = 2^6$$

So the base-2 logarithm of 64 (the second power of 8) is twice the base-2 logarithm of 8:

$$\log_2(8^2) = 2 \cdot \log_2(8)$$

This example suggests the **power rule for logarithms.**

Power Rule for Logarithms

> For $M > 0$ and any real number N,
> $$\log_a(M^N) = N \cdot \log_a(M).$$

The power rule for logarithms says that *the logarithm of a power of a number is equal to the power times the logarithm of the number*, provided that all logarithms are defined and have the same base. The proof is left as an exercise.

You can illustrate the power rule on a calculator, as shown in Fig. 4.34.   □

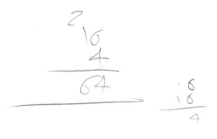

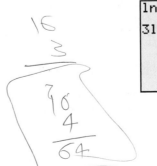

```
ln(17^3)
         8.499640032
3ln(17)
         8.499640032
```

▪ **Figure 4.34**

Example **3** **Using the power rule for logarithms**

Rewrite each expression in terms of $\log(3)$.

a. $\log(3^8)$ **b.** $\log(\sqrt{3})$ **c.** $\log\left(\dfrac{1}{3}\right)$

Solution

a. $\log(3^8) = 8 \cdot \log(3)$ By the power rule for logarithms

b. $\log(\sqrt{3}) = \log(3^{1/2}) = \dfrac{1}{2}\log(3)$ By the power rule for logarithms

c. $\log\left(\dfrac{1}{3}\right) = \log(3^{-1}) = -\log(3)$ By the power rule for logarithms

Try This. Write $\ln(27)$ in terms of $\ln(3)$.   ▪

The Inverse Rules

The definition of logarithms leads to two rules that are useful in solving equations. If $f(x) = a^x$ and $g(x) = \log_a(x)$, then

$$g(f(x)) = g(a^x) = \log_a(a^x) = x$$

for any real number x. The result of this composition is x because the functions are inverses. If we compose in the opposite order we get

$$f(g(x)) = f(\log_a(x)) = a^{\log_a(x)} = x$$

for any positive real number x. The results are called the **inverse rules.**

Inverse Rules

> If $a > 0$ and $a \neq 1$, then
>
> **1.** $\log_a(a^x) = x$ for any real number x
> **2.** $a^{\log_a(x)} = x$ for $x > 0$.

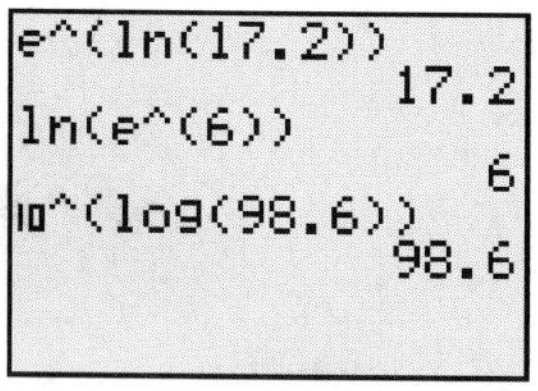

```
e^(ln(17.2))
               17.2
ln(e^(6))
                  6
10^(log(98.6))
               98.6
```

▪ **Figure 4.35**

The inverse rules are easy to use if you remember that $\log_a(x)$ is the power of a that produces x. For example, $\log_2(67)$ is the power of 2 that produces 67. So $2^{\log_2(67)} = 67$. Similarly, $\ln(e^{99})$ is the power of e that produces e^{99}. So $\ln(e^{99}) = 99$. ⬚ You can illustrate the inverse rules with a calculator, as shown in Fig. 4.35.   □

Example **4** **Using the inverse rules**

Simplify each expression.

a. $e^{\ln(x^2)}$ **b.** $\log_7(7^{2x-1})$

Solution

By the inverse rules, $e^{\ln(x^2)} = x^2$ and $\log_7(7^{2x-1}) = 2x - 1$.

Try This. Simplify $10^{\log(5w)}$ and $\log(10^p)$.   ▪

Using the Rules

When simplifying or rewriting expressions, we often apply several rules. In the following box we list all of the available rules of logarithms.

Rules of Logarithms with Base *a*

If M, N, and a are positive real numbers with $a \neq 1$, and x is any real number, then

1. $\log_a(a) = 1$ **2.** $\log_a(1) = 0$

3. $\log_a(a^x) = x$ **4.** $a^{\log_a(N)} = N$

5. $\log_a(MN) = \log_a(M) + \log_a(N)$ **6.** $\log_a(M/N) = \log_a(M) - \log_a(N)$

7. $\log_a(M^x) = x \cdot \log_a(M)$ **8.** $\log_a(1/N) = -\log_a(N)$

Note that rule 1 is a special case of rule 3 with $x = 1$, rule 2 follows from the fact that $a^0 = 1$ for any nonzero base a, and rule 8 is a special case of rule 6 with $M = 1$.

The rules for logarithms with base a in the preceding box apply to all logarithms, including common logarithms (base 10) and natural logarithms (base e). Since natural logarithms are very popular, we list the rules of logarithms again for base e in the following box for easy reference.

Rules of Natural Logarithms

If M and N are positive real numbers and x is any real number, then

1. $\ln(e) = 1$ **2.** $\ln(1) = 0$

3. $\ln(e^x) = x$ **4.** $e^{\ln(N)} = N$

5. $\ln(MN) = \ln(M) + \ln(N)$ **6.** $\ln(M/N) = \ln(M) - \ln(N)$

7. $\ln(M^x) = x \cdot \ln(M)$ **8.** $\ln(1/N) = -\ln(N)$

Example 5   Using the rules of logarithms

Rewrite each expression in terms of $\log(2)$ and $\log(3)$.

a. $\log(6)$ **b.** $\log\left(\dfrac{16}{3}\right)$ **c.** $\log\left(\dfrac{1}{3}\right)$

Solution

a. $\log(6) = \log(2 \cdot 3)$ Factor.

         $= \log(2) + \log(3)$ Rule 5

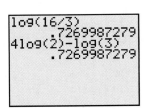

b. $\log\left(\dfrac{16}{3}\right) = \log(16) - \log(3)$ Rule 6

              $= \log(2^4) - \log(3)$ Write 16 as a power of 2.

              $= 4 \cdot \log(2) - \log(3)$ Rule 7

■ **Figure 4.36**

You can check this answer with a calculator, as shown in Fig. 4.36. ☐

c. $\log\left(\dfrac{1}{3}\right) = -\log(3)$ Rule 8

Try This.   Write $\ln(45)$ in terms of $\ln(3)$ and $\ln(5)$.   ■

Be careful to use the rules of logarithms exactly as stated. For example, the logarithm of a product is the sum of the logarithms. The logarithm of a product is generally *not* equal to the product of the logarithms. That is,

$$\log(2 \cdot 3) = \log(2) + \log(3) \qquad \text{but} \qquad \log(2 \cdot 3) \neq \log(2) \cdot \log(3).$$

Example **6** Rewriting a logarithmic expression

Rewrite each expression using a sum or difference of multiples of logarithms.

a. $\ln\left(\dfrac{3x^2}{yz}\right)$ **b.** $\log_3\left(\dfrac{(x-1)^2}{z^{3/2}}\right)$

Solution

a. $\ln\left(\dfrac{3x^2}{yz}\right) = \ln(3x^2) - \ln(yz)$ Quotient rule for logarithms

$\qquad\qquad = \ln(3) + \ln(x^2) - [\ln(y) + \ln(z)]$ Product rule for logarithms

$\qquad\qquad = \ln(3) + 2 \cdot \ln(x) - \ln(y) - \ln(z)$ Power rule for logarithms

Note that $\ln(y) + \ln(z)$ must be in brackets (or parentheses) because of the subtraction symbol preceding it.

b. $\log_3\left(\dfrac{(x-1)^2}{z^{3/2}}\right) = \log_3((x-1)^2) - \log_3(z^{3/2})$ Quotient rule for logarithms

$\qquad\qquad = 2 \cdot \log_3(x-1) - \dfrac{3}{2}\log_3(z)$ Power rule for logarithms

Try This. Write $\log\left(\dfrac{x^2}{5a}\right)$ using sums and/or differences of multiples of logarithms. ■

In Example 7 we use the rules "in reverse" of the way we did in Example 6.

Example **7** Rewriting as a single logarithm

Rewrite each expression as a single logarithm.

a. $\ln(x-1) + \ln(3) - 3 \cdot \ln(x)$ **b.** $\dfrac{1}{2}\log(y) - \dfrac{1}{3}\log(z)$

Solution

a. Use the product rule, the power rule, and then the quotient rule for logarithms:

$\ln(x-1) + \ln(3) - 3 \cdot \ln(x) = \ln[3(x-1)] - \ln(x^3)$

$\qquad\qquad\qquad\qquad\qquad = \ln[3x - 3] - \ln(x^3)$

$\qquad\qquad\qquad\qquad\qquad = \ln\left(\dfrac{3x-3}{x^3}\right)$

b. $\dfrac{1}{2}\log(y) - \dfrac{1}{3}\log(z) = \log(y^{1/2}) - \log(z^{1/3})$

$\qquad\qquad\qquad\qquad = \log(\sqrt{y}) - \log(\sqrt[3]{z})$

$\qquad\qquad\qquad\qquad = \log\left(\dfrac{\sqrt{y}}{\sqrt[3]{z}}\right)$

Try This. Write $\ln(x) - \ln(y) - 2 \cdot \ln(z)$ as a single logarithm. ■

The Base-Change Formula

The exact solution to an exponential equation is often expressed in terms of logarithms. For example, the exact solution to $(1.03)^x = 5$ is $x = \log_{1.03}(5)$. But how do we calculate $\log_{1.03}(5)$? The next example shows how to find a rational approximation for $\log_{1.03}(5)$ using rules of logarithms. In this example we also introduce a new idea in solving equations, *taking the logarithm of each side*. The base-*a* logarithms of two equal quantities are equal because logarithm is a function.

Example 8 A rational approximation for a logarithm

Find an approximate rational solution to $(1.03)^x = 5$. Round to four decimal places.

Solution

Take the logarithm of each side using one of the bases available on a calculator.

$$(1.03)^x = 5$$

$$\log((1.03)^x) = \log(5) \qquad \text{Take the logarithm of each side.}$$

$$x \cdot \log(1.03) = \log(5) \qquad \text{Power rule for logarithms}$$

$$x = \frac{\log(5)}{\log(1.03)} \qquad \text{Divide each side by } \log(1.03).$$

$$\approx 54.4487$$

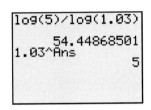

■ **Figure 4.37**

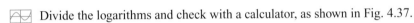 Divide the logarithms and check with a calculator, as shown in Fig. 4.37.

Try This. Solve $5.44^x = 2.3$. Round to four decimal places. ■

The solution to the equation of Example 8 is a base-1.03 logarithm, but we obtained a rational approximation for it using base-10 logarithms. We can use the procedure of Example 8 to write a base-*a* logarithm in terms of a base-*b* logarithm for any bases *a* and *b*:

$$a^x = M \qquad \text{Equivalent equation: } x = \log_a(M)$$

$$\log_b(a^x) = \log_b(M) \qquad \text{Take the base-}b \text{ logarithm of each side.}$$

$$x \cdot \log_b(a) = \log_b(M) \qquad \text{Power rule for logarithms}$$

$$x = \frac{\log_b(M)}{\log_b(a)} \qquad \text{Divide each side by } \log_b(a).$$

Since $x = \log_a(M)$, we have the following **base-change formula.**

Base-Change Formula

> If $a > 0, b > 0, a \neq 1, b \neq 1$, and $M > 0$, then
> $$\log_a(M) = \frac{\log_b(M)}{\log_b(a)}.$$

The base-change formula says that the logarithm of a number in one base is equal to the logarithm of the number in the new base divided by the logarithm of the old base. Using this formula, a logarithm such as $\log_3(7)$ can be easily found with a calculator. Let $a = 3$, $b = 10$, and $M = 7$ in the formula to get

$$\log_3(7) = \frac{\log(7)}{\log(3)} \approx 1.7712.$$

Note that you get the same value for $\log_3(7)$ using natural logarithms, as shown in Fig. 4.38. □

■ **Figure 4.38**

Example **9** **Using the base-change formula with compound interest**

If $2500 is invested at 6% compounded daily, then how long (to the nearest day) would it take for the investment to double in value?

Solution

We want the number of years for $2500 to grow to $5000 at 6% compounded daily. Use $P = \$2500$, $A = \$5000$, $n = 365$, and $r = 0.06$ in the formula for compound interest:

$$A = P\left(1 + \frac{r}{n}\right)^{nt}$$

$$5000 = 2500\left(1 + \frac{0.06}{365}\right)^{365t}$$

$$2 = \left(1 + \frac{0.06}{365}\right)^{365t} \qquad \text{Divide by 2500 to get the form } y = a^x.$$

$$2 \approx (1.000164384)^{365t} \qquad \text{Approximate the base with a decimal number.}$$

$$365t \approx \log_{1.000164384}(2) \qquad \text{Definition of logarithm}$$

$$t \approx \frac{1}{365} \cdot \frac{\ln(2)}{\ln(1.000164384)} \qquad \text{Base-change formula (Use either ln or log.)}$$

$$\approx 11.553 \text{ years}$$

The investment of $2500 will double in about 11 years, 202 days.

Try This. How long (to the nearest day) does it take for $100 to grow to $400 at 5% compounded daily? ■

Note that Example 9 can also be solved by taking the log (or ln) of each side and applying the power rule for logarithms.

$$\log(2) = \log((1 + 0.06/365)^{365t})$$

$$\log(2) = 365t \cdot \log(1 + 0.06/365)$$

$$\frac{\log(2)}{365 \cdot \log(1 + 0.06/365)} = t$$

$$11.553 \approx t$$

Finding the Rate

If a variable is a function of time, we are often interested in finding the rate of growth or decay. In the continuous model, $y = ae^{rt}$, the rate r appears as an exponent. If we know the values of the other variables, we find the rate by solving for r using natural logarithms. If $y = ab^t$ is used instead of $y = ae^{rt}$, the rate does not appear in the formula. However, we can write $y = ab^t$ as $y = ae^{\ln(b) \cdot t}$, because $b = e^{\ln(b)}$. Comparing $e^{\ln(b) \cdot t}$ to e^{rt}, we see that the rate is $\ln(b)$.

In the case of interest compounded periodically, the growth is not continuous and we use the formula $A = P(1 + r/n)^{nt}$. Because the rate appears in the base, it can be found without using logarithms, as shown in the next example.

Example **10** Finding the rate with interest compounded periodically

For what annual percentage rate would $1000 grow to $3500 in 20 years compounded monthly?

Solution

Use $A = 3500$, $P = 1000$, $n = 12$, and $t = 20$ in the compound interest formula.

$$A = P\left(1 + \frac{r}{n}\right)^{nt}$$

$$3500 = 1000\left(1 + \frac{r}{12}\right)^{240}$$

$$3.5 = \left(1 + \frac{r}{12}\right)^{240} \qquad \text{Divide each side by 1000.}$$

Since 240 is even, there is a positive and a negative 240th root of 3.5. Ignore the negative root in this application because it gives a negative interest rate.

$$1 + \frac{r}{12} = (3.5)^{1/240} \qquad \text{Logarithms are not needed because the exponent is a constant.}$$

$$\frac{r}{12} = (3.5)^{1/240} - 1$$

$$r = 12((3.5)^{1/240} - 1)$$

$$\approx 0.063$$

If $1000 earns approximately 6.3% compounded monthly, then it will grow to $3500 in 20 years.

Try This. For what annual rate compounded daily would $100 grow to $300 in 20 years? ∎

For Thought

True or False? Explain.

1. $\dfrac{\log(8)}{\log(3)} = \log(8) - \log(3)$

2. $\ln\left(\sqrt{3}\right) = \dfrac{\ln(3)}{2}$

3. $\dfrac{\log_{19}(8)}{\log_{19}(2)} = \log_3(27)$

4. $\dfrac{\log_2(7)}{\log_2(5)} = \dfrac{\log_3(7)}{\log_3(5)}$

5. $e^{\ln(x)} = x$ for any real number x.

6. The equations $\log(x - 2) = 4$ and $\log(x) - \log(2) = 4$ are equivalent.

7. The equations $x + 1 = 2x + 3$ and $\log(x + 1) = \log(2x + 3)$ are equivalent.

8. $\ln(e^x) = x$ for any real number x.

9. If $30 = x^{50}$, then x is between 1 and 2.

10. If $20 = a^4$, then $\ln(a) = \frac{1}{4}\ln(20)$.

4.3 Exercises

Rewrite each expression as a single logarithm.

1. $\log(5) + \log(3)$

2. $\ln(6) + \ln(2)$

3. $\log_2(x - 1) + \log_2(x)$

4. $\log_3(x + 2) + \log_3(x - 1)$

5. $\log_4(12) - \log_4(2)$

6. $\log_2(25) - \log_2(5)$

7. $\ln(x^8) - \ln(x^3)$

8. $\log(x^2 - 4) - \log(x - 2)$

Rewrite each expression as a sum or difference of logarithms.

9. $\log_2(3x)$

10. $\log_3(xy)$

11. $\log\left(\dfrac{x}{2}\right)$

12. $\log\left(\dfrac{a}{b}\right)$

13. $\log(x^2 - 1)$

14. $\log(a^2 - 9)$

15. $\ln\left(\dfrac{x - 1}{x}\right)$

16. $\ln\left(\dfrac{a + b}{b}\right)$

Rewrite each expression in terms of $\log_a(5)$.

17. $\log_a(5^3)$

18. $\log_a(25)$

19. $\log_a\left(\sqrt{5}\right)$

20. $\log_a\left(\sqrt[3]{5}\right)$

21. $\log_a\left(\dfrac{1}{5}\right)$

22. $\log_a\left(\dfrac{1}{125}\right)$

Simplify each expression.

23. $e^{\ln(\sqrt{y})}$

24. $10^{\log(3x+1)}$

25. $\log(10^{y+1})$

26. $\ln(e^{2k})$

27. $7^{\log_7(999)}$

28. $\log_4(2^{300})$

Rewrite each expression in terms of $\log_a(2)$ and $\log_a(5)$.

29. $\log_a(10)$

30. $\log_a(0.4)$

31. $\log_a(2.5)$

32. $\log_a(250)$

33. $\log_a\left(\sqrt{20}\right)$

34. $\log_a(0.0005)$

35. $\log_a\left(\dfrac{4}{25}\right)$

36. $\log_a(0.1)$

Rewrite each expression as a sum or difference of multiples of logarithms.

37. $\log_3(5x)$

38. $\log_2(xyz)$

39. $\log_2\left(\dfrac{5}{2y}\right)$

40. $\log\left(\dfrac{4a}{3b}\right)$

41. $\log\left(3\sqrt{x}\right)$

42. $\ln\left(\sqrt{x/4}\right)$

43. $\log(3 \cdot 2^{x-1})$

44. $\ln\left(\dfrac{5^{-x}}{2}\right)$

45. $\ln\left(\dfrac{\sqrt[3]{xy}}{t^{4/3}}\right)$

46. $\log\left(\dfrac{3x^2}{(ab)^{2/3}}\right)$

47. $\ln\left(\dfrac{6\sqrt{x-1}}{5x^3}\right)$

48. $\log_4\left(\dfrac{3x\sqrt{y}}{\sqrt[3]{x-1}}\right)$

Rewrite each expression as a single logarithm.

49. $\log_2(5) + 3 \cdot \log_2(x)$

50. $\log(x) + 5 \cdot \log(x)$

51. $\log_7(x^5) - 4 \cdot \log_7(x^2)$

52. $\dfrac{1}{3}\ln(6) - \dfrac{1}{3}\ln(2)$

53. $\log(2) + \log(x) + \log(y) - \log(z)$

54. $\ln(2) + \ln(3) + \ln(5) - \ln(7)$

55. $\dfrac{1}{2}\log(x) - \log(y) + \log(z) - \dfrac{1}{3}\log(w)$

56. $\dfrac{5}{6}\log_2(x) + \dfrac{2}{3}\log_2(y) - \dfrac{1}{2}\log_2(x) - \log_2(y)$

57. $3 \cdot \log_4(x^2) - 4 \cdot \log_4(x^{-3}) + 2 \cdot \log_4(x)$

58. $\dfrac{1}{2}[\log(x) + \log(y)] - \log(z)$

Find an approximate rational solution to each equation. Round answers to four decimal places.

59. $2^x = 9$

60. $3^x = 12$

61. $(0.56)^x = 8$

62. $(0.23)^x = 18.4$

63. $(1.06)^x = 2$

64. $(1.09)^x = 3$

65. $(0.73)^x = 0.5$

66. $(0.62)^x = 0.25$

Use a calculator and the base-change formula to find each logarithm to four decimal places.

67. $\log_4(9)$

68. $\log_3(4.78)$

69. $\log_{9.1}(2.3)$

70. $\log_{1.2}(13.7)$

71. $\log_{1/2}(12)$

72. $\log_{1.05}(3.66)$

Solve each equation. Round answers to four decimal places.

73. $(1.02)^{4t} = 3$

74. $(1.025)^{12t} = 3$

75. $(1.0001)^{365t} = 3.5$

76. $(1.00012)^{365t} = 2.4$

77. $(1 + r)^3 = 2.3$

78. $\left(1 + \dfrac{r}{4}\right)^{20} = 3$

79. $2\left(1 + \dfrac{r}{12}\right)^{360} = 8.4$

80. $5\left(1 + \dfrac{r}{360}\right)^{720} = 12$

Solve each equation. Round answers to four decimal places.

81. $\log_x(33.4) = 5$

82. $\log_x(12.33) = 2.3$

83. $\log_x(0.546) = -1.3$

84. $\log_x(0.915) = -3.2$

For Exercises 85–90 find the time or rate required for each investment given in the table to grow to the specified amount. The letter W represents an unknown principal.

	Principal	Ending Balance	Rate	Compounded	Time
85.	$800	$2000	8%	Daily	?
86.	$10,000	$1,000,000	7.75%	Annually	?
87.	$W	$3W	10%	Quarterly	?
88.	$W	$2W	12%	Monthly	?
89.	$500	$2000	?	Annually	25 yr
90.	$1000	$2500	?	Monthly	8 yr

91. *Ben's Gift to Boston*   Ben Franklin's gift of $4000 to Boston grew to $4.5 million in 200 years. At what interest rate compounded annually would this growth occur?

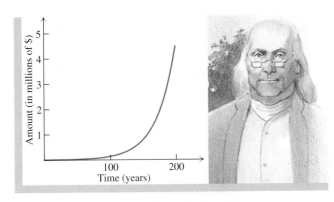

■ **Figure for Exercise 91**

92. *Ben's Gift to Philadelphia*   Ben Franklin's gift of $4000 to Philadelphia grew to $2 million in 200 years. At what interest rate compounded monthly would this growth occur?

93. *Richter Scale*   The common logarithm is used to measure the intensity of an earthquake on the Richter scale. The Richter scale rating of an earthquake of intensity I is given by $\log(I) - \log(I_0)$, where I_0 is the intensity of a small "benchmark" earthquake. Write the Richter scale rating as a single logarithm. What is the Richter scale rating of an earthquake for which $I = 1000 \cdot I_0$?

94. *Colombian Earthquake of 1906*   At 8.6 on the Richter scale, the Colombian earthquake of January 31, 1906, was one of the strongest earthquakes on record. Use the formula from Exercise 93 to write I as a multiple of I_0 for this earthquake.

95. *Time for Growth*   The time in years for a population of size P_0 to grow to size P at the annual growth rate r is given by $t = \ln((P/P_0)^{1/r})$. Use the rules of logarithms to express t in terms of $\ln(P)$ and $\ln(P_0)$.

96. *Formula for pH*   The pH of a solution is given by $\text{pH} = \log(1/H^+)$, where H^+ is the hydrogen ion concentration of the solution. Use the rules of logarithms to express the pH in terms of $\log(H^+)$.

97. *Rollover Time*   The probability that a $1 ticket does not win the Louisiana Lottery is $\frac{7,059,051}{7,059,052}$. The probability p that n independently sold tickets are all losers and the lottery rolls over is given by

$$p = \left(\frac{7,059,051}{7,059,052}\right)^n.$$

a. As n increases is p increasing or decreasing?

b. For what number of tickets sold is the probability of a rollover greater than 50%?

98. *Economic Impact*   An economist estimates that 75% of the money spent in Chattanooga is respent in four days on the average in Chattanooga. So if P dollars are spent, then the economic impact I in dollars after n respendings is given by

$$I = P(0.75)^n.$$

When $I < 0.02P$, then P no longer has an impact on the economy. If the Telephone Psychics Convention brings $1.3 million to the city, then in how many days will that money no longer have an impact on the city?

99. *Marginal Revenue* The revenue in dollars from the sale of x items is given by the function $R(x) = 500 \cdot \log(x + 1)$. The marginal revenue function $MR(x)$ is the difference quotient for $R(x)$ when $h = 1$. Find $MR(x)$ and write it as a single logarithm. What happens to the marginal revenue as x gets larger and larger?

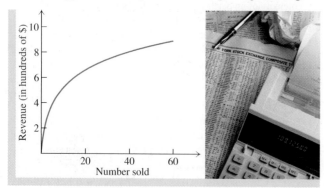

■ **Figure for Exercise 99**

100. *Human Memory Model* A class of college algebra students was given a test on college algebra concepts every month for one year after completing a college algebra course. The mean score for the class t months after completing the course can be modeled by the function $m = \ln[e^{80}/(t + 1)^7]$ for $0 \le t \le 12$. Find the mean score of the class for $t = 0, 5$, and 12. Use the rules of logarithms to rewrite the function.

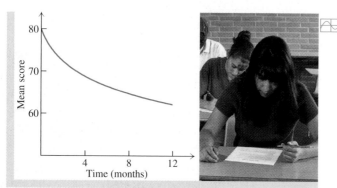

■ **Figure for Exercise 100**

101. *Computers per Capita* The number of computers per 1000 people in the U.S. from 1996 through 2002 is given in the following table (*Computer Industry Almanac*, www.c-i-a.com).
  a. Use exponential regression on a graphing calculator to find the best fitting curve of the form $y = a \cdot b^x$, where $x = 0$ corresponds to 1996.

  b. Write your equation in the form $y = ae^{cx}$.

  c. Assuming that the number of computers per 1000 people is growing continuously, what is the annual percentage rate?

Year	Computers
1996	365
1997	407
1998	452
1999	507
2000	572
2001	625
2002	659

  d. In what year will the number of computers per 1000 people reach 1500?

  e. Judging from the graph of the data and the curve, does the exponential model look like a good model?

For Writing/Discussion

102. *Power Rule* Applying the power rule to $y = \log(x^2)$ yields $y = 2 \cdot \log(x)$, but are these functions the same? What is the domain of each function? Graph the functions. Find another example of a function whose domain changes after application of a rule for logarithms.

103. *Quotient Rule* Write a proof for the quotient rule for logarithms that is similar to the proof given in the text for the product rule for logarithms.

104. *Cooperative Learning* Work in a small group to write a proof for the power rule for logarithms.

105. *Finding Relationships* Graph each of the following pairs of functions on the same screen of a graphing calculator. (Use the base-change formula to graph with bases other than 10 or e.) Explain how the functions in each pair are related.
  a. $y_1 = \log_3\left(\sqrt{3x}\right), y_2 = 0.5 + \log_3(x)$

  b. $y_1 = \log_2(1/x), y_2 = -\log_2(x)$

  c. $y_1 = 3^{x-1}, y_2 = \log_3(x) + 1$

  d. $y_1 = 3 + 2^{x-4}, y_2 = \log_2(x - 3) + 4$

Thinking Outside the Box XLI

Unit Fractions The fractions $\frac{1}{2}, \frac{1}{3}, \frac{1}{4}, \frac{1}{5}, \frac{1}{6}, \frac{1}{7}$, etc., are called *unit fractions*. Some rational numbers can be expressed as sums of unit fractions. For example,

$$\frac{4}{7} = \frac{1}{2} + \frac{1}{14} \quad \text{and} \quad \frac{11}{18} = \frac{1}{3} + \frac{1}{6} + \frac{1}{9}.$$

Write each of the following fractions as a sum of unit fractions using the fewest number of unit fractions, and keep the total of all of the denominators in the unit fractions as small as possible.

  a. $\frac{6}{23}$ **b.** $\frac{14}{15}$ **c.** $\frac{7}{11}$

Handwritten notes at top:

$\ln \sqrt{18}$

$\ln \sqrt{2 \cdot 3^2}$

$\dfrac{2 \cdot 9}{2 \cdot 3^2}$

$\ln 3 \cdot \sqrt{2}^{1/2}$

$\ln 3 + \ln \sqrt{2}$

$\ln 3 + \frac{1}{2} \ln 2$

4.3 Pop Quiz

1. Write $\log(9) + \log(3)$ as a single logarithm.

2. Write $\log(9) - \log(3)$ as a single logarithm.

3. Write $3 \ln(x) + \ln(y)$ as a single logarithm.

4. Write $\ln(5x)$ as a sum of logarithms.

5. Write $\ln\left(\sqrt{18}\right)$ in terms of $\ln(2)$ and $\ln(3)$.

Solve each equation. Round answers to four decimal places.

6. $3^x = 11$

7. $\log_x(22.5) = 3$

Handwritten work:

$x \log 3 = \log 11$

$x = \dfrac{\log 11}{\log 3}$

$x \log 3 = \log 11$

$x = \dfrac{\log 11}{\log 3} =$

$\ln \sqrt{2 \cdot 3^2}$

$\ln 2^{1/2} + \ln 3 = \frac{1}{2} \ln 2 + \ln 3$

$x^3 = 22.5$

$x = \sqrt[3]{22.5}$

Linking Concepts

For Individual or Group Explorations

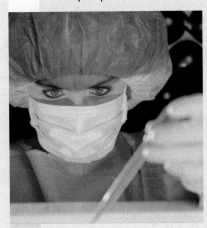

Handwritten: $\sqrt{(8)(9)}$ $\sqrt{8} \cdot \sqrt{9}$

A Logarithmic Model for Water Quality

The United States Geological Survey measures the quality of a water sample by using the diversity index d, given by

$$d = -[p_1 \cdot \log_2(p_1) + p_2 \cdot \log_2(p_2) + \cdots + p_n \cdot \log_2(p_n)],$$

where n is the number of different taxons (biological classifications) represented in the sample and p_1 through p_n are the percentages of organisms in each of the n taxons. For example, if 10% of the organisms in a water sample are E. coli and 90% are fecal coliform, then

$$d = -[0.1 \cdot \log_2(0.1) + 0.9 \cdot \log_2(0.9)] \approx 0.5.$$

a) Find the value of d when the only organism found in a sample is *E. coli* bacteria.

b) Let $n = 3$ in the formula and write the diversity index as a single logarithm.

c) If a water sample is found to contain 20% of its organisms of one type, 30% of another type, and 50% of a third type, then what is the diversity index for the water sample?

d) If the organisms in a water sample are equally distributed among 100 different taxons, then what is the diversity index?

e) If 99% of the organisms in a water sample are from one taxon and the other 1% are equally distributed among 99 other taxons, then what is the diversity index?

f) The diversity index can be found for populations other than organisms in a water sample. Find the diversity index for the dogs in the movie *101 Dalmatians* (cartoon version).

g) Identify a population of your choice and different classifications within the population. (For example, the trees on campus can be classified as pine, maple, spruce, or elm.) Gather real data and calculate the diversity index for your population.

4.4

More Equations and Applications

The rules of Section 4.3 combined with the techniques that we have already used in Sections 4.1 and 4.2 allow us to solve several new types of equations involving exponents and logarithms.

Logarithmic Equations

An equation involving a single logarithm can usually be solved by using the definition of logarithm as we did in Section 4.2.

Example **1** An equation involving a single logarithm

Solve the equation $\log(x - 3) = 4$.

Solution

Write the equivalent equation using the definition of logarithm:

$$\log(x - 3) = 4 \qquad \text{Original equation } \log_a(y) = x \Leftrightarrow y = a^x$$
$$x - 3 = 10^4$$
$$x = 10{,}003$$

Check this number in the original equation. The solution is 10,003.

Try This. Solve $\log(2x + 1) = 3$. ▪

When more than one logarithm is present, we can use the one-to-one property as in Section 4.2 or use the other rules of logarithms to combine logarithms.

Example **2** Equations involving more than one logarithm

Solve each equation.

a. $\log_2(x) + \log_2(x + 2) = \log_2(6x + 1)$ **b.** $\log(x) - \log(x - 1) = 2$
c. $2 \cdot \ln(x) = \ln(x + 3) + \ln(x - 1)$

Solution

a. Since the sum of two logarithms is equal to the logarithm of a product, we can rewrite the left-hand side of the equation.

$$\log_2(x) + \log_2(x + 2) = \log_2(6x + 1)$$
$$\log_2(x(x + 2)) = \log_2(6x + 1) \qquad \text{Product rule for logarithms}$$
$$x^2 + 2x = 6x + 1 \qquad \text{One-to-one property of logarithms}$$
$$x^2 - 4x - 1 = 0 \qquad \text{Solve quadratic equation.}$$
$$x = \frac{4 \pm \sqrt{16 - 4(-1)}}{2} = 2 \pm \sqrt{5}$$

Since $2 - \sqrt{5}$ is a negative number, $\log_2(2 - \sqrt{5})$ is undefined and $2 - \sqrt{5}$ is not a solution. The only solution to the equation is $2 + \sqrt{5}$. Check this solution by using a calculator and the base-change formula.

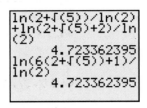 The check is shown with a graphing calculator in Fig. 4.39.   ☐

■ **Figure 4.39**

b. Since the difference of two logarithms is equal to the logarithm of a quotient, we can rewrite the left-hand side of the equation:

$$\log(x) - \log(x - 1) = 2$$

$$\log\!\left(\frac{x}{x - 1}\right) = 2 \qquad \text{Quotient rule for logarithms}$$

$$\frac{x}{x - 1} = 10^2 \qquad \text{Definition of logarithm}$$

$$x = 100x - 100 \qquad \text{Solve for } x.$$

$$-99x = -100$$

$$x = \frac{100}{99}$$

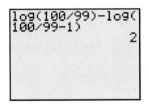

■ **Figure 4.40**

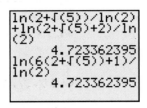 Check 100/99 in the original equation, as shown in Fig. 4.40.   ☐

c. $2 \cdot \ln(x) = \ln(x + 3) + \ln(x - 1)$

$$\ln(x^2) = \ln(x^2 + 2x - 3) \qquad \text{Power rule, product rule}$$

$$x^2 = x^2 + 2x - 3 \qquad \text{One-to-one property}$$

$$0 = 2x - 3$$

$$\frac{3}{2} = x$$

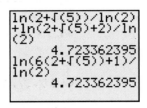 Use a calculator to check 3/2 in the original equation. The only solution to the equation is 3/2.

Try This. Solve $\log(x - 1) + \log(3) = \log(x) - \log(4)$.   ■

All solutions to logarithmic and exponential equations should be checked in the original equation, because extraneous roots can occur, as in Example 2(a). Use your calculator to check every solution and you will increase your proficiency with your calculator.

Exponential Equations

An exponential equation with a single exponential expression can usually be solved by using the definition of logarithm, as in Section 4.2.

Example **3** Equations involving a single exponential expression

Solve the equation $(1.02)^{4t-1} = 5$.

Solution

Write an equivalent equation, using the definition of logarithm.

$$4t - 1 = \log_{1.02}(5)$$

$$t = \frac{1 + \log_{1.02}(5)}{4} \qquad \text{The exact solution}$$

$$= \frac{1 + \dfrac{\ln(5)}{\ln(1.02)}}{4} \qquad \text{Base-change formula}$$

$$\approx 20.5685$$

The approximate solution is 20.5685.

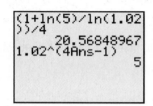

■ **Figure 4.41**

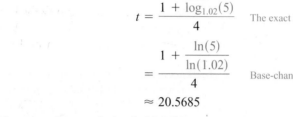 Check with a graphing calculator, as shown in Fig. 4.41.

Try This. Solve $(1.05)^{3t} = 8$. ■

Note that Example 3 can also be solved by taking the ln (or log) of each side and applying the power rule for logarithms:

$$\ln\left(1.02^{4t-1}\right) = \ln(5)$$

$$(4t - 1)\ln(1.02) = \ln(5)$$

$$4t \cdot \ln(1.02) - \ln(1.02) = \ln(5)$$

$$t = \frac{\ln(5) + \ln(1.02)}{4 \cdot \ln(1.02)} \approx 20.5685$$

If an equation has an exponential expression on each side, as in Example 4, then it is best to take the log or ln of each side to solve it.

Example **4** Equations involving two exponential expressions

Find the exact and approximate solutions to $3^{2x-1} = 5^x$.

Solution

$$\ln(3^{2x-1}) = \ln(5^x) \qquad \text{Take the natural logarithm of each side.}$$

$$(2x - 1)\ln(3) = x \cdot \ln(5) \qquad \text{Power rule for logarithms}$$

$$2x \cdot \ln(3) - \ln(3) = x \cdot \ln(5) \qquad \text{Distributive property}$$

$$2x \cdot \ln(3) - x \cdot \ln(5) = \ln(3)$$

$$x[2 \cdot \ln(3) - \ln(5)] = \ln(3)$$

$$x = \frac{\ln(3)}{2 \cdot \ln(3) - \ln(5)} \qquad \text{Exact solution}$$

$$\approx 1.8691$$

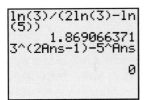

■ **Figure 4.42**

Had we used common logarithms, similar steps would give

$$x = \frac{\log(3)}{2 \cdot \log(3) - \log(5)}$$

$$\approx 1.8691.$$

Check the solution in the original equation, as shown in Fig. 4.42.

Try This. Solve $3^{x-1} = 2^x$.   ■

The technique of Example 4 can be used on any equation of the form $a^M = b^N$. Take the natural logarithm of each side and apply the power rule, to get an equation of the form $M \cdot \ln(a) = N \cdot \ln(b)$. This last equation usually has no exponents and is relatively easy to solve.

Strategy for Solving Equations

We solved equations involving exponential and logarithmic functions in Sections 4.1 through 4.4. There is no formula that will solve every exponential or logarithmic equation, but the following strategy will help you solve exponential and logarithmic equations.

S T R A T E G Y   Solving Exponential and Logarithmic Equations

1. If the equation involves a single logarithm or a single exponential expression, then use the definition of logarithm: $y = \log_a(x)$ if and only if $a^y = x$.

2. Use the one-to-one properties when applicable:
   a) if $a^M = a^N$, then $M = N$.
   b) if $\log_a(M) = \log_a(N)$, then $M = N$.

3. If an equation has several logarithms with the same base, then combine them using the product and quotient rules:
   a) $\log_a(M) + \log_a(N) = \log_a(MN)$
   b) $\log_a(M) - \log_a(N) = \log_a(M/N)$

4. If an equation has only exponential expressions with different bases on each side, then take the natural or common logarithm of each side and use the power rule: $a^M = b^N$ is equivalent to $\ln(a^M) = \ln(b^N)$ or $M \cdot \ln(a) = N \cdot \ln(b)$.

Radioactive Dating

In Section 4.1, we stated that the amount A of a radioactive substance remaining after t years is given by

$$A = A_0 e^{rt},$$

where A_0 is the initial amount present and r is the annual rate of decay for that particular substance. A standard measurement of the speed at which a radioactive substance decays is its **half-life.** The half-life of a radioactive substance is the amount of time that it takes for one-half of the substance to decay. Of course, when one-half has decayed, one-half remains.

Now that we have studied logarithms, we can use the formula for radioactive decay to determine the age of an ancient object that contains a radioactive substance. One such substance is potassium-40, which is found in rocks. Once the rock is formed, the potassium-40 begins to decay. The amount of time that has passed since the formation of the rock can be determined by measuring the amount of potassium-40 that has decayed into argon-40. Dating rocks using potassium-40 is known as **potassium-argon dating.**

Example 5 Finding the age of *Deinonychus* ("terrible claw")

Our chapter-opening case described the 1964 find of *Deinonychus* (*National Geographic*, August 1978). Since dinosaur bones are too old to contain enough organic material for radiocarbon dating, paleontologists often estimate the age of bones by dating volcanic debris in the surrounding rock. The age of *Deinonychus* was determined from the age of surrounding rocks by using potassium-argon dating. The half-life of potassium-40 is 1.31 billion years. If 94.5% of the original amount of potassium-40 is still present in the rock, then how old are the bones of *Deinonychus*?

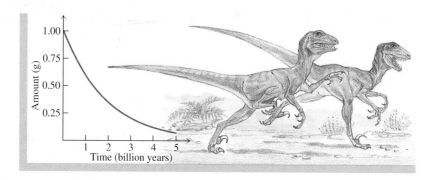

Solution

The half-life is the amount of time that it takes for 1 gram to decay to 0.5 gram. Use $A_0 = 1, A = 0.5$, and $t = 1.31 \times 10^9$ in the formula $A = A_0 e^{rt}$ to find r:

$$0.5 = 1 \cdot e^{(1.31 \times 10^9)(r)}$$

$$(1.31 \times 10^9)(r) = \ln(0.5) \qquad \text{\small Definition of logarithm}$$

$$r = \frac{\ln(0.5)}{1.31 \times 10^9}$$

$$\approx -5.29 \times 10^{-10}$$

Now we can find the amount of time that it takes for 1 gram to decay to 0.945 gram. Use $r \approx -5.29 \times 10^{-10}, A_0 = 1$, and $A = 0.945$ in the formula.

$$0.945 = 1 \cdot e^{(-5.29 \times 10^{-10})(t)}$$

$$(-5.29 \times 10^{-10})(t) = \ln(0.945) \qquad \text{\small Definition of logarithm}$$

$$t = \frac{\ln(0.945)}{-5.29 \times 10^{-10}}$$

$$\approx 107 \text{ million years}$$

The dinosaur *Deinonychus* lived about 107 million years ago.

Try This. If the half-life of a substance is 1 million years and 40% of the original amount is still present in a rock, then how old is the rock? ◼

Newton's Model for Cooling

Newton's law of cooling states that when a warm object is placed in colder surroundings or a cold object is placed in warmer surroundings, then the difference between the two temperatures decreases in an exponential manner. If D_0 is the initial difference in temperature, then the difference D at time t is modeled by the formula

$$D = D_0 e^{kt},$$

where k is a constant that depends on the object and the surroundings. In the next example we use Newton's law of cooling to answer a question that you might have asked yourself as the appetizers were running low.

Example **6** Using Newton's law of cooling

A turkey with a temperature of 40°F is moved to a 350° oven. After 4 hours the internal temperature of the turkey is 170°F. If the turkey is done when its temperature reaches 185°, then how much longer must it cook?

◼ **Figure 4.43**

Solution

The initial difference of 310° (350° − 40°) has dropped to a difference of 180° (350° − 170°) after 4 hours. See Fig. 4.43. With this information we can find k:

$$180 = 310 e^{4k}$$

$$e^{4k} = \frac{180}{310} \qquad \text{Isolate } e^{4k} \text{ by dividing by 310.}$$

$$4k = \ln(18/31) \qquad \text{Definition of logarithm: } e^x = y \Leftrightarrow x = \ln(y)$$

$$k = \frac{\ln(18/31)}{4} \approx -0.1359$$

The turkey is done when the difference in temperature between the turkey and the oven is 165° (the oven temperature 350° minus the turkey temperature 185°). Now find the time for which the difference will be 165°:

$$165 = 310e^{-0.1359t}$$

$$e^{-0.1359t} = \frac{165}{310}$$

$$-0.1359t = \ln(165/310) \qquad \text{Definition of logarithm: } e^x = y \Leftrightarrow x = \ln(y)$$

$$t = \frac{\ln(165/310)}{-0.1359} \approx 4.6404$$

The difference in temperature between the turkey and the oven will be 165° when the turkey has cooked approximately 4.6404 hours. So the turkey must cook about 0.6404 hour (38.4 minutes) longer.

Try This. A cup of boiling water (212°F) is placed outside, where the temperature is 12°F. In 4 minutes the water temperature is 102°F. How much longer will it take for the water temperature to reach 82°F? ■

Paying off a Loan

If n is the number of periods per year, r is the annual percentage rate (APR), t is the number of years, and i is the interest rate per period ($i = r/n$), then the periodic payment R that will pay off a loan of P dollars is given by

$$R = P\frac{i}{1 - (1 + i)^{-nt}}.$$

Homeowners are often interested in how the time will change if they increase the monthly payment. Solving for t requires logarithms.

Example **7** **Finding the time**

A couple still owes $90,000 on a house that is financed at 8% annual percentage rate compounded monthly. If they start paying $1200 per month, then when will the loan be paid off?

Solution

Use $R = 1200$, $P = 90{,}000$, $n = 12$, and $i = 0.08/12$ in the formula for the monthly payment:

$$1200 = 90{,}000\frac{0.08/12}{1 - (1 + 0.08/12)^{-12t}}$$

$$1200\left(1 - (1 + 0.08/12)^{-12t}\right) = 600$$

$$1 - (1 + 0.08/12)^{-12t} = 0.5$$

$$0.5 = (1 + 0.08/12)^{-12t}$$

$$\ln(0.5) = -12t \cdot \ln(1 + 0.08/12)$$

$$\frac{\ln(0.5)}{-12 \cdot \ln(1 + 0.08/12)} = t$$

$$8.6932 \approx t$$

So in approximately 8.6932 years or about 8 years and 8 months the loan will be paid off.

Try This. How long (to the nearest month) does it take to pay off $120,000 financed at 7.5% APR compounded monthly with payments of $2000 per month? ■

For Thought

True or False? Explain.

1. The equation $3(1.02)^x = 21$ is equivalent to $x = \log_{1.02}(7)$.

2. If $x - x \cdot \ln(3) = 8$, then $x = \dfrac{8}{1 - \ln(3)}$.

3. The solution to $\ln(x) - \ln(x - 1) = 6$ is $1 - \sqrt{6}$.

4. If $2^{x-3} = 3^{2x+1}$, then $x - 3 = \log_2(3^{2x+1})$.

5. The exact solution to $3^x = 17$ is 2.5789.

6. The equation $\log(x) + \log(x - 3) = 1$ is equivalent to $\log(x^2 - 3x) = 1$.

7. The equation $4^x = 2^{x-1}$ is equivalent to $2x = x - 1$.

8. The equation $(1.09)^x = 2.3$ is equivalent to $x \cdot \ln(1.09) = \ln(2.3)$.

9. $\ln(2) \cdot \log(7) = \log(2) \cdot \ln(7)$

10. $\log(e) \cdot \ln(10) = 1$

4.4 Exercises

Each of these equations involves a single logarithm. Solve each equation. See the strategy for solving exponential and logarithmic equations on page 399.

1. $\log_2(x) = 3$

2. $\log_3(x) = 0$

3. $\log(x + 20) = 2$

4. $\log(x - 6) = 1$

5. $\log(x^2 - 15) = 1$

6. $\log(x^2 - 5x + 16) = 1$

7. $\log_x(9) = 2$

8. $\log_x(16) = 4$

9. $-2 = \log_x(4)$

10. $-\dfrac{1}{2} = \log_x(9)$

11. $\log_x(10) = 3$

12. $\log_x(5) = 2$

13. $\log_8(x) = -\dfrac{2}{3}$

14. $\log_4(x) = -\dfrac{5}{2}$

Each of these equations involves more than one logarithm. Solve each equation. Give exact solutions.

15. $\log_2(x + 2) + \log_2(x - 2) = 5$

16. $\log(x + 1) - \log(x) = 3$

17. $\log(5) = 2 - \log(x)$

18. $\log(4) = 1 + \log(x - 1)$

19. $\ln(x) + \ln(x + 2) = \ln(8)$

20. $\log_3(x) = \log_3(2) - \log_3(x - 2)$

21. $\log(4) + \log(x) = \log(5) - \log(x)$

22. $\ln(x) - \ln(x + 1) = \ln(x + 3) - \ln(x + 5)$

23. $\log_2(x) - \log_2(3x - 1) = 0$

24. $\log_3(x) + \log_3(1/x) = 0$

25. $x \cdot \ln(3) = 2 - x \cdot \ln(2)$

26. $x \cdot \log(5) + x \cdot \log(7) = \log(9)$

Each of these equations involves a single exponential expression. Solve each equation. Round approximate solutions to four decimal places.

27. $2^{x-1} = 7$

28. $5^{3x} = 29$

29. $(1.09)^{4x} = 3.4$

30. $(1.04)^{2x} = 2.5$

31. $3^{-x} = 30$

32. $10^{-x+3} = 102$

33. $9 = e^{-3x^2}$

34. $25 = 10^{-2x}$

Each of these equations involves more than one exponential expression. Solve each equation. Round approximate solutions to four decimal places.

35. $6^x = 3^{x+1}$

36. $2^x = 3^{x-1}$

37. $e^{x+1} = 10^x$

38. $e^x = 2^{x+1}$

39. $2^{x-1} = 4^{3x}$

40. $3^{3x-4} = 9^x$

41. $6^{x+1} = 12^x$

42. $2^x \cdot 2^{x+1} = 4^{x^2+x}$

Solve each equation. Round approximate solutions to four decimal places.

43. $e^{-\ln(w)} = 3$

44. $10^{2\cdot\log(y)} = 4$

45. $(\log(z))^2 = \log(z^2)$

46. $\ln(e^x) - \ln(e^6) = \ln(e^2)$

47. $4(1.02)^x = 3(1.03)^x$

48. $500(1.06)^x = 400(1.02)^{4x}$

49. $e^{3\cdot\ln(x^2)-2\cdot\ln(x)} = \ln(e^{16})$

50. $\sqrt{\log(x) - 3} = \log(x) - 3$

51. $\left(\frac{1}{2}\right)^{2x-1} = \left(\frac{1}{4}\right)^{3x+2}$

52. $\left(\frac{2}{3}\right)^{x+1} = \left(\frac{9}{4}\right)^{x+2}$

⊞ *Find the approximate solution to each equation by graphing an appropriate function on a graphing calculator and locating the x-intercept. Note that these equations cannot be solved by the techniques that we have learned in this chapter.*

53. $2^x = 3^{x-1} + 5^{-x}$

54. $2^x = \log(x + 4)$

55. $\ln(x + 51) = \log(-48 - x)$ **56.** $2^x = 5 - 3^{x+1}$

57. $x^2 = 2^x$

58. $x^3 = e^x$

Solve each problem.

59. *Finding the Rate* If the half-life of a radioactive substance is 10,000 years, then at what rate is it decaying?
HINT The amount goes from A_0 to $\frac{1}{2}A_0$ in 10,000 years.

60. *Finding the Rate* If the half-life of a drug is 12 hours, then at what rate is it being eliminated from the body?

61. *Dating a Bone* A piece of bone from an organism is found to contain 10% of the carbon-14 that it contained when the organism was living. If the half-life of carbon-14 is 5730 years, then how long ago was the organism alive?
HINT First find the rate of decay for carbon-14.

62. *Old Clothes* If only 15% of the carbon-14 in a remnant of cloth has decayed, then how old is the cloth?
HINT Use the decay rate for carbon-14 from the previous problem.

63. *Dating a Tree* How long does it take for 12 g of carbon-14 in a tree trunk to be reduced to 10 g of carbon-14 by radioactive decay?

64. *Carbon-14 Dating* How long does it take for 2.4 g of carbon-14 to be reduced to 1.3 g of carbon-14 by radioactive decay?

65. *Radioactive Waste* If 25 g of radioactive waste reduces to 20 g of radioactive waste after 8000 years, then what is the half-life for this radioactive element?
HINT First find the rate of decay, then find the time for half of it to decay.

66. *Finding the Half-Life* If 80% of a radioactive element remains radioactive after 250 million years, then what percent remains radioactive after 600 million years? What is the half-life of this element?

67. *Lorazepam* The drug lorazepam, used to relieve anxiety and nervousness, has a half-life of 14 hours. Its chemical structure is shown in the accompanying figure. If a doctor prescribes one 2.5 milligram tablet every 24 hours, then what percentage of the last dosage remains in the patient's body when the next dosage is taken?

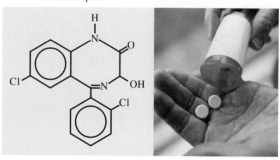

Lorazepam

■ **Figure for Exercise 67**

68. *Drug Build-Up* The level of a prescription drug in the human body over time can be found using the formula

$$L = \frac{D}{1 - (0.5)^{n/h}},$$

where D is the amount taken every n hours and h is the drug's half-life in hours.

a. If 2.5 milligrams of lorazepam with a half-life of 14 hours is taken every 24 hours, then to what level does the drug build up over time?

b. If a doctor wants the level of lorazepam to build up to a level of 5.58 milligrams in a patient taking 2.5 milligram doses, then how often should the doses be taken?

c. What is the difference between taking 2.5 milligrams of lorazepam every 12 hours and taking 5 milligrams every 24 hours?

69. *Dead Sea Scrolls* Willard Libby, a nuclear chemist from the University of Chicago, developed radiocarbon dating in the 1940s. This dating method, effective on specimens up to about 40,000 years old, works best on objects like shells, charred bones, and plants that contain organic matter (carbon). Libby's first great success came in 1951 when he dated the Dead Sea Scrolls. Carbon-14 has a half-life of 5730 years. If Libby found 79.3% of the original carbon-14 still present, then in about what year were the scrolls made?

70. *Leakeys Date Zinjanthropus*   In 1959, archaeologists Louis and Mary Leakey were exploring Olduvai Gorge, Tanzania, when they uncovered the remains of *Zinjanthropus*, an early hominid with traits of both ape and man. Dating of the volcanic rock revealed that 91.2% of the original potassium had not decayed into argon. What age was assigned to the rock and the bones of *Zinjanthropus*? The radioactive decay of potassium-40 to argon-40 occurs with a half-life of 1.31 billion years.

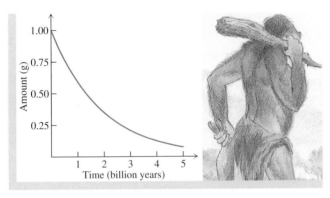

■ **Figure for Exercise 70**

71. *Cooking a Roast*   James knows that to get well-done beef, it should be brought to a temperature of 170°F. He placed a sirloin tip roast with a temperature of 35°F in an oven with a temperature of 325°, and after 3 hr the temperature of the roast was 140°. How much longer must the roast be in the oven to get it well done? If the oven temperature is set at 170°, how long will it take to get the roast well done?

    HINT The difference between the roast temperature and the oven temperature decreases exponentially.

72. *Room Temperature*   Marlene brought a can of polyurethane varnish that was stored at 40°F into her shop, where the temperature was 74°. After 2 hr the temperature of the varnish was 58°. If the varnish must be 68° for best results, then how much longer must Marlene wait until she uses the varnish?

73. *Time of Death*   A detective discovered a body in a vacant lot at 7 A.M. and found that the body temperature was 80°F. The county coroner examined the body at 8 A.M. and found that the body temperature was 72°. Assuming that the body temperature was 98° when the person died and that the air temperature was a constant 40° all night, what was the approximate time of death?

74. *Cooling Hot Steel*   A blacksmith immersed a piece of steel at 600°F into a large bucket of water with a temperature of 65°. After 1 min the temperature of the steel was 200°. How much longer must the steel remain immersed to bring its temperature down to 100°?

75. *Paying off a Loan*   Find the time (to the nearest month) that it takes to pay off a loan of $100,000 at 9% APR compounded monthly with payments of $1250 per month.

76. *Solving for Time*   Solve the formula

$$R = P\frac{i}{1 - (1 + i)^{-nt}}$$

    for t. Then use the result to find the time (to the nearest month) that it takes to pay off a loan of $48,265 at $8\frac{3}{4}$% APR compounded monthly with payments of $700 per month.

77. *Equality of Investments*   Fiona invested $1000 at 6% compounded continuously. At the same time, Maria invested $1100 at 6% compounded daily. How long will it take (to the nearest day) for their investments to be equal in value?

78. *Depreciation and Inflation*   Boris won a $35,000 luxury car on *Wheel of Fortune*. He plans to keep it until he can trade it evenly for a new compact car that currently costs $10,000. If the value of the luxury car decreases by 8% each year and the cost of the compact car increases by 5% each year, then in how many years will he be able to make the trade?

79. *Population of Rabbits*   The population of rabbits in a national forest is modeled by the formula $P = 12,300 + 1000 \cdot \ln(t + 1)$, where t is the number of years from the present.
    a. How many rabbits are now in the forest?

    b. Use the accompanying graph to estimate the number of years that it will take for the rabbit population to reach 15,000.

    c. Use the formula to determine the number of years that it will take for the rabbit population to reach 15,000.

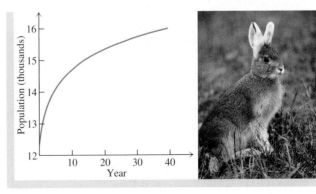

■ **Figure for Exercise 79**

80. *Population of Foxes*   The population of foxes in the forest of Exercise 79 appears to be growing according to the formula $P = 400 + 50 \cdot \ln(90t + 1)$. When the population of foxes is equal to 5% of the population of rabbits, the system is considered to be out of ecological balance. In how many years will the system be out of balance?

81. *Habitat Destruction* Biologists use the species-area curve $n = k \log(A)$ to estimate the number of species n that live in a region of area A, where k is a constant.
a. If 2500 species live in a rain forest of 400 square kilometers, then how many species will be left when half of this rain forest is destroyed by logging?
b. A rain forest of 1200 square kilometers supported 3500 species in 1950. Due to intensive logging, what remains of this forest supported only 1000 species in 2000. What percent of this rain forest has been destroyed?

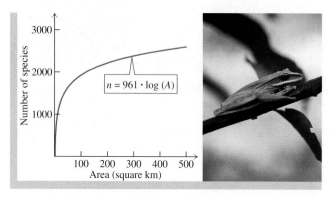

▪ **Figure for Exercise 81**

82. *Extinction of Species* In 1980 an Amazon rain forest of area A contained n species. In the year 2000, after intensive logging, this rain forest was reduced to an area that contained only half as many species. Use the species-area formula from the previous exercise to find the area in 2000.

83. *Visual Magnitude of a Star* If all stars were at the same distance, it would be a simple matter to compare their brightness. However, the brightness that we see, the apparent visual magnitude m, depends on a star's intrinsic brightness, or absolute visual magnitude M_V, and the distance d from the observer in parsecs (1 parsec = 3.262 light years), according to the formula $m = M_V - 5 + 5 \cdot \log(d)$. The values of M_V range from -8 for the intrinsically brightest stars to $+15$ for the intrinsically faintest stars. The nearest star to the sun, Alpha Centauri, has an apparent visual magnitude of 0 and an absolute visual magnitude of 4.39. Find the distance d in parsecs to Alpha Centauri.

84. *Visual Magnitude of Deneb* The star Deneb is 490 parsecs away and has an apparent visual magnitude of 1.26, which means that it is harder to see than Alpha Centauri. Use the formula from Exercise 83 to find the absolute visual magnitude of Deneb. Is Deneb intrinsically very bright or very faint? If Deneb and Alpha Centauri were both 490 parsecs away, then which would appear brighter?

85. *Price Per Gigabyte* According to Moore's Law, the performance of electronic parts grows exponentially while the cost of those parts declines exponentially. The figure on the next page shows the decline in price P for a gigabyte of hard drive storage for the years 1982 through 2005. The relationship looks linear because the price axis has a log scale.
a. Find a formula for P in terms of x, where x is the number of years since 1982. See Exercise 121 of Section 4.2.
b. Find P in 1991.
c. Find the year in which the price will be $0.25.

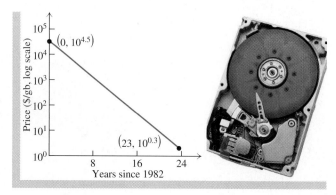

▪ **Figure for Exercise 85**

86. *Hard Drive Capacity* The accompanying figure shows how hard drive capacity C has increased for the years 1982 through 2005. The relationship looks linear because the capacity axis has a log scale.
a. Find a formula for C in terms of x, where x is the number of years since 1982. See the previous exercise.
b. Find C in 1994.
c. Find the year in which the capacity will be 1000 gigabytes.

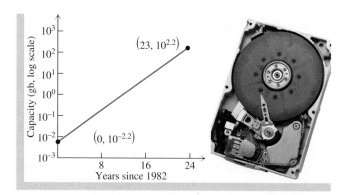

▪ **Figure for Exercise 86**

87. *Noise Pollution* The level of a sound in decibels (db) is determined by the formula

$$\text{sound level} = 10 \cdot \log(I \times 10^{12}) \text{ db},$$

where I is the intensity of the sound in watts per square meter. To combat noise pollution, a city has an ordinance prohibiting sounds above 90 db on a city street. What value of I gives a sound of 90 db?

88. *Doubling the Sound Level* A small stereo amplifier produces a sound of 50 db at a distance of 20 ft from the speakers. Use the formula from Exercise 87 to find the intensity of the sound at this point in the room. If the intensity just found is doubled, what happens to the sound level? What must the intensity be to double the sound level to 100 db?

89. *Present Value of a CD* What amount (present value) must be deposited today in a certificate of deposit so that the investment will grow to $20,000 in 18 years at 6% compounded continuously.

90. *Present Value of a Bond* A $50 U.S. Savings Bond paying 6.22% compounded monthly matures in 11 years 2 months. What is the present value of the bond?

91. *Two-Parent Families* The percentage of households with children that consist of two-parent families is shown in the following table (U.S. Census Bureau, www.census.gov).

Year	Two Parents	
1970	85%	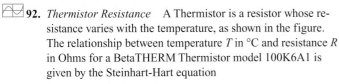
1980	77	
1990	73	
1992	71	
1994	69	
1996	68	
1998	68	
2000	68	

a. Use logarithmic regression on a graphing calculator to find the best-fitting curve of the form $y = a + b \cdot \ln(x)$, where $x = 0$ corresponds to 1960.

b. Use your equation to predict the percentage of two-parent families in 2010.

c. In what year will the percentage of two-parent families reach 50%?

d. Graph your equation and the data on your graphing calculator. Does this logarithmic model look like another model that we have used?

92. *Thermistor Resistance* A Thermistor is a resistor whose resistance varies with the temperature, as shown in the figure. The relationship between temperature T in °C and resistance R in Ohms for a BetaTHERM Thermistor model 100K6A1 is given by the Steinhart-Hart equation

$$T = \frac{1}{A + B \cdot \ln(R) + C \cdot [\ln(R)]^3} - 273.15,$$

where

$$A = 8.27153 \times 10^{-4}$$
$$B = 2.08796 \times 10^{-4}$$
$$C = 8.060985 \times 10^{-8}.$$

At what temperature is the resistance 1×10^5 Ohms? Use a graphing calculator to find the resistance when $T = 35$°C.

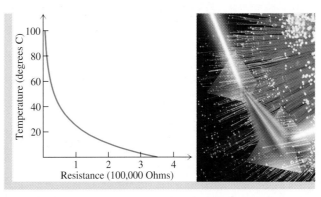

■ **Figure for Exercise 92**

93. *Infinite Series for e^x* The following formula from calculus is used to compute values of e^x:

$$e^x = 1 + x + \frac{x^2}{2!} + \frac{x^3}{3!} + \frac{x^4}{4!} + \cdots + \frac{x^n}{n!} + \cdots,$$

where $n! = 1 \cdot 2 \cdot 3 \cdot \cdots \cdot n$ for any positive integer n. The notation $n!$ is read "n factorial." For example, $3! = 1 \cdot 2 \cdot 3 = 6$. In calculating e^x, the more terms that we use from the formula, the closer we get to the true value of e^x. Use the first five terms of the formula to estimate the value of $e^{0.1}$ and compare your result to the value of $e^{0.1}$ obtained using the e^x-key on your calculator.

94. *Infinite Series for Logarithms* The following formula from calculus can be used to compute values of natural logarithms:

$$\ln(1 + x) = x - \frac{x^2}{2} + \frac{x^3}{3} - \frac{x^4}{4} + \cdots,$$

where $-1 < x < 1$. The more terms that we use from the formula, the closer we get to the true value of $\ln(1 + x)$. Find $\ln(1.4)$ by using the first five terms of the series and compare your result to the calculator value for $\ln(1.4)$.

Thinking Outside the Box XLII

Pile of Pipes Six pipes are placed in a pile as shown in the diagram. The three pipes on the bottom each have a radius of 2 feet. The two pipes on top of those each have a radius of 1 foot. At the top of the pile is a pipe that is tangent to the two smaller pipes and one of the larger pipes. What is its exact radius?

■ **Figure for Thinking Outside the Box XLII**

4.4 Pop Quiz

Solve each equation. Round approximate solutions to four decimal places.

1. $\log_5(x) = 4$

2. $\log(x + 1) = 3$

3. $\ln(x) + \ln(x - 1) = \ln(12)$

4. $3^{x-5} = 10$

5. $8^x = 3^{x+5}$

Linking Concepts

For Individual or Group Explorations

The Logistic Growth Model

*When a virus infects a finite population of size P in which no one is immune, the virus spreads slowly at first, then more rapidly as more people are infected, and finally slows down when nearly everyone has been infected. This situation is modeled by a **logistic curve** of the form*

$$n = \frac{P}{1 + (P - 1) \cdot e^{-ct}},$$

where n is the number of people who have caught the virus on or before day t, and c is a positive constant.

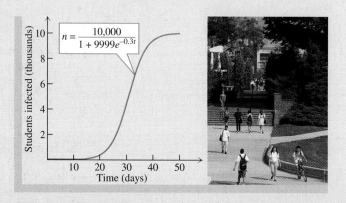

a) According to the model, how many people have caught the virus at time $t = 0$?

b) Now consider what happens when one student carrying a flu virus returns from spring break to a university of 10,000 students. For $c = 0.1, 0.2$, and so on through $c = 0.9$, graph the logistic curve

$$y_1 = 10000/(1 + 9999e^{\wedge}(-cx))$$

and the daily number of new cases

$$y_2 = y_1(x) - y_1(x - 1).$$

For each value of c use the graph of y_2 to find the day on which the flu is spreading most rapidly.

c) The Health Center estimated that the greatest number of new cases of the flu occurred on the 19th day after the students returned from spring break. What value of c should be used to model this situation? How many new cases occurred on the 19th day?

d) Algebraically find the day on which the number of infected students reached 9000.

e) The Health Center has a team of doctors from Atlanta arriving on the 30th day to help with this three-day flu epidemic. What do you think of this plan?

■ ■ ■ Highlights

4.1 Exponential Functions and Their Applications

Exponential Function	$f(x) = a^x$ for $a > 0$ and $a \neq 1$ Domain: $(-\infty, \infty)$, Range: $(0, \infty)$ Horizontal asymptote: $y = 0$	$f(x) = 2^x, g(x) = e^x$
Increasing and Decreasing	$f(x) = a^x$ is increasing on $(-\infty, \infty)$ if $a > 1$, decreasing on $(-\infty, \infty)$ if $0 < a < 1$.	$f(x) = 3^x$ is increasing $g(x) = (0.4)^x$ is decreasing
One-to-One	If $a^{x_1} = a^{x_2}$, then $x_1 = x_2$.	$2^{x-1} = 2^3 \Rightarrow x - 1 = 3$
Amount Formula	P dollars, annual interest rate r, for t years Compounded periodically: n periods per year, $A = P\left(1 + \frac{r}{n}\right)^{nt}$ Compounded continuously: $A = Pe^{rt} (e \approx 2.718)$	$P = \$1000, r = 5\%, t = 10$ yr Compounded monthly: $A = 1000\left(1 + \frac{0.05}{12}\right)^{120}$ Compounded continuously: $A = 1000e^{0.05(10)}$

4.2 Logarithmic Functions and Their Applications

Logarithmic Function	$f(x) = \log_a(x)$ for $a > 0$ and $a \neq 1$ $y = \log_a(x) \Leftrightarrow a^y = x$ Domain: $(0, \infty)$, Range: $(-\infty, \infty)$ Vertical asymptote: $x = 0$ Common log: base 10, $f(x) = \log(x)$ Natural log: base e, $f(x) = \ln(x)$	$f(x) = \log_2(x)$ $f(32) = \log_2(32) = 5$ $f(1) = \log_2(1) = 0$ $f(1/4) = \log_2(1/4) = -2$
Increasing and Decreasing	$f(x) = \log_a(x)$ is increasing on $(0, \infty)$ if $a > 1$, decreasing on $(0, \infty)$ if $0 < a < 1$.	$f(x) = \ln(x)$ is increasing. $g(x) = \log_{1/2}(x)$ is decreasing.
One-to-One	If $\log_a(x_1) = \log_a(x_2)$, then $x_1 = x_2$.	$\log_2(3x) = \log_2(4) \Rightarrow 3x = 4$

4.3 Rules of Logarithms

Product Rule	$\log_a(MN) = \log_a(M) + \log_a(N)$	$\log_2(8x) = 3 + \log_2(x)$
Quotient Rule	$\log_a(M/N) = \log_a(M) - \log_a(N)$	$\log(1/2) = \log(1) - \log(2)$

Power Rule	$\log_a(M^N) = N \cdot \log_a(M)$	$\ln(e^3) = 3 \cdot \ln(e)$
Inverse Rules	$\log_a(a^x) = x$ and $a^{\log_a(x)} = x$	$\ln(e^5) = 5,\ 10^{\log(7)} = 7$
Base-Change	$\log_a(M) = \dfrac{\log_b(M)}{\log_b(a)}$	$\log_3(7) = \dfrac{\ln(7)}{\ln(3)} = \dfrac{\log(7)}{\log(3)}$

4.4 More Equations and Applications

Equation-Solving Strategy

1. Use $y = \log_a(x) \Leftrightarrow a^y = x$ on equations with a single logarithm or single exponential.
2. Use the one-to-one properties to eliminate logarithms or exponential expressions.
3. Combine logarithms using the product and quotient rules.
4. Take a logarithm of each side and use the power rule.

$\log_2(x) = 3 \Leftrightarrow x = 2^3$

$\log(x^2) = \log(5) \Leftrightarrow x^2 = 5$
$e^x = e^{2x-1} \Leftrightarrow x = 2x - 1$
$\ln(x) + \ln(2) = 3$
$\qquad \ln(2x) = 3$
$3^x = 4^{2x}$
$x \cdot \ln(3) = 2x \cdot \ln(4)$

▪▪▪Chapter 4 Review Exercises

Simplify each expression.

1. 2^6 **2.** $\ln(e^2)$ **3.** $\log_2(64)$

4. $3 + 2 \cdot \log(10)$ **5.** $\log_9(1)$ **6.** $5^{\log_5(99)}$

7. $\log_2(2^{17})$ **8.** $\log_2(\log_2(16))$

Let $f(x) = 2^x$, $g(x) = 10^x$, and $h(x) = \log_2(x)$. Simplify each expression.

9. $f(5)$ **10.** $g(-1)$ **11.** $\log(g(3))$

12. $g(\log(5))$ **13.** $(h \circ f)(9)$ **14.** $(f \circ h)(7)$

15. $g^{-1}(1000)$ **16.** $g^{-1}(1)$ **17.** $h(1/8)$

18. $f(1/2)$ **19.** $f^{-1}(8)$ **20.** $(f^{-1} \circ f)(13)$

Rewrite each expression as a single logarithm.

21. $\log(x - 3) + \log(x)$ **22.** $(1/2)\ln(x) - 2 \cdot \ln(y)$

23. $2 \cdot \ln(x) + \ln(y) + \ln(3)$

24. $3 \cdot \log_2(x) - 2 \cdot \log_2(y) + \log_2(z)$

Rewrite each expression as a sum or difference of multiples of logarithms.

25. $\log(3x^4)$ **26.** $\ln\left(\dfrac{x^5}{y^3}\right)$

27. $\log_3\left(\dfrac{5\sqrt{x}}{y^4}\right)$ **28.** $\log_2\left(\sqrt{xy^3}\right)$

Rewrite each expression in terms of $\ln(2)$ and $\ln(5)$.

29. $\ln(10)$ **30.** $\ln(0.4)$ **31.** $\ln(50)$ **32.** $\ln\left(\sqrt{20}\right)$

Find the exact solution to each equation.

33. $\log(x) = 10$ **34.** $\log_3(x + 1) = -1$

35. $\log_x(81) = 4$ **36.** $\log_x(1) = 0$

37. $\log_{1/3}(27) = x + 2$ **38.** $\log_{1/2}(4) = x - 1$

39. $3^{x+2} = \dfrac{1}{9}$ **40.** $2^{x-1} = \dfrac{1}{4}$

41. $e^{x-2} = 9$ **42.** $\dfrac{1}{2^{1-x}} = 3$

43. $4^{x+3} = \dfrac{1}{2^x}$ **44.** $3^{2x-1} \cdot 9^x = 1$

45. $\log(x) + \log(2x) = 5$ **46.** $\log(x + 90) - \log(x) = 1$

47. $\log_2(x) + \log_2(x - 4) = \log_2(x + 24)$

48. $\log_5(x + 18) + \log_5(x - 6) = 2 \cdot \log_5(x)$

49. $2 \cdot \ln(x + 2) = 3 \cdot \ln(4)$

50. $x \cdot \log_2(12) = x \cdot \log_2(3) + 1$

51. $x \cdot \log(4) = 6 - x \cdot \log(25)$

52. $\log(\log(x)) = 1$

Find the missing coordinate so that each ordered pair satisfies the given equation.

53. $y = \left(\dfrac{1}{3}\right)^x$: $(-1, \ \ ), (\ \ , 27), (-1/2, \ \ ), (\ \ , 1)$

54. $y = \log_9(x - 1)$: $(2, \ \ ), (4, \ \ ), (\ \ , 3/2), (\ \ , -1)$

Match each equation to one of the graphs (a)–(h).

55. $y = 2^x$ **56.** $y = 2^{-x}$

57. $y = \log_2(x)$ **58.** $y = \log_{1/2}(x)$

59. $y = 2^{x+2}$ **60.** $y = \log_2(x + 2)$

61. $y = 2^x + 2$ **62.** $y = 2 + \log_2(x)$

a.
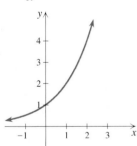

b.
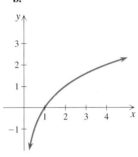

c.
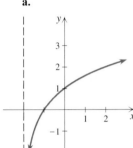

d.
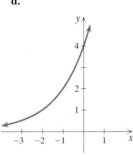

e.
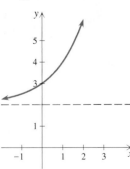

f.
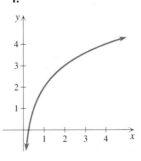

g.
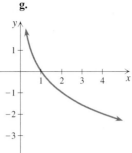

h.
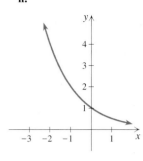

Sketch the graph of each function. State the domain, the range, and whether the function is increasing or decreasing. Identify any asymptotes.

63. $f(x) = 5^x$ **64.** $f(x) = e^x$

65. $f(x) = 10^{-x}$ **66.** $f(x) = (1/2)^x$

67. $y = \log_3(x)$ **68.** $y = \log_5(x)$

69. $y = 1 + \ln(x + 3)$ **70.** $y = 3 - \log_2(x)$

71. $f(x) = 1 + 2^{x-1}$ **72.** $f(x) = 3 - 2^{x+1}$

73. $y = \log_3(-x + 2)$ **74.** $y = 1 + \log_3(x + 2)$

For each function f, find f^{-1}.

75. $f(x) = 7^x$ **76.** $f(x) = 3^x$

77. $f(x) = \log_5(x)$ **78.** $f(x) = \log_8(x)$

79. $f(x) = 3 \cdot \log(x - 1)$ **80.** $f(x) = \log_2(x + 3) - 5$

81. $f(x) = e^{x+2} - 3$ **82.** $f(x) = 2 \cdot 3^x + 1$

Use a calculator to find an approximate solution to each equation. Round answers to four decimal places.

83. $3^x = 10$ **84.** $4^{2x} = 12$

85. $\log_3(x) = 1.876$ **86.** $\log_5(x + 2) = 2.7$

87. $5^x = 8^{x+1}$ **88.** $3^x = e^{x+1}$

Use the rules of logarithms to determine whether each equation is true or false. Explain your answer.

89. $\log_3(81) = \log_3(9) \cdot \log_3(9)$

90. $\log(81) = \log(9) \cdot \log(9)$

91. $\ln(3^2) = (\ln(3))^2$ **92.** $\ln\left(\dfrac{5}{8}\right) = \dfrac{\ln(5)}{\ln(8)}$

93. $\log_2(8^4) = 12$

94. $\log(8.2 \times 10^{-9}) = -9 + \log(8.2)$

95. $\log(1006) = 3 + \log(6)$

96. $\dfrac{\log_2(16)}{\log_2(4)} = \log_2(16) - \log_2(4)$

97. $\dfrac{\log_2(8)}{\log_2(16)} = \log_2(8) - \log_2(16)$

98. $\ln(e^{(e^x)}) = e^x$

99. $\log_2(25) = 2 \cdot \log(5)$

100. $\log_3\!\left(\dfrac{5}{7}\right) = \log(5) - \log(7)$

101. $\log_2(7) = \dfrac{\log(7)}{\log(2)}$ **102.** $\dfrac{\ln(17)}{\ln(3)} = \dfrac{\log(17)}{\log(3)}$

Solve each problem.

103. *Comparing pH* If the hydrogen ion concentration of liquid A is 10 times the hydrogen ion concentration of liquid B, then how does the pH of A compare with the pH of B? See Exercise 123 in Section 4.2 and Exercise 96 in Section 4.3.

104. *Solving a Formula* Solve the formula $A = P + Ce^{-kt}$ for t.

105. *Future Value* If $50,000 is deposited in a bank account paying 5% compounded quarterly, then what will be the value of the account at the end of 18 years?

106. *Future Value* If $30,000 is deposited in First American Savings and Loan in an account paying 6.18% compounded continuously, then what will be the value of the account after 12 years and 3 months?

107. *Doubling Time with Quarterly Compounding* How long (to the nearest quarter) will it take for the investment of Exercise 105 to double?

108. *Doubling Time with Continuous Compounding* How long (to the nearest day) will it take for the investment of Exercise 106 to double?

109. *Finding the Half-Life* The number of grams A of a certain radioactive substance present at time t is given by the formula $A = 25e^{-0.00032t}$, where t is the number of years from the present. How many grams are present initially? How many grams are present after 1000 years? What is the half-life of this substance?

110. *Comparing Investments* Shinichi invested $800 at 6% compounded continuously. At the same time, Toshio invested $1000 at 5% compounded monthly. How long (to the nearest month) will it take for their investments to be equal in value?

111. *Learning Curve* According to an educational psychologist, the number of words learned by a student of a foreign language after t hours in the language laboratory is given by $f(t) = 40{,}000(1 - e^{-0.0001t})$. How many hours would it take to learn 10,000 words?

112. *Pediatric Tuberculosis* Health researchers divided the state of Maryland into regions according to per capita income and found a relationship between per capita income i (in thousands) and the percentage of pediatric TB cases P. The equation $P = 31.5(0.935)^i$ can be used to model the data.
  a. What percentage of TB cases would you expect to find in the region with a per capita income of $8000?
  b. What per capita income would you expect in the region where only 2% of the TB cases occurred?

113. *Sea Worthiness* Three measures of a boat's sea worthiness are given in the accompanying table, where A is the sail area in square feet, d is the displacement in pounds, L is the length at the water line in feet, and b is the beam in feet. The Freedom 40 is a $291,560 sailboat with a sail area of 1026 ft^2, a displacement of 25,005 lb, a beam of 13 ft 6 in., and a length at the water line of 35 ft 1 in. Its sail area-displacement ratio is 19.20, its displacement-length ratio is 258.51, and its capsize screening value is 1.85. Find x, y, and z.

■ **Table for Exercise 113**

Measure	Expression	
Sail area-displacement ratio	$A\!\left(\dfrac{d}{64}\right)^x$	
Displacement-length ratio	$\dfrac{d}{2240}\left(\dfrac{L}{100}\right)^y$	
Capsize-screening value	$b\!\left(\dfrac{d}{64}\right)^z$	

114. *Doubling the Bet* A strategy used by some gamblers when they lose is to play again and double the bet, assuming that the amount won is equal to the amount bet. So if a gambler loses $2, then the gambler plays again and bets $4, then $8, and so on.
  a. Why would the gambler use this strategy?
  b. On the nth bet, the gambler risks 2^n dollars. Estimate the first value of n for which the amount bet is more than $1 million using the accompanying graph of $y = 2^n$.

c. Use logarithms to find the answer to part (b).

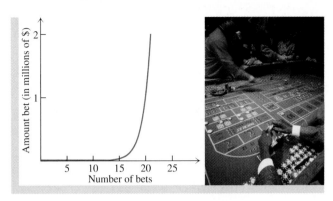

■ **Figure for Exercise 114**

Thinking Outside the Box XLIII

Crescent City Three semicircles are drawn so that their diameters are the three sides of a right triangle as shown in the diagram. What is the ratio of the total area of the two crescents A and B to the area of the triangle T?

■ **Figure for Thinking Outside the Box XLIII**

▪ ▪ ▪ ▪ **Chapter 4 Test**

Simplify each expression.

1. $\log(1000)$

2. $\log_7(1/49)$

3. $3^{\log_3(6.47)}$

4. $\ln\left(e^{\sqrt{2}}\right)$

Find the inverse of each function.

5. $f(x) = \ln(x)$

6. $f(x) = 8^{x+1} - 3$

Write each expression as a single logarithm.

7. $\log(x) + 3 \cdot \log(y)$

8. $\dfrac{1}{2} \cdot \ln(x - 1) - \ln(33)$

Write each expression in terms of $\log_a(2)$ and $\log_a(7)$.

9. $\log_a(28)$

10. $\log_a(3.5)$

Find the exact solution to each equation.

11. $\log_2(x) + \log_2(x - 2) = 3$

12. $\log(10x) - \log(x + 2) = 2 \cdot \log(3)$

Find the exact solution and an approximate solution (to four decimal places) for each equation.

13. $3^x = 5^{x-1}$

14. $\log_3(x - 1) = 5.46$

Graph each equation in the xy-plane. State the domain and range of the function and whether the function is increasing or decreasing. Identify any asymptotes.

15. $f(x) = 2^x + 1$

16. $y = \log_{1/2}(x - 1)$

Solve each problem.

17. What is the only ordered pair that satisfies both $y = \log_2(x)$ and $y = \ln(x)$?

18. A student invested $2000 at 8% annual percentage rate for 20 years. What is the amount of the investment if the interest is compounded quarterly? Compounded continuously?

19. A satellite has a radioisotope power supply. The power supply in watts is given by the formula $P = 50e^{-t/250}$, where t is the time in days. How much power is available at the end of 200 days? What is the half-life of the power supply? If the equipment aboard the satellite requires 9 watts of power to operate properly, then what is the operational life of the satellite?

20. If $4000 is invested at 6% compounded quarterly, then how long will it take for the investment to grow to $10,000?

21. An educational psychologist uses the model $t = -50 \cdot \ln(1 - p)$ for $0 \le p < 1$ to predict the number of hours t that it will take for a child to reach level p in the new video game Mario Goes to Mars (MGM). Level $p = 0$ means that the child knows nothing about MGM. What is the predicted level of a child for 100 hr of playing MGM? If $p = 1$ corresponds to mastery of MGM, then is it possible to master MGM?

Tying it all Together

Chapters 1–4

Solve each equation.

1. $(x - 3)^2 = 4$

2. $2 \cdot \log(x - 3) = \log(4)$

3. $\log_2(x - 3) = 4$

4. $2^{x-3} = 4$

5. $\sqrt{x - 3} = 4$

6. $|x - 3| = 4$

7. $x^2 - 4x = -2$

8. $2^{x-3} = 4^x$

9. $\sqrt[3]{x - 5} = 5$

10. $2^x = 3$

11. $\log(x - 3) + \log(4) = \log(x)$

12. $x^3 - 4x^2 + x + 6 = 0$

Sketch the graph of each function.

13. $y = x^2$

14. $y = (x - 2)^2$

15. $y = 2^x$

16. $y = x^{-2}$

17. $y = \log_2(x - 2)$

18. $y = x - 2$

19. $y = 2x$

20. $y = \log(2^x)$

21. $y = e^2$

22. $y = 2 - x^2$

23. $y = \dfrac{2}{x}$

24. $y = \dfrac{1}{x - 2}$

Find the inverse of each function.

25. $f(x) = \dfrac{1}{3}x$

26. $f(x) = \dfrac{1}{3^x}$

27. $f(x) = \sqrt{x - 2}$

28. $f(x) = 2 + (x - 5)^3$

29. $f(x) = \log(\sqrt{x} - 3)$

30. $f = \{(3, 1), (5, 4)\}$

31. $f(x) = 3 + \dfrac{1}{x - 5}$

32. $f(x) = 3 - e^{\sqrt{x}}$

Find a formula for each composition function given that $p(x) = e^x$, $m(x) = x + 5$, $q(x) = \sqrt{x}$, and $r(x) = \ln(x)$. State the domain and range of the composition function.

33. $p \circ m$

34. $p \circ q$

35. $q \circ p \circ m$

36. $m \circ r \circ q$

37. $p \circ r \circ m$

38. $r \circ q \circ p$

Express each of the following functions as a composition of the functions f, g, and h, where $f(x) = \log_2(x)$, $g(x) = x - 4$, and $h(x) = x^3$.

39. $F(x) = \log_2(x^3 - 4)$

40. $H(x) = (\log_2(x))^3 - 4$

41. $G(x) = (\log_2(x) - 4)^3$

42. $M(x) = (\log_2(x - 4))^3$

Function Gallery: Some Basic Functions of Algebra with Transformations

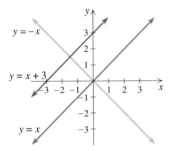

Linear

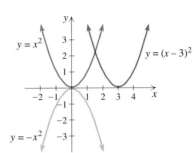

Quadratic

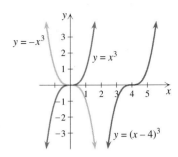

Cubic

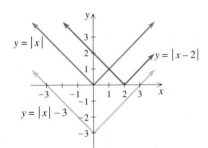

Absolute value

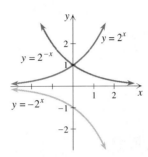

Exponential

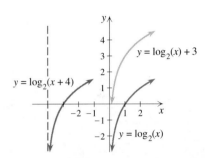

Logarithmic

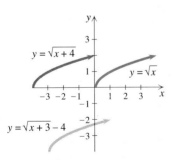

Square root

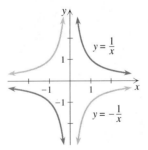

Reciprocal

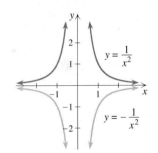

Rational

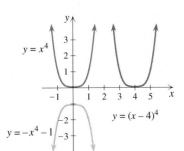

Fourth degree

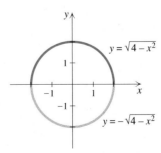

Semicircle

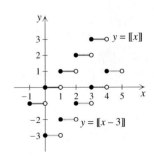

Greatest integer

5 The Trigonometric Functions

SINCE the beginning of human flight, man has been using the air to spy on his neighbors. In 1960 the U-2 spy plane flew at an altitude of 13 miles to photograph the Soviet Union and Cuba. Today, spy planes have been replaced with satellites that orbit the earth at an altitude of 700 miles. The photographs taken are so clear that analysts can determine troop movements as well as the sizes of buildings housing them.

Determining sizes of objects without measuring them physically is one of the principal applications of trigonometry. Long before we dreamed of traveling to the moon, trigonometry was used to calculate the distance to the moon.

WHAT YOU WILL LEARN   In this chapter we explore many applications of trigonometric functions, ranging from finding the velocity of a lawnmower blade to estimating the size of a building in an aerial photograph.

Angles and Their Measurements

Trigonometry was first studied by the Greeks, Egyptians, and Babylonians and used in surveying, navigation, and astronomy. Using trigonometry, they had a powerful tool for finding areas of triangular plots of land, as well as lengths of sides and measures of angles, without physically measuring them. We begin our study of trigonometry by studying angles and their measurements.

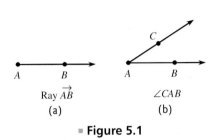

Ray $\overrightarrow{AB}$

(a)

$\angle CAB$

(b)

■ **Figure 5.1**

Degree Measure of Angles

In geometry a **ray** is defined as a point on a line together with all points of the line on one side of that point. Figure 5.1(a) shows ray $\overrightarrow{AB}$. An **angle** is defined as the union of two rays with a common endpoint, the **vertex.** The angle shown in Fig. 5.1(b) is named $\angle A$, $\angle BAC$, or $\angle CAB$. (Read the symbol $\angle$ as "angle.") Angles are also named using Greek letters such as α (alpha), β (beta), γ (gamma), or θ (theta).

An angle is often thought of as being formed by rotating one ray away from a fixed ray as indicated by angle α and the arrow in Fig. 5.2(a). The fixed ray is the **initial side** and the rotated ray is the **terminal side.** An angle whose vertex is the center of a circle, as shown in Fig. 5.2(b), is a **central angle,** and the arc of the circle through which the terminal side moves is the **intercepted arc.** An angle in **standard position** is located in a rectangular coordinate system with the vertex at the origin and the initial side on the positive x-axis as shown in Fig. 5.2(c).

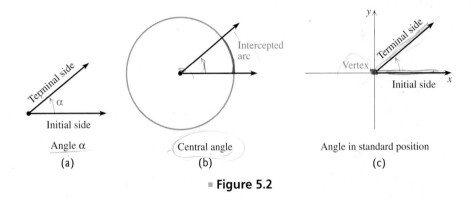

Angle α

(a)

Central angle

(b)

Angle in standard position

(c)

■ **Figure 5.2**

The measure $m(\alpha)$, of an angle α indicates the amount of rotation of the terminal side from the initial position. It is found using any circle centered at the vertex. The circle is divided into 360 equal arcs and each arc is one **degree** (1°).

Definition:
Degree Measure

> The **degree measure of an angle** is the number of degrees in the intercepted arc of a circle centered at the vertex. The degree measure is positive if the rotation is counterclockwise and negative if the rotation is clockwise.

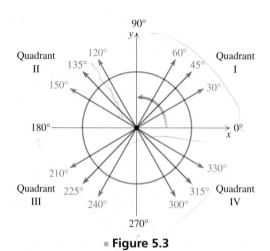

▪ **Figure 5.3**

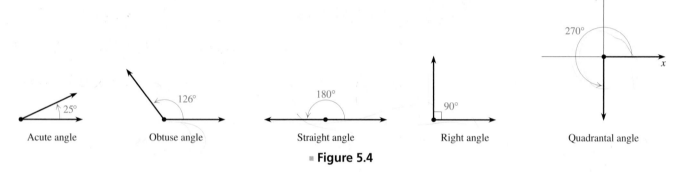

▪ **Figure 5.4**

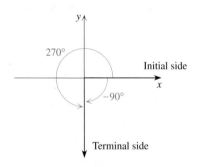

▪ **Figure 5.5**

Figure 5.3 shows the positions of the terminal sides of some angles in standard position with common positive measures between 0° and 360°. An angle with measure between 0° and 90° is an **acute angle.** An angle with measure between 90° and 180° is an **obtuse angle.** An angle of exactly 180° is a **straight angle.** See Fig. 5.4. A 90° angle is a **right angle.** An angle in standard position is said to lie in the quadrant where its terminal side lies. If the terminal side is on an axis, the angle is a **quadrantal angle.** We often think of the degree measure of an angle as the angle itself. For example, we write $m(\alpha) = 60°$ or $\alpha = 60°$, and we say that 60° is an acute angle.

The initial side of an angle may be rotated in a positive or negative direction to get to the position of the terminal side. For example, if the initial side shown in Fig. 5.5 rotates clockwise for one-quarter of a revolution to get to the terminal position, then the measure of the angle is −90°. If the initial side had rotated counterclockwise to get to the position of the terminal side, then the measure of the angle would be 270°. If the initial side had rotated clockwise for one and a quarter revolutions to get to the terminal position, then the angle would be −450°. **Coterminal angles** are angles in standard position that have the same initial side and the same terminal side. The angles −90°, −450°, and 270° are examples of coterminal angles. Any two coterminal angles have degree measures that differ by a multiple of 360° (one complete revolution).

Coterminal Angles

Angles α and β in standard position are coterminal if and only if there is an integer k such that $m(\beta) = m(\alpha) + k360°$.

Note that any angle formed by two rays can be thought of as one angle with infinitely many different measures, or infinitely many different coterminal angles.

Example 1 Finding coterminal angles

Find two positive angles and two negative angles that are coterminal with $-50°$.

Solution

Since any angle of the form $-50° + k360°$ is coterminal with $-50°$, there are infinitely many possible answers. For simplicity, we choose the positive integers 1 and 2 and the negative integers -1 and -2 for k to get the following angles:

$$-50° + 1 \cdot 360° = 310°$$

$$-50° + 2 \cdot 360° = 670°$$

$$-50° + (-1)360° = -410°$$

$$-50° + (-2)360° = -770°$$

The angles $310°$, $670°$, $-410°$, and $-770°$ are coterminal with $-50°$.

Try This. Find two positive and two negative angles coterminal with $10°$. ■

To determine whether two angles are coterminal we must determine whether they differ by a multiple of $360°$.

Example 2 Determining whether angles are coterminal

Determine whether angles in standard position with the given measures are coterminal.

a. $m(\alpha) = 190°, m(\beta) = -170°$
b. $m(\alpha) = 150°, m(\beta) = 880°$

Solution

a. If there is an integer k such that $190 + 360k = -170$, then α and β are coterminal.

$$190 + 360k = -170$$

$$360k = -360$$

$$k = -1$$

Since the equation has an integral solution, α and β are coterminal.

b. If there is an integer k such that $150 + 360k = 880$, then α and β are coterminal.

$$150 + 360k = 880$$

$$360k = 730$$

$$k = \frac{73}{36}$$

Since there is no integral solution to the equation, α and β are not coterminal.

Try This. Determine whether $-690°$ and $390°$ are coterminal. ■

The quadrantal angles, such as 90°, 180°, 270°, and 360°, have terminal sides on an axis and do not lie in any quadrant. Any angle that is not coterminal with a quadrantal angle lies in one of the four quadrants. To determine the quadrant in which an angle lies, add or subtract multiples of 360° (one revolution) to obtain a coterminal angle with a measure between 0° and 360°. A nonquadrantal angle is in the quadrant in which its terminal side lies.

Example **3** **Determining in which quadrant an angle lies**

Name the quadrant in which each angle lies.

a. 230°   **b.** −580°   **c.** 1380°

Solution

a. Since 180° < 230° < 270°, a 230° angle lies in quadrant III.

b. We must add 2(360°) to −580° to get an angle between 0° and 360°:

$$-580° + 2(360°) = 140°$$

So 140° and −580° are coterminal. Since 90° < 140° < 180°, 140° lies in quadrant II and so does a −580° angle.

c. From 1380° we must subtract 3(360°) to obtain an angle between 0° and 360°:

$$1380° - 3(360°) = 300°$$

So 1380° and 300° are coterminal. Since 270° < 300° < 360°, 300° lies in quadrant IV and so does 1380°.

Try This. Name the quadrant in which −890° lies.   ■

Each degree is divided into 60 equal parts called **minutes,** and each minute is divided into 60 equal parts called **seconds.** A minute (min) is 1/60 of a degree (deg), and a second (sec) is 1/60 of a minute or 1/3600 of a degree. An angle with measure 44°12′30″ is an angle with a measure of 44 degrees, 12 minutes, and 30 seconds. Historically, angles were measured by using degrees-minutes-seconds, but with calculators it is convenient to have the fractional parts of a degree written as a decimal number such as 7.218°. Some calculators can handle angles in degrees-minutes-seconds and even convert them to decimal degrees.

Example **4** **Converting degrees-minutes-seconds to decimal degrees**

Convert the measure 44°12′30″ to decimal degrees.

Solution

Since 1 degree = 60 minutes and 1 degree = 3600 seconds, we get

$$12 \text{ min} = 12 \text{ min} \cdot \frac{1 \text{ deg}}{60 \text{ min}} = \frac{12}{60} \text{ deg} \quad \text{and}$$

$$30 \text{ sec} = 30 \text{ sec} \cdot \frac{1 \text{ deg}}{3600 \text{ sec}} = \frac{30}{3600} \text{ deg.}$$

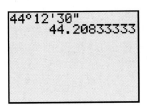

■ **Figure 5.6**

So

$$44°12'30'' = \left(44 + \frac{12}{60} + \frac{30}{3600}\right)° \approx 44.2083°.$$

A graphing calculator can convert degrees-minutes-seconds to decimal degrees as shown in Fig. 5.6.

Try This. Convert $35°15'12''$ to decimal degrees. ■

Note that the conversion of Example 4 was done by "cancellation of units." Minutes in the numerator canceled with minutes in the denominator to give the result in degrees, and seconds in the numerator canceled with seconds in the denominator to give the result in degrees.

Example **5** **Converting decimal degrees to degrees-minutes-seconds**

Convert the measure $44.235°$ to degrees-minutes-seconds.

Solution

First convert $0.235°$ to minutes. Since 1 degree = 60 minutes,

$$0.235 \text{ deg} = 0.235 \text{ deg} \cdot \frac{60 \text{ min}}{1 \text{ deg}} = 14.1 \text{ min}.$$

So $44.235° = 44°14.1'$. Now convert $0.1'$ to seconds. Since 1 minute = 60 seconds,

$$0.1 \text{ min} = 0.1 \text{ min} \cdot \frac{60 \text{ sec}}{1 \text{ min}} = 6 \text{ sec}.$$

So $44.235° = 44°14'6''$.

A graphing calculator can convert to degrees-minutes-seconds as shown in Fig. 5.7.

Try This. Convert $56.321°$ to degrees-minutes-seconds. ■

■ **Figure 5.7**

Note again how the original units canceled in Example 5, giving the result in the desired units of measurement.

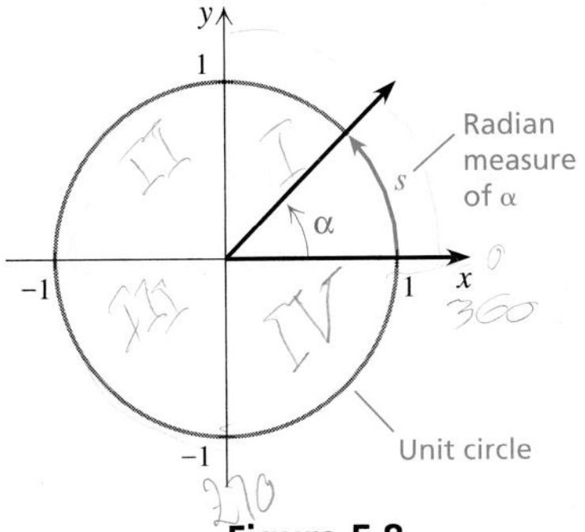

Radian Measure of Angles

Degree measure of angles is used mostly in applied areas such as surveying, navigation, and engineering. Radian measure of angles is used more in scientific fields and results in simpler formulas in trigonometry and calculus.

For radian measures of angles we use a **unit circle** (a circle with radius 1) centered at the origin. The radian measure of an angle in standard position is simply the length of the intercepted arc on the unit circle. See Fig. 5.8. Since the radius of the unit circle is the real number 1 without any dimension (such as feet or inches), the length of an intercepted arc is a real number without any dimension and so the radian measure of an angle is also a real number without any dimension. One **radian** (abbreviated 1 rad) is the real number 1.

■ **Figure 5.8**

Definition:
Radian Measure

> To find the **radian measure** of the angle α in standard position, find the length of the intercepted arc on the unit circle. If the rotation is counterclockwise, the radian measure is the length of the arc. If the rotation is clockwise, the radian measure is the opposite of the length of the arc.

Radian measure is called a **directed length** because it is positive or negative depending on the direction of rotation of the initial side. If s is the length of the intercepted arc on the unit circle for an angle α, as shown in Fig. 5.8, we write $m(\alpha) = s$. To emphasize that s is the length of an arc on a unit circle, we may write $m(\alpha) = s$ radians.

Because the circumference of a circle with radius r is $2\pi r$, the circumference of the unit circle is 2π. If the initial side rotates 360° (one complete revolution), then the length of the intercepted arc is 2π. So an angle of 360° has a radian measure of 2π radians. We express this relationship as $360° = 2\pi$ rad or simply $360° = 2\pi$. Dividing each side by 2 yields $180° = \pi$, which is the basic relationship to remember for conversion of one unit of measurement to the other.

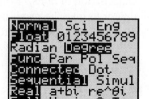

Use the MODE key on a graphing calculator to set the calculator to radian or degree mode as shown in Fig. 5.9. □

■ **Figure 5.9**

Degree-Radian
Conversion

> Conversion from degrees to radians or radians to degrees is based on
>
> 180 degrees = π radians.

To convert degrees to radians or radians to degrees, we use 180 deg = π rad and cancellation of units. For example,

$$1 \text{ deg} = 1 \text{ deg} \cdot \frac{\pi \text{ rad}}{180 \text{ deg}} = \frac{\pi}{180} \text{ rad} \approx 0.01745 \text{ rad}$$

and

$$1 \text{ rad} = 1 \text{ rad} \cdot \frac{180 \text{ deg}}{\pi \text{ rad}} = \frac{180}{\pi} \text{ deg} \approx 57.3 \text{ deg}.$$

If your calculator is in radian mode, as in Fig. 5.10(a), pressing ENTER converts degrees to radians. When the calculator is in degree mode, as in Fig. 5.10(b), pressing ENTER converts radians to degrees. □

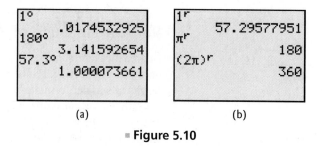

(a)   (b)

■ **Figure 5.10**

Figure 5.11(a) shows an angle of 1°, and Fig. 5.11(b) shows an angle of 1 radian. An angle of 1 radian intercepts an arc on the unit circle equal in length to the radius of the unit circle.

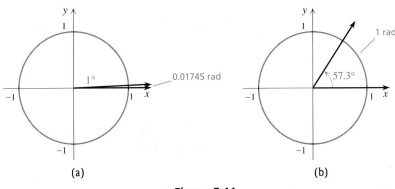

▪ **Figure 5.11**

Example 6 Converting from degrees to radians

Convert the degree measures to radians.

a. 270°   **b.** −23.6°

Solution

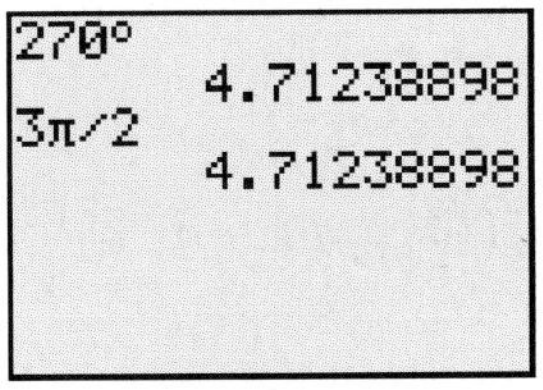

a. To convert degrees to radians, multiply the degree measure by π rad/180 deg:

$$270° = 270 \text{ deg} \cdot \frac{\pi \text{ rad}}{180 \text{ deg}} = \frac{3\pi}{2} \text{ rad}$$

The exact value, $3\pi/2$ rad, is approximately 4.71 rad; but when a measure in radians is a simple multiple of π, we usually write the exact value.

Check this result with a calculator in radian mode as shown in Fig. 5.12.   □

b. $-23.6° = -23.6 \text{ deg} \cdot \dfrac{\pi \text{ rad}}{180 \text{ deg}} \approx -0.412 \text{ rad.}$

```
270°
          4.71238898
3π/2
          4.71238898
```

▪ **Figure 5.12**

Try This. Convert 210° to radians. = 3.665 rad.   ▪

Example 7 Converting from radians to degrees

Convert the radian measures to degrees.

a. $\dfrac{7\pi}{6}$ **b.** 12.3 $210°$ ⓑ $705°$

3.665

Solution

Multiply the radian measure by 180 deg/π rad:

a. $\dfrac{7\pi}{6} = \dfrac{7\pi}{6} \text{ rad} \cdot \dfrac{180 \text{ deg}}{\pi \text{ rad}} = 210°$ **b.** $12.3 = 12.3 \text{ rad} \cdot \dfrac{180 \text{ deg}}{\pi \text{ rad}} \approx 704.7°$

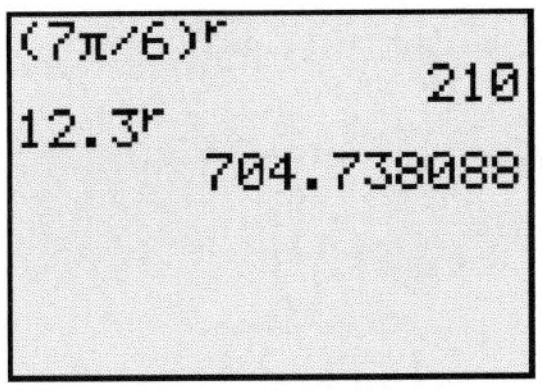

```
(7π/6)ʳ
                210
12.3ʳ
         704.738088
```

▪ **Figure 5.13**

Check these answers in degree mode as shown in Fig. 5.13.

Try This. Convert the radian measure $\frac{5\pi}{3}$ to degrees. = 300   ▪

Figure 5.14 shows angles with common measures in standard position. Coterminal angles in standard position have radian measures that differ by an integral multiple of 2π (their degree measures differ by an integral multiple of $360°$).

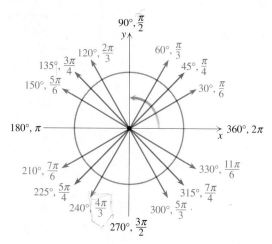

■ **Figure 5.14**

Example 8 Finding coterminal angles using radian measure

Find two positive and two negative angles that are coterminal with $\pi/6$.

Solution

All angles coterminal with $\pi/6$ have a radian measure of the form $\pi/6 + k(2\pi)$, where k is an integer.

$$\frac{\pi}{6} + 1(2\pi) = \frac{\pi}{6} + \frac{12\pi}{6} = \frac{13\pi}{6}$$

$$\frac{\pi}{6} + 2(2\pi) = \frac{\pi}{6} + \frac{24\pi}{6} = \frac{25\pi}{6}$$

$$\frac{\pi}{6} + (-1)(2\pi) = \frac{\pi}{6} - \frac{12\pi}{6} = -\frac{11\pi}{6}$$

$$\frac{\pi}{6} + (-2)(2\pi) = \frac{\pi}{6} - \frac{24\pi}{6} = -\frac{23\pi}{6}$$

The angles $13\pi/6$, $25\pi/6$, $-11\pi/6$, and $-23\pi/6$ are coterminal with $\pi/6$.

Try This. Find two positive and two negative angles coterminal with $-\pi/3$.   ■

Arc Length

Radian measure of a central angle of a circle can be used to easily find the length of the intercepted arc of the circle. For example, an angle of $\pi/2$ radians positioned at the center of the earth, as shown in Fig. 5.15, intercepts an arc on the surface of the

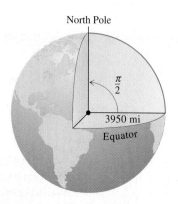

■ **Figure 5.15**

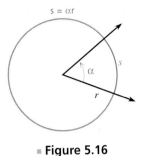

◼ **Figure 5.16**

earth that runs from the equator to the North Pole. Using 3950 miles as the approximate radius of the earth yields a circumference of $2\pi(3950) = 7900\pi$ miles. Since $\pi/2$ radians is $1/4$ of a complete circle, the length of the intercepted arc is $1/4$ of the circumference. So the distance from the equator to the North Pole is $7900\pi/4$ miles, or about 6205 miles.

In general, a central angle α in a circle of radius r intercepts an arc whose length s is a fraction of the circumference of the circle, as shown in Fig. 5.16. Since a complete revolution is 2π radians, that fraction is $\alpha/(2\pi)$. Since the circumference is $2\pi r$, we get

$$s = \frac{\alpha}{2\pi} \cdot 2\pi r = \alpha r.$$

Theorem: Length of an Arc

> The length s of an arc intercepted by a central angle of α *radians* on a circle of radius r is given by
>
> $$s = \alpha r.$$

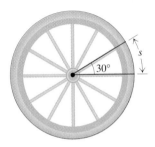

◼ **Figure 5.17**

If α is negative, the formula $s = \alpha r$ gives a negative number for the length of the arc. So s is a directed length. Note that the formula $s = \alpha r$ applies only if α is in radians.

Example ⑨ Finding the length of an arc

The wagon wheel shown in Fig. 5.17 has a diameter of 28 inches and an angle of $30°$ between the spokes. What is the length of the arc s between two adjacent spokes?

Solution

First convert $30°$ to radians:

$$30° = 30 \text{ deg} \cdot \frac{\pi \text{ rad}}{180 \text{ deg}} = \frac{\pi}{6} \text{ rad}$$

Now use $r = 14$ inches and $\alpha = \pi/6$ in the formula for arc length $s = \alpha r$:

$$s = \frac{\pi}{6} \cdot 14 \text{ in.} = \frac{7\pi}{3} \text{ in.} \approx 7.33 \text{ in.}$$

Since the radian measure of an angle is a dimensionless real number, the product of radians and inches is given in inches.

Try This. Find the length of an arc intercepted by a central angle of $45°$ on a circle of radius 8 inches.   ◼

6.2 inch

Example ⑩ Finding the central angle

For four years, U-2 spy planes flew reconnaissance over the Soviet Union, collecting information for the United States. The U-2, designed to fly at 70,000 feet—well out of range of Soviet guns—carried 12,000 feet of film in a camera that could photograph a path 2000 miles long and about 600 miles wide. Find the central angle, to the nearest tenth of a degree, that intercepts an arc of 600 miles on the surface of the earth (radius 3950 miles), as shown in Fig. 5.18.

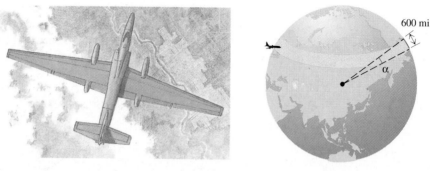

■ **Figure 5.18**

Solution

From the formula $s = \alpha r$ we get $\alpha = s/r$:

$$\alpha = \frac{600}{3950} \approx 0.1519 \text{ rad}$$

Multiply by 180 deg/π rad to find the angle in degrees:

$$\alpha = 0.1519 \text{ rad} \cdot \frac{180 \text{ deg}}{\pi \text{ rad}} \approx 8.7°$$

Try This. Find the central angle in degrees that intercepts an arc of length 3 cm on a circle of radius 50 cm.   ■

Linear and Angular Velocity

Suppose that a point is in motion on a circle. The distance traveled by the point in some unit of time is the length of an arc on the circle. The *linear velocity* of the point is the rate at which that distance is changing. For example, suppose that the helicopter blade of length 10 meters shown in Fig. 5.19 is rotating at 400 revolutions per minute about its center. A point on the tip of the blade (10 meters from the center) travels 20π meters for each revolution of the blade, because $C = 2\pi r$. Since the blade is rotating at 400 revolutions per minute, the point has a linear velocity of 8000π meters per minute.

■ **Figure 5.19**

If we draw a ray from the center of the circle through the point on the tip of the helicopter blade, then an angle is formed by the initial and terminal positions of the ray over some unit of time. The *angular velocity* of the point is the rate at which that angle is changing. For example, the ray shown in Fig. 5.19 rotates through an angle of 2π rad for each revolution of the blade. Since the blade is rotating at 400 revolutions per minute, the angular velocity of the point is 800π radians per minute. The angular velocity does not depend on the length of the blade (or the radius of the circular path), but only on the number of revolutions per unit of time.

We use the letter v to represent linear velocity for a point in circular motion and the Greek letter ω (omega) to represent angular velocity, and define them as follows.

Definition: Linear Velocity and Angular Velocity

If a point is in motion on a circle of radius r through an angle of α radians in time t, then its **linear velocity** is

$$v = \frac{s}{t},$$

where s is the arc length determined by $s = \alpha r$, and its **angular velocity** is

$$\omega = \frac{\alpha}{t}.$$

Since $v = s/t = \alpha r/t$ and $\omega = \alpha/t$, we get $v = r\omega$. We have proved the following theorem.

Theorem: Linear Velocity in Terms of Angular Velocity

If v is the linear velocity of a point on a circle of radius r, and ω is its angular velocity, then

$$v = r\omega.$$

It is important to recognize that both linear and angular velocity can be expressed with many different units. For example, an angular velocity of 240 radians per hour can be expressed in radians per minute by the cancellation of units method:

$$240 \text{ rad/hr} = \frac{240 \text{ rad}}{\cancel{hr}} \cdot \frac{1 \cancel{hr}}{60 \text{ min}} = 4 \text{ rad/min}$$

A linear velocity of 45 ft/sec can be expressed in yards per minute:

$$45 \text{ ft/sec} = \frac{45 \cancel{ft}}{\cancel{sec}} \cdot \frac{60 \cancel{sec}}{1 \text{ min}} \cdot \frac{1 \text{ yd}}{3 \cancel{ft}} = 900 \text{ yd/min}$$

Example **11** Finding angular and linear velocity

What are the angular velocity in radians per second and the linear velocity in miles per hour of the tip of a 22-inch lawnmower blade that is rotating at 2500 revolutions per minute?

Solution

Use the fact that 2π radians $= 1$ revolution and 60 seconds $= 1$ minute to find the angular velocity:

$$\omega = \frac{2500 \text{ rev}}{\text{min}} = \frac{2500 \text{ rev}}{\text{min}} \cdot \frac{2\pi \text{ rad}}{1 \text{ rev}} \cdot \frac{1 \text{ min}}{60 \text{ sec}} \approx 261.799 \text{ rad/sec}$$

To find the linear velocity, use $v = r\omega$ with $r = 11$ inches:

$$v = 11 \text{ in.} \cdot \frac{261.799 \text{ rad}}{\text{sec}} \approx 2879.789 \text{ in./sec}$$

Convert to miles per hour:

$$v = \frac{2879.789 \text{ in.}}{\text{sec}} \cdot \frac{1 \text{ ft}}{12 \text{ in.}} \cdot \frac{1 \text{ mi}}{5280 \text{ ft}} \cdot \frac{3600 \text{ sec}}{1 \text{ hr}} \approx 163.624 \text{ mi/hr}$$

Try This. Find the angular velocity in radians per second and the linear velocity in miles per hour for a particle that is moving in a circular path at 5 revolutions per second on a circle of radius 10 feet. ■

Any point on the surface of the earth (except at the poles) makes one revolution about the axis of the earth in 24 hours. So the angular velocity of a point on the earth is $\pi/12$ radians per hour. The linear velocity of a point on the surface of the earth depends on its distance from the axis of the earth.

Example **12** **Linear velocity on the surface of the earth**

What is the linear velocity in miles per hour of a point on the equator?

Solution

A point on the equator has an angular velocity of $\pi/12$ radians per hour on a circle of radius 3950 miles. Using the formula $v = \omega r$, we get

$$v = \frac{\pi}{12} \text{ rad/hr} \cdot 3950 \text{ mi} = \frac{3950\pi}{12} \text{ mi/hr} \approx 1034 \text{ mi/hr}.$$

Note that we write angular velocity as radians per hour, but radians are omitted from the answer in miles per hour because radians are simply real numbers.

Try This. Find the linear velocity in feet per second for a point on the equator? ■

For Thought

True or False? Explain.

1. An angle is a union of two rays with a common endpoint.

2. The lengths of the rays of $\angle A$ determine the degree measure of $\angle A$.

3. Angles of $5°$ and $-365°$ are coterminal.

4. The radian measure of an angle cannot be negative.

5. An angle of $38\pi/4$ radians is a quadrantal angle.

6. If $m(\angle A) = 210°$, then $m(\angle A) = 5\pi/6$ radians.

7. $25°20'40'' = 25.34°$

8. The angular velocity of Seattle is $\pi/12$ radians per hour.

9. Seattle and Los Angeles have the same linear velocity.

10. A central angle of 1 rad in a circle of radius r intercepts an arc of length r.

5.1 Exercises

Find two positive angles and two negative angles that are coterminal with each given angle.

1. $60°$ **2.** $45°$ **3.** $-16°$ **4.** $-90°$

Determine whether the angles in each given pair are coterminal.

5. $123.4°, -236.6°$ **6.** $744°, -336°$

7. $1055°, 155°$ **8.** $0°, 359.9°$

Name the quadrant in which each angle lies.

9. $85°$ **10.** $110°$ **11.** $-125°$ **12.** $-200°$

13. $300°$ **14.** $205°$ **15.** $750°$ **16.** $-980°$

Match each of the following angles with one of the degree measures: $30°, 45°, 60°, 120°, 135°, 150°$.

17.

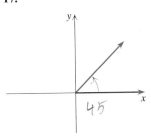

18.

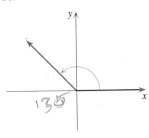

19.

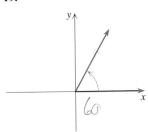

20.

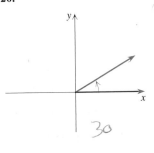

21.

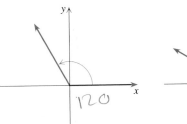

22.

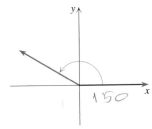

Find the measure in degrees of the least positive angle that is coterminal with each given angle.

23. $400°$ **24.** $540°$ **25.** $-340°$

26. $-180°$ **27.** $-1100°$ **28.** $-840°$

Convert each angle to decimal degrees. When necessary, round to four decimal places.

29. $13°12'$ **30.** $45°6'$ **31.** $-8°30'18''$

32. $-5°45'30''$ **33.** $28°5'9''$ **34.** $44°19'32''$

Convert each angle to degrees-minutes-seconds. Round to the nearest whole number of seconds.

35. $75.5°$ **36.** $39.4°$ **37.** $-17.33°$

38. $-9.12°$ **39.** $18.123°$ **40.** $122.786°$

Convert each degree measure to radian measure. Give exact answers.

41. $30°$ **42.** $45°$

43. $18°$ **44.** $48°$

45. $-67.5°$ **46.** $-105°$

47. $630°$ **48.** $495°$

Convert each degree measure to radian measure. Use the value of π found on a calculator and round answers to three decimal places.

49. 37.4°   **50.** 125.3°   **51.** −13°47′

52. −99°15′   **53.** −53°37′ 6″   **54.** 187°49′ 36″

Convert each radian measure to degree measure. Use the value of π found on a calculator and round approximate answers to three decimal places.

55. $\dfrac{5\pi}{12}$ **56.** $\dfrac{17\pi}{12}$ **57.** $\dfrac{7\pi}{4}$ **58.** $\dfrac{13\pi}{6}$

59. −6π   **60.** −9π   **61.** 2.39   **62.** 0.452

Using radian measure, find two positive angles and two negative angles that are coterminal with each given angle.

63. $\dfrac{\pi}{3}$ **64.** $\dfrac{\pi}{4}$ **65.** $-\dfrac{\pi}{6}$ **66.** $-\dfrac{2\pi}{3}$

Find the measure in radians of the least positive angle that is coterminal with each given angle.

67. 3π   **68.** 6π   **69.** $\dfrac{9\pi}{2}$ **70.** $\dfrac{19\pi}{2}$

71. $-\dfrac{5\pi}{3}$ **72.** $-\dfrac{7\pi}{6}$ **73.** $-\dfrac{13\pi}{3}$ **74.** $-\dfrac{19\pi}{4}$

75. 8.32   **76.** −23.55

Determine whether the angles in each given pair are coterminal.

 77. $\dfrac{3\pi}{4}, \dfrac{29\pi}{4}$ 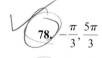 **78.** $-\dfrac{\pi}{3}, \dfrac{5\pi}{3}$

79. $\dfrac{7\pi}{6}, -\dfrac{5\pi}{6}$ **80.** $\dfrac{3\pi}{2}, -\dfrac{9\pi}{2}$

Name the quadrant in which each angle lies.

81. $\dfrac{5\pi}{12}$ **82.** $\dfrac{13\pi}{12}$ **83.** $-\dfrac{6\pi}{7}$ **84.** $-\dfrac{39\pi}{20}$

85. $\dfrac{13\pi}{8}$ **86.** $-\dfrac{11\pi}{8}$ **87.** −7.3   **88.** 23.1

Find the missing degree or radian measure for each position of the terminal side shown. For Exercise 89, use degrees between 0° and 360° and radians between 0 and 2π. For Exercise 90, use degrees between −360° and 0° and radians between −2π and 0. Practice these two exercises until you have memorized the degree and radian measures corresponding to these common angles.

89.

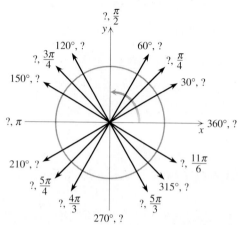

90.

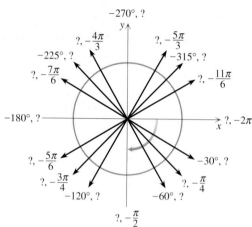

Find the length of the arc intercepted by the given central angle α in a circle of radius r.

91. $\alpha = \dfrac{\pi}{4}, r = 12$ ft   **92.** $\alpha = 1, r = 4$ cm

93. $\alpha = 3°, r = 4000$ mi   **94.** $\alpha = 60°, r = 2$ m

Find the radius of the circle in which the given central angle α intercepts an arc of the given length s.

95. $\alpha = 1, s = 1$ mi   **96.** $\alpha = 0.004, s = 99$ km

97. $\alpha = 180°, s = 10$ km   **98.** $\alpha = 360°, s = 8$ m

Solve each problem.

99. *Distance to the North Pole* Peshtigo, Wisconsin, is on the 45th parallel. This means that an arc from Peshtigo to the North Pole subtends a central angle of 45° as shown in the figure. If the radius of the earth is 3950 mi, then how far is it from Peshtigo to the North Pole?

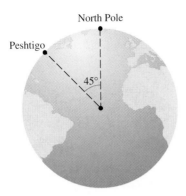

■ Figure for Exercise 99

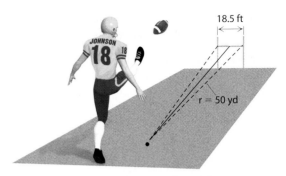

■ Figure for Exercise 102

100. *Distance to the Helper* A surveyor sights her 6-ft 2-in. helper on a nearby hill as shown in the figure. If the angle of sight between the helper's feet and head is 0°37′, then approximately how far away is the helper?

> HINT Assume that the helper is a 6-ft 2-in. arc on a circle of radius *r*.

■ Figure for Exercise 100

101. *Photographing the Earth* From an altitude of 161 mi, the Orbiting Geophysical Observatory (OGO-1) can photograph a path on the surface of the earth that is approximately 2000 mi wide. Find the central angle, to the nearest tenth of a degree, that intercepts an arc of 2000 mi on the surface of the earth (radius 3950 mi).

102. *Margin of Error in a Field Goal* Assume that the distance between the goal posts, 18.5 ft, is the length of an arc on a circle of radius 50 yd as shown in the figure. The kicker aims to kick the ball midway between the uprights. To score a field goal, what is the maximum number of degrees that the actual trajectory can deviate from the intended trajectory? What is the maximum number of degrees for a 20-yd field goal?

103. *Linear Velocity* What is the linear velocity for any edge point of a 12-in.-diameter record spinning at $33\frac{1}{3}$ rev/min?

104. *Angular Velocity* What is the angular velocity in radians per minute for any point on a 45-rpm record?

105. *Lawnmower Blade* What is the linear velocity in miles per hour of the tip of a lawnmower blade spinning at 2800 rev/min for a lawnmower that cuts a 20-in.-wide path?

106. *Router Bit* A router bit makes 45,000 rev/min. What is the linear velocity in miles per hour of the outside edge of a bit that cuts a 1-in.-wide path?

107. *Table Saw* The blade on a table saw rotates at 3450 revolutions per minute. What is the difference in the linear velocity in miles per hour for a point on the edge of a 12-in.-diameter blade and a 10-in.-diameter blade?

108. *Automobile Tire* If a car runs over a nail at 55 mph and the nail is lodged in the tire tread 13 in. from the center of the wheel, then what is the angular velocity of the nail in radians per hour?

109. *Linear Velocity Near the North Pole* Find the linear velocity for a point on the surface of the earth that is one mile from the North Pole.

> HINT First determine how far the point travels in 24 hours.

110. *Linear Velocity of Peshtigo* What are the linear and angular velocities for Peshtigo, Wisconsin (on the 45th parallel) with respect to its rotation around the axis of the earth? (See Exercise 99.)

111. *Eratosthenes Measures the Earth* Over 2200 years ago Eratosthenes read in the Alexandria library that at noon on June 21 a vertical stick in Syene cast no shadow. So on June 21 at noon Eratosthenes set out a vertical stick in Alexandria and found an angle of 7° in the position shown in the drawing on the next page. Eratosthenes reasoned that since the sun is so far away, sunlight must be arriving at the earth in parallel rays. With this assumption he concluded that the earth is round and the central angle in the drawing must also be 7°. He then paid a man to pace off the distance between Syene and Alexandria and found it to be 800 km. From these facts, calculate the circumference of the earth as Eratosthenes did and compare his answer with the circumference calculated by using the currently accepted radius of 6378 km.

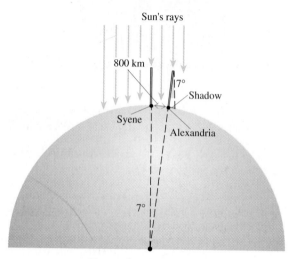

▪ **Figure for Exercise 111**

112. *Achieving Synchronous Orbit*   The space shuttle orbits the earth in about 90 min at an altitude of about 125 mi, but a communication satellite must always remain above a fixed location on the earth, in synchronous orbit with the earth. Since the earth rotates 15° per hour, the angular velocity of a communication satellite must be 15° per hour. To achieve a synchronous orbit with the earth, the radius of the orbit of a satellite must be 6.5 times the radius of the earth (3950 mi). What is the linear velocity in miles per hour of such a satellite?

113. *Area of a Slice of Pizza*   If a 16-in.-diameter pizza is cut into six slices of the same size, then what is the area of each piece?

114. *Area of a Sector of a Circle*   If a slice with central angle α radians is cut from a pizza of radius r, then what is the area of the slice?

115. *Belt and Pulleys*   A belt connects two pulleys with radii 3 in. and 5 in. as shown in the accompanying diagram. The velocity of a point A on the belt is 10 ft/sec. What is the linear velocity (ft/sec) and angular velocity (rad/sec) for point B? What is the linear velocity (ft/sec) and angular velocity (rad/sec) for point C?

116. *Belt and Pulleys*   A belt connects two pulleys with radii 3 in. and 5 in. as shown in the accompanying diagram. Point B is

rotating at 1000 revolutions per minute. What is the linear velocity (ft/sec) for points A, B, and C? How many revolutions per minute is point C making?

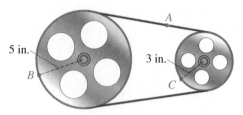

▪ **Figure for Exercises 115 and 116**

117. *Time After Time*   What are the first two times (to the nearest second) after 12 noon for which the minute hand and the hour hand of a clock are perpendicular to each other?

   HINT First find the rates at which the hands are moving.

118. *Perfect Timing*   Tammy starts swimming laps at 4 P.M. exactly. She quits swimming as soon as the minute hand and hour hand of the clock coincide. Find the amount of time she spends swimming to the nearest tenth of a second.

119. *Volume of a Cup*   Sunbelt Paper Products makes conical paper cups by cutting a 4-in. circular piece of paper on the radius and overlapping the paper by an angle α as shown in the accompanying figure.
   a. Find the volume of the cup if $\alpha = 30°$.

   b. Write the volume of the cup as a function of α.

   c. Use the maximum feature of a graphing calculator to find the angle α that gives the maximum volume for the cup.

   d. What is the maximum volume for the cup?

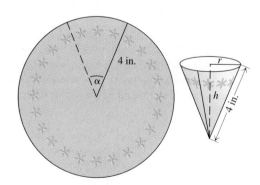

▪ **Figure for Exercise 119**

Thinking Outside the Box XLIV

Triangles and Circles Each of the five circles in the accompanying diagram has radius r. The four right triangles are congruent and each hypotenuse has length 1. Each line segment that appears to intersect a circle intersects it at exactly one point (a point of tangency). Find r.

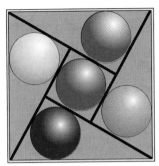

■ **Figure for Thinking Outside the Box XLIV**

5.1 Pop Quiz

1. Find the degree measure of the least positive angle that is coterminal with 1267°.

2. In which quadrant does −120° lie?

3. Convert 70°30′36″ to decimal degrees.

4. Convert 270° to radian measure.

5. Convert $7\pi/4$ to degree measure.

6. Are $-3\pi/4$ and $5\pi/4$ coterminal?

7. Find the exact length of the arc intercepted by a central angle of 60° in a circle with radius 30 feet.

Linking Concepts

For Individual or Group Explorations

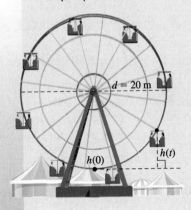

Constructing the Sine Function

Imagine that you are riding on a 20-meter-diameter Ferris wheel that is making three revolutions per minute.

a) What is your linear velocity?

b) What is your angular velocity?

c) If there are eight equally spaced seats on the Ferris wheel, then what is the length of the arc between two adjacent seats?

d) Use a compass to draw a circle with diameter 20 centimeters to represent the Ferris wheel. Let $h(t)$ be your height in meters at time t seconds, where $h(0) = 0$. Locate your position on the Ferris wheel for each time in the following table and find $h(t)$ by measuring your drawing.

t	0	2.5	5	7.5	10	12.5	15	17.5	20
$h(t)$									

e) Sketch the graph of $h(t)$ on graph paper for t ranging from 0 to 60 seconds.

f) How many solutions are there to $h(t) = 18$ in the interval [0, 60]?

g) Find the exact values of $h(2.5)$ and $h(7.5)$.

5.2

The Sine and Cosine Functions

In Section 5.1 we learned that angles can be measured in degrees or radians. Now we define two trigonometric functions whose domain is the set of all angles (measured in degrees or radians). These two functions are unlike any functions defined in algebra, but they form the foundation of trigonometry.

Definition

If α is an angle in standard position whose terminal side intersects the unit circle at point (x, y), as shown in Fig. 5.20, then **sine of α**—abbreviated $\sin(\alpha)$ or $\sin\alpha$—is the y-coordinate of that point, and **cosine of α**—abbreviated $\cos(\alpha)$ or $\cos\alpha$—is the x-coordinate. Sine and cosine are called **trigonometric functions.**

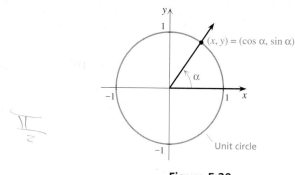

■ **Figure 5.20**

Definition: Sine and Cosine

If α is an angle in standard position and (x, y) is the point of intersection of the terminal side and the unit circle, then

$$\sin\alpha = y \quad \text{and} \quad \cos\alpha = x.$$

The domain of the sine function and the cosine function is the set of angles in standard position, but since each angle has a measure in degrees or radians, we generally use the set of degree measures or the set of radian measures as the domain. If (x, y) is on the unit circle, then $-1 \le x \le 1$ and $-1 \le y \le 1$, so the range of each of these functions is the interval $[-1, 1]$.

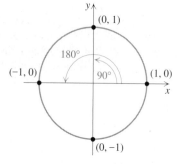

■ **Figure 5.21**

Example **1** **Evaluating sine and cosine at a multiple of 90°**

Find the exact values of the sine and cosine functions for each angle.

a. $90°$ **b.** $\pi/2$ **c.** $180°$ **d.** $-5\pi/2$ **e.** $-720°$

Solution

a. Consider the unit circle shown in Fig. 5.21. Since the terminal side of 90° intersects the unit circle at $(0, 1)$, $\sin(90°) = 1$ and $\cos(90°) = 0$.

b. Since $\pi/2$ is coterminal with $90°$, we have $\sin(\pi/2) = 1$ and $\cos(\pi/2) = 0$.

c. The terminal side for $180°$ intersects the unit circle at $(-1, 0)$. So $\sin(180°) = 0$ and $\cos(180°) = -1$.

d. Since $-5\pi/2$ is coterminal with $3\pi/2$, the terminal side of $-5\pi/2$ lies on the negative y-axis and intersects the unit circle at $(0, -1)$. So $\sin(-5\pi/2) = -1$ and $\cos(-5\pi/2) = 0$.

e. Since $-720°$ is coterminal with $0°$, the terminal side of $-720°$ lies on the positive x-axis and intersects the unit circle at $(1, 0)$. So $\sin(-720°) = 0$ and $\cos(-720°) = 1$.

Try This.   Find the exact values of $\sin(-\pi/2)$ and $\cos(-\pi/2)$.   ■

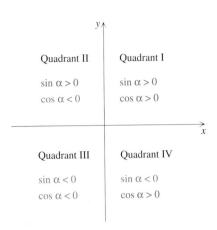

Quadrant II   Quadrant I
$\sin \alpha > 0$ $\sin \alpha > 0$
$\cos \alpha < 0$ $\cos \alpha > 0$

Quadrant III   Quadrant IV
$\sin \alpha < 0$ $\sin \alpha < 0$
$\cos \alpha < 0$ $\cos \alpha > 0$

■ **Figure 5.22**

The *signs* of the sine and cosine functions depend on the quadrant in which the angle lies. For any point (x, y) on the unit circle in quadrant I, the x- and y-coordinates are positive. So if α is an angle in quadrant I, $\sin \alpha > 0$ and $\cos \alpha > 0$. Since the x-coordinate of any point in quadrant II is negative, the cosine of any angle in quadrant II is negative. Figure 5.22 gives the signs of sine and cosine for each of the four quadrants.

Sine and Cosine of a Multiple of 45°

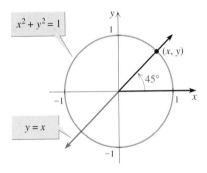

■ **Figure 5.23**

In Example 1, exact values of the sine and cosine functions were found for some angles that were multiples of $90°$ (quadrantal angles). We can also find the exact values of $\sin(45°)$ and $\cos(45°)$ and then use that information to find the exact values of these functions for any multiple of $45°$.

Since the terminal side for $45°$ lies on the line $y = x$, as shown in Fig. 5.23, the x- and y-coordinates at the point of intersection with the unit circle are equal. From Section 1.2, the equation of the unit circle is $x^2 + y^2 = 1$. Because $y = x$, we can write $x^2 + x^2 = 1$ and solve for x:

$$2x^2 = 1$$

$$x^2 = \frac{1}{2}$$

$$x = \pm\sqrt{\frac{1}{2}} = \pm\frac{\sqrt{2}}{2}$$

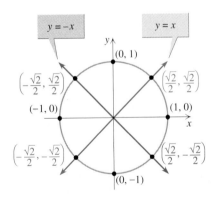

■ **Figure 5.24**

For a $45°$ angle in quadrant I, x and y are both positive numbers. So $\sin(45°) = \cos(45°) = \sqrt{2}/2$.

There are four points where the lines $y = x$ and $y = -x$ intersect the unit circle. The coordinates of these points, shown in Fig. 5.24, can all be found as above or by the symmetry of the unit circle. Figure 5.24 shows the coordinates of the key points for determining the sine and cosine of any angle that is an integral multiple of $45°$.

Example **2**   **Evaluating sine and cosine at a multiple of 45°**

Find the exact value of each expression.

a. $\sin(135°)$ **b.** $\sin\left(\dfrac{5\pi}{4}\right)$ **c.** $2\sin\left(-\dfrac{9\pi}{4}\right)\cos\left(-\dfrac{9\pi}{4}\right)$

Solution

a. The terminal side for 135° lies in quadrant II, halfway between 90° and 180°. As shown in Fig. 5.24, the point on the unit circle halfway between 90° and 180° is $\left(-\sqrt{2}/2,\ \sqrt{2}/2\right)$. So $\sin(135°) = \sqrt{2}/2$.

b. The terminal side for $5\pi/4$ is in quadrant III, halfway between $4\pi/4$ and $6\pi/4$. From Fig. 5.24, $\sin(5\pi/4)$ is the y-coordinate of $\left(-\sqrt{2}/2,\ -\sqrt{2}/2\right)$. So $\sin(5\pi/4) = -\sqrt{2}/2$.

c. Since $-8\pi/4$ is one clockwise revolution, $-9\pi/4$ is coterminal with $-\pi/4$. From Fig. 5.24, we have $\cos(-9\pi/4) = \cos(-\pi/4) = \sqrt{2}/2$ and $\sin(-9\pi/4) = -\sqrt{2}/2$. So

$$2 \sin\left(-\frac{9\pi}{4}\right)\cos\left(-\frac{9\pi}{4}\right) = 2 \cdot \left(-\frac{\sqrt{2}}{2}\right) \cdot \frac{\sqrt{2}}{2} = -1.$$

Try This. Find the exact values of $\sin(-5\pi/4)$ and $\cos(-5\pi/4)$.   ■

Sine and Cosine of a Multiple of 30°

Exact values for the sine and cosine of 60° and the sine and cosine of 30° are found with a little help from geometry. The terminal side of a 60° angle intersects the unit circle at a point (x, y), as shown in Fig. 5.25(a), where x and y are the lengths of the legs of a 30-60-90 triangle whose hypotenuse is length 1.

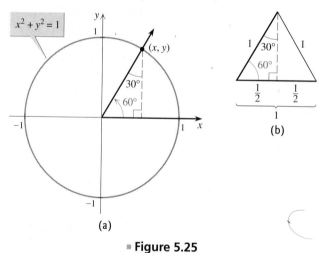

■ **Figure 5.25**

Since two congruent 30-60-90 triangles are formed by the altitude of an equilateral triangle, as shown in Fig. 5.25(b), the side opposite the 30° angle is half the length of the hypotenuse. Since the length of the hypotenuse is 1, the side opposite 30° is 1/2, that is, $x = 1/2$. Use the fact that $x^2 + y^2 = 1$ for the unit circle to find y:

$$\left(\frac{1}{2}\right)^2 + y^2 = 1 \qquad\qquad \text{Replace } x \text{ with } \frac{1}{2} \text{ in } x^2 + y^2 = 1.$$

$$y^2 = \frac{3}{4}$$

$$y = \pm\sqrt{\frac{3}{4}} = \pm\frac{\sqrt{3}}{2}$$

Since $y > 0$ (in quadrant I), $\cos(60°) = 1/2$ and $\sin(60°) = \sqrt{3}/2$.

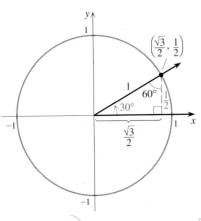

The point of intersection of the terminal side for a 30° angle with the unit circle determines a 30-60-90 triangle exactly the same size as the one described above. In this case the longer leg is on the x-axis, as shown in Fig. 5.26, and the point on the unit circle is $\left(\sqrt{3}/2, 1/2\right)$. So $\cos(30°) = \sqrt{3}/2$, and $\sin(30°) = 1/2$.

Using the symmetry of the unit circle and the values just found for the sine and cosine of 30° and 60°, we can label points on the unit circle, as shown in Fig. 5.27. This figure shows the coordinates of every point on the unit circle where the terminal side of a multiple of 30° intersects the unit circle. For example, the terminal side of a 120° angle in standard position intersects the unit circle at $\left(-1/2, \sqrt{3}/2\right)$, because $120° = 4 \cdot 30°$.

In the next example, we use Fig. 5.27 to evaluate expressions involving multiples of 30° $(\pi/6)$. Note that the square of $\sin\alpha$ could be written as $(\sin\alpha)^2$, but for simplicity it is written as $\sin^2\alpha$. Likewise, $\cos^2\alpha$ is used for $(\cos\alpha)^2$.

■ **Figure 5.26**

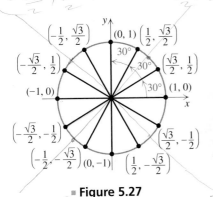

■ **Figure 5.27**

Example 3 Evaluating sine and cosine at a multiple of 30°

Find the exact value of each expression.

a. $\sin\left(\dfrac{7\pi}{6}\right)$ **b.** $\cos^2(-240°) - \sin^2(-240°)$

Solution

a. Since $\pi/6 = 30°$, we have $7\pi/6 = 210°$. To determine the location of the terminal side of $7\pi/6$, notice that $7\pi/6$ is 30° larger than the straight angle 180° and so it lies in quadrant III. From Fig. 5.27, $7\pi/6$ intersects the unit circle at $\left(-\sqrt{3}/2, -1/2\right)$ and $\sin(7\pi/6) = -1/2$.

b. Since $-240°$ is coterminal with 120°, the terminal side for $-240°$ lies in quadrant II and intersects the unit circle at $\left(-1/2, \sqrt{3}/2\right)$, as indicated in Fig. 5.27. So $\cos(-240°) = -1/2$ and $\sin(-240°) = \sqrt{3}/2$. Now use these values in the original expression:

$$\cos^2(-240°) - \sin^2(-240°) = \left(-\frac{1}{2}\right)^2 - \left(\frac{\sqrt{3}}{2}\right)^2 = \frac{1}{4} - \frac{3}{4} = -\frac{1}{2}$$

Try This. Find the exact values of $\sin(-\pi/6)$ and $\cos(-\pi/6)$.   ■

A good way to remember the sines and cosines for the common angles is to note the following patterns:

$$\sin(30°) = \frac{\sqrt{1}}{2}, \quad \sin(45°) = \frac{\sqrt{2}}{2}, \quad \sin(60°) = \frac{\sqrt{3}}{2}$$

$$\cos(30°) = \frac{\sqrt{3}}{2}, \quad \cos(45°) = \frac{\sqrt{2}}{2}, \quad \cos(60°) = \frac{\sqrt{1}}{2}$$

Reference Angles

One way to find $\sin\alpha$ and $\cos\alpha$ is to determine the coordinates of the intersection of the terminal side of α and the unit circle, as we did in Examples 2 and 3. Another way to find $\sin\alpha$ and $\cos\alpha$ is to find the sine and cosine of the corresponding *reference angle,* which is in the interval $[0, \pi/2]$, and attach the appropriate sign. Reference angles are essential for evaluating trigonometric functions if you are using tables that give values only for angles in the interval $[0, \pi/2]$.

Definition:
Reference Angle

If θ is a nonquadrantal angle in standard position, then the **reference angle** for θ is the positive acute angle θ' (read "theta prime") formed by the terminal side of θ and the positive or negative x-axis.

Figure 5.28 shows the reference angle θ' for an angle θ with terminal side in each of the four quadrants. Notice that in each case θ' is the acute angle formed by the terminal side of θ and the x-axis.

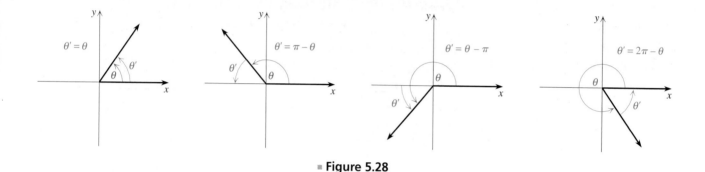

▪ **Figure 5.28**

Example 4 Finding reference angles

For each given angle θ, sketch the reference angle θ' and give the measure of θ' in both radians and degrees.

a. $\theta = 120°$ **b.** $\theta = 7\pi/6$ **c.** $\theta = 690°$ **d.** $\theta = -7\pi/4$

Solution

a. The terminal side of $120°$ is in quadrant II, as shown in Fig. 5.29(a). The reference angle θ' is $180° - 120° = 60°$ or $\pi/3$.

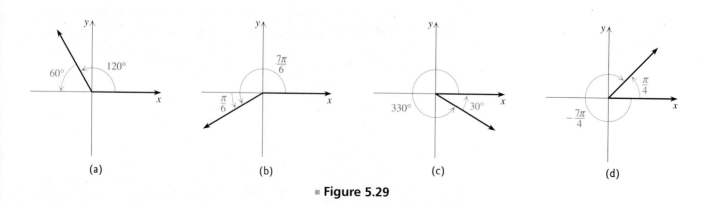

▪ **Figure 5.29**

b. The terminal side of $7\pi/6$ is in quadrant III, as shown in Fig. 5.29(b). The reference angle θ' is $7\pi/6 - \pi = \pi/6$ or $30°$.

c. Since $690°$ is coterminal with $330°$, the terminal side is in quadrant IV, as shown in Fig. 5.29(c). The reference angle θ' is $360° - 330° = 30°$ or $\pi/6$.

d. An angle of $-7\pi/4$ radians has a terminal side in quadrant I, as shown in Fig. 5.29(d), and the reference angle is $\pi/4$ or $45°$.

Try This. Find the reference angle for $135°$ in degrees and radians. ■

Because the measure of any reference angle is between 0 and $\pi/2$, any reference angle can be positioned in the first quadrant. Figure 5.30(a) shows a typical reference angle θ' in standard position with its terminal side intersecting the unit circle at (x, y). So $\cos\theta' = x$ and $\sin\theta' = y$.

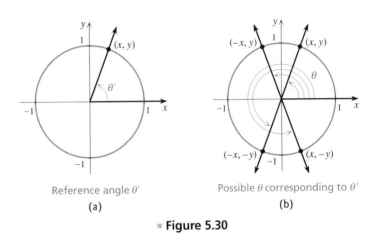

Reference angle θ'
(a)

Possible θ corresponding to θ'
(b)

■ **Figure 5.30**

For what angle θ could θ' be the reference angle? The angle θ (in standard position) that corresponds to θ' could have terminated in quadrant I, II, III, or IV, where it would intersect the unit circle at (x, y), $(-x, y)$, $(-x, -y)$, or $(x, -y)$, as shown in Fig. 5.30(b). So $\cos\theta = \pm x$ and $\sin\theta = \pm y$. Therefore, we can find $\cos\theta'$ and $\sin\theta'$ and prefix the appropriate sign to get $\cos\theta$ and $\sin\theta$. The appropriate sign is determined by the quadrant in which the terminal side of θ lies. The signs of $\sin\theta$ and $\cos\theta$ for the four quadrants were summarized in Fig. 5.22.

Theorem: Evaluating Trigonometric Functions Using Reference Angles

For an angle θ in standard position that is not a quadrantal angle, the value of a trigonometric function of θ can be found by finding the value of the function for its reference angle θ' and prefixing the appropriate sign.

Example **5** **Trigonometric functions using reference angles**

Find the sine and cosine for each angle using reference angles.

a. $\theta = 120°$ **b.** $\theta = 7\pi/6$ **c.** $\theta = 690°$ **d.** $\theta = -7\pi/4$

Solution

a. Figure 5.29(a) shows the terminal side of 120° in quadrant II and its reference angle 60°. Since $\sin\theta > 0$ and $\cos\theta < 0$ for any angle θ in quadrant II, $\sin(120°) > 0$ and $\cos(120°) < 0$. So to get $\cos(120°)$ from $\cos(60°)$, we must prefix a negative sign. We have

$$\sin(120°) = \sin(60°) = \frac{\sqrt{3}}{2} \quad \text{and} \quad \cos(120°) = -\cos(60°) = -\frac{1}{2}.$$

b. Figure 5.29(b) shows the terminal side of $7\pi/6$ in quadrant III and its reference angle $\pi/6$. Since $\sin\theta < 0$ and $\cos\theta < 0$ for any angle θ in quadrant III, we must prefix a negative sign to both $\sin(\pi/6)$ and $\cos(\pi/6)$. So

$$\sin\left(\frac{7\pi}{6}\right) = -\sin\left(\frac{\pi}{6}\right) = -\frac{1}{2} \quad \text{and} \quad \cos\left(\frac{7\pi}{6}\right) = -\cos\left(\frac{\pi}{6}\right) = -\frac{\sqrt{3}}{2}.$$

c. The angle 690° terminates in quadrant IV (where $\sin\theta < 0$ and $\cos\theta > 0$) and its reference angle is 30°, as shown in Fig. 5.29(c). So

$$\sin(690°) = -\sin(30°) = -\frac{1}{2} \quad \text{and} \quad \cos(690°) = \cos(30°) = \frac{\sqrt{3}}{2}.$$

d. The angle $-7\pi/4$ terminates in quadrant I (where $\sin\theta > 0$ and $\cos\theta > 0$) and its reference angle is $\pi/4$, as shown in Fig. 5.29(d). So

$$\sin\left(-\frac{7\pi}{4}\right) = \sin\left(\frac{\pi}{4}\right) = \frac{\sqrt{2}}{2} \quad \text{and} \quad \cos\left(-\frac{7\pi}{4}\right) = \cos\left(\frac{\pi}{4}\right) = \frac{\sqrt{2}}{2}.$$

Try This. Find $\sin(135°)$ and $\cos(135°)$ using reference angles. ■

The Sine and Cosine of a Real Number

The domain of sine and cosine can be thought of as the set of all angles, degree measures of angles, or radian measures of angles. However, since radian measures are just real numbers, the domain of sine and cosine is the set of real numbers. The set of real numbers is often visualized as the domain of sine and cosine by placing a number line (the set of real numbers) next to the unit circle, as shown in Fig. 5.31(a). To find $\sin s$ or $\cos s$ for $s > 0$, wrap the top half of the number line around the unit circle, as shown in Fig. 5.31(b). The real number s then corresponds to an arc on the unit circle of length s, ending at (x, y). Of course, $\sin s = y$ and $\cos s = x$. Arcs corresponding to negative real numbers are found by wrapping the negative half of the number line around the unit circle in a clockwise direction, as shown in Fig. 5.31(c). Because of their relationship to the unit circle, sine and cosine are often called **circular functions.**

Approximate Values for Sine and Cosine

The sine and cosine for any angle that is a multiple of 30° or 45° can be found exactly. These angles are so common that it is important to know these exact values. However, for most other angles or real numbers we use approximate values for the sine and cosine, found with the help of a scientific calculator or a graphing calculator. Calculators can evaluate $\sin\alpha$ or $\cos\alpha$ if α is a real number (radian) or α is the degree measure of an angle. Generally, there is a mode key that sets the calculator to degree mode or radian mode. Consult your calculator manual.

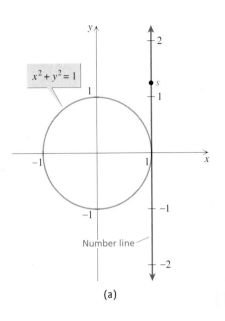
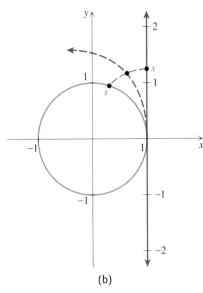
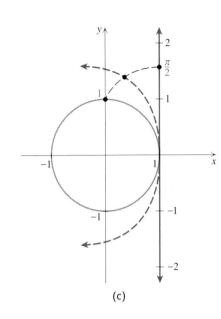

(a)   (b)   (c)

■ **Figure 5.31**

Example **6** **Evaluating sine and cosine with a calculator**

Find each function value rounded to four decimal places.

a. $\sin(4.27)$ **b.** $\cos(-39.46°)$

Solution

a. With the calculator in radian mode, we get $\sin(4.27) \approx -0.9037$.
b. With the calculator in degree mode, we get $\cos(-39.46°) \approx 0.7721$.

If you use the symbol for radians or degrees, as shown in Fig. 5.32, then it is not necessary to change the mode with a graphing calculator.

Try This. Find $\sin(55.6)$ and $\cos(34.2°)$ with a calculator.   ■

■ **Figure 5.32**

The Fundamental Identity

An identity is an equation that is satisfied for all values of the variable for which both sides are defined. The most fundamental identity in trigonometry involves the squares of the sine and cosine functions. For convenience we write $(\sin\alpha)^2$ as $\sin^2\alpha$ and $(\cos\alpha)^2$ as $\cos^2\alpha$. If α is an angle in standard position then $\sin\alpha = y$ and $\cos\alpha = x$, where (x, y) is the point of intersection of the terminal side of α and the unit circle. Since every point on the unit circle satisfies the equation $x^2 + y^2 = 1$, we get the following identity.

The Fundamental Identity of Trigonometry

If α is any angle or real number,

$$\sin^2\alpha + \cos^2\alpha = 1.$$

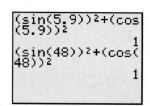

■ **Figure 5.33**

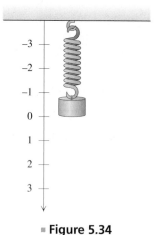

■ **Figure 5.34**

The fundamental identity can be illustrated with a graphing calculator as shown in Fig. 5.33. □

If we know the value of the sine or cosine of an angle, then we can use the fundamental identity to find the value of the other function of the angle.

Example 7 Using the fundamental identity

Find $\cos\alpha$, given that $\sin\alpha = 3/5$ and α is an angle in quadrant II.

Solution

Use the fundamental identity $\sin^2\alpha + \cos^2\alpha = 1$ to find $\cos\alpha$:

$$\left(\frac{3}{5}\right)^2 + \cos^2\alpha = 1 \qquad \text{Replace } \sin\alpha \text{ with } 3/5.$$

$$\cos^2\alpha = \frac{16}{25} \qquad 1 - \frac{9}{25} = \frac{16}{25}$$

$$\cos\alpha = \pm\sqrt{\frac{16}{25}} = \pm\frac{4}{5}$$

Since $\cos\alpha < 0$ for any α in quadrant II, we choose the negative sign and get $\cos\alpha = -4/5$.

Try This. Find $\sin\alpha$ given that $\cos\alpha = 1/4$ and α is in quadrant IV. ■

Modeling the Motion of a Spring

The sine and cosine functions are used in modeling the motion of a spring. If a weight is at rest while hanging from a spring, as shown in Fig. 5.34, then it is at the **equilibrium** position, or 0 on a vertical number line. If the weight is set in motion with an initial velocity v_0 from location x_0, then the location at time t is given by

$$x = \frac{v_0}{\omega} \sin(\omega t) + x_0 \cos(\omega t).$$

The letter ω (omega) is a constant that depends on the stiffness of the spring and the amount of weight on the spring. For positive values of x the weight is below equilibrium, and for negative values it is above equilibrium. The initial velocity is considered to be positive if it is in the downward direction and negative if it is upward. Note that the domain of the sine and cosine functions in this formula is the nonnegative real numbers, and this application has nothing to do with angles.

Example 8 Motion of a spring

A weight on a certain spring is set in motion with an upward velocity of 3 centimeters per second from a position 2 centimeters below equilibrium. Assume that for this spring and weight combination the constant ω has a value of 1. Write a formula that gives the location of the weight in centimeters as a function of the time t in seconds, and find the location of the weight 2 seconds after the weight is set in motion.

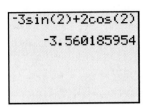

-3sin(2)+2cos(2)
            -3.560185954

■ **Figure 5.35**

Solution

Since upward velocity is negative and locations below equilibrium are positive, use $v_0 = -3$, $x_0 = 2$, and $\omega = 1$ in the formula for the motion of a spring:

$$x = -3 \sin t + 2 \cos t$$

If $t = 2$ seconds, then $x = -3 \sin(2) + 2 \cos(2) \approx -3.6$ centimeters, which is 3.6 centimeters above the equilibrium position.

◨◨ Figure 5.35 shows the calculation for x with a graphing calculator in radian mode.

Try This. If $x = -2 \sin(2t) + 3 \cos(2t)$ gives the location in centimeters of a weight on a spring t seconds after it is set in motion, then what is the location at time $t = 3$?   ■

For Thought

True or False? Explain.

1. $\cos(90°) = 1$

2. $\cos(90) = 0$

3. $\sin(45°) = 1/\sqrt{2}$

4. $\sin(-\pi/3) = \sin(\pi/3)$

5. $\cos(-\pi/3) = -\cos(\pi/3)$ **6.** $\sin(390°) = \sin(30°)$

7. If $\sin(\alpha) < 0$ and $\cos(\alpha) > 0$, then α is an angle in quadrant III.

8. If $\sin^2(\alpha) = 1/4$ and α is in quadrant III, then $\sin(\alpha) = 1/2$.

9. If $\cos^2(\alpha) = 3/4$ and α is an angle in quadrant I, then $\alpha = \pi/6$.

10. For any angle α, $\cos^2(\alpha) = (1 - \sin(\alpha))(1 + \sin(\alpha))$.

5.2 Exercises

Redraw each diagram and label the indicated points with the proper coordinates. Exercise 1 shows the points where the terminal side of every multiple of 45° intersects the unit circle. Exercise 2 shows the points where the terminal side of every multiple of 30° intersects the unit circle. Repeat Exercises 1 and 2 until you can do them from memory.

1.

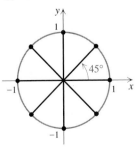

2.

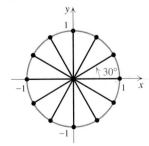

Use the diagrams drawn in Exercises 1 and 2 to help you find the exact value of each function. Do not use a calculator.

3. $\sin(0)$ **4.** $\cos(\pi)$ **5.** $\cos(-90°)$

6. $\sin(0°)$ **7.** $\sin(2\pi)$ **8.** $\cos(-\pi)$

9. $\cos(3\pi/2)$ **10.** $\sin(-5\pi/2)$ **11.** $\sin(135°)$

12. $\cos(-135°)$ **13.** $\sin(-\pi/4)$ **14.** $\sin(3\pi/4)$

15. $\sin(30°)$ **16.** $\sin(120°)$ **17.** $\cos(-60°)$

18. $\cos(-120°)$ **19.** $\cos(7\pi/6)$ **20.** $\cos(-2\pi/3)$

21. $\sin(-4\pi/3)$ **22.** $\sin(5\pi/6)$ **23.** $\sin(390°)$

24. $\sin(765°)$ **25.** $\cos(-420°)$ **26.** $\cos(-450°)$

27. $\cos(13\pi/6)$ **28.** $\cos(-7\pi/3)$

Sketch the given angle in standard position and find its reference angle in degrees and radians.

29. $210°$ **30.** $330°$ **31.** $5\pi/3$ **32.** $7\pi/6$

33. $-300°$ **34.** $-210°$ **35.** $5\pi/6$ **36.** $-13\pi/6$

37. $405°$ **38.** $390°$ **39.** $-3\pi/4$ **40.** $-2\pi/3$

Determine whether each of the following expressions is positive $(+)$ or negative $(-)$ without using a calculator.

41. $\sin(121°)$ **42.** $\cos(157°)$ **43.** $\cos(359°)$

44. $\sin(213°)$ **45.** $\sin(7\pi/6)$ **46.** $\cos(7\pi/3)$

47. $\cos(-3\pi/4)$ **48.** $\sin(-7\pi/4)$

Use reference angles to find the exact value of each expression.

49. $\sin(135°)$ **50.** $\sin(420°)$ **51.** $\cos(5\pi/3)$

52. $\cos(11\pi/6)$ **53.** $\sin(7\pi/4)$ **54.** $\sin(-13\pi/4)$

55. $\cos(-17\pi/6)$ **56.** $\cos(-5\pi/3)$ **57.** $\sin(-45°)$

58. $\cos(-120°)$ **59.** $\cos(-240°)$ **60.** $\sin(-225°)$

Find the exact value of each expression without using a calculator. Check your answer with a calculator.

61. $\dfrac{\cos(\pi/3)}{\sin(\pi/3)}$ **62.** $\dfrac{\sin(-5\pi/6)}{\cos(-5\pi/6)}$

63. $\dfrac{\sin(7\pi/4)}{\cos(7\pi/4)}$ **64.** $\dfrac{\sin(-3\pi/4)}{\cos(-3\pi/4)}$

65. $\sin\left(\dfrac{\pi}{3} + \dfrac{\pi}{6}\right)$ **66.** $\cos\left(\dfrac{\pi}{3} - \dfrac{\pi}{6}\right)$

67. $\dfrac{1 - \cos(5\pi/6)}{\sin(5\pi/6)}$ **68.** $\dfrac{\sin(5\pi/6)}{1 + \cos(5\pi/6)}$

69. $\sin(\pi/4) + \cos(\pi/4)$ **70.** $\sin^2(\pi/6) + \cos^2(\pi/6)$

Use a calculator to find the value of each function. Round answers to four decimal places.

71. $\cos(-359.4°)$ **72.** $\sin(344.1°)$

73. $\sin(23°48')$ **74.** $\cos(49°13')$

75. $\sin(-48°3'12'')$ **76.** $\cos(-9°4'7'')$

77. $\sin(1.57)$ **78.** $\cos(3.14)$

79. $\cos(7\pi/12)$ **80.** $\sin(-13\pi/8)$

Find the exact value of each expression for the given value of θ. Do not use a calculator.

81. $\sin(2\theta)$ if $\theta = \pi/4$

82. $\sin(2\theta)$ if $\theta = \pi/6$

83. $\cos(2\theta)$ if $\theta = \pi/6$

84. $\cos(2\theta)$ if $\theta = \pi/3$

85. $\sin(\theta/2)$ if $\theta = 3\pi/2$

86. $\sin(\theta/2)$ if $\theta = 2\pi/3$

87. $\cos(\theta/2)$ if $\theta = \pi/3$

88. $\cos(\theta/2)$ if $\theta = \pi/2$

Solve each problem.

89. Find $\cos(\alpha)$, given that $\sin(\alpha) = 5/13$ and α is in quadrant II.

90. Find $\sin(\alpha)$, given that $\cos(\alpha) = -4/5$ and α is in quadrant III.

91. Find $\sin(\alpha)$, given that $\cos(\alpha) = 3/5$ and α is in quadrant IV.

92. Find $\cos(\alpha)$, given that $\sin(\alpha) = -12/13$ and α is in quadrant IV.

93. Find $\cos(\alpha)$, given that $\sin(\alpha) = 1/3$ and $\cos(\alpha) > 0$.

94. Find $\sin(\alpha)$, given that $\cos(\alpha) = 2/5$ and $\sin(\alpha) < 0$.

Solve each problem.

95. *Motion of a Spring*   A weight on a vertical spring is given an initial downward velocity of 4 cm/sec from a point 3 cm above equilibrium. Assuming that the constant ω has a value of 1, write the formula for the location of the weight at time t, and find its location 3 sec after it is set in motion.

96. *Motion of a Spring*   A weight on a vertical spring is given an initial upward velocity of 3 in./sec from a point 1 in. below equilibrium. Assuming that the constant ω has a value of $\sqrt{3}$, write the formula for the location of the weight at time t, and find its location 2 sec after it is set in motion.

97. *Spacing Between Teeth*   The length of an arc intercepted by a central angle of θ radians in a circle of radius r is $r\theta$. The length of the chord, c, joining the endpoints of that arc is given by $c = r\sqrt{2 - 2\cos\theta}$. Find the actual distance between the tips of two adjacent teeth on a 12-in.-diameter carbide-tipped circular saw blade with 22 equally spaced teeth. Compare your answer with the length of a circular arc joining two adjacent teeth on a circle 12 in. in diameter. Round to three decimal places.

■ **Figure for Exercise 97**

98. *Shipping Stop Signs*   A manufacturer of stop signs ships 100 signs in a circular drum. Use the formula from Exercise 97 to find the radius of the inside of the circular drum, given that the length of an edge of the stop sign is 10 in. and the signs just fit as shown in the figure.

■ **Figure for Exercise 98**

99. *Cosine in Terms of Sine*   Use the fundamental identity to write two formulas for $\cos\alpha$ in terms of $\sin\alpha$. Indicate which formula to use for a given value of α.

100. *Throwing a Javelin*   The formula

$$d = \frac{1}{32} v_0^2 \sin(2\theta)$$

gives the distance d in feet that a projectile will travel when its launch angle is θ and its initial velocity is v_0 feet per second. Approximately what initial velocity in miles per hour does it take to throw a javelin 367 feet with launch angle 43°, which is a typical launch angle (www.canthrow.com)?

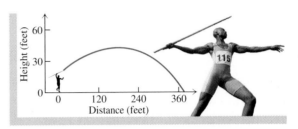

■ **Figure for Exercise 100**

Thinking Outside the Box XLV

Telling Time   At 12 noon the hour hand, minute hand, and second hand of a clock are all pointing straight upward.

a. Find the first time after 12 noon to the nearest tenth of a second at which the angle between the hour hand and minute hand is 120°.

b. Is there a time between 12 noon and 1 P.M. at which the three hands divide the face of the clock into thirds (that is, the angles between the hour hand, minute hand, and second hand are equal and each is 120°)?

c. Does the alignment of the three hands described in part (b) ever occur?

5.2 Pop Quiz

1. What is the reference angle for 120°?

Find the exact value.

2. $\sin(0°)$

3. $\sin(-60°)$

4. $\sin(3\pi/4)$

5. $\cos(90°)$

6. $\cos(-2\pi/3)$

7. $\cos(11\pi/6)$

8. $\sin(\alpha)$, if $\cos(\alpha) = -3/5$ and α is in quadrant III

Linking Concepts

For Individual or
Group Explorations

Domain and Range

The domain for both the sine and cosine functions is $(-\infty, \infty)$ and the range for both functions is $[-1, 1]$. When we combine these trigonometric functions with other functions that we have studied, determining domain and range can get interesting. For example, the range for $f(x) = \ln(\sin(x))$ is $(-\infty, 0]$. For some functions the domain or range can only be approximated. Find the domain and range for each of the following functions. Use a calculator table or graph if necessary.

a) $f(x) = \sin^2(x)$

b) $f(x) = x^2 + \sin(x)$

c) $f(x) = \sin^2(x) + \cos^2(x)$

d) $f(x) = x \cdot \cos(x)$

e) $f(x) = x + \sin(x)$

f) $f(x) = \sin(\cos(x))$

g) $f(x) = \ln(\cos(x))$

h) $f(x) = e^{\sin(x)}$

5.3

The Graphs of the Sine and Cosine Functions

In Section 5.2 we studied the sine and cosine functions. In this section we will study their graphs. The graphs of trigonometric functions are important for understanding their use in modeling physical phenomena such as radio, sound, and light waves, and the motion of a spring or a pendulum.

The Graph of $y = \sin(x)$

Until now, s or α has been used as the independent variable in writing $\sin(s)$ or $\sin(\alpha)$. When graphing functions in an xy-coordinate system, it is customary to use x as the independent variable and y as the dependent variable, as in $y = \sin(x)$. We assume that x is a real number or radian measure unless it is stated that x is the degree measure of an angle. We often omit the parentheses and write $\sin x$ for $\sin(x)$.

Consider some ordered pairs that satisfy $y = \sin x$ for x in $[0, 2\pi]$, as shown in Fig. 5.36. These ordered pairs are best understood by recalling that $\sin x$ is the second coordinate of the terminal point on the unit circle for an arc of length x, as

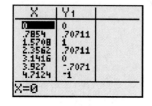

X	Y₁
0	0
.7854	.70711
1.5708	1
2.3562	.70711
3.1416	0
3.927	-.7071
4.7124	-1

X=0

■ **Figure 5.36**

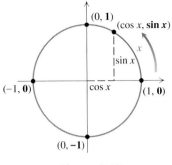

■ **Figure 5.37**

shown in Fig. 5.37. As the arc length x increases from length 0 to $\pi/2$, the second coordinate of the endpoint increases to 1. As the length of x increases from $\pi/2$ to π, the second coordinate of the endpoint decreases to 0. As the arc length x increases from π to 2π, the second coordinate decreases to -1 and then increases to 0. The calculator graph in Fig. 5.38 shows 95 accurately plotted points on the graph of $y = \sin(x)$ between 0 and 2π. The actual graph of $y = \sin(x)$ is a smooth curve through those 95 points, as shown in Fig. 5.39.

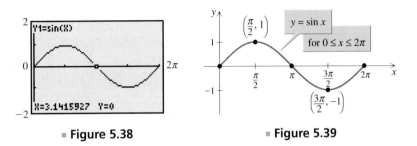

■ **Figure 5.38**   ■ **Figure 5.39**

Since the x-intercepts and the maximum and minimum values of the function occur at multiples of π, we usually label the x-axis with multiples of π, as in Fig. 5.39. There are five key points on the graph of $y = \sin x$ between 0 and 2π. Their exact coordinates are given in the following table.

x	0	$\pi/2$	π	$3\pi/2$	2π
$y = \sin x$	0	1	0	-1	0

Note that these five points divide the interval $[0, 2\pi]$ into four equal parts.

Since the domain of $y = \sin x$ is the set of all real numbers (or radian measures of angles), we must consider values of x outside the interval $[0, 2\pi]$. As x increases from 2π to 4π, the values of $\sin x$ again increase from 0 to 1, decrease to -1, then increase to 0. Because $\sin(x + 2\pi) = \sin x$, the exact shape that we saw in Fig. 5.39 is repeated for x in intervals such as $[2\pi, 4\pi], [-2\pi, 0], [-4\pi, -2\pi]$, and so on. So the curve shown in Fig. 5.40 continues indefinitely to the left and right. The range of $y = \sin x$ is $[-1, 1]$. The graph $y = \sin x$ or any transformation of $y = \sin x$ is called a **sine wave,** a **sinusoidal wave,** or a **sinusoid.**

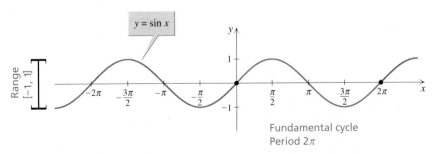

■ **Figure 5.40**

Since the shape of $y = \sin x$ for x in $[0, 2\pi]$ is repeated infinitely often, $y = \sin x$ is called a *periodic function.*

Definition: Periodic Function

> If $y = f(x)$ is a function and a is a nonzero constant such that $f(x) = f(x + a)$ for every x in the domain of f, then f is called a **periodic function.** The smallest such positive constant a is the **period** of the function.

For the sine function, the smallest value of a such that $\sin x = \sin(x + a)$ is $a = 2\pi$, and so the period of $y = \sin x$ is 2π. To prove that 2π is actually the *smallest* value of a for which $\sin x = \sin(x + a)$ is a bit complicated, and we will omit the proof. The graph of $y = \sin x$ over any interval of length 2π is called a **cycle** of the sine wave. The graph of $y = \sin x$ over $[0, 2\pi]$ is called the **fundamental cycle** of $y = \sin x$.

Example **1** Graphing a periodic function

Sketch the graph of $y = 2 \sin x$ for x in the interval $[-2\pi, 2\pi]$.

Solution

We can obtain the graph of $y = 2 \sin x$ from the graph of $y = \sin x$ by stretching the graph of $y = \sin x$ by a factor of 2. In other words, double the y-coordinates of the five key points on the graph of $y = \sin x$ to obtain five key points on the graph of $y = 2 \sin x$ for the interval $[0, 2\pi]$, as shown in the following table.

x	0	$\pi/2$	π	$3\pi/2$	2π
$y = 2 \sin x$	0	2	0	-2	0

The x-intercepts for $y = 2 \sin x$ on $[0, 2\pi]$ are $(0, 0)$, $(\pi, 0)$ and $(2\pi, 0)$. Midway between the intercepts are found the highest point $(\pi/2, 2)$ and the lowest point $(3\pi/2, -2)$. Draw one cycle of $y = 2 \sin x$ through these five points, as shown in Fig. 5.41. Draw another cycle of the sine wave for x in $[-2\pi, 0]$ to complete the graph of $y = 2 \sin x$ for x in $[-2\pi, 2\pi]$.

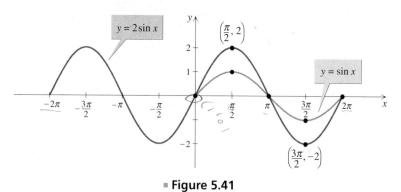

■ **Figure 5.41**

Use $\pi/2$ as the x-scale on your calculator graph as we did in Fig. 5.41. The calculator graph in Fig. 5.42 supports our conclusions.

Try This. Graph $y = 4 \sin x$ for x in the interval $[0, 2\pi]$. ■

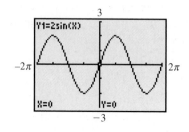

■ **Figure 5.42**

The amplitude of a sine wave is a measure of the "height" of the wave. When an oscilloscope is used to get a picture of the sine wave corresponding to a sound, the amplitude of the sine wave corresponds to the intensity or loudness of the sound.

Definition: Amplitude

> The **amplitude** of a sine wave, or the amplitude of the function, is the absolute value of half the difference between the maximum and minimum y-coordinates on the wave.

Example 2 Finding amplitude

Find the amplitude of the functions $y = \sin x$ and $y = 2 \sin x$.

Solution

For $y = \sin x$, the maximum y-coordinate is 1 and the minimum is -1. So the amplitude is

$$\left| \frac{1}{2} [1 - (-1)] \right| = 1.$$

For $y = 2 \sin x$, the maximum y-coordinate is 2 and the minimum is -2. So the amplitude is

$$\left| \frac{1}{2} [2 - (-2)] \right| = 2.$$

Try This. Find the amplitude for $y = 4 \sin(x)$. $= 4$ ■

X	Y₁	
0	1	
.7854	.70711	
1.5708	0	
2.3562	-.7071	
3.1416	-1	
3.927	-.7071	
4.7124	0	

X=0

■ **Figure 5.43**

The Graph of $y = \cos(x)$

The graph of $y = \cos x$ is best understood by recalling that $\cos x$ is the first coordinate of the terminal point on the unit circle for an arc of length x, as shown in Fig. 5.37. As the arc length x increases from length 0 to $\pi/2$, the first coordinate of the endpoint decreases from 1 to 0. Some ordered pairs that satisfy $y = \cos x$ are shown in Fig. 5.43. As the length of x increases from $\pi/2$ to π, the first coordinate decreases from 0 to -1. As the length increases from π to 2π, the first coordinate increases from -1 to 1. The graph of $y = \cos x$ has exactly the same shape as the graph of $y = \sin x$. If the graph of $y = \sin x$ is shifted a distance of $\pi/2$ to the left, the graphs would coincide. For this reason the graph of $y = \cos x$ is also called a sine wave with amplitude 1 and period 2π. The graph of $y = \cos x$ over $[0, 2\pi]$ is called the **fundamental cycle** of $y = \cos x$. Since $\cos x = \cos(x + 2\pi)$, the fundamental cycle of $y = \cos x$ is repeated on $[2\pi, 4\pi]$, $[-2\pi, 0]$, and so on. The graph of $y = \cos x$ is shown in Fig. 5.44.

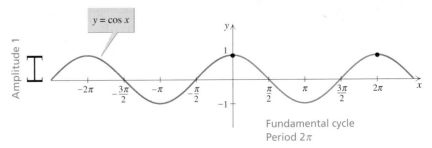

■ **Figure 5.44**

Note that there are five key points on the graph of $y = \cos x$ between 0 and 2π. These points give us the highest and lowest points in the cycle as well as the x-intercepts. The exact coordinates are given in the following table.

x	0	$\pi/2$	π	$3\pi/2$	2π
$y = \cos x$	1	0	-1	0	1

These five points divide the fundamental cycle into four equal parts.

Example 3 Graphing another periodic function

Sketch the graph of $y = -3 \cos x$ for x in the interval $[-2\pi, 2\pi]$ and find its amplitude.

Solution

Make a table of ordered pairs for x in $[0, 2\pi]$ to get one cycle of the graph. Note that the five x-coordinates in the table divide the interval $[0, 2\pi]$ into four equal parts. Multiply the y-coordinates of $y = \cos x$ by -3 to obtain the y-coordinates for $y = -3 \cos x$.

x	0	$\pi/2$	π	$3\pi/2$	2π
$y = -3 \cos x$	-3	0	3	0	-3

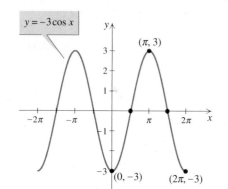

$y = -3 \cos x$

$(\pi, 3)$

$(0, -3)$ $(2\pi, -3)$

▪ **Figure 5.45**

Draw one cycle of $y = -3 \cos x$ through these five points, as shown in Fig. 5.45. Repeat the same shape for x in the interval $[-2\pi, 0]$ to get the graph of $y = -3 \cos x$ for x in $[-2\pi, 2\pi]$. The amplitude is $|0.5(3 - (-3))|$ or 3.

Try This. Graph $y = 4 \cos x$ for x in $[0, 2\pi]$ and find the amplitude. ▪

Transformations of Sine and Cosine

In Section 2.3 we discussed how various changes in a formula affect the graph of the function. We know the changes that cause horizontal or vertical translations, reflections, and stretching or shrinking. The graph of $y = 2 \sin x$ from Example 1 can be obtained by stretching the graph of $y = \sin x$. The graph of $y = -3 \cos x$ in Example 3 can be obtained by stretching and reflecting the graph of $y = \cos x$. The amplitude of $y = 2 \sin x$ is 2 and the amplitude of $y = -3 \sin x$ is 3. In general, the amplitude is determined by the coefficient of the sine or cosine function.

Theorem: Amplitude

The amplitude of $y = A \sin x$ or $y = A \cos x$ is $|A|$.

In Section 2.3 we saw that the graph of $y = f(x - C)$ is a horizontal translation of the graph of $y = f(x)$, to the right if $C > 0$ and to the left if $C < 0$. So the graphs of $y = \sin(x - C)$ and $y = \cos(x - C)$ are horizontal translations of $y = \sin x$ and $y = \cos x$, respectively, to the right if $C > 0$ and to the left if $C < 0$.

Definition: Phase Shift

The **phase shift** of the graph of $y = \sin(x - C)$ or $y = \cos(x - C)$ is C.

Example **4** Horizontal translation

Graph two cycles of $y = \sin(x + \pi/6)$, and determine the phase shift of the graph.

Solution

Since $x + \pi/6 = x - (-\pi/6)$ and $C < 0$, the graph of $y = \sin(x + \pi/6)$ is obtained by moving $y = \sin x$ a distance of $\pi/6$ to the left. Since the phase shift is $-\pi/6$, label the x-axis with multiples of $\pi/6$, as shown in Fig. 5.46. Concentrate on moving the fundamental cycle of $y = \sin x$. The three x-intercepts $(0, 0)$, $(\pi, 0)$, and $(2\pi, 0)$ move to $(-\pi/6, 0)$, $(5\pi/6, 0)$ and $(11\pi/6, 0)$. The high and low points, $(\pi/2, 1)$ and $(3\pi/2, -1)$, move to $(\pi/3, 1)$ and $(4\pi/3, -1)$. Draw one cycle through these five points and continue the pattern for another cycle, as shown in Fig. 5.46. The second cycle could be drawn to the right or left of the first cycle.

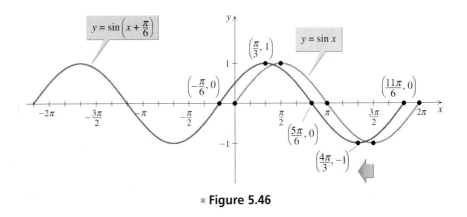

■ **Figure 5.46**

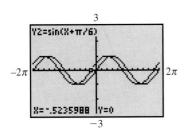

■ **Figure 5.47**

⬳ Use $\pi/6$ as the x-scale on your calculator graph as we did in Fig. 5.46. The calculator graph in Fig. 5.47 supports the conclusion that $y_2 = \sin(x + \pi/6)$ has shifted $\pi/6$ to the left of $y_1 = \sin(x)$.

Try This. Graph one cycle of $y = \cos(x - \pi/6)$ and find the phase shift. ■

The graphs of $y = \sin(x) + D$ and $y = \cos(x) + D$ are vertical translations of $y = \sin x$ and $y = \cos x$, respectively. The translation is upward for $D > 0$ and downward for $D < 0$. The next example combines a phase shift and a vertical translation. Note how we follow the five basic points of the fundamental cycle to see where they go in the transformation.

Example **5** Horizontal and vertical translation

Graph two cycles of $y = \cos(x - \pi/4) + 2$, and determine the phase shift of the graph.

Solution

The graph of $y = \cos(x - \pi/4) + 2$ is obtained by moving $y = \cos x$ a distance of $\pi/4$ to the right and two units upward. Since the phase shift is $\pi/4$, label the x-axis with multiples of $\pi/4$, as shown in Fig. 5.48. Concentrate on moving the fundamental cycle of $y = \cos x$. The points $(0, 1)$, $(\pi, -1)$, and $(2\pi, 1)$ move to $(\pi/4, 3)$, $(5\pi/4, 1)$, and $(9\pi/4, 3)$. The x-intercepts $(\pi/2, 0)$ and $(3\pi/2, 0)$ move to $(3\pi/4, 2)$ and $(7\pi/4, 2)$. Draw one cycle through these five points and continue the pattern for another cycle, as shown in Fig. 5.48.

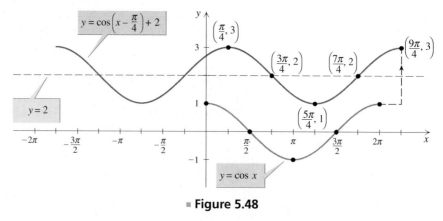

■ **Figure 5.48**

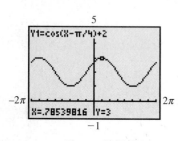

■ **Figure 5.49**

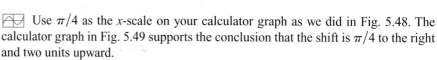 Use $\pi/4$ as the x-scale on your calculator graph as we did in Fig. 5.48. The calculator graph in Fig. 5.49 supports the conclusion that the shift is $\pi/4$ to the right and two units upward.

Try This. Graph one cycle of $y = \sin(x + \pi/4) + 1$ and find the phase shift. ■

Changing the Period

We can alter the sine and cosine functions in a way that we did not alter the functions of algebra in Chapter 2. The period of a periodic function can be changed by replacing x by a multiple of x.

Example **6** **Changing the period**

Graph two cycles of $y = \sin(2x)$ and determine the period of the function.

Solution

The graph of $y = \sin x$ completes its fundamental cycle for $0 \le x \le 2\pi$. So $y = \sin(2x)$ completes one cycle for $0 \le 2x \le 2\pi$, or $0 \le x \le \pi$. So the period is π. For $0 \le x \le \pi$, the x-intercepts are $(0, 0)$, $(\pi/2, 0)$, and $(\pi, 0)$. Midway between 0 and $\pi/2$, at $x = \pi/4$, the function reaches a maximum value of 1, and the function attains its minimum value of -1 at $x = 3\pi/4$. Draw one cycle of the graph through the x-intercepts and through $(\pi/4, 1)$ and $(3\pi/4, -1)$. Then graph another cycle of $y = \sin(2x)$ as shown in Fig. 5.50.

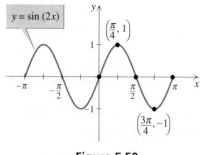

■ **Figure 5.50**

Try This. Graph two cycles of $y = \cos(2x)$ and determine the period. ■

Note that in Example 6 the period of $y = \sin(2x)$ is the period of $y = \sin x$ divided by 2. In general, one complete cycle of $y = \sin(Bx)$ or $y = \cos(Bx)$ for $B > 0$ occurs for $0 \le Bx \le 2\pi$, or $0 \le x \le 2\pi/B$.

Theorem: Period of
$y = \sin(Bx)$ **and** $y = \cos(Bx)$

The period P of $y = \sin(Bx)$ and $y = \cos(Bx)$ is given by

$$P = \frac{2\pi}{B}.$$

Note that the period is a natural number (that is, not a multiple of π) when B is a multiple of π.

Example **7** **A period that is not a multiple of** π

Determine the period of $y = \cos\left(\frac{\pi}{2}x\right)$ and graph two cycles of the function.

Solution

For this function, $B = \pi/2$. To find the period, use $P = 2\pi/B$:

$$P = \frac{2\pi}{\pi/2} = 4$$

So one cycle of $y = \cos\left(\frac{\pi}{2}x\right)$ is completed for $0 \le x \le 4$. The cycle starts at $(0, 1)$ and ends at $(4, 1)$. A minimum point occurs halfway in between, at $(2, -1)$. The x-intercepts are $(1, 0)$ and $(3, 0)$. Draw a curve through these five points to get one cycle of the graph. Continue this pattern from 4 to 8 to get a second cycle, as shown in Fig. 5.51.

Try This. Graph two cycles of $y = \cos(\pi x)$ and determine the period. ▪

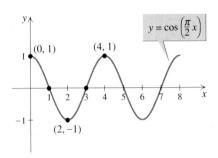

▪ **Figure 5.51**

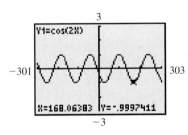

▪ **Figure 5.52**

🔲 A calculator graph for a periodic function can be very misleading. For example, consider $y = \cos(2x)$, shown in Fig. 5.52. From the figure it appears that $y = \cos(2x)$ has a period of about 150 and is a left shift of $y = \cos(x)$. However, we know that the period is π and that there is no shift. What we see in Fig. 5.52 is the pattern formed by choosing 95 equally spaced points on the graph of $y = \cos(2x)$. Equally spaced points on the graph of a periodic function will usually have some kind of pattern, but the pattern may not be a good graph of the function. Because x ranges from -301 to 303 in Fig. 5.52, the spaces between these points are approximately 6 units each, which is enough for about two cycles of $y = \cos(2x)$. So the viewing window is much too large to show the relatively small features of $y = \cos(2x)$. 🔲

The General Sine Wave

We can use any combination of translating, reflecting, phase shifting, stretching, shrinking, or period changing in a single trigonometric function.

The General Sine Wave

The graph of

$$y = A \sin[B(x - C)] + D \qquad \text{or} \qquad y = A \cos[B(x - C)] + D$$

is a sine wave with an amplitude $|A|$, period $2\pi/B$ ($B > 0$), phase shift C, and vertical translation D.

We assume that $B > 0$, because any general sine or cosine function can be rewritten with $B > 0$ using identities from Chapter 6. Notice that A and B affect the shape of the curve, while C and D determine its location.

P R O C E D U R E   Graphing a Sine Wave

To graph $y = A \sin[B(x - C)] + D$ or $y = A \cos[B(x - C)] + D$:

1. Sketch one cycle of $y = \sin Bx$ or $y = \cos Bx$ on $[0, 2\pi/B]$.
2. Change the amplitude of the cycle according to the value of A.
3. If $A < 0$, reflect the curve in the x-axis.
4. Translate the cycle $|C|$ units to the right if $C > 0$ or to the left if $C < 0$.
5. Translate the cycle $|D|$ units upward if $D > 0$ or downward if $D < 0$.

Example 8 A transformation of $y = \sin(x)$

Determine amplitude, period, and phase shift, and sketch two cycles of $y = 2 \sin(3x + \pi) + 1$.

Solution

First we rewrite the function in the form $y = A \sin[B(x - C)] + D$ by factoring 3 out of $3x + \pi$:

$$y = 2 \sin\left[3\left(x + \frac{\pi}{3}\right)\right] + 1$$

From this equation we get $A = 2$, $B = 3$, and $C = -\pi/3$. So the amplitude is 2, the period is $2\pi/3$, and the phase shift is $-\pi/3$. The period change causes the fundamental cycle of $y = \sin x$ on $[0, 2\pi]$ to shrink to the interval $[0, 2\pi/3]$. Now draw one cycle of $y = \sin 3x$ on $[0, 2\pi/3]$, as shown in Fig. 5.53. Stretch the cycle vertically so that it has an amplitude of 2. The numbers $\pi/3$ and 1 shift the cycle a distance of $\pi/3$ to the left and up one unit. So one cycle of the function occurs on $[-\pi/3, \pi/3]$. Check by evaluating the function at the endpoints and midpoint of the

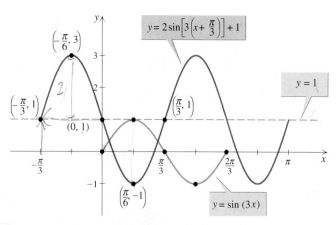

■ **Figure 5.53**

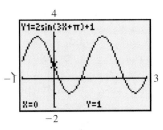

■ **Figure 5.54**

interval $[-\pi/3, \pi/3]$ to get $(-\pi/3, 1)$, $(0, 1)$, and $(\pi/3, 1)$. Since the graph is shifted one unit upward, these points are the points where the curve intersects the line $y = 1$. Evaluate the function midway between these points to get $(-\pi/6, 3)$ and $(\pi/6, -1)$, the highest and lowest points of this cycle. One cycle is drawn through these five points and continued for another cycle, as shown in Fig. 5.53.

The calculator graph in Fig. 5.54 supports our conclusions about amplitude, period, and phase shift. Note that it is easier to obtain the amplitude, period, and phase shift from the equation than from the calculator graph.

Try This. Determine the amplitude, period, and phase shift, and graph one cycle of $y = 3 \sin(2x - \pi) - 1$. ■

Example 9 A transformation of $y = \cos(x)$

Determine amplitude, period, and phase shift, and sketch one cycle of $y = -3 \cos(2x - \pi) - 1$.

Solution

Rewrite the function in the general form as

$$y = -3 \cos[2(x - \pi/2)] - 1.$$

The amplitude is 3. Since the period is $2\pi/B$, the period is $2\pi/2$ or π. The fundamental cycle is shrunk to the interval $[0, \pi]$. Sketch one cycle of $y = \cos 2x$ on $[0, \pi]$, as shown in Fig. 5.55. This cycle is reflected in the x-axis and stretched by a factor of 3. A shift of $\pi/2$ to the right means that one cycle of the original function occurs for $\pi/2 \le x \le 3\pi/2$. Evaluate the function at the endpoints and midpoint of the interval $[\pi/2, 3\pi/2]$ to get $(\pi/2, -4)$, $(\pi, 2)$, and $(3\pi/2, -4)$ for the endpoints and midpoint of this cycle. Since the graph is translated downward one unit, midway between these maximum and minimum points we get points where the graph intersects the line $y = -1$. These points are $(3\pi/4, -1)$ and $(5\pi/4, -1)$. Draw one cycle of the graph through these five points, as shown in Fig. 5.55.

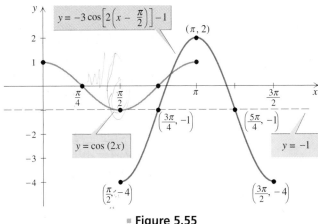

■ **Figure 5.55**

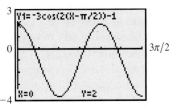

■ **Figure 5.56**

The calculator graph in Fig. 5.56 supports our conclusions.

Try This. Determine the amplitude, period, and phase shift, and graph one cycle of $y = -2 \cos(2x + \pi) + 1$. ■

Frequency

Sine waves are used to model physical phenomena such as radio, sound, or light waves. A high-frequency radio wave is a wave that has a large number of cycles per second. If we think of the *x*-axis as a time axis, then the period of a sine wave is the amount of time required for the wave to complete one cycle, and the reciprocal of the period is the number of cycles per unit of time. For example, the sound wave for middle C on a piano completes 262 cycles per second. The period of the wave is 1/262 second, which means that one cycle is completed in 1/262 second.

Definition: Frequency

> The **frequency** *F* of a sine wave with period *P* is defined by $F = 1/P$.

Example **10** **Frequency of a sine wave** $F = \frac{1}{p}$

Find the frequency of the sine wave given by $y = \sin(524\pi x)$.

Solution

First find the period:

$$P = \frac{2\pi}{524\pi} = \frac{1}{262} \approx 0.004$$

Since $F = 1/P$, the frequency is 262. A sine wave with this frequency completes 262 cycles for *x* in [0, 1] or approximately one cycle in [0, 0.004].

⊡ We cannot draw a graph showing 262 cycles in an interval of length one, but we can see the cycle that occurs in the interval [0, 0.004] by looking at Fig. 5.57.

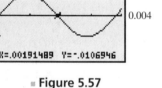

■ **Figure 5.57**

Try This. Find the frequency of the sine wave given by $y = \cos(100\pi x)$.   ■

$P = \frac{2\pi}{100\pi} = \frac{1}{50}$ $F = \frac{1}{p} = \frac{1}{1/50} = 50$

Note that if *B* is a large positive number in $y = \sin(Bx)$ or $y = \cos(Bx)$, then the period is short and the frequency is high.

Sinusoidal Curve Fitting

We now consider the problem of finding an equation for a sinusoid that passes through some given points. If we know the five key points on one cycle of a sinusoid, then we can write an equation for the curve in the form $y = A \sin[B(x - C)] + D$.

Example **11** **Modeling room temperature**

Ten minutes after a furnace is turned on, the temperature in a room reaches 74° and the furnace turns off. It takes two minutes for the room to cool to 70° and two minutes for the furnace to bring it back to 74° as shown in the following table.

Time (min)	10	11	12	13	14	15	16	17	18
Temperature (°F)	74	72	70	72	74	72	70	72	74

Assuming that the temperature (after time 10) is a sine function of the time, find the function and graph it.

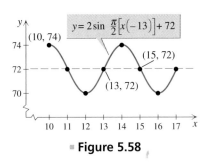

■ **Figure 5.58**

Solution

Because the temperature ranges from 70° to 74°, the amplitude of the sine curve is 2. Because the temperature goes from its maximum of 74° back to 74° in 4 minutes, the period is 4. Since the period is $2\pi/B$, we get $B = \pi/2$. Now concentrate on one cycle of the function. Starting at $(13, 72)$ the temperature increases to its maximum of 74°, decreases to its minimum of 70°, and then ends in the middle at $(17, 72)$. We choose this cycle because it duplicates the behavior of $y = \sin(x)$ on its fundamental cycle $[0, 2\pi]$. So shifting $y = \sin(x)$ to the right 13 and up 72 gives us $y = 2\sin\left[\frac{\pi}{2}(x - 13)\right] + 72$. Its graph is shown in Fig. 5.58. Changing this equation with any right or left shift by a multiple of 4 gives an equivalent equation. For example, $y = 2\sin\left[\frac{\pi}{2}(x - 1)\right] + 72$ is an equivalent equation.

Try This. The points $(1, 2)$, $(2, 5)$, $(3, 2)$, $(4, -1)$, and $(5, 2)$ are on one cycle of a sine wave. Find an equation for the curve. ■

If we have real data, the points will usually not fit exactly on a sine curve as they did in the last example. In this case we can use the sinusoidal regression feature of a graphing calculator to find a sine curve that approximates the data.

$y = \sin$

Function Gallery: **The Sine and Cosine Functions**

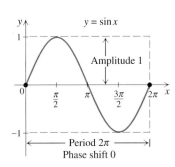

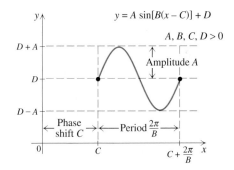

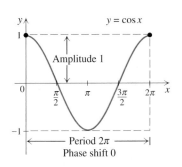

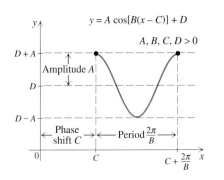

Example **12** **Modeling the time of sunrise**

The times of sunrise in Miami, Florida, on the first of every month for one year are shown in the following table (U.S. Naval Observatory, http://aa.usno.navy.mil). The time is the number of minutes after 5 A.M.

Month	Time	Month	Time
1	127	7	33
2	125	8	47
3	104	9	61
4	72	10	73
5	44	11	90
6	29	12	111

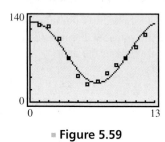

■ **Figure 5.59**

Use the sinusoidal regression feature of a graphing calculator to find an equation that fits the data. Graph the data and the curve on your graphing calculator. Find the period from the equation.

Solution

Enter the data and use the sinusoidal regression feature (SinReg) to get $y = 49.92 \sin(0.47x + 1.42) + 81.54$, where x is the month and y is the number of minutes after 5 A.M. Figure 5.59 shows the data and the sine curve. The period is $2\pi/0.47$ or approximately 13.4 months. Since the period should be 12 months, the sine curve does not fit the data very well. See Exercises 87 and 88 for some data that really look like a sine curve.

Try This. Use sinusoidal regression to find an equation that fits the points $(1, 2.2)$, $(2, 4.9)$, $(3, 1.9)$, $(4, -1.1)$, and $(5, 2.1)$.   ■

For Thought

True or False? Explain.

1. The period of $y = \cos(2\pi x)$ is π.

2. The range of $y = 4 \sin(x) + 3$ is $[-4, 4]$.

3. The graph of $y = \sin(2x + \pi/6)$ has a period of π and a phase shift $\pi/6$.

4. The points $(5\pi/6, 0)$ and $(11\pi/6, 0)$ are on the graph of $y = \cos(x - \pi/3)$.

5. The frequency of the sine wave $y = \sin x$ is $1/(2\pi)$.

6. The period for $y = \sin(0.1\pi x)$ is 20.

7. The graphs of $y = \sin x$ and $y = \cos(x + \pi/2)$ are identical.

8. The period of $y = \cos(4x)$ is $\pi/2$.

9. The maximum value of the function $y = -2 \cos(3x) + 4$ is 6.

10. The range of the function $y = 3 \sin(5x - \pi) + 2$ is $[-1, 5]$.

5.3 Exercises

Match each graph with one of the functions $y = 3 \sin(x)$, $y = 3 \cos(x)$, $y = -2 \sin(x)$, and $y = -2 \cos(x)$. Determine the amplitude of each function.

1.

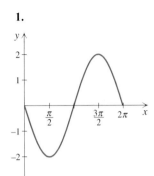

2.

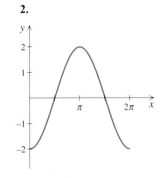

3.

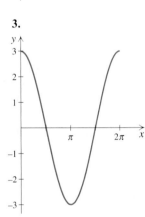

4.

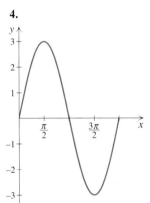

Determine the amplitude, period, and phase shift for each function.

5. $y = -2 \cos x$

6. $y = -4 \cos x$

7. $f(x) = \cos(x - \pi/2)$

8. $f(x) = \sin(x + \pi/2)$

9. $y = -2 \sin(x + \pi/3)$

10. $y = -3 \sin(x - \pi/6)$

Determine the amplitude and phase shift for each function, and sketch at least one cycle of the graph. Label five points as done in the examples.

11. $y = -\sin x$

12. $y = -\cos x$

13. $y = -3 \sin x$

14. $y = 4 \sin x$

15. $y = \dfrac{1}{2} \cos x$

16. $y = \dfrac{1}{3} \cos x$

17. $y = \sin(x + \pi)$

18. $y = \cos(x - \pi)$

19. $y = \cos(x - \pi/3)$

20. $y = \cos(x + \pi/4)$

21. $f(x) = \cos(x) + 2$

22. $f(x) = \cos(x) - 3$

23. $y = -\sin(x) - 1$

24. $y = -\sin(x) + 2$

25. $y = \sin(x + \pi/4) + 2$

26. $y = \sin(x - \pi/2) - 2$

27. $y = 2 \cos(x + \pi/6) + 1$

28. $y = 3 \cos(x + 2\pi/3) - 2$

29. $f(x) = -2 \sin(x - \pi/3) + 1$

30. $f(x) = -3 \cos(x + \pi/3) - 1$

$A = 3 \qquad P = 2\pi \qquad$ Phase $= -\pi/3$

Determine the amplitude, period, and phase shift for each function.

31. $y = 3 \sin(4x)$

32. $y = -2 \cos(3x)$

33. $y = -\cos\left(\dfrac{x}{2}\right) + 3$

34. $y = \sin\left(\dfrac{x}{3}\right) - 5$

35. $y = 2 \sin(x - \pi) + 3$

36. $y = -5 \cos(x + 4) + \pi$

37. $y = -2 \cos\left(2x + \dfrac{\pi}{2}\right) - 1$

38. $y = 4 \cos\left(3x - \dfrac{\pi}{4}\right)$

39. $y = -2 \cos\left(\dfrac{\pi}{2}x + \pi\right)$

40. $y = 8 \sin\left(\dfrac{\pi}{3}x - \dfrac{\pi}{2}\right)$

Find a function of the form $y = A \sin[B(x - C)] + D$ with the given period, phase shift, and range. Answers may vary.

41. $\pi, -\pi/2, [3, 7]$

42. $\pi/2, \pi, [-2, 4]$

43. $2, 2, [-1, 9]$

44. $4, 7, [5, 25]$

45. $\dfrac{1}{2}, -\pi, [-9, 3]$

46. $\dfrac{1}{3}, 2\pi, [-6, 2]$

Find the equation for each curve in its final position.

47. The graph of $y = \sin(x)$ is shifted a distance of $\pi/4$ to the right, reflected in the x-axis, then translated one unit upward.

48. The graph of $y = \cos(x)$ is shifted a distance of $\pi/6$ to the left, reflected in the x-axis, then translated two units downward.

49. The graph of $y = \cos(x)$ is stretched by a factor of 3, shifted a distance of π to the right, translated two units downward, then reflected in the x-axis.

50. The graph of $y = \sin(x)$ is shifted a distance of $\pi/2$ to the left, translated one unit upward, stretched by a factor of 4, then reflected in the x-axis.

Let $f(x) = \sin(x)$, $g(x) = x - \pi/4$, and $h(x) = 3x$. Find a formula for F in each case.

51. $F = f \circ g \circ h$ **52.** $F = h \circ f \circ g$

53. $F = f \circ h \circ g$ **54.** $F = g \circ f \circ h$

Sketch at least one cycle of the graph of each function. Determine the period, phase shift, and range of the function. Label five points on the graph as done in the examples. See the procedure for graphing a sine wave on page 454.

55. $y = \sin(3x)$ **56.** $y = \cos(x/3)$

57. $y = -\sin(2x)$ **58.** $y = -\cos(3x)$

59. $y = \cos(4x) + 2$ **60.** $y = \sin(3x) - 1$

61. $y = 2 - \sin(x/4)$ **62.** $y = 3 - \cos(x/5)$

63. $y = \sin\left(\dfrac{\pi}{3}x\right)$ **64.** $y = \sin\left(\dfrac{\pi}{4}x\right)$

65. $f(x) = \sin\left[2\left(x - \dfrac{\pi}{2}\right)\right]$

66. $f(x) = \sin\left[3\left(x + \dfrac{\pi}{3}\right)\right]$

67. $f(x) = \sin\left(\dfrac{\pi}{2}x + \dfrac{3\pi}{2}\right)$

68. $f(x) = \cos\left(\dfrac{\pi}{3}x - \dfrac{\pi}{3}\right)$

69. $y = 2\cos\left[2\left(x + \dfrac{\pi}{6}\right)\right] + 1$

70. $y = 3\cos\left[4\left(x - \dfrac{\pi}{2}\right)\right] - 1$

71. $y = -\dfrac{1}{2}\sin\left[3\left(x - \dfrac{\pi}{6}\right)\right] - 1$

72. $y = -\dfrac{1}{2}\sin\left[4\left(x + \dfrac{\pi}{4}\right)\right] + 1$

Write an equation of the form $y = A \sin[B(x - C)] + D$ whose graph is the given sine wave.

73.

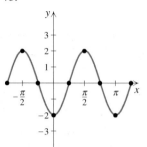

74.

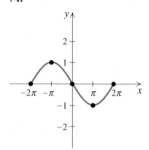

75.

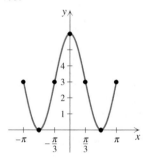

76.

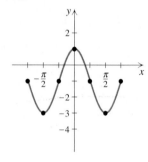

Solve each problem.

77. What is the frequency of the sine wave determined by $y = \sin(200\pi x)$, where x is time in seconds?

78. What is the frequency of the sine wave determined by $y = \cos(0.001\pi x)$, where x is time in seconds?

79. If the period of a sine wave is 0.025 hr, then what is the frequency?

80. If the frequency of a sine wave is 40,000 cycles per second, then what is the period?

81. *Motion of a Spring* A weight hanging on a vertical spring is set in motion with a downward velocity of 6 cm/sec from its equilibrium position. Assume that the constant ω for this particular spring and weight combination is 2. Write the formula that gives the location of the weight in centimeters as a function of the time t in seconds. Find the amplitude and period of the function and sketch its graph for t in the interval $[0, 2\pi]$. (See Section 5.2 for the general formula that describes the motion of a spring.)

82. *Motion of a Spring* A weight hanging on a vertical spring is set in motion with an upward velocity of 4 cm/sec from its equilibrium position. Assume that the constant ω for this particular spring and weight combination is π. Write the formula that gives the location of the weight in centimeters as a function of the time t in seconds. Find the period of the function and sketch its graph for t in the interval $[0, 4]$.

83. *Sun Spots* Astronomers have been recording sunspot activity for over 130 years. The number of sunspots per year varies like a periodic function over time, as shown in the graph. What is the approximate period of this function?

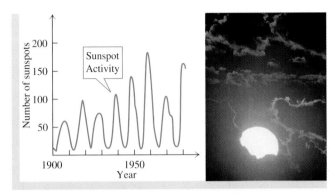

■ **Figure for Exercise 83**

84. *First Pulsar* In 1967, Jocelyn Bell, a graduate student at Cambridge University, England, found the peculiar pattern shown in the graph on a paper chart from a radio telescope. She had made the first discovery of a pulsar, a very small neutron star that emits beams of radiation as it rotates as fast as 1000 times per second. From the graph shown here, estimate the period of the first discovered pulsar, now known as CP 1919.

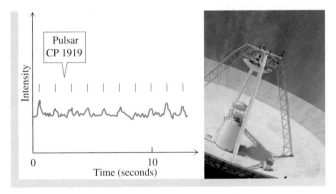

■ **Figure for Exercise 84**

85. *Lung Capacity* The volume of air v in cubic centimeters in the lungs of a certain distance runner is modeled by the equation $v = 400 \sin(60\pi t) + 900$, where t is time in minutes.
  a. What are the maximum and minimum volumes of air in the runner's lungs at any time?

  b. How many breaths does the runner take per minute?

86. *Blood Velocity* The velocity v of blood at a valve in the heart of a certain rodent is modeled by the equation $v = -4\cos(6\pi t) + 4$, where v is in centimeters per second and t is time in seconds.

  a. What are the maximum and minimum velocities of the blood at this valve?

  b. What is the rodent's heart rate in beats per minute?

87. *Periodic Revenue* For the past three years, the manager of The Toggery Shop has observed that revenue reaches a high of about $40,000 in December and a low of about $10,000 in June, and that a graph of the revenue looks like a sinusoid. If the months are numbered 1 through 36 with 1 corresponding to January, then what are the period, amplitude, and phase shift for this sinusoid? What is the vertical translation? Write a formula for the curve and find the approximate revenue for April.

88. *Periodic Cost* For the past three years, the manager of The Toggery Shop has observed that the utility bill reaches a high of about $500 in January and a low of about $200 in July, and the graph of the utility bill looks like a sinusoid. If the months are numbered 1 through 36 with 1 corresponding to January, then what are the period, amplitude, and phase shift for this sinusoid? What is the vertical translation? Write a formula for the curve and find the approximate utility bill for November.

89. *Discovering a Planet* On April 26, 1997, astronomers announced the discovery of a new planet orbiting the star Rho Coronae Borealis (*Sky and Telescope,* July 1997). The astronomers deduced the existence of the planet by measuring the change in line-of-sight velocity of the star over a period of ten months. The measurements appear to fall along a sine wave, as shown in the accompanying figure.
  a. What are the period, amplitude, and equation for the sine wave?

  b. How many earth days does it take for the planet to orbit Rho Coronae Borealis?

  c. Use the equation to estimate the change in line-of-sight velocity on day 36.

  d. On day 36 do you think that the planet is between the earth and Rho or that Rho is between the earth and the planet?

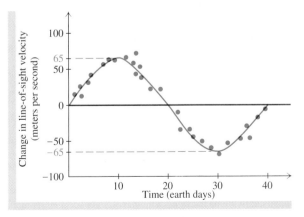

■ **Figure for Exercise 89**

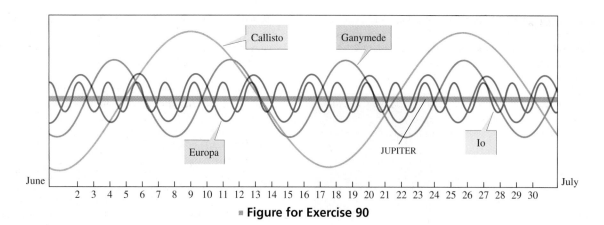

■ **Figure for Exercise 90**

90. *Jupiter's Satellites* The accompanying graph shows the positions in June of Ganymede, Callisto, Io, and Europa, the four bright satellites of Jupiter (*Sky and Telescope*). Jupiter itself is the center horizontal bar. The paths of the satellites are nearly sine waves because the orbits of the satellites are nearly circular. The distances from Jupiter to Io, Europa, Ganymede, and Callisto are 262,000, 417,000, 666,000, and 1,170,000 miles, respectively.

a. From the graph estimate the period of revolution to the nearest hour for each satellite. For which satellite can you obtain the period with the most accuracy?

b. What is the amplitude of each sine wave?

91. *Ocean Waves* Scientists use the same types of terms to describe ocean waves that we use to describe sine waves. The *wave period* is the time between crests and the *wavelength* is the distance between crests. The wave *height* is the vertical distance from the trough to the crest. The accompanying figure shows a *swell* in a coordinate system. Write an equation for the swell, assuming that its shape is that of a sinusoid.

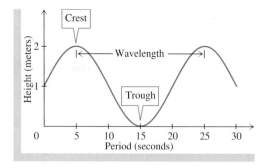

■ **Figure for Exercise 91**

92. *Large Ocean Waves* A *tsunami* is a series of large waves caused by an earthquake. The wavelength for a tsunami can be as long as several hundred kilometers. The accompanying figure shows a tsunami in a coordinate system. Write an equation for the tsunami, assuming that its shape is that of a sinusoid.

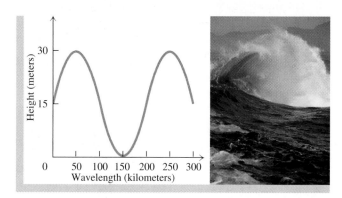

■ **Figure for Exercise 92**

93. *Moon Illumination* The accompanying table shows the percentage of the moon that will be illuminated at midnight for the 31 days of January 2010 (U.S. Naval Observatory, http://aa.usno.navy.mil).

Day	%	Day	%	Day	%
1	100	12	9	23	48
2	97	13	4	24	58
3	92	14	1	25	68
4	84	15	0	26	78
5	74	16	1	27	87
6	64	17	3	28	94
7	53	18	7	29	98
8	42	19	13	30	100
9	32	20	20	31	99
10	23	21	28		
11	15	22	38		

Use the sinusoidal regression feature of a graphing calculator to find the equation for a sine curve that fits the data. Graph the data and the curve on your graphing calculator. Find the period from the equation. Use the equation to predict the percentage of the moon that will be illuminated on February 4, 2010.

94. *Amount of Daylight* The accompanying table gives the number of minutes between sunrise and sunset for the first day of each month in the year 2010 in Miami, Florida.

Month	Time	Month	Time
1	634	7	823
2	660	8	800
3	699	9	759
4	746	10	714
5	788	11	670
6	819	12	639

Use the sinusoidal regression feature of a graphing calculator to find the equation for a sine curve that fits the data. Graph the data and the curve on your graphing calculator. Find the period from the equation. Use the equation to predict the number of minutes between sunrise and sunset on February 1, 2011.

For Writing/Discussion

95. *Periodic Temperature* Air temperature T generally varies in a periodic manner, with highs during the day and lows during the night. Assume that no drastic changes in the weather are expected and let t be time in hours with $t = 0$ at midnight tonight. Graph T for t in the interval $[0, 48]$ and write a function of the form $T = A \sin[B(t - C)] + D$ for your graph. Explain your choices for A, B, C, and D.

96. *Even or Odd* Determine whether $y = \sin x$ and $y = \cos x$ are even or odd functions and explain your answers.

Thinking Outside the Box XLVI

The Survivor There are 13 contestants on a reality television show. They are instructed to each take a seat at a circular table containing 13 chairs that are numbered consecutively with the numbers 1 through 13. The producer then starts at number 1 and tells that contestant that he is a survivor. That contestant then leaves the table. The producer skips a contestant and tells the contestant in chair number 3 that he is a survivor. That contestant then leaves the table. The producer continues around the table skipping a contestant and telling the next contestant that he is a survivor. Each survivor leaves the table. The last person left at the table is *not* a survivor and must leave the show.

a. For $n = 13$ find the unlucky number k, for which the person sitting in chair k must leave the show.

b. Find k for $n = 8$, 16, and 41.

c. Find a formula for k.

5.3 Pop Quiz

1. Determine the amplitude, period, and phase shift for $y = -5 \sin(2x + 2\pi/3)$.

2. List the coordinates for the five key points for one cycle of $y = 3 \sin(2x)$.

3. If $y = \cos(x)$ is shifted $\pi/2$ to the right, reflected in the x-axis, and shifted 3 units upward, then what is the equation of the curve in its final position?

4. What is the range of $f(x) = -4 \sin(x - 3) + 2$?

5. What is the frequency of the sine wave determined by $y = \sin(500\pi x)$, where x is time in minutes?

Linking Concepts

For Individual or Group Explorations

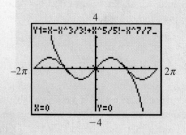

Taylor Polynomials

In calculus it is proved that many functions can be approximated to any degree of accuracy by polynomial functions called Taylor polynomials. The Taylor polynomials for sin(x) are

$$y = x,$$

$$y = x - \frac{x^3}{3!},$$

$$y = x - \frac{x^3}{3!} + \frac{x^5}{5!},$$

$$y = x - \frac{x^3}{3!} + \frac{x^5}{5!} - \frac{x^7}{7!}$$

and so on. As shown in the accompanying figure from a graphing calculator, the seventh degree Taylor polynomial approximates the sine curve very well between $-\pi$ and π.

a) Graph $y_1 = x$ and $y_2 = \sin(x)$ on a graphing calculator and determine approximately the values for x for which y_1 differs from y_2 by less than 0.1.

b) Repeat part (a) using the fifth, ninth, and 19th degree Taylor polynomials for y_1.

c) The Taylor polynomials for $\cos(x)$ are $y = 1$, $y = 1 - x^2/2!$, $y = 1 - x^2/2! + x^4/4!$, $y = 1 - x^2/2! + x^4/4! - x^6/6!$, and so on. Graph the zero, fourth, eighth, and 18th degree Taylor polynomials and determine approximately the intervals on which each of them differs from $\cos(x)$ by less than 0.1. Compare your answers with the answers obtained for parts (a) and (b).

d) Suppose that each operation $(+, -, \times, \div)$ takes one second. Explain how to find $\sin(x)$ to the nearest tenth for any real number x in the shortest amount of time by using Taylor polynomials.

5.4

The Other Trigonometric Functions and Their Graphs

So far we have studied two of the six trigonometric functions. In this section we will define the remaining four functions.

Definitions

The tangent, cotangent, secant, and cosecant functions are all defined in terms of the unit circle. We use the abbreviation tan for tangent, cot for cotangent, sec for secant, and csc for cosecant. As usual, we may think of α as an angle, the measure of an angle in degrees, the measure of an angle in radians, or a real number.

Definition: Tangent, Cotangent, Secant, and Cosecant Functions

If α is an angle in standard position and (x, y) is the point of intersection of the terminal side and the unit circle, we define the **tangent, cotangent, secant,** and **cosecant** functions as

$$\tan \alpha = \frac{y}{x}, \quad \cot \alpha = \frac{x}{y}, \quad \sec \alpha = \frac{1}{x}, \quad \text{and} \quad \csc \alpha = \frac{1}{y}.$$

We exclude from the domain of each function any value of α for which the denominator is 0.

Since $\sin \alpha = y$ and $\cos \alpha = x$, we can rewrite the definitions of tangent, cotangent, secant, and cosecant to get the following identities for these four functions.

Identities from the Definitions

If α is any angle or real number

$$\tan \alpha = \frac{\sin \alpha}{\cos \alpha}, \quad \cot \alpha = \frac{\cos \alpha}{\sin \alpha}, \quad \sec \alpha = \frac{1}{\cos \alpha}, \quad \text{and} \quad \csc \alpha = \frac{1}{\sin \alpha},$$

provided no denominator is zero.

The domain of the tangent and secant functions is the set of angles except those for which $\cos \alpha = 0$. The only points on the unit circle where the first coordinate is 0 are $(0, 1)$ and $(0, -1)$. Angles such as $\pi/2, 3\pi/2, 5\pi/2$, and so on, have terminal sides through either $(0, 1)$ or $(0, -1)$. These angles are of the form $\pi/2 + k\pi$, where k is any integer. So the domain of tangent and secant is

$$\left\{ \alpha \,\middle|\, \alpha \neq \frac{\pi}{2} + k\pi, \text{ where } k \text{ is an integer} \right\}.$$

The only points on the unit circle where the second coordinate is 0 are $(1, 0)$ and $(-1, 0)$. Angles that are multiples of π, such as $0, \pi, 2\pi$, and so on, have terminal sides that go through one of these points. Any angle of the form $k\pi$, where k is any integer, has a terminal side that goes through either $(1, 0)$ or $(-1, 0)$. So $\sin(k\pi) = 0$ for any integer k. By definition, the zeros of the sine function are excluded from the domain of cotangent and cosecant; thus the domain of cotangent and cosecant is

$$\{ \alpha \,|\, \alpha \neq k\pi, \text{ where } k \text{ is an integer} \}.$$

To find the values of the six trigonometric functions for an angle α, first find $\sin \alpha$ and $\cos \alpha$. Then use the identities from the definitions to find values of the other four functions. Of course, $\cot \alpha$ can be found also by using $1/\tan \alpha$, but not if $\tan \alpha$ is undefined like it is for $\alpha = \pm\pi/2$. If $\tan \alpha$ is undefined, then $\cot \alpha = 0$, and if $\tan \alpha = 0$, then $\cot \alpha$ is undefined.

The signs of the six trigonometric functions for angles in each quadrant are shown in Fig. 5.60. It is not necessary to memorize these signs, because they can be easily obtained by knowing the signs of the sine and cosine functions in each quadrant.

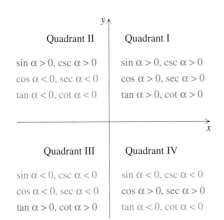

■ **Figure 5.60**

Example **1** Evaluating the trigonometric functions

Find the values of all six trigonometric functions for each angle.

a. $\pi/4$ **b.** $150°$

Solution

a. Use $\sin(\pi/4) = \sqrt{2}/2$ and $\cos(\pi/4) = \sqrt{2}/2$ to find the other values:

$$\tan(\pi/4) = \frac{\sin(\pi/4)}{\cos(\pi/4)} = \frac{\sqrt{2}/2}{\sqrt{2}/2} = 1 \quad \cot(\pi/4) = \frac{\cos(\pi/4)}{\sin(\pi/4)} = 1$$

$$\sec(\pi/4) = \frac{1}{\cos(\pi/4)} = \frac{2}{\sqrt{2}} = \sqrt{2} \quad \csc(\pi/4) = \frac{1}{\sin(\pi/4)} = \frac{2}{\sqrt{2}} = \sqrt{2}$$

b. The reference angle for $150°$ is $30°$. So $\sin(150°) = \sin(30°) = 1/2$ and $\cos(150°) = -\cos(30°) = -\sqrt{3}/2$. Use these values to find the other four values:

$$\tan(150°) = \frac{1/2}{-\sqrt{3}/2} = -\frac{1}{\sqrt{3}} = -\frac{\sqrt{3}}{3} \quad \cot(150°) = -\frac{3}{\sqrt{3}} = -\sqrt{3}$$

$$\sec(150°) = \frac{1}{\cos(150°)} = -\frac{2}{\sqrt{3}} = -\frac{2\sqrt{3}}{3} \quad \csc(150°) = \frac{1}{\sin(150°)} = 2$$

Try This. Find the values of all six trigonometric functions for $-\pi/4$.   ■

Most scientific calculators have keys for the sine, cosine, and tangent functions only. To find values for the other three functions, we use the definitions. Keys labeled $\sin^{-1}$, $\cos^{-1}$, and $\tan^{-1}$ on a calculator are for the inverse trigonometric functions, which we will study in the next section. These keys do *not* give the values of $1/\sin x$, $1/\cos x$, or $1/\tan x$.

Example **2** Evaluating with a calculator

Use a calculator to find approximate values rounded to four decimal places.

a. $\sec(\pi/12)$ **b.** $\csc(123°)$ **c.** $\cot(-12.4)$

Solution

a. $\sec(\pi/12) = \dfrac{1}{\cos(\pi/12)} \approx 1.0353$

b. $\csc(123°) = \dfrac{1}{\sin(123°)} \approx 1.1924$

c. $\cot(-12.4) = \dfrac{1}{\tan(-12.4)} \approx 5.9551$

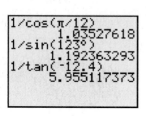

■ **Figure 5.61**

These expressions are evaluated with a graphing calculator in radian mode as shown in Fig. 5.61. Note that in radian mode the degree symbol is used for part (b).

Try This. Find $\sec(\pi/22)$ and $\csc(-14.5)$ to four decimal places.   ■

Graph of $y = \tan(x)$

Consider some ordered pairs that satisfy $y = \tan x$ for x in $(-\pi/2, \pi/2)$. Note that $-\pi/2$ and $\pi/2$ are not in the domain of $y = \tan x$, but x can be chosen close to $\pm \pi/2$.

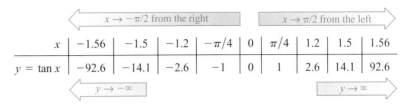

	$x \to -\pi/2$ from the right						$x \to \pi/2$ from the left		
x	-1.56	-1.5	-1.2	$-\pi/4$	0	$\pi/4$	1.2	1.5	1.56
$y = \tan x$	-92.6	-14.1	-2.6	-1	0	1	2.6	14.1	92.6
	$y \to -\infty$						$y \to \infty$		

The graph of $y = \tan x$ includes the points $(-\pi/4, -1)$, $(0, 0)$, and $(\pi/4, 1)$. Since $\tan x = \sin x/\cos x$, the graph of $y = \tan x$ has a vertical asymptote for every zero of the cosine function. So the vertical lines $x = \pi/2 + k\pi$ for any integer k are the **vertical asymptotes.** The behavior of $y = \tan x$ near the asymptotes $x = \pm\pi/2$ can be seen from the table of ordered pairs. As x approaches $\pi/2$ (approximately 1.57) from the left, $\tan x \to \infty$. As x approaches $-\pi/2$ from the right, $\tan x \to -\infty$. In the interval $(-\pi/2, \pi/2)$, the function is increasing. The graph of $y = \tan x$ is shown in Fig. 5.62. The shape of the tangent function between $-\pi/2$ and $\pi/2$ is repeated between each pair of consecutive asymptotes, as is shown in the figure. The period of $y = \tan(x)$ is π, and the fundamental cycle is the portion of the graph between $-\pi/2$ and $\pi/2$. Since the range of $y = \tan x$ is $(-\infty, \infty)$, the concept of amplitude is not defined.

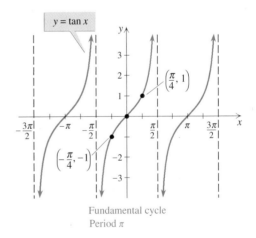

Fundamental cycle
Period π

■ **Figure 5.62**

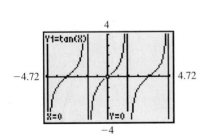

■ **Figure 5.63**

The calculator graph of $y = \tan(x)$ in connected mode is shown in Fig. 5.63. In connected mode the calculator connects two points on opposite sides of each asymptote and appears to draw the vertical asymptotes. ☐

To help us understand the sine and cosine curve, we identified five key points on the fundamental cycle. For the tangent curve, we have three key points and the asymptotes (where the function is undefined). The exact coordinates are given in the following table.

x	$-\pi/2$	$-\pi/4$	0	$\pi/4$	$\pi/2$
$y = \tan x$	undefined	-1	0	1	undefined

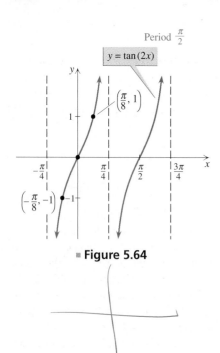

Period $\frac{\pi}{2}$

$y = \tan(2x)$

$\left(\frac{\pi}{8}, 1\right)$

$\left(-\frac{\pi}{8}, -1\right)$

■ **Figure 5.64**

We can transform the graph of the tangent function by using the same techniques that we used for the sine and cosine functions.

Example 3 A tangent function with a transformation

Sketch two cycles of the function $y = \tan(2x)$ and determine the period.

Solution

The graph of $y = \tan x$ completes one cycle for $-\pi/2 < x < \pi/2$. So the graph of $y = \tan(2x)$ completes one cycle for $-\pi/2 < 2x < \pi/2$, or $-\pi/4 < x < \pi/4$. The graph includes the points $(-\pi/8, -1)$, $(0, 0)$, and $(\pi/8, 1)$. The graph is similar to the graph of $y = \tan x$, but the asymptotes are $x = \pi/4 + k\pi/2$ for any integer k, and the period of $y = \tan(2x)$ is $\pi/2$. Two cycles of the graph are shown in Fig. 5.64.

Try This.   Graph two cycles of $y = \tan(2\pi x)$ and determine the period. ■

By understanding what happens to the fundamental cycles of the sine, cosine, and tangent functions when the period changes, we can easily determine the location of the new function and sketch its graph quickly. We start with the fundamental cycles of $y = \sin x$, $y = \cos x$, and $y = \tan x$, which occur over the intervals $[0, 2\pi]$, $[0, 2\pi]$, and $[-\pi/2, \pi/2]$, respectively. For $y = \sin Bx$, $y = \cos Bx$, and $y = \tan Bx$ (for $B > 0$) these fundamental cycles move to the intervals $[0, 2\pi/B]$, $[0, 2\pi/B]$, and $[-\pi/(2B), \pi/(2B)]$, respectively. In every case, divide the old period by B to get the new period. Note that only the nonzero endpoints of the intervals are changed. The following Function Gallery summarizes the period change for the sine, cosine, and tangent functions with $B > 1$.

Function Gallery: Periods of Sine, Cosine, and Tangent ($B > 1$)

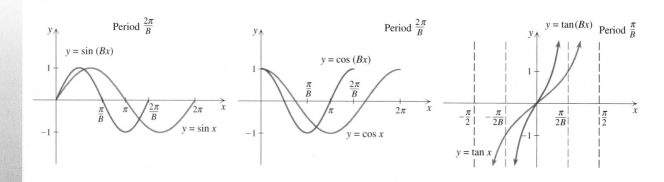

Fundamental cycles

Graph of $y = \cot(x)$

We know that $\cot x = 1/\tan x$, provided $\tan x$ is defined and not 0. So we can use $y = \tan x$ to graph $y = \cot x$. Since $\tan x = 0$ for $x = k\pi$, where k is any integer, the vertical lines $x = k\pi$ for an integer k are the vertical asymptotes of $y = \cot x$. Consider some ordered pairs that satisfy $y = \cot x$ between the asymptotes $x = 0$ and $x = \pi$:

	$x \to 0$ from the right					$x \to \pi$ from the left	
x	0.01	0.02	$\pi/4$	$\pi/2$	$3\pi/4$	3.1	3.14
$y = \cot x$	100.00	49.99	1	0	-1	-24.0	-627.9
	$y \to \infty$					$y \to -\infty$	

As $x \to 0$ from the right, $y \to \infty$, and as $x \to \pi$ from the left, $y \to -\infty$. In the interval $(0, \pi)$, the function is decreasing. The graph of $y = \cot x$ is shown in Fig. 5.65. The shape of the curve between $x = 0$ and $x = \pi$ is repeated between each pair of consecutive asymptotes. The period of $y = \cot x$ is π, and its fundamental cycle is the portion of the graph between 0 and π. The range of $y = \cot x$ is $(-\infty, \infty)$.

Because $\cot x = 1/\tan x$, $\cot x$ is large when $\tan x$ is small, and vice versa. The graph of $y = \cot x$ has an x-intercept wherever $y = \tan x$ has a vertical asymptote, and a vertical asymptote wherever $y = \tan x$ has an x-intercept. So for every integer k, $(\pi/2 + k\pi, 0)$ is an x-intercept of $y = \cot x$, and the vertical line $x = k\pi$ is an asymptote, as shown in Fig. 5.65.

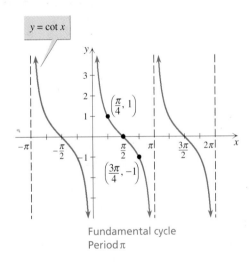

Fundamental cycle
Period π

■ **Figure 5.65**

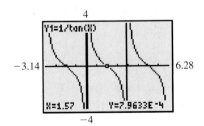

■ **Figure 5.66**

To see the graph of $y = \cot(x)$ on a graphing calculator, you can graph $y = 1/\tan(x)$, as shown in Fig. 5.66. □

For the cotangent curve, note the three key points and the asymptotes (where the function is undefined). The exact coordinates are given in the following table.

x	0	$\pi/4$	$\pi/2$	$3\pi/4$	π
$y = \cot x$	undefined	1	0	-1	undefined

When we graph a transformation of the cotangent function, as in the next example, we must determine what happens to these five features.

Example 4 A cotangent function with a transformation

Sketch two cycles of the function $y = 0.3 \cot(2x + \pi/2)$, and determine the period.

Solution

Factor out 2 to write the function as

$$y = 0.3 \cot\left[2\left(x + \frac{\pi}{4}\right)\right].$$

Since $y = \cot x$ completes one cycle for $0 < x < \pi$, the graph of $y = 0.3 \cot[2(x + \pi/4)]$ completes one cycle for

$$0 < 2\left(x + \frac{\pi}{4}\right) < \pi, \quad \text{or} \quad 0 < x + \frac{\pi}{4} < \frac{\pi}{2}, \quad \text{or} \quad -\frac{\pi}{4} < x < \frac{\pi}{4}.$$

The interval $(-\pi/4, \pi/4)$ for the fundamental cycle can also be obtained by dividing the period π of $y = \cot x$ by 2 to get $\pi/2$ as the period, and then shifting the interval $(0, \pi/2)$ a distance of $\pi/4$ to the left. The factor 0.3 shrinks the y-coordinates. The graph goes through $(-\pi/8, 0.3)$, $(0, 0)$, and $(\pi/8, -0.3)$ as it approaches its vertical asymptotes $x = \pm\pi/4$. The graph for two cycles is shown in Fig. 5.67.

⊟⊟ The calculator graph in Fig. 5.68 supports these conclusions.

Try This. Graph two cycles of $y = 3 \cot(2x + \pi)$ and determine the period. ■

Since the period for $y = \cot x$ is π, the period for $y = \cot(Bx)$ with $B > 0$ is π/B and $y = \cot(Bx)$ completes one cycle on the interval $(0, \pi/B)$. The left-hand asymptote for the fundamental cycle of $y = \cot x$ remains fixed at $x = 0$, while the right-hand asymptote changes to $x = \pi/B$.

Graph of $y = \sec(x)$

Since $\sec x = 1/\cos x$, the values of $\sec x$ are large when the values of $\cos x$ are small. For any x such that $\cos x$ is 0, $\sec x$ is undefined and the graph of $y = \sec x$ has a vertical asymptote. Because of the reciprocal relationship between $\sec x$ and $\cos x$, we first draw the graph of $y = \cos x$ for reference when graphing $y = \sec x$. At every x-intercept of $y = \cos x$, we draw a vertical asymptote, as shown in Fig. 5.69.

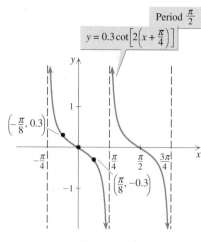

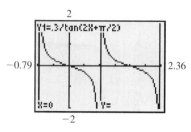

■ **Figure 5.67**

■ **Figure 5.68**

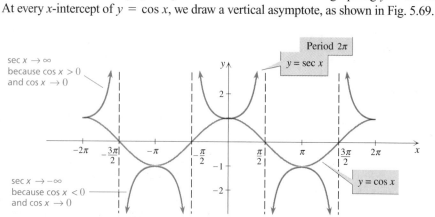

■ **Figure 5.69**

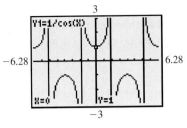

■ **Figure 5.70**

If $\cos x = \pm 1$, then $\sec x = \pm 1$. So every maximum or minimum point on the graph of $y = \cos x$ is also on the graph of $y = \sec x$. If $\cos x > 0$ and $\cos x \to 0$, then $\sec x \to \infty$. If $\cos x < 0$ and $\cos x \to 0$, then $\sec x \to -\infty$. These two facts cause the graph of $y = \sec x$ to approach its asymptotes in the manner shown in Fig. 5.69. The period of $y = \sec x$ is 2π, the same as the period for $y = \cos x$. The range of $y = \sec x$ is $(-\infty, -1] \cup [1, \infty)$.

The calculator graph in Fig. 5.70 supports these conclusions.   □

Example 5   A secant function with a transformation

Sketch two cycles of the function $y = 2 \sec(x - \pi/2)$, and determine the period and the range of the function.

Solution

Since

$$y = 2 \sec\left(x - \frac{\pi}{2}\right) = \frac{2}{\cos(x - \pi/2)},$$

we first graph $y = \cos(x - \pi/2)$, as shown in Fig. 5.71. The function $y = \sec(x - \pi/2)$ goes through the maximum and minimum points on the graph of $y = \cos(x - \pi/2)$, but $y = 2 \sec(x - \pi/2)$ stretches $y = \sec(x - \pi/2)$ by a factor of 2. So the portions of the curve that open up do not go lower than 2, and the portions that open down do not go higher than -2, as shown in the figure. The period is 2π, and the range is $(-\infty, -2] \cup [2, \infty)$.

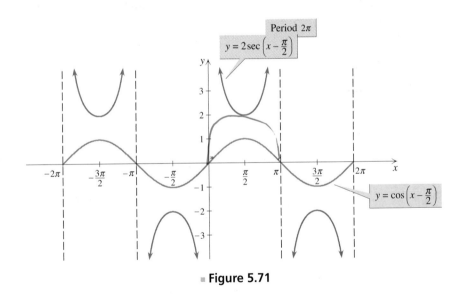

■ **Figure 5.71**

Try This. Graph one cycle of $y = 3 \sec(2x)$ and determine the period and range.   ■

Graph of $y = \csc(x)$

Since $\csc x = 1/\sin x$, the graph of $y = \csc x$ is related to the graph of $y = \sin x$ in the same way that the graphs of $y = \sec x$ and $y = \cos x$ are related. To graph $y = \csc x$, first draw the graph of $y = \sin x$ and a vertical asymptote at each x-intercept.

Since the graph of $y = \sin x$ can be obtained by shifting $y = \cos x$ a distance of $\pi/2$ to the right, the graph of $y = \csc x$ is obtained from $y = \sec x$ by shifting a distance of $\pi/2$ to the right, as shown in Fig. 5.72. The period of $y = \csc x$ is 2π, the same as the period for $y = \sin x$.

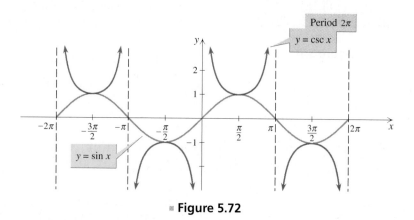

▪ **Figure 5.72**

Example 6   A cosecant function with a transformation

Sketch two cycles of the graph of $y = \csc(2x - 2\pi/3)$ and determine the period and the range of the function.

Solution

Since

$$y = \csc(2x - 2\pi/3)$$

$$= \frac{1}{\sin[2(x - \pi/3)]}$$

we first graph $y = \sin[2(x - \pi/3)]$. The period for $y = \sin[2(x - \pi/3)]$ is π with phase shift of $\pi/3$. So the fundamental cycle of $y = \sin x$ is transformed to occur on the interval $[\pi/3, 4\pi/3]$. Draw at least two cycles of $y = \sin[2(x - \pi/3)]$ with a vertical asymptote (for the cosecant function) at every x-intercept, as shown in Fig. 5.73.

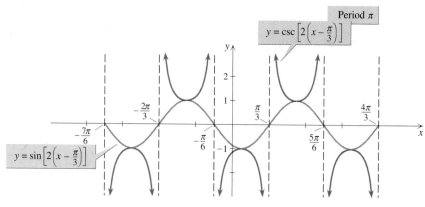

▪ **Figure 5.73**

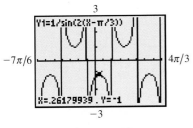

■ **Figure 5.74**

Each portion of $y = \csc[2(x - \pi/3)]$ that opens up has a minimum value of 1, and each portion that opens down has a maximum value of -1. The period of the function is π, and the range is $(-\infty, -1] \cup [1, \infty)$.

The calculator graph in Fig. 5.74 supports these conclusions.

Try This. Graph one cycle of $y = \csc(2x - \pi)$ and determine the period and range. ■

The Function Gallery summarizes some of the facts that we have learned about the six trigonometric functions. Also, one cycle of the graph of each trigonometric function is shown.

Function Gallery: **Trigonometric Functions**

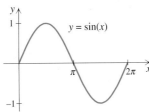

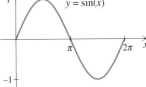

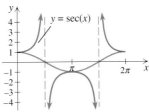

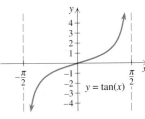

	$y = \sin(x)$	$y = \cos(x)$	$y = \tan(x)$
Domain (*k* any integer)	$(-\infty, \infty)$	$(-\infty, \infty)$	$x \neq \dfrac{\pi}{2} + k\pi$
Range	$[-1, 1]$	$[-1, 1]$	$(-\infty, \infty)$
Period	2π	2π	π
Fundamental cycle	$[0, 2\pi]$	$[0, 2\pi]$	$\left[-\dfrac{\pi}{2}, \dfrac{\pi}{2}\right]$

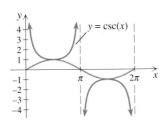

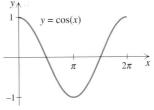

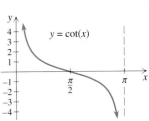

	$y = \csc(x)$	$y = \sec(x)$	$y = \cot(x)$
Domain (*k* any integer)	$x \neq k\pi$	$x \neq \dfrac{\pi}{2} + k\pi$	$x \neq k\pi$
Range	$(-\infty, -1] \cup [1, \infty)$	$(-\infty, -1] \cup [1, \infty)$	$(-\infty, \infty)$
Period	2π	2π	π
Fundamental cycle	$[0, 2\pi]$	$[0, 2\pi]$	$[0, \pi]$

For Thought

True or False? Explain.

1. $\sec(\pi/4) = 1/\sin(\pi/4)$ 2. $\cot(\pi/2) = 1/\tan(\pi/2)$

3. $\csc(60°) = 2\sqrt{3}/3$ 4. $\tan(5\pi/2) = 0$

5. $\sec(95°) = \sqrt{5}$ 6. $\csc(120°) = 2/\sqrt{3}$

7. The graphs of $y = 2 \csc x$ and $y = 1/(2 \sin x)$ are identical.

8. The range of $y = 0.5 \csc(13x - 5\pi)$ is $(-\infty, -0.5] \cup [0.5, \infty)$.

9. The graph of $y = \tan(3x)$ has vertical asymptotes at $x = \pm\pi/6$.

10. The graph of $y = \cot(4x)$ has vertical asymptotes at $x = \pm\pi/4$.

5.4 Exercises

Redraw each unit circle and label each indicated point with the proper value of the tangent function. The points in Exercise 1 are the terminal points for arcs with lengths that are multiples of $\pi/4$. The points in Exercise 2 are the terminal points for arcs with lengths that are multiples of $\pi/6$. Repeat Exercises 1 and 2 until you can do them from memory.

1.

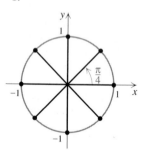

2.

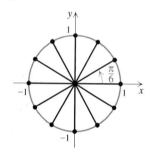

Find the exact value of each of the following expressions without using a calculator.

3. $\tan(\pi/3)$ 4. $\tan(\pi/4)$

5. $\tan(-\pi/4)$ 6. $\tan(\pi/6)$

7. $\cot(\pi/2)$ 8. $\cot(2\pi/3)$

9. $\cot(-\pi/3)$ 10. $\cot(0)$

11. $\sec(\pi/6)$ 12. $\sec(\pi/3)$

13. $\sec(\pi/2)$ 14. $\sec(\pi)$

15. $\csc(-\pi)$ 16. $\csc(\pi/6)$

17. $\csc(3\pi/4)$ 18. $\csc(-\pi/3)$

19. $\tan(135°)$ 20. $\tan(270°)$

21. $\cot(210°)$ 22. $\cot(120°)$

23. $\sec(-120°)$ 24. $\sec(-90°)$

25. $\csc(315°)$ 26. $\csc(240°)$

27. $\cot(-90°)$ 28. $\cot(270°)$

Find the approximate value of each expression. Round to four decimal places.

29. $\tan(1.55)$ 30. $\tan(1.6)$

31. $\cot(-3.48)$ 32. $\cot(22.4)$

33. $\csc(0.002)$ 34. $\csc(1.54)$

35. $\sec(\pi/12)$ 36. $\sec(-\pi/8)$

37. $\cot(0.09°)$ 38. $\cot(179.4°)$

39. $\csc(-44.3°)$ 40. $\csc(-124.5°)$

41. $\sec(89.2°)$ 42. $\sec(-0.024°)$

43. $\tan(-44.6°)$ 44. $\tan(138°)$

Find the exact value of each expression for the given value of θ. Do not use a calculator.

45. $\sec^2(2\theta)$ if $\theta = \pi/6$ 46. $\csc^2(2\theta)$ if $\theta = \pi/8$

47. $\tan(\theta/2)$ if $\theta = \pi/3$ 48. $\csc(\theta/2)$ if $\theta = \pi/2$

49. $\sec(\theta/2)$ if $\theta = 3\pi/2$ 50. $\cot(\theta/2)$ if $\theta = 2\pi/3$

Determine the period and sketch at least one cycle of the graph of each function.

51. $y = \tan(3x)$

52. $y = \tan(4x)$

53. $y = \cot(x + \pi/4)$

54. $y = \cot(x - \pi/6)$

55. $y = \cot(x/2)$

56. $y = \cot(x/3)$

57. $y = \tan(\pi x)$

58. $y = \tan(\pi x/2)$

59. $y = -2 \tan x$

60. $y = -\tan(x - \pi/2)$

61. $y = -\cot(x + \pi/2)$

62. $y = 2 + \cot x$

63. $y = \cot(2x - \pi/2)$

64. $y = \cot(3x + \pi)$

65. $y = \tan\left(\dfrac{\pi}{2}x - \dfrac{\pi}{2}\right)$

66. $y = \tan\left(\dfrac{\pi}{4}x + \dfrac{3\pi}{4}\right)$

Determine the period and sketch at least one cycle of the graph of each function. State the range of each function.

67. $y = \sec(2x)$

68. $y = \sec(3x)$

69. $y = \csc(x - \pi/2)$

70. $y = \csc(x + \pi/4)$

71. $y = \csc(x/2)$

72. $y = \csc(x/4)$

73. $y = \sec(\pi x/2)$

74. $y = \sec(\pi x)$

75. $y = 2 \sec x$

76. $y = \dfrac{1}{2} \sec x$

77. $y = \csc(2x - \pi/2)$

78. $y = \csc(3x + \pi)$

79. $y = -\csc\left(\dfrac{\pi}{2}x + \dfrac{\pi}{2}\right)$

80. $y = -2 \csc(\pi x - \pi)$

81. $y = 2 + 2 \sec(2x)$

82. $y = 2 - 2 \sec\left(\dfrac{x}{2}\right)$

Determine the period and range of each function.

83. $y = \tan(2x - \pi) + 3$

84. $y = 2 \cot(3x + \pi) - 8$

85. $y = 2 \sec(x/2 - 1) - 1$

86. $y = -2 \sec(x/3 - 6) + 3$

87. $y = -3 \csc(2x - \pi) - 4$

88. $y = 4 \csc(3x - \pi) + 5$

Write the equation of each curve in its final position.

89. The graph of $y = \tan(x)$ is shifted $\pi/4$ units to the right, stretched by a factor of 3, then translated 2 units upward.

90. The graph of $y = \cot(x)$ is shifted $\pi/2$ units to the left, reflected in the *x*-axis, then translated 1 unit upward.

91. The graph of $y = \sec(x)$ is shifted π units to the left, reflected in the *x*-axis, then shifted 2 units upward.

92. The graph of $y = \csc(x)$ is shifted 2 units to the right, translated 3 units downward, then reflected in the *x*-axis.

For Writing/Discussion

93. Graph $y = x + \sin x$ for $-100 \le x \le 100$ and $-100 \le y \le 100$. Explain your results.

94. Graph $y = x + \tan x$ for $-6 \le x \le 6$ and $-10 \le y \le 10$. Explain your results.

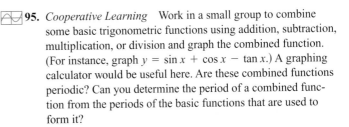 **95.** *Cooperative Learning* Work in a small group to combine some basic trigonometric functions using addition, subtraction, multiplication, or division and graph the combined function. (For instance, graph $y = \sin x + \cos x - \tan x$.) A graphing calculator would be useful here. Are these combined functions periodic? Can you determine the period of a combined function from the periods of the basic functions that are used to form it?

96. *Average Rate of Change* The average rate of change of a function on a short interval $[x, x + h]$ for a fixed value of h is a function itself. Sometimes it is a function that we can recognize by its graph.

a. Graph $y_1 = \sin(x)$ and its average rate of change

$$y_2 = (y_1(x + 0.1) - y_1(x))/0.1$$

for $-2\pi \le x \le 2\pi$. What familiar function does y_2 look like?

b. Repeat part (a) for $y_1 = \cos(x)$, $y_1 = e^x$, $y_1 = \ln(x)$, and $y_1 = x^2$.

97. *Discovering Planets* On October 23, 1996, astronomers announced that they had discovered a planet orbiting the star 16 Cygni B (*Sky and Telescope*, January 1997). They did not see the planet itself. Rather, for 8 years they detected the periodic wobbling motion toward and away from earth that the planet induced in the star as it orbited the star. See the accompanying figure.

a. Estimate the period of the function shown in the figure.

b. Which trigonometric function has a graph similar to the graph in the figure?

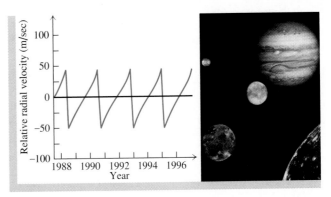

■ **Figure for Exercise 97**

Thinking Outside the Box XLVII

Counting Votes   Fifteen experts are voting to determine the best convertible of the year. The choices are a Porsche Carrera, a Chrysler Crossfire, and a Nissan Roadster. The experts will rank the three cars 1st, 2nd, and 3rd. There are three common ways to determine the winner.

1. *Plurality:* The car with the most first place votes (preferences) is the winner.

2. *Instant runoff:* The car with the least number of preferences is eliminated. Then the ballots where the eliminated car is first are revised so that the second place car is moved to first. Finally, the car with the most preferences is the winner.

3. *The point system:* Two points are given for each time a car occurs in first place on a ballot, one point for each time the car appears in second place on a ballot, and no points for third place.

When the ballots were cast, the Porsche won when plurality was used, the Chrysler won when instant runoff was used, and the Nissan won when the point system was used. Determine 15 actual votes for which this result would occur.

5.4 Pop Quiz

1. What is the period for $y = \tan(3x)$?

2. Find the equations of all asymptotes for $y = \cot(2x)$.

3. Find the equations of all asymptotes for $y = \sec(2x)$.

4. What is the range of $y = 3 \csc(2x)$?

Find the exact value.

5. $\tan(\pi/4)$

6. $\tan(120°)$

7. $\cot(-\pi/3)$

8. $\sec(60°)$

9. $\csc(-3\pi/4)$

Linking Concepts

For Individual or Group Explorations

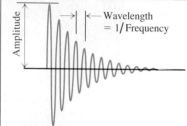

Modeling a Guitar Note

The waveform of a guitar note is characterized by an initial sharp peak that falls off rapidly, as shown in the figure. Guitar effects pedals electronically modify that waveform. We can mathematically modify the oscillations of the sine wave to get the waveform of a guitar note as well as many others. The basic sine wave oscillates between the two horizontal lines $y = 1$ and $y = -1$. If we multiply $\sin(x)$ by any other function $g(x)$ we will get a curve that oscillates between the graphs of $y = g(x)$ and $y = -g(x)$.

a) Graph $y_1 = x \sin(x)$, $y_2 = x$, and $y_3 = -x$. For what exact values of x is $x \sin(x) = x$? For what exact values of x is $x \sin(x) = -x$?

b) For what exact values of x is $x^2 \sin(x) = x^2$? For what exact values of x is $x^2 \sin(x) = -x^2$? Support your conclusions with a graph.

c) Graph $y_1 = \frac{1}{x} \sin(x)$ for $0 \le x \le 10$ and $-2 \le y \le 2$. Is it true that for $x > 0$, $-\frac{1}{x} < \frac{1}{x} \sin(x) < \frac{1}{x}$? Prove your answer.

d) Graph $f(x) = \frac{1}{x} \sin(x)$ for $-2 \le x \le 2$ and $-2 \le y \le 2$. What is $f(0)$? Is it true that for all x in the interval $[-0.1, 0.1]$ for which $f(x)$ is defined, $f(x)$ satisfies $0.99 < f(x) < 1$? Explain.

e) Experiment with functions until you get one that looks like the one in the figure.

The Inverse Trigonometric Functions

We have learned how to find the values of the trigonometric functions for angles or real numbers, but to make the trigonometric functions really useful we must be able to reverse this process. In this section we define the inverses of the trigonometric functions.

The Inverse of the Sine Function

In Chapter 2 we learned that only one-to-one functions are invertible. Since $y = \sin x$ with domain $(-\infty, \infty)$ is a periodic function, it is certainly not one-to-one. However, if we restrict the domain to the interval $[-\pi/2, \pi/2]$, then the restricted function is one-to-one and invertible. Other intervals could be used, but this interval is chosen to keep the inverse function as simple as possible.

The graph of the sine function with domain $[-\pi/2, \pi/2]$ is shown in Fig. 5.75(a). Its range is $[-1, 1]$. We now define the inverse sine function and denote it as $f^{-1}(x) = \sin^{-1} x$ (read "inverse sine of x") or $f^{-1}(x) = \arcsin x$ (read "arc sine of x").

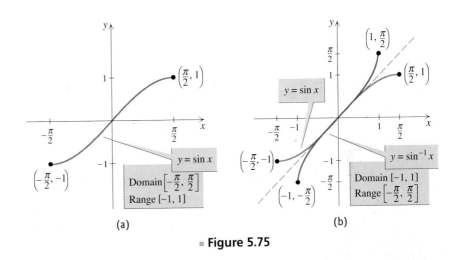

(a)   (b)

■ **Figure 5.75**

Definition: The Inverse Sine Function

The function $y = \sin^{-1} x$ or $y = \arcsin x$ is the inverse of the function $y = \sin x$ restricted to $[-\pi/2, \pi/2]$. The domain of $y = \sin^{-1} x$ is $[-1, 1]$ and its range is $[-\pi/2, \pi/2]$.

The graph of $y = \sin^{-1} x$ is a reflection about the line $y = x$ of the graph of $y = \sin x$ on $[-\pi/2, \pi/2]$, as shown in Fig. 5.75(b).

If $y = \sin^{-1} x$, then y is the real number such that $-\pi/2 \le y \le \pi/2$ and $\sin y = x$. Depending on the context, $\sin^{-1} x$ might also be an angle, a measure of an angle in degrees or radians, or the length of an arc of the unit circle. The expression $\sin^{-1} x$ can be read as "the angle whose sine is x" or "the arc length whose sine is x." The notation $y = \arcsin x$ reminds us that y is the arc length whose sine is x. For example, $\arcsin(1)$ is the arc length in $[-\pi/2, \pi/2]$ whose sine is 1. Since we know that $\sin(\pi/2) = 1$, we have $\arcsin(1) = \pi/2$. We will assume that $\sin^{-1}x$ is a real number unless indicated otherwise.

Note that the nth power of the sine function is usually written as $\sin^n(x)$ as a shorthand notation for $(\sin x)^n$, provided $n \neq -1$. The -1 used in $\sin^{-1}x$ indicates the inverse function and does *not* mean reciprocal. To write $1/\sin x$ using exponents, we must write $(\sin x)^{-1}$.

Example 1 Evaluating the inverse sine function

Find the exact value of each expression without using a table or a calculator.

a. $\sin^{-1}(1/2)$ **b.** $\arcsin(-\sqrt{3}/2)$

Solution

a. The value of $\sin^{-1}(1/2)$ is the number α in the interval $[-\pi/2, \pi/2]$ such that $\sin(\alpha) = 1/2$. We recall that $\sin(\pi/6) = 1/2$, and so $\sin^{-1}(1/2) = \pi/6$. Note that $\pi/6$ is the only value of α in $[-\pi/2, \pi/2]$ for which $\sin(\alpha) = 1/2$.

b. The value of $\arcsin(-\sqrt{3}/2)$ is the number α in $[-\pi/2, \pi/2]$ such that $\sin(\alpha) = -\sqrt{3}/2$. Since $\sin(-\pi/3) = -\sqrt{3}/2$, we have $\arcsin(-\sqrt{3}/2) = -\pi/3$. Note that $-\pi/3$ is the only value of α in $[-\pi/2, \pi/2]$ for which $\sin(\alpha) = -\sqrt{3}/2$.

Try This. Find the exact value of $\sin^{-1}(-\sqrt{2}/2)$. ∎

Example 2 Evaluating the inverse sine function

Find the exact value of each expression in degrees without using a table or a calculator.

a. $\sin^{-1}(\sqrt{2}/2)$ **b.** $\arcsin(0)$

Solution

a. The value of $\sin^{-1}(\sqrt{2}/2)$ in degrees is the angle α in the interval $[-90°, 90°]$ such that $\sin(\alpha) = \sqrt{2}/2$. We recall that $\sin(45°) = \sqrt{2}/2$, and so $\sin^{-1}(\sqrt{2}/2) = 45°$.

b. The value of $\arcsin(0)$ in degrees is the angle α in the interval $[-90°, 90°]$ for which $\sin(\alpha) = 0$. Since $\sin(0°) = 0$, we have $\arcsin(0) = 0°$.

Try This. Find the exact value in degrees of $\sin^{-1}(-\sqrt{3}/2)$. ∎

In the next example, we use a calculator to find the degree measure of an angle whose sine is given. To obtain degree measure, make sure the calculator is in degree mode. Scientific calculators usually have a key labeled $\sin^{-1}$ that gives values for the inverse sine function.

Example 3 Finding an angle given its sine

Let α be an angle such that $-90° < \alpha < 90°$. In each case, find α to the nearest tenth of a degree.

a. $\sin\alpha = 0.88$ **b.** $\sin\alpha = -0.27$

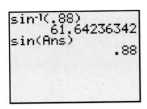

■ **Figure 5.76**

Solution

a. The value of $\sin^{-1}(0.88)$ is the only angle in $[-90°, 90°]$ with a sine of 0.88. Use a calculator in degree mode to get $\alpha = \sin^{-1}(0.88) \approx 61.6°$.

b. Use a calculator to get $\alpha = \sin^{-1}(-0.27) \approx -15.7°$.

Figure 5.76 shows how to find the angle in part (a) on a graphing calculator and how to check. Make sure that the mode is degrees.

Try This. Find α to the nearest tenth of a degree such that $\sin \alpha = 0.3$ and $-90° < \alpha < 90°$. $\approx 17.4°$ ■

The Inverse Cosine Function

Since the cosine function is not one-to-one on $(-\infty, \infty)$, we restrict the domain to $[0, \pi]$, where the cosine function is one-to-one and invertible. The graph of the cosine function with this restricted domain is shown in Fig. 5.77(a). Note that the range of the restricted function is $[-1, 1]$. We now define the inverse of $f(x) = \cos x$ for x in $[0, \pi]$ and denote it as $f^{-1}(x) = \cos^{-1}x$ or $f^{-1}(x) = \arccos x$.

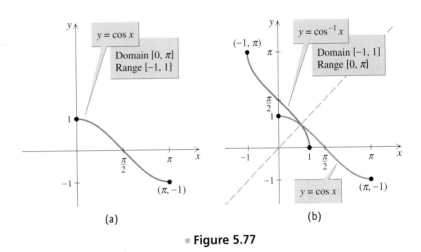

■ **Figure 5.77**

Definition: The Inverse Cosine Function

The function $y = \cos^{-1}x$ or $y = \arccos x$ is the inverse of the function $y = \cos x$ restricted to $[0, \pi]$. The domain of $y = \cos^{-1}x$ is $[-1, 1]$ and its range is $[0, \pi]$.

If $y = \cos^{-1}x$, then y is the real number in $[0, \pi]$ such that $\cos y = x$. The expression $\cos^{-1}x$ can be read as "the angle whose cosine is x" or "the arc length whose cosine is x." The graph of $y = \cos^{-1}x$, shown in Fig. 5.77(b), is obtained by reflecting the graph of $y = \cos x$ (restricted to $[0, \pi]$) about the line $y = x$. We will assume that $\cos^{-1}x$ is a real number unless indicated otherwise.

Example **4** **Evaluating the inverse cosine function**

Find the exact value of each expression without using a table or a calculator.

a. $\cos^{-1}(-1)$ **b.** $\arccos(-1/2)$ **c.** $\cos^{-1}(\sqrt{2}/2)$

Solution

a. The value of $\cos^{-1}(-1)$ is the number α in $[0, \pi]$ such that $\cos(\alpha) = -1$. We recall that $\cos(\pi) = -1$, and so $\cos^{-1}(-1) = \pi$.
b. The value of $\arccos(-1/2)$ is the number α in $[0, \pi]$ such that $\cos(\alpha) = -1/2$.
We recall that $\cos(2\pi/3) = -1/2$, and so $\arccos(-1/2) = 2\pi/3$.
c. Since $\cos(\pi/4) = \sqrt{2}/2$, we have $\cos^{-1}(\sqrt{2}/2) = \pi/4$.

Try This. Find the exact value of $\cos^{-1}(-\sqrt{3}/2)$.   ■

In the next example, we use a calculator to find the degree measure of an angle
whose cosine is given. Most scientific calculators have a key labeled $\cos^{-1}$ that
gives values for the inverse cosine function. To get the degree measure, make sure
the calculator is in degree mode.

Example **5** **Finding an angle given its cosine**

In each case find the angle α to the nearest tenth of a degree, given that
$0° < \alpha < 180°$.

a. $\cos \alpha = 0.23$ **b.** $\cos \alpha = -0.82$

Solution

a. Since $\cos^{-1}(0.23)$ is the unique angle in $[0°, 180°]$ with a cosine of 0.23,
$\alpha = \cos^{-1}(0.23) \approx 76.7°$.
b. Use a calculator to get $\alpha = \cos^{-1}(-0.82) \approx 145.1°$.

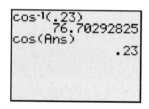

■ **Figure 5.78**

Figure 5.78 shows how to find $\cos^{-1}(0.23)$ on a graphing calculator and how
to check. Make sure that the mode is degrees.

Try This. Find α to the nearest tenth of a degree such that $\cos \alpha = -0.4$ and
$0° < \alpha < 180°$.   ■

Inverses of Tangent, Cotangent, Secant, and Cosecant

Since all of the trigonometric functions are periodic, they must all be restricted to a
domain where they are one-to-one before the inverse functions can be defined.
There is more than one way to choose a domain to get a one-to-one function, but
we will use the most common restrictions. The restricted domain for $y = \tan x$
is $(-\pi/2, \pi/2)$, for $y = \csc x$ it is $[-\pi/2, 0) \cup (0, \pi/2]$, for $y = \sec x$ it is
$[0, \pi/2) \cup (\pi/2, \pi]$, and for $y = \cot x$ it is $(0, \pi)$. The functions $\tan^{-1}$, $\cot^{-1}$,
$\sec^{-1}$, and $\csc^{-1}$ are defined to be the inverses of these restricted functions. The notations arctan, arccot, arcsec, and arccsc are also used for these inverse functions.
The graphs of all six inverse functions, with their domains and ranges, are shown in
the accompanying Function Gallery.

When studying inverse trigonometric functions, you should first learn to evaluate $\sin^{-1}$, $\cos^{-1}$, and $\tan^{-1}$. Then use the identities

$$\csc\alpha = \frac{1}{\sin\alpha}, \qquad \sec\alpha = \frac{1}{\cos\alpha}, \qquad \text{and} \qquad \cot\alpha = \frac{1}{\tan\alpha}$$

to evaluate $\csc^{-1}$, $\sec^{-1}$, and $\cot^{-1}$. For example, $\sin(\pi/6) = 1/2$ and $\csc(\pi/6) = 2$. So the angle whose cosecant is 2 is the same as the angle whose sine is $1/2$. In symbols,

$$\csc^{-1}(2) = \sin^{-1}\left(\frac{1}{2}\right) = \frac{\pi}{6}.$$

In general, $\csc^{-1}x = \sin^{-1}(1/x)$. Likewise, $\sec^{-1}x = \cos^{-1}(1/x)$. For the inverse cotangent, $\cot^{-1}x = \tan^{-1}(1/x)$ only for $x > 0$, because of the choice of $(0, \pi)$ as the range of the inverse cotangent. We have $\cot^{-1}(0) = \pi/2$ and, if $x < 0$, $\cot^{-1}(x) = \tan^{-1}(1/x) + \pi$. These identities are summarized following the Function Gallery.

Function Gallery: **Inverse Trigonometric Functions**

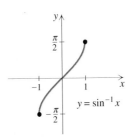

Domain $[-1, 1]$
Range $\left[-\frac{\pi}{2}, \frac{\pi}{2}\right]$

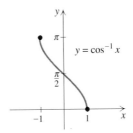

Domain $[-1, 1]$
Range $[0, \pi]$

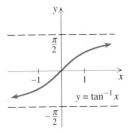

Domain $(-\infty, \infty)$
Range $\left(-\frac{\pi}{2}, \frac{\pi}{2}\right)$

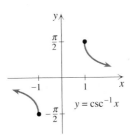

Domain $(-\infty, -1] \cup [1, \infty)$
Range $\left[-\frac{\pi}{2}, 0\right) \cup \left(0, \frac{\pi}{2}\right]$

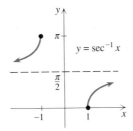

Domain $(-\infty, -1] \cup [1, \infty)$
Range $\left[0, \frac{\pi}{2}\right) \cup \left(\frac{\pi}{2}, \pi\right]$

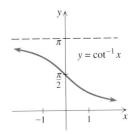

Domain $(-\infty, \infty)$
Range $(0, \pi)$

Identities for the Inverse Functions

$$\csc^{-1}(x) = \sin^{-1}(1/x) \text{ for } |x| \ge 1$$

$$\sec^{-1}(x) = \cos^{-1}(1/x) \text{ for } |x| \ge 1$$

$$\cot^{-1}(x) = \begin{cases} \tan^{-1}(1/x) & \text{for } x > 0 \\ \tan^{-1}(1/x) + \pi & \text{for } x < 0 \\ \pi/2 & \text{for } x = 0 \end{cases}$$

We can see another relationship between $\cot^{-1}$ and $\tan^{-1}$ from their graphs in the Function Gallery. We get the graph of $y = \cot^{-1}(x)$ by reflecting the graph of $y = \tan^{-1}(x)$ with respect to the x-axis and translating it up $\pi/2$ units. Thus,

$$\cot^{-1}(x) = \frac{\pi}{2} - \tan^{-1}(x).$$

Example **6** **Evaluating the inverse functions**

Find the exact value of each expression without using a table or a calculator.

a. $\tan^{-1}(1)$ **b.** $\text{arcsec}(-2)$ **c.** $\csc^{-1}(\sqrt{2})$ **d.** $\text{arccot}(-1/\sqrt{3})$

Solution

a. Since $\tan(\pi/4) = 1$ and since $\pi/4$ is in the range of $\tan^{-1}$, we have

$$\tan^{-1}(1) = \frac{\pi}{4}.$$

b. To evaluate the inverse secant, we use the identity $\sec^{-1}(x) = \cos^{-1}(1/x)$. In this case, the arc whose secant is -2 is the same as the arc whose cosine is $-1/2$. So we must find $\cos^{-1}(-1/2)$. Since $\cos(2\pi/3) = -1/2$ and since $2\pi/3$ is in the range of arccos, we have

$$\text{arcsec}(-2) = \arccos\left(-\frac{1}{2}\right) = \frac{2\pi}{3}.$$

c. To evaluate $\csc^{-1}(\sqrt{2})$, we use the identity $\csc^{-1}(x) = \sin^{-1}(1/x)$ with $x = \sqrt{2}$. So we must find $\sin^{-1}(1/\sqrt{2})$. Since $\sin(\pi/4) = 1/\sqrt{2}$ and since $\pi/4$ is in the range of $\csc^{-1}$,

$$\csc^{-1}(\sqrt{2}) = \sin^{-1}\left(\frac{1}{\sqrt{2}}\right) = \frac{\pi}{4}.$$

d. If x is negative, we use the identity $\cot^{-1}(x) = \tan^{-1}(1/x) + \pi$. Since $x = -1/\sqrt{3}$, we must find $\tan^{-1}(-\sqrt{3})$. Since $\tan^{-1}(-\sqrt{3}) = -\pi/3$, we get

$$\text{arccot}\left(\frac{-1}{\sqrt{3}}\right) = \tan^{-1}(-\sqrt{3}) + \pi = -\frac{\pi}{3} + \pi = \frac{2\pi}{3}.$$

Note that $\cot(-\pi/3) = \cot(2\pi/3) = -1/\sqrt{3}$, but $\text{arccot}(-1/\sqrt{3}) = 2\pi/3$ because $2\pi/3$ is in the range of the function arccot.

Try This. Find the exact value of $\arctan(\sqrt{3})$. ■

The functions $\sin^{-1}$, $\cos^{-1}$, and $\tan^{-1}$ are available on scientific and graphing calculators. The calculator values of the inverse functions are given in degrees or radians, depending on the mode setting. We will assume that the values of the inverse functions are to be in radians unless indicated otherwise. Calculators use the domains and ranges of these functions defined here. The functions $\sec^{-1}$, $\csc^{-1}$, and $\cot^{-1}$ are generally not available on a calculator, and expressions involving these functions must be written in terms of $\sin^{-1}$, $\cos^{-1}$, and $\tan^{-1}$ by using the identities.

Example 7 Evaluating the inverse functions with a calculator

Find the approximate value of each expression rounded to four decimal places.

a. $\sin^{-1}(0.88)$ **b.** $\operatorname{arccot}(2.4)$ **c.** $\csc^{-1}(4)$ **d.** $\cot^{-1}(0)$

Solution

a. Use the inverse sine function in radian mode to get $\sin^{-1}(0.88) \approx 1.0759$.
b. Use the identity $\cot^{-1}(x) = \tan^{-1}(1/x)$ because $2.4 > 0$:

$$\operatorname{arccot}(2.4) = \tan^{-1}\left(\frac{1}{2.4}\right) \approx 0.3948$$

c. $\csc^{-1}(4) = \sin^{-1}(0.25) \approx 0.2527$
d. The value of $\cot^{-1}(0)$ cannot be found on a calculator. But we know that $\cot(\pi/2) = 0$ and $\pi/2$ is in the range $(0, \pi)$ for $\cot^{-1}$, so we have $\cot^{-1}(0) = \pi/2$.

◸◹ The expressions in parts (a), (b), and (c) are shown on a graphing calculator in Fig. 5.79.

Try This. Find $\cot^{-1}(1.3)$ to four decimal places.

0.6557 rad

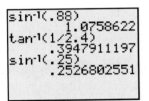

```
sin⁻¹(.88)
          1.0758622
tan⁻¹(1/2.4)
          .3947911197
sin⁻¹(.25)
          .2526802551
```

■ **Figure 5.79**

Compositions of Functions

One trigonometric function can be followed by another to form a composition of functions. We can evaluate an expression such as $\sin(\sin(\alpha))$ because the sine of the real number $\sin(\alpha)$ is defined. However, it is more common to have a composition of a trigonometric function and an inverse trigonometric function. For example, $\tan^{-1}(\alpha)$ is the angle whose tangent is α, and so $\sin(\tan^{-1}(\alpha))$ is the sine of the angle whose tangent is α.

Example 8 Evaluating compositions of functions

Find the exact value of each composition without using a table or a calculator.

a. $\sin(\tan^{-1}(0))$ **b.** $\arcsin(\cos(\pi/6))$ **c.** $\tan\left(\sec^{-1}\left(\sqrt{2}\right)\right)$

Solution

a. Since $\tan(0) = 0$, $\tan^{-1}(0) = 0$. Therefore,

$$\sin(\tan^{-1}(0)) = \sin(0) = 0.$$

b. Since $\cos(\pi/6) = \sqrt{3}/2$, we have

$$\arcsin\left(\cos\left(\frac{\pi}{6}\right)\right) = \arcsin\left(\frac{\sqrt{3}}{2}\right) = \frac{\pi}{3}.$$

c. To find $\sec^{-1}(\sqrt{2})$, we use the identity $\sec^{-1}(x) = \cos^{-1}(1/x)$. Since $\cos(\pi/4) = 1/\sqrt{2}$, we have $\cos^{-1}(1/\sqrt{2}) = \pi/4$ and $\sec^{-1}(\sqrt{2}) = \pi/4$. Therefore,

$$\tan\left(\sec^{-1}(\sqrt{2})\right) = \tan\left(\frac{\pi}{4}\right) = 1.$$

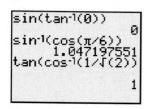 You can check parts (a), (b), and (c) with a graphing calculator, as shown in Fig. 5.80. Note that $\pi/3 \approx 1.047$.

▪ **Figure 5.80**

Try This. Find the exact value of $\sin^{-1}(\sin(5\pi/6))$. ▪

The composition of $y = \sin x$ restricted to $[-\pi/2, \pi/2]$ and $y = \sin^{-1}x$ is the identity function, because they are inverse functions of each other. So

$$\sin^{-1}(\sin x) = x \quad \text{for } x \text{ in } [-\pi/2, \pi/2]$$

and

$$\sin(\sin^{-1} x) = x \quad \text{for } x \text{ in } [-1, 1].$$

Note that if x is not in $[-\pi/2, \pi/2]$, then the sine function followed by the inverse sine function is not the identity function. For example, $\sin^{-1}(\sin(2\pi/3)) = \pi/3$.

The general sine function $f(x) = A \sin[B(x - C)] + D$ is a composition of a trigonometric function and several algebraic functions. Since the sine function has an inverse and the algebraic functions have inverses, we can find the inverse for a general sine function provided it is restricted to a suitable domain.

Example **9** **The inverse of a general sine function**

Find the inverse of $f(x) = 3 \sin(2x) + 5$, where $-\pi/4 \leq x \leq \pi/4$, and determine the domain of f^{-1}.

Solution

Interchange x and y in $y = 3 \sin(2x) + 5$, and then solve for y:

$$x = 3 \sin(2y) + 5 \quad \text{Switch } x \text{ and } y.$$

$$\frac{x - 5}{3} = \sin(2y)$$

$$2y = \sin^{-1}\left(\frac{x - 5}{3}\right) \quad \text{Definition of } \sin^{-1}$$

$$y = \frac{1}{2}\sin^{-1}\left(\frac{x - 5}{3}\right)$$

$$f^{-1}(x) = \frac{1}{2}\sin^{-1}\left(\frac{x - 5}{3}\right)$$

Since $\sin(2x)$ is between -1 and 1, the range of f is $[2, 8]$ and the domain of f^{-1} is $[2, 8]$.

Try This. Find the inverse of $f(x) = 2\cos(3x) - 1$ where $0 \le x \le \pi/3$ and find the domain of f^{-1}. ■

For Thought

True or False? Explain.

1. $\sin^{-1}(0) = \sin(0)$

2. $\sin(3\pi/4) = 1/\sqrt{2}$

3. $\cos^{-1}(0) = 1$

4. $\sin^{-1}(\sqrt{2}/2) = 135°$

5. $\cot^{-1}(5) = \dfrac{1}{\tan^{-1}(5)}$

6. $\sec^{-1}(5) = \cos^{-1}(0.2)$

7. $\sin(\cos^{-1}(\sqrt{2}/2)) = 1/\sqrt{2}$

8. $\sec(\sec^{-1}(2)) = 2$

9. The functions $f(x) = \sin^{-1} x$ and $f^{-1}(x) = \sin x$ are inverse functions.

10. The secant and cosecant functions are inverses of each other.

5.5 Exercises

Find the exact value of each expression without using a calculator or table.

1. $\sin^{-1}(-1/2)$

2. $\sin^{-1}(0)$

3. $\arcsin(1/2)$

4. $\arcsin(\sqrt{3}/2)$

5. $\arcsin(\sqrt{2}/2)$

6. $\arcsin(1)$

Find the exact value of each expression in degrees without using a calculator or table.

7. $\sin^{-1}(-1/\sqrt{2})$

8. $\sin^{-1}(\sqrt{3}/2)$

9. $\arcsin(1/2)$

10. $\arcsin(-1)$

11. $\sin^{-1}(0)$

12. $\sin^{-1}(\sqrt{2}/2)$

In each case find α to the nearest tenth of a degree, where $-90° \le \alpha \le 90°$.

13. $\sin \alpha = -1/3$

14. $\sin \alpha = 0.4138$

15. $\sin \alpha = 0.5682$

16. $\sin \alpha = -0.34$

Find the exact value of each expression without using a calculator or table.

17. $\cos^{-1}(-\sqrt{2}/2)$

18. $\cos^{-1}(1)$

19. $\arccos(1/2)$

20. $\arccos(-\sqrt{3}/2)$

21. $\arccos(-1)$

22. $\arccos(0)$

Find the exact value of each expression in degrees without using a calculator or table.

23. $\cos^{-1}(-\sqrt{2}/2)$

24. $\cos^{-1}(\sqrt{3}/2)$

25. $\arccos(-1)$

26. $\arccos(0)$

27. $\cos^{-1}(-1/2)$

28. $\cos^{-1}(1)$

In each case find α to the nearest tenth of a degree, where $0° \le \alpha \le 180°$.

29. $\cos \alpha = -0.993$

30. $\cos \alpha = 0.7392$

31. $\cos \alpha = 0.001$

32. $\cos \alpha = -0.499$

Find the exact value of each expression without using a calculator or table.

33. $\tan^{-1}(-1)$

34. $\cot^{-1}(1/\sqrt{3})$

35. $\sec^{-1}(2)$

36. $\csc^{-1}(2/\sqrt{3})$

37. $\operatorname{arcsec}(\sqrt{2})$

38. $\arctan(-1/\sqrt{3})$

39. $\operatorname{arccsc}(-2)$

40. $\operatorname{arccot}\left(-\sqrt{3}\right)$

41. $\tan^{-1}(0)$

42. $\sec^{-1}(1)$

43. $\csc^{-1}(1)$

44. $\csc^{-1}(-1)$

45. $\cot^{-1}(-1)$

46. $\cot^{-1}(0)$

47. $\cot^{-1}\left(-\sqrt{3}/3\right)$

48. $\cot^{-1}(1)$

Find the approximate value of each expression with a calculator. Round answers to two decimal places.

49. $\arcsin(0.5682)$

50. $\sin^{-1}(-0.4138)$

51. $\cos^{-1}(-0.993)$

52. $\cos^{-1}(0.7392)$

53. $\tan^{-1}(-0.1396)$

54. $\cot^{-1}(4.32)$

55. $\sec^{-1}(-3.44)$

56. $\csc^{-1}(6.8212)$

57. $\operatorname{arcsec}\left(\sqrt{6}\right)$

58. $\arctan\left(-2\sqrt{7}\right)$

59. $\operatorname{arccsc}\left(-2\sqrt{2}\right)$

60. $\operatorname{arccot}\left(-\sqrt{5}\right)$

61. $\operatorname{arccot}(-12)$

62. $\operatorname{arccot}(0.001)$

63. $\cot^{-1}(15.6)$

64. $\cot^{-1}(-1.01)$

Find the exact value of each composition without using a calculator or table.

65. $\tan(\arccos(1/2))$

66. $\sec\left(\arcsin(1/\sqrt{2})\right)$

67. $\sin^{-1}(\cos(2\pi/3))$

68. $\tan^{-1}(\sin(\pi/2))$

69. $\cot^{-1}(\cot(\pi/6))$

70. $\sec^{-1}(\sec(\pi/3))$

71. $\arcsin(\sin(3\pi/4))$

72. $\arccos(\cos(-\pi/3))$

73. $\tan(\arctan(1))$

74. $\cot(\operatorname{arccot}(0))$

75. $\cos^{-1}(\cos(3\pi/2))$

76. $\sin(\csc^{-1}(-2))$

77. $\cos\left(2\sin^{-1}\left(\sqrt{2}/2\right)\right)$

78. $\tan(2\cos^{-1}(1/2))$

79. $\sin^{-1}(2\sin(\pi/6))$

80. $\cos^{-1}(0.5\tan(\pi/4))$

Use a calculator to find the approximate value of each composition. Round answers to four decimal places. Some of these expressions are undefined.

81. $\sin(\cos^{-1}(0.45))$

82. $\cos(\tan^{-1}(44.33))$

83. $\sec^{-1}(\cos(\pi/9))$

84. $\csc^{-1}(\sec(\pi/8))$

85. $\tan(\sin^{-1}(-0.7))$

86. $\csc^{-1}(\sin(3\pi/7))$

87. $\cot\left(\cos^{-1}\left(-1/\sqrt{7}\right)\right)$

88. $\sin^{-1}(\csc(1.08))$

89. $\cot(\csc^{-1}(3.6))$

90. $\csc(\sec^{-1}(-2.4))$

91. $\sec(\cot^{-1}(5.2))$

92. $\csc(\cot^{-1}(-3.1))$

Find the inverse of each function and state its domain.

93. $f(x) = \sin(2x)$ for $-\dfrac{\pi}{4} \le x \le \dfrac{\pi}{4}$

94. $f(x) = \cos(3x)$ for $0 \le x \le \dfrac{\pi}{3}$

95. $f(x) = 3 + \tan(\pi x)$ for $-\dfrac{1}{2} < x < \dfrac{1}{2}$

96. $f(x) = 2 - \sin(\pi x - \pi)$ for $\dfrac{1}{2} \le x \le \dfrac{3}{2}$

97. $f(x) = \sin^{-1}(x/2) + 3$ for $-2 \le x \le 2$

98. $f(x) = 2\cos^{-1}(5x) + 3$ for $-\dfrac{1}{5} \le x \le \dfrac{1}{5}$

In a circle with radius r, a central angle θ intercepts a chord of length c, where $\theta = \cos^{-1}\left(1 - \dfrac{c^2}{2r^2}\right)$. Use this formula for Exercises 99 and 100.

99. An airplane at 2000 feet flies directly over a gun that has a range of 2400 feet, as shown in the figure. What is the measure in degrees of the angle for which the airplane is within range of the gun?

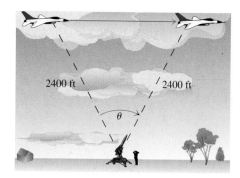

■ **Figure for Exercise 99**

100. A triangle has two sides that are both 5.2 meters long and one side with a length of 1.3 meters. Find the measure in degrees for the smallest angle of the triangle.

For Writing/Discussion

101. Graph the function $y = \sin(\sin^{-1} x)$ for $-2\pi \le x \le 2\pi$ and explain your result.

102. Graph the function $y = \sin^{-1}(\sin x)$ for $-2\pi \le x \le 2\pi$ and explain your result.

103. Graph $y = \sin^{-1}(1/x)$ and explain why the graph looks like the graph of $y = \csc^{-1} x$ shown in the Function Gallery on page 481.

104. Graph $y = \tan^{-1}(1/x)$ and explain why the graph does *not* look like the graph of $y = \cot^{-1} x$ shown in the Function Gallery on page 481.

Thinking Outside the Box XLVIII

Clear Sailing   A sailor plans to install a windshield wiper on a porthole that has radius 1 foot. The wiper blade of length x feet is to be attached to the edge of the porthole as shown in the figure.

The area cleaned by the blade is a sector of a circle centered at the point of attachment.

a. Write the area cleaned by the blade as a function of x.

b. If the blade cleans half of the window, then what is the exact length of the blade?

c. Use a graphing calculator to find the length for the blade that would maximize the area cleaned?

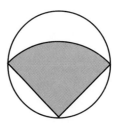

■ **Figure for Thinking Outside the Box XLVIII**

5.5 Pop Quiz

1. Find $f^{-1}(x)$ if $f(x) = \cos(2x)$ for $0 \le x \le \pi/2$.

Find the exact value.

2. $\sin^{-1}(-1)$

3. $\sin^{-1}(1/2)$

4. $\arccos(-1)$

5. $\arctan(-1)$

6. $\tan(\arcsin(1/2))$

7. $\sin^{-1}(\sin(3\pi/4))$

Linking Concepts

For Individual or Group Explorations

Maximizing the Viewing Angle

A billboard that is 30 ft wide is placed 60 ft from a highway, as shown in the figure. The billboard is easiest to read from the highway when the viewing angle α is larger than $7°$.

a) Using results from the next section, we can show that

$$\alpha = \tan^{-1}(90/x) - \tan^{-1}(60/x).$$

Graph this function.

b) For what approximate values of x is α greater than $7°$?

c) For approximately how long does a motorist traveling at 35 mph have a viewing angle larger than $7°$?

5.6

Right Triangle Trigonometry

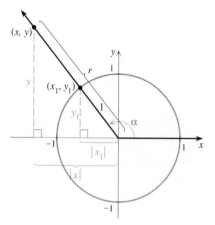

■ **Figure 5.81**

One reason trigonometry was invented was to determine the measures of sides and angles of geometric figures without actually measuring them. In this section we study right triangles (triangles that have a 90° angle) and see what information is needed to determine the measures of all unknown sides and angles of a right triangle.

Trigonometric Ratios

We defined the sine and cosine functions for an angle in standard position in terms of the point at which the terminal side intersects the unit circle. However, it is often the case that we do not know that point, but we do know a different point on the terminal side. When this situation occurs, we can find the values for the sine and cosine of the angle using trigonometric ratios.

Figure 5.81 shows an angle α in quadrant II with the terminal side passing through the point (x, y) and intersecting the unit circle at (x_1, y_1). If we draw vertical line segments down to the x-axis, we form two similar right triangles, as shown in Fig. 5.81. If r is the distance from (x, y) to the origin, then $r = \sqrt{x^2 + y^2}$. The lengths of the legs in the larger triangle are y and $|x|$, and its hypotenuse is r. The lengths of the legs in the smaller triangle are y_1 and $|x_1|$, and its hypotenuse is 1. Since ratios of the lengths of corresponding sides of similar triangles are equal, we have

$$\frac{y_1}{y} = \frac{1}{r} \qquad \text{and} \qquad \frac{|x_1|}{|x|} = \frac{1}{r}.$$

The first equation can be written as $y_1 = y/r$ and the second as $x_1 = x/r$. (Since x_1 and x are both negative in this case, the absolute value symbols can be omitted.) Since $y_1 = \sin \alpha$ and $x_1 = \cos \alpha$, we get

$$\sin \alpha = \frac{y}{r} \qquad \text{and} \qquad \cos \alpha = \frac{x}{r}.$$

This argument can be repeated in each quadrant with (x, y) chosen inside, outside, or on the unit circle. These ratios also give the correct sine or cosine if α is a quadrantal angle. Since all other trigonometric functions are related to the sine and cosine, their values can also be obtained from x, y, and r.

Theorem:
Trigonometric Ratios

> If (x, y) is any point other than the origin on the terminal side of an angle α in standard position and $r = \sqrt{x^2 + y^2}$, then
>
> $$\sin \alpha = \frac{y}{r}, \qquad \cos \alpha = \frac{x}{r}, \qquad \text{and} \qquad \tan \alpha = \frac{y}{x} \ (x \neq 0).$$

Example **1** **Trigonometric ratios**

Find the values of the six trigonometric functions of the angle α in standard position whose terminal side passes through $(4, -2)$.

Solution

Use $x = 4$, $y = -2$, and $r = \sqrt{4^2 + (-2)^2} = \sqrt{20} = 2\sqrt{5}$ to get

$$\sin \alpha = \frac{-2}{2\sqrt{5}} = -\frac{\sqrt{5}}{5}, \qquad \cos \alpha = \frac{4}{2\sqrt{5}} = \frac{2\sqrt{5}}{5}, \qquad \text{and}$$

$$\tan \alpha = \frac{-2}{4} = -\frac{1}{2}.$$

Since cosecant, secant, and cotangent are the reciprocals of sine, cosine, and tangent,

$$\csc \alpha = \frac{1}{\sin \alpha} = -\frac{5}{\sqrt{5}} = -\sqrt{5}, \qquad \sec \alpha = \frac{1}{\cos \alpha} = \frac{5}{2\sqrt{5}} = \frac{\sqrt{5}}{2}, \qquad \text{and}$$

$$\cot \alpha = \frac{1}{\tan \alpha} = -2.$$

Try This. Find $\sin \alpha$, $\cos \alpha$, and $\tan \alpha$ if α is an angle in standard position whose terminal side passes through $(2, 3)$.   ■

Right Triangles

So far the trigonometric functions have been tied to a coordinate system and an angle in standard position. Trigonometric ratios can also be used to evaluate the trigonometric functions for an acute angle of a right triangle without having the angle or the triangle located in a coordinate system.

Consider a right triangle with acute angle α, legs of length x and y, and hypotenuse r, as shown in Fig. 5.82(a). If this triangle is positioned in a coordinate system, as in Fig. 5.82(b), then (x, y) is a point on the terminal side of α and

$$\sin(\alpha) = \frac{y}{r}, \qquad \cos(\alpha) = \frac{x}{r}, \qquad \text{and} \qquad \tan(\alpha) = \frac{y}{x}.$$

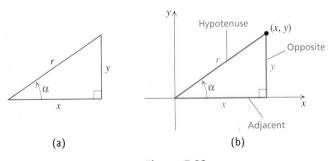

(a)   (b)

■ **Figure 5.82**

However, the values of the trigonometric functions are simply ratios of the lengths of the sides of the right triangle and it is not necessary to move the triangle to a coordinate system to find them. Notice that y is the length of the side **opposite** the angle α, x is the length of the side **adjacent** to α, and r is the length of the **hypotenuse.** We use the abbreviations opp, adj, and hyp to represent the lengths of these sides in the following theorem.

Theorem: Trigonometric Functions of an Acute Angle of a Right Triangle

If α is an acute angle of a right triangle, then

$$\sin \alpha = \frac{\text{opp}}{\text{hyp}}, \qquad \cos \alpha = \frac{\text{adj}}{\text{hyp}}, \qquad \text{and} \qquad \tan \alpha = \frac{\text{opp}}{\text{adj}}.$$

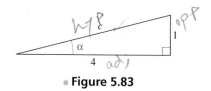

■ **Figure 5.83**

Since the cosecant, secant, and cotangent are the reciprocals of the sine, cosine, and tangent, respectively, the values of all six trigonometric functions can be found for an acute angle of a right triangle.

Example **2** **Trigonometric functions in a right triangle**

Find the values of all six trigonometric functions for the angle α of the right triangle with legs of length 1 and 4, as shown in Fig. 5.83.

Solution

The length of the hypotenuse is $c = \sqrt{4^2 + 1^2} = \sqrt{17}$. Since the length of the side opposite α is 1 and the length of the adjacent side is 4, we have

$$\sin \alpha = \frac{\text{opp}}{\text{hyp}} = \frac{1}{\sqrt{17}} = \frac{\sqrt{17}}{17}, \qquad \cos \alpha = \frac{\text{adj}}{\text{hyp}} = \frac{4}{\sqrt{17}} = \frac{4\sqrt{17}}{17},$$

$$\tan \alpha = \frac{\text{opp}}{\text{adj}} = \frac{1}{4}, \qquad \csc \alpha = \frac{1}{\sin \alpha} = \sqrt{17},$$

$$\sec \alpha = \frac{1}{\cos \alpha} = \frac{\sqrt{17}}{4}, \qquad \cot \alpha = \frac{1}{\tan \alpha} = 4.$$

Try This. A right triangle has legs with lengths 2 and 4. Find $\sin \alpha$, $\cos \alpha$, and $\tan \alpha$ if α is the angle opposite the side of length 4. ■

Solving a Right Triangle

The values of the trigonometric functions for an acute angle of a right triangle are determined by ratios of lengths of sides of the triangle. We can use those ratios along with the inverse trigonometric functions to find missing parts of a right triangle in which some of the measures of angles or lengths of sides are known. Finding all of the unknown lengths of sides or measures of angles is called **solving the triangle.** A triangle can be solved only if enough information is given to determine a unique triangle. For example, the lengths of the sides in a 30-60-90 triangle cannot be found because there are infinitely many such triangles of different sizes. However, the lengths of the missing sides in a 30-60-90 triangle with a hypotenuse of 6 will be found in Example 3.

In solving right triangles, we usually name the acute angles α and β and the lengths of the sides opposite those angles a and b. The 90° angle is γ, and the length of the side opposite γ is c.

Example **3** **Solving a right triangle**

Solve the right triangle in which $\alpha = 30°$ and $c = 6$.

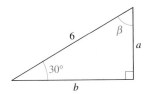

■ **Figure 5.84**

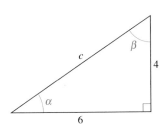

■ **Figure 5.85**

Solution

The triangle is shown in Fig. 5.84. Since $\alpha = 30°, \gamma = 90°$, and the sum of the measures of the angles of any triangle is $180°$, we have $\beta = 60°$. Since $\sin \alpha = \text{opp/hyp}$, we get $\sin 30° = a/6$ and

$$a = 6 \cdot \sin 30° = 6 \cdot \frac{1}{2} = 3.$$

Since $\cos \alpha = \text{adj/hyp}$, we get $\cos 30° = b/6$ and

$$b = 6 \cdot \cos 30° = 6 \cdot \frac{\sqrt{3}}{2} = 3\sqrt{3}.$$

The angles of the right triangle are $30°$, $60°$, and $90°$, and the sides opposite those angles are 3, $3\sqrt{3}$, and 6, respectively.

Try This. Solve the right triangle in which $\alpha = 60°$ and $c = 2$. ■

Example **4** Solving a right triangle

Solve the right triangle in which $a = 4$ and $b = 6$. Find the acute angles to the nearest tenth of a degree.

Solution

The triangle is shown in Fig. 5.85. By the Pythagorean theorem, $c^2 = 4^2 + 6^2$, or $c = \sqrt{52} = 2\sqrt{13}$. To find α, first find $\sin \alpha$:

$$\sin \alpha = \frac{\text{opp}}{\text{hyp}} = \frac{4}{2\sqrt{13}} = \frac{2}{\sqrt{13}}$$

Now, α is the angle whose sine is $2/\sqrt{13}$:

$$\alpha = \sin^{-1}\left(\frac{2}{\sqrt{13}}\right) \approx 33.7°$$

Since $\alpha + \beta = 90°, \beta = 90° - 33.7° = 56.3°$. The angles of the triangle are $33.7°$, $56.3°$, and $90°$, and the sides opposite those angles are 4, 6, and $2\sqrt{13}$, respectively.

Try This. Solve the right triangle in which $a = 2$ and $b = 5$. ■

In Example 4 we could have found α by using $\alpha = \tan^{-1}(4/6) \approx 33.7°$. We could then have found c by using $\cos(33.7°) = 6/c$, or $c = 6/\cos(33.7°) \approx 7.2$. There are many ways to solve a right triangle, but the basic strategy is always the same.

S T R A T E G Y **Solving a Right Triangle**

1. Use the Pythagorean theorem to find the length of a third side when the lengths of two sides are known.

2. Use the trigonometric ratios to find missing sides or angles.

3. Use the fact that the sum of the measures of the angles of a triangle is $180°$ to determine a third angle when two are known.

Significant Digits

In solving right triangles and in most other problems in algebra and trigonometry, we usually assume that the given numbers are exact. Starting with exact numbers, exact answers or approximations to any degree of accuracy desired can be obtained. However, in real applications, lengths of sides in triangles or measures of angles are determined by actual measurement. In such cases, we must be concerned with the accuracy of our measurements and the values obtained from those measurements using trigonometry. For example, an angle of 21.75° at the center of a circle with a one-mile radius intercepts an arc of length 2004.3 feet, but if the same angle is given to the nearest tenth of a degree, 21.8°, then the length of the intercepted arc is 2008.9 feet, a difference of 4.6 feet.

To say that an angle is measured at 21.8° means that the angle is actually between 21.75° and 21.85°. Since all digits in 21.8° are meaningful, we say that 21.8° has three **significant digits.** If a number has no nonzero digits to the left of the decimal point, then zeros that follow the decimal point but precede nonzero digits are not significant. A measurement of 0.00034 centimeters has two significant digits. The number 44.006 has five significant digits. In the measurement 44.0 pounds, the zero means that the measurement was made to the nearest tenth, and so it is significant. Zeros that follow the decimal point but are not followed by nonzero digits are significant. For whole-number measurements without decimal points, we assume that all digits are significant. So measurements of 4782 feet and 8000 miles both have four significant digits. In scientific notation, only significant digits are written. For example, 8×10^3 has only one significant digit, while 8.000×10^3 has four.

When arithmetic operations or trigonometric functions are applied to measurements, the answers are only as accurate as the *least accurate* measurement involved. For example, if it is 23 miles from your house to the airport and 8×10^3 miles (to the nearest thousand) to Hong Kong, then to say that it is 8023 miles to Hong Kong from your house implies an accuracy that is not justified. Since 8×10^3 miles is a less accurate measurement than 23 miles and 8×10^3 miles has one significant digit, any answer obtained from these two measurements has only one significant digit. So only the first digit of 8023 is significant, and it is 8×10^3 miles (to the nearest thousand) from your house to Hong Kong. A "rule of thumb" that works well in most cases is that an answer should be given with the same number of significant digits as is in the least accurate measurement in the computation.

Example **5** **Significant digits**

Assume that the given numbers are measurements. Perform each computation, and give the answer with the appropriate number of significant digits.

a. $\sin(123.4°)$ **b.** $\dfrac{\cos(3°)}{\sin(1.9°)}$ **c.** $544 \cdot \sin(25°)$

Solution

a. Since 123.4° has four significant digits, we round the calculator value to four decimal places. So $\sin(123.4°) \approx 0.8348$.

b. Since the least number of significant digits in 3° and 1.9° is one, round the value obtained on a calculator to one significant digit:

$$\frac{\cos(3°)}{\sin(1.9°)} \approx 3 \times 10^1 \text{ (to the nearest ten)}$$

c. The least number of significant digits in 544 and 25° is two. So

$$544 \cdot \sin(25°) \approx 230 \text{ (to the nearest ten)}.$$

In scientific notation, $544 \cdot \sin(25°) \approx 2.3 \times 10^2$.

Try This. Find $24 \sin(5.461°)$ using the appropriate number of significant digits in the answer. ■

Applications

Using trigonometry, we can find the size of an object without actually measuring the object but by measuring an angle. Two common terms used in this regard are **angle of elevation** and **angle of depression.** The angle of elevation α for a point above a horizontal line is the angle formed by the horizontal line and the observer's line of sight through the point, as shown in Fig. 5.86. The angle of depression β for a point below a horizontal line is the angle formed by the horizontal line and the observer's line of sight through the point, as shown in Fig. 5.86. We use these angles and our skills in solving triangles to find the sizes of objects that would be inconvenient to measure.

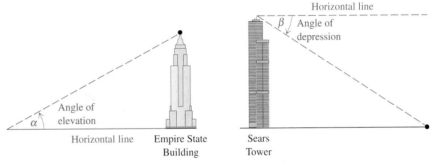

■ **Figure 5.86**

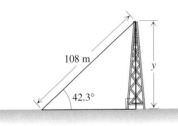

■ **Figure 5.87**

Example **6** **Finding the height of an object**

A guy wire of length 108 meters runs from the top of an antenna to the ground. If the angle of elevation of the top of the antenna, sighting along the guy wire, is 42.3°, then what is the height of the antenna?

Solution

Let y represent the height of the antenna, as shown in Fig. 5.87. Since $\sin(42.3°) = y/108$,

$$y = 108 \cdot \sin(42.3°) \approx 72.7 \text{ meters}.$$

Three significant digits are used in the answer because both of the given measurements contain three significant digits.

Try This. The angle of elevation of the top of a cell phone tower is 38.2° at a distance of 344 feet from the tower. What is the height of the tower? ■

In Example 6 we knew the distance to the top of the antenna, and we found the height of the antenna. If we knew the distance on the ground to the base of the antenna and the angle of elevation of the guy wire, we could still have found the height of the antenna. Both cases involve knowing the distance to the antenna either on the ground or through the air. However, one of the biggest triumphs of trigonometry is being able to find the size of an object or the distance to an object (such as the moon) without going to the object. The next example shows one way to find the height of an object without actually going to it. In Section 7.1, Example 7, we will show another (slightly simpler) solution to the same problem using the law of sines.

Example 7 Finding the height of an object from a distance

The angle of elevation of the top of a water tower from point A on the ground is 19.9°. From point B, 50.0 feet closer to the tower, the angle of elevation is 21.8°. What is the height of the tower?

Solution

Let y represent the height of the tower and x represent the distance from point B to the base of the tower, as shown in Fig. 5.88.

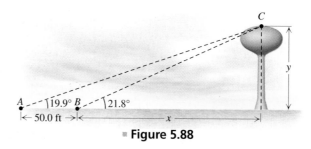

■ **Figure 5.88**

At point B, $\tan 21.8° = y/x$ or

$$x = \frac{y}{\tan 21.8°}.$$

Since the distance to the base of the tower from point A is $x + 50$,

$$\tan 19.9° = \frac{y}{x + 50}$$

or

$$y = (x + 50) \tan 19.9°.$$

To find the value of y we must write an equation that involves only y. Since $x = y/\tan 21.8°$, we can substitute $y/\tan 21.8°$ for x in the last equation:

$$y = \left(\frac{y}{\tan 21.8°} + 50\right) \tan 19.9°$$

$$y = \frac{y \cdot \tan 19.9°}{\tan 21.8°} + 50 \tan 19.9° \quad \text{Distributive property}$$

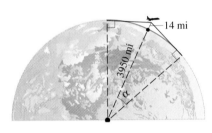

■ **Figure 5.89**

$$y - \frac{y \cdot \tan 19.9°}{\tan 21.8°} = 50 \tan 19.9°$$

$$y\left(1 - \frac{\tan 19.9°}{\tan 21.8°}\right) = 50 \tan 19.9° \qquad \text{Factor out } y.$$

$$y = \frac{50 \tan 19.9°}{1 - \dfrac{\tan 19.9°}{\tan 21.8°}} \approx 191 \text{ feet}$$

This computation is shown on a graphing calculator in Fig. 5.89.

Try This. At one location, the angle of elevation of the top of an antenna is 44.2°. At a point that is 100 feet closer to the antenna, the angle of elevation is 63.1°. What is the height of the antenna? ■

In the next example we combine the solution of a right triangle with the arc length of a circle from Section 5.1 to solve a problem of aerial photography.

Example 8 Photography from a spy plane

In the late 1950s, the Soviets labored to develop a missile that could stop the U-2 spy plane. On May 1, 1960, Nikita S. Khrushchev announced to the world that the Soviets had shot down Francis Gary Powers while Powers was photographing the Soviet Union from a U-2 at an altitude of 14 miles. How wide a path on the earth's surface could Powers see from that altitude? (Use 3950 miles as the earth's radius.)

Solution

Figure 5.90 shows the line of sight to the horizon on the left-hand side and right-hand side of the airplane while flying at the altitude of 14 miles.

Since a line tangent to a circle (the line of sight) is perpendicular to the radius at the point of tangency, the angle α at the center of the earth in Fig. 5.90 is an acute angle of a right triangle with hypotenuse $3950 + 14$ or 3964. So we have

$$\cos \alpha = \frac{3950}{3964}$$

$$\alpha = \cos^{-1}\left(\frac{3950}{3964}\right) \approx 4.8°.$$

The width of the path seen by Powers is the length of the arc intercepted by the central angle 2α or 9.6°. Using the formula $s = \alpha r$ from Section 5.1, where α is in radians, we get

$$s = 9.6 \text{ deg} \cdot \frac{\pi \text{ rad}}{180 \text{ deg}} \cdot 3950 \text{ miles} \approx 661.8 \text{ miles}.$$

From an altitude of 14 miles, Powers could see a path that was 661.8 miles wide. Actually, he photographed a path that was somewhat narrower, because parts of the photographs near the horizon were not usable.

Try This. How wide a path (to the nearest mile) on the earth's surface can be seen from a commercial airliner flying at an altitude of 7 miles? ■

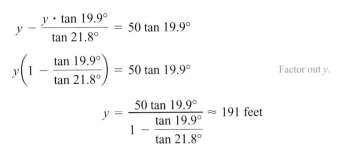

14 mi

3950 mi

α

■ **Figure 5.90**

For Thought

True or False? Explain. For Exercises 1–4, α is an angle in standard position.

1. If the terminal side of α goes through $(5, -10)$, then $\sin \alpha = 10/\sqrt{125}$.

2. If the terminal side of α goes through $(-1, 2)$, then $\sec \alpha = -\sqrt{5}$.

3. If the terminal side of α goes through $(-2, 3)$, then $\alpha = \sin^{-1}(3/\sqrt{13})$.

4. If the terminal side of α goes through $(3, 1)$, then $\alpha = \cos^{-1}(3/\sqrt{10})$.

5. In a right triangle, $\sin \alpha = \cos \beta$, $\sec \alpha = \csc \beta$, and $\tan \alpha = \cot \beta$.

6. If $a = 4$ and $b = 2$ in a right triangle, then $c = \sqrt{6}$.

7. If $a = 6$ and $b = 2$ in a right triangle, then $\beta = \tan^{-1}(3)$.

8. If $a = 8$ and $\alpha = 55°$ in a right triangle, then $b = 8/\tan(55°)$.

9. In a right triangle with sides of length 3, 4, and 5, the smallest angle is $\cos^{-1}(0.8)$.

10. In a right triangle, $\sin(90°) = \text{hyp}/\text{adj}$.

5.6 Exercises

Assume that α is an angle in standard position whose terminal side contains the given point. Find the exact values of $\sin \alpha$, $\cos \alpha$, $\tan \alpha$, $\csc \alpha$, $\sec \alpha$, and $\cot \alpha$.

1. $(3, 4)$

2. $(4, 4)$

3. $(-2, 6)$

4. $(-3, 6)$

5. $(-2, -\sqrt{2})$

6. $(-1, -\sqrt{3})$

7. $(\sqrt{6}, -\sqrt{2})$

8. $(2\sqrt{3}, -2)$

For Exercises 9–14 find exact values of $\sin \alpha$, $\cos \alpha$, $\tan \alpha$, $\sin \beta$, $\cos \beta$, and $\tan \beta$ for the given right triangle.

9.

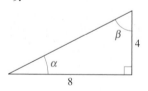

10.

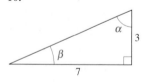

11.

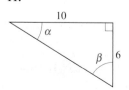

12.

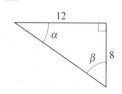

13.

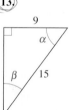

14.

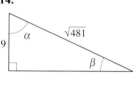

Assume that α is an angle in standard position whose terminal side contains the given point and that $0° < \alpha < 90°$. Find the degree measure of α to the nearest tenth of a degree.

15. $(1.5, 9)$

16. $(4, 5)$

17. $(\sqrt{2}, \sqrt{6})$

18. $(4.3, 6.9)$

Assume that α is an angle in standard position whose terminal side contains the given point and that $0 < \alpha < \pi/2$. Find the radian measure of α to the nearest tenth of a radian.

19. $(4, 6.3)$

20. $(1/3, 1/2)$

21. $(\sqrt{5}, 1)$

22. $(\sqrt{7}, \sqrt{3})$

Solve each right triangle with the given sides and angles. In each case, make a sketch. Note that, α is the acute angle opposite leg a and β is the acute angle opposite leg b. The hypotenuse is c. See the strategy for solving a right triangle on page 491.

23. $\alpha = 60°, c = 20$ **24.** $\beta = 45°, c = 10$

25. $a = 6, b = 8$ **26.** $a = 10, c = 12$

27. $b = 6, c = 8.3$ **28.** $\alpha = 32.4°, b = 10$

29. $\alpha = 16°, c = 20$ **30.** $\beta = 47°, a = 3$

31. $\alpha = 39°9', a = 9$ **32.** $\beta = 19°12', b = 60$

Perform each computation with the given measurements, and give your answers with the appropriate number of significant digits.

33. $456 \cdot \tan 3.2°$ **34.** $123.32 \cdot \tan 36.3°$

35. $\sin^2 65.7°$ **36.** $4 + 6.33 \sin 79.7°$

37. $7.50/\sin 23.559°$ **38.** $12.3/\cot 0.03°$

39. $500/\sec 125.3°$ **40.** $16.3 - 8.31 \cdot \csc 8.76°$

Solve each problem.

41. *Aerial Photography* An aerial photograph from a U-2 spy plane is taken of a building suspected of housing nuclear warheads. The photograph is made when the angle of elevation of the sun is 32°. By comparing the shadow cast by the building to objects of known size in the illustration, analysts determine that the shadow is 80 ft long. How tall is the building?

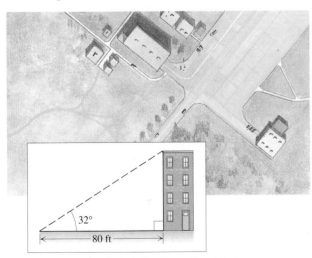

■ **Figure for Exercise 41**

42. *Giant Redwood* A hiker stands 80 feet from a giant redwood tree and sights the top with an angle of elevation of 75°. How tall is the tree to the nearest foot?

43. *Avoiding a Swamp* Muriel was hiking directly toward a long, straight road when she encountered a swamp. She turned 65° to the right and hiked 4 mi in that direction to reach the road. How far was she from the road when she encountered the swamp?

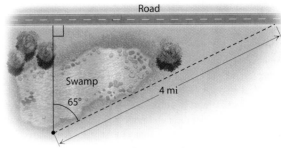

■ **Figure for Exercise 43**

44. *Tall Antenna* A 100 foot guy wire is attached to the top of an antenna. The angle between the guy wire and the ground is 62°. How tall is the antenna to the nearest foot?

45. *Angle of Depression* From a highway overpass, 14.3 m above the road, the angle of depression of an oncoming car is measured at 18.3°. How far is the car from a point on the highway directly below the observer?

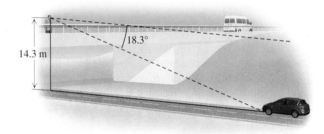

■ **Figure for Exercise 45**

46. *Length of a Tunnel* A tunnel under a river is 196.8 ft below the surface at its lowest point, as shown in the drawing. If the angle of depression of the tunnel is 4.962°, then how far apart on the surface are the entrances to the tunnel? How long is the tunnel?

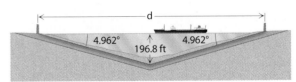

■ **Figure for Exercise 46**

47. *Height of a Crosswalk* The angle of elevation of a pedestrian crosswalk over a busy highway is 8.34°, as shown in the drawing on the next page. If the distance between the ends of the crosswalk measured on the ground is 342 ft, then what is the height h of the crosswalk at the center?

■ **Figure for Exercise 47**

48. *Shortcut to Snyder*   To get from Muleshoe to Snyder, Harry drives 50 mph for 178 mi south on route 214 to Seminole, then goes east on route 180 to Snyder. Harriet leaves Muleshoe one hour later at 55 mph, but takes US 84, which goes straight from Muleshoe to Snyder through Lubbock. If US 84 intersects route 180 at a 50° angle, then how many more miles does Harry drive?

49. *Installing a Guy Wire*   A 41-m guy wire is attached to the top of a 34.6-m antenna and to a point on the ground. How far is the point on the ground from the base of the antenna, and what angle does the guy wire make with the ground?

50. *Robin and Marian*   Robin Hood plans to use a 30-ft ladder to reach the castle window of Maid Marian. Little John, who made the ladder, advised Robin that the angle of elevation of the ladder must be between 55° and 70° for safety. What are the minimum and maximum heights that can safely be reached by the top of the ladder when it is placed against the 50-ft castle wall?

51. *Detecting a Speeder*   A policewoman has positioned herself 500 ft from the intersection of two roads. She has carefully measured the angles of the lines of sight to points *A* and *B* as shown in the drawing. If a car passes from *A* to *B* in 1.75 sec and the speed limit is 55 mph, is the car speeding?
HINT Find the distance from *B* to *A* and use $R = D/T$.

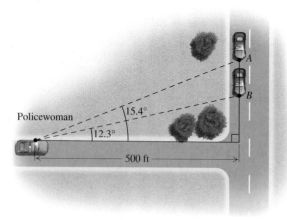

■ **Figure for Exercise 51**

52. *Progress of a Forest Fire*   A forest ranger atop a 3248-ft mesa is watching the progress of a forest fire spreading in her direction. In 5 minutes the angle of depression of the leading edge of the fire changed from 11.34° to 13.51°. At what speed in miles per hour is the fire spreading in the direction of the ranger?

53. *Height of a Rock*   Inscription Rock rises almost straight upward from the valley floor. From one point the angle of elevation of the top of the rock is 16.7°. From a point 168 m closer to the rock, the angle of elevation of the top of the rock is 24.1°. How high is Inscription Rock?

54. *Height of a Balloon*   A hot air balloon is between two spotters who are 1.2 mi apart. One spotter reports that the angle of elevation of the balloon is 76°, and the other reports that it is 68°. What is the altitude of the balloon in miles?

55. *Passing in the Night*   A boat sailing north sights a lighthouse to the east at an angle of 32° from the north, as shown in the drawing. After the boat travels one more kilometer, the angle of the lighthouse from the north is 36°. If the boat continues to sail north, then how close will the boat come to the lighthouse?

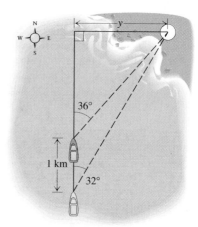

■ **Figure for Exercise 55**

56. *Height of a Skyscraper*   For years the Woolworth skyscraper in New York held the record for the world's tallest office building. If the length of the shadow of the Woolworth building increases by 17.4 m as the angle of elevation of the sun changes from 44° to 42°, then how tall is the building?

57. *Parsecs*   In astronomy the light year (abbreviated ly) and the parsec (abbreviated pc) are the two units used to measure distances. The distance in space at which a line from the earth to the sun subtends an angle of 1 second is 1 parsec. Find the number of miles in 1 parsec by using a right triangle positioned in space with its right angle at the sun as shown in the figure. How many years does it take light to travel 1 parsec? (One light year (ly) is the distance that light travels in one year, one AU is the distance from the earth to the sun, and 1 ly = 63,240 AU.)

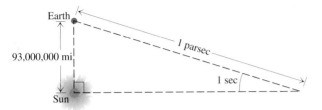

Earth

93,000,000 mi

1 parsec

1 sec

Sun

■ **Figure for Exercise 57**

58. *Catching a Green Flash*   Ed and Diane are on the beach to observe a green flash. As the very last sliver of the sun sinks into the ocean, refraction bends the final ray of sunlight into a rainbow of colors. The red and orange set first, and the green light or *green flash* is the last bit of sunlight seen before sunset.

a. If Diane's eyes are 2 ft above the ocean, then what is the angle α shown in the figure?

b. If Ed's eyes are 6 ft above the ocean, then what is the angle α for Ed?

c. Diane sees the green flash first. The difference between the two alphas is how far the earth must rotate before Ed sees it. How many seconds after Diane sees the green flash does Ed see it?

d. In Exercise 111 of Section 5.1 we saw how Eratosthenes measured the earth. Find the radius of the earth by starting with the time delay for the green flash from part (c). You will need a graphing calculator.

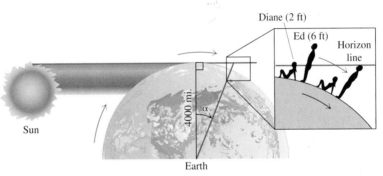

Diane (2 ft)

Ed (6 ft)

Horizon line

4000 mi.

α

Sun

Earth

■ **Figure for Exercise 58**

59. *View from Landsat*   The satellite Landsat orbits the earth at an altitude of 700 mi, as shown in the figure. What is the width of the path on the surface of the earth that can be seen by the cameras of Landsat? Use 3950 mi for the radius of the earth.

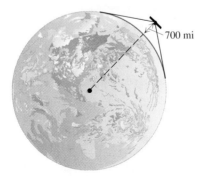

700 mi

■ **Figure for Exercise 59**

60. *Communicating Via Satellite*   A communication satellite is usually put into a synchronous orbit with the earth, which means that it stays above a fixed point on the surface of the earth at all times. The radius of the orbit of such a satellite is 6.5 times the radius of the earth (3950 mi). The satellite is used to relay a signal from one point on the earth to another point on the earth. The sender and receiver of a signal must be in a line of sight with the satellite, as shown in the figure. What is the maximum distance on the surface of the earth between the sender and receiver for this type of satellite?

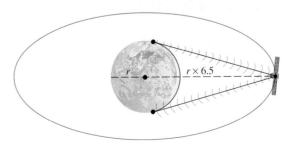

r $r \times 6.5$

■ **Figure for Exercise 60**

61. *Maximizing Area*   A builder wants to build a rectangular house on a triangular lot, as shown in the figure. Find the dimensions of the house that will maximize the area of the house.

> **HINT** Write the length in terms of the width. Then write a quadratic function for the area.

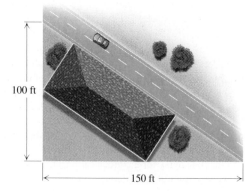

100 ft

150 ft

▪ **Figure for Exercises 61 and 62**

62. *Figuring the Setback*   The builder in the previous exercise just remembered that the house must be set back 10 ft from each side of the property and 40 ft from the street. Now find the dimensions of the house that will maximize the area of the house.

63. *Solving a Lot*   The diagonals of the property shown in the figure are 90 ft and 120 ft. The diagonals cross 30 ft from the street. Find the width w, of the property.
   HINT Write an equation involving w and solve it with a graphing calculator.

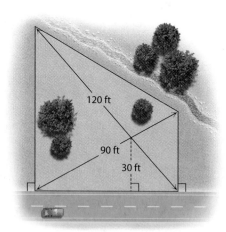

120 ft

90 ft

30 ft

▪ **Figure for Exercise 63**

64. *Hanging a Pipe*   A contractor wants to pick up a 10-ft diameter pipe with a chain of length 40 ft. The chain encircles the pipe and is attached to a hook on a crane. What is the distance between the hook and the pipe? (This is an actual problem that an engineer was asked to solve on the job.)

x ft

40 ft chain

▪ **Figure for Exercise 64**

65. *Blocking a Pipe*   A large pipe is held in place by using a 1-ft high block on one side and a 2-ft high block on the other side. If the length of the arc between the points where the pipe touches the blocks is 6 ft, then what is the radius of the pipe?

> **HINT** Ignore the thickness of the pipe.

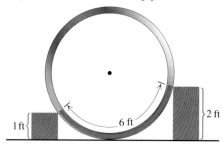

1 ft

6 ft

2 ft

▪ **Figure for Exercise 65**

66. *Pipe on an Inclined Plane*   A large pipe on an inclined plane is held in place by sticks of length 2 ft and 6 ft, as shown in the figure. The sticks are in line with the center of the pipe. The extended lines of the sticks intersect at an 18° angle at the center of the circle. What is the radius of the pipe? Find the radius of the pipe if the angle is 19°.

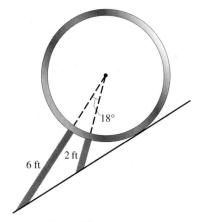

18°

6 ft

2 ft

▪ **Figure for Exercise 66**

For Writing/Discussion

67. *Cooperative Learning*   Construct a device for measuring angle of elevation or angle of depression by attaching a weighted string to a protractor, as shown in the figure. Work with a group to select an object such as a light pole or tree. Measure the angle of elevation to the top of the object and the angle of depression to the bottom of the object while standing at some appropriate distance from the object. Use these angles, the distance to the object, and the distance of your eye from the ground to find the height of the object.

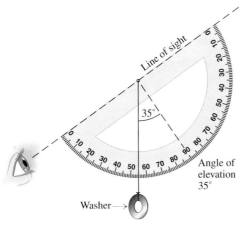

■ **Figure for Exercise 67**

68. *Cooperative Learning*   Work in a group to select a local building and find its height without going up to the building. Use the difference in the lengths of its shadow at two different times of day and the technique of Example 7. Use the protractor and string from the last exercise to measure the angle of elevation of the sun, without looking at the sun.

Thinking Outside the Box XLIX

Kicking a Field Goal   In professional football the ball must be placed between the left and right hash mark when a field goal is to be kicked. The hash marks are 9.25 ft from the center line of the field. The goal post is 30 ft past the goal line. Suppose that θ is the angle between the lines of sight from the ball to the left and right uprights as shown in the figure. The larger the value of θ the easier it is to kick the ball between the uprights. What is the difference (to the nearest thousandth of a degree) between the values of θ at the right hash mark and the value of θ at the center of the field when the ball is 60 ft from the goal line? The two vertical bars on the goal are 18.5 ft apart and the horizontal bar is 10 ft above the ground.

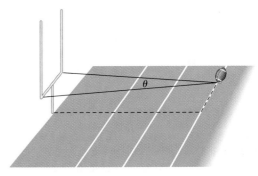

■ **Figure for Thinking Outside the Box XLIX**

5.6 Pop Quiz

1. Find $\sin \alpha$, $\cos \alpha$, and $\tan \alpha$ if the terminal side of α in standard position goes through $(-3, 4)$.

2. A right triangle has legs with lengths 3 and 6, and α is the acute angle opposite the smallest leg. Find exact values for $\sin \alpha$, $\cos \alpha$, and $\tan \alpha$.

3. At a distance of 1000 feet from a building the angle of elevation to the top of the building is 36°. Find the height of the building to the nearest foot.

▪▪▪ Highlights

5.1 Angles and Their Measurements

Degree Measure	Divide the circumference of a circle into 360 equal arcs. The degree measure is the number of degrees through which the initial side of an angle is rotated to get to the terminal side (positive for counterclockwise, negative for clockwise).	90° is 1/4 of a circle. 180° is 1/2 of a circle.
Radian Measure	The length of the arc on the unit circle through which the initial side rotates to get to the terminal side.	$\pi/2$ is 1/4 of the unit circle. π is 1/2 of the unit circle.
Converting	Use π radians $= 180°$ and cancellation of units.	$90° \cdot \dfrac{\pi \text{ rad}}{180°} = \dfrac{\pi}{2} \text{ rad}$
Arc Length	$s = \alpha r$ where s is the arc intercepted by a central angle of α radians on a circle of radius r	$\alpha = 90°, r = 10$ ft $s = \alpha r = \dfrac{\pi}{2} \cdot 10 \text{ ft} = 5\pi \text{ ft}$
Angular and Linear Velocity	A point in motion on a circle of radius r through an angle of α radians or arc length s in time t has angular velocity ω, where $\omega = \alpha/t$, and linear velocity v, where $v = s/t$.	$r = 10$ ft, $\alpha = \pi/2$ $s = 5\pi$ ft, $t = 2$ sec $w = \pi/4$ rad/sec $v = 2.5\pi$ ft/sec

5.2 The Sine and Cosine Functions

Sine and Cosine Functions	If α is an angle in standard position and (x, y) is the point of intersection of the terminal side and the unit circle, then $\sin \alpha = y$ and $\cos \alpha = x$.	$\sin(90°) = 1$ $\cos(\pi/2) = 0$ $\sin(0) = 0$
Domain	The domain for sine or cosine can be the set of angles in standard position, the measures of those angles, or the set of real numbers.	$\sin(30°) = 1/2$ $\sin(\pi/4) = \sqrt{2}/2$ $\sin(72.6) \approx -0.3367$
Fundamental Identity	For any angle α (or real number α), $\sin^2(\alpha) + \cos^2(\alpha) = 1$.	$\sin^2(30°) + \cos^2(30°) = 1$ $\sin^2(\sqrt[3]{2}) + \cos^2(\sqrt[3]{2}) = 1$
Reference Angles	To find the sine (or cosine) of an angle, first find the sine or cosine of its reference angle and prefix the appropriate sign.	$\sin(2\pi/3) = \sin(\pi/3)$ $\cos(2\pi/3) = -\cos(\pi/3)$

5.3 The Graphs of the Sine and Cosine Functions

Graphs	The graphs of $y = \sin x$ and $y = \cos x$ are sine waves, each with period 2π.	
General Sine and Cosine Functions	The graph of $y = A \sin[B(x - C)] + D$ or $y = A \cos[B(x - C)] + D$ is a sine wave with amplitude $|A|$, period $2\pi/|B|$, phase shift C, and vertical translation D.	$y = -3 \sin(2(x - \pi)) + 1$ amplitude 3, period π, phase shift π, vertical translation 1

Graphing a General Sine Function	Start with the five key points on $y = \sin x$: $$(0, 0), \left(\tfrac{\pi}{2}, 1\right), (\pi, 0), \left(\tfrac{3\pi}{2}, -1\right), (2\pi, 0)$$ Divide each x-coordinate by B and add C. Multiply each y-coordinate by A and add D. Sketch one cycle of the general function through the five new points. Graph a general cosine function in the same manner.	$y = 3\sin\!\left(2\!\left(x - \tfrac{\pi}{2}\right)\right) + 1$ Five new points: $\left(\tfrac{\pi}{2}, 1\right), \left(\tfrac{3\pi}{4}, 4\right),$ $(\pi, 1), \left(\tfrac{5\pi}{4}, -2\right), \left(\tfrac{3\pi}{2}, 1\right)$ Draw one cycle through these points.
Frequency	$F = 1/P$, P is the period and F is the frequency	$y = \sin(24\pi x)$, $F = 12$

5.4 The Other Trigonometric Functions and Their Graphs

Tangent, Cotangent, Secant, and Cosecant	If (x, y) is the point of intersection of the unit circle and the terminal side of α, then $\tan\alpha = y/x$, $\cot\alpha = x/y$, $\sec\alpha = 1/x$, and $\csc\alpha = 1/y$. If 0 occurs in a denominator the function is undefined.	$\tan(\pi/4) = 1$ $\cot(\pi/2) = 0$ $\sec(30°) = 2/\sqrt{3}$ $\csc(0)$ is undefined.
Identities	$\tan\alpha = \dfrac{\sin\alpha}{\cos\alpha}, \quad \cot\alpha = \dfrac{\cos\alpha}{\sin\alpha}, \quad \sec\alpha = \dfrac{1}{\cos\alpha}, \quad \csc\alpha = \dfrac{1}{\sin\alpha}$	
Tangent	Period π, fundamental cycle on $(-\pi/2, \pi/2)$, vertical asymptotes $x = \pi/2 + k\pi$	
Cotangent	Period π, fundamental cycle on $(0, \pi)$, vertical asymptotes $x = k\pi$	
Secant	Period 2π, vertical asymptotes $x = \pi/2 + k\pi$	
Cosecant	Period 2π, vertical asymptotes $x = k\pi$	

5.5 The Inverse Trigonometric Functions

Inverse Sine	If $y = \sin^{-1}x$ for x in $[-1, 1]$, then y is the real number in $[-\pi/2, \pi/2]$ such that $\sin y = x$.	$\sin^{-1}(1) = \pi/2$ $\sin^{-1}(-1/2) = -\pi/6$
Inverse Cosine	If $y = \cos^{-1}x$ for x in $[-1, 1]$, then y is the real number in $[0, \pi]$ such that $\cos y = x$.	$\cos^{-1}(-1) = \pi$ $\cos^{-1}(-1/2) = 2\pi/3$
Inverse Tangent	If $y = \tan^{-1}x$ for x in $(-\infty, \infty)$, then y is the real number in $(-\pi/2, \pi/2)$ such that $\tan y = x$.	$\tan^{-1}(1) = \pi/4$ $\tan^{-1}(-1) = -\pi/4$

5.6 Right Triangle Trigonometry

Trigonometric Ratios	If (x, y) is any point other than $(0, 0)$ on the terminal side of α is standard position and $r = \sqrt{x^2 + y^2}$, then $\sin\alpha = y/r$, $\cos\alpha = x/r$, and $\tan\alpha = y/x$ $(x \neq 0)$.	Terminal side of α through $(3, 4)$, gives $r = 5$ and $\sin\alpha = 4/5$, $\cos\alpha = 3/5$, and $\tan\alpha = 4/3$.
Trigonometric Functions in a Right Triangle	If α is an acute angle of a right triangle, then $\sin\alpha = \text{opp/hyp}$, $\cos\alpha = \text{adj/hyp}$, and $\tan\alpha = \text{opp/adj}$.	$\sin\alpha = 5/13$, $\cos\alpha = 12/13$, $\tan\alpha = 5/12$ 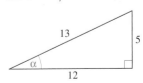

■ ■ ■ Chapter 5 Review Exercises

Find the measure in degrees of the least positive angle that is coterminal with each given angle.

1. 388°   **2.** −840°   **3.** −153°14′27″

4. 455°39′24″   **5.** −π   **6.** −35π/6

7. 13π/5   **8.** 29π/12

Convert each radian measure to degree measure. Do not use a calculator.

9. 5π/3   **10.** −3π/4   **11.** 3π/2   **12.** 5π/6

Convert each degree measure to radian measure. Do not use a calculator.

13. 330°   **14.** 405°   **15.** −300°   **16.** −210°

Fill in the tables. Do not use a calculator.

17.

θ deg	0	30	45	60	90	120	135	150	180
θ rad									
sin θ									
cos θ									
tan θ									

18.

θ rad	0	$\frac{\pi}{6}$	$\frac{\pi}{4}$	$\frac{\pi}{3}$	$\frac{\pi}{2}$	$\frac{2\pi}{3}$	$\frac{3\pi}{4}$	$\frac{5\pi}{6}$	π
θ deg									
sin θ									
cos θ									
tan θ									

Give the exact values of each of the following expressions. Do not use a calculator.

19. $\sin(-\pi/4)$ **20.** $\cos(-2\pi/3)$ **21.** $\tan(\pi/3)$

22. $\sec(\pi/6)$ **23.** $\csc(-120°)$ **24.** $\cot(135°)$

25. $\sin(180°)$ **26.** $\tan(0°)$ **27.** $\cos(3\pi/2)$

28. $\csc(5\pi/6)$ **29.** $\sec(-\pi)$ **30.** $\cot(-4\pi/3)$

31. $\cot(420°)$ **32.** $\sin(390°)$ **33.** $\cos(-135°)$

34. $\tan(225°)$ **35.** $\sec(2\pi/3)$ **36.** $\csc(-3\pi/4)$

37. $\tan(5\pi/6)$ **38.** $\sin(7\pi/6)$

For each triangle shown below, find the exact values of $\sin \alpha$, $\cos \alpha$, $\tan \alpha$, $\csc \alpha$, $\sec \alpha$, and $\cot \alpha$.

39. 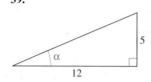   **40.**

Find an approximate value for each expression. Round to four decimal places.

41. $\sin(44°)$ **42.** $\cos(-205°)$ **43.** $\cos(4.62)$

44. $\sin(3.14)$ **45.** $\tan(\pi/17)$ **46.** $\sec(2.33)$

47. $\csc(105°4′)$ **48.** $\cot(55°3′12″)$

Find the exact value of each expression.

49. $\sin^{-1}(-0.5)$ **50.** $\cos^{-1}(-0.5)$

51. $\arctan(-1)$ **52.** $\text{arccot}(1/\sqrt{3})$

53. $\sec^{-1}(\sqrt{2})$ **54.** $\csc^{-1}(\sec(\pi/3))$

55. $\sin^{-1}(\sin(5\pi/6))$ **56.** $\cos(\cos^{-1}(-\sqrt{3}/2))$

Find the exact value of each expression in degrees.

57. $\sin^{-1}(1)$ **58.** $\tan^{-1}(1)$

59. $\arccos(-1/\sqrt{2})$ **60.** $\text{arcsec}(2)$

61. $\cot^{-1}(\sqrt{3})$ **62.** $\cot^{-1}(-\sqrt{3})$

63. $\text{arccot}(0)$ **64.** $\text{arccot}(-\sqrt{3}/3)$

Solve each right triangle with the given parts.

65. $a = 2, b = 3$ **66.** $a = 3, c = 7$

67. $a = 3.2, \alpha = 21.3°$ **68.** $\alpha = 34.6°, c = 9.4$

Sketch at least one cycle of the graph of each function and determine the period and range.

69. $f(x) = 2 \sin(3x)$ **70.** $f(x) = 1 + \cos(x + \pi/4)$

71. $y = \tan(2x + \pi)$ **72.** $y = \cot(x - \pi/4)$

73. $y = \sec\left(\frac{1}{2}x\right)$ **74.** $y = \csc\left(\frac{\pi}{2}x\right)$

75. $y = \dfrac{1}{2}\cos(2x)$

76. $y = 1 - \sin(x - \pi/3)$

77. $y = \cot(2x + \pi/3)$

78. $y = 2\tan(x + \pi/4)$

79. $y = \dfrac{1}{3}\csc(2x + \pi)$

80. $y = 1 + 2\sec(x - \pi/4)$

For each of the following sine curves find an equation of the form
$y = A\sin[B(x - C)] + D$.

81.

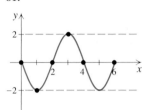

82.

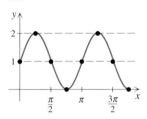

83.

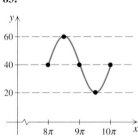

84.

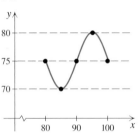

Solve each problem.

85. Find α if α is the angle between 90° and 180° whose sine is 1/2.

86. Find β if β is the angle between 180° and 270° whose cosine is $-\sqrt{3}/2$.

87. Find $\sin\alpha$, given that $\cos\alpha = 1/5$ and α is in quadrant IV.

88. Find $\tan\alpha$, given that $\sin\alpha = 1/3$ and α is in quadrant II.

89. Find x if x is the length of the shortest side of a right triangle that has a 16° angle and a hypotenuse of 24 ft.

90. Find x if x is the length of the hypotenuse of a right triangle that has a 22° angle and a shortest side of length 12 m.

91. Find α if α is the largest acute angle of a right triangle that has legs with lengths 6 cm and 8 cm.

92. Find β if β is the largest acute angle of a right triangle that has a hypotenuse with length 19 yd and a leg with length 8 yd.

93. *Broadcasting the Oldies* If radio station Q92 is broadcasting its oldies at 92.3 FM, then it is broadcasting at a frequency of 92.3 megahertz, or 92.3×10^6 cycles per second. What is the period of a wave with this frequency? (FM stands for frequency modulation.)

94. *AM Radio* If WLS in Chicago is broadcasting at 890 AM (amplitude modulation), then its signal has a frequency of 890 kilohertz, or 890×10^3 cycles per second. What is the period of a wave with this frequency?

95. *Irrigation* A center-pivot irrigation system waters a circular field by rotating about the center of the field. If the system makes one revolution in 8 hr, what is the linear velocity in feet per hour of a nozzle that is 120 ft from the center?

96. *Angular Velocity* If a bicycle with a 26-in.-diameter wheel is traveling 16 mph, then what is the angular velocity of the valve stem in radians per hour?

97. *Crooked Man* A man is standing 1000 ft from a surveyor. The surveyor measures an angle of 0.4° sighting from the man's feet to his head. Assume that the man can be represented by the arc intercepted by a central angle of 0.4° in a circle of radius 1000 ft. Find the height of the man.

98. *Straight Man* Assume that the man of Exercise 97 can be represented by the side a of a right triangle with angle $\alpha = 0.4°$ and $b = 1000$ ft. Find the height of the man, and compare your answer with the answer of the previous exercise.

99. *Oscillating Depth* The depth of water in a tank oscillates between 12 ft and 16 ft. It takes 10 min for the depth to go from 12 to 16 ft and 10 min for the depth to go from 16 to 12 ft. Express the depth as a function of time in the form $y = A\sin[B(x - C)] + D$, where the depth is 16 ft at time 0. Graph one cycle of the function.

100. *Oscillating Temperature* The temperature of the water in a tank oscillates between 100°F and 120°F. It takes 30 min for the temperature to go from 100° to 120° and 30 min for the temperature to go from 120° to 100°. Express the temperature as a function of time in the form $y = A\sin[B(x - C)] + D$, where the temperature is 100° at time 20 min. Graph one cycle of the function.

101. *Shooting a Target* Judy is standing 200 ft from a circular target with a radius of 3 in. To hit the center of the circle, she must hold the gun perfectly level, as shown in the figure. Will she hit the target if her aim is off by one-tenth of a degree in any direction?

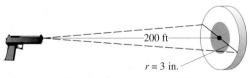

■ **Figure for Exercise 101**

102. *Height of Buildings* Two buildings are 300 ft apart. From the top of the shorter building, the angle of elevation of the top of the taller building is 23°, and the angle of depression of the base of the taller building is 36°, as shown in the figure. How tall is each building?

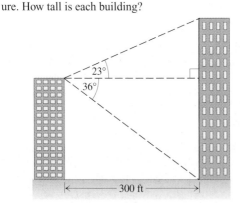

▪ **Figure for Exercise 102**

103. *Rotating Space Station* One design for a space station is a giant rotating "doughnut" where humans live and work on the outside wall, as shown in the figure. The inhabitants would have an artificial gravity equal to the normal gravity on earth when the time T in seconds for one revolution is related to the radius r in meters by the equation $T^2 g = 4\pi^2 r$. If g is acceleration of gravity (9.8 m/sec^2) and the station is built with $r = 90$ m, then what angular velocity in radians per second would be necessary to produce artificial gravity?

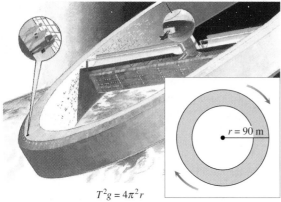

$T^2 g = 4\pi^2 r$

▪ **Figure for Exercise 103**

104. *Cloud Height* Visual Flight Rules require that the height of the clouds be more than 1000 ft for a pilot to fly without instrumentation. At night, cloud height can be determined from the ground by using a searchlight and an observer, as shown in the figure. If the beam of light is aimed straight upward and the observer 500 ft away sights the cloud with an angle of elevation of 55°, then what is the height of the cloud cover?

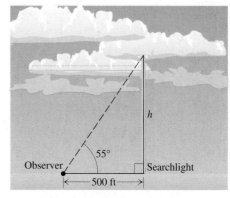

▪ **Figure for Exercise 104**

105. *Radar Range* The radar antenna on a cargo ship is located 90 feet above the water, as shown in the accompanying figure. Find the distance to the horrizon from that height. Use 3950 miles as the radius of the earth.

106. *Increasing Visibility* To increase its radar visibility, a sailboat has a radar reflector mounted on the mast, 25 feet above the water. Find the distance to the horizon from that height. Assuming that radar waves travel in a straight line, at what distance can the cargo ship of the previous exercise get a radar image of the sailboat?

Thinking Outside the Box L

Buckling Bridge A 100 ft bridge expands 1 in. during the heat of the day. Since the ends of the bridge are embedded in rock, the bridge buckles upward and forms an arc of a circle for which the original bridge is a chord. What is the approximate distance moved by the center of the bridge?

▪ **Figure for Exercises 105 and 106**

■■■ Chapter 5 Test

Find the exact value of each expression.

1. $\cos 420°$ **2.** $\sin(-390°)$ **3.** $\tan(3\pi/4)$

4. $\sec(-\pi/3)$ **5.** $\csc(7\pi/6)$ **6.** $\cot(-2\pi/3)$

7. $\sin^{-1}(-1/2)$ **8.** $\cos^{-1}(-1/2)$ **9.** $\arctan(-1)$

10. $\sec^{-1}(\sqrt{3}/2)$ **11.** $\csc^{-1}(-\sqrt{2}/2)$

12. $\sin(\cos^{-1}(-1/3))$

Sketch at least one cycle of the graph of each function. Determine the period, range, and amplitude for each function.

13. $y = \sin(3x) - 2$ **14.** $y = \cos(x + \pi/2)$

15. $y = \tan(\pi x/2)$ **16.** $y = 2\sin(2x + \pi)$

17. $y = 2\sec(x - \pi)$ **18.** $y = \csc(x - \pi/2)$

19. $y = \cot(2x)$ **20.** $y = -\cos(x - \pi/2)$

Solve each problem.

21. Find the arc length intercepted by a central angle of $46°24'6''$ in a circle with a radius of 35.62 m.

22. Find the degree measure of an angle of 2.34 radians.

23. Find the exact value of $\cos \alpha$, given that $\sin \alpha = 1/4$ and α is in quadrant II.

24. Find the exact values of all six trigonometric functions for an angle α in standard position whose terminal side contains the point $(5, -2)$.

25. If a bicycle wheel with a 26-in. diameter is making 103 revolutions per minute, then what is the angular velocity in radians per minute for a point on the tire?

26. At what speed in miles per hour will a bicycle travel if the rider can cause the 26-in.-diameter wheel to rotate 103 revolutions per minute?

27. To estimate the blood pressure of *Brachiosaurus,* Professor Ostrom wanted to estimate the height of the head of the famed *Brachiosaurus* skeleton at Humboldt University in Berlin. From a distance of 11 m from a point directly below the head, the angle of elevation of the head was approximately 48°. Use this information to find the height of the head of *Brachiosaurus.*

28. From a point on the street the angle of elevation of the top of the John Hancock Building is 65.7°. From a point on the street that is 100 ft closer to the building the angle of elevation is 70.1°. Find the height of the building.

29. The pH of a water supply oscillates between 7.2 and 7.8. It takes 2 days for the pH to go from 7.2 to 7.8 and 2 days for the pH to go from 7.8 to 7.2. Express the pH as a function of time in the form $y = A\sin[B(x - C)] + D$ where the pH is 7.2 on day 13. Graph one cycle of the function.

Tying it all Together

Chapters 1–5

Sketch the graph of each function. State the domain and range of each function.

1. $y = 2 + e^x$ **2.** $y = 2 + x^2$ **3.** $y = 2 + \sin x$ **4.** $y = 2 + \ln(x)$

5. $y = \ln(x - \pi/4)$ **6.** $y = \sin(x - \pi/4)$ **7.** $y = \log_2(2x)$ **8.** $y = \cos(2x)$

Determine whether each function is even, odd, or neither.

9. $f(x) = e^x$ **10.** $f(x) = \sin x$ **11.** $f(x) = \cos x$ **12.** $f(x) = x^4 - x^2 + 1$

13. $f(x) = \tan x$ **14.** $f(x) = \sec x$ **15.** $f(x) = \ln(x)$ **16.** $f(x) = x^3 - 3x$

Determine whether each function is increasing or decreasing on the given interval.

17. $f(x) = \sin x,\ (0, \pi/2)$ **18.** $f(x) = 2^x,\ (-\infty, \infty)$ **19.** $f(x) = \tan x,\ (-\pi/2, \pi/2)$

20. $f(x) = \cos x,\ (0, \pi)$ **21.** $f(x) = x^2,\ (-\infty, 0)$ **22.** $f(x) = \sec x,\ (0, \pi/2)$

6 Trigonometric Identities and Conditional Equations

PEOPLE have been hurling missiles of one kind or another since the ice age. Whether a ball, a rock, or a bullet, we've learned through trial and error that a projectile's path is determined by its initial velocity and the angle at which it's launched.

Athletes achieve a balance between accuracy and distance through years of practice. Operators of modern artillery achieve accuracy by calculating velocity and angle of trajectory with mathematical precision.

WHAT YOU WILL LEARN In this chapter, we will use trigonometric equations to analyze the flight of a projectile. Using trigonometry we can find the velocity and angle of trajectory required to launch a projectile a given distance. Then, using a trigonometric identity, we will find the proper angle of trajectory to achieve the maximum distance for a projectile.

Basic Identities

All of the trigonometric functions are related to each other because they are all defined in terms of the coordinates on a unit circle. For this reason, any expression involving them can be written in many different forms. Identities are used to simplify expressions and determine whether expressions are equivalent. Recall that an identity is an equation that is satisfied by *every* number for which both sides are defined. There are infinitely many trigonometric identities, but only the most common identities should be memorized. Here we review the most basic ones.

Identities from Definitions

We saw in Section 5.4 that the following identities follow directly from the unit circle definitions of tangent, cotangent, secant, and cosecant.

Identities from the Definitions

If α is any angle or real number

$$\tan \alpha = \frac{\sin \alpha}{\cos \alpha}, \quad \cot \alpha = \frac{\cos \alpha}{\sin \alpha}, \quad \sec \alpha = \frac{1}{\cos \alpha}, \quad \text{and} \quad \csc \alpha = \frac{1}{\sin \alpha},$$

provided no denominator is zero.

From these identities it follows that tangent and cotangent are reciprocals of each other, secant and cosine are reciprocals of each other, and cosecant and sine are reciprocals of each other.

Reciprocal Identities

$$\sin \alpha = \frac{1}{\csc \alpha} \qquad \cos \alpha = \frac{1}{\sec \alpha} \qquad \tan \alpha = \frac{1}{\cot \alpha}$$

$$\csc \alpha = \frac{1}{\sin \alpha} \qquad \sec \alpha = \frac{1}{\cos \alpha} \qquad \cot \alpha = \frac{1}{\tan \alpha}$$

The fundamental identity $\sin^2 x + \cos^2 x = 1$ from Section 5.2 is based on the definitions of $\sin x$ and $\cos x$ as coordinates of a point on the unit circle. If we divide each side of this identity by $\sin^2 x$, we get a new identity:

$$\frac{\sin^2 x}{\sin^2 x} + \frac{\cos^2 x}{\sin^2 x} = \frac{1}{\sin^2 x}$$

$$1 + \left(\frac{\cos x}{\sin x}\right)^2 = \left(\frac{1}{\sin x}\right)^2$$

$$1 + \cot^2 x = \csc^2 x$$

If we divide each side of the fundamental identity by $\cos^2 x$, we get another new identity:

$$\frac{\sin^2 x}{\cos^2 x} + \frac{\cos^2 x}{\cos^2 x} = \frac{1}{\cos^2 x}$$

$$\tan^2 x + 1 = \sec^2 x$$

Since these three identities come from the Pythagorean theorem, they are called the **Pythagorean identities.**

Pythagorean Identities

$$\sin^2 x + \cos^2 x = 1 \qquad 1 + \cot^2 x = \csc^2 x \qquad \tan^2 x + 1 = \sec^2 x$$

We can support our conclusion that an equation is an identity by examining a graph. For example, consider the graph of $y = \sin^2 x + \cos^2 x$ in Fig. 6.1. The fact that $\sin^2 x + \cos^2 x = 1$ for 95 values of x supports the conclusion that $\sin^2 x + \cos^2 x = 1$ for all real numbers. Because a calculator evaluates a function at a finite number of values only, a calculator graph does not prove that an equation is an identity. Note that $\sin x = 1$ is satisfied for 95 values of x in Fig. 6.2 but it is certainly not an identity. □

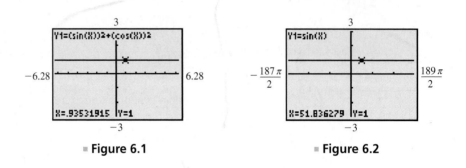

■ **Figure 6.1**   ■ **Figure 6.2**

Using Identities

Identities are used in many ways. One way is to simplify expressions involving trigonometric functions. In the next example we follow the strategy of writing each expression in terms of sines and cosines only and then simplifying the expression.

Example **1** Using identities to simplify

Write each expression in terms of sines and/or cosines, and then simplify.

a. $\dfrac{\tan x}{\sec x}$ **b.** $\sin x + \cot x \cos x$

Solution

a. Rewrite each trigonometric function in terms of sine and cosine:

$$\frac{\tan x}{\sec x} = \frac{\dfrac{\sin x}{\cos x}}{\dfrac{1}{\cos x}} = \frac{\sin x}{\cos x} \cdot \frac{\cos x}{1} \qquad \text{Invert and multiply.}$$

$$= \sin x$$

b. $\sin x + \cot x \cos x = \sin x + \dfrac{\cos x}{\sin x} \cdot \cos x$ Rewrite using sines and cosines.

$$= \sin x + \dfrac{\cos^2 x}{\sin x}$$

$$= \dfrac{\sin^2 x}{\sin x} + \dfrac{\cos^2 x}{\sin x}$$ Multiply $\sin x$ by $\sin x / \sin x$.

$$= \dfrac{\sin^2 x + \cos^2 x}{\sin x}$$

$$= \dfrac{1}{\sin x}$$ Since $\sin^2 x + \cos^2 x = 1$

$$= \csc x$$ Use a basic identity.

Try This. Simplify $\dfrac{\tan x \csc x}{\sec x}$.   ■

Using identities we can write any one of the six trigonometric functions in terms of any other. For example, the identity

$$\sin x = \dfrac{1}{\csc x}$$

expresses the sine function in terms of the cosecant. To write sine in terms of cosine, we rewrite the identity $\sin^2 x + \cos^2 x = 1$:

$$\sin^2 x = 1 - \cos^2 x$$

$$\sin x = \pm \sqrt{1 - \cos^2 x}$$

The $\pm$ symbol in this identity means that if $\sin x > 0$, then the identity is $\sin x = \sqrt{1 - \cos^2 x}$, and if $\sin x < 0$, then the identity is $\sin x = -\sqrt{1 - \cos^2 x}$. Note that the graph of $y = \sqrt{1 - (\cos x)^2}$ in Fig. 6.3 is the same as the graph of $y = \sin x$ only when $\sin x > 0$.   □

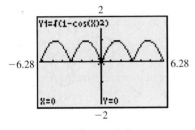

■ **Figure 6.3**

Example **2** **Writing one function in terms of another**

Write an identity that expresses the tangent function in terms of the sine function.

Solution

The Pythagorean identity $\sin^2 x + \cos^2 x = 1$ yields $\cos x = \pm \sqrt{1 - \sin^2 x}$. Since $\tan x = \sin x / \cos x$, we can write

$$\tan x = \pm \dfrac{\sin x}{\sqrt{1 - \sin^2 x}}.$$

We use the positive or negative sign, depending on the value of x.

Try This. Write an identity that expresses the cotangent function in terms of the sine function.   ■

Because all of the trigonometric functions are related by identities, we can find their values for an angle if we know the value of any one of them and the quadrant of the angle.

Example **3** **Using identities to find function values**

Given that $\tan \alpha = -2/3$ and α is in quadrant IV, find the values of the remaining five trigonometric functions at α by using identities.

Solution

Use $\tan \alpha = -2/3$ in the identity $\sec^2 \alpha = 1 + \tan^2 \alpha$:

$$\sec^2 \alpha = 1 + \left(-\frac{2}{3}\right)^2 = \frac{13}{9}$$

$$\sec \alpha = \pm \frac{\sqrt{13}}{3}$$

Since α is in quadrant IV, $\cos \alpha > 0$ and $\sec \alpha > 0$. So $\sec \alpha = \sqrt{13}/3$. Since cosine is the reciprocal of the secant and cotangent is the reciprocal of tangent,

$$\cos \alpha = \frac{3}{\sqrt{13}} = \frac{3\sqrt{13}}{13} \qquad \text{and} \qquad \cot \alpha = -\frac{3}{2}.$$

Since $\sin \alpha$ is negative in quadrant IV, $\sin \alpha = -\sqrt{1 - \cos^2 \alpha}$:

$$\sin \alpha = -\sqrt{1 - \frac{9}{13}} = -\sqrt{\frac{4}{13}} = -\frac{2\sqrt{13}}{13}$$

Since cosecant is the reciprocal of sine, $\csc \alpha = -\sqrt{13}/2$.

Try This. Suppose that $\cot \alpha = -1/3$ and α is in quadrant II. Find $\sin \alpha$ and $\cos \alpha$. ■

In the next example we use the Pythagorean identities to rewrite a composition of trigonometric functions as an algebraic function.

Example **4** **Converting compositions to algebraic functions**

Find an equivalent algebraic expression for $\sin(\arctan(x))$.

Solution

Let $\theta = \arctan(x)$. Since $\tan(\theta) = x$ and $\tan^2(\theta) + 1 = \sec^2(\theta)$, we have $\sec^2(\theta) = x^2 + 1$ and

$$\cos^2(\theta) = \frac{1}{x^2 + 1}.$$

Since $\sin^2(\theta) = 1 - \cos^2(\theta)$, we have

$$\sin^2(\theta) = 1 - \frac{1}{x^2 + 1} = \frac{x^2 + 1}{x^2 + 1} - \frac{1}{x^2 + 1} = \frac{x^2}{x^2 + 1}$$

and

$$\sin(\theta) = \pm\sqrt{\frac{x^2}{x^2 + 1}} = \frac{x}{\sqrt{x^2 + 1}}.$$

We eliminated the radical and the $\pm$ sign in the last equation because the sign of x is the same as the sign of $\sin(\theta)$. Specifically, if $x > 0$, then $0 < \theta < \frac{\pi}{2}$ and $\sin(\theta) > 0$. If $x < 0$, then $-\frac{\pi}{2} < \theta < 0$ and $\sin(\theta) < 0$. So

$$\sin(\arctan(x)) = \frac{x}{\sqrt{x^2 + 1}}.$$

Try This. Find an equivalent algebraic expression for $\sec(\arccos(x))$.   ▪

The identities discussed so far give relationships between the different trigonometric functions. The next identities show how the value of each trigonometric function for $-x$ is related to its value for x.

Odd and Even Identities

In Section 2.3 we defined an **odd function** as one for which $f(-x) = -f(x)$ and an **even function** as one for which $f(-x) = f(x)$. In algebra most of the odd functions have odd exponents and the even functions have even exponents. We can also classify each of the trigonometric functions as either odd or even, but we cannot make the determination based on exponents. Instead, we examine the definitions of the functions in terms of the unit circle.

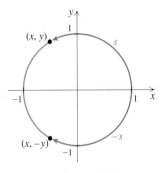

▪ **Figure 6.4**

Figure 6.4 shows the real numbers s and $-s$ as arcs on the unit circle. If the terminal point of s is (x, y), then the terminal point of $-s$ is $(x, -y)$. The y-coordinate for the terminal point of $-s$ is the opposite of the y-coordinate for the terminal point of s. So $\sin(-s) = -\sin(s)$, and sine is an odd function. Since the x-coordinates for s and $-s$ are equal, $\cos(-s) = \cos(s)$, and cosine is an even function. Because sine is an odd function, cosecant is also an odd function:

$$\csc(-s) = \frac{1}{\sin(-s)} = \frac{1}{-\sin(s)} = -\csc(s)$$

Likewise we can establish that secant is an even function and tangent and cotangent are both odd functions. These arguments establish the following identities.

Odd and Even Identities

Odd:	$\sin(-x) = -\sin(x)$	$\csc(-x) = -\csc(x)$
	$\tan(-x) = -\tan(x)$	$\cot(-x) = -\cot(x)$
Even:	$\cos(-x) = \cos(x)$	$\sec(-x) = \sec(x)$

To help you remember which of the trigonometric functions are odd and which are even, remember that the graph of an odd function is symmetric about the origin and the graph of an even function is symmetric about the y-axis. The graphs of $y = \sin x$, $y = \csc x$, $y = \tan x$, and $y = \cot x$ are symmetric about the origin, while $y = \cos x$ and $y = \sec x$ are symmetric about the y-axis.

Example **5** **Using odd and even identities**

Simplify each expression.

a. $\sin(-x)\cot(-x)$ **b.** $\dfrac{1}{1 + \cos(-x)} + \dfrac{1}{1 - \cos x}$

Solution

a. $\sin(-x)\cot(-x) = (-\sin x)(-\cot x)$

$$= \sin x \cdot \cot x = \sin x \cdot \frac{\cos x}{\sin x} = \cos x$$

b. To add these expressions, we need a common denominator.

$$\frac{1}{1 + \cos(-x)} + \frac{1}{1 - \cos x} = \frac{1}{1 + \cos x} + \frac{1}{1 - \cos x} \quad \text{Because } \cos(-x) = \cos x$$

$$= \frac{1(1 - \cos x)}{(1 + \cos x)(1 - \cos x)} + \frac{1(1 + \cos x)}{(1 - \cos x)(1 + \cos x)}$$

$$= \frac{1 - \cos x}{1 - \cos^2 x} + \frac{1 + \cos x}{1 - \cos^2 x}$$

$$= \frac{1 - \cos x}{\sin^2 x} + \frac{1 + \cos x}{\sin^2 x} \quad \text{Pythagorean identity}$$

$$= \frac{1 - \cos x + 1 + \cos x}{\sin^2 x} = \frac{2}{\sin^2 x}$$

$$= 2 \cdot \frac{1}{\sin^2 x} = 2\csc^2 x$$

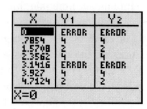

We can check this result by making a table for $y_1 = 1/(1 + \cos(-x)) + 1/(1 - \cos(x))$ and $y_2 = 2/(\sin(x))^2$, as shown in Fig. 6.5. The identical values for y_1 and y_2 support the result.

■ **Figure 6.5**

Try This. Simplify $\csc(-x)\tan(-x)$.   ■

Example **6** **Odd or even functions**

Determine whether each function is odd, even, or neither.

a. $f(x) = \dfrac{\cos(2x)}{x}$ **b.** $g(t) = \sin t + \cos t$

Solution

a. We replace x by $-x$ and see whether $f(-x)$ is equal to $f(x)$ or $-f(x)$:

$$f(-x) = \frac{\cos(2(-x))}{-x} \quad \text{Replace } x \text{ by } -x.$$

$$= \frac{\cos(-2x)}{-x}$$

$$= \frac{\cos(2x)}{-x} \quad \text{Since cosine is an even function}$$

$$= -\frac{\cos(2x)}{x}$$

Since $f(-x) = -f(x)$, the function f is an odd function.

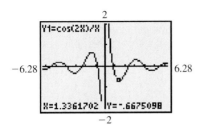

■ **Figure 6.6**

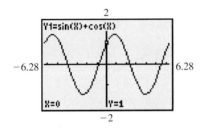

■ **Figure 6.7**

The graph of $y = \cos(2x)/x$ in Fig. 6.6 appears to be symmetric about the origin, which supports the conclusion that the function is odd.   □

b. $g(-t) = \sin(-t) + \cos(-t)$ Replace t by $-t$.

$\qquad\quad = -\sin t + \cos t$ Sine is odd and cosine is even.

Since $g(-t) \neq g(t)$ and $g(-t) \neq -g(t)$, the function is neither odd nor even.

Since the graph of $y = \sin(x) + \cos(x)$ in Fig. 6.7 is neither symmetric about the origin nor symmetric about the y-axis, the graph supports the conclusion that the function is neither odd nor even.

Try This. Determine whether $f(x) = \csc(x) + \tan(x)$ is an odd or even function.   ■

Identity or Not?

In this section we have seen the basic identities, the reciprocal identities, the Pythagorean identities, and the odd-even identities and we are just getting started. With so many identities in trigonometry it might seem difficult to determine whether a given equation is an identity. Remember that an identity is satisfied by every value of the variable for which both sides are defined. If you can find one number for which the left-hand side of the equation has a value different from the right-hand side, then the equation is not an identity.

Example **7** **Proving that an equation is not an identity**

Show that $\sin(2t) = 2 \sin(t)$ is not an identity.

Solution

If $t = \pi/4$, then

$$\sin(2t) = \sin\left(2 \cdot \frac{\pi}{4}\right) = \sin\left(\frac{\pi}{2}\right) = 1$$

and

$$2 \sin(t) = 2 \sin\left(\frac{\pi}{4}\right) = 2 \cdot \frac{\sqrt{2}}{2} = \sqrt{2}.$$

Since the values of $\sin(2t)$ and $2 \sin(t)$ are unequal for $t = \pi/4$, the equation is not an identity.

The graphs of $y_1 = \sin(2x)$ and $y_2 = 2 \sin(x)$ in Fig. 6.8 support the conclusion that $\sin(2t) = 2 \sin(t)$ is not an identity.

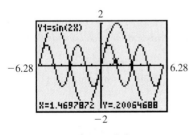

■ **Figure 6.8**

Try This. Show that $\cos(3t) = 3 \cos(t)$ is not an identity.   ■

An equation that is not an identity may have many solutions. So you might have to try more than one value for the variable to find one that fails to satisfy the equation. Section 6.2 will be devoted to proving that given equations are identities.

For Thought

True or False? Explain.

1. The equation $\sin x = \cos x$ is an identity.

2. If we simplify the expression $(\tan x)(\cot x)$, we get 1.

3. The function $f(x) = \sin^2 x$ is an even function.

4. The function $f(x) = \cos^3 x$ is an odd function.

5. If $\sin(2x) = 2\sin(x)\cos(x)$ is an identity, then so is
$\dfrac{\sin(2x)}{\sin(x)} = 2\cos(x)$.

6. The equation $(\sin x + \cos x)^2 = \sin^2 x + \cos^2 x$ is an identity.

7. Tangent is written in terms of secant as
$\tan x = \pm\sqrt{1 - \sec^2 x}$.

8. $\sin(-3)\cos(-3)\tan(-3)\sec(3)\csc(-3)\cot(3) = -1$

9. $\sin^2(-\pi/9) + \cos^2(-\pi/9) = -1$

10. $(1 - \sin(\pi/7))(1 + \sin(\pi/7)) = \cos^2(-\pi/7)$

6.1 Exercises

Write each expression in terms of sines and/or cosines, and then simplify.

1. $\tan x \cos x$

2. $\sin x \cot x$

3. $\sec x \cos x$

4. $\sin x \csc x$

5. $\dfrac{\sec x}{\tan x}$

6. $\dfrac{\cot x}{\csc x}$

7. $\dfrac{\sin x}{\csc x} + \cos^2 x$

8. $\dfrac{\cos x}{\sec x} + \sin^2 x$

9. $\dfrac{\sin x}{\csc x} + \dfrac{\cos x}{\sec x}$

10. $\dfrac{1}{\sin^2 x} - \dfrac{1}{\tan^2 x}$

11. $(1 - \sin\alpha)(1 + \sin\alpha)$

12. $(\sec\alpha - 1)(\sec\alpha + 1)$

13. $(\cos\beta\tan\beta + 1)(\sin\beta - 1)$

14. $(1 + \cos\beta)(1 - \cot\beta\sin\beta)$

15. $\dfrac{1 + \cos\alpha\tan\alpha\csc\alpha}{\csc\alpha}$

16. $\dfrac{(\cos\alpha\tan\alpha + 1)(\sin\alpha - 1)}{\cos^2\alpha}$

Write an identity that expresses the first function in terms of the second.

17. $\cot(x)$, in terms of $\csc(x)$

18. $\sec(x)$, in terms of $\tan(x)$

19. $\sin(x)$, in terms of $\cot(x)$

20. $\cos(x)$, in terms of $\tan(x)$

21. $\tan(x)$, in terms of $\csc(x)$

22. $\cot(x)$, in terms of $\sec(x)$

In each exercise, use identities to find the exact values at α for the remaining five trigonometric functions.

23. $\tan\alpha = 1/2$ and $0 < \alpha < \pi/2$

24. $\sin\alpha = 3/4$ and $\pi/2 < \alpha < \pi$

25. $\cos\alpha = -\sqrt{3}/5$ and α is in quadrant III

26. $\sec\alpha = -4\sqrt{5}/5$ and α is in quadrant II

27. $\cot\alpha = -1/3$ and $-\pi/2 < \alpha < 0$

28. $\csc\alpha = \sqrt{3}$ and $0 < \alpha < \pi/2$

Find an equivalent algebraic expression for each composition.

29. $\sin(\arccos(x))$

30. $\cos(\arcsin(x))$

31. $\cos(\arctan(x))$

32. $\tan(\arccos(x))$

33. $\tan(\arcsin(x))$

34. $\sec(\arcsin(x))$

35. $\sec(\arctan(x))$

36. $\csc(\arcsin(x))$

Simplify each expression.

37. $\sin(-x)\cot(-x)$

38. $\sec(-x) - \sec(x)$

39. $\sin(y) + \sin(-y)$

40. $\cos(y) + \cos(-y)$

41. $\dfrac{\sin(x)}{\cos(-x)} + \dfrac{\sin(-x)}{\cos(x)}$

42. $\dfrac{\cos(-x)}{\sin(-x)} - \dfrac{\cos(-x)}{\sin(x)}$

43. $(1 + \sin(\alpha))(1 + \sin(-\alpha))$

44. $(1 - \cos(-\alpha))(1 + \cos(\alpha))$

45. $\sin(-\beta)\cos(-\beta)\csc(\beta)$

46. $\tan(-\beta)\csc(-\beta)\cos(\beta)$

Determine whether each function is odd, even, or neither.

47. $f(x) = \sin(2x)$ **48.** $f(x) = \cos(2x)$

49. $f(x) = \cos x + \sin x$ **50.** $f(x) = 2\sin x \cos x$

51. $f(t) = \sec^2(t) - 1$ **52.** $f(t) = 2 + \tan(t)$

53. $f(\alpha) = 1 + \sec \alpha$ **54.** $f(\beta) = 1 + \csc \beta$

55. $f(x) = \dfrac{\sin x}{x}$ **56.** $f(x) = x \cos x$

57. $f(x) = x + \sin x$ **58.** $f(x) = \csc(x^2)$

Match each given expression with an equivalent expression (a)–(p).

59. $\sin x$ **60.** $\cos x$ **61.** $\tan x$

62. $\cot x$ **63.** $\sec x$ **64.** $\csc x$

65. $\sin^2 x$ **66.** $\cos^2 x$ **67.** $\tan^2 x$

68. $\sin(-x)$ **69.** $\cos(-x)$ **70.** $\tan(-x)$

71. $\cot(-x)$ **72.** $\sec(-x)$ **73.** $\csc(-x)$

74. $\cot^2 x$ **75.** $\sec^2 x$ **76.** $\csc^2 x$

  a. $-\sin x$ **b.** $-\cot x$ **c.** $-\tan x$

  d. $1 + \tan^2 x$ **e.** $1 + \cot^2 x$ **f.** $-\csc x$

  g. $\dfrac{1}{\sec x}$ **h.** $\dfrac{1}{\csc x}$ **i.** $\dfrac{1}{\sin x}$

  j. $1 - \sin^2 x$ **k.** $1 - \cos^2 x$ **l.** $\sec^2(x) - 1$

  m. $\dfrac{1}{\cos x}$ **n.** $\dfrac{\sin x}{\cos x}$ **o.** $\dfrac{\cos x}{\sin x}$

  p. $\csc^2(x) - 1$

Show that each equation is not an identity. Write your explanation in paragraph form.

77. $(\sin \gamma + \cos \gamma)^2 = \sin^2 \gamma + \cos^2 \gamma$

78. $\tan^2 x - 1 = \sec^2 x$

79. $(1 + \sin \beta)^2 = 1 + \sin^2 \beta$

80. $\sin(2\alpha) = \sin \alpha \cos \alpha$

81. $\sin \alpha = \sqrt{1 - \cos^2 \alpha}$ **82.** $\tan \alpha = \sqrt{\sec^2 \alpha - 1}$

83. $\sin(y) = \sin(-y)$ **84.** $\cos(-y) = -\cos y$

85. $\cos^2 y - \sin^2 y = \sin(2y)$ **86.** $\cos(2x) = 2 \cos x \sin x$

Use identities to simplify each expression.

87. $1 - \dfrac{1}{\cos^2 x}$ **88.** $\dfrac{\sin^4 x - \sin^2 x}{\sec x}$

89. $\dfrac{-\tan^2 t - 1}{\sec^2 t}$ **90.** $\dfrac{\cos w \sin^2 w + \cos^3 w}{\sec w}$

91. $\dfrac{\sin^2 \alpha - \cos^2 \alpha}{1 - 2\cos^2 \alpha}$ **92.** $\dfrac{\sin^3(-\theta)}{\sin^3 \theta - \sin \theta}$

93. $\dfrac{\tan^3 x - \sec^2 x \tan x}{\cot(-x)}$ **94.** $\sin x + \dfrac{\cos^2 x}{\sin x}$

95. $\dfrac{1}{\sin^3 x} - \dfrac{\cot^2 x}{\sin x}$ **96.** $1 - \dfrac{\sec^2 x}{\tan^2 x}$

97. $\sin^4 x - \cos^4 x$ **98.** $\csc^4 x - \cot^4 x$

Solve each problem.

 99. Suppose $\sin \theta = 1/3$. What are the possible values for $\cos \theta$?

100. Suppose $\cos \theta = 4/5$. What are the possible values for $\tan \theta$?

101. Suppose $\sin \theta = u$. Write $\cos \theta$ in terms of u.

102. Suppose $\cos \theta = u$. Write $\tan \theta$ in terms of u.

103. For what values of x is the identity $\tan x = \dfrac{\sin x}{\cos x}$ *not* valid?

104. For what values of x is the identity $\cot x = \dfrac{\cos x}{\sin x}$ *not* valid?

For Writing/Discussion

105. *Algebraic Identities*   List as many algebraic identities as you can and explain why each one is an identity.

106. *Trigonometric Identities*   List as many trigonometric identities as you can and explain why each one is an identity.

Thinking Outside the Box LI

Tangent Circles The three large circles in the accompanying diagram are tangent to each other and each has radius 1. The small circle in the middle is tangent to each of the three large circles. Find its radius.

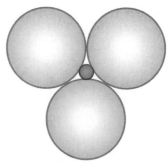

■ **Figure for Thinking Outside the Box LI**

6.1 Pop Quiz

1. Simplify the expression $\cot x \sec x$.

2. Find exact values for $\cos \alpha$ and $\cot \alpha$ if $\sin \alpha = 1/3$ and $0 < \alpha < \pi/2$.

3. Is $f(x) = \cos(3x)$ even or odd?

4. Write an equivalent algebraic expression for $\cos(\arcsin(w))$.

5. Simplify $\dfrac{2}{\sec^2 \alpha} + \dfrac{2}{\csc^2 \alpha}$.

Linking Concepts

For Individual or Group Explorations

Modeling the Motion of a Projectile

A projectile is fired with initial velocity of v_0 feet per second. The projectile can be pictured as being fired from the origin into the first quadrant, making an angle θ with the positive x-axis, as shown in the figure. If there is no air resistance, then at t seconds the coordinates of the projectile (in feet) are $x = v_0 t \cos \theta$ and $y = -16t^2 + v_0 t \sin \theta$. Suppose a projectile leaves the gun at 100 ft/sec and $\theta = 60°$.

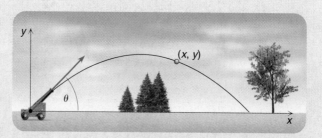

a) What are the coordinates of the projectile at time $t = 4$ sec?

b) For how many seconds is the projectile in the air?

c) How far from the gun does the projectile land?

d) What is the maximum height attained by the projectile?

e) Find expressions in terms of v_0 and θ for the time in the air, the distance from the gun, and the maximum height.

f) Show that $y = -\dfrac{16 \sec^2 \theta}{v_0^2} x^2 + x \tan \theta$.

Verifying Identities

In Section 6.1 we saw that any trigonometric function can be expressed in terms of any other trigonometric function. So an expression involving one or more trigonometric functions could be written in many different equivalent forms. In trigonometry and calculus, where trigonometric functions are used, we must often decide whether two expressions are equivalent. The fact that two expressions are equivalent is expressed as an identity. In this section we concentrate on techniques for verifying that a given equation is an identity.

Developing a Strategy

In simplifying expressions in Section 6.1, we used known identities and properties of algebra to rewrite a complicated expression as a simpler equivalent expression. It may not always be clear that one expression is simpler than another, but an expression that involves fewer symbols is generally considered simpler. To prove or verify that an equation is an identity, we start with the expression on one side of the equation and use known identities and properties of algebra to convert it into the expression on the other side of the equation. It is the same process as simplifying expressions, except that in this case we know exactly what we want as the final expression. If one side appears to be more complicated than the other, we usually start with the more complicated side and simplify it.

Example **1** Simplifying the complicated side

Verify that the following equation is an identity

$$1 + \sec x \sin x \tan x = \sec^2 x$$

Solution

We start with the left-hand side, the more complicated side, and write it in terms of sine and cosine. Our goal is to get $\sec^2 x$ as the final simplified expression.

$$1 + \sec x \sin x \tan x = 1 + \frac{1}{\cos x} \sin x \frac{\sin x}{\cos x}$$

$$= 1 + \frac{\sin^2 x}{\cos^2 x}$$

$$= 1 + \tan^2 x$$

$$= \sec^2 x \qquad \text{Pythagorean identity}$$

Try This. Verify that $1 - \sec(x)\csc(x)\tan(x) = -\tan^2(x)$ is an identity. ■

In verifying identities it is often necessary to multiply binomials or factor trinomials involving trigonometric functions.

Example **2** **Multiplying binomials**

Find each product.

a. $(1 + \tan x)(1 - \tan x)$ **b.** $(2 \sin \alpha + 1)^2$

Solution

a. The product of a sum and a difference is equal to the difference of two squares:

$$(1 + \tan x)(1 - \tan x) = 1 - \tan^2 x$$

b. Use $(a + b)^2 = a^2 + 2ab + b^2$ with $a = 2 \sin \alpha$ and $b = 1$:

$$(2 \sin \alpha + 1)^2 = 4 \sin^2 \alpha + 4 \sin \alpha + 1$$

Try This. Find the product $(1 - 2 \sin x)(1 + 2 \sin x)$. ▪

Example **3** **Factoring with trigonometric functions**

Factor each expression.

a. $\sec^2 x - \tan^2 x$ **b.** $\sin^2 \beta + \sin \beta - 2$

Solution

a. A difference of two squares is equal to a product of a sum and a difference:

$$\sec^2 x - \tan^2 x = (\sec x + \tan x)(\sec x - \tan x)$$

b. Factor this expression as you would factor $x^2 + x - 2 = (x + 2)(x - 1)$:

$$\sin^2 \beta + \sin \beta - 2 = (\sin \beta + 2)(\sin \beta - 1)$$

Try This. Factor $\sin^2 x + 4 \sin x + 4$. ▪

The equations that appear in Examples 2 and 3 are identities. They are not the kind of identities that are memorized, because they can be obtained at any time by the principles of multiplying or factoring. The next example uses the identity $(1 - \sin \alpha)(1 + \sin \alpha) = 1 - \sin^2 \alpha$.

Example **4** **An identity with equal fractions**

Prove that the following equation is an identity:

$$\frac{\cos \alpha}{1 - \sin \alpha} = \frac{1 + \sin \alpha}{\cos \alpha}$$

Solution

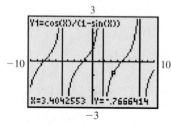 Before attempting to prove an identity, it is a good idea to check the calculator graphs of the opposite sides. The graphs of $y_1 = \cos(x)/(1 - \sin(x))$ and $y_2 = (1 + \sin(x))/\cos(x)$ in Fig. 6.9 appear to coincide. So the equation is probably an identity. □

▪ **Figure 6.9**

To prove that the equation is an identity, multiply the numerator and denominator of the left-hand side by $1 + \sin \alpha$, because $1 + \sin \alpha$ appears in the numerator of the right-hand side.

$$\frac{\cos \alpha}{1 - \sin \alpha} = \frac{\cos \alpha (1 + \sin \alpha)}{(1 - \sin \alpha)(1 + \sin \alpha)}$$

$$= \frac{\cos \alpha (1 + \sin \alpha)}{1 - \sin^2 \alpha} \qquad \text{Leave numerator in factored form.}$$

$$= \frac{\cos \alpha (1 + \sin \alpha)}{\cos^2 \alpha} \qquad \text{Pythagorean identity}$$

$$= \frac{1 + \sin \alpha}{\cos \alpha} \qquad \text{Reduce.}$$

Try This. Prove that $\dfrac{\csc x - 1}{\cot x} = \dfrac{\cot x}{\cot x + 1}$ is an identity.   ■

In Example 4 the numerator and denominator were multiplied by $1 + \sin \alpha$ because the expression on the right-hand side has $1 + \sin \alpha$ in its numerator. We could have multiplied the numerator and denominator of the left-hand side by $\cos \alpha$, because the expression on the right-hand side has $\cos \alpha$ in its denominator. You should prove that the equation of Example 4 is an identity by multiplying the numerator and denominator by $\cos \alpha$.

Another strategy to use when a fraction has a sum or difference in its numerator is to write the fraction as a sum or difference of two fractions, as in the next example.

Example 5 Writing one fraction as two fractions

Prove that the following equation is an identity:

$$\frac{\csc x - \sin x}{\sin x} = \cot^2 x$$

Solution

First rewrite the left-hand side as a difference of two rational expressions.

$$\frac{\csc x - \sin x}{\sin x} = \frac{\csc x}{\sin x} - \frac{\sin x}{\sin x}$$

$$= \csc x \cdot \frac{1}{\sin x} - 1$$

$$= \csc^2 x - 1 \qquad \text{Since } 1/\sin x = \csc x$$

$$= \cot^2 x \qquad \text{Pythagorean identity}$$

Try This. Prove that $\dfrac{\sec x - \cos x}{\cos x} = \tan^2 x$ is an identity.   ■

In verifying an identity, you can start with the expression on either side of the equation and use known identities and properties of algebra to get the expression on the other side. Note that we do *not* start with the given equation and work on it to get simpler and simpler equations as we do when solving an equation, because we do not yet know that the given equation is actually a correct identity. So when an expression involves the sum or difference of fractions, as in the next example, get a common denominator and combine the fractions. Do *not* multiply each side by the LCD as is done when solving an equation involving fractions.

Example 6 Starting with the right-hand side

Prove that the following equation is an identity:

$$2 \tan^2 x = \frac{1}{\csc x - 1} - \frac{1}{\csc x + 1}$$

Solution

In this equation, the right-hand side is the more complicated one. We will simplify the right-hand side by getting a common denominator and combining the fractions, keeping $2 \tan^2 x$ in mind as our goal.

$$\frac{1}{\csc x - 1} - \frac{1}{\csc x + 1} = \frac{1(\csc x + 1)}{(\csc x - 1)(\csc x + 1)} - \frac{1(\csc x - 1)}{(\csc x + 1)(\csc x - 1)}$$

$$= \frac{\csc x + 1}{\csc^2 x - 1} - \frac{\csc x - 1}{\csc^2 x - 1}$$

$$= \frac{2}{\csc^2 x - 1} \qquad \text{Subtract the numerators.}$$

$$= \frac{2}{\cot^2 x} \qquad \text{Pythagorean identity}$$

$$= 2 \cdot \frac{1}{\cot^2 x} = 2 \tan^2 x$$

Try This. Prove that $-2 \cot^2 x = \frac{1}{1 - \sec x} + \frac{1}{1 + \sec x}$ is an identity. ▪

In every example so far, we started with one side and converted it into the other side. Although all identities can be verified by that method, sometimes it is simpler to convert both sides into the same expression. If both sides are shown to be equivalent to a common expression, then the identity is certainly verified. This technique works best when both sides are rather complicated and we simply set out to simplify them.

Example 7 Working on both sides

Prove that the following equation is an identity:

$$\frac{1 - \sin^2 t}{1 - \csc(-t)} = \frac{1 + \sin(-t)}{\csc t}$$

Solution

Use $\csc(-t) = -\csc t$ to simplify the left-hand side:

$$\frac{1 - \sin^2 t}{1 - \csc(-t)} = \frac{1 - \sin^2 t}{1 + \csc t} = \frac{\cos^2 t}{1 + \csc t}$$

Now look for a way to get the right-hand side of the original equation equivalent to

$\cos^2 t/(1 + \csc t)$. Since $\sin(-t) = -\sin t$, we can write $1 + \sin(-t)$ as $1 - \sin t$ and then get $\cos^2 t$ in the numerator by multiplying by $1 + \sin t$.

$$\frac{1 + \sin(-t)}{\csc t} = \frac{1 - \sin t}{\csc t} = \frac{(1 - \sin t)(1 + \sin t)}{\csc t(1 + \sin t)}$$

$$= \frac{1 - \sin^2 t}{\csc t + \csc t \sin t} = \frac{\cos^2 t}{\csc t + \dfrac{1}{\sin t} \sin t}$$

$$= \frac{\cos^2 t}{1 + \csc t}$$

Since both sides of the equation are equivalent to the same expression, the equation is an identity.

Try This. Prove that $\dfrac{1 - \cos^2(-t)}{\sin(-t)} = \tan(-t)\cos(-t)$ is an identity. ■

The Strategy

Verifying identities takes practice. There are many different ways to verify a particular identity, so just go ahead and get started. If you do not seem to be getting anywhere, try another approach. The strategy that we used in the examples is stated below. Other techniques are sometimes needed to prove identities, but you can prove most of the identities in this text using the following strategy.

STRATEGY   Verifying Identities

1. Work on one side of the equation (usually the more complicated side), keeping in mind the expression on the other side as your goal.

2. Some expressions can be simplified quickly if they are rewritten in terms of sines and cosines only. (See Example 1.)

3. To convert one rational expression into another, multiply the numerator and denominator of the first by either the numerator or the denominator of the desired expression. (See Example 4.)

4. If the numerator of a rational expression is a sum or difference, convert the rational expression into a sum or difference of two rational expressions. (See Example 5.)

5. If a sum or difference of two rational expressions occurs on one side of the equation, then find a common denominator and combine them into one rational expression. (See Example 6.)

For Thought

True or False?

Answer true if the equation is an identity and false if it is not. If false, find a value of x for which the two sides have different values.

1. $\dfrac{\sin x}{\csc x} = \sin^2 x$

2. $\dfrac{\cot x}{\tan x} = \tan^2 x$

3. $\dfrac{\sec x}{\csc x} = \tan x$

4. $\sin x \sec x = \tan x$

5. $\dfrac{\cos x + \sin x}{\cos x} = 1 + \tan x$

6. $\sec x + \dfrac{\sin x}{\cos x} = \dfrac{1 + \sin x \cos x}{\cos x}$

7. $\dfrac{1}{1 - \sin x} = \dfrac{1 + \sin x}{\cos^2 x}$

8. $\tan x \cot x = 1$

9. $(1 - \cos x)(1 - \cos x) = \sin^2 x$

10. $(1 - \csc x)(1 + \csc x) = \cot^2 x$

6.2 Exercises

Match each expression on the left with one on the right that completes each equation as an identity.

1. $\cos x \tan x = \,?$

   A. 1

2. $\sec x \cot x = \,?$

   B. $-\tan^2 x$

3. $(\csc x - \cot x)(\csc x + \cot x) = \,?$

   C. $\cot^2 x$

4. $\dfrac{\sin x + \cos x}{\sin x} = \,?$

   D. $\sin x$

5. $(1 - \sec x)(1 + \sec x) = \,?$

   E. $-\sin^2 x$

6. $(\csc x - 1)(\csc x + 1) = \,?$

   F. $\dfrac{1}{\sin x \cos x}$

7. $\dfrac{\csc x - \sin x}{\csc x} = \,?$

   G. $1 + \cot^2 x$

8. $\dfrac{\cos x - \sec x}{\sec x} = \,?$

   H. $\cos^2 x$

9. $\dfrac{\csc x}{\sin x} = \,?$

   I. $\csc x$

10. $\dfrac{\sin x}{\cos x} + \dfrac{\cos x}{\sin x} = \,?$

   J. $1 + \cot x$

Find the products.

11. $(2\cos \beta + 1)(\cos \beta - 1)$ **12.** $(2\csc \beta - 1)(\csc \beta - 3)$

13. $(\csc x + \sin x)^2$ **14.** $(2\cos x - \sec x)^2$

15. $(2\sin \theta - 1)(2\sin \theta + 1)$ **16.** $(3\sec \theta - 2)(3\sec \theta + 2)$

17. $(3\sin \theta + 2)^2$ **18.** $(3\cos \theta - 2)^2$

19. $(2\sin^2 y - \csc^2 y)^2$ **20.** $(\tan^2 y + \cot^2 y)^2$

Find the products and simplify your answers.

21. $(1 - \sin \alpha)(1 + \sin \alpha)$ **22.** $(1 - \cos \alpha)(1 + \cos \alpha)$

23. $(\csc \alpha - 1)(\csc \alpha + 1)$ **24.** $(\sec \alpha - 1)(\sec \alpha + 1)$

25. $(\tan \alpha - \sec \alpha)(\tan \alpha + \sec \alpha)$

26. $(\cot \alpha - \csc \alpha)(\cot \alpha + \csc \alpha)$

Factor each trigonometric expression.

27. $2\sin^2 \gamma - 5\sin \gamma - 3$ **28.** $\cos^2 \gamma - \cos \gamma - 6$

29. $\tan^2 \alpha - 6\tan \alpha + 8$ **30.** $2\cot^2 \alpha + \cot \alpha - 3$

31. $4\sec^2 \beta + 4\sec \beta + 1$ **32.** $9\csc^2 \theta - 12\csc \theta + 4$

33. $\tan^2 \alpha - \sec^2 \beta$ **34.** $\sin^4 y - \cos^4 x$

35. $\sin^2 \beta \cos \beta + \sin \beta \cos \beta - 2\cos \beta$

36. $\cos^2 \theta \tan \theta - 2\cos \theta \tan \theta - 3\tan \theta$

37. $4\sec^4 x - 4\sec^2 x + 1$

38. $\cos^4 x - 2\cos^2 x + 1$

39. $\sin \alpha \cos \alpha + \cos \alpha + \sin \alpha + 1$

40. $2\sin^2 \theta + \sin \theta - 2\sin \theta \cos \theta - \cos \theta$

Simplify each expression.

41. $\dfrac{1}{a} - \dfrac{\cos^2 x}{a}$ **42.** $\dfrac{1}{\cos x} - \dfrac{\sin^2 x}{\cos x}$

43. $\dfrac{\sin(2x)}{2} + \sin(2x)$ **44.** $\cos(2x) - \dfrac{\cos(2x)}{2}$

45. $\dfrac{\tan x}{3} + \dfrac{\tan x}{2}$ **46.** $\dfrac{\sin x}{b} + \dfrac{\sin x}{3b}$

47. $\dfrac{\sin x - \sin^2 x}{\sin x}$ **48.** $\dfrac{\cos^3 x - \cos x}{-\cos x}$

49. $\dfrac{\sin^2 x - \cos^2 x}{\sin x - \cos x}$ **50.** $\dfrac{1 - \cos^2 x}{1 - \cos x}$

51. $\dfrac{\sin^2 x - \sin x - 2}{\sin^2 x - 4}$ **52.** $\dfrac{\tan^2 x - 2\tan x + 1}{1 + \tan(-x)}$

53. $\dfrac{\sin^2(-x) - \sin(-x)}{1 - \sin(-x)}$ **54.** $\dfrac{\cos^2(-x) - \cos(-x)}{1 - \cos(-x)}$

Prove that each of the following equations is an identity. See the strategy for verifying identities on page 523.

55. $\sin(x)\cot(x) = \cos(x)$

56. $\cos^2(x)\tan^2(x) = \sin^2(x)$

57. $1 - \sec(x)\cos^3(x) = \sin^2(x)$

58. $1 - \csc(x)\sin^3(x) = \cos^2(x)$

59. $1 + \sec^2(x)\sin^2(x) = \sec^2(x)$

60. $1 + \csc^2(x)\cos^2(x) = \csc^2(x)$

61. $\dfrac{\sin^3(x) + \sin(x)\cos^2(x)}{\cos(x)} = \tan(x)$

62. $\dfrac{\cos(x)\sin^2(x) + \cos^3(x)}{\sin(x)} = \cot(x)$

63. $\dfrac{\sin(x)}{\csc(x)} + \dfrac{\cos(x)}{\sec(x)} = 1$

64. $\sin^3(x)\csc(x) + \cos^3(x)\sec(x) = 1$

65. $\dfrac{1}{\csc\theta - \cot\theta} = \dfrac{1 + \cos\theta}{\sin\theta}$

66. $\dfrac{-1}{\tan\theta - \sec\theta} = \dfrac{1 + \sin\theta}{\cos\theta}$

67. $\dfrac{\sec x - \cos x}{\sec x} = \sin^2 x$

68. $\dfrac{\sec x - \cos x}{\cos x} = \tan^2 x$

69. $1 - \sin x = \dfrac{1 - \sin^2(-x)}{1 - \sin(-x)}$

70. $\tan^2 x = \dfrac{1 - \sin^2 x\csc^2 x + \sin^2 x}{\cos^2 x}$

71. $\sin^4 w = \dfrac{1 - \cot^2 w + \cos^2 w\cot^2 w}{\csc^2 w}$

72. $\tan^4 z = \dfrac{\sec^2 z - \csc^2 z + \csc^2 z\cos^2 z}{\cot^2 z}$

73. $1 + \csc x\sec x = \dfrac{\cos(-x) - \csc(-x)}{\cos(x)}$

74. $\tan^2(-x) - \dfrac{\sin(-x)}{\sin x} = \sec^2 x$

75. $\tan x\cos x + \csc x\sin^2 x = 2\sin x$

76. $\cot x\sin x - \cos^2 x\sec x = 0$

77. $(1 + \sin\alpha)^2 + \cos^2\alpha = 2 + 2\sin\alpha$

78. $(1 + \cot\alpha)^2 - 2\cot\alpha = \dfrac{1}{(1 - \cos\alpha)(1 + \cos\alpha)}$

79. $\dfrac{\sin^2\beta + \sin\beta - 2}{2\sin\beta - 2} = \dfrac{\sin\beta + 2}{2}$

80. $\dfrac{4\sec^2\beta + 4\sec\beta + 1}{2\sec\beta + 1} = \dfrac{2}{\cos\beta} + 1$

81. $2 - \csc\beta\sin\beta = \sin^2\beta + \cos^2\beta$

82. $(1 - \sin^2\beta)(1 + \sin^2\beta) = 2\cos^2\beta - \cos^4\beta$

83. $\tan x + \cot x = \sec x\csc x$

84. $\dfrac{\csc x}{\cot x} - \dfrac{\cot x}{\csc x} = \dfrac{\tan x}{\csc x}$

85. $\dfrac{\sec x}{\tan x} - \dfrac{\tan x}{\sec x} = \cos x\cot x$

86. $\dfrac{1 - \sin^2 x}{1 - \sin x} = \dfrac{\csc x + 1}{\csc x}$

87. $\sec^2 x = \dfrac{\csc x}{\csc x - \sin x}$

88. $\dfrac{\sin x}{\sin x + 1} = \dfrac{\csc x - 1}{\cot^2 x}$

89. $\dfrac{\csc y + 1}{\csc y - 1} = \dfrac{1 + \sin y}{1 - \sin y}$

90. $\dfrac{1 - 2\cos^2 y}{1 - 2\cos y\sin y} = \dfrac{\sin y + \cos y}{\sin y - \cos y}$

91. $\ln(\sec\theta) = -\ln(\cos\theta)$

92. $\ln(\tan\theta) = \ln(\sin\theta) + \ln(\sec\theta)$

93. $\ln|\sec\alpha + \tan\alpha| = -\ln|\sec\alpha - \tan\alpha|$

94. $\ln|\csc\alpha + \cot\alpha| = -\ln|\csc\alpha - \cot\alpha|$

The equation $f_1(x) = f_2(x)$ is an identity if and only if the graphs of $y = f_1(x)$ and $y = f_2(x)$ coincide at all values of x for which both sides are defined. Graph $y = f_1(x)$ and $y = f_2(x)$ on the same screen of your calculator for each of the following equations. From the graphs, make a conjecture as to whether each equation is an identity, then prove your conjecture.

95. $\dfrac{\sin\theta + \cos\theta}{\sin\theta} = 1 + \cot\theta$

96. $\dfrac{\sin\theta + \cos\theta}{\cos\theta} = \cot\theta + 1$

97. $(\sin x + \csc x)^2 = \sin^2 x + \csc^2 x$

98. $\tan x + \sec x = \dfrac{\sin^2 x + 1}{\cos x}$

99. $\cot x + \sin x = \dfrac{1 + \cos x - \cos^2 x}{\sin x}$

100. $1 - 2\cos^2 x + \cos^4 x = \sin^4 x$

101. $\dfrac{\sin x}{\cos x} - \dfrac{\cos x}{\sin x} = \dfrac{2\cos^2 x - 1}{\sin x\cos x}$

102. $\dfrac{1}{1 - \sin x} + \dfrac{1}{1 + \sin x} = \dfrac{2}{\cos^2 x}$

103. $\dfrac{\cos(-x)}{1 - \sin x} = \dfrac{1 - \sin(-x)}{\cos x}$

104. $\dfrac{\sin^2 x}{1 - \cos x} = 0.99 + \cos x$

For Writing/Discussion

105. Find functions $f_1(x)$ and $f_2(x)$ such that $f_1(x) = f_2(x)$ for infinitely many values of x, but yet $f_1(x) = f_2(x)$ is not an identity. Explain your example.

106. Find functions $f_1(x)$ and $f_2(x)$ such that $f_1(x) \neq f_2(x)$ for infinitely many values of x, but yet $f_1(x) = f_2(x)$ is an identity. Explain your example.

Thinking Outside the Box LII

Cutting Cardboard   A cardboard tube has height 1 and diameter 1. Suppose that it is cut at a 45° angle as shown in the first figure. The tube is then laid flat in the first quadrant as shown in the second figure. What is the equation of the curve? Assume that the curve is a sine wave.

▪ **Figure for Thinking Outside the Box LII**

6.2 Pop Quiz

1. Find the product $(2 \sin x + 1)(\sin x - 1)$.

2. Factor $2 \cos^2 x + \cos x - 1$.

3. Simplify $\dfrac{1}{\cos(-x)} - \dfrac{\sin^2(-x)}{\cos(x)}$.

4. Prove that $\dfrac{\cos(-x) - \sec(-x)}{\sec(x)} = -\sin^2(x)$ is an identity.

Linking Concepts

For Individual or Group Explorations

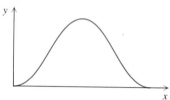

Constructing an Eyebrow Window

A company that manufactures custom aluminum windows makes an eyebrow window that is placed on top of a rectangular window as shown in the diagram. When a customer orders an eyebrow window, the customer gives the width w and height h of the eyebrow. To make the window, the shop needs to know the radius of the circular arc and the length of the circular arc. (This is an actual problem given to the author by a former student who worked for the company.)

a) If the width and height of the eyebrow are 36 in. and 10 in., respectively, then what is the radius of the circular arc?

b) Find the length of the circular arc for a width of 36 in. and a height of 10 in.

c) Find a formula that expresses the radius of the circular arc r in terms of the width of the eyebrow w and the height of the eyebrow h.

d) Find a formula that expresses the length of the circular arc L in terms of w and h.

Sum and Difference Identities

In every identity discussed in Sections 6.1 and 6.2, the trigonometric functions were functions of only a single variable, such as θ or x. In this section we establish several new identities involving sine, cosine, and tangent of a sum or difference of two variables. These identities are used in solving equations and in simplifying expressions. They do not follow from the known identities but rather from the geometry of the unit circle.

The Cosine of a Sum

With so many identities showing relationships between the trigonometric functions, we might be fooled into thinking that almost any equation is an identity. For example, consider the equation

$$\cos(\alpha + \beta) = \cos\alpha + \cos\beta.$$

This equation looks nice, but is it an identity? It is easy to check with a calculator that if $\alpha = 30°$ and $\beta = 45°$, we get

$$\cos(30° + 45°) = \cos(75°) \approx 0.2588 \qquad \text{and} \qquad \cos(30°) + \cos(45°) \approx 1.573.$$

So the equation $\cos(\alpha + \beta) = \cos\alpha + \cos\beta$ is *not* an identity. However, there is an identity for the cosine of a sum of two angles.

To derive an identity for $\cos(\alpha + \beta)$, consider angles α, $\alpha + \beta$, and $-\beta$ in standard position, as shown in Fig. 6.10. Assume that α and β are acute angles, although any angles α and β may be used. The terminal side of α intersects the unit circle at the point $A(\cos\alpha, \sin\alpha)$. The terminal side of $\alpha + \beta$ intersects the unit circle at $B(\cos(\alpha + \beta), \sin(\alpha + \beta))$. The terminal side of $-\beta$ intersects the unit circle at $C(\cos(-\beta), \sin(-\beta))$ or $C(\cos\beta, -\sin\beta)$. Let D be the point $(1, 0)$. Since $\angle BOD = \alpha + \beta$ and $\angle AOC = \alpha + \beta$, the chords $\overline{BD}$ and $\overline{AC}$ are equal in length. Using the distance formula to find the lengths, we get the following equation:

$$\sqrt{(\cos(\alpha + \beta) - 1)^2 + (\sin(\alpha + \beta) - 0)^2} =$$
$$\sqrt{(\cos\alpha - \cos\beta)^2 + (\sin\alpha - (-\sin\beta))^2}$$

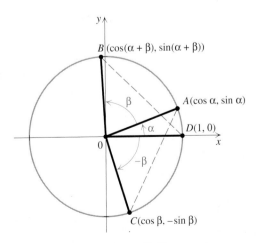

■ **Figure 6.10**

Square each side and simplify:

$$(\cos(\alpha + \beta) - 1)^2 + \sin^2(\alpha + \beta) = (\cos \alpha - \cos \beta)^2 + (\sin \alpha + \sin \beta)^2$$

Square the binomials:

$$\cos^2(\alpha + \beta) - 2\cos(\alpha + \beta) + 1 + \sin^2(\alpha + \beta)$$
$$= \cos^2 \alpha - 2\cos \alpha \cos \beta + \cos^2 \beta + \sin^2 \alpha + 2\sin \alpha \sin \beta + \sin^2 \beta$$

Using the identity $\cos^2 x + \sin^2 x = 1$ once on the left-hand side and twice on the right-hand side, we get the following equation:

$$2 - 2\cos(\alpha + \beta) = 2 - 2\cos \alpha \cos \beta + 2\sin \alpha \sin \beta$$

Subtract 2 from each side and divide each side by -2 to get the identity for the cosine of a sum:

Identity: Cosine of a Sum

$$\cos(\alpha + \beta) = \cos \alpha \cos \beta - \sin \alpha \sin \beta$$

If an angle is the sum of two angles for which we know the exact values of the trigonometric functions, we can use this new identity to find the exact value of the cosine of the sum. For example, we can find the exact values of expressions such as $\cos(75°)$, $\cos(7\pi/12)$, and $\cos(195°)$ because

$$75° = 45° + 30°, \qquad \frac{7\pi}{12} = \frac{\pi}{3} + \frac{\pi}{4}, \qquad \text{and} \qquad 195° = 150° + 45°.$$

Note that we do not need identities to find approximate values for $\cos(75°)$, $\cos(7\pi/12)$, and $\cos(195°)$. All we need is a calculator.

Example **1** **The exact value of a cosine**

Find the exact value of $\cos(75°)$.

Solution

Use $75° = 30° + 45°$ and the identity for the cosine of a sum.

$$\cos 75° = \cos(30° + 45°)$$
$$= \cos(30°)\cos(45°) - \sin(30°)\sin(45°)$$
$$= \frac{\sqrt{3}}{2} \cdot \frac{\sqrt{2}}{2} - \frac{1}{2} \cdot \frac{\sqrt{2}}{2}$$
$$= \frac{\sqrt{6} - \sqrt{2}}{4}$$

To check, evaluate $\cos 75°$ and $\left(\sqrt{6} - \sqrt{2}\right)/4$ using a calculator.

Try This. Find the exact value of $\cos(105°)$ using the cosine of a sum identity.

■

Note that we found the exact value of $\cos(75°)$ only to illustrate an identity. If we need $\cos(75°)$ in an application, we can use the calculator's value for it.

The Cosine of a Difference

To derive an identity for the cosine of a difference, we write $\alpha - \beta = \alpha + (-\beta)$ and use the identity for the cosine of a sum:

$$\cos(\alpha - \beta) = \cos(\alpha + (-\beta))$$
$$= \cos \alpha \cos(-\beta) - \sin \alpha \sin(-\beta)$$

Now use the identities $\cos(-\beta) = \cos \beta$ and $\sin(-\beta) = -\sin \beta$ to get the identity for the cosine of a difference.

Identity: Cosine of a Difference

$$\cos(\alpha - \beta) = \cos \alpha \cos \beta + \sin \alpha \sin \beta$$

If an angle is the difference of two angles for which we know the exact values of sine and cosine, then we can find the exact value for the cosine of the angle.

Example **2** The exact value of a cosine

Find the exact value of $\cos(\pi/12)$.

Solution

Since $\frac{\pi}{3} = \frac{4\pi}{12}$ and $\frac{\pi}{4} = \frac{3\pi}{12}$, we can use $\frac{\pi}{12} = \frac{\pi}{3} - \frac{\pi}{4}$ and the identity for the cosine of a difference:

$$\cos\left(\frac{\pi}{12}\right) = \cos\left(\frac{\pi}{3} - \frac{\pi}{4}\right)$$
$$= \cos \frac{\pi}{3} \cos \frac{\pi}{4} + \sin \frac{\pi}{3} \sin \frac{\pi}{4}$$
$$= \frac{1}{2} \cdot \frac{\sqrt{2}}{2} + \frac{\sqrt{3}}{2} \cdot \frac{\sqrt{2}}{2}$$
$$= \frac{\sqrt{2} + \sqrt{6}}{4}$$

To check, evaluate $\cos(\pi/12)$ and $(\sqrt{2} + \sqrt{6})/4$ using a calculator.

Try This. Find the exact value of $\cos(75°)$ using the cosine of a difference identity. ■

The identities for the cosine of a sum or a difference can be used to find identities for the sine of a sum or difference. To develop those identities, we first find identities that relate each trigonometric function and its **cofunction.** Sine and cosine are cofunctions, tangent and cotangent are cofunctions, and secant and cosecant are cofunctions.

Cofunction Identities

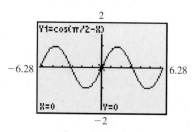

■ **Figure 6.11**

If $\alpha = \pi/2$ is substituted into the identity for $\cos(\alpha - \beta)$, we get another identity:

$$\cos\left(\frac{\pi}{2} - \beta\right) = \cos\frac{\pi}{2}\cos\beta + \sin\frac{\pi}{2}\sin\beta$$

$$= 0 \cdot \cos\beta + 1 \cdot \sin\beta$$

$$= \sin\beta$$

The identity $\cos(\pi/2 - \beta) = \sin\beta$ is a **cofunction identity.**

The graphing calculator graph in Fig. 6.11 supports this conclusion because $y_1 = \cos(\pi/2 - x)$ appears to coincide with $y_2 = \sin(x)$. □

If we let $u = \pi/2 - \beta$ or $\beta = \pi/2 - u$ in the identity $\sin\beta = \cos(\pi/2 - \beta)$, we get the identity

$$\sin\left(\frac{\pi}{2} - u\right) = \cos u.$$

Using these identities, we can establish the cofunction identities for secant, cosecant, tangent, and cotangent. For the cofunction identities, *the value of any trigonometric function at u is equal to the value of its cofunction at $(\pi/2 - u)$.*

Cofunction Identities

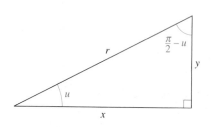

■ **Figure 6.12**

$$\sin\left(\frac{\pi}{2} - u\right) = \cos u \qquad \cos\left(\frac{\pi}{2} - u\right) = \sin u$$

$$\tan\left(\frac{\pi}{2} - u\right) = \cot u \qquad \cot\left(\frac{\pi}{2} - u\right) = \tan u$$

$$\sec\left(\frac{\pi}{2} - u\right) = \csc u \qquad \csc\left(\frac{\pi}{2} - u\right) = \sec u$$

If $0 < u < \pi/2$, then u and $(\pi/2 - u)$ are the measures of the acute angles of a right triangle, as shown in Fig. 6.12. The term *cofunction* comes from the fact that these angles are complementary. The cofunction identities indicate that the value of a trigonometric function of one acute angle of a right triangle is equal to the value of the cofunction of the other. Since the cofunction identities hold also for complementary angles measured in degrees, we can write equations such as

$$\sin(20°) = \cos(70°), \qquad \cot(89°) = \tan(1°), \qquad \text{and} \qquad \sec(50°) = \csc(40°).$$

The cofunction identities are consistent with the values obtained for the trigonometric functions for an acute angle using the ratios of the sides of a right triangle. For example, the cofunction identity $\sin(u) = \cos(\pi/2 - u)$ is correct for angles u and $(\pi/2 - u)$ shown in Fig. 6.12, because

$$\sin(u) = \frac{\text{opp}}{\text{hyp}} = \frac{y}{r}$$

and

$$\cos\left(\frac{\pi}{2} - u\right) = \frac{\text{adj}}{\text{hyp}} = \frac{y}{r}.$$

Example **3** Using the cofunction identities

Use a cofunction identity to find the exact value of $\sin(5\pi/12)$.

Solution

Since $\sin(u) = \cos(\pi/2 - u)$,

$$\sin\left(\frac{5\pi}{12}\right) = \cos\left(\frac{\pi}{2} - \frac{5\pi}{12}\right) = \cos\left(\frac{\pi}{12}\right).$$

From Example 2, $\cos(\pi/12) = \left(\sqrt{2} + \sqrt{6}\right)/4$. So $\sin(5\pi/12) = \left(\sqrt{2} + \sqrt{6}\right)/4$.

Try This. Find the exact value of $\sin(15°)$ using a cofunction identity. ■

Sine of a Sum or a Difference

We now use a cofunction identity to find an identity for $\sin(\alpha + \beta)$:

$$\sin(\alpha + \beta) = \cos\left(\frac{\pi}{2} - (\alpha + \beta)\right)$$

$$= \cos\left(\left(\frac{\pi}{2} - \alpha\right) - \beta\right)$$

$$= \cos\left(\frac{\pi}{2} - \alpha\right)\cos\beta + \sin\left(\frac{\pi}{2} - \alpha\right)\sin\beta$$

$$= \sin\alpha\cos\beta + \cos\alpha\sin\beta$$

To find an identity for $\sin(\alpha - \beta)$, use the identity for $\sin(\alpha + \beta)$:

$$\sin(\alpha - \beta) = \sin(\alpha + (-\beta))$$

$$= \sin\alpha\cos(-\beta) + \cos\alpha\sin(-\beta)$$

$$= \sin\alpha\cos\beta - \cos\alpha\sin\beta$$

The identities for the sine of a sum or difference are stated as follows.

Identities: Sine of a Sum or Difference

$$\sin(\alpha + \beta) = \sin\alpha\cos\beta + \cos\alpha\sin\beta$$
$$\sin(\alpha - \beta) = \sin\alpha\cos\beta - \cos\alpha\sin\beta$$

Example **4** The exact value of a sine

Find the exact value of $\sin(195°)$.

Solution

Use $195° = 150° + 45°$ and the identity for the sine of a sum:

$$\sin(195°) = \sin(150° + 45°)$$

$$= \sin(150°)\cos(45°) + \cos(150°)\sin(45°)$$

$$= \frac{1}{2} \cdot \frac{\sqrt{2}}{2} + \left(-\frac{\sqrt{3}}{2}\right) \cdot \frac{\sqrt{2}}{2}$$

$$= \frac{\sqrt{2} - \sqrt{6}}{4}$$

To check, evaluate $\sin(195°)$ and $\left(\sqrt{2} - \sqrt{6}\right)/4$ using a calculator.

Try This. Find the exact value of $\sin(75°)$ using the sine of a sum identity.   ■

Tangent of a Sum or Difference

We can use the identities for $\sin(\alpha + \beta)$ and $\cos(\alpha + \beta)$ to find an identity for $\tan(\alpha + \beta)$:

$$\tan(\alpha + \beta) = \frac{\sin(\alpha + \beta)}{\cos(\alpha + \beta)}$$

$$= \frac{\sin \alpha \cos \beta + \cos \alpha \sin \beta}{\cos \alpha \cos \beta - \sin \alpha \sin \beta}$$

To express the right-hand side in terms of tangent, multiply the numerator and denominator by $1/(\cos \alpha \cos \beta)$ and use the identity $\tan x = \sin x / \cos x$:

$$\tan(\alpha + \beta) = \frac{\dfrac{\sin \alpha \cos \beta}{\cos \alpha \cos \beta} + \dfrac{\cos \alpha \sin \beta}{\cos \alpha \cos \beta}}{\dfrac{\cos \alpha \cos \beta}{\cos \alpha \cos \beta} - \dfrac{\sin \alpha \sin \beta}{\cos \alpha \cos \beta}}$$

$$= \frac{\tan \alpha + \tan \beta}{1 - \tan \alpha \tan \beta}$$

Use the identity $\tan(-\beta) = -\tan \beta$ to get a similar identity for $\tan(\alpha - \beta)$.

Identities: Tangent
of a Sum or Difference

$$\tan(\alpha + \beta) = \frac{\tan \alpha + \tan \beta}{1 - \tan \alpha \tan \beta} \qquad \tan(\alpha - \beta) = \frac{\tan \alpha - \tan \beta}{1 + \tan \alpha \tan \beta}$$

Example **5** **The exact value of a tangent**

Find the exact value of $\tan(\pi/12)$.

Solution

Use $\pi/12 = \pi/3 - \pi/4$ and the identity for the tangent of a difference:

$$\tan\left(\frac{\pi}{12}\right) = \tan\left(\frac{\pi}{3} - \frac{\pi}{4}\right)$$

$$= \frac{\tan \dfrac{\pi}{3} - \tan \dfrac{\pi}{4}}{1 + \tan \dfrac{\pi}{3} \tan \dfrac{\pi}{4}} = \frac{\sqrt{3} - 1}{1 + \sqrt{3} \cdot 1}$$

$$= \frac{\left(\sqrt{3} - 1\right)\left(\sqrt{3} - 1\right)}{\left(\sqrt{3} + 1\right)\left(\sqrt{3} - 1\right)}$$

$$= \frac{3 - 2\sqrt{3} + 1}{2} = \frac{4 - 2\sqrt{3}}{2} = 2 - \sqrt{3}$$

Try This. Find the exact value of $\tan(75°)$ using the tangent of a sum identity.

■

In the next example we use sum and difference identities in reverse to simplify an expression.

Example ⬛6 Simplifying with sum and difference identities

Use an appropriate identity to simplify each expression.

a. $\cos(47°)\cos(2°) + \sin(47°)\sin(2°)$ **b.** $\sin t \cos 2t - \cos t \sin 2t$

Solution

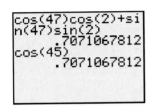

■ **Figure 6.13**

a. This expression is the right-hand side of the identity for the cosine of a difference, $\cos(\alpha - \beta) = \cos \alpha \cos \beta + \sin \alpha \sin \beta$, with $\alpha = 47°$ and $\beta = 2°$:

$$\cos(47°)\cos(2°) + \sin(47°)\sin(2°) = \cos(47° - 2°) = \cos(45°) = \frac{\sqrt{2}}{2}$$

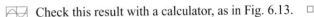   Check this result with a calculator, as in Fig. 6.13.   □

b. We recognize this expression as the right-hand side of the identity for the sine of a difference, $\sin(\alpha - \beta) = \sin \alpha \cos \beta - \cos \alpha \sin \beta$. So

$$\sin t \cos 2t - \cos t \sin 2t = \sin(t - 2t) = \sin(-t) = -\sin t.$$

Try This. Simplify $\sin \alpha \cos 3\alpha + \cos \alpha \sin 3\alpha$.   ■

As shown next, we use identities to find $\sin(\alpha + \beta)$ without knowing α or β.

Example ⬛7 Evaluating a sine of a sum

Find the exact value of $\sin(\alpha + \beta)$, given that $\sin \alpha = -3/5$ and $\cos \beta = -1/3$, with α in quadrant IV and β in quadrant III.

Solution

To use the identity for $\sin(\alpha + \beta)$, we need $\cos \alpha$ and $\sin \beta$ in addition to the given values. Use $\sin \alpha = -3/5$ in the identity $\sin^2 x + \cos^2 x = 1$ to find $\cos \alpha$:

$$\left(-\frac{3}{5}\right)^2 + \cos^2 \alpha = 1, \qquad \cos^2 \alpha = \frac{16}{25}, \qquad \cos \alpha = \pm\frac{4}{5}.$$

Since cosine is positive in quadrant IV, $\cos \alpha = 4/5$. Use $\cos \beta = -1/3$ in $\sin^2 x + \cos^2 x = 1$ to find $\sin \beta$.

$$\sin^2 \beta + \left(-\frac{1}{3}\right)^2 = 1, \qquad \sin^2 \beta = \frac{8}{9}, \qquad \sin \beta = \pm\frac{2\sqrt{2}}{3}.$$

Since sine is negative in quadrant III, $\sin \beta = -2\sqrt{2}/3$. Now use the appropriate values in the identity $\sin(\alpha + \beta) = \sin \alpha \cos \beta + \cos \alpha \sin \beta$:

$$\sin(\alpha + \beta) = -\frac{3}{5}\left(-\frac{1}{3}\right) + \frac{4}{5}\left(-\frac{2\sqrt{2}}{3}\right) = \frac{3}{15} - \frac{8\sqrt{2}}{15} = \frac{3 - 8\sqrt{2}}{15}.$$

To check, we use the inverse functions to find α and β and then $\sin(\alpha + \beta)$ on a graphing calculator. Since $\cos^{-1}(-1/3)$ is in quadrant II, β in degrees is $360 - \cos^{-1}(-1/3)$. The angle α is $\sin^{-1}(-3/5)$, as shown in Fig. 6.14(a). Now use these values for α and β to check that $\sin(\alpha + \beta) = (3 - 8\sqrt{2})/15$, as shown in Fig. 6.14(b).

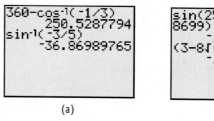

(a)

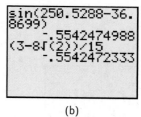

(b)

■ **Figure 6.14**

Try This. Find the exact value of $\cos(\alpha - \beta)$ given that $\sin\alpha = 1/3$, $\cos\beta = -1/3$ and both angles terminate in quadrant II. ■

For Thought

True or False? Explain.

1. $\cos(3°) \cos(2°) - \sin(3°) \sin(2°) = \cos(1°)$

2. $\cos(4) \cos(5) + \sin(4) \sin(5) = \cos(-1)$

3. For any real number t, $\cos(t - \pi/2) = \sin t$.

4. For any radian measure α, $\sin(\alpha - \pi/2) = \cos \alpha$.

5. $\sec(\pi/3) = \csc(\pi/6)$

6. $\sin(5\pi/6) = \sin(2\pi/3) + \sin(\pi/6)$

7. $\sin(5\pi/12) = \sin(\pi/3) \cos(\pi/4) + \cos(\pi/3) \sin(\pi/4)$

8. $\tan(21°30'5'') = \cot(68°29'55'')$

9. For any real number x, $\sec(x) - \csc(\pi/2 - x) = \tan(x) - \cot(\pi/2 - x)$.

10. $\dfrac{\tan(13°) + \tan(-20°)}{1 - \tan(13°)\tan(-20°)} = \dfrac{\tan(45°) + \tan(-52°)}{1 - \tan(45°)\tan(-52°)}$

6.3 Exercises

Use an identity to find the exact value of each expression. Use a calculator to check.

1. $\cos\left(\dfrac{\pi}{3} + \dfrac{\pi}{4}\right)$

2. $\cos\left(\dfrac{2\pi}{3} + \dfrac{\pi}{4}\right)$

3. $\cos(60° - 45°)$

4. $\cos(120° - 45°)$

Use a cofunction identity to fill in the blank.

5. $\sin(20°) = \cos(\quad)$

6. $\cos(15°) = \sin(\quad)$

7. $\tan\left(\dfrac{\pi}{6}\right) = \cot(\quad)$

8. $\cot\left(\dfrac{\pi}{3}\right) = \tan(\quad)$

9. $\sec(90° - 6°) = \csc(\quad)$

10. $\csc(90° - 17°) = \sec(\quad)$

Use an identity to find the exact value of each expression. Use a calculator to check.

11. $\sin\left(\dfrac{\pi}{4} + \dfrac{\pi}{3}\right)$

12. $\sin\left(\dfrac{\pi}{4} + \dfrac{2\pi}{3}\right)$

13. $\sin(60° - 45°)$

14. $\sin(45° - 120°)$

15. $\tan\left(\dfrac{3\pi}{4} + \dfrac{\pi}{3}\right)$

16. $\tan\left(\dfrac{\pi}{4} + \dfrac{\pi}{3}\right)$

17. $\tan(210° - 45°)$

18. $\tan(45° - 150°)$

Find the following sums or differences in terms of π.

19. $\dfrac{\pi}{4} + \dfrac{\pi}{3}$

20. $\dfrac{2\pi}{3} - \dfrac{\pi}{4}$

21. $\dfrac{3\pi}{4} + \dfrac{\pi}{3}$ **22.** $\dfrac{\pi}{6} + \dfrac{\pi}{4}$

Simplify each expression by using sum or difference identities.

23. $\sin(23°)\cos(67°) + \cos(23°)\sin(67°)$

24. $\sin(55°)\cos(10°) - \cos(55°)\sin(10°)$

25. $\cos\left(\dfrac{\pi}{6}\right)\cos\left(\dfrac{\pi}{3}\right) - \sin\left(\dfrac{\pi}{6}\right)\sin\left(\dfrac{\pi}{3}\right)$

26. $\cos\left(\dfrac{7\pi}{12}\right)\cos\left(\dfrac{\pi}{3}\right) + \sin\left(\dfrac{7\pi}{12}\right)\sin\left(\dfrac{\pi}{3}\right)$

27. $\dfrac{\tan(\pi/12) + \tan(\pi/6)}{1 - \tan(\pi/12)\tan(\pi/6)}$

28. $\dfrac{\tan(5\pi/12) - \tan(\pi/12)}{1 + \tan(5\pi/12)\tan(\pi/12)}$

29. $\sin(2k)\cos(k) + \cos(2k)\sin(k)$

30. $\cos(3y)\cos(y) - \sin(3y)\sin(y)$

Express each given angle as $\alpha + \beta$ or $\alpha - \beta$, where $\sin \alpha$ and $\sin \beta$ are known exactly.

31. $75°$ **32.** $15°$ **33.** $165°$ **34.** $195°$

Use appropriate identities to find the exact value of each expression. Do not use a calculator.

35. $\cos(5\pi/12)$ **36.** $\cos(7\pi/12)$

37. $\sin(7\pi/12)$ **38.** $\sin(5\pi/12)$

39. $\tan(75°)$ **40.** $\tan(-15°)$

41. $\sin(-15°)$ **42.** $\sin(165°)$

43. $\cos(195°)$ **44.** $\cos(-75°)$

45. $\tan(-13\pi/12)$ **46.** $\tan(7\pi/12)$

Simplify each expression by using appropriate identities. Do not use a calculator.

47. $\sin(3°)\cos(-87°) + \cos(3°)\sin(87°)$

48. $\sin(34°)\cos(13°) + \cos(-34°)\sin(-13°)$

49. $\cos(-\pi/2)\cos(\pi/5) + \sin(\pi/2)\sin(\pi/5)$

50. $\cos(12°)\cos(-3°) - \sin(12°)\sin(-3°)$

51. $\dfrac{\tan(\pi/7) + \cot(\pi/2 - \pi/6)}{1 + \tan(-\pi/7)\tan(\pi/6)}$

52. $\dfrac{\cot(\pi/2 - \pi/3) + \tan(-\pi/6)}{1 + \tan(\pi/3)\cot(\pi/2 - \pi/6)}$

53. $\sin(14°)\cos(35°) + \cos(14°)\cos(55°)$

54. $\cos(10°)\cos(20°) + \cos(-80°)\sin(-20°)$

Match each expression on the left with an expression on the right that has exactly the same value. Do not use a calculator.

55. $\cos(44°)$ A. $\cos(0)$

56. $\sin(-46°)$ B. $-\cos(44°)$

57. $\cos(46°)$ C. $\cos(36°)$

58. $\sin(136°)$ D. $\cot\left(\dfrac{5\pi}{14}\right)$

59. $\sec(1)$ E. $-\cos(46°)$

60. $\tan\left(\dfrac{\pi}{7}\right)$ F. $\csc\left(\dfrac{\pi - 2}{2}\right)$

61. $\csc\left(\dfrac{\pi}{2}\right)$ G. $\sin(46°)$

62. $\sin(-44°)$ H. $\sin(44°)$

Solve each problem.

63. Find the exact value of $\sin(\alpha + \beta)$, given that $\sin \alpha = 3/5$ and $\sin \beta = 5/13$, with α in quadrant II and β in quadrant I.

64. Find the exact value of $\sin(\alpha - \beta)$, given that $\sin \alpha = -4/5$ and $\cos \beta = 12/13$, with α in quadrant III and β in quadrant IV.

65. Find the exact value of $\cos(\alpha + \beta)$, given that $\sin \alpha = 2/3$ and $\sin \beta = -1/2$, with α in quadrant I and β in quadrant III.

66. Find the exact value of $\cos(\alpha - \beta)$, given that $\cos \alpha = \sqrt{3}/4$ and $\cos \beta = -\sqrt{2}/3$, with α in quadrant I and β in quadrant II.

67. Find the exact value of $\sin(\alpha - \beta)$ if $\sin \alpha = -24/25$ and $\cos \beta = -8/17$ with α in quadrant III and β in quadrant II.

68. Find the exact value of $\sin(\alpha + \beta)$ if $\sin \alpha = 7/25$ and $\sin \beta = -8/17$ with α in quadrant II and β in quadrant III.

69. Find the exact value of $\cos(\alpha - \beta)$ if $\sin \alpha = 24/25$ and $\cos \beta = 8/17$ with α in quadrant II and β in quadrant IV.

70. Find the exact value of $\cos(\alpha + \beta)$ if $\sin \alpha = -7/25$ and $\sin \beta = 8/17$ with α in quadrant IV and β in quadrant II.

Write each expression as a function of α alone.

71. $\cos(\pi/2 + \alpha)$ **72.** $\sin(\alpha - \pi)$

73. $\cos(180° - \alpha)$　　**74.** $\sin(180° - \alpha)$

75. $\sin(360° - \alpha)$　　**76.** $\cos(\alpha - \pi)$

77. $\sin(90° + \alpha)$　　**78.** $\cos(360° - \alpha)$

Verify that each equation is an identity.

79. $\sin(180° - \alpha) = \dfrac{1 - \cos^2 \alpha}{\sin \alpha}$

80. $\cos(x - \pi/2) = \cos x \tan x$

81. $\dfrac{\cos(x + y)}{\cos x \cos y} = 1 - \tan x \tan y$

82. $\dfrac{\sin(x + y)}{\sin x \cos y} = 1 + \cot x \tan y$

83. $\sin(\alpha + \beta) \sin(\alpha - \beta) = \sin^2 \alpha - \sin^2 \beta$

84. $\cos(\alpha + \beta) \cos(\alpha - \beta) = \cos^2 \beta - \sin^2 \alpha$

85. $\cos(2x) = \cos^2 x - \sin^2 x$

86. $\sin(2x) = 2 \sin x \cos x$

87. $\sin(x - y) - \sin(y - x) = 2 \sin x \cos y - 2 \cos x \sin y$

88. $\cos(x - y) + \cos(y - x) = 2 \cos x \cos y + 2 \sin x \sin y$

89. $\tan(s + t) \tan(s - t) = \dfrac{\tan^2 s - \tan^2 t}{1 - \tan^2 s \tan^2 t}$

90. $\tan(\pi/4 + x) = \cot(\pi/4 - x)$

91. $\dfrac{\cos(\alpha + \beta)}{\sin(\alpha - \beta)} = \dfrac{1 - \tan \alpha \tan \beta}{\tan \alpha - \tan \beta}$

92. $\dfrac{\cos(\alpha - \beta)}{\sin(\alpha + \beta)} = \dfrac{1 + \tan \alpha \tan \beta}{\tan \alpha + \tan \beta}$

93. $\sec(v + t) = \dfrac{\cos v \cos t + \sin v \sin t}{\cos^2 v - \sin^2 t}$

94. $\csc(v - t) = \dfrac{\sin v \cos t + \cos v \sin t}{\sin^2 v - \sin^2 t}$

95. $\dfrac{\cos(x + y)}{\cos(x - y)} = \dfrac{\cot y - \tan x}{\cot y + \tan x}$

96. $\dfrac{\sin(x + y)}{\sin(x - y)} = \dfrac{\cot y + \cot x}{\cot y - \cot x}$

97. $\dfrac{\sin(\alpha + \beta)}{\sin \alpha + \sin \beta} = \dfrac{\sin \alpha - \sin \beta}{\sin(\alpha - \beta)}$

98. $\dfrac{\cos(\alpha + \beta)}{\cos \alpha + \sin \beta} = \dfrac{\cos \alpha - \sin \beta}{\cos(\beta - \alpha)}$

For Writing/Discussion

99. Verify the four cofunction identities that were not proved in the text.

100. Verify the identity for the tangent of a difference.

101. Show that $\sin(\alpha + \beta) = \sin \alpha + \sin \beta$ is not an identity.

102. Show that $\cos(\alpha - \beta) = \cos \alpha - \cos \beta$ is not an identity.

103. Without using a calculator, find the exact value of

$$\frac{\sin^2(1°) + \sin^2(2°) + \sin^2(3°) + \cdots + \sin^2(90°)}{\cos^2(1°) + \cos^2(2°) + \cos^2(3°) + \cdots + \cos^2(90°)}.$$

Thinking Outside the Box LIII

Spiral of Triangles　The hypotenuse in the first right triangle shown in the figure has length $\sqrt{2}$. In the second the hypotenuse is $\sqrt{3}$. In the third it is $\sqrt{4}$, and so on. In this sequence of right triangles, which right triangle (after the first one) intersects the first one at more than one point?

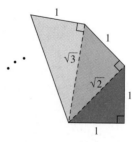

■ **Figure for Thinking Outside the Box LIII**

6.3 Pop Quiz

1. Use an identity to find the exact value of $\cos(135° - 120°)$.

2. Use a cofunction identity to find an angle α for which $\sin(10°) = \cos(\alpha)$.

3. Simplify $\sin(2x)\cos(x) + \cos(2x)\sin(x)$.

4. Find the exact value of $\sin(\alpha - \beta)$, if $\sin \alpha = 4/5$ and $\cos \beta = 1/2$, with α and β in quadrant I.

6.4

Double-Angle and Half-Angle Identities

In Section 6.3 we studied identities for functions of sums and differences of angles. The double-angle and half-angle identities, which we develop next, are special cases of those identities. These special cases occur so often that they are remembered as separate identities.

Double-Angle Identities

To get an identity for $\sin 2x$, replace both α and β by x in the identity for $\sin(\alpha + \beta)$:

$$\sin 2x = \sin(x + x) = \sin x \cos x + \cos x \sin x = 2 \sin x \cos x$$

To find an identity for $\cos 2x$, replace both α and β by x in the identity for $\cos(\alpha + \beta)$:

$$\cos 2x = \cos(x + x) = \cos x \cos x - \sin x \sin x = \cos^2 x - \sin^2 x$$

To get a second form of the identity for $\cos 2x$, replace $\sin^2 x$ with $1 - \cos^2 x$:

$$\cos(2x) = \cos^2 x - \sin^2 x = \cos^2 x - (1 - \cos^2 x) = 2 \cos^2 x - 1$$

Replacing $\cos^2 x$ with $1 - \sin^2 x$ produces a third form of the identity for $\cos 2x$:

$$\cos(2x) = 1 - 2 \sin^2 x$$

To get an identity for $\tan 2x$, we can replace both α and β by x in the identity for $\tan(\alpha + \beta)$:

$$\tan 2x = \tan(x + x) = \frac{\tan x + \tan x}{1 - \tan x \tan x} = \frac{2 \tan x}{1 - \tan^2 x}$$

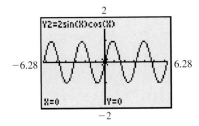

■ **Figure 6.15**

You can visually check identities by graphing both sides. For example, $y_1 = \sin(2x)$ and $y_2 = 2 \sin(x) \cos(x)$ appear to coincide in Fig. 6.15. □
These identities are known as the **double-angle identities.**

Double-Angle Identities

$$\sin 2x = 2 \sin x \cos x \qquad \tan 2x = \frac{2 \tan x}{1 - \tan^2 x}$$

$$\cos 2x = \cos^2 x - \sin^2 x$$

$$\cos 2x = 2 \cos^2 x - 1$$

$$\cos 2x = 1 - 2 \sin^2 x$$

Be careful to learn the double-angle identities exactly as they are written. A "nice looking" equation such as $\cos 2x = 2 \cos x$ could be mistaken for an identity if you are not careful. [Since $\cos(\pi/2) \neq 2 \cos(\pi/4)$, the "nice looking" equation is not an identity.] Remember that an equation is not an identity if at least one permissible value of the variable fails to satisfy the equation.

Example **1** **Evaluating with double-angle identities**

Find $\sin(120°)$, $\cos(120°)$, and $\tan(120°)$ by using double-angle identities.

Solution

Note that $120° = 2 \cdot 60°$ and use the values $\sin(60°) = \sqrt{3}/2$, $\cos(60°) = 1/2$, and $\tan(60°) = \sqrt{3}$ in the appropriate identities.

$$\sin(120°) = 2\sin(60°)\cos(60°) = 2 \cdot \frac{\sqrt{3}}{2} \cdot \frac{1}{2} = \frac{\sqrt{3}}{2}$$

$$\cos(120°) = \cos^2(60°) - \sin^2(60°) = \left(\frac{1}{2}\right)^2 - \left(\frac{\sqrt{3}}{2}\right)^2 = -\frac{1}{2}$$

$$\tan(120°) = \frac{2\tan(60°)}{1 - \tan^2(60°)} = \frac{2 \cdot \sqrt{3}}{1 - (\sqrt{3})^2} = \frac{2\sqrt{3}}{-2} = -\sqrt{3}$$

These results are the well-known values of $\sin(120°)$, $\cos(120°)$, and $\tan(120°)$.

Try This. Find $\sin(60°)$ using a double angle identity. ▪

Half-Angle Identities

The double-angle identities can be used to get identities for $\sin(x/2)$, $\cos(x/2)$, and $\tan(x/2)$. These identities, are known as **half-angle identities.**

To get an identity for $\cos(x/2)$, start by solving the double-angle identity $\cos 2x = 2\cos^2 x - 1$ for $\cos x$:

$$2\cos^2 x - 1 = \cos 2x$$

$$\cos^2 x = \frac{1 + \cos 2x}{2}$$

$$\cos x = \pm\sqrt{\frac{1 + \cos 2x}{2}}$$

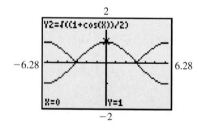

▪ **Figure 6.16**

Because the last equation is correct for any value of x, it is also correct if x is replaced by $x/2$. Replacing x by $x/2$ yields the half-angle identity

$$\cos\frac{x}{2} = \pm\sqrt{\frac{1 + \cos x}{2}}.$$

So for any value of x, $\cos(x/2)$ is equal to either the positive or the negative square root of $(1 + \cos x)/2$.

▱ The graphs of $y_1 = \cos(x/2)$ and $y_2 = \sqrt{(1 + \cos x)/2}$ in Fig. 6.16 illustrate this identity. Graph these curves on your calculator to see that the graph of y_2 does not go below the x-axis. Then use TRACE to see that $y_1 = y_2$ for $y_1 \geq 0$ and $y_1 = -y_2$ for $y_1 < 0$. □

Example **2** **Evaluating with half-angle identities**

Use the half-angle identity to find the exact value of $\cos(\pi/8)$.

Solution

Use $x = \pi/4$ in the half-angle identity for cosine:

$$\cos \frac{\pi}{8} = \cos \frac{\pi/4}{2} = \sqrt{\frac{1 + \cos \pi/4}{2}} = \sqrt{\frac{1 + \dfrac{\sqrt{2}}{2}}{2}}$$

$$= \sqrt{\frac{2 + \sqrt{2}}{4}} = \frac{\sqrt{2 + \sqrt{2}}}{2}$$

The positive square root is used because the angle $\pi/8$ is in the first quadrant, where cosine is positive.

Try This. Find the exact value of $\sin(-22.5°)$ using a half-angle identity.   ■

To get an identity for $\sin(x/2)$, solve $1 - 2\sin^2 x = \cos 2x$ for $\sin x$:

$$1 - 2\sin^2 x = \cos 2x$$

$$-2\sin^2 x = \cos 2x - 1$$

$$\sin^2 x = \frac{1 - \cos 2x}{2}$$

$$\sin x = \pm\sqrt{\frac{1 - \cos 2x}{2}}$$

Replacing x by $x/2$ yields the half-angle identity for sine:

$$\sin \frac{x}{2} = \pm\sqrt{\frac{1 - \cos x}{2}}$$

The half-angle identity for sine can be used to find the exact value for any angle that is one-half of an angle for which we know the exact value of cosine.

Example **3** **Evaluating with half-angle identities**

Use the half-angle identity to find the exact value of $\sin(75°)$.

Solution

Use $x = 150°$ in the half-angle identity for sine:

$$\sin(75°) = \sin\left(\frac{150°}{2}\right) = \pm\sqrt{\frac{1 - \cos(150°)}{2}}$$

Since $75°$ is in quadrant I, $\sin(75°)$ is positive. The reference angle for $150°$ is $30°$ and $\cos(150°)$ is negative, because $150°$ is in quadrant II. So

$$\cos(150°) = -\cos(30°) = -\frac{\sqrt{3}}{2}$$

and

$$\sin(75°) = \sqrt{\frac{1 + \dfrac{\sqrt{3}}{2}}{2}} = \sqrt{\frac{2 + \sqrt{3}}{4}} = \frac{\sqrt{2 + \sqrt{3}}}{2}.$$

√(2+√(3))/2
    .9659258263
sin(75°)
    .9659258263

■ **Figure 6.17**

⊿⊿ Check with a calculator, as shown in Fig. 6.17.

Try This. Find the exact value of cos 75° using the half-angle identity. ■

To get a half-angle identity for tangent, use the half-angle identities for sine and cosine:

$$\tan\frac{x}{2} = \frac{\sin\dfrac{x}{2}}{\cos\dfrac{x}{2}} = \frac{\pm\sqrt{\dfrac{1-\cos x}{2}}}{\pm\sqrt{\dfrac{1+\cos x}{2}}} = \pm\sqrt{\frac{1-\cos x}{1+\cos x}}$$

To get another form of this identity, multiply the numerator and denominator inside the radical by $1 + \cos x$. Use the fact that $\sqrt{\sin^2 x} = |\sin x|$ but $\sqrt{(1+\cos x)^2} = 1 + \cos x$ because $1 + \cos x$ is nonnegative:

$$\tan\frac{x}{2} = \pm\sqrt{\frac{(1-\cos x)(1+\cos x)}{(1+\cos x)^2}} = \pm\sqrt{\frac{\sin^2 x}{(1+\cos x)^2}} = \pm\frac{|\sin x|}{1+\cos x}$$

We use the positive or negative sign depending on the sign of $\tan(x/2)$. It can be shown that $\sin x$ has the same sign as $\tan(x/2)$. (See Exercise 83.) With this fact we can omit the absolute value and the $\pm$ symbol and write the identity as

$$\tan\frac{x}{2} = \frac{\sin x}{1+\cos x}.$$

To get a third form of the identity for $\tan(x/2)$, multiply the numerator and denominator of the right-hand side by $1 - \cos x$:

$$\tan\frac{x}{2} = \frac{\sin x(1-\cos x)}{(1+\cos x)(1-\cos x)} = \frac{\sin x(1-\cos x)}{\sin^2 x} = \frac{1-\cos x}{\sin x}$$

The half-angle identities are summarized as follows.

Half-Angle Identities

$$\sin\frac{x}{2} = \pm\sqrt{\frac{1-\cos x}{2}} \qquad \cos\frac{x}{2} = \pm\sqrt{\frac{1+\cos x}{2}}$$

$$\tan\frac{x}{2} = \pm\sqrt{\frac{1-\cos x}{1+\cos x}} \qquad \tan\frac{x}{2} = \frac{\sin x}{1+\cos x} \qquad \tan\frac{x}{2} = \frac{1-\cos x}{\sin x}$$

Example **4** **Evaluating with half-angle identities**

Use the half-angle identity to find the exact value of $\tan(-15°)$.

Solution

Use $x = -30°$ in the half-angle identity $\tan\frac{x}{2} = \frac{\sin x}{1+\cos x}$.

$$\tan(-15°) = \tan\left(\frac{-30°}{2}\right)$$

$$= \frac{\sin(-30°)}{1 + \cos(-30°)}$$

$$= \frac{-\dfrac{1}{2}}{1 + \dfrac{\sqrt{3}}{2}} = \frac{-1}{2 + \sqrt{3}} = \frac{-1(2 - \sqrt{3})}{(2 + \sqrt{3})(2 - \sqrt{3})}$$

$$= -2 + \sqrt{3}$$

Try This. Find the exact value of $\tan(-22.5°)$ using the half-angle identity. ■

Example 5 Using the identities

Find $\sin(\alpha/2)$, given that $\sin\alpha = 3/5$ and $\pi/2 < \alpha < \pi$.

Solution

Use the identity $\sin^2\alpha + \cos^2\alpha = 1$ to find $\cos\alpha$:

$$\left(\frac{3}{5}\right)^2 + \cos^2\alpha = 1$$

$$\cos^2\alpha = \frac{16}{25}$$

$$\cos\alpha = \pm\frac{4}{5}$$

For $\pi/2 < \alpha < \pi$, we have $\cos\alpha < 0$. So $\cos\alpha = -4/5$. Now use the half-angle identity for sine:

$$\sin\frac{\alpha}{2} = \pm\sqrt{\frac{1 - \cos\alpha}{2}} = \pm\sqrt{\frac{1 + \dfrac{4}{5}}{2}} = \pm\sqrt{\frac{9}{10}} = \pm\frac{\sqrt{90}}{10} = \pm\frac{3\sqrt{10}}{10}$$

If $\pi/2 < \alpha < \pi$, then $\pi/4 < \alpha/2 < \pi/2$. So $\sin(\alpha/2) > 0$ and $\sin(\alpha/2) = 3\sqrt{10}/10$.

Try This. Find $\cos(\alpha/2)$ given that $\sin\alpha = 1/3$ and $\pi/2 < \alpha < \pi$. ■

Verifying Identities

Example 6 A triple-angle identity

Prove that the following equation is an identity:

$$\sin(3x) = \sin x(3\cos^2 x - \sin^2 x)$$

Solution

Write $3x$ as $2x + x$ and use the identity for the sine of a sum:

$$\sin(3x) = \sin(2x + x)$$

$$= \sin 2x \cos x + \cos 2x \sin x \qquad \text{Sine of a sum identity}$$

$$= 2 \sin x \cos x \cos x + (\cos^2 x - \sin^2 x) \sin x \qquad \text{Double-angle identities}$$

$$= 2 \sin x \cos^2 x + \sin x \cos^2 x - \sin^3 x \qquad \text{Distributive property}$$

$$= 3 \sin x \cos^2 x - \sin^3 x$$

$$= \sin x (3 \cos^2 x - \sin^2 x) \qquad \text{Factor.}$$

Try This. Prove $\cos(3x) = \cos^3 x - 3 \cos x \sin^2 x$ is an identity.   ■

Note that in Example 6 we used identities to expand the simpler side of the equation. In this case, simplifying the more complicated side is more difficult.

Example **7** **Verifying identities involving half-angles**

Prove that the following equation is an identity:

$$\tan \frac{x}{2} + \cot \frac{x}{2} = 2 \csc x$$

Solution

Write the left-hand side in terms of $\tan(x/2)$ and then use two different half-angle identities for $\tan(x/2)$:

$$\tan \frac{x}{2} + \cot \frac{x}{2} = \tan \frac{x}{2} + \frac{1}{\tan \dfrac{x}{2}}$$

$$= \frac{\sin x}{1 + \cos x} + \frac{\sin x}{1 - \cos x}$$

$$= \frac{\sin x(1 - \cos x)}{(1 + \cos x)(1 - \cos x)} + \frac{\sin x(1 + \cos x)}{(1 - \cos x)(1 + \cos x)}$$

$$= \frac{2 \sin x}{1 - \cos^2 x}$$

$$= \frac{2 \sin x}{\sin^2 x} = 2 \frac{1}{\sin x} = 2 \csc x$$

Try This. Prove $\sin^2\left(\frac{x}{2}\right) \cos^2\left(\frac{x}{2}\right) = \frac{\sin^2 x}{4}$ is an identity.   ■

For Thought

True or False? Explain.

1. $\dfrac{\sin 42°}{2} = \sin 21° \cos 21°$

2. $\cos\left(\sqrt{8}\right) = 2 \cos^2\left(\sqrt{2}\right) - 1$

3. $\sin 150° = \sqrt{\dfrac{1 - \cos 75°}{2}}$

4. $\sin 200° = -\sqrt{\dfrac{1 - \cos 40°}{2}}$

5. $\tan\dfrac{7\pi}{8} = \sqrt{\dfrac{1 - \cos(7\pi/4)}{1 + \cos(7\pi/4)}}$

6. $\tan(-\pi/8) = \dfrac{1 - \cos(\pi/4)}{\sin(-\pi/4)}$

7. For any real number x, $\dfrac{\sin 2x}{2} = \sin x$.

8. The equation $\cos x = \sqrt{\dfrac{1 + \cos 2x}{2}}$ is an identity.

9. The equation $\sqrt{(1 - \cos x)^2} = 1 - \cos x$ is an identity.

10. If $180° < \alpha < 360°$, then $\sin\alpha < 0$ and $\tan(\alpha/2) < 0$.

6.4 Exercises

Find the exact value of each expression using double-angle identities.

1. $\sin(90°)$ **2.** $\cos(60°)$ **3.** $\tan(60°)$ **4.** $\cos(180°)$

5. $\sin(3\pi/2)$ **6.** $\cos(4\pi/3)$ **7.** $\tan(4\pi/3)$ **8.** $\sin(2\pi/3)$

Find the exact value of each expression using the half-angle identities.

9. $\cos(15°)$ **10.** $\cos(\pi/8)$ **11.** $\sin(15°)$

12. $\sin(-\pi/6)$ **13.** $\tan(15°)$ **14.** $\tan(3\pi/8)$

15. $\sin(22.5°)$ **16.** $\tan(75°)$

For each equation determine whether the positive or negative sign makes the equation correct. Do not use a calculator.

17. $\sin 118.5° = \pm\sqrt{\dfrac{1 - \cos 237°}{2}}$

18. $\sin 222.5° = \pm\sqrt{\dfrac{1 - \cos 445°}{2}}$

19. $\cos 100° = \pm\sqrt{\dfrac{1 + \cos 200°}{2}}$

20. $\cos\dfrac{9\pi}{7} = \pm\sqrt{\dfrac{1 + \cos(18\pi/7)}{2}}$

21. $\tan\dfrac{-5\pi}{12} = \pm\sqrt{\dfrac{1 - \cos(-5\pi/6)}{1 + \cos(-5\pi/6)}}$

22. $\tan\dfrac{17\pi}{12} = \pm\sqrt{\dfrac{1 - \cos(17\pi/6)}{1 + \cos(17\pi/6)}}$

Use identities to simplify each expression. Do not use a calculator.

23. $2\sin 13° \cos 13°$

24. $\sin^2\left(\dfrac{\pi}{5}\right) - \cos^2\left(\dfrac{\pi}{5}\right)$

25. $2\cos^2(22.5°) - 1$

26. $1 - 2\sin^2\left(-\dfrac{\pi}{8}\right)$

27. $\dfrac{\tan 15°}{1 - \tan^2(15°)}$

28. $\dfrac{\tan 30°}{1 - \tan^2(30°)}$

29. $\dfrac{\sin 12°}{1 + \cos 12°}$

30. $\csc 8°(1 - \cos 8°)$

31. $2\sin\left(\dfrac{\pi}{9} - \dfrac{\pi}{2}\right)\cos\left(\dfrac{\pi}{2} - \dfrac{\pi}{9}\right)$

32. $2\cos^2\left(\dfrac{\pi}{5} - \dfrac{\pi}{2}\right) - 1$ **33.** $\cos^2\left(\dfrac{\pi}{9}\right) - \sin^2\left(\dfrac{\pi}{9}\right)$

34. $\dfrac{2}{\cot 5(1 - \tan^2 5)}$

Match each given expression with an equivalent expression (a)–(j).

35. $1 - \cos^2 2$ **36.** $\dfrac{2\tan 2}{1 - \tan^2 2}$ **37.** $\dfrac{1 + \cos 2}{2}$

38. $\dfrac{1 + \cos 2}{\sin 2}$ **39.** $2\sin 2\cos 2$ **40.** $\dfrac{1}{\cos^2 2} - 1$

41. $\dfrac{1 - \cos 2}{1 + \cos 2}$ **42.** $\cos^2 2 - \sin^2 2$

43. $\dfrac{1 - \cos 2}{2}$ **44.** $1 - \sin^2 2$

a. $\sin 4$ **b.** $\cos 4$ **c.** $\sin^2 2$

d. $\cos^2 2$ **e.** $\tan 4$ **f.** $\sin^2 1$

g. $\cos^2 1$ **h.** $\tan^2 1$ **i.** $\cot 1$

j. $\tan^2 2$

In each case, find $\sin\alpha$, $\cos\alpha$, $\tan\alpha$, $\csc\alpha$, $\sec\alpha$, and $\cot\alpha$.

45. $\cos(2\alpha) = 3/5$ and $0° < 2\alpha < 90°$

46. $\cos(2\alpha) = 1/3$ and $360° < 2\alpha < 450°$

47. $\sin(2\alpha) = 5/13$ and $0° < \alpha < 45°$

48. $\sin(2\alpha) = -8/17$ and $180° < 2\alpha < 270°$

49. $\cos(\alpha/2) = -1/4$ and $\pi/2 < \alpha/2 < 3\pi/4$

50. $\sin(\alpha/2) = -1/3$ and $7\pi/4 < \alpha/2 < 2\pi$

51. $\sin(\alpha/2) = 4/5$ and $\alpha/2$ is in quadrant II

52. $\sin(\alpha/2) = 1/5$ and $\alpha/2$ is in quadrant II

Verify that each equation is an identity.

53. $\cos^4 s - \sin^4 s = \cos 2s$

54. $\sin 2s = -2 \sin s \sin(s - \pi/2)$

55. $\cos 3t = \cos^3 t - 3 \sin^2 t \cos t$

56. $\dfrac{\sin 4t}{4} = \cos^3 t \sin t - \sin^3 t \cos t$

57. $\dfrac{\cos 2x + \cos 2y}{\sin x + \cos y} = 2 \cos y - 2 \sin x$

58. $(\sin \alpha - \cos \alpha)^2 = 1 - \sin 2\alpha$

59. $\dfrac{\cos 2x}{\sin^2 x} = \csc^2 x - 2$ 　　**60.** $\dfrac{\cos 2s}{\cos^2 s} = \sec^2 s - 2 \tan^2 s$

61. $2 \sin^2\left(\dfrac{u}{2}\right) = \dfrac{\sin^2 u}{1 + \cos u}$ 　　**62.** $\cos 2y = \dfrac{1 - \tan^2 y}{1 + \tan^2 y}$

63. $\tan^2\left(\dfrac{x}{2}\right) = \dfrac{\sec x + \cos x - 2}{\sec x - \cos x}$

64. $\sec^2\left(\dfrac{x}{2}\right) = \dfrac{2 \sec x + 2}{\sec x + 2 + \cos x}$

65. $\dfrac{1 - \sin^2\left(\dfrac{x}{2}\right)}{1 + \sin^2\left(\dfrac{x}{2}\right)} = \dfrac{1 + \cos x}{3 - \cos x}$

66. $\dfrac{1 - \cos^2\left(\dfrac{x}{2}\right)}{1 - \sin^2\left(\dfrac{x}{2}\right)} = \dfrac{1 - \cos x}{1 + \cos x}$

For each equation, either prove that it is an identity or prove that it is not an identity.

67. $\sin(2x) = 2 \sin x$ 　　**68.** $\dfrac{\cos 2x}{2} = \cos x$

69. $\tan\left(\dfrac{x}{2}\right) = \dfrac{1}{2} \tan x$ 　　**70.** $\tan\left(\dfrac{x}{2}\right) = \sqrt{\dfrac{1 - \cos x}{1 + \cos x}}$

71. $\sin(2x) \cdot \sin\left(\dfrac{x}{2}\right) = \sin^2 x$ 　　**72.** $\tan x + \tan x = \tan 2x$

73. $\cot\dfrac{x}{2} - \tan\dfrac{x}{2} = \dfrac{\sin 2x}{\sin^2 x}$

74. $\csc^2\left(\dfrac{x}{2}\right) + \sec^2\left(\dfrac{x}{2}\right) = 4 \csc^2 x$

Solve each problem.

75. Find the exact value of $\sin(2\alpha)$ given that $\sin(\alpha) = 3/5$ and α is in quadrant II.

76. Find the exact value of $\sin(2\alpha)$ given that $\tan(\alpha) = -8/15$ and α is in quadrant IV.

77. Find the exact value of $\cos(2\alpha)$ given that $\sin(\alpha) = 8/17$ and α is in quadrant II.

78. Find the exact value of $\tan(2\alpha)$ given that $\sin(\alpha) = -4/5$ and α is in quadrant III.

79. In the figure, $\angle B$ is a right angle and the line segment $\overline{AD}$ bisects $\angle A$. If $AB = 5$ and $BC = 3$, what is the exact value of BD?

　　HINT Use the half-angle identity for tangent.

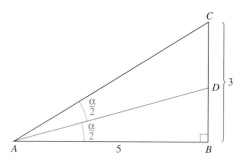

▪ **Figure for Exercise 79**

80. In the figure, $\angle B$ is a right angle and the line segment $\overline{AD}$ bisects $\angle A$. If $AB = 10$ and $BD = 2$, what is the exact value of CD?

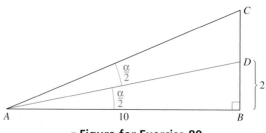

▪ **Figure for Exercise 80**

81. *Viewing Area* Find a formula for the viewing area of a television screen in terms of its diagonal and the angle α shown in the figure. Rewrite the formula using a single trigonometric function.

82. Use the formula from the previous exercise to find the viewing area for a 32 in. diagonal measure television for which $\alpha = 37.2°$.

■ **Figure for Exercises 81 and 82**

For Writing/Discussion

83. Show that $\tan(x/2)$ has the same sign as $\sin x$ for any real number x.

84. Explain why $1 + \cos x \geq 0$ for any real number x.

85. Explain why $\sin 2x = 2 \sin x$ is not an identity by using graphs and by using the definition of the sine function.

Thinking Outside the Box LIV

Completely Saturated   Four lawn sprinklers are positioned at the vertices of a square that is 2 meters on each side. If each sprinkler waters a circular area with radius 2 meters, then what is the exact area of the region that gets watered by all four sprinklers?

6.4 Pop Quiz

1. Find $\sin(2\alpha)$ exactly if $\sin \alpha = 1/4$ and $\pi/2 < \alpha < \pi$.

2. Find $\sin(\alpha/2)$ exactly if $\sin \alpha = -4/5$ and $\pi < \alpha < 3\pi/2$.

3. Prove that $\sin^4 x - \cos^4 x = -\cos(2x)$

Linking Concepts

For Individual or Group Explorations

Modeling the Motion of a Football

A good field-goal kicker must learn through experience how to get the maximum distance in a kick. However, without touching a football, we can find the angle at which the football should be kicked to achieve the maximum distance. Assuming no air resistance, the distance in feet that a projectile (such as a football) travels when launched from the ground with an initial velocity of v_0 ft/sec is given by

$$x = (v_0^2 \sin\theta \cos\theta)/16,$$

where θ is the angle between the trajectory and the ground, as shown in the figure. (See 6.1 Linking Concepts.)

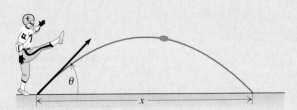

a)   Use an identity to write the distance x as a function of 2θ.

b)   Graph the function that you found in part (a) using an initial velocity of 50 ft/sec.

(continued on next page)

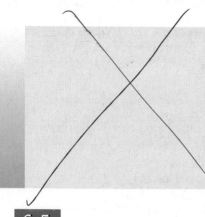

c) From your graph, determine the value of θ that maximizes x. Does this value of θ maximize x for any velocity? Explain.

d) When a kicker kicks a football 55 yd using the angle determined in part (c), what is the initial velocity of the football in miles per hour?

e) Do you think that the actual initial velocity for a 55-yd field goal is larger or smaller than that found in part (d)?

6.5

Product and Sum Identities

In Section 5.2 we saw that the location x at time t for an object on a spring, given an initial velocity v_0 from an initial location x_0, is given by

$$x = \frac{v_0}{\omega} \sin(\omega t) + x_0 \cos(\omega t),$$

where ω is a constant. Even though x is determined by a combination of sine and cosine, an object in motion on a spring oscillates in a "periodic" fashion. In this section we will develop identities that involve sums of trigonometric functions. These identities are used in solving equations, graphing functions, and explaining how the combination of the sine and cosine functions in the spring equation can work together to produce a periodic motion.

Product-to-Sum Identities

In Section 6.3 we learned identities for sine and cosine of a sum or difference:

$$\sin(A + B) = \sin A \cos B + \cos A \sin B$$

$$\sin(A - B) = \sin A \cos B - \cos A \sin B$$

$$\cos(A + B) = \cos A \cos B - \sin A \sin B$$

$$\cos(A - B) = \cos A \cos B + \sin A \sin B$$

Now we can add the first two of these identities to get a new identity:

$$\sin(A + B) = \sin A \cos B + \cos A \sin B$$

$$\underline{\sin(A - B) = \sin A \cos B - \cos A \sin B}$$

$$\sin(A + B) + \sin(A - B) = 2 \sin A \cos B$$

When we divide each side by 2, we get an identity expressing a product in terms of a sum:

$$\sin A \cos B = \frac{1}{2} [\sin(A + B) + \sin(A - B)]$$

We can produce three other, similar identities from the sum and difference identities. The four identities, known as the **product-to-sum identities** are listed below. These

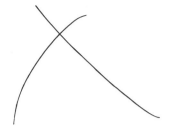

identities are not used as often as the other identities that we study in this chapter. It is not necessary to memorize these identities. Just remember them by name and look them up as necessary.

Product-to-Sum Identities

$$\sin A \cos B = \frac{1}{2}[\sin(A + B) + \sin(A - B)]$$

$$\sin A \sin B = \frac{1}{2}[\cos(A - B) - \cos(A + B)]$$

$$\cos A \sin B = \frac{1}{2}[\sin(A + B) - \sin(A - B)]$$

$$\cos A \cos B = \frac{1}{2}[\cos(A - B) + \cos(A + B)]$$

Example ▮1▮ **Expressing a product as a sum**

Use the product-to-sum identities to rewrite each expression.

a. $\sin 12° \cos 9°$ **b.** $\sin(\pi/12) \sin(\pi/8)$

Solution

a. Use the product-to-sum identity for $\sin A \cos B$:

$$\sin 12° \cos 9° = \frac{1}{2}[\sin(12° + 9°) + \sin(12° - 9°)]$$

$$= \frac{1}{2}[\sin 21° + \sin 3°]$$

b. Use the product-to-sum identity for $\sin A \sin B$:

$$\sin\left(\frac{\pi}{12}\right) \sin\left(\frac{\pi}{8}\right) = \frac{1}{2}\left[\cos\left(\frac{\pi}{12} - \frac{\pi}{8}\right) - \cos\left(\frac{\pi}{12} + \frac{\pi}{8}\right)\right]$$

$$= \frac{1}{2}\left[\cos\left(-\frac{\pi}{24}\right) - \cos\left(\frac{5\pi}{24}\right)\right]$$

$$= \frac{1}{2}\left[\cos\frac{\pi}{24} - \cos\frac{5\pi}{24}\right]$$

```
sin(π/12)sin(π/8
)
      .0990457605
.5(cos(π/24)-cos
(5π/24))
      .0990457605
```

■ **Figure 6.18**

◪⊿ Use a calculator to check, as shown in Fig. 6.18.

Try This. Use a product-to-sum identity to rewrite $\sin(4x) \cos(3x)$. ■

Example ▮2▮ **Evaluating a product**

Use a product-to-sum identity to find the exact value of $\cos(67.5°) \sin(112.5°)$.

Solution

Use the identity $\cos A \sin B = \frac{1}{2}[\sin(A + B) - \sin(A - B)]$:

$$\cos(67.5°)\sin(112.5°) = \frac{1}{2}[\sin(67.5° + 112.5°) - \sin(67.5° - 112.5°)]$$

$$= \frac{1}{2}[\sin(180°) - \sin(-45°)]$$

$$= \frac{1}{2}[0 + \sin(45°)]$$

$$= \frac{1}{2} \cdot \frac{\sqrt{2}}{2} = \frac{\sqrt{2}}{4}$$

Try This. Find the exact value of $\sin(52.5°)\cos(7.5°)$.   ▪

Sum-to-Product Identities

It is sometimes useful to write a sum of two trigonometric functions as a product. Consider the product-to-sum formula

$$\sin A \cos B = \frac{1}{2}[\sin(A + B) + \sin(A - B)].$$

To make the right-hand side look simpler, we let $A + B = x$ and $A - B = y$. From these equations we get $x + y = 2A$ and $x - y = 2B$, or

$$A = \frac{x + y}{2} \qquad \text{and} \qquad B = \frac{x - y}{2}.$$

Substitute these values into the identity for $\sin A \cos B$:

$$\sin\left(\frac{x + y}{2}\right)\cos\left(\frac{x - y}{2}\right) = \frac{1}{2}[\sin x + \sin y]$$

Multiply each side by 2 to get

$$\sin x + \sin y = 2\sin\left(\frac{x + y}{2}\right)\cos\left(\frac{x - y}{2}\right).$$

The other three product-to-sum formulas can also be rewritten as sum-to-product identities by using similar procedures.

Sum-to-Product Identities

$$\sin x + \sin y = 2\sin\left(\frac{x + y}{2}\right)\cos\left(\frac{x - y}{2}\right)$$

$$\sin x - \sin y = 2\cos\left(\frac{x + y}{2}\right)\sin\left(\frac{x - y}{2}\right)$$

$$\cos x + \cos y = 2\cos\left(\frac{x + y}{2}\right)\cos\left(\frac{x - y}{2}\right)$$

$$\cos x - \cos y = -2\sin\left(\frac{x + y}{2}\right)\sin\left(\frac{x - y}{2}\right)$$

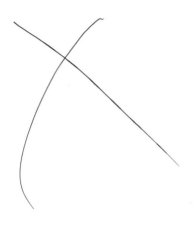

Example **3** Expressing a sum or difference as a product

Use the sum-to-product identities to rewrite each expression.

a. $\sin 8° + \sin 6°$ **b.** $\cos(\pi/5) - \cos(\pi/8)$ **c.** $\sin(6t) - \sin(4t)$

Solution

a. Use the sum-to-product identity for $\sin x + \sin y$:

$$\sin 8° + \sin 6° = 2 \sin\left(\frac{8° + 6°}{2}\right) \cos\left(\frac{8° - 6°}{2}\right) = 2 \sin 7° \cos 1°$$

b. Use the sum-to-product identity for $\cos x - \cos y$:

$$\cos\left(\frac{\pi}{5}\right) - \cos\left(\frac{\pi}{8}\right) = -2 \sin\left(\frac{\pi/5 + \pi/8}{2}\right) \sin\left(\frac{\pi/5 - \pi/8}{2}\right)$$

$$= -2 \sin\left(\frac{13\pi}{80}\right) \sin\left(\frac{3\pi}{80}\right)$$

c. Use the sum-to-product identity for $\sin x - \sin y$:

$$\sin 6t - \sin 4t = 2 \cos\left(\frac{6t + 4t}{2}\right) \sin\left(\frac{6t - 4t}{2}\right)$$

$$= 2 \cos 5t \sin t$$

Try This. Use a sum-to-product identity to rewrite $\cos(4x) + \cos(2x)$.   ■

Example **4** Evaluating a sum

Use a sum-to-product identity to find the exact value of $\cos(112.5°) + \cos(67.5°)$.

Solution

Use the sum-to-product identity $\cos x + \cos y = 2 \cos\left(\frac{x + y}{2}\right) \cos\left(\frac{x - y}{2}\right)$:

$$\cos(112.5°) + \cos(67.5°) = 2 \cos\left(\frac{112.5° + 67.5°}{2}\right) \cos\left(\frac{112.5° - 67.5°}{2}\right)$$

$$= 2 \cos(90°) \cos(22.5°)$$

$$= 0 \quad \text{Because } \cos(90°) = 0.$$

Try This. Use a sum-to-product identity to evaluate $\sin(105°) + \sin(15°)$.   ■

The Function $y = a \sin x + b \cos x$

The function $y = a \sin x + b \cos x$ involves an expression similar to those in the sum-to-product identities, but it is not covered by the sum-to-product identities. Functions of this type occur in applications such as the position of a weight in

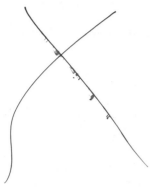

motion due to the force of a spring, the position of a swinging pendulum, and the current in an electrical circuit. In these applications it is important to express this function in terms of a single trigonometric function.

To get an identity for $a \sin x + b \cos x$ for nonzero values of a and b, we let α be an angle in standard position whose terminal side contains the point (a, b). Using trigonometric ratios from Section 5.6, we get

$$\sin \alpha = \frac{b}{r} = \frac{b}{\sqrt{a^2 + b^2}} \quad \text{and} \quad \cos \alpha = \frac{a}{r} = \frac{a}{\sqrt{a^2 + b^2}}.$$

Now write $a \sin x + b \cos x$ with the trigonometric ratios, and then replace the ratios by $\sin \alpha$ and $\cos \alpha$:

$$a \sin x + b \cos x = \sqrt{a^2 + b^2} \left((\sin x) \frac{a}{\sqrt{a^2 + b^2}} + (\cos x) \frac{b}{\sqrt{a^2 + b^2}} \right)$$

$$= \sqrt{a^2 + b^2} \, (\sin x \cos \alpha + \cos x \sin \alpha)$$

$$= \sqrt{a^2 + b^2} \, \sin(x + \alpha) \quad \text{By the identity for the sine of a sum}$$

This identity is called the **reduction formula** because it reduces two trigonometric functions to one.

Theorem:
Reduction Formula

> If α is an angle in standard position whose terminal side contains (a, b), then
>
> $$a \sin x + b \cos x = \sqrt{a^2 + b^2} \sin (x + \alpha) \text{ for any real number } x.$$

To rewrite an expression of the form $a \sin x + b \cos x$ by using the reduction formula, we need to find α so that the terminal side of α goes through (a, b). By using trigonometric ratios, we have

$$\sin \alpha = \frac{b}{\sqrt{a^2 + b^2}}, \quad \cos \alpha = \frac{a}{\sqrt{a^2 + b^2}}, \quad \text{and} \quad \tan \alpha = \frac{b}{a}.$$

Since we know a and b, we can find $\sin \alpha$, $\cos \alpha$, or $\tan \alpha$, and then use an inverse trigonometric function to find α. However, because of the ranges of the inverse functions, the angle obtained from an inverse function might not have its terminal side through (a, b) as required.

Example **5** **Using the reduction formula**

Use the reduction formula to rewrite $-3 \sin x - 3 \cos x$ in the form $A \sin(x + C)$.

Solution

Because $a = -3$ and $b = -3$, we have

$$\sqrt{a^2 + b^2} = \sqrt{18} = 3\sqrt{2}.$$

Since the terminal side of α must go through $(-3, -3)$, we have

$$\cos \alpha = \frac{-3}{3\sqrt{2}} = -\frac{\sqrt{2}}{2}.$$

Now $\cos^{-1}(-\sqrt{2}/2) = 3\pi/4$, but the terminal side of $3\pi/4$ is in quadrant II, as shown in Fig. 6.19. However, we also have $\cos(5\pi/4) = -\sqrt{2}/2$ and the terminal

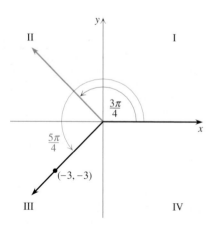

■ **Figure 6.19**

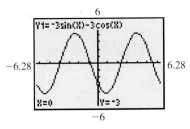

■ **Figure 6.20**

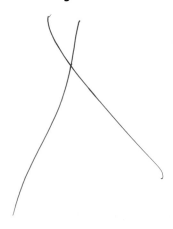

side for $5\pi/4$ does pass through $(-3, -3)$ in quadrant III. So $\alpha = 5\pi/4$. By the reduction formula,

$$-3 \sin x - 3 \cos x = 3\sqrt{2} \sin\left(x + \frac{5\pi}{4}\right).$$

The reduction formula explains why the calculator graph of $y = -3 \sin x - 3 \cos x$ in Fig. 6.20 appears to be a sine wave with amplitude $3\sqrt{2}$.

Try This. Rewrite $-2 \sin x + 2 \cos x$ in the form $A \sin(x + C)$. ■

In the next example, we see that the reduction formula makes it easier to understand functions of the form $y = a \sin x + b \cos x$.

Example **6** **Using the reduction formula in graphing**

Graph one cycle of the function

$$y = \sqrt{3} \sin x + \cos x$$

and state the amplitude, period, and phase shift.

Solution

Because $a = \sqrt{3}$ and $b = 1$, we have $\sqrt{a^2 + b^2} = \sqrt{4} = 2$. The terminal side of α must go through $(\sqrt{3}, 1)$. So $\sin \alpha = 1/2$, and $\sin^{-1}(1/2) = \pi/6$. Since the terminal side of $\pi/6$ is in quadrant I and goes through $(\sqrt{3}, 1)$, we can rewrite the function as

$$y = \sqrt{3} \sin x + \cos x = 2 \sin\left(x + \frac{\pi}{6}\right).$$

From the new form, we see that the graph is a sine wave with amplitude 2, period 2π, and phase shift $-\pi/6$, as shown in Fig. 6.21.

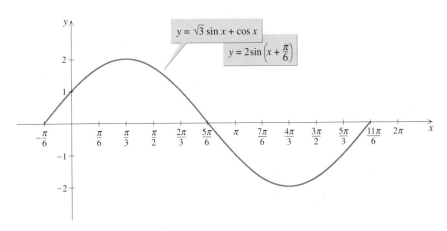

■ **Figure 6.21**

The calculator graph shown in Fig. 6.22 supports these conclusions.

Try This. Determine the amplitude and phase shift for $y = \sin x - \cos x$. ■

3

`Y1=√(3)sin(X)+cos(X)`

−6.28   6.28

X=0   Y=1

−3

■ **Figure 6.22**

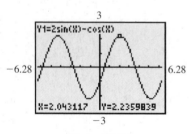

Figure 6.23

Since $y = a \sin x + b \cos x$ and $y = \sqrt{a^2 + b^2} \sin(x + \alpha)$ are the same function, the graph of $y = a \sin x + b \cos x$ is a transformation of $y = \sin x$. So, for nonzero a and b, the graph of $y = a \sin x + b \cos x$ is a sine wave with amplitude $\sqrt{a^2 + b^2}$, period 2π, and phase shift $-\alpha$.

Modeling the Motion of a Spring

In Section 5.2, the function

$$x = \frac{v_0}{\omega} \sin(\omega t) + x_0 \cos(\omega t)$$

was used to give the position x at time t for a weight in motion on a vertical spring. In this formula, v_0 is the initial velocity, x_0 is the initial position, and ω is a constant. The same equation is used for a weight attached to a horizontal spring and set in motion on a frictionless surface, as shown in Fig. 6.23.

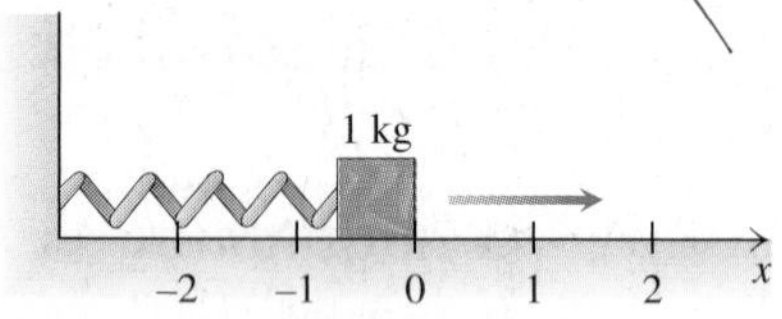

We use the reduction formula on the spring equation to determine how far a spring actually stretches and compresses when the block is set in motion with a particular velocity.

Example **7** **Using the reduction formula with springs**

At time $t = 0$ the 1-kilogram block shown in Fig. 6.23 is moved (and the spring compressed) to a position 1 meter to the left of the resting position. From this position the block is given a velocity of 2 meters per second to the right. Use the reduction formula to find the maximum distance reached by the block from rest. Assume that $\omega = 1$.

Solution

Since $v_0 = 2$, $x_0 = -1$, and $\omega = 1$, the position of the block at time t is given by

$$x = 2 \sin t - \cos t.$$

Use the reduction formula to rewrite this equation. If $a = 2$ and $b = -1$, then

$$\sqrt{a^2 + b^2} = \sqrt{2^2 + (-1)^2} = \sqrt{5}.$$

If the terminal side of α goes through $(2, -1)$, then $\tan \alpha = -1/2$ and

$$\tan^{-1}(-1/2) \approx -0.46.$$

An angle of -0.46 rad lies in quadrant IV and goes through $(2, -1)$. Use $\alpha = -0.46$ in the reduction formula to get

$$x = \sqrt{5} \sin(t - 0.46).$$

Since the amplitude of this function is $\sqrt{5}$, the block oscillates between $x = \sqrt{5}$ meters and $x = -\sqrt{5}$ meters. The maximum distance from $x = 0$ is $\sqrt{5}$ meters. The maximum y value on the calculator graph in Fig. 6.24 supports the conclusion that the amplitude is $\sqrt{5}$.

Try This. The location in centimeters of a block on a spring is given by $x = \sin t - 3 \cos t$. Find the maximum distance reached by the block from rest. ■

Figure 6.24

For Thought

True or False? Explain.

1. $\sin 45° \cos 15° = 0.5[\sin 60° + \sin 30°]$

2. $\cos(\pi/8)\sin(\pi/4) = 0.5[\sin(3\pi/8) - \sin(\pi/8)]$

3. $2\cos 6° \cos 8° = \cos 2° + \cos 14°$

4. $\sin 5° - \sin 9° = 2\cos 7° \sin 2°$

5. $\cos 4 + \cos 12 = 2\cos 8 \cos 4$

6. $\cos(\pi/3) - \cos(\pi/2) = -2\sin(5\pi/12)\sin(\pi/12)$

7. $\sin(\pi/6) + \cos(\pi/6) = \sqrt{2}\sin(\pi/6 + \pi/4)$

8. $\frac{1}{2}\sin(\pi/6) + \frac{\sqrt{3}}{2}\cos(\pi/6) = \sin(\pi/2)$

9. The graph of $y = \frac{1}{2}\sin x + \frac{\sqrt{3}}{2}\cos x$ is a sine wave with amplitude 1.

10. The equation $\frac{1}{\sqrt{2}}\sin x + \frac{1}{\sqrt{2}}\cos x = \sin\left(x + \frac{\pi}{4}\right)$ is an identity.

6.5 Exercises

Use the product-to-sum identities to rewrite each expression.

1. $\sin 13° \sin 9°$

2. $\cos 34° \cos 39°$

3. $\sin 16° \cos 20°$

4. $\cos 9° \sin 8°$

5. $\sin(5°)\cos(10°)$

6. $\cos(6°)\cos(8°)$

7. $\cos\left(\frac{\pi}{6}\right)\cos\left(\frac{\pi}{5}\right)$

8. $\sin\left(\frac{2\pi}{9}\right)\sin\left(\frac{3\pi}{4}\right)$

9. $\cos(5y^2)\cos(7y^2)$

10. $\cos 3t \sin 5t$

11. $\sin(2s - 1)\cos(s + 1)$

12. $\sin(3t - 1)\sin(2t + 3)$

Find the exact value of each product. Do not use a calculator.

13. $\sin(52.5°)\sin(7.5°)$

14. $\cos(105°)\cos(75°)$

15. $\sin\left(\frac{13\pi}{24}\right)\cos\left(\frac{5\pi}{24}\right)$

16. $\cos\left(\frac{5\pi}{24}\right)\sin\left(-\frac{\pi}{24}\right)$

Use the sum-to-product identities to rewrite each expression.

17. $\sin 12° - \sin 8°$

18. $\sin 7° + \sin 11°$

19. $\cos 80° - \cos 87°$

20. $\cos 44° + \cos 31°$

21. $\sin 3.6 - \sin 4.8$

22. $\sin 5.1 + \sin 6.3$

23. $\cos\left(\frac{\pi}{3}\right) - \cos\left(\frac{\pi}{5}\right)$

24. $\cos\left(\frac{1}{2}\right) + \cos\left(\frac{2}{3}\right)$

25. $\cos(5y - 3) - \cos(3y + 9)$

26. $\cos(6t^2 - 1) + \cos(4t^2 - 1)$

27. $\sin 5\alpha - \sin 8\alpha$

28. $\sin 3s + \sin 5s$

Find the exact value of each sum. Do not use a calculator.

29. $\sin(75°) + \sin(15°)$

30. $\sin(285°) - \sin(15°)$

31. $\cos\left(-\frac{\pi}{24}\right) - \cos\left(\frac{7\pi}{24}\right)$

32. $\cos\left(\frac{5\pi}{24}\right) + \cos\left(\frac{\pi}{24}\right)$

Rewrite each expression in the form of $A\sin(x + C)$.

33. $\sin x - \cos x$

34. $2\sin x + 2\cos x$

35. $-\frac{1}{2}\sin x + \frac{\sqrt{3}}{2}\cos x$

36. $\frac{\sqrt{2}}{2}\sin x - \frac{\sqrt{2}}{2}\cos x$

37. $\frac{\sqrt{3}}{2}\sin x - \frac{1}{2}\cos x$

38. $-\frac{\sqrt{3}}{2}\sin x - \frac{1}{2}\cos x$

Write each function in the form $y = A\sin(x + C)$. Then graph at least one cycle and state the amplitude, period, and phase shift.

39. $y = -\sin x + \cos x$

40. $y = \sin x + \sqrt{3}\cos x$

41. $y = \sqrt{2}\sin x - \sqrt{2}\cos x$

42. $y = 2\sin x - 2\cos x$

43. $y = -\sqrt{3}\sin x - \cos x$

44. $y = -\frac{1}{2}\sin x - \frac{\sqrt{3}}{2}\cos x$

For each function, determine the exact amplitude and find the phase shift in radians (to the nearest tenth).

45. $y = 3 \sin x + 4 \cos x$ **46.** $y = \sin x + 5 \cos x$

47. $y = -6 \sin x + \cos x$ **48.** $y = -\sqrt{5} \sin x + 2 \cos x$

49. $y = -3 \sin x - 5 \cos x$

50. $y = -\sqrt{2} \sin x - \sqrt{7} \cos x$

Prove that each equation is an identity.

51. $\dfrac{\sin 3t - \sin t}{\cos 3t + \cos t} = \tan t$

52. $\dfrac{\sin 3x + \sin 5x}{\sin 3x - \sin 5x} = -\dfrac{\tan 4x}{\tan x}$

53. $\dfrac{\cos x - \cos 3x}{\cos x + \cos 3x} = \tan 2x \tan x$

54. $\dfrac{\cos 5y + \cos 3y}{\cos 5y - \cos 3y} = -\cot 4y \cot y$

55. $\cos^2 x - \cos^2 y = -\sin(x + y) \sin(x - y)$

56. $\sin^2 x - \sin^2 y = \sin(x + y) \sin(x - y)$

57. $\left(\sin \dfrac{x + y}{2} + \cos \dfrac{x + y}{2} \right) \left(\sin \dfrac{x - y}{2} + \cos \dfrac{x - y}{2} \right) =$
$\sin x + \cos y$

58. $\sin 2A \sin 2B = \sin^2(A + B) - \sin^2(A - B)$

59. $\cos^2(A - B) - \cos^2(A + B) = \sin^2(A + B) - \sin^2(A - B)$

60. $(\sin A + \cos A)(\sin B + \cos B) =$
$\sin(A + B) + \cos(A - B)$

Solve each problem.

61. *Motion of a Spring*   A block is attached to a spring and set in motion, as in Example 7. For this block and spring, the location on the surface at any time t in seconds is given in meters by $x = \sqrt{3} \sin t + \cos t$. Use the reduction formula to rewrite this function, and find the maximum distance reached by the block from the resting position.

62. *Motion of a Spring*   A block hanging from a spring, as shown in the figure, oscillates in the same manner as the block of Example 7. If a 1-kg block attached to a certain spring is given an upward velocity of 0.3 m/sec from a point 0.5 m below its resting position, then its position at any time t in seconds is given in meters by $x = -0.3 \sin t + 0.5 \cos t$. Use the reduction formula to find the maximum distance that the block travels from the resting position.

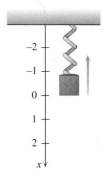

■ **Figure for Exercise 62**

For Writing/Discussion

63. Derive the identity $\cos(2x) = \cos^2 x - \sin^2 x$ from a product-to-sum identity.

64. Derive the identity $\sin(2x) = 2 \sin x \cos x$ from a product-to-sum identity.

65. Prove the three product-to-sum identities that were not proved in the text.

66. Prove the three sum-to-product identities that were not proved in the text.

Thinking Outside the Box LV & LVI

Overlapping Circles   Two lawn sprinklers that each water a circular region with radius a are placed a feet apart. What is the total area watered by the sprinklers?

Marching Ant   An ant starts in the center of the upper left square of an 8 by 8 chessboard. He must crawl to the lower right square of the board, passing through each square exactly once and not passing through a vertex of any square. Either find a path or explain why there is no such path.

6.5 Pop Quiz

1. Use a product to sum identity to simplify $\frac{1}{2}[\sin(2\alpha + \beta) - \sin(2\alpha - \beta)]$.

2. Rewrite $\cos(2\alpha) + \cos(4\alpha)$ as a product using a sum-to-product identity.

3. Write $y = \sqrt{3} \sin x - \cos x$ in the form $y = A \sin(x + C)$.

4. Find the amplitude for $y = 3 \sin x - 5 \cos x$?

Linking Concepts

For Individual or
Group Explorations

Maximizing the Total Profit

Profits at The Christmas Store vary periodically with a high of $50,000 in December and a low of $10,000 in June, as shown in the accompanying graph. The owner of the Christmas Store also owns The Pool Store, where profits reach a high of $80,000 in August and a low of $20,000 in February. Assume that the profit function for each store is a sine wave.

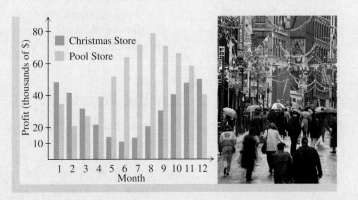

a) Write the profit function for The Christmas Store as a function of the month and sketch its graph.

b) Write the profit function for The Pool Store as a function of the month and sketch its graph.

c) Write the total profit as a function of the month and sketch its graph. What is the period?

d) Use the maximum feature of a graphing calculator to find the owner's maximum total profit and the month in which it occurs.

e) Find the owner's minimum total profit and the month in which it occurs.

f) We know that $y = a \sin x + b \cos x$ is a sine function. However, the sum of two arbitrary sine or cosine functions is not necessarily a sine function. Find an example in which the graph of the sum of two sine functions does not look like a sine curve.

g) Do you think that the sum of any two sine functions is a periodic function? Explain.

6.6

Conditional Trigonometric Equations

An identity is satisfied by *all* values of the variable for which both sides are defined. In this section we investigate **conditional equations,** those equations that are not identities but that have *at least one* solution. For example, the equation $\sin x = 0$ is a conditional equation. Because the trigonometric functions are periodic, conditional equations involving trigonometric functions usually have infinitely many solutions. The equation $\sin x = 0$ is satisfied by $x = 0, \pm\pi, \pm 2\pi, \pm 3\pi$, and so on. All of these solutions are of the form $k\pi$, where k is an integer. In this section we will solve conditional equations and see that identities play a fundamental role in their solution.

Cosine Equations

The most basic conditional equation involving cosine is of the form $\cos x = a$, where a is a number in the interval $[-1, 1]$. If a is not in $[-1, 1]$, $\cos x = a$ has no solution. For a in $[-1, 1]$, the equation $x = \cos^{-1} a$ provides one solution in the interval $[0, \pi]$. From this single solution we can determine all of the solutions because of the periodic nature of the cosine function. We must remember also that the cosine of an arc is the x-coordinate of its terminal point on the unit circle. So arcs that terminate at opposite ends of a vertical chord in the unit circle have the same cosine.

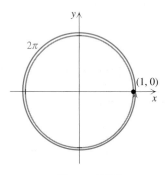

■ **Figure 6.25**

Example **1** Solving a cosine equation

Find all real numbers that satisfy each equation.

a. $\cos x = 1$ **b.** $\cos x = 0$ **c.** $\cos x = -1/2$

Solution

a. One solution to $\cos x = 1$ is

$$x = \cos^{-1}(1) = 0.$$

Since the period of cosine is 2π, any integral multiple of 2π can be added to this solution to get additional solutions. So the equation is satisfied by $0, \pm 2\pi, \pm 4\pi$, and so on. Now $\cos x = 1$ is satisifed only if the arc of length x on the unit circle has terminal point $(1, 0)$, as shown in Fig. 6.25. So there are no more solutions. The solution set is written as

$$\{x \mid x = 2k\pi\},$$

where k is any integer.

b. One solution to $\cos x = 0$ is

$$x = \cos^{-1}(0) = \frac{\pi}{2}.$$

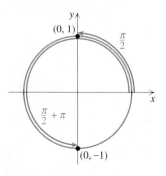

■ **Figure 6.26**

The terminal point for the arc of length $\pi/2$ is $(0, 1)$. So any arc length of the form $\pi/2 + 2k\pi$ is a solution to $\cos x = 0$. However, an arc of length x that terminates at $(0, -1)$ also satisfies $\cos x = 0$, as shown in Fig. 6.26, and these arcs are not included in the form $\pi/2 + 2k\pi$. Since the distance between $(0, 1)$ and and $(0, -1)$ on the unit circle is π, all arcs that terminate at these points are of the form $\pi/2 + k\pi$. So the solution set is

$$\left\{ x \,\middle|\, x = \frac{\pi}{2} + k\pi \right\},$$

where k is any integer.

c. One solution to the equation $\cos x = -1/2$ is

$$x = \cos^{-1}\left(-\frac{1}{2}\right) = \frac{2\pi}{3}.$$

Since the period of cosine is 2π, all arcs of length $2\pi/3 + 2k\pi$ (where k is any integer) satisfy the equation. The arc of length $4\pi/3$ also terminates at a point with x-coordinate $-1/2$, as shown in Fig. 6.27.

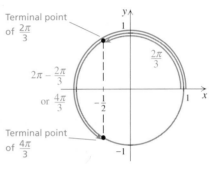

■ **Figure 6.27**

So $4\pi/3$ is also a solution to $\cos x = -1/2$, but it is not included in the form $2\pi/3 + 2k\pi$. So the solution set to $\cos x = -1/2$ is

$$\left\{ x \,\middle|\, x = \frac{2\pi}{3} + 2k\pi \quad \text{or} \quad x = \frac{4\pi}{3} + 2k\pi \right\},$$

where k is any integer. Note that the arcs of length $2\pi/3$ and $4\pi/3$ terminate at opposite ends of a vertical chord in the unit circle.

Note also that the calculator graph of $y_1 = \cos x$ in Fig. 6.28 crosses the horizontal line $y_2 = -1/2$ twice in the interval $[0, 2\pi]$, at $2\pi/3$ and $4\pi/3$.

Try This. Find all real numbers that satisfy $\cos x = \sqrt{3}/2$.   ■

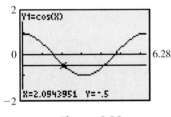

■ **Figure 6.28**

The procedure used in Example 1(c) can be used to solve $\cos x = a$ for any nonzero a between -1 and 1. First find two values of x between 0 and 2π that satisfy the equation. (In general, $s = \cos^{-1}(a)$ and $2\pi - s$ are the values.) Next write all solutions by adding $2k\pi$ to the first two solutions. The solution sets for $a = -1, 0$, and 1 are easier to find and remember. All of the cases for solving $\cos x = a$ are summarized next. The letter k is used to represent any integer. You should not simply memorize this summary, but you should be able to solve $\cos x = a$ for any real number a with the help of a unit circle.

S U M M A R Y   **Solving cos x = a**

1. If $-1 < a < 1$ and $a \neq 0$, then the solution set to $\cos x = a$ is $\{x \mid x = s + 2k\pi \text{ or } x = 2\pi - s + 2k\pi\}$, where $s = \cos^{-1} a$.
2. The solution set to $\cos x = 1$ is $\{x \mid x = 2k\pi\}$.
3. The solution set to $\cos x = 0$ is $\{x \mid x = \pi/2 + k\pi\}$.
4. The solution set to $\cos x = -1$ is $\{x \mid x = \pi + 2k\pi\}$.
5. If $|a| > 1$, then $\cos x = a$ has no solution.

Sine Equations

To solve $\sin x = a$, we use the same strategy that we used to solve $\cos x = a$. We look for the smallest nonnegative solution and then build on it to obtain all solutions. However, the "first" solution obtained from $s = \sin^{-1} a$ might be negative because the range for the function $\sin^{-1}$ is $[-\pi/2, \pi/2]$. In this case $s + 2\pi$ is positive and we build on it to find all solutions. Remember that the sine of an arc on the unit circle is the y-coordinate of the terminal point of the arc. So arcs that terminate at opposite ends of a horizontal chord in the unit circle have the same y-coordinate for the terminal point and the same sine. This fact is the key to finding all solutions of a sine equation.

Example **2** Solving a sine equation

Find all real numbers that satisfy $\sin x = -1/2$.

Solution

One solution to $\sin x = -1/2$ is

$$x = \sin^{-1}\left(-\frac{1}{2}\right) = -\frac{\pi}{6}.$$

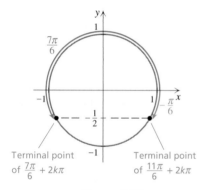

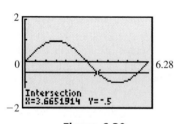

Terminal point
of $\frac{7\pi}{6} + 2k\pi$ Terminal point
of $\frac{11\pi}{6} + 2k\pi$

■ **Figure 6.29**

See Fig. 6.29. The smallest positive arc with the same terminal point as $-\pi/6$ is $-\pi/6 + 2\pi = 11\pi/6$. Since the period of the sine function is 2π, all arcs of the form $11\pi/6 + 2k\pi$ have the same terminal point and satisfy $\sin x = -1/2$. The smallest positive arc that terminates at the other end of the horizontal chord shown in Fig. 6.29 is $\pi - (-\pi/6)$ or $7\pi/6$. So $7\pi/6$ also satisfies the equation, but it is not included in the form $11\pi/6 + 2k\pi$. So the solution set is

$$\left\{x \mid x = \frac{7\pi}{6} + 2k\pi \quad \text{or} \quad x = \frac{11\pi}{6} + 2k\pi\right\},$$

where k is any integer.

Note that the calculator graph of $y_1 = \sin x$ in Fig. 6.30 crosses the horizontal line $y_2 = -1/2$ twice in the interval $[0, 2\pi]$, at $7\pi/6$ and $11\pi/6$.

Try This. Find all real numbers that satisfy $\sin x = 1/2$.   ■

The procedure used in Example 2 can be used to solve $\sin x = a$ for any nonzero a between -1 and 1. First find two values between 0 and 2π that satisfy the equation. (In general, $s = \sin^{-1} a$ and $\pi - s$ work when s is positive; $s + 2\pi$ and

6.28

Intersection
X=3.6651914 Y=-.5

■ **Figure 6.30**

$\pi - s$ work when s is negative.) Next write all solutions by adding $2k\pi$ to the first two solutions. The equations $\sin x = 0$, $\sin x = 1$, and $\sin x = -1$ have solutions that are similar to the corresponding cosine equations. We can summarize the different solution sets to $\sin x = a$ as follows. You should not simply memorize this summary, but you should be able to solve $\sin x = a$ for any real number a with the help of a unit circle.

SUMMARY   Solving sin x = a

1. If $-1 < a < 1$, $a \neq 0$, and $s = \sin^{-1} a$, then the solution set to $\sin x = a$ is
$$\{x \mid x = s + 2k\pi \text{ or } x = \pi - s + 2k\pi\} \text{ for } s > 0,$$
$$\{x \mid x = s + 2\pi + 2k\pi \text{ or } x = \pi - s + 2k\pi\} \text{ for } s < 0.$$

2. The solution set to $\sin x = 1$ is $\{x \mid x = \pi/2 + 2k\pi\}$.

3. The solution set to $\sin x = 0$ is $\{x \mid x = k\pi\}$.

4. The solution set to $\sin x = -1$ is $\{x \mid x = 3\pi/2 + 2k\pi\}$.

5. If $|a| > 1$, then $\sin x = a$ has no solution.

Tangent Equations

Tangent equations are a little simpler than sine and cosine equations, because the tangent function is one-to-one in its fundamental cycle, while sine and cosine are not one-to-one in their fundamental cycles. In the next example, we see that the solution set to $\tan x = a$ consists of any single solution plus multiples of π.

Example 3 Solving a tangent equation

Find all solutions, in degrees.

a. $\tan \alpha = 1$ **b.** $\tan \alpha = -1.34$

Solution

a. Since $\tan^{-1}(1) = 45°$ and the period of tangent is $180°$, all angles of the form $45° + k180°$ satisfy the equation. There are no additional angles that satisfy the equation. The solution set to $\tan \alpha = 1$ is
$$\{\alpha \mid \alpha = 45° + k180°\}.$$

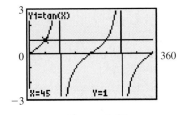
The calculator graphs of $y_1 = \tan x$ and $y_2 = 1$ in Fig. 6.31 support the conclusion that the solutions to $\tan x = 1$ are $180°$ apart. □

b. Since $\tan^{-1}(-1.34) = -53.3°$, one solution is $\alpha = -53.3°$. Since all solutions to $\tan \alpha = -1.34$ differ by a multiple of $180°$, $-53.3° + 180°$, or $126.7°$ is the smallest positive solution. So the solution set is
$$\{\alpha \mid \alpha = 126.7° + k180°\}.$$

■ **Figure 6.31**

Try This. Find all angles α in degrees that satisfy $\tan \alpha = \sqrt{3}$.   ■

To solve $\tan x = a$ for any real number a, we first find the smallest nonnegative solution. (In general, $s = \tan^{-1} a$ works if $s > 0$, and $s + \pi$ works if $s < 0$.) Next, add on all integral multiples of π. The solution to the equation $\tan x = a$ is summarized as follows.

S U M M A R Y   **Solving tan x = a**

If a is any real number and $s = \tan^{-1} a$, then the solution set to $\tan x = a$ is $\{x \,|\, x = s + k\pi\}$ for $s \geq 0$, and $\{x \,|\, x = s + \pi + k\pi\}$ for $s < 0$.

In the summaries for solving sine, cosine, and tangent equations, the domain of x is the set of real numbers. Similar summaries can be made if the domain of x is the set of degree measures. Start by finding $\sin^{-1} a$, $\cos^{-1} a$, or $\tan^{-1} a$ in degrees, then write the solution set in terms of multiples of $180°$ or $360°$.

Equations Involving Multiple Angles

Equations can involve expressions such as $\sin 2x$, $\cos 3\alpha$, or $\tan(s/2)$. These expressions involve a multiple of the variable rather than a single variable such as x, α, or s. In this case we solve for the multiple just as we would solve for a single variable and then find the value of the single variable in the last step of the solution.

Example **4**   **A sine equation involving a double angle**

Find all solutions in degrees to $\sin 2\alpha = 1/\sqrt{2}$.

Solution

The only values for 2α between $0°$ and $360°$ that satisfy the equation are $45°$ and $135°$. (Note that $\sin^{-1}(1/\sqrt{2}) = 45°$ and $135° = 180° - 45°$.) See Fig. 6.32. So proceed as follows:

$$\sin 2\alpha = \frac{1}{\sqrt{2}}$$

$$2\alpha = 45° + k360° \quad \text{or} \quad 2\alpha = 135° + k360°$$

$$\alpha = 22.5° + k180° \quad \text{or} \quad \alpha = 67.5° + k180° \quad \text{\small Divide each side by 2.}$$

The solution set is $\{\alpha \,|\, \alpha = 22.5° + k180° \text{ or } \alpha = 67.5° + k180°\}$, where k is any integer.

Try This. Find all angles α in degrees that satisfy $\sin 2\alpha = 1$. ◾

Note that in Example 4, all possible values for 2α are found and *then* divided by 2 to get all possible values for α. Finding $\alpha = 22.5°$ and then adding on multiples of $360°$ will not produce the same solutions. Observe the same procedure in the next example, where we wait until the final step to divide each side by 3.

Example **5**   **A tangent equation involving angle multiples**

Find all solutions to $\tan 3x = -\sqrt{3}$ in the interval $(0, 2\pi)$.

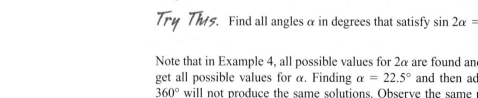

Figure 6.32

Solution

First find the smallest positive value for $3x$ that satisfies the equation. Then form all of the solutions by adding on multiples of π. Since $\tan^{-1}\left(-\sqrt{3}\right) = -\pi/3$, the smallest positive value for $3x$ that satisfies the equation is $-\pi/3 + \pi$, or $2\pi/3$. So proceed as follows:

$$\tan 3x = -\sqrt{3}$$

$$3x = \frac{2\pi}{3} + k\pi$$

$$x = \frac{2\pi}{9} + \frac{k\pi}{3} \qquad \text{Divide each side by 3.}$$

The solutions between 0 and 2π occur if $k = 0, 1, 2, 3, 4,$ and 5. If $k = 6$, then x is greater than 2π. So the solution set is

$$\left\{\frac{2\pi}{9}, \frac{5\pi}{9}, \frac{8\pi}{9}, \frac{11\pi}{9}, \frac{14\pi}{9}, \frac{17\pi}{9}\right\}.$$

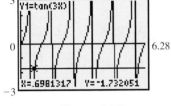

■ **Figure 6.33**

The graphs of $y_1 = \tan(3x)$ and $y_2 = -\sqrt{3}$ in Fig. 6.33 show the six solutions in the interval $(0, 2\pi)$.

Try This. Find all solutions to $\tan(4x) = 1$ in the interval $(0, \pi)$.   ■

More Complicated Equations

More complicated equations involving trigonometric functions are solved by first solving for $\sin x$, $\cos x$, or $\tan x$, and then solving for x. In solving for the trigonometric functions, we may use trigonometric identities or properties of algebra, such as factoring or the quadratic formula. In stating formulas for solutions to equations, we will continue to use the letter k to represent any integer.

Example **6**   **An equation solved by factoring**

Find all solutions in the interval $[0, 2\pi)$ to the equation

$$\sin 2x = \sin x.$$

Solution

First we use the identity $\sin 2x = 2 \sin x \cos x$ to get all of the trigonometric functions written in terms of the variable x alone. Then we rearrange and factor:

$$\sin 2x = \sin x$$

$$2 \sin x \cos x = \sin x \qquad \text{By the double-angle identity}$$

$$2 \sin x \cos x - \sin x = 0 \qquad \text{Subtract } \sin x \text{ from each side.}$$

$$\sin x(2 \cos x - 1) = 0 \qquad \text{Factor.}$$

$$\sin x = 0 \quad \text{or} \quad 2 \cos x - 1 = 0 \qquad \text{Set each factor equal to 0.}$$

$$x = k\pi \quad \text{or} \qquad 2 \cos x = 1$$

$$\cos x = \frac{1}{2}$$

$$x = \frac{\pi}{3} + 2k\pi \quad \text{or} \quad x = \frac{5\pi}{3} + 2k\pi$$

The only solutions in the interval $[0, 2\pi)$ are 0, $\pi/3$, π, and $5\pi/3$.

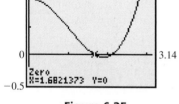

Figure 6.34

⊟ The graphs of $y_1 = \sin(2x)$ and $y_2 = \sin(x)$ in Fig. 6.34 show the four solutions to $\sin 2x = \sin x$ in the interval $[0, 2\pi)$.

Try This. Find all solutions to $\sin(2x) = \cos(x)$ in the interval $(0, 2\pi)$.   ▪

In algebraic equations, we generally do not divide each side by any expression that involves a variable, and the same rule holds for trigonometric equations. In Example 6, we did not divide by $\sin x$ when it appeared on opposite sides of the equation. If we had, the solutions 0 and π would have been lost.

In the next example, an equation of quadratic type is solved by factoring.

Example **7** **An equation of quadratic type**

Find all solutions to the equation

$$6 \cos^2\left(\frac{x}{2}\right) - 7 \cos\left(\frac{x}{2}\right) + 2 = 0$$

in the interval $[0, 2\pi)$. Round approximate answers to four decimal places.

Solution

Let $y = \cos(x/2)$ to get a quadratic equation:

$$6y^2 - 7y + 2 = 0$$

$$(2y - 1)(3y - 2) = 0 \quad \text{Factor.}$$

$$2y - 1 = 0 \quad \text{or} \quad 3y - 2 = 0$$

$$y = \frac{1}{2} \quad \text{or} \quad y = \frac{2}{3}$$

$$\cos \frac{x}{2} = \frac{1}{2} \quad \text{or} \quad \cos \frac{x}{2} = \frac{2}{3} \quad \text{Replace } y \text{ by } \cos(x/2).$$

$$\frac{x}{2} = \frac{\pi}{3} + 2k\pi \text{ or } \frac{5\pi}{3} + 2k\pi \quad \text{or} \quad \frac{x}{2} \approx 0.8411 + 2k\pi \text{ or } 5.4421 + 2k\pi$$

Now multiply by 2 to solve for x:

$$x = \frac{2\pi}{3} + 4k\pi \text{ or } \frac{10\pi}{3} + 4k\pi \quad \text{or} \quad x \approx 1.6821 + 4k\pi \text{ or } 10.844 + 4k\pi$$

For all k, $10\pi/3 + 4k\pi$ and $10.844 + 4k\pi$ are outside the interval $[0, 2\pi)$. The solutions in the interval $[0, 2\pi)$ are $2\pi/3$ and 1.6821.

⊟ The graph of $y = 6(\cos(x/2))^2 - 7 \cos(x/2) + 2$ in Fig. 6.35 appears to cross the x-axis at $2\pi/3$ and 1.6821.

Figure 6.35

Try This. Find all solutions to $2 \sin^2(x) - \sin(x) - 1 = 0$ in the interval $(0, 2\pi)$.
▪

For a trigonometric equation to be of quadratic type, it must be written in terms of a trigonometric function and the square of that function. In the next example, an identity is used to convert an equation involving sine and cosine to one with only sine. This equation is of quadratic type like Example 7, but it does not factor.

Example **8** An equation solved by the quadratic formula

Find all solutions to the equation

$$\cos^2 \alpha - 0.2 \sin \alpha = 0.9$$

in the interval $[0°, 360°)$. Round answers to the nearest tenth of a degree.

Solution

$$\cos^2 \alpha - 0.2 \sin \alpha = 0.9$$

$$1 - \sin^2 \alpha - 0.2 \sin \alpha = 0.9 \qquad \text{Replace } \cos^2 \alpha \text{ with } 1 - \sin^2 \alpha.$$

$$\sin^2 \alpha + 0.2 \sin \alpha - 0.1 = 0 \qquad \text{An equation of quadratic type}$$

$$\sin \alpha = \frac{-0.2 \pm \sqrt{(0.2)^2 - 4(1)(-0.1)}}{2} \qquad \begin{array}{l}\text{Use } a = 1, b = 0.2, \text{ and } c = -0.1 \\ \text{in the quadratic formula.}\end{array}$$

$$\sin \alpha \approx 0.2317 \quad \text{or} \quad \sin \alpha \approx -0.4317$$

Find two positive solutions to $\sin \alpha = 0.2317$ in $[0, 360°)$ by using a calculator to get $\alpha = \sin^{-1}(0.2317) \approx 13.4°$ and $180° - \alpha \approx 166.6°$. Now find two positive solutions to $\sin \alpha = -0.4317$ in $[0, 360°)$. Using a calculator, we get $\alpha = \sin^{-1}(-0.4317) \approx -25.6°$ and $180° - \alpha \approx 205.6°$. Since $-25.6°$ is negative, we use $-25.6° + 360°$ or $334.4°$ along with $205.6°$ as the two solutions. We list all possible solutions as

$$\alpha \approx 13.4°, 166.6°, 205.6°, \text{ or } 334.4° \quad (+k360° \text{ in each case}).$$

The solutions in $[0°, 360°)$ are $13.4°$, $166.6°$, $205.6°$, and $334.4°$.

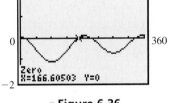 The graph of $y = \cos^2 x - 0.2 \sin x - 0.9$ in Fig. 6.36 appears to cross the x-axis four times in the interval $[0°, 360°)$.

■ **Figure 6.36**

Try This. Find all solutions to $\cos \alpha - \sin^2 \alpha = 0$ in the interval $[0°, 360°)$. ■

The Pythagorean identities $\sin^2 x = 1 - \cos^2 x$, $\csc^2 x = 1 + \cot^2 x$, and $\sec^2 x = 1 + \tan^2 x$ are frequently used to replace one function by the other. If an equation involves sine and cosine, cosecant and cotangent, or secant and tangent, we might be able to square each side and then use these identities.

Example **9** Square each side of the equation

Find all values of y in the interval $[0, 360°)$ that satisfy the equation

$$\tan 3y + 1 = \sqrt{2} \sec 3y.$$

Solution

$$(\tan 3y + 1)^2 = \left(\sqrt{2} \sec 3y\right)^2 \qquad \text{Square each side.}$$

$$\tan^2 3y + 2 \tan 3y + 1 = 2 \sec^2 3y$$

$$\tan^2 3y + 2 \tan 3y + 1 = 2(\tan^2 3y + 1) \qquad \text{Since } \sec^2 x = \tan^2 x + 1$$

$$-\tan^2 3y + 2 \tan 3y - 1 = 0 \qquad \text{Subtract } 2 \tan^2 3y + 2 \text{ from both sides.}$$

$$\tan^2 3y - 2 \tan 3y + 1 = 0$$

$$(\tan 3y - 1)^2 = 0$$

$$\tan 3y - 1 = 0$$

$$\tan 3y = 1$$

$$3y = 45° + k180°$$

Because we squared each side, we must check for extraneous roots. First check $3y = 45°$. If $3y = 45°$, then $\tan 3y = 1$ and $\sec 3y = \sqrt{2}$. Substituting these values into $\tan 3y + 1 = \sqrt{2} \sec 3y$ gives us

$$1 + 1 = \sqrt{2} \cdot \sqrt{2}. \quad \text{Correct.}$$

If we add any *even* multiple of 180°, or any multiple of 360°, to 45°, we get the same values for $\tan 3y$ and $\sec 3y$. So for any k, $3y = 45° + k360°$ satisfies the original equation. Now check 45° plus *odd* multiples of 180°. If $3y = 225°$ ($k = 1$), then

$$\tan 3y = 1 \qquad \text{and} \qquad \sec 3y = -\sqrt{2}.$$

These values do not satisfy the original equation. Since $\tan 3y$ and $\sec 3y$ have these same values for $3y = 45° + k180°$ for any odd k, the only solutions are of the form $3y = 45° + k360°$, or $y = 15° + k120°$. The solutions in the interval $[0°, 360°)$ are 15°, 135°, and 255°.

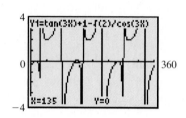 The graph of $y = \tan(3x) + 1 - \sqrt{2}/\cos(3x)$ in Fig. 6.37 appears to touch the x-axis at these three locations.

Try This. Find all α in $[0°, 360°)$ that satisfy $\sin\alpha - \cos\alpha = \dfrac{1}{\sqrt{2}}$. ■

■ **Figure 6.37**

There is no single method that applies to all trigonometric equations, but the following strategy will help you to solve them.

S T R A T E G Y **Solving Trigonometric Equations**

1. Know the solutions to $\sin x = a$, $\cos x = a$, and $\tan x = a$.

2. Solve an equation involving only multiple angles as if the equation had a single variable.

3. Simplify complicated equations by using identities. Try to get an equation involving only one trigonometric function.

4. If possible, factor to get different trigonometric functions into separate factors.

5. For equations of quadratic type, solve by factoring or the quadratic formula.

6. Square each side of the equation, if necessary, so that identities involving squares can be applied. (Check for extraneous roots.)

Solving a Spring Equation

Next, we consider the motion of a weight on a spring, discussed in Section 6.5.

Example **10** Solving a spring equation

In Example 7 of Section 6.5, a weight in motion attached to a spring had location x given by $x = 2 \sin t - \cos t$. For what values of t is the weight at position $x = 0$?

Solution

To solve $2 \sin t - \cos t = 0$, divide each side by $2 \cos t$. We can divide by $2 \cos t$ because the values of t for which $2 \cos t$ is zero do not satisfy the original equation.

$$2 \sin t = \cos t$$

$$\frac{\sin t}{\cos t} = \frac{1}{2}$$

$$\tan t = \frac{1}{2}$$

Since the weight is set in motion at time $t = 0$, the values of t are positive. Since $\tan^{-1}(1/2) = 0.46$, the weight is at position $x = 0$ for $t = 0.46 + k\pi$ for k a nonnegative integer.

Try This. The location x for a weight on a spring is given by $x = 3 \sin t - 2 \cos t$ where t is time in seconds. Find all t for which $x = 0$. ■

Modeling Projectile Motion

The distance d (in feet) traveled by a projectile fired from the ground with an angle of elevation θ is related to the initial velocity v_0 (in feet per second) by the equation $v_0^2 \sin 2\theta = 32d$. If the projectile is pictured as being fired from the origin into the first quadrant, then the x- and y-coordinates (in feet) of the projectile at time t (in seconds) are given by $x = v_0 t \cos \theta$ and $y = -16t^2 + v_0 t \sin \theta$.

Example **11** The path of a projectile

A catapult is placed 100 feet from the castle wall, which is 35 feet high. The soldier wants the burning bale of hay to clear the top of the wall and land 50 feet inside the castle wall. If the initial velocity of the bale is 70 feet per second, then at what angle should the bale of hay be launched so that it will travel 150 feet and pass over the castle wall?

Solution

Use the equation $v_0^2 \sin 2\theta = 32d$ with $v_0 = 70$ feet per second and $d = 150$ feet to find θ:

$$70^2 \sin 2\theta = 32(150)$$

$$\sin 2\theta = \frac{32(150)}{70^2} \approx 0.97959$$

The launch angle θ must be in the interval $(0°, 90°)$, so we look for values of 2θ in the interval $(0°, 180°)$. Since $\sin^{-1}(0.97959) \approx 78.4°$, both $78.4°$ and $180° - 78.4° = 101.6°$ are possible values for 2θ. So possible values for θ are $39.2°$ and $50.8°$. See Fig. 6.38 on the next page.

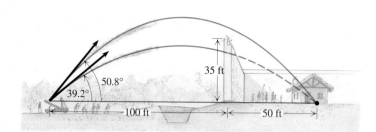

■ **Figure 6.38**

Use the equation $x = v_0 t \cos \theta$ to find the time at which the bale is 100 feet from the catapult (measured horizontally) by using each of the possible values for θ:

$$100 = 70t \cos(39.2°) \qquad\qquad 100 = 70t \cos(50.8°)$$

$$t = \frac{100}{70 \cos(39.2°)} \approx 1.84 \text{ seconds} \qquad t = \frac{100}{70 \cos(50.8°)} \approx 2.26 \text{ seconds}$$

Use the equation $y = -16t^2 + v_0 t \sin \theta$ to find the altitude of the bale at time $t = 1.84$ seconds and $t = 2.26$ seconds:

$$y = -16(1.84)^2 + 70(1.84) \sin 39.2° \qquad y = -16(2.26)^2 + 70(2.26) \sin 50.8°$$

$$\approx 27.2 \text{ feet} \qquad\qquad\qquad \approx 40.9 \text{ feet}$$

If the burning bale is launched on a trajectory with an angle of 39.2°, then it will have an altitude of only 27.2 feet when it reaches the castle wall. If it is launched with an angle of 50.8°, then it will have an altitude of 40.9 feet when it reaches the castle wall. Since the castle wall is 35 feet tall, the 50.8° angle must be used for the bale to reach its intended target.

Try This.   Find the launch angle so that a projectile with initial velocity of 100 ft per second will travel 200 feet.   ■

For Thought

True or False? Explain.

1. The only solutions to $\cos \alpha = 1/\sqrt{2}$ in $[0°, 360°)$ are 45° and 135°.

2. The only solution to $\sin x = -0.55$ in $[0, \pi)$ is $\sin^{-1}(-0.55)$.

3. $\{x \mid x = -29° + k360°\} = \{x \mid x = 331° + k360°\}$, where k is any integer.

4. The solution set to $\tan x = -1$ is $\left\{x \mid x = \frac{7\pi}{4} + k\pi\right\}$, where k is any integer.

5. $2 \cos^2 x + \cos x - 1 = (2 \cos x - 1)(\cos x + 1)$ is an identity.

6. The equation $\sin^2 x = \sin x \cos x$ is equivalent to $\sin x = \cos x$.

7. One solution to $\sec x = 2$ is $\dfrac{1}{\cos^{-1}(2)}$.

8. The solution set to $\cot x = 3$ for x in $[0, \pi)$ is $\{\tan^{-1}(1/3)\}$.

9. The equation $\sin x = \cos x$ is equivalent to $\sin^2 x = \cos^2 x$.

10. $\left\{x \mid 3x = \frac{\pi}{2} + 2k\pi\right\} = \left\{x \mid x = \frac{\pi}{6} + 2k\pi\right\}$, where k is any integer.

6.6 Exercises

Find all real numbers that satisfy each equation. Do not use a calculator. See the summaries for solving $\cos x = a$, $\sin x = a$, *and* $\tan x = a$ *on pages 558, 559, and 560.*

1. $\cos x = -1$ **2.** $\cos x = 0$

3. $\sin x = 0$ **4.** $\sin x = 1$

5. $\sin x = -1$ **6.** $\cos x = 1$

7. $\cos x = 1/2$ **8.** $\cos x = \sqrt{2}/2$

9. $\sin x = \sqrt{2}/2$ **10.** $\sin x = \sqrt{3}/2$

11. $\tan x = 1$ **12.** $\tan x = \sqrt{3}/3$

13. $\cos x = -\sqrt{3}/2$ **14.** $\cos x = -\sqrt{2}/2$

15. $\sin x = -\sqrt{2}/2$ **16.** $\sin x = -\sqrt{3}/2$

17. $\tan x = -1$ **18.** $\tan x = -\sqrt{3}$

Find all angles in degrees that satisfy each equation. Round approximate answers to the nearest tenth of a degree.

19. $\cos \alpha = 0$ **20.** $\cos \alpha = -1$ **21.** $\sin \alpha = 1$

22. $\sin \alpha = -1$ **23.** $\tan \alpha = 0$ **24.** $\tan \alpha = -1$

25. $\cos \alpha = 0.873$ **26.** $\cos \alpha = -0.158$

27. $\sin \alpha = -0.244$ **28.** $\sin \alpha = 0.551$

29. $\tan \alpha = 5.42$ **30.** $\tan \alpha = -2.31$

Find all real numbers that satisfy each equation.

31. $\cos(x/2) = 1/2$ **32.** $2 \cos 2x = -\sqrt{2}$

33. $\cos 3x = 1$ **34.** $\cos 2x = 0$

35. $2 \sin(x/2) - 1 = 0$ **36.** $\sin 2x = 0$

37. $2 \sin 2x = -\sqrt{2}$ **38.** $\sin(x/3) + 1 = 0$

39. $\tan 2x = \sqrt{3}$ **40.** $\sqrt{3} \tan(3x) + 1 = 0$

41. $\tan 4x = 0$ **42.** $\tan 3x = -1$

43. $\sin(\pi x) = 1/2$ **44.** $\tan(\pi x/4) = 1$

45. $\cos(2\pi x) = 0$ **46.** $\sin(3\pi x) = 1$

Find all values of α *in* $[0°, 360°)$ *that satisfy each equation.*

47. $2 \sin \alpha = -\sqrt{3}$ **48.** $\tan \alpha = -\sqrt{3}$

49. $\sqrt{2} \cos 2\alpha - 1 = 0$ **50.** $\sin 6\alpha = 1$

51. $\sec 3\alpha = -\sqrt{2}$ **52.** $\csc(5\alpha) + 2 = 0$

53. $\cot(\alpha/2) = \sqrt{3}$ **54.** $\sec(\alpha/2) = \sqrt{2}$

Find all values of α *in degrees that satisfy each equation. Round approximate answers to the nearest tenth of a degree.*

55. $\sin 3\alpha = 0.34$ **56.** $\cos 2\alpha = -0.22$

57. $\sin 3\alpha = -0.6$ **58.** $\tan 4\alpha = -3.2$

59. $\sec 2\alpha = 4.5$ **60.** $\csc 3\alpha = -1.4$

61. $\csc(\alpha/2) = -2.3$ **62.** $\cot(\alpha/2) = 4.7$

Find all real numbers in the interval $[0, 2\pi)$ *that satisfy each equation. Round approximate answers to the nearest tenth. See the strategy for solving trigonometric equations on page 564.*

63. $3 \sin^2 x = \sin x$ **64.** $2 \tan^2 x = \tan x$

65. $2 \cos^2 x + 3 \cos x = -1$ **66.** $2 \sin^2 x + \sin x = 1$

67. $5 \sin^2 x - 2 \sin x = \cos^2 x$ **68.** $\sin^2 x - \cos^2 x = 0$

69. $\tan x = \sec x - \sqrt{3}$ **70.** $\csc x - \sqrt{3} = \cot x$

71. $\sin x + \sqrt{3} = 3\sqrt{3} \cos x$

72. $6 \sin^2 x - 2 \cos x = 5$

73. $\tan x \sin 2x = 0$

74. $3 \sec^2 x \tan x = 4 \tan x$

75. $\sin 2x - \sin x \cos x = \cos x$

76. $2 \cos^2 2x - 8 \sin^2 x \cos^2 x = -1$

77. $\sin x \cos(\pi/4) + \cos x \sin(\pi/4) = 1/2$

78. $\sin(\pi/6) \cos x - \cos(\pi/6) \sin x = -1/2$

79. $\sin 2x \cos x - \cos 2x \sin x = -1/2$

80. $\cos 2x \cos x - \sin 2x \sin x = 1/2$

Find all values of θ *in the interval* $[0°, 360°)$ *that satisfy each equation. Round approximate answers to the nearest tenth of a degree.*

81. $\cos^2\left(\dfrac{\theta}{2}\right) = \sec \theta$ **82.** $2 \sin^2\left(\dfrac{\theta}{2}\right) = \cos \theta$

83. $2 \sin \theta = \cos \theta$

84. $3 \sin 2\theta = \cos 2\theta$

85. $\sin 3\theta = \csc 3\theta$

86. $\tan^2 \theta - \cot^2 \theta = 0$

87. $\tan^2 \theta - 2 \tan \theta - 1 = 0$

88. $\cot^2 \theta - 4 \cot \theta + 2 = 0$

89. $9 \sin^2 \theta + 12 \sin \theta + 4 = 0$

90. $12 \cos^2 \theta + \cos \theta - 6 = 0$

91. $\dfrac{\tan 3\theta - \tan \theta}{1 + \tan 3\theta \tan \theta} = \sqrt{3}$ **92.** $\dfrac{\tan 3\theta + \tan 2\theta}{1 - \tan 3\theta \tan 2\theta} = 1$

93. $8 \cos^4 \theta - 10 \cos^2 \theta + 3 = 0$

94. $4 \sin^4 \theta - 5 \sin^2 \theta + 1 = 0$

95. $\sec^4 \theta - 5 \sec^2 \theta + 4 = 0$

96. $\cot^4 \theta - 4 \cot^2 \theta + 3 = 0$

Solve each equation. (These equations are types that will arise in Chapter 7.)

97. $\dfrac{\sin 33.2°}{a} = \dfrac{\sin 45.6°}{13.7}$ **98.** $\dfrac{\sin 49.6°}{55.1} = \dfrac{\sin 88.2°}{b}$

99. $\dfrac{\sin \alpha}{23.4} = \dfrac{\sin 67.2°}{25.9}$ for $0° < \alpha < 90°$

100. $\dfrac{\sin 9.7°}{15.4} = \dfrac{\sin \beta}{52.9}$ for $90° < \beta < 180°$

101. $(3.6)^2 = (5.4)^2 + (8.2)^2 - 2(5.4)(8.2) \cos \alpha$
for $0° < \alpha < 90°$

102. $(6.8)^2 = (3.2)^2 + (4.6)^2 - 2(3.2)(4.6) \cos \alpha$
for $90° < \alpha < 180°$

⎓   *One way to solve an equation with a graphing calculator is to rewrite the equation with 0 on the right-hand side, then graph the function that is on the left-hand side. The x-coordinate of each x-intercept of the graph is a solution to the original equation. For each equation, find all real solutions (to the nearest tenth) in the interval* $[0, 2\pi)$.

103. $\sin(x/2) = \cos 3x$ **104.** $2 \sin x = \csc(x + 0.2)$

105. $\dfrac{x}{2} - \dfrac{\pi}{6} + \dfrac{\sqrt{3}}{2} = \sin x$ **106.** $x^2 = \sin x$

Solve each problem.

107. *Motion of a Spring*   A block is attached to a spring and set in motion on a frictionless plane. Its location on the surface at any time t in seconds is given in meters by

$x = \sqrt{3} \sin 2t + \cos 2t$. For what values of t is the block at its resting position $x = 0$?

108. *Motion of a Spring*   A block is set in motion hanging from a spring and oscillates about its resting position $x = 0$ according to the function $x = -0.3 \sin 3t + 0.5 \cos 3t$. For what values of t is the block at its resting position $x = 0$?

109. *Wave Action*   The vertical position of a floating ball in an experimental wave tank is given by the equation $x = 2 \sin(\pi t/3)$, where x is the number of feet above sea level and t is the time in seconds. For what values of t is the ball $\sqrt{3}$ ft above sea level?

110. *Periodic Sales*   The number of car stereos sold by a national department store chain varies seasonally and is a function of the month of the year. The function

$$x = 6.2 + 3.1 \sin\left(\frac{\pi}{6}(t - 9)\right)$$

gives the anticipated sales (in thousands of units) as a function of the number of the month ($t = 1, 2, \ldots, 12$). In what month does the store anticipate selling 9300 units?

111. *Firing an M-16*   A soldier is accused of breaking a window 3300 ft away during target practice. If the muzzle velocity for an M-16 is 325 ft/sec, then at what angle would it have to be aimed for the bullet to travel 3300 ft? The distance d (in feet) traveled by a projectile fired at an angle θ is related to the initial velocity v_0 (in feet per second) by the equation $v_0^2 \sin 2\theta = 32d$.

112. *Firing an M-16*   If you were accused of firing an M-16 into the air and breaking a window 4000 ft away, what would be your defense?

113. *Choosing the Right Angle*   Cincinnati Reds centerfielder Ken Griffey, Jr., fields a ground ball and attempts to make a 230-ft throw to home plate. Given that Griffey commonly makes long throws at 90 mph, find the two possible angles at which he can throw the ball to home plate. Find the time saved by choosing the smaller angle.

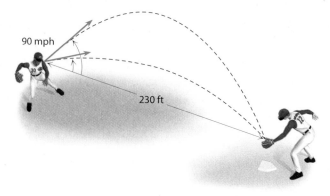

90 mph

230 ft

■ **Figure for Exercise 113**

114. *Muzzle Velocity* The 8-in. (diameter) howitzer on the U.S. Army's M-110 can propel a projectile a distance of 18,500 yd. If the angle of elevation of the barrel is 45°, then what muzzle velocity (in feet per second) is required to achieve this distance?

Thinking Outside the Box LVII

Two Common Triangles An equilateral triangle with sides of length 1 and an isosceles right triangle with legs of length 1 are positioned as shown in the accompanying diagram. Find the exact area of the shaded triangle.

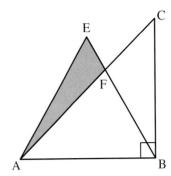

■ **Figure for Thinking Outside the Box LVII**

6.6 Pop Quiz

Find all angles α in degrees that satisfy each equation.

1. $\sin\alpha = \sqrt{2}/2$

2. $\cos\alpha = 1/2$

3. $\tan\alpha = -1$

Find all real numbers in $[0, 2\pi]$ that satisfy each equation.

4. $\sin(x/2) = 1/2$

5. $\cos(x) = 1$

6. $\tan(2x) = 1$

Linking Concepts

For Individual or Group Explorations

Modeling Baseball Strategy

An outfielder picks up a ground ball and wants to make a quick 150 ft throw to first base. To save time, he does not throw the ball directly to the first baseman. He skips the ball off the artificial surface at 75 ft so that the ball reaches first base after a bounce, as shown in the figure. Let's examine this strategy. The equations

$$x = v_0 t \cos \theta \quad and \quad y = -16t^2 + v_0 t \sin \theta + h_0$$

give the coordinates of the ball at time t seconds for an initial angle of θ degrees, initial velocity v_0 ft/sec, and initial height h_0 ft. Assume that the ball is thrown at 130 ft/sec from a height of 5 ft and the ball is caught at a height of 5 ft.

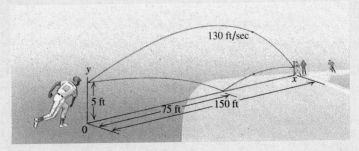

(continued on next page)

a) At what angle from horizontal must the ball be thrown to reach first base without a skip?

b) How long does it take the ball to get to first base without a skip?

c) At what angle from horizontal must the ball be thrown so that it strikes the ground at 75 ft?

d) How long does it take for the ball to reach the skip point at 75 ft?

e) Assume that the path of the ball is symmetric with respect to the skip point and double your answer in part (d) to get the time that it takes the ball to reach first base.

f) How much time is saved by skipping the ball?

g) How long would it take for the ball to reach first base if it could be thrown in a straight line?

h) Discuss the assumptions made for this model. Did we ignore anything that might affect our results?

▪ ▪ ▪ Highlights

6.1 Basic Identities

Identity

An equation that is satisfied by every number for which both sides are defined.

$$\sin^2 x + \cos^2 x = 1$$
$$\tan x = \sin(x)/\cos(x)$$

Reciprocal Identities

$$\sin \alpha = \frac{1}{\csc \alpha} \qquad \cos \alpha = \frac{1}{\sec \alpha} \qquad \tan \alpha = \frac{1}{\cot \alpha}$$

$$\csc \alpha = \frac{1}{\sin \alpha} \qquad \sec \alpha = \frac{1}{\cos \alpha} \qquad \cot \alpha = \frac{1}{\tan \alpha}$$

Pythagorean Identities

$$\sin^2 x + \cos^2 x = 1 \qquad 1 + \cot^2 x = \csc^2 x \qquad \tan^2 x + 1 = \sec^2 x$$

Odd Identities

$$\sin(-x) = -\sin(x) \qquad \csc(-x) = -\csc(x)$$
$$\tan(-x) = -\tan(x) \qquad \cot(-x) = -\cot(x)$$

Even Identities

$$\cos(-x) = \cos(x) \qquad \sec(-x) = \sec(x)$$

6.2 Verifying Identities

Strategy

1. Convert one side of the identity into the expression on the other side.
2. Write all functions in terms of sines and cosines, then simplify.
3. Multiply the numerator and denominator of a rational expression by either the numerator or denominator of the desired rational expression.

4. If the numerator of a rational expression is a sum or difference, separate it into two rational expressions.

5. Combine a sum or difference of two rational expressions into a single rational expression.

6.3 Sum and Difference Identities

Cosine of a Sum or Difference

$$\cos(\alpha + \beta) = \cos \alpha \cos \beta - \sin \alpha \sin \beta$$
$$\cos(\alpha - \beta) = \cos \alpha \cos \beta + \sin \alpha \sin \beta$$

Sine of a Sum or Difference

$$\sin(\alpha + \beta) = \sin \alpha \cos \beta + \cos \alpha \sin \beta$$
$$\sin(\alpha - \beta) = \sin \alpha \cos \beta - \cos \alpha \sin \beta$$

Tangent of a Sum or Difference

$$\tan(\alpha \pm \beta) = \frac{\tan \alpha \pm \tan \beta}{1 \mp \tan \alpha \tan \beta}$$

Cofunction Identities

The value of any trigonometric function at u equals its value at $\pi/2 - u$.

6.4 Double-Angle and Half-Angle Identities

Double-Angle Identities

$$\sin 2x = 2 \sin x \cos x$$

$$\cos 2x = \cos^2 x - \sin^2 x$$

$$= 2 \cos^2 x - 1$$

$$= 1 - 2 \sin^2 x$$

$$\tan 2x = \frac{2 \tan x}{1 - \tan^2 x}$$

Half-Angle Identities

$$\sin \frac{x}{2} = \pm \sqrt{\frac{1 - \cos x}{2}}$$

$$\cos \frac{x}{2} = \pm \sqrt{\frac{1 + \cos x}{2}}$$

$$\tan \frac{x}{2} = \pm \sqrt{\frac{1 - \cos x}{1 + \cos x}}$$

$$= \frac{\sin x}{1 + \cos x}$$

$$= \frac{1 - \cos x}{\sin x}$$

6.5 Product and Sum Identities

Product-to-Sum Identities

$$\sin A \cos B = \frac{1}{2}[\sin(A + B) + \sin(A - B)]$$

$$\sin A \sin B = \frac{1}{2}[\cos(A - B) - \cos(A + B)]$$

$$\cos A \sin B = \frac{1}{2}[\sin(A + B) - \sin(A - B)]$$

$$\cos A \cos B = \frac{1}{2}[\cos(A - B) + \cos(A + B)]$$

Sum-to-Product Identities

$$\sin x + \sin y = 2 \sin\left(\frac{x + y}{2}\right) \cos\left(\frac{x - y}{2}\right)$$

$$\sin x - \sin y = 2 \cos\left(\frac{x + y}{2}\right) \sin\left(\frac{x - y}{2}\right)$$

$$\cos x + \cos y = 2 \cos\left(\frac{x + y}{2}\right) \cos\left(\frac{x - y}{2}\right)$$

$$\cos x - \cos y = -2 \sin\left(\frac{x + y}{2}\right) \sin\left(\frac{x - y}{2}\right)$$

Reduction Formula	If α is an angle in standard position whose terminal side contains (a, b) and x is a real number, then $a \sin x + b \cos x = \sqrt{a^2 + b^2} \sin(x + \alpha)$.	$\alpha = \tan^{-1}(4/3)$ $3 \sin x + 4 \cos x$ $= 5 \sin(x + \tan^{-1}(4/3))$

6.6 Conditional Trigonometric Equations

Solving $\cos x = a$	If $\lvert a \rvert \leq 1$, find all solutions in $[0, 2\pi]$, then add all multiples of 2π to them and simplify. If $\lvert a \rvert > 1$ there are no solutions.	$\cos x = 0, x = \dfrac{\pi}{2} + k\pi$ $\cos x = \dfrac{1}{2}, x = \dfrac{\pi}{3} + 2k\pi$ or $x = \dfrac{5\pi}{3} + 2k\pi$
Solving $\sin x = a$	If $\lvert a \rvert \leq 1$, find all solutions in $[0, 2\pi]$, then add all multiples of 2π to them and simplify. If $\lvert a \rvert > 1$ there are no solutions.	$\sin x = 0, x = k\pi$ $\sin x = \dfrac{1}{2}, x = \dfrac{\pi}{6} + 2k\pi$ or $x = \dfrac{5\pi}{6} + 2k\pi$
Solving $\tan x = a$	If a is a real number, find all solutions in $[-\pi/2, \pi/2]$, then add multiples of π to them and simplify.	$\tan x = 1, x = \dfrac{\pi}{4} + k\pi$

■■■ Chapter 6 Review Exercises

Simplify each expression.

1. $(1 - \sin \alpha)(1 + \sin \alpha)$

2. $\csc x \tan x + \sec(-x)$

3. $(1 - \csc x)(1 - \csc(-x))$

4. $\dfrac{\cos^2 x - \sin^2 x}{\sin 2x}$

5. $\dfrac{1}{1 + \sin \alpha} - \dfrac{\sin(-\alpha)}{\cos^2 \alpha}$

6. $2 \sin\left(\dfrac{\pi}{2} - \alpha\right) \cos\left(\dfrac{\pi}{2} - \alpha\right)$

7. $\dfrac{2 \tan 2s}{1 - \tan^2 2s}$

8. $\dfrac{\tan 2w - \tan 4w}{1 + \tan 2w \tan 4w}$

9. $\sin 3\theta \cos 6\theta - \cos 3\theta \sin 6\theta$

10. $\dfrac{\sin 2y}{1 + \cos 2y}$

11. $\dfrac{1 - \cos 2z}{\sin 2z}$

12. $\cos^2\left(\dfrac{x}{2}\right) - \sin^2\left(\dfrac{x}{2}\right)$

Match each expression with an equivalent expression (a)–(h).

13. $\sin(20°)$

14. $\tan(85°)$

15. $\cos(90°)$

16. $\cot(40°)$

17. $\sec(\pi/6)$

18. $\csc(\pi/8)$

19. $\sin(5\pi/12)$

20. $\cot(\pi/3)$

a. $\csc(\pi/3)$ **b.** $\sec(3\pi/8)$ **c.** $\sin(0°)$

d. $\tan(50°)$ **e.** $\cos(70°)$ **f.** $\tan(\pi/6)$

g. $\cos(\pi/12)$ **h.** $\cot(5°)$

Use identities to find the exact values of the remaining five trigonometric functions at α.

21. $\cos \alpha = -5/13$ and $\pi/2 < \alpha < \pi$

22. $\tan \alpha = 5/12$ and $\pi < \alpha < 3\pi/2$

Find the exact values of the six trigonometric functions at α.

23. $\sin\left(\dfrac{\pi}{2} - \alpha\right) = -3/5$ and $\pi < \alpha < 3\pi/2$

24. $\csc\left(\dfrac{\pi}{2} - \alpha\right) = 3$ and $0 < \alpha < \pi/2$

25. $\sin(\alpha/2) = 3/5$ and $3\pi/4 < \alpha/2 < \pi$

26. $\cos(\alpha/2) = -1/3$ and $\pi/2 < \alpha/2 < 3\pi/4$

Determine whether each equation is an identity. Prove your answer.

27. $(\sin x + \cos x)^2 = 1 + \sin 2x$

28. $\cos(A - B) = \cos A \cos B - \sin A \sin B$

29. $\csc^2 x - \cot^2 x = \tan^2 x - \sec^2 x$

30. $\sin^2\left(\dfrac{x}{2}\right) = \dfrac{\sin^2 x}{2 + \sin 2x \csc x}$

Determine whether each function is odd, even, or neither.

31. $f(x) = \dfrac{\sin x - \tan x}{\cos x}$

32. $f(x) = 1 + \sin^2 x$

33. $f(x) = \dfrac{\cos x - \sin x}{\sec x}$

34. $f(x) = \csc^3 x - \tan^3 x$

35. $f(x) = \dfrac{\sin x \tan x}{\cos x + \sec x}$

36. $f(x) = \sin x + \cos x$

Match each given expression with an equivalent expression (a)–(j).

37. $\cos 4$

38. $\sin 4$

39. $\sin 2 + \sin 4$

40. $\sin 4 - \sin 2$

41. $2 \sin 4 \cos 4$

42. $\cos^2 4 - \sin^2 4$

43. $\cos 4 + \cos 2$

44. $\cos 4 - \cos 2$

45. $\dfrac{\sin 4}{1 + \cos 4}$

46. $\dfrac{1 + \cos 4}{2}$

a. $2 \cos 3 \sin 1$

b. $\sin 8$

c. $\tan 2$

d. $\cos 8$

e. $2 \sin 3 \cos 1$

f. $\sin(\pi/2 - 4)$

g. $-\sin(-4)$

h. $2 \cos 3 \cos 1$

i. $-2 \sin 3 \sin 1$

j. $\cos^2 2$

Prove that each of the following equations is an identity.

47. $\sec 2\theta = \dfrac{1 + \tan^2 \theta}{1 - \tan^2 \theta}$

48. $\tan^2 \theta = \dfrac{1 - \cos 2\theta}{1 + \cos 2\theta}$

49. $\sin^2\left(\dfrac{x}{2}\right) = \dfrac{\csc^2 x - \cot^2 x}{2 \csc^2 x + 2 \csc x \cot x}$

50. $\cot(-x) = \dfrac{1 - \sin^2 x}{\cos(-x) \sin(-x)}$

51. $\cot(\alpha - 45°) = \dfrac{1 + \tan \alpha}{\tan \alpha - 1}$

52. $\cos(\alpha + 45°) = \dfrac{\cos \alpha - \sin \alpha}{\sqrt{2}}$

53. $\dfrac{\sin 2\beta}{2 \csc \beta} = \sin^2 \beta \cos \beta$

54. $\sin(45° - \beta) = \dfrac{\cos 2\beta}{\sqrt{2}(\cos \beta + \sin \beta)}$

55. $\dfrac{\cot^3 y - \tan^3 y}{\sec^2 y + \cot^2 y} = 2 \cot 2y$

56. $\dfrac{\sin^3 y - \cos^3 y}{\sin y - \cos y} = \dfrac{2 + \sin 2y}{2}$

57. $\cos 4x = 8 \sin^4 x - 8 \sin^2 x + 1$

58. $\cos 3x = \cos x(1 - 4 \sin^2 x)$

59. $\sin^4 2x = 16 \sin^4 x - 32 \sin^6 x + 16 \sin^8 x$

60. $1 - \cos^6 x = 3 \sin^2 x - 3 \sin^4 x + \sin^6 x$

Use an appropriate identity to find the exact value of each expression.

61. $\tan(-\pi/12)$

62. $\sin(-\pi/8)$

63. $\sin(-75°)$

64. $\cos(105°)$

Write each function in the form $y = A \sin(x + C)$, and graph the function for $-2\pi \le x \le 2\pi$. Determine the amplitude and phase shift.

65. $y = 4 \sin x + 4 \cos x$

66. $y = \sqrt{3} \sin x + 3 \cos x$

67. $y = -2 \sin x + \cos x$

68. $y = -2 \sin x - \cos x$

Find all real numbers that satisfy each equation.

69. $2 \cos 2x + 1 = 0$

70. $2 \sin 2x + \sqrt{3} = 0$

71. $\left(\sqrt{3} \csc x - 2\right)(\csc x - 2) = 0$

72. $\left(\sec x - \sqrt{2}\right)\left(\sqrt{3} \sec x + 2\right) = 0$

73. $2 \sin^2 x + 1 = 3 \sin x$

74. $4 \sin^2 x = \sin x + 3$

75. $-8\sqrt{3} \sin\dfrac{x}{2} = -12$

76. $-\cos\dfrac{x}{2} = \sqrt{2} + \cos\dfrac{x}{2}$

77. $\cos \dfrac{x}{2} - \sin x = 0$

78. $\sin 2x = \tan x$

79. $\cos 2x + \sin^2 x = 0$

80. $\tan \dfrac{x}{2} = \sin x$

81. $\sin x \cos x + \sin x + \cos x + 1 = 0$

82. $\sin 2x \cos 2x - \cos 2x + \sin 2x - 1 = 0$

Find all angles α in $[0°, 360°)$ that satisfy each equation.

83. $\sin \alpha \cos \alpha = \dfrac{1}{2}$

84. $\cos 2\alpha = \cos \alpha$

85. $\sin \alpha = \cos \alpha + 1$

86. $\cos \alpha \csc \alpha = \cot^2 \alpha$

87. $\sin^2 \alpha + \cos^2 \alpha = \dfrac{1}{2}$

88. $\sec^2 \alpha - \tan^2 \alpha = 0$

89. $4 \sin^4 2\alpha = 1$

90. $\sin 2\alpha = \tan \alpha$

91. $\tan 2\alpha = \tan \alpha$

92. $\tan \alpha = \cot \alpha$

93. $\sin 2\alpha \cos \alpha + \cos 2\alpha \sin \alpha = \cos 3\alpha$

94. $\cos 2\alpha \cos \alpha - \sin 2\alpha \sin \alpha = \cot 3\alpha$

Use the sum-to-product identities to rewrite each expression as a product.

95. $\cos 15° + \cos 19°$

96. $\cos 4° - \cos 6°$

97. $\sin(\pi/4) - \sin(-\pi/8)$

98. $\sin(-\pi/6) + \sin(\pi/12)$

Use the product-to-sum identities to write each expression as a sum or difference.

99. $2 \sin 11° \cos 13°$

100. $2 \sin 8° \sin 12°$

101. $2 \cos(x/4) \cos(x/3)$

102. $2 \cos s \sin 3s$

Solve each problem.

103. Find the exact value of $\sin(\alpha - \beta)$ if $\sin \alpha = \sqrt{3}/2$ and $\cos \beta = \sqrt{2}/2$ with α in quadrant II and β in quadrant I.

104. Find the exact value of $\sin(\alpha + \beta)$ if $\sin \alpha = -\sqrt{2}/2$ and $\sin \beta = 1/2$ with α in quadrant IV and β in quadrant II.

105. Find the exact value of $\cos(\alpha - \beta)$ if $\sin \alpha = \sqrt{3}/2$ and $\cos \beta = -\sqrt{2}/2$ with α in quadrant I and β in quadrant II.

106. Find the exact value of $\cos(\alpha + \beta)$ if $\sin \alpha = \sqrt{2}/2$ and $\sin \beta = 1/2$ with α in quadrant II and β in quadrant I.

107. *Motion of a Spring* A block is set in motion hanging from a spring and oscillates about its resting position $x = 0$ according to the function $x = 0.6 \sin 2t + 0.4 \cos 2t$, where x is in centimeters and t is in seconds. For what values of t in the interval $[0, 3]$ is the block at its resting position $x = 0$?

108. *Battle of Gettysburg* The Confederates had at least one 24-lb mortar at the battle of Gettysburg in the Civil War. If the muzzle velocity of a projectile was 400 ft/sec, then at what angles could the cannon be aimed to hit the Union Army 3000 ft away? The distance d (in feet) traveled by a projectile fired at an angle θ is related to the initial velocity v_0 (in feet per second) by the equation $v_0^2 \sin 2\theta = 32d$.

Thinking Outside the Box LVIII

Two Wrongs Make Right In the following addition problem each letter represents a different digit from 1 through 9 and zero is not allowed. Find digits that make the addition correct.

$$
\begin{array}{r}
\text{WRONG} \\
+ \text{WRONG} \\
\hline
\text{RIGHT}
\end{array}
$$

■■■ Chapter 6 Test

Use identities to simplify each expression.

1. $\sec x \cot x \sin 2x$

2. $\sin 2t \cos 5t + \cos 2t \sin 5t$

3. $\dfrac{1}{1 - \cos y} + \dfrac{1}{1 + \cos y}$

4. $\dfrac{\tan(\pi/5) + \tan(\pi/10)}{1 - \tan(\pi/5) \tan(\pi/10)}$

Prove that each of the following equations is an identity.

5. $\dfrac{\sin \beta \cos \beta}{\tan \beta} = 1 - \sin^2 \beta$

6. $\dfrac{1}{\sec \theta - 1} - \dfrac{1}{\sec \theta + 1} = 2 \cot^2 \theta$

7. $\cos\left(\dfrac{\pi}{2} - x\right) \cos(-x) = \dfrac{\sin(2x)}{2}$

8. $\tan \dfrac{t}{2} \cos^2 t - \tan \dfrac{t}{2} = \dfrac{\sin t}{\sec t} - \sin t$

Find all solutions to each equation.

9. $\sin(-\theta) = 1$

10. $\cos 3s = \dfrac{1}{2}$

11. $\tan 2t = -\sqrt{3}$ **12.** $\sin 2\theta = \cos \theta$

Find all values of α in $[0°, 360°)$ that satisfy each equation. Round approximate answers to the nearest tenth of a degree.

13. $3 \sin^2 \alpha - 4 \sin \alpha + 1 = 0$

14. $\dfrac{\tan 2\alpha - \tan 7\alpha}{1 + \tan 2\alpha \tan 7\alpha} = 1$

Solve each problem.

15. Write $y = \sin x - \sqrt{3} \cos x$ in the form $y = A \sin(x + C)$ and graph the function. Determine the period, amplitude, and phase shift.

16. Given that $\csc \alpha = 2$ and α is in quadrant II, find the exact values of α for the remaining five trigonometric functions.

17. Determine whether the function $f(x) = x \sin x$ is odd, even, or neither.

18. Use an appropriate identity to find the exact value of $\sin(-\pi/12)$.

19. Prove that $\tan x + \tan y = \tan(x + y)$ is not an identity.

20. A car with worn shock absorbers hits a pothole and oscillates about its normal riding position. At time t (in seconds) the front bumper is distance d (in inches) above or below its normal position, where $d = 2 \sin 3t - 4 \cos 3t$. For what values of t (to the nearest tenth of a second) in the interval $[0, 4]$ is the front bumper at its normal position $d = 0$?

Tying it all Together

Chapters 1–6

Determine whether each function is even or odd.

1. $f(x) = 3x^3 - 2x$ **2.** $f(x) = 2|x|$ **3.** $f(x) = x^3 + \sin x$ **4.** $f(x) = x^3 \sin x$

5. $f(x) = x^4 - x^2 + 1$ **6.** $f(x) = 1/x$ **7.** $f(x) = \dfrac{\sin x}{x}$ **8.** $f(x) = |\sin x|$

Determine whether each equation is an identity. Prove your answer.

9. $\sin(\alpha + \beta) = \sin \alpha + \sin \beta$ **10.** $(\alpha + \beta)^2 = \alpha^2 + \beta^2$ **11.** $3x + 5x = 8x$

12. $(x + 3)^2 = x^2 + 6x + 9$ **13.** $\sin^{-1}(x) = \dfrac{1}{\sin x}$ **14.** $\sin^2 x = \sin(x^2)$

Solve each right triangle that has the given parts.

15. $\alpha = 30°$ and $a = 4$ **16.** $a = \sqrt{3}$ and $b = 1$ **17.** $\cos \beta = 0.3$ and $b = 5$ **18.** $\sin \alpha = 0.6$ and $a = 2$

In Chapter 7 we will study expressions like the following, which combine trigonometric functions and complex numbers. Write each expression in the form $a + bi$, where a and b are real numbers.

19. $\cos(\pi/3) + i \sin(\pi/3)$ **20.** $2\left(\cos \dfrac{2\pi}{3} + i \sin \dfrac{2\pi}{3}\right)$ **21.** $(\cos 225° + i \sin 225°)^2$

22. $(\cos 3° + i \sin 3°)(\cos 3° - i \sin 3°)$ **23.** $(2 + i)^3$ **24.** $\left(\sqrt{2} - \sqrt{2}i\right)^4$

7 Applications of Trigonometry

MECHANICAL devices that perform manipulative tasks under their own power (robots) are as old as recorded history. As early as 3000 B.C., Egyptians built waterclocks and articulated figures.

The first industrial robot joined GM's production line in 1961. In the 21st century, robotic "steel collar" workers will increasingly perform boring, repetitive, and dangerous jobs associated with a high degree of human error. They will assemble parts, mix chemicals, and work in areas that would be deadly to their human counterparts.

WHAT YOU WILL LEARN   In this chapter, we will see that trigonometry is the mathematics you need to model repetitive motion. We will learn how to use trigonometry to position a robot's arm. We'll also learn how trigonometry is used in designing machines and finding forces acting on parts.

7.1

The Law of Sines

In Chapter 5 we used trigonometry to solve right triangles. In this section and the next we will learn to solve any triangle for which we have enough information to determine the shape of the triangle.

Oblique Triangles

Any triangle without a right angle is called an **oblique triangle.** As usual, we use α, β, and γ for the angles of a triangle and a, b, and c, respectively, for the lengths of the sides opposite those angles, as shown in Fig. 7.1. The vertices at angles α, β, and γ are labeled A, B, and C, respectively. To solve an oblique triangle, we must know at least three parts of the triangle, at least one of which must be the length of a side. We can classify the different cases for the three known parts as follows:

1. One side and any two angles (ASA or AAS)
2. Two sides and a nonincluded angle (SSA)
3. Two sides and an included angle (SAS)
4. Three sides (SSS)

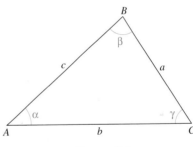

■ **Figure 7.1**

We can actually solve all of these cases by dividing the triangles into right triangles and using right triangle trigonometry. Since that method is quite tedious, we develop the *law of sines* and the *law of cosines.* The first two cases can be handled with the law of sines. We will discuss the last two cases in Section 7.2, when we develop the law of cosines.

The Law of Sines

The **law of sines** says that *the ratio of the sine of an angle and the length of the side opposite the angle is the same for each angle of a triangle.*

Theorem: The Law of Sines

In any triangle,

$$\frac{\sin \alpha}{a} = \frac{\sin \beta}{b} = \frac{\sin \gamma}{c}.$$

■ **PROOF**   Either the triangle is an acute triangle (all acute angles) or it is an obtuse triangle (one obtuse angle). We consider the case of the obtuse triangle here and leave the case of the acute triangle as Exercise 43. Triangle ABC is shown in Fig. 7.2 with an altitude of length h_1 drawn from point C to the opposite side and an altitude of length h_2 drawn from point B to the extension of the opposite side.

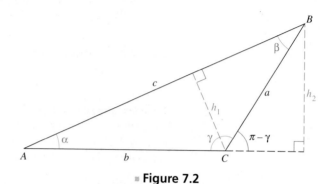

■ **Figure 7.2**

Since α and β are now in right triangles, we have

$$\sin \alpha = \frac{h_1}{b} \qquad \text{or} \qquad h_1 = b \sin \alpha$$

and

$$\sin \beta = \frac{h_1}{a} \qquad \text{or} \qquad h_1 = a \sin \beta.$$

Replace h_1 by $b \sin \alpha$ in the equation $h_1 = a \sin \beta$ to get

$$b \sin \alpha = a \sin \beta.$$

Dividing each side of the last equation by ab yields

$$\frac{\sin \alpha}{a} = \frac{\sin \beta}{b}.$$

Using the largest right triangle in Fig. 7.2, we have

$$\sin \alpha = \frac{h_2}{c} \qquad \text{or} \qquad h_2 = c \sin \alpha.$$

Note that the acute angle $\pi - \gamma$ in Fig. 7.2 is the reference angle for the obtuse angle γ. We know that sine is positive for both acute and obtuse angles, so $\sin \gamma = \sin(\pi - \gamma)$. Since $\pi - \gamma$ is an angle of a right triangle, we can write

$$\sin \gamma = \sin(\pi - \gamma) = \frac{h_2}{a} \qquad \text{or} \qquad h_2 = a \sin \gamma.$$

From $h_2 = c \sin \alpha$ and $h_2 = a \sin \gamma$ we get $c \sin \alpha = a \sin \gamma$ or

$$\frac{\sin \alpha}{a} = \frac{\sin \gamma}{c}$$

and we have proved the law of sines.   ■

The law of sines can also be written in the form

$$\frac{a}{\sin \alpha} = \frac{b}{\sin \beta} = \frac{c}{\sin \gamma}.$$

In solving triangles, it is usually simplest to use the form in which the unknown quantity appears in the numerator.

In our first example we use the law of sines to solve a triangle for which we are given two angles and an included side. Remember that if the dimensions given for triangles are measurements, then any answers obtained from those measurements are only as accurate as the least accurate of the measurements. (See the discussion of significant digits in Chapter 5.)

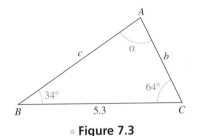

■ **Figure 7.3**

Example **1** Given two angles and an included side (ASA)

Given $\beta = 34°$, $\gamma = 64°$, and $a = 5.3$, solve the triangle.

Solution

To sketch the triangle, first draw side a, then draw angles of approximately 34° and 64° on opposite ends of a. Label all parts (see Fig. 7.3). Since the sum of the three angles of a triangle is 180°, the third angle α is 82°. By the law of sines,

$$\frac{5.3}{\sin 82°} = \frac{b}{\sin 34°}$$

$$b = \frac{5.3 \sin 34°}{\sin 82°} \approx 3.0.$$

Again, by the law of sines,

$$\frac{c}{\sin 64°} = \frac{5.3}{\sin 82°}$$

$$c = \frac{5.3 \sin 64°}{\sin 82°} \approx 4.8.$$

So $\alpha = 82°$, $b \approx 3.0$, and $c \approx 4.8$ solves the triangle.

Try This. Given $\alpha = 28°$, $\beta = 66°$, and $c = 8.2$, solve the triangle. ■

The AAS case is similar to the ASA case. If two angles are known, then the third can be found by using the fact that the sum of all angles is 180°. If we know all angles and any side, then we can proceed to find the remaining sides with the law of sines as in Example 1.

The Ambiguous Case (SSA)

In the AAS and ASA cases we are given any two angles with positive measures and the length of any side. If the total measure of the two angles is less than 180°, then a unique triangle is determined. However, for two sides and a *nonincluded* angle (SSA), there are several possibilities. So the SSA case is called the **ambiguous case.** Drawing the diagram in the proper order will help you in understanding the ambiguous case.

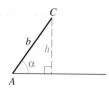

Draw α and b: find h

■ **Figure 7.4**

Suppose we are given an acute angle α ($0° < \alpha < 90°$) and sides a and b. Side a is opposite the angle α, and side b is adjacent to it. Draw an angle of approximate size α in standard position with terminal side of length b, as shown in Fig. 7.4. Don't draw in side a yet. Let h be the distance from C to the initial side of α. Since $\sin \alpha = h/b$, we have $h = b \sin \alpha$. Now we are ready to draw in side a, but there are four possibilities for its location.

1. If $a < h$, then no triangle can be formed, because a cannot reach from point C to the initial side of α. This situation is shown in Fig. 7.5(a).
2. If $a = h$, then exactly one right triangle is formed, as in Fig. 7.5(b).
3. If $h < a < b$, then exactly two triangles are formed, because a reaches to the initial side in two places, as in Fig. 7.5(c).
4. If $a \geq b$, then only one triangle is formed, as in Fig. 7.5(d).

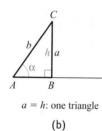

$a < h$: no triangle

(a)

$a = h$: one triangle

(b)

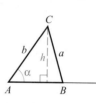

$h < a < b$: two triangles

(c)

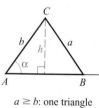

$a \geq b$: one triangle

(d)

■ **Figure 7.5**

If we start with α, a, and b, where $90° \leq \alpha < 180°$, then there are only two possibilities for the number of triangles determined. If $a \leq b$, then no triangle is formed, as in Fig. 7.6(a). If $a > b$, one triangle is formed, as in Fig. 7.6(b).

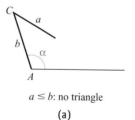

$a \leq b$: no triangle

(a)

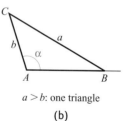

$a > b$: one triangle

(b)

■ **Figure 7.6**

It is not necessary to memorize all of the SSA cases shown in Figs. 7.5 and 7.6. If you draw the triangle for a given problem in the order suggested, then it will be clear how many triangles are possible with the given parts. We must decide how many triangles are possible, before we can solve the triangle(s).

Example 2 SSA with no triangle

Given $\alpha = 41°$, $a = 3.3$, and $b = 5.4$, solve the triangle.

Solution

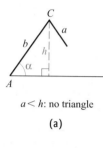

■ **Figure 7.7**

Draw the 41° angle and label its terminal side 5.4, as shown in Fig. 7.7. Side a must go opposite α, but do not put it in yet. Find the length of the altitude h from C. Using a trigonometric ratio, we get $\sin 41° = h/5.4$, or $h = 5.4 \sin 41° \approx 3.5$. Since

$a = 3.3$, a is *shorter* than the altitude h, and side a will not reach from point C to the initial side of α. So there is no triangle with the given parts.

Try This.   Given $\beta = 38°$, $b = 2.9$, and $c = 5.9$, solve the triangle.   ▪

Example **3** SSA with one triangle

Given $\gamma = 125°$, $b = 5.7$, and $c = 8.6$, solve the triangle.

Solution

Draw the $125°$ angle and label its terminal side 5.7. Since $8.6 > 5.7$, side c will reach from point A to the initial side of γ, as shown in Fig. 7.8, and form a single triangle. To solve this triangle, we use the law of sines:

$$\frac{\sin 125°}{8.6} = \frac{\sin \beta}{5.7}$$

$$\sin \beta = \frac{5.7 \sin 125°}{8.6}$$

$$\beta = \sin^{-1}\left(\frac{5.7 \sin 125°}{8.6}\right) \approx 32.9°$$

Since the sum of α, β, and γ is $180°$, $\alpha = 22.1°$. Now use α and the law of sines to find a:

$$\frac{a}{\sin 22.1°} = \frac{8.6}{\sin 125°}$$

$$a = \frac{8.6 \sin 22.1°}{\sin 125°} \approx 3.9$$

Try This.   Given $\beta = 38°$, $b = 6.4$, and $c = 5.9$, solve the triangle.   ▪

Example **4** SSA with two triangles

Given $\beta = 56.3°$, $a = 8.3$, and $b = 7.6$, solve the triangle.

Solution

Draw an angle of approximately $56.3°$, and label its terminal side 8.3. Side b must go opposite β, but do not put it in yet. Find the length of the altitude h from point C to the initial side of β, as shown in Fig. 7.9(a). Since $\sin 56.3° = h/8.3$, we get

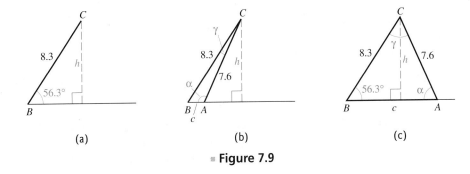

▪ **Figure 7.9**

A

α

8.6

5.7

$125°$

β

C a B

▪ **Figure 7.8**

$h = 8.3 \sin 56.3° \approx 6.9$. Because b is longer than h but shorter than a, there are two triangles that satisfy the given conditions, as shown in parts (b) and (c) of Fig. 7.9. In either triangle we have

$$\frac{\sin \alpha}{8.3} = \frac{\sin 56.3°}{7.6}$$

$$\sin \alpha = 0.9086.$$

This equation has two solutions in $[0°, 180°]$. For part (c) we get $\alpha = \sin^{-1}(0.9086) = 65.3°$, and for part (b) we get $\alpha = 180° - 65.3° = 114.7°$. Using the law of sines, we get $\gamma = 9.0°$ and $c = 1.4$ for part (b), and we get $\gamma = 58.4°$ and $c = 7.8$ for part (c).

Try This. Given $\beta = 38°$, $b = 4.7$, and $c = 5.9$, solve the triangle. ■

The remaining cases for solving triangles are presented in the next section, which also contains a summary of how to proceed in each case. You may wish to refer to it now.

Area of a Triangle

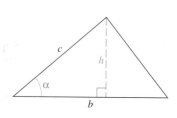

■ **Figure 7.10**

The formula $A = \frac{1}{2}bh$ gives the area of a triangle in terms of a side and the altitude to that side. We can rewrite this formula in terms of two sides and their included angle. Consider the triangle shown in Fig. 7.10, in which α is an acute angle and h is inside the triangle. Since $\sin \alpha = h/c$, we get $h = c \sin \alpha$. Substitute $c \sin \alpha$ for h in the formula to get

$$A = \frac{1}{2} bc \sin \alpha.$$

This formula is also correct in all other possible cases: α is acute and h is the side of the triangle opposite α, α is acute and h lies outside the triangle, α is a right angle, and α is obtuse. (See Exercise 44.) Likewise, the formula can be written using the angles β or γ. We have the following theorem.

Theorem: Area of a Triangle

> The area of a triangle is given by
>
> $$A = \frac{1}{2} bc \sin \alpha, \qquad A = \frac{1}{2} ac \sin \beta, \qquad \text{and} \qquad A = \frac{1}{2} ab \sin \gamma.$$

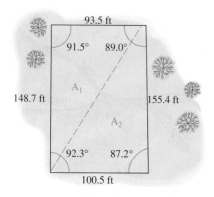

■ **Figure 7.11**

Surveyors describe property in terms of the lengths of the sides and the angles between those sides. In the next example we find the area of a piece of property, using the new area formula.

Example **5** Area of a quadrilateral

The town of Hammond is considering the purchase of a four-sided piece of property to be used for a playground. The dimensions of the property are given in Fig. 7.11. Find the area of the property in square feet.

Solution

Divide the property into two triangles, as shown in Fig. 7.11, and find the area of each:

$$A_1 = \frac{1}{2}(148.7)(93.5)\sin 91.5° \approx 6949.3 \text{ square feet}$$

$$A_2 = \frac{1}{2}(100.5)(155.4)\sin 87.2° \approx 7799.5 \text{ square feet}$$

The total area of the property is 14,748.8 square feet.

Try This. Two sides of a triangular piece of property are 244 feet and 206 feet, and the angle between these sides is 87.4°. Find the area to the nearest square foot. ▪

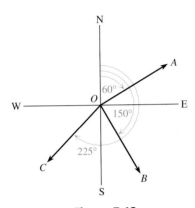

▪ **Figure 7.12**

Bearing

The measure of an angle that describes the direction of a ray is called the **bearing** of the ray. In air navigation, bearing is given as a nonnegative angle less than 360° measured in a clockwise direction from a ray pointing due north. So in Fig. 7.12 the bearing of ray $\overrightarrow{OA}$ is 60°, the bearing of ray $\overrightarrow{OB}$ is 150°, and the bearing of $\overrightarrow{OC}$ is 225°.

Example **6** **Using bearing in solving triangles**

A bush pilot left the Fairbanks Airport in a light plane and flew 100 miles toward Fort Yukon in still air on a course with a bearing of 18°. She then flew due east (bearing 90°) for some time to drop supplies to a snowbound family. After the drop, her course to return to Fairbanks had a bearing of 225°. What was her maximum distance from Fairbanks?

Solution

Figure 7.13 shows the course of her flight. To change course from bearing 18° to bearing 90° at point B, the pilot must add 72° to the bearing. So $\angle ABC$ is 108°. A bearing of 225° at point C means that $\angle BCA$ is 45°. Finally, we obtain $\angle BAC = 27°$ by subtracting 18° and 45° from 90°.

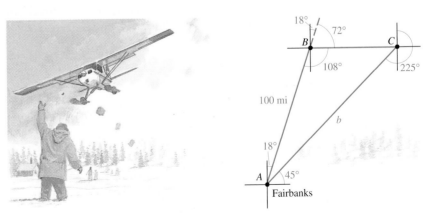

▪ **Figure 7.13**

We can find the length of $\overline{AC}$ (the maximum distance from Fairbanks) by using the law of sines:

$$\frac{b}{\sin 108°} = \frac{100}{\sin 45°}$$

$$b = \frac{100 \cdot \sin 108°}{\sin 45°} \approx 134.5$$

So the pilot's maximum distance from Fairbanks was 134.5 miles.

Try This. A pilot flew north from the airport for 88 miles, then east until he headed back to the airport on a course with bearing 210°. What was his maximum distance from the airport?   ■

In marine navigation and surveying, the bearing of a ray is the acute angle the ray makes with a ray pointing due north or due south. Along with the acute angle, directions are given that indicate in which quadrant the ray lies. For example, in Fig. 7.12, $\overrightarrow{OA}$ has a bearing 60° east of north (N60°E), $\overrightarrow{OB}$ has a bearing 30° east of south (S30°E), and $\overrightarrow{OC}$ has a bearing 45° west of south (S45°W).

One More Application

In the next example we find the height of an object from a distance. This same problem was solved in Section 5.6 Example 7. Note how much easier the solution is here using the law of sines.

Example **7** **Finding the height of an object from a distance**

The angle of elevation of the top of a water tower from point A on the ground is 19.9°. From point B, 50.0 feet closer to the tower, the angle of elevation is 21.8°. What is the height of the tower?

Solution

Let y represent the height of the tower. All angles of triangle ABC can be determined as shown in Fig. 7.14.

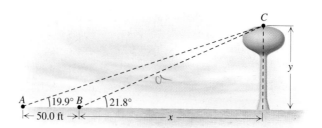

■ **Figure 7.14**

Apply the law of sines in triangle *ABC*:

$$\frac{a}{\sin 19.9°} = \frac{50}{\sin 1.9°}$$

$$a = \frac{50 \sin 19.9°}{\sin 1.9°} \approx 513.3 \text{ feet}$$

Now, using the smaller right triangle, we have $\sin 21.8° = y/513.3$ or $y = 513.3 \sin 21.8° \approx 191$. So the height of the tower is 191 feet.

Try This. The angle of elevation of the top of a building from point *A* on the ground is 24.2°. From point *B*, which is 44.5 feet closer, the angle of elevation is 38.1°. What is the height of the building?   ▪

For Thought

True or False? Explain.

1. If we know the measures of two angles of a triangle, then the measure of the third angle is determined.

2. If $\frac{\sin 9°}{a} = \frac{\sin 17°}{88}$, then $a = \frac{\sin 17°}{88 \sin 9°}$.

3. The equation $\frac{\sin \alpha}{5} = \frac{\sin 44°}{18}$ has exactly one solution in $[0, 180°]$.

4. One solution to $\frac{\sin \beta}{2.3} = \frac{\sin 39°}{1.6}$ is $\beta = \sin^{-1}\left(\frac{2.3 \sin 39°}{1.6}\right)$.

5. $\frac{\sin 60°}{\sqrt{3}} = \frac{\sin 30°}{1}$

6. No triangle exists with $\alpha = 60°$, $b = 10$ feet, and $a = 500$ feet.

7. There is exactly one triangle with $\beta = 30°$, $c = 20$, and $b = 10$.

8. There are two triangles with $\alpha = 135°$, $b = 17$, and $a = 19$.

9. The area of a triangle is determined by the lengths of two sides and the measure of the included angle.

10. A right triangle's area is one-half the product of the lengths of its legs.

7.1 Exercises

Solve each triangle.

1.

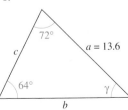

2.

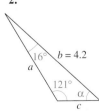

3.

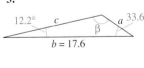

4.

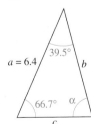

Solve each triangle with the given parts.

5. $\alpha = 10.3°, \gamma = 143.7°, c = 48.3$

6. $\beta = 94.7°, \alpha = 30.6°, b = 3.9$

7. $\beta = 120.7°, \gamma = 13.6°, a = 489.3$

8. $\alpha = 39.7°, \gamma = 91.6°, b = 16.4$

Determine the number of triangles with the given parts and solve each triangle.

9. $\alpha = 39.6°, c = 18.4, a = 3.7$

10. $\beta = 28.6°, a = 40.7, b = 52.5$

11. $\gamma = 60°, b = 20, c = 10\sqrt{3}$

12. $\alpha = 41.2°, a = 8.1, b = 10.6$

13. $\beta = 138.1°, c = 6.3, b = 15.6$

14. $\gamma = 128.6°, a = 9.6, c = 8.2$

15. $\beta = 32.7°, a = 37.5, b = 28.6$

16. $\alpha = 30°, c = 40, a = 20$

17. $\gamma = 99.6°, b = 10.3, c = 12.4$

18. $\alpha = 75.3, a = 12.4, b = 9.8$

Find the area of each triangle with the given parts.

19. $a = 12.9, b = 6.4, \gamma = 13.7°$

20. $b = 42.7, c = 64.1, \alpha = 74.2°$

21. $\alpha = 39.4°, b = 12.6, a = 13.7$

22. $\beta = 74.2°, c = 19.7, b = 23.5$

23. $\alpha = 42.3°, \beta = 62.1°, c = 14.7$

24. $\gamma = 98.6°, \beta = 32.4°, a = 24.2$

25. $\alpha = 56.3°, \beta = 41.2°, a = 9.8$

26. $\beta = 25.6°, \gamma = 74.3°, b = 17.3$

Find the area of each region. Give the exact area for Exercise 27 and round to the nearest 10 square feet for Exercise 28.

27.

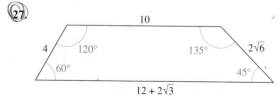

28.

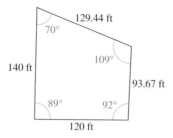

Solve each problem.

29. *Observing Traffic* A traffic report helicopter left the WKPR studios on a course with a bearing of 210°. After flying 12 mi to reach interstate highway 20, the helicopter flew due east along I-20 for some time. The helicopter headed back to WKPR on a course with a bearing of 310° and reported no accidents along I-20. For how many miles did the helicopter fly along I-20?

30. *Course of a Fighter Plane* During an important NATO exercise, an F-14 Tomcat left the carrier Nimitz on a course with a bearing of 34° and flew 400 mi. Then the F-14 flew for some distance on a course with a bearing of 162°. Finally, the plane flew back to its starting point on a course with a bearing of 308°. What distance did the plane fly on the final leg of the journey?

31. *Surveying Property* A surveyor locating the corners of a triangular piece of property started at one corner and walked 480 ft in the direction of N36°W to reach the next corner. The surveyor turned and walked S21°W to get to the next corner of the property. Finally, the surveyor walked in the direction N82°E to get back to the starting point. What is the area of the property in square feet?

32. *Sailing* Joe and Jill set sail from the same point, with Joe sailing in the direction of S4°E and Jill sailing in the direction S9°W. After 4 hr, Jill was 2 mi due west of Joe. How far had Jill sailed?

33. *Cellular One* The angle of elevation of the top of a cellular telephone tower from point *A* on the ground is 18.1°. From point *B*, 32.5 ft closer to the tower, the angle of elevation is 19.3°. What is the height of the tower?

34. *Moving Back* A surveyor determines that the angle of elevation of the top of a building from a point on the ground is 30.4°. He then moves back 55.4 ft and determines that the angle of elevation is 23.2°. What is the height of the building?

35. *Designing an Addition* A 40-ft-wide house has a roof with a 6-12 pitch (the roof rises 6 ft for a run of 12 ft). The owner plans a 14-ft-wide addition that will have a 3-12 pitch to its roof. Find the lengths of $\overline{AB}$ and $\overline{BC}$ in the accompanying figure.

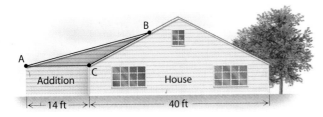

- **Figure for Exercise 35**

36. *Saving an Endangered Hawk*   A hill has an angle of inclination of 36°, as shown in the figure. A study completed by a state's highway commission showed that the placement of a highway requires that 400 ft of the hill, measured horizontally, be removed. The engineers plan to leave a slope alongside the highway with an angle of inclination of 62°, as shown in the figure. Located 750 ft up the hill measured from the base is a tree containing the nest of an endangered hawk. Will this tree be removed in the excavation?

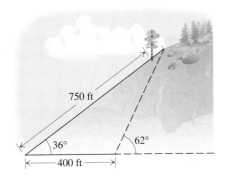

- **Figure for Exercise 36**

37. *Making a Kite*   A kite is made in the shape shown in the figure. Find the surface area of the kite in square inches.

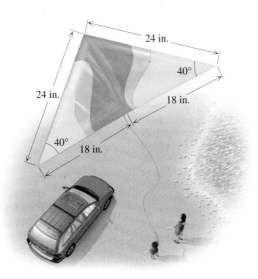

- **Figure for Exercise 37**

38. *Area of a Wing*   The F-106 Delta Dart once held a world speed record of Mach 2.3. Its sweptback triangular wings have the dimensions given in the figure. Find the area of one wing in square feet.

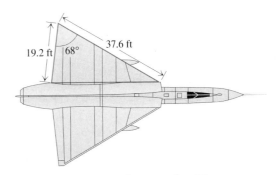

- **Figure for Exercise 38**

39. *Filtering the Sun*   When the sun is directly overhead, the sun's light is filtered by approximately 10 miles of atmosphere. As the sun sets, the intensity decreases because the light must pass a greater distance through the atmosphere, as shown in the figure.

a. Find the distance d in the figure when the angle of elevation of the sun is 30°. Use 3950 miles as the radius of the earth.

b. If the sun is directly overhead at 12 noon, then at what time (to the nearest tenth of a second) is the angle of elevation of the sun 30°? The distance from the earth to the sun is 93 million miles.

c. Find the distance that the sunlight passes through the atmosphere at sunset.

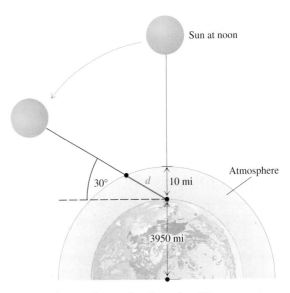

- **Figure for Exercise 39**

40. *Flat Earth* Assume that the earth is flat with a 10-mile thick layer of atmosphere. Find the distance that the sunlight passes through the atmosphere when the angle of elevation of the sun is 30°. Compare your answer to that obtained in part (a) of the previous exercise.

41. *Shot Down* A cruise missile is traveling straight across the desert at 548 mph at an altitude of 1 mile, as shown in the figure. A gunner spots the missile coming in his direction and fires a projectile at the missile when the angle of elevation of the missile is 35°. If the speed of the projectile is 688 mph, then for what angle of elevation of the gun will the projectile hit the missile?

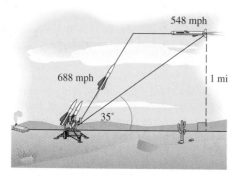

▪ **Figure for Exercise 41**

42. *Angle of Completion* When the ball is snapped, Smith starts running at a 50° angle to the line of scrimmage. At the moment when Smith is at a 60° angle from Jones, Smith is running at 17 ft/sec and Jones passes the ball at 60 ft/sec to Smith. However, to complete the pass, Jones must lead Smith by the angle θ shown in the figure. Find θ. (Of course, Jones finds θ in his head. Note that θ can be found without knowing any distances.)

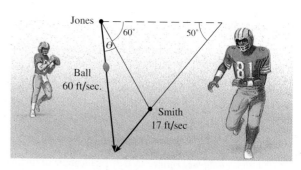

▪ **Figure for Exercise 42**

For Writing/Discussion

43. *Law of Sines* Prove the law of sines for the case in which the triangle is an acute triangle.

44. *Area of a Triangle* Prove the trigonometric formula for the area of a triangle in the cases mentioned in the text but not proved in the text.

45. *Cooperative Learning* Find a map or a description of a piece of property in terms of sides and angles. Show it to your class and explain how to find the area of the property in square feet.

Thinking Outside the Box LIX & LX

Packing Beans A can of green beans has a 3-in. diameter and 4-in. height. How many cans of beans will fit in a box that measures 24 in. long, 24 in. wide, and 4 in. high on the inside?

Working Together Big Red can build a brick wall by himself in 10 hr, whereas Slim would take 12 hr to build the same wall by himself. If they work together they talk sports, and each of them lays 10 fewer bricks per hour. To get the job done in a hurry, the boss assigns both men and they finish the wall in 6 hr. How many bricks are in the wall?

7.1 Pop Quiz

1. If $\alpha = 8°$, $\beta = 121°$, and $c = 12$, then what is γ?

2. If $\alpha = 20.4°$, $\beta = 27.3°$, and $c = 38.5$, then what is a?

3. If $\alpha = 33.5°$, $a = 7.4$, and $b = 10.6$, then what is β?

4. Find the area of the triangle in which $a = 6$ ft, $b = 15$ ft, and $\gamma = 66.7°$.

Linking Concepts

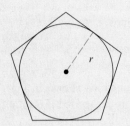

Discovering Area and Circumference Formulas

In these exercises you will discover the area and circumference formulas for the circle. You will work with regular polygons that circumscribe a circle of radius r, and you will use the new formula for the area of a triangle:

$$A = \frac{1}{2} bc \sin \alpha$$

a) Suppose a regular pentagon circumscribes a circle of radius r, as shown in the figure. Show that the area of the pentagon is $5r^2 \tan(36°)$.

b) Now suppose that a regular polygon of n sides circumscribes a circle of radius r. Show that the area of the n-gon is given by $A = nr^2 \tan(180°/n)$.

c) Express the result of part (b), using the language of variation discussed in Section 2.6.

d) Find the constant of proportionality for a decagon, kilogon, and megagon.

e) What happens to the shape of the n-gon as n increases? So what can you conclude is the formula for the area of a circle of radius r?

f) Use trigonometry to find a formula for the perimeter of an n-gon that circumscribes a circle of radius r.

g) Use the result of part (f) to find a formula for the circumference of a circle of radius r.

h) Graph the function $y = x \tan(\pi/x)$, where x is a real number (radian mode). Identify all vertical and horizontal asymptotes.

7.2

The Law of Cosines

In Section 7.1 we discussed the AAS, ASA, and SSA cases for solving triangles using the law of sines. For the SAS and SSS cases, which cannot be handled with the law of sines, we develop the law of cosines. This law is a generalization of the Pythagorean theorem and applies to any triangle. However, it is usually stated and used only for oblique triangles because it is not needed for right triangles.

The Law of Cosines

The **law of cosines** gives a formula for the square of any side of an oblique triangle in terms of the other two sides and their included angle.

Theorem: Law of Cosines

If triangle ABC is an oblique triangle with sides a, b, and c and angles α, β, and γ, then

$$a^2 = b^2 + c^2 - 2bc \cos \alpha,$$

$$b^2 = a^2 + c^2 - 2ac \cos \beta,$$

$$c^2 = a^2 + b^2 - 2ab \cos \gamma.$$

▪ **PROOF** Given triangle ABC, position the triangle as shown in Fig. 7.15. The vertex C is in the first quadrant if α is acute and in the second if α is obtuse. Both cases are shown in Fig. 7.15.

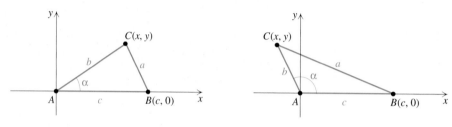

▪ **Figure 7.15**

In either case, the x-coordinate of C is $x = b \cos \alpha$, and the y-coordinate of C is $y = b \sin \alpha$. The distance from C to B is a, but we can also find that distance by using the distance formula:

$$a = \sqrt{(b \cos \alpha - c)^2 + (b \sin \alpha - 0)^2}$$

$$a^2 = (b \cos \alpha - c)^2 + (b \sin \alpha)^2$$

$$= b^2 \cos^2 \alpha - 2bc \cos \alpha + c^2 + b^2 \sin^2 \alpha$$

$$= b^2(\cos^2 \alpha + \sin^2 \alpha) + c^2 - 2bc \cos \alpha$$

Using $\cos^2 x + \sin^2 x = 1$, we get the first equation of the theorem:

$$a^2 = b^2 + c^2 - 2bc \cos \alpha$$

Similar arguments with B and C at $(0, 0)$ produce the other two equations. ▪

In any triangle with unequal sides, the largest angle is opposite the largest side, and the smallest angle is opposite the smallest side. We need this fact in the first example.

Example **1** Given three sides of a triangle (SSS)

Given $a = 8.2$, $b = 3.7$, and $c = 10.8$, solve the triangle.

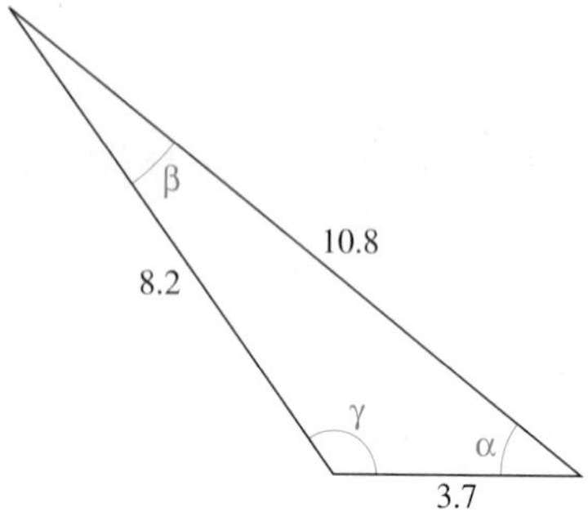

Solution

Draw the triangle and label it as in Fig. 7.16. Since c is the longest side, use

$$c^2 = a^2 + b^2 - 2ab \cos \gamma$$

▪ **Figure 7.16**

to find the largest angle γ:

$$-2ab \cos \gamma = c^2 - a^2 - b^2$$

$$\cos \gamma = \frac{c^2 - a^2 - b^2}{-2ab} = \frac{(10.8)^2 - (8.2)^2 - (3.7)^2}{-2(8.2)(3.7)} \approx -0.5885$$

$$\gamma = \cos^{-1}(-0.5885) \approx 126.1°$$

We could finish with the law of cosines, but in this case it is simpler to use the law of sines, which involves fewer computations:

$$\frac{\sin \beta}{b} = \frac{\sin \gamma}{c}$$

$$\frac{\sin \beta}{3.7} = \frac{\sin 126.1°}{10.8}$$

$$\sin \beta \approx 0.2768$$

There are two solutions to $\sin \beta = 0.2768$ in $[0°, 180°]$. However, β must be less than 90°, because γ is 126.1°. So $\beta = \sin^{-1}(0.2768) \approx 16.1°$. Finally, $\alpha = 180° - 126.1° - 16.1° = 37.8°$.

Try This. Given $a = 3.8$, $b = 9.6$, and $c = 7.7$, solve the triangle. ▪

In solving the SSS case, we always *find the largest angle first,* using the law of cosines. The remaining two angles must be acute angles. So when using the law of sines to find another angle, we need only find an acute solution to the equation.

In Section 7.1 we learned that two adjacent sides and a nonincluded angle (SSA) might not determine a triangle. To determine a triangle with three given sides (SSS), the sum of the lengths of any two must be greater than the length of the third side. This fact is called the **triangle inequality.** To understand the triangle inequality, try to draw a triangle with sides of lengths 1 in., 2 in., and 5 in. Two sides and the included angle (SAS) will determine a triangle provided the angle is between 0° and 180°.

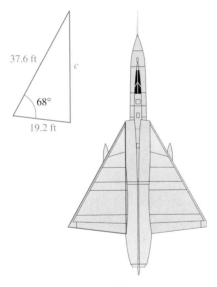

▪ **Figure 7.17**

Example **2** Given two sides and an included angle (SAS)

The wing of the F-106 Delta Dart is triangular in shape, with the dimensions given in Fig. 7.17. Find the length of the side labeled c in Fig. 7.17.

Solution

Using the law of cosines, we find c as follows:

$$c^2 = (19.2)^2 + (37.6)^2 - 2(19.2)(37.6) \cos 68° \approx 1241.5$$

$$c \approx \sqrt{1241.5} \approx 35.2 \text{ feet}$$

Try This. Given $b = 5.8$, $c = 3.6$, and $\alpha = 39.5°$, find a. ▪

| PROCEDURE | Solving Triangles (In all cases draw pictures.) |

ASA (For example α, c, β)

1. Find γ using $\gamma = 180° - \alpha - \beta$.
2. Find a and c using the law of sines.

SSA (For example a, b, α)

1. Find $h = b \sin \alpha$. If $h > a$, then there is no triangle.
2. If $h = a$, then there is one right triangle ($\beta = 90°$ and b is the hypotenuse). Solve it using right triangle trigonometry.
3. If $h < a < b$ there are two triangles, one with β acute and one with β obtuse. Find the acute β using the law of sines. Subtract it from $180°$ to get the obtuse β. In each of the two triangles, find γ using $\gamma = 180° - \alpha - \beta$ and c using the law of sines.
4. If $a \geq b$ there is only one triangle and β is acute (α or γ might be obtuse). Find β using the law of sines. Then find γ using $\gamma = 180° - \alpha - \beta$. Find c using the law of sines.

SSS (For example a, b, c)

1. Find the largest angle using the law of cosines. The largest angle is opposite the largest side.
2. Find another angle using the law of sines.
3. Find the third angle by subtracting the first two from $180°$.

SAS (For example b, α, c)

1. Find a using the law of cosines.
2. Use the law of sines to find β if $b < c$ or γ if $c < b$. If $b = c$, then find either β or γ.
3. Find the last angle by subtracting the first two from $180°$.

Area of a Triangle by Heron's Formula

In Section 7.1 we saw the formulas $A = \frac{1}{2}bh$ and $A = \frac{1}{2}bc \sin \alpha$ for the area of a triangle. For one formula we must know a side and the altitude to that side, and for the other we need two sides and the measure of their included angle. Using the law of cosines, we can get a formula for the area of a triangle that involves only the lengths of the sides. This formula is known as **Heron's area formula**, named after Heron of Alexandria, who is believed to have discovered it in about A.D. 75.

Heron's Area Formula

The area of a triangle with sides a, b, and c is given by the formula

$$A = \sqrt{S(S - a)(S - b)(S - c)},$$

where $S = (a + b + c)/2$.

▪ **PROOF** First rewrite the equation $a^2 = b^2 + c^2 - 2bc \cos \alpha$ as follows:

$$2bc \cos \alpha = b^2 + c^2 - a^2$$

$$4b^2c^2 \cos^2 \alpha = (b^2 + c^2 - a^2)^2$$

Now write the area formula $A = \frac{1}{2}bc \sin \alpha$ in terms of $4b^2c^2 \cos \alpha$:

$$A = \frac{1}{2}bc \sin \alpha$$

$$4A = 2bc \sin \alpha \qquad \text{Multiply each side by 4.}$$

$$16A^2 = 4b^2c^2 \sin^2 \alpha \qquad \text{Square both sides.}$$

$$16A^2 = 4b^2c^2(1 - \cos^2 \alpha) \qquad \text{Pythagorean identity}$$

$$16A^2 = 4b^2c^2 - 4b^2c^2 \cos^2 \alpha \qquad \text{Distributive property}$$

Using the expression for $4b^2c^2 \cos^2 \alpha$ obtained above, we get

$$16A^2 = 4b^2c^2 - (b^2 + c^2 - a^2)^2.$$

The last equation could be solved for A in terms of the lengths of the three sides, but it would not be Heron's formula. To get Heron's formula, let

$$S = \frac{(a + b + c)}{2} \qquad \text{or} \qquad 2S = a + b + c.$$

It can be shown (with some effort) that

$$4b^2c^2 - (b^2 + c^2 - a^2)^2 = 2S(2S - 2a)(2S - 2b)(2S - 2c).$$

Using this fact in the above equation yields Heron's formula:

$$16A^2 = 16S(S - a)(S - b)(S - c)$$

$$A^2 = S(S - a)(S - b)(S - c)$$

$$A = \sqrt{S(S - a)(S - b)(S - c)}$$ ■

Example **3** **Area of a triangle using only the sides**

A piece of property in downtown Houston is advertised for sale at $45 per square foot. If the lengths of the sides of the triangular lot are 220 feet, 234 feet, and 160 feet, then what is the asking price for the lot?

Solution

Using Heron's formula with $S = \frac{220 + 234 + 160}{2} = 307$, we get

$$A = \sqrt{307(307 - 220)(307 - 234)(307 - 160)} \approx 16{,}929.69 \text{ square feet.}$$

At $45 per square foot, the asking price is $761,836.

Try This. Given $a = 12$, $b = 8$, and $c = 6$, find the area of the triangle.   ■

Applications

In the next example we use the law of cosines to find a formula for the length of a chord in terms of the radius of the circle and the central angle subtended by the chord.

Example **4** **A formula for the length of a chord**

A central angle α in a circle of radius r intercepts a chord of length a. Write a formula for a in terms of α and r.

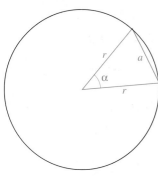

▪ **Figure 7.18**

Solution

We can apply the law of cosines to the triangle formed by the chord and the radii at the endpoints of the chord, as shown in Fig. 7.18.

$$a^2 = r^2 + r^2 - 2r^2 \cos \alpha$$
$$= r^2(2 - 2 \cos \alpha)$$
$$a = r\sqrt{2 - 2 \cos \alpha}$$

Try This.   Find the length of the chord intercepted by a central angle of 33.8° in a circle of radius 22.4 feet.   ▪

When you reach for a light switch, your brain controls the angles at your elbow and your shoulder that put your hand at the location of the switch. An arm on a robot works in much the same manner. In the next example we see how the law of sines and the law of cosines are used to determine the proper angles at the joints so that a robot's hand can be moved to a position given in the coordinates of the workspace.

Example 5 Positioning a robotic arm

A robotic arm with a 0.5-meter segment and a 0.3-meter segment is attached at the origin, as shown in Fig. 7.19. The computer-controlled arm is positioned by rotating each segment through angles θ_1 and θ_2, as shown in Fig. 7.19. Given that we want to have the end of the arm at the point (0.7, 0.2), find θ_1 and θ_2 to the nearest tenth of a degree.

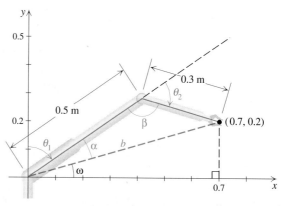

▪ **Figure 7.19**

Solution

In the right triangle shown in Fig. 7.19, we have $\tan \omega = 0.2/0.7$ and

$$\omega = \tan^{-1}\left(\frac{0.2}{0.7}\right) \approx 15.95°.$$

To get the angles to the nearest tenth, we use more accuracy along the way and round to the nearest tenth only on the final answer. Find b using the Pythagorean theorem:

$$b = \sqrt{0.7^2 + 0.2^2} \approx 0.73$$

Find β using the law of cosines:

$$(0.73)^2 = 0.5^2 + 0.3^2 - 2(0.5)(0.3) \cos \beta$$

$$\cos \beta = \frac{0.73^2 - 0.5^2 - 0.3^2}{-2(0.5)(0.3)} = -0.643$$

$$\beta = \cos^{-1}(-0.643) \approx 130.02°$$

Find α using the law of sines:

$$\frac{\sin \alpha}{0.3} = \frac{\sin 130.02°}{0.73}$$

$$\sin \alpha = \frac{0.3 \cdot \sin 130.02°}{0.73} = 0.3147$$

$$\alpha = \sin^{-1}(0.3147) \approx 18.34°$$

Since $\theta_1 = 90° - \alpha - \omega$,

$$\theta_1 = 90° - 18.34° - 15.95° = 55.71°.$$

Since β and θ_2 are supplementary,

$$\theta_2 = 180° - 130.02° = 49.98°.$$

So the longer segment of the arm is rotated 55.7° and the shorter segment is rotated 50.0°.

Try This. Solve the triangle that has vertices $A(0, 0)$, $B(1, 5)$, and $C(8, 3)$.   ■

Since the angles in Example 5 describe a clockwise rotation, the angles could be given negative signs to indicate the direction of rotation. In robotics, the direction of rotation is important, because there may be more than one way to position a robotic arm at a desired location. In fact, Example 5 has infinitely many solutions. See if you can find another one.

For Thought

True or False? Explain.

1. If $\gamma = 90°$ in triangle ABC, then $c^2 = a^2 + b^2$.

2. If a, b, and c are the sides of a triangle, then $a = \sqrt{c^2 + b^2 - 2bc \cos \gamma}$.

3. If a, b, and c are the sides of any triangle, then $c^2 = a^2 + b^2$.

4. The smallest angle of a triangle lies opposite the shortest side.

5. The equation $\cos \alpha = -0.3421$ has two solutions in $[0°, 180°]$.

6. If the largest angle of a triangle is obtuse, then the other two are acute.

7. The equation $\sin \beta = 0.1235$ has two solutions in $[0°, 180°]$.

8. In the SSS case of solving a triangle it is best to find the largest angle first.

9. There is no triangle with sides $a = 3.4$, $b = 4.2$, and $c = 8.1$.

10. There is no triangle with $\alpha = 179.9°$, $b = 1$ inch, and $c = 1$ mile.

7.2 Exercises

Solve each triangle.

1.

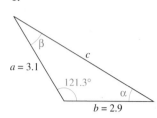

2.

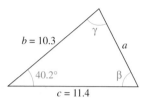

3.

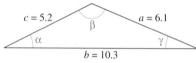

4.

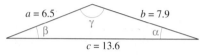

Solve each triangle with the given information. See the procedure for solving triangles on page 592.

5. $a = 6.8, c = 2.4, \beta = 10.5°$

6. $a = 1.3, b = 14.9, \gamma = 9.8°$

7. $a = 18.5, b = 12.2, c = 8.1$

8. $a = 30.4, b = 28.9, c = 31.6$

9. $b = 9.3, c = 12.2, \alpha = 30°$

10. $a = 10.3, c = 8.4, \beta = 88°$

11. $a = 6.3, b = 7.1, c = 6.8$

12. $a = 4.1, b = 9.8, c = 6.2$

13. $a = 7.2, \beta = 25°, \gamma = 35°$

14. $b = 12.3, \alpha = 20°, \gamma = 120°$

Determine the number of triangles with the given parts.

15. $a = 3, b = 4, c = 7$

16. $a = 2, b = 9, c = 5$

17. $a = 10, b = 5, c = 8$

18. $a = 3, b = 15, c = 16$

19. $c = 10, \alpha = 40°, \beta = 60°, \gamma = 90°$

20. $b = 6, \alpha = 62°, \gamma = 120°$

21. $b = 10, c = 1, \alpha = 179°$

22. $a = 10, c = 4, \beta = 2°$

23. $b = 8, c = 2, \gamma = 45°$

24. $a = \sqrt{3}/2, b = 1, \alpha = 60°$

Find the area of each triangle using Heron's formula.

25. $a = 16, b = 9, c = 10$

26. $a = 12, b = 8, c = 17$

27. $a = 3.6, b = 9.8, c = 8.1$

28. $a = 5.4, b = 8.2, c = 12.0$

29. $a = 346, b = 234, c = 422$

30. $a = 124.8, b = 86.4, c = 154.2$

Use the most appropriate formula for the area of a triangle to find the area of each triangle.

31.

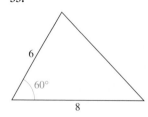

32.

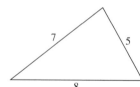

33.

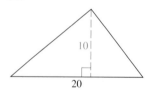

34.

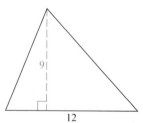

35.

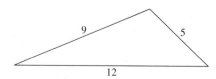

36.

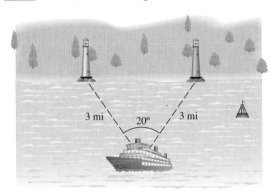

Solve each problem.

37. *Length of a Chord*   What is the length of the chord intercepted by a central angle of 19° in a circle of radius 30 ft?

38. *Boating*   The boat shown in the accompanying figure is 3 mi from both lighthouses and the angle between the line of sight to the lighthouses is 20°. Find the distance between the lighthouses.
HINT The distance is the length of a chord.

■ **Figure for Exercise 38**

39. *The Pentagon*   The Pentagon in Washington D.C. is 921 ft on each side, as shown in the accompanying figure. What is the distance from a vertex to the center of the Pentagon?
HINT The sides of the Pentagon can be viewed as chords in a circle.

■ **Figure for Exercise 39**

40. *A Hexagon*   If the length of each side of a regular hexagon is 10 ft, then what is the distance from a vertex to the center?

41. *Hiking*   Jan and Dean started hiking from the same location at the same time. Jan hiked at 4 mph with bearing N12°E, and Dean hiked at 5 mph with bearing N31°W. How far apart were they after 6 hr?

42. *Flying*   Andrea and Carlos left the airport at the same time. Andrea flew at 180 mph on a course with bearing 80°, and

Carlos flew at 240 mph on a course with bearing 210°. How far apart were they after 3 hr?

43. *Positioning a Solar Panel*   A solar panel with a width of 1.2 m is positioned on a flat roof, as shown in the figure. What is the angle of elevation α of the solar panel?

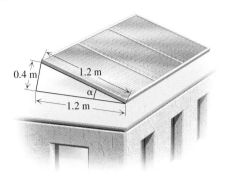

■ **Figure for Exercise 43**

44. *Installing an Antenna*   A 6-ft antenna is installed at the top of a roof, as shown in the figure. A guy wire is to be attached to the top of the antenna and to a point 10 ft down the roof. If the angle of elevation of the roof is 28°, then what length guy wire is needed?

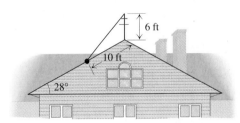

■ **Figure for Exercise 44**

45. *Adjacent Pipes*   An engineer wants to position three pipes at the vertices of a triangle, as shown in the figure. If the pipes *A*, *B*, and *C* have radii 2 in., 3 in., and 4 in., respectively, then what are the measures of the angles of the triangle *ABC*?

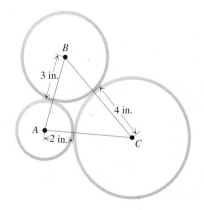

■ **Figure for Exercise 45**

46. *Firing a Torpedo* A submarine sights a moving target at a distance of 820 m. A torpedo is fired 9° ahead of the target, as shown in the drawing, and travels 924 m in a straight line to hit the target. How far has the target moved from the time the torpedo is fired to the time of the hit?

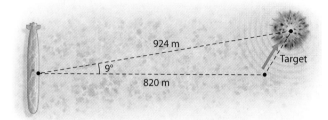

■ **Figure for Exercise 46**

47. *Planning a Tunnel* A tunnel is planned through a mountain to connect points A and B on two existing roads, as shown in the figure. If the angle between the roads at point C is 28°, what is the distance from point A to B? Find $\angle CBA$ and $\angle CAB$ to the nearest tenth of a degree.

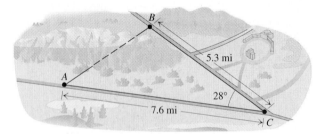

■ **Figure for Exercise 47**

48. *Scattering Angle* On June 30, 1861, Comet Tebutt, one of the greatest comets, was visible even before sunset. One of the factors that causes a comet to be extra bright is a small *scattering angle* θ, shown in the accompanying figure. When Comet Tebutt was at its brightest, it was 0.133 a.u. from the earth, 0.894 a.u. from the sun, and the earth was 1.017 a.u. from the sun. Find the *phase angle* α and the scattering angle θ for Comet Tebutt on June 30, 1861. (One astronomical unit (a.u.) is the average distance between the earth and the sun.)

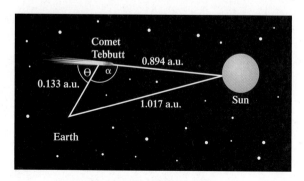

Figure for Exercise 48

49. *Side of a Pentagon* A regular pentagon is inscribed in a circle of radius 10 m. Find the length of a side of the pentagon.

50. *Central Angle* A central angle α in a circle of radius 5 m intercepts a chord of length 1 m. What is the measure of α?

51. *Positioning a Human Arm* A human arm consists of an upper arm of 30 cm and a lower arm of 30 cm, as shown in the figure. To move the hand to the point (36, 8), the human brain chooses angle θ_1 and θ_2, as shown in the figure. Find θ_1 and θ_2 to the nearest tenth of a degree.

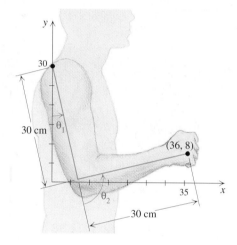

■ **Figure for Exercise 51**

52. *Making a Shaft* The end of a steel shaft for an electric motor is to be machined so that it has three flat sides of equal widths, as shown in the figure. Given that the radius of the shaft is 10 mm and the length of the arc between each pair of flat sides must be 2.5 mm, find the width s of each flat side.

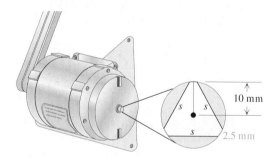

■ **Figure for Exercise 52**

53. *Total Eclipse* Astronomers refer to the angles alpha and beta (α and β) in the accompanying figure as the "diameters" of the sun and the moon, respectively. These angles/diameters vary as the sun, moon, and earth travel in their orbits. When the moon moves between the earth and the sun, a total eclipse occurs, provided the moon's apparent diameter is larger than that of the sun, or if beta > alpha. The actual diameters of the sun and moon are 865,000 miles and 2163 miles, respectively, and those distances do not vary.

a. The distance from the earth to the sun varies from 91,400,000 miles to 94,500,000 miles. Find the minimum and maximum α to the nearest hundredth of a degree.
HINT Use the law of cosines.

b. The distance from the earth to the moon varies from 225,800 miles to 252,000 miles. Find the minimum and maximum β to the nearest hundredth of a degree.
HINT Use the law of cosines.

c. Is it possible for the earth, the moon, and the sun to be in perfect alignment without a total eclipse occurring?

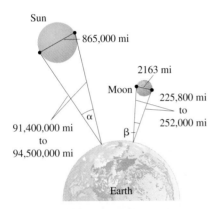

■ **Figure for Exercise 53**

54. *Total Eclipse on Jupiter* The distance from Jupiter to the sun varies from 7.406×10^8 km to 8.160×10^8 km. The diameter of the sun is 1.39×10^6 km.

a. Determine the minimum and maximum diameter of the sun to the nearest hundredth of a degree as seen from Jupiter.
HINT See the previous exercise and use the law of cosines.

b. The distance from Jupiter to its moon Callisto is 1.884×10^6 km and Callisto's diameter is 2420 km. Determine the diameter of Callisto as seen from Jupiter.
HINT Use the law of cosines.

c. Is it possible for Callisto to produce a total eclipse of the sun on Jupiter?

55. *Attack of the Grizzly* A forest ranger is 150 ft above the ground in a fire tower when she spots an angry grizzly bear east of the tower with an angle of depression of 10°, as shown in the figure. Southeast of the tower she spots a hiker with an angle of depression of 15°. Find the distance between the hiker and the angry bear.

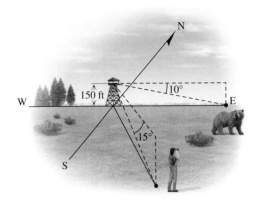

■ **Figure for Exercise 55**

56. *Smuggler's Blues* A smuggler's boat sets out at midnight at 20 mph on a course that makes a 40° angle with the shore, as shown in the figure. One hour later a DEA boat sets out also at 20 mph from a dock 80 miles up the coast to intercept the smuggler's boat. Find the angle θ (in the figure) for which the DEA boat will intercept the smuggler's boat. At what time will the interception occur?

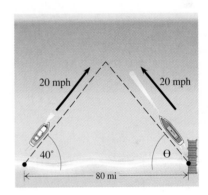

■ **Figure for Exercise 56**

For Writing/Discussion

57. A central angle θ in a circle of radius r intercepts a chord of length a, where $0° \leq \theta \leq 180°$. Show that $a = 2r \sin(\theta/2)$.

58. Explain why the second largest side in a triangle with unequal sides is opposite an acute angle.

59. Explain why the Pythagorean theorem is a special case of the law of cosines.

60. Find the area of the triangle with sides 37, 48, and 86, using Heron's formula. Explain your result.

61. Find the area of the triangle with sides 31, 87, and 56, using Heron's formula. Explain your result.

62. Find the area of the triangle with sides of length 6 ft, 9 ft, and 13 ft by using the formula

$$A = \frac{1}{4} \sqrt{4b^2c^2 - (b^2 + c^2 - a^2)^2},$$

and check your result using a different formula for the area of a triangle. Prove that this formula gives the area of any triangle with sides a, b, and c.

Thinking Outside the Box LXI

Watering the Lawn   Josie places her lawn sprinklers at the vertices of a triangle that has sides of 9 m, 10 m, and 11 m. The sprinklers water in circular patterns with radii of 4, 5, and 6 m. No area is watered by more than one sprinkler. What amount of area inside the triangle is not watered by any of the three sprinklers? Give your answer to the nearest thousandth of a square meter.

7.2 Pop Quiz

1. If $\alpha = 12.3°$, $b = 10.4$, and $c = 8.1$, then what is a?

2. If $a = 6$, $b = 7$, and $c = 12$, then what is γ?

3. Find the exact area of the triangle with sides of lengths 7, 8, and 9 using Heron's formula.

Linking Concepts

For Individual or Group Explorations

Modeling the Best View

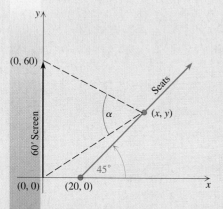

A large-screen theater has a screen that is 60 *feet high, as shown in the figure. The seats are placed on a 45° incline so that the seats are close to the screen.*

a) Find the distance to the top of the screen and to the bottom of the screen for a person sitting at coordinates (30, 10).

b) Find the viewing angle α for a person sitting at (30, 10).

c) Write the viewing angle α as a function of x and graph the function using a graphing calculator.

d) If the good seats are those for which the viewing angle is greater than 60°, then what are the x-coordinates of the good seats?

e) If the best seat has the largest viewing angle, then what are the coordinates of the best seat?

7.3

Vectors

By using the law of sines and the law of cosines, we can find all of the missing parts of any triangle for which we have enough information to determine its shape. In this section we will use these and many other tools of trigonometry in the study of vectors.

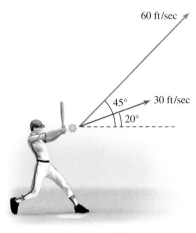

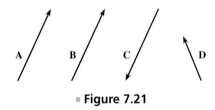

■ **Figure 7.20**

■ **Figure 7.21**

Definitions

Quantities such as length, area, volume, temperature, and time have magnitude only and are completely characterized by a single real number with appropriate units (such as feet, degrees, or hours). Such quantities are called **scalar quantities,** and the corresponding real numbers are **scalars.** Quantities that involve both a magnitude and a direction, such as velocity, acceleration, and force, are **vector quantities,** and they can be represented by **directed line segments.** These directed line segments are called **vectors.**

For an example of vectors, consider two baseballs hit into the air, as shown in Fig. 7.20. One is hit with an initial velocity of 30 feet per second at an angle of 20° from the horizontal, and the other is hit with an initial velocity of 60 feet per second at an angle of 45° from the horizontal. The length of a vector represents the **magnitude** of the vector quantity. So the vector representing the faster baseball is drawn twice as long as the other vector. The **direction** is indicated by the position of the vector and the arrowhead at one end.

We have used the notation $\overline{AB}$ to name a line segment with endpoints A and B and $\overrightarrow{AB}$ to name a ray with initial point A and passing through B. When studying vectors, the notation $\overrightarrow{AB}$ is used to name a vector with **initial point** A and **terminal point** B. The vector $\overrightarrow{AB}$ terminates at B, while the ray $\overrightarrow{AB}$ goes beyond B. Since the direction of the vector is the same as the order of the letters, to indicate a vector in print we will omit the arrow and use boldface type. In handwritten work, an arrow over the letter or letters indicates a vector. Thus **AB** and $\overrightarrow{AB}$ represent the same vector. The vector **BA** is a vector with initial point B and terminal point A, and is not the same as **AB**. A vector whose endpoints are not specified is named a single uppercase or lowercase letter. For example, **b**, **B**, $\vec{b}$, and $\vec{B}$ are names of vectors. The magnitude of vector **A** is written as $|\mathbf{A}|$.

Two vectors are **equal** if they have the same magnitude and the same direction. Equal vectors may be in different locations. They do not necessarily coincide. In Fig. 7.21, **A** = **B** because they have the same direction and the same magnitude. Vector **B** is *not* equal to vector **C** because they have the same magnitude but opposite directions. **C** ≠ **D** because they have different magnitudes and different directions. The **zero vector** is a vector that has no magnitude and no direction. It is denoted by a boldface zero, **0.**

Scalar Multiplication and Addition

Suppose two tugboats are pulling on a barge that has run aground in the Mississippi River as shown in Fig. 7.22.

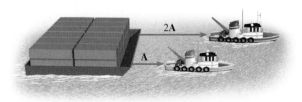

■ **Figure 7.22**

The smaller boat exerts a force of 2000 pounds in an easterly direction and its force is represented by the vector **A.** If the larger boat exerts a force of 4000 pounds in the same direction, then its force can be represented by the vector 2**A.** This example illustrates the operation of **scalar multiplication.**

Definition:
Scalar Multiplication

> For any scalar k and vector **A,** k**A** is a vector with magnitude $|k|$ times the magnitude of **A.** If $k > 0$, then the direction of k**A** is the same as the direction of **A.** If $k < 0$, the direction of k**A** is opposite to the direction of **A.** If $k = 0$, then k**A** = **0.**

The vector 2**A** has twice the magnitude of **A,** and $-\frac{2}{3}$**A** has two-thirds of the magnitude. For -1**A,** we write $-$**A.** Some examples of scalar multiplication are shown in Fig. 7.23.

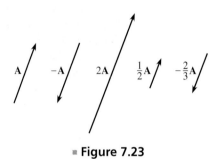

■ **Figure 7.23**

Suppose two draft horses are pulling on a tree stump with forces of 200 pounds and 300 pounds, as shown in Fig. 7.24, with an angle of 65° between the forces.

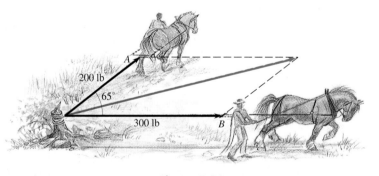

■ **Figure 7.24**

The two forces are represented by vectors **A** and **B.** If **A** and **B** had the same direction, then there would be a total force of 500 pounds acting on the stump. However, a total force of 500 pounds is not achieved because of the angle between the forces. It can be shown by experimentation that one force acting along the diagonal of the parallelogram shown in Fig. 7.24, with a magnitude equal to the length of the diagonal, has the same effect on the stump as the two forces **A** and **B.** In physics, this result is known as the **parallelogram law.** The single force **A** + **B** acting along the diagonal is called the **sum** or **resultant** of **A** and **B.** The parallelogram law motivates our definition of vector addition.

Definition:
Vector Addition

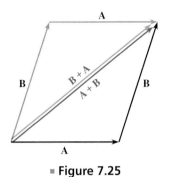

■ **Figure 7.25**

To find the resultant or sum **A** + **B** of any vectors **A** and **B**, position **B** (without changing its magnitude or direction) so that the initial point of **B** coincides with the terminal point of **A**, as in Fig. 7.25. The vector that begins at the initial point of **A** and terminates at the terminal point of **B** is the vector **A** + **B**.

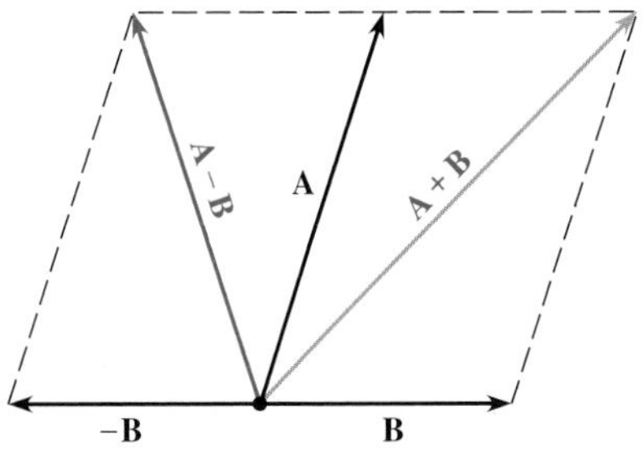

Note that the vector **A** + **B** in Fig. 7.25 coincides with the diagonal of a parallelogram whose adjacent sides are **A** and **B**. We can see from Fig. 7.25 that **A** + **B** = **B** + **A**, and vector addition is commutative. If **A** and **B** have the same direction or opposite directions, then no parallelogram is formed, but **A** + **B** can be found by the procedure given in the definition. Each vector in the sum **A** + **B** is called a **component** of the sum.

For every vector **A**, there is a vector −**A**, the **opposite** of **A**, having the same magnitude as **A** but opposite direction. The sum of a vector and its opposite is the zero vector, **A** + (−**A**) = **0**. Vector subtraction is defined like subtraction of real numbers. For any two vectors **A** and **B**, **A** − **B** = **A** + (−**B**). Fig. 7.26 shows **A** − **B**.

The law of sines and the law of cosines can be used to find the magnitude and direction of a resultant vector. Remember that in any parallelogram the opposite sides are equal and parallel, and adjacent angles are supplementary. The diagonals of a parallelogram do not bisect the angles of a parallelogram unless the adjacent sides of the parallelogram are equal in length.

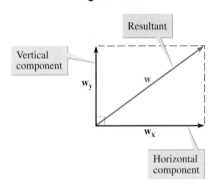

■ **Figure 7.26**

Horizontal and Vertical Components

Any nonzero vector **w** is the sum of a **vertical component** and a **horizontal component**, as shown in Fig. 7.27. The horizontal component is denoted **w**$_x$ and the vertical component is denoted **w**$_y$. The vector **w** is the diagonal of the rectangle formed by the vertical and horizontal components.

If a vector **w** is placed in a rectangular coordinate system so that its initial point is the origin (as in Fig. 7.28), then **w** is called a **position vector** or **radius vector**. A position vector is a convenient representative to focus on when considering all vectors that are equal to a certain vector. The angle θ ($0° \leq \theta < 360°$) formed by the positive x-axis and a position vector is the **direction angle** for the position vector (or any other vector that is equal to the position vector).

If the vector **w** has magnitude r, direction angle θ, horizontal component **w**$_x$, and vertical component **w**$_y$, as shown in Fig. 7.28, then by using trigonometric ratios we get

$$\cos \theta = \frac{|\mathbf{w_x}|}{r} \quad \text{and} \quad \sin \theta = \frac{|\mathbf{w_y}|}{r}$$

or

$$|\mathbf{w_x}| = r \cos \theta \quad \text{and} \quad |\mathbf{w_y}| = r \sin \theta.$$

If the direction of **w** is such that $\sin \theta$ or $\cos \theta$ is negative, then we can write

$$|\mathbf{w_x}| = |r \cos \theta| \quad \text{and} \quad |\mathbf{w_y}| = |r \sin \theta|.$$

■ **Figure 7.27**

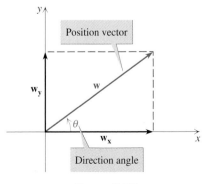

■ **Figure 7.28**

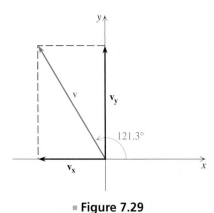

■ **Figure 7.29**

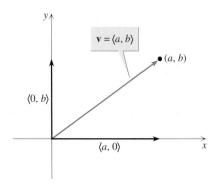

■ **Figure 7.30**

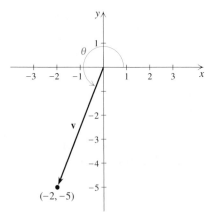

■ **Figure 7.31**

Example **1** **Finding horizontal and vertical components**

Find the magnitude of the horizontal and vertical components for a vector **v** with magnitude 8.3 and direction angle 121.3°.

Solution

The vector **v** and its horizontal and vertical components $\mathbf{v_x}$ and $\mathbf{v_y}$ are shown in Fig. 7.29. The magnitudes of $\mathbf{v_x}$ and $\mathbf{v_y}$ are found as follows:

$$|\mathbf{v_x}| = |8.3 \cos 121.3°| = 4.3$$
$$|\mathbf{v_y}| = |8.3 \sin 121.3°| = 7.1$$

The direction angle for $\mathbf{v_x}$ is 180°, and the direction angle for $\mathbf{v_y}$ is 90°.

Try This. Find the magnitude of the horizontal and vertical components for a vector **v** with magnitude 5.6 and direction angle 22°. ■

Component Form of a Vector

Any vector is the resultant of its horizontal and vertical components. Since the horizontal and vertical components of a vector determine the vector, it is convenient to use a notation for vectors that involves them. The notation $\langle a, b \rangle$ is used for the position vector with terminal point (a, b), as shown in Fig. 7.30. The form $\langle a, b \rangle$ is called **component form** because its horizontal component is $\langle a, 0 \rangle$ and its vertical component is $\langle 0, b \rangle$. Since the vector $\mathbf{v} = \langle a, b \rangle$ extends from $(0, 0)$ to (a, b), its magnitude is the distance between these points:

$$|\mathbf{v}| = \sqrt{a^2 + b^2}$$

In Example 2 we find the magnitude and direction for a vector given in component form, and in Example 3 we find the component form for a vector given as a directed line segment.

Example **2** **Finding magnitude and direction from component form**

Find the magnitude and direction angle of the vector $\mathbf{v} = \langle -2, -5 \rangle$.

Solution

The vector **v** shown in Fig. 7.31 has magnitude

$$|\mathbf{v}| = \sqrt{(-2)^2 + (-5)^2} = \sqrt{29}.$$

To find the direction angle θ, use trigonometric ratios to get $\sin \theta = -5/\sqrt{29}$. Since $\sin^{-1}(-5/\sqrt{29}) = -68.2°$, we have $\theta = 180° - (-68.2°) = 248.2°$.

Try This. Find the magnitude and direction angle for $\mathbf{v} = \langle 2, -6 \rangle$. ■

Example **3** **Finding the component form given magnitude and direction**

Find the component form for a vector of magnitude 40 mph with direction angle 330°.

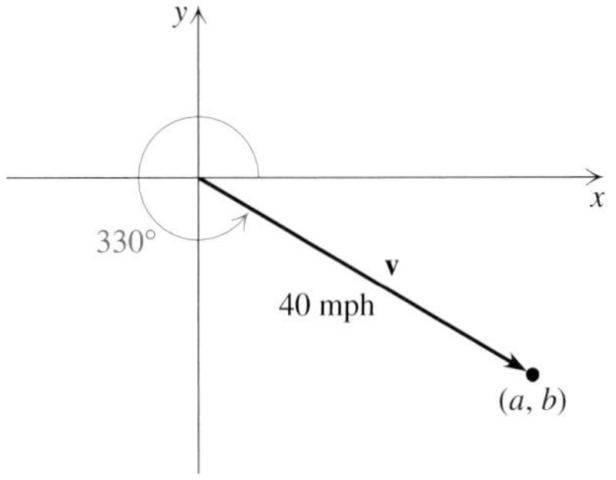

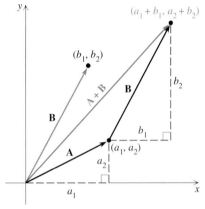

■ **Figure 7.32**

■ **Figure 7.33**

Rules for Scalar Product, Vector Sum, Vector Difference, and Dot Product

Solution

To find the component form, we need the terminal point (a, b) for a vector of magnitude 40 mph positioned as shown in Fig. 7.32. Since $a = r \cos \theta$ and $b = r \sin \theta$, we have

$$a = 40 \cos 330° = 40 \frac{\sqrt{3}}{2} = 20 \sqrt{3} \quad \text{and}$$

$$b = 40 \sin 330° = 40\left(-\frac{1}{2}\right) = -20.$$

So the component form of the vector is $\langle 20\sqrt{3}, -20 \rangle$.

Try This. Find the component form for a vector of magnitude 50 mph with direction angle 120°. ■

We originally defined vectors as directed line segments and performed operations with these directed line segments. When vectors are written in component form, operations with vectors are easier to perform. Figure 7.33 illustrates the addition of $\mathbf{A} = \langle a_1, a_2 \rangle$ and $\mathbf{B} = \langle b_1, b_2 \rangle$, where (a_1, a_2) and (b_1, b_2) are in the first quadrant. It is easy to see that the endpoint of $\mathbf{A} + \mathbf{B}$ is $(a_1 + b_1, a_2 + b_2)$ and so $\mathbf{A} + \mathbf{B} = \langle a_1 + b_1, a_2 + b_2 \rangle$. The sum can be found in component form by adding the components instead of drawing directed line segments. The component forms for the three operations that we have already studied and a new operation called **dot product** are given in the following box.

If $\mathbf{A} = \langle a_1, a_2 \rangle$, $\mathbf{B} = \langle b_1, b_2 \rangle$, and k is a scalar, then

1. $k\mathbf{A} = \langle ka_1, ka_2 \rangle$ **Scalar product**
2. $\mathbf{A} + \mathbf{B} = \langle a_1 + b_1, a_2 + b_2 \rangle$ **Vector sum**
3. $\mathbf{A} - \mathbf{B} = \langle a_1 - b_1, a_2 - b_2 \rangle$ **Vector difference**
4. $\mathbf{A} \cdot \mathbf{B} = a_1 b_1 + a_2 b_2$ **Dot product**

Example **4** **Operations with vectors in component form**

Let $\mathbf{w} = \langle -3, 2 \rangle$ and $\mathbf{z} = \langle 5, -1 \rangle$. Perform the operations indicated.

a. $\mathbf{w} - \mathbf{z}$ **b.** $-8\mathbf{z}$ **c.** $3\mathbf{w} + 4\mathbf{z}$ **d.** $\mathbf{w} \cdot \mathbf{z}$

Solution

a. $\mathbf{w} - \mathbf{z} = \langle -3, 2 \rangle - \langle 5, -1 \rangle = \langle -8, 3 \rangle$
b. $-8\mathbf{z} = -8\langle 5, -1 \rangle = \langle -40, 8 \rangle$
c. $3\mathbf{w} + 4\mathbf{z} = 3\langle -3, 2 \rangle + 4\langle 5, -1 \rangle = \langle -9, 6 \rangle + \langle 20, -4 \rangle = \langle 11, 2 \rangle$
d. $\mathbf{w} \cdot \mathbf{z} = \langle -3, 2 \rangle \cdot \langle 5, -1 \rangle = -15 + (-2) = -17$

Try This. Let $\mathbf{u} = \langle -1, -3 \rangle$ and $\mathbf{v} = \langle 3, -4 \rangle$. Find $\mathbf{u} + 3\mathbf{v}$ and $\mathbf{u} \cdot \mathbf{v}$. ■

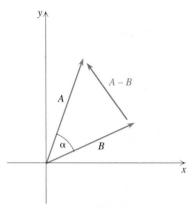

■ **Figure 7.34**

The Angle Between Two Vectors

If $\mathbf{A} = k\mathbf{B}$ for a nonzero scalar k, then $\mathbf{A}$ and $\mathbf{B}$ are **parallel** vectors. If $\mathbf{A}$ and $\mathbf{B}$ have the same direction ($k > 0$), the angle between $\mathbf{A}$ and $\mathbf{B}$ is $0°$. If they have opposite directions ($k < 0$), the angle between them is $180°$. If $\mathbf{A}$ and $\mathbf{B}$ are nonparallel vectors with the same initial point, then the vectors $\mathbf{A}$, $\mathbf{B}$, and $\mathbf{A} - \mathbf{B}$ form a triangle, as in Fig. 7.34. The **angle between the vectors $\mathbf{A}$ and $\mathbf{B}$** is the angle α shown in Fig. 7.34. The angle between two vectors is in the interval $[0°, 180°]$. If the angle between $\mathbf{A}$ and $\mathbf{B}$ is $90°$, then the vectors are **perpendicular** or **orthogonal.**

The following theorem relates the angle between two vectors, the magnitudes of the vectors, and the dot product of the vectors. The theorem is simply a vector version of the law of cosines.

Theorem:
Dot Product

If $\mathbf{A}$ and $\mathbf{B}$ are nonzero vectors and α is the angle between them, then

$$\cos \alpha = \frac{\mathbf{A} \cdot \mathbf{B}}{|\mathbf{A}| \, |\mathbf{B}|}.$$

■ **PROOF**   Let $\mathbf{A} = \langle a_1, a_2 \rangle$, $\mathbf{B} = \langle b_1, b_2 \rangle$, and $\mathbf{A} - \mathbf{B} = \langle a_1 - b_1, a_2 - b_2 \rangle$. The vectors $\mathbf{A}$, $\mathbf{B}$, and $\mathbf{A} - \mathbf{B}$ form a triangle, as shown in Fig. 7.34. Apply the law of cosines to this triangle and simplify as follows.

$$|\mathbf{A}|^2 + |\mathbf{B}|^2 - 2|\mathbf{A}| \, |\mathbf{B}| \cos \alpha = |\mathbf{A} - \mathbf{B}|^2$$

$$(a_1)^2 + (a_2)^2 + (b_1)^2 + (b_2)^2 - 2|\mathbf{A}| \, |\mathbf{B}| \cos \alpha = (a_1 - b_1)^2 + (a_2 - b_2)^2$$

$$-2|\mathbf{A}| \, |\mathbf{B}| \cos \alpha = -2a_1 b_1 - 2a_2 b_2$$

$$|\mathbf{A}| \, |\mathbf{B}| \cos \alpha = a_1 b_1 + a_2 b_2$$

$$|\mathbf{A}| \, |\mathbf{B}| \cos \alpha = \mathbf{A} \cdot \mathbf{B}$$

$$\cos \alpha = \frac{\mathbf{A} \cdot \mathbf{B}}{|\mathbf{A}| \, |\mathbf{B}|}$$ ■

Note that $\cos \alpha = 0$ if and only if $\mathbf{A} \cdot \mathbf{B} = 0$. So two vectors are perpendicular if and only if their dot product is zero. Two vectors are parallel if and only if $\cos \alpha = \pm 1$.

Example 5   Angle between two vectors

Find the smallest positive angle between each pair of vectors.

a. $\langle -3, 2 \rangle$, $\langle 4, 5 \rangle$
b. $\langle -5, 9 \rangle$, $\langle 9, 5 \rangle$

Solution

a. Use the dot product theorem to find the cosine of the angle and then a calculator to find the angle:

$$\cos \alpha = \frac{\langle -3, 2 \rangle \cdot \langle 4, 5 \rangle}{|\langle -3, 2 \rangle| \, |\langle 4, 5 \rangle|} = \frac{-2}{\sqrt{13} \sqrt{41}}$$

$$\alpha = \cos^{-1}\left(-2/\left(\sqrt{13}\sqrt{41}\right)\right) \approx 94.97°$$

b. Because $\langle -5, 9 \rangle \cdot \langle 9, 5 \rangle = 0$, the vectors are perpendicular and the angle between them is 90°.

Try This. Find the smallest positive angle between $\langle 1, 3 \rangle$ and $\langle 5, 2 \rangle$. ■

Unit Vectors

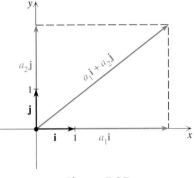

■ **Figure 7.35**

The vectors $\mathbf{i} = \langle 1, 0 \rangle$ and $\mathbf{j} = \langle 0, 1 \rangle$ are called **unit vectors** because each has magnitude one. For any vector $\langle a_1, a_2 \rangle$, as shown in Fig. 7.35, we have

$$\langle a_1, a_2 \rangle = a_1 \langle 1, 0 \rangle + a_2 \langle 0, 1 \rangle = a_1 \mathbf{i} + a_2 \mathbf{j}.$$

The form $a_1 \mathbf{i} + a_2 \mathbf{j}$ is called a **linear combination** of the vectors $\mathbf{i}$ and $\mathbf{j}.$ These unit vectors are thought of as fundamental vectors, because any vector can be expressed as a linear combination of them.

Example **6** Unit vectors

Write each vector as a linear combination of the unit vectors $\mathbf{i}$ and $\mathbf{j}.$

a. $\mathbf{A} = \langle -2, 6 \rangle$ **b.** $\mathbf{B} = \langle -4, -1 \rangle$

Solution

a. $\mathbf{A} = \langle -2, 6 \rangle = -2\mathbf{i} + 6\mathbf{j}$
b. $\mathbf{B} = \langle -4, -1 \rangle = -4\mathbf{i} - \mathbf{j}$

Try This. Write $\langle -1, 7 \rangle$ as a linear combination of $\mathbf{i}$ and $\mathbf{j}.$ ■

Applications of Vectors

When we use vectors to model forces acting on an object, we usually ignore friction and assume that all forces are acting on a single point. In the next example, we find the force that results when the two horses mentioned earlier in this section are pulling on a tree stump.

Example **7** Magnitude and direction of a resultant

Two draft horses are pulling on a tree stump with forces of 200 pounds and 300 pounds, as shown in Fig. 7.36. If the angle between the forces is 65°, then what is the magnitude of the resultant force? What is the angle between the resultant and the 300-pound force?

Solution

The resultant coincides with the diagonal of the parallelogram shown in Fig. 7.36. We can view the resultant as a side of a triangle in which the other two sides are 200 and 300, and their included angle is 115°. The law of cosines can be used to find the magnitude of the resultant vector **v**:

$$|\mathbf{v}|^2 = 300^2 + 200^2 - 2(300)(200)\cos 115° = 180,714.19$$

$$|\mathbf{v}| = 425.1$$

We can use the law of sines to find the angle α between the resultant and the 300-pound force:

$$\frac{\sin \alpha}{200} = \frac{\sin 115°}{425.1}$$

$$\sin \alpha = 0.4264$$

Since $\sin^{-1}(0.4264) = 25.2°$, both 25.2° and 154.8° are solutions to $\sin \alpha = 0.4264$. However, α must be smaller than 65°, so the angle between the resultant and the 300-pound force is 25.2°. So one horse pulling in the direction of the resultant with a force of 425.1 pounds would have the same effect on the stump as the two horses pulling at an angle of 65°.

Try This. Forces of 100 pounds and 200 pounds are acting on a point. If the angle between the forces is 30°, then what is the magnitude of the resultant? ■

Example 7 can be solved with horizontal and vertical components. The horizontal component of the 200-pound force is 200 cos 65° or 84.5 pounds. The vertical component is 200 sin 65° or 181.3. The 300-pound force and the 84.5-pound force are in the same direction. So the original problem is equivalent to forces of 181.3 pounds and 384.5 pounds acting at a right angle to each other, as in Fig. 7.37. Now the resultant force is the diagonal of a rectangle and its magnitude is $\sqrt{181.3^2 + 384.5^2}$ or 425.1 pounds.

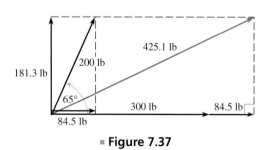

■ **Figure 7.37**

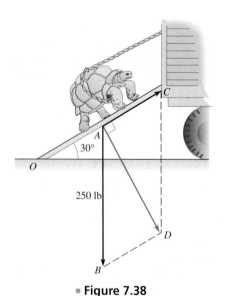

■ **Figure 7.38**

Example 8 Finding a force

Workers at the Audubon Zoo must move a giant tortoise to his new home. Find the amount of force required to pull a 250-pound tortoise up a ramp leading into a truck. The angle of elevation of the ramp is 30°. See Fig. 7.38.

Solution

The weight of the tortoise is a 250-pound force in a downward direction, shown as vector **AB** in Fig. 7.38. The tortoise exerts a force against the ramp at a 90° angle with the ramp, shown as vector **AD**. The magnitude of **AD** is less than the magnitude of **AB** because of the incline of the ramp. The force required to pull the tortoise up the ramp is vector **AC** in Fig. 7.38. The force against the ramp, **AD,** is the resultant of **AB** and **AC.** Since $\angle O = 30°$, we have $\angle OAB = 60°$ and $\angle BAD = 30°$. Because $\angle CAD = 90°$, we have $\angle ADB = 90°$. We can find any of the missing parts to the right triangle *ABD.* Since

$$\sin 30° = \frac{\text{opposite}}{\text{hypotenuse}} = \frac{|\mathbf{BD}|}{|\mathbf{AB}|},$$

we have

$$|\mathbf{BD}| = |\mathbf{AB}| \sin 30° = 250 \cdot \frac{1}{2} = 125 \text{ pounds}.$$

Since the opposite sides of a parallelogram are equal, the magnitude of **AC** is also 125 pounds.

Try This. Find the amount of force required to push an 800-pound block of ice up a ramp that is inclined 10°.   ■

If an airplane heads directly against the wind or with the wind, the wind speed is subtracted from or added to the air speed of the plane to get the ground speed of the plane. When the wind is at some other angle to the direction of the plane, some portion of the wind speed will be subtracted from or added to the air speed to determine the ground speed. In addition, the wind causes the plane to travel on a course different from where it is headed. We can use the vector $\mathbf{v}_1$ to represent the heading and air speed of the plane, as shown in Fig. 7.39.

The vector $\mathbf{v}_2$ represents the wind direction and speed. The resultant of $\mathbf{v}_1$ and $\mathbf{v}_2$ is the vector $\mathbf{v}_3$, where $\mathbf{v}_3$ represents the course and ground speed of the plane. The angle between the heading and the course is the **drift angle.** Recall that the bearing of a vector used to describe direction in air navigation is a nonnegative angle smaller than 360° measured in a clockwise direction from due north. For example, the wind in Fig. 7.39 is out of the southwest and has a bearing of 45°. The vector $\mathbf{v}_1$, describing the heading of the plane, points due west and has a bearing of 270°.

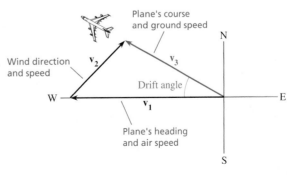

■ **Figure 7.39**

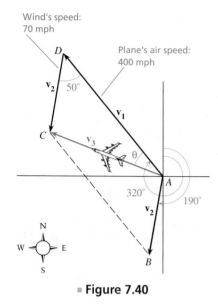

Wind's speed:
70 mph

Plane's air speed:
400 mph

■ **Figure 7.40**

Example 9 Finding course and ground speed of an airplane

The heading of an executive's Lear jet has a bearing of 320°. The wind is 70 mph with a bearing of 190°. Given that the air speed of the plane is 400 mph, find the drift angle, the ground speed, and the course of the airplane.

Solution

Figure 7.40 shows vectors for the heading v_1, the wind v_2, and the course v_3. Subtract the bearing of the wind from that of the heading to get $\angle DAB = 320° - 190° = 130°$. Since $ABCD$ is a parallelogram, $\angle CDA = 50°$. Apply the law of cosines to triangle ACD to get

$$|v_3|^2 = 70^2 + 400^2 - 2(70)(400)\cos 50° = 128{,}903.9$$

$$|v_3| \approx 359.0.$$

The ground speed is approximately 359.0 mph. The drift angle θ is found by using the law of sines:

$$\frac{\sin \theta}{70} = \frac{\sin 50°}{359.0}$$

$$\sin \theta = 0.1494$$

Since θ is an acute angle, $\theta = \sin^{-1}(0.1494) = 8.6°$. The course v_3 has a bearing of $320° - 8.6° = 311.4°$.

Try This. A jet is headed northwest with an airspeed of 500 mph. The wind is 100 mph with a bearing of 200°. Find the drift angle, the ground speed, and the course of the jet. ■

For Thought

True or False? Explain.

1. The vector 2**v** has the same direction as **v** but twice the magnitude of **v**.

2. The magnitude of **A** + **B** is the sum of the magnitudes of **A** and **B**.

3. The magnitude of −**A** is equal to the magnitude of **A**.

4. For any vector **A**, **A** + (−**A**) = **0**.

5. The parallelogram law says that the opposite sides of a parallelogram are equal in length.

6. In the coordinate plane, the direction angle is the angle formed by the *y*-axis and a vector with initial point at the origin.

7. If **v** has magnitude *r* and direction angle θ, then the horizontal component of **v** is a vector with direction angle 0° and magnitude $r \cos \theta$.

8. The magnitude of $\langle 3, -4 \rangle$ is 5.

9. The vectors $\langle -2, 3 \rangle$ and $\langle -6, 9 \rangle$ have the same direction angle.

10. The direction angle of $\langle -2, 2 \rangle$ is $\cos^{-1}\left(-2/\sqrt{8}\right)$.

7.3 Exercises

For each given pair of vectors **A** *and* **B**, *draw the vectors* **A** + **B** *and* **A** − **B**.

1.

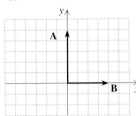

2.

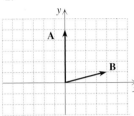

3.

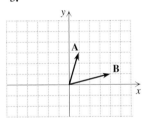

4.

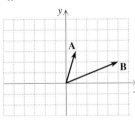

5.

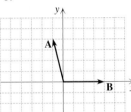

6.

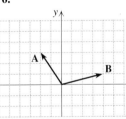

Determine which of the position vectors shown in the accompanying figure has the given magnitude and direction angle.

7. Magnitude 2 and direction angle 90°

8. Magnitude 3 and direction angle 180°

9. Magnitude $3\sqrt{2}$ and direction angle 45°

10. Magnitude 4 and direction angle 30°

11. Magnitude $3\sqrt{2}$ and direction angle 135°

12. Magnitude 5 and direction angle 120°

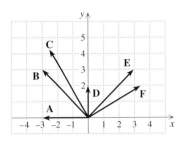

■ **Figure for Exercises 7–12**

Find the magnitude of the horizontal and vertical components for each vector **v** *with the given magnitude and given direction angle* θ.

13. $|\mathbf{v}| = 4.5, \theta = 65.2°$ **14.** $|\mathbf{v}| = 6000, \theta = 13.1°$

15. $|\mathbf{v}| = 8000, \theta = 155.1°$ **16.** $|\mathbf{v}| = 445, \theta = 211.1°$

17. $|\mathbf{v}| = 234, \theta = 248°$ **18.** $|\mathbf{v}| = 48.3, \theta = 349°$

Find the magnitude and direction angle of each vector.

19. $\langle \sqrt{3}, 1 \rangle$ **20.** $\langle -1, \sqrt{3} \rangle$

21. $\langle -\sqrt{2}, \sqrt{2} \rangle$ **22.** $\langle \sqrt{2}, -\sqrt{2} \rangle$

23. $\langle 8, -8\sqrt{3} \rangle$ **24.** $\langle -1/2, -\sqrt{3}/2 \rangle$

25. $\langle 5, 0 \rangle$ **26.** $\langle 0, -6 \rangle$

27. $\langle -3, 2 \rangle$ **28.** $\langle -4, -2 \rangle$

29. $\langle 3, -1 \rangle$ **30.** $\langle 2, -6 \rangle$

Find the component form for each vector **v** *with the given magnitude and direction angle* θ.

31. $|\mathbf{v}| = 8, \theta = 45°$ **32.** $|\mathbf{v}| = 12, \theta = 120°$

33. $|\mathbf{v}| = 290, \theta = 145°$ **34.** $|\mathbf{v}| = 5.3, \theta = 321°$

35. $|\mathbf{v}| = 18, \theta = 347°$ **36.** $|\mathbf{v}| = 3000, \theta = 209.1°$

Let $\mathbf{r} = \langle 3, -2 \rangle$, $\mathbf{s} = \langle -1, 5 \rangle$, *and* $\mathbf{t} = \langle 4, -6 \rangle$. *Perform the operations indicated. Write the vector answers in the form* $\langle a, b \rangle$.

37. 5**r**   **38.** −4**s**   **39.** 2**r** + 3**t**

40. **r** − **t**   **41.** **s** + 3**t**   **42.** $\dfrac{\mathbf{r} + \mathbf{s}}{2}$

43. $r - (s + t)$ **44.** $r - s - t$ **45.** $r \cdot s$

46. $s \cdot t$

Find the angle to the nearest tenth of a degree between each given pair of vectors.

47. $\langle 2, 1 \rangle, \langle 3, 5 \rangle$ **48.** $\langle 2, 3 \rangle, \langle 1, 5 \rangle$

49. $\langle -1, 5 \rangle, \langle 2, 7 \rangle$ **50.** $\langle -2, -5 \rangle, \langle 1, -9 \rangle$

51. $\langle -6, 5 \rangle, \langle 5, 6 \rangle$ **52.** $\langle 2, 7 \rangle, \langle 7, -2 \rangle$

Determine whether each pair of vectors is parallel, perpendicular, or neither.

53. $\langle -2, 3 \rangle, \langle 6, 4 \rangle$ **54.** $\langle 2, 3 \rangle, \langle 8, 12 \rangle$

55. $\langle 1, 7 \rangle, \langle -2, -14 \rangle$ **56.** $\langle 2, -4 \rangle, \langle 2, 1 \rangle$

57. $\langle 5, 3 \rangle, \langle 2, 5 \rangle$ **58.** $\langle 2, 6 \rangle, \langle 6, 2 \rangle$

Write each vector as a linear combination of the unit vectors **i** *and* **j**.

59. $\langle 2, 1 \rangle$ **60.** $\langle 1, 5 \rangle$

61. $\left\langle -3, \sqrt{2} \right\rangle$ **62.** $\left\langle \sqrt{2}, -5 \right\rangle$

63. $\langle 0, -9 \rangle$ **64.** $\langle -1/2, 0 \rangle$

65. $\langle -7, -1 \rangle$ **66.** $\langle 1, 1 \rangle$

Given that **A** $= \langle 3, 1 \rangle$ *and* **B** $= \langle -2, 3 \rangle$, *find the magnitude and direction angle for each of the following vectors.*

67. A + B   **68. A − B**

69. −3A   **70. 5B**

71. B − A   **72. B + A**

73. $-\mathbf{A} + \dfrac{1}{2}\mathbf{B}$ **74.** $\dfrac{1}{2}\mathbf{A} - 2\mathbf{B}$

For each given pair of vectors, find the magnitude and direction angle of the resultant.

75.

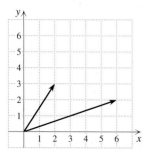

76.

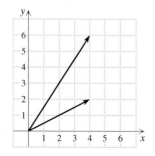

77.

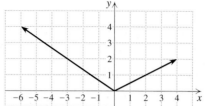

78.

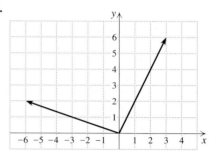

79.

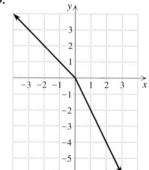

80.

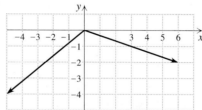

In each case, find the magnitude of the resultant force and the angle between the resultant and each force.

81. Forces of 3 lb and 8 lb act at an angle of 90° to each other.

82. Forces of 2 lb and 12 lb act at an angle of 60° to each other.

83. Forces of 4.2 newtons (a unit of force from physics) and 10.3 newtons act at an angle of 130° to each other.

84. Forces of 34 newtons and 23 newtons act at an angle of 100° to each other.

Solve each problem.

85. *Magnitude of a Force*   The resultant of a 10-lb force and another force has a magnitude of 12.3 lb at an angle of 23.4° with the 10-lb force. Find the magnitude of the other force and the angle between the two forces.

86. *Magnitude of a Force*   The resultant of a 15-lb force and another force has a magnitude of 9.8 lb at an angle of 31° with the 15-lb force. Find the magnitude of the other force and the angle between the other force and the resultant.

87. *Moving a Donkey*   Two prospectors are pulling on ropes attached around the neck of a donkey that does not want to move. One prospector pulls with a force of 55 lb, and the other pulls with a force of 75 lb. If the angle between the ropes is 25°, as shown in the figure, then how much force must the donkey use in order to stay put? (The donkey knows the proper direction in which to apply his force.)

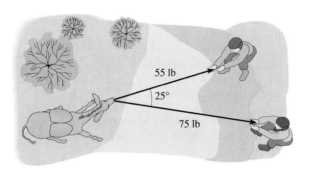

■ **Figure for Exercise 87**

88. *Pushing a Shopping Cart*   Ronnie, Phyllis, and Ted are conducting a vector experiment in a Wal-Mart parking lot. Ronnie is pushing a cart containing Phyllis to the east at 5 mph while Ted is pushing it to the north at 3 mph. What is Phyllis's speed and in what direction (measured from north) is she moving?

89. *Gaining Altitude*   An airplane with an air speed of 520 mph is climbing at an angle of 30° from the horizontal. What are the magnitudes of the horizontal and vertical components of the speed vector?

90. *Acceleration of a Missile*   A missile is fired with an angle of elevation of 22°, with an acceleration of 30 m/sec². What are the magnitudes of the horizontal and vertical components of the acceleration vector?

91. *Rock and Roll*   In Roman mythology, Sisyphus, king of Corinth, revealed a secret of Zeus and thus incurred the god's wrath. As punishment, Zeus banished him to Hades, where he was doomed for eternity to roll a rock uphill, only to have it roll back on him. If Sisyphus stands in front of a 4000-lb spherical rock on a 20° incline, as shown in the figure, then

what force applied in the direction of the incline would keep the rock from rolling down the incline?

HINT   The result of the 4000-lb force and the force of Sisyphus is perpendicular to the incline.

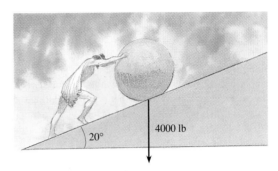

Figure for Exercise 91

92. *Weight of a Ball*   A solid steel ball is placed on a 10° incline. If a force of 3.2 lb in the direction of the incline is required to keep the ball in place, then what is the weight of the ball?

93. *Winch Force*   Find the amount of force required for a winch to pull a 3000-lb car up a ramp that is inclined at 20°.

94. *Ice Force*   If the amount of force required to push a block of ice up an ice-covered driveway that is inclined at 25° is 100 lb, then what is the weight of the block?

95. *Super Force*   If Superman exerts 1000 lb of force to prevent a 5000-lb boulder from rolling down a hill and crushing a bus full of children, then what is the angle of inclination of the hill?

96. *Sisy's Slope*   If Sisyphus exerts a 500-lb force in rolling his 4000-lb spherical boulder uphill, then what is the angle of inclination of the hill?

97. *Due East*   A plane is headed due east with an air speed of 240 mph. The wind is from the north at 30 mph. Find the bearing for the course and the ground speed of the plane.

98. *Due West*   A plane is headed due west with an air speed of 300 mph. The wind is from the north at 80 mph. Find the bearing for the course and the ground speed of the plane.

99. *Ultralight*   An ultralight is flying northeast at 50 mph. The wind is from the north at 20 mph. Find the bearing for the course and the ground speed of the ultralight.

100. *Superlight*   A superlight is flying northwest at 75 mph. The wind is from the south at 40 mph. Find the bearing for the course and the ground speed of the superlight.

101. *Course of an Airplane*   An airplane is heading on a bearing of 102° with an air speed of 480 mph. If the wind is out of the northeast (bearing 225°) at 58 mph, then what are the bearing of the course and the ground speed of the airplane?

102. *Course of a Helicopter*   The heading of a helicopter has a bearing of 240°. If the 70-mph wind has a bearing of 185° and the air speed of the helicopter is 195 mph, then what are the bearing of the course and the ground speed of the helicopter?

103. *Going with the Flow*   A river is 2000 ft wide and flowing at 6 mph from north to south. A woman in a canoe starts on the eastern shore and heads west at her normal paddling speed of 2 mph. In what direction (measured clockwise from north) will she actually be traveling? How far downstream from a point directly across the river will she land?

104. *Crossing a River*   If the woman in Exercise 103 wants to go directly across the river and she paddles at 8 mph, then in what direction (measured clockwise from north) must she aim her canoe? How long will it take her to go directly across the river?

105. *Distance and Rate*   A trigonometry student wants to cross a river that is 0.2 mi wide and has a current of 1 mph. The boat goes 3 mph in still water.
   a. Write the distance the boat travels as a function of the angle β shown in the figure.

   b. Write the actual speed of the boat as a function of α and β.

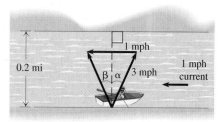

■ **Figure for Exercises 105 and 106**

106. *Minimizing the Time*   Write the time for the trip in the previous exercise as a function of α. Find the angle α for which the student will cross the river in the shortest amount of time.

107. *My Three Elephants*   Amal uses three elephants to pull a very large log out of the jungle. The papa elephant pulls with 800 lb of force, the mama elephant pulls with 500 lb of force, and the baby elephant pulls with 200 lb of force. The angles between the forces are shown in the figure. What is the magnitude of the resultant of all three forces? If mama is pulling due east, then in what direction will the log move?

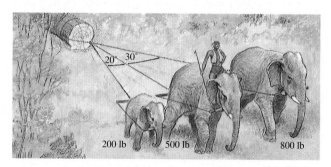

■ **Figure for Exercise 107**

For Writing/Discussion

108. *Distributive*   Prove that scalar multiplication is distributive over vector addition, first using the component form and then using a geometric argument.

109. *Associative*   Prove that vector addition is associative, first using the component form and then using a geometric argument.

Thinking Outside the Box LXII

Related Angles
a. Find the triangle with the smallest perimeter and whole-number sides such that the measure of one of the angles is twice the measure of another angle.

b. Find general expressions that could be used to determine infinitely many such triangles.

c. Find the next two larger such triangles.

7.3 Pop Quiz

1. Find $|v_x|$ and $|v_y|$ if $|v| = 5.6$ and the direction angle of **v** is 33.9°.

2. Find the magnitude and direction angle of the vector $\mathbf{v} = \langle -2, 6 \rangle$.

3. Find $\mathbf{v} - \mathbf{w}$, $3\mathbf{v}$, and $\mathbf{v} \cdot \mathbf{w}$, if $\mathbf{v} = \langle -1, 3 \rangle$ and $\mathbf{w} = \langle 2, 6 \rangle$.

4. Find the smallest positive angle between $\langle 1, 4 \rangle$ and $\langle 2, 6 \rangle$.

5. An airplane is headed due east with an air speed of 300 mph. The wind is from the north at 60 mph. What is the ground speed and heading of the airplane?

Linking Concepts

Minimizing the Total Time

Suk wants to go from point A to point B, as shown in the figure. Point B is 1 mile up-stream and on the north side of the 0.4-mi-wide river. The speed of the current is 0.5 mph. Her boat will travel 4 mph in still water. When she gets to the north side she will travel by bicycle at 6 mph. Let α represent the bearing of her course and β represent the drift angle, as shown in the figure.

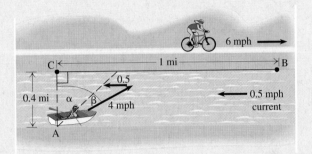

a) When $\alpha = 12°$, find β and the actual speed of the boat.

b) Find the total time for the trip when $\alpha = 12°$.

c) Write the total time for the trip as a function of α and graph the function.

d) Find the approximate angle α that minimizes the total time for the trip and find the corresponding β.

7.4

Trigonometric Form of Complex Numbers

Until now we have concentrated on applying the trigonometric functions to real-life situations. So the trigonometric form of complex numbers, introduced in this section, may be somewhat surprising. However, with this form we can expand our knowledge about complex numbers and find powers and roots of complex numbers that we could not find without it.

The Complex Plane

A complex number $a + bi$ can be thought of as an ordered pair (a, b), where a is the real part and b is the imaginary part of the complex number. Ordered pairs representing complex numbers are found in a coordinate system just like ordered pairs of real numbers. The horizontal axis is called the **real axis,** and the vertical axis is called the **imaginary axis,** as shown in Fig. 7.41. This coordinate system is called the **complex plane.** The complex plane provides an order to the complex number system and allows us to treat complex numbers like vectors.

The absolute value of a real number is the distance on the number line between the number and the origin. The **absolute value of a complex number** is the distance between the complex number and the origin in the complex plane.

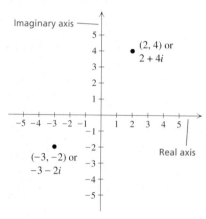

The Complex Plane

■ **Figure 7.41**

Definition: Absolute Value of *a + bi*

The absolute value of the complex number $a + bi$ is defined by

$$|a + bi| = \sqrt{a^2 + b^2}.$$

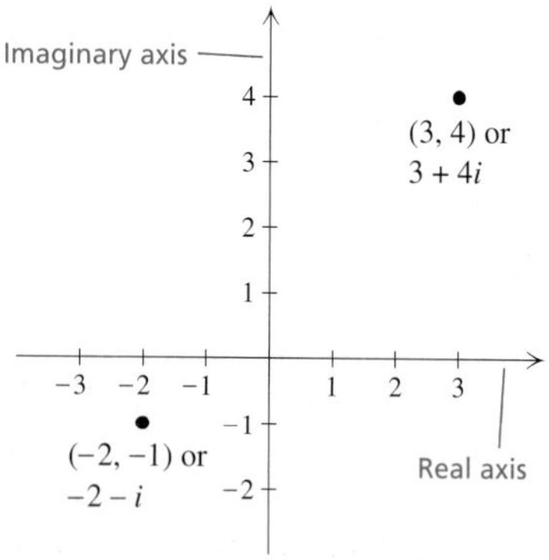

The absolute value of $a + bi$ is the same as the magnitude of the vector $\langle a, b \rangle$.

Example 1 Graphing complex numbers

Graph each complex number and find its absolute value.

a. $3 + 4i$ **b.** $-2 - i$

Solution

a. The complex number $3 + 4i$ is located in the first quadrant, as shown in Fig. 7.42, three units to the right of the origin and four units upward. Using the definition of absolute value of a complex number, we have

$$|3 + 4i| = \sqrt{3^2 + 4^2} = 5.$$

b. The complex number $-2 - i$ is located in the third quadrant (Fig. 7.42) and

$$|-2 - i| = \sqrt{(-2)^2 + (-1)^2} = \sqrt{5}.$$

Try This. Find the absolute value of $5 - i$. ■

■ **Figure 7.42**

Trigonometric Form of a Complex Number

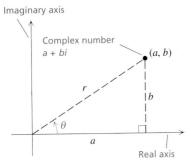

Imaginary axis

Complex number
$a + bi$

(a, b)

r

b

θ

a

Real axis

■ **Figure 7.43**

Multiplication and division of complex numbers in the standard form $a + bi$ is rather complicated, as is finding powers and roots, but in trigonometric form these operations are simpler and more natural. A complex number is written in **trigonometric form** by a process similar to finding the horizontal and vertical components of a vector.

Consider the complex number $a + bi$ shown in Fig. 7.43, where

$$r = |a + bi| = \sqrt{a^2 + b^2}.$$

Since (a, b) is a point on the terminal side of θ in standard position, we have

$$a = r \cos \theta \quad \text{and} \quad b = r \sin \theta.$$

Substituting for a and b, we get

$$a + bi = (r \cos \theta) + (r \sin \theta)i = r(\cos \theta + i \sin \theta).$$

Definition: Trigonometric Form of a Complex Number

If $z = a + bi$ is a complex number, then the **trigonometric form** of z is

$$z = r(\cos \theta + i \sin \theta),$$

where $r = \sqrt{a^2 + b^2}$ and θ is an angle in standard position whose terminal side contains the point (a, b).

The number r is called the **modulus** of $a + bi$, and θ is called the **argument** of $a + bi$. Since there are infinitely many angles whose terminal side contains (a, b), the trigonometric form of a complex number is not unique. However, we usually use the smallest nonnegative value for θ that satisfies

$$a = r \cos \theta \quad \text{or} \quad b = r \sin \theta.$$

In the next two examples, we convert from standard to trigonometric form and vice versa.

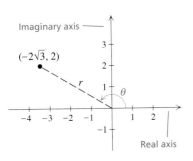

Imaginary axis

$(-2\sqrt{3}, 2)$

r

θ

Real axis

■ **Figure 7.44**

Example 2 Writing a complex number in trigonometric form

Write each complex number in trigonometric form.

a. $-2\sqrt{3} + 2i$ **b.** $3 - 2i$

Solution

a. First we locate $-2\sqrt{3} + 2i$ in the complex plane, as shown in Fig. 7.44. Find r using $a = -2\sqrt{3}$, $b = 2$, and $r = \sqrt{a^2 + b^2}$:

$$r = \sqrt{(-2\sqrt{3})^2 + 2^2} = 4$$

Since $a = r \cos \theta$, we have $-2\sqrt{3} = 4 \cos \theta$ or

$$\cos \theta = -\frac{\sqrt{3}}{2}.$$

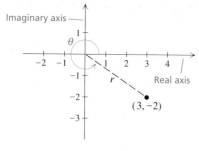

Imaginary axis

θ

-2 -1   1   2   3   4

-1 r Real axis

-2

(3, -2)

-3

■ **Figure 7.45**

The smallest nonnegative solution to this equation is $\theta = 5\pi/6$. Since the terminal side of $5\pi/6$ goes through $\left(-2\sqrt{3}, 2\right)$, we get

$$-2\sqrt{3} + 2i = 4\left(\cos\frac{5\pi}{6} + i\sin\frac{5\pi}{6}\right).$$

b. Locate $3 - 2i$ in the complex plane, as shown in Fig. 7.45. Next, find r by using $a = 3$, $b = -2$, and $r = \sqrt{a^2 + b^2}$:

$$r = \sqrt{(3)^2 + (-2)^2} = \sqrt{13}$$

Since $a = r\cos\theta$, we have $3 = \sqrt{13}\cos\theta$ or

$$\cos\theta = \frac{3}{\sqrt{13}}.$$

Use a calculator to get

$$\cos^{-1}\left(\frac{3}{\sqrt{13}}\right) = 33.7°.$$

Since θ is an angle whose terminal side contains $(3, -2)$, choose $\theta = 360° - 33.7° = 326.3°$. So

$$3 - 2i = \sqrt{13}(\cos 326.3° + i\sin 326.3°).$$

Try This. Write $1 + 2i$ in trigonometric form using degree measure.   ■

Example **3** **Writing a complex number in standard form**

Write each complex number in the form $a + bi$.

a. $\sqrt{2}\left(\cos\frac{\pi}{4} + i\sin\frac{\pi}{4}\right)$ **b.** $3.6(\cos 143° + i\sin 143°)$

Solution

a. Since $\cos(\pi/4) = \sqrt{2}/2$ and $\sin(\pi/4) = \sqrt{2}/2$, we have

$$\sqrt{2}\left(\cos\frac{\pi}{4} + i\sin\frac{\pi}{4}\right) = \sqrt{2}\left(\frac{\sqrt{2}}{2} + i\frac{\sqrt{2}}{2}\right)$$

$$= 1 + i.$$

b. Use a calculator to find $\cos 143°$ and $\sin 143°$:

$$3.6(\cos 143° + i\sin 143°) \approx -2.875 + 2.167i$$

Try This. Write $12\left(\cos\frac{\pi}{6} + i\sin\frac{\pi}{6}\right)$ in standard form.   ■

Products and Quotients in Trigonometric Form

In standard form, multiplication and division of complex numbers are rather complicated whereas addition and subtraction are simple. In trigonometric form addition and subtraction are complicated whereas multiplication and division are simple. We multiply the complex numbers

$$z_1 = r_1(\cos \theta_1 + i \sin \theta_1) \qquad \text{and} \qquad z_2 = r_2(\cos \theta_2 + i \sin \theta_2)$$

as follows:

$$\begin{aligned} z_1 z_2 &= r_1(\cos \theta_1 + i \sin \theta_1) \cdot r_2(\cos \theta_2 + i \sin \theta_2) \\ &= r_1 r_2 (\cos \theta_1 + i \sin \theta_1)(\cos \theta_2 + i \sin \theta_2) \\ &= r_1 r_2 [(\cos \theta_1 \cos \theta_2 - \sin \theta_1 \sin \theta_2) + i(\sin \theta_1 \cos \theta_2 + \cos \theta_1 \sin \theta_2)] \end{aligned}$$

Using the identities for cosine and sine of a sum, we get

$$z_1 z_2 = r_1 r_2 [\cos(\theta_1 + \theta_2) + i \sin(\theta_1 + \theta_2)].$$

The product of z_1 and z_2 can be found by multiplying their moduli r_1 and r_2 and adding their arguments θ_1 and θ_2.

We find the quotient of z_1 and z_2 as follows:

$$\begin{aligned} \frac{z_1}{z_2} &= \frac{r_1(\cos \theta_1 + i \sin \theta_1)}{r_2(\cos \theta_2 + i \sin \theta_2)} \\ \\ &= \frac{r_1(\cos \theta_1 + i \sin \theta_1)(\cos \theta_2 - i \sin \theta_2)}{r_2(\cos \theta_2 + i \sin \theta_2)(\cos \theta_2 - i \sin \theta_2)} \\ \\ &= \frac{r_1}{r_2}[(\cos \theta_1 \cos \theta_2 + \sin \theta_1 \sin \theta_2) + i(\sin \theta_1 \cos \theta_2 - \cos \theta_1 \sin \theta_2)] \\ \\ &= \frac{r_1}{r_2}[\cos(\theta_1 - \theta_2) + i \sin(\theta_1 - \theta_2)] \end{aligned}$$

The quotient of z_1 and z_2 can be found by dividing their moduli r_1 and r_2 and subtracting their arguments θ_1 and θ_2.

These two results are stated in the following theorem.

Theorem: Product and Quotient of Complex Numbers

If $z_1 = r_1(\cos \theta_1 + i \sin \theta_1)$ and $z_2 = r_2(\cos \theta_2 + i \sin \theta_2)$, then

$$z_1 z_2 = r_1 r_2 [\cos(\theta_1 + \theta_2) + i \sin(\theta_1 + \theta_2)]$$

and

$$\frac{z_1}{z_2} = \frac{r_1}{r_2}[\cos(\theta_1 - \theta_2) + i \sin(\theta_1 - \theta_2)].$$

Example **4** **A product and quotient using trigonometric form**

Find $z_1 z_2$ and z_1/z_2 for $z_1 = 6(\cos 2.4 + i \sin 2.4)$ and $z_2 = 2(\cos 1.8 + i \sin 1.8)$. Express the answers in the form $a + bi$.

Solution

Find the product by multiplying the moduli and adding the arguments:

$$z_1 z_2 = 6 \cdot 2(\cos(2.4 + 1.8) + i \sin(2.4 + 1.8))$$
$$= 12(\cos 4.2 + i \sin 4.2)$$
$$= 12 \cos 4.2 + 12i \sin 4.2$$
$$\approx -5.88 - 10.46i$$

Find the quotient by dividing the moduli and subtracting the arguments:

$$\frac{z_1}{z_2} = \frac{6}{2}(\cos(2.4 - 1.8) + i \sin(2.4 - 1.8))$$
$$= 3(\cos(0.6) + i \sin(0.6))$$
$$= 3 \cos 0.6 + 3i \sin 0.6$$
$$\approx 2.48 + 1.69i$$

Try This. Find $z_1 z_2$ for $z_1 = 4\left(\cos \frac{\pi}{12} + i \sin \frac{\pi}{12}\right)$ and $z_2 = 8\left(\cos \frac{\pi}{12} + i \sin \frac{\pi}{12}\right)$. ■

Example **5** **A product and quotient using trigonometric form**

Use trigonometric form to find $z_1 z_2$ and z_1/z_2, given that $z_1 = -2 + 2i\sqrt{3}$ and $z_2 = \sqrt{3} + i$.

Solution

First convert z_1 and z_2 into trigonometric form. Since $\sqrt{(-2)^2 + (2\sqrt{3})^2} = 4$ and $120°$ is a positive angle whose terminal side contains $(-2, 2\sqrt{3})$, we have

$$z_1 = 4(\cos 120° + i \sin 120°).$$

Since $\sqrt{(\sqrt{3})^2 + 1^2} = 2$ and $30°$ is a positive angle whose terminal side contains $(\sqrt{3}, 1)$, we have

$$z_2 = 2(\cos 30° + i \sin 30°).$$

To find the product, we multiply the moduli and add the arguments:

$$z_1 z_2 = 2 \cdot 4[\cos(120° + 30°) + i \sin(120° + 30°)]$$
$$= 8(\cos 150° + i \sin 150°)$$

Using $\cos 150° = -\sqrt{3}/2$ and $\sin 150° = 1/2$, we get

$$z_1 z_2 = 8\left(-\frac{\sqrt{3}}{2} + i \frac{1}{2}\right) = -4\sqrt{3} + 4i.$$

To find the quotient, we divide the moduli and subtract the arguments:

$$\frac{z_1}{z_2} = \frac{4}{2}[\cos(120° - 30°) + i\sin(120° - 30°)]$$

$$= 2(\cos 90° + i\sin 90°)$$

$$= 2(0 + i) = 2i$$

Check by performing the operations with z_1 and z_2 in standard form.

Try This. Find $z_1 z_2$ using trigonometric form for $z_1 = 2\sqrt{3} + 2i$ and $z_2 = 3 + 3i\sqrt{3}$. ■

Complex Conjugates

In standard form, $a + bi$ and $a - bi$ are complex conjugates. Their product is the real number $a^2 + b^2$. In trigonometric form, $a + bi = r(\cos\theta + i\sin\theta)$, and we could write $a - bi = r(\cos\theta - i\sin\theta)$. But trigonometric form has a plus sign in it. To multiply in trigonometric form (by multiplying the moduli and adding the arguments), the plus sign is necessary. It can be shown that $a - bi = r(\cos(-\theta) + i\sin(-\theta))$. (See Exercise 75.) Now if you multiply the moduli and add the arguments, the product is a real number. (See Exercise 76.)

Example **6**   **Complex conjugates in trigonometric form**

Find the product of $2(\cos(\pi/3) + i\sin(\pi/3))$ and its conjugate using multiplication in trigonometric form.

Solution

The conjugate is $2(\cos(-\pi/3) + i\sin(-\pi/3))$. We find the product by multiplying the moduli and adding the arguments:

$$2\left(\cos\frac{\pi}{3} + i\sin\frac{\pi}{3}\right) \cdot 2\left(\cos\left(-\frac{\pi}{3}\right) + i\sin\left(-\frac{\pi}{3}\right)\right) = 4(\cos 0 + i\sin 0) = 4.$$

Try This. Find the product of $6(\cos 30° + i\sin 30°)$ and its conjugate using multiplication in trigonometric form. ■

For Thought

True or False? Explain.

1. The complex number $3 - 4i$ lies in quadrant IV of the complex plane.

2. The absolute value of $-2 - 5i$ is $2 + 5i$.

3. If θ is an angle whose terminal side contains $(1, -3)$, then $\cos\theta = 1/\sqrt{10}$.

4. If θ is an angle whose terminal side contains $(-2, -3)$, then $\tan\theta = 2/3$.

5. $i = 1(\cos 0° + i\sin 0°)$

6. The smallest positive solution to $\sqrt{3} = 2\cos\theta$ is $\theta = 30°$.

7. The modulus of $2 - 5i$ is $\sqrt{29}$.

8. An argument for $2 - 4i$ is $\theta = 360° - \cos^{-1}(1/\sqrt{5})$.

9. $3(\cos\pi/4 + i\sin\pi/4) \cdot 2(\cos\pi/2 + i\sin\pi/2) = 6(\cos 3\pi/4 + i\sin 3\pi/4)$

10. $\dfrac{3(\cos\pi/4 + i\sin\pi/4)}{2(\cos\pi/2 + i\sin\pi/2)} = 1.5(\cos\pi/4 + i\sin\pi/4)$

7.4 Exercises

Graph each complex number, and find its absolute value.

1. $8i$ 　　　　**2.** $-3i$ 　　　　**3.** -9

4. $-\sqrt{6}$ 　　　**5.** $2 - 6i$ 　　**6.** $-1 - i$

7. $-2 + 2i\sqrt{3}$ 　**8.** $-\sqrt{3} - i$ 　**9.** $\dfrac{1}{\sqrt{2}} - \dfrac{i}{\sqrt{2}}$

10. $\dfrac{\sqrt{3}}{2} + \dfrac{i}{2}$ 　**11.** $3 + 3i$ 　**12.** $-4 - 4i$

Write each complex number in trigonometric form, using degree measure for the argument.

13. 8 　　　　　**14.** -7 　　　　**15.** $i\sqrt{3}$

16. $-5i$ 　　　　**17.** $-3 + 3i$ 　　**18.** $4 - 4i$

19. $-\dfrac{3}{\sqrt{2}} + \dfrac{3i}{\sqrt{2}}$ 　**20.** $\dfrac{\sqrt{3}}{6} + \dfrac{i}{6}$ 　**21.** $-\sqrt{3} + i$

22. $-2 - 2i\sqrt{3}$ 　**23.** $3 + 4i$ 　　**24.** $-2 + i$

25. $-3 + 5i$ 　　**26.** $-2 - 4i$ 　　**27.** $3 - 6i$

28. $5 - 10i$

Write each complex number in the form $a + bi$.

29. $\sqrt{2}(\cos 45° + i \sin 45°)$ 　　**30.** $6(\cos 30° + i \sin 30°)$

31. $\dfrac{\sqrt{3}}{2}(\cos 150° + i \sin 150°)$

32. $12(\cos (\pi/10) + i \sin(\pi/10))$

33. $\dfrac{1}{2}(\cos 3.7 + i \sin 3.7)$

34. $4.3(\cos(\pi/9) + i \sin(\pi/9))$

35. $3(\cos 90° + i \sin 90°)$ 　　**36.** $4(\cos 180° + i \sin 180°)$

37. $\sqrt{3}(\cos(3\pi/2) + i \sin(3\pi/2))$

38. $8.1(\cos \pi + i \sin \pi)$ 　　**39.** $\sqrt{6}(\cos 60° + i \sin 60°)$

40. $0.5(\cos(5\pi/6) + i \sin(5\pi/6))$

Perform the indicated operations. Write the answer in the form $a + bi$.

41. $2(\cos 150° + i \sin 150°) \cdot 3(\cos 300° + i \sin 300°)$

42. $\sqrt{3}(\cos 45° + i \sin 45°) \cdot \sqrt{2}(\cos 315° + i \sin 315°)$

43. $\sqrt{3}(\cos 10° + i \sin 10°) \cdot \sqrt{2}(\cos 20° + i \sin 20°)$

44. $8(\cos 100° + i \sin 100°) \cdot 3(\cos 35° + i \sin 35°)$

45. $[3(\cos 45° + i \sin 45°)]^2$

46. $\left[\sqrt{5}(\cos(\pi/12) + i \sin(\pi/12))\right]^2$

47. $\dfrac{4(\cos(\pi/3) + i \sin(\pi/3))}{2(\cos(\pi/6) + i \sin(\pi/6))}$

48. $\dfrac{9(\cos(\pi/4) + i \sin(\pi/4))}{3(\cos(5\pi/4) + i \sin(5\pi/4))}$

49. $\dfrac{4.1(\cos 36.7° + i \sin 36.7°)}{8.2(\cos 84.2° + i \sin 84.2°)}$

50. $\dfrac{18(\cos 121.9° + i \sin 121.9°)}{2(\cos 325.6° + i \sin 325.6°)}$

Find $z_1 z_2$ and z_1/z_2 for each pair of complex numbers, using trigonometric form. Write the answer in the form $a + bi$.

51. $z_1 = 4 + 4i, z_2 = -5 - 5i$

52. $z_1 = -3 + 3i, z_2 = -2 - 2i$

53. $z_1 = \sqrt{3} + i, z_2 = 2 + 2i\sqrt{3}$

54. $z_1 = -\sqrt{3} + i, z_2 = 4\sqrt{3} - 4i$

55. $z_1 = 2 + 2i, z_2 = \sqrt{2} - i\sqrt{2}$

56. $z_1 = \sqrt{5} + i\sqrt{5}, z_2 = -\sqrt{6} - i\sqrt{6}$

57. $z_1 = 3 + 4i, z_2 = -5 - 2i$

58. $z_1 = 3 - 4i, z_2 = -1 + 3i$

59. $z_1 = 2 - 6i, z_2 = -3 - 2i$

60. $z_1 = 1 + 4i, z_2 = -4 - 2i$

61. $z_1 = 3i, z_2 = 1 + i$

62. $z_1 = 4, z_2 = -3 + i$

For each given complex number, determine its complex conjugate in trigonometric form.

63. $3\left(\cos \dfrac{\pi}{4} + i \sin \dfrac{\pi}{4}\right)$ 　　**64.** $\sqrt{2}\left(\cos \dfrac{\pi}{8} + i \sin \dfrac{\pi}{8}\right)$

65. $2\sqrt{3}(\cos(-20°) + i \sin(-20°))$

66. $9(\cos 14° + i \sin 14°)$

Find the product of the given complex number and its complex conjugate in trigonometric form.

67. $3\left(\cos \dfrac{\pi}{6} + i \sin \dfrac{\pi}{6}\right)$

68. $5\left(\cos \dfrac{\pi}{7} + i \sin \dfrac{\pi}{7}\right)$

69. $2(\cos 7° + i \sin 7°)$

70. $6(\cos 5° + i \sin 5°)$

Solve each problem.

71. Given that $z = 4 + 4i$, find z^2 by writing z in trigonometric form and computing $z \cdot z$.

72. Given that $z = -3 + 3i$, find z^2 by writing z in trigonometric form and computing $z \cdot z$.

73. Given that $z = 3 + 3i$, find z^3 by writing z in trigonometric form and computing the product $z \cdot z \cdot z$.

74. Given that $z = \sqrt{3} + i$, find z^4 by writing z in trigonometric form and computing the product $z \cdot z \cdot z \cdot z$.

For Writing/Discussion

75. Show that the complex conjugate of $z = r(\cos \theta + i \sin \theta)$ is $\bar{z} = r(\cos(-\theta) + i \sin(-\theta))$.

76. Find the product of z and $\bar{z}$, using the trigonometric forms from Exercise 75.

77. Show that the reciprocal of $z = r(\cos \theta + i \sin \theta)$ is $z^{-1} = r^{-1}(\cos \theta - i \sin \theta)$, provided $r \neq 0$.

78. Given that $z = r(\cos \theta + i \sin \theta)$, find z^2 and z^{-2}, provided $r \neq 0$.

79. Find the sum of $6(\cos 9° + i \sin 9°)$ and $3(\cos 5° + i \sin 5°)$. Find the sum of $1 + 3i$ and $5 - 7i$. Is it easier to add complex numbers in trigonometric form or standard form?

80. Find the quotient when $6(\cos 9° + i \sin 9°)$ is divided by $3(\cos 5° + i \sin 5°)$. Find the quotient when $1 + 3i$ is divided by $5 - 7i$. Is it easier to divide complex numbers in trigonometric form or standard form?

Thinking Outside the Box LXIII

Two Squares and Two Triangles   The two squares with areas of 25 and 36 shown in the accompanying figure are positioned so that $AB = 7$. Find the exact area of triangle *TSC*.

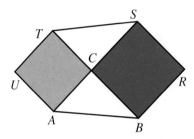

▪ **Figure for Thinking Outside the Box LXIII**

7.4 Pop Quiz

1. Find the absolute value of $3 - i$.

2. Write $5 + 5i$ in trigonometric form.

3. Write $2\sqrt{3}(\cos(5\pi/6) + i \sin(5\pi/6))$ in standard form.

4. Find zw if $z = \sqrt{2}(\cos(\pi/12) + i\sin(\pi/12))$ and $w = 3\sqrt{2}(\cos(\pi/12) + i \sin(\pi/12))$.

5. Find the product of $\sqrt{2}(\cos(\pi/4) + i \sin(\pi/4))$ and its complex conjugate.

7.5

Powers and Roots of Complex Numbers

We know that the square root of a negative number is an imaginary number, but we have not yet encountered roots of imaginary numbers. In this section we will use the trigonometric form of a complex number to find any power or root of a complex number.

De Moivre's Theorem

Powers of a complex number can be found by repeated multiplication. Repeated multiplication can be done with standard form or trigonometric form. However, in trigonometric form, a product is found by simply multiplying the moduli and adding the arguments. So, if $z = r(\cos \theta + i \sin \theta)$, then

$$z^2 = r(\cos \theta + i \sin \theta) \cdot r(\cos \theta + i \sin \theta) = r^2(\cos 2\theta + i \sin 2\theta).$$

We can easily find z^3 because $z^3 = z^2 \cdot z$:

$$z^3 = r^2(\cos 2\theta + i \sin 2\theta) \cdot r(\cos \theta + i \sin \theta) = r^3(\cos 3\theta + i \sin 3\theta)$$

Multiplying z^3 by z gives

$$z^4 = r^4(\cos 4\theta + i \sin 4\theta).$$

If you examine the results of z^2, z^3, and z^4, you will see a pattern, which can be stated as follows.

De Moivre's Theorem

> If $z = r(\cos \theta + i \sin \theta)$ is a complex number and n is any positive integer, then
> $$z^n = r^n(\cos n\theta + i \sin n\theta).$$

De Moivre's theorem, named after the French mathematician Abraham De Moivre (1667–1754), can be proved by using mathematical induction, which is presented in Chapter 11.

Example 1 A power of a complex number

Use De Moivre's theorem to simplify $\left(-\sqrt{3} + i\right)^8$. Write the answer in the form $a + bi$.

Solution

First write $-\sqrt{3} + i$ in trigonometric form. The modulus is 2, and the argument is 150°. So

$$-\sqrt{3} + i = 2(\cos 150° + i \sin 150°).$$

Use De Moivre's theorem to find the eighth power:

$$\left(-\sqrt{3} + i\right)^8 = [2(\cos 150° + i \sin 150°)]^8$$
$$= 2^8[\cos(8 \cdot 150°) + i \sin(8 \cdot 150°)]$$
$$= 256[\cos 1200° + i \sin 1200°]$$

Since $1200° = 3 \cdot 360° + 120°$,

$$\cos 1200° = \cos 120° = -\frac{1}{2} \quad \text{and} \quad \sin 1200° = \sin 120° = \frac{\sqrt{3}}{2}.$$

Use these values to simplify the trigonometric form of $\left(-\sqrt{3} + i\right)^8$:

$$\left(-\sqrt{3} + i\right)^8 = 256\left(-\frac{1}{2} + i\frac{\sqrt{3}}{2}\right) = -128 + 128i\sqrt{3}$$

Try This. Simplify $(1 + i)^6$.   ■

We have seen how De Moivre's theorem is used to find a positive integral power of a complex number. It is also used for finding roots of complex numbers.

Roots of a Complex Number

In the real number system, a is an nth root of b if $a^n = b$. We have a similar definition of nth root in the complex number system.

Definition: nth Root of a Complex Number

> The complex number $w = a + bi$ is an nth root of the complex number z if
>
> $$(a + bi)^n = z.$$

In Example 1, we saw that $\left(-\sqrt{3} + i\right)^8 = -128 + 128i\sqrt{3}$. So $-\sqrt{3} + i$ is an eighth root of $-128 + 128i\sqrt{3}$.

The definition of nth root does not indicate how to find an nth root, but a formula for all nth roots can be found by using trigonometric form. Suppose that the complex number w is an nth root of the complex number z, where the trigonometric forms of w and z are

$$w = s(\cos\alpha + i\sin\alpha) \qquad \text{and} \qquad z = r(\cos\theta + i\sin\theta).$$

By De Moivre's theorem, $w^n = s^n(\cos n\alpha + i\sin n\alpha)$. Since $w^n = z$, we have

$$w^n = s^n(\cos n\alpha + i\sin n\alpha) = r(\cos\theta + i\sin\theta).$$

This last equation gives two different trigonometric forms for w^n. The absolute value of w^n in one form is s^n and in the other it is r. So $s^n = r$, or $s = r^{1/n}$. Because the argument of a complex number is any angle in the complex plane whose terminal side goes through the point representing the complex number, any two angles used for the argument must differ by a multiple of 2π. Since the argument for w^n is either $n\alpha$ or θ, we have $n\alpha = \theta + 2k\pi$ or

$$\alpha = \frac{\theta + 2k\pi}{n}.$$

We can get n different values for α from this formula by using $k = 0, 1, 2, \ldots, n - 1$. These results are summarized in the following theorem.

Theorem: nth Roots of a Complex Number

> For any positive integer n, the complex number $z = r(\cos\theta + i\sin\theta)$ has exactly n distinct nth roots, given by the expression
>
> $$r^{1/n}\left[\cos\left(\frac{\theta + 2k\pi}{n}\right) + i\sin\left(\frac{\theta + 2k\pi}{n}\right)\right]$$
>
> for $k = 0, 1, 2, \ldots, n - 1$.

The n distinct nth roots of a complex number all have the same modulus, $r^{1/n}$. What distinguishes between these roots is the n different arguments determined by

$$\alpha = \frac{\theta + 2k\pi}{n}$$

for $k = 0, 1, 2, \ldots, n - 1$. If θ is measured in degrees, then the n different arguments are determined by

$$\alpha = \frac{\theta + k360°}{n}$$

for $k = 0, 1, 2, \ldots, n - 1$. Note that if $k \geq n$, we get new values for α but no new nth roots, because the values of $\sin \alpha$ and $\cos \alpha$ have already occurred for some $k < n$. For example, if $k = n$, then $\alpha = \theta/n + 2\pi$, and the values for $\cos \alpha$ and $\sin \alpha$ are the same as they were for $\alpha = \theta/n$, which corresponds to $k = 0$.

Example 2 Finding nth roots

Find all of the fourth roots of the complex number $-8 + 8i\sqrt{3}$.

Solution

First find the modulus of $-8 + 8i\sqrt{3}$:

$$\sqrt{(-8)^2 + (8\sqrt{3})^2} = 16$$

Since the terminal side of θ contains $(-8, 8\sqrt{3})$, we have $\cos \theta = -8/16 = -1/2$. Choose $\theta = 120°$. So the fourth roots are generated by the expression

$$16^{1/4}\left[\cos\left(\frac{120° + k360°}{4}\right) + i\sin\left(\frac{120° + k360°}{4}\right)\right].$$

Evaluating this expression for $k = 0, 1, 2,$ and 3 (because $n = 4, n - 1 = 3$), we get

$$2[\cos 30° + i\sin 30°] = \sqrt{3} + i$$
$$2[\cos 120° + i\sin 120°] = -1 + i\sqrt{3}$$
$$2[\cos 210° + i\sin 210°] = -\sqrt{3} - i$$
$$2[\cos 300° + i\sin 300°] = 1 - i\sqrt{3}.$$

Try This. Find all fourth roots of $-8 - 8i\sqrt{3}$. ■

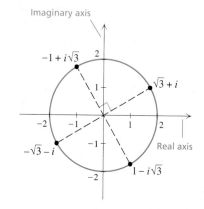

Fourth Roots of $-8 + 8i\sqrt{3}$

■ **Figure 7.46**

The graph of the original complex number in Example 2 and its four fourth roots reveals an interesting pattern. Since the modulus of $-8 + 8i\sqrt{3}$ is 16, $-8 + 8i\sqrt{3}$ lies on a circle of radius 16 centered at the origin. Since the modulus of each fourth root is $16^{1/4} = 2$, the fourth roots lie on a circle of radius 2 centered at the origin, as shown in Fig. 7.46. Note how the fourth roots in Fig. 7.46 divide the circumference of the circle into four equal parts. In general, the nth roots of $z = r(\cos \theta + i\sin \theta)$ divide the circumference of a circle of radius $r^{1/n}$ into n equal parts. This symmetric arrangement of the nth roots on a circle provides a simple check on whether the nth roots are correct.

Consider the equation $x^6 - 1 = 0$. By the fundamental theorem of algebra, this equation has six solutions in the complex number system if multiplicity is considered. Since this equation is equivalent to $x^6 = 1$, each solution is a sixth root of 1. The roots of 1 are called the **roots of unity.** The fundamental theorem of algebra does not guarantee that all of the roots are unique, but the theorem on the nth roots of a complex number does.

Example **3** **The six sixth roots of unity**

Solve the equation $x^6 - 1 = 0$.

Solution

The solutions to the equation are the sixth roots of 1. Since $1 = 1(\cos 0° + i \sin 0°)$, the expression for the sixth roots is

$$1^{1/6}\left[\cos\left(\frac{0° + k360°}{6}\right) + i \sin\left(\frac{0° + k360°}{6}\right)\right].$$

Evaluating this expression for $k = 0, 1, 2, \ldots, 5$ gives the six roots:

$$\cos 0° + i \sin 0° = 1 \qquad\qquad \cos 60° + i \sin 60° = \frac{1}{2} + i\frac{\sqrt{3}}{2}$$

$$\cos 120° + i \sin 120° = -\frac{1}{2} + i\frac{\sqrt{3}}{2} \qquad \cos 180° + i \sin 180° = -1$$

$$\cos 240° + i \sin 240° = -\frac{1}{2} - i\frac{\sqrt{3}}{2} \qquad \cos 300° + i \sin 300° = \frac{1}{2} - i\frac{\sqrt{3}}{2}$$

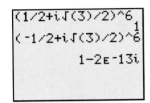

■ **Figure 7.47**

You can check the six roots using a graphing calculator that handles complex numbers, as shown in Fig. 7.47. Note that the calculator does not always get the exact answer.

Try This. Find all complex solutions to $x^3 - 27 = 0$.   ■

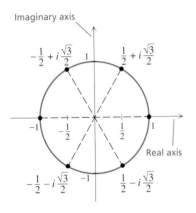

Sixth Roots of Unity

■ **Figure 7.48**

A graph of the sixth roots of unity found in Example 3 is shown in Fig. 7.48. Notice that they divide the circle of radius 1 into six equal parts. According to the conjugate pairs theorem of Chapter 3, the complex solutions to a polynomial equation with real coefficients occur in conjugate pairs. In Example 3, the complex solutions of $x^6 - 1 = 0$ did occur in conjugate pairs. This fact gives us another method of checking whether the six roots are correct. In general, the nth roots of any real number occur in conjugate pairs. Note that the fourth roots of the imaginary number in Example 2 did not occur in conjugate pairs.

For Thought

True or False? Explain.

1. $(2 + 3i)^2 = 4 + 9i^2$

2. If $z = 2(\cos 120° + i \sin 120°)$, then $z^3 = 8i$.

3. If z is a complex number with modulus r, then the modulus of z^4 is r^4.

4. If the argument of z is θ, then the argument of z^4 is $\cos 4\theta$.

5. $[\cos(\pi/3) + i \sin(\pi/6)]^2 = \cos(2\pi/3) + i \sin(\pi/3)$

6. If $\cos \alpha = \cos \beta$, then $\alpha = \beta + 2k\pi$ for some integer k.

7. One of the fifth roots of 32 is $2(\cos 72° + i \sin 72°)$.

8. All solutions to $x^8 = 1$ in the complex plane lie on the unit circle.

9. All solutions to $x^8 = 1$ lie on the axes or the lines $y = \pm x$.

10. The equation $x^4 + 81 = 0$ has two real and two imaginary solutions.

7.5 Exercises

Use De Moivre's theorem to simplify each expression. Write the answer in the form $a + bi$.

1. $[3(\cos 30° + i \sin 30°)]^3$

2. $[2(\cos 45° + i \sin 45°)]^5$

3. $\left[\sqrt{2}(\cos 120° + i \sin 120°)\right]^4$

4. $\left[\sqrt{3}(\cos 150° + i \sin 150°)\right]^3$

5. $[\cos(\pi/12) + i \sin(\pi/12)]^8$

6. $[\cos(\pi/6) + i \sin(\pi/6)]^9$

7. $\left[\sqrt{6}(\cos(2\pi/3) + i \sin(2\pi/3))\right]^4$

8. $\left[\sqrt{18}(\cos(5\pi/6) + i \sin(5\pi/6))\right]^3$

9. $[4.3(\cos 12.3° + i \sin 12.3°)]^5$

10. $[4.9(\cos 37.4° + i \sin 37.4°)]^6$

Simplify each expression, by using trigonometric form and De Moivre's theorem. Write the answer in the form $a + bi$.

11. $(2 + 2i)^3$

12. $(1 - i)^3$

13. $\left(\sqrt{3} - i\right)^4$

14. $\left(-2 + 2i\sqrt{3}\right)^4$

15. $\left(-3 - 3i\sqrt{3}\right)^5$

16. $\left(2\sqrt{3} - 2i\right)^5$

17. $(2 + 3i)^4$

18. $(4 - i)^5$

19. $(2 - i)^4$

20. $(-1 - 2i)^6$

21. $(1.2 + 3.6i)^3$

22. $(-2.3 - i)^3$

Find the indicated roots. Express answers in trigonometric form.

23. The square roots of $4(\cos 90° + i \sin 90°)$

24. The cube roots of $8(\cos 30° + i \sin 30°)$

25. The fourth roots of $\cos 120° + i \sin 120°$

26. The fifth roots of $32(\cos 300° + i \sin 300°)$

27. The sixth roots of $64(\cos \pi + i \sin \pi)$

28. The fourth roots of $16[\cos(3\pi/2) + i \sin(3\pi/2)]$

Find the indicated roots in the form $a + bi$. Check by graphing the roots in the complex plane.

29. The cube roots of 1

30. The cube roots of 8

31. The fourth roots of 16

32. The fourth roots of 1

33. The fourth roots of -1

34. The fourth roots of -16

35. The cube roots of i

36. The cube roots of $-8i$

37. The square roots of $-2 + 2i\sqrt{3}$

38. The square roots of $-4i$

39. The square roots of $1 + 2i$

40. The cube roots of $-1 + 3i$

Solve each equation. Express answers in the form $a + bi$.

41. $x^3 + 1 = 0$

42. $x^3 + 125 = 0$

43. $x^4 - 81 = 0$

44. $x^4 + 81 = 0$

45. $x^2 + 2i = 0$

46. $ix^2 + 3 = 0$

47. $x^7 - 64x = 0$

48. $x^9 - x = 0$

49. $x^5 + 5x^3 + 8x^2 + 40 = 0$

50. $x^5 + x^3 - 27x^2 - 27 = 0$

Solve each equation. Express answers in trigonometric form.

51. $x^5 - 2 = 0$

52. $x^5 + 3 = 0$

53. $x^4 + 3 - i = 0$

54. $ix^3 + 2 - i = 0$

Solve each problem.

55. Write the expression $[\cos(\pi/3) + i \sin(\pi/6)]^3$ in the form $a + bi$.

56. Solve the equation $x^6 - 1 = 0$ by factoring and the quadratic formula. Compare your answers to Example 3 of this section.

57. Solve $x^2 + (-1 + i)x - i = 0$.
 HINT Use the quadratic formula and De Moivre's theorem.

58. Solve $x^2 + (-1 - 3i)x + (-2 + 2i) = 0$.
 HINT Use the quadratic formula and De Moivre's theorem.

For Writing/Discussion

59. Explain why $x^6 - 2x^3 + 1 = 0$ has three distinct solutions, $x^6 - 2x^3 = 0$ has four distinct solutions, and $x^6 - 2x = 0$ has six distinct solutions.

60. Find all real numbers a and b for which it is true that $\sqrt{a + bi} = \sqrt{a} + i\sqrt{b}$.

Thinking Outside the Box LXIV

Double-Boxed A rectangular box contains a delicate statue. The shipping department places the box containing the statue inside a 3 ft by 4 ft rectangular box as shown from above in the accompanying figure. If the box containing the statue is 1 ft wide, then what is its length? Find a four-decimal place approximation.

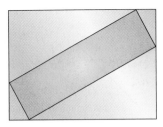

■ **Figure for Thinking Outside the Box LXIV**

7.5 Pop Quiz

1. Use De Moivre's theorem to simplify $(1 + i)^8$.

2. Find all fourth roots of -16.

7.6

Polar Equations

The Cartesian coordinate system is a system for describing the location of points in a plane. It is not the only coordinate system available. In this section we will study the **polar coordinate system.** In this coordinate system, a point is located by using a *directed distance* and an *angle* in a manner that will remind you of the magnitude and direction angle of a vector and the modulus and argument for a complex number.

Polar Coordinates

In the rectangular coordinate system, points are named according to their position with respect to an x-axis and a y-axis. In the polar coordinate system, we have a fixed point called the **pole** and a fixed ray called the **polar axis.** A point P has coordinates (r, θ), where r is the directed distance from the pole to P and θ is an angle whose initial side is the polar axis and whose terminal side contains the point. See Fig. 7.49.

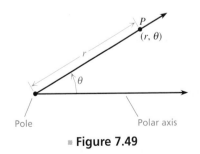

■ **Figure 7.49**

Since we are so familiar with rectangular coordinates, we retain the x- and y-axes when using polar coordinates. The pole is placed at the origin, and the polar

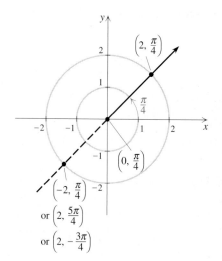

■ **Figure 7.50**

axis is placed along the positive *x*-axis. The angle θ is any angle (in degrees or radians) in standard position whose terminal side contains the point. As usual, θ is positive for a counterclockwise rotation and negative for a clockwise rotation. In polar coordinates, *r* can be any real number. For example, the ordered pair $(2, \pi/4)$ represents the point that lies two units from the origin on the terminal side of the angle $\pi/4$. The point $(0, \pi/4)$ is at the origin. The point $(-2, \pi/4)$ lies two units from the origin on the line through the terminal side of $\pi/4$ but in the direction opposite to $(2, \pi/4)$. See Fig. 7.50. Any ordered pair in polar coordinates names a single point, but the coordinates of a point in polar coordinates are not unique. For example, $(-2, \pi/4)$, $(2, 5\pi/4)$, and $(2, -3\pi/4)$ all name the same point.

Example 1 Plotting points in polar coordinates

Plot the points whose polar coordinates are $(2, 5\pi/6)$, $(-3, \pi)$, $(1, -\pi/2)$, and $(-1, 450°)$.

Solution

The terminal side of $5\pi/6$ lies in the second quadrant, so $(2, 5\pi/6)$ is two units from the origin along this ray. The terminal side of the angle π points in the direction of the negative *x*-axis, but since the first coordinate of $(-3, \pi)$ is negative, the point is located three units in the direction opposite to the direction of the ray. So $(-3, \pi)$ lies three units from the origin on the positive *x*-axis. The point $(-3, \pi)$ is the same point as $(3, 0)$. The terminal side of $-\pi/2$ lies on the negative *y*-axis. So $(1, -\pi/2)$ lies one unit from the origin on the negative *y*-axis. The terminal side of $450°$ lies on the positive *y*-axis. Since the first coordinate of $(-1, 450°)$ is negative, the point is located one unit in the opposite direction. The point $(-1, 450°)$ is the same as $(1, -\pi/2)$. All points are shown in Fig. 7.51.

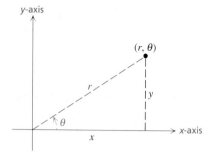

■ **Figure 7.51**

Try This. Plot $(2, 3\pi/4)$ and $(-1, \pi/2)$ in polar coordinates. ■

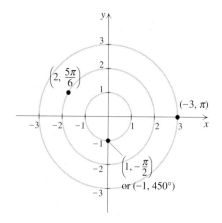

■ **Figure 7.52**

Polar-Rectangular Conversions

To graph equations in polar coordinates, we must be able to switch the coordinates of a point in either system to the other. Suppose that (r, θ) is a point in the first quadrant with $r > 0$ and θ acute, as shown in Fig. 7.52. If (x, y) is the same point in rectangular coordinates, then *x* and *y* are the lengths of the legs of the right triangle shown in Fig. 7.52. Using the Pythagorean theorem and trigonometric ratios, we have

$$x^2 + y^2 = r^2, \quad x = r \cos \theta, \quad y = r \sin \theta, \quad \text{and} \quad \tan \theta = \frac{y}{x}.$$

It can be shown that these equations hold for any point (r, θ), except that $\tan \theta$ is undefined if $x = 0$. The following rules for converting from one system to the other follow from these relationships.

Polar-Rectangular Conversion Rules

To convert (r, θ) to rectangular coordinates (x, y), use

$$x = r \cos \theta \quad \text{and} \quad y = r \sin \theta.$$

To convert (x, y) to polar coordinates (r, θ), use

$$r = \sqrt{x^2 + y^2}$$

and any angle θ in standard position whose terminal side contains (x, y).

There are several ways to find an angle θ whose terminal side contains (x, y). One way to find θ (provided $x \neq 0$) is to find an angle that satisfies $\tan \theta = y/x$ and goes through (x, y). Remember that the angle $\tan^{-1}(y/x)$ is between $-\pi/2$ and $\pi/2$ and will go through (x, y) only if $x > 0$. If $x = 0$, then the point (x, y) is on the y-axis, and you can use $\theta = \pi/2$ or $\theta = -\pi/2$ as appropriate.

Example **2** **Polar-rectangular conversion**

a. Convert $(6, 210°)$ to rectangular coordinates.
b. Convert the rectangular coordinates $(-3, 6)$ to polar coordinates.

Solution

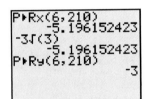

■ **Figure 7.53**

a. Use $r = 6$, $\cos 210° = -\sqrt{3}/2$, and $\sin 210° = -1/2$ in the formulas $x = r \cos \theta$ and $y = r \sin \theta$:

$$x = 6\left(-\frac{\sqrt{3}}{2}\right) = -3\sqrt{3} \quad \text{and} \quad y = 6\left(-\frac{1}{2}\right) = -3$$

So $(6, 210°)$ in polar coordinates is $\left(-3\sqrt{3}, -3\right)$ in rectangular coordinates.

You can check this result with a graphing calculator, as shown in Fig. 7.53. □

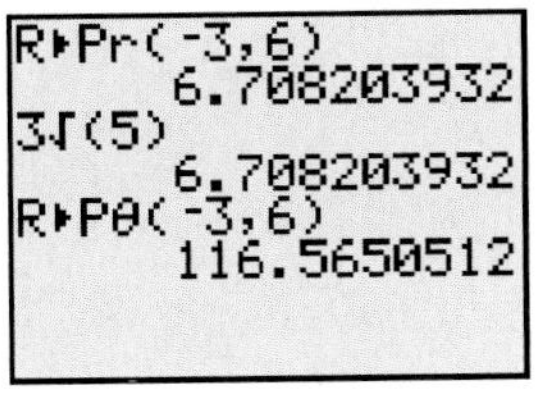

b. To convert $(-3, 6)$ to polar coordinates, find r:

$$r = \sqrt{(-3)^2 + 6^2} = 3\sqrt{5}$$

■ **Figure 7.54**

Use a calculator to find $\tan^{-1}(-2) \approx -63.4°$. To get an angle whose terminal side contains $(-3, 6)$, use $\theta = 180° - 63.4° = 116.6°$. So $(-3, 6)$ in polar coordinates is $\left(3\sqrt{5}, 116.6°\right)$. Since there are infinitely many representations for any point in polar coordinates, this answer is not unique. In fact, another possibility is $\left(-3\sqrt{5}, -63.4°\right)$.

You can check this result with a graphing calculator, as shown in Fig. 7.54.

Try This. Convert $(3, 45°)$ to rectangular coordinates and convert $\left(-2, 2\sqrt{3}\right)$ to polar coordinates. ■

Polar Equations

An equation in two variables (typically x and y) that is graphed in the rectangular coordinate system is called a **rectangular** or **Cartesian equation.** An equation in two variables (typically r and θ) that is graphed in the polar coordinate system is called a **polar equation.** Certain polar equations are easier to graph than the equivalent

Cartesian equations. We can graph a polar equation in the same way that we graph a rectangular equation, that is, we can simply plot enough points to get the shape of the graph. However, since most of our polar equations involve trigonometric functions, finding points on these curves can be tedious. A graphing calculator can be used to great advantage here.

Example 3 Graphing a polar equation

Sketch the graph of the polar equation $r = 2 \cos \theta$.

Solution

If $\theta = 0°$, then $r = 2 \cos 0° = 2$. So, the ordered pair $(2, 0°)$ is on the graph. If $\theta = 30°$, then $r = 2 \cos 30° = \sqrt{3}$, and $\left(\sqrt{3}, 30°\right)$ is on the graph. These ordered pairs and several others that satisfy the equation are listed in the following table:

θ	0°	30°	45°	60°	90°	120°	135°	150°	180°
r	2	$\sqrt{3}$	$\sqrt{2}$	1	0	-1	$-\sqrt{2}$	$-\sqrt{3}$	-2

Plot these points and draw a smooth curve through them to get the graph shown in Fig. 7.55. If θ is larger than $180°$ or smaller than $0°$, we get different ordered pairs, but they are all located on the curve drawn in Fig. 7.55. For example, $\left(-\sqrt{3}, 210°\right)$ satisfies $r = 2 \cos \theta$, but it has the same location as $\left(\sqrt{3}, 30°\right)$.

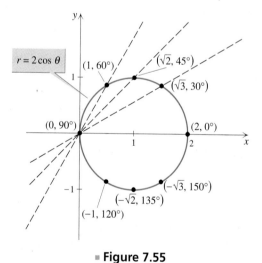

■ **Figure 7.55**

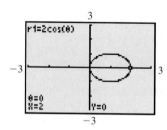

■ **Figure 7.56**

⊞ To check this graph with a calculator, set the mode to polar and enter $r = 2 \cos \theta$. The graph in Fig.7.56 supports the graph in Fig. 7.55.

Try This. Graph $r = 4 \sin \theta$ in polar coordinates. ■

The graph $r = 2 \cos \theta$ in Fig. 7.55 looks like a circle. To verify that it is a circle, we can convert the polar equation to an equivalent rectangular equation because we know the form of the equation of a circle in rectangular coordinates. This conversion is done in Example 6.

In the next example, a simple polar equation produces a curve that is not usually graphed when studying rectangular equations because the equivalent rectangular equation is quite complicated.

Example 4 Graphing a polar equation

Sketch the graph of the polar equation $r = 3 \sin 2\theta$.

Solution

The ordered pairs in the following table satisfy the equation $r = 3 \sin 2\theta$. The values of r are rounded to the nearest tenth.

θ	0°	15°	30°	45°	60°	90°	135°	180°	225°	270°	315°	360°
r	0	1.5	2.6	3	2.6	0	−3	0	3	0	−3	0

As θ varies from 0° to 90°, the value of r goes from 0 to 3, then back to 0, creating a loop in quadrant I. As θ varies from 90° to 180°, the value of r goes from 0 to −3 then back to 0, creating a loop in quadrant IV. As θ varies from 180° to 270°, a loop in quadrant III is formed; and as θ varies from 270° to 360°, a loop in quadrant II is formed. If θ is chosen greater than 360° or less than 0°, we get the same points over and over because the sine function is periodic. The graph of $r = 3 \sin 2\theta$ is shown in Fig. 7.57. The graph is called a **four-leaf rose.**

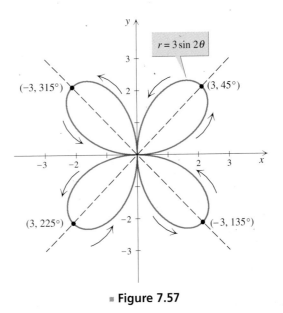

■ **Figure 7.57**

With a graphing calculator you can make a table of ordered pairs, as shown in Fig. 7.58. Make this table yourself and scroll through the table to see how the radius oscillates between 3 and −3 as the angle varies. The calculator graph of $r = 3 \sin(2\theta)$

in polar mode with θ between $0°$ and $360°$ is shown in Fig. 7.59. This graph supports the graph shown in Fig. 7.57.

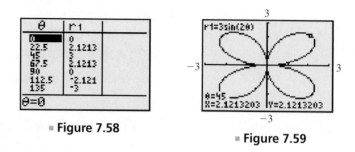

■ **Figure 7.58**

■ **Figure 7.59**

Try This. Graph $r = \cos 2\theta$ in polar coordinates. ■

Example 5 Graphing a polar equation

Sketch the graph of the polar equation $r = \theta$, where θ is in radians and $\theta \geq 0$.

Solution

The ordered pairs in the following table satisfy $r = \theta$. The values of r are rounded to the nearest tenth.

θ	0	$\pi/4$	$\pi/2$	$3\pi/4$	π	$3\pi/2$	2π	3π
r	0	0.8	1.6	2.4	3.1	4.7	6.3	9.4

The graph of $r = \theta$, called **the spiral of Archimedes,** is shown as a smooth curve in Fig. 7.60. As the value of θ increases, the value of r increases, causing the graph to

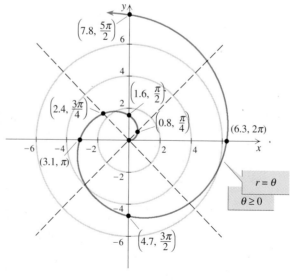

■ **Figure 7.60**

spiral out from the pole. There is no repetition of points as there was in Examples 3 and 4, because no periodic function is involved.

With a graphing calculator you can make a table of ordered pairs for $r = \theta$, as shown in Fig. 7.61. The calculator graph of $r = \theta$ in polar coordinates for θ ranging from 0 through 40 radians is shown in Fig. 7.62. This graph shows much more of the spiral than Fig. 7.60. It supports the conclusion that the graph of $r = \theta$ is a spiral.

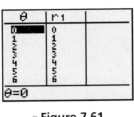

■ **Figure 7.61**

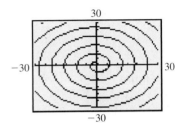

■ **Figure 7.62**

Try This. Graph $r = -\theta$ for θ in radians and $\theta \geq 0$. ■

The Function Gallery on page 637 shows the graphs of several types of polar equations.

Converting Equations

We know that certain types of rectangular equations have graphs that are particular geometric shapes, such as lines, circles, and parabolas. We can use our knowledge of equations in rectangular coordinates with equations in polar coordinates (and vice versa) by converting the equations from one system to the other. For example, we can determine whether the graph of $r = 2 \cos \theta$ in Example 3 is a circle by finding the equivalent Cartesian equation and deciding whether it is the equation of a circle. When converting from one system to another, we use the relationships

$$x^2 + y^2 = r^2, \qquad x = r \cos \theta, \qquad \text{and} \qquad y = r \sin \theta.$$

Example **6** **Converting a polar equation to a rectangular equation**

Write an equivalent rectangular equation for the polar equation $r = 2 \cos \theta$.

Solution

First multiply each side of $r = 2 \cos \theta$ by r to get $r^2 = 2r \cos \theta$. Now eliminate r and θ by making substitutions using $r^2 = x^2 + y^2$ and $x = r \cos \theta$:

$$r^2 = 2r \cos \theta$$

$$x^2 + y^2 = 2x$$

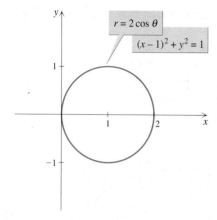

- **Figure 7.63**

From Section 1.3, the standard equation of a circle with center (h, k) and radius r is $(x - h)^2 + (y - k)^2 = r^2$. Complete the square to get $x^2 + y^2 = 2x$ into the standard form of the equation of a circle:

$$x^2 - 2x + y^2 = 0$$
$$x^2 - 2x + 1 + y^2 = 0 + 1$$
$$(x - 1)^2 + y^2 = 1$$

Since we recognize the rectangular equation as the equation of a circle centered at $(1, 0)$ with radius 1, the graph of $r = 2 \cos \theta$ shown in Fig. 7.63 is a circle centered at $(1, 0)$ with radius 1.

Try This. Convert $r = 3 \sin \theta$ to a rectangular equation. ■

In the next example we convert the rectangular equation of a line and circle into polar coordinates.

Example **7** **Converting a rectangular equation to a polar equation**

For each rectangular equation, write an equivalent polar equation.

a. $y = 3x - 2$ **b.** $x^2 + y^2 = 9$

Solution

a. Substitute $x = r \cos \theta$ and $y = r \sin \theta$, and then solve for r:

$$y = 3x - 2$$
$$r \sin \theta = 3r \cos \theta - 2$$
$$r \sin \theta - 3r \cos \theta = -2$$
$$r(\sin \theta - 3 \cos \theta) = -2$$
$$r = \frac{-2}{\sin \theta - 3 \cos \theta}$$

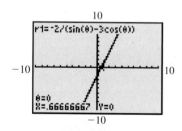

- **Figure 7.64**

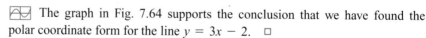 The graph in Fig. 7.64 supports the conclusion that we have found the polar coordinate form for the line $y = 3x - 2$. □

b. Substitute $r^2 = x^2 + y^2$ to get polar coordinates:

$$x^2 + y^2 = 9$$
$$r^2 = 9$$
$$r = \pm 3$$

A polar equation for a circle of radius 3 centered at the origin is $r = 3$. The equation $r = -3$ is also a polar equation for the same circle.

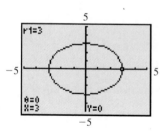 The graph in Fig. 7.65 supports the conclusion that $r = 3$ is a circle in polar coordinates. Graph $r = 3$ on your calculator, using a viewing window that makes the circle look round.

- **Figure 7.65**

Try This. Convert $y = -2x + 5$ into a polar equation. ■

In Example 7 we saw that a straight line has a rather simple equation in rectangular coordinates but a more complicated equation in polar coordinates. A circle centered at the origin has a very simple equation in polar coordinates but a more complicated equation in rectangular coordinates. The graphs of simple polar equations are typically circular or somehow "centered" at the origin.

Function Gallery: Functions in Polar Coordinates

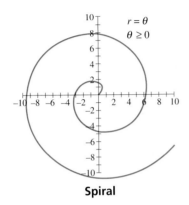

$r = \theta$
$\theta \geq 0$

Spiral

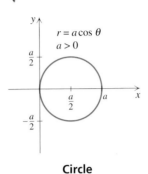

$r = a\cos\theta$
$a > 0$

Circle

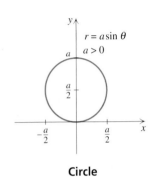

$r = a\sin\theta$
$a > 0$

Circle

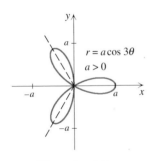

$r = a\cos 3\theta$
$a > 0$

Three-Leaf Rose

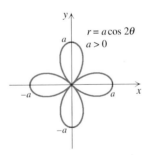

$r = a\cos 2\theta$
$a > 0$

Four-Leaf Rose

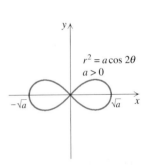

$r^2 = a\cos 2\theta$
$a > 0$

Lemniscate

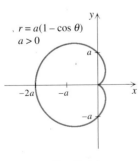

$r = a(1 - \cos\theta)$
$a > 0$

Cardioid

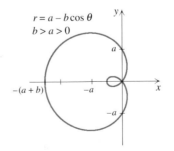

$r = a - b\cos\theta$
$b > a > 0$

Limaçon

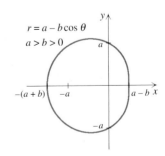

$r = a - b\cos\theta$
$a > b > 0$

Limaçon

For Thought

True or False? Explain.

1. The distance of the point (r, θ) from the origin depends only on r.

2. The distance of the point (r, θ) from the origin is r.

3. The ordered pairs $(2, \pi/4)$, $(2, -3\pi/4)$, and $(-2, 5\pi/4)$ all represent the same point in polar coordinates.

4. The equations relating rectangular and polar coordinates are $x = r \sin \theta$, $y = r \cos \theta$, and $x^2 + y^2 = r^2$.

5. The point $(-4, 225°)$ in polar coordinates is $(2\sqrt{2}, 2\sqrt{2})$ in rectangular coordinates.

6. The graph of $0 \cdot r + \theta = \pi/4$ in polar coordinates is a straight line.

7. The graphs of $r = 5$ and $r = -5$ are identical.

8. The ordered pairs $(-\sqrt{2}/2, \pi/3)$ and $(\sqrt{2}/2, \pi/3)$ satisfy $r^2 = \cos 2\theta$.

9. The graph of $r = 1/\sin \theta$ is a vertical line.

10. The graphs of $r = \theta$ and $r = -\theta$ are identical.

7.6 Exercises

Find polar coordinates for each given point, using radian measure for the angle.

1.

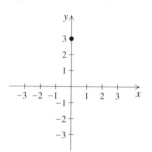

2.

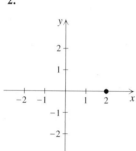

3.

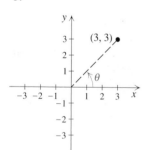

4.

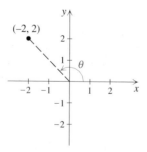

Plot the points whose polar coordinates are given.

5. $(2, 0°)$ **6.** $(-3, 0°)$ **7.** $(0, 35°)$

8. $(0, 90°)$ **9.** $(3, \pi/6)$ **10.** $(2, \pi/4)$

11. $(-2, 2\pi/3)$ **12.** $(-1, \pi/6)$ **13.** $(2, -\pi/4)$

14. $(1, -2\pi/3)$ **15.** $(3, -225°)$ **16.** $(2, -180°)$

17. $(-2, 45°)$ **18.** $(-3, 30°)$ **19.** $(4, 390°)$

20. $(3, 13\pi/6)$

Convert the polar coordinates of each point to rectangular coordinates.

21. $(4, 0°)$ **22.** $(-5, 0°)$ **23.** $(0, \pi/4)$

24. $(0, -\pi)$ **25.** $(1, \pi/6)$ **26.** $(2, \pi/4)$

27. $(-3, 3\pi/2)$ **28.** $(-2, 2\pi)$

29. $(\sqrt{2}, 135°)$ **30.** $(\sqrt{3}, 150°)$

31. $(-\sqrt{6}, -60°)$ **32.** $(-\sqrt{2}/2, -45°)$

Convert the rectangular coordinates of each point to polar coordinates. Use degrees for θ.

33. $(\sqrt{3}, 3)$ **34.** $(4, 4)$ **35.** $(-2, 2)$

36. $(-2, 2\sqrt{3})$ **37.** $(0, 2)$ **38.** $(-2, 0)$

39. $(-3, -3)$ **40.** $(2, -2)$ **41.** $(1, 4)$

42. $(-2, 3)$ **43.** $(\sqrt{2}, -2)$ **44.** $(-2, -\sqrt{3})$

Convert the polar coordinates of each point to rectangular coordinates rounded to the nearest hundredth.

45. $(4, 26°)$ **46.** $(-5, 33°)$ **47.** $(2, \pi/7)$

48. $(3, 2\pi/9)$ **49.** $(-2, 1.1)$ **50.** $(6, 2.3)$

Convert the rectangular coordinates of each point to polar coordinates. Round r to the nearest tenth and θ to the nearest tenth of a degree.

51. $(4, 5)$ **52.** $(-5, 3)$

53. $(-2, -7)$ **54.** $(3, -8)$

Sketch the graph of each polar equation.

55. $r = 2 \sin \theta$ **56.** $r = 3 \cos \theta$

57. $r = 3 \cos 2\theta$ **58.** $r = -2 \sin 2\theta$

59. $r = 2\theta$ for θ in radians

60. $r = \theta$ for $\theta \le 0$ and θ in radians

61. $r = 1 + \cos \theta$ (cardioid)

62. $r = 1 - \cos \theta$ (cardioid)

63. $r^2 = 9 \cos 2\theta$ (lemniscate)

64. $r^2 = 4 \sin 2\theta$ (lemniscate)

65. $r = 4 \cos 2\theta$ (four-leaf rose)

66. $r = 3 \sin 2\theta$ (four-leaf rose)

67. $r = 2 \sin 3\theta$ (three-leaf rose)

68. $r = 4 \cos 3\theta$ (three-leaf rose)

69. $r = 1 + 2 \cos \theta$ (limaçon) **70.** $r = 2 + \cos \theta$ (limaçon)

71. $r = 3.5$ **72.** $r = -5$

73. $\theta = 30°$ **74.** $\theta = 3\pi/4$

For each polar equation, write an equivalent rectangular equation.

75. $r = 4 \cos \theta$ **76.** $r = 2 \sin \theta$ **77.** $r = \dfrac{3}{\sin \theta}$

78. $r = \dfrac{-2}{\cos \theta}$ **79.** $r = 3 \sec \theta$ **80.** $r = 2 \csc \theta$

81. $r = 5$ **82.** $r = -3$ **83.** $\theta = \dfrac{\pi}{4}$

84. $\theta = 0$ **85.** $r = \dfrac{2}{1 - \sin \theta}$ **86.** $r = \dfrac{3}{1 + \cos \theta}$

For each rectangular equation, write an equivalent polar equation.

87. $x = 4$ **88.** $y = -6$

89. $y = -x$ **90.** $y = x\sqrt{3}$

91. $x^2 = 4y$ **92.** $y^2 = 2x$

93. $x^2 + y^2 = 4$ **94.** $2x^2 + y^2 = 1$

95. $y = 2x - 1$ **96.** $y = -3x + 5$

97. $x^2 + (y - 1)^2 = 1$ **98.** $(x + 1)^2 + y^2 = 4$

Graph each pair of polar equations on the same screen of your calculator and use the trace feature to estimate the polar coordinates of all points of intersection of the curves. Check your calculator manual to see how to graph polar equations on your calculator.

99. $r = 1, r = 2 \sin 3\theta$ **100.** $r = \sin \theta, r = \sin 2\theta$

101. $r = 3 \sin 2\theta, r = 1 - \cos \theta$

102. $r = 3 \sin 4\theta, r = 2$

For Writing/Discussion

103. Explain why θ must be radian measure for the equation $r = 2\theta$, but θ can be in radians or degrees for the equation $r = 2 \cos \theta$.

104. Show that a polar equation for the straight line $y = mx$ is $\theta = \tan^{-1} m$.

Thinking Outside the Box LXV

Laying Pipe A circular pipe with radius 1 is placed in a v-shaped trench whose sides form an angle of θ radians. In the cross section shown here, the pipe touches the sides of the trench at points A and B.

a. Find the area inside the circle and below the line segment AB in terms of θ (the blue area).

b. Find the area below the circle and inside the trench in terms of θ (the red area).

c. If θ is nearly equal to π, then the blue area and the red area are both very small. If θ is equal to π both areas are zero. As θ approaches π there is a limit to the ratio of the blue area to the red area. Use a calculator to determine this limit.

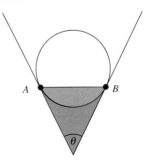

∎ **Figure for Thinking Outside the Box LXV**

7.6 Pop Quiz

1. If the polar coordinates of a point are $(-1, 15\pi/4)$, then in which quadrant does the point lie?

2. Convert $(4, 150°)$ to rectangular coordinates.

3. Convert the rectangular coordinates $(-2, 2)$ to polar coordinates using radians for the angle.

4. Convert the polar equation $r = 4\cos\theta$ into a rectangular equation.

5. Find the center and radius of the circle $r = -16\sin\theta$.

6. Convert $y = x + 1$ into polar coordinates.

7.7

Parametric Equations

We know how to graph points in the plane using the rectangular coordinate system and using polar coordinates. Points in the plane can also be located using parametric equations. We have already used parametric equations when we gave the x- and y-coordinates of a projectile as functions of time in Chapter 6. In this section we will study parametric equations in more detail.

Graphs of Parametric Equations

If $f(t)$ and $g(t)$ are functions of t, where t is in some interval of real numbers, then the equations $x = f(t)$ and $y = g(t)$ are called **parametric equations.** The variable t is called the **parameter** and the graph of the parametric equations is said to be defined **parametrically.** If the parameter is thought of as time, then we know when each point of the graph is plotted. If no interval is specified for t, then t is assumed to be any real number for which both $f(t)$ and $g(t)$ are defined.

Example **1** Graphing a line segment

Graph the parametric equations $x = 3t - 2$ and $y = t + 1$ for t in the interval $[0, 3]$. Determine the domain (the set of x-coordinates) and the range (the set of y-coordinates) for the function or relation that you graphed.

Solution

Make a table of ordered pairs corresponding to values of t between 0 and 3:

t	x	y
0	-2	1
1	1	2
2	4	3
3	7	4

Since $x = 3t - 2$, we have $t = \frac{x + 2}{3}$. Since t is in the interval $[0, 3]$, we have $0 \le \frac{x + 2}{3} \le 3$. Solving for x, we get $-2 \le x \le 7$. So x is in the interval $[-2, 7]$. The graph is not the whole line. It is the line segment with endpoints $(-2, 1)$ and

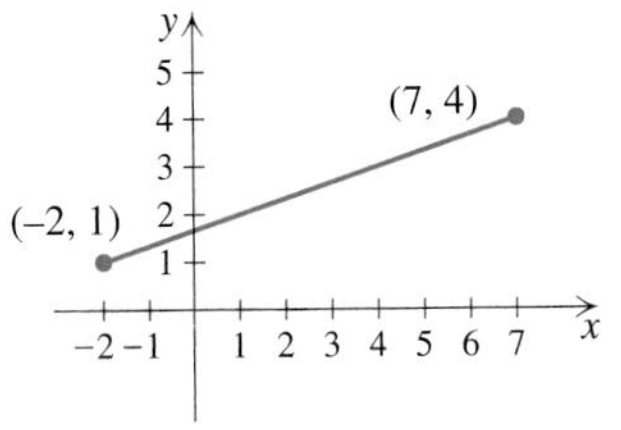

(7, 4) shown in Fig. 7.66. We use solid dots at the end points to show that they are included in the graph. The domain is the interval $[-2, 7]$ and the range is the interval $[1, 4]$.

To check Example 1 with a graphing calculator, set your calculator to parametric mode and enter the parametric equations, as shown in Fig. 7.67(a). Set the limits on the viewing window and the parameter as in Fig. 7.67(b). The graph is shown in Fig. 7.67(c). □

■ **Figure 7.66**

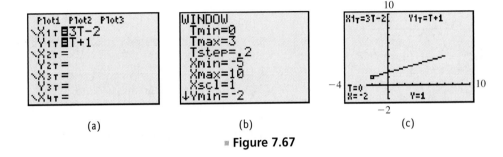

(a)   (b)   (c)

■ **Figure 7.67**

Try This. Graph $x = t + 5$ and $y = 2t - 1$ for t in $[0, 5]$.   ■

Eliminating the Parameter

In rectangular coordinates, we know that $y = mx + b$ is a line, $(x - h)^2 + (y - k)^2 = r^2$ is a circle, and $y = ax^2 + bx + c$ is a parabola. Since we have little experience with parametric equations, it may not be obvious when a system of parametric equations has a familiar graph. However, it is often possible to identify a graph by eliminating the parameter and writing an equation involving only x and y.

Example **2** **Eliminating the parameter**

Eliminate the parameter and identify the graph of the parametric equations. Determine the domain (the set of x-coordinates) and the range (the set of y-coordinates).

a. $x = 3t - 2, y = t + 1, -\infty < t < \infty$
b. $x = 7 \sin t, y = 7 \cos t, -\infty < t < \infty$

Solution

a. Solve $y = t + 1$ for t to get $t = y - 1$. Now replace t in the other equation by $y - 1$:

$$x = 3(y - 1) - 2$$
$$x = 3y - 5$$
$$3y = x + 5$$
$$y = \frac{1}{3}x + \frac{5}{3}$$

After eliminating the parameter, we get $y = \frac{1}{3}x + \frac{5}{3}$, which is the equation of a line with slope 1/3 and y-intercept $(0, 5/3)$. Because $-\infty < t < \infty$ and both x and y are linear functions of t, the doman of $y = \frac{1}{3}x + \frac{5}{3}$ is $(-\infty, \infty)$ and its range is $(-\infty, \infty)$. So the graph is the entire line. Note that in Example 1 these same parametric equations with a different interval for t determined a line segment.

b. The simplest way to eliminate the parameter in this case is to use the trigonometric identity $\sin^2(\theta) + \cos^2(\theta) = 1$. Because $\sin t = x/7$ and $\cos t = y/7$, we have $(x/7)^2 + (y/7)^2 = 1$ or $x^2 + y^2 = 49$. So the graph is a circle centered at the origin with radius 7. The domain of this relation is $[-7, 7]$ and the range is $[-7, 7]$.

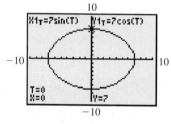

■ **Figure 7.68**

You can check this conclusion with a graphing calculator, as shown in Fig. 7.68.

Try This. Eliminate the parameter and identify the graph $x = 4t - 9$ and $y = -t + 1$ for t in $(-\infty, \infty)$. Determine the domain and range. ■

Writing Parametric Equations

Because a nonvertical straight line has a unique slope and y-intercept, it has a unique equation in slope-intercept form. However, a polar equation for a curve is not unique and neither are parametric equations for a curve. For example, consider the line $y = 2x + 1$. For parametric equations we could let $x = t$ and $y = 2t + 1$. We could also let $x = 4t$ and $y = 8t + 1$. We could even write $x = t^3 + 7$ and $y = 2t^3 + 15$. Each of these pairs of parametric equations produces the line $y = 2x + 1$.

Example **3** **Writing parametric equations for a line segment**

Write parametric equations for the line segment between $(1, 3)$ and $(5, 8)$ for t in the interval $[0, 2]$, with $t = 0$ corresponding to $(1, 3)$ and $t = 2$ corresponding to $(5, 8)$.

Solution

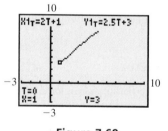

■ **Figure 7.69**

We can make both parametric equations linear functions of t. If $x = mt + b$ and $t = 0$ corresponds to $x = 1$, then $1 = m \cdot 0 + b$ and $b = 1$. So $x = mt + 1$. If $t = 2$ corresponds to $x = 5$, then $5 = m \cdot 2 + 1$ or $m = 2$. So we have $x = 2t + 1$. Using similar reasoning for the y-coordinates we get $y = 2.5t + 3$.

You can use a graphing calculator to check, as shown in Fig. 7.69.

Try This. Write parametric equations for the line segment between $(1, 2)$ and $(8, 10)$ with t in the interval $[0, 1]$, where $t = 0$ corresponds to $(1, 2)$ and $t = 1$ corresponds to $(8, 10)$. ■

While it may not seem obvious how to write parametric equations for a particular rectangular equation, there is a simple way to find parametric equations for a polar equation $r = f(\theta)$. Because $x = r \cos \theta$ and $y = r \sin \theta$, we can substitute $f(\theta)$ for r and write $x = f(\theta) \cos \theta$ and $y = f(\theta) \sin \theta$. In this case the parameter is θ and we have parametric equations for the polar curve. The parametric equations for a polar curve can be used to graph a polar curve on a calculator that is capable of handling parametric equations but not polar equations.

Example **4** **Converting a polar equation to parametric equations**

Write parametric equations for the polar equation $r = 1 - \cos\theta$.

Solution

Replace r by $1 - \cos\theta$ in the equations $x = r\cos\theta$ and $y = r\sin\theta$ to get $x = (1 - \cos\theta)\cos\theta$ and $y = (1 - \cos\theta)\sin\theta$. We know from Section 7.6 that the graph of $r = 1 - \cos\theta$ is the cardioid shown in Fig. 7.70. So the graph of these parametric equations is the same cardioid.

The calculator graph of the parametric equations in Fig. 7.71 appears to be a cardioid and supports the graph shown in Fig. 7.70.

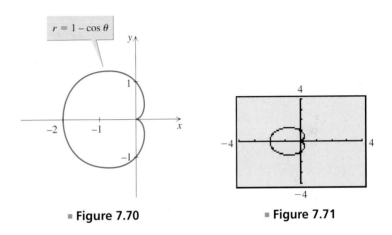

■ **Figure 7.70**   ■ **Figure 7.71**

Try This. Write parametric equations for the polar equation $r = 3\cos\theta$.   ■

For Thought

True or False? Explain.

1. If $x = 3t + 1$ and $y = 4t - 2$, then t is the variable and x and y are the parameters.

2. Parametric equations are graphed in the rectangular coordinate system.

3. The graph of $x = 0.5t$ and $y = 2t + 1$ is a straight line with slope 4.

4. The graph of $x = \cos t$ and $y = \sin t$ is a sine wave.

5. The graph of $x = 3t + 1$ and $y = 6t - 1$ for $0 \le t \le 3$ includes the point $(2, 1)$.

6. The graph of $x = w^2 - 3$ and $y = w + 5$ for $-2 < w < 2$ includes the point $(1, 7)$.

7. The graph of $x = e^t$ and $y = e^t$ lies entirely within the first quadrant.

8. The graph of $x = -\sin t$ and $y = \cos t$ for $0 < t < \pi/2$ lies entirely within the second quadrant.

9. The parametric equations $x = e^t$ and $y = e^t$ have the same graph as $x = \ln t$ and $y = \ln t$.

10. The polar equation $r = \cos\theta$ can be graphed using the parametric equations $x = \cos^2\theta$ and $y = \cos\theta\sin\theta$.

7.7 Exercises

Complete the table that accompanies each pair of parametric equations.

1. $x = 4t + 1$, $y = t - 2$, for $0 \le t \le 3$

2. $x = 3 - t$, $y = 2t + 5$, for $2 \le t \le 7$

t	x	y
0		
1		
	7	
		1

t	x	y
2		
3		
	-2	
		19

3. $x = t^2$, $y = 3t - 1$, for $1 \le t \le 5$

4. $x = \sqrt{t}$, $y = t + 4$, for $0 \le t \le 9$

t	x	y
1		
2.5		
	5	
		11
	25	

t	x	y
0		
2		
4		
		12
	3	

Graph each pair of parametric equations in the rectangular coordinate system. Determine the domain (the set of x-coordinates) and the range (the set of y-coordinates).

5. $x = 3t - 2$, $y = t + 3$, for $0 \le t \le 4$

6. $x = 4 - 3t$, $y = 3 - t$, for $1 \le t \le 3$

7. $x = t - 1$, $y = t^2$, for t in $(-\infty, \infty)$

8. $x = t - 3$, $y = 1/t$, for t in $(-\infty, \infty)$

9. $x = \sqrt{w}$, $y = \sqrt{1 - w}$, for $0 < w < 1$

10. $x = \ln t$, $y = t + 3$, for $-2 < t < 2$

11. $x = \cos t$, $y = \sin t$

12. $x = 0.5t$, $y = \sin t$

Eliminate the parameter and identify the graph of each pair of parametric equations. Determine the domain (the set of x-coordinates) and the range (the set of y-coordinates).

13. $x = 4t - 5$, $y = 3 - 4t$

14. $x = 5t - 1$, $y = 4t + 6$

15. $x = -4 \sin 3t$, $y = 4 \cos 3t$

16. $x = 2 \sin t \cos t$, $y = 3 \sin 2t$

17. $x = t/4$, $y = e^t$

18. $x = t - 5$, $y = t^2 - 10t + 25$

19. $x = \tan t$, $y = 2 \tan t + 3$

20. $x = \tan t$, $y = -\tan^2 t + 3$

Write a pair of parametric equations that will produce the indicated graph. Answers may vary.

21. The line segment starting at $(2, 3)$ with $t = 0$ and ending at $(5, 9)$ with $t = 2$

22. The line segment starting at $(-2, 4)$ with $t = 3$ and ending at $(5, -9)$ with $t = 7$

23. That portion of the circle $x^2 + y^2 = 4$ that lies in the third quadrant

24. That portion of the circle $x^2 + y^2 = 9$ that lies below the x-axis.

25. The vertical line through $(3, 1)$

26. The horizontal line through $(5, 2)$

27. The circle whose polar equation is $r = 2 \sin \theta$

28. The four-leaf rose whose polar equation is $r = 5 \sin(2\theta)$

Graph the following pairs of parametric equations with the aid of a graphing calculator. These are uncommon curves that would be difficult to describe in rectangular or polar coordinates.

29. $x = \cos 3t$, $y = \sin t$

30. $x = \sin t$, $y = t^2$

31. $x = t - \sin t$, $y = 1 - \cos t$ (cycloid)

32. $x = t - \sin t$, $y = -1 + \cos t$ (inverted cycloid)

33. $x = 4 \cos t - \cos 4t$, $y = 4 \sin t - \sin 4t$ (epicycloid)

34. $x = \sin^3 t$, $y = \cos^3 t$ (hypocycloid)

The following problems involve the parametric equations for the path of a projectile

$$x = v_0(\cos \theta)t \quad \text{and} \quad y = -16t^2 + v_0(\sin \theta)t + h_0,$$

where θ is the angle of inclination of the projectile at the launch, v_0 is the initial velocity of the projectile in feet per second, and h_0 is the initial height of the projectile in feet.

35. An archer shoots an arrow from a height of 5 ft at an angle of inclination of 30° with a velocity of 300 ft/sec. Write the parametric equations for the path of the projectile and sketch the graph of the parametric equations.

36. If the arrow of Exercise 35 strikes a target at a height of 5 ft, then how far is the target from the archer?

37. For how many seconds is the arrow of Exercise 35 in flight?

38. What is the maximum height reached by the arrow in Exercise 35?

Thinking Outside the Box LXVI

Lakefront Property A man-made lake in the shape of a triangle is bounded on each of its sides by a square lot as shown in the figure. The square lots are 8, 13, and 17 acres, respectively. What is the area of the lake in square feet?

HINT One acre is 43,560 square feet.

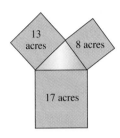

▪ **Figure for Thinking Outside the Box LXVI**

7.7 Pop Quiz

1. The graph of $x = 2t + 5$ and $y = 3t - 7$ for t in [3, 5] is a line segment. What are the endpoints?

2. Eliminate the parameter and identify the graph of $x = 3\cos t$ and $y = 3\sin t$ for $-\infty < t < \infty$.

3. Write parametric equations for the line segment between (0, 1) and (3, 5), where $t = 0$ corresponds to (0, 1) and $t = 4$ corresponds to (3, 5).

Linking Concepts

For Individual or Group Explorations

Distance

We know that the distance d between (x_1, y_1) and (x_2, y_2) in rectangular coordinates is given by the formula

$$d = \sqrt{(x_2 - x_1)^2 + (y_2 - y_1)^2}.$$

In the following exercises you will investigate distance in other coordinate systems.

a) Show that the distance between (r_1, θ_1) and (r_2, θ_2) in polar coordinates is given by

$$d = \sqrt{r_1^2 + r_2^2 - 2r_1r_2\cos(\theta_2 - \theta_1)}.$$

by converting the distance formula in rectangular coordinates into polar coordinates.

b) Find the distance formula given in part (a) without using the distance formula in rectangular coordinates.

(continued on next page)

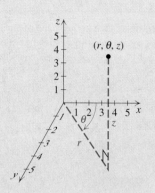

Cylindrical coordinates

c) The *cylindrical coordinate system* for three-dimensional space is (r, θ, z) where r and θ are polar coordinates in the xy-plane and z is the altitude above or below the xy-plane, as shown in the accompanying figure. Show that the distance d between (r_1, θ_1, z_1) and (r_2, θ_2, z_2) is given by

$$d = \sqrt{r_1^2 + r_2^2 - 2r_1r_2\cos(\theta_2 - \theta_1) + (z_1 - z_2)^2}.$$

d) Points on the surface of the earth (a sphere of radius r) are located by two coordinates (α, β), where α is longitude and β is latitude. The distance S between two points on the surface of the earth is the length of an arc through the two points. Show that the distance S between (α_1, β_1) and (α_2, β_2) is given by

$$S = r \cdot \cos^{-1}(\sin \beta_1 \sin \beta_2 + \cos \beta_1 \cos \beta_2 \cos(\alpha_2 - \alpha_1)),$$

assuming that α is the angle west of the prime meridian and β is the angle from the equator (positive in the northern hemisphere and negative in the southern hemisphere).

e) Find the longitude and latitude for Paris, France, and Chicago, Illinois, and use the formula in part (d) to calculate the distance between them. Using the radius of the earth as 3963 miles, you should get a distance of 4140 miles. Explain any discrepancies.

■ ■ ■ Highlights

7.1 The Law of Sines

Law of Sines

In any triangle $\dfrac{\sin \alpha}{a} = \dfrac{\sin \beta}{b} = \dfrac{\sin \gamma}{c}$.

$a = \sqrt{3}, b = 1, c = 2$
$\alpha = \pi/3, \beta = \pi/6, \gamma = \pi/2$
$\dfrac{\sin\left(\frac{\pi}{3}\right)}{\sqrt{3}} = \dfrac{\sin\left(\frac{\pi}{6}\right)}{1} = \dfrac{\sin\left(\frac{\pi}{2}\right)}{2}$

Area of a Triangle

The area of a triangle is one-half the product of any two sides and the sine of the angle between them: $A = \dfrac{1}{2} bc \sin \alpha$

Equilateral triangle with sides of length 4: $A =$
$\dfrac{1}{2} \cdot 4 \cdot 4 \cdot \sin(60°) = 4\sqrt{3}$

7.2 The Law of Cosines

Law of Cosines

If a is the side opposite angle α in any triangle then $a^2 = b^2 + c^2 - 2bc \cos \alpha$.

$a = 4, b = 5, c = 6$
$4^2 = 5^2 + 6^2 - 60\cos \alpha$

Heron's Area Formula

If a, b, and c are the sides of a triangle and $S = (a + b + c)/2$, then

$$A = \sqrt{S(S - a)(S - b)(S - c)}.$$

$S = (4 + 6 + 8)/2 = 9$
$A = \sqrt{9 \cdot 5 \cdot 3 \cdot 1} = \sqrt{135}$

7.3 Vectors

Component Form

The vector with initial point $(0, 0)$ and terminal point (a, b) is denoted $\langle a, b \rangle$.

$\langle 2, 3 \rangle$ has initial point $(0, 0)$ and terminal point $(2, 3)$.

Magnitude	$\lvert \langle a, b \rangle \rvert = \sqrt{a^2 + b^2}$	$\lvert \langle 2, 3 \rangle \rvert = \sqrt{13}$
Scalar Product	If k is a scalar and $\mathbf{A}$ is a vector, then $\lvert k\mathbf{A} \rvert = \lvert k \rvert \cdot \lvert \mathbf{A} \rvert$. If $k > 0$, $k\mathbf{A}$ has same direction as $\mathbf{A}$. If $k < 0$, $k\mathbf{A}$ has opposite direction as $\mathbf{A}$.	$4\langle 1, 2 \rangle = \langle 4, 8 \rangle$ $-4\langle 1, 2 \rangle = \langle -4, -8 \rangle$
Resultant of A and B	If $\mathbf{B}$ is placed so that its initial point coincides with the terminal point of $\mathbf{A}$, then $\mathbf{A} + \mathbf{B}$ is the vector from the initial point of $\mathbf{A}$ to the terminal point of $\mathbf{B}$.	$\mathbf{A} = \langle 1, 4 \rangle, \mathbf{B} = \langle 2, 5 \rangle$ $\mathbf{A} + \mathbf{B} = \langle 3, 9 \rangle$
Dot Product	$\langle a, b \rangle \cdot \langle c, d \rangle = ac + bd$	$\langle 1, 4 \rangle \cdot \langle 2, 5 \rangle = 22$
Angle Between Two Vectors	If α is the angle between nozero vectors $\mathbf{A}$ and $\mathbf{B}$, then $\cos \alpha = \dfrac{\mathbf{A} \cdot \mathbf{B}}{\lvert \mathbf{A} \rvert \lvert \mathbf{B} \rvert}$.	$\mathbf{A} = \langle 1, 4 \rangle, \mathbf{B} = \langle 2, 5 \rangle$ $\cos \alpha = \dfrac{22}{\sqrt{17}\sqrt{29}}$

7.4 Trigonometric Form of Complex Numbers

Absolute Value	$\lvert a + bi \rvert = \sqrt{a^2 + b^2}$	$\lvert 2 + 3i \rvert = \sqrt{13}$
Trigonometric Form	$z = a + bi = r(\cos \theta + i \sin \theta)$, where $a = r \cos \theta$, $b = r \sin \theta$, $r = \sqrt{a^2 + b^2}$ and θ is an angle in standard position whose terminal side contains (a, b).	$z = \sqrt{3} + i$ $= 2\left(\cos \dfrac{\pi}{6} + i \sin \dfrac{\pi}{6}\right)$
Multiplying and Dividing	$z_1 = r_1(\cos \theta_1 + i \sin \theta_1)$ $z_2 = r_2(\cos \theta_2 + i \sin \theta_2)$ $z_1 z_2 = r_1 r_2 [\cos(\theta_1 + \theta_2) + i \sin(\theta_1 + \theta_2)]$ $\dfrac{z_1}{z_2} = \dfrac{r_1}{r_2}[\cos(\theta_1 - \theta_2) + i \sin(\theta_1 - \theta_2)]$	$z_1 = 8\left(\cos \dfrac{\pi}{3} + i \sin \dfrac{\pi}{3}\right)$ $z_2 = 2\left(\cos \dfrac{\pi}{6} + i \sin \dfrac{\pi}{6}\right)$ $z_1 z_2 = 16\left(\cos \dfrac{\pi}{2} + i \sin \dfrac{\pi}{2}\right)$ $\dfrac{z_1}{z_2} = 4\left(\cos \dfrac{\pi}{6} + i \sin \dfrac{\pi}{6}\right)$

7.5 Powers and Roots of Complex Numbers

De Moivre's Theorem	If $z = r(\cos \theta + i \sin \theta)$ and n is a positive integer, then $z^n = r^n(\cos n\theta + i \sin n\theta)$.	$z = 2\left(\cos \dfrac{\pi}{6} + i \sin \dfrac{\pi}{6}\right)$ $z^3 = 8\left(\cos \dfrac{\pi}{2} + i \sin \dfrac{\pi}{2}\right)$
Roots	The n distinct nth roots of $r(\cos \theta + i \sin \theta)$ are $r^{1/n}\left[\cos\left(\dfrac{\theta + 2k\pi}{n}\right) + i \sin\left(\dfrac{\theta + 2k\pi}{n}\right)\right]$ for $k = 0, 1, 2, \ldots, n - 1$.	Square roots of z: $\sqrt{2}\left(\cos \dfrac{\pi}{12} + i \sin \dfrac{\pi}{12}\right)$ $\sqrt{2}\left(\cos \dfrac{13\pi}{12} + i \sin \dfrac{13\pi}{12}\right)$

7.6 Polar Equations

Polar Coordinates	If $r > 0$, then (r, θ) is r units from the origin on the terminal side of θ in standard position. If $r < 0$, then (r, θ) is $\lvert r \rvert$ units from the origin on the extension of the terminal side of θ.	$(2, \pi/4)$ $(-2, 5\pi/4)$
Converting	Polar to rectangular: $x = r \cos \theta$, $y = r \sin \theta$. Rectangular to polar: $r = \sqrt{x^2 + y^2}$ and the terminal side of θ contains (x, y).	Polar: $(2, \pi/4)$ Rectangular: $\left(\sqrt{2}, \sqrt{2}\right)$

7.7 Parametric Equations

| **Parametric Equations** | $x = f(t)$ and $y = g(t)$ where t is a parameter in some interval of real numbers. | $x = 2t, y = t^2$ for t in $(0, 5)$ |
| **Converting to Rectangular** | Eliminate the parameter. | $y = (x/2)^2$ for x in $(0, 10)$ |

■ ■ ■ Chapter 7 Review Exercises

Solve each triangle that exists with the given parts. If there is more than one triangle with the given parts, then solve each one.

1. $\gamma = 48°, a = 3.4, b = 2.6$

2. $a = 6, b = 8, c = 10$

3. $\alpha = 13°, \beta = 64°, c = 20$

4. $\alpha = 50°, a = 3.2, b = 8.4$

5. $a = 3.6, b = 10.2, c = 5.9$

6. $\beta = 36.2°, \gamma = 48.1°, a = 10.6$

7. $a = 30.6, b = 12.9, c = 24.1$

8. $\alpha = 30°, a = \sqrt{3}, b = 2\sqrt{3}$

9. $\beta = 22°, c = 4.9, b = 2.5$

10. $\beta = 121°, a = 5.2, c = 7.1$

Find the area of each triangle.

11.

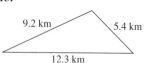

12.

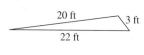

13.

14.

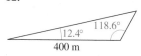

*Find the magnitude of the horizontal and vertical components for each vector **v** with the given magnitude and given direction angle θ.*

15. $|\mathbf{v}| = 6, \theta = 23.3°$

16. $|\mathbf{v}| = 4.5, \theta = 156°$

17. $|\mathbf{v}| = 3.2, \theta = 231.4°$

18. $|\mathbf{v}| = 7.3, \theta = 344°$

Find the magnitude and direction for each vector.

19. $\langle 2, 3 \rangle$

20. $\langle -4, 3 \rangle$

21. $\langle -3.2, -5.1 \rangle$

22. $\langle 2.1, -3.8 \rangle$

*Find the component form for each vector **v** with the given magnitude and direction angle.*

23. $|\mathbf{v}| = \sqrt{2}, \theta = 45°$

24. $|\mathbf{v}| = 6, \theta = 60°$

25. $|\mathbf{v}| = 9.1, \theta = 109.3°$

26. $|\mathbf{v}| = 5.5, \theta = 344.6°$

Perform the vector operations. Write your answer in the form $\langle a, b \rangle$ if the answer is a vector.

27. $2\langle -3, 4 \rangle$

28. $-3\langle 4, -1 \rangle$

29. $\langle 2, -5 \rangle - 2\langle 1, 6 \rangle$

30. $3\langle 1, 2 \rangle + 4\langle -1, -2 \rangle$

31. $\langle -1, 5 \rangle \cdot \langle 4, 2 \rangle$

32. $\langle -4, 7 \rangle \cdot \langle 7, 4 \rangle$

*Rewrite each vector **v** in the form $a_1\mathbf{i} + a_2\mathbf{j}$, where $\mathbf{i} = \langle 1, 0 \rangle$ and $\mathbf{j} = \langle 0, 1 \rangle$.*

33. In component form, $\mathbf{v} = \langle -4, 8 \rangle$.

34. In component form, $\mathbf{v} = \langle 3.2, -4.1 \rangle$.

35. The direction angle for **v** is 30° and its magnitude is 7.2.

36. The magnitude of **v** is 6 and it has the same direction as the vector $\langle 2, 5 \rangle$.

Find the absolute value of each complex number.

37. $3 - 5i$

38. $3.6 + 4.8i$

39. $\sqrt{5} + i\sqrt{3}$

40. $-2\sqrt{2} + 3i\sqrt{5}$

Write each complex number in trigonometric form, using degree measure for the argument.

41. $-4.2 + 4.2i$

42. $3 - i\sqrt{3}$

43. $-2.3 - 7.2i$

44. $4 + 9.2i$

Write each complex number in the form $a + bi$.

45. $\sqrt{3}(\cos 150° + i \sin 150°)$

46. $\sqrt{2}(\cos 225° + i \sin 225°)$

47. $6.5(\cos 33.1° + i \sin 33.1°)$

48. $14.9(\cos 289.4° + i \sin 289.4°)$

Find the product and quotient of each pair of complex numbers, using trigonometric form.

49. $z_1 = 2.5 + 2.5i, z_2 = -3 - 3i$

50. $z_1 = -\sqrt{3} + i, z_2 = -2 - 2i\sqrt{3}$

51. $z_1 = 2 + i, z_2 = 3 - 2i$

52. $z_1 = -3 + i, z_2 = 2 - i$

Use De Moivre's theorem to simplify each expression. Write the answer in the form $a + bi$.

53. $[2(\cos 45° + i \sin 45°)]^3$

54. $\left[\sqrt{3}(\cos 210° + i \sin 210°)\right]^4$

55. $(4 + 4i)^3$ **56.** $\left(1 - i\sqrt{3}\right)^4$

Find the indicated roots. Express answers in the form $a + bi$.

57. The square roots of i **58.** The cube roots of $-i$

59. The cube roots of $\sqrt{3} + i$ **60.** The square roots of $3 + 3i$

61. The cube roots of $2 + i$ **62.** The cube roots of $3 - i$

63. The fourth roots of $625i$ **64.** The fourth roots of $-625i$

Convert the polar coordinates of each point to rectangular coordinates.

65. $(5, 60°)$ **66.** $(-4, 30°)$

67. $\left(\sqrt{3}, 100°\right)$ **68.** $\left(\sqrt{5}, 230°\right)$

Convert the rectangular coordinates of each point to polar coordinates. Use radians for θ.

69. $\left(-2, -2\sqrt{3}\right)$ **70.** $\left(-3\sqrt{2}, 3\sqrt{2}\right)$

71. $(2, -3)$ **72.** $(-4, -5)$

Sketch the graph of each polar equation.

73. $r = -2 \sin \theta$ **74.** $r = 5 \sin 3\theta$

75. $r = 2 \cos 2\theta$ **76.** $r = 1.1 - \cos \theta$

77. $r = 500 + \cos \theta$ **78.** $r = 500$

79. $r = \dfrac{1}{\sin \theta}$ **80.** $r = \dfrac{-2}{\cos \theta}$

For each polar equation, write an equivalent rectangular equation.

81. $r = \dfrac{1}{\sin \theta + \cos \theta}$ **82.** $r = -6 \cos \theta$

83. $r = -5$ **84.** $r = \dfrac{1}{1 + \sin \theta}$

For each rectangular equation, write an equivalent polar equation.

85. $y = 3$ **86.** $x^2 + (y + 1)^2 = 1$

87. $x^2 + y^2 = 49$ **88.** $2x + 3y = 6$

Sketch the graph of each pair of parametric equations.

89. $x = 3t, y = 3 - t$, for t in $(0, 1)$

90. $x = t - 3, y = t^2$, for t in $(-\infty, \infty)$

91. $x = -\sin t, y = -\cos t$, for t in $[0, \pi/2]$

92. $x = -\cos t, y = \sin t$, for t in $[0, \pi]$

Solve each problem.

93. *Resultant Force*   Forces of 12 lb and 7 lb act at a 30° angle to each other. Find the magnitude of the resultant force and the angle that the resultant makes with each force.

94. *Course of a Cessna*   A twin-engine Cessna is heading on a bearing of 35° with an air speed of 180 mph. If the wind is out of the west (bearing 90°) at 40 mph, then what are the bearing of the course and the ground speed of the airplane?

95. *Dividing Property*   Mrs. White Eagle gave each of her children approximately half of her four-sided lot in Gallup by dividing it on a diagonal. If Susan's piece is 482 ft by 364 ft by 241 ft and Seth's piece is 482 ft by 369 ft by 238 ft, then which child got the larger piece?

96. *Area, Area, Area*   A surveyor found that two sides of a triangular lot were 135.4 ft and 164.1 ft, with an included angle of 86.4°. Find the area of this lot, using each of the three area formulas.

97. *Pipeline Detour*   A pipeline was planned to go from A to B, as shown in the figure. However, Mr. Smith would not give permission for the pipeline to cross his property. The pipeline was laid 431 ft from A to C and then 562 ft from C to B. If $\angle C$ is 122° and the cost of the pipeline was \$21.60/ft, then how much extra was spent to go around Mr. Smith's property?

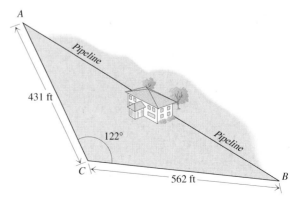

■ **Figure for Exercise 97**

98. *In the Wrong Place* In a lawsuit filed against a crane operator, a pedestrian of average height claims that he was struck by
a wrecking ball. At the time of the accident, the operator had
the ball extended 40 ft from the end of the 60-ft boom as
shown in the figure, and the angle of elevation of the boom
was 53°. How far from the crane would the pedestrian have to
stand to be struck by this wrecking ball?

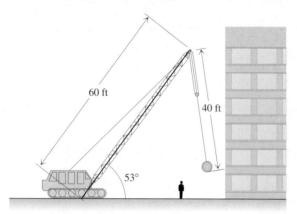

■ **Figure for Exercise 98**

99. *Detroit Pistons* The pistons in a gasoline engine are connected to a crankshaft, as shown in the figure. If the length of
the connecting rod is c and the radius of revolution around the
center of the crankshaft is r, then the distance from the center
of the crankshaft to the center of the piston, a, varies from
$c + r$ to $c - r$.
a. Show that $a = \sqrt{c^2 - r^2 \sin^2 \theta} + r \cos \theta$.

b. Suppose that the crankshaft is rotating at 426 rpm and time
$t = 0$ min corresponds to $\theta = 0$ and $a = c + r$. If $c = 12$
in., $r = 2$ in., and $t = 0.1$ min, then what is a?

■ **Figure for Exercise 99**

Thinking Outside the Box LXVII

Tiling a Room Tile Mart is selling new T-shaped ceramic tiles as
shown in the figure. Each tile is 3 ft long and 2 ft wide, and covers
four square feet.
a. Is it possible to tile completely an 8 ft by 8 ft room with these
T-shaped tiles? No cutting, breaking, or overlapping of the tiles
is allowed. Explain.

b. Is it possible to tile a 6 ft by 6 ft room with these tiles?

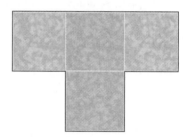

■ **Figure for Thinking Outside the Box LXVII**

■■■ Chapter 7 Test

Determine the number of triangles with the given parts and solve each triangle.

1. $\alpha = 30°, b = 4, a = 2$

2. $\alpha = 60°, b = 4.2, a = 3.9$

3. $a = 3.6, \alpha = 20.3°, \beta = 14.1°$

4. $a = 2.8, b = 3.9, \gamma = 17°$

5. $a = 4.1, b = 8.6, c = 7.3$

Given $\mathbf{A} = \langle -3, 2 \rangle$ *and* $\mathbf{B} = \langle 1, 4 \rangle$, *find the magnitude and direction angle for each of the following vectors.*

6. $\mathbf{A} + \mathbf{B}$ **7.** $\mathbf{A} - \mathbf{B}$ **8.** $3\mathbf{B}$

Write each complex number in trigonometric form, using degree measure for the argument.

9. $3 + 3i$ **10.** $-1 + i\sqrt{3}$ **11.** $-4 - 2i$

Perform the indicated operations. Write the answer in the form $a + bi$.

12. $3(\cos 20° + i \sin 20°) \cdot 2(\cos 25° + i \sin 25°)$

13. $[2(\cos 10° + i \sin 10°)]^9$ **14.** $\dfrac{3(\cos 63° + i \sin 63°)}{2(\cos 18° + i \sin 18°)}$

Give the rectangular coordinates for each of the following points in the polar coordinate system.

15. $(5, 30°)$ **16.** $(-3, -\pi/4)$ **17.** $(33, 217°)$

Sketch the graph of each equation in polar coordinates.

18. $r = 5 \cos \theta$ **19.** $r = 3 \cos 2\theta$

Solve each problem.

20. Find the area of the triangle in which $a = 4.1$ m, $b = 6.8$ m, and $c = 9.5$ m.

21. A vector $\mathbf{v}$ in the coordinate plane has direction angle $\theta = 37.2°$ and $|\mathbf{v}| = 4.6$. Find real numbers a_1 and a_2 such that $\mathbf{v} = a_1\mathbf{i} + a_2\mathbf{j}$, where $\mathbf{i} = \langle 1, 0 \rangle$ and $\mathbf{j} = \langle 0, 1 \rangle$.

22. Find all of the fourth roots of -81.

23. Write an equation equivalent to $x^2 + y^2 + 5y = 0$ in polar coordinates.

24. Write an equation equivalent to $r = 5 \sin 2\theta$ in rectangular coordinates.

25. Find a pair of parametric equations whose graph is the line segment joining the points $(-2, -3)$ and $(4, 5)$.

26. The bearing of an airplane is $40°$ with an air speed of 240 mph. If the wind is out of the northwest (bearing $135°$) at 30 mph, then what are the bearing of the course and the ground speed of the airplane?

Tying it all Together

Chapters 1–7

Find all real and imaginary solutions to each equation.

1. $x^4 - x = 0$

2. $x^3 - 2x^2 - 5x + 6 = 0$

3. $x^6 + 2x^5 - x - 2 = 0$

4. $x^7 - x^4 + 2x^3 - 2 = 0$

5. $2 \sin 2x - 2 \cos x + 2 \sin x = 1$

6. $4x \sin x + 2 \sin x - 2x = 1$

7. $e^{\sin x} = 1$

8. $\sin(e^x) = 1/2$

9. $2^{2x-3} = 32$

10. $\log(x - 1) - \log(x + 2) = 2$

Sketch the graph of each function using rectangular or polar coordinates as appropriate.

11. $y = \sin x$

12. $y = e^x$

13. $r = \sin \theta$

14. $r = \theta$

15. $y = \sqrt{\sin x}$

16. $y = \ln(\sin x)$

17. $r = \sin(\pi/3)$

18. $y = x^{1/3}$

Evaluate each expression.

19. $\log(\sin(\pi/2))$

20. $\sin(\log(1))$

21. $\cos(\ln(e^\pi))$

22. $\ln(\cos(2\pi))$

23. $\sin^{-1}\left(\log_2\left(\sqrt{2}\right)\right)$

24. $\cos^{-1}\left(\ln\left(\sqrt{e}\right)\right)$

25. $\tan^{-1}(\log(0.1))$

26. $\tan^{-1}(\ln(e))$

Perform the indicated operations and simplify.

27. $\dfrac{1}{x + 2} + \dfrac{1}{x - 2}$

28. $\dfrac{1}{1 - \sin x} + \dfrac{1}{1 + \sin x}$

29. $\dfrac{1}{2 + \sqrt{3}} + \dfrac{1}{2 - \sqrt{3}}$

30. $\log\left(\dfrac{1}{x + 2}\right) + \log\left(\dfrac{1}{x - 2}\right)$

31. $\dfrac{x^2 - 9}{2x - 6} \cdot \dfrac{4x + 12}{x^2 + 6x + 9}$

32. $\dfrac{1 - \sin^2(x)}{2 \cos^2(x) - \sin(2x)} \cdot \dfrac{\cos^2(x) - \sin^2(x)}{2 \cos^2(x) + \sin(2x)}$

33. $\dfrac{\sqrt{8}}{2 - \sqrt{3}} \cdot \dfrac{\sqrt{2}}{4 + \sqrt{12}}$

34. $\log\left(\dfrac{x^2 + 3x + 2}{x^2 + 5x + 6}\right) - \log\left(\dfrac{x + 1}{x + 3}\right)$

35. $\dfrac{\dfrac{1}{x - 2} + \dfrac{1}{x + 2}}{\dfrac{1}{x^2 - 4} - \dfrac{1}{x + 2}}$

36. $\dfrac{\dfrac{1}{\sin(2x)} - \tan(x)}{\dfrac{\cot(x)}{2} - \dfrac{\tan(x)}{2}}$

Function Gallery: Some Basic Functions of Trigonometry

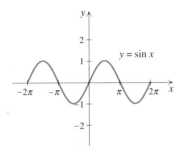

Period 2π
Amplitude 1

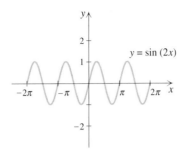

Period π
Amplitude 1

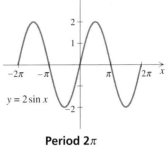

Period 2π
Amplitude 2

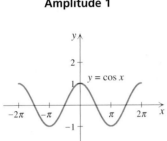

Period 2π
Amplitude 1

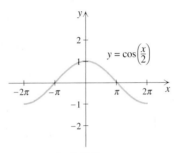

Period 4π
Amplitude 1

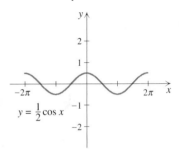

Period 2π
Amplitude $\frac{1}{2}$

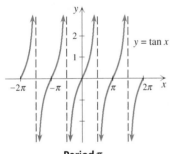

Period π

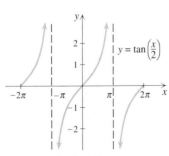

Period 2π

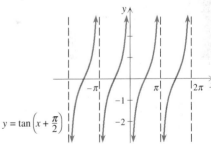

Period π

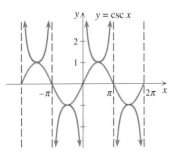

Period 2π

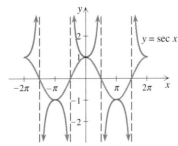

Period 2π

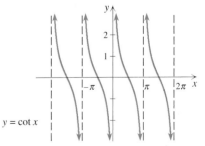

Period π

8 Systems of Equations and Inequalities

THROUGHOUT history humans have been directly or indirectly influenced by the world's oceans. Ocean waters serve as a source of food and valuable minerals, as a vast highway for commerce, and as a place for both recreation and waste disposal.

The world ocean has an area of 139 million square miles and occupies 70% of the surface of the earth. Yet, it has been said that we know more about outer space than we do about the oceans. The oceans hold the answers to many important questions about the development of the earth and the history of life on earth.

WHAT YOU WILL LEARN   In this chapter we will use systems of equations to solve problems involving two or more variables. In particular, we will learn how geophysicists use systems of equations in their effort to map the ocean floor and expand their knowledge of this important resource.

Systems of Linear Equations in Two Variables

In Section 1.2 we defined a linear equation in two variables as an equation of the form

$$Ax + By = C,$$

where A and B are not both zero, and we discussed numerous applications of linear equations. There are infinitely many ordered pairs that satisfy a single linear equation. In applications, however, we are often interested in finding a single ordered pair that satisfies a *pair* of linear equations. In this section we discuss several methods for solving that problem.

Solving a System by Graphing

Any collection of two or more equations is called a **system of equations.** For example, the system of equations consisting of $x + 2y = 6$ and $2x - y = -8$ is written as follows:

$$x + 2y = 6$$
$$2x - y = -8$$

The **solution set** of a system of two linear equations in two variables is the set of all ordered pairs that satisfy *both* equations of the system. The graph of an equation shows all ordered pairs that satisfy it, so we can solve some systems by graphing the equations and observing which points (if any) satisfy all of the equations.

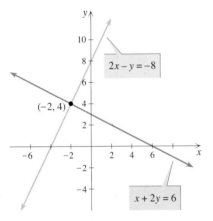

■ **Figure 8.1**

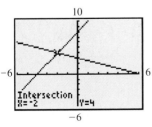

■ **Figure 8.2**

Example **1** Solving a system by graphing

Solve each system by graphing.

a. $x + 2y = 6$
 $2x - y = -8$

b. $y = \dfrac{1}{2}x + 2$
 $x - 2y = 4$

Solution

a. Graph the straight line

$$x + 2y = 6$$

by using its intercepts, $(0, 3)$ and $(6, 0)$. Graph the straight line

$$2x - y = -8$$

by using its intercepts, $(0, 8)$ and $(-4, 0)$. The graphs are shown in Fig. 8.1. The lines appear to intersect at $(-2, 4)$. Check $(-2, 4)$ in both equations. Since

$$-2 + 2(4) = 6 \qquad \text{and} \qquad 2(-2) - 4 = -8$$

are both correct, we can be certain that $(-2, 4)$ satisfies both equations. The solution set of the system is $\{(-2, 4)\}$.

You can check by graphing the equations on a calculator and finding the intersection, as shown in Fig. 8.2. ☐

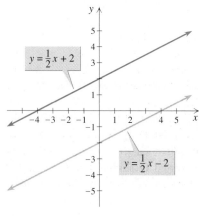

■ **Figure 8.3**

b. Use the y-intercept $(0, 2)$ and the slope $1/2$ to graph

$$y = \frac{1}{2}x + 2,$$

as shown in Fig. 8.3. Since $x - 2y = 4$ is equivalent to

$$y = \frac{1}{2}x - 2,$$

its graph has y-intercept $(0, -2)$ and is parallel to the first line. Since the lines are parallel, there is no point that satisfies both equations of the system. In fact, if you substitute any value of x in the two equations, the corresponding y-values will differ by 4.

Try This. Solve the system $y = x - 3$ and $x + y = 7$ by graphing.   ■

Independent, Inconsistent, and Dependent Equations

Most of the systems of equations that occur in applications correspond to pairs of lines that intersect in a single point, as in Fig. 8.1. In this case the equations are called **independent** or the system is called an independent system. If the two lines corresponding to the equations are parallel, as in Fig. 8.3, then there is no solution to the system and the equations are called **inconsistent** or the system is called inconsistent. The third possibility is that two equations are equivalent and have the same graph. In this case the equations are called **dependent.** For example, the equations of the system

$$\begin{aligned} x + y &= 5 \\ 2x + 2y &= 10 \end{aligned}$$

have the same graph (a line) and the system is dependent. Any ordered pair that satisfies one of these equations satisfies both of these equations. So the solution set to the system consists of all ordered pairs that satisfy one of the equations, described in set notation as

$$\{(x, y) \mid x + y = 5\}.$$

Since the equations are equivalent, we could use either equation in this set notation, but we usually use the simpler one. Figure 8.4 illustrates typical graphs for independent, inconsistent, and dependent systems.

Independent system:
one solution

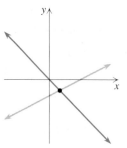

Lines have different slopes.

Inconsistent system:
no solution

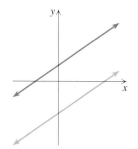

Lines have same slope,
different y-intercepts.

Dependent system:
infinitely many solutions

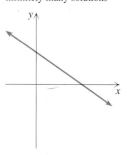

Lines have same slope,
same y-intercept.

■ **Figure 8.4**

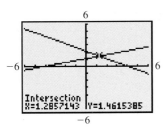

■ **Figure 8.5**

The Substitution Method

Graphing the equations of a system helps us to visualize the system and determine how many solutions it has. However, solving systems of linear equations by graphing is not very accurate unless the solution is fairly simple. The accuracy of graphing can be improved with a graphing calculator, but even with a graphing calculator, we generally get only approximate solutions. For example, the solution to $y_1 = (28 - 7x)/13$ and $y_2 = (29 + 7x)/26$ is $(9/7, 19/13)$, but the graphing calculator solution in Fig. 8.5 does not give this exact answer. However, by using an algebraic technique such as the substitution method, we can get exact solutions quickly. In this method, shown in Example 2, we eliminate a variable from one equation by substituting an expression for that variable from the other equation.

Example **2** **Solving a system by substitution**

Solve each system by substitution.

a. $3x - y = 6$ **b.** $y = 2x + 1000$
      $6x + 5y = -23$ $0.05x + 0.06y = 400$

Solution

a. Since y occurs with coefficient -1 in $3x - y = 6$, it is simpler to isolate y in this equation than to isolate any other variable in the system.

$$-y = -3x + 6$$

$$y = 3x - 6$$

Use $3x - 6$ in place of y in the equation $6x + 5y = -23$:

$$6x + 5(3x - 6) = -23 \quad \text{Substitution}$$

$$6x + 15x - 30 = -23$$

$$21x = 7$$

$$x = \frac{1}{3}$$

The x-coordinate of the solution is $\frac{1}{3}$. To find y, use $x = \frac{1}{3}$ in $y = 3x - 6$:

$$y = 3\left(\frac{1}{3}\right) - 6$$

$$y = -5 \cdot$$

Check that $\left(\frac{1}{3}, -5\right)$ satisfies both of the original equations. The solution set is $\left\{\left(\frac{1}{3}, -5\right)\right\}$.

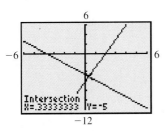

■ **Figure 8.6**

   The graphs of $y_1 = 3x - 6$ and $y_2 = (-23 - 6x)/5$ in Fig. 8.6 support this solution. ☐

b. The first equation, $y = 2x + 1000$, already has one variable isolated. So we can replace y by $2x + 1000$ in $0.05x + 0.06y = 400$:

$$0.05x + 0.06(2x + 1000) = 400 \qquad \text{Substitution}$$

$$0.05x + 0.12x + 60 = 400$$

$$0.17x = 340$$

$$x = 2000$$

Use $x = 2000$ in $y = 2x + 1000$ to find y:

$$y = 2(2000) + 1000$$

$$y = 5000$$

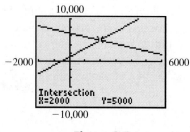

10,000

−2000   6000

Intersection
X=2000   Y=5000

−10,000

■ **Figure 8.7**

Check $(2000, 5000)$ in the original equations. The solution set is $\{(2000, 5000)\}$. The graphs of $y_1 = 2x + 1000$ and $y_2 = (400 - 0.05x)/0.06$ in Fig. 8.7 support this solution.

Try This. Solve $y = 2x - 3$ and $x + 2y = -1$ by substitution. ■

If substitution results in a false statement, then the system is inconsistent. If substitution results in an identity, then the system is dependent. In the next example we solve an inconsistent system and a dependent system by substitution.

Example **3** Inconsistent and dependent systems

Solve each system by substitution.

a. $3x - y = 9$ **b.** $\dfrac{1}{2}x - \dfrac{2}{3}y = -2$

  $2y - 6x = 7$ $4y = 3x + 12$

Solution

a. Solve $3x - y = 9$ for y to get $y = 3x - 9$. Replace y by $3x - 9$ in the equation $2y - 6x = 7$:

$$2(3x - 9) - 6x = 7$$

$$6x - 18 - 6x = 7$$

$$-18 = 7$$

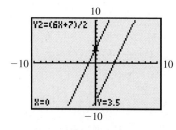

10

Y2=(6X+7)/2

−10   10

X=0   Y=3.5

−10

■ **Figure 8.8**

Since the last statement is false, the system is inconsistent and has *no solution*. The graphs of $y_1 = 3x - 9$ and $y_2 = (6x + 7)/2$ in Fig. 8.8 appear to be parallel lines and support the conclusion that the system is inconsistent.   □

b. Solve $4y = 3x + 12$ for y to get $y = \dfrac{3}{4}x + 3$. Replace y by $\dfrac{3}{4}x + 3$ in the first equation:

$$\frac{1}{2}x - \frac{2}{3}\left(\frac{3}{4}x + 3\right) = -2$$

$$\frac{1}{2}x - \frac{1}{2}x - 2 = -2$$

$$-2 = -2$$

Since the last statement is an identity, the system is dependent. The solution set is $\{(x, y) \mid 4y = 3x + 12\}$.

Try This. Solve $y = 3x - 5$ and $6x - 2y = 1$ by substitution. ▪

Note that there are many ways of writing the solution set in Example 3(b). Since $4y = 3x + 12$ is equivalent to $y = \frac{3}{4}x + 3$ or to $x = \frac{4}{3}y - 4$, we could write the solution set as $\left\{ \left(x, \frac{3}{4}x + 3\right) \mid x \text{ is any real number} \right\}$ or $\left\{ \left(\frac{4}{3}y - 4, y\right) \mid y \text{ is any real number} \right\}$. The variable used in describing the solution set does not matter. We could even use another variable and write $\left\{ \left(t, \frac{3}{4}t + 3\right) \mid t \text{ is any real number} \right\}$. All of these sets contain exactly the same ordered pairs.

The Addition Method

In the substitution method we eliminate a variable in one equation by substituting from the other equation. In the addition method we eliminate a variable by adding the two equations. It might be necessary to multiply each equation by an appropriate number so that a variable will be eliminated by the addition.

Example **4** **Solving systems by addition**

Solve each system by addition.

a. $3x - y = 9$ **b.** $2x - 3y = -2$

$\quad 2x + y = 1$ $\quad\quad 3x - 2y = 12$

Solution

a. Add the equations to eliminate the y-variable:

$$3x - y = 9$$
$$\underline{2x + y = 1}$$
$$5x \quad\quad = 10$$
$$x \quad\quad = 2$$

Use $x = 2$ in $2x + y = 1$ to find y:

$$2(2) + y = 1$$
$$y = -3$$

Substituting the values $x = 2$ and $y = -3$ in the original equation yields $3(2) - (-3) = 9$ and $2(2) + (-3) = 1$, which are both correct. So $(2, -3)$ satisfies both equations, and the solution set to the system is $\{(2, -3)\}$.

b. To eliminate x upon addition, we multiply the first equation by 3 and the second equation by -2:

$$3(2x - 3y) = 3(-2)$$
$$-2(3x - 2y) = -2(12)$$

This multiplication produces $6x$ in one equation and $-6x$ in the other. So the x-variable is eliminated upon addition of the equations.

$$
\begin{aligned}
6x - 9y &= -6 \\
\underline{-6x + 4y} &= \underline{-24} \\
-5y &= -30 \\
y &= 6
\end{aligned}
$$

Use $y = 6$ in $2x - 3y = -2$ to find x:

$$
\begin{aligned}
2x - 3(6) &= -2 \\
2x - 18 &= -2 \\
2x &= 16 \\
x &= 8
\end{aligned}
$$

If $y = 6$ is used in the other equation, $3x - 2y = 12$, we would also get $x = 8$. Substituting $x = 8$ and $y = 6$ in both of the original equations yields $2(8) - 3(6) = -2$ and $3(8) - 2(6) = 12$, which are both correct. So the solution set to the system is $\{(8, 6)\}$.

Try This. Solve $x + y = 3$ and $3x - 2y = 4$ by addition. ■

In Example 4(b), we started with the given system and multiplied the first equation by 3 and the second equation by -2 to get

$$
\begin{aligned}
6x - 9y &= -6 \\
-6x + 4y &= -24.
\end{aligned}
$$

Since each equation of the new system is equivalent to an equation of the old system, the solution sets to these systems are identical. Two systems with the same solution set are **equivalent systems.** If we had multiplied the first equation by 2 and the second by -3, we would have obtained the equivalent system

$$
\begin{aligned}
4x - 6y &= -4 \\
-9x + 6y &= -36
\end{aligned}
$$

and we would have eliminated y by adding the equations.

When we have a choice of which method to use for solving a system, we generally avoid graphing because it is often inaccurate. Substitution or addition both yield exact solutions, but sometimes one method is easier to apply than the other. Substitution is usually used when one equation gives one variable in terms of the other, as in Example 2. Addition is usually used when both equations are in the form $Ax + By = C$, as in Example 4. By doing the exercises, you will soon discover which method works best on a given system.

When a system is solved by the addition method, an inconsistent system results in a false statement and a dependent system results in an identity, just as they did for the substitution method.

Example **5** Inconsistent and dependent systems

Solve each system by addition.

a. $0.2x - 0.4y = 0.5$ **b.** $\dfrac{1}{2}x - \dfrac{2}{3}y = -2$

  $x - 2y = 1.3$ $-3x + 4y = 12$

Solution

a. It is usually a good idea to eliminate the decimals in the coefficients, so we multiply the first equation by 10:

$$2x - 4y = 5 \qquad \text{First equation multiplied by 10}$$

$$x - 2y = 1.3$$

Now multiply the second equation by -2 and add to eliminate x:

$$2x - 4y = 5$$
$$\underline{-2x + 4y = -2.6}$$
$$0 = 2.4$$

Since $0 = 2.4$ is false, there is no solution to the system.

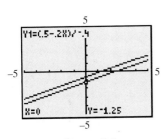 The graphs of $y_1 = (0.5 - 0.2x)/(-0.4)$ and $y_2 = (1.3 - x)/(-2)$ in Fig. 8.9 appear to be parallel lines and support the conclusion that there is no solution to the system.   □

b. To eliminate fractions in the coefficients, multiply the first equation by the LCD 6:

$$3x - 4y = -12$$
$$\underline{-3x + 4y = 12}$$
$$0 = 0$$

Since $0 = 0$ is an identity, the solution set is $\{(x, y) \mid -3x + 4y = 12\}$.

Try This. Solve $\frac{1}{2}x - \frac{1}{4}y = 1$ and $2x - y = 3$ by addition.   ■

Note that there are many ways to solve a system by addition. In Example 5(a), we could have multiplied the first equation by -5 or the second equation by -0.2. In either case, x would be eliminated upon addition. Try this for yourself.

Modeling with a System of Equations

We solved many problems involving linear equations in the past, but we always wrote all unknown quantities in terms of a single variable. Now that we can solve systems of equations, we can model situations involving two unknown quantities by using two variables and a system of equations.

Example **6** Modeling using a system of equations

While on a stakeout protecting earth from the scum of the universe, Agent K observed five men in black enter a coffee shop and get three doughnuts and five coffees for $3.30. Next, three aliens in disguise entered the shop and got four doughnuts and three coffees for $2.75. Finally, Agent Zed from the home office entered and got a doughnut and a cup of coffee. How much was Zed's bill?

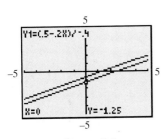

■ **Figure 8.9**

Solution

Let x be the cost of one doughnut and y be the cost of one cup of coffee. We can write a system of equations about x and y:

$$3x + 5y = 3.30$$
$$4x + 3y = 2.75$$

To eliminate x, multiply the first equation by -4 and the second by 3:

$$-4(3x + 5y) = -4(3.30)$$
$$3(4x + 3y) = 3(2.75)$$

Add the two resulting equations:

$$-12x - 20y = -13.20$$
$$\underline{12x + 9y = 8.25}$$
$$-11y = -4.95$$
$$y = 0.45$$

Use $y = 0.45$ in $3x + 5y = 3.30$ to find x:

$$3x + 5(0.45) = 3.30$$
$$3x + 2.25 = 3.30$$
$$3x = 1.05$$
$$x = 0.35$$

Check that 35 cents for a doughnut and 45 cents for a cup of coffee satisfy the statements in the original problem. Given these prices, Agent Zed's bill is 80 cents. The graphs of $y_1 = (3.30 - 3x)/5$ and $y_2 = (2.75 - 4x)/3$ in Fig. 8.10 support this conclusion.

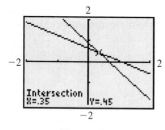

■ **Figure 8.10**

Try This. Two DVDs and three CDs cost $78. One DVD and four CDs cost $74. What is the cost of a DVD? ■

For Thought

True or False? Explain.

The following systems are referenced in these statements.
(a) $x + y = 5$ (b) $x - 2y = 4$ (c) $x = 5 + 3y$
    $x - y = 1$ $3x - 6y = 8$ $9y - 3x = -15$

1. The ordered pair $(2, 3)$ is in the solution set to $x + y = 5$.

2. The ordered pair $(2, 3)$ is in the solution set to system (a).

3. System (a) is inconsistent.

4. There is no solution to system (b).

5. Adding the equations in system (a) would eliminate y.

6. To solve system (c), we could substitute $5 + 3y$ for x in $9y - 3x = -15$.

7. System (c) is inconsistent.

8. The solution set to system (c) is the set of all real numbers.

9. The graphs of the equations of system (c) intersect at a single point.

10. The graphs of the equations of system (b) are parallel.

8.1 Exercises

Determine whether the given point is in the solution set to the given system.

1. $(1, 3)$
$$x + y = 4$$
$$x - y = -2$$

2. $(-1, 2)$
$$x + y = 1$$
$$2x - 3y = -8$$

3. $(-1, 5)$
$$2x + y = 3$$
$$x - 2y = -9$$

4. $(3, 2)$
$$3x - y = 7$$
$$2x + 4y = 16$$

Solve each system by inspecting the graphs of the equations.

5. $2x - 3y = -4$
$$y = -2x + 4$$

6. $x + 2y = -1$
$$2x + 3y = -3$$

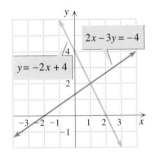

7. $3x - 4y = 0$
$$y = \frac{3}{4}x + 2$$

8. $x - 2y = -3$
$$y = \frac{1}{2}x + \frac{3}{2}$$

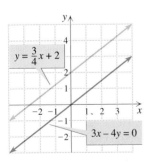

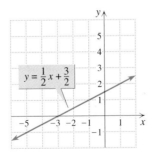

Solve each system by graphing.

9. $x + y = 5$
$$x - y = 1$$

10. $2x + y = -1$
$$y - x = 5$$

11. $y = x - 2$
$$y = -x + 4$$

12. $y = -3x$
$$x - 2y = 7$$

13. $3y + 2x = 6$
$$y = -\frac{2}{3}x - 1$$

14. $2x + 4y = 12$
$$2y = 6 - x$$

15. $y = \frac{1}{2}x - 3$
$$2x - 4y = 12$$

16. $y = -2x + 6$
$$4x + 2y = 8$$

Solve each system by substitution. Determine whether each system is independent, inconsistent, or dependent.

17. $y = 2x + 1$
$$3x - 4y = 1$$

18. $5x - 6y = 23$
$$x = 6 - 3y$$

19. $x + y = 1$
$$2x - 3y = 8$$

20. $x + 2y = 3$
$$2x + y = 5$$

21. $y - 3x = 5$
$$3(x + 1) = y - 2$$

22. $2y = 1 - 4x$
$$2x + y = 0$$

23. $2y = 6 - 3x$
$$\frac{1}{2}x + \frac{1}{3}y = 3$$

24. $2x = 10 - 5y$
$$\frac{1}{5}x + \frac{1}{2}y = 1$$

25. $x + y = 200$
$$0.05x + 0.06y = 10.50$$

26. $2x + y = 300$
$$\frac{1}{2}x + \frac{1}{3}y = 80$$

27. $y = 3x + 1$
$$y = 3x - 7$$

28. $2x + y = 9$
$$4x + 2y = 10$$

29. $\frac{1}{2}x - \frac{1}{3}y = 12$
$$\frac{1}{4}x - \frac{1}{2}y = 1$$

30. $0.05x + 0.1y = 10$
$$0.06x + 0.2y = 16$$

Solve each system by addition. Determine whether each system is independent, inconsistent, or dependent.

31. $x + y = 20$
$$x - y = 6$$

32. $3x - 2y = 7$
$$-3x + y = 5$$

33. $x - y = 5$
$$3x + 2y = 10$$

34. $x - 4y = -3$
$$-3x + 5y = 2$$

35. $x - y = 7$
$$y - x = 5$$

36. $2x - y = 6$
$$-4x + 2y = 9$$

37. $2x + 3y = 1$
$$3x - 5y = -8$$

38. $-2x + 5y = 14$
$$7x + 6y = -2$$

39. $0.05x + 0.1y = 0.6$
$$x + 2y = 12$$

40. $0.02x - 0.04y = 0.08$
$$x - 2y = 4$$

41. $\dfrac{x}{2} + \dfrac{y}{2} = 5$

$\dfrac{3x}{2} - \dfrac{2y}{3} = 2$

42. $\dfrac{x}{4} + \dfrac{y}{3} = 0$

$\dfrac{x}{8} - \dfrac{y}{6} = 2$

43. $3x - 2.5y = -4.2$

$0.12x + 0.09y = 0.4932$

44. $1.5x - 2y = 8.5$

$3x + 1.5y = 6$

Classify each system as independent, inconsistent, or dependent without doing any written work.

45. $y = 5x - 6$

$y = -5x - 6$

46. $y = 5x - 6$

$y = 5x + 4$

47. $5x - y = 6$

$y = 5x - 6$

48. $5x - y = 6$

$y = -5x + 6$

📉 *Solve each system by graphing the equations on a graphing calculator and estimating the point of intersection.*

49. $y = 0.5x + 3$

$y = 0.499x + 2$

50. $y = 2x - 3$

$y = 1.9999x - 2$

51. $0.23x + 0.32y = 1.25$

$0.47x - 1.26y = 3.58$

52. $342x - 78y = 474$

$123x + 145y = 397$

Solve each problem using two variables and a system of two equations. Solve the system by the method of your choice. Note that some of these problems lead to dependent or inconsistent systems.

53. *Two-Income Family* Althea has a higher income than Vaughn and their total income is $82,000. If their salaries differ by $16,000, then what is the income of each?

54. *Males and Females* A total of 76 young Republicans attended a strategy meeting. The number of females exceeded the number of males by 2. How many of each gender were at the meeting?

55. *Income on Investments* Carmen made $25,000 profit on the sale of her condominium. She lent part of the profit to Jim's Orange Grove at 10% interest and the remainder to Ricky's Used Cars at 8% interest. If she received $2200 in interest after one year, then how much did she lend to each business?

56. *Stock Market Losses* In 2005 Gerhart lost twice as much in the futures market as he did in the stock market. If his losses totaled $18,630, then how much did he lose in each market?

57. *Zoo Admission Prices* The Springfield Zoo has different admission prices for adults and children. When Mr. and Mrs. Weaver went with their five children, the bill was $33. If Mrs. Wong and her three children got in for $18.50, then what is the price of an adult's ticket and what is the price of a child's ticket?

58. *Book Prices* At the Book Exchange, all paperbacks sell for one price and all hardbacks sell for another price. Tanya got six paperbacks and three hardbacks for $8.25, while Gretta got four paperbacks and five hardbacks for $9.25. What was Todd's bill for seven paperbacks and nine hardbacks?

59. *Getting Fit* The Valley Health Club sold a dozen memberships in one week for a total of $6000. If male memberships cost $500 and female memberships cost $500, then how many male memberships and how many female memberships were sold?

60. *Quality Time* Mr. Thomas and his three children paid a total of $65.75 for admission to Water World. Mr. and Mrs. Li and their six children paid a total of $131.50. What is the price of an adult's ticket and what is the price of a child's ticket?

61. *Cows and Ostriches* A farmer has some cows and ostriches. One day he observed that his animals, which are normal, have 84 eyes and 122 legs. How many animals of each type does he have?

62. *Snakes and Iguanas* The farmer's wife collects snakes and iguanas. One day she observed that her reptiles, which are normal, have a total of 60 eyes and 68 feet. How many reptiles of each type does she have?

63. *Cows and Horses* A rancher has some normal cows and horses. One day he observed that his animals have a total of 96 legs and 24 tails. How many animals of each type does he have?

64. *Snakes and Mice* The rancher's wife raises snakes and white mice. One day she observed that her animals have a total of 78 eyes and 38 tails. How many animals of each type does she have?

65. *Coffee and Muffins* On Monday the office staff paid a total of $7.77 including tax for 3 coffees and 7 muffins. On Tuesday the bill was $14.80 including tax for 6 coffees and 14 muffins. If the sales tax rate is 7%, then what is the price of a coffee and what is the price of a muffin?

66. *Graduating Seniors* In Sociology 410 there are 55 more males than there are females. Two-thirds of the males and two-thirds of the females are graduating seniors. If there are 30 more graduating senior males than graduating senior females, then how many males and how many females are in the class?

67. *Political Party Preference* The results of a survey of students at Central High School concerning political party preference are given in the accompanying table. If 230 students preferred the Democratic party and 260 students preferred the Republican party, then how many students are there at CHS?

■ **Table for Exercise 67**

	M	F	
Democrat	50%	30%	
Republican	20%	60%	
Other	30%	10%	

68. *Protein and Carbohydrates*   Nutritional information for Rice Krispies and Grape-nuts is given in the accompanying table. How many servings of each would it take to get exactly 23 g of protein and 215 g of carbohydrates?

   HINT Write an equation for protein and another for carbohydrates.

▪ **Table for Exercise 68**

	Rice Krispies	Grape-nuts
Protein (g/serving)	2	3
Carbohydrates (g/serving)	25	23

69. *Distribution of Coin Types*   Isabelle paid for her $1.75 lunch with 87 coins. If all of the coins were nickels and pennies, then how many were there of each type?

70. *Coin Collecting*   Theodore has a collection of 166 old coins consisting of quarters and dimes. If he figures that each coin is worth two and a half times its face value, then his collection is worth $61.75. How many of each type of coin does he have?

71. *Bird Mobile*   A wood carver is making a bird mobile, as shown in the accompanying figure. The weights of the horizontal bars and strings are negligible. The mobile will balance if the product of the weight and distance on one side of the balance point is equal to the product of the weight and distance on the other side. For what values of x and y will the mobile be balanced?

   HINT Write an equation for each balancing point.

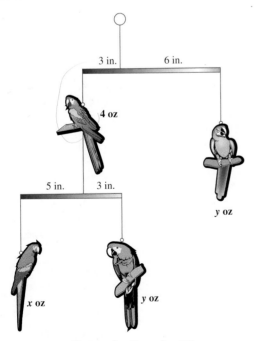

▪ **Figure for Exercise 71**

72. *Doubles and Singles*   The Executive Inn rents a double room for $10 more per night than a single. One night the motel took in $2159 by renting 15 doubles and 26 singles. What is the rental price for each type of room?

73. *Furniture Rental*   A civil engineer has a choice of two plans for renting furniture for her new office. Under Plan A she pays $800 plus $150 per month, while under Plan B she pays $200 plus $200 per month. For each plan, write the cost as a function of the number of months. Which plan is cheaper in the long run? For what number of months do the two plans cost the same?

74. *Flat Tax*   The 2004 tax rate schedule for a single taxpayer is given in the accompanying table. Suppose the federal tax were simplified to be $100 plus 25% of taxable income. Use a system of equations to find the taxable income at which a single taxpayer would pay the same amount of tax under the simplified plan as under the 2004 schedule.

▪ **Table for Exercise 74**   2004 Tax rate schedule– single taxpayers

If Taxable Income is Over	But not Over	Your Tax is	of the Amount Over
$0	$7150	10%	$0
7150	29,050	715 + 15%	7150
29,050	70,350	4000 + 25%	29,050
70,350	146,750	14,325 + 28%	70,350
146,750	319,100	35,717 + 33%	146,750
319,100	no limit	92,593 + 35%	319,100

75. *Prescribing Drugs*   Doctors often prescribe the same drugs for children as they do for adults. If a is the age of a child and D is the adult dosage, then to find the child's dosage d, doctors can use the formula $d = 0.08aD$ (Fried's rule) or $d = D(a + 1)/24$ (Cowling's rule). For what age do the two formulas give the same child's dosage?

76. *Tax Reform*   One plan for federal income tax reform is to tax an individual's income in excess of $15,000 at a 17% rate. Another plan is to institute a national retail sales tax of 15%. If an individual spends 75% of his or her income in retail stores where it is taxed at 15%, then for what income would the amount of tax be the same under either plan?

77. *Approaching Trucks*   A 60-ft truck doing 40 mph is approaching a 40-ft truck doing 50 mph on a two-lane road. How long (in seconds) does it take them to pass each other?

78. *Passing Trucks*   A 40-ft truck doing 50 mph and a 60-ft truck doing 40 mph are traveling in the same direction in adjacent lanes on an interstate highway. How long (in seconds) does it take the faster truck to pass the slower truck?

A system of equations can be used to find the equation of a line that goes through two points. For example, if $y = ax + b$ goes through $(3, 5)$, then a and b must satisfy $3a + b = 5$. For each given pair of points, find the equation of the line $y = ax + b$ that goes through the points by solving a system of equations.

79. $(-3, 9), (2, -1)$ **80.** $(1, -1), (3, 7)$

81. $(-2, 3), (4, -7)$ **82.** $(-3, -1), (4, 9)$

For Writing/Discussion

83. *Number of Solutions*   Explain how you can tell (without graphing) whether a system of linear equations has one solution, no solutions, or infinitely many solutions. Be sure to account for linear equations that are not functions.

84. *Cooperative Learning*   Write a step-by-step procedure (or algorithm) based on the addition method that will solve any system of two equations of the form $Ax + By = C$. Ask a classmate to solve a system using your procedure.

85. *Cooperative Learning*   Write an independent system of two linear equations for which $(2, -3)$ is the solution. Ask a classmate to solve your system.

86. *Cooperative Learning*   Write a dependent system of two linear equations for which $\{(t, t + 5) \mid t$ is any real number$\}$ is the solution set. Ask a classmate to solve your system.

Thinking Outside the Box LXVIII & LXIX

Many Means   The mean score for those who passed the last test was 65, whereas the mean score for those who failed that test was 35. The mean for the entire class was 53. What percentage of the students in the class passed the test?

Cubic Power   Find all real solutions to the equation

$$(x^2 + 2x - 24)^{x^3 - 9x^2 + 20x} = 1.$$

8.1 Pop Quiz

Solve each system and classify each system as independent, inconsistent, or dependent.

1. $7x - 3y = 4$
   $y = 2x$

2. $3x - 5y = 11$
   $7x + 5y = 19$

3. $5x - 2y = -1$
   $4x + 3y = 13$

4. $3x = 1 - 9y$
   $3y + x = 8$

5. $y = x + 1$
   $5x - 5y + 5 = 0$

Linking Concepts

For Individual or Group Explorations

Modeling Life Expectancy

The accompanying table gives the life expectancy at birth for U.S. men and women (Center for Disease Control, www.cdc.gov).

Year of Birth	Life Expectancy (Male)	Life Expectancy (Female)
1930	58.1	61.6
1940	60.8	65.2
1950	65.6	71.1
1960	66.6	73.1
1970	67.1	74.7
1980	70.0	77.4
1990	71.1	78.6
2000	74.4	79.7

a) Use linear regression on your graphing calculator to find the life expectancy for men as a function of the year of birth.

b) Use linear regression on your graphing calculator to find the life expectancy for women as a function of the year of birth.

c) Graph the functions that you found in parts (a) and (b) on the same coordinate system.

d) According to this model, will men ever catch up to women in life expectancy?

e) For what year of birth did men and women have the same life expectancy?

f) Interpret the slope of these two lines.

g) Why do you think that life expectancy for women is increasing at a greater rate than life expectancy for men?

8.2

Systems of Linear Equations in Three Variables

Systems of many linear equations in many variables are used to model a variety of situations ranging from airline scheduling to allocating resources in manufacturing. The same techniques that we are studying with small systems can be extended to much larger systems. In this section we use the techniques of substitution and addition from Section 8.1 to solve systems of linear equations in three variables.

Definitions

A **linear equation in three variables** x, y, and z is an equation of the form

$$Ax + By + Cz = D,$$

where A, B, C, and D are real numbers with A, B, and C not all equal to zero. For example,

$$x + y + 2z = 9$$

is a linear equation in three variables. The equation is called *linear* because its form is similar to that of a linear equation in two variables. A solution to a linear equation in three variables is an **ordered triple** of real numbers in the form (x, y, z) that satisfies the equation. For instance, the ordered triple $(1, 2, 3)$ is a solution to

$$x + y + 2z = 9$$

because $1 + 2 + 2(3) = 9$. Other ordered triples, such as $(4, 5, 0)$ or $(3, 4, 1)$, are also in the solution set to $x + y + 2z = 9$. In fact, there are infinitely many ordered triples in the solution set to a linear equation in three variables.

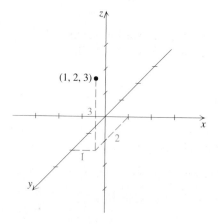

▪ **Figure 8.11**

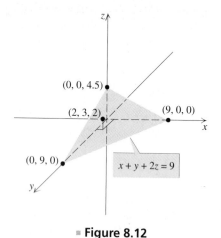

▪ **Figure 8.12**

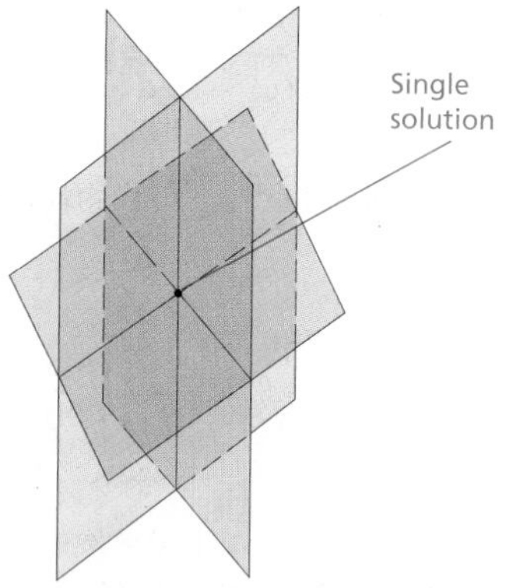

Independent System of Three Equations

▪ **Figure 8.13**

The graph of the solution set of a linear equation in three variables requires a three-dimensional coordinate system. A three-dimensional coordinate system has a z-axis through the origin of the xy-plane, as shown in Fig. 8.11. The third coordinate of a point indicates its distance above or below the xy-plane. The point $(1, 2, 3)$ is shown in Fig. 8.11.

The graph of a linear equation in three variables is a plane and not a line as the name might suggest. We think of a plane as an infinite sheet of paper (with no edges), but that is difficult to draw. One way to draw a representation of a plane in a three-dimensional coordinate system is to draw a triangle whose vertices are the points of intersection of the plane and the axes.

Example **1** Graphing a plane

Sketch the graph of $x + y + 2z = 9$ in a three-dimensional coordinate system by locating the three intercepts. Find one additional point that satisfies the equation and plot it.

Solution

Let x and y both be zero in the equation:

$$0 + 0 + 2z = 9$$

$$z = 4.5$$

So $(0, 0, 4.5)$ is the z-intercept of the plane. If x and z are both zero, then we get $y = 9$. So $(0, 9, 0)$ is the y-intercept. If y and z are both zero, then we get $x = 9$. So $(9, 0, 0)$ is the x-intercept. Plot the three intercepts and draw a triangle, as shown in Fig. 8.12. If $x = 2$ and $y = 3$, then $z = 2$. So $(2, 3, 2)$ satisfies the equation and appears to be on the plane when it is plotted as in Fig. 8.12. Of course there is more to a plane than the triangle in Fig. 8.12, but the triangle gives us an idea of the location of the plane.

Try This. Graph $x + 2y + z = 6$ in a three-dimensional coordinate system. ▪

It is fairly easy to draw a triangle through the intercepts to represent a plane, as we did in Example 1. However, if we were to draw two or three planes in this manner, it could be very difficult to see how they intersect. So we will often refer to the graphs of the equations to aid in understanding a system, but we will not attempt to solve a system by graphing. We will solve systems by using the algebraic methods of substitution, addition, or a combination of both.

Independent Systems

As with linear systems in two variables, a linear system in three variables can have one, infinitely many, or zero solutions. If the three equations correspond to three planes that intersect at a single point, as in Fig. 8.13, then the solution to the system is a single ordered triple. In this case, the system is called **independent.** Note that for simplicity, Fig. 8.13 is drawn without showing the coordinate axes.

Example **2** An independent system of equations

Solve the system.

(1) $x + y - z = 0$ (2) $3x - y + 3z = -2$ (3) $x + 2y - 3z = -1$

Solution

Look for a variable that is easy to eliminate by addition. Since y occurs in Eq. (1) and $-y$ occurs in Eq. (2), we can eliminate y by adding Eqs. (1) and (2):

$$x + y - z = 0$$
$$\underline{3x - y + 3z = -2}$$
$$(4) \quad 4x \qquad + 2z = -2$$

Now repeat the process to eliminate y from Eqs. (1) and (3). Multiply Eq. (1) by -2 and add the result to Eq. (3):

$$-2x - 2y + 2z = 0 \qquad \text{Eq. (1) multiplied by } -2$$
$$\underline{x + 2y - 3z = -1} \qquad \text{Eq. (3)}$$
$$(5) \quad -x \qquad - z = -1$$

Equations (4) and (5) are a system of two linear equations in two variables. We could solve this system by substitution or addition. To solve by addition, multiply Eq. (5) by 2 and add to Eq. (4) to eliminate z:

$$4x + 2z = -2 \qquad \text{Eq. (4)}$$
$$\underline{-2x - 2z = -2} \qquad \text{Eq. (5) multiplied by 2}$$
$$2x \qquad = -4$$
$$x = -2$$

Use $x = -2$ in $4x + 2z = -2$ to find z:

$$4(-2) + 2z = -2$$
$$2z = 6$$
$$z = 3$$

To find y, use $x = -2$ and $z = 3$ in $x + y - z = 0$ (Eq. 1):

$$-2 + y - 3 = 0$$
$$y = 5$$

Check that the ordered triple $(-2, 5, 3)$ satisfies all three of the original equations:

(1) $\qquad -2 + 5 - 3 = 0 \qquad$ Correct.

(2) $\quad 3(-2) - 5 + 3(3) = -2 \qquad$ Correct.

(3) $\quad -2 + 2(5) - 3(3) = -1 \qquad$ Correct.

The solution set is $\{(-2, 5, 3)\}$.

Try This. Solve $x + y + z = 9$, $x - y + 2z = 1$, and $x + y - z = 5$.   ■

When solving a system of three equations in three variables, we try to get a system of two equations in two variables by eliminating one of the variables. Any one of the three variables can be eliminated first, but we usually look for the easiest one. We may eliminate that variable from the first and second, the first and third, or the second and third equations. For example, we could have started Example 3 by

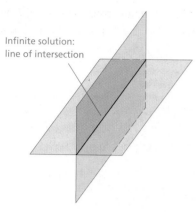

Infinite solution:
line of intersection

Dependent System of Two Equations

■ **Figure 8.14**

adding Eqs. (2) and (3) to eliminate z. We then repeat the process, eliminating the same variable from another pair of equations in the system. After getting two equations in two unknowns, we solve that system using methods for systems involving two variables.

Systems with Infinite Solution Sets

Two planes in three-dimensional space either are parallel or intersect along a line, as shown in Fig. 8.14. If the planes are parallel, there is no common point and no solution to the system. If the two planes intersect along a line, then there are infinitely many points on that line that satisfy both equations of the system. In our next example we solve a system of two equations whose graphs are planes that intersect along a line.

Example **3** **Two linear equations in three variables**

Solve the system.   (1) $-2x + 3y - z = -1$

(2) $x - 2y + z = 3$

Solution

Add the equations to eliminate z:

$$-2x + 3y - z = -1$$
$$\underline{x - 2y + z = 3}$$
$$-x + y \quad\ = 2$$

(3) $y = x + 2$

Equation (3) indicates that (x, y, z) satisfies both equations if and only if $y = x + 2$. Write Eq. (2) as $z = 3 - x + 2y$ and substitute $y = x + 2$ into this equation:

$$z = 3 - x + 2(x + 2)$$
$$z = x + 7$$

Now (x, y, z) satisfies (1) and (2) if and only if $y = x + 2$ and $z = x + 7$. So the solution set to the system could be written as

$$\{(x, y, z) \mid y = x + 2 \quad\text{and}\quad z = x + 7\}$$

or more simply

$$\{(x, x + 2, x + 7) \mid x \text{ is any real number}\}.$$

We write the solution set in this manner, because the system has infinitely many solutions. Every real number corresponds to a solution. For example, if $x = 1, 2$, or 3 in $(x, x + 2, x + 7)$ we get the solutions $(1, 3, 8)$, $(2, 4, 9)$, and $(3, 5, 10)$. Note that we can write the solution set in terms of x, y, or z. Since $z = x + 7$ and $y = x + 2$, we have $x = z - 7$ and $y = z - 7 + 2 = z - 5$. So the solution set can also be written as

$$\{(z - 7, z - 5, z) \mid z \text{ is any real number}\}.$$

Now if $z = 8, 9$, or 10 we get the solutions $(1, 3, 8)$, $(2, 4, 9)$, and $(3, 5, 10)$.

Try This. Solve $x + y + z = 2$ and $x - 2y - z = 4$.   ■

A line in three-dimensional space does not have a simple equation like a line in a two-dimensional coordinate system. In Example 3, the triple $(x, x + 2, x + 7)$ is a point on the line of intersection of the two planes, for any real number x. For example, the points $(0, 2, 7)$, $(1, 3, 8)$, and $(-2, 0, 5)$ all satisfy both equations of the system and lie on the line of intersection of the planes.

In the next example we solve a system consisting of three planes that intersect along a single line, as shown in Fig. 8.15. This example is similar to Example 3.

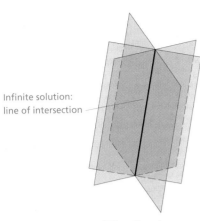

Infinite solution:
line of intersection

Dependent System of Three Equations

▪ **Figure 8.15**

Example 4 A dependent system of three equations

Solve the system.

$$(1) \qquad 2x + y - z = 1$$

$$(2) \quad -3x - 3y + 2z = 1$$

$$(3) \quad -10x - 14y + 8z = 10$$

Solution

Examine the system and decide which variable to eliminate. If Eq. (1) is multiplied by 2, then $-2z$ will appear in Eq. (1) and $2z$ in Eq. (2). So, multiply Eq. (1) by 2 and add the result to Eq. (2) to eliminate z:

$$4x + 2y - 2z = 2 \qquad \text{Eq. (1) multiplied by 2}$$
$$\underline{-3x - 3y + 2z = 1} \qquad \text{Eq. (2)}$$
$$(4) \qquad x - y \qquad\;\; = 3$$

Now eliminate z from Eqs. (2) and (3) by multiplying Eq. (2) by -4 and adding the result to Eq. (3):

$$12x + 12y - 8z = -4 \qquad \text{Eq. (2) multiplied by } -4$$
$$\underline{-10x - 14y + 8z = 10} \qquad \text{Eq. (3)}$$
$$(5) \qquad 2x - 2y \qquad\;\; = 6$$

Note that Eq. (5) is a multiple of Eq. (4). Since Eqs. (4) and (5) are dependent, the original system has infinitely many solutions. We can describe all solutions in terms of the single variable x. To do this, get $y = x - 3$ from Eq. (4). Then substitute $x - 3$ for y in Eq. (1) to find z in terms of x:

$$z = 2x + y - 1 \qquad \text{Eq. (1) solved for } z.$$
$$z = 2x + (x - 3) - 1 \qquad \text{Replace } y \text{ with } x - 3.$$
$$z = 3x - 4$$

An ordered triple (x, y, z) satisfies the system provided $y = x - 3$ and $z = 3x - 4$. So the solution set to the system is $\{(x, x - 3, 3x - 4) \,|\, x \text{ is any real number}\}$.

Try This. Solve $x + y + z = 1$, $x - y - z = 3$, and $3x + y + z = 5$. ▪

If all of the original equations are equivalent, then the solution set to the system is the set of all points that satisfy one of the equations. For example, the system

$$x + y + z = 1$$
$$2x + 2y + 2z = 2$$
$$3x + 3y + 3z = 3$$

has solution set $\{(x, y, z) \,|\, x + y + z = 1\}$.

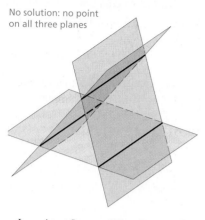

No solution: no point
on all three planes

Inconsistent System of Three Equations

■ **Figure 8.16**

Inconsistent Systems

There are several ways that three planes can be positioned so that there is no single point that is on all three planes. For example, Fig. 8.16 corresponds to a system where there are ordered triples that satisfy two equations, but no ordered triple that satisfies all three equations. Whatever the configuration of the planes, if there are no points in common to all three, the corresponding system is called **inconsistent** and has no solution. It is easy to identify a system that has no solution because a false statement will occur when we try to solve the system.

Example 5 A system with no solution

Solve the system.

$$(1) \quad x + y - z = 5$$
$$(2) \quad x + 2y - 3z = 9$$
$$(3) \quad x - y + 3z = 3$$

Solution

Multiply Eq. (1) by -2 and add the result to Eq. (2):

$$-2x - 2y + 2z = -10 \quad \text{Eq. (1) multiplied by } -2$$
$$\underline{x + 2y - 3z = 9} \quad\quad \text{Eq. (2)}$$
$$(4) \quad -x - z = -1$$

Add Eq. (1) and Eq. (3) to eliminate y:

$$x + y - z = 5 \quad \text{Eq. (1)}$$
$$\underline{x - y + 3z = 3} \quad \text{Eq. (3)}$$
$$2x + 2z = 8$$
$$(5) \quad\quad x + z = 4$$

Now add Eq. (4) and Eq. (5):

$$-x - z = -1 \quad \text{Eq. (4)}$$
$$\underline{x + z = 4} \quad\quad \text{Eq. (5)}$$
$$0 = 3$$

Since $0 = 3$ is false, the system is inconsistent. There is no solution.

Try This. Solve $x + y + z = 1$, $x - y - z = 3$, and $3x + y + z = 7$.   ■

Just as we find the equation of a line from two given points, we can find the equation of a parabola of the form $y = ax^2 + bx + c$ from three given points. Each of the three points determines a linear equation in the three variables a, b, and c. As long as the three points are not colinear, this system of three equations in three unknowns will have a unique solution as demonstrated in the next example.

Example 6 Finding the equation of a parabola given three points

Find the equation of the parabola opening upward ′ or ′ downward through $(-1, 10)$, $(1, 4)$, and $(2, 7)$.

Solution

Letting $x = -1$ and $y = 10$ in $y = ax^2 + bx + c$, we get $10 = a(-1)^2 + b(-1) + c$ or $10 = a - b + c$. For $x = 1$ and $y = 4$ we get $4 = a + b + c$, and for $x = 2$ and $y = 7$ we get $7 = 4a + 2b + c$. The three points determine the following system.

$$(1) \quad a - b + c = 10$$

$$(2) \quad a + b + c = 4$$

$$(3) \quad 4a + 2b + c = 7$$

Multiply (1) by -1 and add the result to (2) and to (3) to eliminate c:

$$(4) \qquad 2b = -6$$

$$(5) \quad 3a + 3b = -3$$

Equation (4) yields $b = -3$. Substitute $b = -3$ into (5) to get $a = 2$. Substitute $a = 2$ and $b = -3$ into (2) to get $2 - 3 + c = 4$ or $c = 5$. So the equation of the parabola through these three points is $y = 2x^2 - 3x + 5$.

Try This. Find the equation of the parabola that goes through $(-1, 1)$, $(1, 3)$, and $(2, 7)$.   ■

Modeling with a System of Three Equations

Problems that involve three unknown quantities can often be modeled with a system of three linear equations in three variables.

Example **7** **A problem involving three unknowns**

Lionel delivers milk, bread, and eggs to Marcie's Camp Store. On Monday the bill for eight half-gallons of milk, four loaves of bread, and six dozen eggs was $14.20. On Tuesday the bill for five half-gallons of milk, ten loaves of bread, and three dozen eggs was $14.50. On Wednesday the bill for two half-gallons of milk, five loaves of bread, and seven dozen eggs was $9.50. How much is Thursday's bill for one half-gallon of milk, two loaves of bread, and one dozen eggs?

Solution

Let x represent the price of a half-gallon of milk, y represent the price of a loaf of bread, and z represent the price of a dozen eggs. We can write an equation for the bill on each of the three days:

$$(1) \quad 8x + 4y + 6z = 14.20$$

$$(2) \quad 5x + 10y + 3z = 14.50$$

$$(3) \quad 2x + 5y + 7z = 9.50$$

Multiply Eq. (2) by -2 and add the result to Eq. (1):

$$\begin{array}{ll} 8x + 4y + 6z = 14.20 & \text{Eq. (1)} \\ \underline{-10x - 20y - 6z = -29} & \text{Eq. (2) multiplied by } -2 \\ -2x - 16y = -14.80 & \\ (4) \qquad x + 8y = 7.40 & \end{array}$$

Multiply Eq. (2) by -7, multiply Eq. (3) by 3, and add the results:

$$-35x - 70y - 21z = -101.50 \quad \text{Eq. (2) multiplied by } -7$$
$$\underline{6x + 15y + 21z = 28.50} \quad \text{Eq. (3) multiplied by 3}$$
$$(5) \quad -29x - 55y \quad\quad = -73$$

Multiply Eq. (4) by 29 and add the result to Eq. (5) to eliminate x:

$$29x + 232y = 214.60 \quad \text{Eq. (4) multiplied by 29}$$
$$\underline{-29x - 55y = -73} \quad \text{Eq. (5)}$$
$$177y = 141.60$$
$$y = 0.80$$

Use $y = 0.80$ in $x + 8y = 7.40$ (Eq. 4):

$$x + 8(0.80) = 7.40$$
$$x + 6.40 = 7.40$$
$$x = 1$$

Use $x = 1$ and $y = 0.80$ in $8x + 4y + 6z = 14.20$ (Eq. 1) to find z:

$$8(1) + 4(0.80) + 6z = 14.20$$
$$6z = 3$$
$$z = 0.50$$

Milk is \$1.00 per half-gallon, bread is 80 cents per loaf, and eggs are 50 cents per dozen. Thursday's bill should be \$3.10.

Try This. Two adults with one student and one child paid a total of \$14 for admission to the zoo. One adult with two students and one child paid a total of \$12. Two adults with two students and three children paid a total of \$19. Find the admissions prices for adults, students, and children. ■

For Thought

True or False? Explain.

The following systems are referenced in statements 1–7:

(a) $\quad x + y - z = 2$
$\quad\quad -x - y + z = 4$
$\quad\quad x - 2y + 3z = 9$

(b) $\ x + y - z = 6$
$\quad\quad x + y + z = 4$
$\quad\quad x - y - z = 8$

(c) $\quad x - y + z = 1$
$\quad\quad -x + y - z = -1$
$\quad\quad 2x - 2y + 2z = 2$

1. The point $(1, 1, 0)$ is in the solution set to $x + y - z = 2$.

2. The point $(1, 1, 0)$ is in the solution set to system (a).

3. System (a) is inconsistent.

4. The point $(2, 3, -1)$ satisfies all equations of system (b).

5. The point $(6, -1, -1)$ satisfies all equations of system (b).

6. The solution set to system (c) is
$\{(x, y, z) \mid x - y + z = 1\}$.

7. System (c) is dependent.

8. The solution set to $y = 2x + 3$ is $\{(x, 2x + 3) \mid x$ is any real number$\}$.

9. $(3, 1, 0) \in \{(x + 2, x, x - 1) \mid x$ is any real number$\}$.

10. x nickels, y dimes, and z quarters are worth $5x + 10y + 25z$ dollars.

8.2 Exercises

Sketch the graph of each equation in a three-dimensional coordinate system.

1. $x + y + z = 5$ **2.** $x + 2y + z = 6$

3. $x + y - z = 3$ **4.** $2x + y - z = 6$

Determine whether the given point is in the solution set to the given system.

5. $(1, 3, 2)$

$x + y + z = 6$
$x - y - z = -4$
$2x + y - z = 3$

6. $(-1, 2, 4)$

$x + y - z = -3$
$2x - 3y + z = -4$
$x - y + 3z = 9$

7. $(-1, 5, 2)$

$2x + y - z = 1$
$x - 2y + z = -9$
$x - y - 2z = -8$

8. $(3, 2, 1)$

$3x - y + z = 8$
$2x + y - z = 9$
$x - 3y + z = -2$

Solve each system of equations.

9.
$x + y + z = 6$
$2x - 2y - z = -5$
$3x + y - z = 2$

10.
$3x - y + 2z = 14$
$x + y - z = 0$
$2x - y + 3z = 18$

11.
$3x + 2y + z = 1$
$x + y - 2z = -4$
$2x - 3y + 3z = 1$

12.
$4x - 2y + z = 13$
$3x - y + 2z = 13$
$x + 3y - 3z = -10$

13.
$2x + y - 2z = -15$
$4x - 2y + z = 15$
$x + 3y + 2z = -5$

14.
$x - 2y - 3z = 4$
$2x - 4y + 5z = -3$
$5x - 6y + 4z = -7$

Find three ordered triples that belong to each of the following sets. Answers may vary.

15. $\{(x, x + 3, x - 5) \mid x \text{ is any real number}\}$

16. $\{(x, 2x - 4, x - 9) \mid x \text{ is any real number}\}$

17. $\{(2y, y, y - 7) \mid y \text{ is any real number}\}$

18. $\{(3 - z, 2 - z, z) \mid z \text{ is any real number}\}$

Fill in the blanks so that the two sets are equal.

19. $\{(x, x + 3, x - 5) \mid x \text{ is any real number}\}$
$= \{(\underline{\quad}, y, \underline{\quad}) \mid y \text{ is any real number}\}$

20. $\{(x, 2x, 3x) \mid x \text{ is any real number}\}$
$\left\{\left(\underline{\quad}, y, \underline{\quad}\right) \mid y \text{ is any real number}\right\}$

21. $\{(x, x + 1, x - 1) \mid x \text{ is any real number}\}$
$= \{(\underline{\quad}, \underline{\quad}, z) \mid z \text{ is any real number}\}$

22. $\{(x, x - 1, x + 5) \mid x \text{ is any real number}\}$
$= \{(\underline{\quad}, \underline{\quad}, z) \mid z \text{ is any real number}\}$

23. $\{(x, 2x + 1, 3x - 1) \mid x \text{ is any real number}\}$
$= \left\{\left(\underline{\quad}, y, \underline{\quad}\right) \mid y \text{ is any real number}\right\}$

24. $\{(x, 3x, 2x - 4) \mid x \text{ is any real number}\}$
$= \left\{\left(\underline{\quad}, \underline{\quad}, z\right) \mid z \text{ is any real number}\right\}$

Solve each system.

25.
$x + 2y - 3z = -17$
$3x - 2y - z = -3$

26.
$x + 2y + z = 4$
$2x - y - z = 3$

27.
$x + 2y - 3z = 5$
$-x - 2y + 3z = -5$
$2x + 4y - 6z = 10$

28.
$2x - 6y + 4z = 8$
$3x - 9y + 6z = 12$
$5x - 15y + 10z = 20$

29.
$x - 2y + 3z = 5$
$2x - 4y + 6z = 3$
$2x - 3y + z = 9$

30.
$-2x + y - 3z = 6$
$4x - y + z = 2$
$2x - y + 3z = 1$

31.
$x + y - z = 2$
$2x - y + z = 4$

32.
$-2x + 2y - z = 4$
$2x - y + z = 1$

33.
$x + y = 5$
$y - z = 2$
$x + z = 3$

34.
$2x - y = -1$
$-2x + z = 1$
$y - z = 0$

35.
$x - y + z = 7$
$2y - 3z = -13$
$3x - 2z = -3$

36.
$2x + y - z = 5$
$2y + 3z = -14$
$-3y - 2z = 11$

37.
$x + y + 2z = 7.5$
$3x + 4y + z = 12$
$5x + 2y + 5z = 21$

38. $100x + 200y + 500z = 47$

$350x + 5y + 250z = 33.9$

$200x + 80y + 100z = 23.4$

39. $x + y + z = 9000$

$0.05x + 0.06y + 0.09z = 710$

$z = 3y$

40. $x + y + z = 200{,}000$

$0.09x + 0.08y + 0.12z = 20{,}200$

$z = x + y$

41. $x = 2y - 1$

$y = 3z + 2$

$z = 2x - 3$

42. $x + 2y - 3z = 0$

$2x - y + z = 0$

$3x + y - 4z = 0$

Use a system of equations to find the parabola of the form
$y = ax^2 + bx + c$ *that goes through the three given points.*

43. $(-1, -2), (2, 1), (-2, 1)$ **44.** $(1, 2), (2, 3), (3, 6)$

45. $(0, 0), (1, 3), (2, 2)$

46. $(0, -6), (1, -3), (2, 6)$

47. $(0, 4), (-2, 0), (-3, 1)$

48. $(0, 6), (3, 0), (-1, 12)$

Write a linear equation in three variables that is satisfied by all three of the given ordered triples.

49. $(0, 0, 1), (0, 1, 0), (1, 0, 0)$

50. $(0, 0, 2), (0, 1, 0), (1, 0, 0)$

51. $(1, 1, 1), (0, 2, 0), (1, 0, 0)$

52. $(1, 0, 1), (2, 1, 0), (0, 2, 1)$

Solve each problem by using a system of three linear equations in three variables.

53. *Just Numbers* The sum of three numbers is 40. The difference between the largest and the smallest is 12, and the largest is equal to the sum of the two smaller numbers. Find the numbers.

54. *Perimeter* The perimeter of a triangle is 40 meters. The sum of the two shorter sides is 2 meters more than the longest side, and the longest side is 11 meters longer than the shortest side. Find the sides.

55. *Quizzes* The Rabbit had an average (mean) score of 7 on the first three College Algebra 101 quizzes. His second quiz score was one point higher than the first quiz score and the third was 4 points higher than the second. What were the three scores?

HINT Mean is the total of the scores divided by the number of scores.

56. *More Quizzes* Dr. M had an average (mean) score of 8 on the first two Chemistry 302 quizzes. After she took the third quiz her average was 12. If her score on the third quiz was 13 points higher than her score on the first quiz, then what were the three quiz scores?

57. *Stocks, Bonds, and a Mutual Fund* Marita invested a total of $25,000 in stocks, bonds, and a mutual fund. In one year she earned 8% on her stock investment, 10% on her bond investment, and 6% on her mutual fund, with a total return of $1860. Unfortunately, the amount invested in the mutual fund was twice as large as the amount she invested in the bonds. How much did she invest in each?

58. *Age-Groups* In 1980 the population of Springfield was 1911. In 1990 the number of people under 20 years old increased 10%, while the number of people in the 20 to 60 category decreased by 8%, and the number of people over 60 increased by one-third. If the 1990 population was 2136 and in 1990 the number of people over 60 was equal to the number of people 60 and under, then how many were in each age group in 1980?

59. *Fast Food Inflation* Last year you could get a hamburger, fries, and a Coke at Francisco's Drive-In for $1.90. Since the price of a hamburger has increased 10%, the price of fries has increased 20%, and the price of a Coke has increased 50%, the same meal now costs $2.37. If the price of a Coke is now 9 cents more than that of a hamburger, then what was the price of each item last year?

60. *Misplaced House Numbers* Angelo on Elm Street removed his three-digit house number for painting and noticed that the sum of the digits was 9 and that the units digit was three times as large as the hundreds digit. When the painters put the house number back up, they reversed the digits. The new house number was now 396 larger than the correct house number. What is Angelo's correct address?

61. *Weight Distribution*   A driver of a 1200-pound race car wants to have 51% of the car's weight on the left front and left rear tires and 48% of the car's weight on the left rear and right rear tires. If there must be at least 280 pounds on every tire, then find three possible weight distributions. Answers may vary.

62. *Burgers, Fries, and Cokes*   Jennifer bought 5 burgers, 7 orders of fries, and 6 Cokes for $11.25. Marylin bought 6 burgers, 8 orders of fries, and 7 Cokes for $13.20. John wants to buy an order of fries from Marylin for $0.80, but Marylin says that she paid more than that for the fries. What do you think? Find three possibilities for the prices of the burgers, fries, and Cokes. Answers may vary.

63. *Distribution of Coins*   Emma paid the $10.36 bill for her lunch with 232 coins consisting of pennies, nickels, and dimes. If the number of nickels plus the number of dimes was equal to the number of pennies, then how many coins of each type did she use?

64. *Students, Teachers, and Pickup Trucks*   Among the 564 students and teachers at Jefferson High School, 128 drive to school each day. One-fourth of the male students, one-sixth of the female students, and three-fourths of the teachers drive. Among those who drive to school, there are 41 who drive pickup trucks. If one-half of the driving male students, one-tenth of the driving female students, and one-third of the driving teachers drive pickups, then how many male students, female students, and teachers are there?

65. *Milk, Coffee, and Doughnuts*   The employees from maintenance go for coffee together every day at 9 A.M. On Monday, Hector paid $5.45 for three cartons of milk, four cups of coffee, and seven doughnuts. On Tuesday, Guillermo paid $5.30 for four milks, two coffees, and eight doughnuts. On Wednesday, Anna paid $5.15 for two milks, five coffees, and six doughnuts. On Thursday, Alphonse had to pay for five milks, two coffees, and nine doughnuts. How much change did he get back from his $10 bill?

66. *Average Age of Vehicles*   The average age of the Johnsons' cars is eight years. Three years ago the Toyota was twice as old as the Ford. Two years ago the sum of the Buick's and the Ford's ages was equal to the age of the Toyota. How old is each car now?

67. *Fish Mobile*   A sculptor is designing a fish mobile, as shown in the accompanying figure. The weights of the horizontal bars and strings are negligible compared to the cast-iron fish. The mobile will balance if the product of the weight and distance on one side of the balance point is equal to the product of the weight and distance on the other side. How much must the bottom three fish weigh for the mobile to be balanced?
HINT   Write an equation for each balancing point.

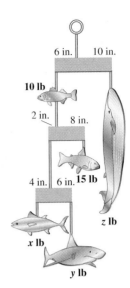

■ **Figure for Exercise 67**

68. *Efficiency for Descending Flight*   In studying the flight of birds, Vance Tucker measured the efficiency (the relationship between power input and power output) for parakeets flying at various speeds in a descending flight pattern. He recorded an efficiency of 0.18 at 12 mph, 0.23 at 22 mph, and 0.14 at 30 mph. Tucker's measurements suggest that efficiency E is a quadratic function of the speed s. Find the quadratic function whose graph goes through the three given ordered pairs, and find the speed that gives the maximum efficiency for descending flight.

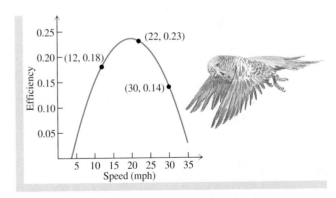

■ **Figure for Exercise 68**

69. *Path of an Arrow*   An arrow shot into the air follows the parabolic path shown in the figure on the next page. When the arrow is 10 meters from the origin, its altitude is 40 meters. When the arrow is 20 meters from the origin, its altitude is 70 meters.

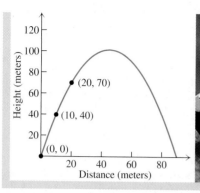

■ **Figure for Exercise 69**

a. Find the equation of the parabola.

b. What is the highest altitude reached by the arrow?

c. How many meters from the origin will the arrow land?

 70. *Quadratic Regression*   Use the data from Exercise 68 and quadratic regression on a graphing calculator to find the equation of the parabola that passes through the given points. Then use your calculator to find the *x*-coordinate corresponding to the maximum *y*-coordinate on the graph of the parabola. Compare your results to those in Exercise 68.

For Writing/Discussion

71. *Cooperative Learning*   Write a system of three linear equations in three unknowns for which $(1/2, 1/3, 1/4)$ is the only solution. Ask a classmate to solve the system.

72. *Cooperative Learning*   Write a word problem for which a system of three linear equations in three unknowns can be used to find the solution. Ask a classmate to solve the problem.

Thinking Outside the Box LXX

Hungry Workers   Two bricklayers and a contractor went to a fast food restaurant for lunch. The first bricklayer paid $14.25 for 8 hamburgers, 5 orders of fries, and 2 Cokes. The second bricklayer paid $8.51 for 5 hamburgers, 3 orders of fries, and 1 Coke. What did the contractor pay for 1 hamburger, 1 order of fries, and 1 Coke?

8.2 Pop Quiz

Solve each system.

1. $x + y + z = 22$
  $x + 2y - z = 11$
  $x - y + z = 8$

2. $2x - y + 2z = 4$
  $x + y - z = 2$

Linking Concepts

For Individual or Group Explorations

Accounting Problems

For Class C corporations in Louisiana the amount of state income tax is deductible on the federal income tax return and the amount of federal income tax is deductible on the state return. Assume that the state tax rate is 5%, the federal tax rate is 30%, and the corporation has an income of $200,000 before taxes.

a)   Write the amount of state tax as a function of the amount of federal tax.

b)   Write the amount of federal tax as a function of the amount of state tax.

c)   Solve your system to find the amount of state tax and the amount of federal tax.

d)   If the corporation wants to give a 20% bonus to employees, the accountant deducts the amount of the bonus, the state tax, and the federal tax from the $200,000 income to get the amount on which a 20% bonus is computed. The bonus and state tax are deductible before the federal tax is computed, and the federal tax and bonus are deductible before the state tax is computed. Find the amount of the bonus, the federal tax, and the state tax for this corporation.

Nonlinear Systems of Equations

In Sections 8.1 and 8.2 we solved linear systems of equations, but we have seen many equations in two variables that are not linear. Equations such as

$$y = 3x^2, \qquad y = \sqrt{x}, \qquad y = |x|, \qquad y = 10^x, \qquad \text{and} \qquad y = \log(x)$$

are called *nonlinear equations* because their graphs are not straight lines. If a system has at least one nonlinear equation, it is called a **nonlinear system.** Systems of non-linear equations arise in applications just as systems of linear equations do. In this section we will use the techniques that we learned for linear systems to solve nonlin-ear systems. We are seeking only the real solutions to the systems in this section.

Solving by Elimination of Variables

To solve nonlinear systems, we combine equations to eliminate variables just as we did for linear systems. However, since the graphs of nonlinear equations are not straight lines, the graphs might intersect at more than one point, and the solution set might contain more than one point. The next examples show systems whose solution sets contain two points, four points, and one point.

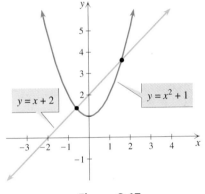

$y = x + 2$

$y = x^2 + 1$

■ **Figure 8.17**

Example **1** A parabola and a line

Solve the system of equations and sketch the graph of each equation on the same coordinate plane.

$$y = x^2 + 1$$
$$y - x = 2$$

Solution

The graph of $y = x^2 + 1$ is a parabola opening upward. The graph of $y = x + 2$ is a line with y-intercept $(0, 2)$ and slope 1. The graphs are shown in Fig. 8.17. To find the exact coordinates of the points of intersection of the graphs, we solve the system by substitution. Substitute $y = x^2 + 1$ into $y - x = 2$:

$$x^2 + 1 - x = 2$$
$$x^2 - x - 1 = 0$$

Solve this quadratic equation using the quadratic formula:

$$x = \frac{1 \pm \sqrt{1 - 4(1)(-1)}}{2(1)} = \frac{1 \pm \sqrt{5}}{2}$$

Use $x = \left(1 \pm \sqrt{5}\right)/2$ in $y = x + 2$ to find y:

$$y = \frac{1 \pm \sqrt{5}}{2} + 2 = \frac{5 \pm \sqrt{5}}{2}$$

The solution set to the system is

$$\left\{ \left(\frac{1 + \sqrt{5}}{2}, \frac{5 + \sqrt{5}}{2} \right), \left(\frac{1 - \sqrt{5}}{2}, \frac{5 - \sqrt{5}}{2} \right) \right\}.$$

To check the solution, use a calculator to find the approximations (1.62, 3.62) and (−0.62, 1.38). These points of intersection are consistent with Fig. 8.17. You should also check the decimal approximations in the original equations. Note that we could have solved this system by solving the second equation for x (or for y) and then substituting into the first.

Try This. Solve $y = x^2 - 1$ and $x + y = 5$. ■

In the next example we find the points of intersection for the graph of an absolute value function and a parabola.

Example **2** An absolute value function and a parabola

Solve the system of equations and sketch the graph of each equation on the same coordinate plane.

$$y = |3x|$$
$$y = x^2 + 2$$

Solution

The graph of $y = x^2 + 2$ is a parabola opening upward with vertex at (0, 2). The graph of $y = |3x|$ is V-shaped and passes through (0, 0) and (± 1, 3). Both graphs are shown in Fig. 8.18. To eliminate y, we can substitute $|3x|$ for y in the second equation:

$$|3x| = x^2 + 2$$

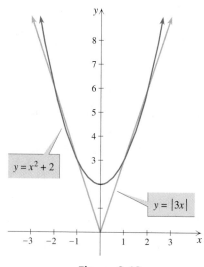

■ **Figure 8.18**

Write an equivalent compound equation without absolute value and solve:

$$3x = x^2 + 2 \quad \text{or} \qquad\qquad 3x = -(x^2 + 2)$$

$$x^2 - 3x + 2 = 0 \qquad \text{or} \qquad x^2 + 3x + 2 = 0$$

$$(x - 2)(x - 1) = 0 \qquad \text{or} \quad (x + 2)(x + 1) = 0$$

$$x = 2 \quad \text{or} \quad x = 1 \qquad \text{or} \quad x = -2 \quad \text{or} \quad x = -1$$

Using each of these values for x in the equation $y = |3x|$ yields the solution set $\{(-2, 6), (-1, 3), (1, 3), (2, 6)\}$. This solution set is consistent with what we see on the graphs in Fig. 8.18.

Try This. Solve $y = |2x|$ and $y = -x^2 + 2x + 5$.   ■

It is not necessary to draw the graphs of the equations of a nonlinear system to solve it. However, the graphs give us an idea of how many solutions to expect for the system. So graphs can be used to support our solutions.

Example **3**   Solving a nonlinear system

Solve the system of equations.

$$(1) \quad x^2 + y^2 = 25$$

$$(2) \quad \frac{x^2}{18} + \frac{y^2}{32} = 1$$

Solution

Write Eq. (1) as $y^2 = 25 - x^2$ and substitute into Eq. (2):

$$\frac{x^2}{18} + \frac{25 - x^2}{32} = 1$$

$$288\left(\frac{x^2}{18} + \frac{25 - x^2}{32}\right) = 288 \cdot 1 \qquad \text{The LCD for 18 and 32 is 288.}$$

$$16x^2 + 9(25 - x^2) = 288$$

$$7x^2 + 225 = 288$$

$$7x^2 = 63$$

$$x^2 = 9$$

$$x = \pm 3$$

Use $x = 3$ in $y^2 = 25 - x^2$: $y^2 = 25 - 3^2$

$$y^2 = 16$$

$$y = \pm 4$$

Use $x = -3$ in $y^2 = 25 - x^2$: $y^2 = 25 - (-3)^2$

$$y^2 = 16$$

$$y = \pm 4$$

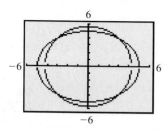

■ **Figure 8.19**

The solution set is $\{(-3, 4), (-3, -4), (3, 4), (3, -4)\}$. Check that all four ordered pairs satisfy both equations of the original system.

⊡ The graphs of $y = \pm\sqrt{25 - x^2}$ and $y = \pm\sqrt{32 - 32x^2/18}$ in Fig. 8.19 intersect at four points and support the conclusion that there are four solutions to the system.

Try This. Solve $x^2 + y^2 = 5$ and $4x^2 + 9y^2 = 35$.   ■

Example 4 Solving a nonlinear system

Solve the system of equations.

$$(1) \quad \frac{2}{x} + \frac{3}{y} = \frac{1}{2}$$

$$(2) \quad \frac{4}{x} + \frac{1}{y} = \frac{2}{3}$$

Solution

Since these equations have the same form, we can use addition to eliminate a variable. Multiply Eq. (2) by -3 and add the result to Eq. (1):

$$\frac{2}{x} + \frac{3}{y} = \frac{1}{2}$$

$$\frac{-12}{x} + \frac{-3}{y} = -2$$

$$\overline{\phantom{\frac{-12}{x}}}$$

$$\frac{-10}{x} \qquad = -\frac{3}{2}$$

$$3x = 20$$

$$x = \frac{20}{3}$$

Use $x = 20/3$ in Eq. (1) to find y:

$$\frac{2}{20/3} + \frac{3}{y} = \frac{1}{2}$$

$$\frac{3}{10} + \frac{3}{y} = \frac{1}{2}$$

$$3y + 30 = 5y \qquad \text{Multiply each side by } 10y.$$

$$30 = 2y$$

$$15 = y$$

The solution set is $\{(20/3, 15)\}$. Check this solution in the original system.

Try This. Solve $\frac{1}{x} + \frac{1}{y} = 2$ and $\frac{1}{x} - \frac{1}{y} = 4$.   ■

Modeling with a Nonlinear System

In the next example we solve a problem using a nonlinear system.

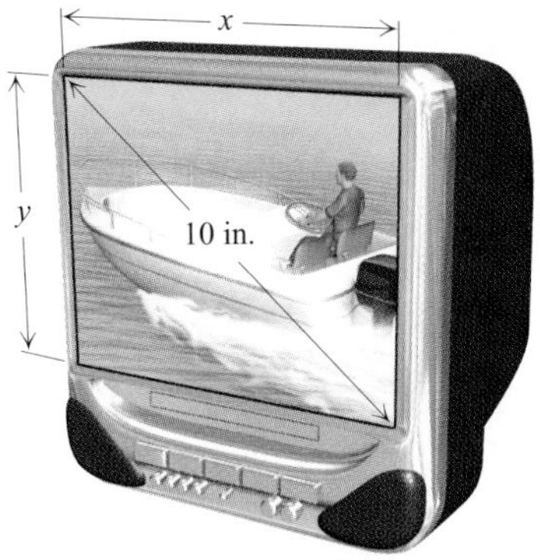

Example 5 Application of a nonlinear system

A 10-inch diagonal measure television is advertised as having a viewing area of 48 square inches. What are the width and height of the screen?

Solution

Let x be the width and y be the height, as shown in Fig. 8.20. Use the Pythagorean theorem to write the first equation and the formula $A = LW$ to write the second:

$$x^2 + y^2 = 10^2$$

$$xy = 48$$

Write $xy = 48$ as $y = 48/x$ and replace y by $48/x$ in the first equation:

$$x^2 + \left(\frac{48}{x}\right)^2 = 100$$

$$x^2 + \frac{2304}{x^2} = 100$$

$$x^4 + 2304 = 100x^2$$

$$x^4 - 100x^2 + 2304 = 0 \qquad \text{An equation of quadratic type}$$

$$(x^2 - 36)(x^2 - 64) = 0 \qquad \text{Solve by factoring.}$$

$$x^2 = 36 \quad \text{or} \quad x^2 = 64$$

$$x = \pm 6 \quad \text{or} \quad x = \pm 8$$

Since x cannot be negative in this situation, x is either 6 or 8. If $x = 6$ then $y = 8$; and if $x = 8$, then $y = 6$. The width of a television screen is usually larger than the height, so we conclude that the screen is 8 inches wide and 6 inches high.

Try This. The area of a rectangle with a 13-foot diagonal is 60 square feet. Find the length and width. ■

■ **Figure 8.20**

For Thought

True or False? Explain.

1. The line $y = x$ intersects the circle $x^2 + y^2 = 1$ at two points.

2. A line and a circle intersect at two points or not at all.

3. A parabola and a circle can intersect at three points.

4. The parabolas $y = x^2 - 1$ and $y = 1 - x^2$ do not intersect.

5. The line $y = x$ intersects $x^2 + y^2 = 2$ at $(1, 1)$, $(1, -1)$, $(-1, 1)$, and $(-1, -1)$.

6. Two distinct circles can intersect at more than two points.

7. The area of a right triangle is half the product of the lengths of its legs.

8. The surface area of a rectangular solid with length L, width W, and height H is $2LW + 2LH + 2WH$.

9. Two numbers with a sum of 6 and a product of 7 are $3 - \sqrt{2}$ and $3 + \sqrt{2}$.

10. It is impossible to find two numbers with a sum of 7 and a product of 1.

8.3 Exercises

Determine whether the given point is in the solution set to the given system.

1. $(-1, 4)$
$y - 2x = 6$
$y = x^2 + 3$

2. $(2, -3)$
$y = x - 5$
$y = |x - 3| - 4$

3. $(4, -5)$
$y = \sqrt{x} - 7$
$x - y = 1$

4. $(-3, -4)$
$x^2 + y^2 = 25$
$2x - 3y = 18$

Solve each system algebraically. Then graph both equations on the same coordinate system to support your solution.

5. $y = x^2$
$y = x$

6. $y = -x^2$
$y = -2x$

7. $5x - y = 6$
$y = x^2$

8. $2x^2 - y = 8$
$7x + y = -4$

9. $y - x = 3$
$y = |x|$

10. $y = |x| - 1$
$2y - x = 1$

11. $y = |x|$
$y = x^2$

12. $y = 2|x| - 1$
$y = x^2$

13. $y = \sqrt{x}$
$y = 2x$

14. $y = \sqrt{x + 3}$
$x - 4y = -7$

15. $y = x^3$
$y = 4x$

16. $y = x^2$
$x = y^2$

17. $y = x^3 - x$
$y = x$

18. $y = x^4 - x^2$
$y = x^2$

19. $x^2 + y^2 = 1$
$y = x$

20. $x^2 + y^2 = 25$
$2x - 3y = -6$

Solve each system.

21. $x + y = -4$
$xy = 1$

22. $x + y = 10$
$xy = 21$

23. $2x^2 - y^2 = 1$
$x^2 - 2y^2 = -1$

24. $x^2 + y^2 = 5$
$x^2 + 4y^2 = 14$

25. $xy - 2x = 2$
$2x - y = 1$

26. $y - xy = -10$
$x - 2y = -7$

27. $\dfrac{3}{x} - \dfrac{1}{y} = \dfrac{13}{10}$
$\dfrac{1}{x} + \dfrac{2}{y} = \dfrac{9}{10}$

28. $\dfrac{2}{x} + \dfrac{3}{2y} = \dfrac{11}{4}$
$\dfrac{5}{2x} - \dfrac{2}{y} = \dfrac{3}{2}$

29. $x^2 + xy - y^2 = -5$
$x + y = 1$

30. $x^2 + xy + y^2 = 12$
$x + y = 2$

31. $x^2 + 2xy - 2y^2 = -11$
$-x^2 - xy + 2y^2 = 9$

32. $-3x^2 + 2xy - y^2 = -9$
$3x^2 - xy + y^2 = 15$

Solve each system of exponential or logarithmic equations.

33. $y = 2^{x+1}$
$y = 4^{-x}$

34. $y = 3^{2x+1}$
$y = 9^{-x}$

35. $y = \log_2(x)$
$y = \log_4(x + 2)$

36. $y = \log_2(-x)$
$y = \log_2(x + 4)$

37. $y = \log_2(x + 2)$
$y = 3 - \log_2(x)$

38. $y = \log(2x + 4)$
$y = 1 + \log(x - 2)$

39. $y = 3^x$
$y = 2^x$

40. $y = 6^{x-1}$
$y = 2^{x+1}$

Using a graphing calculator, we can solve systems that are too difficult to solve algebraically. Solve each system by graphing its equations on a graphing calculator and finding points of intersection to the nearest tenth.

41. $y = \log_2(x)$
$y = 2^x - 3$

42. $x^2 - y^2 = 1$
$y = 2^{x-3}$

43. $x^2 + y^2 = 4$
$y = \log_3(x)$

44. $y = \dfrac{x^3}{6} + \dfrac{x^2}{2} + x + 1$
$y = e^x$

45. $y = x^2$
$y = 2^x$

46. $y = e^x$
$y = x^3 + 1$

Solve each problem using a system of two equations in two unknowns.

47. *Unknown Numbers* Find two numbers whose sum is 6 and whose product is -16.

48. *More Unknown Numbers* Find two numbers whose sum is -8 and whose product is -20.

49. *Legs of a Right Triangle* Find the lengths of the legs of a right triangle whose hypotenuse is 15 m and whose area is 54 m^2.

50. *Sides of a Rectangle*   What are the length and width of a rectangle that has a perimeter of 98 cm and a diagonal of 35 cm?

51. *Sides of a Triangle*   Find the lengths of the sides of a triangle whose perimeter is 12 ft and whose angles are 30°, 60°, and 90°.

> HINT Use a, $a/2$, and b to represent the lengths of the sides.

52. *Size of a Vent*   Kwan is constructing a triangular vent in the gable end of a house, as shown in the accompanying diagram. If the pitch of the roof is 6–12 (run 12 ft and rise 6 ft) and the vent must have an area of 4.5 ft², then what size should he make the base and height of the triangle?

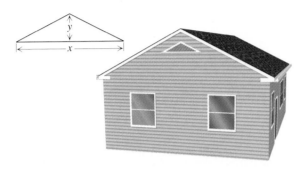

■ **Figure for Exercise 52**

53. *Air Mobile*   A hobbyist is building a mobile out of model airplanes, as shown in the accompanying figure. Find x and y so that the mobile will balance. Ignore the weights of the horizontal bars and the strings. The mobile will be in balance if the product of the weight and distance on one side of the balance point is equal to the product of the weight and distance on the other side.

> HINT Write an equation for each balancing point.

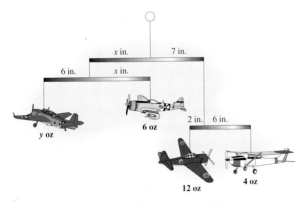

■ **Figure for Exercise 53**

54. *Making an Arch*   A bricklayer is constructing a circular arch with a radius of 9 ft, as shown in the figure. If the height h must be 1.5 times as large as the width w, then what are h and w?

> HINT The center of the circle is the midpoint of w.

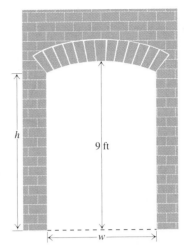

■ **Figure for Exercise 54**

55. *Pumping Tomato Soup*   At the Acme Soup Company a large vat of tomato soup can be filled in 8 min by pump A and pump B working together. Pump B can be reversed so that it empties the vat at the same rate at which it fills the vat. One day a worker filled the vat in 12 min using pumps A and B, but accidentally ran pump B in reverse. How long does it take each pump to fill the vat working alone?

56. *Planting Strawberries*   Blanche and Morris can plant an acre of strawberries in 8 hr working together. Morris takes 2 hr longer to plant an acre of strawberries working alone than it takes Blanche working alone. How long does it take Morris to plant an acre by himself?

57. *Lost Numbers*   Find two complex numbers whose sum is 6 and whose product is 10.

58. *More Lost Numbers*   Find two complex numbers whose sum is 1 and whose product is 5.

59. *Voyage of the Whales*   In one of Captain James Kirk's most challenging missions, he returned to late-twentieth-century San Francisco to bring back a pair of humpback whales. Chief Engineer Scotty built a tank of transparent aluminum to hold the time-traveling cetaceans. If Scotty's 20-ft-high tank had a volume of 36,000 ft³, and it took 7200 ft² of transparent aluminum to cover all six sides, then what were the length and width of the tank?

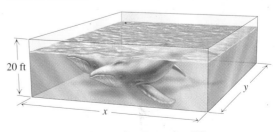

■ **Figure for Exercise 59**

60. *Dimensions of a Laundry Room* The plans call for a rectangular laundry room in the Wilson's new house. Connie wants to increase the width by 1 ft and the length by 2 ft, which will increase the area by 30 ft². Christopher wants to increase the width by 2 ft and decrease the length by 3 ft, which will decrease the area by 6 ft². What are the original dimensions for the laundry room?

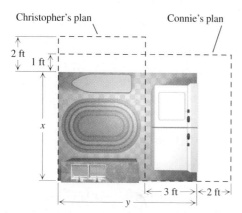

▪ **Figure for Exercise 60**

61. *Two Models* One demographer believes that the population growth of a certain country is best modeled by the function $P(t) = 20e^{0.07t}$, while a second demographer believes that the population growth of that same country is best modeled by the function $P(t) = 20 + 2t$. In each case t is the number of years from the present and $P(t)$ is given in millions of people. For what values of t do these two models give the same population? In how many years is the population predicted by the exponential model twice as large as the population predicted by the linear model?

62. *Simple or Compound* Harvey borrowed $2000 from Chet (his brother-in-law) and made no payments. After t years Harvey won the lottery and decided to pay up. However, Chet figured the value of the debt at twice what Harvey figured it. They both used 5% annual interest, but Harvey used simple interest and Chet used interest compounded continuously. Find t to the nearest day.

63. *Mixed Signals* Sally and Bob live quite far from each other on Interstate 75 and decide to get together on Saturday. However, they get their signals crossed and they each start out at sunrise for the other's house. At noon they pass each other on the freeway. At 4 P.M. Sally arrives at Bob's house and at 9 P.M. Bob arrives at Sally's house. At what time did the sun rise? See the figure.

▪ **Figure for Exercise 63**

64. *Westward Ho* A wagon train that is one mile long advances one mile at a constant rate. During the same time period, the wagon master rides his horse at a constant rate from the front of the wagon train to the rear, and then back to the front. How far did the wagon master ride?

65. *Traveling Mouse* A rectangular train car that is 40 feet long and 10 feet wide is traveling north at a constant rate. A mouse starts from the left rear corner (southwest corner) of the car and runs around the perimeter of the car in a clockwise direction (first going north) at a constant rate. The mouse is back where he started when the train has advanced 40 feet. On the train car the mouse has traveled 100 feet, but how far did he travel relative to the ground?

For Writing/Discussion

66. *Cooperative Learning* Write a nonlinear system of two equations that has three points in its solution set and a nonlinear system of two equations that has no solution. Ask a classmate to solve your systems.

67. *How Many?* Explain why the system

$$y = 3 + \log_2(x - 1)$$
$$8x - 8 = 2^y$$

has infinitely many solutions.

68. *Line and Circle* For what values of b does the solution set to $y = x + b$ and $x^2 + y^2 = 1$ consist of one point, two points, and no points? Explain.

Thinking Outside the Box LXXI & LXXII

Perfect Squares   Solve the system:

$$x + \sqrt{y} = 32$$
$$y + \sqrt{x} = 54$$

From Left to Right   Find the smallest positive integer whose first digit on the left is 1, such that multiplying the number by 3 simply moves the 1 to the first position on the right.

HINT   If 147 times 3 were 471, we would have the required number.

8.3 Pop Quiz

Solve each system.

1. $y = x^2$

$y = x$

2. $y = |x| + 1$

$y = \dfrac{1}{2}x + 4$

3. $x^2 - y^2 = 5$

$x^2 + y^2 = 13$

Linking Concepts

For Individual or Group Explorations

Measuring Ocean Depth

Geophysicists who map the ocean floor use sound reflection to measure the depth of the ocean. The problem is complicated by the fact that the speed of sound in water is not constant, but depends on the temperature and other conditions of the water. Let v be the velocity of sound through the water and d_1 be the depth of the ocean below the ship, as shown in the figure.

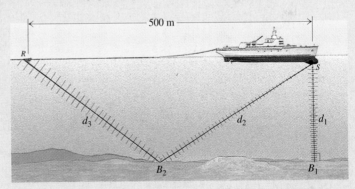

a)   The time that it takes for sound to travel to the ocean floor at point B_1 and back to the ship at point S is measured as 0.280 sec. Write d_1 as a function of v.

b)   It takes 0.446 sec for sound to travel from point S to point B_2 and then to a receiver at R, which is towed 500 meters behind the ship. Assuming that $d_2 = d_3$, write d_2 as a function of v.

c)   Write d_1 as a function of d_2.

d)   Solve your system to find the ocean depth d_1.

8.4

Partial Fractions

In algebra we usually learn a process and then learn to reverse it. For example, we multiply two binomials, and then we factor trinomials into a product of two binomials. We solve polynomial equations, and we write polynomial equations with given solutions. After we studied functions, we learned about their inverses. We have learned how to add rational expressions. Now we will reverse the process of addition. We start with a rational expression and write it as a sum of two or more simpler rational expressions. This technique shows a nice application of systems of equations.

The Basic Idea

Before trying to reverse addition of rational expressions, recall how to add them.

Example **1** Adding rational expressions

Perform the indicated operation.

$$\frac{3}{x-3} + \frac{-2}{x+1}$$

Solution

The least common denominator (LCD) for $x - 3$ and $x + 1$ is $(x - 3)(x + 1)$. We convert each rational expression or fraction into an equivalent fraction with this denominator:

$$\frac{3}{x-3} + \frac{-2}{x+1} = \frac{3(x+1)}{(x-3)(x+1)} + \frac{-2(x-3)}{(x+1)(x-3)}$$

$$= \frac{3x+3}{(x-3)(x+1)} + \frac{-2x+6}{(x-3)(x+1)}$$

$$= \frac{x+9}{(x-3)(x+1)}$$

Try This. Find the sum $\dfrac{4}{x-2} + \dfrac{3}{x+5}$.　　　　　　　　■

The following rational expression is similar to the result in Example 1.

$$\frac{8x-7}{(x+1)(x-2)}$$

Thus, it is possible that this fraction is the sum of two fractions with denominators $x + 1$ and $x - 2$. In the next example, we will find those two fractions.

Example **2** Reversing the addition of rational expressions

Write the following rational expression as a sum of two rational expressions.

$$\frac{8x-7}{(x+1)(x-2)}$$

Solution

To write the given expression as a sum, we need numbers A and B such that

$$\frac{8x - 7}{(x + 1)(x - 2)} = \frac{A}{x + 1} + \frac{B}{x - 2}.$$

Simplify this equation by multiplying each side by the LCD, $(x + 1)(x - 2)$:

$$(x + 1)(x - 2)\frac{8x - 7}{(x + 1)(x - 2)} = (x + 1)(x - 2)\left(\frac{A}{x + 1} + \frac{B}{x - 2}\right)$$

$$8x - 7 = A(x - 2) + B(x + 1)$$

$$8x - 7 = Ax - 2A + Bx + B$$

$$8x - 7 = (A + B)x - 2A + B \qquad \text{Combine like terms.}$$

Since the last equation is an identity, the coefficient of x on one side equals the coefficient of x on the other, and the constant on one side equals the constant on the other. So A and B satisfy the following two equations.

$$A + B = 8$$

$$-2A + B = -7$$

We can solve this system of two linear equations in two unknowns by addition:

$$A + B = 8$$

$$\underline{2A - B = 7} \qquad \text{Second equation multiplied by } -1$$

$$3A \qquad = 15$$

$$A = 5$$

If $A = 5$, then $B = 3$, and we have

$$\frac{8x - 7}{(x + 1)(x - 2)} = \frac{5}{x + 1} + \frac{3}{x - 2}.$$

Check by adding the fractions on the right-hand side of the equation.

Try This. Write $\dfrac{2x - 3}{(x - 1)(x + 3)}$ as a sum of two rational expressions. ▪

Each of the two fractions on the right-hand side of the equation

$$\frac{8x - 7}{(x + 1)(x - 2)} = \frac{5}{x + 1} + \frac{3}{x - 2}$$

is called a **partial fraction**. This equation shows the **partial fraction decomposition** of the rational expression on the left-hand side.

General Decomposition

In general, let $N(x)$ be the polynomial in the numerator and $D(x)$ be the polynomial in the denominator of the fraction that is to be decomposed. *We will decompose only fractions for which the degree of the numerator is smaller than the degree of the denominator.* If the degree of $N(x)$ is not smaller than the degree of $D(x)$, we can use long division to write the rational expression as quotient + remainder/divisor. For example,

$$\frac{x^3 + x^2 - x + 5}{x^2 + x - 6} = x + \frac{5x + 5}{x^2 + x - 6} = x + \frac{A}{x + 3} + \frac{B}{x - 2}.$$

You should find the values of A and B in the above equation as we did in Example 2, and check.

If a factor of D(x) is repeated n times, then all powers of the factor from 1 through n might occur as denominators in the partial fractions. To understand the reason for this statement, look at

$$\frac{7}{8} = \frac{1}{2} + \frac{1}{4} + \frac{1}{8} = \frac{1}{2} + \frac{1}{2^2} + \frac{1}{2^3}.$$

Here the factor 2 is repeated three times in $D(x) = 8$. Notice that each of the powers of 2 $(2^1, 2^2, 2^3)$ occurs in the denominators of the partial fractions.

Now consider a fraction $N(x)/D(x)$, where $D(x) = (x - 1)(x + 3)^2$. Since the factor $x + 3$ occurs to the second power, both $x + 3$ and $(x + 3)^2$ occur in the partial fraction decomposition of $N(x)/D(x)$. For example, we write the partial fraction decomposition for $(3x^2 + 17x + 12)/D(x)$ as follows:

$$\frac{3x^2 + 17x + 12}{(x - 1)(x + 3)^2} = \frac{A}{x - 1} + \frac{B}{x + 3} + \frac{C}{(x + 3)^2}$$

It is possible that we do not need the fraction $B/(x + 3)$, but we do not know this until we find the value of B. If $B = 0$, then the decomposition does not include a fraction with denominator $x + 3$. This decomposition is completed in Example 3.

To find the partial fraction decomposition of a rational expression, the denominator must be factored into a product of prime polynomials. *If a quadratic prime polynomial occurs in the denominator, then the numerator of the partial fraction for that polynomial is of the form Ax + B.* For example, if $D(x) = x^3 + x^2 + 4x + 4$, then $D(x) = (x^2 + 4)(x + 1)$. The partial fraction decomposition for $(5x^2 + 3x + 13)/D(x)$ is written as follows:

$$\frac{5x^2 + 3x + 13}{x^3 + x^2 + 4x + 4} = \frac{Ax + B}{x^2 + 4} + \frac{C}{x + 1}$$

This decomposition is completed in Example 4.

The main points to remember for partial fraction decomposition are summarized as follows.

S T R A T E G Y   Decomposition into Partial Fractions

To decompose a rational expression $N(x)/D(x)$ into partial fractions, use the following strategies:

1. If the degree of the numerator $N(x)$ is greater than or equal to the degree of the denominator $D(x)$, use division to express $N(x)/D(x)$ as quotient + remainder/divisor and decompose the resulting fraction.

2. If the degree of $N(x)$ is less than the degree of $D(x)$, factor the denominator completely into prime factors that are either linear $(ax + b)$ or quadratic $(ax^2 + bx + c)$.

3. For each linear factor of the form $(ax + b)^n$, the partial fraction decomposition must include the following fractions:

$$\frac{A_1}{ax + b} + \frac{A_2}{(ax + b)^2} + \cdots + \frac{A_n}{(ax + b)^n}$$

4. For each quadratic factor of the form $(ax^2 + bx + c)^m$, the partial fraction decomposition must include the following fractions:

$$\frac{B_1x + C_1}{ax^2 + bx + c} + \frac{B_2x + C_2}{(ax^2 + bx + c)^2} + \cdots + \frac{B_mx + C_m}{(ax^2 + bx + c)^m}$$

5. Set up and solve a system of equations involving the As, Bs, and/or Cs.

The strategy for decomposition applies to very complicated rational expressions. To actually carry out the decomposition, we must be able to solve the system of equations that arises. Theoretically, we can solve systems of many equations in many unknowns. Practically, we are limited to fairly simple systems of equations. Large systems of equations are generally solved by computers or calculators using techniques that we will develop in the next chapter.

Example 3 Repeated linear factor

Find the partial fraction decomposition for the rational expression

$$\frac{3x^2 + 17x + 12}{(x - 1)(x + 3)^2}.$$

Solution

Since the factor $x + 3$ occurs twice in the original denominator, it might occur in the partial fractions with powers 1 and 2:

$$\frac{3x^2 + 17x + 12}{(x - 1)(x + 3)^2} = \frac{A}{x - 1} + \frac{B}{x + 3} + \frac{C}{(x + 3)^2}$$

Multiply each side of the equation by the LCD, $(x - 1)(x + 3)^2$.

$$3x^2 + 17x + 12 = A(x + 3)^2 + B(x - 1)(x + 3) + C(x - 1)$$
$$= Ax^2 + 6Ax + 9A + Bx^2 + 2Bx - 3B + Cx - C$$
$$= (A + B)x^2 + (6A + 2B + C)x + 9A - 3B - C$$

Next, write a system of equations by equating the coefficients of like terms from opposite sides of the equation. The corresponding coefficients are highlighted above.

$$A + B \qquad = 3$$
$$6A + 2B + C = 17$$
$$9A - 3B - C = 12$$

We can solve this system of three equations in the variables A, B, and C by first eliminating C. Add the last two equations to get $15A - B = 29$. Add this equation to $A + B = 3$:

$$15A - B = 29$$
$$\underline{A + B = 3}$$
$$16A = 32$$
$$A = 2$$

If $A = 2$ and $A + B = 3$, then $B = 1$. Use $A = 2$ and $B = 1$ in the equation $6A + 2B + C = 17$:

$$6(2) + 2(1) + C = 17$$
$$C = 3$$

The partial fraction decomposition is written as follows:

$$\frac{3x^2 + 17x + 12}{(x - 1)(x + 3)^2} = \frac{2}{x - 1} + \frac{1}{x + 3} + \frac{3}{(x + 3)^2}$$

Try This. Find the partial fraction decomposition for $\dfrac{4x^2 + 4x - 4}{(x + 1)^2(x - 1)}$. ■

Example 4 Single prime quadratic factor

Find the partial fraction decomposition for the rational expression

$$\frac{5x^2 + 3x + 13}{x^3 + x^2 + 4x + 4}.$$

Solution

Factor the denominator by grouping:

$$x^3 + x^2 + 4x + 4 = x^2(x + 1) + 4(x + 1) = (x^2 + 4)(x + 1)$$

Write the partial fraction decomposition:

$$\frac{5x^2 + 3x + 13}{x^3 + x^2 + 4x + 4} = \frac{Ax + B}{x^2 + 4} + \frac{C}{x + 1}$$

Note that $Ax + B$ is used over the prime quadratic polynomial $x^2 + 4$. Multiply each side of this equation by the LCD, $(x^2 + 4)(x + 1)$:

$$5x^2 + 3x + 13 = (Ax + B)(x + 1) + C(x^2 + 4)$$
$$= Ax^2 + Bx + Ax + B + Cx^2 + 4C$$
$$= (A + C)x^2 + (A + B)x + B + 4C$$

Write a system of equations by equating the coefficients of like terms from opposite sides of the last equation:

$$A + C = 5$$
$$A + B = 3$$
$$B + 4C = 13$$

One way to solve the system is to substitute $C = 5 - A$ and $B = 3 - A$ into $B + 4C = 13$:

$$3 - A + 4(5 - A) = 13$$
$$3 - A + 20 - 4A = 13$$
$$23 - 5A = 13$$
$$-5A = -10$$
$$A = 2$$

Since $A = 2$, we get $C = 5 - 2 = 3$ and $B = 3 - 2 = 1$. So the partial fraction decomposition is written as follows:

$$\frac{5x^2 + 3x + 13}{x^3 + x^2 + 4x + 4} = \frac{2x + 1}{x^2 + 4} + \frac{3}{x + 1}$$

Try This. Find the partial fraction decomposition for $\dfrac{5x^2 - 4x + 11}{(x^2 + 2)(x - 1)}$. ■

Example 5 Repeated prime quadratic factor

Find the partial fraction decomposition for the rational expression

$$\frac{4x^3 - 2x^2 + 7x - 6}{4x^4 + 12x^2 + 9}.$$

Solution

The denominator factors as $(2x^2 + 3)^2$, and $2x^2 + 3$ is prime. Write the partial fractions using denominators $2x^2 + 3$ and $(2x^2 + 3)^2$.

$$\frac{4x^3 - 2x^2 + 7x - 6}{(2x^2 + 3)^2} = \frac{Ax + B}{2x^2 + 3} + \frac{Cx + D}{(2x^2 + 3)^2}$$

Multiply each side of the equation by the LCD, $(2x^2 + 3)^2$, to get the following equation:

$$4x^3 - 2x^2 + 7x - 6 = (Ax + B)(2x^2 + 3) + Cx + D$$
$$= 2Ax^3 + 2Bx^2 + 3Ax + 3B + Cx + D$$
$$= 2Ax^3 + 2Bx^2 + (3A + C)x + 3B + D$$

Equating the coefficients produces the following system of equations:

$$2A = 4$$
$$2B = -2$$
$$3A + C = 7$$
$$3B + D = -6$$

From the first two equations $2A = 4$ and $2B = -2$, we get $A = 2$ and $B = -1$. Using $A = 2$ in $3A + C = 7$ gives $C = 1$. Using $B = -1$ in $3B + D = -6$ gives $D = -3$. So the partial fraction decomposition is written as follows:

$$\frac{4x^3 - 2x^2 + 7x - 6}{(2x^2 + 3)^2} = \frac{2x - 1}{2x^2 + 3} + \frac{x - 3}{(2x^2 + 3)^2}$$

Try This. Find the partial fraction decomposition for $\dfrac{3x^3 + 3x^2 + x - 2}{(3x^2 - 1)^2}$.   ▪

For Thought

True or False? Explain.

1. $\dfrac{1}{x} + \dfrac{3}{x + 1} = \dfrac{4x + 1}{x^2 + x}$ for any real number x except 0 and -1.

2. $x + \dfrac{3x}{x^2 - 1} = \dfrac{x^3 + 2x}{x^2 - 1}$ for any real number x except -1 and 1.

3. The partial fraction decomposition of $\dfrac{x^2}{x^2 - 9}$ is

$\dfrac{x^2}{x^2 - 9} = \dfrac{A}{x - 3} + \dfrac{B}{x + 3}$.

4. In the decomposition $\dfrac{5}{8} = \dfrac{A}{2} + \dfrac{B}{2^2} + \dfrac{C}{2^3}$, $A = 1$, $B = 0$, and $C = 1$.

5. The partial fraction decomposition of $\dfrac{3x - 1}{x^3 + x}$ is

$\dfrac{3x - 1}{x^3 + x} = \dfrac{A}{x} + \dfrac{B}{x^2 + 1}$.

6. $\dfrac{1}{x^2 - 1} = \dfrac{1}{x - 1} + \dfrac{1}{x + 1}$ for any real number except 1 and -1.

7. $\dfrac{x^3 + 1}{x^2 + x - 2} = x - 1 + \dfrac{3x - 1}{x^2 + x - 2}$ for any real number except 1 and -2.

8. $x^3 - 8 = (x - 2)(x^2 + 4x + 4)$ for any real number x.

9. $\dfrac{x^2 + 2x}{x^3 - 1} = \dfrac{1}{x - 1} + \dfrac{1}{x^2 + x + 1}$ for any real number except 1.

10. There is no partial fraction decomposition for $\dfrac{2x}{x^2 + 9}$.

8.4 Exercises

Perform the indicated operations.

1. $\dfrac{3}{x - 2} + \dfrac{4}{x + 1}$

2. $\dfrac{-1}{x + 5} + \dfrac{-3}{x - 4}$

3. $\dfrac{1}{x - 1} + \dfrac{-3}{x^2 + 2}$

4. $\dfrac{x + 3}{x^2 + x + 1} + \dfrac{1}{x - 1}$

5. $\dfrac{2x + 1}{x^2 + 3} + \dfrac{x^3 + 2x + 2}{(x^2 + 3)^2}$

6. $\dfrac{3x - 1}{x^2 + x - 3} + \dfrac{x^3 + x - 1}{(x^2 + x - 3)^2}$

7. $\dfrac{1}{x - 1} + \dfrac{2x + 3}{(x - 1)^2} + \dfrac{x^2 + 1}{(x - 1)^3}$

8. $\dfrac{3}{x + 2} + \dfrac{x - 1}{(x + 2)^2} + \dfrac{1}{x^2 + 2}$

Find A and B for each partial fraction decomposition.

9. $\dfrac{12}{x^2 - 9} = \dfrac{A}{x - 3} + \dfrac{B}{x + 3}$

10. $\dfrac{5x + 2}{x^2 - 4} = \dfrac{A}{x - 2} + \dfrac{B}{x + 2}$

Find the partial fraction decomposition for each rational expression. See the strategy for decomposition into partial fractions on page 690.

11. $\dfrac{5x - 1}{(x + 1)(x - 2)}$

12. $\dfrac{-3x - 5}{(x + 3)(x - 1)}$

13. $\dfrac{2x + 5}{x^2 + 6x + 8}$

14. $\dfrac{x + 2}{x^2 + 12x + 32}$

15. $\dfrac{2}{x^2 - 9}$

16. $\dfrac{1}{9x^2 - 1}$

17. $\dfrac{1}{x^2 - x}$

18. $\dfrac{2}{x^2 - 2x}$

Find A, B, and C for each partial fraction decomposition.

19. $\dfrac{x^2 + x - 31}{(x + 3)^2(x - 2)} = \dfrac{A}{x + 3} + \dfrac{B}{(x + 3)^2} + \dfrac{C}{x - 2}$

20. $\dfrac{-x^2 - 9x - 12}{(x + 1)^2(x - 3)} = \dfrac{A}{x + 1} + \dfrac{B}{(x + 1)^2} + \dfrac{C}{x - 3}$

Find each partial fraction decomposition.

21. $\dfrac{4x - 1}{(x - 1)^2(x + 2)}$

22. $\dfrac{5x^2 - 15x + 7}{(x - 2)^2(x + 1)}$

23. $\dfrac{20 - 4x}{(x + 4)(x - 2)^2}$

24. $\dfrac{x^2 - 3x + 14}{(x + 3)(x - 1)^2}$

25. $\dfrac{3x^2 + 3x - 2}{x^3 + x^2 - x - 1}$

26. $\dfrac{-2x^2 + 8x + 6}{x^3 - 3x^2 - 9x + 27}$

Find A, B, and C for each partial fraction decomposition.

27. $\dfrac{x^2 - x - 7}{(x + 1)(x^2 + 4)} = \dfrac{A}{x + 1} + \dfrac{Bx + C}{x^2 + 4}$

28. $\dfrac{4x^2 - 3x + 7}{(x - 1)(x^2 + 3)} = \dfrac{A}{x - 1} + \dfrac{Bx + C}{x^2 + 3}$

Find each partial fraction decomposition.

29. $\dfrac{5x^2 + 5x}{(x + 2)(x^2 + 1)}$

30. $\dfrac{3x^2 - 2x + 11}{(x^2 + 5)(x - 1)}$

31. $\dfrac{x^2 - 2}{(x + 1)(x^2 + x + 1)}$

32. $\dfrac{3x^2 + 7x - 4}{(x^2 + 3x + 1)(x - 2)}$

Find each partial fraction decomposition.

33. $\dfrac{-2x - 7}{x^2 + 4x + 4}$

34. $\dfrac{-3x + 2}{x^2 - 2x + 1}$

35. $\dfrac{6x^2 - x + 1}{x^3 + x^2 + x + 1}$

36. $\dfrac{3x^2 - 2x + 8}{x^3 + 2x^2 + 4x + 8}$

37. $\dfrac{3x^3 - x^2 + 19x - 9}{x^4 + 18x^2 + 81}$

38. $\dfrac{-x^3 - 10x - 3}{x^4 + 10x^2 + 25}$

39. $\dfrac{3x^2 + 17x + 14}{x^3 - 8}$

40. $\dfrac{2x^2 + 17x - 21}{x^3 + 27}$

41. $\dfrac{2x^3 + x^2 + 3x - 2}{x^2 - 1}$

42. $\dfrac{2x^3 - 19x - 9}{x^2 - 9}$

43. $\dfrac{3x^3 - 2x^2 + x - 2}{(x^2 + x + 1)^2}$

44. $\dfrac{x^3 - 8x^2 - 5x - 33}{(x^2 + x + 4)^2}$

45. $\dfrac{3x^3 + 4x^2 - 12x + 16}{x^4 - 16}$

46. $\dfrac{9x^2 + 3}{81x^4 - 1}$

47. $\dfrac{5x^3 + x^2 + x - 3}{x^4 - x^3}$

48. $\dfrac{2x^3 + 3x^2 - 8x + 4}{x^4 - 4x^2}$

49. $\dfrac{6x^2 - 28x + 33}{(x - 2)^2(x - 3)}$

50. $\dfrac{7x^2 + 45x + 58}{(x + 3)^2(x + 1)}$

51. $\dfrac{9x^2 + 21x - 24}{x^3 + 4x^2 - 11x - 30}$

52. $\dfrac{3x^2 + 24x + 6}{2x^3 - x^2 - 13x - 6}$

53. $\dfrac{x^2 - 2}{x^3 - 3x^2 + 3x - 1}$

54. $\dfrac{2x^2}{x^3 + 3x^2 + 3x + 1}$

Find the partial fraction decomposition for each rational expression. Assume that a, b, and c are nonzero constants.

55. $\dfrac{x}{(ax + b)^2}$

56. $\dfrac{x^2}{(ax + b)^3}$

57. $\dfrac{x + c}{ax^2 + bx}$

58. $\dfrac{1}{x^3(ax + b)}$

59. $\dfrac{1}{x^2(ax + b)}$

60. $\dfrac{1}{x^2(ax + b)^2}$

Thinking Outside the Box LXXIII

The Lizard and the Fly   A cylindrical garbage can has a height of 4 ft and a circumference of 6 ft. On the inside of the can, 1 ft from the top, is a fly. On the opposite side of the can, 1 ft from the bottom and on the outside, is a lizard. What is the shortest distance that the lizard must walk to reach the fly?

▪ **Figure for Thinking Outside the Box LXXIII**

8.4 Pop Quiz

Find the partial fraction decomposition.

1. $\dfrac{6x - 10}{x^2 - 25}$

2. $\dfrac{4x^2 + 2x + 6}{(x + 1)(x^2 + 3)}$

Linking Concepts

For Individual or Group Explorations

Modeling Work

⊞ *Suppose that Andrew can paint a garage by himself in a hours and Betty can paint the same garage by herself in b hours. Working together they can paint the garage in 2 hr 24 min.*

a)   What is the sum of $1/a$ and $1/b$?

b)   Write a as a function of b and graph it.

c)   What are the horizontal and vertical asymptotes? Interpret the asymptotes in the context of this situation.

d)   Use a graphing calculator to make a table for a and b. If a and b are positive integers, find all possible values for a and b from the table.

e)   Now suppose that the time it takes for Andrew, Betty, or Carl to paint a garage working alone is a, b, or c hours, respectively, where a, b, and c are positive integers. If they paint the garage together in 1 hr 15 min, then what are the possible values of a, b, and c? Explain your solution.

8.5

Inequalities and Systems of Inequalities in Two Variables

Earlier in this chapter we solved systems of equations in two variables. In this section we turn to linear and nonlinear inequalities in two variables and systems of inequalities in two variables. In the next section systems of linear inequalities will be used to solve linear programming problems.

Linear Inequalities

Three hamburgers and two Cokes cost at least $5.50. If a hamburger costs x dollars and a Coke costs y dollars, then this sentence can be written as the linear inequality

$$3x + 2y \geq 5.50.$$

A linear inequality in two variables is simply a linear equation in two variables with the equal sign replaced by an inequality symbol.

Definition: Linear Inequality

> If A, B, and C are real numbers with A and B not both zero, then
>
> $$Ax + By < C$$
>
> is called a **linear inequality in two variables.** In place of $<$ we can also use the symbols $\leq$, $>$, or $\geq$.

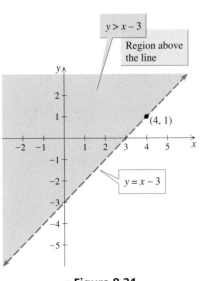

■ **Figure 8.21**

Some examples of linear inequalities are

$$y > x - 3, \qquad 2x - y \geq 1, \qquad x + y < 0, \qquad \text{and} \qquad y \leq 3.$$

The inequality $y \leq 3$ is considered an inequality in two variables because it can be written in the form $0 \cdot x + y \leq 3$.

An ordered pair (a, b) is a **solution to an inequality** if the inequality is true when x is replaced by a and y is replaced by b. Ordered pairs such as $(9, 1)$ and $(7, 3)$ satisfy $y \leq 3$. Ordered pairs such as $(4, 2)$ and $(-1, -2)$ satisfy $y > x - 3$.

The ordered pairs satisfying $y = x - 3$ form a line in the rectangular coordinate system, but what does the solution set to $y > x - 3$ look like? An ordered pair satisfies $y > x - 3$ whenever the y-coordinate is *greater than* the x-coordinate minus 3. For example, $(4, 1)$ satisfies the equation $y = x - 3$ and is on the line, while $(4, 1.001)$ satisfies $y > x - 3$ and is above the line. In fact, any point in the coordinate plane directly above $(4, 1)$ satisfies $y > x - 3$ (and any point directly below $(4, 1)$ satisfies $y < x - 3$). Since this fact holds true for every point on the line $y = x - 3$, the solution set to the inequality is the set of all points *above* the line. The graph of $y > x - 3$ is indicated by shading the region above the line, as shown in Fig. 8.21. We draw a dashed line for $y = x - 3$ because it is not part of the solution set to $y > x - 3$. However, the graph of $y \geq x - 3$ in Fig. 8.22 on the next page has a solid boundary line because the line is included in its solution set. Notice also that the region below the line is the solution set to $y < x - 3$.

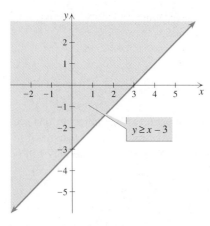

■ **Figure 8.22**

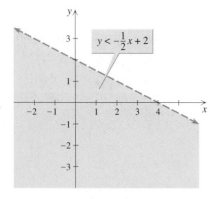

■ **Figure 8.23**

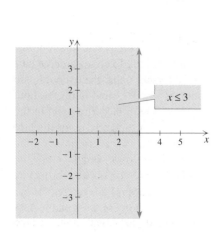

■ **Figure 8.26**

Example **1** **Graphing linear inequalities**

Graph the solution set to each inequality.

a. $y < -\dfrac{1}{2}x + 2$ **b.** $2x - y \geq 1$ **c.** $y > 2$ **d.** $x \leq 3$

Solution

a. Graph the corresponding equation

$$y = -\frac{1}{2}x + 2$$

as a dashed line by using its intercept $(0, 2)$ and slope $-\dfrac{1}{2}$. Since the inequality symbol is $<$, the solution set to the inequality is the region below this line, as shown in Fig. 8.23.

b. Solve the inequality for y:

$$2x - y \geq 1$$

$$-y \geq -2x + 1$$

$$y \leq 2x - 1 \qquad \text{Multiply by } -1 \text{ and reverse the inequality.}$$

Graph the line $y = 2x - 1$ by using its intercept $(0, -1)$ and its slope 2. Because of the symbol $\leq$, we use a solid line and shade the region below it, as shown in Fig. 8.24.

c. Every point in the region above the horizontal line $y = 2$ satisfies $y > 2$. See Fig. 8.25.

d. Every point on or to the left of the vertical line $x = 3$ satisfies $x \leq 3$. See Fig. 8.26.

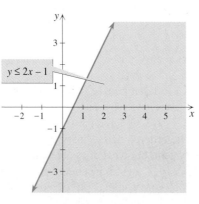

■ **Figure 8.24**

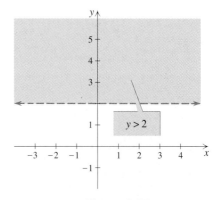

■ **Figure 8.25**

Try This. Graph the solution set to $y \leq -2x + 4$. ■

The graph in Example 1(b) is the region below the line because $2x - y \geq 1$ is equivalent to $y \leq 2x - 1$. The graph of $y > mx + b$ is above the line and

$y < mx + b$ is below the line. The symbol $>$ corresponds to "above the line" and $<$ corresponds to "below the line" *only if the inequality is solved for y.*

With the **test point method** it is not necessary to solve the inequality for y. The graph of $Ax + By = C$ divides the plane into two regions. On one side, $Ax + By > C$; and on the other, $Ax + By < C$. We can simply test a single point in one of the regions to see which is which.

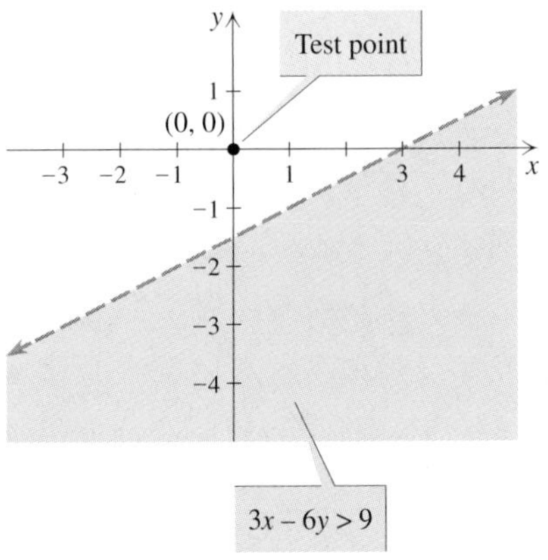

Example 2 Graphing an inequality using test points

Use the test point method to graph $3x - 6y > 9$.

Solution

First graph $3x - 6y = 9$, using a dashed line going through its intercepts $\left(0, -\frac{3}{2}\right)$ and $(3, 0)$. Select a test point that is not on the line, say, $(0, 0)$. Test $(0, 0)$ in $3x - 6y > 9$:

$$3 \cdot 0 - 6 \cdot 0 > 9$$

$$0 > 9 \quad \text{Incorrect.}$$

Since $(0, 0)$ does not satisfy $3x - 6y > 9$, any point on the *other* side of the line satisfies $3x - 6y > 9$. So we shade the region that does not contain $(0, 0)$, as shown in Fig. 8.27.

Try This. Graph the solution set to $2x - 5y < 20$. ■

■ **Figure 8.27**

Nonlinear Inequalities

Any nonlinear equation in two variables becomes a nonlinear inequality in two variables when the equal sign is replaced by an inequality symbol. The test point method is generally the easiest to use for graphing nonlinear inequalities in two variables.

Example 3 Graphing nonlinear inequalities in two variables

Graph the solution set to each nonlinear inequality.
a. $y > x^2$ **b.** $x^2 + y^2 \leq 4$ **c.** $y < \log_2(x)$

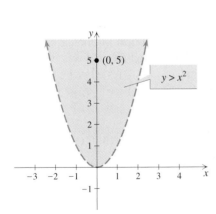

■ **Figure 8.28**

Solution

a. The graph of $y = x^2$ is a parabola opening upward with vertex at $(0, 0)$. Draw the parabola dashed, as shown in Fig. 8.28, and select a point that is not on the parabola, say, $(0, 5)$. Since $5 > 0^2$ is correct, the region of the plane containing $(0, 5)$ is shaded.

b. The graph of $x^2 + y^2 = 4$ is a circle of radius 2 centered at $(0, 0)$. Select $(0, 0)$ as a test point. Since $0^2 + 0^2 \leq 4$ is correct, we shade the region inside the circle, as shown in Fig. 8.29 on the next page.

c. The graph of $y = \log_2(x)$ is a curve through the points $(1, 0)$, $(2, 1)$, and $(4, 2)$, as shown in Fig. 8.30 on the next page. Select $(4, 0)$ as a test point. Since $0 < \log_2(4)$ is correct, shade the region shown in Fig. 8.30.

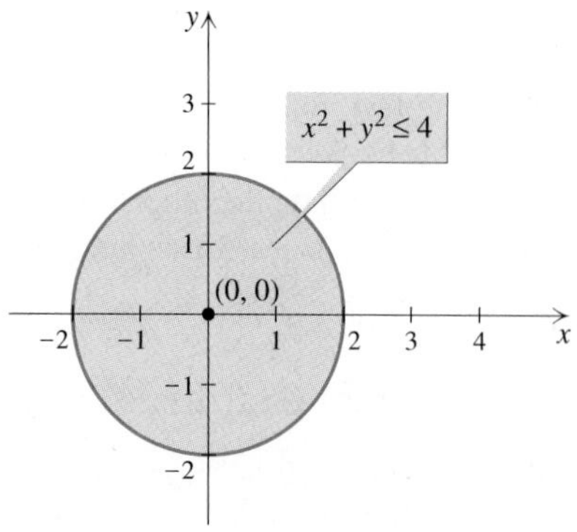

■ **Figure 8.29**

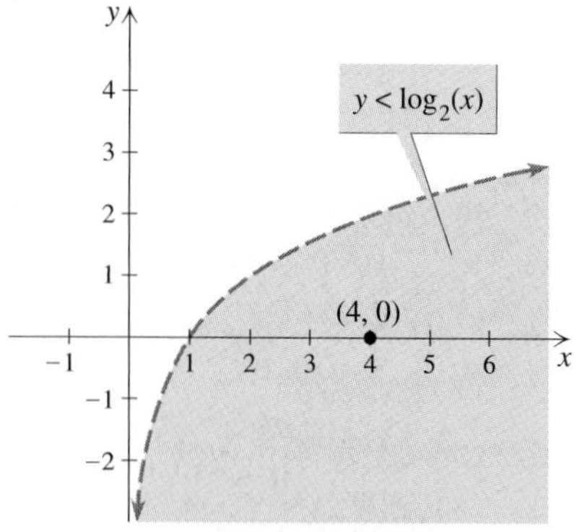

■ **Figure 8.30**

Try This. Graph the solution set to $y < |x|$.   ■

Systems of Inequalities

The solution set to a system of inequalities in two variables consists of all ordered pairs that satisfy *all* of the inequalities in the system. For example, the system

$$x + y > 5$$
$$x - y < 9$$

has (4, 2) as a solution because $4 + 2 > 5$ and $4 - 2 < 9$. There are infinitely many solutions to this system.

The solution set to a system is generally a region of the coordinate plane. It is the intersection of the solution sets to the individual inequalities. To find the solution set to a system, we graph the equation corresponding to each inequality in the system and then test a point in each region to see whether it satisfies all inequalities of the system.

Example 4 Solving a system of linear inequalities

Graph the solution set to the system.

$$x + 2y \leq 4$$
$$y \geq x - 3$$

Solution

The graph of $x + 2y = 4$ is a line through (0, 2) and (4, 0). The graph of $y = x - 3$ is a line through (0, −3) with slope 1. These two lines divide the plane into four regions, as shown in Fig. 8.31. Select a test point in each of the four regions. We use (0, 0), (0, 5), (0, −5), and (5, 0) as test points. Only (0, 0) satisfies both of the inequalities of the system. So we shade the region containing (0, 0) including its boundaries, as shown in Fig. 8.32.

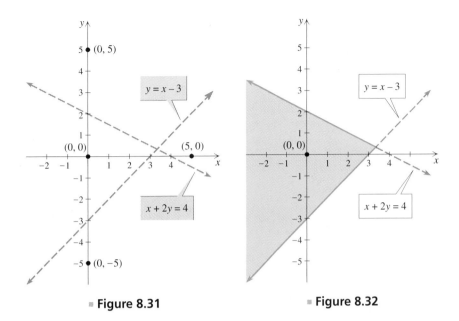

■ **Figure 8.31**   ■ **Figure 8.32**

Try This. Graph the solution set to $y < x$ and $y > 3 - x$.   ■

Note that the set of points indicated in Fig. 8.32 is the intersection of the set of points on or below the line $x + 2y = 4$ with the set of points on or above the line $y = x - 3$. If you can visualize how these regions intersect, then you can find the solution set without test points.

Example **5** **Solving a system of nonlinear inequalities**

Graph the solution set to the system.

$$x^2 + y^2 \leq 16$$
$$y > 2^x$$

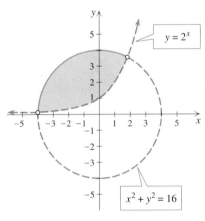

■ **Figure 8.33**

Solution

Points that satisfy $x^2 + y^2 \leq 16$ are on or inside the circle of radius 4 centered at $(0, 0)$. Points that satisfy $y > 2^x$ are above the curve $y = 2^x$. Points that satisfy both inequalities are on or inside the circle and above the curve $y = 2^x$, as shown in Fig. 8.33. Note that the circular boundary of the solution set is drawn as a solid curve because of the $\leq$ symbol. We could also find the solution set by using test points.

Try This. Graph the solution set to $y < 1 - |x|$ and $y > x^2 - 1$.   ■

Example **6** **Solving a system of three inequalities**

Graph the solution set to the system.

$$y > x^2$$
$$y < x + 6$$
$$y < -x + 6$$

Solution

Points that satisfy $y > x^2$ are above the parabola $y = x^2$. Points that satisfy $y < x + 6$ are below the line $y = x + 6$. Points that satisfy $y < -x + 6$ are below the line $y = -x + 6$. Points that satisfy all three inequalities lie in the region shown in Fig. 8.34.

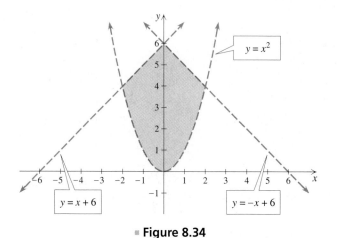

■ **Figure 8.34**

Try This. Graph the solution set to $y < 3$, $x < 2$, and $y > -x$. ■

For Thought

True or False? Explain.

1. The point $(1, 3)$ satisfies the inequality $y > x + 2$.

2. The graph of $x - y < 2$ is the region below the line $x - y = 2$.

3. The graph of $x + y > 2$ is the region above the line $x + y = 2$.

4. The graph of $x^2 + y^2 > 5$ is the region outside the circle of radius 5.

The following systems are referenced in statements 5–10:

(a) $x - 2y < 3$
   $y - 3x > 5$

(b) $x^2 + y^2 > 9$ (c) $y \geq x^2 - 4$
   $y < x + 2$ $y < x + 2$

5. The point $(-2, 1)$ is in the solution set to system (a).

6. The solution to system (b) consists of points outside a circle and below a line.

7. The point $(-2, 0)$ is in the solution set to system (c).

8. The point $(-1, 2)$ is a test point for system (a).

9. The origin is in the solution set to system (c).

10. The point $(2, 3)$ is in the solution set to system (b).

8.5 Exercises

Match each inequality with one of the graphs (a)–(d).

1. $y > x - 2$

2. $x < 2 - y$

3. $x - y > 2$

4. $x + y > 2$

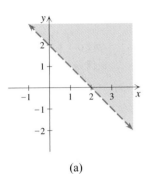

(a)

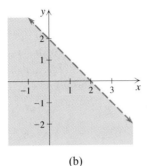

(b)

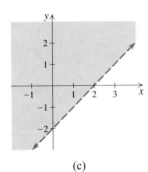

(c)

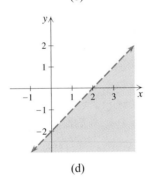

(d)

Sketch the graph of the solution set to each linear inequality in the rectangular coordinate system.

5. $y < 2x$

6. $x > y$

7. $x + y > 3$

8. $2x + y < 1$

9. $2x - y \le 4$

10. $x - 2y \ge 6$

11. $y < -3x - 4$

12. $y > \frac{2}{3}x - 3$

13. $x - 3 \ge 0$

14. $y + 1 \le 0$

15. $20x - 30y \le 6000$

16. $30y + 40x > 1200$

17. $y < 3$

18. $x > 0$

Sketch the graph of each nonlinear inequality.

19. $y > -x^2$

20. $y < 4 - x^2$

21. $x^2 + y^2 \ge 1$

22. $x^2 + y^2 < 36$

23. $x > |y|$

24. $y < x^{1/3}$

25. $x \ge y^2$

26. $y^2 > x - 1$

27. $y \ge x^3$

28. $y < |x - 1|$

29. $y > 2^x$

30. $y < \log_2(x)$

Sketch the graph of the solution set to each system of inequalities.

31. $y > x - 4$
$y < -x - 2$

32. $y < \frac{1}{2}x + 1$
$y < -\frac{1}{3}x + 1$

33. $3x - 4y \le 12$
$x + y \ge -3$

34. $2x + y \ge -1$
$y - 2x \le -3$

35. $3x - y < 4$
$y < 3x + 5$

36. $x - y > 0$
$y + 4 > x$

37. $y + x < 0$
$y > 3 - x$

38. $3x - 2y \le 6$
$2y - 3x \le -8$

39. $x + y < 5$
$y \ge 2$

40. $x \le 2$
$y > -2$

41. $y < x - 3$
$x \le 4$

42. $y > 0$
$y \le x$

Sketch the graph of the solution set to each nonlinear system of inequalities.

43. $y > x^2 - 3$
$y < x + 1$

44. $y < 5 - x^2$
$y > (x - 1)^2$

45. $x^2 + y^2 \ge 4$
$x^2 + y^2 \le 16$

46. $x^2 + y^2 \le 9$
$y \ge x - 1$

47. $(x - 3)^2 + y^2 \le 25$
$(x + 3)^2 + y^2 \le 25$

48. $x^2 + y^2 \ge 64$
$x^2 + y^2 \le 16$

49. $x^2 + y^2 > 4$
$|x| \le 4$

50. $x^2 + y^2 < 36$
$|y| < 3$

51. $y > |2x| - 4$
$y \le \sqrt{4 - x^2}$

52. $y < 4 - |x|$
$y \ge |x| - 4$

53. $|x - 1| < 2$
$|y - 1| < 4$

54. $x \ge y^2 - 1$
$(x + 1)^2 + y^2 \ge 4$

Solve each system of inequalities.

55. $x \geq 0$

$y \geq 0$

$x + y \leq 4$

56. $x \geq 0$

$y \geq 0$

$y \geq -\dfrac{1}{2}x + 2$

57. $x \geq 0, y \geq 0$

$x + y \geq 4$

$y \geq -2x + 6$

58. $x \geq 0, y \geq 0$

$4x + 3y \leq 12$

$3x + 4y \leq 12$

59. $x^2 + y^2 \geq 9$

$x^2 + y^2 \leq 25$

$y \geq |x|$

60. $x - 2 < y < x + 2$

$x^2 + y^2 < 16$

$x > 0, y > 0$

61. $y > (x - 1)^3$

$y > 1$

$x + y > -2$

62. $x \geq |y|$

$y \geq -3$

$2y - x \leq 4$

63. $y > 2^x$

$y < 6 - x^2$

$x + y > 0$

64. $x^2 + y \leq 5$

$y \geq x^3 - x$

$y \leq 4$

Write a system of inequalities whose solution set is the region shown.

65.

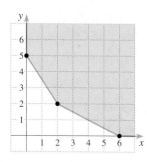

66.

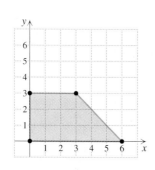

67.

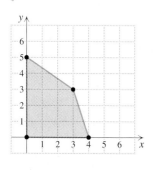

68.

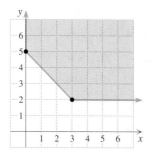

Find a system of inequalities to describe the given region.

69. Points inside the square that has vertices $(2, 2)$, $(-2, 2)$, $(-2, -2)$, and $(2, -2)$.

70. Points inside the triangle that has vertices $(0, 0)$, $(0, 6)$, and $(3, 0)$.

71. Points in the first quadrant less than nine units from the origin.

72. Points that are closer to the *x*-axis than they are to the *y*-axis.

Use a graphing calculator to graph the equation corresponding to each inequality. From the display of the graphing calculator, locate one ordered pair in the solution set to the system and check that it satisfies all inequalities of the system.

73. $y > 2^{x+2}$

$y < x^3 - 3x$

74. $y > e^{x-0.8}$

$y < \log(x + 2.5)$

75. $y < -0.5x^2 + 150x - 11{,}226.6$

$y > -0.11x + 38$

76. $y < x^2$

$y > 2^x$

$y < 5x - 6$

Write a system of inequalities that describes the possible solutions to each problem and graph the solution set to the system.

77. *Size Restrictions*   United Parcel Service defines the girth of a box as the sum of the length, twice the width, and twice the height. The maximum girth that UPS will accept is 130 in. If the length of a box is 50 in., then what inequality must be satisfied by the width and height? Draw a graph showing the acceptable widths and heights for a length of 50 in.

78. *More Restrictions*   United Parcel Service defines the girth of a box as the sum of the length, twice the width, and twice the height. The maximum girth that UPS will accept is 130 in. A shipping clerk wants to ship parts in a box that has a height of 24 in. For easy handling, he wants the box to have a width that is less than or equal to two-thirds of the length. Write a system of inequalities that the box must satisfy and draw a graph showing the possible lengths and widths.

79. *Inventory Control*   A car dealer stocks mid-size and full-size cars on her lot, which cannot hold more than 110 cars. On the average, she borrows $10,000 to purchase a mid-size car and $15,000 to purchase a full-size car. How many cars of each type could she stock if her total debt cannot exceed $1.5 million?

80. *Delicate Balance*   A fast food restaurant must have a minimum of 30 employees and a maximum of 50. To avoid charges of sexual bias, the company has a policy that the number of employees of one sex must never exceed the number of employees of the other sex by more than six. How many persons of each sex could be employed at this restaurant?

81. *Political Correctness*   A political party is selling $50 tickets and $100 tickets for a fund-raising banquet in a hall that cannot hold more than 500 people. To show that the party represents the people, the number of $100 tickets must not be greater than 20% of the total number of tickets sold. How many tickets of each type can be sold?

82. *Mixing Alloys*   A metallurgist has two alloys available. Alloy A is 20% zinc and 80% copper, while alloy B is 60% zinc and 40% copper. He wants to melt down and mix x ounces of alloy A with y ounces of alloy B to get a metal that is at most 50% zinc and at most 60% copper. How many ounces of each alloy should he use if the new piece of metal must not weigh more than 20 ounces?

Thinking Outside the Box LXXIV

Changing Places   Five men and five women are rowing a long boat with 11 seats in a row. The five men are in front, the five women are in the back, and there is an empty seat in the middle. The five men in the front of the boat want to exchange seats with the five women in back. A rower can move from his/her seat to the next empty seat or he/she can step over one person without capsizing the boat. What is the minimum number of moves needed for the five men in front to change places with the five women in back?

8.5 Pop Quiz

1. Does the graph of $x - 2y > 0$ consist of the region above or below the line $x - 2y = 0$?

2. Does the graph of $x^2 + y^2 < 9$ consist of the region inside or outside of the circle $x^2 + y^2 = 9$?

3. Does $(0, 1)$ satisfy the system $x + 6y \geq 2$ and $x + 2y \leq -5$?

4. Write a system of inequalities whose solution set consists of points in the second quadrant that are less than 5 units from the origin.

5. What is the area of the graph of the solution set to $|x - 2| \leq 5$ and $|y - 3| \leq 2$?

Linking Concepts

For Individual or Group Explorations

Describing Regular Polygons

The equilateral triangle in the figure is inscribed in the circle $x^2 + y^2 = 1$.

a) Write a system of inequalities whose graph consists of all points on and inside the equilateral triangle.

b) Write a system of inequalities whose graph consists of all points on and inside a square inscribed in $x^2 + y^2 = 1$. Position the square in any manner that you choose.

c) Write a system of inequalities whose graph consists of all points on and inside a regular pentagon inscribed in $x^2 + y^2 = 1$. You will need trigonometry for this one.

d) Write a system of inequalities whose graph consists of all points on and inside a regular hexagon inscribed in $x^2 + y^2 = 1$.

8.6

The Linear Programming Model

In this section we apply our knowledge of systems of linear inequalities to solving linear programming problems. Linear programming is a method that can be used to solve many practical business problems. Linear programming can tell us how to allocate resources to achieve a maximum profit, minimum labor cost, or a most nutritious meal.

Graphing the Constraints

In the simplest linear programming applications we have two variables that must satisfy several linear inequalities. These inequalities are called the **constraints** because they restrict the variables to only certain values. The graph of the solution set to the system is used to indicate the points that satisfy all of the constraints. Any point that satisfies all of the constraints is called a **feasible solution** to the problem.

Example **1** **Graphing the constraints**

Graph the solution set to the system of inequalities and identify each vertex of the region.

$$x \geq 0, \quad y \geq 0$$
$$2x + y \leq 6$$
$$x + y \leq 4$$

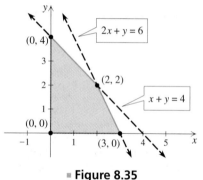

■ **Figure 8.35**

Solution

The points on or to the right of the y-axis satisfy $x \geq 0$. The points on or above the x-axis satisfy $y \geq 0$. The points on or below the line $2x + y = 6$ satisfy $2x + y \leq 6$. The points on or below the line $x + y = 4$ satisfy $x + y \leq 4$. Graph each straight line and shade the region that satisfies all four inequalities, as shown in Fig. 8.35. Three of the vertices are easily identified as $(0, 0)$, $(0, 4)$, and $(3, 0)$. The fourth vertex is at the intersection of $x + y = 4$ and $2x + y = 6$. Multiply $x + y = 4$ by -1 and add the result to $2x + y = 6$:

$$-x - y = -4$$
$$\underline{2x + y = 6}$$
$$x \qquad = 2$$

If $x = 2$ and $x + y = 4$, then $y = 2$. So the fourth vertex is $(2, 2)$.

Try This.   Graph the solution set to $x \geq 0, y \geq 0, x + y \leq 6$, and $x + 2y \leq 8$ and identify each vertex of the region.   ■

In linear programming, the constraints usually come from physical limitations of resources described within the specific problem. Constraints that are always satisfied are called **natural constraints.** For example, the requirement that the number of employees be greater than or equal to zero is a natural constraint. In the next example we write the constraints and then graph the points in the coordinate plane that satisfy all of the constraints.

Example 2 Finding and graphing the constraints

Bruce builds portable storage buildings. He uses 10 sheets of plywood and 15 studs in a small building, and he uses 15 sheets of plywood and 45 studs in a large building. Bruce has available only 60 sheets of plywood and 135 studs. Write the constraints on the number of small and large portable buildings that he can build with the available supplies, and graph the solution set to the system of constraints.

Solution

Let x represent the number of small buildings and y represent the number of large buildings. The natural constraints are

$$x \geq 0 \quad \text{and} \quad y \geq 0$$

because he cannot build a negative number of buildings. Since he has only 60 sheets of plywood available, we must have

$$10x + 15y \leq 60$$

$$2x + 3y \leq 12 \qquad \text{Divide each side by 5.}$$

Since he has only 135 studs available, we must have

$$15x + 45y \leq 135$$

$$x + 3y \leq 9 \qquad \text{Divide each side by 15.}$$

The conditions stated lead to the following system of constraints:

$$x \geq 0, \quad y \geq 0 \qquad \text{Natural constraints}$$

$$2x + 3y \leq 12 \qquad \text{Constraint on amount of plywood}$$

$$x + 3y \leq 9 \qquad \text{Constraint on number of studs}$$

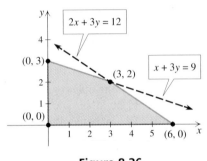

■ **Figure 8.36**

The graph of the solution set to this system is shown in Fig. 8.36.

Try This. A small shop assembles electronic components. Component X uses 4 transistors and 6 diodes. Component Y uses 8 transistors and 2 diodes. The shop has on hand only 40 transistors and 30 diodes. Write the constraints on the number of components that can be assembled and graph the system. ■

Maximizing or Minimizing a Linear Function

In Example 2, any ordered pair within the shaded region of Fig. 8.36 is a feasible solution to the problem of deciding how many buildings of each type could be built. For each feasible solution within the shaded region, Bruce makes some amount of profit. Of course, Bruce wants to find a feasible solution that will yield the maximum possible profit. In general, the function that we wish to maximize or minimize, subject to the constraints, is called the **objective function.**

If Bruce makes a profit of $400 on a small building and $500 on a large building, then the total profit from x small and y large buildings is

$$P = 400x + 500y.$$

Since the profit is a function of x and y, we write the objective function as

$$P(x, y) = 400x + 500y.$$

The function P is a linear function of x and y. The domain of P is the region graphed in Fig. 8.36.

Definition: Linear Function in Two Variables

A **linear function in two variables** is a function of the form

$$f(x, y) = Ax + By + C,$$

where A, B, and C are real numbers such that A and B are not both zero.

Bruce is interested in the maximum profit, subject to the constraints on x and y. Suppose $x = 1$ and $y = 1$; then the profit is

$$P(1, 1) = 400(1) + 500(1) = \$900.$$

In fact, the profit is $900 for any x and y that satisfy

$$400x + 500y = 900.$$

If he wants $1300 profit, then x and y must satisfy

$$400x + 500y = 1300,$$

and the profit is $1800 if x and y satisfy

$$400x + 500y = 1800.$$

The graphs of these lines are shown in Fig. 8.37. Notice that the larger profit is found on the higher *profit line* and all of the profit lines are parallel. Bruce wants the highest profit line that intersects the region of feasible solutions. You can see in Fig. 8.37 that the highest line that intersects the region and is parallel to the other profit lines will intersect the region at the vertex (6, 0). So he should build six small buildings and no large buildings to maximize the profit. The maximum profit is $P(6, 0) = 400(6) + 500(0) = \$2400.$

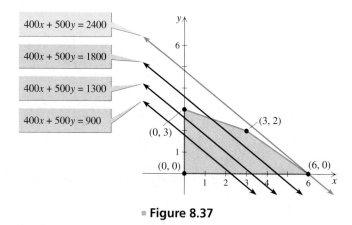

■ **Figure 8.37**

In another case we may be looking for the point that would *minimize* a linear function of the two variables. *In general, if the maximum or minimum value exists, then the maximum or minimum value of a linear function subject to linear constraints occurs at a vertex of the region determined by the constraints.* The minimum profit for the portable buildings occurs when no buildings are built, at the vertex (0, 0). It is possible that the maximum or minimum value occurs at two adjacent vertices and at every point along the line segment joining them.

It is a bit cumbersome and possibly inaccurate to graph parallel lines and then find the highest (or lowest) one that intersects the region determined by the constraints. Instead, we can use the following procedure for linear programming, which does not depend as much on graphing.

PROCEDURE   **Linear Programming**

Use the following steps to find the maximum or minimum value of a linear function subject to linear constraints.

1. Graph the region that satisfies all of the constraints.
2. Determine the coordinates of each vertex of the region.
3. Evaluate the function at each vertex of the region.
4. Identify which vertex gives the maximum or minimum value of the function.

To use the new procedure on Bruce's buildings, note that in Fig. 8.37 the vertices are $(0, 0)$, $(6, 0)$, $(0, 3)$, and $(3, 2)$. Compute the profit at each vertex:

$$P(0, 0) = 400(0) + 500(0) = \$0 \qquad \text{Minimum profit}$$

$$P(6, 0) = 400(6) + 500(0) = \$2400 \qquad \text{Maximum profit}$$

$$P(0, 3) = 400(0) + 500(3) = \$1500$$

$$P(3, 2) = 400(3) + 500(2) = \$2200$$

From this list, we see that the maximum profit is $2400, when six small buildings and no large buildings are built, and the minimum profit is $0, when no buildings of either type are built.

In the next example we use the linear programming technique to find the minimum value of a linear function subject to a system of constraints.

Example **3** **Finding the minimum value of a linear function**

One serving of Muesli breakfast cereal contains 4 grams of protein and 30 grams of carbohydrates. One serving of Multi Bran Chex contains 2 grams of protein and 25 grams of carbohydrates. A dietitian wants to mix these two cereals to make a batch that contains at least 44 grams of protein and at least 450 grams of carbohydrates. If the cost of Muesli is 21 cents per serving and the cost of Multi Bran Chex is 14 cents per serving, then how many servings of each cereal would minimize the cost and satisfy the constraints?

Solution

Let x = the number of servings of Muesli and y = the number of servings of Multi Bran Chex. If the batch is to contain at least 44 grams of protein, then

$$4x + 2y \geq 44.$$

If the batch is to contain at least 450 grams of carbohydrates, then

$$30x + 25y \geq 450.$$

Simplify each inequality and use the two natural constraints to get the following system:

$$x \geq 0, \quad y \geq 0$$

$$2x + y \geq 22$$

$$6x + 5y \geq 90$$

The graph of the constraints is shown in Fig. 8.38. The vertices are (0, 22), (5, 12), and (15, 0). The cost in dollars for x servings of Muesli and y servings of Multi Bran Chex is $C(x, y) = 0.21x + 0.14y$. Evaluate the cost at each vertex.

$$C(0, 22) = 0.21(0) + 0.14(22) = \$3.08$$

$$C(5, 12) = 0.21(5) + 0.14(12) = \$2.73 \quad \text{Minimum cost}$$

$$C(15, 0) = 0.21(15) + 0.14(0) = \$3.15$$

The minimum cost of $2.73 is attained by using 5 servings of Muesli and 12 servings of Multi Bran Chex. Note that $C(x, y)$ does not have a maximum value on this region. The cost increases without bound as x and y increase.

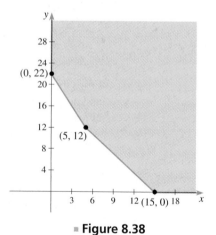

■ **Figure 8.38**

Try This. The shop described in the Try This for Example 2 gets $2 for each X-component assembled and $3 for each Y-component assembled. How many components of each type would maximize the revenue and satisfy the constraints? ■

The examples of linear programming given in this text are simple examples. Problems in linear programming in business can involve a hundred or more variables subject to as many inequalities. These problems are solved by computers using matrix methods, but the basic idea is the same as we have seen in this section.

For Thought

True or False? Explain.

1. The graph of $x \geq 0$ in the coordinate plane consists only of points on the x-axis that are at or to the right of the origin.

2. The graph of $y \geq 2$ in the coordinate plane consists of the points on or to the right of the line $y = 2$.

3. The graph of $x + y \leq 5$ does not include the origin.

4. The graph of $2x + 3y = 12$ has x-intercept $(0, 4)$ and y-intercept $(6, 0)$.

5. The graph of a system of inequalities is the intersection of their individual solution sets.

6. In linear programming, constraints are inequalities that restrict the values of the variables.

7. The function $f(x, y) = 4x^2 + 9y^2 + 36$ is a linear function of x and y.

8. The value of $R(x, y) = 30x + 15y$ at the point $(1, 3)$ is 75.

9. If $C(x, y) = 7x + 9y + 3$, then $C(0, 5) = 45$.

10. To solve a linear programming problem, we evaluate the objective function at the vertices of the region determined by the constraints.

8.6 Exercises

Graph the solution set to each system of inequalities and identify each vertex of the region.

1. $x \geq 0, y \geq 0$
   $x + y \leq 4$

2. $x \geq 0, y \geq 0$
   $2x + y \leq 4$

3. $x \geq 0, y \geq 0$
   $x \leq 1, y \leq 3$

4. $x \geq 0, y \geq 0$
   $y \leq x, x \leq 3$

5. $x \geq 0, y \geq 0$
   $x + y \leq 4$
   $2x + y \leq 6$

6. $x \geq 0, y \geq 0$
   $50x + 40y \leq 200$
   $10x + 20y \leq 60$

7. $x \geq 0, y \geq 0$
   $2x + y \geq 4$
   $x + y \geq 3$

8. $x \geq 0, y \geq 0$
   $20x + 10y \geq 40$
   $5x + 5y \geq 15$

9. $x \geq 0, y \geq 0$
   $3x + y \geq 6$
   $x + y \geq 4$

10. $x \geq 0, y \geq 0$
    $2x + y \geq 6$
    $x + 2y \geq 6$

11. $x \geq 0, y \geq 0$
    $3x + y \geq 8$
    $x + y \geq 6$

12. $x \geq 0, y \geq 0$
    $x + 4y \leq 20$
    $4x + y \leq 64$

Find the maximum value of the objective function $T(x, y) = 2x + 3y$ on each given region.

13.

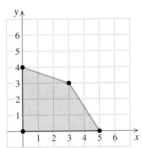

14.

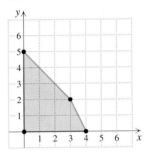

Find the minimum value of the objective function $H(x, y) = 2x + 2y$ on each given region.

15.

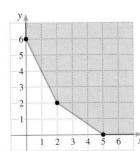

16.

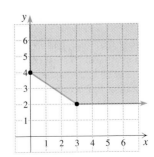

Find the maximum or minimum value of each objective function subject to the given constraints.

17. Maximize $P(x, y) = 5x + 9y$ subject to $x \geq 0$, $y \geq 0$, and $x + 2y \leq 6$.

18. Maximize $P(x, y) = 25x + 31y$ subject to $x \geq 0$, $y \geq 0$, and $5x + 6y \leq 30$.

19. Minimize $C(x, y) = 3x + 2y$ subject to $x \geq 0$, $y \geq 0$, $x + y \geq 4$.

20. Minimize $C(x, y) = 2x + 5y$ subject to $x \geq 0$, $y \geq 0$, and $2x + y \geq 8$.

21. Minimize $C(x, y) = 10x + 20y$ subject to $x \geq 0$, $y \geq 0$, $x + y \geq 8$, and $3x + 5y \geq 30$.

22. Maximize $R(x, y) = 50x + 20y$ subject to $x \geq 0$, $y \geq 0$, $3x + y \leq 18$, and $2x + y \leq 14$.

Solve each problem. See the procedure for linear programming on page 709.

23. *Maximizing Revenue*   Bob and Betty make bird houses and mailboxes in their craft shop near Gatlinburg. Each bird house requires 3 hr of work from Bob and 1 hr from Betty. Each mailbox requires 4 hr of work from Bob and 2 hr of work from Betty. Bob cannot work more than 48 hr per week and Betty cannot work more than 20 hr per week. If each bird house sells for $12 and each mailbox sells for $20, then how many of each should they make to maximize their revenue?
   HINT Write an inequality about the amount of time Bob can work and another for Betty.

24. *Maximizing Revenue*   At Taco Town a taco contains 2 oz of ground beef and 1 oz of chopped tomatoes. A burrito contains 1 oz of ground beef and 3 oz of chopped tomatoes. Near closing time the cook discovers that they have only 22 oz of ground beef and 36 oz of tomatoes left. The manager directs the cook to use the available resources to maximize their revenue for the remainder of the shift. If a taco sells for 40 cents and a burrito sells for 65 cents, then how many of each should they make to maximize their revenue?
   HINT Write an inequality about the amount of available beef and another for the tomatoes.

25. *Bird Houses and Mailboxes*   If a bird house sells for $18 and a mailbox for $20, then how many of each should Bob and Betty build to maximize their revenue, subject to the constraints of Exercise 23?

26. *Tacos and Burritos*   If a taco sells for 20 cents and a burrito for 65 cents, then how many of each should be made to maximize the revenue, subject to the constraints of Exercise 24?

27. *Minimizing Operating Costs*   Kimo's Material Company hauls gravel to a construction site using a small truck and a large truck. The carrying capacity and operating cost per load are given in the accompanying table. Kimo must deliver a minimum of 120 yd³ per day to satisfy his contract with the builder. The union contract with his drivers requires that the total number of loads per day be a minimum of 8. How many loads should be made in each truck per day to minimize the total cost?

■ **Table for Exercise 27**

	Small Truck	Large Truck	
Capacity (yd³)	12	20	
Cost per load	$70	$60	

28. *Minimizing Labor Costs*   Tina's Telemarketing employs part-time and full-time workers. The number of hours worked per week and the pay per hour for each is given in the accompanying table. Tina needs at least 1200 hr of work done per week. To qualify for certain tax breaks, she must have at least 45 employees. How many part-time and full-time employees should be hired to minimize Tina's weekly labor cost?

■ **Table for Exercise 28**

	Part-time	Full-time	
Hours per week	20	40	
Pay per hour	$6	$8	

29. *Small Trucks and Large Trucks*   If it costs $70 per load to operate the small truck and $75 per load to operate the large truck, then how many loads should be made in each truck per day to minimize the total cost, subject to the constraints of Exercise 27?

30. *Part-Time and Full-Time Workers*   If the labor cost for a part-timer is $9/hr and the labor cost for a full-timer is $8/hr, then how many of each should be employed to minimize the weekly labor cost, subject to the constraints of Exercise 28?

Thinking Outside the Box LXXV

Ten Tangents   In the accompanying figure, $AB = 7$, $AC = 12$, and $BC = 10$. There is a point D on BC such that the circles inscribed in triangles ABD and ACD are both tangent to line AD at a common point E. Find the length of BD.
   HINT Two tangent segments from a point to a circle are equal in length.

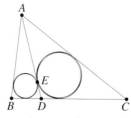

■ **Figure for Thinking Outside the Box LXXV**

8.6 Pop Quiz

1. Find all vertices of the region $x \geq 0$, $y \geq 0$, and $x + 2y \leq 6$.

2. On the region in the previous problem, find the maximum value of $P(x, y) = 20x + 50y - 100$.

3. Find all vertices of the region $x \geq 0$, $y \geq 0$, $x + 2y \geq 6$, and $2x + y \geq 9$.

4. Find the minimum value of $C(x, y) = 5x + 4y$ on the region of the previous problem.

Linking Concepts

For Individual or Group Explorations

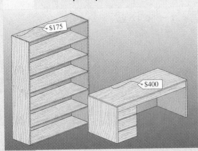

Modeling Numerous Constraints

Lucy's Woodworks makes desks and bookcases for furniture stores. Lucy sells the desks for $400 and the bookcases for $175. Each desk takes 30 ft² of oak plywood, 12 ft of molding, 1 quart of stain, 3 pints of lacquer, 9 drawer pulls, 7 drawer glides, and 20 hours of labor. Each bookcase requires 20 ft² of oak plywood, 15 ft of molding, 1 pint of stain, 1 quart of lacquer, 20 shelf pins, and 12 hours of labor. Lucy has available only 960 ft² of oak plywood, 480 ft of molding, 8 gallons of stain, 15 gallons of lacquer, 270 drawer pulls, 350 drawer glides, 580 shelf pins, and 720 hours of labor.

a) How many desks and how many bookcases should Lucy make to maximize her revenue?

b) Suppose Lucy could increase her supply of one item. Which item would have the most significant impact on her revenue? Explain your answer.

■ ■ ■ Highlights

8.1 Systems of Linear Equations in Two Variables

System	Two equation of the form $Ax + By = C$ where A and B are not both zero	$3x - y = 5$ $3x + y = 13$
Solution Set	The set of ordered pairs that satisfy all equations of the system	$\{(3, 4)\}$
Addition Method	Add the equations (or appropriate multiples) of the equations to eliminate a variable.	$3x - y = 5$ $\underline{3x + y = 13}$ $6x \quad\quad = 18$
Substitution Method	Substitute one equation into the other to eliminate a variable.	$y = 3x - 5$ $3x + (3x - 5) = 13$

Independent	A system with one solution.	$y = 2x - 3, y = -2x - 3$
Inconsistent	A system with no solutions.	$y = 2x - 3, y = 2x - 4$
Dependent	A system with infinitely many solutions.	$y = 2x - 3, 2y = 4x - 6$

8.2 Systems of Linear Equations in Three Variables

System	Two or three equations of the form $Ax + By + Cz = D$, where A, B, and C are not all zero	$x + y - z = 0$ $x - y + z = 2$ $x - 2y + z = 1$
Solution Set	Single-ordered triple, no solution, or infinitely many solutions	$\{(1, 1, 2)\}$
Methods	Use a combination of addition and/or substitution to eliminate variables.	

8.3 Nonlinear Systems of Equations

| **Nonlinear System** | Two equations in two variables, where at least one is not linear | $x + y = 12$
 $y = x^2$ |
| **Solving** | Use a combination of addition and/or substitution to eliminate variables. | $x + x^2 = 12$ |

8.4 Partial Fractions

| **Partial Fraction Decomposition** | Reverses the addition of rational expressions | $\dfrac{3x - 2}{x(x - 1)} = \dfrac{2}{x} + \dfrac{1}{x - 1}$ |

8.5 Inequalities and Systems of Inequalities in Two Variables

Linear Inequalities	$y > mx + b$ is the region above $y = mx + b$. $y < mx + b$ is the region below $y = mx + b$. $x > k$ is the region to the right of $x = k$. $x < k$ is the region to the left of $x = k$.	$y > x$ is above $y = x$. $y < x$ is below $y = x$. $x > 2$ is to the right of $x = 2$. $x < 2$ is to the left of $x = 2$.
Including Boundaries	For $\leq$ or $\geq$ the boundary lines are included and drawn with solid lines.	$x \geq 2$ includes the line $x = 2$.
System of Inequalities	Test a point in each region determined by the boundary lines or curves.	$y > 2$ and $y < 3$ is the region between $y = 2$ and $y = 3$.

8.6 The Linear Programming Model

| **Constraints** | Inequalities concerning the variables in a linear programming problem. | |
| **Principle of Linear Programming** | The maximum or minimum of a linear function subject to linear constraints occurs at a vertex of the region determined by the constraints. | |

■ ■ ■ Chapter 8 Review Exercises

Solve each system by graphing.

1. $2x - 3y = -9$
$3x + y = 14$

2. $3x - 2y = 0$
$y = -2x - 7$

3. $x + y = 2$
$2y - 3x = 9$

4. $x - y = 30$
$2x + 3y = 10$

Solve each system by the method of your choice. Indicate whether each system is independent, dependent, or inconsistent.

5. $3x - 5y = 19$
$y = x$

6. $x + y = 9$
$y = x - 3$

7. $4x - 3y = 6$
$3x + 2y = 9$

8. $3x - 2y = 4$
$5x + 7y = 1$

9. $6x + 2y = 2$
$y = -3x + 1$

10. $x - y = 9$
$2y - 2x = -18$

11. $3x - 4y = 12$
$8y - 6x = 9$

12. $y = -5x + 3$
$5x + y = 6$

Solve each system.

13. $x + y - z = 8$
$2x + y + z = 1$
$x + 2y + 3z = -5$

14. $2x + 3y - 2z = 8$
$3x - y + 4z = -20$
$x + y - z = 3$

15. $x + y + z = 1$
$2x - y + 2z = 2$
$2x + 2y + 2z = 2$

16. $x - y - z = 9$
$x + y + 2z = -9$
$-2x + 2y + 2z = -18$

17. $x + y + z = 1$
$2x - y + 3z = 5$
$x + y + z = 4$

18. $2x - y + z = 4$
$x - y + z = -1$
$-x + y - z = 0$

Solve each nonlinear system of equations. Find real solutions only.

19. $x^2 + y^2 = 4$
$x = y^2$

20. $x^2 - y^2 = 9$
$x^2 + y^2 = 7$

21. $y = |x|$
$y = x^2$

22. $y = 2x^2 + x - 3$
$6x + y = 12$

Find the partial fraction decomposition for each rational expression.

23. $\dfrac{7x - 7}{(x - 3)(x + 4)}$

24. $\dfrac{x - 13}{x^2 - 6x + 5}$

25. $\dfrac{7x^2 - 7x + 23}{x^3 - 3x^2 + 4x - 12}$

26. $\dfrac{10x^2 - 6x + 2}{(x - 1)^2(x + 2)}$

Graph the solution set to each inequality.

27. $x^2 + (y - 3)^2 < 9$

28. $2x - 9y \le 18$

29. $x \le (y - 1)^2$

30. $y < 6 - 2x^2$

Graph the solution set to each system of inequalities.

31. $2x - 3y \ge 6$
$x \le 2$

32. $x \le 3, y \ge 1$
$x - y \ge -5$

33. $y \ge 2x^2 - 6$
$x^2 + y^2 \le 9$

34. $x^2 + y^2 \ge 16$
$2y \ge x^2 - 16$

35. $x \geq 0, y \geq 1$

$x + 2y \leq 10$

$3x + 4y \leq 24$

36. $x \geq 0, y \geq 0$

$30x + 60y \leq 1200$

$x + y \leq 30$

37. $x \geq 0, y \geq 0$

$x + 6y \geq 60$

$x + y \geq 35$

38. $x \geq 0$

$y \geq x + 1$

$x + y \leq 5$

Solve each problem, using a system of equations.

39. Find the equation of the line through $(-2, 3)$ and $(4, -1)$.

40. Find the equation of the line through $(4, 7)$ and $(-2, -3)$.

41. Find the equation of the parabola through $(1, 4)$, $(3, 20)$, and $(-2, 25)$.

42. Find the equation of the parabola through $(-1, 10)$, $(2, -5)$, and $(3, -18)$.

43. *Tacos and Burritos*   At Taco Town a taco contains 1 oz of meat and 2 oz of cheese, while a burrito contains 2 oz of meat and 3 oz of cheese. In 1 hr the cook used 181 oz of meat and 300 oz of cheese making tacos and burritos. How many of each were made?

44. *Imported and Domestic Cars*   Nicholas had 10% imports on his used car lot, and Seymour had 30% imports on his used car lot. After Nicholas bought Seymour's entire stock, Nicholas had 300 cars, of which 22% were imports. How many cars were on each lot originally?

45. *Daisies, Carnations, and Roses*   Esther's Flower Shop sells a bouquet containing five daisies, three carnations, and two roses for $3.05. Esther also sells a bouquet containing three daisies, one carnation, and four roses for $2.75. Her Valentine's Day Special contains four daisies, two carnations, and one rose for $2.10. How much should she charge for her Economy Special, which contains one daisy, one carnation, and one rose?

46. *Peppers, Tomatoes, and Eggplants*   Ngan planted 81 plants in his garden at a total cost of $23.85. The 81 plants consisted of peppers at 20 cents each, tomatoes at 35 cents each, and eggplants at 30 cents each. If the total number of peppers and tomatoes was only half the number of eggplants, then how many of each did he plant?

Solve each linear programming problem.

47. *Minimum*   Find the minimum value of the function $C(x, y) = 0.42x + 0.84y$ subject to the constraints $x \geq 0, y \geq 0, x + 6y \geq 60$, and $x + y \geq 35$.

48. *Maximum*   Find the maximum value of the function $P(x, y) = 1.23x + 1.64y$ subject to the constraints $x \geq 0, y \geq 0, x + 2y \leq 40$, and $x + y \leq 30$.

49. *Pipeline or Barge*   A refinery gets its oil from a pipeline or from barges. The refinery needs at least 12 million barrels per day. The maximum capacity of the pipeline is 12 million barrels per day, and the maximum that can be delivered by barge is 8 million barrels per day. The refinery has a contract to buy at least 6 million barrels per day through the pipeline. If the cost of oil by barge is $18 per barrel and the cost of oil from the pipeline is $20 per barrel, then how much oil should be purchased from each source to minimize the cost and satisfy the constraints?

50. *Fluctuating Costs*   If the cost of oil by barge goes up to $21 per barrel while the cost of oil by pipeline stays at $20 per barrel, then how much oil should be purchased from each source to minimize the cost and satisfy the constraints of Exercise 49?

Thinking Outside the Box LXXVI & LXXVII

Maximizing Mileage   Dan knows that tires on the front of his SUV will wear out in 20,000 miles, whereas tires on the rear of his SUV will wear out in 30,000 miles. If Dan buys five new tires for his SUV, then what is the maximum mileage that he can get out of the set of tires by rotating them? Describe how and when they should be rotated.

Tree Farming   A tree farmer is planting pine seedlings in a rectangular field that is 1000 ft by 3000 ft. The trees can be planted on the very edge of the field, but the trees must be at least 10 ft apart. What is the maximum number of trees that can be planted in this field?

▪▪▪ Chapter 8 Test

Solve the system by the indicated method.

1. Graphing:

$2x + 3y = 6$

$y = \dfrac{1}{3}x + 5$

2. Substitution:

$2x + y = 4$

$3x - 4y = 9$

3. Addition:

$10x - 3y = 22$

$7x + 2y = 40$

Determine whether each of the following systems is independent, inconsistent, or dependent.

4. $x = 6 - y$

$3x + 3y = 4$

5. $y = \dfrac{1}{2}x + 3$

$x - 2y = -6$

6. $y = 2x - 1$

$y = 3x + 20$

7. $y = -x + 2$

$y = -x + 5$

Find the solution set to each system of equations in three variables.

8. $2x - y + z = 4$

$-x + 2y - z = 6$

9. $x - 2y - z = 2$

$2x + 3y + z = -1$

$3x - y - 3z = -4$

10. $x + y + z = 1$

$x + y - z = 4$

$-x - y + z = 2$

Solve each system.

11. $x^2 + y^2 = 16$

$x^2 - 4y^2 = 16$

12. $x + y = -2$

$y = x^2 - 5x$

Find the partial fraction decomposition for each rational expression.

13. $\dfrac{2x + 10}{x^2 - 2x - 8}$

14. $\dfrac{4x^2 + x - 2}{x^3 - x^2}$

Graph the solution set to each inequality or system of inequalities.

15. $2x - y < 8$

16. $x + y \le 5$

$x - y < 0$

17. $x^2 + y^2 \le 9$

$y \le 1 - x^2$

Solve each problem.

18. In a survey of 52 students in the cafeteria, it was found that 15 were commuters. If one-quarter of the female students and one-third of the male students in the survey were commuters, then how many of each sex were surveyed?

19. General Hospital is planning an aggressive advertising campaign to bolster the hospital's image in the community. Each television commercial reaches 14,000 people and costs $9000, while each newspaper ad reaches 6000 people and costs $3000. The advertising budget for the campaign is limited to $99,000, and the advertising agency has the capability of producing a maximum of 23 ads and/or commercials during the time allotted for the campaign. What mix of television commercials and newspaper ads will maximize the audience exposure, subject to the given constraints?

Tying it all Together

Chapters 1–8

Solve each equation.

1. $\dfrac{x-2}{x+5} = \dfrac{11}{24}$

2. $\dfrac{1}{x} + \dfrac{x-2}{x+5} = \dfrac{11}{24}$

3. $5 - 3(x+2) - 2(x-2) = 7$

4. $|3 - 2x| = 5$

5. $\sqrt{3-2x} = 5$

6. $3x^2 - 4 = 0$

7. $\dfrac{(x-2)^2}{x^2} = 1$

8. $2^{x-1} = 9$

9. $\log(x+1) + \log(x+4) = 1$

10. $x^{-2/3} = 0.25$

11. $x^2 - 3x = 6$

12. $2(x-3)^2 - 1 = 0$

Solve each inequality and graph the solution set on the number line.

13. $3 - 2x > 0$

14. $|3 - 2x| > 0$

15. $x^2 \geq 9$

16. $(x-2)(x+4) \leq 27$

Graph the solution set to each inequality in the coordinate plane.

17. $3 - 2x > y$

18. $|3 - 2x| > y$

19. $x^2 \geq 9$

20. $(x-2)(x+4) \leq y$

9 Matrices and Determinants

N the year 2020, traffic gridlocks in our major cities will be a thing of the past. At least that is the vision of urban planners, who see mathematics and computers playing an ever increasing role in traffic management.

In the future, powerful computers will pinpoint drivers' locations, select routes and speeds, display maps, prevent collisions, and even issue weather reports. Drivers will simply relax and let the computer do the work.

WHAT YOU WILL LEARN To perform such feats, computer systems must instantly solve huge linear programming problems involving hundreds of thousands of variables. However, the techniques that we learned in Chapter 8 are not readily adaptable to computers. In this chapter we will study matrices and several methods for solving linear systems that are used on computers.

719

9.1

Solving Linear Systems Using Matrices

In this section we learn a method for solving systems of linear equations that is an improved version of the addition method of Section 8.1. The new method requires some new terminology.

Matrices

Twenty-six female students and twenty-four male students responded to a survey on income in a college algebra class. Among the female students, 5 classified themselves as low-income, 10 as middle-income, and 11 as high-income. Among the male students, 9 were low-income, 2 were middle-income, and 13 were high-income. Each student is classified in two ways, according to gender and income. This information can be written in a *matrix*:

$$\begin{array}{c} \\ \text{Female} \\ \text{Male} \end{array} \begin{array}{ccc} \text{L} & \text{M} & \text{H} \\ \begin{bmatrix} 5 & 10 & 11 \\ 9 & 2 & 13 \end{bmatrix} \end{array}$$

In this matrix we can see the class makeup according to gender and income. A matrix provides a convenient way to organize a two-way classification of data.

A **matrix** is a rectangular array of real numbers. The **rows** of a matrix run horizontally, and the **columns** run vertically. A matrix with only one row is a **row matrix** or **row vector,** and a matrix with only one column is a **column matrix** or **column vector.** A matrix with m rows and n columns has **size** $m \times n$ (read "m by n"). The number of rows is always given first. For example, the matrix used to classify the students is a 2×3 matrix.

Example 1 Finding the size of a matrix

Determine the size of each matrix.

a. $[-4 \quad 3 \quad -2]$ **b.** $\begin{bmatrix} 3 & -1 \\ 4 & 2 \end{bmatrix}$ **c.** $\begin{bmatrix} -5 & 19 \\ 14 & 2 \\ 0 & -1 \end{bmatrix}$ **d.** $\begin{bmatrix} -4 & 46 & 8 \\ 1 & 0 & 13 \\ -12 & -5 & 2 \end{bmatrix}$

Solution

Matrix (a) is a row matrix with size 1×3. Matrix (b) has size 2×2, matrix (c) has size 3×2, and matrix (d) has size 3×3.

Try This. Determine the size of the matrix $\begin{bmatrix} 1 & 3 & 5 & 7 \\ 2 & 4 & 6 & 8 \end{bmatrix}$. ■

A **square matrix** has an equal number of rows and columns. Matrices (b) and (d) of Example 1 are square matrices. Each number in a matrix is called an **entry** or an **element.** The matrix [5] is a 1×1 matrix with only one entry, 5.

The Augmented Matrix

We now see how matrices are used to represent systems of linear equations. The solution to a system of linear equations such as

$$x - 3y = 11$$
$$2x + y = 1$$

depends on the coefficients of x and y and the constants on the right-hand side of the equation. The **coefficient matrix** for this system is the matrix

$$\begin{bmatrix} 1 & -3 \\ 2 & 1 \end{bmatrix},$$

whose entries are the coefficients of the variables. (The coefficient of y in $x - 3y = 11$ is -3.) The constants from the right-hand side of the system are attached to the matrix of coefficients, to form the **augmented matrix** of the system:

$$\begin{bmatrix} 1 & -3 & \bigm| & 11 \\ 2 & 1 & \bigm| & 1 \end{bmatrix}$$

Each row of the augmented matrix represents an equation of the system, while the columns represent the coefficients of x, the coefficients of y, and the constants, respectively. The vertical line represents the equal signs.

Example **2** **Determining the augmented matrix**

Write the augmented matrix for each system of equations.

a. $x = y + 3$ **b.** $x + 2y - z = 1$ **c.** $x - y = 2$
     $y = 4 - x$ $2x \quad + 3z = 5$ $y - z = 3$
                      $3x - 2y + z = 0$

Solution

a. To write the augmented matrix, the equations must be in standard form with the variables on the left-hand side and constants on the right-hand side:

$$x - y = 3$$
$$x + y = 4$$

We write the augmented matrix using the coefficients of the variables and the constants:

$$\begin{bmatrix} 1 & -1 & \bigm| & 3 \\ 1 & 1 & \bigm| & 4 \end{bmatrix}$$

b. Use the coefficient 0 for each variable that is missing:

$$\begin{bmatrix} 1 & 2 & -1 & \bigm| & 1 \\ 2 & 0 & 3 & \bigm| & 5 \\ 3 & -2 & 1 & \bigm| & 0 \end{bmatrix}$$

c. The augmented matrix for this system is a 2×4 matrix:

$$\begin{bmatrix} 1 & -1 & 0 & \bigm| & 2 \\ 0 & 1 & -1 & \bigm| & 3 \end{bmatrix}$$

Try This. Write the augmented matrix for the system $\begin{array}{l} x - y = -2 \\ 2x + y = 3 \end{array}$.   ■

Example 3 Writing a system for an augmented matrix

Write the system of equations represented by each augmented matrix.

a. $\begin{bmatrix} 1 & 3 & \bigm| & -5 \\ 2 & -3 & \bigm| & 1 \end{bmatrix}$ **b.** $\begin{bmatrix} 1 & 0 & \bigm| & 7 \\ 0 & 1 & \bigm| & 3 \end{bmatrix}$ **c.** $\begin{bmatrix} 2 & 5 & 1 & \bigm| & 2 \\ -3 & 0 & 4 & \bigm| & -1 \\ 4 & -5 & 2 & \bigm| & 3 \end{bmatrix}$

Solution

a. Use the first two numbers in each row as the coefficients of x and y and the last number as the constant to get the following system:

$$x + 3y = -5$$
$$2x - 3y = 1$$

b. The augmented matrix represents the following system:

$$x = 7$$
$$y = 3$$

c. Use the first three numbers in each row as the coefficients of x, y, and z and the last number as the constant to get the following system:

$$2x + 5y + z = 2$$
$$-3x + 4z = -1$$
$$4x - 5y + 2z = 3$$

Try This. Write the system for the matrix $\begin{bmatrix} 2 & 1 & \bigm| & 5 \\ -1 & 4 & \bigm| & 6 \end{bmatrix}$.   ■

Two systems of linear equations are **equivalent** if they have the same solution set, whereas two augmented matrices are **equivalent** if the systems they represent are equivalent. The augmented matrices

$$\begin{bmatrix} 1 & 1 & \bigm| & 5 \\ 2 & 3 & \bigm| & 13 \end{bmatrix} \quad \text{and} \quad \begin{bmatrix} 1 & 1 & \bigm| & 5 \\ 0 & 1 & \bigm| & 3 \end{bmatrix}$$

are equivalent because their corresponding systems

$$\begin{array}{l} x + y = 5 \\ 2x + 3y = 13 \end{array} \quad \text{and} \quad \begin{array}{l} x + y = 5 \\ y = 3 \end{array}$$

are equivalent. Each system has solution set $\{(2, 3)\}$.

Recall that to solve a single equation, we write simpler and simpler equivalent equations to get an equation whose solution is obvious. Similarly, to solve a system of equations, we write simpler and simpler equivalent systems to get a system whose solution is obvious. We now look at operations that can be performed on augmented matrices to obtain simpler equivalent augmented matrices.

The Gaussian Elimination Method

The rows of an augmented matrix represent the equations of a system. Since the equations of a system can be written in any order, two rows of an augmented matrix can be interchanged if necessary. Since multiplication of both sides of an equation by the same nonzero number produces an equivalent equation, multiplying each entry in a row of the augmented matrix by a nonzero number produces an equivalent augmented matrix. In Section 8.1, two equations were added to eliminate a variable. In the augmented matrix, elimination of variables is accomplished by adding the entries in one row to the corresponding entries in another row. These two row operations can be combined to add a multiple of one row to another, just as was done in solving systems by addition. The three **row operations** for an augmented matrix are summarized as follows.

S U M M A R Y **Row Operations**

Any of the following row operations on an augmented matrix gives an equivalent augmented matrix:

1. Interchanging two rows of the matrix.
2. Multiplying every entry in a row by the same nonzero real number.
3. Adding to a row a nonzero multiple of another row.

To solve a system of two linear equations in two variables using the **Gaussian elimination method,** we use row operations to obtain simpler and simpler augmented matrices. We want to get an augmented matrix that corresponds to a system whose solution is obvious. An augmented matrix of the following form is the simplest:

$$\left[\begin{array}{cc|c} 1 & 0 & a \\ 0 & 1 & b \end{array}\right]$$

Notice that this augmented matrix corresponds to the system $x = a$ and $y = b$, for which the solution set is $\{(a, b)\}$.

The **diagonal** of a matrix consists of the entries in the first row first column, second row second column, third row third column, and so on. A square matrix with ones on the diagonal and zeros elsewhere is an **identity matrix:**

$$\left[\begin{array}{cc} a_{11} & a_{12} \\ a_{21} & a_{22} \end{array}\right] \qquad \left[\begin{array}{cc} 1 & 0 \\ 0 & 1 \end{array}\right] \qquad \left[\begin{array}{ccc} 1 & 0 & 0 \\ 0 & 1 & 0 \\ 0 & 0 & 1 \end{array}\right]$$

The diagonal of a 2×2 matrix   The 2×2 identity matrix   The 3×3 identity matrix

The goal of the Gaussian elimination method is to convert the coefficient matrix (in the augmented matrix) into an identity matrix using row operations. If the system has a unique solution, then it will appear in the right-most column of the final augmented matrix.

Example **4** **Using the Gaussian elimination method**

Use row operations to solve the system.

$$2x - 4y = 16$$
$$3x + y = 3$$

Solution

Start with the augmented matrix:

$$\begin{bmatrix} 2 & -4 & | & 16 \\ 3 & 1 & | & 3 \end{bmatrix}$$

The first step is to multiply the first row R_1 by $\frac{1}{2}$ to get a 1 in the first position on the diagonal. Think of this step as replacing R_1 by $\frac{1}{2}R_1$. We show this in symbols as $\frac{1}{2}R_1 \rightarrow R_1$. Read the arrow as "replaces."

$$\begin{bmatrix} 1 & -2 & | & 8 \\ 3 & 1 & | & 3 \end{bmatrix} \quad \frac{1}{2}R_1 \rightarrow R_1$$

To get a 0 in the first position of R_2, multiply R_1 by -3 and add the result to R_2. Since $-3R_1 = [-3, 6, -24]$ and $R_2 = [3, 1, 3]$, $-3R_1 + R_2 = [0, 7, -21]$. We are replacing R_2 with $-3R_1 + R_2$:

$$\begin{bmatrix} 1 & -2 & | & 8 \\ 0 & 7 & | & -21 \end{bmatrix} \quad -3R_1 + R_2 \rightarrow R_2$$

To get a 1 in the second position on the diagonal, multiply R_2 by $\frac{1}{7}$:

$$\begin{bmatrix} 1 & -2 & | & 8 \\ 0 & 1 & | & -3 \end{bmatrix} \quad \frac{1}{7}R_2 \rightarrow R_2$$

Now row 2 is in the form needed to solve the system. We next get a 0 as the second entry in R_1. Multiply row 2 by 2 and add the result to row 1. Since $2R_2 = [0, 2, -6]$ and $R_1 = [1, -2, 8]$, $2R_2 + R_1 = [1, 0, 2]$. We get the following matrix.

$$\begin{bmatrix} 1 & 0 & | & 2 \\ 0 & 1 & | & -3 \end{bmatrix} \quad 2R_2 + R_1 \rightarrow R_1$$

Note that the coefficient of y in the first equation is now 0. The system associated with the last augmented matrix is $x = 2$ and $y = -3$. So the solution set to the system is $\{(2, -3)\}$. Check in the original system.

Try This. Use row operations to solve the system $\begin{array}{r} 2x - y = 7 \\ x + 3y = 14 \end{array}$. ■

The procedure used in Example 4 to solve a system that has a unique solution is summarized below. For consistent and dependent systems, see Examples 5 and 6.

PROCEDURE

The Gaussian Elimination Method for an Independent System of Two Equations

To solve a system of two linear equations in two variables using Gaussian elimination, perform the following row operations on the augmented matrix.

1. If necessary, interchange R_1 and R_2 so that R_1 begins with a nonzero entry.
2. Get a 1 in the first position on the diagonal by multiplying R_1 by the reciprocal of the first entry in R_1.
3. Add an appropriate multiple of R_1 to R_2 to get 0 below the first 1.
4. Get a 1 in the second position on the diagonal by multiplying R_2 by the reciprocal of the second entry in R_2.
5. Add an appropriate multiple of R_2 to R_1 to get 0 above the second 1.
6. Read the unique solution from the last column of the final augmented matrix.

Inconsistent and Dependent Equations in Two Variables

A system is independent if it has a single solution. The coefficient matrix of an independent system is equivalent to an identity matrix. A system is inconsistent if it has no solution and dependent if it has infinitely many solutions. The coefficient matrix for an inconsistent or dependent system is not equivalent to an identity matrix. However, we can still use Gaussian elimination to simplify the coefficient matrix and determine the solution.

Example **5** An inconsistent system in two variables

Solve the system.

$$x = y + 3$$
$$2y = 2x + 5$$

Solution

Write both equations in the form $Ax + By = C$:

$$x - y = 3$$
$$-2x + 2y = 5$$

Start with the augmented matrix:

$$\begin{bmatrix} 1 & -1 & | & 3 \\ -2 & 2 & | & 5 \end{bmatrix}$$

To get a 0 in the first position of R_2, multiply R_1 by 2 and add the result to R_2:

$$\begin{bmatrix} 1 & -1 & | & 3 \\ 0 & 0 & | & 11 \end{bmatrix} \quad 2R_1 + R_2 \rightarrow R_2$$

It is impossible to convert this augmented matrix to the desired form of the Gaussian elimination method. However, we can obtain the solution by observing that the second row corresponds to the equation $0 = 11$. So the system is inconsistent, and there is no solution.

Try This. Use row operations to solve the system $\begin{aligned} 3x - y &= 1 \\ -6x + 2y &= 4 \end{aligned}$.   ■

Applying Gaussian elimination to an inconsistent system (as in Example 5) causes a row to appear with 0 as the entry for each coefficient but a nonzero entry for the constant. For a dependent system of two equations in two variables (as in the next example), a 0 will appear in every entry of some row.

Example 6 A dependent system in two variables

Solve the system.

$$2x + y = 4$$
$$4x + 2y = 8$$

Solution

Start with the augmented matrix:

$$\left[\begin{array}{cc|c} 2 & 1 & 4 \\ 4 & 2 & 8 \end{array}\right]$$

Notice that the second row is twice the first row. So instead of getting a 1 in the first position on the diagonal, we multiply R_1 by -2 and add the result to R_2:

$$\left[\begin{array}{cc|c} 2 & 1 & 4 \\ 0 & 0 & 0 \end{array}\right] \quad -2R_1 + R_2 \rightarrow R_2$$

The second row of this augmented matrix gives us the equation $0 = 0$. So the system is dependent. Every ordered pair that satisfies the first equation satisfies both equations. The solution set is $\{(x, y) \mid 2x + y = 4\}$. Since $y = 4 - 2x$, every ordered pair of the form $(x, 4 - 2x)$ is a solution. So the solution set can be written also as $\{(x, 4 - 2x)\}$.

Try This. Use row operations to solve the system $\begin{aligned} 4x + y &= 5 \\ 8x + 2y &= 10 \end{aligned}$.   ■

Gaussian Elimination with Three Variables

In the next example, Gaussian elimination is used on a system involving three variables. For three linear equations in three variables, x, y, and z, we try to get the augmented matrix into the form

$$\left[\begin{array}{ccc|c} 1 & 0 & 0 & a \\ 0 & 1 & 0 & b \\ 0 & 0 & 1 & c \end{array}\right],$$

from which we conclude that $x = a$, $y = b$, and $z = c$.

Example **7** **Gaussian elimination with three variables**

Use the Gaussian elimination method to solve the following system:

$$2x - y + z = 1$$
$$x + y - 2z = 5$$
$$3x - y - z = 8$$

Solution

Write the augmented matrix:

$$\left[\begin{array}{ccc|c} 2 & -1 & 1 & 1 \\ 1 & 1 & -2 & 5 \\ 3 & -1 & -1 & 8 \end{array}\right]$$

To get the first 1 on the diagonal, we could multiply R_1 by $\frac{1}{2}$ or interchange R_1 and R_2. Interchanging R_1 and R_2 is simpler because it avoids getting fractions in the entries:

$$\left[\begin{array}{ccc|c} 1 & 1 & -2 & 5 \\ 2 & -1 & 1 & 1 \\ 3 & -1 & -1 & 8 \end{array}\right] \quad R_1 \leftrightarrow R_2$$

Now use the first row to get 0's below the first 1 on the diagonal. First, to get a 0 in the first position of the second row, multiply the first row by -2 and add the result to the second row. Then, to get a 0 in the first position of the third row, multiply the first row by -3 and add the result to the third row. These two steps eliminate the variable x from the second and third rows.

$$\left[\begin{array}{ccc|c} 1 & 1 & -2 & 5 \\ 0 & -3 & 5 & -9 \\ 0 & -4 & 5 & -7 \end{array}\right] \quad \begin{array}{l} -2R_1 + R_2 \to R_2 \\ -3R_1 + R_3 \to R_3 \end{array}$$

To get a 1 in the second position on the diagonal, multiply the second row by $-\frac{1}{3}$:

$$\left[\begin{array}{ccc|c} 1 & 1 & -2 & 5 \\ 0 & 1 & -\frac{5}{3} & 3 \\ 0 & -4 & 5 & -7 \end{array}\right] \quad -\frac{1}{3}R_2 \to R_2$$

To get a 0 above the second 1 on the diagonal, multiply R_2 by -1 and add the result to R_1. To get a 0 below the second 1 on the diagonal, multiply R_2 by 4 and add the result to the third row:

$$\left[\begin{array}{ccc|c} 1 & 0 & -\frac{1}{3} & 2 \\ 0 & 1 & -\frac{5}{3} & 3 \\ 0 & 0 & -\frac{5}{3} & 5 \end{array}\right] \quad \begin{array}{l} -1R_2 + R_1 \to R_1 \\ \\ 4R_2 + R_3 \to R_3 \end{array}$$

To get a 1 in the third position on the diagonal, multiply R_3 by $-\frac{3}{5}$:

$$\left[\begin{array}{ccc|c} 1 & 0 & -\frac{1}{3} & 2 \\ 0 & 1 & -\frac{5}{3} & 3 \\ 0 & 0 & 1 & -3 \end{array}\right] \quad -\frac{3}{5}R_3 \to R_3$$

Use the third row to get 0's above the third 1 on the diagonal:

$$\begin{bmatrix} 1 & 0 & 0 & | & 1 \\ 0 & 1 & 0 & | & -2 \\ 0 & 0 & 1 & | & -3 \end{bmatrix} \quad \begin{matrix} \frac{1}{3}R_3 + R_1 \rightarrow R_1 \\ \frac{5}{3}R_3 + R_2 \rightarrow R_2 \end{matrix}$$

This last matrix corresponds to $x = 1, y = -2,$ and $z = -3$. So the solution set to the system is $\{(1, -2, -3)\}$. Check in the original system.

Try This. Use Gaussian elimination to solve $\begin{aligned} x + y - z &= 0 \\ -x + y \phantom{{}+z} &= 4 \\ -x + y + z &= 10 \end{aligned}$. ▪

In Example 7, there were two different row operations that would produce the first 1 on the diagonal. When doing Gaussian elimination by hand, you should choose the row operations that make the computations the simplest. When this method is performed by a computer, the same sequence of steps is used on every system. Any system of three equations that has a unique solution can be solved with the following sequence of steps.

PROCEDURE

The Gaussian Elimination Method for an Independent System of Three Equations

To solve a system of three linear equations in three variables using Gaussian elimination, perform the following row operations on the augmented matrix.

1. Get a 1 in the first position on the diagonal by multiplying R_1 by the reciprocal of the first entry in R_1. (First interchange rows if necessary.)

2. Add appropriate multiples of R_1 to R_2 and R_3 to get 0's below the first 1.

3. Get a 1 in the second position on the diagonal by multiplying R_2 by the reciprocal of the second entry in R_2. (First interchange rows R_2 and R_3 if necessary.)

4. Add appropriate multiples of R_2 to R_1 and R_3 to get 0's above and below the second 1.

5. Get a 1 in the third position on the diagonal by multiplying R_3 by the reciprocal of the third entry in R_3.

6. Add appropriate multiples of R_3 to R_1 and R_2 to get 0's above the third 1.

7. Read the unique solution from the last column of the final augmented matrix.

Inconsistent and Dependent Systems in Three Variables

Applying Gaussian elimination to inconsistent or dependent systems in two or three variables is similar. If a system is inconsistent, then a row will appear with 0 as the entry for each coefficient but a nonzero entry for the constant. If a system of three equations in three variables is dependent then a row will appear in which all entries are 0. The next example shows a system with fewer equations than variables. In this case we do not necessarily get a row in which all entries are 0.

Example **8** **A dependent system in three variables**

Solve the system.

$$x - y + z = 2$$
$$-2x + y + 2z = 5$$

Solution

Start with the augmented matrix:

$$\begin{bmatrix} 1 & -1 & 1 & | & 2 \\ -2 & 1 & 2 & | & 5 \end{bmatrix}$$

Perform the following row operations to get ones and zeros in the first two columns, as you would for a 2 × 2 matrix:

$$\begin{bmatrix} 1 & -1 & 1 & | & 2 \\ 0 & -1 & 4 & | & 9 \end{bmatrix} \qquad 2R_1 + R_2 \to R_2$$

$$\begin{bmatrix} 1 & -1 & 1 & | & 2 \\ 0 & 1 & -4 & | & -9 \end{bmatrix} \qquad -1R_2 \to R_2$$

$$\begin{bmatrix} 1 & 0 & -3 & | & -7 \\ 0 & 1 & -4 & | & -9 \end{bmatrix} \qquad R_2 + R_1 \to R_1$$

The last matrix corresponds to the system.

$$x - 3z = -7$$
$$y - 4z = -9$$

or $x = 3z - 7$ and $y = 4z - 9$. The system is dependent, and the solution set is

$$\{(3z - 7, 4z - 9, z) \mid z \text{ is any real number}\}.$$

Replace x by $3z - 7$ and y by $4z - 9$ in the original equations to check.

Try This. Use Gaussian elimination to solve $\begin{array}{l} x - y + 2z = 3 \\ -x + 2y + z = 2 \end{array}$. ■

The Gaussian elimination method can be applied to a system of n linear equations in m unknowns. However, it is a rather tedious method to perform when n and m are greater than 2, especially when fractions are involved. Computers can be programmed to work with matrices, and Gaussian elimination is frequently used for computer solutions.

You can enter matrices into a graphing calculator and perform row operations with the calculator. However, performing row operations with a calculator is still rather tedious. In Section 9.4 we will see a much simpler method for solving systems with a calculator. □

Applications

There is much discussion among city planners and transportation experts about automated highways (*Scientific American*, www.sciamarchive.com). An automated highway system would use a central computer and computers within vehicles to manage traffic flow by controlling traffic lights, rerouting traffic away from congested areas, and perhaps even driving the vehicle for you. It is easy to say that computers will control the complex traffic system of the future, but a computer does only what it is programmed to do. The next example shows one type of problem that a computer would be continually solving in order to control traffic flow.

Example 9 Traffic control in the future

Figure 9.1 shows the intersections of four one-way streets. The numbers on the arrows represent the number of cars per hour that desire to enter each intersection and leave each intersection. For example, 400 cars per hour want to enter intersection P from the north on First Avenue while 300 cars per hour want to head east from intersection Q on Elm Street. The letters w, x, y, and z represent the number of cars per hour passing the four points between these four intersections, as shown in Fig. 9.1.

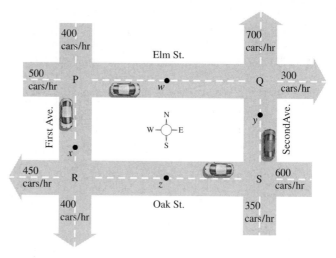

■ **Figure 9.1**

a. Find values for w, x, y, and z that would realize this desired traffic flow.
b. If construction on Oak Street limits z to 300 cars per hour, then how many cars per hour would have to pass w, x, and y?

Solution

a. The solution to the problem is based on the fact that the number of cars entering an intersection per hour must equal the number leaving that intersection per hour, if the traffic is to keep flowing. Since 900 cars (400 + 500) enter intersection P, 900 must leave, $x + w = 900$. Writing a similar equation for each intersection yields the system shown next.

$$w + x = 900$$

$$w + y = 1000$$

$$x + z = 850$$

$$y + z = 950$$

For this system, the augmented matrix is

$$\left[\begin{array}{cccc|c} 1 & 1 & 0 & 0 & 900 \\ 1 & 0 & 1 & 0 & 1000 \\ 0 & 1 & 0 & 1 & 850 \\ 0 & 0 & 1 & 1 & 950 \end{array}\right].$$

Use row operations to get the equivalent matrix:

$$\left[\begin{array}{cccc|c} 1 & 1 & 0 & 0 & 900 \\ 0 & 1 & 0 & 1 & 850 \\ 0 & 0 & 1 & 1 & 950 \\ 0 & 0 & 0 & 0 & 0 \end{array}\right].$$

The system is a dependent system and does not have a unique solution. From this matrix we get $w = 50 + z$, $x = 850 - z$, and $y = 950 - z$. The problem is solved by any ordered 4-tuple of the form $(50 + z, 850 - z, 950 - z, z)$ where z is a nonnegative integer. (In general, n numbers or expressions separated by commas within parentheses is called an n-tuple.)

b. If z is limited to 300 because of construction, then the solution is (350, 550, 650, 300). To keep traffic flowing when $z = 300$, the system must route 350 cars past w, 550 past x, and 650 past y.

Try This. A small gift basket contains 3 apples and 5 pears. A large gift basket contains 8 apples and 7 pears. If 30 apples and 31 pears were used, then how many baskets of each type were made? Use Gaussian elimination. ■

One-way streets were used in Example 9 to keep the problem simple, but you can imagine two-way streets where at each intersection there are three ways to route traffic. You can also imagine many streets and many intersections all subject to systems of equations, with computers continually controlling flow to keep all traffic moving. Computerized traffic control may be a few years away, but similar problems are being solved today at AT&T to route long-distance calls and at American Airlines to schedule flight crews and equipment.

For Thought

True or False? Explain.

1. The augmented matrix for a system of two linear equations in two unknowns is a 2×2 matrix.

2. The augmented matrix
for $\begin{array}{r} x - y = 4 \\ 3x + y = 5 \end{array}$ is $\left[\begin{array}{cc|c} x & -y & 4 \\ 3x & y & 5 \end{array}\right].$

3. The augmented matrix
for $\begin{array}{r} 3x - 2y = 4 \\ x + y = 6 \end{array}$ is $\left[\begin{array}{cc|c} 3 & -2 & 4 \\ 1 & 1 & 6 \end{array}\right].$

4. The matrix $\left[\begin{array}{cc|c} 1 & 0 & 2 \\ 1 & -1 & -3 \end{array}\right]$ corresponds to the
system $\begin{array}{r} x = 2 \\ x - y = -3. \end{array}$

5. The matrix $\left[\begin{array}{cc|c} 1 & 3 & -2 \\ -1 & -5 & 4 \end{array}\right]$ is equivalent
to $\left[\begin{array}{cc|c} 1 & 3 & -2 \\ 0 & -2 & 2 \end{array}\right].$

6. The matrix $\begin{bmatrix} 1 & 2 & | & 3 \\ 1 & -3 & | & 2 \end{bmatrix}$ is equivalent

to $\begin{bmatrix} 1 & 2 & | & 3 \\ 0 & -5 & | & -1 \end{bmatrix}$.

7. The matrix $\begin{bmatrix} 1 & 0 & | & 2 \\ 0 & 1 & | & 7 \end{bmatrix}$ corresponds to the

system $\begin{aligned} x + y &= 2 \\ x + y &= 7. \end{aligned}$

8. The system corresponding to $\begin{bmatrix} 1 & 3 & | & 5 \\ 0 & 0 & | & 7 \end{bmatrix}$ is inconsistent.

9. The system corresponding to $\begin{bmatrix} -1 & 2 & | & -3 \\ 0 & 0 & | & 0 \end{bmatrix}$ is inconsistent.

10. The notation $2R_1 + R_3 \to R_3$ means to replace R_3 by $2R_1 + R_3$.

9.1 Exercises

Determine the size of each matrix.

1. $[1 \quad 5 \quad 8]$

2. $\begin{bmatrix} -3 \\ y \\ 5 \end{bmatrix}$

3. $[7]$

4. $\begin{bmatrix} x & y \\ z & w \end{bmatrix}$

5. $\begin{bmatrix} -5 & 12 \\ 99 & 6 \\ 0 & 0 \end{bmatrix}$

6. $\begin{bmatrix} 1 & 5 & 7 \\ 3 & 0 & 5 \\ 2 & -6 & -3 \end{bmatrix}$

15. $\begin{bmatrix} 5 & 0 & 0 & | & 6 \\ -4 & 0 & 2 & | & -1 \\ 4 & 4 & 0 & | & 7 \end{bmatrix}$

16. $\begin{bmatrix} 1 & 0 & 1 & | & 2 \\ 0 & 1 & -1 & | & -6 \\ 1 & -1 & 1 & | & 5 \end{bmatrix}$

17. $\begin{bmatrix} 1 & -1 & 2 & | & 1 \\ 0 & 1 & 4 & | & 3 \\ 0 & 0 & 0 & | & 0 \end{bmatrix}$

18. $\begin{bmatrix} 1 & 1 & 1 & | & 3 \\ 0 & 1 & 2 & | & 7 \\ 0 & 0 & 0 & | & 0 \end{bmatrix}$

Write the augmented matrix for each system of equations.

7. $\begin{aligned} x - 2y &= 4 \\ 3x + 2y &= -5 \end{aligned}$

8. $\begin{aligned} 4x - y &= 1 \\ x + 3y &= 5 \end{aligned}$

9. $\begin{aligned} x - y - z &= 4 \\ x + 3y - z &= 1 \\ 2y - 5z &= -6 \end{aligned}$

10. $\begin{aligned} x + 3y &= 5 \\ y - 4z &= 8 \\ -2x + 5z &= 7 \end{aligned}$

11. $\begin{aligned} x + 3y - z &= 5 \\ x + z &= 0 \end{aligned}$

12. $\begin{aligned} x - y &= 6 \\ x + z &= 7 \end{aligned}$

Write the system of equations represented by each augmented matrix.

13. $\begin{bmatrix} 3 & 4 & | & -2 \\ 3 & -5 & | & 0 \end{bmatrix}$

14. $\begin{bmatrix} 1 & 0 & | & -7 \\ 0 & 1 & | & 5 \end{bmatrix}$

Perform the indicated row operation on the given augmented matrix. See the summary of row operations on page 723.

19. $R_1 \leftrightarrow R_2$: $\begin{bmatrix} -2 & 4 & | & 1 \\ 1 & 2 & | & 0 \end{bmatrix}$

20. $R_1 \leftrightarrow R_2$: $\begin{bmatrix} 4 & 9 & | & 3 \\ 1 & 0 & | & 5 \end{bmatrix}$

21. $\frac{1}{2}R_1 \to R_1$: $\begin{bmatrix} 2 & 8 & | & 2 \\ 0 & 3 & | & 6 \end{bmatrix}$

22. $-\frac{1}{3}R_1 \to R_1$: $\begin{bmatrix} -3 & 6 & | & 12 \\ 0 & 9 & | & 3 \end{bmatrix}$

23. $3R_1 + R_2 \to R_2$: $\begin{bmatrix} 1 & -2 & | & 1 \\ -3 & 5 & | & 0 \end{bmatrix}$

24. $-2R_2 + R_1 \to R_1$: $\begin{bmatrix} 1 & 2 & | & 7 \\ 0 & 1 & | & 4 \end{bmatrix}$

For each given sequence of augmented matrices determine the system that has been solved, the solution, and the row operation that was used on each matrix to obtain the next matrix in the sequence. For the row operations use the notation that was introduced in the examples.

25. $\begin{bmatrix} 2 & 4 & | & 14 \\ 5 & 4 & | & 5 \end{bmatrix}$
$\begin{bmatrix} 1 & 2 & | & 7 \\ 5 & 4 & | & 5 \end{bmatrix}$
$\begin{bmatrix} 1 & 2 & | & 7 \\ 0 & -6 & | & -30 \end{bmatrix}$
$\begin{bmatrix} 1 & 2 & | & 7 \\ 0 & 1 & | & 5 \end{bmatrix}$
$\begin{bmatrix} 1 & 0 & | & -3 \\ 0 & 1 & | & 5 \end{bmatrix}$

26. $\begin{bmatrix} 3 & 5 & | & -2 \\ 1 & 2 & | & -1 \end{bmatrix}$
$\begin{bmatrix} 1 & 2 & | & -1 \\ 3 & 5 & | & -2 \end{bmatrix}$
$\begin{bmatrix} 1 & 2 & | & -1 \\ 0 & -1 & | & 1 \end{bmatrix}$
$\begin{bmatrix} 1 & 2 & | & -1 \\ 0 & 1 & | & -1 \end{bmatrix}$
$\begin{bmatrix} 1 & 0 & | & 1 \\ 0 & 1 & | & -1 \end{bmatrix}$

Solve each system using Gaussian elimination. State whether each system is independent, dependent, or inconsistent. See the procedure for the Gaussian elimination method for two equations on page 725.

27. $x + y = 5$
$-2x + y = -1$

28. $x - y = 2$
$3x - y = 12$

29. $2x + 2y = 8$
$-3x - y = -6$

30. $3x - 6y = 9$
$2x + y = -4$

31. $2x - y = 3$
$3x + 2y = 15$

32. $2x - 3y = -1$
$3x - 2y = 1$

33. $0.4x - 0.2y = 0$
$x + 1.5y = 2$

34. $0.2x + 0.6y = 0.7$
$0.5x - y = 0.5$

35. $3a - 5b = 7$
$-3a + 5b = 4$

36. $2s - 3t = 9$
$4s - 6t = 1$

37. $0.5u + 1.5v = 2$
$3u + 9v = 12$

38. $m - 2.5n = 0.5$
$-4m + 10n = -2$

39. $y = 4 - 2x$
$x = 8 + y$

40. $3x = 1 + 2y$
$y = 2 - x$

Solve each system using Gaussian elimination. State whether each system is independent, dependent, or inconsistent. See the procedure for the Gaussian elimination method for three equations on page 728.

41. $x + y + z = 6$
$x - y - z = 0$
$2y - z = 3$

42. $x - y + z = 2$
$-x + y + z = 4$
$-x + z = 2$

43. $2x + y = 2 + z$
$x + 2y = 2 + z$
$x + 2z = 2 + y$

44. $3x = 4 + y$
$x + y = z - 1$
$2z = 3 - x$

45. $2a - 2b + c = -2$
$a + b - 3c = 3$
$a - 3b + c = -5$

46. $r - 3s - t = -3$
$-r - s + 2t = 1$
$-r + 2s - t = -2$

47. $3y = x + z$
$x - y - 3z = 4$
$x + y + 2z = -1$

48. $z = 2 + x$
$2x - y = 1$
$y + 3z = 15$

49. $x - 2y + 3z = 1$
$2x - 4y + 6z = 2$
$-3x + 6y - 9z = -3$

50. $4x - 2y + 6z = 4$
$2x - y + 3z = 2$
$-2x + y - 3z = -2$

51. $x - y + z = 2$
$2x + y - z = 1$
$2x - 2y + 2z = 5$

52. $x - y + z = 4$
$x + y - z = 1$
$x + y - z = 3$

53. $x + y - z = 3$
$3x + y + z = 7$
$x - y + 3z = 1$

54. $x + 2y + 2z = 4$
$2x + y + z = 1$
$-x + y + z = 3$

55. $2x - y + 3z = 1$
$x + y - z = 4$

56. $x + 3y + z = 6$
$-x + y - z = 2$

57. $x - y + z - w = 2$
$-x + 2y - z - w = -1$
$2x - y - z + w = 4$
$x + 3y - 2z - 3w = 6$

58. $3a - 2b + c + d = 0$
$a - b + c - d = -4$
$-2a + b + 3c - 2d = 5$
$2a + 3b - c - d = -3$

Write a system of linear equations for each problem and solve the system using Gaussian elimination.

59. *Wages from Two Jobs*   Mike works a total of 60 hr per week at his two jobs. He makes $8 per hour at Burgers-R-Us and $9 per hour at the Soap Opera Laundromat. If his total pay for one week is $502 before taxes, then how many hours does he work at each job?

60. *Postal Rates*   Noriko spent $19.80 on postage inviting a total of 60 guests to her promotion party. Each woman was sent a picture postcard showing the company headquarters in Tokyo while each man was invited with a letter. If she put a 25-cent stamp on each postcard and a 37-cent stamp on each letter, then how many guests of each gender were invited?

61. *Investment Portfolio* Petula invested a total of $40,000 in a no-load mutual fund, treasury bills, and municipal bonds. Her total return of $3660 came from an 8% return on her investment in the no-load mutual fund, a 9% return on the treasury bills, and 12% return on the municipal bonds. If her total investment in treasury bills and municipal bonds was equal to her investment in the no-load mutual fund, then how much did she invest in each?

62. *Nutrition* The accompanying table shows the percentage of U.S. Recommended Daily Allowances (RDA) for phosphorous (P), magnesium (Mg), and calcium (Ca) in one ounce of three breakfast cereals (without milk). If Hulk Hogan got 98% of the RDA of phosphorous, 84% of the RDA of magnesium, and 38% of the RDA of calcium by eating a large bowl of each (without milk), then how many ounces of each cereal did he eat?

 HINT Write an equation for P, one for Mg, and one for Ca.

■ **Table for Exercise 62**

	P	Mg	Ca	
Kix	4%	2%	4%	
Quick Oats	10%	10%	0%	
Muesli	8%	8%	2%	

63. *Cubic Curve Fitting* Find a, b, and c such that the graph of $y = ax^3 + bx + c$ goes through the points $(-1, 4)$, $(1, 2)$, and $(2, 7)$.

64. *Quadratic Curve Fitting* Find a, b, and c such that the graph of $y = ax^2 + bx + c$ goes through the points $(-1, 0)$, $(1, 0)$, and $(3, 0)$.

65. *Traffic Control I* The diagram shows the number of cars that desire to enter and leave each of three intersections on three one-way streets in a 60-minute period. The letters x, y, and z represent the number of cars passing the three points between these three intersections, as shown in the diagram. Find values for x, y, and z that would realize this desired traffic flow. If construction on JFK Boulevard limits the value of

z to 50, then what values for x and y would keep the traffic flowing?

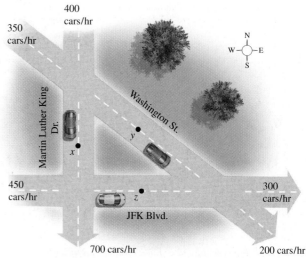

■ **Figure for Exercises 65 and 66**

66. *Traffic Control II* Southbound M. L. King Drive in the figure leads to another intersection that cannot always handle 700 cars in a 60-minute period. Change 700 to 600 in the figure and write a system of three equations in x, y, and z. What is the solution to this system? If you had control over the 400 cars coming from the north into the first intersection on M. L. King Drive, what number would you use in place of 400 to get the system flowing again?

Thinking Outside the Box LXXVIII

The Gigantic The ocean liner Gigantic began taking on water after hitting an iceberg. Water was coming in at a uniform rate and some amount had already accumulated. The captain looked in some tables and found that 12 identical pumps could pump out all of the water in 3 hours, while 5 of those same pumps could do it in 10 hours. To calm the passengers, the captain wanted all of the water out in 2 hours. How many pumps are needed?

9.1 Pop Quiz

1. Determine the size of $\begin{bmatrix} 1 & 2 & 3 \\ 4 & 5 & 6 \end{bmatrix}$.

2. Write the augmented matrix for $\begin{array}{c} 2x - 3y = -9 \\ x + 4y = 23 \end{array}$.

Solve by Gaussian elimination.

3. $\begin{array}{c} 2x - 3y = -9 \\ x + 4y = 23 \end{array}$

4. $\begin{array}{c} x - y + z = 4 \\ -x + 2y + z = -1 \\ -x + y + 4z = 6 \end{array}$

Linking Concepts

For Individual or Group Explorations

Modeling Traffic Flow

The accompanying figure shows two one-way streets and two two-way streets. The numbers and arrows represent the number of cars per hour that desire to enter or leave each intersection. The variables a, b, c, d, e, and f represent the number of cars per hour that pass the six points between the four intersections. To keep traffic flowing, the number of cars per hour that enter an intersection must equal the number per hour that leave an intersection.

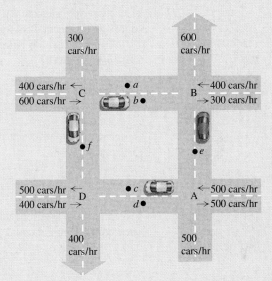

a) Write four equations (one for each intersection) with the six variables.

b) Does the system have a unique solution?

c) If $a = 300$, $c = 400$, and $f = 200$ cars/hr, find b, d, and e.

d) If the number of cars entering intersection A from the south is changed from 500 to 800 cars/hr, then what happens to the system and why?

9.2

Operations with Matrices

In Section 9.1, matrices were used to keep track of the coefficients and constants in systems of equations. Matrices are also useful for simplifying and organizing information ranging from inventories to win-loss records of sports teams. In this section we study matrices in more detail and learn how operations with matrices are used in applications.

Notation

In Section 9.1 a matrix was defined as a rectangular array of numbers. Capital letters are used to name matrices and lowercase letters to name their entries. A general $m \times n$ matrix with m rows and n columns is given as follows:

$$A = \begin{bmatrix} a_{11} & a_{12} & a_{13} & \cdots & a_{1n} \\ a_{21} & a_{22} & a_{23} & \cdots & a_{2n} \\ a_{31} & a_{32} & a_{33} & \cdots & a_{3n} \\ \vdots & \vdots & \vdots & & \vdots \\ a_{m1} & a_{m2} & a_{m3} & \cdots & a_{mn} \end{bmatrix}$$

The subscripts indicate the position of each entry. For example, a_{32} is the entry in the third row and second column. The entry in the ith row and jth column is denoted by a_{ij}.

Two matrices are **equal** if they have the same size and the corresponding entries are equal. We write

$$\begin{bmatrix} 0.5 \\ 0.25 \end{bmatrix} = \begin{bmatrix} \frac{1}{2} \\ \frac{1}{4} \end{bmatrix}$$

because these matrices have the same size and their corresponding entries are equal. The matrices

$$\begin{bmatrix} 4 & 3 \\ 2 & 1 \end{bmatrix} \quad \text{and} \quad \begin{bmatrix} 1 & 3 \\ 2 & 1 \end{bmatrix}$$

have the same size, but they are not equal because the entries in the first row and first column are not equal. The matrices

$$[3 \quad 5] \quad \text{and} \quad \begin{bmatrix} 3 \\ 5 \end{bmatrix}$$

are not equal because the first has size 1×2 (a row matrix) and the second has size 2×1 (a column matrix).

Example **1** **Equal matrices**

Determine the values of x, y, and z that make the following matrix equation true:

$$\begin{bmatrix} x & y - 1 \\ z & 7 \end{bmatrix} = \begin{bmatrix} 1 & 3 \\ 2 & 7 \end{bmatrix}$$

Solution

If these matrices are equal, then the corresponding entries are equal. So $x = 1$, $y = 4$, and $z = 2$.

Try This. If $\begin{bmatrix} x & 2 \\ 3 & z + 1 \end{bmatrix} = \begin{bmatrix} 1 & y + 1 \\ 3 & 2z - 1 \end{bmatrix}$, then what are x, y, and z? ■

Addition and Subtraction of Matrices

Matrices are rectangular arrays of real numbers, and in many ways they behave like real numbers. We can define matrix operations that have many properties similar to the properties of the real numbers.

Definition: Matrix Addition

> The sum of two $m \times n$ matrices A and B is the $m \times n$ matrix denoted $A + B$ whose entries are the sums of the corresponding entries of A and B.

Note that only matrices that have the same size can be added. There is no definition for the sum of matrices of different sizes.

Example **2** **Sum of matrices**

Find $A + B$ given that $A = \begin{bmatrix} -4 & 3 \\ 5 & -2 \end{bmatrix}$ and $B = \begin{bmatrix} 7 & -3 \\ 2 & -5 \end{bmatrix}$.

Solution

To find $A + B$, add the corresponding entries of A and B:

$$A + B = \begin{bmatrix} -4 & 3 \\ 5 & -2 \end{bmatrix} + \begin{bmatrix} 7 & -3 \\ 2 & -5 \end{bmatrix} = \begin{bmatrix} -4 + 7 & 3 + (-3) \\ 5 + 2 & -2 + (-5) \end{bmatrix} = \begin{bmatrix} 3 & 0 \\ 7 & -7 \end{bmatrix}$$

To check, define matrices A and B on your graphing calculator using the matrix edit feature. Then use the matrix names feature to display $A + B$ and find the sum as in Fig. 9.2(a), (b), and (c).

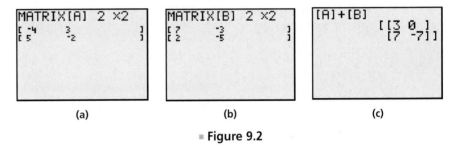

(a)   (b)   (c)

■ **Figure 9.2**

Try This. Find $A + B$ if $A = \begin{bmatrix} 3 & 4 \\ -1 & 5 \end{bmatrix}$ and $B = \begin{bmatrix} 5 & -4 \\ 1 & 4 \end{bmatrix}$.   ■

If all of the entries of a matrix are zero, the matrix is called a **zero matrix.** There is a zero matrix for every size. In matrix addition, the zero matrix behaves just like the additive identity 0 in the set of real numbers. For example,

$$\begin{bmatrix} 5 & -2 \\ 3 & -4 \end{bmatrix} + \begin{bmatrix} 0 & 0 \\ 0 & 0 \end{bmatrix} = \begin{bmatrix} 5 & -2 \\ 3 & -4 \end{bmatrix}.$$

In general, an $n \times n$ zero matrix is called the **additive identity** for $n \times n$ matrices.

For any matrix A, the **additive inverse** of A, denoted $-A$, is the matrix of the same size as A such that each entry of $-A$ is the opposite of the corresponding entry of A. Since corresponding entries are added in matrix addition, $A + (-A)$ is a zero matrix.

Example **3** **Additive inverses of matrices**

Find $-A$ and $A + (-A)$ for $A = \begin{bmatrix} -1 & 2 & 0 \\ 4 & -3 & 5 \\ 2 & 0 & -9 \end{bmatrix}$.

Solution

To find $-A$, find the opposite of every entry of A:

$$-A = \begin{bmatrix} 1 & -2 & 0 \\ -4 & 3 & -5 \\ -2 & 0 & 9 \end{bmatrix}$$

$$A + (-A) = \begin{bmatrix} -1 & 2 & 0 \\ 4 & -3 & 5 \\ 2 & 0 & -9 \end{bmatrix} + \begin{bmatrix} 1 & -2 & 0 \\ -4 & 3 & -5 \\ -2 & 0 & 9 \end{bmatrix} = \begin{bmatrix} 0 & 0 & 0 \\ 0 & 0 & 0 \\ 0 & 0 & 0 \end{bmatrix}$$

Therefore, the sum of A and $-A$ is the additive identity for 3×3 matrices.

Try This. Find $-A$ and $-A + A$ if $A = \begin{bmatrix} 2 & 3 \\ -4 & 1 \end{bmatrix}$.   ▪

The difference of two real numbers a and b is defined by $a - b = a + (-b)$. The difference of two matrices of the same size is defined similarly.

Definition:
Matrix Subtraction

The difference of two $m \times n$ matrices A and B is the $m \times n$ matrix denoted $A - B$, where $A - B = A + (-B)$.

Even though subtraction is defined as addition of the additive inverse, we can certainly find the difference for two matrices by subtracting their corresponding entries. Note that we can subtract corresponding entries only if the matrices have the same size.

Example **4** **Subtraction of matrices**

Let $A = [3 \quad 5 \quad 8]$, $B = [3 \quad -1 \quad 6]$, $C = \begin{bmatrix} -3 \\ 5 \\ 6 \end{bmatrix}$, and $D = \begin{bmatrix} 4 \\ 7 \\ 2 \end{bmatrix}$. Find the following matrices.

a. $A - B$ **b.** $C - D$ **c.** $A - C$

Solution

a. To find $A - B$, subtract the corresponding entries of the matrices:

$$A - B = [3 \quad 5 \quad 8] - [3 \quad -1 \quad 6] = [0 \quad 6 \quad 2]$$

b. To find $C - D$, subtract the corresponding entries:

$$C - D = \begin{bmatrix} -3 \\ 5 \\ 6 \end{bmatrix} - \begin{bmatrix} 4 \\ 7 \\ 2 \end{bmatrix} = \begin{bmatrix} -7 \\ -2 \\ 4 \end{bmatrix}$$

c. Since A and C do not have the same size, $A - C$ is not defined.

Try This. Find $A - B$ if $A = [1 \quad -2 \quad 4]$ and $B = [1 \quad 2 \quad 3]$.   ■

Scalar Multiplication

A matrix of size 1×1 is a matrix with only one entry. To distinguish a 1×1 matrix from a real number, a real number is called a **scalar** when we are dealing with matrices. We define multiplication of a matrix by a scalar as follows.

Definition:
Scalar Multiplication

> If A is an $m \times n$ matrix and b is a scalar, then the matrix bA is the $m \times n$ matrix obtained by multiplying each entry of A by the real number b.

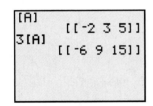

■ **Figure 9.3**

Example **5** Scalar multiplication

Given that $A = [-2 \quad 3 \quad 5]$ and $B = \begin{bmatrix} -3 & 4 \\ 2 & -6 \end{bmatrix}$, find the following matrices.

a. $3A$ **b.** $-2B$ **c.** $-1A$

Solution

a. The matrix $3A$ is the 1×3 matrix formed by multiplying each entry of A by 3:

$$3A = [-6 \quad 9 \quad 15]$$

To check, define A on your calculator and perform scalar multiplication as in Fig. 9.3. □

b. Multiply each entry of B by -2:

$$-2B = \begin{bmatrix} 6 & -8 \\ -4 & 12 \end{bmatrix}$$

c. Multiply each entry of A by -1:

$$-1A = [2 \quad -3 \quad -5]$$

The scalar product of A and -1 is the additive inverse of A, $-1A = -A$.

Try This. Find $4C$ if $C = [1 \quad -2 \quad 9]$.   ■

In Section 9.3 multiplication of matrices will be defined in a manner that is very different from scalar multiplication.

Modeling with Matrices

In Section 9.1 we saw how a matrix is used to represent a system of equations. Just the essential parts of the system, the coefficients, are listed in the matrix. Matrices are also very useful in two-way classifications of data. The matrix just contains the essentials, and we must remember what the entries represent.

Example **6** **Applications of Matrices**

The following table shows the number of transistors and resistors purchased by a manufacturer from suppliers A and B for the first week of January.

	A	B
Transistors	400	800
Resistors	600	500

a. Write the data in the table as a 2 × 2 matrix S_1.
b. Use scalar multiplication to find a matrix S_2 whose entries are all 10% larger than the corresponding entries of S_1.
c. Suppose that S_2 is the supply matrix for the second week of January. Find $S_1 + S_2$ and explain what its entries represent.

Solution

a. The supply table can be written as the following 2 × 2 matrix:

$$S_1 = \begin{bmatrix} 400 & 800 \\ 600 & 500 \end{bmatrix}$$

b. If the entries of S_2 are 10% larger than the entries of S_1, then $S_2 = S_1 + 0.10S_1 = 1.1S_1$:

$$S_2 = 1.1S_1 = 1.1\begin{bmatrix} 400 & 800 \\ 600 & 500 \end{bmatrix} = \begin{bmatrix} 440 & 880 \\ 660 & 550 \end{bmatrix}$$

c. The entries of $S_1 + S_2$ give the total number of transistors and resistors purchased from suppliers A and B for the first two weeks of January.

$$S_1 + S_2 = \begin{bmatrix} 400 & 800 \\ 600 & 500 \end{bmatrix} + \begin{bmatrix} 440 & 880 \\ 660 & 550 \end{bmatrix} = \begin{bmatrix} 840 & 1680 \\ 1260 & 1050 \end{bmatrix}$$

Try This. A small post office delivered 11,000 pieces of first class mail, 20,000 pieces of second class mail, and 16,000 pieces of third class mail in 2005. Write this data as a 3 × 1 matrix. For 2015 the post office expects a 50% increase in each class. Use scalar multiplication to find the matrix whose entries give the expected number in each class. ■

Example 6 is a simple example of how matrices are used with real data. You can imagine the same example with a matrix showing many different items from many different suppliers. Of course, computers, and even graphing calculators, are used to perform the operations when the matrices are large.

For Thought

True or False? Explain.

The following statements refer to the matrices

$$A = \begin{bmatrix} 1 \\ 3 \end{bmatrix}, \quad B = \begin{bmatrix} 1 \\ 3 \end{bmatrix}, \quad C = \begin{bmatrix} 1 & 1 \\ 3 & 3 \end{bmatrix},$$

$$D = \begin{bmatrix} -3 & 5 \\ 1 & -2 \end{bmatrix} \quad \text{and} \quad E = \begin{bmatrix} -2 & 6 \\ 4 & 1 \end{bmatrix}.$$

1. $A = B$ **2.** $A = C$ **3.** $A + B = C$

4. $C + D = E$

5. $A - B = \begin{bmatrix} 0 \\ 0 \end{bmatrix}$ **6.** $3B = \begin{bmatrix} 3 \\ 3 \end{bmatrix}$

7. $-A = \begin{bmatrix} 3 \\ 1 \end{bmatrix}$ **8.** $A + C = \begin{bmatrix} 2 & 1 \\ 6 & 3 \end{bmatrix}$

9. $C - A = \begin{bmatrix} 1 & 0 \\ 3 & 0 \end{bmatrix}$ **10.** $C + 2D = \begin{bmatrix} -5 & 11 \\ 4 & 1 \end{bmatrix}$

9.2 Exercises

Determine the values of x, y, and z that make each matrix equation true.

1. $\begin{bmatrix} x \\ 5 \end{bmatrix} = \begin{bmatrix} 2 \\ y \end{bmatrix}$ **2.** $\begin{bmatrix} 2z \\ 5x \end{bmatrix} = \begin{bmatrix} -1 \\ 2 \end{bmatrix}$

3. $\begin{bmatrix} 2x & 4y \\ 3z & 8 \end{bmatrix} = \begin{bmatrix} 6 & 16 \\ z+y & 8 \end{bmatrix}$

4. $\begin{bmatrix} -x & 2y \\ 3 & x+y \end{bmatrix} = \begin{bmatrix} 3 & -6 \\ 3 & 4z \end{bmatrix}$

Find the following sums.

5. $\begin{bmatrix} 3 \\ 5 \end{bmatrix} + \begin{bmatrix} 2 \\ 1 \end{bmatrix}$

6. $\begin{bmatrix} -1 \\ 2 \end{bmatrix} + \begin{bmatrix} 0 \\ 3 \end{bmatrix}$

7. $\begin{bmatrix} -0.5 & -0.03 \\ 2 & -0.33 \end{bmatrix} + \begin{bmatrix} 2 & 1 \\ -0.05 & 1 \end{bmatrix}$

8. $\begin{bmatrix} -0.05 & -0.1 \\ 0.2 & -1 \end{bmatrix} + \begin{bmatrix} 1 & -2 \\ -3 & 0.01 \end{bmatrix}$

9. $\begin{bmatrix} 2 & -3 & 4 \\ 4 & -6 & 8 \\ 6 & -3 & 1 \end{bmatrix} + \begin{bmatrix} 1 & -1 & 1 \\ 0 & 1 & -1 \\ 0 & 0 & 1 \end{bmatrix}$

10. $\begin{bmatrix} -3 & 5 & 1 \\ -8 & 2 & 4 \\ 4 & 5 & -3 \end{bmatrix} + \begin{bmatrix} -4 & -8 & -3 \\ 5 & 0 & -6 \\ -3 & 4 & -1 \end{bmatrix}$

For each matrix A find $-A$ and $A + (-A)$.

11. $A = \begin{bmatrix} 1 & -4 \\ -5 & 6 \end{bmatrix}$ **12.** $A = \begin{bmatrix} -3 & 5 \\ 0 & 2 \end{bmatrix}$

13. $A = \begin{bmatrix} 3 & 0 & -1 \\ 8 & -2 & 1 \\ -3 & 6 & 3 \end{bmatrix}$ **14.** $A = \begin{bmatrix} 4 & 8 & 3 \\ -5 & 0 & 6 \\ -1 & 1 & -9 \end{bmatrix}$

Let $A = \begin{bmatrix} -4 & 1 \\ 3 & 0 \end{bmatrix}, B = \begin{bmatrix} -1 & -2 \\ 7 & 4 \end{bmatrix}, C = \begin{bmatrix} -3 & -4 \\ 2 & -5 \end{bmatrix},$

$D = \begin{bmatrix} -4 \\ 5 \end{bmatrix},$ *and* $E = \begin{bmatrix} -1 \\ 2 \end{bmatrix}.$ *Find each of the following matrices, if possible.*

15. $B - A$ **16.** $A - B$

17. $B - C$ **18.** $C - B$

19. $B - E$ **20.** $A - D$

21. $3A$ **22.** $5B$

23. $-1D$ **24.** $-3E$

25. $3A + 3C$ **26.** $3(A + C)$

27. $2A - B$ **28.** $-C - 3B$

29. $2D - 3E$ **30.** $4D + 0E$

31. $D + A$ **32.** $E + C$

33. $(A + B) + C$ **34.** $A + (B + C)$

Perform the following operations. If it is not possible to perform an operation, explain.

35. $\begin{bmatrix} 0.2 & 0.1 \\ 0.4 & 0.3 \end{bmatrix} + \begin{bmatrix} 0.2 & 0.05 \\ 0.3 & 0.8 \end{bmatrix}$

36. $\begin{bmatrix} \frac{1}{2} & \frac{1}{3} \\ \frac{1}{4} & 1 \end{bmatrix} - \begin{bmatrix} -\frac{1}{2} & \frac{1}{6} \\ 1 & \frac{1}{3} \end{bmatrix}$

37. $3\begin{bmatrix} \frac{1}{6} & \frac{1}{2} \\ 1 & -4 \end{bmatrix}$

38. $-2\begin{bmatrix} -1 & 4 \\ 3 & 1 \end{bmatrix}$

39. $\begin{bmatrix} -2 & 4 \\ 6 & 8 \end{bmatrix} - 4\begin{bmatrix} -3 & 1 \\ 2 & -2 \end{bmatrix}$

40. $2\begin{bmatrix} \frac{1}{4} \\ \frac{1}{3} \end{bmatrix} + 3\begin{bmatrix} \frac{1}{4} \\ \frac{1}{6} \end{bmatrix}$

41. $\begin{bmatrix} 1 & 3 & 7 \end{bmatrix} + \begin{bmatrix} -1 \\ 2 \\ 3 \end{bmatrix}$

42. $\begin{bmatrix} -1 & 3 \\ 2 & 5 \end{bmatrix} + 5\begin{bmatrix} -2 \\ 4 \end{bmatrix}$

43. $\begin{bmatrix} -5 & 4 \\ -2 & 3 \\ 0 & 1 \end{bmatrix} - \begin{bmatrix} -4 & -9 \\ 7 & 0 \\ -6 & 3 \end{bmatrix}$

44. $\begin{bmatrix} 2 & 3 & 4 \\ 4 & 6 & 8 \\ 6 & 3 & 1 \end{bmatrix} + \begin{bmatrix} 1 & 0 & 0 \\ 0 & 1 & 0 \\ 0 & 0 & 1 \end{bmatrix}$

45. $\begin{bmatrix} \sqrt{2} & 4 & \sqrt{12} \end{bmatrix} + \begin{bmatrix} \sqrt{8} & -2 & \sqrt{3} \end{bmatrix}$

46. $2\begin{bmatrix} -\sqrt{2} \\ \sqrt{5} \\ \sqrt{27} \end{bmatrix} - \begin{bmatrix} -\sqrt{8} \\ \sqrt{20} \\ -\sqrt{3} \end{bmatrix}$

47. $2\begin{bmatrix} a \\ b \end{bmatrix} + 3\begin{bmatrix} 2a \\ 4b \end{bmatrix} - 5\begin{bmatrix} -a \\ 3b \end{bmatrix}$

48. $2\begin{bmatrix} -a \\ b \\ c \end{bmatrix} - \begin{bmatrix} -3a \\ 4b \\ -c \end{bmatrix} - 6\begin{bmatrix} a \\ -b \\ 2c \end{bmatrix}$

49. $0.4\begin{bmatrix} -x & y \\ 2x & 8y \end{bmatrix} - 0.3\begin{bmatrix} 2x & 3y \\ 5x & -y \end{bmatrix}$

50. $a\begin{bmatrix} a & b \\ c & 2b \end{bmatrix} - b\begin{bmatrix} b & a \\ 0 & 3a \end{bmatrix}$

51. $2\begin{bmatrix} x & y & z \\ -x & 2y & 3z \\ x & -y & -3z \end{bmatrix} - \begin{bmatrix} -x & 0 & 3z \\ 4x & y & -z \\ 2x & 5y & z \end{bmatrix}$

52. $\frac{1}{2}\begin{bmatrix} 2 & -8 & -4 \\ 14 & 16 & 10 \\ 4 & -6 & 2 \end{bmatrix} + \frac{1}{3}\begin{bmatrix} 6 & -9 & 0 \\ 0 & 12 & -6 \\ 21 & -18 & -3 \end{bmatrix}$

Each of the following matrix equations corresponds to a system of linear equations. Write the system of equations and solve it by the method of your choice.

53. $\begin{bmatrix} x + y \\ x - y \end{bmatrix} = \begin{bmatrix} 5 \\ 1 \end{bmatrix}$

54. $\begin{bmatrix} x - y \\ 2x + y \end{bmatrix} = \begin{bmatrix} -1 \\ 4 \end{bmatrix}$

55. $\begin{bmatrix} 2x + 3y \\ x - 4y \end{bmatrix} = \begin{bmatrix} 7 \\ -13 \end{bmatrix}$

56. $\begin{bmatrix} x - 3y \\ 2x + y \end{bmatrix} = \begin{bmatrix} 1 \\ -5 \end{bmatrix}$

57. $\begin{bmatrix} x + y + z \\ x - y - z \\ x - y + z \end{bmatrix} = \begin{bmatrix} 8 \\ -7 \\ 2 \end{bmatrix}$

58. $\begin{bmatrix} 2x + y + z \\ x - 2y - z \\ x - y + z \end{bmatrix} = \begin{bmatrix} 7 \\ -6 \\ 2 \end{bmatrix}$

Solve each problem.

59. *Budgeting* In January, Terry spent $120 on food, $30 on clothing, and $40 on utilities. In February she spent $130 on food, $70 on clothing, and $50 on utilities. In March she spent $140 on food, $60 on clothing, and $45 on utilities. Write a 3×1 matrix for each month's expenditures and find the sum of the three matrices. What do the entries in the sum represent?

60. *Nutritional Content* According to manufacturers' labels, one serving of Kix contains 110 calories, 2 g of protein, and 40 mg of potassium. One serving of Quick Oats contains 100 calories, 4 g of protein, and 100 mg of potassium. One serving of Muesli contains 120 calories, 3 g of protein, and 115 mg of potassium. Write 1×3 matrices K, Q, and M, which express the nutritional content of each cereal. Find $2K + 2Q + 3M$ and indicate what the entries represent.

61. *Arming the Villagers* In preparation for an attack by rampaging warlords, Xena accumulated 40 swords, 30 longbows, and 80 arrows to arm the villagers. Her companion Gabrielle obtained 80 swords, 90 longbows, and 200 arrows from a passing arms dealer. Write this information as a 3×2 matrix. One week later both Xena and Gabrielle managed to increase their supplies in each category by 50%. Write a 3×2 matrix for the supply of armaments in the second week.

62. *Recommended Daily Allowances* The percentages of the U.S. Recommended Daily Allowances for phosphorus, magnesium, and calcium for a 1-oz serving of Kix are 4, 2, and 4, respectively. The percentages of the U.S. Recommended Daily Allowances for phosphorus, magnesium, and calcium in 1/2 cup of milk are 11, 4, and 16, respectively. Write a 1×3 matrix K giving the percentages for 1 oz of Kix and a 1×3 matrix M giving the percentages for 1/2 cup of milk. Find the matrix $K + 2M$; indicate what its entries represent.

For Writing/Discussion

Let $A = \begin{bmatrix} a_{11} & a_{12} \\ a_{21} & a_{22} \end{bmatrix}$, $B = \begin{bmatrix} b_{11} & b_{12} \\ b_{21} & b_{22} \end{bmatrix}$, and $C = \begin{bmatrix} c_{11} & c_{12} \\ c_{21} & c_{22} \end{bmatrix}$ for the following problems.

63. Is $A + B = B + A$? Is addition of 2×2 matrices commutative? Explain.

64. Is addition of 3×3 matrices commutative? Explain.

65. Is $(A + B) + C = A + (B + C)$? Is addition of 2×2 matrices associative? Explain.

66. Is addition of 3×3 matrices associative? Explain.

67. Is $k(A + B) = kA + kB$ for any constant k? Is scalar multiplication distributive over addition of 2×2 matrices?

68. Is scalar multiplication distributive over addition of $n \times n$ matrices for each natural number n?

69. In the set of 2×2 matrices, which matrix is the additive identity?

70. Does every 2×2 matrix have an additive inverse with respect to the appropriate additive identity?

Thinking Outside the Box LXXIX

Shade-Tree Mechanic   Bubba has filled his 8-quart radiator with antifreeze, but he should have put in only 4 quarts of antifreeze and 4 quarts of water. He has an empty 5-quart container and an empty 3-quart container, but no other way of measuring the antifreeze. How can he get only 4 quarts of antifreeze in his radiator?

9.2 Pop Quiz

$A = \begin{bmatrix} 1 \\ 3 \end{bmatrix}$, $B = \begin{bmatrix} -1 \\ 5 \end{bmatrix}$, $C = \begin{bmatrix} 1 & 2 \\ 3 & 4 \end{bmatrix}$, $D = \begin{bmatrix} 1 & 0 \\ 0 & 1 \end{bmatrix}$. Perform the matrix operations if possible.

1. $A + B$

2. $3A - B$

3. $-A + A$

4. $A + C$

5. $C + 2D$

9.3

Multiplication of Matrices

In Sections 9.1 and 9.2 we saw how matrices are used for solving equations and for representing two-way classifications of data. We saw how addition and scalar multiplication of matrices could be useful in applications. In this section you will learn to find the product of two matrices. Matrix multiplication is more complicated than addition or subtraction, but it is also very useful in applications.

An Application

Before presenting the general definition of multiplication, let us look at an example where multiplication of matrices is useful. Table 9.1 shows the number of economy, mid-size, and large cars rented by individuals and corporations at a rental agency in a

■ Table 9.1

	Econo	Mid	Large
Individuals	3	2	6
Corporations	5	2	4

■ **Table 9.2**

	Bonus Points	**Free Miles**
Econo	20	50
Mid	30	100
Large	40	150

single day. Table 9.2 shows the number of bonus points and free miles given in a promotional program for each of the three car types.

The 1×3 row matrix $[3 \quad 2 \quad 6]$ from Table 9.1 represents the number of economy, mid-size, and large cars rented by individuals. The 3×1 column matrix $\begin{bmatrix} 20 \\ 30 \\ 40 \end{bmatrix}$ from Table 9.2 represents the bonus points given for each economy, mid-size, and large car that is rented. The product of these two matrices is a 1×1 matrix whose entry is the sum of the products of the corresponding entries:

$$[3 \quad 2 \quad 6] \begin{bmatrix} 20 \\ 30 \\ 40 \end{bmatrix} = [3(20) + 2(30) + 6(40)] = [360]$$

The product of this row matrix and this column matrix gives the total number of bonus points given to individuals on the rental of the 11 cars.

Now write Table 9.1 as a 2×3 matrix giving the number of cars of each type rented by individuals and corporations and write Table 9.2 as a 3×2 matrix giving the number of bonus points and free miles for each type of car rented.

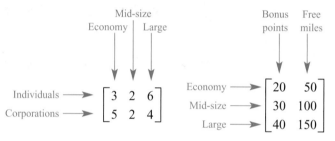

The product of these two matrices is a 2×2 matrix that gives the total bonus points and free miles both for individuals and corporations.

$$\begin{bmatrix} 3 & 2 & 6 \\ 5 & 2 & 4 \end{bmatrix} \begin{bmatrix} 20 & 50 \\ 30 & 100 \\ 40 & 150 \end{bmatrix} = \begin{bmatrix} 360 & 1250 \\ 320 & 1050 \end{bmatrix} \begin{array}{l} \longleftarrow \text{Individuals} \\ \longleftarrow \text{Corporations} \end{array}$$

The product of the 2×3 matrix and the 3×2 matrix is a 2×2 matrix. Take a careful look at where the entries in the 2×2 matrix come from:

$$3(20) + 2(30) + 6(40) = 360 \quad \text{Total bonus points for individuals}$$

$$3(50) + 2(100) + 6(150) = 1250 \quad \text{Total free miles for individuals}$$

$$5(20) + 2(30) + 4(40) = 320 \quad \text{Total bonus points for corporations}$$

$$5(50) + 2(100) + 4(150) = 1050 \quad \text{Total free miles for corporations}$$

Each entry in the 2×2 matrix is found by multiplying the entries of a *row* of the first matrix by the corresponding entries of a *column* of the second matrix and adding the results. To multiply any matrices, the number of entries in a row of the first matrix must equal the number of entries in a column of the second matrix.

Matrix Multiplication

The product of two matrices is illustrated by the previous example of the rental cars. The general definition of matrix multiplication follows.

Definition:
Matrix Multiplication

> The product of an $m \times n$ matrix A and an $n \times p$ matrix B is an $m \times p$ matrix AB whose entries are found as follows. The entry in the ith row and jth column of AB is found by multiplying each entry in the ith row of A by the corresponding entry in the jth column of B and adding the results.

Note that by the definition we can multiply an ***m*** $\times$ *n* matrix and an *n* $\times$ ***p*** matrix to get an ***m*** $\times$ ***p*** matrix. To find a product AB, *each row of A must have the same number of entries as each column of B.* The entry c_{ij} in AB comes from the ith row of A and the jth column of B as shown below.

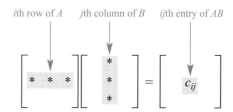

Example **1** **Multiplying matrices**

Find the following products.

a. $[2 \quad 3]\begin{bmatrix} -3 \\ 4 \end{bmatrix}$

b. $\begin{bmatrix} 1 & 2 \\ 3 & 4 \end{bmatrix}\begin{bmatrix} -2 \\ 5 \end{bmatrix}$

c. $\begin{bmatrix} 1 & 3 \\ 5 & 7 \end{bmatrix}\begin{bmatrix} 2 & 4 \\ 6 & 8 \end{bmatrix}$

Solution

a. The product of a **1** $\times$ **2** matrix and a **2** $\times$ **1** matrix is a **1** $\times$ **1** matrix. The only entry in the product is found by multiplying the first row of the first matrix by the first column of the second matrix:

$$[2 \quad 3]\begin{bmatrix} -3 \\ 4 \end{bmatrix} = [2(-3) + 3(4)] = [6]$$

b. The product of a **2** $\times$ **2** matrix and **2** $\times$ **1** matrix is a **2** $\times$ **1** matrix. Multiply the corresponding entries in each row of the first matrix and the only column of the second matrix:

$$\begin{bmatrix} 1 & 2 \\ 3 & 4 \end{bmatrix}\begin{bmatrix} -2 \\ 5 \end{bmatrix} = \begin{bmatrix} 8 \\ 14 \end{bmatrix} \qquad \begin{matrix} 1(-2) + 2(5) = 8 \\ 3(-2) + 4(5) = 14 \end{matrix}$$

c. The product of a **2 × 2** matrix and a 2 × 2 matrix is a **2 × 2** matrix. Multiply the corresponding entries in each row of the first matrix and each column of the second matrix.

$$\begin{bmatrix} 1 & 3 \\ 5 & 7 \end{bmatrix}\begin{bmatrix} 2 & 4 \\ 6 & 8 \end{bmatrix} = \begin{bmatrix} 20 & 28 \\ 52 & 76 \end{bmatrix}$$

$$1 \cdot 2 + 3 \cdot 6 = 20$$
$$1 \cdot 4 + 3 \cdot 8 = 28$$
$$5 \cdot 2 + 7 \cdot 6 = 52$$
$$5 \cdot 4 + 7 \cdot 8 = 76$$

Try This. Find AB if $A = \begin{bmatrix} 1 & 2 \\ 3 & 4 \end{bmatrix}$ and $B = \begin{bmatrix} 5 & 6 \\ 7 & 8 \end{bmatrix}$. ■

Example **2** **Multiplying matrices**

Find AB and BA in each case.

a. $A = \begin{bmatrix} 1 & 3 \\ 5 & 7 \\ 8 & 2 \end{bmatrix}$, $B = \begin{bmatrix} 2 & 4 & -1 \\ 6 & -3 & 2 \end{bmatrix}$ **b.** $A = \begin{bmatrix} 1 & 3 & 4 \\ 2 & 5 & 6 \\ 7 & 8 & 9 \end{bmatrix}$, $B = \begin{bmatrix} 1 & 0 & 1 \\ 0 & 1 & 0 \\ 0 & 1 & 1 \end{bmatrix}$

Solution

a. The product of 3 × 2 matrix A and 2 × 3 matrix B is the 3 × 3 matrix AB. The first row of AB is found by multiplying the corresponding entries in the first row of A and each column of B:

$$1 \cdot 2 + 3 \cdot 6 = 20, \qquad 1 \cdot 4 + 3(-3) = -5, \qquad \text{and} \qquad 1(-1) + 3 \cdot 2 = 5$$

So 20, −5, and 5 form the first row of AB. The second row of AB is formed from multiplying the second row of A and each column from B:

$$5 \cdot 2 + 7 \cdot 6 = 52, \qquad 5 \cdot 4 + 7(-3) = -1, \qquad \text{and} \qquad 5(-1) + 7 \cdot 2 = 9$$

So 52, −1, and 9 form the second row of AB. The third row of AB is formed by multiplying corresponding entries in the third row of A and each column of B.

$$AB = \begin{bmatrix} 1 & 3 \\ 5 & 7 \\ 8 & 2 \end{bmatrix}\begin{bmatrix} 2 & 4 & -1 \\ 6 & -3 & 2 \end{bmatrix} = \begin{bmatrix} 20 & -5 & 5 \\ 52 & -1 & 9 \\ 28 & 26 & -4 \end{bmatrix}$$

The product of 2 × 3 matrix B and 3 × 2 matrix A is the 2 × 2 matrix BA:

$$BA = \begin{bmatrix} 2 & 4 & -1 \\ 6 & -3 & 2 \end{bmatrix}\begin{bmatrix} 1 & 3 \\ 5 & 7 \\ 8 & 2 \end{bmatrix}$$

$$= \begin{bmatrix} 14 & 32 \\ 7 & 1 \end{bmatrix}$$

$$2(1) + 4(5) + (-1)(8) = 14$$
$$2(3) + 4(7) + (-1)(2) = 32$$
$$6(1) + (-3)(5) + 2(8) = 7$$
$$6(3) + (-3)(7) + 2(2) = 1$$

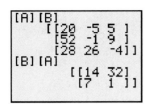

■ **Figure 9.4**

To check, enter A and B into a graphing calculator and find the products, as shown in Fig. 9.4. □

b. The product of 3×3 matrix A and 3×3 matrix B is the 3×3 matrix AB:

$$AB = \begin{bmatrix} 1 & 3 & 4 \\ 2 & 5 & 6 \\ 7 & 8 & 9 \end{bmatrix} \begin{bmatrix} 1 & 0 & 1 \\ 0 & 1 & 0 \\ 0 & 1 & 1 \end{bmatrix} = \begin{bmatrix} 1 & 7 & 5 \\ 2 & 11 & 8 \\ 7 & 17 & 16 \end{bmatrix}$$

The product of 3×3 matrix B and 3×3 matrix A is the 3×3 matrix BA:

$$BA = \begin{bmatrix} 1 & 0 & 1 \\ 0 & 1 & 0 \\ 0 & 1 & 1 \end{bmatrix} \begin{bmatrix} 1 & 3 & 4 \\ 2 & 5 & 6 \\ 7 & 8 & 9 \end{bmatrix} = \begin{bmatrix} 8 & 11 & 13 \\ 2 & 5 & 6 \\ 9 & 13 & 15 \end{bmatrix}$$

Try This. Find AB and BA if $A = \begin{bmatrix} 2 \\ 3 \end{bmatrix}$ and $B = [7 \quad 4]$. ■

Example 2(a) shows matrices A and B where $AB \neq BA$. In Example 2(a), AB and BA are not even the same size. It can also happen that AB is defined but BA is undefined because of the sizes of A and B. So multiplication of matrices is generally not commutative. Operations with matrices have some properties that are similar to the properties of operations with real numbers. We will not study the properties of matrices extensively in this text, but some properties for 2×2 matrices are discussed in the exercises.

Matrix Equations

Recall that two matrices are equal provided they are the same size and their corresponding entries are equal. This definition is used to solve matrix equations.

Example **3** **Solving a matrix equation**

Find the values of x and y that satisfy the matrix equation

$$\begin{bmatrix} 3 & 2 \\ 4 & -1 \end{bmatrix} \begin{bmatrix} x \\ y \end{bmatrix} = \begin{bmatrix} 1 \\ -6 \end{bmatrix}.$$

Solution

Multiply the matrices on the left-hand side to get the matrix equation

$$\begin{bmatrix} 3x + 2y \\ 4x - y \end{bmatrix} = \begin{bmatrix} 1 \\ -6 \end{bmatrix}.$$

Since matrices of the same size are equal only when all of their corresponding entries are equal, we have the following system of equations:

(1) $3x + 2y = 1$

(2) $4x - y = -6$

Solve by eliminating y:

$$3x + 2y = 1$$

$$\underline{8x - 2y = -12} \quad \text{Eq. (2) multiplied by 2.}$$

$$11x \quad = -11$$

$$x = -1$$

Now use $x = -1$ in Eq. (1):

$$3(-1) + 2y = 1$$

$$2y = 4$$

$$y = 2$$

You should check that the matrix equation is satisfied if $x = -1$ and $y = 2$.

Try This. Solve $\begin{bmatrix} 1 & 1 \\ 1 & -1 \end{bmatrix} \begin{bmatrix} x \\ y \end{bmatrix} = \begin{bmatrix} 15 \\ -3 \end{bmatrix}$. ■

In Example 3, we rewrote a matrix equation as a system of equations. We will now write a system of equations as a matrix equation. In fact, any system of linear equations can be written as a matrix equation in the form $AX = B$, where A is a matrix of coefficients, X is a column matrix of variables, and B is a column matrix of constants. In the next example we write a system of two equations as a matrix equation. In Section 9.4 we will solve systems of equations by using matrix operations on the corresponding matrix equations.

Example **4** **Writing a matrix equation**

Write the following system as an equivalent matrix equation in the form of $AX = B$.

$$2x + y = 5$$

$$x - y = 4$$

Solution

Let $A = \begin{bmatrix} 2 & 1 \\ 1 & -1 \end{bmatrix}$, $X = \begin{bmatrix} x \\ y \end{bmatrix}$, and $B = \begin{bmatrix} 5 \\ 4 \end{bmatrix}$. The system of equations is equivalent to the matrix equation

$$\begin{bmatrix} 2 & 1 \\ 1 & -1 \end{bmatrix} \begin{bmatrix} x \\ y \end{bmatrix} = \begin{bmatrix} 5 \\ 4 \end{bmatrix},$$

which is of the form $AX = B$. Check by multiplying the two matrices on the left-hand side to get

$$\begin{bmatrix} 2x + y \\ x - y \end{bmatrix} = \begin{bmatrix} 5 \\ 4 \end{bmatrix}.$$

By the definition of equal matrices, this matrix equation is correct provided that $2x + y = 5$ and $x - y = 4$, which is the original system.

Try This. Write $\begin{aligned} x - y &= 3 \\ x + 2y &= 5 \end{aligned}$ as the matrix equation $AX = B$. ■

For Thought

True or False? Explain.

The following statements refer to the matrices

$$A = \begin{bmatrix} 1 \\ 6 \end{bmatrix}, B = [7 \quad 9], C = \begin{bmatrix} 2 & 3 \\ 4 & 5 \end{bmatrix},$$

$$D = \begin{bmatrix} 1 & 0 \\ 0 & 1 \end{bmatrix}, \text{ and } E = \begin{bmatrix} 2 & -1 \\ 0 & 3 \end{bmatrix}.$$

1. The size of AB is 2×2.

2. The size of BA is 1×1.

3. The size of AC is 1×2.

4. The size of CA is 2×2.

5. $DC = C$ and $CD = C$.

6. $BC = [50 \quad 66]$

7. $AB = \begin{bmatrix} 7 & 9 \\ 42 & 54 \end{bmatrix}$ **8.** $CE = \begin{bmatrix} 4 & 7 \\ 8 & 11 \end{bmatrix}$

9. $BA = [61]$ **10.** $CE = EC$

9.3 Exercises

Find the size of AB in each case if the matrices can be multiplied.

1. A has size 3×2, B has size 2×5

2. A has size 3×1, B has size 1×3

3. A has size 1×4, B has size 4×1

4. A has size 4×2, B has size 2×5

5. A has size 5×1, B has size 1×5

6. A has size 2×1, B has size 1×6

7. A has size 3×3, B has size 3×3

8. A has size 4×4, B has size 4×1

9. A has size 3×4, B has size 3×4

10. A has size 4×2, B has size 3×4

Find the following products.

11. $[-3 \quad 2]\begin{bmatrix} 4 \\ 1 \end{bmatrix}$ **12.** $[-1 \quad 2]\begin{bmatrix} 4 \\ 2 \end{bmatrix}$

13. $\begin{bmatrix} 1 & 3 \\ 2 & -4 \end{bmatrix}\begin{bmatrix} 1 \\ 3 \end{bmatrix}$ **14.** $\begin{bmatrix} 0 & 2 \\ -3 & 1 \end{bmatrix}\begin{bmatrix} 2 \\ 5 \end{bmatrix}$

15. $\begin{bmatrix} 5 & 1 \\ 2 & 1 \end{bmatrix}\begin{bmatrix} 1 & 2 \\ 3 & 1 \end{bmatrix}$ **16.** $\begin{bmatrix} 3 & 1 \\ 4 & 0 \end{bmatrix}\begin{bmatrix} 0 & 3 \\ 1 & 6 \end{bmatrix}$

17. $\begin{bmatrix} 3 \\ 1 \end{bmatrix}[5 \quad 6]$ **18.** $\begin{bmatrix} -2 \\ 2 \end{bmatrix}[4 \quad 3]$

Find AB and BA in each case.

19. $A = \begin{bmatrix} 1 & 3 \\ 2 & 4 \\ 5 & 6 \end{bmatrix}, B = \begin{bmatrix} 1 & 0 & 1 \\ -1 & 1 & 0 \end{bmatrix}$

20. $A = \begin{bmatrix} 1 & 1 \\ -1 & 0 \\ 1 & -1 \end{bmatrix}, B = \begin{bmatrix} 2 & 1 & 0 \\ 3 & 2 & 0 \end{bmatrix}$

21. $A = \begin{bmatrix} 1 & 2 & 3 \\ 2 & 1 & 3 \\ 3 & 2 & 1 \end{bmatrix}, B = \begin{bmatrix} 1 & 1 & 1 \\ 0 & 1 & 1 \\ 0 & 0 & 1 \end{bmatrix}$

22. $A = \begin{bmatrix} 2 & 0 & 0 \\ 2 & 2 & 0 \\ 2 & 2 & 2 \end{bmatrix}, B = \begin{bmatrix} 0 & 0 & 1 \\ 2 & 0 & 0 \\ 0 & 3 & 0 \end{bmatrix}$

Let $A = \begin{bmatrix} 2 \\ -3 \\ 1 \end{bmatrix}, B = [2 \quad 3 \quad 4], C = \begin{bmatrix} 2 & 3 \\ 4 & 5 \\ 1 & 0 \end{bmatrix},$

$D = \begin{bmatrix} 2 & -1 & 1 \\ 0 & 3 & 2 \end{bmatrix}, \text{ and } E = \begin{bmatrix} 1 & 1 & 1 \\ 0 & 1 & 1 \\ 0 & 0 & 1 \end{bmatrix}.$ *Find the following if*

possible.

23. AB **24.** BC **25.** BE **26.** BA

27. CE **28.** DE **29.** EC **30.** BD

31. DC **32.** CD **33.** ED **34** EE

35. EA **36.** DA **37.** AE **38.** EB

39. $AB + 2E$ **40.** $EA - 2A$

If A is a square matrix then $A^2 = AA$, $A^3 = AAA$, and so on. Let $A = \begin{bmatrix} 1 & 0 \\ 1 & 1 \end{bmatrix}$. Find the following.

41. A^2　　**42.** A^3　　**43.** A^4　　**44.** A^5

Find each product if possible.

45. $\begin{bmatrix} 2 & 0 \\ 3 & 1 \end{bmatrix}\begin{bmatrix} 1 & 1 \\ 0 & 1 \end{bmatrix}$

46. $\begin{bmatrix} -1 & 2 \\ 3 & 4 \end{bmatrix}\begin{bmatrix} 2 & -1 \\ 5 & 2 \end{bmatrix}$

47. $\begin{bmatrix} 7 & 4 \\ 5 & 3 \end{bmatrix}\begin{bmatrix} 3 & -4 \\ -5 & 7 \end{bmatrix}$

48. $\begin{bmatrix} -2 & -3 \\ 5 & 8 \end{bmatrix}\begin{bmatrix} -8 & -3 \\ 5 & 2 \end{bmatrix}$

49. $\begin{bmatrix} -0.5 & 4 \\ 9 & 0.7 \end{bmatrix}\begin{bmatrix} 1 & 0 \\ 0 & 1 \end{bmatrix}$

50. $\begin{bmatrix} 1 & 0 \\ 0 & 1 \end{bmatrix}\begin{bmatrix} -0.7 & 1.2 \\ 3 & 1.1 \end{bmatrix}$

51. $[-2 \quad 3]\begin{bmatrix} a & 3b \\ 2a & b \end{bmatrix}$

52. $\begin{bmatrix} -2 & 5 \\ 6 & 4 \end{bmatrix}\begin{bmatrix} x \\ y \end{bmatrix}$

53. $\begin{bmatrix} a & 0 \\ 0 & b \end{bmatrix}\begin{bmatrix} -2 & 5 & 3 \\ 1 & 4 & 6 \end{bmatrix}$

54. $\begin{bmatrix} 1 & 1 \\ 1 & -1 \end{bmatrix}\begin{bmatrix} x \\ y \end{bmatrix}$

55. $[1 \quad 2 \quad 3]\begin{bmatrix} 1 & 0 & 1 \\ 0 & 1 & 1 \\ 1 & 0 & 1 \end{bmatrix}$

56. $\begin{bmatrix} 1 & 1 & 0 \\ 1 & 0 & 1 \\ 0 & 1 & 1 \end{bmatrix}\begin{bmatrix} -2 \\ 3 \\ 5 \end{bmatrix}$

57. $[-1 \quad 0 \quad 3]\begin{bmatrix} -5 \\ 1 \\ 4 \end{bmatrix}$

58. $\begin{bmatrix} -5 \\ 1 \\ 4 \end{bmatrix}[-1 \quad 0 \quad 3]$

59. $\begin{bmatrix} x \\ y \end{bmatrix}[x \quad y]$

60. $[x \quad y]\begin{bmatrix} x \\ y \end{bmatrix}$

61. $\begin{bmatrix} -1 & 2 & 3 \\ 3 & 4 & 4 \end{bmatrix}\begin{bmatrix} \sqrt{2} \\ 0 \\ \sqrt{2} \end{bmatrix}$

62. $[2 \quad 3]\begin{bmatrix} 0 & \sqrt{2} & 5 \\ -1 & \sqrt{8} & 0 \end{bmatrix}$

63. $\begin{bmatrix} \frac{1}{2} & \frac{1}{3} \\ \frac{1}{4} & \frac{1}{5} \end{bmatrix}\begin{bmatrix} -8 & 12 \\ -5 & 15 \end{bmatrix}$

64. $\begin{bmatrix} \frac{1}{4} & \frac{1}{2} \\ \frac{1}{8} & -\frac{1}{2} \end{bmatrix}\begin{bmatrix} -\frac{1}{2} & \frac{1}{4} \\ \frac{1}{2} & \frac{1}{4} \end{bmatrix}$

65. $[3 \quad 0 \quad 3][2 \quad 4 \quad 6]$

66. $\begin{bmatrix} 2 \\ 5 \end{bmatrix}\begin{bmatrix} 7 & 2 \\ 3 & 1 \end{bmatrix}$

67. $\begin{bmatrix} 1 & 0 & -1 \\ 0 & 1 & 0 \\ 1 & 1 & 1 \end{bmatrix}\begin{bmatrix} 9 & 8 & 10 \\ 3 & 5 & 2 \\ 7 & 8 & 4 \end{bmatrix}$

68. $\begin{bmatrix} 1 & 1 & 1 \\ 0 & 1 & 1 \\ 0 & 0 & 1 \end{bmatrix}\begin{bmatrix} -2 & 3 & -4 \\ 2 & 5 & 7 \\ -3 & 0 & -6 \end{bmatrix}$

69. $\begin{bmatrix} 5 & -3 & -2 \\ 4 & 2 & 6 \\ 2 & 3 & -8 \end{bmatrix}\begin{bmatrix} 0.2 & 0.3 \\ 0.2 & -0.4 \\ 0.3 & 0.5 \end{bmatrix}$

70. $\begin{bmatrix} 0.2 & 0.1 & 0.7 \\ 0.3 & 0.3 & 0.4 \end{bmatrix}\begin{bmatrix} 20 & 30 & 40 \\ 10 & 20 & 50 \\ 60 & 50 & 40 \end{bmatrix}$

Write each matrix equation as a system of equations and solve the system by the method of your choice.

71. $\begin{bmatrix} 2 & -3 \\ 1 & 2 \end{bmatrix}\begin{bmatrix} x \\ y \end{bmatrix} = \begin{bmatrix} 0 \\ 7 \end{bmatrix}$

72. $\begin{bmatrix} 1 & 5 \\ -2 & 4 \end{bmatrix}\begin{bmatrix} x \\ y \end{bmatrix} = \begin{bmatrix} 2 \\ 10 \end{bmatrix}$

73. $\begin{bmatrix} 2 & 3 \\ 4 & 6 \end{bmatrix}\begin{bmatrix} x \\ y \end{bmatrix} = \begin{bmatrix} 5 \\ 9 \end{bmatrix}$

74. $\begin{bmatrix} 1 & -3 \\ -2 & 6 \end{bmatrix}\begin{bmatrix} x \\ y \end{bmatrix} = \begin{bmatrix} 1 \\ -2 \end{bmatrix}$

75. $\begin{bmatrix} 1 & 1 & 1 \\ 0 & 1 & 1 \\ 0 & 0 & 1 \end{bmatrix}\begin{bmatrix} x \\ y \\ z \end{bmatrix} = \begin{bmatrix} 4 \\ 5 \\ 6 \end{bmatrix}$

76. $\begin{bmatrix} 2 & 3 & 1 \\ 0 & 1 & 4 \\ 0 & 0 & 2 \end{bmatrix}\begin{bmatrix} x \\ y \\ z \end{bmatrix} = \begin{bmatrix} 0 \\ 3 \\ 6 \end{bmatrix}$

Write a matrix equation of the form $AX = B$ that corresponds to each system of equations.

77. $2x + 3y = 9$
$4x - y = 6$

78. $x - y = -7$
$x + 2y = 8$

79. $x + 2y - z = 3$
$3x - y + 3z = 1$
$2x + y - 4z = 0$

80. $x + y + z = 1$
$2x + y - z = 4$
$x - y - 3z = 2$

Solve each problem.

81. *Building Costs* A contractor builds two types of houses. The costs for labor and materials for the economy model and the deluxe model in thousands of dollars are shown in the table. Write the information in the table as a matrix A. Suppose that the contractor built four economy models and seven deluxe models. Write a matrix Q of the appropriate size containing the quantity of each type. Find the product matrix AQ. What do the entries of AQ represent?

■ **Table for Exercise 81**

	Economy	Deluxe	
Labor Cost	$24	$40	
Material Cost	$38	$70	

82. *Nutritional Information* According to the manufacturers, the breakfast cereals Almond Delight and Basic 4 contain the grams of protein, carbohydrates, and fat per serving listed in the table. Write the information in the table as a matrix A. In one week Julia ate four servings of Almond Delight and three servings of Basic 4. Write a matrix Q of the appropriate size expressing the quantity of each type that Julia ate. Find the product AQ. What do the entries of AQ represent?

▪ **Table for Exercise 82**

	Almond Delight	Basic 4
Protein	2	3
Carbohydrates	23	28
Fat	2	2

For Writing/Discussion

Let $A = \begin{bmatrix} a_{11} & a_{12} \\ a_{21} & a_{22} \end{bmatrix}$, $B = \begin{bmatrix} b_{11} & b_{12} \\ b_{21} & b_{22} \end{bmatrix}$, and $C = \begin{bmatrix} c_{11} & c_{12} \\ c_{21} & c_{22} \end{bmatrix}$.

Determine whether each of the following statements is true, and explain your answer.

83. $AB = BA$ (commutative)

84. $(AB)C = A(BC)$ (associative)

85. For any real number k, $k(A + B) = kA + kB$.

86. $A(B + C) = AB + AC$ (distributive)

87. For any real numbers s and t, $sA + tA = (s + t)A$.

88. Multiplication of 1×1 matrices is commutative.

Thinking Outside the Box LXXX

Statewide Play-Offs Twenty-four teams competed in the statewide soccer play-offs. To reduce the amount of travel, the teams were divided into a north section and a south section, with the winner from each section to meet in a final match. Within each section, each team played every other team once, getting 1 point for a win, 1/2 point for a tie, and no points for a loss. There were 69 more games played in the north than in the south. In the south, Springville scored 5.5 points in the play-offs. How many games did Springville win?

9.3 **Pop Quiz**

$A = \begin{bmatrix} 2 \\ 4 \end{bmatrix}$, $B = \begin{bmatrix} 1 \\ 3 \end{bmatrix}$, $C = \begin{bmatrix} 1 & 3 \\ 5 & 7 \end{bmatrix}$, $D = \begin{bmatrix} 2 & 4 \\ 0 & 8 \end{bmatrix}$.

Perform the matrix operations if possible.

1. AB

2. AC

3. CA

4. CD

5. DC

Linking Concepts

For Individual or Group Explorations

Using Matrices to Rank Teams

The table shown here gives the records of all four teams in a soccer league. An entry of 1 indicates that the row team has defeated the column team. (There are no ties.) The teams are ranked, not by their percentage of victories, but by the number of points received under a ranking scheme that gives a team credit for the quality of the team it defeats. Since team A defeated B and C, A gets two points. Since B defeated C and D, and C defeated D, A gets 3 more secondary points for a total of 5 points. Since D's only victory is over A, D gets 1 point for that victory plus 2 secondary points for A's defeats of B and C, giving D a total of 3 points.

a) Write the table as a 4×4 matrix M and find M^2.

b) Explain what the entries of M^2 represent.

(continued on next page)

c) Now let T be a 4×1 matrix with a 1 in every entry. Find $(M + M^2)T$.

d) Explain what the entries of $(M + M^2)T$ represent.

e) Is it possible for one team to have a better win-loss record than another, but end up ranked lower than the other because of this point scheme? Give an example to support your answer.

f) Make up a win-loss table (with no ties) for a six-team soccer league like the given table. Use a graphing calculator to find $(M + M^2)T$. Compare the percentage of games won by each team with its ranking by this scheme. Is it possible for a team to have a higher percentage of wins but still be ranked lower than another team?

	A	B	C	D
A	0	1	1	0
B	0	0	1	1
C	0	0	0	1
D	1	0	0	0

g) Find $(2M + M^2)T$ for the matrix M from part (f) and explain its entries. What is the significance of the number 2?

h) Compare the ranking of the six teams using $(2M + M^2)T$ and $(M + M^2)T$. Is it possible that $(2M + M^2)T$ could change the order of the teams?

9.4

Inverses of Matrices

In previous sections we learned to add, subtract, and multiply matrices. There is no definition for division of matrices. In Section 9.1 we defined the identity matrix to be a square matrix with ones on the diagonal and zeros elsewhere. In this section we will see that the identity matrix behaves like the multiplicative identity 1 in the real number system ($1 \cdot a = a$ and $a \cdot 1 = a$ for any real number a). That is why it is called the identity matrix. We will also see that for certain matrices there are inverse matrices such that the product of a matrix and its inverse matrix is the identity matrix.

The Identity Matrix

If $A = \begin{bmatrix} a_{11} & a_{12} \\ a_{21} & a_{22} \end{bmatrix}$ and $I = \begin{bmatrix} 1 & 0 \\ 0 & 1 \end{bmatrix}$, then

$$\begin{bmatrix} a_{11} & a_{12} \\ a_{21} & a_{22} \end{bmatrix}\begin{bmatrix} 1 & 0 \\ 0 & 1 \end{bmatrix} = \begin{bmatrix} a_{11} & a_{12} \\ a_{21} & a_{22} \end{bmatrix} \text{ and } \begin{bmatrix} 1 & 0 \\ 0 & 1 \end{bmatrix}\begin{bmatrix} a_{11} & a_{12} \\ a_{21} & a_{22} \end{bmatrix} = \begin{bmatrix} a_{11} & a_{12} \\ a_{21} & a_{22} \end{bmatrix}.$$

So $AI = A$ and $IA = A$ for any 2×2 matrix A. Note that the 2×2 identity matrix is not an identity matrix for 3×3 matrices, but a 3×3 matrix with ones on the diagonal and zeros elsewhere is the identity matrix for 3×3 matrices.

Definition: Identity Matrix

For each positive integer n, the **$n \times n$ identity matrix I** is an $n \times n$ matrix with ones on the diagonal and zeros elsewhere. In symbols,

$$I = \begin{bmatrix} 1 & 0 & 0 & \cdots & 0 \\ 0 & 1 & 0 & \cdots & 0 \\ 0 & 0 & 1 & \cdots & 0 \\ \vdots & \vdots & \vdots & & \vdots \\ 0 & 0 & 0 & \cdots & 1 \end{bmatrix}.$$

We use the letter I for the identity matrix for any size, but the size of I should be clear from the context.

Example **1** **Using an identity matrix**

Show that $BI = B$ and $IB = B$ where I is the 3×3 identity matrix and

$$B = \begin{bmatrix} 2 & 3 & 5 \\ 1 & 0 & 4 \\ 5 & 7 & 2 \end{bmatrix}.$$

Solution

Check that $BI = B$:

$$\begin{bmatrix} 2 & 3 & 5 \\ 1 & 0 & 4 \\ 5 & 7 & 2 \end{bmatrix} \begin{bmatrix} 1 & 0 & 0 \\ 0 & 1 & 0 \\ 0 & 0 & 1 \end{bmatrix} = \begin{bmatrix} 2 & 3 & 5 \\ 1 & 0 & 4 \\ 5 & 7 & 2 \end{bmatrix} \qquad \begin{aligned} 2 \cdot 1 + 3 \cdot 0 + 5 \cdot 0 &= 2 \\ 2 \cdot 0 + 3 \cdot 1 + 5 \cdot 0 &= 3 \\ 2 \cdot 0 + 3 \cdot 0 + 5 \cdot 1 &= 5 \end{aligned}$$

The computations at the right show that 2, 3, and 5 form the first row of BI. The second and third rows are found similarly. Now check that $IB = B$:

$$\begin{bmatrix} 1 & 0 & 0 \\ 0 & 1 & 0 \\ 0 & 0 & 1 \end{bmatrix} \begin{bmatrix} 2 & 3 & 5 \\ 1 & 0 & 4 \\ 5 & 7 & 2 \end{bmatrix} = \begin{bmatrix} 2 & 3 & 5 \\ 1 & 0 & 4 \\ 5 & 7 & 2 \end{bmatrix}$$

Try This. Find AB and BA if $A = \begin{bmatrix} 1 & 3 \\ 8 & 9 \end{bmatrix}$ and $B = \begin{bmatrix} 1 & 0 \\ 0 & 1 \end{bmatrix}$. ■

The Inverse of a Matrix

If the product of two $n \times n$ matrices is the $n \times n$ identity matrix, then the two matrices are *multiplicative inverses* of each other. Matrices also have additive inverses. Since we are discussing only multiplicative inverses in this section, we will simply call them inverses.

Definition: Inverse of a Matrix

The **inverse** of an $n \times n$ matrix A is an $n \times n$ matrix A^{-1} (if it exists) such that $AA^{-1} = I$ and $A^{-1}A = I$. (Read A^{-1} as "A inverse.")

If A has an inverse, then A is **invertible.** Before we learn how to find the inverse of a matrix, we use the definition to determine whether two given matrices are inverses.

Example 2 Using the definition of inverse matrices

Determine whether $A = \begin{bmatrix} 3 & 4 \\ 5 & 7 \end{bmatrix}$ and $B = \begin{bmatrix} 7 & -4 \\ -5 & 3 \end{bmatrix}$ are inverses of each other.

Solution

Find the products AB and BA:

$$AB = \begin{bmatrix} 3 & 4 \\ 5 & 7 \end{bmatrix} \begin{bmatrix} 7 & -4 \\ -5 & 3 \end{bmatrix} = \begin{bmatrix} 1 & 0 \\ 0 & 1 \end{bmatrix}$$

$$BA = \begin{bmatrix} 7 & -4 \\ -5 & 3 \end{bmatrix} \begin{bmatrix} 3 & 4 \\ 5 & 7 \end{bmatrix} = \begin{bmatrix} 1 & 0 \\ 0 & 1 \end{bmatrix}$$

A and B are inverses because $AB = BA = I$, where I is the 2×2 identity matrix.

Try This. Find AB and BA if $A = \begin{bmatrix} 3 & 5 \\ 1 & 2 \end{bmatrix}$ and $B = \begin{bmatrix} 2 & -5 \\ -1 & 3 \end{bmatrix}$.   ■

If we are given the matrix

$$A = \begin{bmatrix} 3 & 4 \\ 5 & 7 \end{bmatrix}$$

from Example 2, how do we find its inverse if it is not already known? According to the definition, A^{-1} is a 2×2 matrix such that $AA^{-1} = I$ and $A^{-1}A = I$. So if

$$A^{-1} = \begin{bmatrix} x & y \\ z & w \end{bmatrix},$$

then

$$\begin{bmatrix} 3 & 4 \\ 5 & 7 \end{bmatrix} \begin{bmatrix} x & y \\ z & w \end{bmatrix} = \begin{bmatrix} 1 & 0 \\ 0 & 1 \end{bmatrix} \quad \text{and} \quad \begin{bmatrix} x & y \\ z & w \end{bmatrix} \begin{bmatrix} 3 & 4 \\ 5 & 7 \end{bmatrix} = \begin{bmatrix} 1 & 0 \\ 0 & 1 \end{bmatrix}.$$

To find A^{-1} we solve these matrix equations. If A is invertible, both equations will have the same solution. We will work with the first one. Multiply the two matrices on the left-hand side of the first equation to get the following equation:

$$\begin{bmatrix} 3x + 4z & 3y + 4w \\ 5x + 7z & 5y + 7w \end{bmatrix} = \begin{bmatrix} 1 & 0 \\ 0 & 1 \end{bmatrix}$$

Equate the corresponding terms from these equal matrices to get the following two systems:

$$3x + 4z = 1 \qquad 3y + 4w = 0$$
$$5x + 7z = 0 \qquad 5y + 7w = 1$$

We can solve these two systems by using the Gaussian elimination method from Section 9.1. The augmented matrices for these systems are

$$\begin{bmatrix} 3 & 4 & | & 1 \\ 5 & 7 & | & 0 \end{bmatrix} \quad \text{and} \quad \begin{bmatrix} 3 & 4 & | & 0 \\ 5 & 7 & | & 1 \end{bmatrix}.$$

Note that the two augmented matrices have the same coefficient matrix. Since we would use the same row operations on each of them, we can solve the systems simultaneously by combining the two systems into one augmented matrix denoted $[A\,|\,I]$:

$$[A\,|\,I] = \begin{bmatrix} 3 & 4 & | & 1 & 0 \\ 5 & 7 & | & 0 & 1 \end{bmatrix}$$

So the problem of finding A^{-1} is equivalent to the problem of solving two systems by Gaussian elimination. A is invertible if and only if these systems have a solution. Multiply the first row of $[A\,|\,I]$ by $\frac{1}{3}$ to get a 1 in the first row, first column:

$$\begin{bmatrix} 1 & \frac{4}{3} & | & \frac{1}{3} & 0 \\ 5 & 7 & | & 0 & 1 \end{bmatrix} \quad \frac{1}{3}R_1 \to R_1$$

Now multiply row 1 by -5 and add the result to row 2:

$$\begin{bmatrix} 1 & \frac{4}{3} & | & \frac{1}{3} & 0 \\ 0 & \frac{1}{3} & | & -\frac{5}{3} & 1 \end{bmatrix} \quad -5R_1 + R_2 \to R_2$$

Multiply row 2 by 3:

$$\begin{bmatrix} 1 & \frac{4}{3} & | & \frac{1}{3} & 0 \\ 0 & 1 & | & -5 & 3 \end{bmatrix} \quad 3R_2 \to R_2$$

Multiply row 2 by $-\frac{4}{3}$ and add the result to row 1:

$$\begin{bmatrix} 1 & 0 & | & 7 & -4 \\ 0 & 1 & | & -5 & 3 \end{bmatrix} \quad -\frac{4}{3}R_2 + R_1 \to R_1$$

The numbers in the first column to the right of the bar give the values of x and z, while the numbers in the second column give the values of y and w. So $x = 7$, $y = -4, z = -5$, and $w = 3$ give the solutions to the two systems, and

$$A^{-1} = \begin{bmatrix} 7 & -4 \\ -5 & 3 \end{bmatrix}.$$

Since A^{-1} is the same matrix that was called B in Example 2, we can be sure that $AA^{-1} = I$ and $A^{-1}A = I$. Note that the matrix A^{-1} actually appeared on the right-hand side of the final augmented matrix, while the 2×2 identity matrix I appeared on the left-hand side. So A^{-1} is found by simply using row operations to convert the matrix $[A\,|\,I]$ into the matrix $[I\,|\,A^{-1}]$.

The essential steps for finding the inverse of a matrix are listed as follows.

PROCEDURE   Finding A^{-1}

Use the following steps to find the inverse of a square matrix A.

1. Write the augmented matrix $[A\,|\,I]$, where I is the identity matrix of the same size as A.

2. Use row operations (the Gaussian elimination method) to convert the left-hand side of the augmented matrix into I.

3. If the left-hand side can be converted to I, then $[A\,|\,I]$ becomes $[I\,|\,A^{-1}]$, and A^{-1} appears on the right-hand side of the augmented matrix.

4. If the left-hand side cannot be converted to I, then A is not invertible.

Example 3 Finding the inverse of a matrix

Find the inverse of the matrix

$$A = \begin{bmatrix} 2 & -3 \\ 1 & 1 \end{bmatrix}.$$

Solution

Write the augmented matrix $[A\,|\,I]$:

$$\begin{bmatrix} 2 & -3 & | & 1 & 0 \\ 1 & 1 & | & 0 & 1 \end{bmatrix}$$

Use row operations to convert $[A\,|\,I]$ into $[I\,|\,A^{-1}]$:

$$\begin{bmatrix} 1 & 1 & | & 0 & 1 \\ 2 & -3 & | & 1 & 0 \end{bmatrix} \quad R_1 \leftrightarrow R_2$$

$$\begin{bmatrix} 1 & 1 & | & 0 & 1 \\ 0 & -5 & | & 1 & -2 \end{bmatrix} \quad -2R_1 + R_2 \to R_2$$

$$\begin{bmatrix} 1 & 1 & | & 0 & 1 \\ 0 & 1 & | & -\frac{1}{5} & \frac{2}{5} \end{bmatrix} \quad -\frac{1}{5}R_2 \to R_2$$

$$\begin{bmatrix} 1 & 0 & | & \frac{1}{5} & \frac{3}{5} \\ 0 & 1 & | & -\frac{1}{5} & \frac{2}{5} \end{bmatrix} \quad -R_2 + R_1 \to R_1$$

Since the last matrix is in the form $[I\,|\,A^{-1}]$, we get

$$A^{-1} = \begin{bmatrix} \frac{1}{5} & \frac{3}{5} \\ -\frac{1}{5} & \frac{2}{5} \end{bmatrix}.$$

Check that $AA^{-1} = A^{-1}A = I$.

The inverse of a matrix can be found with a graphing calculator. Enter the matrix A from Example 3 into a calculator. Then use the x^{-1} key to find A^{-1}, as shown in Fig. 9.5.

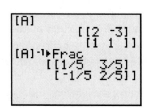

■ **Figure 9.5**

Try This. Find the inverse of $\begin{bmatrix} 3 & 2 \\ 7 & 5 \end{bmatrix}$. ■

Example 4 A noninvertible matrix

Find the inverse of the matrix

$$A = \begin{bmatrix} 1 & -3 \\ -1 & 3 \end{bmatrix}.$$

Solution

Use row operations on the augmented matrix $[A\,|\,I]$:

$$\left[\begin{array}{rr|rr} 1 & -3 & 1 & 0 \\ -1 & 3 & 0 & 1 \end{array}\right]$$

$$\left[\begin{array}{rr|rr} 1 & -3 & 1 & 0 \\ 0 & 0 & 1 & 1 \end{array}\right] \quad R_1 + R_2 \to R_2$$

The row containing all zeros on the left-hand side of the augmented matrix indicates that the left-hand side (the matrix A) cannot be converted to I using row operations. So A is not invertible. Note that the row of zeros in the augmented matrix means that there is no solution to the systems that must be solved to find A^{-1}.

Try This. Find the inverse of $\begin{bmatrix} 3 & -1 \\ -3 & 1 \end{bmatrix}$. ■

Example **5** The inverse of a 3 × 3 matrix

Find the inverse of the matrix

$$A = \begin{bmatrix} 0 & 1 & 2 \\ 1 & 0 & 3 \\ 0 & 1 & 4 \end{bmatrix}.$$

Solution

Perform row operations to convert $[A\,|\,I]$ into $[I\,|\,A^{-1}]$, where I is the 3×3 identity matrix.

$$\left[\begin{array}{rrr|rrr} 0 & 1 & 2 & 1 & 0 & 0 \\ 1 & 0 & 3 & 0 & 1 & 0 \\ 0 & 1 & 4 & 0 & 0 & 1 \end{array}\right] \quad \text{The augmented matrix}$$

Interchange the first and second rows to get the first 1 on the diagonal:

$$\left[\begin{array}{rrr|rrr} 1 & 0 & 3 & 0 & 1 & 0 \\ 0 & 1 & 2 & 1 & 0 & 0 \\ 0 & 1 & 4 & 0 & 0 & 1 \end{array}\right] \quad R_1 \leftrightarrow R_2$$

Since the first column is now in the desired form, we work on the second column:

$$\left[\begin{array}{rrr|rrr} 1 & 0 & 3 & 0 & 1 & 0 \\ 0 & 1 & 2 & 1 & 0 & 0 \\ 0 & 0 & 2 & -1 & 0 & 1 \end{array}\right] \quad -R_2 + R_3 \to R_3$$

We now get a 1 in the last position on the diagonal and zeros above it:

$$\left[\begin{array}{rrr|rrr} 1 & 0 & 3 & 0 & 1 & 0 \\ 0 & 1 & 2 & 1 & 0 & 0 \\ 0 & 0 & 1 & -0.5 & 0 & 0.5 \end{array}\right] \quad \tfrac{1}{2}R_3 \to R_3$$

$$\left[\begin{array}{rrr|rrr} 1 & 0 & 0 & 1.5 & 1 & -1.5 \\ 0 & 1 & 0 & 2 & 0 & -1 \\ 0 & 0 & 1 & -0.5 & 0 & 0.5 \end{array}\right] \quad -3R_3 + R_1 \to R_1 \text{ and } -2R_3 + R_2 \to R_2$$

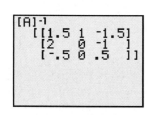

■ **Figure 9.6**

So

$$A^{-1} = \begin{bmatrix} 1.5 & 1 & -1.5 \\ 2 & 0 & -1 \\ -0.5 & 0 & 0.5 \end{bmatrix}.$$

Check that $AA^{-1} = I$ and $A^{-1}A = I$.

Of course it is a lot simpler to find A^{-1} with a calculator, as shown in Fig. 9.6.

Try This. Find the inverse of $\begin{bmatrix} 1 & 1 & 3 \\ -1 & 0 & 0 \\ -1 & -1 & -2 \end{bmatrix}$. ■

Finding the inverse of matrices larger than 2×2 is rather tedious, but technology can be used to great advantage here. We will now see why inverse matrices are so important.

Solving Systems of Equations Using Matrix Inverses

In Section 9.3 we saw that a system of n linear equations in n unknowns could be written as a matrix equation of the form $AX = B$, where A is a matrix of coefficients, X is a matrix of variables, and B is a matrix of constants. If A^{-1} exists, then we can multiply each side of this equation by A^{-1}. Since matrix multiplication is not commutative in general, A^{-1} is placed to the left of the matrices on each side:

$$AX = B$$

$$A^{-1}(AX) = A^{-1}B \quad \text{Multiply each side by } A^{-1}.$$

$$(A^{-1}A)X = A^{-1}B \quad \text{Matrix multiplication is associative.}$$

$$IX = A^{-1}B \quad \text{Since } A^{-1}A = I$$

$$X = A^{-1}B \quad \text{Since } I \text{ is the identity matrix}$$

The last equation indicates that the values of the variables in the matrix X are equal to the entries in the matrix $A^{-1}B$. So solving the system is equivalent to finding $A^{-1}B$. This result is summarized in the following theorem.

Theorem: Solving a System Using A^{-1}

> If a system of n linear equations in n variables has a unique solution, then the solution is given by
>
> $$X = A^{-1}B,$$
>
> where A is the matrix of coefficients, B is the matrix of constants, and X is the matrix of variables.

If there is no solution or there are infinitely many solutions, the matrix of coefficients is not invertible.

Example **6** **Using the inverse of a matrix to solve a system**

Solve the system by using A^{-1}.

$$2x - 3y = 1$$
$$x + y = 8$$

Solution

For this system,

$$A = \begin{bmatrix} 2 & -3 \\ 1 & 1 \end{bmatrix}, \qquad X = \begin{bmatrix} x \\ y \end{bmatrix}, \qquad \text{and} \qquad B = \begin{bmatrix} 1 \\ 8 \end{bmatrix}.$$

Since the matrix A is the same as in Example 3, use A^{-1} from Example 3. Multiply A^{-1} and B to obtain

$$\begin{bmatrix} \frac{1}{5} & \frac{3}{5} \\ -\frac{1}{5} & \frac{2}{5} \end{bmatrix} \begin{bmatrix} 1 \\ 8 \end{bmatrix} = \begin{bmatrix} 5 \\ 3 \end{bmatrix}.$$

Since $X = A^{-1}B$, we have

$$X = \begin{bmatrix} x \\ y \end{bmatrix} = \begin{bmatrix} 5 \\ 3 \end{bmatrix}.$$

So $x = 5$ and $y = 3$. Check this solution in the original system.

With a calculator you can enter A and B, then find $A^{-1}B$, as shown in Fig. 9.7.

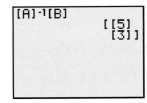

■ **Figure 9.7**

Try This. Solve $\begin{matrix} 3x + 2y = 4 \\ 7x + 5y = 1 \end{matrix}$ by using A^{-1}.   ■

Since computers and even pocket calculators can find inverses of matrices, the method of solving an equation by first finding A^{-1} is a popular one for use with machines. By hand, this method may seem somewhat tedious. However, it is useful when there are many systems to solve with the same coefficients. The following system has the same coefficients as the system of Example 6:

$$2x - 3y = -1$$
$$x + y = -3$$

So

$$X = A^{-1}B = \begin{bmatrix} \frac{1}{5} & \frac{3}{5} \\ -\frac{1}{5} & \frac{2}{5} \end{bmatrix} \begin{bmatrix} -1 \\ -3 \end{bmatrix} = \begin{bmatrix} -2 \\ -1 \end{bmatrix}$$

or $x = -2$ and $y = -1$. As long as the coefficients of x and y are unchanged, the same A^{-1} is used to solve the system.

Application of Matrices to Secret Codes

There are many ways of encoding a message so that no one other than the intended recipient can understand the message. One way is to use a matrix to encode the message and the inverse matrix to decode the message. Assume that a space is 0, A is 1, B is 2, C is 3, and so on. The numerical equivalent of the word HELP is **8, 5, 12, 16.** List these numbers in two 2 × 1 matrices. Then multiply the matrices by a coding matrix. We will use the 2 × 2 matrix A from Example 6:

$$\begin{bmatrix} 2 & -3 \\ 1 & 1 \end{bmatrix}\begin{bmatrix} 8 \\ 5 \end{bmatrix} = \begin{bmatrix} 1 \\ 13 \end{bmatrix} \quad \text{and} \quad \begin{bmatrix} 2 & -3 \\ 1 & 1 \end{bmatrix}\begin{bmatrix} 12 \\ 16 \end{bmatrix} = \begin{bmatrix} -24 \\ 28 \end{bmatrix}.$$

So the encoded message sent is 1, 13, −24, 28. The person receiving the message must know that

$$A^{-1} = \begin{bmatrix} \frac{1}{5} & \frac{3}{5} \\ -\frac{1}{5} & \frac{2}{5} \end{bmatrix}.$$

To decode this message, we find

$$\begin{bmatrix} \frac{1}{5} & \frac{3}{5} \\ -\frac{1}{5} & \frac{2}{5} \end{bmatrix}\begin{bmatrix} 1 \\ 13 \end{bmatrix} = \begin{bmatrix} 8 \\ 5 \end{bmatrix} \quad \text{and} \quad \begin{bmatrix} \frac{1}{5} & \frac{3}{5} \\ -\frac{1}{5} & \frac{2}{5} \end{bmatrix}\begin{bmatrix} -24 \\ 28 \end{bmatrix} = \begin{bmatrix} 12 \\ 16 \end{bmatrix}.$$

So the message is 8, 5, 12, 16, or HELP. Of course, any invertible matrix of any size and its inverse could be used for this coding scheme, and all of the encoding and decoding could be done by a computer. If a person or computer didn't know A or A^{-1} or even their size, then how hard do you think it would be to break this code?

For Thought

True or False? Explain.

The following statements refer to the matrices

$$A = \begin{bmatrix} 2 & 3 \\ 3 & 5 \end{bmatrix}, \quad B = \begin{bmatrix} 5 & -3 \\ -3 & 2 \end{bmatrix}, \quad C = \begin{bmatrix} 4 & 6 \\ 3 & 5 \\ 2 & 1 \end{bmatrix},$$

$$D = \begin{bmatrix} 11 \\ 19 \end{bmatrix}, \quad \text{and} \quad I = \begin{bmatrix} 1 & 0 \\ 0 & 1 \end{bmatrix}.$$

1. $AB = BA = I$

2. $B = A^{-1}$

3. B is an invertible matrix.

4. $AC = CA$

5. $CI = C$

6. C is an invertible matrix.

7. The system $\begin{array}{l} 2x + 3y = 11 \\ 3x + y = 19 \end{array}$ is equivalent to $A\begin{bmatrix} x \\ y \end{bmatrix} = D.$

8. $A^{-1}D = \begin{bmatrix} -2 \\ 5 \end{bmatrix}$

9. The solution set to the system $\begin{array}{l} 2x + 3y = 11 \\ 3x + y = 19 \end{array}$ is $\{(-2, 5)\}.$

10. The solution set to the system $\begin{array}{l} 2x + 3y = 3 \\ 3x + y = -7 \end{array}$ is obtained from $A^{-1}\begin{bmatrix} 3 \\ -7 \end{bmatrix}.$

9.4 Exercises

For each given matrix A, show that $AI = A$ and $IA = A$ where I is the identity matrix of the appropriate size.

1. $A = \begin{bmatrix} 1 & 3 \\ 4 & 6 \end{bmatrix}$

2. $A = \begin{bmatrix} 3 & 2 \\ 5 & 9 \end{bmatrix}$

3. $A = \begin{bmatrix} 3 & 2 & 1 \\ 5 & 6 & 2 \\ 7 & 8 & 3 \end{bmatrix}$

4. $A = \begin{bmatrix} 3 & 4 & 1 \\ 2 & 2 & 5 \\ 0 & 3 & 6 \end{bmatrix}$

Find the following products.

5. $\begin{bmatrix} 1 & 0 \\ 0 & 1 \end{bmatrix}\begin{bmatrix} -3 & 5 \\ 12 & 6 \end{bmatrix}$

6. $\begin{bmatrix} 4 & 8 \\ 9 & -2 \end{bmatrix}\begin{bmatrix} 1 & 0 \\ 0 & 1 \end{bmatrix}$

7. $\begin{bmatrix} -2 & -3 \\ 3 & 4 \end{bmatrix}\begin{bmatrix} 4 & 3 \\ -3 & -2 \end{bmatrix}$

8. $\begin{bmatrix} 3 & 2 \\ 3 & 3 \end{bmatrix}\begin{bmatrix} 1 & -\frac{2}{3} \\ -1 & 1 \end{bmatrix}$

9. $\begin{bmatrix} 3 & 4 \\ 3 & 5 \end{bmatrix}\begin{bmatrix} \frac{5}{3} & -\frac{4}{3} \\ -1 & 1 \end{bmatrix}$

10. $\begin{bmatrix} 1 & 2 \\ 4 & 5 \end{bmatrix}\begin{bmatrix} -\frac{5}{3} & \frac{2}{3} \\ \frac{4}{3} & -\frac{1}{3} \end{bmatrix}$

11. $\begin{bmatrix} 3 & 5 & 1 \\ 4 & 5 & 7 \\ 4 & 9 & 2 \end{bmatrix}\begin{bmatrix} 1 & 0 & 0 \\ 0 & 1 & 0 \\ 0 & 0 & 1 \end{bmatrix}$

12. $\begin{bmatrix} 1 & 0 & 0 \\ 0 & 1 & 0 \\ 0 & 0 & 1 \end{bmatrix}\begin{bmatrix} 4 & 0 & 5 \\ 0 & 7 & 9 \\ 3 & 1 & 2 \end{bmatrix}$

13. $\begin{bmatrix} 1 & 0 & 2 \\ 1 & 3 & 0 \\ 0 & 1 & 0 \end{bmatrix}\begin{bmatrix} 0 & 1 & -3 \\ 0 & 0 & 1 \\ 0.5 & -0.5 & 1.5 \end{bmatrix}$

14. $\begin{bmatrix} 2 & 1 & 0 \\ 1 & 0 & 2 \\ 0 & 1 & 1 \end{bmatrix}\begin{bmatrix} 0.4 & 0.2 & -0.4 \\ 0.2 & -0.4 & 0.8 \\ -0.2 & 0.4 & 0.2 \end{bmatrix}$

15. $\begin{bmatrix} 1 & 1 & 0 \\ 0 & 1 & 1 \\ 1 & 0 & 1 \end{bmatrix}\begin{bmatrix} 0.5 & -0.5 & 0.5 \\ 0.5 & 0.5 & -0.5 \\ -0.5 & 0.5 & 0.5 \end{bmatrix}$

16. $\begin{bmatrix} 0.4 & 0.2 & -0.4 \\ 0.2 & -0.4 & 0.8 \\ -0.2 & 0.4 & 0.2 \end{bmatrix}\begin{bmatrix} 2 & 1 & 0 \\ 1 & 0 & 2 \\ 0 & 1 & 1 \end{bmatrix}$

Determine whether the matrices in each pair are inverses of each other.

17. $\begin{bmatrix} 3 & 1 \\ 11 & 4 \end{bmatrix}, \begin{bmatrix} 4 & -1 \\ -11 & 3 \end{bmatrix}$

18. $\begin{bmatrix} \frac{1}{2} & 0 \\ 0 & \frac{1}{2} \end{bmatrix}, \begin{bmatrix} 2 & 0 \\ 0 & 2 \end{bmatrix}$

19. $\begin{bmatrix} \frac{1}{2} & -1 \\ 3 & -12 \end{bmatrix}, \begin{bmatrix} 4 & 2 \\ 1 & 1 \end{bmatrix}$

20. $\begin{bmatrix} 1 & 2 & 3 \\ 0 & 1 & 2 \\ 0 & 0 & 1 \end{bmatrix}, \begin{bmatrix} 1 & -2 & 1 \\ 0 & 1 & -2 \\ 0 & 0 & 1 \end{bmatrix}$

21. $\begin{bmatrix} 1 & 0 & 0 \\ 0 & \frac{1}{2} & 0 \\ & & \end{bmatrix}, \begin{bmatrix} 1 & 0 \\ 0 & 2 \\ 3 & 4 \end{bmatrix}$

22. $\begin{bmatrix} 1 & 2 \\ 3 & 4 \\ 5 & 6 \end{bmatrix}, \begin{bmatrix} 1 & \frac{1}{2} \\ \frac{1}{3} & \frac{1}{4} \\ \frac{1}{5} & \frac{1}{6} \end{bmatrix}$

Find the inverse of each matrix A if possible. Check that $AA^{-1} = I$ and $A^{-1}A = I$. See the procedure for finding A^{-1} on page 755.

23. $\begin{bmatrix} 1 & 4 \\ 0 & 2 \end{bmatrix}$

24. $\begin{bmatrix} 1 & 3 \\ 0 & -1 \end{bmatrix}$

25. $\begin{bmatrix} 1 & 6 \\ 1 & 9 \end{bmatrix}$

26. $\begin{bmatrix} 1 & 4 \\ 3 & 8 \end{bmatrix}$

27. $\begin{bmatrix} -2 & -3 \\ 3 & 4 \end{bmatrix}$

28. $\begin{bmatrix} 3 & 4 \\ 4 & 5 \end{bmatrix}$

29. $\begin{bmatrix} 1 & -5 \\ -1 & 3 \end{bmatrix}$

30. $\begin{bmatrix} 4 & 3 \\ -3 & -2 \end{bmatrix}$

31. $\begin{bmatrix} -1 & 5 \\ 2 & -10 \end{bmatrix}$

32. $\begin{bmatrix} 2 & 6 \\ 1 & 3 \end{bmatrix}$

33. $\begin{bmatrix} 1 & 1 & 0 \\ 0 & -1 & -1 \\ 1 & 0 & -1 \end{bmatrix}$

34. $\begin{bmatrix} 1 & -1 & 2 \\ 1 & 2 & 3 \\ 2 & 1 & 5 \end{bmatrix}$

35. $\begin{bmatrix} 1 & 1 & 1 \\ 1 & -1 & -1 \\ 1 & -1 & 1 \end{bmatrix}$

36. $\begin{bmatrix} 1 & 0 & 2 \\ 0 & 2 & 0 \\ 1 & 3 & 0 \end{bmatrix}$

37. $\begin{bmatrix} 0 & 2 & 0 \\ 3 & 3 & 2 \\ 2 & 5 & 1 \end{bmatrix}$

38. $\begin{bmatrix} 4 & 1 & -3 \\ 0 & 1 & 0 \\ -3 & 1 & 2 \end{bmatrix}$

39. $\begin{bmatrix} 1 & 0 & 1 \\ 0 & 2 & 2 \\ 2 & 1 & 0 \end{bmatrix}$

40. $\begin{bmatrix} 1 & 3 & 0 \\ 0 & 3 & -2 \\ 0 & -5 & 3 \end{bmatrix}$

41. $\begin{bmatrix} 0 & 4 & 2 \\ 0 & 3 & 2 \\ 1 & -1 & 1 \end{bmatrix}$ **42.** $\begin{bmatrix} 1 & 2 & 0 \\ 2 & 3 & -1 \\ 0 & 1 & 2 \end{bmatrix}$

43. $\begin{bmatrix} 1 & 2 & 3 & 4 \\ 0 & 1 & 2 & 3 \\ 0 & 0 & 1 & 2 \\ 0 & 0 & 0 & 1 \end{bmatrix}$ **44.** $\begin{bmatrix} 1 & 0 & 0 & 0 \\ -2 & 1 & 0 & 0 \\ 3 & -2 & 1 & 0 \\ 5 & 3 & -2 & 1 \end{bmatrix}$

Solve each system of equations by using A^{-1}. Note that the matrix of coefficients in each system is a matrix from Exercises 23–44.

45. $x + 6y = -3$
$x + 9y = -6$

46. $x + 4y = 5$
$3x + 8y = 7$

47. $x + 6y = 4$
$x + 9y = 5$

48. $x + 4y = 1$
$3x + 8y = 5$

49. $-2x - 3y = 1$
$3x + 4y = -1$

50. $3x + 4y = 1$
$4x + 5y = 2$

51. $x - 5y = -5$
$-x + 3y = 1$

52. $4x + 3y = 2$
$-3x - 2y = -1$

53. $x + y + z = 3$
$x - y - z = -1$
$x - y + z = 5$

54. $x + 2z = -4$
$2y = 6$
$x + 3y = 7$

55. $2y = 6$
$3x + 3y + 2z = 16$
$2x + 5y + z = 19$

56. $4x + y - 3z = 3$
$y = -2$
$-3x + y + 2z = -5$

Solve each system of equations by using A^{-1} if possible.

57. $0.3x = 3 - 0.1y$
$4y = 7 - 2x$

58. $2x = 3y - 7$
$y = x + 4$

59. $x - y + z = 5$
$2x - y + 3z = 1$
$y + z = -9$

60. $x + y - z = 4$
$2x - 3y + z = 2$
$4x - y - z = 6$

61. $x + y + z = 1$
$2x + 4y + z = 2$
$x + 3y + 6z = 3$

62. $0.5x - 0.25y + 0.1z = 3$
$0.2x - 0.5y + 0.2z = -2$
$0.1x + 0.3y - 0.5z = -8$

⊠ *Most graphing calculators can perform operations with matrices, including matrix inversion and multiplication. Solve the following systems, using a graphing calculator to find A^{-1} and the product $A^{-1}B$.*

63. $0.1x + 0.2y + 0.1z = 27$
$0.5x + 0.2y + 0.3z = 9$
$0.4x + 0.8y + 0.1z = 36$

64. $3x + 6y + 4z = 9$
$x + 2y - 2z = -18$
$-x + 4y + 3z = 54$

65. $1.5x - 5y + 3z = 16$
$2.25x - 4y + z = 24$
$2x + 3.5y - 3z = -8$

66. $2.1x - 3.4y + 5z = 100$
$1.3x + 2y - 8z = 250$
$2.5x + 3y - 9.1z = 300$

Solve each problem.

67. Find all matrices A such that $A = \begin{bmatrix} a & 7 \\ 3 & b \end{bmatrix}$,

$A^{-1} = \begin{bmatrix} -b & 7 \\ 3 & -a \end{bmatrix}$, and a and b are positive integers.

68. Find all matrices of the form $A = \begin{bmatrix} a & a \\ 0 & c \end{bmatrix}$ such that $A^2 = I$.

Write a system of equations for each problem. Solve the system using an inverse matrix.

69. *Eggs and Magazines* Stephanie bought a dozen eggs and a magazine at the Handy Mart. Her bill including tax was $2.70. If groceries are taxed at 8% and magazines at 5% and she paid 15 cents in tax on the purchase, then what was the price of each item?

70. *Dogs and Suds* The French Club sold 48 hot dogs and 120 soft drinks at the game on Saturday for a total of $103.20. If the price of a hot dog was 40 cents more than the price of a soft drink, then what was the price of each item?

71. *Plywood and Insulation* A contractor purchased four loads of plywood and six loads of insulation on Monday for $2500, and three loads of plywood and five loads of insulation on Tuesday for $1950. Find the cost of one load of plywood and the cost of one load of insulation.

72. *Virtual Pets* One Monday KZ Toys received a shipment of 12 Nano Puppies and 6 Giga Pets for a total cost of $138. Tuesday's shipment contained 4 Nano Puppies and 8 Giga Pets for a total cost of $76. Wednesday's shipment of 25 Nano Puppies and 33 Giga Pets did not include an invoice. What was the cost of Wednesday's shipment?

The following messages were encoded by using the matrix

$A = \begin{bmatrix} 3 & 1 \\ 5 & 2 \end{bmatrix}$ *and the coding scheme described in this section.*

Find A^{-1} and use it to decode the messages.

73. 36, 65, 49, 83, 12, 24, 66, 111, 33, 55

74. 15, 29, 26, 45, 24, 46, 3, 5, 46, 83, 6, 12, 77, 133

Write a system of equations for each of the following problems and solve the system using matrix inversion and matrix multiplication on a graphing calculator.

75. *Mixing Investments* The Asset Manager Fund keeps 76% of its money in stocks while the Magellan Fund keeps 90% of its money in stocks. How should an investor divide $60,000 between these two mutual funds so that 86% of the money is in stocks?

76. *Mixing Investments* The Asset Manager Mutual Fund investment mix is 76% stocks, 20% bonds, and 4% cash. The Magellan Fund mix is 90% stocks, 9% bonds, and 1% cash. The Puritan Fund mix is 60% stocks, 33% bonds, and 7% cash. How should an investor divide $50,000 between these

three funds so that 74% of the money is in stocks, 21.7% is in bonds, and 4.3% is in cash?

77. *Stocking Supplies* Fernando purchases supplies for an import store. His first shipment on Monday was for 24 animal totems, 33 trade-bead necklaces, and 12 tribal masks for a total price of $202.23. His second shipment was for 19 animal totems, 40 trade-bead necklaces, and 22 tribal masks for a total price of $209.38. His third shipment was for 30 animal totems, 9 trade-bead necklaces, and 19 tribal masks for a total price of $167.66. For the fourth shipment the computer was down, and Fernando did not know the price of each item. What is the price of each item?

78. *On the Bayou* A-Bear's Catering Service charges its customers according to the number of servings of each item that is supplied at the party. The table shows the number of servings of jambalaya, crawfish pie, filé gumbo, iced tea, and dessert for the last five customers, along with the total cost of each party. What amount does A-Bear's charge per serving of each item?

HINT Write a system of five equations in five unknowns and use a graphing calculator to solve it.

■ **Table for Exercise 78**

	Jambalaya	Crawfish Pie	Filé Gumbo	Iced Tea	Dessert	Cost
Boudreaux	36	28	35	90	68	$344.35
Thibodeaux	37	19	56	84	75	$369.10
Fontenot	49	55	70	150	125	$588.90
Arceneaux	58	34	52	122	132	$529.50
Gautreaux	44	65	39	133	120	$521.65

Thinking Outside the Box LXXXI

Maximizing Products

a) What is the maximum product for two whole numbers whose sum is 10? What are the numbers?

b) What is the maximum product for any number of whole numbers whose sum is 10? What are the numbers?

c) What is the maximum product for any number of whole numbers whose sum is 18? What are the numbers?

9.4 Pop Quiz

1. Find the product $\begin{bmatrix} 1 & 0 \\ 0 & 1 \end{bmatrix}\begin{bmatrix} 2 & 4 \\ 6 & 8 \end{bmatrix}$.

2. Find the inverse of $\begin{bmatrix} 2 & 5 \\ 3 & 8 \end{bmatrix}$.

3. Solve $\begin{array}{l} 2x + 5y = 4 \\ 3x + 8y = 3 \end{array}$ using an inverse matrix.

Linking Concepts

For Individual or
Group Explorations

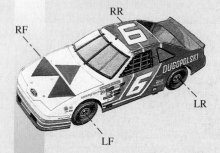

Weight Distribution of a Race Car

Race car drivers adjust the weight distribution of their cars according to track conditions. However, according to a NASCAR rule, no more than 52% of a car's weight can be on any pair of tires.

a)   A driver of a 1250-pound car wants to have 50% of its weight on the left rear and left front tires and 48% of its weight on the left rear and right front tires. If the right front weight is fixed at 288 pounds, then what amount of weight should be on the other three tires?

b)   Is the NASCAR rule satisfied with the weight distribution found in part (a)?

c)   A driver of a 1300-pound car wants to have 50% of the car's weight on the left front and left rear tires, 48% on the left rear and right front tires, and 51% on the left rear and right rear tires. How much weight should be on each of the four tires?

d)   The driver of the 1300-pound car wants to satisfy the NASCAR 52% rule and have as much weight as possible on the left front tire. What is the maximum amount of weight that can be on that tire?

9.5

Solution of Linear Systems in Two Variables Using Determinants

We have solved linear systems of equations by graphing, substitution, addition, Gaussian elimination, and inverse matrices. Graphing, substitution, and addition are feasible only with relatively simple systems. By contrast, the Gaussian elimination and inverse matrix methods are readily performed by computers or even hand-held calculators. With a machine doing the work, they can be applied to complicated systems such as those in Exercises 77 and 78 of Section 9.4. Determinants, which we now discuss, can also be used by computers and calculators and give us another method that is not limited to simple systems.

The Determinant of a 2 × 2 Matrix

Before we can solve a system of equations by using determinants, we need to learn what a determinant is and how to find it. The **determinant** of a square matrix is a real number associated with the matrix. Every square matrix has a determinant. The determinant of a 1 × 1 matrix is the single entry of the matrix. For a 2 × 2 matrix the determinant is defined as follows.

Definition: Determinant of a 2 × 2 Matrix

The **determinant** of the matrix $\begin{bmatrix} a_{11} & a_{12} \\ a_{21} & a_{22} \end{bmatrix}$ is the real number $a_{11}a_{22} - a_{21}a_{12}$.

In symbols,

$$\begin{vmatrix} a_{11} & a_{12} \\ a_{21} & a_{22} \end{vmatrix} = a_{11}a_{22} - a_{21}a_{12}.$$

If a matrix is named A, then the determinant of that matrix is denoted as $|A|$ or $\det(A)$. Even though the symbol for determinant looks like the absolute value symbol, the value of a determinant may be any real number. For a 2 × 2 matrix, that number is found by subtracting the products of the diagonal entries:

$$\begin{vmatrix} a_{11} & a_{12} \\ a_{21} & a_{22} \end{vmatrix} = a_{11}a_{22} - a_{21}a_{12}$$

Example 1 The determinant of a 2 × 2 matrix

Find the determinant of each matrix.

a. $\begin{bmatrix} 3 & -1 \\ 4 & -5 \end{bmatrix}$ **b.** $\begin{bmatrix} 4 & -6 \\ 2 & -3 \end{bmatrix}$

Solution

a. $\begin{vmatrix} 3 & -1 \\ 4 & -5 \end{vmatrix} = 3(-5) - (4)(-1) = -15 + 4 = -11$

b. $\begin{vmatrix} 4 & -6 \\ 2 & -3 \end{vmatrix} = 4(-3) - (2)(-6) = -12 + 12 = 0$

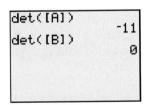

To find these determinants with a calculator, enter the matrices A and B, then use the determinant function, as in Fig. 9.8.

■ **Figure 9.8**

Try This. Find the determinant of $\begin{bmatrix} 2 & 4 \\ -3 & 1 \end{bmatrix}$. ■

Cramer's Rule for Systems in Two Variables

We will now see how determinants arise in the solution of a system of two linear equations in two unknowns. Consider a general system of two linear equations in two unknowns, x and y,

$$(1) \quad a_1x + b_1y = c_1$$

$$(2) \quad a_2x + b_2y = c_2$$

where a_1, a_2, b_1, b_2, c_1 and c_2 are real numbers.

To eliminate y, multiply Eq. (1) by b_2 and Eq. (2) by $-b_1$:

$$b_2 a_1 x + b_2 b_1 y = b_2 c_1 \qquad \text{Eq. (1) is multiplied by } b_2.$$

$$\underline{-b_1 a_2 x - b_1 b_2 y = -b_1 c_2} \qquad \text{Eq. (2) is multiplied by } -b_1.$$

$$a_1 b_2 x - a_2 b_1 x \qquad\quad = c_1 b_2 - c_2 b_1 \qquad \text{Add.}$$

$$(a_1 b_2 - a_2 b_1)x = c_1 b_2 - c_2 b_1 \qquad \text{Factor out } x.$$

$$x = \frac{c_1 b_2 - c_2 b_1}{a_1 b_2 - a_2 b_1} \qquad \text{Provided that } a_1 b_2 - a_2 b_1 \neq 0$$

This formula for x can be written using determinants as

$$x = \frac{\begin{vmatrix} c_1 & b_1 \\ c_2 & b_2 \end{vmatrix}}{\begin{vmatrix} a_1 & b_1 \\ a_2 & b_2 \end{vmatrix}}, \qquad \text{provided } a_1 b_2 - a_2 b_1 \neq 0.$$

The same procedure is used to eliminate x and get the following formula for y in terms of determinants:

$$y = \frac{\begin{vmatrix} a_1 & c_1 \\ a_2 & c_2 \end{vmatrix}}{\begin{vmatrix} a_1 & b_1 \\ a_2 & b_2 \end{vmatrix}}, \qquad \text{provided } a_1 b_2 - a_2 b_1 \neq 0$$

Notice that there are three determinants involved in solving for x and y. Let

$$D = \begin{vmatrix} a_1 & b_1 \\ a_2 & b_2 \end{vmatrix}, \qquad D_x = \begin{vmatrix} c_1 & b_1 \\ c_2 & b_2 \end{vmatrix}, \qquad \text{and} \qquad D_y = \begin{vmatrix} a_1 & c_1 \\ a_2 & c_2 \end{vmatrix}.$$

Note that D is the determinant of the original matrix of coefficients of x and y. D appears in the denominator for both x and y. D_x is the determinant D with the constants c_1 and c_2 replacing the first column of D. D_y is the determinant D with the constants c_1 and c_2 replacing the second column of D. These formulas for solving a system of two linear equations in two variables are known as **Cramer's rule.**

Cramer's Rule for Systems in Two Variables

For the system of equations

$$a_1 x + b_1 y = c_1$$
$$a_2 x + b_2 y = c_2$$

let

$$D = \begin{vmatrix} a_1 & b_1 \\ a_2 & b_2 \end{vmatrix}, \qquad D_x = \begin{vmatrix} c_1 & b_1 \\ c_2 & b_2 \end{vmatrix}, \qquad \text{and} \qquad D_y = \begin{vmatrix} a_1 & c_1 \\ a_2 & c_2 \end{vmatrix}.$$

If $D \neq 0$, then the solution to the system is given by

$$x = \frac{D_x}{D} \qquad \text{and} \qquad y = \frac{D_y}{D}.$$

Note that D_x is obtained from D by replacing the x-column with the constants c_1 and c_2, whereas D_y is obtained from D by replacing the y-column with the constants c_1 and c_2.

Example **2** **Applying Cramer's rule**

Use Cramer's rule to solve the system.

$$3x = 2y + 9$$

$$3y = x + 3$$

Solution

To apply Cramer's rule, rewrite both equations in the form $Ax + By = C$:

$$3x - 2y = 9$$

$$-x + 3y = 3$$

First find the determinant of the coefficient matrix using the coefficients of x and y:

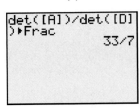

$$D = \begin{vmatrix} 3 & -2 \\ -1 & 3 \end{vmatrix} = 3(3) - (-1)(-2) = 7$$

Next we find the determinants D_x and D_y. For D_x, use 9 and 3 in the x-column, and for D_y, use 9 and 3 in the y-column:

$$D_x = \begin{vmatrix} 9 & -2 \\ 3 & 3 \end{vmatrix} = 33 \quad \text{and} \quad D_y = \begin{vmatrix} 3 & 9 \\ -1 & 3 \end{vmatrix} = 18.$$

By Cramer's rule,

$$x = \frac{D_x}{D} = \frac{33}{7} \quad \text{and} \quad y = \frac{D_y}{D} = \frac{18}{7}.$$

Check that $x = 33/7$ and $y = 18/7$ satisfy the original system.

To check this result with a calculator, define matrices D and A, as in Fig. 9.9(a). Find x with Cramer's rule, as in Fig. 9.9(b). You can find y in a similar manner.

(a)

(b)

▪ **Figure 9.9**

Try This. Solve $\begin{matrix} 2x + 5y = 17 \\ -3x + y = 0 \end{matrix}$ by using Cramer's rule.   ▪

A system of two linear equations in two unknowns, may have a unique solution, no solution, or infinitely many solutions. Cramer's rule works only on systems that have a unique solution. For inconsistent or dependent systems, $D = 0$ and Cramer's rule will not give the solution. If $D = 0$, then another method must be used to determine the solution set.

Example **3** **Inconsistent and dependent systems**

Use Cramer's rule to solve each system if possible.

a. $\begin{matrix} 2x - 4y = 8 \\ -x + 2y = -4 \end{matrix}$ **b.** $\begin{matrix} 2x - 4y = 8 \\ -x + 2y = 6 \end{matrix}$

Solution

The coefficient matrix is the same for both systems:

$$D = \begin{vmatrix} 2 & -4 \\ -1 & 2 \end{vmatrix} = 0$$

So Cramer's rule does not apply to either system. Multiply the second equation in each system by 2 and add the equations:

a. $2x - 4y = 8$ **b.** $2x - 4y = 8$
    $-2x + 4y = -8$ $-2x + 4y = 12$
        $0 = 0$ $0 = 20$

System (a) is dependent, and the solution set is $\{(x, y) \mid -x + 2y = -4\}$. System (b) is inconsistent and has no solution.

Try This. Solve $\begin{matrix} x + y = 6 \\ -2x - 2y = 5 \end{matrix}$ by Cramer's rule if possible. ▪

A system of two linear equations in two variables has a unique solution if and only if the determinant of the matrix of coefficients is nonzero. We have not yet seen how to find a determinant of a larger matrix, but the same result is true for a system of n linear equations in n variables. In Section 9.4 we learned that a system of n linear equations in n variables has a unique solution if and only if the matrix of coefficients is invertible. These two results are combined in the following theorem to give a means of identifying whether a matrix is invertible.

Theorem: Invertible Matrices

> A matrix is invertible if and only if it has a nonzero determinant.

Example **4** **Determinants and inverse matrices**

Are the matrices $A = \begin{bmatrix} 2 & -3 \\ 4 & 5 \end{bmatrix}$ and $B = \begin{bmatrix} 3 & 5 \\ 6 & 10 \end{bmatrix}$ invertible?

Solution

Since $|A| = 10 - (-12) = 22$, A is an invertible matrix. However, because $|B| = 30 - 30 = 0$, B is not an invertible matrix.

Try This. Determine whether $\begin{bmatrix} 1 & 6 \\ -2 & 9 \end{bmatrix}$ has an inverse. ▪

For Thought

True or False? Explain.

The following statements refer to the matrices

$A = \begin{bmatrix} 3 & -5 \\ 1 & 4 \end{bmatrix}$, $B = \begin{bmatrix} 4 & -2 \\ -10 & 5 \end{bmatrix}$,

$C = \begin{bmatrix} 2 & -5 \\ 6 & 4 \end{bmatrix}$,   and $E = \begin{bmatrix} 3 & 2 \\ 1 & 6 \end{bmatrix}$.

1. $|A| = 7$

2. A is invertible.

3. $|B| = 0$

4. B is invertible.

5. The system $\begin{matrix} 3x - 5y = 2 \\ x + 4y = 6 \end{matrix}$ is independent.

6. $|CE| = |C| \cdot |E|$

7. The solution to the system $\begin{array}{l} 3x^2 - 5y^2 = 2 \\ x^2 + 4y = 6 \end{array}$ is $x = \dfrac{|C|}{|A|}$ and $y = \dfrac{|E|}{|A|}$.

8. The determinant of the 2×2 identity matrix I is 1.

9. The matrix $\begin{bmatrix} 2 & 0.1 \\ 100 & 5 \end{bmatrix}$ is invertible.

10. $\begin{bmatrix} 5 & 3 \\ 1 & 6 \end{bmatrix} = 27$

9.5 Exercises

Find the determinant of each matrix.

1. $\begin{bmatrix} 1 & 3 \\ 0 & 2 \end{bmatrix}$

2. $\begin{bmatrix} 0 & 4 \\ 2 & -1 \end{bmatrix}$

3. $\begin{bmatrix} 3 & 4 \\ 2 & 9 \end{bmatrix}$

4. $\begin{bmatrix} 7 & 2 \\ 3 & -2 \end{bmatrix}$

5. $\begin{bmatrix} -0.3 & -0.5 \\ -0.7 & 0.2 \end{bmatrix}$

6. $\begin{bmatrix} -\frac{1}{3} & \frac{4}{3} \\ -3 & \frac{2}{3} \end{bmatrix}$

7. $\begin{bmatrix} \frac{1}{8} & -\frac{3}{8} \\ 2 & -\frac{1}{4} \end{bmatrix}$

8. $\begin{bmatrix} -1 & -3 \\ -5 & -8 \end{bmatrix}$

9. $\begin{bmatrix} 0.02 & 0.4 \\ 1 & 20 \end{bmatrix}$

10. $\begin{bmatrix} -0.3 & 0.4 \\ 3 & -4 \end{bmatrix}$

11. $\begin{bmatrix} 3 & -5 \\ -9 & 15 \end{bmatrix}$

12. $\begin{bmatrix} -6 & 2 \\ 3 & -1 \end{bmatrix}$

Find all values for a that make each equation correct.

13. $\begin{vmatrix} a & 2 \\ 3 & 4 \end{vmatrix} = 10$

14. $\begin{vmatrix} 1 & 7 \\ 3 & a \end{vmatrix} = 5$

15. $\begin{vmatrix} a & 8 \\ 2 & a \end{vmatrix} = 0$

16. $\begin{vmatrix} a & 1 \\ a & a \end{vmatrix} = 0$

Solve each system, using Cramer's rule when possible.

17. $\begin{array}{l} 2x + y = 5 \\ x + 2y = 7 \end{array}$

18. $\begin{array}{l} 3x + y = 10 \\ x + y = 6 \end{array}$

19. $\begin{array}{l} x - 2y = 7 \\ x + 2y = -5 \end{array}$

20. $\begin{array}{l} x - y = 1 \\ 3x + y = 7 \end{array}$

21. $\begin{array}{l} 2x - y = -11 \\ x + 3y = 12 \end{array}$

22. $\begin{array}{l} 3x - 2y = -4 \\ -5x + 4y = -1 \end{array}$

23. $\begin{array}{l} x = y + 6 \\ x + y = 5 \end{array}$

24. $\begin{array}{l} 3x + y = 7 \\ 4x = y - 4 \end{array}$

25. $\begin{array}{l} \frac{1}{2}x - \frac{1}{3}y = 4 \\ \frac{1}{4}x + \frac{1}{2}y = 6 \end{array}$

26. $\begin{array}{l} \frac{1}{4}x + \frac{2}{3}y = 25 \\ \frac{3}{5}x - \frac{1}{10}y = 12 \end{array}$

27. $\begin{array}{l} 0.2x + 0.12y = 148 \\ x + \quad y = 900 \end{array}$

28. $\begin{array}{l} 0.08x + 0.05y = 72 \\ 2x - \quad y = 0 \end{array}$

29. $\begin{array}{l} 3x + y = 6 \\ -6x - 2y = -12 \end{array}$

30. $\begin{array}{l} 8x - 4y = 2 \\ 4x - 2y = 1 \end{array}$

31. $\begin{array}{l} 8x - y = 9 \\ -8x + y = 10 \end{array}$

32. $\begin{array}{l} 12x + 3y = 9 \\ 4x + y = 6 \end{array}$

33. $\begin{array}{l} y = x - 3 \\ y = 3x + 9 \end{array}$

34. $\begin{array}{l} y = \dfrac{x - 3}{2} \\ x + 2y = 15 \end{array}$

35. $\begin{array}{l} \sqrt{2}x + \sqrt{3}y = 4 \\ \sqrt{18}x - \sqrt{12}y = -3 \end{array}$

36. $\begin{array}{l} \dfrac{\sqrt{3}x}{3} + y = 1 \\ x - \sqrt{3}y = 0 \end{array}$

37. $\begin{array}{l} x^2 + y^2 = 25 \\ x^2 - y = 5 \end{array}$

38. $\begin{array}{l} x^2 + y = 8 \\ x^2 - y = 4 \end{array}$

39. $\begin{array}{l} x - 2y = y^2 \\ \frac{1}{2}x - y = 2 \end{array}$

40. $\begin{array}{l} y = x^2 \\ x + y = 30 \end{array}$

Determine whether each matrix is invertible by finding the determinant of the matrix.

41. $\begin{bmatrix} 4 & 0.5 \\ 2 & 3 \end{bmatrix}$

42. $\begin{bmatrix} -5 & 2 \\ 4 & -1 \end{bmatrix}$

43. $\begin{bmatrix} 3 & -4 \\ 9 & -12 \end{bmatrix}$

44. $\begin{bmatrix} \frac{1}{2} & 12 \\ \frac{1}{3} & 8 \end{bmatrix}$

Solve the following systems using Cramer's rule and a graphing calculator.

45. $3.47x + 23.09y = 5978.95$

$\quad 12.48x + 3.98y = 2765.34$

46. $0.0875x + 0.1625y = 564.40$

$\quad\quad x + \quad\quad y = 4232$

Solve each problem, using two linear equations in two variables and Cramer's rule.

47. *The Survey Says*   A survey of 615 teenagers found that 44% of the boys and 35% of the girls would like to be taller. If altogether 231 teenagers in the survey wished they were taller, how many boys and how many girls were in the survey?

48. *Consumer Confidence*   A survey of 900 Americans found that 680 of them had confidence in the economy. If 80% of the women and 70% of the men surveyed expressed confidence in the economy, then how many men and how many women were surveyed.

49. *Acute Angles*   One acute angle of a right triangle is 1° larger than twice the other acute angle. What are the measures of the acute angles?

50. *Isosceles Triangle*   If the smallest angle of an isosceles triangle is 2° smaller than any other angle, then what is the measure of each angle?

51. *Average Salary*   The average salary for the president and vice-president of Intermax Office Supply is $200,000. If the president's salary is $100,000 more than the vice president's, then what is the salary of each?

52. *A Losing Situation*   Morton Motor Express lost a full truckload of TVs and VCRs. The truck carrying the shipment had a capacity of 2350 ft³. On the insurance claim the TVs were valued at $400 each and the VCRs were valued at $225 each,

for a total value of $147,500. If each TV was in a carton with a volume of 8 ft³ and each VCR was in a carton with a volume of 2.5 ft³, then how many TVs and VCRs were in the shipment?

For Writing/Discussion

The following exercises investigate some of the properties of determinants. For these exercises let $M = \begin{bmatrix} 3 & 2 \\ 5 & 4 \end{bmatrix}$ and $N = \begin{bmatrix} 2 & 7 \\ 1 & 5 \end{bmatrix}$.

53. Find $|M|$, $|N|$, and $|MN|$. Is $|MN| = |M| \cdot |N|$?

54. Find M^{-1} and $|M^{-1}|$. Is $|M^{-1}| = 1/|M|$?

55. Prove that the determinant of a product of two 2×2 matrices is equal to the product of their determinants.

56. Prove that if A is any 2×2 invertible matrix, then the determinant of A^{-1} is the reciprocal of the determinant of A.

57. Find $|-2M|$. Is $|-2M| = -2 \cdot |M|$?

58. Prove that if k is any scalar and A is any 2×2 matrix, then $|kA| = k^2 \cdot |A|$.

Thinking Outside the Box LXXXII

Ponder These Pills   A blind man must take two pills every morning; one is type A and the other is type B. Since the pills are identical to the blind man, he keeps them in bottles of different sizes to tell them apart. One day he places one pill of type A on the counter and accidentally drops two pills of type B next to it. Now he has three pills on the counter that he cannot tell apart. They are identical in size, shape, texture, and weight. He is poor and cannot waste these pills, he cannot put them back in the bottles, and he must take his medications. What can he do?

9.5 Pop Quiz

1. Find the determinant of $\begin{bmatrix} 4 & 2 \\ 3 & -1 \end{bmatrix}$.

2. Solve $\begin{array}{l} 4x + 2y = 3 \\ 3x - \ y = 1 \end{array}$ using Cramer's rule.

3. Is $\begin{bmatrix} 9 & 2 \\ 6 & 8 \end{bmatrix}$ invertible?

Solution of Linear Systems in Three Variables Using Determinants

The determinant can be defined for any square matrix. In this section we define the determinant of a 3×3 matrix by extending the definition of the determinant for 2×2 matrices. We can then solve linear systems of three equations in three unknowns, using an extended version of Cramer's rule. The first step is to define a determinant of a certain part of a matrix, a *minor*.

Minors

To each entry of a 3×3 matrix there corresponds a 2×2 matrix, which is obtained by deleting the row and column in which that entry appears. The determinant of this 2×2 matrix is called the **minor** of that entry.

Example **1** Finding the minor of an entry

Find the minors for the entries $-2, 5$, and 6 in the 3×3 matrix $\begin{bmatrix} -2 & -3 & -1 \\ -7 & 4 & 5 \\ 0 & 6 & 1 \end{bmatrix}$.

Solution
To find the minor for the entry -2, delete the first row and first column.

$$\begin{bmatrix} -2 & -3 & -1 \\ -7 & 4 & 5 \\ 0 & 6 & 1 \end{bmatrix}$$

The minor for -2 is $\begin{vmatrix} 4 & 5 \\ 6 & 1 \end{vmatrix} = 4 - (30) = -26$. To find the minor for the entry 5, delete the second row and third column.

$$\begin{bmatrix} -2 & -3 & -1 \\ -7 & 4 & 5 \\ 0 & 6 & 1 \end{bmatrix}$$

The minor for 5 is $\begin{vmatrix} -2 & -3 \\ 0 & 6 \end{vmatrix} = -12 - (0) = -12$. To find the minor for the entry 6, delete the third row and second column.

$$\begin{bmatrix} -2 & -3 & -1 \\ -7 & 4 & 5 \\ 0 & 6 & 1 \end{bmatrix}$$

The minor for 6 is $\begin{vmatrix} -2 & -1 \\ -7 & 5 \end{vmatrix} = -10 - (7) = -17$.

Try This. Find the minor for 7 in the matrix $\begin{bmatrix} 1 & 2 & 3 \\ 4 & 5 & 6 \\ 7 & 8 & 9 \end{bmatrix}$. ■

The Determinant of a 3 × 3 Matrix

The determinant of a 3 × 3 matrix is defined in terms of minors. Let M_{ij} be the 2 × 2 matrix obtained from M by deleting the ith row and jth column. The determinant of M_{ij}, $|M_{ij}|$, is the minor for a_{ij}.

Definition: Determinant of a 3 × 3 Matrix

If $A = \begin{bmatrix} a_{11} & a_{12} & a_{13} \\ a_{21} & a_{22} & a_{23} \\ a_{31} & a_{32} & a_{33} \end{bmatrix}$, then the determinant of A, $|A|$, is defined as

$$|A| = a_{11}|M_{11}| - a_{21}|M_{21}| + a_{31}|M_{31}|.$$

To find the determinant of A, each entry in the first column of A is multiplied by its minor. This process is referred to as **expansion by minors** about the first column. Note the sign change on the middle term in the expansion.

Example **2** **The determinant of a 3 × 3 matrix**

Find $|A|$, given that $A = \begin{bmatrix} -2 & -3 & -1 \\ -7 & 4 & 5 \\ 0 & 6 & 1 \end{bmatrix}$.

Solution

Use the definition to expand by minors about the first column.

$$|A| = -2 \cdot \begin{vmatrix} 4 & 5 \\ 6 & 1 \end{vmatrix} - (-7) \cdot \begin{vmatrix} -3 & -1 \\ 6 & 1 \end{vmatrix} + 0 \cdot \begin{vmatrix} -3 & -1 \\ 4 & 5 \end{vmatrix}$$

$$= -2(-26) + 7(3) + 0(-11)$$

$$= 73$$

Try This. Find $|A|$ if $A = \begin{bmatrix} 1 & 2 & 3 \\ 4 & 5 & 6 \\ 7 & 8 & 9 \end{bmatrix}$. ■

The value of the determinant of a 3 × 3 matrix can be found by expansion by minors about any row or column. However, you must use alternating plus and minus signs to precede the coefficients of the minors according to the following **sign array**:

$$\begin{bmatrix} + & - & + \\ - & + & - \\ + & - & + \end{bmatrix}$$

The signs in the sign array are used for the determinant of *any* 3 × 3 matrix and they are independent of the signs of the entries in the matrix. Notice that in Example 2, when we expanded by minors about the first column, we used the signs "+ − +" from the first column of the sign array. These signs were used in addition to the signs that appear on the entries themselves (-2, -7, and 0).

Example **3** **Expansion by minors about the second column**

Expand by minors using the second column to find $|A|$, given that

$$A = \begin{bmatrix} -2 & -3 & -1 \\ -7 & 4 & 5 \\ 0 & 6 & 1 \end{bmatrix}.$$

Solution

Use the signs "$- + -$" from the second column of the sign array:

$$\begin{bmatrix} + & - & + \\ - & + & - \\ + & - & + \end{bmatrix}$$

The coefficients $-3, 4$, and 6 from the second column of A are preceded by the signs from the second column of the sign array:

From the sign array

$$|A| = -(-3) \cdot \begin{vmatrix} -7 & 5 \\ 0 & 1 \end{vmatrix} + (4) \cdot \begin{vmatrix} -2 & -1 \\ 0 & 1 \end{vmatrix} - (6) \cdot \begin{vmatrix} -2 & -1 \\ -7 & 5 \end{vmatrix}$$

From second column of A

$$= 3(-7) + 4(-2) - 6(-17)$$

$$= 73$$

To check, define matrix A on a calculator and find the determinant, as in Fig. 9.10.

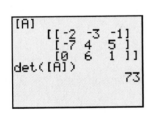

```
[A]
   [[-2  -3  -1]
    [-7   4   5]
    [ 0   6   1]]
det([A])
                73
```

■ **Figure 9.10**

Try This. Expand by minors using the second column to find $|A|$ if

$$A = \begin{bmatrix} 1 & 2 & 3 \\ 5 & -7 & 4 \\ -6 & 0 & 8 \end{bmatrix}.$$ ■

In Examples 2 and 3 we got the same value for $|A|$ by using two different expansions. Expanding about any row or column gives the same result, but the computations can be easier if we examine the matrix and choose the row or column that contains the most zeros. Using zeros for the coefficients of one or more minors simplifies the work, because we do not have to evaluate the minors that are multiplied by zero. If a row or column of a matrix contains all zeros, then the determinant of the matrix is 0.

Example **4** **Expansion by minors using the simplest row or column**

Find $|B|$, given that $B = \begin{bmatrix} 4 & 2 & 1 \\ -6 & 3 & 5 \\ 0 & 0 & -7 \end{bmatrix}.$

Solution

Since the third row has two zeros, we expand by minors about the third row. Use the signs "+ − +" from the third row of the sign array and the coefficients $0, 0,$ and -7 from the third row of B:

$$|B| = 0 \cdot \begin{vmatrix} 2 & 1 \\ 3 & 5 \end{vmatrix} - 0 \cdot \begin{vmatrix} 4 & 1 \\ -6 & 5 \end{vmatrix} + (-7) \cdot \begin{vmatrix} 4 & 2 \\ -6 & 3 \end{vmatrix}$$

$$= -7(24)$$

$$= -168$$

Try This. Find $|A|$ if $A = \begin{bmatrix} 3 & 5 & 2 \\ -1 & 7 & 0 \\ 4 & 9 & 0 \end{bmatrix}.$ ■

Determinant of a 4 × 4 Matrix

The determinant of a 4 × 4 matrix is also found by expanding by minors about a row or column. The following 4 × 4 sign array of alternating + and − signs (starting with + in the upper-left position) is used for the signs in the expansion:

$$\begin{bmatrix} + & - & + & - \\ - & + & - & + \\ + & - & + & - \\ - & + & - & + \end{bmatrix}$$

The minor for an entry of a 4 × 4 matrix is the determinant of the 3 × 3 matrix found by deleting the row and column of that entry. In general, the determinant of an $n \times n$ matrix is defined in terms of determinants of $(n-1) \times (n-1)$ matrices in the same manner.

Example **5** Determinant of a 4 × 4 matrix

Find $|A|$, given that $A = \begin{bmatrix} -2 & -3 & 0 & 4 \\ 1 & -6 & 1 & -1 \\ 2 & 0 & 1 & 5 \\ 4 & 0 & 3 & 1 \end{bmatrix}.$

Solution

Since the second column has two zeros, we expand by minors about the second column, using the signs "− + − +" from the second column of the sign array:

$$|A| = -(-3)\begin{vmatrix} 1 & 1 & -1 \\ 2 & 1 & 5 \\ 4 & 3 & 1 \end{vmatrix} + (-6)\begin{vmatrix} -2 & 0 & 4 \\ 2 & 1 & 5 \\ 4 & 3 & 1 \end{vmatrix}$$

$$- 0\begin{vmatrix} -2 & 0 & 4 \\ 1 & 1 & -1 \\ 4 & 3 & 1 \end{vmatrix} + 0\begin{vmatrix} -2 & 0 & 4 \\ 1 & 1 & -1 \\ 2 & 1 & 5 \end{vmatrix}$$

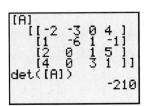

■ **Figure 9.11**

Evaluate the determinant of the first two 3×3 matrices to get

$$|A| = 3(2) - 6(36) = -210.$$

To check, enter A into your calculator and find its determinant, as in Fig. 9.11.

Try This. Find $|A|$ if $A = \begin{bmatrix} 1 & 6 & 0 & 0 \\ 0 & 1 & 7 & 9 \\ 0 & 0 & 1 & 0 \\ 0 & 0 & 0 & 1 \end{bmatrix}$. ■

Cramer's Rule for Systems in Three Variables

Cramer's rule for solving a system of three linear equations in three variables consists of formulas for finding x, y, and z in terms of determinants.

Cramer's Rule for Systems in Three Variables

For the system of equations

$$a_1x + b_1y + c_1z = d_1$$
$$a_2x + b_2y + c_2z = d_2$$
$$a_3x + b_3y + c_3z = d_3$$

let

$$D = \begin{vmatrix} a_1 & b_1 & c_1 \\ a_2 & b_2 & c_2 \\ a_3 & b_3 & c_3 \end{vmatrix}, \qquad D_x = \begin{vmatrix} d_1 & b_1 & c_1 \\ d_2 & b_2 & c_2 \\ d_3 & b_3 & c_3 \end{vmatrix}$$

$$D_y = \begin{vmatrix} a_1 & d_1 & c_1 \\ a_2 & d_2 & c_2 \\ a_3 & d_3 & c_3 \end{vmatrix}, \qquad \text{and} \qquad D_z = \begin{vmatrix} a_1 & b_1 & d_1 \\ a_2 & b_2 & d_2 \\ a_3 & b_3 & d_3 \end{vmatrix}.$$

If $D \neq 0$, then the solution to the system is given by

$$x = \frac{D_x}{D}, \qquad y = \frac{D_y}{D}, \qquad \text{and} \qquad z = \frac{D_z}{D}.$$

The development of this rule is similar to the development for Cramer's rule in two variables in Section 9.5, and so we will omit it.

Note that D_x, D_y, and D_z are obtained by replacing, respectively, the first, second, and third columns of D by the constants d_1, d_2, and d_3.

Example **6** Solving a system using Cramer's rule

Use Cramer's rule to solve the system

$$x + y + z = 0$$
$$2x - y + z = -1$$
$$-x + 3y - z = -8.$$

Solution

To use Cramer's rule, we first evaluate D, D_x, D_y, and D_z:

$$D = \begin{vmatrix} 1 & 1 & 1 \\ 2 & -1 & 1 \\ -1 & 3 & -1 \end{vmatrix} = 1 \cdot \begin{vmatrix} -1 & 1 \\ 3 & -1 \end{vmatrix} - 2 \cdot \begin{vmatrix} 1 & 1 \\ 3 & -1 \end{vmatrix} + (-1) \cdot \begin{vmatrix} 1 & 1 \\ -1 & 1 \end{vmatrix}$$

$$= 1(-2) - 2(-4) - 1(2)$$

$$= 4$$

To find D_x, D_y, or D_z, expand by minors about the first row because the first row contains a zero in each case:

$$D_x = \begin{vmatrix} 0 & 1 & 1 \\ -1 & -1 & 1 \\ -8 & 3 & -1 \end{vmatrix} = 0 \cdot \begin{vmatrix} -1 & 1 \\ 3 & -1 \end{vmatrix} - (1) \cdot \begin{vmatrix} -1 & 1 \\ -8 & -1 \end{vmatrix} + (1) \cdot \begin{vmatrix} -1 & -1 \\ -8 & 3 \end{vmatrix}$$

$$= -1(9) + 1(-11)$$

$$= -20$$

$$D_y = \begin{vmatrix} 1 & 0 & 1 \\ 2 & -1 & 1 \\ -1 & -8 & -1 \end{vmatrix} = 1 \cdot \begin{vmatrix} -1 & 1 \\ -8 & -1 \end{vmatrix} - (0) \cdot \begin{vmatrix} 2 & 1 \\ -1 & -1 \end{vmatrix} + (1) \cdot \begin{vmatrix} 2 & -1 \\ -1 & -8 \end{vmatrix}$$

$$= 1(9) + 1(-17)$$

$$= -8$$

$$D_z = \begin{vmatrix} 1 & 1 & 0 \\ 2 & -1 & -1 \\ -1 & 3 & -8 \end{vmatrix} = 1 \cdot \begin{vmatrix} -1 & -1 \\ 3 & -8 \end{vmatrix} - (1) \cdot \begin{vmatrix} 2 & -1 \\ -1 & -8 \end{vmatrix} + (0) \cdot \begin{vmatrix} 2 & -1 \\ -1 & 3 \end{vmatrix}$$

$$= 1(11) - 1(-17)$$

$$= 28$$

Now, by Cramer's rule,

$$x = \frac{D_x}{D} = \frac{-20}{4} = -5, \quad y = \frac{D_y}{D} = \frac{-8}{4} = -2, \quad \text{and} \quad z = \frac{D_z}{D} = \frac{28}{4} = 7.$$

Check that the ordered triple $(-5, -2, 7)$ satisfies all three equations. The solution set to the system is $\{(-5, -2, 7)\}$.

Try This. Use Cramer's rule to solve $\begin{aligned} x + y - z &= 0 \\ -x + y \quad &= 4 \\ -x + y + z &= 10 \end{aligned}$. ■

Cramer's rule can provide the solution to any system of three linear equations in three variables that has a unique solution. Its advantage is that it can give the value of any one of the variables without having to solve for the others. If $D = 0$, then Cramer's rule does not give the solution to the system, but it does indicate that the system is either dependent or inconsistent. If $D = 0$, then we must use another method to complete the solution to the system.

Example **7** **Solving a system with $D = 0$**

Use Cramer's rule to solve the system.

$$(1) \quad 2x + y - z = 3$$

$$(2) \quad 4x + 2y - 2z = 6$$

$$(3) \quad 6x + 3y - 3z = 9$$

Solution

To use Cramer's rule, we first evaluate D:

$$D = \begin{vmatrix} 2 & 1 & -1 \\ 4 & 2 & -2 \\ 6 & 3 & -3 \end{vmatrix} = 2 \cdot \begin{vmatrix} 2 & -2 \\ 3 & -3 \end{vmatrix} - 4 \cdot \begin{vmatrix} 1 & -1 \\ 3 & -3 \end{vmatrix} + 6 \cdot \begin{vmatrix} 1 & -1 \\ 2 & -2 \end{vmatrix}$$

$$= 2(0) - 4(0) + 6(0)$$

$$= 0$$

Because $D = 0$, Cramer's rule cannot be used to solve the system. We could use the Gaussian elimination method, but note that Eqs. (2) and (3) are obtained by multiplying Eq. (1) by 2 and 3, respectively. Since all three equations are equivalent, the solution set to the system is $\{(x, y, z) \mid 2x + y - z = 3\}$.

Try This. Use Cramer's rule to solve $\begin{aligned} x - y + z &= 2 \\ 2x - 2y + 2z &= 4 \\ 3x - 3y + 3z &= 6 \end{aligned}$ if possible.   ■

In the last two chapters we have discussed several different methods for solving systems of linear equations. Studying different methods increases our understanding of systems of equations. A small system can usually be solved by any of these methods, but for large systems that are solved with computers, the most efficient and popular method is probably the Gaussian elimination method or a variation of it. Since you can find determinants, invert matrices, and perform operations with them on a graphing calculator, you can use either Cramer's rule or inverse matrices with a graphing calculator.

For Thought

True or False? Explain.

Statements 1–5 reference the matrix $A = \begin{bmatrix} 2 & -3 & 1 \\ 3 & 4 & 2 \\ 0 & 0 & 1 \end{bmatrix}$.

1. The sign array is used to determine whether $|A|$ is positive or negative.

2. $|A| = 2 \cdot \begin{vmatrix} 4 & 2 \\ 0 & 1 \end{vmatrix} - (-3) \cdot \begin{vmatrix} 3 & 2 \\ 0 & 1 \end{vmatrix} + 1 \cdot \begin{vmatrix} 2 & -3 \\ 3 & 4 \end{vmatrix}$

3. We can find $|A|$ by expanding about any row or column.

4. $|A| = \begin{vmatrix} 2 & -3 \\ 3 & 4 \end{vmatrix}$

5. We can find $|A|$ by expanding by minors about the diagonal.

6. A minor is a 2×2 matrix.

7. By Cramer's rule, the value of x is D/D_x.

8. If a matrix has a row in which all entries are zero, then the determinant of the matrix is 0.

9. By Cramer's rule, there is no solution to a system for which $D = 0$.

10. Cramer's rule works on any system of nonlinear equations.

9.6 Exercises

Find the indicated minors, using the matrix $\begin{bmatrix} 2 & -3 & 1 \\ 4 & 5 & -6 \\ 7 & 9 & -8 \end{bmatrix}$.

1. Minor for 2   **2.** Minor for -3 **3.** Minor for 1

4. Minor for 4   **5.** Minor for 5   **6.** Minor for -6

7. Minor for 9   **8.** Minor for -8

Find the determinant of each 3×3 *matrix, using expansion by minors about the first column.*

9. $\begin{bmatrix} 1 & -4 & 0 \\ -3 & 1 & -2 \\ 3 & -1 & 5 \end{bmatrix}$ **10.** $\begin{bmatrix} 1 & -3 & 2 \\ 3 & 1 & -4 \\ 2 & 3 & 6 \end{bmatrix}$

11. $\begin{bmatrix} 3 & -1 & 2 \\ 0 & 4 & -1 \\ 5 & 1 & -2 \end{bmatrix}$ **12.** $\begin{bmatrix} -1 & 3 & -1 \\ 0 & 2 & -3 \\ 2 & 6 & -9 \end{bmatrix}$

13. $\begin{bmatrix} -2 & 5 & 1 \\ -3 & 0 & -1 \\ 0 & 2 & -7 \end{bmatrix}$ **14.** $\begin{bmatrix} 0 & -6 & 2 \\ -1 & 4 & -2 \\ 5 & 3 & -1 \end{bmatrix}$

15. $\begin{bmatrix} 0.1 & 30 & 1 \\ 0.4 & 20 & 6 \\ 0.7 & 90 & 8 \end{bmatrix}$ **16.** $\begin{bmatrix} 3 & 0.3 & 10 \\ 5 & 0.5 & 30 \\ 8 & 0.1 & 80 \end{bmatrix}$

Evaluate the following determinants, using expansion by minors about the row or column of your choice.

17. $\begin{vmatrix} -1 & 3 & 5 \\ -2 & 0 & 0 \\ 4 & 3 & -4 \end{vmatrix}$ **18.** $\begin{vmatrix} 8 & -9 & 1 \\ 3 & 4 & 0 \\ -2 & 1 & 0 \end{vmatrix}$

19. $\begin{vmatrix} 1 & 1 & 1 \\ 2 & 2 & 2 \\ 4 & 4 & 4 \end{vmatrix}$ **20.** $\begin{vmatrix} 4 & -1 & 3 \\ 4 & -1 & 3 \\ 4 & -1 & 3 \end{vmatrix}$

21. $\begin{vmatrix} 0 & -1 & 0 \\ 3 & 4 & 6 \\ -2 & 3 & -5 \end{vmatrix}$ **22.** $\begin{vmatrix} 2 & 0 & 0 \\ 56 & 3 & -4 \\ 88 & 5 & -2 \end{vmatrix}$

23. $\begin{vmatrix} 2 & 0 & 1 \\ 4 & 0 & 6 \\ -7 & 9 & -8 \end{vmatrix}$ **24.** $\begin{vmatrix} 2 & -3 & 1 \\ -2 & 5 & -6 \\ 0 & 0 & 0 \end{vmatrix}$

25. $\begin{vmatrix} 3 & 0 & 1 & 5 \\ 2 & -3 & 2 & 0 \\ -2 & 3 & 1 & 2 \\ 2 & -4 & 1 & 3 \end{vmatrix}$ **26.** $\begin{vmatrix} 1 & -4 & 2 & 0 \\ -2 & -1 & 0 & -3 \\ 2 & 2 & 4 & 1 \\ 3 & 0 & -3 & 1 \end{vmatrix}$

27. $\begin{vmatrix} 2 & -3 & 4 & 6 \\ 1 & -5 & 0 & 0 \\ 1 & 3 & 1 & -3 \\ -2 & 0 & 2 & 1 \end{vmatrix}$ **28.** $\begin{vmatrix} -2 & 4 & 0 & 5 \\ 2 & -1 & 0 & 7 \\ 3 & 2 & 0 & -1 \\ 2 & 2 & -3 & 4 \end{vmatrix}$

Solve each system, using Cramer's rule where possible.

29. $\begin{aligned} x + y + z &= 6 \\ x - y + z &= 2 \\ 2x + y + z &= 7 \end{aligned}$ **30.** $\begin{aligned} 2x - 2y + 3z &= 7 \\ x + y - z &= -2 \\ 3x + y - 2z &= 5 \end{aligned}$

31. $\begin{aligned} x + 2y \phantom{{}+ z} &= 8 \\ x - 3y + z &= -2 \\ 2x - y \phantom{{}+ z} &= 1 \end{aligned}$ **32.** $\begin{aligned} 2x + y \phantom{{}+ 3z} &= -4 \\ 3y - z &= -1 \\ x \phantom{{}+ 3y} + 3z &= -16 \end{aligned}$

33. $\begin{aligned} 2x - 3y + z &= 1 \\ x + 4y - z &= 0 \\ 3x - y + 2z &= 0 \end{aligned}$ **34.** $\begin{aligned} -2x + y - z &= 0 \\ x - y + 3z &= 1 \\ 3x + 3y + 2z &= 0 \end{aligned}$

35. $\begin{aligned} x + y + z &= 2 \\ 2x - y + 3z &= 0 \\ 3x + y - z &= 0 \end{aligned}$ **36.** $\begin{aligned} x - 2y - z &= 0 \\ -x + y + 3z &= 0 \\ x + 3y + z &= 3 \end{aligned}$

37. $\begin{aligned} x + y - 2z &= 1 \\ x - 2y + z &= 2 \\ 2x - y - z &= 3 \end{aligned}$ **38.** $\begin{aligned} x + y + z &= 4 \\ -2x - y + 3z &= 1 \\ y + 5z &= 9 \end{aligned}$

39.
$$x - y + z = 5$$
$$x + 2y + 3z = 8$$
$$2x - 2y + 2z = 16$$

40.
$$3x + 6y + 9z = 12$$
$$x + 2y + 3z = 0$$
$$x - y - 3z = 0$$

46.
$$3.6x + 4.5y + 6.8z = 45{,}300$$
$$0.09x + 0.05y + 0.04z = 474$$
$$x + y - z = 0$$

Solve each problem, using a system of three equations in three unknowns and Cramer's rule.

41. *Age Disclosure*   Jackie, Rochelle, and Alisha will not disclose their ages. However, the average of the ages of Jackie and Rochelle is 33, the average for Rochelle and Alisha is 25, and the average for Jackie and Alisha is 19. How old is each?

42. *Bennie's Coins*   Bennie emptied his pocket of 49 coins to pay for his $5.50 lunch. He used only nickels, dimes, and quarters, and the total number of dimes and quarters was one more than the number of nickels. How many of each type of coin did he use?

43. *What a Difference a Weight Makes*   A sociology professor gave two one-hour exams and a final exam. Ian was distressed with his average score of 60 for the three tests and went to see the professor. Because of Ian's improvement during the semester, the professor offered to count the final exam as 60% of the grade and the two tests equally, giving Ian a weighted average of 76. Ian countered that since he improved steadily during the semester, the first test should count 10%, the second 20%, and the final 70% of the grade, giving a weighted average of 83. What were Ian's actual scores on the two tests and the final exam?

44. *Cookie Time*   Cheryl, of Cheryl's Famous Cookies, set out 18 cups of flour, 14 cups of sugar, and 13 cups of shortening for her employees to use to make some batches of chocolate chip, oatmeal, and peanut butter cookies. She left for the day without telling them how many batches of each type to bake. The table gives the number of cups of each ingredient required for one batch of each type of cookie. How many batches of each were they supposed to bake?

■ **Table for Exercise 44**

	Flour	Sugar	Shortening	
Chocolate chip	2	2	1	
Oatmeal	1	1	2	
Peanut butter	4	2	1	

Use the determinant feature of a graphing calculator to solve each system by Cramer's rule.

45.
$$0.2x - 0.3y + 1.2z = 13.11$$
$$0.25x + 0.35y - 0.9z = -1.575$$
$$2.4x - y + 1.25z = 42.02$$

Solve each problem, using Cramer's rule and a graphing calculator.

47. *Gasoline Sales*   The Runway Deli sells regular unleaded, plus unleaded, and supreme unleaded Shell gasoline. The number of gallons of each grade and the total receipts for gasoline are shown in the table for the first three weeks of February. What was the price per gallon for each grade?

■ **Table for Exercise 47**

	Regular	Plus	Supreme	Receipts
Week 1	1270	980	890	$5924.86
Week 2	1450	1280	1050	$7138.22
Week 3	1340	1190	1060	$6789.41

48. *Gasoline Sales*   John's Curb Market sells regular, plus, and supreme unleaded Citgo gasoline. John sold the same amount of each grade of gasoline each week for the first three weeks of February. The prices that he charged and the receipts for each week are shown in the accompanying table. How many gallons of each grade did he sell each week?

■ **Table for Exercise 48**

	Regular	Plus	Supreme	Receipts
Week 1	$1.799	$1.899	$1.999	$8736.37
Week 2	$1.749	$1.849	$1.949	$8504.87
Week 3	$1.759	$1.899	$1.949	$8596.07

Extend Cramer's rule to four linear equations in four unknowns. Solve each system, using the extended Cramer's rule.

49.
$$w + x + y + z = 4$$
$$2w - x + y + 3z = 13$$
$$w + 2x - y + 2z = -2$$
$$w - x - y + 4z = 8$$

50.
$$2w + 2x - 2y + z = 11$$
$$w + x + y + z = 10$$
$$4w - 3x + 2y - 5z = 6$$
$$w + 3x - y + 9z = 20$$

The equation of a line through two points can be expressed as an equation involving a determinant.

51. Show that the following equation is equivalent to the equation of the line through $(3, -5)$ and $(-2, 6)$.

$$\begin{vmatrix} x & y & 1 \\ 3 & -5 & 1 \\ -2 & 6 & 1 \end{vmatrix} = 0$$

52. Show that the following equation is equivalent to the equation of the line through (x_1, y_1) and (x_2, y_2).

$$\begin{vmatrix} x & y & 1 \\ x_1 & y_1 & 1 \\ x_2 & y_2 & 1 \end{vmatrix} = 0$$

For Writing/Discussion

Prove each of the following statements for any 3×3 matrix A.

53. If all entries in any row or column of A are zero, then $|A| = 0$.

54. If A has two identical rows (or columns), then $|A| = 0$.

55. If all entries in a row (or column) of A are multiplied by a constant k, then the determinant of the new matrix is $k \cdot |A|$.

56. If two rows (or columns) of A are interchanged, then the determinant of the new matrix is $-|A|$.

Thinking Outside the Box LXXXIII

The Missing Dollar Three students each pay $10 for a $30 room at the Magnolia Inn. After the students are in their room, the night clerk realizes that the student rate is actually only $25. She gives the bellboy five singles to give back to the students. The bellboy can't decide how to split five singles among three students, so he gives each student one dollar and keeps the other two for himself. Now the students have paid $9 apiece and the bellboy has $2. Three times 9 is 27 plus 2 is 29. So where is the other dollar? Explain.

9.6 Pop Quiz

1. Find the determinant of $\begin{bmatrix} 1 & 2 & 3 \\ 2 & 1 & 1 \\ 3 & 3 & 4 \end{bmatrix}$.

2. Solve the system using Cramer's rule.

$$\begin{aligned} x + y - z &= 1 \\ x - y + 2z &= 9 \\ 2x + y + z &= 12 \end{aligned}$$

3. Is $\begin{bmatrix} 1 & 2 & 3 \\ 0 & 2 & 6 \\ 0 & 5 & 15 \end{bmatrix}$ invertible?

■ ■ ■ Highlights

9.1 Solving Linear Systems Using Matrices

Matrix	An $m \times n$ matrix is a rectangular array of numbers with m rows and n columns.	$\begin{bmatrix} 1 & 2 \\ 3 & 4 \end{bmatrix}$
Augmented Matrix	The coefficients of the variables together with the constants from a linear system	$\begin{aligned} x + 2y &= 4 \\ 3x - y &= 5 \end{aligned}$ $\begin{bmatrix} 1 & 2 & \vert & 4 \\ 3 & -1 & \vert & 5 \end{bmatrix}$

Gaussian Elimination	Row operations are used to simplify the augmented matrix.	$\begin{bmatrix} 1 & 0 & \vert & 2 \\ 0 & 1 & \vert & 1 \end{bmatrix}$ $x = 2, y = 1$

9.2 Operations with Matrices

Equal Matrices	The same size and all corresponding entries are equal.	$\begin{bmatrix} 1 & 2 \\ 3 & 4 \end{bmatrix} = \begin{bmatrix} 1 & 2 \\ 3 & 4 \end{bmatrix}$
Addition or Subtraction	Add or subtract the corresponding entries for matrices of the same size.	$\begin{bmatrix} 1 & 2 \\ 3 & 4 \end{bmatrix} + \begin{bmatrix} 2 & 3 \\ 4 & 5 \end{bmatrix} = \begin{bmatrix} 3 & 5 \\ 7 & 9 \end{bmatrix}$
Scalar Multiple	To find the product of a scalar and a matrix multiply each entry by the scalar.	$3\begin{bmatrix} 1 & 2 \\ 3 & 4 \end{bmatrix} = \begin{bmatrix} 3 & 6 \\ 9 & 12 \end{bmatrix}$

9.3 Multiplication of Matrices

Multiplication	The ijth entry of AB is the sum of the products of the corresponding entries in the ith row of A and the jth column of B.	$\begin{bmatrix} 1 & 0 \\ 2 & 1 \end{bmatrix}\begin{bmatrix} 0 & 1 \\ 3 & 4 \end{bmatrix} = \begin{bmatrix} 0 & 1 \\ 3 & 6 \end{bmatrix}$

9.4 Inverses of Matrices

Identity Matrix	An $n \times n$ matrix with ones on the diagonal and zeros elsewhere	$I = \begin{bmatrix} 1 & 0 \\ 0 & 1 \end{bmatrix}$
Inverse Matrix	The inverse of A is a matrix A^{-1} such that $AA^{-1} = I$.	$\begin{bmatrix} 1 & 1 \\ 5 & 4 \end{bmatrix}\begin{bmatrix} -4 & 1 \\ 5 & -1 \end{bmatrix} = \begin{bmatrix} 1 & 0 \\ 0 & 1 \end{bmatrix}$
Solution to Systems	The solution to $AX = B$ is $X = A^{-1}B$.	$\begin{bmatrix} -4 & 1 \\ 5 & -1 \end{bmatrix}\begin{bmatrix} x \\ y \end{bmatrix} = \begin{bmatrix} 5 \\ 6 \end{bmatrix}$ $\begin{bmatrix} x \\ y \end{bmatrix} = \begin{bmatrix} 1 & 1 \\ 5 & 4 \end{bmatrix}\begin{bmatrix} 5 \\ 6 \end{bmatrix} = \begin{bmatrix} 11 \\ 49 \end{bmatrix}$

9.5 Solution of Linear Systems in Two Variables Using Determinants

Determinant	$\begin{vmatrix} a & b \\ c & d \end{vmatrix} = ad - bc$	$\begin{vmatrix} 1 & 2 \\ 3 & -1 \end{vmatrix} = -1 - 6 = -7$ $\begin{vmatrix} 4 & 2 \\ 5 & -1 \end{vmatrix} = -14, \begin{vmatrix} 1 & 4 \\ 3 & 5 \end{vmatrix} = -7$

Cramer's Rule   If $\dfrac{a_1x_1 + b_1y_1 = c_1}{a_2x_2 + b_2y_2 = c_2}$ is an independent system,

$$x + 2y = 4$$
$$3x - y = 5$$

then $x = \dfrac{\begin{vmatrix} c_1 & b_1 \\ c_2 & b_2 \end{vmatrix}}{\begin{vmatrix} a_1 & b_1 \\ a_2 & b_2 \end{vmatrix}}$ and $y = \dfrac{\begin{vmatrix} a_1 & c_1 \\ a_2 & c_2 \end{vmatrix}}{\begin{vmatrix} a_1 & b_1 \\ a_2 & b_2 \end{vmatrix}}.$

$$x = \frac{-14}{-7} = 2, \quad y = \frac{-7}{-7} = 1$$

9.6 Solution of Linear Systems in Three Variables Using Determinants

Cramer's Rule   If $\begin{array}{l} a_1x_1 + b_1y_1 + c_1z_1 = d_1 \\ a_2x_2 + b_2y_2 + c_2z_2 = d_2 \\ a_3x_3 + b_3y_3 + c_3z_3 = d_3 \end{array}$ is an independent system and

$$D = \begin{vmatrix} a_1 & b_1 & c_1 \\ a_2 & b_2 & c_2 \\ a_3 & b_3 & c_3 \end{vmatrix}, \quad D_x = \begin{vmatrix} d_1 & b_1 & c_1 \\ d_2 & b_2 & c_2 \\ d_3 & b_3 & c_3 \end{vmatrix}, \quad D_y = \begin{vmatrix} a_1 & d_1 & c_1 \\ a_2 & d_2 & c_2 \\ a_3 & d_3 & c_3 \end{vmatrix}, \quad \text{and} \quad D_z = \begin{vmatrix} a_1 & b_1 & d_1 \\ a_2 & b_2 & d_2 \\ a_3 & b_3 & d_3 \end{vmatrix},$$

then $x = D_x/D, \quad y = D_y/D, \quad$ and $\quad z = D_z/D.$

▪ ▪ ▪ ▪ Chapter 9 Review Exercises

Let $A = \begin{bmatrix} 2 & -3 \\ -2 & 4 \end{bmatrix}, B = \begin{bmatrix} 3 & 7 \\ 1 & 2 \end{bmatrix}, C = \begin{bmatrix} -1 \\ 3 \end{bmatrix},$

$D = \begin{bmatrix} 5 \\ -3 \end{bmatrix}, E = \begin{bmatrix} 1 \\ -4 \\ 3 \end{bmatrix}, F = [3 \quad 2 \quad -1],$ and

$G = \begin{bmatrix} -1 & 0 & 0 \\ 1 & 1 & 0 \\ -2 & 3 & 1 \end{bmatrix}.$ Find each of the following matrices or

determinants if possible.

1. $A + B$ **2.** $A - B$ **3.** $2A - B$ **4.** $2A + 3B$

5. AB **6.** BA **7.** $D + E$ **8.** $F + G$

9. AC **10.** BD **11.** EF **12.** FE

13. FG **14.** GE **15.** GF **16.** EG

17. A^{-1} **18.** B^{-1} **19.** G^{-1} **20.** $A^{-1}C$

21. $(AB)^{-1}$ **22.** $A^{-1}B^{-1}$ **23.** AA^{-1} **24.** GG^{-1}

25. $|A|$ **26.** $|B|$ **27.** $|G|$ **28.** $|C|$

Solve each of the following systems by all three methods: Gaussian elimination, matrix inversion, and Cramer's rule.

29. $x + y = 9$
   $2x - y = 1$

30. $x - 2y = 3$
   $x + 2y = 2$

31. $2x + y = -1$
   $3x + 2y = 0$

32. $3x - y = 1$
   $-2x + y = 1$

33. $x - 5y = 9$
   $-2x + 10y = -18$

34. $3x - y = 4$
   $6x - 2y = 6$

35. $0.05x + 0.1y = 1$
   $10x + 20y = 20$

36. $0.04x - 0.2y = 3$
   $2x - 10y = 150$

37. $x + y - 2z = -3$
   $-x + 2y - z = 0$
   $-x - y + 3z = 6$

38. $x - y + z = 5$
   $x + y + 3z = 11$
   $-x + 2y - z = -5$

39. $y - 3z = 1$
   $x + 2y = 5$
   $x + 4z = 1$

40. $3x + y - 2z = 0$
   $-2y + z = 0$
   $y + 3z = 14$

41. $x - y + z = 2$
$x - 2y - z = 1$
$2x - 3y = 3$

42. $x + 2y + z = 1$
$2x + 4y + 2z = 0$
$-x - 2y - z = 2$

43. $x - 3y - z = 2$
$x - 3y - z = 1$
$x - 3y - z = 0$

44. $2x - y - z = 0$
$x + y + z = 3$
$3x = 3$

Find the values of x, y, and z that make each of the equations true.

45. $\begin{bmatrix} x \\ x + y \end{bmatrix} = \begin{bmatrix} 9 \\ -3 \end{bmatrix}$

46. $\begin{bmatrix} x^2 \\ x - y \end{bmatrix} = \begin{bmatrix} 4 \\ 1 \end{bmatrix}$

47. $\begin{bmatrix} 1 & 1 \\ 2 & 1 \end{bmatrix} \begin{bmatrix} x \\ y \end{bmatrix} = \begin{bmatrix} 6 \\ 8 \end{bmatrix}$

48. $\begin{bmatrix} 0 & 0.5 \\ 1 & 1 \end{bmatrix} \begin{bmatrix} x \\ y \end{bmatrix} = \begin{bmatrix} 7 \\ 9 \end{bmatrix}$

49. $\begin{bmatrix} x \\ y \end{bmatrix} + \begin{bmatrix} y \\ -x \end{bmatrix} = \begin{bmatrix} -3 \\ y \end{bmatrix}$

50. $\begin{bmatrix} y \\ x \end{bmatrix} - \begin{bmatrix} x \\ y \end{bmatrix} = \begin{bmatrix} 4 \\ 5 \end{bmatrix}$

51. $\begin{bmatrix} x + y & 0 & 0 \\ 0 & y + z & 0 \\ 0 & 0 & x + z \end{bmatrix} = \begin{bmatrix} 1 & 0 & 0 \\ 0 & 1 & 0 \\ 0 & 0 & 1 \end{bmatrix}$

52. $\begin{bmatrix} 0 & x - y & 2z \\ 0 & 0 & y - z \\ 0 & 0 & 0 \end{bmatrix} = \begin{bmatrix} 0 & 2 & 5 \\ 0 & 0 & 3 \\ 0 & 0 & 0 \end{bmatrix}$

53. $\begin{bmatrix} 1 & 1 & 0 \\ 0 & 1 & 2 \\ 1 & 0 & 3 \end{bmatrix} \begin{bmatrix} x \\ y \\ z \end{bmatrix} = \begin{bmatrix} -1 \\ 7 \\ 17 \end{bmatrix}$

54. $\begin{bmatrix} 1 & 1 & 1 \\ -1 & 1 & -1 \\ 1 & 0 & 2 \end{bmatrix} \begin{bmatrix} x \\ y \\ z \end{bmatrix} = \begin{bmatrix} 0 \\ 0 \\ 0 \end{bmatrix}$

Solve each problem, using a system of linear equations in two or three variables. Use the method of your choice from this chapter.

55. *Fine for Polluting* A small manufacturing plant must pay a fine of $10 for each gallon of pollutant A and $6 for each gallon of pollutant B per day that it discharges into a nearby stream. If the manufacturing process produces three gallons of pollutant A for every four gallons of pollutant B and the daily fine is $4060, then how many gallons of each are discharged each day?

56. *Friends* Ross, Joey, and Chandler spent a total of $216 on coffee and pastry last month at the Central Perk coffee shop. Joey and Ross's expenses totaled only half as much as Chandler's. If Joey spent $12 more than Ross, then how much did each spend?

57. *Utility Bills* Bette's total expense for water, gas, and electricity for one month including tax was $189.83. There is a 6% state tax on electricity, a 5% city tax on gas, and a 4% county tax on water. If her total expenses included $9.83 in taxes and her electric bill including tax was twice the gas bill including tax, then how much was each bill including tax?

58. *Predicting Car Sales* In the first three months of the year, West Coast Cadillac sold 38, 42, and 49 new cars, respectively. Find the equation of the parabola of the form $y = ax^2 + bx + c$ that passes through $(1, 38)$, $(2, 42)$, and $(3, 49)$, as shown in the accompanying figure. (The fact that each point satisfies $y = ax^2 + bx + c$ gives three linear equations in a, b, and c.) Assuming that the fourth month's sales will fall on that same parabola, what would be the predicted sales for the fourth month?

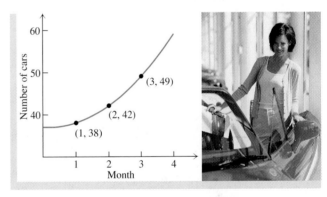

■ **Figure for Exercise 58**

Thinking Outside the Box LXXXIV

Disappearing Dogs Someone opened up the cages at Pet Depot and more than 100 puppies got away! There were exactly 300 puppies to begin with. The Daily Mixup reported: "Of the pups that remained, a third were Dobermans, a quarter were schnauzers, a fifth were beagles, a seventh were poodles, and a ninth were dachshunds. The original number of beagles was three times the number of dachshunds that stayed." The Daily Mixup got just one of the fractions wrong. How many beagles escaped?

■ ■ ■ Chapter 9 Test

Solve each system, using Gaussian elimination.

1. $2x - 3y = 1$
$x + 9y = 4$

2. $2x - y + z = 5$
$x - 2y - z = -2$
$3x - y - z = 6$

3. $x - y - z = 1$
$2x + y - z = 0$
$5x - 2y - 4z = 3$

Let $A = \begin{bmatrix} 1 & -1 \\ -2 & 4 \end{bmatrix}$, $B = \begin{bmatrix} 2 & -3 \\ -4 & 6 \end{bmatrix}$, $C = \begin{bmatrix} -2 \\ 1 \end{bmatrix}$,

$D = \begin{bmatrix} 3 \\ -2 \end{bmatrix}$, $E = \begin{bmatrix} 2 \\ 3 \\ -1 \end{bmatrix}$, $F = [1 \quad 0 \quad -1]$, and

$G = \begin{bmatrix} -2 & 3 & 1 \\ -3 & 1 & 3 \\ 0 & 2 & -1 \end{bmatrix}$. Find each of the following matrices or

determinants if possible.

4. $A + B$ **5.** $2A - B$ **6.** AB **7.** AC

8. CB **9.** FG **10.** EF **11.** A^{-1}

12. G^{-1} **13.** $|A|$ **14.** $|B|$ **15.** $|G|$

Solve each system, using Cramer's rule.

16. $x - y = 2$
    $-2x + 4y = 2$

17. $2x - 3y = 6$
    $-4x + 6y = 1$

18. $-2x + 3y + z = -2$
    $-3x + y + 3z = -4$
    $2y - z = 0$

Solve each system by using inverse matrices.

19. $x - y = 1$
    $-2x + 4y = -8$

20. $-2x + 3y + z = 1$
    $-3x + y + 3z = 0$
    $2y - z = -1$

Solve by using a method from this chapter.

21. The manager of a computer store bought x copies of the program Math Skillbuilder for $10 each at the beginning of the year and sold y copies of the program for $35 each. At the end of the year, the program was obsolete and she destroyed 12 unsold copies. If her net profit for the year was $730, then how many were bought and how many were sold?

22. Find a, b, and c such that the graph of $y = ax^2 + b\sqrt{x} + c$ goes through the points $(0, 3)$, $(1, -1/2)$, and $(4, 3)$.

Tying it all Together

Chapters 1–9

Solve each equation.

1. $2(x + 3) - 5x = 7$ **2.** $\frac{1}{2}\left(x - \frac{1}{3}\right) + \frac{1}{5} = 1$ **3.** $\frac{1}{2}(2x - 2)(6x - 8) = 4$ **4.** $1 - \frac{1}{2}(8x - 4) = 9$

Solve each system of equations by the specified method.

5. Graphing:
    $2x + y = 6$
    $x - 2y = 8$

6. Substitution:
    $y + 2x = 1$
    $2x + 6y = 2$

7. Addition:
    $2x - 0.06y = 20$
    $3x + 0.01y = 20$

8. Gaussian elimination:
    $2x - y = -1$
    $x + 3y = -11$

9. Matrix inversion:
    $3x - 5y = -7$
    $-x + y = 1$

10. Cramer's rule:
    $4x - 3y = 5$
    $3x - 5y = 1$

11. Your choice:
    $3x - 5y = 6$
    $3x - 5y = 1$

12. Your choice:
    $x - 2y = 3$
    $-2x + 4y = -6$

Solve each nonlinear system of equations.

13. $x^2 + y^2 = 25$
    $x - y = -1$

14. $x^2 - y = 1$
    $x + y = 1$

15. $x^2 - y^2 = 1$
    $x^2 + y^2 = 3$

16. $x^2 - y^2 = 1$
    $2x^2 + 3y^2 = 2$

10 The Conic Sections

THE International Space Station is now the brightest star in the night sky. The one-million pound structure is 356 feet across and 290 feet long. The station houses seven people in orbit 250 miles above Earth at an angle of inclination of 51.6° to the equator.

The station is a laboratory in a realm where gravity, temperature, and pressure can be maintained in ways impossible on Earth. Experiments involving advanced industrial materials, communications technology, medical techniques, and more are conducted in the station.

WHAT YOU WILL LEARN   A great deal of mathematics is involved in every aspect of this project. The conic sections, which we study in this chapter, are the basic curves that are used for modeling orbits and making antennae that receive radio signals.

10.1

The Parabola

The parabola, circle, ellipse, and hyperbola can be defined as the four curves that are obtained by intersecting a double right circular cone and a plane, as shown in Fig. 10.1. That is why these curves are known as **conic sections.** If the plane passes through the vertex of the cone, then the intersection of the cone and the plane is called a **degenerate conic.** The conic sections can also be defined as the graphs of certain equations (as we did for the parabola in Section 3.1). However, the useful properties of the curves are not apparent in either of these approaches. In this chapter we give geometric definitions from which we derive equations for the conic sections. This approach allows us to better understand the properties of the conic sections.

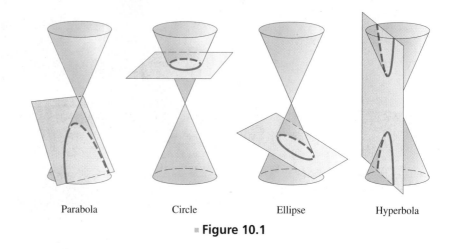

| Parabola | Circle | Ellipse | Hyperbola |

■ **Figure 10.1**

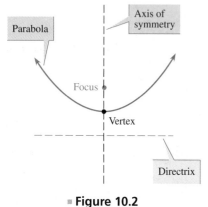

■ **Figure 10.2**

Definition

Previously in Section 3.1, we defined a parabola algebraically as the graph of $y = ax^2 + bx + c$ for $a \neq 0$. This equation can be expressed also in the form $y = a(x - h)^2 + k$. No equation is mentioned in our geometric definition of a parabola.

Definition: Parabola

> A **parabola** is the set of all points in the plane that are equidistant from a fixed line (the **directrix**) and a fixed point not on the line (the **focus**).

Figure 10.2 shows a parabola with its directrix, focus, axis of symmetry, and vertex. In terms of the directrix and focus, the **axis of symmetry** can be described as the line perpendicular to the directrix and containing the focus. The **vertex** is the point on the axis of symmetry that is equidistant from the focus and directrix. If we position the directrix and focus in a coordinate plane with the directrix horizontal, we can find an equation that is satisfied by all points of the parabola.

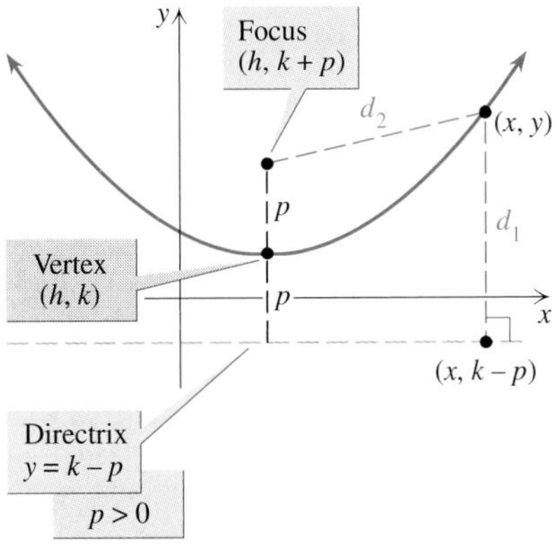

■ **Figure 10.3**

Developing the Equation

Start with a focus and a horizontal directrix, as shown in Fig. 10.3. If we use the coordinates (h, k) for the vertex, then the focus is $(h, k + p)$ and the directrix is $y = k - p$, where p (the **focal length**) is the directed distance from the vertex to the focus. If the focus is above the vertex, then $p > 0$, and if the focus is below the vertex, then $p < 0$. The distance from the vertex to the focus or the vertex to the directrix is $|p|$.

The distance d_1 from an arbitrary point (x, y) on the parabola to the directrix is the distance from (x, y) to $(x, k - p)$, as shown in Fig. 10.3. We use the distance formula from Section 1.3 to find d_1:

$$d_1 = \sqrt{(x - x)^2 + (y - (k - p))^2} = \sqrt{y^2 - 2(k - p)y + (k - p)^2}$$

Now we find the distance d_2 between (x, y) and the focus $(h, k + p)$:

$$d_2 = \sqrt{(x - h)^2 + (y - (k + p))^2} = \sqrt{(x - h)^2 + y^2 - 2(k + p)y + (k + p)^2}$$

Since $d_1 = d_2$ for every point (x, y) on the parabola, we have the following equation.

$$\sqrt{y^2 - 2(k - p)y + (k - p)^2} = \sqrt{(x - h)^2 + y^2 - 2(k + p)y + (k + p)^2}$$

You should verify that squaring each side and simplifying yields

$$y = \frac{1}{4p}(x - h)^2 + k.$$

This equation is of the form $y = a(x - h)^2 + k$, where $a = 1/(4p)$. So the curve determined by the geometric definition has an equation that is an equation of a parabola according to the algebraic definition. We state these results as follows.

Theorem: The Equation of a Parabola

■ **Figure 10.4**

> The equation of a parabola with focus $(h, k + p)$ and directrix $y = k - p$ is
> $$y = a(x - h)^2 + k,$$
> where $a = 1/(4p)$ and (h, k) is the vertex.

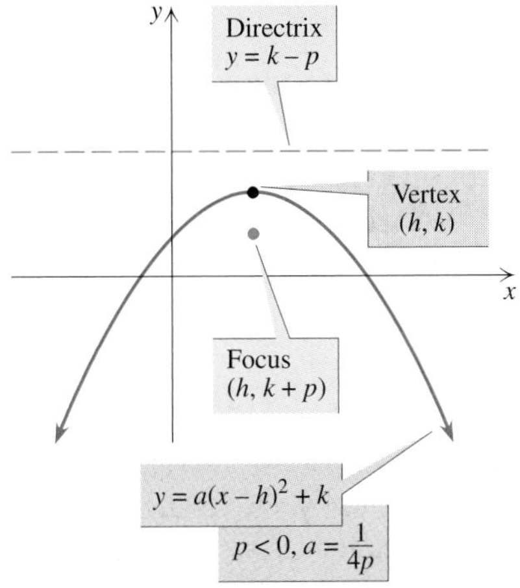

The link between the geometric definition and the equation of a parabola is

$$a = \frac{1}{4p}.$$

For any particular parabola, a and p have the same sign. If they are both positive, the parabola opens upward and the focus is above the directrix. If they are both negative, the parabola opens downward and the focus is below the directrix. Figure 10.4 shows the positions of the focus, directrix, and vertex for a parabola with $p < 0$. Since a is inversely proportional to p, smaller values of $|p|$ correspond to larger values of $|a|$ and to "narrower" parabolas.

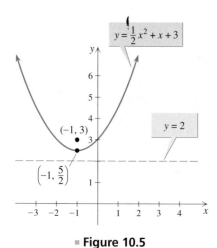

■ Figure 10.5

Example **1** **Writing the equation from the focus and directrix**

Find the equation of the parabola with focus $(-1, 3)$ and directrix $y = 2$.

Solution

The focus is one unit above the directrix, as shown in Fig. 10.5. So $p = 1/2$. Therefore $a = 1/(4p) = 1/2$. Since the vertex is halfway between the focus and directrix, the y-coordinate of the vertex is $(3 + 2)/2$ and the vertex is $(-1, 5/2)$. Use $a = 1/2, h = -1$, and $k = 5/2$ in the formula $y = a(x - h)^2 + k$ to get the equation

$$y = \frac{1}{2}(x - (-1))^2 + \frac{5}{2}.$$

Simplify to get the equation $y = \frac{1}{2}x^2 + x + 3$.

Try This. Find the equation of the parabola with focus $(2, 5)$ and directrix $y = 6$.
■

The Standard Equation of a Parabola

If we start with the standard equation of a parabola, $y = ax^2 + bx + c$, we can identify the vertex, focus, and directrix by rewriting it in the form $y = a(x - h)^2 + k$.

Example **2** **Finding the vertex, focus, and directrix**

Find the vertex, focus, and directrix of the graph of $y = -3x^2 - 6x + 2$.

Solution

Use completing the square to write the equation in the form $y = a(x - h)^2 + k$:

$$y = -3(x^2 + 2x) + 2$$
$$= -3(x^2 + 2x + 1 - 1) + 2$$
$$= -3(x^2 + 2x + 1) + 2 + 3$$
$$= -3(x + 1)^2 + 5$$

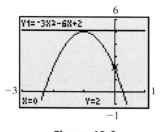

■ Figure 10.6

The vertex is $(-1, 5)$, and the parabola opens downward because $a = -3$. Since $a = 1/(4p)$, we have $1/(4p) = -3$, or $p = -1/12$. Because the parabola opens downward, the focus is $1/12$ unit below the vertex $(-1, 5)$ and the directrix is a horizontal line $1/12$ unit above the vertex. The focus is $(-1, 59/12)$, and the directrix is $y = 61/12$.

The graphs of $y_1 = -3x^2 - 6x + 2$ and $y_2 = 61/12$ in Fig. 10.6 show how close the directrix is to the vertex for this parabola.

Try This. Find the vertex, focus, and directrix for $y = 2x^2 + 8x + 11$. ■

In Section 3.1 we learned that the x-coordinate of the vertex of the parabola $y = ax^2 + bx + c$ is $-b/(2a)$. We can use $x = -b/(2a)$ and $a = 1/(4p)$ to determine the focus and directrix without completing the square.

Example **3** Finding the vertex, focus, and directrix

Find the vertex, focus, and directrix of the parabola $y = 2x^2 + 6x - 7$ without completing the square, and determine whether the parabola opens upward or downward.

Solution

First use $x = -b/(2a)$ to find the x-coordinate of the vertex:

$$x = \frac{-b}{2a} = \frac{-6}{2 \cdot 2} = -\frac{3}{2}$$

To find the y-coordinate of the vertex, let $x = -3/2$ in $y = 2x^2 + 6x - 7$:

$$y = 2\left(-\frac{3}{2}\right)^2 + 6\left(-\frac{3}{2}\right) - 7 = \frac{9}{2} - 9 - 7 = -\frac{23}{2}$$

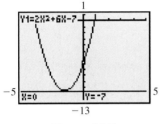

■ **Figure 10.7**

The vertex is $(-3/2, -23/2)$. Since $a = 2$, the parabola opens upward. Use $2 = 1/(4p)$ to get $p = 1/8$. Since the parabola opens upward, the directrix is $1/8$ unit below the vertex and the focus is $1/8$ unit above the vertex. The directrix is $y = -93/8$, and the focus is $(-3/2, -91/8)$.

You can graph $y_1 = -93/8$ and $y_2 = 2x^2 + 6x - 7$, as shown in Fig. 10.7, to check the position of the parabola and its directrix.

Try This. Find the vertex, focus, and directrix for $y = x^2 + 4x - 1$ without completing the square. ■

Graphing a Parabola

According to the geometric definition of a parabola, every point on the parabola is equidistant from its focus and directrix. However, it is not easy to find points satisfying that condition. The German mathematician Johannes Kepler (1571–1630) devised a method for drawing a parabola with a given focus and directrix: A piece of string, with length equal to the length of the T-square, is attached to the end of the T-square and the focus, as shown in Fig. 10.8. A pencil is moved down the edge of

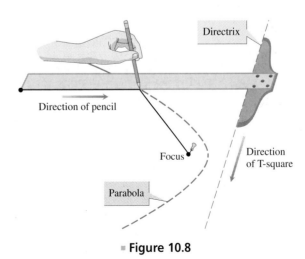

■ **Figure 10.8**

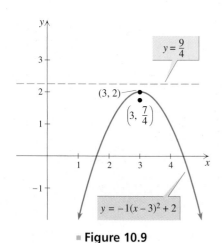

the T-square, holding the string against it, while the T-square is moved along the directrix toward the focus. While the pencil is aligning the string with the edge of the T-square, it remains equidistant from the focus and directrix.

Although Kepler's method does give the graph of the parabola from the focus and directrix, the most accurate graphs are now drawn with the equation and a computer. In the next example we start with the focus and directrix, find the equation, and then draw the graph.

Example **4** **Graphing a parabola given its focus and directrix**

Find the vertex, axis of symmetry, x-intercepts, and y-intercept of the parabola that has focus $(3, 7/4)$ and directrix $y = 9/4$. Sketch the graph, showing the focus and directrix.

Solution

First draw the focus and directrix on the graph, as shown in Fig. 10.9. Since the vertex is midway between the focus and directrix, the vertex is $(3, 2)$. The distance between the focus and vertex is $1/4$. Since the directrix is above the focus, the parabola opens downward, and we have $p = -1/4$. Use $a = 1/(4p)$ to get $a = -1$. Use $a = -1$ and the vertex $(3, 2)$ in the equation $y = a(x - h)^2 + k$ to get

$$y = -1(x - 3)^2 + 2.$$

The axis of symmetry is the vertical line $x = 3$. If $x = 0$, then $y = -1(0 - 3)^2 + 2 = -7$. So the y-intercept is $(0, -7)$. Find the x-intercepts by setting y equal to 0 in the equation:

$$-1(x - 3)^2 + 2 = 0$$
$$(x - 3)^2 = 2$$
$$x - 3 = \pm\sqrt{2}$$
$$x = 3 \pm \sqrt{2}$$

The x-intercepts are $\left(3 - \sqrt{2}, 0\right)$ and $\left(3 + \sqrt{2}, 0\right)$. Two additional points that satisfy $y = -1(x - 3)^2 + 2$ are $(1, -2)$ and $(4, 1)$. Using all of this information, we get the graph shown in Fig. 10.9.

Try This. Find the x- and y-intercepts for a parabola with focus $(2, -3/2)$ and directrix $y = -2$.   ▪

Parabolas Opening to the Left or Right

The graphs of $y = 2x^2$ and $x = 2y^2$ are both parabolas. Interchanging the variables simply changes the roles of the x- and y-axes. The parabola $y = 2x^2$ opens upward, whereas the parabola $x = 2y^2$ opens to the right. For parabolas opening right or left, the directrix is a vertical line. If the focus is to the right of the directrix, then the parabola opens to the right, and if the focus is to the left of the directrix, then the parabola opens to the left. Figure 10.10 shows the relative locations of the vertex, focus, and directrix for parabolas of the form $x = a(y - k)^2 + h$.

▪ **Figure 10.9**

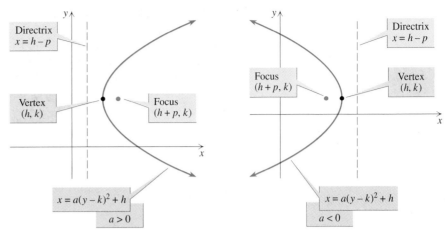

■ **Figure 10.10**

The equation $x = a(y - k)^2 + h$ can be written as $x = ay^2 + by + c$. So the graph of $x = ay^2 + by + c$ is also a parabola opening to the right for $a > 0$ and to the left for $a < 0$. Because the roles of x and y are interchanged, $-b/(2a)$ is now the y-coordinate of the vertex, and the axis of symmetry is the horizontal line $y = -b/(2a)$.

Example 5 Graphing a parabola with a vertical directrix

Find the vertex, axis of symmetry, y-intercepts, focus, and directrix for the parabola $x = y^2 - 2y$. Find several other points on the parabola and sketch the graph.

Solution

Because $a = 1$, the parabola opens to the right. The y-coordinate of the vertex is

$$y = \frac{-b}{2a} = \frac{-(-2)}{2(1)} = 1.$$

If $y = 1$, then $x = (1)^2 - 2(1) = -1$ and the vertex is $(-1, 1)$. The axis of symmetry is the horizontal line $y = 1$. To find the y-intercepts, we solve $y^2 - 2y = 0$ by factoring:

$$y(y - 2) = 0$$

$$y = 0 \quad \text{or} \quad y = 2$$

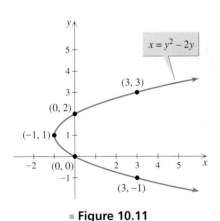

■ **Figure 10.11**

The y-intercepts are $(0, 0)$ and $(0, 2)$. Using all of this information and the additional points $(3, 3)$ and $(3, -1)$, we get the graph shown in Fig. 10.11. Because $a = 1$ and $a = 1/(4p)$, we have $p = 1/4$. So the focus is $1/4$ unit to the right of the vertex at $(-3/4, 1)$. The directrix is the vertical line $1/4$ unit to the left of the vertex, $x = -5/4$.

Try This. Find the vertex, axis of symmetry, y-intercepts, focus, and directrix for $x = y^2 + 4y$ and graph the parabola. ∎

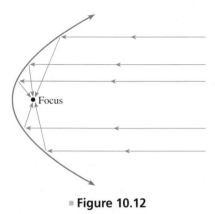

■ **Figure 10.12**

Applications

In Section 3.1 we saw an important application of a parabola. Because of the shape of a parabola, a quadratic function has a maximum value or a minimum value at the vertex of the parabola. However, parabolas are important for another totally different reason. When a ray of light, traveling parallel to the axis of symmetry, hits a parabolic reflector, it is reflected toward the focus of the parabola. See Fig. 10.12. This property is used in telescopes to magnify the light from distant stars. For spotlights, in which the light source is at the focus, the reflecting property is used in reverse. Light originating at the focus is reflected off the parabolic reflector and projected outward in a narrow beam. The reflecting property is used also in telephoto camera lenses, radio antennas, satellite dishes, eavesdropping devices, and flashlights.

Example 6 **The Hubble telescope**

The Hubble space telescope uses a glass mirror with a parabolic cross section and a diameter of 2.4 meters, as shown in Fig. 10.13. If the focus of the parabola is 57.6 meters from the vertex, then what is the equation of the parabola used for the mirror? How much thicker is the mirror at the edge than at its center?

Solution

If we position the parabola with the vertex at the origin and opening upward, as shown in Fig. 10.13, then its equation is of the form $y = ax^2$. Since the distance between the focus and vertex is 57.6 meters, $p = 57.6$. Using $a = 1/(4p)$, we get $a = 1/(4 \cdot 57.6) \approx 0.004340$. The equation of the parabola is

$$y = 0.004340x^2.$$

To find the difference in thickness at the edge, let $x = 1.2$ in the equation of the parabola:

$$y = 0.004340(1.2)^2 \approx 0.006250 \text{ meter}$$

The mirror is 0.006250 meter thicker at the edge than at the center.

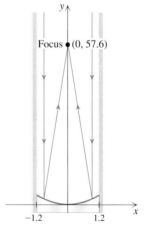

■ **Figure 10.13**

Try This. Find the equation of the parabola with vertex (0, 0) and focus (0, 10).

■

For a telescope to work properly, the glass mirror must be ground with great precision. Prior to the 1993 repair, the Hubble space telescope did not work as well as hoped (Space Telescope Science Institute, www.stsci.edu), because the mirror was actually ground to be only 0.006248 meter thicker at the outside edge, two millionths of a meter smaller than it should have been!

For Thought

True or False? Explain.

1. A parabola with focus $(0, 0)$ and directrix $y = -1$ has vertex $(0, 1)$.

2. A parabola with focus $(3, 0)$ and directrix $x = -1$ opens to the right.

3. A parabola with focus $(4, 5)$ and directrix $x = 1$ has vertex $(5/2, 5)$.

4. For $y = x^2$, the focus is $(0, -1/4)$.

5. For $x = y^2$, the focus is $(1/4, 0)$.

6. A parabola with focus $(2, 3)$ and directrix $y = -5$ has no x-intercepts.

7. The parabola $y = 4x^2 - 9x$ has no y-intercept.

8. The vertex of the parabola $x = 2(y + 3)^2 - 5$ is $(-5, -3)$.

9. The parabola $y = (x - 5)^2 + 4$ has its focus at $(5, 4)$.

10. The parabola $x = -3y^2 + 7y - 5$ opens downward.

10.1 Exercises

Each of the following graphs shows a parabola along with its vertex, focus, and directrix. Determine the coordinates of the vertex and focus, and the equation of the directrix.

1.

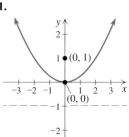

2.

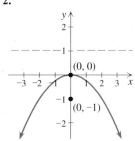

3.

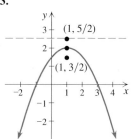

4.

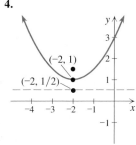

5.

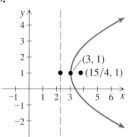

6.

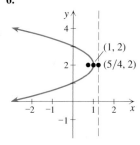

Find the equation of the parabola with the given focus and directrix.

7. Focus $(0, 2)$, directrix $y = -2$

8. Focus $(0, 1)$, directrix $y = -1$

9. Focus $(0, -3)$, directrix $y = 3$

10. Focus $(0, -1)$, directrix $y = 1$

11. Focus $(3, 5)$, directrix $y = 2$

12. Focus $(-1, 5)$, directrix $y = 3$

13. Focus $(1, -3)$, directrix $y = 2$

14. Focus $(1, -4)$, directrix $y = 0$

15. Focus $(-2, 1.2)$, directrix $y = 0.8$

16. Focus $(3, 9/8)$, directrix $y = 7/8$

Find the equation of the parabola with the given focus and directrix.

17.

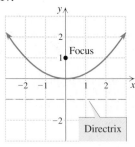

18.

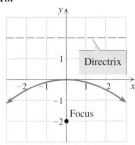

19.

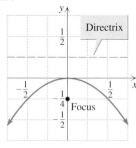

20.

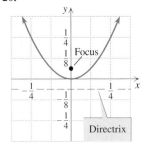

Determine the vertex, focus, and directrix for each parabola.

21. $y = (x - 1)^2$

22. $y = (x + 2)^2$

23. $y = \frac{1}{4}(x - 3)^2$

24. $y = \frac{1}{2}(x + 5)^2$

25. $y = -2(x - 3)^2 + 4$

26. $y = -4(x - 1)^2 + 3$

Use completing the square to write each equation in the form $y = a(x - h)^2 + k$. Identify the vertex, focus, and directrix.

27. $y = x^2 - 8x + 3$

28. $y = x^2 + 2x - 5$

29. $y = 2x^2 + 12x + 5$

30. $y = 3x^2 + 12x + 1$

31. $y = -2x^2 + 6x + 1$

32. $y = -3x^2 - 6x + 5$

33. $y = 5x^2 + 30x$

34. $y = -2x^2 + 12x$

35. $y = \frac{1}{8}x^2 - \frac{1}{2}x + \frac{9}{2}$

36. $y = \frac{1}{4}x^2 + \frac{1}{2}x - \frac{7}{4}$

Find the vertex, focus, and directrix of each parabola without completing the square, and determine whether the parabola opens upward or downward.

37. $y = x^2 - 4x + 3$

38. $y = x^2 - 6x - 7$

39. $y = -x^2 + 2x - 5$

40. $y = -x^2 + 4x + 3$

41. $y = 3x^2 - 6x + 1$

42. $y = 2x^2 + 4x - 1$

43. $y = -\frac{1}{2}x^2 - 3x + 2$

44. $y = -\frac{1}{2}x^2 + 3x - 1$

45. $y = \frac{1}{4}x^2 + 5$

46. $y = -\frac{1}{8}x^2 - 6$

Find the vertex, axis of symmetry, x-intercepts, and y-intercept of the parabola that has the given focus and directrix. Sketch the graph, showing the focus and directrix.

47. Focus $(1/2, -2)$, directrix $y = -5/2$

48. Focus $(1, -35/4)$, directrix $y = -37/4$

49. Focus $(-1/2, 6)$, directrix $y = 13/2$

50. Focus $(-1, 35/4)$, directrix $y = 37/4$

Find the vertex, axis of symmetry, x-intercepts, y-intercept, focus, and directrix for each parabola. Sketch the graph, showing the focus and directrix.

51. $y = \frac{1}{2}(x + 2)^2 + 2$

52. $y = \frac{1}{2}(x - 4)^2 + 1$

53. $y = -\frac{1}{4}(x + 4)^2 + 2$

54. $y = -\frac{1}{4}(x - 2)^2 + 4$

55. $y = \frac{1}{2}x^2 - 2$

56. $y = -\frac{1}{4}x^2 + 4$

57. $y = x^2 - 4x + 4$

58. $y = (x - 4)^2$

59. $y = \frac{1}{3}x^2 - x$

60. $y = \frac{1}{5}x^2 + x$

Find the vertex, axis of symmetry, x-intercept, y-intercepts, focus, and directrix for each parabola. Sketch the graph, showing the focus and directrix.

61. $x = -y^2$

62. $x = y^2 - 2$

63. $x = -\frac{1}{4}y^2 + 1$

64. $x = \frac{1}{2}(y - 1)^2$

65. $x = y^2 + y - 6$ **66.** $x = y^2 + y - 2$

67. $x = -\dfrac{1}{2}y^2 - y - 4$ **68.** $x = -\dfrac{1}{2}y^2 + 3y + 4$

69. $x = 2(y - 1)^2 + 3$ **70.** $x = 3(y + 1)^2 - 2$

71. $x = -\dfrac{1}{2}(y + 2)^2 + 1$ **72.** $x = -\dfrac{1}{4}(y - 2)^2 - 1$

Find the equation of the parabola determined by the given information.

73. Focus $(1, 5)$, vertex $(1, 4)$

74. Directrix $y = 5$, vertex $(2, 3)$

75. Vertex $(0, 0)$, directrix $x = -2$

76. Focus $(-2, 3)$, vertex $(-9/4, 3)$

Solve each problem.

77. *The Hale Telescope* The focus of the Hale telescope, in the accompanying figure, on Palomar Mountain in California is 55 ft above the mirror (at the vertex). The Pyrex glass mirror is 200 in. in diameter and 23 in. thick at the center. Find the equation for the parabola that was used to shape the glass. How thick is the glass on the outside edge?

HINT Find the equation of the parabola with focus $(0, 55)$ that passes through the origin.

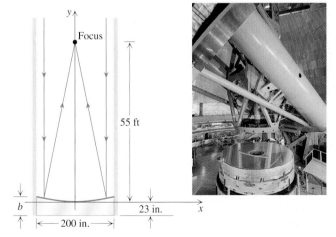

▪ **Figure for Exercise 77**

78. *Eavesdropping* From the Edmund Scientific catalog you can buy a device that will "pull in voices up to three-quarters of a mile away with our electronic parabolic microphone." The 18.75-in.-diameter plastic shield reflects sound waves to a microphone located at the focus. Given that the microphone is located 6 in. from the vertex of the parabolic shield, find the equation for a cross section of the shield. What is the depth of the shield?

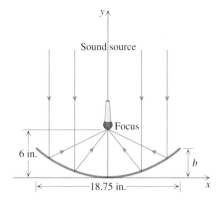

▪ **Figure for Exercise 78**

Use a graphing calculator to solve each problem.

79. Graph $y = x^2$ using the viewing window with $-1 \leq x \leq 1$ and $0 \leq y \leq 1$. Graph $y = 2x^2 - 4x + 5$ using the viewing window with $-1 \leq x \leq 3$ and $3 \leq y \leq 11$. What can you say about the two graphs?

80. Find two different viewing windows in which the graph of $y = 3x^2 + 30x + 71$ looks just like the graph of $y = x^2$ in the viewing window with $-1 \leq x \leq 1$ and $0 \leq y \leq 1$.

81. The graph of $x = -y^2$ is a parabola opening to the left with vertex at the origin. Find two functions whose graphs will together form this parabola and graph them on your calculator.

82. You can illustrate the reflective property of the parabola $x = -y^2$ on the screen of your calculator. First graph $x = -y^2$ as in the previous exercise. Next, by using the graphs of two line segments, make it appear that a particle coming in horizontally from the left toward $(-1, 1)$ is reflected off the parabola and heads toward the focus $(-1/4, 0)$. Consult your manual to see how to graph line segments.

For Writing/Discussion

83. *Derive the Equation* In this section we derived the equation of a parabola with a given focus and a horizontal directrix. Use the same technique to derive the equation of a parabola with a given focus and a vertical directrix.

84. *Cooperative Learning* Working in groups, write a summary of everything that we have learned about parabolas. Be sure to include parabolas opening up and down and parabolas opening left and right.

Thinking Outside the Box LXXXV

Stacking Pipes   Pipes with radii of 2 ft and 3 ft are placed next to each other and anchored so that they cannot move, as shown in the diagram. What is the radius of the largest pipe that can be placed on top of these two pipes?

■ **Figure for Thinking Outside the Box LXXXV**

10.1 Pop Quiz

1. Find the equation of the parabola with focus $(0, 3)$ and directrix $y = 1$.

2. Find the vertex, focus, and directrix for
$$y = -\frac{1}{16}(x - 2)^2 + 3.$$

3. Find the vertex, axis of symmetry, x-intercept, y-intercepts, focus, and directrix for $x = y^2 - 4y + 3$.

Linking Concepts

For Individual or Group Explorations

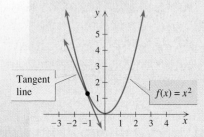

Tangent Lines

A line tangent to a parabola must intersect the parabola at exactly one point, as shown in the accompanying figure. Of course it is impossible to draw a graph that really looks like there is one point of intersection. However, if we find a tangent line algebraically, then we can be sure that it intersects only once.

a) Consider $y = x^2$ and the point $(3, 9)$. If $y - 9 = m(x - 3)$ is tangent to the parabola at $(3, 9)$, then the system

$$y = x^2$$
$$y - 9 = m(x - 3)$$

must have exactly one solution. For what value of m does the system have exactly one solution?

b) Graph the parabola and the tangent line found in part (a) on a graphing calculator and then use the intersect feature of the calculator to find the point of intersection.

c) There is another line that intersects $y = x^2$ at $(3, 9)$ and only at $(3, 9)$. What is it? Why did it not appear in part (a)?

d) Use the same reasoning as used in part (a) to find the slope of the tangent line to $y = x^2$ at (x_1, y_1).

e) Use the result of part (d) to find the equation of the tangent line to $y = x^2$ at $(-2.5, 6.25)$.

The Ellipse and the Circle

Ellipses and circles can be obtained by intersecting planes and cones, as shown in Fig. 10.1. Here we will develop general equations for ellipses, and then circles, by starting with geometric definitions, as we did with the parabola.

Definition of Ellipse

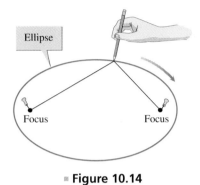

An easy way to draw an ellipse is illustrated in Fig. 10.14. A string is attached at two fixed points, and a pencil is used to take up the slack. As the pencil is moved around the paper, the sum of the distances of the pencil point from the two fixed points remains constant, because the length of the string is constant. This method of drawing an ellipse is used in construction, and it illustrates the geometric definition of an ellipse.

■ **Figure 10.14**

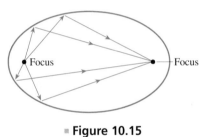

■ **Figure 10.15**

Definition: Ellipse

An **ellipse** is the set of points in a plane such that the sum of their distances from two fixed points is a constant. Each fixed point is called a **focus** (plural: foci) of the ellipse.

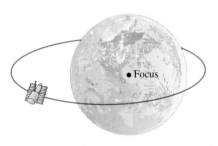

■ **Figure 10.16**

The ellipse, like the parabola, has interesting reflecting properties. All light or sound waves emitted from one focus are reflected off the ellipse to concentrate at the other focus, as shown in Fig. 10.15. This property is used in light fixtures such as a dentist's light, for which a concentration of light at a point is desired, and in a whispering gallery like the one in the U.S. Capitol Building. In a whispering gallery, a whisper emitted at one focus is reflected off the elliptical ceiling and is amplified so that it can be heard at the other focus, but not anywhere in between.

The orbits of the planets around the sun and satellites around Earth are elliptical. For the orbit of Earth, the sun is at one focus of the elliptical path. For the orbit of a satellite such as the Hubble space telescope, the center of Earth is one focus. See Fig. 10.16.

The Equation of the Ellipse

The ellipse shown in Fig. 10.17 has foci at $(c, 0)$ and $(-c, 0)$, and y-intercepts $(0, b)$ and $(0, -b)$, where $c > 0$ and $b > 0$. The line segment $\overline{V_1 V_2}$ is the **major axis,** and the line segment $\overline{B_1 B_2}$ is the **minor axis.** For any ellipse, the major axis is longer than the minor axis, and the foci are on the major axis. The **center** of an ellipse is the midpoint of the major (or minor) axis. The ellipse in Fig. 10.17 is centered at the origin. The **vertices** of an ellipse are the endpoints of the major axis. The vertices of the ellipse in Fig. 10.17 are the x-intercepts.

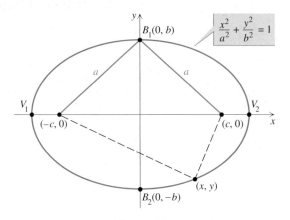

■ **Figure 10.17**

Let a be the distance between $(c, 0)$ and the y-intercept $(0, b)$, as shown in Fig. 10.17. The sum of the distances from the two foci to $(0, b)$ is $2a$. So for any point (x, y) on the ellipse, the distance from (x, y) to $(c, 0)$ plus the distance from (x, y) to $(-c, 0)$ is equal to $2a$. Writing this last statement as an equation (using the distance formula) gives the equation of the ellipse shown in Fig. 10.17:

$$\sqrt{(x - c)^2 + (y - 0)^2} + \sqrt{(x - (-c))^2 + (y - 0)^2} = 2a$$

With some effort, this equation can be greatly simplified. We will provide the major steps in simplifying it, and leave the details as an exercise. First simplify inside the radicals and isolate them to get

$$\sqrt{x^2 - 2xc + c^2 + y^2} = 2a - \sqrt{x^2 + 2xc + c^2 + y^2}.$$

Next, square each side and simplify again to get

$$a\sqrt{x^2 + 2xc + c^2 + y^2} = a^2 + xc.$$

Squaring each side again yields

$$a^2 x^2 - c^2 x^2 + a^2 y^2 = a^4 - a^2 c^2$$

$$(a^2 - c^2)x^2 + a^2 y^2 = a^2(a^2 - c^2) \quad \text{Factor.}$$

Since $a^2 = b^2 + c^2$, or $a^2 - c^2 = b^2$ (from Fig. 10.17), replace $a^2 - c^2$ by b^2:

$$b^2 x^2 + a^2 y^2 = a^2 b^2$$

$$\frac{x^2}{a^2} + \frac{y^2}{b^2} = 1 \quad \text{Divide each side by } a^2 b^2.$$

We have proved the following theorem.

Theorem: Equation of an Ellipse with Center (0, 0) and Horizontal Major Axis

The equation of an ellipse centered at the origin with foci $(c, 0)$ and $(-c, 0)$ and y-intercepts $(0, b)$ and $(0, -b)$ is

$$\frac{x^2}{a^2} + \frac{y^2}{b^2} = 1, \quad \text{where } a^2 = b^2 + c^2.$$

If we start with the foci on the y-axis and x-intercepts $(b, 0)$ and $(-b, 0)$, then we can develop a similar equation, which is stated in the following theorem. If the foci are on the y-axis, then the y-intercepts are the vertices and the major axis is vertical.

Theorem: Equation of an Ellipse with Center (0, 0) and Vertical Major Axis

The equation of an ellipse centered at the origin with foci $(0, c)$ and $(0, -c)$ and x-intercepts $(b, 0)$ and $(-b, 0)$ is

$$\frac{x^2}{b^2} + \frac{y^2}{a^2} = 1, \quad \text{where } a^2 = b^2 + c^2.$$

Consider the ellipse in Fig. 10.18 with foci $(c, 0)$ and $(-c, 0)$ and equation

$$\frac{x^2}{a^2} + \frac{y^2}{b^2} = 1,$$

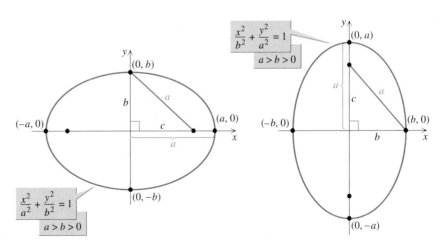

▪ **Figure 10.18**

where $a > b > 0$. If $y = 0$ in this equation, then $x = \pm a$. So the vertices (or x-intercepts) of the ellipse are $(a, 0)$ and $(-a, 0)$, and a is the distance from the center to a vertex. The distance from a focus to an endpoint of the minor axis is a also. *So in any ellipse the distance from the focus to an endpoint of the minor axis is the same as the distance from the center to a vertex.* The graphs in Fig. 10.18 will help you remember the relationship between a, b, and c.

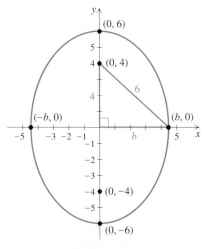

■ **Figure 10.19**

Example **1** **Writing the equation of an ellipse**

Sketch an ellipse with foci at $(0, 4)$ and $(0, -4)$ and vertices $(0, 6)$ and $(0, -6)$, and find the equation for this ellipse.

Solution

Since the vertices are on the y-axis, the major axis is vertical and the ellipse is elongated in the direction of the y-axis. A sketch of the ellipse appears in Fig. 10.19. Because a is the distance from the center to a vertex, $a = 6$. Because c is the distance from the center to a focus, $c = 4$. To write the equation of the ellipse, we need the value of b^2. Use $a = 6$ and $c = 4$ in $a^2 = b^2 + c^2$ to get

$$6^2 = b^2 + 4^2$$

$$b^2 = 20$$

So the x-intercepts are $\left(\sqrt{20}, 0\right)$ and $\left(-\sqrt{20}, 0\right)$, and the equation of the ellipse is

$$\frac{x^2}{20} + \frac{y^2}{36} = 1.$$

Try This. Find the equation and y-intercepts for the ellipse with foci $(\pm 3, 0)$ and vertices $(\pm 8, 0)$. ■

Graphing an Ellipse Centered at the Origin

For $a > b > 0$, the graph of

$$\frac{x^2}{a^2} + \frac{y^2}{b^2} = 1$$

is an ellipse centered at the origin with a horizontal major axis, x-intercepts $(a, 0)$ and $(-a, 0)$, and y-intercepts $(0, b)$ and $(0, -b)$, as shown in Fig. 10.18. The foci are on the major axis and are determined by $a^2 = b^2 + c^2$ or $c^2 = a^2 - b^2$. Remember that when the denominator for x^2 is larger than the denominator for y^2, the major axis is horizontal. *To sketch the graph of an ellipse centered at the origin, simply locate the four intercepts and draw an ellipse through them.*

Example **2** **Graphing an ellipse with foci on the x-axis**

Sketch the graph and identify the foci of the ellipse

$$\frac{x^2}{9} + \frac{y^2}{4} = 1.$$

Solution

To sketch the ellipse, we find the x-intercepts and the y-intercepts. If $x = 0$, then $y^2 = 4$ or $y = \pm 2$. So the y-intercepts are $(0, 2)$ and $(0, -2)$. If $y = 0$, then $x = \pm 3$. So the x-intercepts are $(3, 0)$ and $(-3, 0)$. To make a rough sketch of an ellipse, plot only the intercepts and draw an ellipse through them, as shown in Fig. 10.20. Since this ellipse is elongated in the direction of the x-axis, the foci are on the x-axis. Use $a = 3$ and $b = 2$ in $c^2 = a^2 - b^2$, to get $c^2 = 9 - 4 = 5$. So $c = \pm\sqrt{5}$, and the foci are $\left(\sqrt{5}, 0\right)$ and $\left(-\sqrt{5}, 0\right)$.

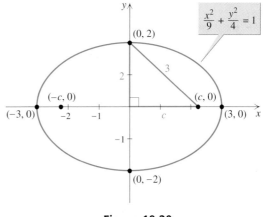

■ **Figure 10.20**

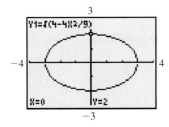

■ **Figure 10.21**

To check, solve for y to get $y = \pm\sqrt{4 - 4x^2/9}$. Then graph $y_1 = \sqrt{4 - 4x^2/9}$ and $y_2 = -y_1$, as shown in Fig. 10.21.

Try This. Graph $\dfrac{x^2}{81} + \dfrac{y^2}{25} = 1$ and identify the foci.   ■

For $a > b > 0$, the graph of

$$\frac{x^2}{b^2} + \frac{y^2}{a^2} = 1$$

is an ellipse with a vertical major axis, x-intercepts $(b, 0)$ and $(-b, 0)$, and y-intercepts $(0, a)$ and $(0, -a)$, as shown in Fig. 10.18. When the denominator for y^2 is larger than the denominator for x^2, the major axis is vertical. The foci are always on the major axis and are determined by $a^2 = b^2 + c^2$ or $c^2 = a^2 - b^2$. Remember that this relationship between a, b, and c is determined by the Pythagorean theorem, because a, b, and c are the lengths of sides of a right triangle, as shown in Fig. 10.18.

Example **3** **Graphing an ellipse with foci on the y-axis**

Sketch the graph of $11x^2 + 3y^2 = 66$ and identify the foci.

Solution

We first divide each side of the equation by 66 to get the standard equation:

$$\frac{x^2}{6} + \frac{y^2}{22} = 1$$

If $x = 0$, then $y^2 = 22$ or $y = \pm\sqrt{22}$. So the y-intercepts are $\left(0, \sqrt{22}\right)$ and $\left(0, -\sqrt{22}\right)$. If $y = 0$, then $x = \pm\sqrt{6}$ and the x-intercepts are $\left(\sqrt{6}, 0\right)$ and $\left(-\sqrt{6}, 0\right)$. Plot these four points and draw an ellipse through them, as shown in Fig. 10.22. Since this ellipse is elongated in the direction of the y-axis, the foci are on the y-axis. Use $a^2 = 22$ and $b^2 = 6$ in $c^2 = a^2 - b^2$ to get $c^2 = 22 - 6 = 16$. So $c = \pm 4$, and the foci are $(0, 4)$ and $(0, -4)$.

Try This. Graph $25x^2 + 4y^2 = 100$ and identify the foci.   ■

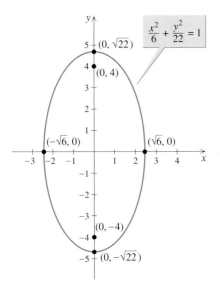

■ **Figure 10.22**

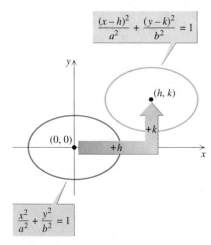

$$\frac{(x-h)^2}{a^2} + \frac{(y-k)^2}{b^2} = 1$$

(h, k)

$+k$

$(0,0)$

$+h$

$$\frac{x^2}{a^2} + \frac{y^2}{b^2} = 1$$

▪ **Figure 10.23**

Translations of Ellipses

Although an ellipse is not the graph of a function, its graph can be translated in the same manner. Figure 10.23 shows the graphs of

$$\frac{(x-h)^2}{a^2} + \frac{(y-k)^2}{b^2} = 1 \quad \text{and} \quad \frac{x^2}{a^2} + \frac{y^2}{b^2} = 1.$$

They have the same size and shape, but the graph of the first equation is centered at (h, k) rather than at the origin. So the graph of the first equation is obtained by translating the graph of the second horizontally h units and vertically k units.

Example **4** Graphing an ellipse centered at (h, k)

Sketch the graph and identify the foci of the ellipse

$$\frac{(x-3)^2}{25} + \frac{(y+1)^2}{9} = 1.$$

Solution

The graph of this equation is a translation of the graph of

$$\frac{x^2}{25} + \frac{y^2}{9} = 1,$$

three units to the right and one unit downward. The center of the ellipse is $(3, -1)$. Since $a^2 = 25$, the vertices lie five units to the right and five units to the left of $(3, -1)$. So the ellipse goes through $(8, -1)$ and $(-2, -1)$. Since $b^2 = 9$, the graph includes points three units above and three units below the center. So the ellipse goes through $(3, 2)$ and $(3, -4)$, as shown in Fig. 10.24. Since $c^2 = 25 - 9 = 16$, $c = \pm 4$. The major axis is on the horizontal line $y = -1$. So the foci are found four units to the right and four units to the left of the center at $(7, -1)$ and $(-1, -1)$.

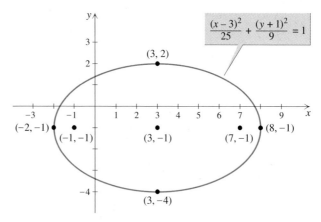

▪ **Figure 10.24**

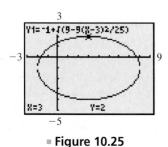

▪ **Figure 10.25**

To check the location of the ellipse, graph

$$y_1 = -1 + \sqrt{9 - 9(x-3)^2/25} \quad \text{and} \quad y_2 = -1 - \sqrt{9 - 9(x-3)^2/25}$$

on a graphing calculator, as in Fig. 10.25.

Try This. Graph $\dfrac{(x-1)^2}{81} + \dfrac{(y+4)^2}{25} = 1$ and identify the foci. ■

The Circle

The circle is the simplest curve of the four conic sections, and it is a special case of the ellipse. In keeping with our approach to the conic sections, we give a geometric definition of a circle and then use the distance formula to derive the standard equation for a circle. The standard equation was derived in this way in Section 1.3. For completeness, we now repeat the definition and derivation.

Definition: Circle

A **circle** is a set of all points in a plane such that their distance from a fixed point (the **center**) is a constant (the **radius**).

As shown in Fig. 10.26, a point (x, y) is on a circle with center (h, k) and radius r if and only if

$$\sqrt{(x-h)^2 + (y-k)^2} = r.$$

If we square both sides of this equation, we get the standard equation of a circle.

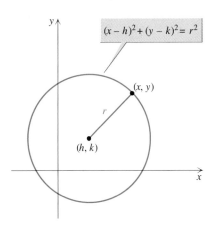

$$(x-h)^2 + (y-k)^2 = r^2$$

(x, y)

r

(h, k)

■ **Figure 10.26**

Theorem: Standard Equation of a Circle

The standard equation of a circle with center (h, k) and radius r ($r > 0$) is

$$(x-h)^2 + (y-k)^2 = r^2.$$

The equation $(x-h)^2 + (y-k)^2 = a$ is a circle of radius $\sqrt{a}$ if $a > 0$. If $a = 0$, only (h, k) satisfies $(x-h)^2 + (y-k)^2 = 0$ and the point (h, k) is a degenerate circle. If $a < 0$, then no ordered pair satisfies $(x-h)^2 + (y-k)^2 = a$. If h and k are zero, then we get the standard equation of a circle centered at the origin, $x^2 + y^2 = r^2$.

A circle is an ellipse in which the two foci coincide at the center. If the foci are identical, then $a = b$ and the equation for an ellipse becomes the equation for a circle with radius a.

Example 5 Finding the equation for a circle

Write the equation for the circle that has center $(4, 5)$ and passes through $(-1, 2)$.

Solution

The radius is the distance from $(4, 5)$ to $(-1, 2)$:

$$r = \sqrt{(4 - (-1))^2 + (5 - 2)^2} = \sqrt{25 + 9} = \sqrt{34}$$

Use $h = 4$, $k = 5$, and $r = \sqrt{34}$ in $(x - h)^2 + (y - k)^2 = r^2$ to get the equation

$$(x - 4)^2 + (y - 5)^2 = 34.$$

Try This. Write the equation for the circle that has center $(-2, 3)$ and passes through $(2, 0)$.   ▪

To graph a circle, we must know the center and radius. A compass or a string can be used to keep the pencil at a fixed distance from the center.

Example 6 Finding the center and radius

Determine the center and radius, and sketch the graph of $x^2 + 4x + y^2 - 2y = 11$.

Solution

Use completing the square to get the equation into the standard form:

$$x^2 + 4x + 4 + y^2 - 2y + 1 = 11 + 4 + 1$$

$$(x + 2)^2 + (y - 1)^2 = 16$$

From the standard form, we recognize that the center is $(-2, 1)$ and the radius is 4. The graph is shown in Fig. 10.27.

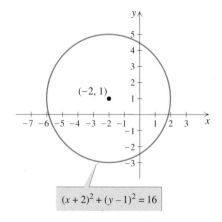

$(x + 2)^2 + (y - 1)^2 = 16$

▪ **Figure 10.27**

Try This. Find the center and radius for $x^2 + 6x + y^2 - 8y = 0$.   ▪

Applications

The **eccentricity** e of an ellipse is defined by $e = c/a$, where c is the distance from the center to a focus and a is one-half the length of the major axis. Since $0 < c < a$, we have $0 < e < 1$. For an ellipse that appears circular, the foci are close to the center and c is small compared with a. So its eccentricity is near 0. For an ellipse that is very elongated, the foci are close to the vertices and c is nearly equal to a. So its eccentricity is near 1. See Fig. 10.28. A satellite that orbits the earth has an elliptical orbit that is nearly circular, with eccentricity close to 0. On the other hand, Halley's comet is in a very elongated elliptical orbit around the sun, with eccentricity close to 1.

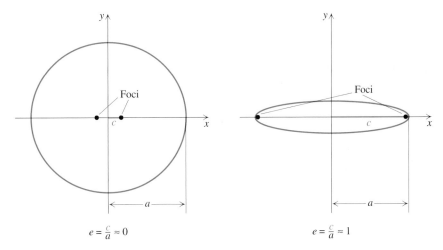

$$e = \frac{c}{a} \approx 0 \qquad\qquad e = \frac{c}{a} \approx 1$$

▪ **Figure 10.28**

Example **7** **Eccentricity of an orbit**

The first artificial satellite to orbit Earth was Sputnik I, launched by the Soviet Union in 1957. The altitude of Sputnik varied from 132 miles to 583 miles above the surface of Earth. If the center of Earth is one focus of its elliptical orbit and the radius of Earth is 3950 miles, then what was the eccentricity of the orbit?

Solution

The center of Earth is one focus F_1 of the elliptical orbit, as shown in Fig. 10.29. When Sputnik I was 132 miles above Earth, it was at a vertex V_1, 4082 miles from

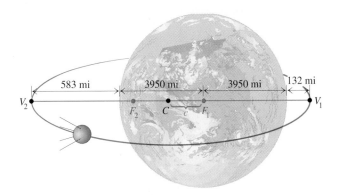

▪ **Figure 10.29**

F_1 (4082 = 3950 + 132). When Sputnik I was 583 miles above Earth, it was at the other vertex V_2, 4533 miles from F_1 (4533 = 3950 + 583). So the length of the major axis is 4082 + 4533, or 8615 miles. Since the length of the major axis is $2a$, we get $a = 4307.5$ miles. Since c is the distance from the center of the ellipse C to F_1, we get $c = 4307.5 - 4082 = 225.5$. Use $e = c/a$ to get

$$e = \frac{225.5}{4307.5} \approx 0.052.$$

So the eccentricity of the orbit was approximately 0.052.

Try This. Find the eccentricity for the ellipse $\frac{x^2}{81} + \frac{y^2}{25} = 1$.   ■

For Thought

True or False? Explain.

1. The x-intercepts for $\frac{x^2}{9} + \frac{y^2}{4} = 1$ are $(9, 0)$ and $(-9, 0)$.

2. The graph of $2x^2 + y^2 = 1$ is an ellipse.

3. The ellipse $\frac{x^2}{16} + \frac{y^2}{25} = 1$ has a major axis of length 10.

4. The x-intercepts for $0.5x^2 + y^2 = 1$ are $(\sqrt{2}, 0)$ and $(-\sqrt{2}, 0)$.

5. The y-intercepts for $x^2 + \frac{y^2}{3} = 1$ are $(0, \sqrt{3})$ and $(0, -\sqrt{3})$.

6. A circle is a set of points, and the center is one of those points.

7. If the foci of an ellipse coincide, then the ellipse is a circle.

8. No ordered pair satisfies the equation $(x - 3)^2 + (y + 1)^2 = 0$.

9. The graph of $(x - 1)^2 + (y + 2)^2 + 9 = 0$ is a circle of radius 3.

10. The radius of the circle $x^2 - 4x + y^2 + y = 9$ is 3.

10.2 Exercises

Each of the following graphs shows an ellipse along with its foci, vertices, and center. Determine the coordinates of the foci, vertices, and center.

1.   **2.**

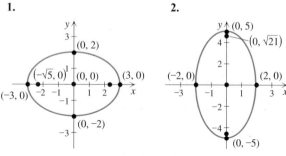

3.   **4.**

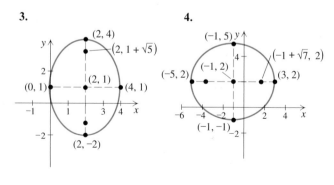

Find the equation of each ellipse described below and sketch its graph.

5. Foci $(-2, 0)$ and $(2, 0)$, and y-intercepts $(0, -3)$ and $(0, 3)$.

6. Foci $(-3, 0)$ and $(3, 0)$, and y-intercepts $(0, -4)$ and $(0, 4)$.

7. Foci $(-4, 0)$ and $(4, 0)$, and x-intercepts $(-5, 0)$ and $(5, 0)$.

8. Foci $(-1, 0)$ and $(1, 0)$, and x-intercepts $(-3, 0)$ and $(3, 0)$.

9. Foci $(0, 2)$ and $(0, -2)$, and x-intercepts $(2, 0)$ and $(-2, 0)$.

10. Foci $(0, 6)$ and $(0, -6)$, and x-intercepts $(2, 0)$ and $(-2, 0)$.

11. Foci $(0, 4)$ and $(0, -4)$, and y-intercepts $(0, 7)$ and $(0, -7)$.

12. Foci $(0, 3)$ and $(0, -3)$, and y-intercepts $(0, 4)$ and $(0, -4)$.

Sketch the graph of each ellipse and identify the foci.

13. $\dfrac{x^2}{16} + \dfrac{y^2}{4} = 1$

14. $\dfrac{x^2}{16} + \dfrac{y^2}{9} = 1$

15. $\dfrac{x^2}{9} + \dfrac{y^2}{36} = 1$

16. $x^2 + \dfrac{y^2}{4} = 1$

17. $\dfrac{x^2}{25} + y^2 = 1$

18. $\dfrac{x^2}{6} + \dfrac{y^2}{10} = 1$

19. $\dfrac{y^2}{25} + \dfrac{x^2}{9} = 1$

20. $\dfrac{y^2}{9} + \dfrac{x^2}{4} = 1$

21. $9x^2 + y^2 = 9$

22. $x^2 + 4y^2 = 4$

23. $4x^2 + 9y^2 = 36$

24. $9x^2 + 25y^2 = 225$

Sketch the graph of each ellipse and identify the foci.

25. $\dfrac{(x-1)^2}{16} + \dfrac{(y+3)^2}{9} = 1$ **26.** $\dfrac{(x+2)^2}{16} + \dfrac{(y+1)^2}{4} = 1$

27. $\dfrac{(x-3)^2}{9} + \dfrac{(y+2)^2}{25} = 1$ **28.** $(x-5)^2 + \dfrac{(y-3)^2}{9} = 1$

29. $(x+4)^2 + 36(y+3)^2 = 36$

30. $9(x-1)^2 + 4(y+3)^2 = 36$

31. $9x^2 - 18x + 4y^2 + 16y = 11$

32. $4x^2 + 16x + y^2 - 6y = -21$

33. $9x^2 - 54x + 4y^2 + 16y = -61$

34. $9x^2 + 90x + 25y^2 - 50y = -25$

Find the equation of each ellipse and identify its foci.

35.

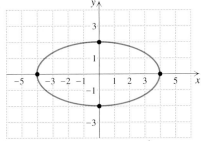

36.

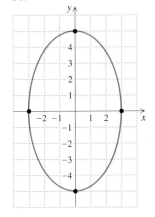

37.

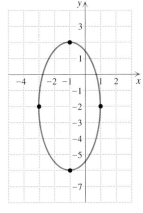

38.

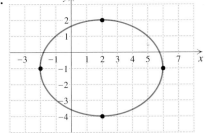

Write the equation for each circle described.

39. Center $(0, 0)$ and radius 2

40. Center $(0, 0)$ and radius $\sqrt{5}$

41. Center $(0, 0)$ and passing through $(4, 5)$

42. Center $(0, 0)$ and passing through $(-3, -4)$

43. Center $(2, -3)$ and passing through $(4, 1)$

44. Center $(-2, -4)$ and passing through $(1, -1)$

45. Diameter has endpoints $(3, 4)$ and $(-1, 2)$.

46. Diameter has endpoints $(3, -1)$ and $(-4, 2)$.

Determine the center and radius of each circle and sketch its graph.

47. $x^2 + y^2 = 100$ **48.** $x^2 + y^2 = 25$

49. $(x - 1)^2 + (y - 2)^2 = 4$ **50.** $(x + 2)^2 + (y - 3)^2 = 9$

51. $(x + 2)^2 + (y + 2)^2 = 8$ **52.** $x^2 + (y - 3)^2 = 9$

Find the center and radius of each circle.

53. $x^2 + y^2 + 2y = 8$ **54.** $x^2 - 6x + y^2 = 1$

55. $x^2 + 8x + y^2 = 10y$ **56.** $x^2 + y^2 = 12x - 12y$

57. $x^2 + 4x + y^2 = 5$ **58.** $x^2 + y^2 - 6y = 0$

59. $x^2 - x + y^2 + y = \dfrac{1}{2}$ **60.** $x^2 + 5x + y^2 + 3y = \dfrac{1}{2}$

61. $x^2 + \dfrac{2}{3}x + y^2 + \dfrac{1}{3}y = \dfrac{1}{9}$ **62.** $x^2 + \dfrac{1}{2}x + y^2 + \dfrac{1}{2}y = \dfrac{1}{8}$

63. $2x^2 + 4x + 2y^2 = 1$ **64.** $x^2 + y^2 = \dfrac{3}{2}y$

Write each of the following equations in one of the forms:

$$y = a(x - h)^2 + k, \quad x = a(y - h)^2 + k,$$

$$\frac{(x - h)^2}{a^2} + \frac{(y - k)^2}{b^2} = 1, \quad \text{or} \quad (x - h)^2 + (y - k)^2 = r^2.$$

Then identify each equation as the equation of a parabola, an ellipse, or a circle.

65. $y = x^2 + y^2$ **66.** $4x = x^2 + y^2$

67. $4x^2 + 12y^2 = 4$ **68.** $2x^2 + 2y^2 = 4 - y$

69. $2x^2 + 4x = 4 - y$ **70.** $2x^2 + 4y^2 = 4 - y$

71. $2 - x = (2 - y)^2$ **72.** $3(3 - x)^2 = 9 - y$

73. $2(4 - x)^2 = 4 - y^2$ **74.** $\dfrac{x^2}{4} + \dfrac{y^2}{4} = 1$

75. $9x^2 = 1 - 9y^2$ **76.** $9x^2 = 1 - 9y$

Solve each problem.

77. *Foci and a Point* Find the equation of the ellipse that goes through the point (2, 3) and has foci (2, 0) and (−2, 0).

78. *Foci and Eccentricity* Find the equation of an ellipse with foci (±5, 0) and eccentricity 1/2.

79. *Focus of Elliptical Reflector* An elliptical reflector is 10 in. in diameter and 3 in. deep. A light source is at one focus (0, 0), 3 in. from the vertex (−3, 0), as shown in the accompanying figure. Find the location of the other focus.

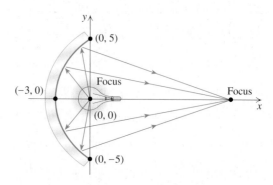

■ **Figure for Exercise 79**

80. *Constructing an Elliptical Arch* A mason is constructing an elliptical form for an arch out of a 4-ft by 12-ft sheet of plywood, as shown in the accompanying figure. What length string is needed to draw the ellipse on the plywood (by attaching the ends at the foci as discussed at the beginning of this section)? Where are the foci for this ellipse?

HINT Find the equation of the ellipse with x-intercepts (±6, 0) and y-intercepts (0, ±4).

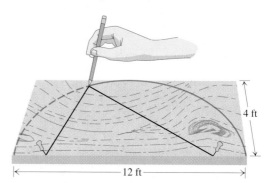

■ **Figure for Exercise 80**

81. *Comet Hale-Bopp* Comet Hale-Bopp, which was clearly seen in April of 1997, orbits the sun in an elliptical orbit every 4200 years. At *perihelion,* the closest point to the sun, the comet is approximately 1 AU from the sun, as shown in the accompanying figure (*Sky and Telescope,* April 1997, used with permission. Guy Ottewell, Astronomical Calendar, 1997, p. 59). At *aphelion,* the farthest point from the sun, the comet is 520 AU from the sun. Find the equation of the ellipse. What is the eccentricity of the orbit?

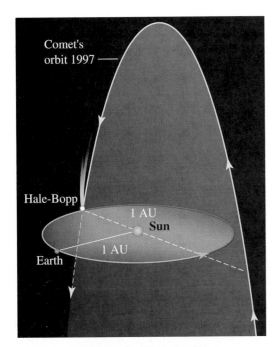

■ **Figure for Exercise 81**

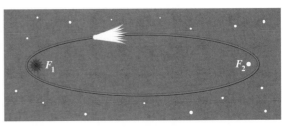

■ **Figure for Exercise 83**

84. *Adjacent Circles*   A 13-in.-diameter mag wheel and a 16-in.-diameter mag wheel are placed in the first quadrant, as shown in the figure. Write an equation for the circular boundary of each wheel.

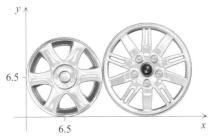

■ **Figure for Exercise 84**

82. *Orbit of the Moon*   The moon travels on an elliptical path with Earth at one focus. If the maximum distance from the moon to Earth is 405,500 km and the minimum distance is 363,300 km, then what is the eccentricity of the orbit?

83. *Halley's Comet*   The comet Halley, last seen in 1986, travels in an elliptical orbit with the sun at one focus. Its minimum distance to the sun is 8×10^7 km, and the eccentricity of its orbit is 0.97. What is its maximum distance from the sun? Comet Halley will next be visible from Earth in 2062.

85. *Picturing Earth*   Solve $x^2 + y^2 = 6360^2$ for y and graph the resulting two equations on a graphing calculator to get a picture of Earth (a circle of radius 6360 km) as seen from space. The center of Earth is at the origin.

86. *Orbit of Mir*   The elliptical orbit of the Mir space station had the equation

$$\frac{(x - 5)^2}{6735^2} + \frac{y^2}{6734.998^2} = 1,$$

where one of the foci is at the center of Earth. Solve this equation for y and graph the resulting two functions on the same screen as the graph obtained in the previous exercise. The

■ **Figure for Exercise 86**

graph of the circle and the ellipse together will give you an idea of the altitude of Mir compared with the size of Earth. What was the eccentricity of the orbit of Mir? (Mir was terminated in 2001.)

87. *Tangent to an Ellipse*   It can be shown that the line tangent to the ellipse $x^2/a^2 + y^2/b^2 = 1$ at the point (x_1, y_1) has the equation

$$\frac{x_1 x}{a^2} + \frac{y_1 y}{b^2} = 1.$$

a. Find the equation of the line tangent to the ellipse $x^2/25 + y^2/9 = 1$ at $(-4, 9/5)$.

b. Graph the ellipse and the tangent line on a graphing calculator and use the intersect feature to find the point of intersection.

88. *Points on an Ellipse*   Find all points on the ellipse $9x^2 + 25y^2 = 225$ that are twice as far from one focus as they are from the other focus.

For Writing/Discussion

89. *Best Reflector*   Both parabolic and elliptical reflectors reflect sound waves to a focal point where they are amplified. To eavesdrop on the conversation of the quarterback on a football field, which type of reflector is preferable and why?

90. *Details*   Fill in the details in the development of the equation of the ellipse that is outlined in this section.

91. *Development*   Develop the equation of the ellipse with foci at $(0, \pm c)$ and x-intercepts $(\pm b, 0)$ from the definition of ellipse.

92. *Cooperative Learning*   To make drapes gather properly, they must be made 2.5 times as wide as the window. Discuss with your classmates the problem of making drapes for a semicircular window. Explain in detail how to cut the material. Of course these drapes do not open and close, they just hang there. This problem was given to the author by a friend in the drapery business.

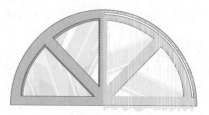

■ **Figure for Exercise 92**

Thinking Outside the Box LXXXVI

Three Circles   A circle of radius 1 is centered at the origin and a circle of radius $1/2$ is centered at $(1/2, 0)$. A third circle is positioned so that it is tangent to the other two circles and the y-axis as shown in the figure. Find the center and radius of the third circle.

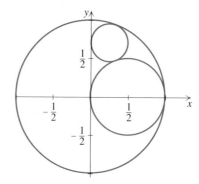

■ **Figure for Thinking Outside the Box LXXXVI**

10.2 Pop Quiz

1. Find the equation of the ellipse with foci $(\pm 3, 0)$ and y-intercepts $(0, \pm 5)$.

2. Find the foci for $\dfrac{(x - 1)^2}{9} + \dfrac{(y - 3)^2}{25} = 1$.

3. Find the center and radius for the circle $x^2 + 4x + y^2 - 10y = 0$.

Linking Concepts

For Individual or
Group Explorations

Apogee, Perigee, and Eccentricity

*For a planet or satellite in an elliptical orbit around a focus of the ellipse, perigee
(P) is defined to be its closest distance to the focus and apogee (A) is defined to be
its greatest distance from the focus.*

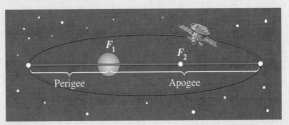

a)   Show that $(A - P)/(A + P)$ is equal to the eccentricity of the orbit.

b)   Find the apogee and perigee for the satellite Sputnik I discussed in Example 7
of this section.

c)   Use the apogee and perigee to find the eccentricity of the orbit of Sputnik I.

d)   Pluto orbits the sun in an elliptical orbit with eccentricity 0.2484. The perigee
for Pluto is 29.64 AU (astronomical units). Find the apogee for Pluto.

e)   Mars orbits the sun in an elliptical orbit with eccentricity 0.0934. The apogee
for Mars is 2.492×10^8 km. Find the perigee for Mars.

10.3

The Hyperbola

The last of the four conic sections, the hyperbola, has two branches and each one has
a focus, as shown in Fig. 10.30. The hyperbola also has a useful reflecting property.

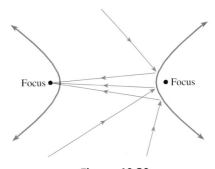

▪ **Figure 10.30**

A light ray aimed at one focus is reflected toward the other focus, as shown in Fig. 10.30. This reflecting property is used in telescopes, as shown in Fig. 10.31. Within the telescope, a small hyperbolic mirror with the same focus as the large parabolic mirror reflects the light to a more convenient location for viewing.

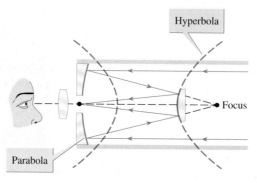

■ **Figure 10.31**

Hyperbolas also occur in the context of supersonic noise pollution. The sudden change in air pressure from an aircraft traveling at supersonic speed creates a cone-shaped wave through the air. The plane of the ground and this cone intersect along one branch of a hyperbola, as shown in Fig. 10.32. A sonic boom is heard on the ground along this branch of the hyperbola. Since this curve where the sonic boom is heard travels along the ground with the aircraft, supersonic planes are restricted from flying across the continental United States.

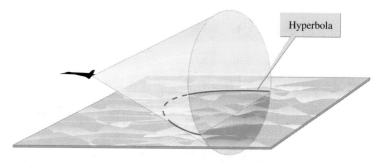

■ **Figure 10.32**

A hyperbola may also occur as the path of a moving object such as a spacecraft traveling past the moon on its way toward Venus, a comet passing in the neighborhood of the sun, or an alpha particle passing by the nucleus of an atom.

The Definition

A hyperbola can be defined as the intersection of a cone and a plane, as shown in Fig. 10.1. As we did for the other conic sections, we will give a geometric definition of a hyperbola and use the distance formula to derive its equation.

Definition: Hyperbola

A **hyperbola** is the set of points in a plane such that the difference between the distances from two fixed points (foci) is constant.

For a point on a hyperbola, the *difference* between the distances from two fixed points is constant, and for a point on an ellipse, the *sum* of the distances from two fixed points is constant. The definitions of a hyperbola and an ellipse are similar, and we will see their equations are similar also. Their graphs, however, are very different. In the hyperbola shown in Fig. 10.33, the branches look like parabolas, *but they are not parabolas* because they do not satisfy the geometric definition of a parabola. A hyperbola with foci on the x-axis, as in Fig. 10.33, is said to open to the left and right.

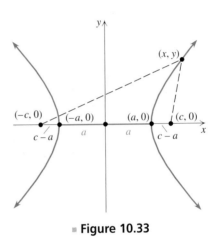

■ **Figure 10.33**

Developing the Equation

The hyperbola shown in Fig. 10.33 has foci at $(c, 0)$ and $(-c, 0)$ and x-intercepts or **vertices** at $(a, 0)$ and $(-a, 0)$, where $a > 0$ and $c > 0$. The line segment between the vertices is the **transverse axis.** The point $(0, 0)$, halfway between the foci, is the **center.** The point $(a, 0)$ is on the hyperbola. The distance from $(a, 0)$ to the focus $(c, 0)$ is $c - a$. The distance from $(a, 0)$ to $(-c, 0)$ is $c + a$. So the constant difference is $2a$. For an arbitrary point (x, y), the distance to $(c, 0)$ is subtracted from the distance to $(-c, 0)$ to get the constant $2a$:

$$\sqrt{(x - (-c))^2 + (y - 0)^2} - \sqrt{(x - c)^2 + (y - 0)^2} = 2a$$

For (x, y) on the other branch, we would subtract in the opposite order, but we get the same simplified form. The simplified form for this equation is similar to that for the ellipse. We will provide the major steps for simplifying it and leave the details as an exercise. First simplify inside the radicals and isolate them to get

$$\sqrt{x^2 + 2xc + c^2 + y^2} = 2a + \sqrt{x^2 - 2xc + c^2 + y^2}.$$

Squaring each side and simplifying yields

$$xc - a^2 = a\sqrt{x^2 - 2xc + c^2 + y^2}.$$

Square each side again and simplify to get

$$c^2x^2 - a^2x^2 - a^2y^2 = c^2a^2 - a^4$$
$$(c^2 - a^2)x^2 - a^2y^2 = a^2(c^2 - a^2) \quad \text{Factor.}$$

Now $c^2 - a^2$ is positive, because $c > a$. So let $b^2 = c^2 - a^2$ and substitute:

$$b^2x^2 - a^2y^2 = a^2b^2$$
$$\frac{x^2}{a^2} - \frac{y^2}{b^2} = 1 \qquad \text{Divide each side by } a^2b^2.$$

We have proved the following theorem.

Theorem: Equation of a Hyperbola Centered at (0, 0) Opening Left and Right	The equation of a hyperbola centered at the origin with foci $(c, 0)$ and $(-c, 0)$ and x-intercepts $(a, 0)$ and $(-a, 0)$ is $$\frac{x^2}{a^2} - \frac{y^2}{b^2} = 1,$$ where $b^2 = c^2 - a^2$. 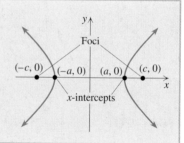

If the foci are positioned on the y-axis, then we say that the hyperbola opens up and down. For hyperbolas that open up and down, we have the following theorem.

Theorem: Equation of a Hyperbola Centered at (0, 0) Opening Up and Down	The equation of a hyperbola centered at the origin with foci $(0, c)$ and $(0, -c)$ and y-intercepts $(0, a)$ and $(0, -a)$ is $$\frac{y^2}{a^2} - \frac{x^2}{b^2} = 1,$$ where $b^2 = c^2 - a^2$. 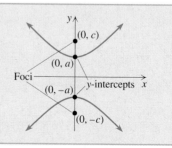

Graphing a Hyperbola Centered at (0, 0)

If we solve the equation $\frac{x^2}{a^2} - \frac{y^2}{b^2} = 1$, for y, we get the following:

$$\frac{y^2}{b^2} = \frac{x^2}{a^2} - 1$$

$$y^2 = \frac{b^2 x^2}{a^2} - b^2 \qquad \text{Multiply each side by } b^2.$$

$$y^2 = \frac{b^2 x^2}{a^2} - \frac{a^2 b^2 x^2}{a^2 x^2} \qquad \text{Write } b^2 \text{ as } \frac{a^2 b^2 x^2}{a^2 x^2}.$$

$$y^2 = \frac{b^2 x^2}{a^2}\left(1 - \frac{a^2}{x^2}\right) \qquad \text{Factor out } \frac{b^2 x^2}{a^2}.$$

$$y = \pm\frac{b}{a}x\sqrt{1 - \frac{a^2}{x^2}}$$

As $x \to \infty$, the value of $(a^2/x^2) \to 0$ and the value of y can be approximated by $y = \pm(b/a)x$. So the lines

$$y = \frac{b}{a}x \qquad \text{and} \qquad y = -\frac{b}{a}x$$

are oblique asymptotes for the graph of the hyperbola. The graph of this hyperbola is shown with its asymptotes in Fig. 10.34. The asymptotes are essential for determining the proper shape of the hyperbola. The asymptotes go through the points (a, b), $(a, -b)$, $(-a, b)$, and $(-a, -b)$. The rectangle with these four points as vertices is called the **fundamental rectangle.** The line segment with endpoints $(0, b)$ and $(0, -b)$ is called the **conjugate axis.** The location of the fundamental rectangle is determined by the conjugate axis and the transverse axis.

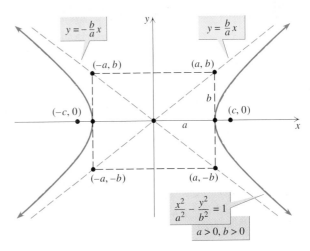

■ **Figure 10.34**

Note that since $c^2 = a^2 + b^2$, the distance c from the center to a focus is equal to the distance from the center to (a, b), as shown in Fig. 10.35 on the following page.

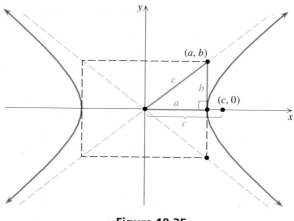

■ **Figure 10.35**

The following steps will help you graph hyperbolas opening left and right.

P R O C E D U R E   **Graphing the Hyperbola $\dfrac{x^2}{a^2} - \dfrac{y^2}{b^2} = 1$**

To graph $\dfrac{x^2}{a^2} - \dfrac{y^2}{b^2} = 1$ for $a > 0$ and $b > 0$, do the following:

1. Locate the x-intercepts $(a, 0)$ and $(-a, 0)$.
2. Draw the rectangle through $(\pm a, 0)$ and through $(0, \pm b)$.
3. Extend the diagonals of the rectangle to get the asymptotes.
4. Draw a hyperbola opening to the left and right from the x-intercepts approaching the asymptotes.

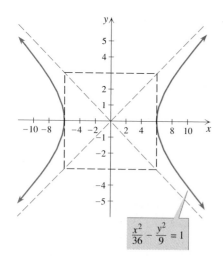

■ **Figure 10.36**

Example **1** **Graphing a hyperbola opening left and right**

Determine the foci and the equations of the asymptotes, and sketch the graph of

$$\frac{x^2}{36} - \frac{y^2}{9} = 1.$$

Solution

If $y = 0$, then $x = \pm 6$. The x-intercepts are $(6, 0)$ and $(-6, 0)$. Since $b^2 = 9$, the fundamental rectangle goes through $(0, 3)$ and $(0, -3)$ and through the x-intercepts. Next draw the fundamental rectangle and extend its diagonals to get the asymptotes. Draw the hyperbola opening to the left and right, as shown in Fig. 10.36. To find the foci, use $c^2 = a^2 + b^2$:

$$c^2 = 36 + 9 = 45$$

$$c = \pm \sqrt{45}$$

So the foci are $\left(\sqrt{45}, 0\right)$ and $\left(-\sqrt{45}, 0\right)$. From the graph we get the asymptotes

$$y = \frac{1}{2}x \qquad \text{and} \qquad y = -\frac{1}{2}x.$$

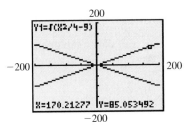

■ **Figure 10.37**

Try This. Determine the foci and equations of the asymptotes, and graph $\dfrac{x^2}{49} - \dfrac{y^2}{25} = 1$. ■

The graph of a hyperbola gets closer and closer to its asymptotes. So in a large viewing window you cannot tell the difference between the hyperbola and its asymptotes. For example, the hyperbola of Example 1, $y = \pm\sqrt{x^2/4 - 9}$, looks like its asymptotes $y = \pm(1/2)x$ in Fig. 10.37. □

We can show that hyperbolas opening up and down have asymptotes just as hyperbolas opening right and left have. The lines

$$y = \frac{a}{b}x \qquad \text{and} \qquad y = -\frac{a}{b}x$$

are asymptotes for the graph of

$$\frac{y^2}{a^2} - \frac{x^2}{b^2} = 1,$$

as shown in Fig. 10.38. Note that the asymptotes are essential for determining the shape of the hyperbola.

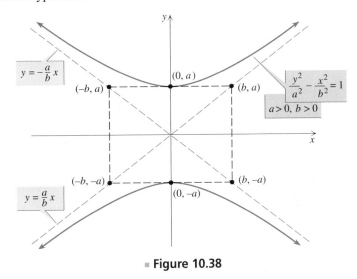

■ **Figure 10.38**

The following steps will help you graph hyperbolas opening up and down.

P R O C E D U R E   **Graphing the Hyperbola $\dfrac{y^2}{a^2} - \dfrac{x^2}{b^2} = 1$**

To graph $\dfrac{y^2}{a^2} - \dfrac{x^2}{b^2} = 1$ for $a > 0$ and $b > 0$, do the following:

1. Locate the y-intercepts $(0, a)$ and $(0, -a)$.
2. Draw the rectangle through $(0, \pm a)$ and through $(b, 0)$ and $(-b, 0)$.
3. Extend the diagonals of the rectangle to get the asymptotes.
4. Draw a hyperbola opening up and down from the y-intercepts approaching the asymptotes.

Example **2** **Graphing a hyperbola opening up and down**

Determine the foci and the equations of the asymptotes, and sketch the graph of

$$4y^2 - 9x^2 = 36.$$

Solution

Divide each side of the equation by 36 to get the equation into the standard form for the equation of the hyperbola:

$$\frac{y^2}{9} - \frac{x^2}{4} = 1$$

If $x = 0$, then $y = \pm 3$. The y-intercepts are $(0, 3)$ and $(0, -3)$. Since $b^2 = 4$, the fundamental rectangle goes through the y-intercepts and through $(2, 0)$ and $(-2, 0)$. Draw the fundamental rectangle and extend its diagonals to get the asymptotes. Draw a hyperbola opening up and down from the y-intercepts approaching the asymptotes, as shown in Fig. 10.39. To find the foci, use $c^2 = a^2 + b^2$:

$$c^2 = 9 + 4 = 13$$
$$c = \pm \sqrt{13}$$

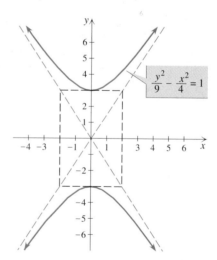

$$\frac{y^2}{9} - \frac{x^2}{4} = 1$$

■ **Figure 10.39**

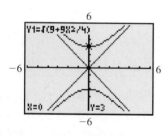

■ **Figure 10.40**

The foci are $\left(0, \sqrt{13}\right)$ and $\left(0, -\sqrt{13}\right)$. From the fundamental rectangle, we can see that the equations of the asymptotes are

$$y = \frac{3}{2}x \qquad \text{and} \qquad y = -\frac{3}{2}x.$$

◩ Check by graphing $y = \pm\sqrt{9 + 9x^2/4}$ and $y = \pm 1.5x$ with a calculator, as shown in Fig. 10.40.

Try This. Determine the foci and equations of the asymptotes, and graph $\frac{y^2}{9} - \frac{x^2}{81} = 1$. ∎ ■

Hyperbolas Centered at (h, k)

The graph of a hyperbola can be translated horizontally and vertically by replacing x by $x - h$ and y by $y - k$.

Theorem: Hyperbolas Centered at (h, k)

A hyperbola centered at (h, k), opening left and right, has a horizontal transverse axis and equation

$$\frac{(x - h)^2}{a^2} - \frac{(y - k)^2}{b^2} = 1.$$

A hyperbola centered at (h, k), opening up and down, has a vertical transverse axis and equation

$$\frac{(y - k)^2}{a^2} - \frac{(x - h)^2}{b^2} = 1.$$

The foci and vertices are on the transverse axis. The distance from center to the vertices is a and the distance from center to foci is c where $c^2 = a^2 + b^2$, $a > 0$, and $b > 0$.

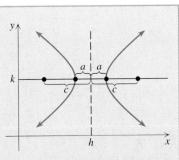

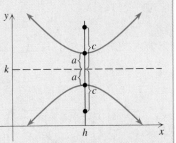

Example **3** Graphing a hyperbola centered at (h, k)

Determine the foci and equations of the asymptotes, and sketch the graph of

$$\frac{(x - 2)^2}{9} - \frac{(y + 1)^2}{4} = 1.$$

Solution

The graph that we seek is the graph of

$$\frac{x^2}{9} - \frac{y^2}{4} = 1$$

translated so that its center is $(2, -1)$. Since $a = 3$, the vertices are three units from $(2, -1)$ at $(5, -1)$ and $(-1, -1)$. Because $b = 2$, the fundamental rectangle passes through the vertices and points that are two units above and below $(2, -1)$. Draw the fundamental rectangle through the vertices and through $(2, 1)$ and $(2, -3)$. Extend the diagonals of the fundamental rectangle for the asymptotes, and draw the hyperbola opening to the left and right, as shown in Fig. 10.41 on the following page. Use $c^2 = a^2 + b^2$ to get $c = \sqrt{13}$. The foci are on the transverse axis, $\sqrt{13}$ units from the center $(2, -1)$. So the foci are $\left(2 + \sqrt{13}, -1\right)$ and $\left(2 - \sqrt{13}, -1\right)$.

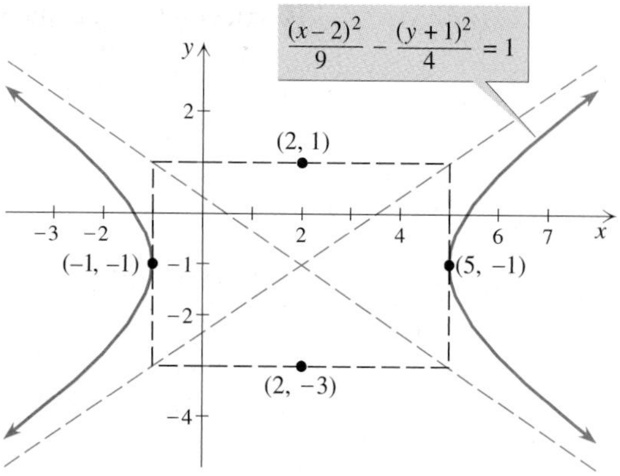

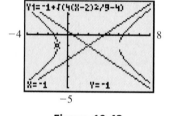

$$\frac{(x-2)^2}{9} - \frac{(y+1)^2}{4} = 1$$

■ **Figure 10.41**

The asymptotes have slopes $\pm 2/3$ and pass through $(2, -1)$. Using the point-slope form for the equation of the line, we get the equations

$$y = \frac{2}{3}x - \frac{7}{3} \quad \text{and} \quad y = -\frac{2}{3}x + \frac{1}{3}$$

as the equations of the asymptotes.

▱▱ Check by graphing

$$y_1 = -1 + \sqrt{4(x-2)^2/9 - 4},$$

$$y_2 = -1 - \sqrt{4(x-2)^2/9 - 4},$$

$$y_3 = (2x - 7)/3, \text{ and}$$

$$y_4 = (-2x + 1)/3$$

■ **Figure 10.42**

on a graphing calculator, as in Fig. 10.42.

Try This. Determine the foci and equations of the asymptotes, and graph

$$\frac{(x+2)^2}{49} - \frac{(y-1)^2}{25} = 1.$$ ■

Finding the Equation of a Hyperbola

The equation of a hyperbola depends on the location of the foci, center, vertices, transverse axis, conjugate axis, and asymptotes. However, it is not necessary to have all of this information to write the equation. In the next example, we find the equation of a hyperbola given only its transverse axis and conjugate axis.

Example **4** **Writing the equation of a hyperbola**

Find the equation of a hyperbola whose transverse axis has endpoints $(0, \pm 4)$ and whose conjugate axis has endpoints $(\pm 2, 0)$.

Solution

Since the vertices are the endpoints of the transverse axis, the vertices are $(0, \pm 4)$ and the hyperbola opens up and down. Since the fundamental rectangle goes through

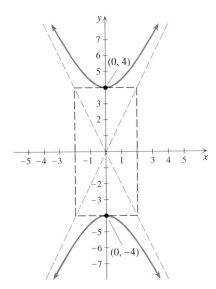

■ **Figure 10.43**

the endpoints of the transverse axis and the endpoints of the conjugate axis, we can sketch the hyperbola shown in Fig. 10.43. Since the hyperbola opens up and down and is centered at the origin, its equation is of the form

$$\frac{y^2}{a^2} - \frac{x^2}{b^2} = 1,$$

for $a > 0$ and $b > 0$. From the fundamental rectangle we get $a = 4$ and $b = 2$. So the equation of the hyperbola is

$$\frac{y^2}{16} - \frac{x^2}{4} = 1.$$

Try This. Find the equation of the hyperbola whose transverse axis has endpoints $(\pm 3, 0)$ and whose conjugate axis has endpoints $(0, \pm 8)$. ■

Example **5** **Writing the equation of a hyperbola**

Find the equation of the hyperbola with asymptotes $y = \pm 4x$ and vertices $(\pm 3, 0)$.

Solution

The vertices of the hyperbola and the vertices of the fundamental rectangle have the same x-coordinates, ± 3. Since the asymptotes go through the vertices of the fundamental rectangle, the vertices of the fundamental rectangle are $(\pm 3, \pm 12)$. Since the parabola opens right and left, its equation is $\frac{x^2}{3^2} - \frac{y^2}{12^2} = 1$ or $\frac{x^2}{9} - \frac{y^2}{144} = 1$.

Try This. Find the equation of the hyperbola with asymptotes $y = \pm 2x$ and vertices $(0, \pm 6)$. ■

For Thought

True or False? Explain.

1. The graph of $\frac{x^2}{4} - \frac{y}{9} = 1$ is a hyperbola.

2. The graph of $\frac{x^2}{16} - \frac{y^2}{9} = 1$ has y-intercepts $(0, 3)$ and $(0, -3)$.

3. The hyperbola $y^2 - x^2 = 1$ opens up and down.

4. The graph of $y^2 = 4 + 16x^2$ is a hyperbola.

5. Every point that satisfies $y = \frac{b}{a}x$ must satisfy $\frac{x^2}{a^2} - \frac{y^2}{b^2} = 1$.

6. The asymptotes for $x^2 - \frac{y^2}{4} = 1$ are $y = 2x$ and $y = -2x$.

7. The foci for $\frac{x^2}{16} - \frac{y^2}{9} = 1$ are $(5, 0)$ and $(-5, 0)$.

8. The points $(0, \sqrt{8})$ and $(0, -\sqrt{8})$ are the foci for $\frac{y^2}{3} - \frac{x^2}{5} = 1$.

9. The graph of $\frac{x^2}{9} - \frac{y^2}{4} = 1$ intersects the line $y = \frac{2}{3}x$.

10. The graph of $y^2 = 1 - x^2$ is a hyperbola centered at the origin.

10.3 Exercises

Each of the following graphs shows a hyperbola along with its vertices, foci, and asymptotes. Determine the coordinates of the vertices and foci, and the equations of the asymptotes.

1.

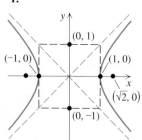

2.

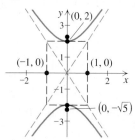

3.

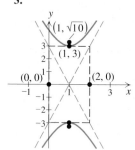

4.

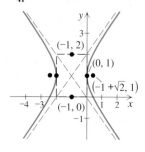

Determine the foci and the equations of the asymptotes, and sketch the graph of each hyperbola. See the procedures for graphing hyperbolas on pages 816 and 817.

5. $\dfrac{x^2}{4} - \dfrac{y^2}{9} = 1$

6. $\dfrac{x^2}{16} - \dfrac{y^2}{9} = 1$

7. $\dfrac{y^2}{4} - \dfrac{x^2}{25} = 1$

8. $\dfrac{y^2}{9} - \dfrac{x^2}{16} = 1$

9. $\dfrac{x^2}{4} - y^2 = 1$

10. $x^2 - \dfrac{y^2}{4} = 1$

11. $x^2 - \dfrac{y^2}{9} = 1$

12. $\dfrac{x^2}{9} - y^2 = 1$

13. $16x^2 - 9y^2 = 144$

14. $9x^2 - 25y^2 = 225$

15. $x^2 - y^2 = 1$

16. $y^2 - x^2 = 1$

Sketch the graph of each hyperbola. Determine the foci and the equations of the asymptotes.

17. $\dfrac{(x+1)^2}{4} - \dfrac{(y-2)^2}{9} = 1$

18. $\dfrac{(x+3)^2}{16} - \dfrac{(y+2)^2}{25} = 1$

19. $\dfrac{(y-1)^2}{4} - (x+2)^2 = 1$

20. $\dfrac{(y-2)^2}{4} - \dfrac{(x-1)^2}{9} = 1$

21. $\dfrac{(x+2)^2}{16} - \dfrac{(y-3)^2}{9} = 1$

22. $\dfrac{(x+1)^2}{16} - \dfrac{(y+2)^2}{25} = 1$

23. $(y-3)^2 - (x-3)^2 = 1$

24. $(y+2)^2 - (x+2)^2 = 1$

Find the equation of each hyperbola described below.

25. Asymptotes $y = \frac{1}{2}x$ and $y = -\frac{1}{2}x$, and x-intercepts $(6, 0)$ and $(-6, 0)$

26. Asymptotes $y = x$ and $y = -x$, and y-intercepts $(0, 2)$ and $(0, -2)$

27. Foci $(5, 0)$ and $(-5, 0)$ and x-intercepts $(3, 0)$ and $(-3, 0)$

28. Foci $(0, 5)$ and $(0, -5)$, and y-intercepts $(0, 4)$ and $(0, -4)$

29. Vertices of the fundamental rectangle $(3, \pm 5)$ and $(-3, \pm 5)$, and opening left and right

30. Vertices of the fundamental rectangle $(1, \pm 7)$ and $(-1, \pm 7)$, and opening up and down

Find the equation of each hyperbola.

31.

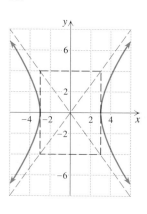

32.

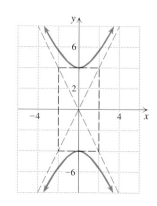

33.

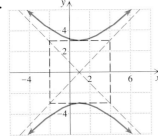

34.

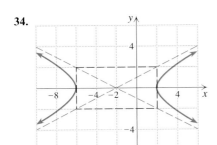

Rewrite each equation in one of the standard forms of the conic sections and identify the conic section.

35. $y^2 - x^2 + 2x = 2$

36. $4y^2 + x^2 - 2x = 15$

37. $y - x^2 = 2x$

38. $y^2 + x^2 - 2x = 0$

39. $25x^2 = 2500 - 25y^2$

40. $100x^2 = 25y^2 + 2500$

41. $100y^2 = 2500 - 25x$

42. $100y^2 = 2500 - 25x^2$

43. $2x^2 - 4x + 2y^2 - 8y = -9$

44. $2x^2 + 4x + y = -7$

45. $2x^2 + 4x + y^2 + 6y = -7$

46. $9x^2 - 18x + 4y^2 + 16y = 11$

47. $25x^2 - 150x - 8y = 4y^2 - 121$

48. $100y^2 + 4x = x^2 + 104$

Solve each problem.

49. *Telephoto Lens* The focus of the main parabolic mirror of a telephoto lens is 10 in. above the vertex, as shown in the drawing. A hyperbolic mirror is placed so that its vertex is at (0, 8). If the hyperbola has center (0, 0) and foci (0, ±10), then what is the equation of the hyperbola?

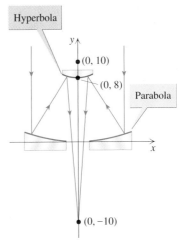

■ **Figure for Exercise 49**

50. *Parabolic Mirror* Find the equation of the cross section of the parabolic mirror described in Exercise 49.

51. *Marine Navigation* In 1990 the loran (long range navigation) system had about 500,000 users (International Loran Association, www.loran.org). A loran unit measures the difference in time that it takes for radio signals from pairs of fixed points to reach a ship. The unit then finds the equations of two hyperbolas that pass through the location of the ship and determines the location of the ship. Suppose that the hyperbolas $9x^2 - 4y^2 = 36$ and $16y^2 - x^2 = 16$ pass through the location of a ship in the first quadrant. Find the exact location of the ship.

52. *Air Navigation* A pilot is flying in the coordinate system shown in the accompanying figure. Using radio signals emitted from (4, 0) and (−4, 0), he knows that he is four units closer to (4, 0) than he is to (−4, 0). Find the equation of the hyperbola with foci (±4, 0) that passes through his location. He also knows that he is two units closer to (0, 4) than he is to (0, −4). Find the equation of the hyperbola with foci (0, ±4) that passes through his location. If the pilot is in the first quadrant, then what is his exact location?

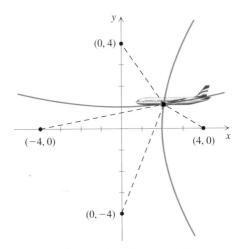

■ **Figure for Exercise 52**

53. *Points on a Hyperbola* Find all points of the hyperbola $x^2 - y^2 = 1$ that are twice as far from one focus as they are from the other focus.

54. *Perpendicular Asymptotes* For what values of a and b are the asymptotes of the hyperbola $x^2/a^2 - y^2/b^2 = 1$ perpendicular?

Graph each hyperbola on a graphing calculator along with its asymptotes. Observe how close the hyperbola gets to its asymptotes as x gets larger and larger. If $x = 50$, what is the difference between the y-value on the asymptotes and the y-value on the hyperbola?

55. $x^2 - y^2 = 1$

56. $\dfrac{y^2}{4} - \dfrac{x^2}{9} = 1$

For Writing/Discussion

57. *Details* Fill in the details in the development of the equation of a hyperbola that is outlined at the beginning of this section.

58. *Cooperative Learning* Work in groups to find conditions on $A, B, C, D,$ and E that will determine whether $Ax^2 + Bx + Cy^2 + Dy = E$ represents a parabola, an ellipse, a circle, or a hyperbola (without rewriting the equation). Test your conditions on the equations of Exercises 35–48.

59. *DeAlwis's Theorem* Let (x_1, y_1) and (x_2, y_2) be any two points on the hyperbola $x^2 - y^2 = 1$. Consider the parallelogram that has these two points as opposite vertices and sides that are parallel to the asymptotes of the hyperbola. Show that the other two opposite vertices of the parallelogram lie on a line through the origin.

Thinking Outside the Box LXXXVII & LXXXVIII

Transporting a Pipe Two corridors are 8 ft wide and 8 ft high and meet at a right angle. What is the longest length of iron pipe that can be transported around this corner without bending the pipe?

A Little Lagniappe A regulation basketball in the NBA has a circumference of 30 in. A manufacturer packages a basketball in a cubic box so that it just fits. As lagniappe, the manufacturer includes a small rubber ball in each corner of the box. The small rubber balls also fit exactly into the corners. Find the exact radius of each of the small rubber balls.

10.3 Pop Quiz

1. Find the foci and the equations of the asymptotes for the hyperbola $\dfrac{x^2}{9} - \dfrac{y^2}{4} = 1$.

2. Find the equation of the hyperbola that has vertices $(0, \pm 6)$ and asymptotes $y = \pm 3x$.

Linking Concepts

For Individual or Group Explorations

Eccentricity and Focal Radii

For the hyperbola $x^2/a^2 - y^2/b^2 = 1$, the eccentricity e is defined as $e = c/a$, where $a^2 + b^2 = c^2$. The line segments joining a point on the hyperbola to the foci, as shown in the figure, are called the focal radii.

a) What is the eccentricity of a hyperbola if the asymptotes are perpendicular?

b) Let e_1 and e_2 be the eccentricities of the hyperbolas $x^2/a^2 - y^2/b^2 = 1$ and $y^2/b^2 - x^2/a^2 = 1$, respectively. Show that $e_1^2 e_2^2 = e_1^2 + e_2^2$.

c) Show that for a point (x_1, y_1) on the right-hand branch of $x^2/a^2 - y^2/b^2 = 1$, the length of the shorter focal radius is $x_1 e - a$ and the length of the longer focal radius is $x_1 e + a$.

HINT Start with the definition of the hyperbola.

■ ■ ■ Highlights

10.1 The Parabola

Parabola: Geometric Definition	The set of all points in the plane that are equidistant from a fixed line (directrix) and a fixed point (focus) not on the line.	All points on $y = x^2$ are equidistant from $y = -1/4$ and $(0, 1/4)$.
Parabolas Opening Up and Down	$y = ax^2 + bx + c$ opens up if $a > 0$ or down if $a < 0$. The x-coordinate of the vertex is $-b/(2a)$.	$y = 2x^2 + 4x - 1$ Opens up, vertex $(-1, -3)$
	$y = a(x - h)^2 + k$ opens up if $a > 0$ or down if $a < 0$, with vertex (h, k), focus $(h, k + p)$, and directrix $y = k - p$, where $a = 1/(4p)$.	$y = 2(x + 1)^2 - 3$ Opens up, vertex $(-1, -3)$, focus $(-1, -23/8)$, directrix $y = -25/8$
Parabolas Opening Left and Right	$x = ay^2 + by + c$ opens right if $a > 0$ or left if $a < 0$. They y-coordinate of the vertex is $-b/(2a)$.	$x = y^2 - 4y + 3$ Opens right, vertex $(-1, 2)$
	$x = a(y - k)^2 + h$ opens right if $a > 0$ or left if $a < 0$, with vertex (h, k), focus $(h + p, k)$, and directrix $x = h - p$, where $a = 1/(4p)$.	$x = (y - 2)^2 - 1$ Opens right, vertex $(-1, 2)$, focus $(-3/4, 2)$, directrix $y = -5/4$

10.2 The Ellipse and the Circle

Ellipse: Geometric Definition	The set of points in the plane such that the sum of their distances from two fixed points (foci) is constant	
Horizontal Major Axis	If $a > b > 0$, the ellipse $\dfrac{x^2}{a^2} + \dfrac{y^2}{b^2} = 1$ has intercepts $(\pm a, 0)$ and $(0, \pm b)$ and foci $(\pm c, 0)$ where $c^2 = a^2 - b^2$.	$\dfrac{x^2}{25} + \dfrac{y^2}{9} = 1$ intercepts $(\pm 5, 0)$, $(0, \pm 3)$, foci $(\pm 4, 0)$
Vertical Major Axis	If $a > b > 0$, and ellipse $\dfrac{x^2}{b^2} + \dfrac{y^2}{a^2} = 1$ has intercepts $(\pm b, 0)$ and $(0, \pm a)$ and foci $(0, \pm c)$ where $c^2 = a^2 - b^2$.	$\dfrac{x^2}{9} + \dfrac{y^2}{25} = 1$ intercepts $(\pm 3, 0)$, $(0, \pm 5)$, foci $(0, \pm 4)$
Circle: Geometric Definition	The set of points in a plane such that their distance from a fixed point (the center) is constant (the radius)	
Centered at Origin	$x^2 + y^2 = r^2$ for $r > 0$ has center $(0, 0)$ and radius r.	$x^2 + y^2 = 9$ center $(0, 0)$, radius 3
Centered at (h, k)	$(x - h)^2 + (y - k)^2 = r^2$ has center (h, k) and radius r.	$(x - 2)^2 + (y + 3)^2 = 25$ center $(2, -3)$, radius 5

10.3 The Hyperbola

Hyperbola: Geometric Definition	The set of points in the plane such that the difference between the distances from two fixed points (foci) is constant	
Centered at Origin, Opening Left and Right	$\dfrac{x^2}{a^2} - \dfrac{y^2}{b^2} = 1$ opens left and right, x-intercepts $(\pm a, 0)$ and foci $(\pm c, 0)$ where $c^2 = a^2 + b^2$.	$\dfrac{x^2}{16} - \dfrac{y^2}{9} = 1$ x-intercepts $(\pm 4, 0)$, foci $(\pm 5, 0)$
Fundamental Rectangle	Goes through $(\pm a, 0)$ and $(0, \pm b)$, asymptotes $y = \pm(b/a)x$	Asymptotes $y = \pm\dfrac{3}{4}x$
Centered at Origin, Opening Up and Down	$\dfrac{y^2}{a^2} - \dfrac{x^2}{b^2} = 1$ opens up and down, intercepts $(0, \pm a)$ and foci $(0, \pm c)$ where $c^2 = a^2 + b^2$.	$\dfrac{y^2}{16} - \dfrac{x^2}{9} = 1$ y-intercepts $(\pm 4, 0)$, foci $(0, \pm 5)$
Fundamental Rectangle	Goes through $(0, \pm a)$ and $(\pm b, 0)$, asymptotes $y = \pm(a/b)x$	Asymptotes $y = \pm\dfrac{4}{3}x$
Centered at (h, k)	$\dfrac{(x - h)^2}{a^2} - \dfrac{(y - k)^2}{b^2} = 1$ opens left and right, center (h, k), foci $(h \pm c, k)$ where $c^2 = a^2 + b^2$.	$\dfrac{(x - 1)^2}{16} - \dfrac{(y + 3)^2}{9} = 1$ center $(1, -3)$, foci $(6, -3), (-4, -3)$
	$\dfrac{(y - k)^2}{a^2} - \dfrac{(x - h)^2}{b^2} = 1$ opens up and down, center (h, k), foci $(h, k \pm c)$ where $c^2 = a^2 + b^2$.	$\dfrac{(y - 5)^2}{16} - \dfrac{(x + 2)^2}{9} = 1$ center $(-2, 5)$, foci $(-2, 0), (-2, 10)$

■■■Chapter 10 Review Exercises

For Exercises 1–6, sketch the graph of each parabola. Determine the x- and y-intercepts, vertex, axis of symmetry, focus, and directrix for each.

1. $y = x^2 + 4x - 12$

2. $y = 4x - x^2$

3. $y = 6x - 2x^2$

4. $y = 2x^2 - 4x + 2$

5. $x = y^2 + 4y - 6$

6. $x = -y^2 + 6y - 9$

Sketch the graph of each ellipse, and determine its foci.

7. $\dfrac{x^2}{16} + \dfrac{y^2}{36} = 1$

8. $\dfrac{x^2}{64} + \dfrac{y^2}{16} = 1$

9. $\dfrac{(x - 1)^2}{8} + \dfrac{(y - 1)^2}{24} = 1$

10. $\dfrac{(x + 2)^2}{16} + \dfrac{(y - 1)^2}{7} = 1$

11. $5x^2 - 10x + 4y^2 + 24y = -1$

12. $16x^2 - 16x + 4y^2 + 4y = 59$

Determine the center and radius of each circle, and sketch its graph.

13. $x^2 + y^2 = 81$

14. $6x^2 + 6y^2 = 36$

15. $(x + 1)^2 + y^2 = 4$

16. $(x - 2)^2 + (y + 3)^2 = 9$

17. $x^2 + 5x + y^2 + \dfrac{1}{4} = 0$

18. $x^2 + 3x + y^2 + 5y = \dfrac{1}{2}$

Write the standard equation for each circle with the given center and radius.

19. Center $(0, -4)$, radius 3

20. Center $(-2, -5)$, radius 1

21. Center $(-2, -7)$, radius $\sqrt{6}$

22. Center $\left(\dfrac{1}{2}, -\dfrac{1}{4}\right)$, radius $\dfrac{\sqrt{2}}{2}$

Sketch the graph of each hyperbola. Determine the foci and the equations of the asymptotes.

23. $\dfrac{x^2}{64} - \dfrac{y^2}{36} = 1$

24. $\dfrac{y^2}{100} - \dfrac{x^2}{64} = 1$

25. $\dfrac{(y - 2)^2}{64} - \dfrac{(x - 4)^2}{16} = 1$

26. $\dfrac{(x - 5)^2}{100} - \dfrac{(y - 10)^2}{225} = 1$

27. $x^2 - 4x - 4y^2 + 32y = 64$

28. $y^2 - 6y - 4x^2 + 48x = 279$

Identify each equation as the equation of a parabola, ellipse, circle, or hyperbola. Try to do these problems without rewriting the equations.

29. $x^2 = y^2 + 1$

30. $x^2 + y^2 = 1$

31. $x^2 = 1 - 4y^2$

32. $x^2 + 4x + y^2 = 0$

33. $x^2 + y = 1$

34. $y^2 = 1 - x$

35. $x^2 + 4x = y^2$

36. $9x^2 + 7y^2 = 63$

Write each equation in standard form, then sketch the graph of the equation.

37. $x^2 = 4 - y^2$

38. $x^2 = 4y^2 + 4$

39. $x^2 = 4y + 4$

40. $y^2 = 4x - 4$

41. $x^2 = 4 - 4y^2$

42. $x^2 = 4y - y^2$

43. $4y^2 = 4x - x^2$

44. $x^2 - 4x = y^2 + 4y + 4$

Determine the equation of each conic section described below.

45. A parabola with focus $(1, 3)$ and directrix $x = 1/2$

46. A circle centered at the origin and passing through $(2, 8)$

47. An ellipse with foci $(\pm 4, 0)$ and vertices $(\pm 6, 0)$

48. A hyperbola with asymptotes $y = \pm 3x$ and y-intercepts $(0, \pm 3)$

49. A circle with center $(1, 3)$ and passing through $(-1, -1)$

50. An ellipse with foci $(3, 2)$ and $(-1, 2)$, and vertices $(5, 2)$ and $(-3, 2)$

51. A hyperbola with foci $(\pm 3, 0)$ and x-intercepts $(\pm 2, 0)$

52. A parabola with focus $(0, 3)$ and directrix $y = 1$

Assume that each of the following graphs is the graph of a parabola, ellipse, circle, or hyperbola. Find the equation for each graph.

53.

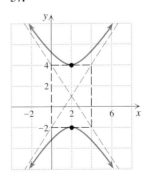

54.

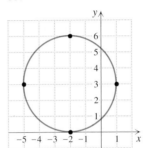

55.

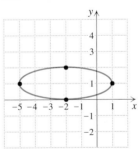

56.

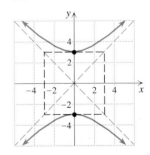

57.

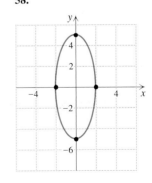

58.

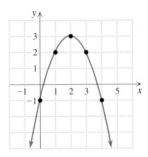

Set each problem.

59. *Nuclear Power*   A cooling tower for a nuclear power plant has a hyperbolic cross section, as shown in the accompanying figure. The diameter of the tower at the top and bottom is 240 ft, while the diameter at the middle is 200 ft. The height of the tower is $48\sqrt{11}$ ft. Find the equation of the hyperbola, using the coordinate system shown in the figure.

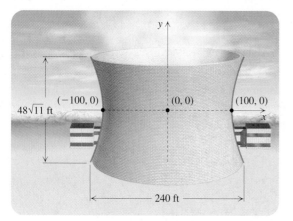

▪ **Figure for Exercise 59**

60. *Searchlight*   The bulb in a searchlight is positioned 10 in. above the vertex of its parabolic reflector, as shown in the accompanying figure. The width of the reflector is 30 in. Using the coordinate system given in the figure, find the equation of the parabola and the thickness t of the reflector at its outside edge.

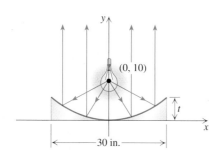

▪ **Figure for Exercise 60**

61. *Whispering Gallery*   In the whispering gallery shown in the accompanying figure, the foci of the ellipse are 60 ft apart.

Each focus is 4 ft from the vertex of an elliptical reflector. Using the coordinate system given in the figure, find the equation of the ellipse that is used to make the elliptical reflectors and the dimension marked h in the figure.

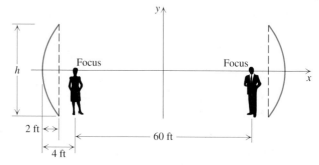

▪ **Figure for Exercise 61**

Thinking Outside the Box IXC

Falling Painter   A painter has her brush positioned a ft from the top of the ladder and b ft from the bottom as shown in the figure. Unfortunately, the ladder was placed on a frictionless surface and the bottom starts sliding away from the wall on which the ladder is leaning. As the painter falls to the ground, she keeps the brush in the same position so that it paints an arc of a curve on the adjacent wall.
a. What kind of curve is it?
b. Find an equation for the curve in an appropriate coordinate system.

▪ **Figure for Thinking Outside the Box IXC**

▪▪▪ Chapter 10 Test

Sketch the graph of each equation.

1. $x^2 + y^2 = 8$

2. $100x^2 + 9y^2 = 900$

3. $x^2 + 6x - y = -8$

4. $\dfrac{y^2}{25} - \dfrac{x^2}{9} = 1$

5. $x^2 + 6x + y^2 - 2y = 0$

6. $\dfrac{(x-2)^2}{9} - \dfrac{(y+3)^2}{4} = 1$

Determine whether each equation is the equation of a parabola, an ellipse, or a hyperbola.

7. $x^2 - 8x = y^2$

8. $x^2 - 8x = y$

9. $8x - x^2 = y^2$

10. $8x - x^2 = 8y^2$

Determine the equation of each conic section.

11. A circle with center $(-3, 4)$ and radius $2\sqrt{3}$

12. A parabola with focus $(2, 0)$ and directrix $x = -2$

13. An ellipse with foci $(0, \pm\sqrt{6})$ and x-intercepts $(\pm 2, 0)$

14. A hyperbola with foci $(\pm 8, 0)$ and vertices $(\pm 6, 0)$

Solve each problem.

15. Find the focus, directrix, vertex, and axis of symmetry for the parabola $y = x^2 - 4x$.

16. Find the foci, length of the major axis, and length of the minor axis for the ellipse $x^2 + 4y^2 = 16$.

17. Find the foci, vertices, equations of the asymptotes, length of the transverse axis, and length of the conjugate axis for the hyperbola $16y^2 - x^2 = 16$.

18. Find the center and radius of the circle
$$x^2 + x + y^2 - 3y = -\frac{1}{4}.$$

19. A lithotripter is used to disintegrate kidney stones by bombarding them with high-energy shock waves generated at one focus of an elliptical reflector. The lithotripter is positioned so that the kidney stone is at the other focus of the reflector. If the equation $81x^2 + 225y^2 = 18{,}225$ is used for the cross section of the elliptical reflector, with centimeters as the unit of measurement, then how far from the point of generation will the waves be focused?

Tying it all Together

Chapter 1–10

Sketch the graph of each equation.

1. $y = 6x - x^2$

2. $y = 6x$

3. $y = 6 - x^2$

4. $y^2 = 6 - x^2$

5. $y = 6 + x$

6. $y^2 = 6 - x$

7. $y = (6 - x)^2$

8. $y = |x + 6|$

9. $y = 6^x$

10. $y = \log_6(x)$

11. $y = \dfrac{1}{x^2 - 6}$

12. $4x^2 + 9y^2 = 36$

13. $4x^2 - 9y^2 = 36$

14. $y = 6x - x^3$

15. $2x + 3y = 6$

16. $x^2 - 6x = y^2 - 6$

Solve each equation.

17. $3(x - 3) + 5 = -9$

18. $5x - 4(2x - 3) = 17$

19. $2\left(x - \dfrac{1}{3}\right) - 3\left(\dfrac{1}{2} - x\right) = \dfrac{3}{2}$

20. $-4\left(x - \dfrac{1}{2}\right) = 3\left(x + \dfrac{2}{3}\right)$

21. $\dfrac{1}{2}x - \dfrac{1}{3} = \dfrac{1}{4}x + \dfrac{3}{2}$

22. $0.05(x - 20) + 0.02(x + 10) = 2.7$

23. $2x^2 + 31x = 51$

24. $2x^2 + 31x = 0$

25. $x^2 - 34x + 286 = 0$

26. $x^2 - 34x + 290 = 0$

11 Sequences, Series, and Probability

GAMBLERS and gambling are as old as recorded history. Homer described how Agamemnon had his soldiers cast lots to see who would face Hector. In the Bible, Moses divided the lands among the tribes by casting lots.

It wasn't until 1654 that a mathematical theory of probability was created by French mathematicians Blaise Pascal and Pierre de Fermat. In 1657 the Dutch scientist Christian Huygens published the first book on probability. It was a treatise on problems associated with gambling.

WHAT YOU WILL LEARN   In this chapter we will see how discrete mathematics can help us calculate accumulated monthly savings, various ways to select lottery numbers, and the economic impact of a $2 million payroll. We will also study the basic rules of probability and even compute the probability of winning a lottery.

11.1

Sequences

In everyday life, we hear the term *sequence* in many contexts. We describe a sequence of events, we make a sequence of car payments, or we get something out of sequence. In this section we will give a mathematical definition of the term and explore several applications.

Definition

We can think of a sequence of numbers as *an ordered list* of numbers. For example, your grades on the first four algebra tests can be listed to form a finite sequence. The sequence

$$10, 20, 30, 40, 50, \ldots$$

is an infinite sequence that lists the positive multiples of 10.

We can think of a sequence as a list, but saying that a sequence is a list is too vague for mathematics. The definition of sequence can be clearly stated by using the terminology of functions.

Definition: Sequence

> A **finite sequence** is a function whose domain is $\{1, 2, 3, \ldots, n\}$, the positive integers less than or equal to a fixed positive integer n.
>
> An **infinite sequence** is a function whose domain is the set of all positive integers.

When the domain is apparent, we will refer to either a finite or an infinite sequence as a sequence.

The function $f(n) = n^2$ with domain $\{1, 2, 3, 4, 5\}$ is a finite sequence. For the independent variable of a sequence, we usually use n (for natural number) rather than x and assume that only natural numbers can be used in place of n. For the dependent variable $f(n)$, we generally write a_n (read "a sub n"). So this finite sequence is also defined by

$$a_n = n^2 \qquad \text{for } 1 \le n \le 5.$$

The **terms** of the sequence are the values of the dependent variable a_n. We call a_n the **nth term** or the **general term** of the sequence. The equation $a_n = n^2$ provides a formula for finding the nth term. The five terms of this sequence are $a_1 = 1$, $a_2 = 4$, $a_3 = 9$, $a_4 = 16$, and $a_5 = 25$. We refer to a listing of the terms as the sequence. So 1, 4, 9, 16, 25 is a finite sequence with five terms.

Example **1** Listing terms of a finite sequence

Find all terms of the sequence

$$a_n = n^2 - n + 2 \qquad \text{for } 1 \le n \le 4.$$

Solution

Replace n in the formula by each integer from 1 through 4:

$$a_1 = 1^2 - 1 + 2 = 2, \qquad a_2 = 2^2 - 2 + 2 = 4, \qquad a_3 = 8, \qquad a_4 = 14$$

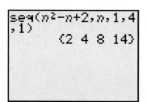

■ **Figure 11.1**

The four terms of the sequence are 2, 4, 8, and 14.

⊞ On a calculator the terms of the sequence $n^2 - n + 2$ are found for $n = 1$ through $n = 4$ in increments of 1, as shown in Fig. 11.1.

Try This. Find all terms of the sequence $a_n = n^2 - 3$ for $1 \le n \le 3$. ■

Example 2 Listing terms of an infinite sequence

Find the first four terms of the infinite sequence

$$a_n = \frac{(-1)^{n-1} 2^n}{n}.$$

Solution

Replace n in the formula by each integer from 1 through 4:

$$a_1 = \frac{(-1)^{1-1} 2^1}{1} = 2 \qquad a_2 = \frac{(-1)^1 2^2}{2} = -2$$

$$a_3 = \frac{(-1)^2 2^3}{3} = \frac{8}{3} \qquad a_4 = \frac{(-1)^3 2^4}{4} = -4$$

The first four terms of the infinite sequence are 2, -2, $8/3$, and -4.

⊞ This sequence is defined on a calculator, and the first four terms are listed in Fig. 11.2.

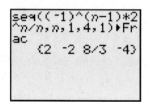

■ **Figure 11.2**

Try This. Find the first three terms of the sequence $a_n = (-1)^n 3^{n-1}$. ■

Factorial Notation

Products of consecutive positive integers occur often in sequences and other functions discussed in this chapter. For example, consider the product

$$5 \cdot 4 \cdot 3 \cdot 2 \cdot 1 = 120.$$

The notation 5! (read "five factorial") is used to represent the product of the positive integers from 1 through 5. So $5! = 5 \cdot 4 \cdot 3 \cdot 2 \cdot 1 = 120$. In general, $n!$ is the product of the positive integers from 1 through n. We will find it convenient when writing formulas to have a meaning for 0! even though it does not represent a product of positive integers. The value given to 0! is 1.

Definition:
Factorial Notation

> For any positive integer n, the notation $n!$ (read **"n factorial"**) is defined by
> $$n! = n \cdot (n - 1) \cdot \cdots \cdot 3 \cdot 2 \cdot 1.$$
> The symbol 0! is defined to be 1, $0! = 1$.

Example 3 A sequence involving factorial notation

Find the first five terms of the infinite sequence whose nth term is

$$a_n = \frac{(-1)^n}{(n-1)!}.$$

Solution

Replace n in the formula by each integer from 1 through 5. Use $0! = 1$, $1! = 1$, $2! = 2 \cdot 1 = 2$, $3! = 3 \cdot 2 \cdot 1 = 6$, and $4! = 4 \cdot 3 \cdot 2 \cdot 1 = 24$.

$$a_1 = \frac{(-1)^1}{(1-1)!} = \frac{-1}{0!} = -1 \qquad a_2 = \frac{(-1)^2}{1!} = 1$$

$$a_3 = \frac{(-1)^3}{2!} = -\frac{1}{2} \qquad a_4 = \frac{(-1)^4}{3!} = \frac{1}{6} \qquad a_5 = \frac{(-1)^5}{4!} = -\frac{1}{24}$$

Using these five terms, we write the infinite sequence as follows:

$$-1, 1, -\frac{1}{2}, \frac{1}{6}, -\frac{1}{24}, \ldots$$

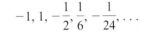 You can use the factorial function and the fraction feature to find a_3, a_4, and a_5 with a calculator, as shown in Fig. 11.3.

Try This. Find the first four terms of the sequence $a_n = (-1)^{n-1}/n!$. ■

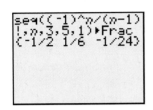

■ **Figure 11.3**

Most scientific calculators have a factorial key, which might be labeled $x!$ or $n!$. The factorial key is used to find values such as $12!$, which is $479,001,600$. The value of $69!$ is the largest factorial that many calculators can calculate, because $70!$ is larger than 10^{100}. Try your calculator to find the largest factorial that it can calculate.

Finding a Formula for the nth Term

We often have the terms of a sequence and want to write a formula for the sequence. For example, consider your two parents, four grandparents, eight great-grandparents, and so on. The sequence $2, 4, 8, \ldots$ lists the number of ancestors you have in each generation going backward in time. Continuing the sequence, the number of great-great-grandparents is 16. Therefore, we want a formula in which each term is twice the one preceding it. The formula

$$a_n = 2^n$$

is the correct formula for the sequence of ancestors. Note that the formula of Example 1,

$$a_n = n^2 - n + 2,$$

gives the same first three terms as this formula but does not have 16 as the fourth term. So if a sequence such as $2, 4, 8, \ldots$ is given out of context, many formulas might be found that will produce the same given terms. When attempting to write a formula for a given sequence of terms, look for an "obvious" pattern. Of course, some sequences do not have an obvious pattern. For example, there is no known pattern to the infinite sequence $2, 3, 5, 7, 11, 13, 17, \ldots$, the sequence of prime numbers.

Example **4** **Finding a formula for a sequence**

Write a formula for the general term of each infinite sequence.

a. $6, 8, 10, 12, \ldots$ **b.** $3, 5, 7, 9, \ldots$ **c.** $1, -\frac{1}{4}, \frac{1}{9}, -\frac{1}{16}, \ldots$

Solution

a. Assuming that all terms of the sequence are multiples of 2, we might try the expression $2n$ for the general term. Since the domain of a sequence is the set of positive integers, the expression $2n$ gives the values 2, 4, 6, 8, and so on. But 2 and 4 are not part of this sequence. To get the desired sequence, use the formula $a_n = 2n + 4$. Check by using $a_n = 2n + 4$ to find

$$a_1 = 2(1) + 4 = 6, \qquad a_2 = 2(2) + 4 = 8,$$
$$a_3 = 2(3) + 4 = 10, \qquad a_4 = 2(4) + 4 = 12.$$

b. Since every odd number is an even number plus 1, let $a_n = 2n + 1$. Using $a_n = 2n + 1$, we get $a_1 = 3, a_2 = 5, a_3 = 7$, and so on. So the general term of the given sequence is $a_n = 2n + 1$.

c. To obtain alternating signs, use a power of -1. The expression $(-1)^{n+1}$ for the numerator will give a value of 1 in the numerator when n is odd and a value of -1 in the numerator when n is even. Since the denominators are the squares of the positive integers, we use n^2 to produce the denominators. So the nth term of the sequence is

$$a_n = \frac{(-1)^{n+1}}{n^2}.$$

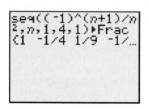

■ **Figure 11.4**

⊞ Check the formula by finding a_1, a_2, a_3, and a_4 with a calculator, as shown in Fig. 11.4.

Try This. Find the nth term of the sequence $-\dfrac{1}{2}, \dfrac{1}{4}, -\dfrac{1}{6}, \dfrac{1}{8}, \ldots$. ■

Recursion Formulas

So far, the formulas used for the nth term of a sequence have expressed the nth term as a function of n, the number of the term. In another approach, a **recursion formula** gives the nth term as a function *of the previous term*. If the first term is known, then a recursion formula determines the remaining terms of the sequence.

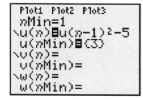

■ **Figure 11.5**

Example **5** A recursion formula

Find the first four terms of the infinite sequence in which $a_1 = 3$ and $a_n = (a_{n-1})^2 - 5$ for $n \geq 2$.

Solution

We are given $a_1 = 3$. If $n = 2$, the recursion formula is $a_2 = (a_1)^2 - 5$. Since $a_1 = 3$, we get $a_2 = 3^2 - 5 = 4$. To find the next two terms, we let $n = 3$ and $n = 4$ in the recursion formula:

$$a_3 = (a_2)^2 - 5 = 4^2 - 5 = 11 \quad \text{Since } a_2 = 4$$
$$a_4 = (a_3)^2 - 5 = 11^2 - 5 = 116 \quad \text{Since } a_3 = 11$$

So the first four terms of the infinite sequence are 3, 4, 11, and 116.

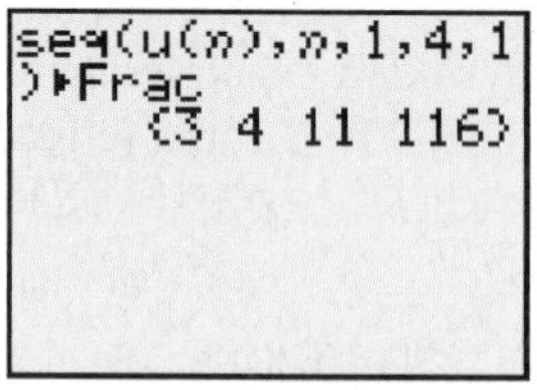

⊞ On a calculator a recursion formula is defined in the sequence mode using the Y = key, as shown in Fig. 11.5. The first four terms are shown in Fig. 11.6.

■ **Figure 11.6**

Try This. Find the first four terms of the infinite sequence in which $a_1 = 3$ and $a_n = 2a_{n-1} - 4$ for $n \geq 2$. ■

Arithmetic Sequences

An **arithmetic sequence** can be defined as a sequence in which there is a **common difference** d between consecutive terms or it can be defined (as follows) by giving a general formula that will produce such a sequence.

Definition: Arithmetic Sequence

> A sequence that has an nth term of the form
>
> $$a_n = a_1 + (n - 1)d,$$
>
> where a_1 and d are any real numbers, is called an **arithmetic sequence.**

Note that $a_n = a_1 + (n - 1)d$ can be written as $a_n = dn + (a_1 - d)$. So the terms of an arithmetic sequence can also be described as a multiple of the term number plus a constant, $(a_1 - d)$. The next example illustrates arithmetic sequences in these two forms.

Example **6** **Finding the terms of an arithmetic sequence**

Find the first four terms and the 30th term of each arithmetic sequence.

a. $a_n = -7 + (n - 1)6$ **b.** $a_n = -\dfrac{1}{2}n + 8$

Solution

a. Let n take the values from 1 through 4:

$$a_1 = -7 + (1 - 1)6 = -7$$
$$a_2 = -7 + (2 - 1)6 = -1$$
$$a_3 = -7 + (3 - 1)6 = 5$$
$$a_4 = -7 + (4 - 1)6 = 11$$

The first four terms of the sequence are $-7, -1, 5, 11$. The 30th term is

$$a_{30} = -7 + (30 - 1)6 = 167.$$

b. Let n take the values from 1 through 4:

$$a_1 = -\frac{1}{2}(1) + 8 = \frac{15}{2}$$

$$a_2 = -\frac{1}{2}(2) + 8 = 7$$

$$a_3 = -\frac{1}{2}(3) + 8 = \frac{13}{2}$$

$$a_4 = -\frac{1}{2}(4) + 8 = 6$$

The first four terms of the sequence are $15/2, 7, 13/2, 6$. The 30th term is

$$a_{30} = -\frac{1}{2}(30) + 8 = -7.$$

Try This. Find the first four terms of the sequence $a_n = 2n - 4$.   ■

A sequence is an arithmetic sequence if and only if there is a common difference between consecutive terms. The common difference can be positive or negative. In Example 6(a) the common difference is 6 and in Example 6(b) the common difference is $-\frac{1}{2}$. If you know the common difference and the first term, then you can write a formula for the general term of an arithmetic sequence.

Example **7** **Finding a formula for an arithmetic sequence**

Determine whether each sequence is arithmetic. If it is, then write a formula for the general term of the sequence.

a. 3, 7, 11, 15, 19, . . .   **b.** 5, 2, −1, −4, . . .   **c.** 2, 6, 18, 54, . . .

Solution

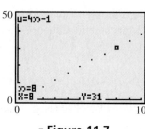

■ **Figure 11.7**

a. Since each term of the sequence is 4 larger than the previous term, the sequence is arithmetic and $d = 4$. Since $a_1 = 3$, the formula for the nth term is

$$a_n = 3 + (n - 1)4.$$

This formula can be simplified to

$$a_n = 4n - 1.$$

Points that satisfy $u_n = 4n - 1$ lie in a straight line in Fig. 11.7 because an arithmetic sequence has the same form as a linear function.   □

b. Since each term of this sequence is 3 smaller than the previous term, the sequence is arithmetic and $d = -3$. Since $a_1 = 5$, we have

$$a_n = 5 + (n - 1)(-3).$$

This formula can be simplified to

$$a_n = -3n + 8.$$

c. For 2, 6, 18, 54, . . . , we get $6 - 2 = 4$ and $18 - 6 = 12$. Since the difference between consecutive terms is not constant, the sequence is not arithmetic.

Try This. Find the nth term of the sequence 5, 11, 17, 23, ...   ■

The formula $a_n = a_1 + (n - 1)d$ involves four quantities, a_n, a_1, n, and d. If any three of them are known, the fourth can be found.

Example **8** **Finding a term of an arithmetic sequence**

An insurance representative made $30,000 her first year and $60,000 her seventh year. Assume that her annual salary figures form an arithmetic sequence and predict what she will make in her tenth year.

Solution

The seventh term is $a_7 = a_1 + (7 - 1)d$. Use $a_7 = 60,000$ and $a_1 = 30,000$ in this equation to find d:

$$60,000 = 30,000 + (7 - 1)d$$

$$30,000 = 6d$$

$$5000 = d$$

Now use the formula $a_n = 30{,}000 + (n - 1)(5000)$ to find a_{10}:

$$a_{10} = 30{,}000 + (10 - 1)(5000)$$

$$a_{10} = 75{,}000$$

So the predicted salary for her tenth year is $75,000.

Try This. Find the 30th term of an arithmetic sequence in which $a_1 = 12$ and $a_5 = 24$.   ▪

Using a recursion formula, an arithmetic sequence with first term a_1 and constant difference d is defined by $a_n = a_{n-1} + d$ for $n \geq 2$.

Example **9** A recursion formula for an arithmetic sequence

Find the first four terms of the sequence in which $a_1 = -7$ and $a_n = a_{n-1} + 6$ for $n \geq 2$.

Solution

The recursion formula indicates that each term after the first is obtained by adding 6 to the previous term:

$$a_1 = -7$$

$$a_2 = a_1 + 6 = -7 + 6 = -1$$

$$a_3 = a_2 + 6 = -1 + 6 = 5$$

$$a_4 = a_3 + 6 = 5 + 6 = 11$$

The first four terms are $-7, -1, 5$, and 11. Note that this recursion formula produces the same sequence as the formula in Example 6(a).

On a calculator define u_n to be $u_{n-1} + 6$ and get the first four terms, as shown in Fig. 11.8.

Try This. Find the first four terms of the sequence in which $a_1 = 50$ and $a_n = a_{n-1} - 5$ for $n \geq 2$.   ▪

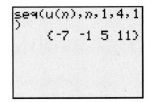

■ **Figure 11.8**

For Thought

True or False? Explain.

1. The equation $a_n = e^n$ for n a natural number defines a sequence.

2. The domain of a finite sequence is the set of positive integers.

3. We can think of a sequence as a list of the values of the dependent variable.

4. The letter n is used to represent the dependent variable.

5. The first four terms of $a_n = (-1)^{n-1}n^3$ are $-1, 8, -27, 81$.

6. The fifth term of $a_n = -3 + (n - 1)6$ is 5.

7. The common difference in the arithmetic sequence $7, 4, 1, -2, \ldots$ is 3.

8. The sequence $1, 4, 9, 16, 25, 36, \ldots$ is an arithmetic sequence.

9. If the first term of an arithmetic sequence is 4 and the third term is 14, then the fourth term is 24.

10. The sequence $a_n = 5 + 2n$ is an arithmetic sequence.

11.1 Exercises

Find all terms of each finite sequence.

1. $a_n = n^2, 1 \leq n \leq 7$

2. $a_n = (n-1)^2, 1 \leq n \leq 5$

3. $b_n = \dfrac{(-1)^{n+1}}{n+1}, 1 \leq n \leq 8$

4. $b_n = (-1)^n 3n, 1 \leq n \leq 4$

5. $c_n = (-2)^{n-1}, 1 \leq n \leq 6$

6. $c_n = (-3)^{n-2}, 1 \leq n \leq 6$

7. $a_n = 2^{2-n}, 1 \leq n \leq 5$

8. $a_n = \left(\dfrac{1}{2}\right)^{3-n}, 1 \leq n \leq 7$

9. $a_n = -6 + (n-1)(-4), 1 \leq n \leq 5$

10. $a_n = -2 + (n-1)4, 1 \leq n \leq 8$

11. $b_n = 5 + (n-1)(0.5), 1 \leq n \leq 7$

12. $b_n = \dfrac{1}{4} + (n-1)\left(-\dfrac{1}{2}\right), 1 \leq n \leq 5$

Find the first four terms and the 10th term of each infinite sequence whose nth term is given.

13. $a_n = -0.1n + 9$

14. $a_n = 0.3n - 0.4$

15. $a_n = 8 + (n-1)(-3)$

16. $a_n = -7 + (n-1)(0.5)$

17. $a_n = \dfrac{4}{2n+1}$

18. $a_n = \dfrac{2}{n^2+1}$

19. $a_n = \dfrac{(-1)^n}{(n+1)(n+2)}$

20. $a_n = \dfrac{(-1)^{n+1}}{(n+1)^2}$

Find the first five terms of the infinite sequence whose nth term is given.

21. $a_n = (2n)!$

22. $a_n = (n-1)!$

23. $a_n = \dfrac{2^n}{n!}$

24. $a_n = \dfrac{n^2}{n!}$

25. $b_n = \dfrac{n!}{(n-1)!}$

26. $b_n = \dfrac{(n+2)!}{n!}$

27. $c_n = \dfrac{(-1)^n}{(n+1)!}$

28. $c_n = \dfrac{(-2)^{2n-1}}{(n-1)!}$

29. $t_n = \dfrac{e^{-n}}{(n+2)!}$

30. $t_n = \dfrac{e^n}{n!}$

Write a formula for the nth term of each infinite sequence. Do not use a recursion formula.

31. $2, 4, 6, 8, \ldots$

32. $1, 3, 5, 7, \ldots$

33. $9, 11, 13, 15, \ldots$

34. $14, 16, 18, 20, \ldots$

35. $1, -1, 1, -1, \ldots$

36. $-\dfrac{1}{2}, \dfrac{1}{2}, -\dfrac{1}{2}, \dfrac{1}{2}, \ldots$

37. $1, 8, 27, 64, \ldots$

38. $1, -\dfrac{1}{8}, \dfrac{1}{27}, -\dfrac{1}{64}, \ldots$

39. $e, e^2, e^3, e^4, \ldots$

40. $\pi, 4\pi, 9\pi, 16\pi, \ldots$

41. $1, \dfrac{1}{2}, \dfrac{1}{4}, \dfrac{1}{8}, \ldots$

42. $1, -3, 9, -27, \ldots$

Find the first four terms and the eighth term of each infinite sequence given by a recursion formula.

43. $a_n = 3a_{n-1} + 2, a_1 = -4$

44. $a_n = 1 - \dfrac{1}{a_{n-1}}, a_1 = 2$

45. $a_n = (a_{n-1})^2 - 3, a_1 = 2$

46. $a_n = (a_{n-1})^2 - 2, a_1 = -2$

47. $a_n = a_{n-1} + 7, a_1 = -15$

48. $a_n = \dfrac{1}{2}a_{n-1}, a_1 = 8$

Find the first four terms and the 10th term of each arithmetic sequence.

49. $a_n = 6 + (n-1)(-3)$

50. $b_n = -12 + (n-1)4$

51. $c_n = 1 + (n-1)(-0.1)$

52. $q_n = 10 - 5n$

53. $w_n = -\dfrac{1}{3}n + 5$

54. $t_n = \dfrac{1}{2}n + \dfrac{1}{2}$

Determine whether each given sequence could be an arithmetic sequence.

55. $2, 3, 4, 5, \ldots$

56. $-7, -4, -2, 0, \ldots$

57. $1, 0.5, 1, 0.5, \ldots$

58. $3, 0, -3, -6, \ldots$

59. $2, 4, 8, 16, \ldots$

60. $1, \dfrac{5}{4}, \dfrac{3}{2}, \dfrac{7}{4}, \ldots$

61. $\dfrac{\pi}{4}, \dfrac{\pi}{2}, \dfrac{3\pi}{4}, \pi, \ldots$

62. $1, 2, 3, 2, 3, \ldots$

Write a formula for the nth term of each arithmetic sequence. Do not use a recursion formula.

63. 1, 6, 11, 16, . . .

64. 2, 5, 8, 11, . . .

65. 0, 2, 4, 6, . . .

66. −3, 3, 9, 15, . . .

67. 5, 1, −3, −7, . . .

68. 1, −1, −3, −5, . . .

69. 1, 1.1, 1.2, 1.3, . . .

70. 2, 2.75, 3.5, 4.25, . . .

71. $\dfrac{\pi}{6}, \dfrac{\pi}{3}, \dfrac{\pi}{2}, \dfrac{2\pi}{3}, \ldots$

72. $\dfrac{\pi}{12}, \dfrac{\pi}{6}, \dfrac{\pi}{4}, \dfrac{\pi}{3}, \ldots$

73. 20, 35, 50, 65, . . .

74. 70, 60, 50, 40, . . .

Find the indicated part of each arithmetic sequence.

75. Find the eighth term of the sequence that has a first term of −3 and a common difference of 5.

76. Find the 11th term of the sequence that has a first term of 4 and a common difference of −0.8.

77. Find the 10th term of the sequence whose third term is 6 and whose seventh term is 18.

78. Find the eighth term of the sequence whose second term is 20 and whose fifth term is 10.

79. Find the common difference of the sequence in which the first term is 12 and the 21st term is 96.

80. Find the common difference of the sequence in which the first term is 5 and the 11th term is −10.

81. Find a formula for a_n, given that $a_3 = 10$ and $a_7 = 20$.

82. Find a formula for a_n, given that $a_5 = 30$ and $a_{10} = -5$.

Write a recursion formula for each sequence.

83. 3, 12, 21, 30, . . .

84. 30, 25, 20, 15, . . .

85. $\dfrac{1}{3}, 1, 3, 9, \ldots$

86. $4, -1, -\dfrac{1}{4}, -\dfrac{1}{16}, \ldots$

87. 16, 4, 2, $\sqrt{2}$, . . .

88. $t^2, t^4, t^8, t^{16}, \ldots$

Solve each problem.

89. *Recursive Pricing*   The MSRP for a 2005 Jeep Grand Cherokee was $35,265 (Edmund's, www.edmunds.com). Analysts estimate that prices will increase 6% per year for the next five years. Find the price to the nearest dollar for this model for the years 2006 through 2010. Write a formula for this sequence.

90. *Rising Salary*   Suppose that you made $35,265 in 2005 and your boss promised that you will get a $2116 raise each year

for the next five years. Find your salary for the years 2006 through 2010. Write a formula for this sequence. Is your salary in 2010 equal to the price of the 2010 Jeep Grand Cherokee from the previous exercise?

91. *Reading Marathon*   On November 1 an English teacher had his class read five pages of a long novel. He then told them to increase their daily reading by three pages each day. For example, on November 2 they should read eight pages. Write a formula for the number of pages that they will read on the nth day of November. If they follow the teacher's instructions, then how many pages will they be reading on the last day of November?

92. *Stiff Penalty*   If a contractor does not complete a multimillion-dollar construction project on time, he must pay a penalty of $500 for the first day that he is late, $700 for the second day, $900 for the third day, and so on. Each day the penalty is $200 larger than the previous day. Write a formula for the penalty on the nth day. What is the penalty for the 10th day?

93. *Nursing Home Care*   The average annual cost for nursing home care in 2005 was $51,000 (www.nursinghomereports.com). If the average increases by $1800 each year, then what will the average annual cost be in 2014?

94. *Good Planning*   Sam's retirement plan gives her a fixed raise of d dollars each year. If her retirement income was $24,500 in her fifth year of retirement and $25,700 in her ninth year, then what was her income her first year? What will her income be in her 13th year of retirement?

95. *Countertops*   It takes C_n corner tiles, E_n edge tiles, and I_n interior tiles to cover an n-ft by n-ft island, as shown in the accompanying figure. Assuming all tiles are 6 in. by 6 in., write expressions for C_n, E_n, and I_n.

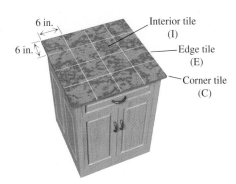

■ **Figure for Exercise 95**

96. *Countertop Pricing*   In the previous exercise, corner tiles cost $0.89 each, edge tiles cost $0.79 each, and interior tiles cost $0.69 each. Write an expression for K_n, the cost of the tiles to cover an n-ft by n-ft countertop.

For Writing/Discussion

97. Explain the difference between a function and a sequence.

98. *Cooperative Learning* Write your own formula for the *n*th term of a sequence on a piece of paper and list the first five terms. Disclose the terms one at a time to your classmates, giving them the opportunity to guess the formula after each disclosed term.

Thinking Outside the Box XC & XCI

Some Factorials Find the ones digit and the tens digit in the sum

$$0! + 1! + 2! + 3! + 4! + \cdots + 10000!.$$

Moving the 6 Find the least positive integer whose last (units) digit is 6 such that moving the 6 to the first digit produces a number that is 4 times as large as the original number. For example, if 123456 multiplied by 4 were 612345, then we would have found such a number.

11.1 Pop Quiz

1. List the first four terms of the sequence $a_n = (n - 1)^2$.

2. Find a formula for the *n*th term of the sequence $5, -10, 15, -20, \ldots$.

3. Find the first four terms of $a_n = -2a_{n-1}$ where $a_1 = -1$.

4. Find the ninth term of the arithmetic sequence that has first term 2 and fifth term 18.

Linking Concepts

For Individual or Group Explorations

Converging Sequences

If the terms of a sequence a_n get closer and closer to some number L as n gets larger and larger without bound, then we say that the sequence converges to L and L is the limit of the sequence. If the sequence does not converge, then it diverges. These terms are defined with greater precision and studied extensively in calculus, but with a graphing calculator we can gain a good understanding of these ideas.

a) Let $a_n = (0.99999)^n$ and find a_n for $n = 100, 1000,$ and $1,000,000$. What do you think is the limit of this sequence?

b) Let $a_n = (1.00001)^n$ and find a_n for $n = 100, 1000,$ and $1,000,000$. What do you think is the limit of this sequence?

c) Let $a_n = (1 + 1/n)^n$ and find a_n for $n = 1000, 10,000$ and $100,000$. This sequence converges to a number that you have seen earlier in this course. What is the limit? What is the difference between this sequence and the one in part (b)?

d) Graph the functions $f(x) = (0.99999)^x, g(x) = (1.00001)^x,$ and $h(x) = (1 + 1/x)^x$. Identify any horizontal asymptotes for these graphs. What is the relationship between horizontal asymptotes and limits of sequences?

e) Consider the sequence $a_n = k^n$ where k is a fixed real number. For what values of k do you think the sequence converges and for what values do you think it diverges?

11.2

Series

In this section we continue the study of sequences, but here we concentrate on finding the sum of the terms of a sequence. For example, if an employee starts at $20,000 per year and gets a $1200 raise each year for the next 39 years, then the total pay for 40 years of work is the sum of 40 terms of an arithmetic sequence:

$$20{,}000 + 21{,}200 + 22{,}400 + \cdots + 66{,}800$$

This sum is an example of a *series*. We could add the 40 terms to find the total pay, but we will soon discover a formula for this sum that involves only the first term, the last term, and the number of terms. (The total investment in an employee over a lifetime of work is often cited as a reason for selecting personnel carefully.)

Summation Notation

As a shorthand way to indicate the sum of the terms of a sequence, we adopt a new notation called **summation notation.** We use the Greek letter Σ (sigma) in summation notation. For example, the sum of the annual salaries for 40 years of work is written as

$$\sum_{n=1}^{40} [20{,}000 + (n - 1)(1200)].$$

Following the letter sigma is the formula for the nth term of an arithmetic sequence in which the first term is 20,000 and the common difference is 1200. The numbers below and above the letter sigma indicate that this is an expression for the sum of the first through fortieth terms of this sequence.

As another example, the sum of the squares of the first five positive integers can be written as

$$\sum_{i=1}^{5} i^2.$$

To evaluate this sum, let i take the integral values from 1 though 5 in the expression i^2. Thus

$$\sum_{i=1}^{5} i^2 = 1^2 + 2^2 + 3^2 + 4^2 + 5^2 = 1 + 4 + 9 + 16 + 25 = 55.$$

The letter i in the summation notation is called the **index of summation.** Although we usually use i or n for the index of summation, any letter may be used. For example, the expressions

$$\sum_{n=1}^{5} n^2, \quad \sum_{j=1}^{5} j^2, \quad \text{and} \quad \sum_{i=1}^{5} i^2$$

all have the same value.

In the summation notation, the expression following the letter sigma is the *general term* of a sequence. The numbers below and above sigma indicate which terms of the sequence are to be added.

Example **1** Evaluating summations

Find the sum in each case.

a. $\displaystyle\sum_{i=1}^{6} (-1)^i 2^{i-1}$ **b.** $\displaystyle\sum_{n=3}^{7} (2n - 1)$ **c.** $\displaystyle\sum_{i=1}^{5} 4$

Solution

a. Evaluate $(-1)^i 2^{i-1}$ for $i = 1$ through 6 and add the resulting terms:

$$\sum_{i=1}^{6} (-1)^i 2^{i-1} = (-1)^1 2^0 + (-1)^2 2^1 + (-1)^3 2^2 + (-1)^4 2^3 + (-1)^5 2^4 + (-1)^6 2^5$$

$$= -1 + 2 - 4 + 8 - 16 + 32 = 21$$

b. Find the third through seventh terms of the sequence whose general term is $2n - 1$ and add the results:

$$\sum_{n=3}^{7} (2n - 1) = 5 + 7 + 9 + 11 + 13 = 45$$

c. Every term of this series is 4. The notation $i = 1$ through 5 means that we add the first five 4's from a sequence in which every term is 4:

$$\sum_{i=1}^{5} 4 = 4 + 4 + 4 + 4 + 4 = 20$$

Try This. Find the sum $\displaystyle\sum_{i=2}^{5} (-1)^i (2i)$. ■

An expression in summation notation or an expression such as $1 + 2 + 3$, in which we have not actually performed the addition, is called an **indicated sum.**

Definition: Series

> The indicated sum of the terms of a sequence is called a **series.**

Just as a sequence may be finite or infinite, a series may be finite or infinite. In summation notation we use the infinity symbol to indicate that there is no end to the terms of the series. For example,

$$1 + \frac{1}{2} + \frac{1}{3} + \frac{1}{4} + \cdots = \sum_{n=1}^{\infty} \frac{1}{n}.$$

To write a series in summation notation, a formula must be found for the nth term of the corresponding sequence.

Example **2** Writing a series in summation notation

Write each series using summation notation.

a. $2 + 4 + 6 + 8 + 10$ **b.** $\dfrac{1}{5} - \dfrac{1}{7} + \dfrac{1}{9} - \dfrac{1}{11} + \dfrac{1}{13}$

c. $1 + \dfrac{1}{4} + \dfrac{1}{9} + \dfrac{1}{16} + \cdots$

Solution

a. The series consists of a sequence of even integers. The nth term for this sequence is $a_n = 2n$. This series consists of five terms of this sequence.

$$2 + 4 + 6 + 8 + 10 = \sum_{i=1}^{5} 2i$$

b. This series has two features of interest: The denominators of the fractions are odd integers and the signs alternate. If we use $a_n = 2n + 1$ as the general term for odd integers, we get $a_1 = 3$ and $a_2 = 5$. We do not always have to use $i = 1$ in the summation notation. In this series it is easier to use $i = 2$ through 6. Use $(-1)^i$ to get the alternating signs:

$$\frac{1}{5} - \frac{1}{7} + \frac{1}{9} - \frac{1}{11} + \frac{1}{13} = \sum_{i=2}^{6} \frac{(-1)^i}{2i + 1}$$

c. Notice that the denominators are the squares of the positive integers and the series is infinite. So

$$1 + \frac{1}{4} + \frac{1}{9} + \frac{1}{16} + \cdots = \sum_{n=1}^{\infty} \frac{1}{n^2}.$$

Try This. Write the series $-\frac{1}{2} + \frac{1}{4} - \frac{1}{6} + \frac{1}{8}$ using summation notation. ■

Changing the Index of Summation

In Example 2(b) the index of summation i ranged from 2 through 6, but the starting point of the index is arbitrary. The notation

$$\sum_{j=3}^{7} \frac{(-1)^{j-1}}{2j - 1}$$

is another notation for the same sum. The summation notation for a given series can be written so that the index begins at any given number.

Example 3 Adjusting the index of summation

Rewrite the series so that instead of index i, it has index j, where j starts at 1.

a. $\displaystyle\sum_{i=2}^{6} \frac{(-1)^i}{2i + 1}$ **b.** $\displaystyle\sum_{i=3}^{\infty} \frac{1}{i^2}$

Solution

In this series, i takes the values 2 through 6. If j starts at 1, then $j = i - 1$, and j takes the values 1 through 5. If $j = i - 1$, then $i = j + 1$. To change the formula for the general term of the series, replace i by $j + 1$:

$$\sum_{i=2}^{6} \frac{(-1)^i}{2i + 1} = \sum_{j=1}^{5} \frac{(-1)^{j+1}}{2(j + 1) + 1} = \sum_{j=1}^{5} \frac{(-1)^{j+1}}{2j + 3}$$

Check that these two series have exactly the same five terms.

b. In this series, i takes the values 3 through ∞. If j starts at 1, then $j = i - 2$ and $i = j + 2$. Since the series is infinite, j ranges from 1 through ∞:

$$\sum_{i=3}^{\infty} \frac{1}{i^2} = \sum_{j=1}^{\infty} \frac{1}{(j+2)^2}$$

Check that these two series have exactly the same terms.

Try This. Rewrite the series $\sum_{i=2}^{7}(-1)^{i-1}i^3$ with index j starting at 1.   ■

The Mean

If you take a sequence of three tests, then your "average" is the sum of the three test scores divided by 3. What is commonly called the "average" is called the *mean* or *arithmetic mean* in mathematics. The mean can be defined using summation notation.

Definition: Mean

> The **mean** of the numbers $x_1, x_2, x_3, \ldots, x_n$ is the number $\bar{x}$, given by
>
> $$\bar{x} = \frac{\sum\limits_{i=1}^{n} x_i}{n}.$$

Example **4** **Finding the mean of a sequence of numbers**

Find the mean of the numbers $-12, 3, 0, 5, -2, 9$.

Solution

To find the mean, divide the total of the six numbers by 6:

$$\bar{x} = \frac{-12 + 3 + 0 + 5 + (-2) + 9}{6} = \frac{1}{2}$$

The mean is $1/2$.

Try This. Find the mean of the numbers $5, 12, -17, 23$.   ■

Arithmetic Series

The indicated sum of an arithmetic sequence is an **arithmetic series.** The sum of a finite arithmetic series can be found without actually adding all of the terms. Let S represent the sum of the even integers from 2 through 50. We can find S by using the following procedure:

$$S = \ 2 + \ 4 + \ 6 + \ 8 + \cdots + 48 + 50$$

$$\underline{S = 50 + 48 + 46 + 44 + \cdots + \ 4 + \ 2} \qquad \text{Write terms in reverse order.}$$

$$2S = 52 + 52 + 52 + 52 + \cdots + 52 + 52 \qquad \text{Add corresponding terms.}$$

Since there are 25 numbers in the series of even integers from 2 through 50, the number 52 appears 25 times on the right-hand side of the last equation.

$$2S = 25(52) = 1300$$

$$S = 650$$

So the sum of the even integers from 2 through 50 is 650.

We can use the same idea to develop a formula for the sum of n terms of any arithmetic series. Let $S_n = a_1 + a_2 + a_3 + \cdots + a_n$ be a finite arithmetic series. Since there is a constant difference between the terms, S_n can be written forwards and backwards as follows:

$$
\begin{aligned}
S_n &= a_1 && + (a_1 + d) + (a_1 + 2d) + \cdots + a_n \\
S_n &= a_n && + (a_n - d) + (a_n - 2d) + \cdots + a_1 \\
\hline
2S_n &= (a_1 + a_n) + (a_1 + a_n) + (a_1 + a_n) + \cdots + (a_1 + a_n) && \text{Add.}
\end{aligned}
$$

Now, there are n terms of the type $a_1 + a_n$ on the right-hand side of the last equation, so the right-hand side can be simplified:

$$2S_n = n(a_1 + a_n)$$

$$S_n = \frac{n}{2}(a_1 + a_n)$$

This result is summarized as follows.

Theorem: Sum of an Arithmetic Series

The sum S_n of the first n terms of an arithmetic series with first term a_1 and nth term a_n is given by the formula

$$S_n = \frac{n}{2}(a_1 + a_n).$$

The mean of n numbers is their sum divided by n. To find the mean $\bar{a}$ for the first n terms of an arithmetic series, divide S_n by n to get

$$\bar{a} = \frac{a_1 + a_n}{2}.$$

So $S_n = n\bar{a}$. The sum of the first n terms is n times the "average" term.

Example **5** Finding the sum of an arithmetic series

Find the sum of each arithmetic series.

a. $\displaystyle\sum_{i=1}^{15} (3i - 5)$ **b.** $36 + 41 + 46 + 51 + \cdots + 91$

Solution

a. To find the sum, we need to know the first term, the last term, and the number of terms. For this series, $a_1 = -2$, $a_{15} = 40$, and $n = 15$. So

$$\sum_{i=1}^{15} (3i - 5) = \frac{15}{2}(-2 + 40) = 285.$$

b. For this series, $a_1 = 36$ and $a_n = 91$, but to find S_n, the number of terms n must be known. We can find n from the formula for the general term of the arithmetic sequence $a_n = a_1 + (n - 1)d$ by using $d = 5$:

$$91 = 36 + (n - 1)5$$

$$55 = (n - 1)5$$

$$11 = n - 1$$

$$12 = n$$

We can now use $n = 12$, $a_1 = 36$, and $a_{12} = 91$ in the formula to get the sum of the 12 terms of this arithmetic series:

$$S_{12} = \frac{12}{2}(36 + 91) = 762$$

Try This. Find the sum $\sum_{i=1}^{24}(2i - 8)$. ■

In the next example we return to the problem of finding the sum of the 40 annual salaries presented at the beginning of this section.

Example **6** **Total salary for 40 years of work**

Find the total salary for an employee who is paid \$20,000 for the first year and receives a \$1200 raise each year for the next 39 years. Find the mean of the 40 annual salaries.

Solution

The salaries form an arithmetic sequence whose nth term is $20{,}000 + (n - 1)1200$. The total of the first 40 terms of the sequence of salaries is given by

$$S_{40} = \sum_{n=1}^{40}[20{,}000 + (n - 1)1200] = \frac{40}{2}(20{,}000 + 66{,}800) = 1{,}736{,}000.$$

The total salary for 40 years of work is \$1,736,000. Divide the total salary by 40 to get a mean salary of \$43,400.

Try This. Find the mean of the numbers 20, 40, 60, 80, . . . , 960. ■

For Thought

True or False? Explain.

1. $\sum_{i=1}^{3}(-2)^i = -6$

2. $\sum_{i=1}^{6}(0 \cdot i + 5) = 30$

3. $\sum_{i=1}^{k} 5i = 5\left(\sum_{i=1}^{k} i\right)$

4. $\sum_{i=1}^{k}(i^2 + 1) = \left(\sum_{i=1}^{k} i^2\right) + k$

5. There are nine terms in the series $\sum_{i=5}^{14} 3i^2$.

6. $\sum_{i=2}^{8}(-1)^i 3i^2 = \sum_{j=1}^{7}(-1)^{j-1}3(j + 1)^2$

7. The series $\displaystyle\sum_{n=1}^{9} (3n - 5)$ is an arithmetic series.

8. The sum of the first n counting numbers is $\dfrac{n(n + 1)}{2}$.

9. The sum of the even integers from 8 through 68 inclusive is $\dfrac{60}{2}(8 + 68)$.

10. $\displaystyle\sum_{i=1}^{10} i^2 = \dfrac{10}{2}(1 + 100)$

11.2 Exercises

Find the sum of each series.

1. $\displaystyle\sum_{i=1}^{5} i^2$

2. $\displaystyle\sum_{i=0}^{4} i^2$

3. $\displaystyle\sum_{j=0}^{3} 2^j$

4. $\displaystyle\sum_{j=0}^{4} (2j)$

5. $\displaystyle\sum_{n=0}^{4} n!$

6. $\displaystyle\sum_{n=0}^{3} \dfrac{1}{n!}$

7. $\displaystyle\sum_{i=1}^{10} 3$

8. $\displaystyle\sum_{i=0}^{5} 4$

9. $\displaystyle\sum_{j=1}^{7} (2j + 5)$

10. $\displaystyle\sum_{j=1}^{6} (2j - 1)$

11. $\displaystyle\sum_{n=7}^{44} (-1)^n$

12. $\displaystyle\sum_{n=3}^{47} (-1)^{n+1}$

13. $\displaystyle\sum_{i=0}^{5} i(i - 1)(i - 2)$

14. $\displaystyle\sum_{i=1}^{6} (i - 2)(i - 3)$

Write each series in summation notation. Use the index i and let i begin at 1 in each summation.

15. $1 + 2 + 3 + 4 + 5 + 6$

16. $2 + 4 + 6 + 8 + 10 + 12$

17. $-1 + 3 - 5 + 7 - 9$

18. $3 - 6 + 9 - 12 + 15 - 18$

19. $1 + 4 + 9 + 16 + 25$

20. $1 + 3 + 9 + 27 + 81$

21. $1 - \dfrac{1}{2} + \dfrac{1}{4} - \dfrac{1}{8} + \dfrac{1}{16} - \cdots$

22. $-1 + \dfrac{1}{2} - \dfrac{1}{3} + \dfrac{1}{4} - \cdots$

23. $\ln(x_1) + \ln(x_2) + \ln(x_3) + \cdots$

24. $x^3 + x^4 + x^5 + x^6 + x^7 + \cdots$

25. $a + ar + ar^2 + \cdots + ar^{10}$

26. $b^2 + b^3 + b^4 + \cdots + b^{12}$

Determine whether each equation is true or false.

27. $\displaystyle\sum_{i=1}^{3} i^2 = \sum_{j=2}^{4} (j - 1)^2$

28. $\displaystyle\sum_{i=0}^{4} i! = \sum_{j=2}^{6} (j - 2)!$

29. $\displaystyle\sum_{x=1}^{5} (2x - 1) = \sum_{y=3}^{7} (2y - 1)$

30. $\displaystyle\sum_{x=1}^{8} (2x) = \sum_{y=3}^{9} (2y - 4)$

31. $\displaystyle\sum_{x=1}^{5} (x + 5) = \sum_{j=0}^{4} (j + 6)$

32. $\displaystyle\sum_{i=0}^{3} x^i = \sum_{j=1}^{4} x^{j-1}$

Rewrite each series using the new index j as indicated.

33. $\displaystyle\sum_{i=1}^{32} (-1)^i = \sum_{j=0}$

34. $\displaystyle\sum_{i=1}^{10} 2^i = \sum_{j=0}$

35. $\displaystyle\sum_{i=4}^{13} (2i + 1) = \sum_{j=1}$

36. $\displaystyle\sum_{i=7}^{12} (3i - 4) = \sum_{j=1}$

37. $\displaystyle\sum_{x=2}^{10} \dfrac{10!}{x!(10 - x)!} = \sum_{j=0}$

38. $\displaystyle\sum_{i=2}^{9} \dfrac{x^i}{i!} = \sum_{j=0}$

39. $\displaystyle\sum_{n=2}^{\infty} \dfrac{5^n e^{-5}}{n!} = \sum_{j=5}$

40. $\displaystyle\sum_{i=0}^{\infty} 3^{2i-1} = \sum_{j=3}$

Write out all of the terms of each series.

41. $\sum_{i=0}^{5} 0.5r^i$

42. $\sum_{i=1}^{6} i^n$

43. $\sum_{j=0}^{4} a^{4-j}b^j$

44. $\sum_{j=0}^{3} (-1)^j x^{3-j} y^j$

45. $\sum_{i=0}^{2} \frac{2}{i!(2-i)!} a^{2-i}b^i$

46. $\sum_{j=0}^{3} \frac{6}{j!(3-j)!} a^{3-j}b^j$

Find the mean of each sequence of numbers. Round approximate answers to three decimal places.

47. 6, 23, 45

48. 33, 42, 78, 19

49. −6, 0, 3, 4, 3, 92

50. 12, 20, 12, 30, 28, 28, 10

51. $\sqrt{2}, \pi, 33.6, -19.4, 52$

52. $\sqrt{5}, -3\sqrt{3}, \pi/2, e, 98.6$

Find the sum of each arithmetic series.

53. $\sum_{i=1}^{12} (6i - 9)$

54. $\sum_{i=1}^{11} (0.5i + 4)$

55. $\sum_{n=3}^{15} (-0.1n + 1)$

56. $\sum_{n=4}^{20} (-0.3n + 2)$

57. $1 + 2 + 3 + \cdots + 47$

58. $2 + 4 + 6 + \cdots + 88$

59. $5 + 10 + 15 + \cdots + 95$

60. $10 + 20 + 30 + \cdots + 450$

61. $10 + 9 + 8 + \cdots + (-238)$

62. $1000 + 998 + 996 + \cdots + 888$

63. $8 + 5 + 2 + (-1) + \cdots + (-16)$

64. $5 + 1 + (-3) + (-7) + \cdots + (-27)$

65. $3 + 7 + 11 + 15 + \cdots + 55$

66. $-6 + 1 + 8 + 15 + \cdots + 50$

67. $\frac{1}{2} + \frac{3}{4} + 1 + \frac{5}{4} + \cdots + 5$

68. $1 + \frac{4}{3} + \frac{5}{3} + 2 + \cdots + \frac{22}{3}$

Find the mean of each arithmetic sequence.

69. 2, 4, 6, 8, 10, 12, 14, 16

70. 3, 6, 9, 12, 15, 18, 21, 24, 27

71. 2, 4, 6, . . . , 142

72. 3, 6, 9, . . . , 150

Solve each problem using the ideas of series.

73. *Total Salary*   If a graphic artist makes $30,000 his first year and gets a $1000 raise each year, then what will be his total salary for 30 years of work? What is his mean annual salary for 30 years of work?

74. *Assigned Reading*   An English teacher with a minor in mathematics told her students that if they read $2n + 1$ pages of a long novel on the nth day of October, for each day of October, then they will exactly finish the novel in October. How many pages are there in this novel? What is the mean number of pages read per day?

75. *Mount of Cans I*   A grocer wants to build a "mountain" out of cans of mountain-grown coffee, as shown in the accompanying figure. The first level is to be a rectangle containing 9 rows of 12 cans in each row. Each level after the first is to contain one less row with 12 cans in each row. Finally, the top level is to contain one row of 12 cans. Write a sequence whose terms are the number of cans at each level. Write the sum of the terms of this sequence in summation notation and find the number of cans in the mountain.

■ **Figure for Exercise 75**

76. *Mount of Cans II*   Suppose that the grocer in Exercise 75 builds the "mountain" so that each level after the first contains one less row and one less can in each row. Write a sequence whose terms are the number of cans at each level. Write the sum of the terms of this sequence in summation notation and find the number of cans in the mountain.

77. *Annual Payments*   Wilma deposited $1000 into an account paying 5% compounded annually each January 1 for ten consecutive years. Given that the first deposit was made January 1, 1990, and the last was made January 1, 1999, write a series in summation notation whose sum is the amount in the account on January 1, 2000.

78. *Compounded Quarterly*   Duane deposited $100 into an account paying 4% compounded quarterly each January 1 for eight consecutive years. Given that the first deposit was made

January 1, 1990, and the last was made January 1, 1997, write a series in summation notation whose sum is the amount in the account on January 1, 2000.

79. *Drug Therapy*   A doctor instructed a patient to start taking 200 mg of Dilantin every 8 hours to control seizures. If the half-life of Dilantin for this patient is 12 hours, then he still has 63% of the last dose in his body when he takes the next dose. Immediately after taking the fourth pill, the amount of Dilantin in the patient's body is

$$200 + 200(0.63) + 200(0.63)^2 + 200(0.63)^3.$$

Find the sum of this series.

80. *Extended Drug Therapy*   Use summation notation to write a series for the amount of Dilantin in the body of the patient in the previous exercise after one week on Dilantin. Find the sum of the series.

Find the indicated mean.

81. Find the mean of the 9th through the 60th terms inclusive of the sequence $a_n = 5n + 56$.

82. Find the mean of the 15th through the 55th terms inclusive of the sequence $a_n = 7 - 4n$.

83. Find the mean of the 7th through the 10th terms inclusive of the sequence in which $a_1 = -2$ and $a_n = (a_{n-1})^2 - 3$ for $n \geq 2$.

84. Find the mean of the 5th through the 8th terms inclusive of the sequence in which $a_n = (-1/2)^n$.

For Writing/Discussion

85. Explain the difference between a sequence and a series.

86. *Cooperative Learning*   Write your own formula for the *n*th term a_n of a sequence on a piece of paper. Find the *partial sums* $S_1 = a_1$, $S_2 = a_1 + a_2$, $S_3 = a_1 + a_2 + a_3$, and so on. Disclose the first five partial sums one at a time to your classmates, giving them the opportunity to guess the formula for a_n after each disclosed partial sum.

Thinking Outside the Box XCII

Supersize Sum   Find the exact value of the following expression:

$$\sum_{i=10^7}^{10^8-1} i + \sum_{i=10^9}^{10^{10}-1} i$$

11.2 Pop Quiz

1. Find the sum of the series $\sum_{n=1}^{5} n!$.

2. Write the series $1 + \dfrac{1}{2} + \dfrac{1}{4} + \dfrac{1}{8} + \cdots$ in summation

notation beginning with $i = 1$.

3. Find the sum of the series $\sum_{i=5}^{24} (3i - 5)$.

Linking Concepts

For Individual or Group Explorations

Various Means

When we find the mean of a set of numbers, we are attempting to find the "middle" or "center" of the set of numbers. The arithmetic mean that we defined in this section is used so frequently that it is hard to believe that there could be any other way to find the middle. However, there are several other ways to find the middle that produce approximately the same results. The wealth (in today's dollars) of the top five richest Americans of all time is shown in the table (Infoplease, www.infoplease.com).

(continued on next page)

Name	Wealth ($billions)
John D. Rockefeller	189.6
Andrew Carnegie	100.5
Cornelius Vanderbilt	95.9
John Jacob Astor	78.0
William H. Gates III	61.7

a) Find the *median*, the score for which approximately one-half of the scores are lower and one-half of the scores are higher.

b) Find the *geometric mean*, the nth root of the product of the n scores:

$$GM = \sqrt[n]{x_1 \cdot x_2 \cdot x_3 \cdot \cdots \cdot x_n}$$

c) Find the *harmonic mean*, the number of scores divided by the sum of the reciprocals of the scores:

$$HM = \frac{n}{\sum \frac{1}{x}}$$

d) Find the *quadratic mean* by finding the sum of the squares of the scores, divide by n, then take the square root:

$$QM = \sqrt{\frac{\sum x^2}{n}}$$

11.3

Geometric Sequences and Series

An arithmetic sequence has a constant difference between consecutive terms. That simple relationship allowed us to find a formula for the sum of n terms of an arithmetic sequence. In this section we study another type of sequence in which there is a simple relationship between consecutive terms.

Geometric Sequences

A **geometric sequence** is defined as a sequence in which there is a **constant ratio** r between consecutive terms. The formula that follows will produce such a sequence.

Definition:
Geometric Sequence

> A sequence with general term $a_n = ar^{n-1}$ is called a **geometric sequence** with **common ratio r**, where $r \neq 1$ and $r \neq 0$.

According to the definition, every geometric sequence has the following form:

$$a, \quad ar, \quad ar^2, \quad ar^3, \quad ar^4, \ldots$$

Every term (after the first) is a constant multiple of the term preceding it. Of course, that constant multiple is the constant ratio r. Note that if $a = r$ then $a_n = ar^{n-1} = rr^{n-1} = r^n$.

Example 1 Listing the terms of a geometric sequence

Find the first four terms of the geometric sequence.

a. $a_n = 3^n$ **b.** $a_n = 100 \left(\frac{1}{2} \right)^{n-1}$

Solution

a. Replace n with the integers from 1 through 4 to get $3^1, 3^2, 3^3$, and 3^4. So the first four terms are 3, 9, 27, and 81.

b. Replace n with the integers from 1 through 4 to get

$$100 \left(\frac{1}{2} \right)^0, 100 \left(\frac{1}{2} \right)^1, 100 \left(\frac{1}{2} \right)^2, 100 \left(\frac{1}{2} \right)^3.$$

So the first four terms are 100, 50, 25, and 12.5.

Try This. Find the first four terms of the sequence $a_n = 3 \cdot 2^n$. ■

To write a formula for the nth term of a geometric sequence, we need the first term and the constant ratio.

Example 2 Finding a formula for the nth term

Write a formula for the nth term of each geometric sequence.

a. $0.3, 0.03, 0.003, 0.0003, \ldots$ **b.** $2, -6, 18, -54, \ldots$

Solution

a. Since each term after the first is one-tenth of the term preceding it, we use $r = 0.1$ and $a = 0.3$ in the formula $a_n = ar^{n-1}$:

$$a_n = 0.3(0.1)^{n-1}$$

b. Choose any two consecutive terms and divide the second by the first to obtain the common ratio. So $r = -6/2 = -3$. Since the first term is 2,

$$a_n = 2(-3)^{n-1}.$$

Try This. Find the nth term of the geometric sequence $5, 10, 20, 40, \ldots$. ■

In Section 11.2 we learned some interesting facts about arithmetic sequences. However, this information is useful only if we can determine whether a given sequence is arithmetic. In this section we will learn some facts about geometric sequences. Likewise, we must be able to determine whether a given sequence is geometric even when it is not given in exactly the same form as the definition.

Example **3** **Identifying a geometric sequence**

Find the first four terms of each sequence and determine whether the sequence is geometric.

a. $b_n = (-2)^{3n}$
b. $a_1 = 1.25$ and $a_n = -2a_{n-1}$ for $n \geq 2$
c. $c_n = 3n$

Solution

a. Use $n = 1, 2, 3$, and 4 in the formula $b_n = (-2)^{3n}$ to find the first four terms:

$$-8, \quad 64, \quad -512, \quad 4096, \ldots$$

The ratio of any term and the preceding term can be found from the formula

$$\frac{b_n}{b_{n-1}} = \frac{(-2)^{3n}}{(-2)^{3(n-1)}} = (-2)^3 = -8.$$

Since there is a common ratio of -8, the sequence is geometric.

b. Use the recursion formula $a_n = -2a_{n-1}$ to obtain each term after the first:

$$a_1 = 1.25$$
$$a_2 = -2a_1 = -2.5$$
$$a_3 = -2a_2 = 5$$
$$a_4 = -2a_3 = -10$$

This recursion formula defines the sequence

$$1.25, -2.5, 5, -10, \ldots$$

The formula $a_n = -2a_{n-1}$ means that each term is a constant multiple of the term preceding it. The constant ratio is -2 and the sequence is geometric.

c. Use $c_n = 3n$ to find $c_1 = 3, c_2 = 6, c_3 = 9$, and $c_4 = 12$. Since $6/3 = 2$ and $9/6 = 1.5$, there is no common ratio for consecutive terms. The sequence is not geometric. Because each term is 3 larger than the preceding term, the sequence is arithmetic.

Try This. Find the first four terms of $a_n = 2^n \cdot 3^{n-1}$ and determine whether the sequence is geometric.   ■

The formula for the general term of a geometric sequence involves a_n, a, n, and r. If we know the value of any three of these quantities, then we can find the value of the fourth.

Example **4** **Finding the number of terms in a geometric sequence**

A certain ball always rebounds 2/3 of the distance from which it falls. If the ball is dropped from a height of 9 feet, and later it is observed rebounding to a height of 64/81 feet, then how many times did it bounce?

Solution

After the first bounce the ball rebounds to $9(2/3) = 6$ feet. After the second bounce the ball rebounds to a height of $9(2/3)^2 = 4$ feet. The first term of this sequence is 6 and the common ratio is $2/3$. So after the nth bounce the ball rebounds to a height h_n given by

$$h_n = 6\left(\frac{2}{3}\right)^{n-1}.$$

To find the number of bounces, solve the following equation, which states that the nth bounce rebounds to $64/81$ feet.

$$6\left(\frac{2}{3}\right)^{n-1} = \frac{64}{81}$$

$$\left(\frac{2}{3}\right)^{n-1} = \frac{32}{243} = \left(\frac{2}{3}\right)^5$$

$$n - 1 = 5$$

$$n = 6$$

The ball bounced six times.

Try This. How many terms are in the geometric sequence 8, 16, 32, . . . , 32768?

■

Geometric Series

The indicated sum of the terms of a geometric sequence is called a **geometric series.** To find the actual sum of a geometric series, we can use a procedure similar to that used for finding the sum of an arithmetic series.

Consider the geometric sequence $a_n = 2^{n-1}$. Let S_{10} represent the sum of the first ten terms of this geometric sequence:

$$S_{10} = 1 + 2 + 4 + 8 + \cdots + 512$$

If we multiply each side of this equation by -2, the opposite of the common ratio, we get

$$-2S_{10} = -2 - 4 - 8 - 16 - \cdots - 512 - 1024.$$

Adding S_{10} and $-2S_{10}$ eliminates most of the terms:

$$S_{10} = 1 + 2 + 4 + 8 + 16 + \cdots + 512$$
$$\underline{-2S_{10} = \quad - 2 - 4 - 8 - 16 - \cdots - 512 - 1024}$$
$$-S_{10} = 1 \qquad\qquad\qquad\qquad\quad - 1024 \quad \text{Add.}$$
$$-S_{10} = -1023$$
$$S_{10} = 1023$$

The "trick" to finding the sum of n terms of a geometric series is to change the signs and shift the terms so that most terms "cancel out" when the two equations are added. This method can also be used to find a general formula for the sum of a geometric series.

Let S_n represent the sum of the first n terms of the geometric sequence $a_n = ar^{n-1}$.

$$S_n = a + ar + ar^2 + \cdots + ar^{n-1}$$

Adding S_n and $-rS_n$ eliminates most of the terms:

$$
\begin{aligned}
S_n &= a + ar + ar^2 + \qquad\quad \cdots + ar^{n-1} \\
-rS_n &= \quad - ar - ar^2 - ar^3 - \cdots - ar^{n-1} - ar^n \\
\hline
S_n - rS_n &= a \qquad\qquad\qquad\qquad\qquad\qquad - ar^n \quad \text{Add.}\\
(1-r)S_n &= a(1 - r^n) \\
S_n &= \frac{a(1-r^n)}{1-r} \qquad\qquad\qquad\qquad \text{Provided that } r \neq 1
\end{aligned}
$$

So the sum of a geometric series can be found if we know the first term, the constant ratio, and the number of terms. This result is summarized in the following theorem.

Theorem: Sum of a Finite Geometric Series

> If S_n represents the sum of the first n terms of a geometric series with first term a and common ratio r $(r \neq 1)$, then
> $$S_n = \frac{a(1-r^n)}{1-r}.$$

Example 5 Finding the sum of a geometric series

Find the sum of each geometric series.

a. $1 + \dfrac{1}{3} + \dfrac{1}{9} + \cdots + \dfrac{1}{243}$ **b.** $\displaystyle\sum_{j=0}^{10} 100(1.05)^j$

Solution

a. To find the sum of a finite geometric series, we need the first term a, the ratio r, and the number of terms n. To find n, use $a = 1$, $a_n = 1/243$, and $r = 1/3$ in the formula $a_n = ar^{n-1}$:

$$1\left(\frac{1}{3}\right)^{n-1} = \frac{1}{243}$$

$$n - 1 = 5 \qquad \text{Because } \left(\frac{1}{3}\right)^5 = \frac{1}{243}$$

$$n = 6$$

Now use $n = 6$, $a = 1$, $r = 1/3$ in the formula $S_n = \dfrac{a(1-r^n)}{1-r}$:

$$S_6 = \frac{1\left(1 - \left(\dfrac{1}{3}\right)^6\right)}{1 - \dfrac{1}{3}} = \frac{\dfrac{728}{729}}{\dfrac{2}{3}} = \frac{364}{243}$$

b. First write out some terms of the series:

$$\sum_{j=0}^{10} 100(1.05)^j = 100 + 100(1.05) + 100(1.05)^2 + \cdots + 100(1.05)^{10}$$

In this geometric series, $a = 100$, $r = 1.05$, and $n = 11$:

$$\sum_{j=0}^{10} 100(1.05)^j = S_{11} = \frac{100(1 - (1.05)^{11})}{1 - 1.05} \approx 1420.68$$

Try This. Find the sum $\sum_{n=1}^{20} 3(2)^{n-1}$.   ■

Infinite Geometric Series

In the geometric series

$$2 + 4 + 8 + 16 + \cdots,$$

in which $r = 2$, the terms get larger and larger. So the sum of the first n terms increases without bound as n increases. In the geometric series

$$\frac{1}{2} + \frac{1}{4} + \frac{1}{8} + \frac{1}{16} + \cdots,$$

in which $r = 1/2$, the terms get smaller and smaller. The sum of n terms of this series is less than 1 no matter how large n is. (To see this, add some terms on your calculator.) We can explain the different behavior of these series by examining the term r^n in the formula

$$S_n = \frac{a(1 - r^n)}{1 - r}.$$

If $r = 2$, the values of r^n increase without bound as n gets larger, causing S_n to increase without bound. If $r = 1/2$, the values of $(1/2)^n$ approach 0 as n gets larger. In symbols, $(1/2)^n \to 0$ as $n \to \infty$. If $r^n \to 0$, then $1 - r^n$ is approximately 1. If we replace $1 - r^n$ by 1 in the formula for S_n, we get

$$S_n \approx \frac{a}{1 - r} \quad \text{for large values of } n.$$

In the above series $r = 1/2$ and $a = 1/2$. So if n is large, we have

$$S_n \approx \frac{\dfrac{1}{2}}{1 - \dfrac{1}{2}} = 1.$$

```
.99^100
       .3660323413
.99^500
       .006570483
1.01^500
       144.7727724
```

■ **Figure 11.9**

In general, it can be proved that $r^n \to 0$ as $n \to \infty$ provided that $|r| < 1$, and r^n does not get close to 0 as $n \to \infty$ for $|r| \geq 1$. You will better understand these ideas if you use a calculator to find some large powers of r for various values of r as in Fig. 11.9. So, if $|r| < 1$ and n is large, then S_n is approximately $a/(1 - r)$. Furthermore, by using more terms in the sum we can get a sum that is arbitrarily close to the number $a/(1 - r)$. In this sense we say that the sum of all terms of the infinite geometric series is $a/(1 - r)$.

Theorem: Sum of an Infinite Geometric Series

If $a + ar + ar^2 + \cdots$ is an infinite geometric series with $|r| < 1$, then the sum S of all of the terms is given by

$$S = \frac{a}{1 - r}.$$

We can use the infinity symbol ∞ to indicate the sum of infinitely many terms of an infinite geometric series as follows:

$$a + ar + ar^2 + \cdots = \sum_{i=1}^{\infty} ar^{i-1}$$

Example **6** **Finding the sum of an infinite geometric series**

Find the sum of each infinite geometric series.

a. $1 + \dfrac{1}{3} + \dfrac{1}{9} + \cdots$ **b.** $\displaystyle\sum_{j=1}^{\infty} 100(-0.99)^j$ **c.** $\displaystyle\sum_{i=0}^{\infty} 3(1.01)^i$

Solution

a. The first term is 1, and the common ratio is $1/3$. So the sum of the infinite series is

$$S = \frac{1}{1 - \dfrac{1}{3}} = \frac{3}{2}.$$

b. The first term is -99, and the common ratio is -0.99. So the sum of the infinite geometric series is

$$S = \frac{-99}{1 - (-0.99)} = \frac{-99}{1.99} = -\frac{9900}{199}.$$

c. The first term is 3, and the common ratio is 1.01. Since the absolute value of the ratio is greater than 1, this infinite geometric series has no sum.

Try This. Find the sum $\displaystyle\sum_{n=1}^{\infty} 5(0.3)^n$. ■

Applications

In the next example we convert an infinite repeating decimal number into its rational form using the formula for the sum of an infinite geometric series.

Example **7** **Repeating decimals to fractions**

Convert each repeating decimal into a fraction.

a. $0.3333\ldots$ **b.** $1.2417417417\ldots$

Solution

a. Write the repeating decimal as an infinite geometric series:

$$0.3333\ldots = 0.3 + 0.3 \cdot 10^{-1} + 0.3 \cdot 10^{-2} + 0.3 \cdot 10^{-3} + \cdots$$

The first term is 0.3 and the ratio is 10^{-1}. Use these numbers in the formula for the sum of an infinite geometric series $S = \dfrac{a}{1 - r}$:

$$0.3333\ldots = \frac{0.3}{1 - 10^{-1}} = \frac{0.3}{0.9} = \frac{3}{9} = \frac{1}{3}$$

b. First separate the repeating part from the nonrepeating part and then convert the repeating part into a fraction using the formula for the sum of an infinite geometric series:

$$1.2417417417417\ldots = 1.2 + 0.0417417417417\ldots$$

$$= 1.2 + 417 \cdot 10^{-4} + 417 \cdot 10^{-7} + 417 \cdot 10^{-10} + \cdots$$

$$= \frac{12}{10} + \frac{417 \cdot 10^{-4}}{1 - 10^{-3}} \qquad a = 417 \cdot 10^{-4} \text{ and } r = 10^{-3}$$

$$= \frac{12}{10} + \frac{417}{10{,}000 - 10}$$

$$= \frac{12}{10} + \frac{417}{9990}$$

$$= \frac{12405}{9990} = \frac{827}{666}$$

Try This. Convert $0.0626262\ldots$ to a fraction. ■

In the next example we use the formula for the sum of an infinite geometric series in a physical situation. Since physical processes do not continue infinitely, the formula for the sum of infinitely many terms is used as an approximation for the sum of a large finite number of terms of a geometric series in which $|r| < 1$.

Example **8** Total distance traveled in bungee jumping

A man jumping from a bridge with a bungee cord tied to his legs falls 120 feet before being pulled back upward by the bungee cord. If he always rebounds 1/3 of the distance that he has fallen and then falls 2/3 of the distance of his last rebound, then approximately how far does the man travel before coming to rest?

■ **Figure 11.10**

Solution

In actual practice, the man does not go up and down infinitely on the bungee cord. However, to get an approximate answer, we can model this situation shown in Fig. 11.10 with two infinite geometric sequences, the sequence of falls and the sequence of rises. The man falls 120 feet, then rises 40 feet. He falls 80/3 feet, then rises 80/9 feet. He falls 160/27 feet, then rises 160/81 feet. The total distance the man falls is given by the following series:

$$F = 120 + \frac{80}{3} + \frac{160}{27} + \cdots = \frac{120}{1 - \dfrac{2}{9}} = \frac{1080}{7} \text{ feet}$$

The total distance the man rises is given by the following series:

$$R = 40 + \frac{80}{9} + \frac{160}{81} + \cdots = \frac{40}{1 - \dfrac{2}{9}} = \frac{360}{7} \text{ feet}$$

The total distance he travels before coming to rest is the sum of these distances, 1440/7 feet, or approximately 205.7 feet.

Try This. Start at 0 on a number line. Move 8 units right, 4 units left, 2 units right, 1 unit left, and so on (forever). Where do you end up? ■

In the next example we see that a geometric sequence occurs in an investment earning compound interest. If we have an initial deposit earning compound interest, then the amounts in the account at the end of consecutive years form a geometric sequence.

Example 9 A geometric sequence in investment

The parents of a newborn decide to start saving early for her college education. On the day of her birth, they invest $3000 at 6% compounded annually. Find the amount of the investment at the end of each of the first four years and find a formula for the amount at the end of the nth year. Find the amount at the end of the 18th year.

Solution

At the end of the first year the amount is $3000(1.06). At the end of the second year the amount is $3000(1.06)^2$. Use a calculator to find the amounts at the ends of the first four years.

$$a_1 = 3000(1.06)^1 = \$3180$$

$$a_2 = 3000(1.06)^2 = \$3370.80$$

$$a_3 = 3000(1.06)^3 = \$3573.05$$

$$a_4 = 3000(1.06)^4 = \$3787.43$$

A formula for the nth term of this geometric sequence is $a_n = 3000(1.06)^n$. The amount at the end of the 18th year is

$$a_{18} = 3000(1.06)^{18} = \$8563.02.$$

Try This. A student invests $2000 at 5% compounded annually. Find a formula for the amount of the investment at the end of the nth year. ■

One of the most important applications of the sum of a finite geometric series is in projecting the value of an annuity. An **annuity** is a sequence of equal periodic payments. If each payment earns the same rate of compound interest, then the total value of the annuity can be found by using the formula for the sum of a finite geometric series.

Example 10 Finding the value of an annuity

To maintain his customary style of living after retirement, a single man earning $50,000 per year at age 65 must have saved a minimum of $250,000 (Fidelity Investments, www.fidelity.com). If Chad invests $1000 at the beginning of each year for 40 years in an investment paying 6% compounded annually, then what is the value of this annuity at the end of the 40th year?

Solution

The last deposit earns interest for only one year and amounts to 1000(1.06). The second to last deposit earns interest for two years and amounts to $1000(1.06)^2$. This pattern continues down to the first deposit, which earns interest for 40 years and amounts to $1000(1.06)^{40}$. The value of the annuity is the sum of a finite geometric series:

$$S_{40} = 1000(1.06) + 1000(1.06)^2 + \cdots + 1000(1.06)^{40}$$

Since the first term is 1000(1.06), use $a = 1000(1.06)$, $r = 1.06$, and $n = 40$:

$$S_{40} = \frac{1000(1.06)(1 - (1.06)^{40})}{1 - 1.06} \approx \$164{,}047.68$$

Try This. A student invests \$2000 at the beginning of each year for 20 years at 5% compounded annually. Find the amount of the investment at the end of the 20th year.

■

For Thought

True or False? Explain.

1. The sequence 2, 6, 24, . . . is a geometric sequence.

2. The sequence $a_n = 3(2)^{3-n}$ is a geometric sequence.

3. The first term of the geometric sequence $a_n = 5(0.3)^n$ is 1.5.

4. The common ratio in the geometric sequence $a_n = 5^{-n}$ is 5.

5. A geometric series is the indicated sum of a geometric sequence.

6. $3 + 6 + 12 + 24 + \cdots = \dfrac{3}{1 - 2}$

7. $\displaystyle\sum_{i=1}^{9} 3(0.6)^i = \dfrac{1.8(1 - 0.6)^9}{1 - 0.6}$

8. $\displaystyle\sum_{i=0}^{4} 2(10)^i = 22{,}222$

9. $\displaystyle\sum_{i=1}^{\infty} 3(0.1)^i = \dfrac{1}{3}$

10. $\displaystyle\sum_{i=1}^{\infty} \left(\dfrac{1}{2}\right)^i = 1$

11.3 Exercises

Find the first four terms of each geometric sequence. What is the common ratio?

1. $a_n = 3 \cdot 2^{n-1}$

2. $a_n = 2 \cdot (3)^{n-1}$

3. $b_n = 800 \cdot \left(\dfrac{1}{2}\right)^n$

4. $b_n = 27 \cdot \left(\dfrac{1}{3}\right)^n$

5. $c_n = \left(-\dfrac{2}{3}\right)^{n-1}$

6. $c_n = \left(-\dfrac{3}{2}\right)^{n-2}$

Find the common ratio in each geometric sequence.

7. $4, 2, 1, \dfrac{1}{2}, \ldots$

8. $1, 5, 25, 125, \ldots$

9. $10^2, 10^3, 10^4, \ldots$

10. $10^{-1}, 10^{-2}, 10^{-3}, \ldots$

11. $-1, 2, -4, 8, \ldots$

12. $81, -27, 9, -3, \ldots$

13. $1, -1, 1, -1, \ldots$

14. $-5, 5, -5, 5, \ldots$

Write a formula for the nth term of each geometric sequence. Do not use a recursion formula.

15. $\dfrac{1}{6}, \dfrac{1}{3}, \dfrac{2}{3}, \dfrac{4}{3}, \ldots$

16. $\dfrac{1}{6}, 0.5, 1.5, 4.5, \ldots$

17. $0.9, 0.09, 0.009, 0.0009, \ldots$

18. $3, 9, 27, 81, \ldots$

19. $4, -12, 36, -108, \ldots$

20. $5, -1, \dfrac{1}{5}, -\dfrac{1}{25}, \ldots$

Identify each sequence as arithmetic, geometric, or neither.

21. $2, 4, 6, 8, \ldots$

22. $2, 4, 8, 16, \ldots$

23. $1, 2, 4, 6, 8, \ldots$

24. $0, 2, 4, 8, 16, \ldots$

25. $2, -4, 8, -16, \ldots$

26. $0, 2, 4, 6, 8, \ldots$

27. $\dfrac{1}{6}, \dfrac{1}{3}, \dfrac{1}{2}, \dfrac{2}{3}, \ldots$

28. $\dfrac{1}{6}, \dfrac{1}{3}, 1, 4, \ldots$

29. $\dfrac{1}{6}, \dfrac{1}{3}, \dfrac{2}{3}, \dfrac{4}{3}, \ldots$

30. $3, 2, 1, 0, -1, \ldots$

31. $1, 4, 9, 16, 25, \ldots$

32. $5, 1, \dfrac{1}{5}, \dfrac{1}{25}, \ldots$

Find the first four terms of each sequence and identify each sequence as arithmetic, geometric, or neither.

33. $a_n = 2n$

34. $a_n = 2^n$

35. $a_n = n^2$

36. $a_n = n!$

37. $a_n = 2^{-n}$

38. $a_n = n + 2$

39. $b_n = 2^{2n+1}$

40. $d_n = \dfrac{1}{3^{n/2}}$

41. $c_1 = 3, c_n = -3c_{n-1}$ for $n \geq 2$

42. $h_1 = \sqrt{2}, h_n = \sqrt{3}h_{n-1}$ for $n \geq 2$

Find the required part of each geometric sequence.

43. Find the number of terms of a geometric sequence with first term 3, common ratio 1/2, and last term 3/1024.

44. Find the number of terms of a geometric sequence with first term 1/64, common ratio −2, and last term −512.

45. Find the first term of a geometric sequence with sixth term 1/81 and common ratio of 1/3.

46. Find the first term of a geometric sequence with seventh term 1/16 and common ratio 1/2.

47. Find the common ratio for a geometric sequence with first term 2/3 and third term 6.

48. Find the common ratio for a geometric sequence with second term −1 and fifth term −27.

49. Find a formula for the *n*th term of a geometric sequence with third term −12 and sixth term 96.

50. Find a formula for the *n*th term of a geometric sequence with second term −40 and fifth term 0.04.

Find the sum of each finite geometric series by using the formula for S_n. Check your answer by actually adding up all of the terms. Round approximate answers to four decimal places.

51. $1 + 2 + 4 + 8 + 16$

52. $100 + 10 + 1 + 0.1 + 0.01$

53. $2 + 4 + 8 + 16 + 32 + 64$

54. $1 + (-1) + 1 + (-1) + 1 + (-1) + 1$

55. $9 + 3 + 1 + \dfrac{1}{3} + \dfrac{1}{9} + \dfrac{1}{27}$

56. $8 + 4 + 2 + 1 + \dfrac{1}{2} + \dfrac{1}{4} + \dfrac{1}{8}$

57. $6 + 2 + \dfrac{2}{3} + \dfrac{2}{9} + \dfrac{2}{27}$

58. $2 + 10 + 50 + 250 + 1250$

59. $1.5 - 3 + 6 - 12 + 24 - 48 + 96 - 192$

60. $1 - \dfrac{1}{2} + \dfrac{1}{4} - \dfrac{1}{8} + \dfrac{1}{16} - \dfrac{1}{32} + \dfrac{1}{64}$

61. $\displaystyle\sum_{i=1}^{12} 2(1.05)^{i-1}$

62. $\displaystyle\sum_{i=0}^{30} 300(1.08)^{i}$

63. $\displaystyle\sum_{i=0}^{7} 200(1.01)^{i}$

64. $\displaystyle\sum_{i=1}^{20} 421(1.09)^{i-1}$

Write each geometric series in summation notation.

65. $3 - 1 + \dfrac{1}{3} - \dfrac{1}{9} + \dfrac{1}{27}$

66. $2 + 1 + \dfrac{1}{2} + \dfrac{1}{4} + \dfrac{1}{8} + \dfrac{1}{16}$

67. $0.6 + 0.06 + 0.006 + \cdots$

68. $4 - 1 + \dfrac{1}{4} - \dfrac{1}{16} + \cdots$

69. $-4.5 + 1.5 - 0.5 + \dfrac{1}{6} - \cdots$

70. $a + ab + ab^2 + \cdots + ab^{37}$

Find the sum of each infinite geometric series where possible.

71. $3 - 1 + \dfrac{1}{3} - \dfrac{1}{9} + \dfrac{1}{27} - \cdots$

72. $1 + \dfrac{1}{2} + \dfrac{1}{4} + \dfrac{1}{8} + \dfrac{1}{16} + \cdots$

73. $0.9 + 0.09 + 0.009 + \cdots$

74. $-1 + \dfrac{1}{4} - \dfrac{1}{16} + \dfrac{1}{64} - \cdots$

75. $-9.9 + 3.3 - 1.1 + \cdots$

76. $1.2 - 2.4 + 4.8 - 9.6 + \cdots$

77. $\displaystyle\sum_{i=1}^{\infty} 34(0.01)^i$ **78.** $\displaystyle\sum_{i=0}^{\infty} 300(0.99)^i$

79. $\displaystyle\sum_{i=0}^{\infty} 300(-1.06)^i$ **80.** $\displaystyle\sum_{i=0}^{\infty} (0.98)^i$

81. $\displaystyle\sum_{i=1}^{\infty} 6(0.1)^i$ **82.** $\displaystyle\sum_{i=0}^{\infty} (0.1)^i$

83. $\displaystyle\sum_{i=0}^{\infty} 34(-0.7)^i$ **84.** $\displaystyle\sum_{i=1}^{\infty} 123(0.001)^i$

Use the formula for the sum of an infinite geometric series to write each repeating decimal number as a fraction.

85. $0.04444\ldots$ **86.** $0.0121212\ldots$

87. $8.2545454\ldots$ **88.** $3.65176176176\ldots$

Use the ideas of geometric series to solve each problem.

89. *Drug Therapy*   If a patient starts taking 100 mg of Lamictal every 8 hours to control seizures, then the amount of Lamictal in the patient's body after taking the twenty-fifth pill is given by

$$\sum_{n=1}^{25} 100(0.69)^{n-1}.$$

Find this sum.

HINT Use the formula for the sum of a geometric series.

90. *Long-Range Therapy*   The long-range build-up of Lamictal in the body of a patient taking 100 mg three times a day is given by

$$\sum_{n=1}^{\infty} 100(0.69)^{n-1}.$$

Find the sum of this series. What is the difference between the long-range build-up and the amount in the person after 25 pills?

91. *Sales Goals*   A group of college students selling magazine subscriptions during the summer sold one subscription on June 1. The sales manager was encouraged by this and said that their daily goal for each day of June is to double the sales of the previous day. If the students work every day during June and meet this goal, then what is the total number of magazine subscriptions that they will sell during June?

92. *Family Tree*   Consider yourself, your parents, your grandparents, your great-grandparents, your great-great-grandparents, and so on, back to your grandparents with the word "great" used in front 40 times. What is the total number of people that you are considering?

93. *Compound Interest*   Given that $4000 is deposited at the beginning of a quarter into an account earning 8% annual interest compounded quarterly, write a formula for the amount in the account at the end of the nth quarter. How much is in the account at the end of 37 quarters?

94. *Compound Interest*   Given that $8000 is deposited at the beginning of a month into an account earning 6% annual interest compounded monthly, write a formula for the amount in the account at the end of the nth month. How much is in the account at the end of the 56th month?

95. *Value of an Annuity*   If you deposit $200 on the first of each month for 12 months into an account paying 12% annual interest compounded monthly, then how much is in the account at the end of the 12th month? Note that each deposit earns 1% per month for a different number of months.

HINT Write out the terms of this finite geometric series.

96. *Saving for Retirement*   If you deposit $9000 on the first of each year for 40 years into an account paying 8% compounded annually, then how much is in the account at the end of the 40th year?

HINT Write out the terms of this finite geometric series.

97. *Saving for Retirement*   If $100 is deposited at the end of each month for 30 years in a retirement account earning 9% compounded monthly, then what is the value of this annuity immediately after the last payment?

98. *Down Payment* To get a down payment for a house, a couple plans to deposit $800 at the end of each quarter for 36 quarters in an account paying 6% compounded quarterly. What is the value of this annuity immediately after the last deposit?

99. *Bouncing Ball* Suppose that a ball always rebounds 2/3 of the distance from which it falls. If this ball is dropped from a height of 9 ft, then approximately how far does it travel before coming to rest?

100. *Saturating the Market* A sales manager has set a team goal of $1 million in sales. The team plans to get 1/2 of the goal the first week of the campaign. Every week after the first they will sell only 1/2 as much as the previous week because the market will become saturated. How close will the team be to the sales goal after 15 weeks of selling?

101. *Economic Impact* The anticipated payroll of the new General Dynamics plant in Hammond, Louisiana, is $2 million annually. It is estimated that 75% of this money is spent in Hammond by its recipients. The people who receive that money again spend 75% of it in Hammond, and so on. The total of all this spending is the *total economic impact* of the plant on Hammond. What is the total economic impact of the plant?

> **HINT** Write out some terms of this infinite geometric series.

102. *Disaster Relief* If the federal government provides $300 million in disaster relief to the people of Louisiana after a hurricane, then 80% of that money is spent in Louisiana. If the money is respent over and over at a rate of 80% in Louisiana, then what is the total economic impact of the $300 million on the state?

For Writing/Discussion

103. Can an arithmetic sequence and a geometric sequence have the same first three terms? Explain your answer.

104. *Cooperative Learning* Get a "super ball" and work in a small group to measure the distance that it rebounds from falls of 8 ft, 6 ft, 4 ft, and 2 ft. Is it reasonable to assume that the rebound distance is a constant percentage of the fall distance? Use an infinite series to find the total distance that your ball travels vertically before coming to rest when it is dropped from a height of 8 ft. Discuss the applicability of the infinite series model to this physical experiment.

105. Consider the functions $y = r^x$ and $y = x^r$ for $x > 0$. Graph these functions for some values of r with $|r| < 1$ and for some values of r with $|r| \geq 1$. Make a conjecture about the relationship between the value of r and the values of r^x and x^r as $x \to \infty$.

106. Consider the function $y = 5(1 - r^x)/(1 - r)$ for $x > 0$. Graph this function for some values of r with $|r| < 1$ and for some values of r with $|r| > 1$. Make a conjecture about the relationship between the value of r and the value of y as $x \to \infty$. Explain how these graphs are related to series.

Thinking Outside the Box XCIII

Crawling Ant In a right triangle with sides 5, 12, and 13 ft, an ant starts at the vertex of the right angle and crawls straight toward the hypotenuse. Its path is perpendicular to the hypotenuse and creates two more right triangles. The ant then crawls straight toward the hypotenuse of the smaller of the two new right triangles. The path keeps creating two new right triangles and the ant forever crawls toward the hypotenuse of the smaller of the two new right triangles. What is the total distance traveled by the ant?

11.3 Pop Quiz

1. What is the common ratio in $9, 3, 1, \frac{1}{3}, \ldots$?

2. Write a formula for the nth term of the sequence $2, -4, 8, -16, \ldots$.

3. Find the first term of a geometric sequence in which $r = -\frac{1}{2}$ and $a_6 = 3$.

4. Find the sum $\sum_{i=1}^{8} 5(-2)^i$.

5. Find the sum $\sum_{n=2}^{\infty} (0.1)^n$.

Linking Concepts

For Individual or
Group Explorations

Annuities

*An annuity consists of periodic payments into an account paying compound interest.
Suppose that R dollars is deposited at the beginning of each period for n periods and
the compound interest rate is i per period.*

a) What is the amount of the first deposit at the time of the last deposit?

b) Write a series for the total amount of all *n* deposits at the time of the last deposit.

c) Use the formula for the sum of a finite geometric series to show that the sum of
this series is $R((1 + i)^n - 1)/i$.

d) Find the amount at the time of the last deposit for an annuity of $2000 per year
for 30 years at 12% compounded annually.

e) Find the amount 7 months after the last deposit for an annuity of $800 per
month for 13 years with a rate of 9% compounded monthly.

11.4

Counting and Permutations

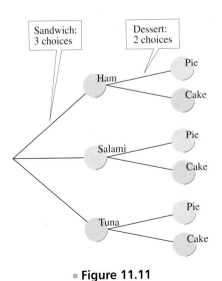

■ **Figure 11.11**

In the first three sections of this chapter we studied sequences and series. In this section we will look at sequences of events and count the number of ways that a sequence of events can occur. But in this instance, we actually try to avoid counting in the usual sense because this method of counting would probably be too cumbersome or tedious. In this new context, counting means finding the number of ways in which something can be done without actually listing all of the ways and counting them.

The Fundamental Counting Principle

Let's say that the cafeteria lunch special includes a choice of sandwich and dessert. In the terminology of counting, choosing a sandwich is an **event** and choosing a dessert is another event. Suppose that there are three outcomes, ham, salami, or tuna, for the first event and two outcomes, pie or cake, for the second event. How many different lunches are available using these choices? We can make a diagram showing all of the possibilities as in Fig. 11.11. This diagram is called a **tree diagram.** Considering only the types of sandwich and dessert, the tree diagram shows six different lunches. Of course, 6 can be obtained by multiplying 3 and 2. This example illustrates the fundamental counting principle.

**Fundamental
Counting Principle**

> If event *A* has *m* different outcomes and event *B* has *n* different outcomes, then there are *mn* different ways for events *A* and *B* to occur together.

The fundamental counting principle can also be used for more than two events, as is illustrated in the next example.

Example 1 Applying the fundamental counting principle

When ordering a new car, you are given a choice of three engines, two transmissions, six colors, three interior designs, and whether or not to get air conditioning. How many different cars can be ordered considering these choices?

Solution

There are three outcomes to the event of choosing the engine, two outcomes to choosing the transmission, six outcomes to choosing the color, three outcomes to choosing the interior, and two outcomes to choosing the air conditioning (to have it or not). So the number of different cars available is $3 \cdot 2 \cdot 6 \cdot 3 \cdot 2$ or 216.

Try This. A computer can be ordered with 3 different monitors, 4 different hard drives, and 2 different amounts of RAM. How many configurations are available for this computer? ■

Example 2 Applying the fundamental counting principle

How many different license plates are possible if each plate consists of three letters followed by a three-digit number? Assume that repetitions in the letters or numbers are allowed and that any of the ten digits may be used.

Solution

Since there are 26 choices for each of the three letters and ten choices for each of the three numbers, by the fundamental counting principle, the number of license plates is $26 \cdot 26 \cdot 26 \cdot 10 \cdot 10 \cdot 10 = 26^3 \cdot 10^3 = 17,576,000$.

Try This. How many different license plates are possible if each plate consists of two letters followed by four digits, with repetitions allowed? ■

Permutations

In Examples 1 and 2, each choice was independent of the previous choices. But the number of ways in which an event can occur often depends on what has already occurred. For example, in arranging three students in a row, the choice of the student for the second seat depends on which student was placed first. Consider the following six different sequential arrangements (or permutations) of three students, Ann, Bob, and Carol:

Ann, Bob, Carol	Bob, Ann, Carol	Carol, Ann, Bob
Ann, Carol, Bob	Bob, Carol, Ann	Carol, Bob, Ann

These six permutations can also be shown in a tree diagram, as we did in Fig. 11.12. Since the students are distinct objects, there can be no repetition of students. An arrangement such as Ann, Ann, Bob is not allowed. A **permutation** is an ordering or arrangement of distinct objects in a sequential manner. There is a first, a second, a

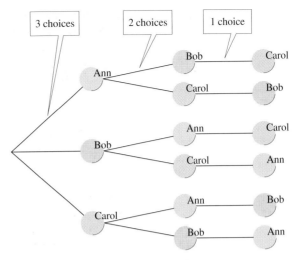

■ **Figure 11.12**

third, and so on. High-level diplomats are usually *not* seated so that they are arranged in a sequential order. They are seated at round tables so that no one is first, or second, or third.

Example **3** **Finding permutations**

A Federal Express driver must make ten deliveries to ten different addresses. In how many ways can she make those deliveries?

Solution

For the event of choosing the first address there are ten outcomes. For the event of choosing the second address, there are nine outcomes (since one has already been chosen). For the third address, there are eight outcomes, and so on. So, according to the fundamental counting principle, the number of permutations of the ten addresses is

$$10 \cdot 9 \cdot 8 \cdot 7 \cdot 6 \cdot 5 \cdot 4 \cdot 3 \cdot 2 \cdot 1 = 10! = 3,628,800.$$

Try This. In how many ways can 13 students stand in a single cafeteria line? ■

   In general, the number of arrangements of any n distinct objects in a sequential manner is referred to as the number of permutations of n things taken n at a time, and the notation $P(n, n)$ is used to represent this number. The phrase "taken n at a time" indicates that all of the n objects are used in the arrangements. (We will soon discuss permutations in which not all of the n objects are used.) Since there are n choices for the first object, $n - 1$ choices for the second object, $n - 2$ choices for the third object, and so on, we have $P(n, n) = n!$.

Theorem: Permutations of *n* Things Taken *n* at a Time

The notation $P(n, n)$ represents the number of permutations of n things taken n at a time, and $P(n, n) = n!$.

Sometimes we are interested in permutations in which not all of the objects are used. For example, if the Federal Express driver wants to deliver three of the ten packages before lunch, then how many ways are there to make those three deliveries? There are ten possibilities for the first delivery, nine for the second, and eight for the third. So the number of ways to make three stops before lunch, or the number of permutations of 10 things taken 3 at a time, is $10 \cdot 9 \cdot 8 = 720$. The notation $P(10, 3)$ is used to represent the number of permutations of 10 things taken 3 at a time. Notice that

$$P(10, 3) = \frac{10!}{7!} = \frac{10 \cdot 9 \cdot 8 \cdot 7 \cdot 6 \cdot 5 \cdot 4 \cdot 3 \cdot 2 \cdot 1}{7 \cdot 6 \cdot 5 \cdot 4 \cdot 3 \cdot 2 \cdot 1} = 10 \cdot 9 \cdot 8 = 720.$$

In general, we have the following theorem.

Theorem: Permutations of *n* Things Taken *r* at a Time

> The notation $P(n, r)$ represents the number of permutations of n things taken r at a time, and
>
> $$P(n, r) = \frac{n!}{(n - r)!} \quad \text{for } 0 \leq r \leq n.$$

Even though n things are usually not taken 0 at a time, 0 is allowed in the formula. Recall that by definition, $0! = 1$. For example, $P(8, 0) = 8!/8! = 1$ and $P(0, 0) = 0!/0! = 1/1 = 1$. The one way to choose 0 objects from 8 objects (or no objects) is to do nothing.

Many calculators have a key that gives the value of $P(n, r)$ when given n and r. The notation nPr is often used on calculators and elsewhere for $P(n, r)$.

Example **4** Finding permutations of *n* things taken *r* at a time

The Beau Chene Garden Club has 12 members. They plan to elect a president, a vice-president, and a treasurer. How many different outcomes are possible for the election if each member is eligible for each office and no one can hold two offices?

Solution

The number of ways in which these offices can be filled is precisely the number of permutations of 12 things taken 3 at a time:

$$P(12, 3) = \frac{12!}{(12 - 3)!} = \frac{12!}{9!} = 12 \cdot 11 \cdot 10 = 1320$$

Use the permutation function or factorials on a calculator to check, as in Fig. 11.13.

Try This. In how many ways can a group of 10 friends on a rented sailboat elect a captain and a first mate?

■

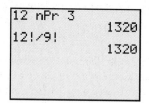

```
12 nPr 3
              1320
12!/9!
              1320
```

■ **Figure 11.13**

For Thought

True or False? Explain.

1. If a product code consists of a single letter followed by a two-digit number from 10 through 99, then $89 \cdot 26$ different codes are available.

2. If a fraternity name consists of three Greek letters chosen from the 24 letters in the Greek alphabet with repetitions allowed, then $23 \cdot 22 \cdot 21$ different fraternity names are possible.

3. If an outfit consists of a tie, a shirt, a pair of pants, and a coat, and John has three ties, five shirts, and three coats that all match his only pair of pants, then John has 11 outfits available to wear.

4. The number of ways in which five people can line up to buy tickets is 120.

5. The number of permutations of 10 things taken 2 at a time is 90.

6. The number of different ways to mark the answers to a 20-question multiple-choice test with each question having four choices is $P(20, 4)$.

7. The number of different ways to mark the answers to this sequence of 10 "For Thought" questions is 2^{10}.

8. $\dfrac{1000!}{998!} = 999{,}000$

9. $P(10, 9) = P(10, 1)$

10. $P(29, 1) = 1$

11.4 Exercises

Solve each problem using the fundamental counting principle.

1. *Tossing a Coin*   A coin is tossed and either heads (H) or tails (T) is recorded. If the coin is tossed twice, then how many outcomes are possible?

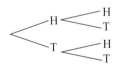

2. *Blue Plate Special*   For $4.99 you can choose from two entrees and three desserts. How many different meals are possible?

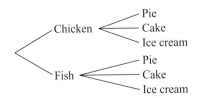

3. *Rams Record*   If the Rams win or lose each game, then how many outcomes are possible for the first three games of the season?

4. *Planning Outfits*   If you can choose from three shirts and four pairs of pants, then how many outfits are possible?

5. *Traveling Sales Representative*   A sales representative can take either of two different routes from Sacramento to Stockton and any one of four different routes from Stockton to San Francisco. How many different routes can she take from Sacramento to San Francisco, going through Stockton?

6. *Optional Equipment*   A new car can be ordered in any one of nine different colors, with three different engines, two different transmissions, three different body styles, two different interior designs, and four different stereo systems. How many different cars are available?

7. *Sleepless Night*   The ghosts of Christmas Past, Present, and Future plan on visiting Scrooge at 1:00, 2:00, and 3:00 in the morning. All three are available for haunting at all three times. Make a tree diagram showing all of the different orders in which the three ghosts can visit Scrooge. How many different arrangements are possible for this haunting schedule?

8. *Track Competition*   Juan, Felix, Ronnie, and Ted are in a 100-m race. Make a tree diagram showing all possible orders in which they can finish the race, assuming that there are no ties. In how many ways can they finish the race?

9. *Poker Hands*   A poker hand consists of five cards drawn from a deck of 52. How many different poker hands are there consisting of an ace, king, queen, jack, and ten?

10. *Drawing Cards from a Deck* In a certain card game, four cards are drawn from a deck of 52. How many different hands are there containing one heart, one spade, one club, and one diamond?

11. *Have It Your Way* Wendy's Old Fashioned Hamburgers once advertised that 256 different hamburgers were available at Wendy's. This number was obtained by using the fundamental counting principle and considering whether or not to include each one of several different options on the burger. How many different optional items were used to get this number?

12. *Choosing a Pizza* A pizza can be ordered in three sizes with either thick crust or thin crust. You have to decide whether to include each of four different meats on your pizza. You must also decide whether to include green peppers, onions, mushrooms, anchovies, and/or black olives. How many different pizzas can you order?

Evaluate each expression.

13. $\dfrac{7!}{4!}$

14. $\dfrac{9!}{6!}$

15. $\dfrac{5!}{0!}$

16. $\dfrac{6!}{0!}$

17. $\dfrac{78!}{77!}$

18. $\dfrac{56!}{55!}$

19. $P(4, 2)$

20. $P(5, 2)$

21. $P(5, 3)$

22. $P(6, 4)$

23. $P(7, 3)$

24. $P(16, 4)$

25. $P(9, 5)$

26. $P(7, 2)$

27. $P(5, 5)$

28. $P(4, 4)$

29. $P(11, 3)$

30. $P(15, 5)$

31. $P(99, 0)$

32. $P(44, 0)$

33. $P(105, 2)$

34. $P(120, 2)$

Solve each problem using the idea of permutations.

35. *Atonement of Hercules* The King of Tiryens ordered the Greek hero Hercules to atone for murdering his own family by performing 12 difficult and dangerous tasks. How many different orders are there for Hercules to perform the 12 tasks?

36. *Parading in Order* A small Mardi Gras parade consists of eight floats and three marching bands. In how many different orders can they line up to parade?

37. *Inspecting Restaurants* A health inspector must visit 3 of 15 restaurants on Monday. In how many ways can she pick a first, second, and third restaurant to visit?

38. *Assigned Reading* In how may ways can an English professor randomly give out one copy each of *War and Peace, The Grapes of Wrath, Moby Dick,* and *Gone with the Wind,* and four copies of *Jurassic Park* to a class of eight students?

39. *Scheduling Radio Shows* The program director for a public radio station has 26 half-hour shows available for Sunday evening. How many different schedules are possible for the 6:00 to 10:00 P.M. time period?

40. *Choosing Songs* A disc jockey must choose eight songs from the top 20 to play in the next 30-minute segment of her show. How many different arrangements are possible for this segment?

Solve each counting problem.

41. *Multiple-Choice Test* How many different ways are there to mark the answers to a six-question multiple-choice test in which each question has four possible answers?

42. *Choosing a Name* A novelist has decided on four possible first names and three possible last names for the main character in his next book. In how many ways can he name the main character?

43. *Choosing a Prize* A committee has four different VCRs, three different CD players, and six different CDs available. The person chosen as Outstanding Freshman will receive one item from each category. How many different prizes are possible?

44. *Phone Extensions* How many different four-digit extensions are available for a company phone system if the first digit cannot be 0?

45. *Computer Passwords* How many different three-letter computer passwords are available if any letters can be used but repetition of letters is not allowed?

46. *Company Cars* A new Cadillac, a new Dodge, and a used Taurus are to be assigned randomly to three of ten real estate salespersons. In how many ways can the assignment be made?

47. *Phone Numbers* How many different seven-digit phone numbers are available in Jamestown if the first three digits are either 345, 286, or 329?

48. *Electronic Mail* Bob Smith's e-mail address at International Plumbing Supply is bobs@ips.com. If all e-mail addresses at IPS consist of four letters followed by @ips.com, then how many possible addresses are there?

49. *Clark to the Rescue* The archvillain Lex Luther fires a nuclear missile into California's fault line, causing a tremendous earthquake. During the resulting chaos, Superman has to save Lois from a rock slide, rescue Jimmy from a bursting dam, stop a train from derailing, and catch a school bus that is plummeting from the Golden Gate Bridge. How many different ways are there for the Man of Steel to arrange these four rescues?

50. *Bus Routes* A bus picks up passengers from five hotels in downtown Seattle before heading to the Seattle-Tacoma Airport. How many different ways are there to arrange these stops?

51. *Taking a Test*   How many ways are there to choose the answers to a test that consists of five true-false questions followed by six multiple-choice questions with four options each?

52. *Granting Tenure*   A faculty committee votes on whether or not to grant tenure to each of four candidates. How many different possible outcomes are there to the vote?

53. *Possible Words*   Ciara is entering a contest (sponsored by a detergent maker) that requires finding all three-letter words that can be made from the word WASHING. No letter may be used more than once. To make sure that she does not miss any, Ciara plans to write down all possible three-letter words and then look up each one in a dictionary. How many possible three-letter words will be on her list?

54. *Listing Permutations*   Make a list of all of the permutations of the letters A, B, C, D, and E taken three at a time. How many permutations should be in your list?

For Writing/Discussion

55. *Listing Subsets*   List all of the subsets of each of the sets $\{A\}, \{A, B\}, \{A, B, C\}$, and $\{A, B, C, D\}$. Find a formula for the number of subsets of a set of n elements.

56. *Number of Subsets*   Explain how the fundamental counting principle can be used to find the number of subsets of a set of n elements.

Thinking Outside the Box XCIV

Perfect Paths   In Perfect City the avenues run east and west, the streets run north and south, and all of the blocks are square. Ms. Peabody lives at the corner of First Ave. and First Street. Every day she walks 16 blocks to the post office at the corner of Ninth Ave. and Ninth Street. To keep from getting bored she returns home using a different 16-block route. If she never repeats a route, then how long will it take for her to walk all possible routes? (Answer in terms of years and days, using 365 days for every year.)

11.4 Pop Quiz

1. Evaluate $\dfrac{102!}{100!}$ and $P(10, 4)$.

2. In how many ways can a class of 40 students choose a president, vice-president, secretary, and treasurer?

3. How many ways can a student mark the answers to a 6-question multiple choice test in which each question has 5 possible answers?

4. If a student chooses one of his 20 pairs of shorts and one of his 30 shirts to wear, then how many ways are there for him to select his outfit?

Linking Concepts

For Individual or Group Explorations

Minimizing the Distance

A supply boat must travel from Cameron, Louisiana, to four oil rigs in the Gulf of Mexico and return to Cameron. In a rectangular coordinate system Cameron is at $(0, 0)$. The coordinates of the rigs are $(-10, 5), (0, 6), (8, 20)$, and $(12, 4)$ where the units are miles.

a)   How many routes are possible?

b)   List all of the possible routes.

c)   Find the distance to the nearest tenth of a mile from each location to every other location.

d)   Find the length of each of the listed routes. Are there any duplications?

e)   What is the shortest possible route?

f)   If the supply boat had to stop at 40 oil rigs, then how many routes are possible? How long would it take you to find the shortest route?

11.5

Combinations, Labeling, and the Binomial Theorem

In Section 11.4 we learned the fundamental counting principle, and we applied it to finding the number of permutations of n objects taken r at a time. In permutations, the r objects are arranged in a sequential manner. In this section, we will find the number of ways to choose r objects from n distinct objects when the order in which the objects are chosen or placed is unimportant.

Combinations of n Things Taken r at a Time

In how many ways can two students be selected to go to the board from a class of four students: Adams, Baird, Campbell, and Dalton? Assuming that the two selected are treated identically, the choice of Adams and Baird is no different from the choice of Baird and Adams. Set notation provides a convenient way of listing all possible choices of two students from the four available students because in set notation $\{A, B\}$ is the same as $\{B, A\}$. We can easily list all subsets or **combinations** of two elements taken from the set $\{A, B, C, D\}$:

$$\{A, B\} \quad \{A, C\} \quad \{A, D\} \quad \{B, C\} \quad \{B, D\} \quad \{C, D\}$$

The number of these subsets is the number of combinations of four things taken two at a time, denoted by $C(4, 2)$. Since there are six subsets, $C(4, 2) = 6$.

If we had a first and a second prize to give to two of the four students, then $P(4, 2) = 4 \cdot 3 = 12$ is the number of ways to award the prizes. Since the prizes are different, AB is different from BA, AC is different from CA, and so on. The 12 permutations are listed here:

$$AB \quad AC \quad AD \quad BC \quad BD \quad CD$$
$$BA \quad CA \quad DA \quad CB \quad DB \quad DC$$

From the list of combinations of four things taken two at a time, we made the list of permutations of four things two at a time by rearranging each combination. So $P(4, 2) = 2 \cdot C(4, 2)$.

In general, we can list all combinations of n things taken r at a time, then rearrange each of those subsets of r things in $r!$ ways to obtain all of the permutations of n things taken r at a time. So $P(n, r) = r! \, C(n, r)$, or

$$C(n, r) = \frac{P(n, r)}{r!}.$$

Since $P(n, r) = n!/(n - r)!$, we have

$$C(n, r) = \frac{n!}{(n - r)! \, r!}.$$

These results are summarized in the following theorem.

Theorem: Combinations of n Things Taken r at a Time

The number of combinations of n things taken r at a time (or the number of subsets of size r from a set of n elements) is given by the formula

$$C(n, r) = \frac{n!}{(n - r)! \, r!} \quad \text{for } 0 \leq r \leq n.$$

Many calculators can calculate the value of $C(n, r)$ when given the value of n and r. The notations $\binom{n}{r}$ or nCr may be used on your calculator or elsewhere for $C(n, r)$. Note that

$$C(n, n) = \frac{n!}{0!\, n!} = 1 \quad \text{and} \quad C(n, 0) = \frac{n!}{n!\, 0!} = 1.$$

There is only one way to choose all n objects from a group of n objects if the order does not matter, and there is only one way to choose no object ($r = 0$) from a group of n objects.

Example 1 Combinations

To raise money, an alumni association prints lottery tickets with the numbers 1 through 11 printed on each ticket, as shown in Fig. 11.14. To play the lottery, one must circle three numbers on the ticket. In how many ways can three numbers be chosen out of 11 numbers on the ticket?

ALUMNI ASSOCIATION LOTTERY

Circle three numbers:

1 ② 3 4 5 6 ⑦ 8 ⑨ 10 11

■ **Figure 11.14**

Solution

Choosing three numbers from a list of 11 numbers is the same as choosing a subset of size 3 from a set of 11 elements. So the number of ways to choose the three numbers in playing the lottery is the number of combinations of 11 things taken 3 at a time:

$$C(11, 3) = \frac{11!}{8!\, 3!}$$

$$= \frac{11 \cdot 10 \cdot 9 \cdot 8 \cdot 7 \cdot 6 \cdot 5 \cdot 4 \cdot 3 \cdot 2 \cdot 1}{8 \cdot 7 \cdot 6 \cdot 5 \cdot 4 \cdot 3 \cdot 2 \cdot 1 \cdot 3 \cdot 2 \cdot 1}$$

$$= \frac{11 \cdot 10 \cdot 9}{3 \cdot 2 \cdot 1} \quad \text{\small Divide numerator and denominator by 8!.}$$

$$= 165$$

```
11 nCr 3
                165
11!/(8!3!)
                165
```

■ **Figure 11.15**

Use the combination function or factorials on a calculator to check, as in Fig. 11.15.

Try This. In how many ways can a group of 10 friends on a rented sailboat select 3 of their group as deck hands? ■

Labeling

In a **labeling problem,** n distinct objects are to be given labels, each object getting exactly one label. For example, each person in a class is "labeled" with a letter grade at the end of the semester. Each student living on campus is "labeled" with the name of the dormitory in which the student resides. In a labeling problem, each distinct object gets one label, but there may be several types of labels and many labels of each type.

Example 2 A labeling problem

Twelve students have volunteered to help with a political campaign. The campaign director needs three telephone solicitors, four door-to-door solicitors, and five envelope stuffers. In how many ways can these jobs (labels) be assigned to these 12 students?

Solution

Since the three telephone solicitors all get the same type of label, the number of ways to select the three students is $C(12, 3)$. The number of ways to select four door-to-door solicitors from the remaining nine students is $C(9, 4)$. The number of ways to select the five envelope stuffers from the remaining five students is $C(5, 5)$. By the fundamental counting principle, the number of ways to make all three selections is

$$C(12, 3) \cdot C(9, 4) \cdot C(5, 5) = \frac{12!}{9!\,3!} \cdot \frac{9!}{5!\,4!} \cdot \frac{5!}{0!\,5!}$$

$$= \frac{12!}{3!\,4!\,5!}$$

$$= 27{,}720.$$

Try This. In how many ways can a group of 10 friends on a rented sailboat select 3 of their group as deck hands, 3 as officers, and 4 as galley workers?   ■

Note that in Example 2 there were 12 distinct objects to be labeled with three labels of one type, four labels of another type, and five labels of a third type, and the number of ways to assign those labels was found to be $12!/(3!\,4!\,5!)$. Instead of using combinations and the fundamental counting principle, as in Example 2, we can use the following theorem.

Labeling Theorem

> If each of n distinct objects is to be assigned one label and there are r_1 labels of the first type, r_2 labels of the second type,..., and r_k labels of the kth type, where $r_1 + r_2 + \cdots + r_k = n$, then the number of ways to assign the labels is
>
> $$\frac{n!}{r_1!\,r_2! \cdot\,\cdots\,\cdot r_k!}.$$

Example 3 Rearrangement of letters in a word

How many different arrangements are there for the 11 letters in the word MISSISSIPPI?

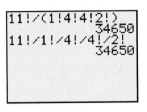

■ **Figure 11.16**

Solution

This problem is a labeling problem if we think of the 11 positions for the letters as 11 distinct objects to be labeled. There is one M-label, and there are four S-labels, four I-labels, and two P-labels. So the number of ways to arrange the letters in MISSISSIPPI is

$$\frac{11!}{1!\,4!\,4!\,2!} = 34{,}650.$$

To calculate this value, either enclose the denominator in parentheses or divide by each factorial in the denominator, as in Fig. 11.16.

Try This. How many different arrangements are there for the letters in the word CALCULUS? ■

You may think of permutation and combination problems as being very different, but they both are labeling problems in actuality. For example, to find the number of subsets of size 3 from a set of size 5, we are assigning three I-labels and two N-labels (I for "in the subset" and N for "not in the subset") to the five distinct objects of the set. Note that

$$\frac{5!}{3!\,2!} = C(5,3).$$

To find the number of ways to give a first, second, and third prize to three of ten people, we are assigning one F-label, one S-label, one T-label, and seven N-labels (N for "no prize") to the ten distinct people. Note that

$$\frac{10!}{1!\,1!\,1!\,7!} = P(10,3).$$

Combinations, Permutations, and Counting

Note that the combination formula counts the number of subsets of n objects taken r at a time. The objects are not necessarily placed in a subset, but they are all treated alike, as in Example 1. In counting the combinations in Example 1, the only important thing is *which* numbers are circled; the order in which the numbers are circled is not important. By contrast, the permutation formula counts the number of ways to select r objects from n, *in order.* For example, the number of ways to award a first, second, and third prize to three of five people is $P(5,3) = 60$, because the order matters. The number of ways to give three identical prizes to three of five people is $C(5,3) = 10$, because the order of the awards doesn't matter.

Do not forget that we also have the fundamental counting principle to count the number of ways in which a sequence of events can occur. In the next example we use both the fundamental counting principle and the combination formula.

Example **4** **Combinations and the counting principle**

A company employs nine male welders and seven female welders. A committee of three male welders and two female welders is to be chosen to represent the welders in negotiations with management. How many different committees can be chosen?

Solution

First observe that three male welders can be chosen from the nine available in $C(9, 3) = 84$ ways. Next observe that two female welders can be chosen from seven available in $C(7, 2) = 21$ ways. Now use the fundamental counting principle to get $84 \cdot 21 = 1764$ ways to choose the males and then the females.

Try This. A student has 20 rock and 15 country CDs. How many ways are there to select 3 rock and 2 country CDs to play in his five-disc CD player? ■

Binomial Expansion

One of the first facts we learn in algebra is

$$(a + b)^2 = a^2 + 2ab + b^2.$$

Our new labeling technique can be used to find the *coefficients* for the square of $a + b$ and any higher power of $a + b$. But before using counting techniques, let's look for patterns in the higher powers of $a + b$ and learn some shortcuts that can be applied in the simple cases.

We can find $(a + b)^3$ by multiplying $(a + b)^2$ by $a + b$. The powers of $a + b$ from the zero power to the fourth power are listed here.

$$(a + b)^0 = 1$$

$$(a + b)^1 = a + b$$

$$(a + b)^2 = a^2 + 2ab + b^2$$

$$(a + b)^3 = a^3 + 3a^2b + 3ab^2 + b^3$$

$$(a + b)^4 = a^4 + 4a^3b + 6a^2b^2 + 4ab^3 + b^4$$

The right-hand sides of these equations are referred to as **binomial expansions.** Each one after the first one is obtained by multiplying the previous expansion by $a + b$. To find the expansion for $(a + b)^5$, multiply the expansion for $(a + b)^4$ by $a + b$:

$$
\begin{array}{r}
a^4 \;+\; 4a^3b \;+\; 6a^2b^2 + 4ab^3 + b^4 \\
a \;+\; b \\
\hline
a^4b \;+\; 4a^3b^2 \;+\; 6a^2b^3 + 4ab^4 + b^5 \\
a^5 + 4a^4b \;+\; 6a^3b^2 \;+\; 4a^2b^3 \;+\; ab^4 \\
\hline
a^5 + 5a^4b + 10a^3b^2 + 10a^2b^3 + 5ab^4 + b^5
\end{array}
$$

Now examine the expansion of $(a + b)^5$. There are six terms and the exponents in each term have a sum of 5. The powers of a decrease from left to right, while the powers of b increase. The **binomial coefficients** 1, 5, 10, 10, 5, 1 can be obtained from the coefficients of $(a + b)^4$. The coefficients of a^5 and b^5 are 1 because a^5 and b^5 are obtained only from $a \cdot a^4$ and $b \cdot b^4$. The coefficient in $5a^4b$ is the sum of the coefficients of the first two terms of $(a + b)^4$, and the coefficient in $10a^3b^2$ is the sum of the coefficients of the second and third terms of $(a + b)^4$. This pattern continues as follows:

$$(a + b)^4 = \qquad 1a^4 + 4a^3b + 6a^2b^2 + 4ab^3 + 1b^4$$

$$(a + b)^5 = 1a^5 + 5a^4b^1 + 10a^3b^2 + 10a^2b^3 + 5ab^4 + 1b^5$$

It is easy to remember the coefficients for the first few powers of $a + b$ by using **Pascal's triangle.** Pascal's triangle is a triangular array of binomial coefficients:

1					Coefficient of $(a + b)^0$
1	1				Coefficients of $(a + b)^1$
1	2	1			Coefficients of $(a + b)^2$
1	3	3	1		Coefficients of $(a + b)^3$
1	4	6	4	1	Coefficients of $(a + b)^4$
1	5	10	10	5	1 Coefficients of $(a + b)^5$

Pascal's triangle

Each row in Pascal's triangle starts and ends with 1 and the coefficients between the 1's are obtained from the previous row by adding consecutive entries.

Example **5** **Using Pascal's triangle**

Use Pascal's triangle to find each binomial expansion.

a. $(2x - 3y)^4$ **b.** $(a + b)^6$

Solution

a. Think of $(2x - 3y)^4$ as $(2x + (-3y))^4$ and use the coefficients 1, 4, 6, 4, 1 from Pascal's triangle:

$$(2x - 3y)^4 = 1(2x)^4(-3y)^0 + 4(2x)^3(-3y)^1 + 6(2x)^2(-3y)^2$$
$$+ 4(2x)^1(-3y)^3 + 1(2x)^0(-3y)^4$$
$$= 16x^4 - 96x^3y + 216x^2y^2 - 216xy^3 + 81y^4$$

b. Use the coefficients 1, 5, 10, 10, 5, 1 from Pascal's triangle to obtain the coefficients 1, 6, 15, 20, 15, 6, 1. Use these coefficients with decreasing powers of a and increasing powers of b to get

$$(a + b)^6 = a^6 + 6a^5b + 15a^4b^2 + 20a^3b^3 + 15a^2b^4 + 6ab^5 + b^6.$$

Try This. Use Pascal's triangle to expand $(a - 2b)^4$.   ■

The Binomial Theorem

Pascal's triangle provides an easy way to find a binomial expansion for small powers of a binomial, but it is not practical for large powers. To get the binomial coefficients for larger powers, we count the number of like terms of each type using the idea of labeling. For example, consider

$$(a + b)^3 = (a + b)(a + b)(a + b) = a^3 + 3a^2b + 3ab^2 + b^3.$$

The terms of the product come from all of the different ways there are to select either a or b from each of the three factors and multiply the selections. The coefficient of a^3 is 1 because we get a^3 only from aaa, where a is chosen from each factor. The coefficient of a^2b is 3 because a^2b is obtained from aab, aba, and baa. From the labeling theorem, the number of ways to rearrange the letters aab is $3!/(2!\,1!) = 3$.

So the coefficient of a^2b is the number of ways to label the three factors with two a-labels and one b-label.

In general, the expansion of $(a + b)^n$ contains $n + 1$ terms in which the exponents have a sum of n. The coefficient of $a^{n-r}b^r$ is the number of ways to label n factors with $(n - r)$ of the a-labels and r of the b-labels:

$$\frac{n!}{r!(n - r)!}$$

The *binomial theorem* expresses this result using summation notation.

The Binomial Theorem

If n is a positive integer, then for any real numbers a and b,

$$(a + b)^n = \sum_{r=0}^{n} \binom{n}{r} a^{n-r} b^r, \quad \text{where} \quad \binom{n}{r} = \frac{n!}{r!(n - r)!}.$$

Example 6 Using the binomial theorem

Write out the first three terms in the expansion of $(x - 2y)^{10}$.

Solution

Write $(x - 2y)^{10}$ as $(x + (-2y))^{10}$. According to the binomial theorem,

$$(x + (-2y))^{10} = \sum_{r=0}^{10} \binom{10}{r} x^{10-r} (-2y)^r.$$

To find the first three terms, let $r = 0, 1$, and 2:

$$(x - 2y)^{10} = \binom{10}{0} x^{10}(-2y)^0 + \binom{10}{1} x^9(-2y)^1 + \binom{10}{2} x^8(-2y)^2 + \cdots$$

$$= x^{10} - 20x^9y + 180x^8y^2 + \cdots$$

Try This. Find the first three terms $(a + 3b)^{12}$. ■

Example 7 Using the binomial theorem

What is the coefficient of a^7b^2 in the binomial expansion of $(a + b)^9$?

Solution

Compare a^7b^2 to $a^{n-r}b^r$ to see that $r = 2$ and $n = 9$. According to the binomial theorem, the coefficient of a^7b^2 is

$$\binom{9}{2} = \frac{9!}{2! \, 7!} = \frac{9 \cdot 8}{2} = 36.$$

The term $36a^7b^2$ occurs in the expansion of $(a + b)^9$.

Try This. What is the coefficient of x^3y^9 in the expansion of $(x + y)^{12}$? ■

In the next example, labeling is used to find the coefficient of a term in a power of a trinomial.

Example 8 Trinomial coefficients

What is the coefficient of a^3b^2c in the expansion of $(a + b + c)^6$?

Solution

The terms of the product $(a + b + c)^6$ come from all of the different ways there are to select a, b, or c from each of the six distinct factors and multiply the selections. The number of times that a^3b^2c occurs is the same as the number of rearrangements of $aaabbc$, which is a labeling problem. The number of rearrangements of $aaabbc$ is

$$\frac{6!}{3!\,2!\,1!} = \frac{6 \cdot 5 \cdot 4 \cdot 3 \cdot 2 \cdot 1}{3 \cdot 2 \cdot 1 \cdot 2 \cdot 1 \cdot 1} = 60.$$

So the term $60a^3b^2c$ occurs in the expansion of $(a + b + c)^6$.

Try This. What is the coefficient of $x^3y^4z^2$ in the expansion of $(x + y + z)^9$? ■

For Thought

True or False? Explain.

1. The number of ways to choose three questions to answer out of five questions on an essay test is $C(5, 2)$.

2. The number of ways to answer a five-question multiple-choice test in which each question has three choices is $P(5, 3)$.

3. The number of ways to pick a Miss America and the first runner-up from the five finalists is 20.

4. The binomial expansion for $(x + y)^n$ contains n terms.

5. For any real numbers a and b, $(a + b)^5 = \displaystyle\sum_{i=0}^{5} \binom{5}{i} a^i b^{5-i}$.

6. The sum of the binomial coefficients in $(a + b)^n$ is 2^n.

7. $P(8, 3) < C(8, 3)$

8. $P(7, 3) = (3!) \cdot C(7, 3)$

9. $P(8, 3) = P(8, 5)$

10. $C(1, 1) = 1$

11.5 Exercises

Evaluate each expression.

1. $\dfrac{5!}{3!\,2!}$

2. $\dfrac{6!}{4!\,2!}$

3. $\dfrac{7!}{6!\,1!}$

4. $\dfrac{8!}{1!\,7!}$

5. $C(5, 1)$

6. $C(9, 1)$

7. $C(8, 4)$

8. $C(7, 3)$

9. $C(7, 4)$

10. $C(6, 3)$

11. $C(10, 0)$

12. $C(9, 0)$

13. $C(12, 12)$

14. $C(11, 11)$

Solve each problem.

15. *Selecting* The vice-president of Southern Insurance will select two of his four secretaries to attend the board meeting. How many selections are possible? List all possible selections from the secretaries Alice, Brenda, Carol, and Dolores.

16. *Choosing* Northwest Distributors will buy two pickup trucks, but has found three that meet its needs, a Ford, a Chevrolet, and a Toyota. In how many ways can Northwest choose two out of the three trucks? List all possible choices.

17. *Surviving* Five people are struggling to survive in the wilderness. In this week's episode, the producers will send two of the five back to civilization. In how many ways can the two be selected?

18. *Scheduling* Tiffany has the prerequisites for six different computer science courses. In how many ways can she select two for her schedule next semester.

19. *Job Candidates* The search committee has narrowed the applicants to five unranked candidates. In how many ways can three be chosen for an in-depth interview?

20. *Spreading the Flu* In how many ways can nature select five students out of a class of 20 students to get the flu?

21. *Playing a Lottery* In a certain lottery the player chooses six numbers from the numbers 1 through 49. In how many ways can the six numbers be chosen?

22. *Fantasy Five* In a different lottery the player chooses five numbers from the numbers 1 through 39. In how many ways can the five numbers be chosen?

23. *Poker Hands* How many five-card poker hands are there if you draw five cards from a deck of 52?

24. *Bridge Hands* How many 13-card bridge hands are there if you draw 13 cards from a deck of 52?

Solve each problem using the idea of labeling.

25. *Assigning Topics* An instructor in a history class of ten students wants term papers written on World War II, World War I, and the Civil War. If he randomly assigns World War II to five students, World War I to three students, and the Civil War to two students, then in how many ways can these assignments be made?

26. *Parking Tickets* Officer O'Reilly is in charge of 12 officers working a sporting event. In how many ways can she choose six to control parking, four to work security, and two to handle emergencies?

27. *Rearranging Letters* How many permutations are possible using the letters in the word ALABAMA?

28. *Spelling Mistakes* How many incorrect spellings of the word FLORIDA are there, using all of the correct letters?

29. *Determining Chords* How many distinct chords (line segments with endpoints on the circle) are determined by three points lying on a circle? By four points? By five points? By n points?

30. *Determining Triangles* How many distinct triangles are determined by six points lying on a circle, where the vertices of each triangle are chosen from the six points?

31. *Assigning Volunteers* Ten students volunteered to work in the governor's reelection campaign. Three will be assigned to making phone calls, two will be assigned to stuffing envelopes, and five will be assigned to making signs. In how many ways can the assignments be made?

32. *Assigning Vehicles* Three identical Buicks, four identical Fords, and three identical Toyotas are to be assigned to ten traveling salespeople. In how many ways can the assignments be made?

The following problems may involve combinations, permutations, or the fundamental counting principle.

33. *Television Schedule* Arnold plans to spend the evening watching television. Each of the three networks runs one-hour shows starting at 7:00, 8:00, and 9:00 P.M. In how many ways can he watch three complete shows from 7:00 to 10:00 P.M.?

34. *Cafeteria Meals* For $3.98 you can get a salad, main course, and dessert at the cafeteria. If you have a choice of five different salads, six different main courses, and four different desserts, then how many different $3.98 meals are there?

35. *Fire Code Inspections* The fire inspector in Cincinnati must select three night clubs from a list of eight for an inspection of their compliance with the fire code. In how many ways can she select the three night clubs?

36. *Prize-Winning Pigs* In how many ways can a red ribbon, a blue ribbon, and a green ribbon be awarded to three of six pigs at the county fair?

37. *Choosing a Team* From the nine male and six female sales representatives for an insurance company, a team of three men and two women will be selected to attend a national conference on insurance fraud. In how many ways can the team of five be selected?
    HINT Select the men and the women, then use the fundamental counting principle.

38. *Poker Hands* How many five-card poker hands are there containing three hearts and two spades?
    HINT Select the hearts and the spades, then use the fundamental counting principle.

39. *Returning Exam Papers* In how many different orders can Professor Pereira return 12 exam papers to 12 students?

40. *Saving the Best Till Last* In how many ways can Professor Yang return examination papers to 12 students if she always returns the worst paper first and the best paper last?

41. *Rolling Dice* A die is to be rolled twice, and each time the number of dots showing on the top face is to be recorded. If we think of the outcome of two rolls as an ordered pair of numbers, then how many outcomes are there?

42. *Having Children* A couple plans to have three children. Considering the sex and order of birth of each of the three children, how many outcomes are possible to this plan?

43. *Marching Bands* In how many ways can four marching bands and three floats line up for a parade if a marching band must lead the parade and two bands cannot march next to one another?

44. *Marching Bands* In how many ways can four marching bands and four floats line up for a parade if a marching band must lead the parade and two bands cannot march next to one another?

Write the complete binomial expansion for each of the following powers of a binomial.

45. $(x + y)^2$

46. $(x + 5)^2$

47. $(2a - 3)^2$

48. $(2x - 3y)^2$

49. $(a - 2)^3$

50. $(b^2 - 3)^3$

51. $(2a + b^2)^3$

52. $(x + 3)^3$

53. $(x - 2y)^4$

54. $(y - 2)^4$

55. $(x^2 + 1)^4$

56. $(2r + 3t^2)^4$

57. $(a - 3)^5$

58. $(b + 2y)^5$

59. $(x + 2a)^6$

60. $(2b - 1)^6$

Write the first three terms of each binomial expansion.

61. $(x + y)^9$

62. $(a - b)^{10}$

63. $(2x - y)^{12}$

64. $(a + 2b)^{11}$

65. $(2s - 0.5t)^8$

66. $(3y^2 + a)^{10}$

67. $(m^2 - 2w^3)^9$

68. $(ab^2 - 5c)^8$

Solve each problem.

69. What is the coefficient of w^3y^5 in the expansion of $(w + y)^8$?

70. What is the coefficient of a^3z^9 in the expansion of $(a + z)^{12}$?

71. What is the coefficient of a^5b^8 in the expansion of $(b - 2a)^{13}$?

72. What is the coefficient of x^6y^5 in the expansion of $(0.5x - y)^{11}$?

73. What is the coefficient of $a^2b^4c^6$ in the expansion of $(a + b + c)^{12}$?

74. What is the coefficient of $x^3y^2z^3$ in the expansion of $(x - y - 2z)^8$?

75. What is the coefficient of a^3b^7 in the expansion of $(a + b + 2c)^{10}$?

76. What is the coefficient of $w^2xy^3z^9$ in the expansion of $(w + x + y + z)^{15}$?

For Writing/Discussion

77. Explain why $C(n, r) = C(n, n - r)$.

78. Is $P(n, r) = P(n, n - r)$? Explain your answer.

79. Explain the difference between permutations and combinations.

80. Prove that $\sum_{i=0}^{n} \binom{n}{i} = 2^n$ for any positive integer n.

81. *Cooperative Learning* Make up three counting problems: a combination problem, a permutation problem, and a labeling problem. Give them to a classmate to solve.

Thinking Outside the Box XCV

Strange Sequence What is the one-millionth term in the sequence

$$1,2,2,3,3,3,4,4,4,4,5,5,5,5,5, \ldots$$

in which each positive integer n occurs n times?

11.5 Pop Quiz

1. Evaluate $C(160, 4)$.

2. In how many ways can a class of 40 students choose three to represent the class at a convention?

3. In how many ways can 30 male employees and 25 female employees choose two males and two females to represent the employees in talks with management?

4. Expand $(x + 2y)^5$.

Linking Concepts

For Individual or Group Explorations

Poker Hands

In the game of poker, a hand consists of five cards drawn from a deck of 52. The best hand is the royal flush. It consists of the ace, king, queen, jack, and ten of one suit. There are only four possible royal flushes. Find the number of possible hands of each of the following types. Explain your reasoning. The hands are listed in order from the least to the most common. Of course, a hand in this list beats any hand that is below it in this list.

a) Straight flush (five cards in a sequence in the same suit, but not a royal flush)

b) Four of a kind (for example, four kings or four twos)

c) Full house (three of a kind together with a pair)

d) Flush (five cards in a single suit, but not a straight)

e) Three of a kind

f) Two pairs

g) One pair

11.6

Probability

We often hear statements such as "The probability of rain today is 70%," "I have a 50-50 chance of passing English," or "I have one chance in a million of winning the lottery." It is clear that we have a better chance of getting rained on than we have of winning a lottery, but what exactly is probability? In this section we will study probability and use the counting techniques of Sections 11.4 and 11.5 to make precise statements concerning probabilities.

The Probability of an Event

An **experiment** is any process for which the outcome is uncertain. For example, if we toss a coin, roll a die, draw a poker hand from a deck, or arrange people in a line in a manner that makes the outcome uncertain, then these processes are experiments. A **sample space** is the set of all possible outcomes to an experiment. If each outcome occurs with about the same frequency when the experiment is repeated many times, then the outcomes are called **equally likely.**

The simplest experiments for determining probabilities are ones in which the outcomes are equally likely. For example, if a fair coin is tossed, then the sample space S consists of two equally likely outcomes, heads and tails:

$$S = \{H, T\}$$

An **event** is a subset of a sample space. The subset $E = \{H\}$ is the event of getting heads when the coin is tossed. If $n(S)$ is the number of equally likely outcomes in S, then $n(S) = 2$. If $n(E)$ is the number of equally likely outcomes in E, then $n(E) = 1$. The probability of the event E, $P(E)$, is the ratio of $n(E)$ to $n(S)$:

$$P(E) = \frac{n(E)}{n(S)} = \frac{1}{2}$$

Definition: Probability of an Event

If S is a sample space of equally likely outcomes to an experiment and the event E is a subset of S, then the **probability of E, $P(E)$**, is defined by

$$P(E) = \frac{n(E)}{n(S)}.$$

Since E is a subset of S, $0 \le n(E) \le n(S)$. So $P(E)$ is a number between 0 and 1, inclusive. If there are no outcomes in the event E, then $P(E) = 0$ and it is impossible for E to occur. If $n(E) = n(S)$, then $P(E) = 1$ and the event E is certain to occur. For example, if E is the event of getting two heads on a single toss of a coin, then $n(E) = 0$ and $P(E) = 0/2 = 0$. If E is the event of getting fewer than two heads on a single toss of a coin, then for either outcome H or T there are fewer than two heads. So $E = \{H, T\}$, $n(E) = 2$, and $P(E) = 2/2 = 1$.

Probability predicts the future, but not exactly. The probability that heads will occur when a coin is tossed is $1/2$, but there is no way to know whether heads or tails will occur on any future toss. However, the probability $1/2$ indicates that if the coin is tossed many times, then *about* $1/2$ of the tosses will be heads.

Example 1 Tossing a pair of coins

What is the probability of getting at least one head when a pair of coins is tossed?

Solution

Since there are two equally like outcomes for the first coin and two equally likely outcomes for the second coin, by the fundamental counting principle there are four equally likely outcomes to the experiment of tossing a pair of coins. We can list the outcomes as ordered pairs:

$$S = \{(H, H), (H, T), (T, H), (T, T)\}.$$

The ordered pairs (H, T) and (T, H) must be listed as different outcomes, because the first coordinate is the result from the first coin and the second coordinate is the result from the second coin. Since three of these outcomes result in at least one head,

$$E = \{(H, H), (H, T), (T, H)\}$$

and $n(E) = 3$. So

$$P(E) = \frac{n(E)}{n(S)} = \frac{3}{4}.$$

Try This. Find the probability of getting fewer than two heads when a pair of coins is tossed. ■

Example 2 Rolling a single die

What is the probability of getting a number larger than 4 when a single six-sided die is rolled?

Solution

When a die is rolled, the number of dots showing on the upper face of the die is counted. So the sample space of equally likely outcomes is

$$S = \{1, 2, 3, 4, 5, 6\} \qquad \text{and} \qquad n(S) = 6.$$

Since only 5 and 6 are larger than 4,

$$E = \{5, 6\} \qquad \text{and} \qquad n(E) = 2.$$

According to the definition of probability,

$$P(E) = \frac{n(E)}{n(S)} = \frac{2}{6} = \frac{1}{3}.$$

Try This. Find the probability of getting a number larger than five when a single die is rolled. ■

Example 3 Rolling a pair of dice

What is the probability of getting a sum of 5 when a pair of dice is rolled?

Solution

Since there are six equally likely outcomes for each die, by the fundamental counting principle there are 6 · 6 or 36 equally likely outcomes to rolling the pair. See Fig. 11.17. We can list the 36 outcomes as ordered pairs as shown.

$$
\begin{aligned}
S = \{ &(1, 1), (1, 2), (1, 3), (1, 4), (1, 5), (1, 6), \\
&(2, 1), (2, 2), (2, 3), (2, 4), (2, 5), (2, 6), \\
&(3, 1), (3, 2), (3, 3), (3, 4), (3, 5), (3, 6), \\
&(4, 1), (4, 2), (4, 3), (4, 4), (4, 5), (4, 6), \\
&(5, 1), (5, 2), (5, 3), (5, 4), (5, 5), (5, 6), \\
&(6, 1), (6, 2), (6, 3), (6, 4), (6, 5), (6, 6)\}
\end{aligned}
$$

The ordered pairs in color in the sample space are the ones in which the sum of the entries is 5. So the phrase "sum of the numbers is 5" describes the event

$$E = \{(4, 1), (3, 2), (2, 3), (1, 4)\},$$

and

$$P(E) = \frac{n(E)}{n(S)} = \frac{4}{36} = \frac{1}{9}.$$

Try This. Find the probability of getting a sum of three when a pair of dice is rolled. ■

■ **Figure 11.17**

Example **4** **Probability of winning a lottery**

To play the Alumni Association Lottery, you pick three numbers from the integers from 1 through 11 inclusive. What is the probability that you win the $50 prize by picking the same three numbers as the ones chosen by the Alumni Association?

Solution

The number of ways to select three numbers out of 11 numbers is $C(11, 3) = 165$. Since only one of the 165 equally likely outcomes is the winning combination, the probability of winning is $1/165$.

Try This. Two friends each randomly predict the five finalists from the 50 candidates for Miss America. What is the probability that the friends have made exactly the same prediction? ■

The Addition Rule

One way to find the probability of an event is to list all of the equally likely outcomes to the experiment and count those outcomes that make up the event. The addition rule provides a relationship between probabilities of events that can be used to find probabilities.

In rolling a pair of dice, let A be the event that doubles occur and let B be the event that the sum is 4. We have

$$A = \{(1, 1), (2, 2), (3, 3), (4, 4), (5, 5), (6, 6)\}$$

and

$$B = \{(3, 1), (2, 2), (1, 3)\}.$$

The event that doubles *and* a sum of 4 occur is the event $A \cap B$:

$$A \cap B = \{(2, 2)\}$$

Count the ordered pairs to get $P(A) = 6/36$, $P(B) = 3/36$, and $P(A \cap B) = 1/36$. The event that either doubles *or* a sum of 4 occurs is the event $A \cup B$:

$$A \cup B = \{(1, 1), (2, 2), (3, 3), (4, 4), (5, 5), (6, 6), (3, 1), (1, 3)\}$$

Even though $(2, 2)$ belongs to both A and B, it is listed only once in $A \cup B$. Note that $P(A \cup B)$ is $8/36$ and

$$\frac{8}{36} = \frac{6}{36} + \frac{3}{36} - \frac{1}{36}.$$

Since $(2, 2)$ is counted in both A and B, $P(A \cap B)$ is subtracted from $P(A) + P(B)$. This example illustrates the addition rule.

The Addition Rule

If A and B are any events in a sample space, then

$$P(A \cup B) = P(A) + P(B) - P(A \cap B).$$

If $P(A \cap B) = 0$, then A and B are called **mutually exclusive** events and

$$P(A \cup B) = P(A) + P(B).$$

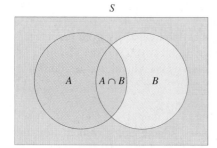

S

A   *A ∩ B*   *B*

■ **Figure 11.18**

Two events are mutually exclusive if it is impossible for both to occur. For example, when a pair of dice is rolled once, getting a sum of 4 and getting a sum of 5 are mutually exclusive events.

The diagram in Fig. 11.18 illustrates the addition rule for two sets A and B whose intersection is not empty. To find $P(A \cup B)$ we add $P(A)$ and $P(B)$, but we must subtract $P(A \cap B)$ because it is counted in both $P(A)$ and $P(B)$. Since $P(A \cap B) = 0$ for mutually exclusive events, the addition rule for mutually exclusive events is a special case of the general addition rule.

Example **5**   **Using the addition rule**

At Hillside Community College, 60% of the students are commuters (C), 50% are female (F), and 30% are female commuters. If a student is selected at random, what is the probability that the student is either a female or a commuter?

Solution

By the addition rule, the probability of selecting either a female or a commuter is

$$P(F \cup C) = P(F) + P(C) - P(F \cap C)$$
$$= 0.50 + 0.60 - 0.30$$
$$= 0.80$$

Try This.   At a country club, 40% of the members are retired, 60% are over 55, and 30% are over 55 and retired. If a member is selected at random, then what is the probability that the member is either retired or over 55?   ■

Example **6**   **The addition rule with mutually exclusive events**

In rolling a pair of dice, what is the probability that the sum is 12 or at least one die shows a 2?

Solution

Let A be the event that the sum is 12 and B be the event that at least one die shows a 2. Since A occurs on only one of the 36 equally likely outcomes, $(6, 6)$ (see Example 3), $P(A) = 1/36$. B occurs on 11 of the equally likely outcomes, so $P(B) = 11/36$. Since A and B are mutually exclusive, we have

$$P(A \cup B) = P(A) + P(B) = \frac{1}{36} + \frac{11}{36} = \frac{12}{36} = \frac{1}{3}.$$

Try This.   What is the probability that the sum is 3 or 5 when a pair of dice is rolled?   ■

Complementary Events

If the probability of rain today is 60%, then the probability that it does not rain is 40%. Rain and not rain are called complementary events. For complementary events there is no possibility that both occur, and one of them must occur. If A is an event, then A' (read "A prime" or "A complement") represents the complement of the event A. Note that complementary events are mutually exclusive, but mutually exclusive events are not necessarily complementary.

Definition: Complementary Events

Two events A and A' are called **complementary events** if $A \cap A' = \varnothing$ and $A \cup A' = S$.

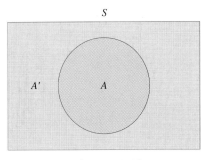

■ **Figure 11.19**

The diagram in Fig. 11.19 illustrates complementary events. The region inside the rectangle represents the sample space S, while the region inside the circle represents the event A. The event A' is the region inside the rectangle, but outside the circle.

We can show that the probability of any event and the probability of its complement have a sum of 1 (Exercise 74). In symbols:

Theorem: Complementary Events

If A and A' are complementary events, then $P(A) + P(A') = 1$.

We often use this property without thinking about it. For example, if the Rams have a 90% chance of winning Sunday's game, then the Rams have a 10% chance of losing the game.

Example **7** **The probability of complementary events**

What is the probability of getting a sum that is not equal to 5 when rolling a pair of dice?

Solution

From Example 3, the probability that the sum is 5 is $1/9$. Getting a sum that is not equal to 5 is the complement of getting a sum that is equal to 5. So

$$P(\text{sum is } 5) + P(\text{sum is not } 5) = 1$$

$$P(\text{sum is not } 5) = 1 - P(\text{sum is } 5).$$

So the probability that the sum is not 5 is $1 - 1/9$, or $8/9$.

Try This. What is the probability that the sum is not 3 when a pair of dice is rolled? ■

Odds

If the probability is $4/5$ that the Braves will win the World Series and $1/5$ that they will lose, then they are four times as likely to win as they are to lose. We say that the *odds* in favor of the Braves winning the World Series are 4 to 1. Notice that odds are *not* probabilities. Odds are ratios of probabilities. Odds are usually written as ratios of whole numbers.

Definition: Odds

> If A is any event, then the **odds in favor of** A are defined as the ratio $P(A)$ to $P(A')$ and the **odds against** A are defined as the ratio of $P(A')$ to $P(A)$.

Example 8 Odds for and against an event

What are the odds in favor of getting a sum of 5 when rolling a pair of dice? What are the odds against a sum of 5?

Solution

From Examples 3 and 5, $P(\text{sum is 5}) = 1/9$ and $P(\text{sum is not 5}) = 8/9$. The odds in favor of getting a sum of 5 are 1/9 to 8/9. Multiply each fraction by 9 to get the odds 1 to 8. The odds against a sum of 5 are 8 to 1.

Try This. What are the odds in favor of getting a sum of 3 when a pair of dice is rolled? ■

The odds in favor of A are found from the ratio $P(A)$ to $P(A')$, but the equation $P(A) + P(A') = 1$ can be used to find $P(A)$ and $P(A')$.

Example 9 Finding probabilities from odds

If the odds in favor of the Rams going to the Super Bowl are 1 to 5, then what is the probability that the Rams will go to the Super Bowl?

Solution

Since 1 to 5 is the ratio of the probability of going to that of not going, the probability of not going is 5 times as large as the probability of going. Let $P(G) = x$ and $P(G') = 5x$. Since

$$P(G) + P(G') = 1,$$

we have

$$x + 5x = 1$$
$$6x = 1$$
$$x = \frac{1}{6}.$$

So the probability that the Rams will go to the Super Bowl is 1/6.

Try This. If the odds in favor of the Eagles winning the playoff game are 4 to 5, then what is the probability that the Eagles win the game? ■

We can express the idea found in Example 9 as a theorem relating odds to probabilities.

Theorem: Converting from Odds to Probability	If the odds in favor of event E are a to b, then $$P(E) = \frac{a}{a+b} \quad \text{and} \quad P(E') = \frac{b}{a+b}.$$

For Thought

True or False? Explain.

1. If S is a sample space of equally likely outcomes and E is a subset of S, then $P(E) = n(E)$.

2. If an experiment consists of tossing four coins, then the sample space consists of eight equally likely outcomes.

3. If a single coin is tossed twice, then P(at least one tail) $= 0.75$.

4. If a pair of dice is rolled, then P(at least one 4) $= 11/36$.

5. If four coins are tossed, then P(at least one head) $= 4/16$.

6. If two coins are tossed, then the complement of getting exactly two heads is getting exactly two tails.

7. If the probability of getting exactly three tails in a toss of three coins is $1/8$, then the probability of getting at least one head is $7/8$.

8. If P(snow today) $= 0.7$, then the odds in favor of snow are 7 to 10.

9. If the odds in favor of an event E are 3 to 4, then $P(E) = 3/4$.

10. The ratio of $1/5$ to $4/5$ is equivalent to the ratio of 4 to 1.

11.6 Exercises

List all equally likely outcomes in the sample space for each of the following experiments.

1. A pair of coins is tossed (4 outcomes).

2. A pair of dice is rolled (36 outcomes).

3. A coin and a six-sided die are tossed simultaneously (12 outcomes).

4. A coin is tossed three times (8 outcomes).

Use the sample spaces in Exercises 1–4 to find the following probabilities.

5. What is the probability of getting exactly one head when a pair of coins is tossed?

6. What is the probability of getting exactly two tails when a pair of coins is tossed?

7. What is the probability of getting two sixes when a pair of dice is rolled?

8. What is the probability of getting the same number on both dies when a pair of dice is rolled?

9. A coin and a six-sided die are tossed. What is the probability of getting heads and a five?

10. A coin and a six-sided die are tossed. What is the probability of getting tails and an even number?

11. What is the probability of getting all tails when a coin is tossed three times?

12. What is the probability of getting at least one head when a coin is tossed three times?

Solve each probability problem.

13. *Rolling a Die* If a single die is rolled, then what is the probability of getting
    a. a number larger than 2?
    b. a number less than or equal to 6?
    c. a number other than 4?
    d. a number larger than 8?
    e. a number smaller than 2?

14. *Tossing a Coin*   If a single coin is tossed, then what is the probability of getting
   a. heads?
   b. fewer than two tails?
   c. exactly three tails?

15. *Tossing Two Coins Once*   If a pair of coins is tossed, then what is the probability of getting
   a. exactly two tails?
   b. at least one head?
   c. exactly two heads?
   d. at most one head?

16. *Tossing One Coin Twice*   If a single coin is tossed twice, then what is the probability of getting
   a. heads followed by tails?
   b. two tails in a row?
   c. heads on the second toss?
   d. exactly one head?

17. *Rolling a Pair of Dice*   If a pair of dice is rolled, then what is the probability of getting
   a. a pair of 3's?
   b. at least one 3?
   c. a sum of 6?
   d. a sum greater than 2?
   e. a sum less than 3?

18. *Rolling a Die Twice*   If a single die is rolled twice, then what is the probability of getting
   a. a 1 followed by a 6?
   b. a sum of 4?
   c. a 5 on the second roll?
   d. no more than two 4's?
   e. an even number followed by an odd number?

19. *Business Expansion*   The board of directors for a major corporation cannot decide whether to build its new assembly plant in Dallas, Memphis, or Chicago. If one of these cities is chosen at random, then what is the probability that
   a. Dallas is chosen?
   b. Memphis is not chosen?
   c. Topeka is chosen?

20. *Scratch and Win*   A batch of 100,000 scratch-and-win tickets contains 10,000 that are redeemable for a free order of french fries. If you randomly select one of these tickets, then what is the probability that
   a. you win an order of french fries?
   b. you do not win an order of french fries?

21. *Colored Marbles*   A marble is selected at random from a jar containing three red marbles, four yellow marbles, and six green marbles. What is the probability that

   a. the marble is red?
   b. the marble is not yellow?
   c. the marble is either red or green?
   d. the marble is neither red nor green?

22. *Choosing a Chairperson*   A committee consists of one Democrat, six Republicans, and seven Independents. If one person is randomly selected from the committee to be the chairperson, then what is the probability that
   a. the person is a Democrat?
   b. the person is either a Democrat or a Republican?
   c. the person is not a Republican?

23. *Numbered Marbles*   A jar contains nine marbles numbered 1 through 9. Two marbles are randomly selected one at a time without replacement. What is the probability that
   a. 1 is selected first and 9 is selected second?
   b. the sum of the numbers selected is 4?
   c. the sum of the numbers selected is 5?

24. *Foul Play*   A company consists of a president, a vice-president, and 10 salespeople. If 2 of the 12 people are randomly selected to win a Hawaiian vacation, then what is the probability that none of the salespeople is a winner?

25. *Poker Hands*   If a five-card poker hand is drawn from a deck of 52, then what is the probability that
   a. the hand contains the ace, king, queen, jack, and ten of hearts?
   b. the hand contains 1 three, 1 four, 1 five, 1 six, and 1 seven?

26. *Lineup*   If four people with different names and different weights randomly line up to buy concert tickets, then what is the probability that
   a. they line up in alphabetical order?
   b. they line up in order of increasing weight?

Use the addition rule to solve each problem.

27. If $P(A) = 0.8$, $P(B) = 0.6$, and $P(A \cap B) = 0.5$, then what is $P(A \cup B)$?

28. If $P(C) = 0.3$, $P(D) = 0.5$, and $P(C \cap D) = 0.2$, then what is $P(C \cup D)$?

29. If $P(E) = 0.2$, $P(F) = 0.7$, and $P(E \cup F) = 0.8$, then what is $P(E \cap F)$?

30. If $P(A) = 0.4$, $P(B) = 0.7$, and $P(A \cup B) = 0.9$, then what is $P(A \cap B)$?

31. If $P(A) = 0.2$, $P(B) = 0.3$, and $P(A \cap B) = 0$, then what is $P(A \cup B)$?

32. If $P(C) = 0.3$, $P(D) = 0.6$, and $P(C \cap D) = 0$, then what is $P(C \cup D)$?

Consider the sample space of 36 equally likely outcomes to the experiment in which a pair of dice is rolled. In each case determine whether the events A and B are mutually exclusive.

33. *A*: The sum is four.
B: The sum is five.

34. *A*: The sum is odd.
B: The sum is even.

35. *A*: The sum is less than five.
B: The sum is even.

36. *A*: The sum is ten.
B: The numbers are the same.

37. *A*: One of the numbers is two.
B: The sum is greater than nine.

38. *A*: The sum is twelve.
B: The numbers are different.

Solve each problem.

39. *Insurance Categories*   Among the drivers insured by American Insurance, 64% are women, 38% of the drivers are in a high-risk category, and 24% of the drivers are high-risk women. If a driver is randomly selected from that company, what is the probability that the driver is either high-risk or a woman?

40. *Six or Four*   What is the probability of getting either a sum of 6 or at least one 4 in the roll of a pair of dice?

41. *Family of Five*   A couple plan to have three children. Assuming that males and females are equally likely, what is the probability that they have either three boys or three girls?

42. *Ten or Four*   What is the probability of getting either a sum of 10 or a sum of 4 in the roll of a pair of dice?

43. *Pick a Card*   What is the probability of getting either a heart or a king when drawing a single card from a deck of 52 cards?

44. *Any Card*   What is the probability of getting either a heart or a diamond when drawing a single card from a deck of 52 cards?

45. *Selecting Students*   The accompanying pie chart shows the percentages of freshmen, sophomores, juniors, and seniors at Washington High School. Find the probability that a randomly selected student is
a. a freshman.
b. a junior or a senior.
c. not a sophomore.

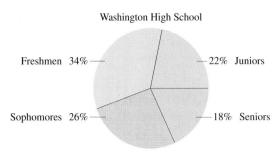

Washington High School

Freshmen 34% — 22% Juniors

Sophomores 26% — 18% Seniors

■ **Figure for Exercise 45**

46. *Earth-Crossing Asteroids*   To reduce the risk of being hit by a killer asteroid, scientists want to locate and track all of the asteroids that are visible through telescopes. Each asteroid would be classified according to whether its diameter *D* is less than 1 km and whether its orbit crossed the orbit of Earth (an Earth-crossing orbit). The table with the figure shows a hypothetical classification of 900 asteroids visible to an amateur astronomer. If an amateur astronomer randomly spots an asteroid one night, then what is the probability that
a. it is an Earth-crossing asteroid and its diameter is greater than or equal to 1 km?
b. it is either an Earth-crossing asteroid or its diameter is greater than or equal to 1 km?

	$D < 1$ km	$D \geq 1$ km	
Earth-crossing	210	90	
Not Earth-crossing	340	260	

■ **Figure for Exercise 46**

Solve each problem.

47. *Drive Defensively*   If the probability of surviving a head-on car accident at 55 mph is 0.001, then what is the probability of not surviving?

48. *Tax Time*   If the probability of a tax return not being audited by the IRS is 0.91, then what is the probability of a tax return being audited?

49. *Rolling Fours*   A pair of dice is rolled. What is the probability of
    a. getting a pair of fours?
    b. not getting a pair of fours?
    c. getting at least one number that is not a four?

50. *Tossing Triplets*   Three coins are tossed. What is the probability of
    a. getting three heads?
    b. not getting three heads?
    c. getting at least one head?

Solve each problem.

51. The probability that it rains today is 4/5 and the probability that it does not rain today is 1/5. What are the odds in favor of rain?

52. The probability that the Yankees win the World Series is 9/10 and the probability that the Yankees do not win the World Series is 1/10. What are the odds in favor of the Yankees winning the World Series?

53. *Hurricane Alley*   If the probability is 80% that the eye of hurricane Zelda comes ashore within 30 mi of Biloxi, then what are the odds in favor of the eye coming ashore within 30 mi of Biloxi?

54. *On Target*   If the probability that an arrow hits its target is 7/9, then what are the odds
    a. in favor of the arrow hitting its target?
    b. against the arrow hitting its target?

55. *Stock Market Rally*   If the probability that the stock market goes up tomorrow is 1/4, then what are the odds
    a. in favor of the stock market going up tomorrow?
    b. against the stock market going up tomorrow?

56. *Read My Lips*   If the probability of new taxes this year is 4/5, then what are the odds
    a. in favor of new taxes?
    b. against new taxes?

57. *Weather Forecast*   If the odds in favor of rain today are 4 to 1, then what is the probability of rain today?

58. *Checkmate*   If the odds are 2 to 1 in favor of Big Blue (the IBM computer) beating the Russian grand master, then what is the probability that Big Blue wins?

59. *Morning Line*   If the Las Vegas odds makers set the odds at 9 to 1 in favor of the Tigers winning their next game, then
    a. what are the odds against the Tigers winning their next game?
    b. what is the probability that the Tigers win their next game?

60. *Public Opinion*   If a pollster says that the odds are 7 to 1 against the reelection of the mayor, then
    a. what are the odds in favor of the reelection of the mayor?
    b. what is the probability that the mayor will be reelected?

61. *Two Out of Four*   What are the odds in favor of getting exactly two heads in four tosses of a coin?

62. *Rolling Once*   What are the odds in favor of getting a 5 in a single roll of a die?

63. *Lucky Seven*   What are the odds in favor of getting a sum of 7 when rolling a pair of dice?

64. *Four Out of Two*   What are the odds in favor of getting at least one 4 when rolling a pair of dice?

65. *Only One Winner*   If 2 million lottery tickets are sold and only one of them is the winning ticket, then what are the odds in favor of winning if you hold a single ticket?

66. *Pick Six*   What are the odds in favor of winning a lottery in which you must choose six numbers from the numbers 1 through 49?

67. *Five in Five*   If the odds in favor of getting five heads in five tosses of a coin are 1 to 31, then what is the probability of getting five heads in five tosses of a coin?

68. *Electing Jones*   If the odds against Jones winning the election are 3 to 5, then what is the probability that Jones will win the election?

For Writing/Discussion

69. Explain the difference between mutually exclusive events and complementary events.

70. Explain the difference between probability and odds.

71. *Cooperative Learning*   Put three pennies into a can. Shake and toss the pennies 100 times. After each toss have your helper record whether 0, 1, 2, or 3 heads are showing. On what percent of the tosses did 0, 1, 2, and 3 heads occur? What are the theoretical probabilities of obtaining 0, 1, 2, and 3 heads in a toss of three coins? How well does the theoretical model fit the actual coin toss?

72. Use a calculator to generate 100 random numbers between 0 and 1 to simulate tossing three pennies 100 times. Record 0 heads if the random number x satisfies $0 < x < 0.125$, record 1 head for $0.125 < x < 0.5$, record 2 heads for $0.5 < x < 0.875$, and record 3 heads for $0.875 < x < 1$. On what percent of the simulated tosses did 0, 1, 2, and 3 heads occur? Compare your results with those of Exercise 71.

73. Explain how you could simulate tossing three coins 100 times using a random number generator, knowing only that heads and tails are equally likely on each toss. Use your method and compare the results with those of the previous exercise.

74. Use the addition rule to prove that $P(A) + P(A') = 1$ for any complementary events A and A'.

Thinking Outside the Box XCVI

Tennis Date   Two friends agree to meet at the court for tennis between 12 noon and 1 P.M. If each arrival time is randomly chosen between 12:00 and 1:00, then what is the probability that neither friends must wait more than 10 minutes for the other?

11.6 Pop Quiz

1. What is the probability of getting all heads when a coin is tossed three times?

2. What is the probability of getting a sum of nine when a pair of dice is rolled?

3. If $P(A) = 0.6$, $P(B) = 0.4$, and $P(A \cup B) = 0.8$, then what is $P(A \cap B)$?

4. If the odds in favor of the Gators winning the game are 7 to 5, then what is the probability that the Gators win the game?

5. What is the probability of drawing either a spade or a face card when drawing a single card from a deck of 52 cards?

Linking Concepts

For Individual or Group Explorations

Dumping Pennies

Place 100 *pennies into a can, shake well, then dump them onto the floor. Set aside all of the pennies that show heads and place the remaining pennies into the can and repeat this process until there are no pennies left to be placed back into the can. Let* a_n *be the number of pennies that are in the can on the nth dump.*

a) Enter the ordered pairs (n, a_n) into a graphing calculator and use regression to find the equations of the best line and the best exponential curve that fits the data.

b) Graph the line and the exponential curve along with the ordered pairs (n, a_n). Which model looks best with the data?

c) Based on probability, what fraction of the pennies dumped would you expect to place back into the can after each dump? Does a number close to this number appear in your exponential function?

d) What does the exponential curve $y = 200(1/2)^x$ have to do with all of this?

e) Based on probability, about how many dumps would you expect to make before you have no pennies left?

11.7

Mathematical Induction

A **statement** is a sentence or equation that is either true or false. Statements involving positive integers are common in the study of sequences and series and in other areas of mathematics. For example, the statement

$$1^3 + 2^3 + 3^3 + \cdots + n^3 = \frac{n^2(n + 1)^2}{4}$$

is true for every positive integer n. This statement claims that

$$1^3 = \frac{1^2(1 + 1)^2}{4}, \qquad 1^3 + 2^3 = \frac{2^2(2 + 1)^2}{4}, \qquad 1^3 + 2^3 + 3^3 = \frac{3^2(3 + 1)^2}{4},$$

and so on. You can easily verify that the statement is true in these first three cases, but that does not prove that the statement is true for *every* positive integer n. In this section we will learn how to prove that a statement is true for every positive integer without performing infinitely many verifications.

Statements Involving Positive Integers

Suppose that S_n is used to denote the statement that $2(n + 3)$ is equal to $2n + 6$ for any positive integer n. This statement is written in symbols as follows:

$$S_n: \quad 2(n + 3) = 2n + 6$$

Is S_n true for every positive integer n? We can check that $2(1 + 3) = 2(1) + 6$, $2(2 + 3) = 2(2) + 6$, and $2(3 + 3) = 2(3) + 6$ are all correct, but no matter how many of these equations are checked, it will not prove that S_n is true for every positive integer n. However, we do know that $a(b + c) = ab + ac$ for any real numbers a, b, and c. Since positive integers are real numbers, S_n is true for every positive integer n because of the distributive property of the real numbers.

In Sections 11.2 and 11.3, statements involving positive integers occurred in the study of series. Consider the statement that the sum of the first n positive odd integers is n^2. In symbols,

$$S_n: \quad 1 + 3 + 5 + 7 + \cdots + (2n - 1) = n^2.$$

Note that S_n is a statement about the sum of an arithmetic series. However, we do know that the sum of n terms of an arithmetic series is the sum of the first and last terms multiplied by $n/2$:

$$S = \frac{n}{2}(a_1 + a_n) = \frac{n}{2}(1 + 2n - 1) = n^2$$

So S_n is true for every positive integer n.

The Principle of Mathematical Induction

So far we have proved two statements true for every positive integer n. One statement was proved by using a property of the real numbers, and the other followed from the formula for the sum of an arithmetic series. The techniques used so far will not always work. For this reason, mathematicians have developed another technique, which is called the **principle of mathematical induction** or simply **mathematical induction.**

Suppose you want to prove that you can make a very long (possibly infinite) journey on foot. You could argue that you can take the first step. Next you could argue that after every step you will use your forward momentum to take another step. Therefore, you can make the journey. Note that both parts of the argument are necessary. If every step taken leads to another step, but it is not possible to make the first step, then the journey can't be made. Likewise, if all you can make is the first step, then you will certainly not make the journey. A proof by mathematical induction is like this example of proving you can make a long journey.

The first step in mathematical induction is to prove that a statement S_n is true in the very first case. S_1 must be proved true. The second step is to prove that the truth of S_k implies the truth of S_{k+1} for any positive integer k. Note that S_k is not proved true. We assume that S_k is true and show that this assumption leads to the truth of S_{k+1}. Mathematicians agree that these two steps are necessary and sufficient to prove that S_n is true for all positive integers n, and so the principle of mathematical induction is accepted as an axiom in mathematics.

Principle of Mathematical Induction

> Let S_n be a statement for every positive integer n. If
>
> 1. S_1 is true, and
> 2. the truth of S_k implies the truth of S_{k+1} for every positive integer k, then S_n is true for every positive integer n.

To prove that a statement S_n is true for every positive integer n, the only statements that we work with are S_1, S_k, and S_{k+1}. So in the first example, we will practice simply writing those statements (correctly).

Example 1 Writing S_1, S_k, and S_{k+1}

For the given statement S_n, write the three statements S_1, S_k, and S_{k+1}.

$$S_n: \quad 1^2 + 2^2 + 3^2 + \cdots + n^2 = \frac{n(n+1)(2n+1)}{6}$$

Solution

Write S_1 by replacing n by 1 in the statement S_n. For $n = 1$ the left-hand side contains only one term:

$$1^2 = \frac{1(1+1)(2 \cdot 1 + 1)}{6}$$

S_k can be written by replacing n by k in the statement S_n:

$$1^2 + 2^2 + 3^2 + \cdots + k^2 = \frac{k(k+1)(2k+1)}{6}$$

S_{k+1} can be written by replacing n by $k+1$ in the statement S_n:

$$1^2 + 2^2 + 3^2 + \cdots + (k+1)^2 = \frac{(k+1)(k+1+1)(2(k+1)+1)}{6}$$

$$1^2 + 2^2 + 3^2 + \cdots + (k+1)^2 = \frac{(k+1)(k+2)(2k+3)}{6}$$

Try This. Write S_1, S_k, and S_{k+1} if S_n is the following statement:

$$S_n: \quad 3 + 6 + 9 + \cdots + 3n = \frac{3n(n+1)}{2}.$$ ■

Earlier we proved that the sum of the first n odd integers was n^2 by using the formula for the sum of an arithmetic series. In the next example we prove the statement again, this time using mathematical induction. Note that we name the statements in this example T_n to distinguish them from the statements S_n in the last example.

Example 2 Proof by mathematical induction

Use mathematical induction to prove that the statement

$$T_n: \quad 1 + 3 + 5 + \cdots + (2n - 1) = n^2$$

is true for every positive integer n.

Solution
Step 1: Write T_1 by replacing n by 1 in the given equation.

$$T_1: \quad 1 = 1^2$$

Since $1 = 1^2$ is correct, T_1 is true.
Step 2: Assume that T_k is true. That is, we assume the equation

$$1 + 3 + 5 + \cdots + (2k - 1) = k^2$$

is correct. Now we show that the truth of T_k implies that T_{k+1} is true. The next odd integer after $2k - 1$ is $2(k + 1) - 1$. Since T_k is true, we can add $2(k + 1) - 1$ to each side of the equation:

$$1 + 3 + 5 + \cdots + (2k - 1) + (2(k + 1) - 1) = k^2 + 2(k + 1) - 1$$

It is not necessary to write $2k - 1$ on the left-hand side because $2k - 1$ is the odd integer that precedes the odd integer $2(k + 1) - 1$:

$$1 + 3 + 5 + \cdots + 2(k + 1) - 1 = k^2 + 2k + 1 \quad \text{Simplify.}$$
$$1 + 3 + 5 + \cdots + 2(k + 1) - 1 = (k + 1)^2 \quad \text{This statement is } T_{k+1}.$$

We have shown that the truth of T_k implies that T_{k+1} is true. So by the principle of mathematical induction, the statement T_n is true for every positive integer n.

Try This. Prove $3 + 6 + 9 + \cdots + 3n = \frac{3n(n+1)}{2}$ for every positive integer n.   ■

Dominoes can be used to illustrate mathematical induction. Imagine an infinite sequence of dominoes, arranged so that when one falls over, it knocks over the one next to it, which knocks over the next one, and so on. If we can topple the first domino, then all dominoes in the infinite sequence will (theoretically) fall over.

In the next example, mathematical induction is used to prove a statement that was presented earlier in this section.

Example 3 Proof by mathematical induction

Use mathematical induction to prove that the statement

$$T_n: \quad 1^3 + 2^3 + 3^3 + \cdots + n^3 = \frac{n^2(n+1)^2}{4}$$

is true for every positive integer n.

Solution

Step 1: If $n = 1$, then the statement T_1 is

$$1^3 = \frac{1^2(1+1)^2}{4}.$$

This equation is correct by arithmetic, so T_1 is true.

Step 2: If we assume that T_k is true, then

$$1^3 + 2^3 + 3^3 + \cdots + k^3 = \frac{k^2(k+1)^2}{4}.$$

Add $(k + 1)^3$ to each side of the equation:

$$1^3 + 2^3 + 3^3 + \cdots + k^3 + (k+1)^3 = \frac{k^2(k+1)^2}{4} + (k+1)^3$$

$$1^3 + 2^3 + 3^3 + \cdots + (k+1)^3 = \frac{k^2(k+1)^2}{4} + \frac{4(k+1)^3}{4} \qquad \text{Find the LCD.}$$

$$1^3 + 2^3 + 3^3 + \cdots + (k+1)^3 = \frac{(k+1)^2(k^2 + 4(k+1))}{4} \qquad \text{Factor out } (k+1)^2.$$

$$1^3 + 2^3 + 3^3 + \cdots + (k+1)^3 = \frac{(k+1)^2(k+2)^2}{4}$$

Since the last equation is T_{k+1}, we have shown that the truth of T_k implies the truth of T_{k+1} for every positive integer k. By the principle of mathematical induction, T_n is true for every positive integer n.

Try This. Prove $9 + 8 + 7 + \cdots + (10 - n) = \frac{n(19 - n)}{2}$ for every positive integer n. ■

Mathematical induction can be applied to the situation of a new college graduate who seeks a lifelong career. The graduate believes he or she can get a first job. The graduate also believes that every job will provide some experience that will guarantee getting a next job. If these two beliefs are really correct, then the graduate will have a lifelong career.

In the next example, we prove a statement involving inequality.

Example 4 Proof by mathematical induction

Use mathematical induction to prove that the statement

$$W_n: \quad 3^n < (n+2)!$$

is true for every positive integer n.

Solution

Step 1: If $n = 1$, then the statement W_1 is

$$3^1 < (1 + 2)!.$$

This inequality is correct and W_1 is true because $(1 + 2)! = 3! = 6$.

Step 2: If W_k is assumed to be true for a positive integer k, then

$$3^k < (k + 2)!.$$

Multiply each side by 3 to get

$$3^{k+1} < 3 \cdot (k + 2)!.$$

Since $3 < k + 3$ for $k \geq 1$, we have

$$3^{k+1} < (k + 3) \cdot (k + 2)!$$

$$3^{k+1} < (k + 3)!$$

$$3^{k+1} < ((k + 1) + 2)!.$$

Since the last inequality is W_{k+1}, we have shown that the truth of W_k implies that W_{k+1} is true. By the principle of mathematical induction, W_n is true for all positive integers n.

Try This. Prove $4^n < (n + 3)!$ for every positive integer n. ∎

When writing a mathematical proof, we try to convince the reader of the truth of a statement. Most proofs that are included in this text are fairly "mechanical," in that an obvious calculation or simplification gives the desired result. Mathematical induction provides a framework for a certain kind of proof, but within that framework we may still need some human ingenuity (as in the second step of Example 4). How much ingenuity is required in a proof is not necessarily related to the complexity of the statement. The story of Fermat's last theorem provides a classic example of how difficult it can be to prove a simple statement. The French mathematician Pierre de Fermat studied the problem of finding all positive integers that satisfy $a^n + b^n = c^n$. Of course, if $n = 2$ there are solutions such as $3^2 + 4^2 = 5^2$ and $5^2 + 12^2 = 13^2$. In 1637, Fermat stated that there are no solutions with $n \geq 3$. The proof of this simple statement (Fermat's last theorem) stumped mathematicians for over 350 years. Finally, in 1993 Dr. Andrew Wiles of Princeton University announced that he had proved it. Dr. Wiles estimated that the written details of his proof, which he worked on for seven years, would take over 200 pages! Errors were soon found in Wiles's proof and it took him two more years to correct it. To this day mathematicians continue to work on the problem, looking to simplify Wiles's proof or to prove it using a different approach.

For Thought

True or False? Explain.

1. The equation $\sum_{i=1}^{n} (4i - 2) = 2n^2$ is true if $n = 1$.

2. The inequality $n^3 < 4n + 15$ is true for $n = 1, 2,$ and 3.

3. For each positive integer n, $\dfrac{n - 1}{n + 1} < 0.9$.

4. Mathematical induction can be used to prove that $3(x + 1) = 3x + 3$ for every real number x.

5. If S_0 is true and the truth of S_{k-1} implies the truth of S_k for every positive integer k, then S_n is true for every nonnegative integer n.

6. If $n = k + 1$, then $\sum\limits_{i=1}^{n} \dfrac{1}{i(i + 1)} = \dfrac{n}{n + 1}$ becomes

$$\sum\limits_{i=1}^{k} \dfrac{1}{i(i + 1)} = \dfrac{k + 1}{k + 2}.$$

7. Mathematical induction can be used to prove that

$$\sum\limits_{i=1}^{\infty} 2^{-i} = 1.$$

8. The statement $n^2 - n > 0$ is true for $n = 1$.

9. The statement $n^2 - n > 0$ is true for $n > 1$.

10. The statement $n^2 - n > 0$ is true for every positive integer n.

11.7 Exercises

Determine whether each statement is true for $n = 1, 2,$ and 3.

1. $\sum\limits_{i=1}^{n} (3i - 1) = \dfrac{3n^2 + n}{2}$

2. $\sum\limits_{i=1}^{n} \left(\dfrac{1}{2}\right)^i = 1 - 2^{-n}$

3. $\sum\limits_{i=1}^{n} \dfrac{1}{i(i + 1)} = \dfrac{n}{n + 1}$

4. $\sum\limits_{i=1}^{n} 3^i = \dfrac{3(3^n - 1)}{2}$

5. $\sum\limits_{i=1}^{n} i^2 = 4n - 3$

6. $\sum\limits_{i=1}^{n} 4i = 8n - 4$

7. $n^2 < n^3$

8. $(0.5)^{n-1} > 0.5$

For each given statement S_n, write the statements $S_1, S_2, S_3,$ and S_4 and verify that they are true.

9. S_n: $\sum\limits_{i=1}^{n} i = \dfrac{n(n + 1)}{2}$

10. S_n: $\sum\limits_{i=1}^{n} i^2 = \dfrac{n(n + 1)(2n + 1)}{6}$

11. S_n: $\sum\limits_{i=1}^{n} i^3 = \dfrac{n^2(n + 1)^2}{4}$

12. S_n: $\sum\limits_{i=1}^{n} i^4 = \dfrac{n(n + 1)(2n + 1)(3n^2 + 3n - 1)}{30}$

13. S_n: $7^n - 1$ is divisible by 6

14. S_n: $3^n > 2n$

For each given statement S_n, write the statements $S_1, S_k,$ and S_{k+1}.

15. S_n: $\sum\limits_{i=1}^{n} 2i = n(n + 1)$

16. S_n: $\sum\limits_{i=1}^{n} 5i = \dfrac{5n(n + 1)}{2}$

17. S_n: $2 + 6 + 10 + \cdots + (4n - 2) = 2n^2$

18. S_n: $3 + 8 + 13 + \cdots + (5n - 2) = \dfrac{n(5n + 1)}{2}$

19. S_n: $\sum\limits_{i=1}^{n} 2^i = 2^{n+1} - 2$

20. S_n: $\sum\limits_{i=1}^{n} 5^{i+1} = \dfrac{5^{n+2} - 25}{4}$

21. S_n: $(ab)^n = a^n b^n$ **22.** S_n: $(a + b)^n = a^n + b^n$

23. S_n: If $0 < a < 1$, then $0 < a^n < 1$.

24. S_n: If $a > 1$, then $a^n > 1$.

Use mathematical induction to prove that each statement is true for each positive integer n.

25. $1 + 2 + 3 + \cdots + n = \dfrac{n(n + 1)}{2}$

26. $2 + 4 + 6 + \cdots + 2n = n(n + 1)$

27. $3 + 7 + 11 + \cdots + (4n - 1) = n(2n + 1)$

28. $2 + 7 + 12 + \cdots + (5n - 3) = \dfrac{n(5n - 1)}{2}$

29. $\sum\limits_{i=1}^{n} 2^i = 2^{n+1} - 2$ **30.** $\sum\limits_{i=1}^{n} 5^{i+1} = \dfrac{5^{n+2} - 25}{4}$

31. $\sum\limits_{i=1}^{n} (3i - 1) = \dfrac{3n^2 + n}{2}$ **32.** $\sum\limits_{i=1}^{n} \left(\dfrac{1}{2}\right)^i = 1 - 2^{-n}$

33. $1^2 + 2^2 + 3^2 + \cdots + n^2 = \dfrac{n(n + 1)(2n + 1)}{6}$

34. $\dfrac{1}{1 \cdot 2} + \dfrac{1}{2 \cdot 3} + \dfrac{1}{3 \cdot 4} + \cdots + \dfrac{1}{n(n + 1)} = \dfrac{n}{n + 1}$

35. $1 \cdot 3 + 2 \cdot 4 + 3 \cdot 5 + \cdots + n(n + 2) = \dfrac{n}{6}(n + 1)(2n + 7)$

36. $\dfrac{1}{1 \cdot 4} + \dfrac{1}{4 \cdot 7} + \dfrac{1}{7 \cdot 10} + \cdots + \dfrac{1}{(3n - 2)(3n + 1)} = \dfrac{n}{3n + 1}$

37. $\dfrac{1}{1 \cdot 3} + \dfrac{1}{3 \cdot 5} + \dfrac{1}{5 \cdot 7} + \cdots + \dfrac{1}{(2n - 1)(2n + 1)} = \dfrac{n}{2n + 1}$

38. $\dfrac{1}{1 \cdot 2 \cdot 3} + \dfrac{1}{2 \cdot 3 \cdot 4} + \dfrac{1}{3 \cdot 4 \cdot 5} + \cdots + \dfrac{1}{n(n + 1)(n + 2)} = \dfrac{n(n + 3)}{4(n + 1)(n + 2)}$

39. If $0 < a < 1$, then $0 < a^n < 1$.

40. If $a > 1$, then $a^n > 1$.

41. $n < 2^n$ **42.** $2^{n-1} \leq n!$

43. The integer $5^n - 1$ is divisible by 4 for every positive integer n.

44. The integer $7^n - 1$ is divisible by 6 for every positive integer n.

45. If a and b are constants, then $(ab)^n = a^n b^n$.

46. If a and m are constants, then $(a^m)^n = a^{mn}$.

47. If x is any real number with $x \neq 1$, then
$$1 + x + x^2 + x^3 + \cdots + x^n = \dfrac{x^{n+1} - 1}{x - 1}.$$

48. The number of subsets from a set with n elements is 2^n.
    HINT If you add a new element to a set, every subset either contains the new element or not.

For Writing/Discussion

49. Write a paragraph describing mathematical induction in your own words.

50. Mathematical induction is used to prove that a statement is true for every positive integer. What steps do you think are necessary to prove that a statement is true for every integer?

Thinking Outside the Box XCVII

Ones and Tens   Determine the units digit and the tens digit of the expression
$$(1!)^3 + (2!)^3 + (3!)^3 + (4!)^3 + \cdots + (101!)^3.$$

11.7 Pop Quiz

1. Prove $\sum\limits_{i=1}^{n}(2i - 1) = n^2$ for every positive integer n by using mathematical induction.

▪▪▪ Highlights

11.1 Sequences

Finite Sequence	A function whose domain is the set of positive integers less than or equal to some fixed positive integer	$a_n = n^2$ for $1 \leq n \leq 5$ $1, 4, 9, 16, 25$
Infinite Sequence	A function whose domain is the set of all positive integers	$a_n = 2n$ for $n \geq 1$ $2, 4, 6, 8, \ldots$
Arithmetic Sequence	nth term is $a_n = a_1 + (n - 1)d$, where a_1 is the first term and d is the common difference	$a_1 = 2 + (n - 1)5$ $2, 7, 12, 17, \ldots$

11.2 Series

Series	The indicated sum of a sequence	$\sum\limits_{n=1}^{5} n^2 = 1 + 4 + 9 + 16 + 25$

Arithmetic Series	The sum of an arithmetic sequence	

$$S_n = \sum_{i=1}^{n}[a_1 + (n-1)d] = \frac{n}{2}(a_1 + a_n)$$

$$\sum_{i=1}^{10} 2i = \frac{10}{2}(2 + 20)$$

11.3 Geometric Sequences and Series

Geometric Sequence	$a_n = ar^{n-1}$, where $r \neq 1$ and $r \neq 0$, and r is the common ratio	$a_n = 8(1/2)^{n-1}$ $8, 4, 2, 1, 1/2, 1/4, \ldots$
Geometric Series	The indicated sum of a geometric sequence	$8 + 4 + 2 + 1 + \cdots$

Finite Sum

$$S_n = \sum_{i=1}^{n} ar^{n-1} = \frac{a(1 - r^n)}{1 - r}$$

$$\sum_{i=1}^{5} 8(0.5)^{n-1} = \frac{8(1 - (0.5)^5)}{1 - 0.5}$$

Infinite Sum

$$S = \sum_{i=1}^{\infty} ar^{n-1} = \frac{a}{1 - r} \text{ provided } |r| < 1$$

$$\sum_{i=1}^{\infty} 8(0.5)^{n-1} = \frac{8}{1 - 0.5} = 16$$

11.4 Counting and Permutations

Fundamental Counting Principle	If event A has m outcomes and event B has n outcomes, then there are mn ways for A and B to occur.	There are $2 \cdot 6$ outcomes to tossing a coin and rolling a die.
Permutation	An arrangement of distinct objects in a line	Permutations of a, b, and c: abc, acb, bac, bca, cab, cba
Counting Permutations	The number of permutations of n things taken r at a time is denoted by $P(n, r)$ and given by	$P(3, 2) = 6$ ab, ba, ac, ca, bc, cb

$$P(n, r) = \frac{n!}{(n - r)!} \text{ for } 0 \leq r \leq n.$$

11.5 Combinations, Labeling, and the Binomial Theorem

Combination	An unordered choice of objects from a set of distinct objects	$\{a, b, c\}$ is a combination taken from the set $\{a, b, c, d\}$.
Counting Combinations	The number of combinations of n things taken r at a time is denoted by $C(n, r)$ or $\binom{n}{r}$ and given by $C(n, r) = \frac{n!}{(n - r)!r!}$ for $0 \leq r \leq n$.	$C(4, 3) = \frac{4!}{(4 - 3)!3!} = 4$
Labeling	The number of ways to label n objects with r_1 labels of type 1, r_2 labels of type 2, $\ldots$, and r_k labels of type k, where $r_1 + r_2 + \cdots + r_k = n$, is $\dfrac{n!}{r_1!r_2! \cdot \cdots \cdot r_k!}$.	The number of ways to rearrange "rearrange" is $\dfrac{9!}{3!2!2!1!1!}$.
Binomial Theorem	$(a + b)^n = \sum_{r=0}^{n} \binom{n}{r} a^{n-r} b^r$, where n is a positive integer	$(a + b)^3 = \binom{3}{0} a^3 b^0 + \binom{3}{1} a^2 b^1 + \binom{3}{2} a^1 b^2 + \binom{3}{3} a^0 b^3$

11.6 Probability

Probability	If S is a sample space of equally likely outcomes to an experiment and E is a subset of S, then $P(E) = n(E)/n(S)$.	Tossing a coin, $S = \{H, T\}$, $E = \{H\}, P(E) = 1/2$
Addition Rule	Any events: $P(A \cup B) = P(A) + P(B) - P(A \cap B)$ Mutually exclusive events: $P(A \cup B) = P(A) + P(B)$	$P(E \text{ or } S) =$ $P(E) + P(S) - P(E \cap S)$ $P(H \text{ or } T) = P(H) + P(T)$
Complementary Events	A and A' where $A \cap A' = \varnothing$ and $P(A) + P(A') = 1$	$\{H\} \cap \{T\} = \varnothing$ $P(H) + P(T) = 1$
Odds	In favor of A: the ratio of $P(A)$ to $P(A')$ Against A: the ratio of $P(A')$ to $P(A)$	Odds in favor of H are 1 to 1.

11.7 Mathematical Induction

Principle of Mathematical Induction	To prove that the statement S_n is true for every positive integer n, prove that S_1 is true and prove that for any positive integer k, assuming that S_k is true implies that S_{k+1} is true.

■ ■ ■ Chapter 11 Review Exercises

List all terms of each finite sequence.

1. $a_n = 2^{n-1}$ for $1 \le n \le 5$

2. $a_n = 3n - 2$ for $1 \le n \le 4$

3. $a_n = \dfrac{(-1)^n}{n!}$ for $1 \le n \le 4$

4. $a_n = (n - 2)^2$ for $1 \le n \le 6$

List the first three terms of each infinite sequence.

5. $a_n = 3(0.5)^{n-1}$

6. $b_n = \dfrac{1}{n(n + 1)}$

7. $c_n = -3n + 6$

8. $d_n = \dfrac{(-1)^n}{n^2}$

Find the sum of each series.

9. $\displaystyle\sum_{i=1}^{4} (0.5)^i$

10. $\displaystyle\sum_{i=1}^{3} (5i - 1)$

11. $\displaystyle\sum_{i=1}^{50} (4i + 7)$

12. $\displaystyle\sum_{i=1}^{4} 6$

13. $\displaystyle\sum_{i=1}^{\infty} 0.3(0.1)^{i-1}$

14. $\displaystyle\sum_{i=1}^{\infty} 5(-0.8)^i$

15. $\displaystyle\sum_{i=1}^{20} 1000(1.05)^{i-1}$

16. $\displaystyle\sum_{i=1}^{10} 6\left(\dfrac{1}{3}\right)^i$

Write a formula for the nth term of each sequence.

17. $-\dfrac{1}{3}, \dfrac{1}{4}, -\dfrac{1}{5}, \ldots$

18. $20, 17, 14, \ldots$

19. $6, 1, \dfrac{1}{6}, \ldots$

20. $1, -4, 9, -16, \ldots$

Write each series in summation notation. Use the index i and let i begin at 1.

21. $\dfrac{1}{2} - \dfrac{1}{3} + \dfrac{1}{4} - \cdots$

22. $5 + \dfrac{5}{2} + \dfrac{5}{4} + \dfrac{5}{8} + \cdots$

23. $2 + 4 + 6 + \cdots + 28$

24. $1 + 4 + 9 + \cdots + n^2$

Solve each problem.

25. *Common Ratio*   Find the common ratio for a geometric sequence that has a first term of 4 and a seventh term of 256.

26. *Common Difference*   Find the common difference for an arithmetic sequence that has a first term of 5 and a seventh term of 29.

27. *Compounded Quarterly*   If $100 is deposited in an account paying 9% compounded quarterly, then how much will be in the account at the end of 10 years?

28. *Compounded Monthly*   If $40,000 is deposited in an account paying 6% compounded monthly, then how much will be in the account at the end of nine years?

29. *Annual Payments*   If $1000 is deposited at the beginning of each year for 10 years in an account paying 6% compounded annually, then what is the total value of the 10 deposits at the end of the 10th year?

30. *Monthly Payments*   If $50 is deposited at the beginning of each month for 20 years in an account paying 6% compounded monthly, then what is the total value of the 240 deposits at the end of the 20th year?

31. *Tummy Masters*   TV Specialities sold 100,000 plastic Tummy Masters for $99.95 each during the first month that the product was advertised. Each month thereafter, sales levels were about 90% of the sales in the previous month. If this pattern continues, then what is the approximate total number of Tummy Masters that could be sold?

32. *Bouncing Ball*   A ball made from a new synthetic rubber will rebound 97% of the distance from which it is dropped. If the ball is dropped from a height of 6 ft, then approximately how far will it travel before coming to rest?

Write the complete binomial expansion for each of the following powers of a binomial.

33. $(a + 2b)^4$

34. $(x - 5)^3$

35. $(2a - b)^5$

36. $(w + 2)^6$

Write the first three terms of each binomial expansion.

37. $(a + b)^{10}$

38. $(x - 2y)^9$

39. $\left(2x + \dfrac{y}{2}\right)^8$

40. $(2a - 3b)^7$

Solve each problem.

41. What is the coefficient of a^4b^9 in the expansion of $(a + b)^{13}$?

42. What is the coefficient of x^8y^7 in the expansion of $(2x - y)^{15}$?

43. What is the coefficient of $w^2x^3y^6$ in the expansion of $(w + 2x + y)^{11}$?

44. What is the coefficient of x^5y^7 in the expansion of $(2w + x + y)^{12}$?

45. How many terms are there in the expansion of $(a + b)^{23}$?

46. Write out the first seven rows of Pascal's triangle.

Solve each counting problem.

47. *Multiple-Choice Test*   How many different ways are there to mark the answers to a nine-question multiple-choice test in which each question has five possible answers?

48. *Scheduling Departures*   Six airplanes are scheduled to depart at 2:00 on the same runway. In how many ways can they line up for departure?

49. *Three-Letter Words*   John is trying to find all three-letter English words that can be formed without repetition using the letters in the word FLORIDA. He plans to write down all possible three-letter "words" and then check each one with a dictionary to see if it is actually a word in the English language. How many possible three-letter "words" are there?

50. *Selecting a Team*   The sales manager for an insurance company must select a team of two agents to give a presentation. If the manager has five male agents and six female agents available, and the team must consist of one man and one woman, then how many different teams are possible?

51. *Placing Advertisements*   A candidate for city council is going to place one newspaper advertisement, one radio advertisement, and one television advertisement. If there are three newspapers, five radio stations, and four television stations available, then in how many ways can the advertisements be placed?

52. *Signal Flags*   A ship has nine different flags available. A signal consists of three flags displayed on a vertical pole. How many different signals are possible?

53. *Choosing a Vacation*   A travel agent offers a vacation in which you can visit any five cities, chosen from Paris, Rome, London, Istanbul, Monte Carlo, Vienna, Madrid, and Berlin. How many different vacations are possible, not counting the order in which the cities are visited?

54. *Counting Subsets*   How many four-element subsets are there for the set $\{a, b, c, d, e, f, g\}$?

55. *Choosing a Committee*   A city council consists of five Democrats and three Republicans. In how many ways can four council members be selected by the mayor to go to a convention in San Francisco if the mayor
   a. may choose any four?
   b. must choose four Democrats?
   c. must choose two Democrats and two Republicans?

56. *Full House* In a five-card poker hand, a full house is three cards of one kind and two cards of another. How many full houses are there consisting of three queens and two 10's?

57. *Arranging Letters* How many different arrangements are there for the letters in the word KANSAS? In TEXAS?

58. *Marking Pickup Trucks* Eight Mazda pickups of different colors are on sale through Saturday only. How many ways are there for the dealer to mark two of them $7000, two of them $8000, and four of them $9000?

59. *Possible Families* A couple plan to have seven children. How many different families are possible, considering the sex of each child and the order of birth?

60. *Triple Feature* The Galaxy Theater has three horror films to show on Saturday night. In how many different ways can the program for a triple feature be arranged?

Solve each probability problem.

61. *Just Guessing* If Miriam randomly marks the answers to a ten-question true-false test, then what is the probability that she gets all ten correct? What is the probability that she gets all ten wrong?

62. *Large Family* If a couple plan to have six children, then what is the probability that they get six girls?

63. *Jelly Beans* There are six red, five green, and two yellow jelly beans in a jar. If a jelly bean is selected at random, then what is the probability that
  a. the selected bean is green?
  b. the selected bean is either yellow or red?
  c. the selected bean is blue?
  d. the selected bean is not purple?

64. *Rolling Dice* A pair of dice is rolled. What is the probability that
  a. at least one die shows an even number?
  b. the sum is an even number?
  c. the sum is 4?
  d. the sum is 4 and at least one die shows an even number?
  e. the sum is 4 or at least one die shows an even number?

65. *My Three Sons* Suppose a couple plan to have three children. What are the odds in favor of getting three boys?

66. *Sum of Six* Suppose a pair of dice is rolled. What are the odds in favor of the sum being 6?

67. *Gone Fishing* If 90% of the fish in Lake Louise are perch and a single fish will be caught at random, then what are the odds in favor of catching a perch?

68. *Future Plans* In a survey of high school students, 80% said they planned to attend college. If a student is selected at random from this group, then what are the odds against getting one who plans to attend college?

69. *Enrollment Data* At MSU, 60% of the students are enrolled in an English class, 70% are enrolled in a mathematics class, and 40% are enrolled in both. If a student is selected at random, then what is the probability that the student is enrolled in either mathematics or English?

70. *Rolling Again* If a pair of dice is rolled, then what is the probability that the sum is either 5 or 6?

Evaluate each expression.

71. $8!$

72. $1!$

73. $\dfrac{5!}{3!}$

74. $\dfrac{7!}{(7-3)!}$

75. $\dfrac{9!}{3!\,3!\,3!}$

76. $\dfrac{10!}{0!\,2!\,3!\,5!}$

77. $C(8, 6)$

78. $\dbinom{12}{3}$

79. $P(8, 4)$

80. $P(4, 4)$

81. $C(8, 1)$

82. $C(12, 0)$

Use mathematical induction to prove that each statement is true for each positive integer n.

83. $3 + 6 + 9 + \cdots + 3n = \dfrac{3}{2}(n^2 + n)$

84. $\displaystyle\sum_{i=1}^{n} 2(3)^{i-1} = 3^n - 1$

Thinking Outside the Box XCVIII

Counting Bees Every male bee has only a female parent, whereas every female bee has a male and a female parent. Assume that every male or female bee mates with only one other bee. How many ancestors does a male bee have going back 10 generations?

▪▪▪ Chapter 11 Test

List all terms of each finite sequence.

1. $a_n = 2.3 + (n - 1)(0.5),\ 1 \le n \le 4$

2. $c_1 = 20$ and $c_n = \dfrac{1}{2}c_{n-1}$ for $2 \le n \le 4$

Write a formula for the nth term of each infinite sequence. Do not use a recursion formula.

3. $0, -1, 4, -9, 16, \ldots$

4. $7, 10, 13, 16, \ldots$

5. $\dfrac{1}{3}, -\dfrac{1}{6}, \dfrac{1}{12}, -\dfrac{1}{24}, \ldots$

Find the sum of each series.

6. $\displaystyle\sum_{j=0}^{53} (3j - 5)$

7. $\displaystyle\sum_{i=1}^{23} 300(1.05)^i$

8. $\displaystyle\sum_{n=1}^{\infty} (0.98)^n$

Solve each problem.

9. Find a formula for the nth term of an arithmetic sequence whose first term is -3 and whose ninth term is 9.

10. How many different nine-letter "words" can be made from the nine letters in TENNESSEE?

11. On the first day of April, $300 worth of lottery tickets were sold at the Quick Mart. If sales increased by $10 each day, then what was the mean of the daily sales amounts for April?

12. A customer at an automatic bank teller must enter a four-digit secret number to use the machine. If any of the integers from 0 through 9 can be used for each of the four digits, then how many secret numbers are there? If the machine gives a customer three tries to enter the secret number, then what is the probability that an unauthorized person can guess the secret number and gain access to the account?

13. Desmond and Molly Jones can save $700 per month toward their $120,000 dream house by living with Desmond's parents. If $700 is deposited at the beginning of each month for 300 months into an account paying 6% compounded monthly, then what is the value of this annuity at the end of the 300th month? If the price of a $120,000 house increases 6% each year, then what will it cost at the end of the 25th year? After 25 years of saving, will they have enough to buy the house and move out of his parents' house?

14. Write all the terms of the binomial expansion for $(a - 2x)^5$.

15. Write the first three terms of the binomial expansion for $(x + y^2)^{24}$.

16. Write the binomial expansion for $(m + y)^{30}$ using summation notation.

17. If a pair of fair dice is rolled, then what is the probability that the sum of the numbers showing is 7? What are the odds in favor of rolling a 7?

18. An employee randomly selects three of the 12 months of the year in which to take a vacation. In how many ways can this selection be made?

19. In a seventh grade class of 12 boys and 10 girls, the teacher randomly selects two boys and two girls to be crossing guards. How many outcomes are there to this process?

20. In a race of eight horses, a bettor randomly selects three horses for the categories of win, place, and show. What is the probability that the bettor gets the horses and the order of finish correct?

21. Use mathematical induction to prove that $\displaystyle\sum_{i=1}^{n} \left(\dfrac{1}{2}\right)^i < 1$ for every positive integer n.

APPENDIX

Solutions to Try This Exercises

Chapter P

Section 1

P.1.1 a. Because I is the set of irrational numbers and 0 is a rational number, $0 \in I$ is false.

b. Because the real numbers (R) include the rational numbers and the irrational numbers, $I \subseteq R$ is true.

c. Because $\sqrt{5}$ is an irrational number, $\sqrt{5} \in I$ is true.

P.1.2 Since $ab = ba$ for all real numbers, we can write $x \cdot 3 = 3x$.

P.1.3 $-(1 - w) = -1 + w = w - 1$

P.1.4 If $a < 0$ then $|a| = -a$. So $|-9| = -(-9) = 9$.

P.1.5 $d(-5, -9) = |-5 - (-9)|$
$$= |-5 + 9| = |4| = 4$$

P.1.6 a. $5^2 = 5 \cdot 5 = 25$

b. $-5^2 = -(5^2) = -25$

P.1.7 a. $(-1 + 3)(5 - 6) = (2)(-1) = -2$

b. $2 - |3 - 9| = 2 - |-6| = 2 - 6 = -4$

P.1.8. a. $3 - 6 \cdot 2 = 3 - 12 = -9$

b. $4 - 5 \cdot 2^3 = 4 - 5 \cdot 8 = 4 - 40 = -36$

P.1.9 $a^2 - b^2 = (-2)^2 - (-3)^2 = 4 - 9$
$$= -5$$

P.1.10 $-2(x - 3) - 3(1 - x) = -2x + 6 - 3 + 3x = x + 3$

Section 2

P.2.1 a. $2^{-2} \cdot 4^3 = \dfrac{1}{2^2} \cdot 64 = \dfrac{1}{4} \cdot 64 = 16$

b. $\left(\dfrac{1}{2}\right)^{-3} \cdot 12^{-2} = 2^3 \cdot \dfrac{1}{12^2} = \dfrac{8}{144} = \dfrac{1}{18}$

P.2.2 $-2a^4 b^3(-3a^5 b^6) = (-2)(-3)a^{4+5} b^{3+6} = 6a^9 b^9$

P.2.3 $-5a^{-7} b^{-5}(9a^{-2} b^8) = -45a^{-9} b^3$
$$= -\dfrac{45b^3}{a^9}$$

P.2.4 $(2a^{m-2})^2(-2a^{4m})^3 = 4a^{2m-4}(-8a^{12m}) = -32a^{14m-4}$

P.2.5 Move the decimal point two places to the left to get $3.78 \times 10^{-2} = 0.0378$.

P.2.6 Move the decimal point six places to the left to get $5,480,000 = 5.48 \times 10^6$.

P.2.7 $(7 \times 10^{14})(5 \times 10^{-3}) = 35 \times 10^{11} = 3.5 \times 10^1 \times 10^{11}$
$$= 3.5 \times 10^{12}$$

P.2.8 Use $r = 2.4 \times 10^{-3}$ in $V = \dfrac{4}{3}\pi r^3$:

$$V = \dfrac{4}{3}\pi(2.4 \times 10^{-3})^3 \approx 5.8 \times 10^{-8} \text{ in.}^3$$

Section 3

P.3.1 a. Since $3^2 = 9$, $9^{1/2} = 3$.

b. Since $2^4 = 16$, $16^{1/4} = 2$.

P.3.2 a. $9^{3/2} = (9^{1/2})^3 = 3^3 = 27$

b. $16^{-5/4} = \dfrac{1}{(16^{1/4})^5} = \dfrac{1}{2^5} = \dfrac{1}{32}$

P.3.3 For any real number w, $(w^4)^{1/4} = |w|$.

P.3.4 $(a^{1/3}a^{1/2})^{12} = (a^{1/3+1/2})^{12}$
$$= (a^{5/6})^{12} = a^{10}$$

P.3.5 a. $\sqrt{100} = 100^{1/2} = 10$

b. $\sqrt[3]{-27} = (-27)^{1/3} = -3$

P.3.6 $5^{2/3} = \sqrt[3]{5^2} = \sqrt[3]{25}$

P.3.7 $\sqrt[3]{-8m^9} = \sqrt[3]{-8} \cdot \sqrt[3]{m^9} = -2m^3$

P.3.8 $\sqrt{8y^7} = \sqrt{4y^6} \cdot \sqrt{2y} = 2y^3 \sqrt{2y}$

P.3.9 $\sqrt{50} - \sqrt{8} = 5\sqrt{2} - 2\sqrt{2} = 3\sqrt{2}$

P.3.10 $\sqrt[3]{5} \cdot \sqrt{2} = 5^{1/3} \cdot 2^{1/2} = 5^{2/6} \cdot 2^{3/6}$
$$= \sqrt[6]{5^2 \cdot 2^3} = \sqrt[6]{200}$$

Section 4

P.4.1 The complex number $i - 5$ is imaginary because the coefficient of i is nonzero. In standard form it is written as $-5 + i$.

P.4.2 $(4 - 3i)(1 + 2i) = 4 - 3i + 8i - 6i^2$
$$= 4 + 5i - 6(-1) = 10 + 5i$$

P.4.3 $i^{35} = (i^4)^8 i^3 = 1^8(-i) = -i$

P.4.4 $(3 - 5i)(3 + 5i) = 9 - 25i^2 = 34$

P.4.5 $\dfrac{4}{1 + i} = \dfrac{4(1 - i)}{(1 + i)(1 - i)}$
$$= \dfrac{4 - 4i}{2} = 2 - 2i$$

P.4.6 $\dfrac{2 - \sqrt{-12}}{2} = \dfrac{2 - 2i\sqrt{3}}{2} = 1 - i\sqrt{3}$

Section 5

P.5.1 The degree of $-x^3 + 6x^2$ is 3 and the leading coefficient is -1.

P.5.2 $P(-2) = -(-2)^2 - 4(-2) + 9$
$$= -4 + 8 + 9 = 13$$

P.5.3 $(-x^2 + 3x) - (x^2 - 5x + 1) = -x^2 + 3x - x^2 + 5x - 1$
$$= -2x^2 + 8x - 1$$

P.5.4 Subtract: $\begin{array}{r} -3x^2 - x + 1 \\ \underline{x^2 \qquad - 9} \\ -4x^2 - x + 10 \end{array}$

P.5.5 $(x^2 - 2)(x^2 - 3) = x^2(x^2 - 3) - 2(x^2 - 3)$
$$= x^4 - 3x^2 - 2x^2 + 6 = x^4 - 5x^2 + 6$$

P.5.6 $(x - 2)(x + 9) = x^2 - 2x + 9x - 18$
$$= x^2 + 7x - 18$$

P.5.7 $(2\sqrt{3} - 1)(\sqrt{3} + 2) = 2\sqrt{3} \cdot \sqrt{3} - \sqrt{3} + 4\sqrt{3} - 2$
$$= 6 + 3\sqrt{3} - 2 = 4 + 3\sqrt{3}$$

P.5.8 $(2a - 5)^2 = (2a)^2 - 2 \cdot 2a \cdot 5 + 5^2$
$$= 4a^2 - 20a + 25$$

P.5.9 $\dfrac{2}{2 - \sqrt{2}} = \dfrac{2(2 + \sqrt{2})}{(2 - \sqrt{2})(2 + \sqrt{2})}$
$$= \dfrac{4 + 2\sqrt{2}}{2} = 2 + \sqrt{2}$$

P.5.10
$$\begin{array}{r} x - 2 \\ x - 1 \overline{)\, x^2 - 3x + 2} \\ \underline{x^2 - x} \\ -2x + 2 \\ \underline{-2x + 2} \\ 0 \end{array}$$

The quotient is $x - 2$.

Section 6

P.6.1 Factor out $4a$ or $-4a$:
$$-12a^3 + 8a^2 - 16a = 4a(-3a^2 + 2a - 4)$$
$$-12a^3 + 8a^2 - 16a = -4a(3a^2 - 2a + 4)$$
P.6.2 $w^3 + w^2 - 3w - 3 = w^2(w + 1) - 3(w + 1)$
$$= (w^2 - 3)(w + 1)$$
P.6.3 Two numbers that have a product of -70 and a sum of -3 are 7 and -10:
$$x^2 - 3x - 70 = (x + 7)(x - 10)$$
P.6.4 Two numbers that have a product of -36 and a sum of 5 are 9 and -4:
$$12x^2 + 5x - 3 = 12x^2 + 9x - 4x - 3$$
$$= 3x(4x + 3) - 1(4x + 3)$$
$$= (3x - 1)(4x + 3)$$
P.6.5 $3x^2 - 20x - 7 = (3x + 1)(x - 7)$
P.6.6 $4x^2 - 28x + 49 = (2x)^2 - 2(2x)(7) + 7^2 = (2x - 7)^2$
P.6.7 $8x^3 + 125 = (2x)^3 + 5^3$
$$= (2x + 5)(4x^2 - 10x + 25)$$
P.6.8 Let $a = w^2 + 1$:
$$(w^2 + 1)^2 - 3(w^2 + 1) - 18$$
$$= a^2 - 3a - 18 = (a - 6)(a + 3)$$
$$= (w^2 + 1 - 6)(w^2 + 1 + 3)$$
$$= (w^2 - 5)(w^2 + 4)$$
P.6.9 $2y^4 - 162 = 2(y^4 - 81)$
$$= 2(y^2 - 9)(y^2 + 9)$$
$$= 2(y - 3)(y + 3)(y^2 + 9)$$
P.6.10
$$
\begin{array}{r}
y^2 - 5y + 6 \\
y - 1{\overline{\smash{\big)}\,y^3 - 6y^2 + 11y - 6}} \\
\underline{y^3 - y^2} \\
-5y^2 + 11y \\
\underline{-5y^2 + 5y} \\
6y - 6 \\
\underline{6y - 6} \\
0
\end{array}
$$
$$y^3 - 6y^2 + 11y - 6 = (y - 1)(y^2 - 5y + 6)$$
$$= (y - 1)(y - 2)(y - 3)$$

Section 7

P.7.1 Since the denominator cannot have a value of zero, the domain of $\dfrac{x - 2}{x^2 - 9}$ is the set of all real numbers except -3 and 3 or $\{x \mid x \neq -3 \text{ and } x \neq 3\}$.

P.7.2 $\dfrac{3x - 9}{x^2 - 9} = \dfrac{3(x - 3)}{(x - 3)(x + 3)} = \dfrac{3}{x + 3}$

P.7.3 $\dfrac{3x + 3}{x^2 - 1} \cdot \dfrac{2x - 2}{9} = \dfrac{3(x + 1)}{(x - 1)(x + 1)} \cdot \dfrac{2(x - 1)}{3 \cdot 3} = \dfrac{2}{3}$

P.7.4 $\dfrac{25 - a^2}{6} \div \dfrac{2a - 10}{8} = \dfrac{-1(a - 5)(a + 5)}{2 \cdot 3} \cdot \dfrac{2 \cdot 2 \cdot 2}{2(a - 5)}$
$$= \dfrac{(-1)(a + 5)2}{3} = \dfrac{-2a - 10}{3}$$

P.7.5 $\dfrac{y}{y + 2} = \dfrac{y(y + 1)}{(y + 2)(y + 1)}$
$$= \dfrac{y^2 + y}{y^2 + 3y + 2}$$

P.7.6 $\dfrac{y}{y + 2} + \dfrac{2}{y + 1} = \dfrac{y(y + 1)}{(y + 2)(y + 1)} + \dfrac{2(y + 2)}{(y + 1)(y + 2)}$
$$= \dfrac{y^2 + y + 2y + 4}{(y + 2)(y + 1)} = \dfrac{y^2 + 3y + 4}{(y + 2)(y + 1)}$$

P.7.7 $\dfrac{\dfrac{1}{2a} - \dfrac{1}{4}}{\dfrac{3}{a} + \dfrac{1}{6}} = \dfrac{12a\left(\dfrac{1}{2a} - \dfrac{1}{4}\right)}{12a\left(\dfrac{3}{a} + \dfrac{1}{6}\right)}$
$$= \dfrac{6 - 3a}{36 + 2a}$$

P.7.8 $\dfrac{x^{-2}}{y^{-3} - x^{-3}} = \dfrac{x^3 y^3 (x^{-2})}{x^3 y^3 (y^{-3} - x^{-3})}$
$$= \dfrac{xy^3}{x^3 - y^3}$$

Chapter 1

Section 1

1.1.1 $5(3x - 2) = 5 - 7(x - 1)$
$$15x - 10 = 5 - 7x + 7$$
$$22x = 22$$
$$x = 1$$
Check: $5(3 \cdot 1 - 2) = 5 - 7(1 - 1)$
$$5 = 5$$
The solution set is $\{1\}$.
1.1.2 $x(x - 1) - 6 = (x - 3)(x + 2)$
$$x^2 - x - 6 = x^2 - x - 6$$
Since both sides are identical, it is an identity.

1.1.3 $\dfrac{2}{x - 3} - \dfrac{3}{x + 3} = \dfrac{4}{x^2 - 9}$
$$(x - 3)(x + 3)\left(\dfrac{2}{x - 3} - \dfrac{3}{x + 3}\right) = (x^2 - 9)\dfrac{4}{x^2 - 9}$$
$$2(x + 3) - 3(x - 3) = 4$$
$$-x + 15 = 4$$
$$-x = -11$$
$$x = 11$$
Check: $\dfrac{2}{11 - 3} - \dfrac{3}{11 + 3} = \dfrac{1}{28}, \dfrac{4}{11^2 - 9} = \dfrac{1}{28}$
The solution set to this conditional equation is $\{11\}$.

1.1.4 $\dfrac{2}{3.4x} - \dfrac{1}{8.9} = \dfrac{4}{4.7}$
$$\dfrac{2}{3.4x} = \dfrac{4}{4.7} + \dfrac{1}{8.9}$$
$$\dfrac{2}{3.4} = x\left(\dfrac{4}{4.7} + \dfrac{1}{8.9}\right)$$
$$\dfrac{\dfrac{2}{3.4}}{\dfrac{4}{4.7} + \dfrac{1}{8.9}} = x$$
$$x \approx 0.611$$
Check: $\dfrac{2}{3.4(0.611)} - \dfrac{1}{8.9} \approx 0.851, \dfrac{4}{4.7} \approx 0.851$

The solution set is $\{0.611\}$.

1.1.5 $|2x - 3| = 5$

$2x - 3 = 5$ or $2x - 3 = -5$

$\quad 2x = 8$ or $\quad 2x = -2$

$\quad\quad x = 4$ or $\quad\quad x = -1$

Check: $|2 \cdot 4 - 3| = 5$ and $|2 \cdot (-1) - 3| = 5$

The solution set is $\{-1, 4\}$.

1.1.6 Solve the following equation.

$$25,000 = 355.9x + 11,075.3$$
$$13,924.7 = 355.9x$$
$$x \approx 39$$

So 39 years after 1990, or in 2029 the median income will reach $25,000.

Section 2

1.2.1 $A = \dfrac{1}{2}hb_1 + \dfrac{1}{2}hb_2$

$2A = hb_1 + hb_2$

$2A = h(b_1 + b_2)$

$h = \dfrac{2A}{b_1 + b_2}$

1.2.2 $A = \dfrac{1}{2}h(b_1 + b_2)$

$20 = \dfrac{1}{2} \cdot 2(b_1 + 3)$

$20 = b_1 + 3$

$17 = b_1$

1.2.3 Let x represent the price of the computer and $0.05x$ the amount of tax.

$$x + 0.05x = 1506.75$$
$$1.05x = 1506.75$$
$$x = 1435$$
$$0.05x = 71.75$$

The amount of tax was $71.75.

1.2.4 Let x represent the width and $5x - 20$ represent the length.

$$2W + 2L = P$$
$$2x + 2(5x - 20) = 800$$
$$12x - 40 = 800$$
$$12x = 840$$
$$x = 70$$
$$5x - 20 = 330$$

So the length is 330 cm.

1.2.5 Let x represent her rate uphill and $x + 2$ her rate downhill. The distance is $6x$ or $3(x + 2)$.

$$6x = 3(x + 2)$$
$$6x = 3x + 6$$
$$3x = 6$$
$$x = 2$$
$$6x = 12$$

The total distance hiked is 24 miles.

1.2.6 Let x represent her speed on the return trip. Her time to work is $1/3$ hr and her time for the return trip is $20/x$. Her average speed of 50 mph is the total distance divided by the total time:

$$\frac{20 + 20}{\dfrac{1}{3} + \dfrac{20}{x}} = 50$$

$$40 = 50\left(\frac{1}{3} + \frac{20}{x}\right)$$

$$\frac{4}{5} = \frac{1}{3} + \frac{20}{x}$$

$$15x \cdot \frac{4}{5} = 15x\left(\frac{1}{3} + \frac{20}{x}\right)$$

$$12x = 5x + 300$$

$$7x = 300$$

$$x = 42\frac{6}{7}$$

Her average speed on the return trip was $42\frac{6}{7}$ mph.

1.2.7 Let x represent the number of gallons of 40% acid solution.

$$0.40x + 0.20(30) = 0.35(x + 30)$$
$$0.40x + 6 = 0.35x + 10.5$$
$$0.05x = 4.5$$
$$x = 90$$

Use 90 gallons of 40% acid solution.

1.2.8 Let x represent the number of hours the small pipe is used and $x - 2$ represent the number of hours the large pipe is used.

$$\frac{1}{12}x + \frac{1}{8}(x - 2) = 1$$

$$\frac{1}{12}x + \frac{1}{8}x - \frac{1}{4} = 1$$

$$24\left(\frac{1}{12}x + \frac{1}{8}x - \frac{1}{4}\right) = 24 \cdot 1$$

$$2x + 3x - 6 = 24$$

$$5x = 30$$

$$x = 6$$

It will take 6 hours to fill the tank.

Section 3

1.3.1

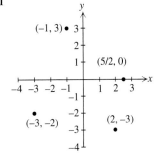

1.3.2 $\sqrt{(-3 - (-1))^2 + (-2 - 4)^2} = \sqrt{4 + 36} = \sqrt{40} = 2\sqrt{10}$

1.3.3 The midpoint of the diagonal with endpoints $(0, 0)$ and $(3, 5)$ is $(3/2, 5/2)$. The midpoint of the diagonal with endpoints $(4, 1)$ and $(-1, 4)$ is also $(3/2, 5/2)$. So the diagonals bisect each other.

1.3.4 The circle has center $(-2, 4)$ and radius 5.

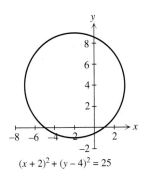

$(x + 2)^2 + (y - 4)^2 = 25$

1.3.5 The radius is the distance from the center $(2, -1)$ to $(3, 6)$:

$$\sqrt{(2-3)^2 + (-1-6)^2} = \sqrt{50}$$

The equation is $(x - 2)^2 + (y + 1)^2 = 50$.

1.3.6 Complete the squares:

$$x^2 + 3x + \frac{9}{4} + y^2 - 2y + 1 = 0 + \frac{9}{4} + 1$$

$$\left(x + \frac{3}{2}\right)^2 + (y - 1)^2 = \frac{13}{4}$$

The graph is a circle with center $(-3/2, 1)$ and radius $\sqrt{13}/2$ or about 1.8.

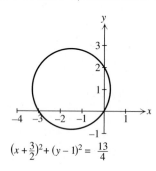

$$\left(x + \frac{3}{2}\right)^2 + (y - 1)^2 = \frac{13}{4}$$

1.3.7 If $x = 0$, then $5y = 10$ or $y = 2$. If $y = 0$, then $2x = 10$ or $x = 5$. Draw a line through the intercepts $(0, 2)$ and $(5, 0)$.

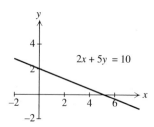

1.3.8 The ordered pairs $(-2, 5)$, $(0, 5)$, and $(2, 5)$ satisfy $y = 5$. So the graph is a horizontal line.

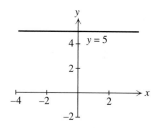

1.3.9 Graph $y = 0.34(x - 2.3) + 4.5$ and find the x-intercept. The x-intercept is $(-10.93529, 0)$ and the solution to the equation is approximately -10.93529.

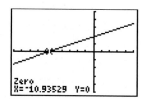

Section 4

1.4.1 slope $= \dfrac{-3 - 5}{-1 - (-2)} = \dfrac{-8}{1} = -8$

1.4.2 Use the slope -8 from the preceding answer and the point $(-2, 5)$ in point-slope form.

$$y - 5 = -8(x - (-2))$$
$$y - 5 = -8x - 16$$
$$y = -8x - 11$$

1.4.3 Solve the equation for y:

$$3x + 5y = 15$$
$$5y = -3x + 15$$
$$y = -\frac{3}{5}x + 3$$

The slope is $-\frac{3}{5}$ and the y-intercept is $(0, 3)$.

1.4.4 Start at the y-intercept $(0, 1)$ and use the slope $-3/2$ to locate a second point $(2, -2)$ which is down 3 and 2 to the right from $(0, 1)$:

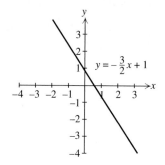

1.4.5 Use slope-intercept form:

$$y = \frac{3}{4}x + \frac{1}{2}$$
$$4y = 3x + 2$$
$$-3x + 4y = 2$$
$$3x - 4y = -2$$

1.4.6 The slope of any line parallel to $y = -\frac{1}{2}x + 9$ is $-\frac{1}{2}$. Use slope $-\frac{1}{2}$ and $(2, 4)$ in the point-slope form:

$$y - 4 = -\frac{1}{2}(x - 2)$$
$$y - 4 = -\frac{1}{2}x + 1$$
$$y = -\frac{1}{2}x + 5$$

1.4.7 Write $2x - y = 8$ as $y = 2x - 8$. The slope of any line perpendicular to $y = 2x - 8$ is $-\frac{1}{2}$. Use $-\frac{1}{2}$ and $(-2, 1)$ in point-slope form:

$$y - 1 = -\frac{1}{2}(x - (-2))$$
$$y - 1 = -\frac{1}{2}x - 1$$
$$y = -\frac{1}{2}x$$

1.4.8 The slope of the side with endpoints $(0, 0)$ and $(5, 2)$ is $\dfrac{2 - 0}{5 - 0}$ or $\dfrac{2}{5}$. The slope of the side with endpoints $(5, 2)$ and $(1, 12)$ is $\dfrac{12 - 2}{1 - 5}$ or $-\dfrac{5}{2}$. So the sides are perpendicular and the triangle is a right triangle.

1.4.9 Find the equation of the line through (100, 70) and (200, 90). The slope is $\frac{90-70}{200-100}$ or 0.20. Use the point-slope formula with C for cost in dollars and m for miles:

$$C - 70 = 0.20(m - 100)$$
$$C - 70 = 0.20m - 20$$
$$C = 0.20m + 50$$

1.4.10 Let F be the number of file cabinets and B be the number of bookshelves. Solve $100F + 150B = 3000$ for F:

$$100F = -150B + 3000$$
$$F = -1.5B + 30$$

The slope -1.5 means that increasing B by 1 causes F to decrease by 1.5.

Section 5

1.5.1 a. Since the points appear to be approximately in line, the relationship is linear.
b. Since the points appear to be in an approximate parabolic shape, the relationship is nonlinear.
1.5.2 Graph the data and draw a line that approximately fits the data. At 2010 the second coordinate should be about 70. So the cost in 2010 should be about $70.
1.5.3 Enter the data into your calculator and find the regression equation $C = 1.926x + 12.093$ where x is the number of years after 1980. $C(30) \approx \$69.87$.

Section 6

1.6.1 $(x - 3)^2 = 16$

$$x - 3 = \pm\sqrt{16}$$
$$x = 3 \pm 4$$
$$x = 7 \text{ or } -1$$

The solution set is $\{-1, 7\}$.
1.6.2 $x^2 - 7x - 18 = 0$

$$(x - 9)(x + 2) = 0$$
$$x - 9 = 0 \text{ or } x + 2 = 0$$
$$x = 9 \text{ or } \qquad x = -2$$

The solution set is $\{-2, 9\}$.
1.6.3 $2x^2 - 4x - 1 = 0$

$$x^2 - 2x - \frac{1}{2} = 0$$
$$x^2 - 2x + 1 = \frac{1}{2} + 1$$
$$(x - 1)^2 = \frac{3}{2}$$
$$x - 1 = \pm\sqrt{\frac{3}{2}}$$
$$x = 1 \pm \frac{\sqrt{6}}{2} = \frac{2 \pm \sqrt{6}}{2}$$

The solution set is $\left\{\frac{2 - \sqrt{6}}{2}, \frac{2 + \sqrt{6}}{2}\right\}$.
1.6.4 $2x^2 - 3x - 2 = 0$

$$x = \frac{3 \pm \sqrt{(-3)^2 - 4(2)(-2)}}{2(2)}$$
$$= \frac{3 \pm \sqrt{25}}{4} = \frac{3 \pm 5}{4}$$

The solution set is $\left\{-\frac{1}{2}, 2\right\}$.
1.6.5 $b^2 - 4ac = (-7)^2 - 4(5)(9) = -131$
Since the discriminant is negative, the equation has no real solutions.
1.6.6 Let x be Josh's average speed and $x + 5$ be Bree's average speed. Their times differ by one-half hour:

$$\frac{100}{x} - \frac{90}{x + 5} = \frac{1}{2}$$
$$2x(x + 5)\left(\frac{100}{x} - \frac{90}{x + 5}\right) = 2x(x + 5)\frac{1}{2}$$
$$200x + 1000 - 180x = x^2 + 5x$$
$$-x^2 + 15x + 1000 = 0$$
$$x^2 - 15x - 1000 = 0$$
$$(x - 40)(x + 25) = 0$$
$$x = 40 \text{ or } x = -25$$

Josh averaged 40 mph and Bree 45 mph.
1.6.7 The ball is back on the earth when $h = 0$:

$$-16t^2 + 40t + 6 = 0$$
$$t = \frac{-40 \pm \sqrt{(40)^2 - 4(-16)(6)}}{2(-16)}$$
$$t = \frac{-40 \pm \sqrt{1984}}{-32} = \frac{-40 \pm 8\sqrt{31}}{-32}$$
$$t = \frac{5 \pm \sqrt{31}}{4} \approx 2.64 \quad \text{or} \quad -0.14$$

The ball is in the air for $(5 + \sqrt{31})/4$ or about 2.64 seconds.
1.6.8 Let x and $x + 2$ represent the lengths of the legs.

$$x^2 + (x + 2)^2 = 6^2$$
$$2x^2 + 4x + 4 = 36$$
$$x^2 + 2x - 16 = 0$$
$$x = \frac{-2 \pm \sqrt{2^2 - 4(1)(-16)}}{2(1)} = \frac{-2 \pm \sqrt{68}}{2}$$
$$= \frac{-2 \pm 2\sqrt{17}}{2} = -1 \pm \sqrt{17}$$

The short leg is $-1 + \sqrt{17}$ or about 3.1 ft and the long leg is $1 + \sqrt{17}$ or about 5.1 ft.
1.6.9 Let x be the number of years after 1980. Quadratic regression yields

$$C = 0.0495x^2 - 0.973x + 11.850.$$

So $C(30) \approx \$27.21$.

Section 7

1.7.1 The interval $(-\infty, 5]$ consists of all real numbers that are less than or equal to 5, and that is the solution set to $x \leq 5$.
1.7.2 $2 - 5x \leq 7$

$$-5x \leq 5$$
$$x \geq -1$$

The solution set is $[-1, \infty)$ and it is graphed as follows.

$$\xleftarrow{\hspace{2em}} \begin{array}{ccccccc} + & + & [& + & + & + & + \\ -3 & -2 & -1 & 0 & 1 & 2 & 3 \end{array} \xrightarrow{\hspace{2em}}$$

1.7.3 $\frac{1}{2}x + \frac{1}{3} \leq \frac{1}{3}x + 1$

$$6\left(\frac{1}{2}x + \frac{1}{3}\right) \leq 6\left(\frac{1}{3}x + 1\right)$$
$$3x + 2 \leq 2x + 6$$
$$x \leq 4$$

The solution set is $(-\infty, 4]$ and it is graphed as follows.

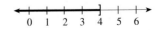

1.7.4 $A \cup B$ consists of all real numbers between 1 and 9. So $A \cup B = (1, 9)$. $A \cap B$ consists of all real numbers that belong to both A and B. So $A \cap B = [4, 6)$.

1.7.5 $2x > -4$ and $4 - x \geq 0$
$\qquad x > -2$ and $\quad -x \geq -4$
$\qquad x > -2$ and $\qquad x \leq 4$
The solution set is $(-2, 4]$.

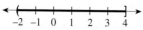

1.7.6 $3x + 2 > -1$ and $5 < -3 - 4x$
$\qquad 3x > -3$ and $4x < -8$
$\qquad x > -1$ and $\quad x < -2$
Since $(-\infty, -2) \cap (-1, \infty) = \varnothing$, the solution set is the empty set, $\varnothing$.

1.7.7 $|x - 6| - 3 \leq -2$
$\qquad |x - 6| \leq 1$
$\quad -1 \leq x - 6 \leq 1$
$\qquad 5 \leq x \leq 7$
The solution set is $[5, 7]$.

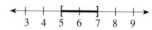

1.7.8 Let x be the third commission.

$$100 < \frac{80 + 90 + x}{3} < 110$$
$$300 < 170 + x < 330$$
$$130 < x < 160$$

The third commission must be in the interval $(130, 160)$ or between \$130 and \$160.

1.7.9 $\dfrac{|x - 10|}{10} < 0.01$

$\qquad |x - 10| < 0.1$
$\quad -0.1 < x - 10 < 0.1$
$\qquad 9.9 < x < 10.1$

If the actual amount dispensed is between 9.9 and 10.1 gallons, then the pump is certified as accurate.

Chapter 2

Section 1

2.1.1 Tax is determined by finding 5% of p and rounding to the nearest cent. So t is a function of p. One cannot determine p from t because 10 cents in tax is paid on an item that costs \$2 or \$2.01. So p is not a function of t.

2.1.2 Since no vertical line can be drawn so that it crosses this graph more than once, y is a function of x.

2.1.3 **a.** Since $(5, 5)$ and $(5, 7)$ have the same first coordinate and different second coordinates, the relation is not a function.
b. Since no ordered pair in the table has the same first coordinate and different second coordinates, the relation is a function.

2.1.4 Since $(-1, 1)$ and $(-1, -1)$ both satisfy $x^3 + y^2 = 0$, the equation does not define y as a function of x.

2.1.5 Since $\sqrt{x + 3}$ is a real number only if $x + 3 \geq 0$ or $x \geq -3$, the domain is $[-3, \infty)$. Since $\sqrt{x + 3} \geq 0$, the range is $[0, \infty)$. Since y is uniquely determined by $y = \sqrt{x + 3}$, this relation is a function.

2.1.6 **a.** $f(4) = 4 - 3 = 1$

b. If $f(x) = 9$, then $x - 3 = 9$ and $x = 12$.

2.1.7 Replace x with $x + 2$ to get

$$f(x + 2) = (x + 2)^2 - 4$$
$$= x^2 + 4x + 4 - 4$$
$$= x^2 + 4x$$

2.1.8 $\dfrac{28{,}645 - 13{,}837}{2000 - 2006} = -2468$

The average rate of change was $-\$2468$ per year.

2.1.9 $\dfrac{f(x + h) - f(x)}{h} = \dfrac{(x + h)^2 - (x + h) - (x^2 - x)}{h}$

$\qquad = \dfrac{x^2 + 2xh + h^2 - x - h - x^2 + x}{h}$

$\qquad = \dfrac{2xh + h^2 - h}{h} = 2x + h - 1$

2.1.10 For a square, $P = 4s$. So $s = P/4$ expresses the side as a function of the perimeter.

Section 2

2.2.1 Since no vertical line crosses the graph more than once, y is a function of x.

x	-4	-2	0	2	4
$y = \frac{1}{2}x^2$	8	2	0	2	8

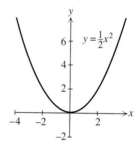

Since any real number can be used for x, the domain is $(-\infty, \infty)$. Since $y \geq 0$, the range is $[0, \infty)$.

2.2.2 Since no vertical line crosses the graph more than once, y is a function of x.

x	1	0	-3	-8
$y = \sqrt{1 - x}$	0	1	2	3

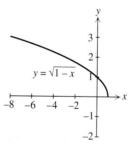

Since $1 - x \geq 0$ or $x \leq 1$ the domain is $(-\infty, 1]$. Since $y \geq 0$, the range is $[0, \infty)$.

2.2.3 Since $(-1, 1)$ and $(-1, -1)$ both satisfy $x = -y^2$, y is not uniquely determined by x and the equation is not a function. Also, y is not a function of x by the vertical line test.

$x = -y^2$	-4	-1	0	-1	-4
y	-2	-1	0	1	2

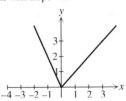

Since x can be any real number, the domain is $(-\infty, \infty)$. Since $y \geq 2$, the range is $[2, \infty)$.

2.2.7 For $x \geq 0$ the graph is a line with slope 1 starting at the origin. For $x < 0$ the graph is a line with slope -2.

Since any real number can be used for y, the range is $(-\infty, \infty)$. Since $x \leq 0$, the domain is $(-\infty, 0]$.

2.2.4 Since no vertical line crosses the graph more than once, y is a function of x.

x	0	1	8	27
$y = -\sqrt[3]{x}$	0	-1	-2	-3

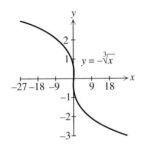

Since any real number can be used for x, the domain is $(-\infty, \infty)$. Since any real number can occur for y, the range is $(-\infty, \infty)$.

2.2.5

x	-3	0	3
$y = -\sqrt{9 - x^2}$	0	-3	0

The graph is a semicircle.

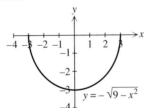

Since x must be between -3 and 3 inclusive, the domain is $[-3, 3]$. Since y is between -3 and 0 inclusive, the range is $[-3, 0]$.

2.2.6 Since no vertical line crosses the graph more than once, y is a function of x.

x	0	± 1	± 2
$y = |x| + 2$	2	3	4

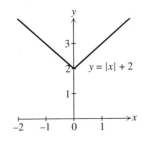

Since x can be any real number, the domain is $(-\infty, \infty)$. Since $y \geq 0$, the range is $[0, \infty)$.

2.2.8

x	$[0, 1)$	$[1, 2)$	$[2, 3)$
$y = -[\![x]\!]$	0	-1	-2

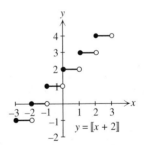

Since any real number can be used for x, the domain is $(-\infty, \infty)$. The range is the set of integers.

2.2.9

x	$[0, 1)$	$[1, 2)$	$[2, 3)$
$y = [\![x + 2]\!]$	2	3	4

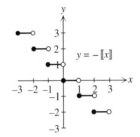

Since any real number can be used for x, the domain is $(-\infty, \infty)$. The range is the set of integers.

2.2.10 Graph $f(x) = -3x$ as follows.

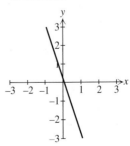

Since the y-coordinates are decreasing as we move from left to right on the graph, the function is decreasing.

2.2.11 Graph $f(x)$ as follows.

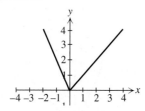

The function is decreasing on $(-\infty, 0)$ and increasing on $(0, \infty)$.

Section 3

2.3.1 Note that g lies one unit above f and h lies two units below f.

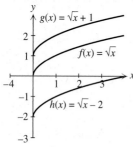

2.3.2 Note that g lies two units to the right of f and h lies one unit to the left of f.

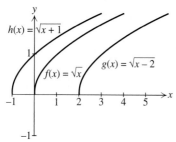

2.3.3 Note that $f(x) = (x - 2)^2$ goes through $(1, 1)$, $(2, 0)$, and $(3, 1)$ and lies two units to the right of $y = x^2$.

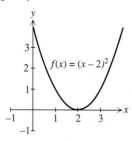

2.3.4 The graph of $g(x) = -\sqrt{x}$ is a reflection of the graph of $f(x) = \sqrt{x}$. Note that g lies below the x-axis and f lies above the x-axis.

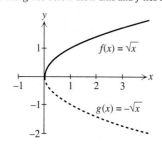

2.3.5 The graph of $g(x) = 3|x|$ is obtained by stretching the graph of $f(x) = |x|$. The graph of $h(x) = \frac{1}{3}|x|$ is obtained by shrinking the graph of $f(x)$.

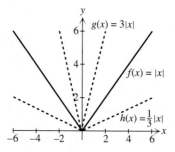

2.3.6 The graph of $y = |x|$ is translated 1 unit to the left, stretched by a factor of 2, reflected in the x-axis, and finally translated 4 units upward to obtain the graph of $y = 4 - 2|x + 1|$. The graph is v-shaped and extends downward from $(-1, 4)$.

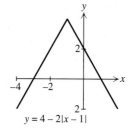

2.3.7 The graph of $y = x$ is stretched by a factor of 2, reflected in the x-axis, and translated 5 units upward to obtain the graph of $y = -2x + 5$.

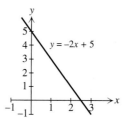

2.3.8 Since $f(-x) = -2(-x)^2 + 5 = -2x^2 + 5$, we have $f(-x) = f(x)$. So the graph is symmetric about the y-axis.

2.3.9 The graph of $y = 2 - |x - 1|$ is shown here.

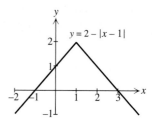

Since the graph is below the x-axis when $x \leq -1$ or when $x \geq 3$, the solution set to the inequality is $(-\infty, -1] \cup [3, \infty)$.

Section 4

2.4.1 $(h + j)(5) = h(5) + j(5)$
$= 5^2 + 3 \cdot 5 = 40,$
$(h \cdot j)(2) = h(2) \cdot j(2)$
$= 2^2 \cdot 3(2) = 24,$
$(h/j)(a) = h(a)/j(a)$
$= a^2/(3a) = a/3$

2.4.2 Note that $(h + j)(2) = h(2) + j(2) = 10 + 5 = 15,$
$(h + j)(6) = h(6) + j(6) = 8 + 0 = 8,$ and $(h + j)(4)$ is undefined because 4 is not in the domain of j. So $h + j = \{(2, 15), (6, 8)\}$ and the domain is $\{2, 6\}$. Note that $(h/j)(2) = h(2)/j(2) = 10/5 = 2, (h/j)(4)$ is undefined because 4 is not in the domain of j, and $(h/j)(6)$ is undefined because $j(6) = 0$. So $h/j = \{(2, 2)\}$ and the domain is $\{2\}$.

2.4.3 The domain of $h + j$ is the intersection of the domains of h and j. So $(h + j)(x) = \sqrt{x} + x$ and the domain is $[0, \infty)$. Since $j(0) = 0$, the domain of h/j is $(0, \infty)$ and $(h/j)(x) = \sqrt{x}/x$.

2.4.4 Note that $(j \circ h)(2) = j(h(2)) = j(0) = 7, (j \circ h)(4) = j(h(4)) = j(0) = 7,$ and $(j \circ h)(6)$ is undefined. So $j \circ h = \{(2, 7), (4, 7)\}$ and the domain is $\{2, 4\}$.

2.4.5 Note that $h(5) = \sqrt{5 - 1} = 2$ and $j(5) = 2 \cdot 5 = 10$. So $(h \circ j)(5) = h(j(5)) = h(10) = \sqrt{10 - 1} = 3$ and $(j \circ h)(5) = j(h(5)) = j(2) = 2(2) = 4.$

2.4.6 Since $(h \circ j)(x) = h(j(x)) = h(3x) = \sqrt{3x + 3}$ we must have $3x + 3 \geq 0$ or $x \geq -1$. So the domain is $[-1, \infty)$.

2.4.7 The function K is the composition of subtracting 3, taking the square root, and then multiplying by 2, in that order. So $K = h \circ f \circ g$. Check: $h(f(g(x))) = h(f(x - 3)) = h(\sqrt{x - 3}) = 2\sqrt{x - 3}$

2.4.8 Substitute $r = C/(2\pi)$ into $d = 2r$ to get $d = 2 \cdot \dfrac{C}{2\pi}$ or $d = \dfrac{C}{\pi}$.

2.4.9 In $C(x) = 0.10x$, x is revenue. Since $R(x) = 80x$, replace x with $80x$ to get $C(x) = 0.10(80x) = 8x$ where x is the number of books.

Section 5

2.5.1 Since there are no ordered pairs with the same second coordinate and different first coordinates, the function is invertible and $h^{-1} = \{(1, 2), (4, 3), (0, 4)\}$.

2.5.2 Since $h^{-1} = \{(1, 2), (4, 3), (0, 4)\}, h(3) = 4, h^{-1}(4) = 3,$ and $(h \circ h^{-1})(1) = h(h^{-1}(1)) = h(2) = 1.$

2.5.3 Since no horizontal line can cross the graph in (a) more than once, the function is one-to-one. Since the horizontal line $y = -2$ crosses the graph in (b) more than once, the function is not one-to-one.

2.5.4 If $h(x_1) = h(x_2)$, then $5x_1^2 = 5x_2^2$ or $x_1^2 = x_2^2$. But $x_1^2 = x_2^2$ does not imply that $x_1 = x_2$, because $2^2 = (-2)^2$. So the function is not one-to-one.

2.5.5 First switch x and y in $y = x^3 - 5$ then solve for y:
$$x = y^3 - 5$$
$$y^3 = x + 5$$
$$y = \sqrt[3]{x + 5}$$
$$h^{-1}(x) = \sqrt[3]{x + 5}$$

2.5.6 Find $f(g(x))$ and $g(f(x))$:
$$f(g(x)) = 2\left(\frac{x - 1}{2}\right) + 1 = x - 1 + 1 = x$$
$$g(f(x)) = \frac{(2x + 1) - 1}{2} = \frac{2x}{2} = x$$

Since these equations are correct for any real number x, the functions are inverses of each other.

2.5.7 The domain of f is $[-2, \infty)$ and the range is $[0, \infty)$. So $f^{-1}(x) = x^2 - 2$ for $x \geq 0$.

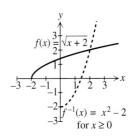

$f(x) = \sqrt{x + 2}$

$f^{-1}(x) = x^2 - 2$ for $x \geq 0$

2.5.8 Since f is a composition of multiplying x by 2/3 and then adding 6, f^{-1} is a composition of subtracting 6 and then dividing by 2/3 (or multiplying by 3/2). So $f^{-1}(x) = \frac{3}{2}(x - 6) = \frac{3}{2}x - 9.$

Section 6

2.6.1 Since $c = ks$ and $3.60 = k(12)$, we have $k = 0.30$. So if $s = 16$, then $c = 0.30(16) = 4.80$. The cost of a 16-ounce smoothie is \$4.80.

2.6.2 Since $t = k/n$ and $12 = k/4$, we have $k = 48$. So if $n = 6$, then $t = 48/6 = 8$. So it takes 6 rakers 8 hours to complete the job.

2.6.3 Since $c = kLH$ and $3000 = k(200 \cdot 5)$, we have $k = 3$. So if $L = 250$ and $H = 6$, then $c = 3 \cdot 250 \cdot 6 = 4500$. So the cost of a 6-foot fence that is 250 feet long is \$4500.

2.6.4 Since the variation is direct, we use multiplication: $M = kwz^3$.

2.6.5 Use $W = 480, m = 5,$ and $t = 4$ in $W = km/t^2$ to find k:
$$480 = \frac{k(5)}{4^2}$$
$$k = 1536$$

Now use $k = 1536, m = 3,$ and $t = 6$ to find W:
$$W = \frac{1536(3)}{6^2} = 128$$

Chapter 3

Section 1

3.1.1 $f(x) = 2x^2 - 8x + 9$
$f(x) = 2(x^2 - 4x) + 9$
$f(x) = 2(x^2 - 4x + 4 - 4) + 9$
$f(x) = 2(x^2 - 4x + 4) - 8 + 9$
$f(x) = 2(x - 2)^2 + 1$

3.1.2 $f(x) = -2x^2 - 4x + 1$
$f(x) = -2(x^2 + 2x) + 1$
$f(x) = -2(x^2 + 2x + 1 - 1) + 1$
$f(x) = -2(x^2 + 2x + 1) + 2 + 1$
$f(x) = -2(x + 1)^2 + 3$

The graph is a parabola opening downward from $(-1, 3)$. The graph goes through $(0, 1)$ and $(-2, 1)$.

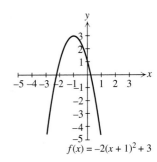

$f(x) = -2(x + 1)^2 + 3$

3.1.3 $x = \dfrac{-b}{2a} = \dfrac{-(-12)}{2(3)} = 2$

$f(2) = 3(2^2) - 12(2) + 5 = -7$

The vertex is $(2, -7)$.

3.1.4 The vertex is $(5, -4)$ and the axis of symmetry is $x = 5$. Since the parabola opens upward from the vertex, the range is $[-4, \infty)$ and -4 is the minimum value of the function. The function is decreasing on $(-\infty, 5)$ and increasing on $(5, \infty)$.

3.1.5 If $x = 0$, then $y = 3(0 - 5)^2 - 4 = 71$. If $y = 0$, then

$3(x - 5)^2 - 4 = 0$ or $x = 5 \pm \sqrt{\dfrac{4}{3}} = 5 \pm \dfrac{2\sqrt{3}}{3} = \dfrac{15 \pm 2\sqrt{3}}{3}$.

The intercepts are $(0, 71)$ and $\left(\dfrac{15 \pm 2\sqrt{3}}{3}, 0\right)$.

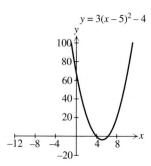

$y = 3(x - 5)^2 - 4$

3.1.6

$x^2 - 2x > 8$

$x^2 - 2x - 8 > 0$

$(x - 4)(x + 2) > 0$

$x - 4 \quad -\,-\,-\,-\,-\,-\,0 + +$

$x + 2 \quad -\,-\,0 + + + + + +$

$\qquad\qquad -2 \qquad 4$

The solution set is $(-\infty, -2) \cup (4, \infty)$.

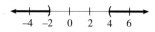

3.1.7 Solve $x^2 - 2x - 4 = 0$:

$x = \dfrac{2 \pm \sqrt{4 - 4(1)(-4)}}{2(1)} = 1 \pm \sqrt{5}$

The numbers $1 - \sqrt{5}$ and $1 + \sqrt{5}$ divide the number line into three intervals, $(-\infty, 1 - \sqrt{5})$, $(1 - \sqrt{5}, 1 + \sqrt{5})$, and $(1 + \sqrt{5}, \infty)$. Test one point from each interval in $x^2 - 2x - 4 < 0$ to see that the solution set is $(1 - \sqrt{5}, 1 + \sqrt{5})$.

3.1.8 Since $L + W = 25$, $W = 25 - L$.

$A = L(25 - L) = -L^2 + 25L$

$L = \dfrac{-b}{2a} = \dfrac{-25}{2(-1)} = 12.5$

If $L = 12.5$, then $W = 12.5$ and the rectangle with the largest area is a 12.5-m by 12.5-m square.

Section 2

3.2.1

$$
\begin{array}{r}
x - 6 \\
x - 1\overline{)x^2 - 7x + 9} \\
\underline{x^2 - x} \\
-6x + 9 \\
\underline{-6x + 6} \\
3
\end{array}
$$

Since the remainder is 3, $P(1) = 3$.

3.2.2

$$
\begin{array}{r|rrrr}
-2 & 1 & 0 & -7 & 5 \\
 & & -2 & 4 & 6 \\
\hline
 & 1 & -2 & -3 & 11
\end{array}
$$

The quotient is $x^2 - 2x - 3$ and the remainder is 11.

3.2.3

$$
\begin{array}{r|rrrr}
3 & 1 & 1 & 0 & -9 \\
 & & 3 & 12 & 36 \\
\hline
 & 1 & 4 & 12 & 27
\end{array}
$$

$P(3) = 27$

3.2.4

$$
\begin{array}{r|rrrr}
1 & 1 & 0 & -3 & 2 \\
 & & 1 & 1 & -2 \\
\hline
 & 1 & 1 & -2 & 0
\end{array}
$$

Since the remainder is 0, $x - 1$ is a factor.

$P(x) = (x - 1)(x^2 + x - 2)$

$\quad\quad = (x - 1)^2(x + 2)$

3.2.5 The factors of 2 are 1 and 2. The factors of 3 are 1 and 3. All possible factors of 3 over factors of 2 are ± 1, ± 3, $\pm\frac{1}{2}$, and $\pm\frac{3}{2}$.

3.2.6 Try the possible rational zeros.

$$
\begin{array}{r|rrrr}
-1 & 2 & 1 & -4 & -3 \\
 & & -2 & 1 & 3 \\
\hline
 & 2 & -1 & -3 & 0
\end{array}
$$

$h(x) = (x + 1)(2x^2 - x - 3)$

$\quad\quad = (x + 1)(x + 1)(2x - 3)$

The zeros are -1 and $3/2$.

Section 3

3.3.1 Set each factor equal to zero to get 0 as a root with multiplicity 3, -2 as a root with multiplicity 2, and $5/2$ as a root.

3.3.2 If the coefficients are to be real, then both $-i$ and its conjugate i must be roots.

$$(x - 3)(x - i)(x + i) = 0$$
$$(x - 3)(x^2 + 1) = 0$$
$$x^3 - 3x^2 + x - 3 = 0$$

3.3.3 The number of variations in sign for $P(x) = x^3 - 5x^2 + 4x + 3$ is 2. So there are either 0 or 2 positive roots. The number of variations in sign for $P(-x) = -x^3 - 5x^2 - 4x + 3$ is 1. So there is exactly 1 negative root. Since there must be a total of 3 roots, there is one negative root and 2 positive roots or one negative root and 2 imaginary roots.

3.3.4 The number of variations in sign for $P(x) = x^4 - 6x^2 + 10$ is 2 and for $P(-x) = x^4 - 6x^2 + 10$ it is also 2. So there are either 0 or 2 positive roots and 0 or 2 negative roots. So the possibilities are 0 positive, 0 negative, 4 imaginary; 0 positive, 2 negative, 2 imaginary; 2 positive, 0 negative, 2 imaginary; 2 positive, 2 negative, 0 imaginary.

3.3.5 The first positive integer for which all terms of the bottom row of synthetic division are nonnegative is 3. So 3 is the best upper bound for the roots. The first negative integer for which the terms of the bottom row alternate in sign is -5. So -5 is the best lower bound for the roots by the theorem on bounds.

3.3.6 Try possible rational roots with synthetic division:

$$
\begin{array}{r|rrrr}
\frac{1}{2} & 2 & -5 & -8 & 5 \\
 & & 1 & -2 & -5 \\
\hline
 & 2 & -4 & -10 & 0
\end{array}
$$

$2x^2 - 4x - 10 = 0$

$x^2 - 2x - 5 = 0$

$$x = \frac{2 \pm \sqrt{4 - 4(1)(-5)}}{2(1)} = 1 \pm \sqrt{6}$$

The roots are $1/2, 1 - \sqrt{6}$, and $1 + \sqrt{6}$.

Section 4

3.4.1 $x^3 - 2x^2 + 5x - 10 = 0$
$x^2(x - 2) + 5(x - 2) = 0$
$(x^2 + 5)(x - 2) = 0$
$x^2 + 5 = 0 \quad$ or $\quad x - 2 = 0$
$x = \pm i\sqrt{5} \quad$ or $\qquad x = 2$
The solution set is $\left\{2, -i\sqrt{5}, i\sqrt{5}\right\}$.

3.4.2 $\qquad\qquad x^5 = 27x^2$
$\qquad\qquad x^5 - 27x^2 = 0$
$x^2(x - 3)(x^2 + 3x + 9) = 0$
$x^2 = 0 \quad$ or $\quad x = 3 \quad$ or $\quad x = \dfrac{-3 \pm \sqrt{-27}}{2}$
$\qquad\qquad\qquad\qquad\qquad = -\dfrac{3}{2} \pm \dfrac{3}{2}i\sqrt{3}$
The solution set is $\left\{0, 3, -\frac{3}{2} \pm \frac{3}{2}i\sqrt{3}\right\}$.

3.4.3 $\sqrt{x} + 12 = x$
$\sqrt{x} = x - 12$
$x = x^2 - 24x + 144$
$0 = x^2 - 25x + 144$
$0 = (x - 9)(x - 16)$
$x = 9 \quad$ or $\quad x = 16$
Since 9 does not satisfy $\sqrt{x} + 12 = x$ the solution set is $\{16\}$.

3.4.4 $\sqrt{3x - 2} - \sqrt{x} = 2$
$\sqrt{3x - 2} = 2 + \sqrt{x}$
$3x - 2 = 4 + 4\sqrt{x} + x$
$2x - 6 = 4\sqrt{x}$
$x - 3 = 2\sqrt{x}$
$x^2 - 6x + 9 = 4x$
$x^2 - 10x + 9 = 0$
$(x - 9)(x - 1) = 0$
$x = 9 \quad$ or $\quad x = 1$
Since $\sqrt{3x - 2} - \sqrt{x} = 2$ is not satisfied by 1, the solution set is $\{9\}$.

3.4.5 $x^{-4/5} = 16$
$(x^{-4/5})^{-5/4} = \pm 16^{-5/4}$
$x = \pm\dfrac{1}{32}$
The solution set is $\{\pm 1/32\}$.

3.4.6 $x^4 - 9x^2 + 20 = 0$
$(x^2 - 5)(x^2 - 4) = 0$
$x^2 - 5 = 0 \quad$ or $\quad x^2 - 4 = 0$
$x = \pm\sqrt{5} \quad$ or $\qquad x = \pm 2$
The solution set is $\left\{\pm\sqrt{5}, \pm 2\right\}$.

3.4.7 $(x^2 + x)^2 - 8(x^2 + x) + 12 = 0$
$u^2 - 8u + 12 = 0$
$(u - 6)(u - 2) = 0$
$u - 6 = 0 \quad$ or $\qquad u - 2 = 0$
$x^2 + x - 6 = 0 \quad$ or $\quad x^2 + x - 2 = 0$
$(x + 3)(x - 2) = 0 \quad$ or $\quad (x + 2)(x - 1) = 0$
$x = -3 \quad$ or $\quad x = 2 \quad$ or $\quad x = -2 \quad$ or $\quad x = 1$
The solution set is $\{-3, -2, 1, 2\}$.

3.4.8 $x^{2/3} - x^{1/3} - 6 = 0$
$(x^{1/3} - 3)(x^{1/3} + 2) = 0$
$x^{1/3} = 3 \quad$ or $\quad x^{1/3} = -2$
$x = 27 \quad$ or $\qquad x = -8$
The solution set is $\{-8, 27\}$.

3.4.9 $|x^2 - x - 4| = 2$
$x^2 - x - 4 = 2 \qquad$ or $\quad x^2 - x - 4 = -2$
$x^2 - x - 6 = 0 \qquad$ or $\quad x^2 - x - 2 = 0$
$(x - 3)(x + 2) = 0 \quad$ or $\quad (x - 2)(x + 1) = 0$
$x = 3 \quad$ or $\quad x = -2 \quad$ or $\quad x = 2 \quad$ or $\quad x = -1$
The solution set is $\{-2, -1, 2, 3\}$.

3.4.10 $|x| = |x - 1|$
$x = x - 1 \quad$ or $\qquad x = -(x - 1)$
$0 = -1 \qquad$ or $\quad 2x = 1$
$\qquad\qquad\qquad$ or $\qquad x = \dfrac{1}{2}$
The solution set is $\{1/2\}$.

3.4.11 Let $D = 10$ and solve for P:
$10 = \sqrt{P} - \sqrt{P - 400}$
$\sqrt{P - 400} = \sqrt{P} - 10$
$P - 400 = P - 20\sqrt{P} + 100$
$-500 = -20\sqrt{P}$
$25 = \sqrt{P}$
$625 = P$
At \$625 the demand is 10 suits per month.

Section 5

3.5.1 Since $f(-x) = x^3 + 4x$, we have $f(-x) = -f(x)$ and the graph is symmetric about the origin.

3.5.2 Since $x - 1$ occurs with an odd power, the graph crosses the x-axis at $(1, 0)$. Since $x + 5$ occurs with an even power, the graph does not cross the x-axis at $(-5, 0)$.

3.5.3 Since the leading coefficient is negative and the degree is odd, as $x \to \infty, y \to -\infty$. As $x \to -\infty, y \to \infty$.

3.5.4 Since $f(-x) = f(x)$ the graph is symmetric about the y-axis. Since $f(x) = x^2(x - 2)(x + 2)$ the graph crosses the x-axis at $(2, 0)$ and $(-2, 0)$ and touches but does not cross at $(0, 0)$. As $x \to \pm\infty, y \to \infty$.

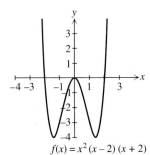

$$f(x) = x^2(x - 2)(x + 2)$$

3.5.5 The solutions to $x^4 - 4x^2 = 0$ are $-2, 0$, and 2. They divide the number line into four intervals $(-\infty, -2), (-2, 0), (0, 2)$, and $(2, \infty)$. Testing one number from each interval in $x^4 - 4x^2 < 0$ reveals that the solution set to the inequality is $(-2, 0) \cup (0, 2)$.

Section 6

3.6.1 Since $x^2 - 9 = 0$ only if $x = \pm 3$, the domain is $(-\infty, -3) \cup (-3, 3) \cup (3, \infty)$.

3.6.2 The vertical lines $x = -3$ and $x = 3$ are vertical asymptotes. Since the degree of the denominator is larger than the degree of the numerator, the x-axis is a horizontal asymptote.

3.6.3

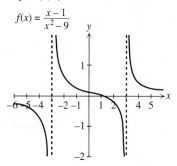

$$x - 1 \overline{)3x^2 + 0x - 4} \quad \begin{array}{c} 3x + 3 \end{array}$$

$$\frac{3x^2 - 3x}{3x - 4}$$

$$\frac{3x - 3}{-1}$$

$$\frac{3x^2 - 4}{x - 1} = 3x + 3 + \frac{-1}{x - 1}$$

The line $y = 3x + 3$ is an oblique asymptote and the line $x = 1$ is a vertical asymptote.

3.6.4 The vertical asymptotes are $x = \pm 3$ and the horizontal asymptote is $y = 0$. The x-intercept is $(1, 0)$.

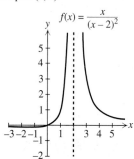

$$f(x) = \frac{x - 1}{x^2 - 9}$$

3.6.5 The line $x = 2$ is a vertical asymptote and the x-axis is a horizontal asymptote. The x-intercept is $(0, 0)$.

$$f(x) = \frac{x}{(x - 2)^2}$$

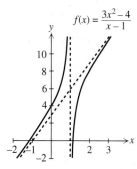

3.6.6 From 3.6.3, the oblique asymptote is $y = 3x + 3$ and the vertical asymptote is $x = 1$.

$$f(x) = \frac{3x^2 - 4}{x - 1}$$

3.6.7 If $x \neq 1$ and $x \neq -1$, then $f(x) = \frac{1}{x + 1}$. So there is a vertical asymptote at $x = -1$ and a hole at $(1, 1/2)$.

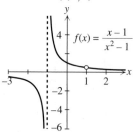

$$f(x) = \frac{x - 1}{x^2 - 1}$$

3.6.8

$$\frac{1}{x - 1} > 1$$

$$\frac{1}{x - 1} - 1 > 0$$

$$\frac{1}{x - 1} - \frac{x - 1}{x - 1} > 0$$

$$\frac{2 - x}{x - 1} > 0$$

$$\begin{array}{lllllllll} 2 - x & + & + & + & + & + & 0 & - & - & - \\ x - 1 & - & - & - & 0 & + & + & + & + & + \end{array}$$

$$\begin{array}{cccc} 0 & 1 & 2 & 3 \end{array}$$

The quotient is positive if x is in the interval $(1, 2)$.

3.6.9 The expression is undefined if $x = 1$ and has a value of 0 if $x = -3$. These numbers divide the number line into three intervals, $(-\infty, -3)$, $(-3, 1)$, and $(1, \infty)$. Test one number in each interval to see that $\frac{x + 3}{x - 1} > 0$ is satisfied for x in $(-\infty, -3) \cup (1, \infty)$.

3.6.10 The average cost per mile is given by $C = \frac{0.80x + 100}{x}$ where x is the number of miles. As x increases, C approaches the horizontal asymptote, which is $0.80 per mile.

Chapter 4

Section 1

4.1.1 If $f(x) = 9^x$, then $f(-2) = 9^{-2} = 1/81$ and $f(1/2) = 9^{1/2} = 3$.

4.1.2 The ordered pairs $(-2, 1/81)$, $(0, 1)$, and $(2, 81)$ are on the graph. The domain is $(-\infty, \infty)$ and the range is $(0, \infty)$.

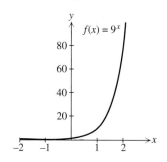

$$f(x) = 9^x$$

4.1.3 The ordered pairs $(-2, 81)$, $(0, 1)$, and $(2, 1/81)$ are on the graph. The domain is $(-\infty, \infty)$ and the range is $(0, \infty)$.

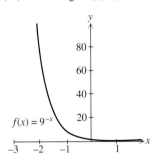

4.1.4 The ordered pairs $(1, 3/2)$, $(2, 2)$, and $(3, 3)$ are on the graph. The domain is $(-\infty, \infty)$ and the range is $(1, \infty)$.

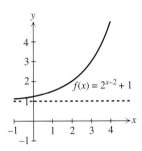

4.1.5 The ordered pairs $(-1, -4)$, $(0, -1)$, and $(1, -1/4)$ are on the graph. The domain is $(-\infty, \infty)$ and the range is $(-\infty, 0)$.

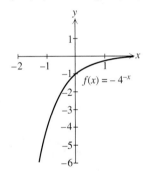

4.1.6 $2^{-x} = \dfrac{1}{8}$

$2^{-x} = 2^{-3}$

$-x = -3$

$x = 3$

4.1.7 $3^{4-x} = \dfrac{1}{9}$

$3^{4-x} = 3^{-2}$

$4 - x = -2$

$-x = -6$

$x = 6$

4.1.8 $A = 9000\left(1 + \dfrac{0.054}{12}\right)^{12\cdot 6}$

$\approx \$12{,}434.78$

4.1.9 $A = 8000e^{0.063(7.25)}$

$\approx \$12{,}631.47$

4.1.10 $A = 2.9e^{-1.21\times 10^{-4}(0)} = 2.9$
The initial amount is 2.9 grams.
$A = 2.9e^{-1.21\times 10^{-4}(6500)} \approx 1.3$ grams

Section 2

4.2.1 Since $4^{-2} = 1/16$, $\log_4(1/16) = -2$.

4.2.2 Since $10^2 = 100$, $\log(100) = 2$.

4.2.3 The ordered pairs $(1/6, -1)$, $(1, 0)$, $(6, 1)$, and $(36, 2)$ are on the graph. The domain is $(0, \infty)$ and the range is $(-\infty, \infty)$.

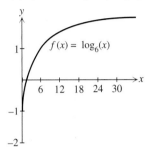

4.2.4 The ordered pairs $(36, -2)$, $(6, -1)$, $(1, 0)$, $(1/6, 1)$, and $(1/36, 2)$ are on the graph. The domain is $(0, \infty)$ and the range is $(-\infty, \infty)$.

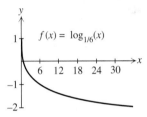

4.2.5 The ordered pairs $(0, 0)$, $(1, -1)$, $(3, -2)$, $(7, -3)$, and $(-1/2, 1)$ are on the graph. The domain is $(-1, \infty)$ and the range is $(-\infty, \infty)$.

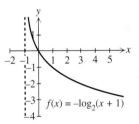

4.2.6 Since $y = a^x$ is equivalent to $\log_a(y) = x$, $\log_4(w) = 7$ is equivalent to $w = 4^7$.

4.2.7 $f(x) = 3 \cdot e^{5x-1}$

$y = 3 \cdot e^{5x-1}$

$x = 3 \cdot e^{5y-1}$

$\dfrac{x}{3} = e^{5y-1}$

$\ln\left(\dfrac{x}{3}\right) = 5y - 1$

$5y = \ln\left(\dfrac{x}{3}\right) + 1$

$y = \dfrac{1}{5}\ln\left(\dfrac{x}{3}\right) + \dfrac{1}{5}$

$f^{-1}(x) = \dfrac{1}{5}\ln\left(\dfrac{x}{3}\right) + \dfrac{1}{5}$

4.2.8 $\log_5(2x) = -3$

$$2x = 5^{-3}$$
$$2x = \frac{1}{125}$$
$$x = \frac{1}{250}$$

4.2.9 $10^{3x} = 70$

$$3x = \log(70)$$
$$x = \frac{\log(70)}{3} \approx 0.6150$$

4.2.10 $12 = 4 \cdot e^{0.05t}$

$$3 = e^{0.05t}$$
$$0.05t = \ln(3)$$
$$t = \frac{\ln(3)}{0.05} \approx 21.9722$$

The time is 21.9722 years or 21 years and 355 days (to the nearest day).

Section 3

4.3.1 $\log(x) + \log(y) = \log(xy)$

4.3.2 $\log(5x) - \log(5) = \log(5x/5)$
$$= \log(x)$$

4.3.3 $\ln(27) = \ln(3^3) = 3 \cdot \ln(3)$

4.3.4 $10^{\log(5w)} = 5w, \log(10^p) = p$

4.3.5 $\ln(45) = \ln(3^2 \cdot 5)$
$$= 2 \cdot \ln(3) + \ln(5).$$

4.3.6 $\log\left(\dfrac{x^2}{5a}\right) = \log(x^2) - \log(5a)$

$$= 2 \cdot \log(x) - [\log(5) + \log(a)]$$
$$= 2 \cdot \log(x) - \log(5) - \log(a)$$

4.3.7 $\ln(x) - \ln(y) - 2 \cdot \ln(z)$

$$= \ln\left(\frac{x}{y}\right) - \ln(z^2)$$
$$= \ln\left(\frac{x}{yz^2}\right)$$

4.3.8 $5.44^x = 2.3$

$$\log(5.44^x) = \log(2.3)$$
$$x \cdot \log(5.44) = \log(2.3)$$
$$x = \frac{\log(2.3)}{\log(5.44)}$$
$$x \approx 0.4917$$

4.3.9 $400 = 100\left(1 + \dfrac{0.05}{365}\right)^{365t}$

$$4 = \left(1 + \frac{0.05}{365}\right)^{365t}$$
$$4 \approx 1.000136986^{365t}$$
$$365t \approx \log_{1.000136986}(4)$$
$$t \approx \frac{1}{365} \cdot \frac{\ln(4)}{\ln(1.000136986)}$$
$$\approx 27.72778621 \text{ yr}$$
$$\approx 27 \text{ years } 266 \text{ days}$$

4.3.10 $300 = 100\left(1 + \dfrac{r}{365}\right)^{365(20)}$

$$3 = \left(1 + \frac{r}{365}\right)^{7300}$$
$$1 + \frac{r}{365} = 3^{1/7300}$$
$$r = 365(3^{1/7300} - 1)$$
$$r \approx 5.5\%$$

Section 4

4.4.1 $\log(2x + 1) = 3$

$$2x + 1 = 10^3$$
$$2x = 999$$
$$x = 499.5$$

4.4.2 $\log(x - 1) + \log(3) = \log(x) - \log(4)$

$$\log(3x - 3) = \log(x/4)$$
$$3x - 3 = \frac{x}{4}$$
$$12x - 12 = x$$
$$11x = 12$$
$$x = \frac{12}{11}$$

4.4.3 $(1.05)^{3t} = 8$

$$3t = \log_{1.05}(8)$$
$$t = \frac{1}{3} \cdot \frac{\ln(8)}{\ln(1.05)}$$
$$t \approx 14.2067$$

4.4.4
$$3^{x-1} = 2^x$$
$$\ln(3^{x-1}) = \ln(2^x)$$
$$(x - 1)\ln(3) = x \cdot \ln(2)$$
$$x \cdot \ln(3) - \ln(3) = x \cdot \ln(2)$$
$$x \cdot \ln(3) - x \cdot \ln(2) = \ln(3)$$
$$x(\ln(3) - \ln(2)) = \ln(3)$$
$$x \cdot \ln(3/2) = \ln(3)$$
$$x = \frac{\ln(3)}{\ln(3/2)} \approx 2.7095$$

4.4.5 First find r:

$$1 = 2e^{1\times 10^6 r}$$
$$0.5 = e^{1\times 10^6 r}$$
$$1 \times 10^6 r = \ln(0.5)$$
$$r = \frac{\ln(0.5)}{1 \times 10^6} \approx -6.93 \times 10^{-7}$$

Next find t:

$$0.40 = 1e^{-6.93\times 10^{-7} t}$$
$$-6.93 \times 10^{-7} t = \ln(0.40)$$
$$t = \frac{\ln(0.40)}{-6.93 \times 10^{-7}}$$
$$\approx 1.32 \times 10^6 \text{ years}$$

4.4.6 The difference in temperature goes from 200° to 90° in 4 minutes.

$$90 = 200e^{4r}$$
$$e^{4r} = 0.45$$
$$4r = \ln(0.45)$$
$$r = \frac{\ln(0.45)}{4} \approx -0.1996$$

Let t be the time required for the difference to go from 90° to 70°.

$$70 = 90e^{-0.1996t}$$
$$e^{-0.1996t} = 7/9$$
$$-0.1996t = \ln(7/9)$$
$$t = \frac{\ln(7/9)}{-0.1996}$$
$$\approx 1.2589 \text{ minutes}$$

4.4.7 $2000 = 120{,}000 \dfrac{0.075/12}{1 - (1 + 0.075/12)^{-12t}}$

$$1 - (1 + 0.075/12)^{-12t} = 0.375$$
$$0.625 = (1 + 0.075/12)^{-12t}$$
$$-12t \cdot \ln(1 + 0.075/12) = \ln(0.625)$$
$$t = \frac{\ln(0.625)}{-12 \cdot \ln(1 + 0.075/12)}$$
$$\approx 6.2863 \text{ yr}$$
$$\approx 6 \text{ years } 3 \text{ mo}$$

Chapter 5

Section 1

5.1.1 $10° + 1 \cdot 360° = 370°$
$10° + 2 \cdot 360° = 730°$
$10° + (-1) \cdot 360° = -350°$
$10° + (-2) \cdot 360° = -710°$

5.1.2 $-690 + 360k = 390$
$360k = 1080$
$k = 3$
Since $-690°$ and $390°$ differ by a multiple of $360°$, they are coterminal.

5.1.3 $-890° + 3(360°) = 190°$
Since $-890°$ is coterminal with $190°$ (which is in quadrant III), $-890°$ lies in quadrant III.

5.1.4 $15' = 15 \text{ min} \cdot \dfrac{1 \text{ deg}}{60 \text{ min}} = \dfrac{1}{4} \text{ deg}$

$12'' = 12 \text{ sec} \cdot \dfrac{1 \text{ deg}}{3600 \text{ sec}} = \dfrac{1}{300} \text{ deg}$

$35° \, 15' \, 12'' = \left(35 + \dfrac{1}{4} + \dfrac{1}{300}\right) \text{ deg}$
$\approx 35.2533°$

5.1.5 $0.321 \text{ deg} \cdot \dfrac{60 \text{ min}}{1 \text{ deg}} = 19.26 \text{ min}$

$0.26 \text{ min} \cdot \dfrac{60 \text{ sec}}{1 \text{ min}} = 15.6 \text{ sec}$

$56.321° = 56°19'15.6''$

5.1.6 $210° = 210 \text{ deg} \cdot \dfrac{\pi \text{ rad}}{180 \text{ deg}}$

$= \dfrac{7\pi}{6} \text{ rad}$

5.1.7 $\dfrac{5\pi}{3} \text{ rad} = \dfrac{5\pi}{3} \text{ rad} \cdot \dfrac{180 \text{ deg}}{\pi \text{ rad}}$

$= 300°$

5.1.8 $-\dfrac{\pi}{3} + 1 \cdot 2\pi = -\dfrac{\pi}{3} + \dfrac{6\pi}{3} = \dfrac{5\pi}{3}$

$-\dfrac{\pi}{3} + 2 \cdot 2\pi = -\dfrac{\pi}{3} + \dfrac{12\pi}{3} = \dfrac{11\pi}{3}$

$-\dfrac{\pi}{3} + (-1) \cdot 2\pi = -\dfrac{\pi}{3} - \dfrac{6\pi}{3} = -\dfrac{7\pi}{3}$

$-\dfrac{\pi}{3} + (-2) \cdot 2\pi = -\dfrac{\pi}{3} - \dfrac{12\pi}{3} = -\dfrac{13\pi}{3}$

5.1.9 Since $45° = \pi/4$ radians,

$s = \alpha r = \dfrac{\pi}{4} \cdot 8 \text{ in.} = 2\pi \text{ in.} \approx 6.28 \text{ in.}$

5.1.10 Since $\alpha = s/r$, $\alpha = 3/50$ rad.
So $\alpha = \dfrac{3}{50} \text{ rad} \cdot \dfrac{180 \text{ deg}}{\pi \text{ rad}} \approx 3.4°.$

5.1.11 $\omega = \dfrac{5 \text{ rev}}{\text{sec}} \cdot \dfrac{2\pi \text{ rad}}{\text{rev}} = 10\pi \text{ rad/sec}$

$\approx 31.416 \text{ rad/sec}$

$v = r\omega$

$= 10 \text{ ft} \cdot \dfrac{10\pi \text{ rad}}{\text{sec}} \cdot \dfrac{1 \text{ mi}}{5280 \text{ ft}} \cdot \dfrac{3600 \text{ sec}}{1 \text{ hr}}$

$\approx 214.199 \text{ mi/hr}$

5.1.12 $v = \omega r$

$= \dfrac{2\pi \text{ rad}}{24 \text{ hr}} \cdot 3950 \text{ mi} \cdot \dfrac{5280 \text{ ft}}{1 \text{ mi}} \cdot \dfrac{1 \text{ hr}}{3600 \text{ sec}}$

$\approx 1516.691 \text{ ft/sec}$

Section 2

5.2.1 The terminal side of $-\pi/2$ intersects the unit circle at $(0, -1)$. So $\sin(-\pi/2) = -1$ and $\cos(-\pi/2) = 0$.

5.2.2 The terminal side of $-5\pi/4$ intersects the unit circle at $\left(-\sqrt{2}/2, \sqrt{2}/2\right)$. So $\sin(-5\pi/4) = \sqrt{2}/2$ and $\cos(-5\pi/4) = -\sqrt{2}/2$.

5.2.3 The terminal side of $-\pi/6$ intersects the unit circle at $\left(\sqrt{3}/2, -1/2\right)$. So $\sin(-\pi/6) = -1/2$ and $\cos(-\pi/6) = \sqrt{3}/2$.

5.2.4 Since the terminal side of $135°$ is in quadrant II, the reference angle is $180° - 135°$ or $45°$. In radians it is $\pi/4$.

5.2.5 The reference angle for $135°$ is $45°$. So $\sin(135°) = \sin(45°) = \sqrt{2}/2$ and $\cos(135°) = -\cos(45°) = -\sqrt{2}/2$.

5.2.6 With a calculator in radian mode, $\sin(55.6) \approx -0.8126$. With a calculator in degree mode, $\cos(34.2°) \approx 0.8271$.

5.2.7 $\sin^2\alpha + \left(\dfrac{1}{4}\right)^2 = 1$

$\sin^2\alpha = \dfrac{15}{16}$

$\sin\alpha = \pm\sqrt{\dfrac{15}{16}} = \pm\dfrac{\sqrt{15}}{4}$

Since $\sin\alpha < 0$ in quadrant IV, $\sin\alpha = -\sqrt{15}/4$.

5.2.8 Let $t = 3$ to get
$x = -2\sin(2 \cdot 3) + 3\cos(2 \cdot 3)$
≈ 3.4
So the weight is 3.4 cm below equilibrium.

Section 3

5.3.1 The curve goes through $(0, 0)$, $(\pi/2, 4)$, $(\pi, 0)$, $(3\pi/2, -4)$, and $(2\pi, 0)$.

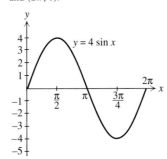

5.3.2 The maximum y-coordinate is 4 and the minimum is -4. The amplitude is $\left|\dfrac{1}{2}(4 - (-4))\right|$ or 4.

5.3.3 The curve goes through $(0, 4)$, $(\pi/2, 0)$, $(\pi, -4)$, $(3\pi/2, 0)$, and $(2\pi, 4)$. The amplitude is 4.

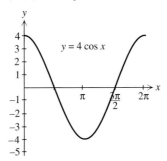

5.3.4 The curve goes through $(\pi/6, 1)$, $(2\pi/3, 0)$, $(7\pi/6, -1)$, $(5\pi/3, 0)$, and $(13\pi/6, 1)$. The phase shift is $\pi/6$.

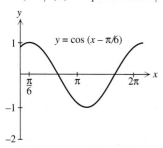

5.3.5 The curve goes through $(-\pi/4, 1)$, $(\pi/4, 2)$, $(3\pi/4, 1)$, $(5\pi/4, 0)$, and $(7\pi/4, 1)$. The phase shift is $-\pi/4$.

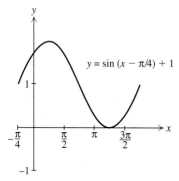

5.3.6 The period is π. So two cycles occur in the interval $[0, 2\pi]$. The curve goes through $(0, 1)$, $(\pi/4, 0)$, $(\pi/2, -1)$, $(3\pi/4, 0)$, and $(\pi, 1)$.

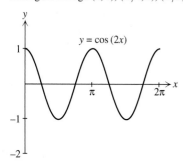

5.3.7 The period is $2\pi/\pi$ or 2. So two cycles occur in the interval $[0, 4]$. The curve goes through $(0, 1)$, $(1/2, 0)$, $(1, -1)$, $(3/2, 0)$, and $(2, 1)$.

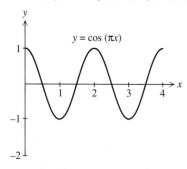

5.3.8 Rewrite the equation as $y = 3\sin\left(2\left(x - \dfrac{\pi}{2}\right)\right) - 1$. The period is $2\pi/2$ or π. The amplitude is 3 and the phase shift is $\pi/2$. So one cycle occurs in the interval $[-\pi/2, \pi/2]$. The curve goes through $(-\pi/2, -1)$, $(-\pi/4, 2)$, $(0, -1)$, $(\pi/4, -4)$, and $(\pi/2, -1)$.

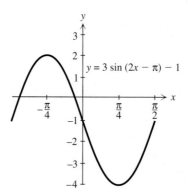

5.3.9 Rewrite the equation as $y = -2\cos\left(2\left(x + \dfrac{\pi}{2}\right)\right) + 1$. The period is $2\pi/2$ or π. The amplitude is 2 and the phase shift is $-\pi/2$. So one cycle occurs in the interval $[-\pi/2, \pi/2]$. The curve goes through $(-\pi/2, -1)$, $(-\pi/4, 1)$, $(0, 3)$, $(\pi/4, 1)$, and $(\pi/2, -1)$.

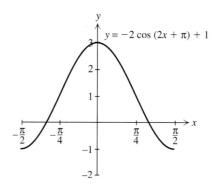

5.3.10 The period is $2\pi/(100\pi)$ or $1/50$. Since $F = 1/P, F = 50$.

5.3.11 Since the range is $[-1, 5]$ the amplitude is 3. Since one cycle occurs on $[1, 5]$ the period is 4 and the phase shift is 1.

So $y = 3\sin\left(\dfrac{\pi}{2}(x - 1)\right) + 2$.

5.3.12 Enter the five points into your calculator and use sinsusoidal regression to get $y = 3.003\sin(1.562x - 1.497) + 1.972$.

Section 4

5.4.1 In quadrant IV, only cosine and secant are positive. So
$\sin(-\pi/4) = -\sqrt{2}/2$, $\cos(-\pi/4) = \sqrt{2}/2$, $\tan(-\pi/4) = -1$,
$\sec(-\pi/4) = 1/\cos(-\pi/4) = \sqrt{2}$,
$\csc(-\pi/4) = 1/\sin(-\pi/4) = -\sqrt{2}$, and
$\cot(-\pi/4) = 1/\tan(-\pi/4) = -1$.

5.4.2 $\sec(\pi/22) = 1/\cos(\pi/22) \approx 1.0103$
$\csc(-14.5) = 1/\sin(-14.5) \approx -1.0696$

5.4.3 The period is $\pi/(2\pi)$ or $1/2$. So two cycles occur on the interval $[-1/4, 3/4]$.

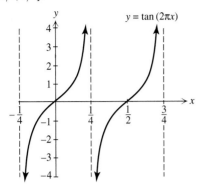

$y = \tan(2\pi x)$

5.4.4 Rewrite the equation as $y = 3\cot\left(2\left(x + \dfrac{\pi}{2}\right)\right)$. The period is $\pi/2$. So two cycles occur on the interval $[-\pi/2, \pi/2]$. The vertical asymptotes are at 0 and $\pm\pi/2$.

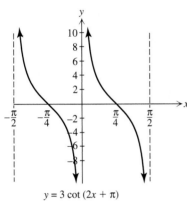

$y = 3\cot(2x + \pi)$

5.4.5 The period is $2\pi/2$ or π and the range is $(-\infty, -3] \cup [3, \infty)$. First graph one cycle of $y = 3\cos(2x)$ on $[0, \pi]$ as a dashed curve.

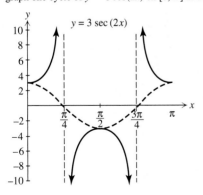

$y = 3\sec(2x)$

5.4.6 The period is $2\pi/2$ or π and the range is $(-\infty, -1] \cup [1, \infty)$. First graph one cycle of $y = \cos(2x - \pi)$ on $[0, \pi]$ as a dashed curve.

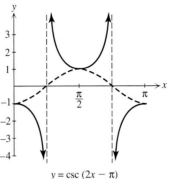

$y = \csc(2x - \pi)$

Section 5

5.5.1 Since $-\pi/4$ is in the interval $[-\pi/2, \pi/2]$ and $\sin(-\pi/4) = -\sqrt{2}/2$, $\sin^{-1}\left(-\sqrt{2}/2\right) = -\pi/4$.

5.5.2 Since $-60°$ is in the interval $[-90°, 90°]$ and $\sin(-60°) = -\sqrt{3}/2$, $\sin^{-1}\left(-\sqrt{3}/2\right) = -60°$.

5.5.3 Use a calculator in degree mode to get $\alpha = \sin^{-1}(0.3) \approx 17.5°$.

5.5.4 Since $5\pi/6$ is in the interval $[0, 2\pi]$ and $\cos(5\pi/6) = -\sqrt{3}/2$, $\cos^{-1}\left(-\sqrt{3}/2\right) = 5\pi/6$.

5.5.5 Use a calculator in degree mode to get $\alpha = \cos^{-1}(-0.4) \approx 113.6°$.

5.5.6 Since $\pi/3$ is in the interval $[-\pi/2, \pi/2]$ and $\tan(\pi/3) = \sqrt{3}$, $\arctan\left(\sqrt{3}\right) = \pi/3$.

5.5.7 $\cot^{-1}(1.3) = \tan^{-1}(1/1.3) \approx 0.6557$

5.5.8 Note that $\sin(5\pi/6) = \sin(\pi/6) = 1/2$. So $\sin^{-1}(\sin(5\pi/6)) = \sin^{-1}(1/2) = \pi/6$ because the range of $\sin^{-1}$ is $[-\pi/2, \pi/2]$.

5.5.9 $f(x) = 2\cos(3x) - 1$
$$y = 2\cos(3x) - 1$$
$$x = 2\cos(3y) - 1$$
$$\cos(3y) = \frac{x+1}{2}$$
$$3y = \cos^{-1}\left(\frac{x+1}{2}\right)$$
$$y = \frac{1}{3}\cos^{-1}\left(\frac{x+1}{2}\right)$$
$$f^{-1}(x) = \frac{1}{3}\cos^{-1}\left(\frac{x+1}{2}\right)$$

Since $-1 \le \cos(3x) \le 1$, the range of f is the interval $[-3, 1]$ and the domain of f^{-1} is $[-3, 1]$.

Section 6

5.6.1 Since $r = \sqrt{2^2 + 3^2} = \sqrt{13}$, $\sin\alpha = 3/\sqrt{13} = 3\sqrt{13}/13$, $\cos\alpha = 2/\sqrt{13} = 2\sqrt{13}/13$, and $\tan\alpha = 3/2$.

5.6.2 $c = \sqrt{2^2 + 4^2} = 2\sqrt{5}$
$$\sin\alpha = 4/\left(2\sqrt{5}\right) = 2\sqrt{5}/5$$
$$\cos\alpha = 2/\left(2\sqrt{5}\right) = \sqrt{5}/5$$
$$\tan\alpha = 4/2 = 2$$

5.6.3 Since $\alpha = 60°$, $\beta = 30°$ and $\gamma = 90°$. Since $c = 2$, $\sin(60°) = a/2$ and $a = 2 \cdot \sqrt{3}/2 = \sqrt{3}$. Since $c = 2$, $\sin(30°) = b/2$ and $b = 2 \cdot 1/2 = 1$. So the angles are $30°$, $60°$, and $90°$, and the sides opposite those angles are 1, $\sqrt{3}$, and 2, respectively.

5.6.4 $c = \sqrt{2^2 + 5^2} = \sqrt{29}$
$\alpha = \sin^{-1}(2/\sqrt{29}) \approx 21.8°$
$\beta = \sin^{-1}(5/\sqrt{29}) \approx 68.2°$
$\gamma = 90°$

5.6.5 The least number of significant digits in 24 and 5.461 is 2.
So $24\sin(5.461°) \approx 2.3$.

5.6.6 The height of the tower a is the side opposite 38.2° in a right triangle in which the other leg b is 344 feet. So $\tan(38.2°) = a/344$ or $a = 344\tan(38.2°) \approx 271$ feet.

5.6.7 Let y be the height of the antenna and x be the distance from the second location to the antenna. So $\tan 63.1° = y/x$ and $\tan 44.2° = y/(x + 100)$. Substitute.

$$\tan 44.2° = \frac{y}{\dfrac{y}{\tan 63.1°} + 100}$$

$$\tan 44.2° \left(\frac{y}{\tan 63.1°} + 100\right) = y$$

$$\left(\frac{\tan 44.2°}{\tan 63.1°} - 1\right)y = -100 \cdot \tan 44.2°$$

$$y \approx 192\,\text{feet}$$

5.6.8 The line of sight to the horizon is perpendicular to the radius at the horizon. So the acute angle α at the center of the earth in this right triangle is $\cos^{-1}(3950/3957)$. The width of the path is the length of the arc intercepted by a central angle of 2α in a circle of radius 3950 miles. So using the arc-length formula and radian mode, the width is $2 \cdot \cos^{-1}(3950/3957) \cdot 3950$ or about 470 miles.

Chapter 6

Section 1

6.1.1 $\dfrac{\tan x \csc x}{\sec x} = \dfrac{\dfrac{\sin x}{\cos x} \cdot \dfrac{1}{\sin x}}{\dfrac{1}{\cos x}}$

$= \dfrac{\sin x}{\cos x} \cdot \dfrac{1}{\sin x} \cdot \dfrac{\cos x}{1} = 1$

6.1.2 $1 + \cot^2 x = \csc^2 x$
$\cot^2 x = \csc^2 x - 1$
$\cot x = \pm\sqrt{\dfrac{1}{\sin^2 x} - 1}$

6.1.3 $\csc^2 x = 1 + \cot^2 x$
$\csc x = \pm\sqrt{1 + \cot^2 x}$
$\sin x = \pm\dfrac{1}{\sqrt{1 + \cot^2 x}}$

In quadrant II sine is positive and cosine is negative.

$\sin \alpha = \dfrac{1}{\sqrt{1 + (-1/3)^2}}$

$= \dfrac{1}{\sqrt{10/9}} = \dfrac{3\sqrt{10}}{10}$

$\cos \alpha = -\sqrt{1 - \sin^2\alpha}$
$= -\sqrt{1 - 9/10} = -\sqrt{10}/10$

6.1.4 Let $\theta = \arccos x$. Since $\cos\theta = x$
$\sec(\arccos(x)) = \sec(\theta) = 1/\cos\theta = 1/x$.

6.1.5 $\csc(-x)\tan(-x) = (-\csc x)(-\tan x)$

$= \csc(x)\tan(x) = \dfrac{1}{\sin x} \cdot \dfrac{\sin x}{\cos x}$

$= \sec x$

6.1.6 $f(-x) = \csc(-x) + \tan(-x)$
$= -\csc(x) - \tan(x)$
$= -(\csc x + \tan x) = -f(x)$
Since $f(-x) = -f(x)$ the function is an odd function.

6.1.7 If $t = \pi/3$, then
$\cos(3t) = \cos(\pi) = -1$ and
$3\cos(t) = 3\cos(\pi/3) = 3/2$.
So the equation is not an identity.

Section 2

6.2.1 $1 - \sec(x)\csc(x)\tan(x)$

$= 1 - \dfrac{1}{\cos x} \dfrac{1}{\sin x} \dfrac{\sin x}{\cos x}$

$= 1 - \sec^2 x$

$= -\tan^2 x$

6.2.2 $(1 - 2\sin x)(1 + 2\sin x)$
$= 1 - 4\sin^2 x$

6.2.3 $\sin^2 x + 4\sin x + 4 = (\sin x + 2)^2$

6.2.4 $\dfrac{\csc x - 1}{\cot x} = \dfrac{(\csc x - 1)(\csc x + 1)}{(\cot x)(\csc x + 1)}$

$= \dfrac{\csc^2 x - 1}{(\cot x)(\csc x + 1)}$

$= \dfrac{\cot^2 x}{(\cot x)(\csc x + 1)} = \dfrac{\cot x}{\csc x + 1}$

6.2.5 $\dfrac{\sec x - \cos x}{\cos x} = \dfrac{\sec x}{\cos x} - \dfrac{\cos x}{\cos x}$

$= \sec^2 x - 1 = \tan^2 x$

6.2.6 $\dfrac{1}{1 - \sec x} + \dfrac{1}{1 + \sec x}$

$= \dfrac{1(1 + \sec x) + 1(1 - \sec x)}{(1 - \sec x)(1 + \sec x)}$

$= \dfrac{2}{1 - \sec^2 x} = \dfrac{2}{-\tan^2 x}$

$= -2\cot^2 x$

6.2.7 $\dfrac{1 - \cos^2(-t)}{\sin(-t)} = \dfrac{1 - \cos^2(t)}{-\sin(t)}$

$= \dfrac{\sin^2(t)}{-\sin(t)} = -\sin t$

$\tan(-t)\cos(-t) = -\tan(t)\cos(t)$

$= -\dfrac{\sin t}{\cos t}\cos t = -\sin t$

Since both sides simplify to the same expression, the equation is an identity.

Section 3

6.3.1 $\cos(105°) = \cos(60° + 45°)$
$= \cos 60° \cos 45° - \sin 60° \sin 45°$
$= \dfrac{1}{2} \cdot \dfrac{\sqrt{2}}{2} - \dfrac{\sqrt{3}}{2} \cdot \dfrac{\sqrt{2}}{2} = \dfrac{\sqrt{2} - \sqrt{6}}{4}$

6.3.2 $\cos(75°) = \cos(135° - 60°)$
$= \cos 135° \cos 60° + \sin 135° \sin 60°$
$= -\dfrac{\sqrt{2}}{2} \cdot \dfrac{1}{2} + \dfrac{\sqrt{2}}{2} \cdot \dfrac{\sqrt{3}}{2} = \dfrac{\sqrt{6} - \sqrt{2}}{4}$

6.3.3 Use the result of 6.3.2:
$\sin(15°) = \cos(90° - 15°)$
$= \cos(75°) = \dfrac{\sqrt{6} - \sqrt{2}}{4}$

6.3.4 $\sin(75°) = \sin(30° + 45°)$
$$= \sin 30° \cos 45° + \cos 30° \sin 45°$$
$$= \frac{1}{2} \cdot \frac{\sqrt{2}}{2} + \frac{\sqrt{3}}{2} \cdot \frac{\sqrt{2}}{2} = \frac{\sqrt{2} + \sqrt{6}}{4}$$

6.3.5 $\tan(75°) = \tan(30° + 45°)$
$$= \frac{\tan 30° + \tan 45°}{1 - \tan 30° \tan 45°}$$
$$= \frac{\dfrac{\sqrt{3}}{3} + 1}{1 - \dfrac{\sqrt{3}}{3} \cdot 1} = \frac{\sqrt{3} + 3}{3 - \sqrt{3}}$$
$$= \frac{12 + 6\sqrt{3}}{6} = 2 + \sqrt{3}$$

6.3.6 $\sin \alpha \cos 3\alpha + \cos \alpha \sin 3\alpha$
$$= \sin(\alpha + 3\alpha) = \sin(4\alpha)$$

6.3.7 Since $\sin \alpha = 1/3$,
$$\left(\frac{1}{3}\right)^2 + \cos^2 \alpha = 1 \text{ or } \cos \alpha = -\frac{2\sqrt{2}}{3}.$$
Since $\cos \beta = -1/3$,
$$\sin^2 \beta + \left(-\frac{1}{3}\right)^2 = 1 \text{ or } \sin \beta = \frac{2\sqrt{2}}{3}.$$
$$\cos(\alpha - \beta) = \cos \alpha \cos \beta + \sin \alpha \sin \beta$$
$$= -\frac{2\sqrt{2}}{3}\left(-\frac{1}{3}\right) + \frac{1}{3} \cdot \frac{2\sqrt{2}}{3}$$
$$= \frac{4\sqrt{2}}{9}$$

Section 4

6.4.1 $\sin(60°) = 2\sin 30° \cos 30°$
$$= 2 \cdot \frac{1}{2} \cdot \frac{\sqrt{3}}{2} = \frac{\sqrt{3}}{2}$$

6.4.2 $\sin(-22.5°) = -\sqrt{\dfrac{1 - \cos(-45°)}{2}}$
$$= -\sqrt{\frac{1 - \dfrac{\sqrt{2}}{2}}{2}}$$
$$= -\frac{\sqrt{2 - \sqrt{2}}}{2}$$

6.4.3 $\cos 75° = \cos \dfrac{150°}{2}$
$$= \sqrt{\frac{1 + \cos 150°}{2}}$$
$$= \sqrt{\frac{1 + \left(-\dfrac{\sqrt{3}}{2}\right)}{2}}$$
$$= \sqrt{\frac{2 - \sqrt{3}}{4}} = \frac{\sqrt{2 - \sqrt{3}}}{2}$$

6.4.4 $\tan(-22.5°) = \dfrac{\sin(-45°)}{1 + \cos(-45°)}$
$$= \frac{-\dfrac{\sqrt{2}}{2}}{1 + \dfrac{\sqrt{2}}{2}} = -\frac{\sqrt{2}}{2 + \sqrt{2}}$$
$$= -\frac{2\sqrt{2} - 2}{2} = 1 - \sqrt{2}$$

6.4.5 Cosine is negative in quadrant II and positive in quadrant I.
Since $\sin \alpha = 1/3$, $\left(\dfrac{1}{3}\right)^2 + \cos^2 \alpha = 1$ or $\cos \alpha = -\dfrac{2\sqrt{2}}{3}$.
$$\cos \frac{\alpha}{2} = \sqrt{\frac{1 + \cos \alpha}{2}} = \sqrt{\frac{1 - \dfrac{2\sqrt{2}}{3}}{2}}$$
$$= \sqrt{\frac{3 - 2\sqrt{2}}{6}}$$

6.4.6 $\cos(3x) = \cos(x + 2x)$
$$= \cos x \cos 2x - \sin x \sin 2x$$
$$= \cos x (\cos^2 x - \sin^2 x) - \sin x(2\sin x \cos x)$$
$$= \cos^3 x - \cos x \sin^2 x - 2\sin^2 x \cos x$$
$$= \cos^3 x - 3\cos x \sin^2 x$$

6.4.7 $\sin^2\left(\dfrac{x}{2}\right) \cos^2\left(\dfrac{x}{2}\right)$
$$= \frac{1 - \cos x}{2} \cdot \frac{1 + \cos x}{2}$$
$$= \frac{1 - \cos^2 x}{4} = \frac{\sin^2 x}{4}$$

Section 5

6.5.1 $\sin(4x) \cos(3x) = \dfrac{1}{2}[\sin(7x) + \sin(x)]$

6.5.2 $\sin(52.5°) \cos(7.5°)$
$$= \frac{1}{2}[\sin(60°) + \sin(45°)]$$
$$= \frac{1}{2}\left(\frac{\sqrt{3}}{2} + \frac{\sqrt{2}}{2}\right) = \frac{\sqrt{3} + \sqrt{2}}{4}$$

6.5.3 $\cos(4x) + \cos(2x) = 2\cos 3x \cos x$

6.5.4 $\sin(105°) + \sin(15°)$
$$= 2\sin(60°) \cos(45°)$$
$$= 2 \cdot \frac{\sqrt{3}}{2} \cdot \frac{\sqrt{2}}{2} = \frac{\sqrt{6}}{2}$$

6.5.5 The terminal side of $3\pi/4$ goes through $(-2, 2)$. So
$$-2 \sin x + 2\cos x$$
$$= \sqrt{(-2)^2 + 2^2} \sin(x + 3\pi/4)$$
$$= 2\sqrt{2} \sin(x + 3\pi/4)$$

6.5.6 The terminal side of $-\pi/4$ contains $(1, -1)$.
So $y = \sin x - \cos x = \sqrt{2} \sin(x - \pi/4)$. The amplitude is $\sqrt{2}$ and the phase shift is $\pi/4$.

6.5.7 The amplitude is $\sqrt{a^2 + b^2}$ or $\sqrt{10}$. So the maximum distance that the block travels from rest is $\sqrt{10}$ cm.

Section 6

6.6.1 One solution is $x = \cos^{-1}\left(\sqrt{3}/2\right) = \pi/6$. Also, $\cos(11\pi/6) = \sqrt{3}/2$. So the solution set is $\{x | x = \pi/6 + 2k\pi \text{ or } x = 11\pi/6 + 2k\pi\}$, where k is any integer.

6.6.2 One solution is $x = \sin^{-1}(1/2) = \pi/6$. Also, $\sin(5\pi/6) = 1/2$. So the solution set is $\{x | x = \pi/6 + 2k\pi \text{ or } x = 5\pi/6 + 2k\pi\}$, where k is any integer.

6.6.3 Since $\alpha = \tan^{-1}\left(\sqrt{3}\right) = 60°$, the solution set is $\{x | x = 60° + k180°\}$, where k is any integer.

6.6.4 Since $2\alpha = \sin^{-1}(1) = 90°$, we have $2\alpha = 90° + k360°$. So the solution set is $\{\alpha | \alpha = 45° + k180°\}$, where k is any integer.

6.6.5 Since $4x = \tan^{-1}(1) = \pi/4$, $4x = \pi/4 + k\pi$ and $x = \pi/16 + k\pi/4$. The solutions between 0 and π occur if $k = 0, 1, 2, 3$. So the solution set is $\left\{\frac{\pi}{16}, \frac{5\pi}{16}, \frac{9\pi}{16}, \frac{13\pi}{16}\right\}$.

6.6.6
$$\sin(2x) = \cos(x)$$
$$2\sin x \cos x - \cos x = 0$$
$$\cos x(2\sin x - 1) = 0$$
$$\cos x = 0 \text{ or } \sin x = \frac{1}{2}$$
$$x = \frac{\pi}{2} + k\pi \text{ or } x = \frac{\pi}{6} + 2k\pi \text{ or } x = \frac{5\pi}{6} + 2k\pi$$

The only solutions in the interval $(0, 2\pi)$ are $\pi/6$, $\pi/2$, $5\pi/6$ and $3\pi/2$.

6.6.7 $2\sin^2(x) - \sin(x) - 1 = 0$
$$(\sin x - 1)(2\sin x + 1) = 0$$
$$\sin x = 1 \text{ or } \sin x = -\frac{1}{2}$$
$$x = \frac{\pi}{2} + 2k\pi \text{ or } x = \frac{7\pi}{6} + 2k\pi \text{ or } x = \frac{11\pi}{6} + 2k\pi$$

The only solutions in the interval $(0, 2\pi)$ are $\pi/2$, $7\pi/6$, and $11\pi/6$.

6.6.8 $\cos \alpha - \sin^2 \alpha = 0$
$$\cos \alpha - (1 - \cos^2 \alpha) = 0$$
$$\cos^2 \alpha + \cos \alpha - 1 = 0$$

Apply the quadratic formula: $\cos \alpha = \dfrac{-1 \pm \sqrt{5}}{2} \approx 0.6180 \text{ or } -1.6180$
Note that $\cos \alpha = -1.6180$ has no solutions. So $\alpha \approx 51.8° + k360°$ or $308.2° + k360°$. The only solutions in the interval $[0°, 360°)$ are $51.8°$ and $308.2°$.

6.6.9
$$\sin \alpha - \cos \alpha = \frac{1}{\sqrt{2}}$$
$$\sin^2 \alpha - 2\sin \alpha \cos \alpha + \cos^2 \alpha = \frac{1}{2}$$
$$1 - 2\sin \alpha \cos \alpha = \frac{1}{2}$$
$$2\sin \alpha \cos \alpha = \frac{1}{2}$$
$$\sin 2\alpha = \frac{1}{2}$$
$$2\alpha = 30° + k360° \text{ or } 150° + k360°$$
$$\alpha = 15° + k180° \text{ or } 75° + k180°$$

Only $15°$, $75°$, $195°$, and $255°$ are in the interval $[0°, 360°)$. Checking reveals that only $75°$ and $195°$ satisfy the original equation.

6.6.10 $3\sin t - 2\cos t = 0$
$$\frac{\sin t}{\cos t} = \frac{2}{3}$$
$$\tan t = \frac{2}{3}$$
$$t = \tan^{-1}\left(\frac{2}{3}\right) + k\pi$$

So $x = 0$ when $t \approx 0.588 + k\pi$, where k is a nonnegative integer.

6.6.11 Use $v_0^2 \sin 2\theta = 32d$
$$100^2 \sin 2\theta = 32(200)$$
$$\sin 2\theta = 0.64$$
Since $\sin^{-1}(0.64) \approx 39.8°$, $2\theta \approx 39.8° + k360°$ or $140.2° + k360°$ and $\theta \approx 19.9° + k180°$ or $70.1° + k180°$. Since the launch angle must be acute, the only possibilities are $19.9°$ and $70.1°$.

Chapter 7

Section 1

7.1.1 Note that c is the side between α and β and that $\gamma = 86°$.
$$\frac{a}{\sin 28°} = \frac{b}{\sin 66°} = \frac{8.2}{\sin 86°}$$
$$a = \sin 28° \cdot \frac{8.2}{\sin 86°} \approx 3.9$$
$$b = \sin 66° \cdot \frac{8.2}{\sin 86°} \approx 7.5$$

7.1.2 First draw angle β with vertex B and mark off side c on one ray of the angle. The other endpoint of c is vertex A. Next find the altitude from A to the opposite side.
$$h = 5.9 \sin 38° \approx 3.6$$
Since $b = 2.9$ and that is less than 3.6, side b cannot reach from A to the opposite side. So there is no triangle with these parts.

7.1.3 First draw angle β with vertex B and mark off side c on one ray of the angle. The other endpoint of c is vertex A. Since b is larger than c, there is one obtuse triangle that is formed.
$$\frac{a}{\sin \alpha} = \frac{6.4}{\sin 38°} = \frac{5.9}{\sin \gamma}$$
$$\sin \gamma = \sin 38° \cdot \frac{5.9}{6.4}$$
$$\gamma = \sin^{-1}\left(\sin 38° \cdot \frac{5.9}{6.4}\right) \approx 34.6°$$
Now $\alpha \approx 107.4°$.
$$a = \sin 107.4° \cdot \frac{6.4}{\sin 38°} \approx 9.9$$

7.1.4 First draw angle β with vertex B and mark off side c on one ray of the angle. The other endpoint of c is vertex A. Next find the altitude from A to the opposite side.
$$h = 5.9 \sin 38° \approx 3.6$$
Since $b = 4.7$ and that is larger than 3.6 and smaller than 5.9, it will reach from A to the opposite side in two places. So there are two triangles.
$$\frac{a}{\sin \alpha} = \frac{4.7}{\sin 38°} = \frac{5.9}{\sin \gamma}$$
$$\sin \gamma = \sin 38° \cdot \frac{5.9}{4.7}$$
Since $\sin^{-1}\left(\sin 38° \cdot \frac{5.9}{4.7}\right) \approx 50.6°$, γ is $50.6°$ or $129.4°$. Now if $\gamma = 50.6°$, then $\alpha \approx 91.4°$ and
$$a = \sin 91.4° \cdot \frac{4.7}{\sin 38°} \approx 7.6.$$
For the obtuse triangle in which $\gamma = 129.4°$, we have $\alpha \approx 12.6°$ and
$$a = \sin 12.6° \cdot \frac{4.7}{\sin 38°} \approx 1.7.$$

7.1.5 A $= \dfrac{1}{2} \cdot 244 \cdot 206 \cdot \sin 87.4°$
$$\approx 25{,}106 \text{ square feet}$$

7.1.6 The triangle is a 30-60-90 triangle in which the side opposite 60° is 88. The maximum distance from the airport is side c, the hypotenuse. Using right triangle ratios, $\sin 60° = 88/c$ or $c \approx 101.6$ mi.

7.1.7 Let C be the point at the top of the building and D be the point on the ground directly below C. In ΔABC, $m\angle A = 24.2°$, $m\angle B = 141.9°$, and $m\angle C = 13.9°$. Find a:

$$\frac{a}{\sin 24.2°} = \frac{44.5}{\sin 13.9°}$$
$$a \approx 75.9$$

In right triangle BCD, $\sin 38.1° = CD/a$ or $CD = a \sin 38.1 \approx 46.9$ feet.

Section 2

7.2.1 Use the law of cosines to find the largest angle:
$$9.6^2 = 3.8^2 + 7.7^2 - 2(3.8)(7.7)\cos\beta$$
$$\beta = \cos^{-1}\left(\frac{9.6^2 - 3.8^2 - 7.7^2}{-2(3.8)(7.7)}\right) \approx 108.4°$$

Use the law of sines to find the next angle:
$$\frac{3.8}{\sin\alpha} = \frac{9.6}{\sin 108.4°}$$
$$\alpha = \sin^{-1}\left(\frac{3.8\sin 108.4°}{9.6}\right) \approx 22.1°$$

Now subtract α and β from $180°$ to get $\gamma \approx 49.5°$.

7.2.2 Use the law of cosines to find a:
$$a^2 = 5.8^2 + 3.6^2 - 2(5.8)(3.6)\cos 39.5°$$
$$a \approx 3.8$$

7.2.3 $S = \dfrac{12 + 8 + 6}{2} = 13$
$$A = \sqrt{13(13-12)(13-8)(13-6)}$$
$$= \sqrt{455} \approx 21.3$$

7.2.4 $a = r\sqrt{2 - 2\cos\alpha}$
$$= 22.4\sqrt{2 - 2\cos 33.8°}$$
$$\approx 13.0\,\text{feet}$$

7.2.5 Use the distance formula to find the lengths of the sides:
$AB = \sqrt{26}$, $BC = \sqrt{53}$, and $AC = \sqrt{73}$.

Use the law of cosines to find the largest angle:
$$73 = 53 + 26 - 2\sqrt{53}\sqrt{26}\cos\beta$$
$$\beta \approx 85.4°$$

Use the law of sines to find α:
$$\frac{\sqrt{53}}{\sin\alpha} = \frac{\sqrt{73}}{\sin 85.4°}$$
$$\alpha \approx 58.1°$$
$$\gamma \approx 36.5°$$

Section 3

7.3.1 $|\mathbf{v_x}| = |5.6\cos 22°| \approx 5.2$
$|\mathbf{v_y}| = |5.6\sin 22°| \approx 2.1$

7.3.2 First find the magnitude:
$$|\mathbf{v}| = \sqrt{2^2 + (-6)^2} = 2\sqrt{10}$$

Use trigonometric ratios to find the direction angle θ:
$$\sin^{-1}\left(\frac{-6}{2\sqrt{10}}\right) \approx -71.6°$$

Since θ must be in $[0°, 360°)$, $\theta \approx -71.6° + 360° = 288.4°$

7.3.3 For terminal point (a, b)
$$a = r\cos\theta = 50\cos(120°)$$
$$= 50(-1/2) = -25$$
$$b = r\sin\theta = 50\sin(120°)$$
$$= 50(\sqrt{3}/2) = 25\sqrt{3}.$$

So the component form is $\langle -25, 25\sqrt{3}\rangle$.

7.3.4 $\mathbf{u} + 3\mathbf{v} = \langle -1, -3\rangle + 3\langle 3, -4\rangle = \langle 8, -15\rangle$
$\mathbf{u} \cdot \mathbf{v} = \langle -1, -3\rangle \cdot \langle 3, -4\rangle = -3 + 12 = 9$

7.3.5 $\cos\alpha = \dfrac{\langle 1, 3\rangle \cdot \langle 5, 2\rangle}{|\langle 1, 3\rangle| \cdot |\langle 5, 2\rangle|}$
$$= \frac{11}{\sqrt{10}\sqrt{29}}$$
$$\alpha \approx 49.8°$$

7.3.6 $\langle -1, 7\rangle = -1\langle 1, 0\rangle + 7\langle 0, 1\rangle$
$$= -1\mathbf{i} + 7\mathbf{j}$$

7.3.7 Let $\mathbf{v}$ be the resultant.
$$|\mathbf{v}|^2 = 100^2 + 200^2 - 2 \cdot 100 \cdot 200\cos 150°$$
$$|\mathbf{v}| = 290.9 \text{ pounds}$$

7.3.8 Draw a figure and label it as in Example 8.
$$|\mathbf{BD}| = |\mathbf{AB}|\sin 10° = 800\sin 10°$$
$$\approx 138.9 \text{ pounds}$$

7.3.9 Draw a figure and label it as in Example 9. Heading northwest is $315°$. The vector $\mathbf{v_3}$ is the course. The angle between $\mathbf{v_1}$ and $\mathbf{v_2}$ is $315° - 200°$ or $115°$. The angle at point D is $180° - 115°$ or $65°$.
$$|\mathbf{v_3}|^2 = 100^2 + 500^2 - 2 \cdot 100 \cdot 500\cos 65°$$
$$|\mathbf{v_3}| = 466.6 \text{ mph (the ground speed)}$$

If θ is the drift angle, then $\dfrac{\sin\theta}{100} = \dfrac{\sin 65°}{466.6}$

or $\theta \approx 11.2°$. So the bearing of the course is $315° - 11.2°$ or $303.8°$.

Section 4

7.4.1 $|5 - i| = \sqrt{5^2 + (-1)^2} = \sqrt{26}$

7.4.2 For $z = 1 + 2i$, $r = \sqrt{1^2 + 2^2} = \sqrt{5}$. From $a = r\cos\theta$ we get $1 = \sqrt{5}\cos\theta$ or $\theta \approx 63.4°$. Since the terminal side of $63.4°$ goes through $(1, 2)$, $z \approx \sqrt{5}(\cos 63.4° + i\sin 63.4°)$.

7.4.3 $12\left(\cos\dfrac{\pi}{6} + i\sin\dfrac{\pi}{6}\right)$
$$= 12\left(\frac{\sqrt{3}}{2} + i\frac{1}{2}\right) = 6\sqrt{3} + 6i$$

7.4.4 Multiply the moduli and add the arguments:
$$z_1 z_2 = 32\left(\cos\frac{\pi}{6} + i\sin\frac{\pi}{6}\right)$$
$$= 32\left(\frac{\sqrt{3}}{2} + i\frac{1}{2}\right) = 16\sqrt{3} + 16i$$

7.4.5 For $z_1 = 2\sqrt{3} + 2i$, $r = \sqrt{12 + 4} = 4$. So $a = r\cos\theta$ yields $\cos\theta = \sqrt{3}/2$ and $\theta = 30°$. For $z_2 = 3 + 3i\sqrt{3}$, $r = \sqrt{9 + 27} = 6$. So $a = r\cos\theta$ yields $\cos\theta = 1/2$ and $\theta = 60°$. So
$$z_1 = 4(\cos 30° + i\sin 30°),$$
$$z_2 = 6(\cos 60° + i\sin 60°),$$
and
$$z_1 z_2 = 24(\cos 90° + i\sin 90°)$$
$$= 24(0 + i) = 24i$$

7.4.6 The conjugate of $6(\cos 30° + i\sin 30°)$ is $6(\cos(-30°) + i\sin(-30°))$. The product is $36(\cos 0° + i\sin 0°)$ or $36(1 + 0i) = 36$.

Section 5

7.5.1 $1 + i = \sqrt{2}(\cos 45° + i\sin 45°)$
$$(1 + i)^6 = (\sqrt{2})^6(\cos(270°) + i\sin(270°))$$
$$= 8(0 + (-1)i) = -8i$$

7.5.2 Write $-8 - 8i\sqrt{3}$ in trigonometric form as $16(\cos 240° + i \sin 240°)$. So

$$2\left(\cos\left(\frac{240° + k360°}{4}\right) + i \sin\left(\frac{240° + k360°}{4}\right)\right)$$

generates the fourth roots for $k = 0, 1, 2, 3$:

$$2(\cos(60°) + i \sin(60°)) = 1 + i\sqrt{3}$$
$$2(\cos(150°) + i \sin(150°)) = -\sqrt{3} + i$$
$$2(\cos(240°) + i \sin(240°)) = -1 - i\sqrt{3}$$
$$2(\cos(330°) + i \sin(330°)) = \sqrt{3} - i$$

7.5.3 The solutions are the three cube roots of 27, where $27 = 27(\cos 0° + i \sin 0°)$. The roots are given by

$$27^{1/3}\left(\cos\left(\frac{0° + k360°}{3}\right) + i\sin\left(\frac{0° + k360°}{3}\right)\right)$$

where $k = 0, 1, 2$:

$$3(\cos 0° + i \sin 0°) = 3$$
$$3(\cos 120° + i \sin 120°) = -\frac{3}{2} + \frac{3\sqrt{3}}{2}i$$
$$3(\cos 240° + i \sin 240°) = -\frac{3}{2} - \frac{3\sqrt{3}}{2}i$$

Section 6

7.6.1

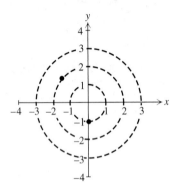

7.6.2 $x = 3\cos 45° = 3\sqrt{2}/2$
$y = 3\sin 45° = 3\sqrt{2}/2$

So $(3, 45°)$ in rectangular coordinates is $(3\sqrt{2}/2, 3\sqrt{2}/2)$.
For $(-2, 2\sqrt{3})$, $r = \sqrt{4 + 12} = 4$ and $\theta = \cos^{-1}(-2/4) = 120°$.
So $(-2, 2\sqrt{3})$ in polar coordinates is $(4, 120°)$.

7.6.3 Some pairs that satisfy $r = 4\sin\theta$ are $(0°, 0)$, $(45°, 2\sqrt{2})$, $(90°, 4)$, $(135°, 2\sqrt{2})$, and $(180°, 0)$.

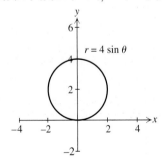

7.6.4 Some pairs that satisfy $r = \cos 2\theta$ are $(0°, 1)$, $(22.5°, \sqrt{2}/2)$, $(45°, 0)$, $(67.5°, -\sqrt{2}/2)$, and $(90°, -1)$.

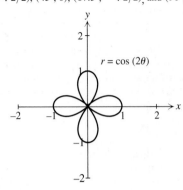

7.6.5 Some pairs that satisfy $r = -\theta$ are $(0, 0)$, $(\pi/4, -\pi/4)$, $(\pi/2, -\pi/2)$, $(3\pi/4, -3\pi/4)$, and $(\pi, -\pi)$.

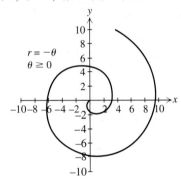

7.6.6 $r = 3\sin\theta$
$r^2 = 3r\sin\theta$
$x^2 + y^2 = 3y$

This rectangular equation can be rewritten in the standard form for a circle:
$$x^2 + y^2 - 3y = 0$$
$$x^2 + (y - 3/2)^2 = 9/4$$

7.6.7
$$y = -2x + 5$$
$$r\sin\theta = -2r\cos\theta + 5$$
$$r\sin\theta + 2r\cos\theta = 5$$
$$r(\sin\theta + 2\cos\theta) = 5$$
$$r = \frac{5}{\sin\theta + 2\cos\theta}$$

Section 7

7.7.1 The graph is a line segment with endpoints $(5, -1)$ corresponding to $t = 0$ and $(10, 9)$ corresponding to $t = 5$.

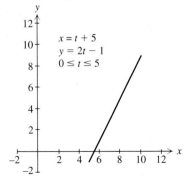

7.7.2 If $y = -t + 1$, then $t = 1 - y$. Substitute $t = 1 - y$ into $x = 4t - 9$ to get $x = 4(1 - y) - 9$ or $x + 4y = -5$ which is a straight line. The domain is $(-\infty, \infty)$ and the range is $(-\infty, \infty)$.

7.7.3 If $x = mt + b$, then $1 = m \cdot 0 + b$ and $8 = m \cdot 1 + b$ yields $b = 1$ and $m = 7$. So $x = 7t + 1$. If $y = mt + b$, then $2 = m \cdot 0 + b$ and $10 = m \cdot 1 + b$ yields $b = 2$ and $m = 8$. So $y = 8t + 2$.

7.7.4 Let $r = 3\cos\theta$ in $x = r\cos\theta$ and $y = r\sin\theta$ to get $x = 3\cos^2\theta$ and $y = 3\cos\theta\sin\theta$. Graph the parametric equations to see that the graph is a circle with diameter 3.

Chapter 8

Section 1

8.1.1 Graph $y = x - 3$ through $(0, -3)$ and $(1, -2)$. Graph $x + y = 7$ through $(0, 7)$ and $(7, 0)$.

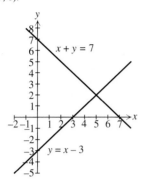

Check $(5, 2)$ in both equations to be sure that the solution set is $\{(5, 2)\}$.

8.1.2 Substitute $y = 2x - 3$ into $x + 2y = -1$:
$$x + 2(2x - 3) = -1$$
$$5x - 6 = -1$$
$$5x = 5$$
$$x = 1$$
$$y = 2(1) - 3 = -1$$
The solution set is $\{(1, -1)\}$.

8.1.3 Substitute $y = 3x - 5$ into $6x - 2y = 1$:
$$6x - 2(3x - 5) = 1$$
$$10 = 1$$
The system is inconsistent and has no solution.

8.1.4 Multiply $x + y = 3$ by 2 to get $2x + 2y = 6$, then add:
$$\begin{array}{r} 2x + 2y = 6 \\ 3x - 2y = 4 \\ \hline 5x = 10 \\ x = 2 \end{array}$$
If $x = 2$ and $x + y = 3$, then $2 + y = 3$ and $y = 1$. The solution set is $\{(2, 1)\}$.

8.1.5 Multiply $\frac{1}{2}x - \frac{1}{4}y = 1$ by 4 to get $2x - y = 4$. Multiply $2x - y = 3$ by -1, then add:
$$\begin{array}{r} 2x - y = 4 \\ -2x + y = -3 \\ \hline 0 = 1 \end{array}$$
The system is inconsistent and has no solution.

8.1.6 If d is the cost of a DVD and c is the cost of a CD, then $2d + 3c = 78$ and $d + 4c = 74$. Substitute $d = 74 - 4c$ into the first equation:
$$2(74 - 4c) + 3c = 78$$
$$148 - 8c + 3c = 78$$
$$-5c = -70$$
$$c = 14$$
$$d = 74 - 4(14) = 18$$
The cost of a DVD is $18.

Section 2

8.2.1 The intercepts are $(0, 0, 6)$, $(0, 3, 0)$, and $(6, 0, 0)$.

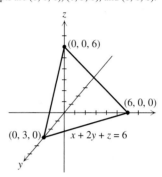

8.2.2 Eliminate y from the first and second and then the second and third equations:
$$\begin{array}{r} x + y + z = 9 \\ x - y + 2z = 1 \\ \hline 2x + 3z = 10 \end{array} \qquad \begin{array}{r} x - y + 2z = 1 \\ x + y - z = 5 \\ \hline 2x + z = 6 \end{array}$$

$$\begin{array}{r} 2x + 3z = 10 \\ -2x - z = -6 \\ \hline 2z = 4 \\ z = 2 \end{array}$$
If $z = 2$, then $2x + 2 = 6$ and $x = 2$. If $z = 2$ and $x = 2$, then $2 + y + 2 = 9$ and $y = 5$. The solution set is $\{(2, 5, 2)\}$.

8.2.3
$$\begin{array}{r} x + y + z = 2 \\ x - 2y - z = 4 \\ \hline 2x - y = 6 \\ y = 2x - 6 \end{array}$$

$$\begin{aligned} z &= 2 - x - y \\ &= 2 - x - (2x - 6) \\ &= -3x + 8 \end{aligned}$$
The solution set is $\{(x, 2x - 6, -3x + 8) \mid x \text{ is any real number}\}$.

8.2.4
$$\begin{array}{r} x + y + z = 1 \\ x - y - z = 3 \\ \hline 2x = 4 \\ x = 2 \end{array}$$
Replacing x with 2 in any of the three equations yields $y + z = -1$ or $z = -y - 1$. So the solution set is $\{(2, y, -y - 1) \mid y \text{ is any real number}\}$.

8.2.5
$$\begin{array}{r} x + y + z = 1 \\ x - y - z = 3 \\ \hline 2x = 4 \\ x = 2 \end{array}$$
Replacing x with 2 in the first equation yields $y + z = -1$ and in the third equation $y + z = 1$. Substitution yields $-1 = 1$. So there is no solution to the system.

8.2.6 Using the form $y = ax^2 + bx + c$ and the three points yields the following system.

$$a - b + c = 1 \quad (1)$$
$$a + b + c = 3 \quad (2)$$
$$4a + 2b + c = 7 \quad (3)$$

Subtracting (1) from (2) yields $2b = 2$ or $b = 1$. Use $b = 1$ in (1) and (3) to get $a + c = 2$ and $4a + c = 5$. Subtracting yields $3a = 3$ or $a = 1$. If $a = 1$, then $c = 1$ and $y = x^2 + x + 1$.

8.2.7
$$2a + s + c = 14 \quad (1)$$
$$a + 2s + c = 12 \quad (2)$$
$$2a + 2s + 3c = 19 \quad (3)$$

Subtract twice (2) from (1) and (3) from (1):

$$
\begin{array}{r}
2a + s + c = 14 \\
2a + 4s + 2c = 24 \\
\hline
-3s - c = -10
\end{array}
\qquad
\begin{array}{r}
2a + s + c = 14 \\
2a + 2s + 3c = 19 \\
\hline
-s - 2c = -5
\end{array}
$$

Multiply $-3s - c = -10$ by -2 and add:

$$
\begin{array}{r}
6s + 2c = 20 \\
-s - 2c = -5 \\
\hline
5s \quad\quad = 15 \\
s = 3
\end{array}
$$

If $s = 3$, then $6(3) + 2c = 20$ or $c = 1$. If $s = 3$ and $c = 1$, then $2a + 3 + 1 = 14$ or $a = 5$. So the admissions are adults \$5, students \$3, and children \$1.

Section 3

8.3.1 Substitute $y = x^2 - 1$ into $x + y = 5$:

$$x + x^2 - 1 = 5$$
$$x^2 + x - 6 = 0$$
$$(x + 3)(x - 2) = 0$$
$$x = -3 \quad \text{or} \quad x = 2$$
$$y = 8 \quad \text{or} \quad y = 3$$

The solution set to the system is $\{(-3, 8), (2, 3)\}$.

8.3.2 Substitution yields $|2x| = -x^2 + 2x + 5$, which is equivalent to $2x = -x^2 + 2x + 5$ or $2x = -(-x^2 + 2x + 5)$. Solve each of these equations:

$$2x = -x^2 + 2x + 5$$
$$x^2 = 5$$
$$x = \pm\sqrt{5}$$

$$2x = -(-x^2 + 2x + 5)$$
$$x^2 - 4x - 5 = 0$$
$$(x - 5)(x + 1) = 0$$
$$x = 5 \text{ or } x = -1$$

Only $\sqrt{5}$ and -1 yield points that satisfy both systems. So the solution set is $\left\{(-1, 2), \left(\sqrt{5}, 2\sqrt{5}\right)\right\}$.

8.3.3 Multiplying the first equation by -4 and adding yields $5y^2 = 15$ or $y = \pm\sqrt{3}$. If $y = \pm\sqrt{3}$, then $x^2 + 3 = 5$ or $x = \pm\sqrt{2}$. There are four ordered pairs in the solution set: $\left(\sqrt{2}, \sqrt{3}\right)$, $\left(\sqrt{2}, -\sqrt{3}\right)$, $\left(-\sqrt{2}, \sqrt{3}\right)$, and $\left(-\sqrt{2}, -\sqrt{3}\right)$.

8.3.4 Adding the equations yields $\frac{2}{x} = 6$ or $x = 1/3$. If $x = 1/3$, then $3 + \frac{1}{y} = 2$ or $y = -1$. So the solutions set is $\{(1/3, -1)\}$.

8.3.5 $L^2 + W^2 = 169$ and $LW = 60$. Substitute $W = 60/L$ into the first equation:

$$L^2 + \frac{3600}{L^2} = 169$$
$$L^4 + 3600 = 169L^2$$
$$L^4 - 169L^2 + 3600 = 0$$
$$(L^2 - 25)(L^2 - 144) = 0$$
$$L^2 = 25 \text{ or } L^2 = 144$$
$$L = \pm5 \text{ or } L = \pm12$$

So the length is 12 ft and the width is 5 ft.

Section 4

8.4.1
$$\frac{4}{x - 2} + \frac{3}{x + 5}$$
$$= \frac{4(x + 5)}{(x - 2)(x + 5)} + \frac{3(x - 2)}{(x + 5)(x - 2)}$$
$$= \frac{7x + 14}{(x + 5)(x - 2)}$$

8.4.2 $\dfrac{2x - 3}{(x - 1)(x + 3)} = \dfrac{A}{x - 1} + \dfrac{B}{x + 3}$

Multiply each side by $(x - 1)(x + 3)$:

$$2x - 3 = A(x + 3) + B(x - 1)$$
$$2x - 3 = (A + B)x + 3A - B$$

So $A + B = 2$ and $3A - B = -3$.

$$
\begin{array}{r}
A + B = 2 \\
3A - B = -3 \\
\hline
4A \quad\quad = -1 \\
A = -1/4
\end{array}
$$

If $A = -1/4$, then $-1/4 + B = 2$ and $B = 9/4$.

$$\frac{2x - 3}{(x - 1)(x + 3)} = \frac{-1/4}{x - 1} + \frac{9/4}{x + 3}$$

8.4.3 $\dfrac{4x^2 + 4x - 4}{(x + 1)^2(x - 1)}$
$$= \frac{A}{x + 1} + \frac{B}{(x + 1)^2} + \frac{C}{x - 1}$$

Multiply each side by $(x + 1)^2(x - 1)$:

$$4x^2 + 4x - 4$$
$$= A(x^2 - 1) + B(x - 1) + C(x + 1)^2$$
$$= (A + C)x^2 + (B + 2C)x + (-A - B + C)$$

$$A + C = 4$$
$$B + 2C = 4$$
$$-A - B + C = -4$$

Substitute $A = 4 - C$ and $B = 4 - 2C$ into the last equation:

$$-(4 - C) - (4 - 2C) + C = -4$$
$$4C = 4$$
$$C = 1$$

If $C = 1$, then $B = 2$ and $A = 3$ and

$$\frac{4x^2 + 4x - 4}{(x + 1)^2(x - 1)}$$
$$= \frac{3}{x + 1} + \frac{2}{(x + 1)^2} + \frac{1}{x - 1}.$$

8.4.4 $\dfrac{5x^2 - 4x + 11}{(x^2 + 2)(x - 1)}$

$= \dfrac{Ax + B}{x^2 + 2} + \dfrac{C}{x - 1}$

Multiply each side by $(x^2 + 2)(x - 1)$:

$5x^2 - 4x + 11$
$= (Ax + B)(x - 1) + C(x^2 + 2)$
$= (A + C)x^2 + (-A + B)x + (-B + 2C)$

$A + C = 5 \quad (1)$
$-A + B = -4 \quad (2)$
$-B + 2C = 11 \quad (3)$

Adding (1) and (2) yields $B + C = 1$. Add this result and (3) to get $3C = 12$ or $C = 4$. If $C = 4$, then $A = 1$ and $B = -3$. So

$$\frac{5x^2 - 4x + 11}{(x^2 + 2)(x - 1)} = \frac{x - 3}{x^2 + 2} + \frac{4}{x - 1}.$$

8.4.5 $\dfrac{3x^3 + 3x^2 + x - 2}{(3x^2 - 1)^2}$

$= \dfrac{Ax + B}{3x^2 - 1} + \dfrac{Cx + D}{(3x^2 - 1)^2}$

Multiply each side by $(3x^2 - 1)^2$:

$3x^3 + 3x^2 + x - 2$
$= (Ax + B)(3x^2 - 1) + Cx + D$
$= 3Ax^3 + 3Bx^2 + (-A + C)x + (-B + D)$

$3A = 3 \quad (1)$
$3B = 3 \quad (2)$
$-A + C = 1 \quad (3)$
$-B + D = -2 \quad (4)$

Since $3A = 3$, we have $A = 1$ and $C = 2$. Since $3B = 3$, we have $B = 1$ and $D = -1$. So

$$\frac{3x^3 + 3x^2 + x - 2}{(3x^2 - 1)^2} = \frac{x + 1}{3x^2 - 1} + \frac{2x - 1}{(3x^2 - 1)^2}.$$

Section 5

8.5.1 Graph the solid line $y = -2x + 4$ through $(0, 4)$ and $(2, 0)$. Shade below the line for $y \le -2x + 4$.

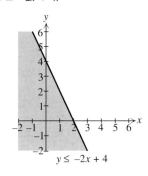

8.5.2 Graph the dashed line $2x - 5y = 20$ through $(0, -4)$ and $(10, 0)$. Since $(0, 0)$ satisfies $2x - 5y < 20$, shade the region containing $(0, 0)$.

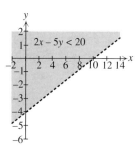

8.5.3 Draw the v-shaped graph of $y = |x|$. Then shade below for $y < |x|$.

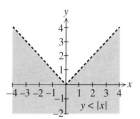

8.5.4 Graph dashed lines for $y = x$ and $y = 3 - x$. Test a point in each of the four regions. The only region that satisfies both inequalities is the region containing $(5, 0)$.

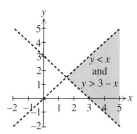

8.5.5 Draw a v-shaped graph through $(\pm 1, 0)$ and $(0, 1)$ for $y = 1 - |x|$. Draw a parabola through $(\pm 1, 0)$ and $(0, -1)$ for $y = x^2 - 1$. The inequality is satisfied only in the region containing $(0, 0)$.

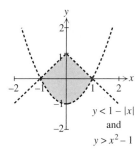

8.5.6 Graph the horizontal dashed line $y = 3$, the vertical dashed line $x = 2$, and the dashed line $y = -x$ through $(0, 0)$. The only region that satisfies all three inequalities is the region containing $(0, 2)$.

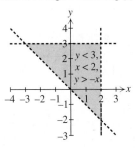

Section 6

8.6.1 Graph $x + y = 6$ through $(0, 6)$ and $(6, 0)$. Graph $x + 2y = 8$ through $(0, 4)$ and $(8, 0)$. The region that satisfies all inequalities is below both of these lines and in the first quadrant. The vertices are $(0, 0)$, $(0, 4)$, $(4, 2)$, and $(6, 0)$.

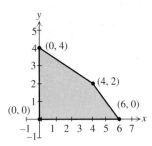

8.6.2 Let x be the number of X-components and y be the number of Y-components. The inequalities are
$x \geq 0, y \geq 0, 4x + 8y \leq 40$, and $6x + 2y \leq 30$, or
$x \geq 0, y \geq 0, x + 2y \leq 10$, and $3x + y \leq 15$.

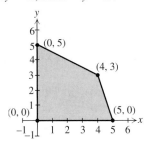

8.6.3 The revenue in dollars is given by $R(x, y) = 2x + 3y$. Find R at each vertex of the region graphed in the solution to 8.6.2. Since $R(0, 0) = 0$, $R(0, 5) = 15$, $R(4, 3) = 17$, and $R(5, 0) = 10$, the maximum revenue occurs when four X-components and three Y-components are assembled.

Chapter 9

Section 1

9.1.1 Since the matrix has 2 rows and 4 columns its size is 2×4.

9.1.2 Write the coefficients and the constants in a matrix as
$$\begin{bmatrix} 1 & -1 & | & -2 \\ 2 & 1 & | & 3 \end{bmatrix}.$$

9.1.3 Using the first two columns for the coefficients and the last column for the constants, the system is $\begin{array}{r} 2x + y = 5 \\ -x + 4y = 6 \end{array}$.

9.1.4 $\begin{bmatrix} 2 & -1 & | & 7 \\ 1 & 3 & | & 14 \end{bmatrix}$

$\begin{bmatrix} 1 & 3 & | & 14 \\ 2 & -1 & | & 7 \end{bmatrix} \quad R_1 \leftrightarrow R_2$

$\begin{bmatrix} 1 & 3 & | & 14 \\ 0 & -7 & | & -21 \end{bmatrix} \quad -2R_1 + R_2 \to R_2$

$\begin{bmatrix} 1 & 3 & | & 14 \\ 0 & 1 & | & 3 \end{bmatrix} \quad -\frac{1}{7}R_2 \to R_2$

$\begin{bmatrix} 1 & 0 & | & 5 \\ 0 & 1 & | & 3 \end{bmatrix} \quad -3R_2 + R_1 \to R_1$

The solution set to the system is $\{(5, 3)\}$.

9.1.5 Multiply the first row by 2 and add to the second.
$$\begin{bmatrix} 3 & -1 & | & 1 \\ -6 & 2 & | & 4 \end{bmatrix} \quad \begin{bmatrix} 3 & -1 & | & 1 \\ 0 & 0 & | & 6 \end{bmatrix}$$
The second row corresponds to the equation $0 = 6$. So the system is inconsistent and there is no solution.

9.1.6 Multiply the first row by -2 and add to the second.
$$\begin{bmatrix} 4 & 1 & | & 5 \\ 8 & 2 & | & 10 \end{bmatrix} \quad \begin{bmatrix} 4 & 1 & | & 5 \\ 0 & 0 & | & 0 \end{bmatrix}$$
The second row corresponds to the equation $0 = 0$. So the system is dependent. The solution set is $\{(x, y) \,|\, 4x + y = 5\}$ or $\{(x, 5 - 4x) \,|\, x$ is any real number$\}$.

9.1.7 $\begin{bmatrix} 1 & 1 & -1 & | & 0 \\ -1 & 1 & 0 & | & 4 \\ -1 & 1 & 1 & | & 10 \end{bmatrix}$

$\begin{bmatrix} 1 & 1 & -1 & | & 0 \\ 0 & 2 & -1 & | & 4 \\ 0 & 2 & 0 & | & 10 \end{bmatrix} \quad \begin{array}{l} R_1 + R_2 \to R_2 \\ R_1 + R_3 \to R_3 \end{array}$

$\begin{bmatrix} 1 & 1 & -1 & | & 0 \\ 0 & 2 & 0 & | & 10 \\ 0 & 2 & -1 & | & 4 \end{bmatrix} \quad R_2 \leftrightarrow R_3$

$\begin{bmatrix} 1 & 1 & -1 & | & 0 \\ 0 & 1 & 0 & | & 5 \\ 0 & 2 & -1 & | & 4 \end{bmatrix} \quad \frac{1}{2}R_2 \to R_2$

$\begin{bmatrix} 1 & 0 & -1 & | & -5 \\ 0 & 1 & 0 & | & 5 \\ 0 & 0 & -1 & | & -6 \end{bmatrix} \quad \begin{array}{l} -R_2 + R_1 \to R_1 \\ -2R_2 + R_3 \to R_3 \end{array}$

$\begin{bmatrix} 1 & 0 & 0 & | & 1 \\ 0 & 1 & 0 & | & 5 \\ 0 & 0 & 1 & | & 6 \end{bmatrix} \quad \begin{array}{l} -R_3 + R_1 \to R_1 \\ -R_3 \to R_3 \end{array}$

The solution set is $\{(1, 5, 6)\}$.

9.1.8 $\begin{bmatrix} 1 & -1 & 2 & | & 3 \\ -1 & 2 & 1 & | & 2 \end{bmatrix}$

$\begin{bmatrix} 1 & -1 & 2 & | & 3 \\ 0 & 1 & 3 & | & 5 \end{bmatrix} \quad R_1 + R_2 \to R_2$

$\begin{bmatrix} 1 & 0 & 5 & | & 8 \\ 0 & 1 & 3 & | & 5 \end{bmatrix} \quad R_2 + R_1 \to R_1$

So $x + 5z = 8$ and $y + 3z = 5$. So the solution set is
$\{(8 - 5z, 5 - 3z, z) \mid z \text{ is any real number}\}$. In terms of x the solution set
is $\left\{\left(x, \frac{3x + 1}{5}, \frac{8 - x}{5}\right) \mid x \text{ is any real number}\right\}$.

9.1.9 $3S + 8L = 30$
$\quad\quad 5S + 7L = 31$

$$\begin{bmatrix} 3 & 8 & | & 30 \\ 5 & 7 & | & 31 \end{bmatrix}$$

Use Gaussian elimination to get the following equivalent augmented
matrix.

$$\begin{bmatrix} 1 & 0 & | & 2 \\ 0 & 1 & | & 3 \end{bmatrix}$$

So there were 2 small and 3 large baskets made.

Section 2

9.2.1 If the matrices are equal, then the corresponding entries are equal. So
$x = 1, y + 1 = 2$ or $y = 1$, and $z + 1 = 2z - 1$ or $z = 2$.

9.2.2 Add the corresponding entries to get

$$A + B = \begin{bmatrix} 8 & 0 \\ 0 & 9 \end{bmatrix}.$$

9.2.3 If $A = \begin{bmatrix} 2 & 3 \\ -4 & 1 \end{bmatrix}$, then

$$-A = \begin{bmatrix} -2 & -3 \\ 4 & -1 \end{bmatrix} \text{ and } -A + A = \begin{bmatrix} 0 & 0 \\ 0 & 0 \end{bmatrix}.$$

9.2.4 Subtract the corresponding entries to get $A - B = \begin{bmatrix} 0 & -4 & 1 \end{bmatrix}$.

9.2.5 Multiply each entry by 4 to get
$4C = \begin{bmatrix} 4 & -8 & 36 \end{bmatrix}$.

9.2.6 The matrix $\begin{bmatrix} 11{,}000 \\ 20{,}000 \\ 16{,}000 \end{bmatrix}$ represents the amount of mail in each class.

Multiply by 1.5 to get a 50% increase:

$$1.5 \begin{bmatrix} 11{,}000 \\ 20{,}000 \\ 16{,}000 \end{bmatrix} = \begin{bmatrix} 16{,}500 \\ 30{,}000 \\ 24{,}000 \end{bmatrix}$$

Section 3

9.3.1 Multiply each row of A by each column of B. For example,
$1 \cdot 5 + 2 \cdot 7 = 19$ and $1 \cdot 6 + 2 \cdot 8 = 22$.

$$AB = \begin{bmatrix} 19 & 22 \\ 43 & 50 \end{bmatrix}$$

9.3.2 $AB = \begin{bmatrix} 2 \\ 3 \end{bmatrix} \begin{bmatrix} 7 & 4 \end{bmatrix} = \begin{bmatrix} 14 & 8 \\ 21 & 12 \end{bmatrix}$

$$BA = \begin{bmatrix} 7 & 4 \end{bmatrix} \begin{bmatrix} 2 \\ 3 \end{bmatrix} = \begin{bmatrix} 26 \end{bmatrix}$$

9.3.3 Multiply and set the corresponding entries equal to get $x + y = 15$
and $x - y = -3$. Solving this system by addition yields $x = 6$ and $y = 9$.

9.3.4 The system $\begin{matrix} x - y = 3 \\ x + 2y = 5 \end{matrix}$ can be written as $\begin{bmatrix} 1 & -1 \\ 1 & 2 \end{bmatrix} \begin{bmatrix} x \\ y \end{bmatrix} = \begin{bmatrix} 3 \\ 5 \end{bmatrix}$.

Section 4

9.4.1 Since B is the 2×2 identity matrix, $AB = A$ and $BA = A$.

9.4.2 $AB = \begin{bmatrix} 3 & 5 \\ 1 & 2 \end{bmatrix} \begin{bmatrix} 2 & -5 \\ -1 & 3 \end{bmatrix} = \begin{bmatrix} 1 & 0 \\ 0 & 1 \end{bmatrix}$

$$BA = \begin{bmatrix} 2 & -5 \\ -1 & 3 \end{bmatrix} \begin{bmatrix} 3 & 5 \\ 1 & 2 \end{bmatrix} = \begin{bmatrix} 1 & 0 \\ 0 & 1 \end{bmatrix}$$

9.4.3 $\begin{bmatrix} 3 & 2 & | & 1 & 0 \\ 7 & 5 & | & 0 & 1 \end{bmatrix}$

$$\begin{bmatrix} 3 & 2 & | & 1 & 0 \\ 1 & 1 & | & -2 & 1 \end{bmatrix} \quad -2R_1 + R_2 \to R_2$$

$$\begin{bmatrix} 1 & 1 & | & -2 & 1 \\ 3 & 2 & | & 1 & 0 \end{bmatrix} \quad R_1 \leftrightarrow R_2$$

$$\begin{bmatrix} 1 & 1 & | & -2 & 1 \\ 0 & -1 & | & 7 & -3 \end{bmatrix} \quad -3R_1 + R_2 \to R_2$$

$$\begin{bmatrix} 1 & 0 & | & 5 & -2 \\ 0 & 1 & | & -7 & 3 \end{bmatrix} \quad \begin{matrix} R_1 + R_2 \to R_1 \\ -R_2 \to R_2 \end{matrix}$$

The inverse matrix is $\begin{bmatrix} 5 & -2 \\ -7 & 3 \end{bmatrix}$.

9.4.4 $\begin{bmatrix} 3 & -1 & | & 1 & 0 \\ -3 & 1 & | & 0 & 1 \end{bmatrix}$

$$\begin{bmatrix} 3 & -1 & | & 1 & 0 \\ 0 & 0 & | & 1 & 1 \end{bmatrix} \quad R_1 + R_2 \to R_2$$

Since the left hand side of the augmented matrix has a row of zeros, it cannot be converted to the identity and there is no inverse to the matrix.

9.4.5 $\begin{bmatrix} 1 & 1 & 3 & | & 1 & 0 & 0 \\ -1 & 0 & 0 & | & 0 & 1 & 0 \\ -1 & -1 & -2 & | & 0 & 0 & 1 \end{bmatrix}$

$$\begin{bmatrix} 1 & 1 & 3 & | & 1 & 0 & 0 \\ 0 & 1 & 3 & | & 1 & 1 & 0 \\ 0 & 0 & 1 & | & 1 & 0 & 1 \end{bmatrix} \quad \begin{matrix} R_1 + R_2 \to R_2 \\ R_1 + R_3 \to R_3 \end{matrix}$$

$$\begin{bmatrix} 1 & 0 & 0 & | & 0 & -1 & 0 \\ 0 & 1 & 3 & | & 1 & 1 & 0 \\ 0 & 0 & 1 & | & 1 & 0 & 1 \end{bmatrix} \quad -R_2 + R_1 \to R_1$$

$$\begin{bmatrix} 1 & 0 & 0 & | & 0 & -1 & 0 \\ 0 & 1 & 0 & | & -2 & 1 & -3 \\ 0 & 0 & 1 & | & 1 & 0 & 1 \end{bmatrix} \quad -3R_3 + R_2 \to R_2$$

The inverse matrix is $\begin{bmatrix} 0 & -1 & 0 \\ -2 & 1 & -3 \\ 1 & 0 & 1 \end{bmatrix}$.

9.4.6 The inverse of the matrix of coefficients was found in 9.4.3. So

$$\begin{bmatrix} x \\ y \end{bmatrix} = A^{-1}B = \begin{bmatrix} 5 & -2 \\ -7 & 3 \end{bmatrix} \begin{bmatrix} 4 \\ 1 \end{bmatrix}$$

$$= \begin{bmatrix} 18 \\ -25 \end{bmatrix}$$

Section 5

9.5.1 $\begin{vmatrix} 2 & 4 \\ -3 & 1 \end{vmatrix} = 2 \cdot 1 - (-3)(4) = 14$

9.5.2 $D = \begin{vmatrix} 2 & 5 \\ -3 & 1 \end{vmatrix} = 2 \cdot 1 - (-3)(5) = 17$

$$D_x = \begin{vmatrix} 17 & 5 \\ 0 & 1 \end{vmatrix} = 17 \cdot 1 - (0)(5) = 17$$

$$D_y = \begin{vmatrix} 2 & 17 \\ -3 & 0 \end{vmatrix} = 2 \cdot 0 - (-3)(17) = 51$$

$$x = D_x/D = 1, y = D_y/D = 3$$

So the solution set to the system is $\{(1, 3)\}$.

9.5.3 Since $D = \begin{vmatrix} 1 & 1 \\ -2 & -2 \end{vmatrix} = 1(-2) - (-2)(1) = 0$, the system cannot

be solved by Cramer's rule. Multiplying the first equation by 2 and adding to the second yields $0 = 17$. So there is no solution to the system.

9.5.4 A matrix has an inverse if and only if its determinant is nonzero.

Since $\begin{vmatrix} 1 & 6 \\ -2 & 9 \end{vmatrix} = 1(9) - (-2)(6) = 21$, the matrix has an inverse.

Section 6

9.6.1 The minor for 7 is $\begin{vmatrix} 2 & 3 \\ 5 & 6 \end{vmatrix}$ or -3.

9.6.2 Expand by minors about the first column:

$$|A| = 1\begin{vmatrix} 5 & 6 \\ 8 & 9 \end{vmatrix} - 4\begin{vmatrix} 2 & 3 \\ 8 & 9 \end{vmatrix} + 7\begin{vmatrix} 2 & 3 \\ 5 & 6 \end{vmatrix}$$
$$= 1(-3) - 4(-6) + 7(-3) = 0$$

9.6.3 Expand by minors about the second column:

$$|A| = -(2)\begin{vmatrix} 5 & 4 \\ -6 & 8 \end{vmatrix} + (-7)\begin{vmatrix} 1 & 3 \\ -6 & 8 \end{vmatrix} - 0\begin{vmatrix} 1 & 3 \\ 5 & 4 \end{vmatrix}$$
$$= -2(64) - 7(26) - 0(-11) = -310$$

9.6.4 Expand by minors about the third column:

$$|A| = 2\begin{vmatrix} -1 & 7 \\ 4 & 9 \end{vmatrix} - 0\begin{vmatrix} 3 & 5 \\ 4 & 9 \end{vmatrix} + 0\begin{vmatrix} 3 & 5 \\ -1 & 7 \end{vmatrix}$$
$$= 2(-37) = -74$$

9.6.5 Expand by minors about the first column:

$$|A| = 1\begin{vmatrix} 1 & 7 & 9 \\ 0 & 1 & 0 \\ 0 & 0 & 1 \end{vmatrix} = 1 \cdot 1\begin{vmatrix} 1 & 0 \\ 0 & 1 \end{vmatrix} = 1$$

9.6.6 First find the determinants $D = 2, D_x = 2, D_y = 10, D_z = 12$. Then find $x = D_x/D = 1, y = D_y/D = 5$, and $z = D_z/D = 6$.

9.6.7 Since $D = 0$, the system cannot be solved by Cramer's rule. Since the second and third equations are multiples of the first, the equations are dependent. The solution set is $\{(x, y, z) \mid x - y + z = 2\}$.

Chapter 10

Section 1

10.1.1 Since the focus is one unit below the directrix, $p = -1/2$ and $a = 1/(4p) = -1/2$. The y-coordinate of the vertex is $(5 + 6)/2$ or $11/2$ and the vertex is $(2, 11/2)$. So the equation is

$$y = -\frac{1}{2}(x - 2)^2 + \frac{11}{2} \text{ or } y = -\frac{1}{2}x^2 + 2x + \frac{7}{2}.$$

10.1.2 $y = 2x^2 + 8x + 11$
$\qquad y = 2(x^2 + 4x) + 11$
$\qquad y = 2(x^2 + 4x + 4) + 11 - 8$
$\qquad y = 2(x + 2)^2 + 3$

The vertex is $(-2, 3)$. Since $a = 2, p = 1/8$. Because the parabola opens upward the focus is $1/8$ unit above the vertex at $(-2, 25/8)$ and the directrix is $y = 23/8$.

10.1.3 $x = -b/(2a) = -4/(2 \cdot 1) = -2$
$\qquad y = (-2)^2 + 4(-2) - 1 = -5$

The vertex is $(-2, -5)$. Since $a = 1$ the parabola opens upward and $p = 1/4$. The focus is $1/4$ unit above the vertex at $(-2, -19/4)$ and the directrix is $y = -21/4$.

10.1.4 The vertex is halfway between the focus and directrix at $(2, -7/4)$. The parabola opens upward, $p = 1/4$, and $a = 1$. The equation is $y = (x - 2)^2 - \frac{7}{4}$. If $x = 0$, then $y = 9/4$ and the y-intercept is $(0, 9/4)$.

$$(x - 2)^2 - \frac{7}{4} = 0$$
$$x - 2 = \pm\frac{\sqrt{7}}{2}$$
$$x = \frac{4 \pm \sqrt{7}}{2}$$

The x-intercepts are $\left(\frac{4 \pm \sqrt{7}}{2}, 0\right)$.

10.1.5 Since $y = -b/(2a) = -2, x = (-2)^2 + 4(-2) = -4$. The vertex is $(-4, -2)$ and the axis of symmetry is $y = -2$. If $x = 0$, then $y^2 + 4y = 0$. Solve for y to find the y-intercepts $(0, 0)$ and $(0, -4)$. Since $a = 1, p = 1/4$, and the focus is $(-15/4, -2)$. The directrix is $x = -17/4$.

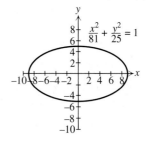

10.1.6 Since the focus is $(0, 10)$ and it is above the vertex $(0, 0)$, $p = 10$ and $a = 1/40$. Since the vertex is $(0, 0)$, the equation is $y = \frac{1}{40}x^2$.

Section 2

10.2.1 Since $a = 8$ and $c = 3, a^2 = b^2 + c^2$ yields $b^2 = 55$. So the equation is $\frac{x^2}{64} + \frac{y^2}{55} = 1$. The y-intercepts are $(0, \pm\sqrt{55})$.

10.2.2 The ellipse goes through $(0, \pm 5)$ and $(\pm 9, 0)$. Since $c^2 = 81 - 25, c = \pm 2\sqrt{14}$. The foci are $(\pm 2\sqrt{14}, 0)$.

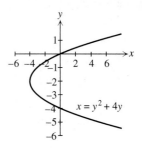

10.2.3 The ellipse $\dfrac{x^2}{4} + \dfrac{y^2}{25} = 1$ goes through $(0, \pm 5)$ and $(\pm 2, 0)$. Since $c^2 = 25 - 4$, $c = \pm\sqrt{21}$. The foci are $(0, \pm\sqrt{21})$.

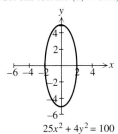

$$25x^2 + 4y^2 = 100$$

10.2.4 The foci for $\dfrac{x^2}{81} + \dfrac{y^2}{25} = 1$ are $(\pm 2\sqrt{14}, 0)$. Move them 1 unit to the right and 4 units downward to get the foci for the transformed graph $(1 \pm 2\sqrt{14}, -4)$. The ellipse passes through points that are 9 units to the left or right of its center $(1, -4)$ and 5 units above or below its center.

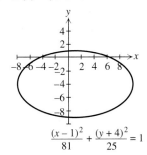

$$\dfrac{(x-1)^2}{81} + \dfrac{(y+4)^2}{25} = 1$$

10.2.5 Find the radius:
$$r = \sqrt{(-2-2)^2 + (3-0)^2} = 5$$
The equation is $(x+2)^2 + (y-3)^2 = 25$.

10.2.6
$$x^2 + 6x + y^2 - 8y = 0$$
$$x^2 + 6x + 9 + y^2 - 8y + 16 = 9 + 16$$
$$(x+3)^2 + (y-4)^2 = 25$$
The center is $(-3, 4)$ and the radius is 5.

10.2.7 Since the foci are $(\pm 2\sqrt{14}, 0)$, $c = 2\sqrt{14}$. Since $a = 9$, we have $e = 2\sqrt{14}/9 \approx 0.83$.

Section 3

10.3.1 If $y = 0$, $x = \pm 7$. Since $c^2 = 49 + 25$, $c = \pm\sqrt{74}$.
The foci are $(\pm\sqrt{74}, 0)$. The asymptotes are $y = \pm\dfrac{5}{7}x$.

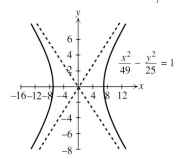

10.3.2 If $x = 0$, $y = \pm 3$. Since $c^2 = 9 + 81$, $c = \pm 3\sqrt{10}$. The foci are $(0, \pm 3\sqrt{10})$. The asymptotes are $y = \pm\dfrac{1}{3}x$.

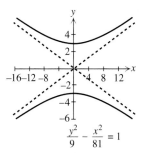

$$\dfrac{y^2}{9} - \dfrac{x^2}{81} = 1$$

10.3.3 This hyperbola is centered at $(-2, 1)$. Move the graph of $\dfrac{x^2}{49} - \dfrac{y^2}{25} = 1$ two units to the left and one unit upward. Draw the fundamental rectangle through $(5, 1)$, $(-9, 1)$, $(-2, 6)$, and $(-2, -4)$. The foci are $(-2 + \sqrt{74}, 1)$ and $(-2 - \sqrt{74}, 1)$. The asymptotes have slopes $\pm 5/7$ and pass through $(-2, 1)$. So their equations are $y = \dfrac{5}{7}x + \dfrac{17}{7}$ and $y = -\dfrac{5}{7}x - \dfrac{3}{7}$.

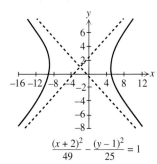

$$\dfrac{(x+2)^2}{49} - \dfrac{(y-1)^2}{25} = 1$$

10.3.4 The vertices are $(\pm 3, 0)$ and the hyperbola opens left and right. Since it is centered at the origin, its equation is $\dfrac{x^2}{9} - \dfrac{y^2}{64} = 1$.

10.3.5 If $y = \pm 6$, then $x = \pm 3$. Since the asymptotes pass through the vertices of the fundamental rectangle, the vertices are $(\pm 3, \pm 6)$. Since this hyperbola opens up and down, its equation is $\dfrac{y^2}{36} - \dfrac{x^2}{9} = 1$.

Chapter 11

Section 1

11.1.1 If $n = 1$, $a_1 = 1^2 - 3 = -2$. If $n = 2$, $a_2 = 2^2 - 3 = 1$. If $n = 3$, $a_3 = 3^2 - 3 = 6$. So the terms are -2, 1, and 6.

11.1.2 $a_1 = (-1)^1 3^{1-1} = -1$
$a_2 = (-1)^2 3^{2-1} = 3$
$a_3 = (-1)^3 3^{3-1} = -9$
The first three terms are -1, 3, and -9.

11.1.3 $a_1 = (-1)^{1-1}/1! = 1$
$a_2 = (-1)^{2-1}/2! = -1/2$
$a_3 = (-1)^{3-1}/3! = 1/6$
$a_4 = (-1)^{4-1}/4! = -1/24$
The first four terms are 1, $-1/2$, $1/6$, and $-1/24$.

11.1.4 Use a power of -1 to get the alternating signs and $2n$ to get the even integers. The nth term is $a_n = \dfrac{(-1)^n}{2n}$.

11.1.5 $a_1 = 3$
$a_2 = 2a_1 - 4 = 2(3) - 4 = 2$
$a_3 = 2a_2 - 4 = 2(2) - 4 = 0$
$a_4 = 2a_3 - 4 = 2(0) - 4 = -4$
The first four terms are 3, 2, 0, and -4.

11.1.6 $a_1 = 2(1) - 4 = -2$
$a_2 = 2(2) - 4 = 0$
$a_3 = 2(3) - 4 = 2$
$a_4 = 2(4) - 4 = 4$
The first four terms are -2, 0, 2, and 4.

11.1.7 The first term is 5 and the constant difference is 6 in this arithmetic sequence. So $a_n = 5 + (n-1)6$ or $a_n = 6n - 1$.

11.1.8 Since $a_1 = 12$ and $a_5 = 24$, $a_5 = 12 + (5-1)d = 24$. Solve this equation to get $d = 3$ and $a_n = 12 + (n-1)3$. So $a_{30} = 12 + (30-1)3 = 99$.

11.1.9 Since $a_1 = 50$, $a_2 = 50 - 5 = 45$, $a_3 = 45 - 5 = 40$, and $a_4 = 40 - 5 = 35$. So the first four terms are 50, 45, 40, and 35.

Section 2

11.2.1 $\sum\limits_{i=2}^{5}(-1)^i(2i) = (-1)^2 4 + (-1)^3 6 + (-1)^4 8 + (-1)^5 10 = -4$

11.2.2 A power of -1 gives the alternating signs and $2n$ gives the even numbers in the summation notation $\sum\limits_{n=1}^{4} \dfrac{(-1)^n}{2n}$.

11.2.3 Since $i = j + 1$, the series is written as
$$\sum_{j=1}^{6}(-1)^j(j+1)^3.$$

11.2.4 The mean is $\dfrac{5 + 12 + (-17) + 23}{4}$ or 5.75.

11.2.5 The sum of this arithmetic series is $\dfrac{24}{2}(-6 + 40)$ or 408.

11.2.6 Find the number of terms by solving $a_n = 20 + (n-1)20 = 960$. The solution is 48. So the sum of the arithmetic sequence is $\dfrac{48}{2}(20 + 960)$, or 23,520, and the mean is 23,520/48, or 490.

Section 3

11.3.1 $a_1 = 3 \cdot 2^1 = 6$
$a_2 = 3 \cdot 2^2 = 12$
$a_3 = 3 \cdot 2^3 = 24$
$a_4 = 3 \cdot 2^4 = 48$
The first four terms are 6, 12, 24, and 48.

11.3.2 The first term is 5 and the common ratio is 2. So $a_n = 5 \cdot 2^{n-1}$.

11.3.3 $a_1 = 2^1 \cdot 3^0 = 2$
$a_2 = 2^2 \cdot 3^1 = 12$
$a_3 = 2^3 \cdot 3^2 = 72$
$a_4 = 2^4 \cdot 3^3 = 432$
The first four terms are 2, 12, 72, and 432. Since there is a common ratio of 6, the sequence is geometric.

11.3.4 The common ratio is 2 and $a_n = 8 \cdot 2^{n-1}$. Solve $8 \cdot 2^{n-1} = 32768$ to find the number of terms.
$$2^{n-1} = 4096 = 2^{12}$$
$$n - 1 = 12$$
$$n = 13$$
There are 13 terms in the sequence.

11.3.5 $S_{20} = \dfrac{3(1 - 2^{20})}{1 - 2} = 3{,}145{,}725$

11.3.6 $S = \dfrac{5(0.3)}{1 - 0.3} = \dfrac{15}{7}$

11.3.7 $0.062626262\ldots$
$= \dfrac{62}{10^3} + \dfrac{62}{10^5} + \dfrac{62}{10^7} + \cdots = \dfrac{62/10^3}{1 - 1/10^2}$
$= \dfrac{62}{10^3 - 10} = \dfrac{62}{990} = \dfrac{31}{495}$

11.3.8 Find the total of the right movements:
$$S = \dfrac{8}{1 - 1/4} = \dfrac{32}{3}$$
Find the total of the left movements:
$$S = \dfrac{4}{1 - 1/4} = \dfrac{16}{3}$$
You are getting closer and closer to $\dfrac{32}{3} - \dfrac{16}{3}$ or $\dfrac{16}{3}$.

11.3.9 $a_n = 2000(1.05)^n$

11.3.10 $S_{20} = 2000(1.05)^1 + 2000(1.05)^2 + 2000(1.05)^3 + \cdots + 2000(1.05)^{20}$
$= \dfrac{2000(1.05)(1 - 1.05^{20})}{1 - 1.05} \approx \$69{,}438.50$

Section 4

11.4.1 By the fundamental counting principle there are $3 \cdot 4 \cdot 2$ or 24 different configurations.

11.4.2 By the fundamental counting principle there are $26^2 \cdot 10^4$ or 6,760,000 possible license plates.

11.4.3 The number of ways for 13 students to line up is 13! or 6,227,020,800.

11.4.4 The number of ways is $P(10, 2)$ or $10 \cdot 9$, or 90.

Section 5

11.5.1 The number of ways is $C(10, 3)$ or 120.

11.5.2 The number of ways is $C(10, 3) \cdot C(7, 3) \cdot C(4, 4)$ or 4200.

11.5.3 The number of arrangements is $\dfrac{8!}{2!2!2!1!1!}$ or 5040.

11.5.4 The number of ways is $C(20, 3) \cdot C(15, 2)$ or 119,700.

11.5.5 $(a - 2b)^4 = 1a^4(-2b)^0 + 4a^3(-2b)^1 + 6a^2(-2b)^2 + 4a^1(-2b)^3 + 1a^0(-2b)^4 = a^4 - 8a^3b + 24a^2b^2 - 32ab^3 + 16b^4$

11.5.6 $(a + 3b)^{12} = \binom{12}{0}a^{12}(3b)^0$
$+ \binom{12}{1}a^{11}(3b)^1 + \binom{12}{2}a^{10}(3b)^2 + \cdots$
$= a^{12} + 36a^{11}b + 594a^{10}b^2 + \cdots$

11.5.7 The coefficient of x^3y^9 is either $\begin{pmatrix} 12 \\ 3 \end{pmatrix}$ or $\begin{pmatrix} 12 \\ 9 \end{pmatrix}$. In either case the term is $220x^3y^9$.

11.5.8 The coefficient of $x^3y^4z^2$ is $\dfrac{9!}{3!4!2!}$ or 1260. The term is $1260x^3y^4z^2$.

Section 6

11.6.1 Fewer than two heads occurs on three of the four outcomes for tossing a pair of coins. So the probability is 3/4.

11.6.2 A number larger than five occurs on only one of the six outcomes for rolling a single die. So the probability is 1/6.

11.6.3 A sum of three occurs on $(1, 2)$ and $(2, 1)$. So the probability is 2/36 or 1/18.

11.6.4 The number of choices is $C(50, 5)$ or 2,118,760. So the probability is $\dfrac{1}{2{,}118{,}760}$ that they have made the same choice.

11.6.5 $P(R \cup F)$
$$= P(R) + P(F) - P(R \cap F)$$
$$= 0.40 + 0.60 - 0.30 = 0.70$$

11.6.6 Since $P(3) = 2/36$ and $P(5) = 4/36$ and these events are mutually exclusive, $P(3 \text{ or } 5) = 6/36$ or 1/6.

11.6.7 Since $P(3) = 2/36$, the probability that the sum is not 3 is 34/36 or 17/18.

11.6.8 Since $P(3) = 2/36$ and $P(\text{not } 3)$ is 34/36, the odds in favor of the sum being 3 are 2 to 34 or 1 to 17.

11.6.9 The probability that the Eagles win is $4/(4 + 5)$ or 4/9.

Section 7

11.7.1 $S_1: 3 = \dfrac{3 \cdot 1(1 + 1)}{2}$

$S_k: 3 + 6 + 9 + \cdots + 3k = \dfrac{3k(k + 1)}{2}$

$S_{k+1}: 3 + 6 + 9 + \cdots + 3(k + 1) = \dfrac{3(k + 1)(k + 2)}{2}$

11.7.2 If $n = 1$ we have $3 = \dfrac{3 \cdot 1(1 + 1)}{2}$, which is true.

$$3 + 6 + 9 + \cdots + 3k + 3(k + 1)$$
$$= \frac{3k(k + 1)}{2} + 3(k + 1)$$
$$= \frac{3k(k + 1)}{2} + \frac{2 \cdot 3(k + 1)}{2}$$
$$= \frac{3k(k + 1) + 6(k + 1)}{2}$$
$$= \frac{3(k + 2)(k + 1)}{2}$$

So the equation is true for $n = k + 1$, assuming that it is true for $n = k$. And that proves the statement is true for every positive integer n.

11.7.3 If $n = 1$ we have $9 = \dfrac{1(19 - 1)}{2}$, which is true.

$$9 + 8 + 7 + \cdots + (10 - k) + (10 - (k + 1))$$
$$= \frac{k(19 - k)}{2} + (10 - (k + 1))$$
$$= \frac{k(19 - k) + 2(9 - k)}{2}$$
$$= \frac{19k - k^2 + 18 - 2k}{2}$$
$$= \frac{18 + 17k - k^2}{2}$$
$$= \frac{(18 - k)(1 + k)}{2}$$
$$= \frac{(k + 1)(19 - (k + 1))}{2}$$

So the equation is true for $n = k + 1$, assuming that it is true for $n = k$. And that proves the statement is true for every positive integer n.

11.7.4 If $n = 1$, the inequality is $4^1 < 4!$, which is true. Assume that $4^k < (k + 3)!$:

$$4 \cdot 4^k < 4 \cdot (k + 3)!$$

Since $4 < k + 4$ for every positive integer k, $4^{k+1} < (k + 4)(k + 3)!$ or $4^{k+1} < (k + 4)!$. So by mathematical induction the inequality is true for every positive integer n.

Answers to Exercises

CHAPTER P

Section P.1

For Thought: **1.** F **2.** T **3.** T **4.** F **5.** T **6.** F
7. F **8.** F **9.** T **10.** F

Exercises: **1.** e, true **3.** h, true **5.** g, true **7.** c, true
9. All **11.** $\{-\sqrt{2}, \sqrt{3}, \pi, 5.090090009\ldots\}$ **13.** $\{0, 1\}$
15. $x + 7$ **17.** $5x + 15$ **19.** $5(x + 1)$ **21.** $(-13 + 4) + x$
23. 8 **25.** $\sqrt{3}$ **27.** $y^2 - x^2$ **29.** 7.2 **31.** $\sqrt{5}$ **33.** 5
35. 22 **37.** 12 **39.** 3/4 **41.** 8 **43.** -49 **45.** 16
47. $-1/64$ **49.** -8 **51.** 12 **53.** -3 **55.** 1 **57.** -41
59. -7 **61.** 26 **63.** 52 **65.** 99 **67.** -61 **69.** 13
71. -1 **73.** 2 **75.** 41 **77.** 6 **79.** -5 **81.** -5
83. -35 **85.** 56 **87.** $-2x$ **89.** $0.85x$ **91.** $-6xy$
93. $3 - 2x$ **95.** $3x - y$ **97.** $-3x - 6$ **99.** $11x - 17$
101. a. $0.40r - 0.60a + 132$ **b.** 148 beats/min
c. $0.40r - 0.30a + 123$
103. Smallest to largest: $-\dfrac{1}{2}, -\dfrac{5}{12}, -\dfrac{1}{3}, 0, \dfrac{1}{3}, \dfrac{5}{12}, \dfrac{1}{2}$

Section P.2

For Thought: **1.** T **2.** T **3.** T **4.** T **5.** F **6.** F
7. T **8.** T **9.** F **10.** F

Exercises: **1.** 1/3 **3.** $-1/16$ **5.** 8 **7.** 8/27 **9.** 4
11. 4/5 **13.** 24 **15.** 3/2 **17.** $-1/500$ **19.** $-6x^{11}y^{11}$
21. $2x^6$ **23.** $-21b^5$ **25.** $2a^{11}$ **27.** $4m^3 + m^2$ **29.** 225
31. $\dfrac{1}{6y^3}$ **33.** $3x^4$ **35.** $\dfrac{n^2}{2}$ **37.** $17a^6$ **39.** $-4x^6$ **41.** $-\dfrac{8x^6}{27}$
43. $\dfrac{25}{y^4}$ **45.** $\dfrac{64y^3}{27x^{15}}$ **47.** x^{b+5} **49.** $\dfrac{-125a^{6t}}{b^{9t}}$ **51.** $\dfrac{-3y^{6v}}{2x^{5w}}$
53. a^{-4s+20} **55.** 43,000 **57.** 0.0000356 **59.** 5×10^6
61. 6.72×10^{-5} **63.** 0.000000007 **65.** 2×10^{10}
67. 2×10^{16} **69.** 2×10^{-3} **71.** 4×10^{-29} **73.** 1×10^{23}
75. 9.936×10^{-5} **77.** 4.78×10^{-41} **79.** BMI = 37.7
81. $D = 366.5$ **83.** \$25,825 **85.** 1.577×10^{24} tons
87. 6379 km **89.** 3.3×10^5 **91. a.** 5365.4 **b.** 9702.38
d. $9 \times 10^3 + 6 \times 10^1 + 3 \times 10^0 + 2 \times 10^{-1} + 4 \times 10^{-2} + 1 \times 10^{-3}$, not unique
e. $4 \times 10^4 + 3 \times 10^3 + 2 \times 10^0 + 1 \times 10^{-1} + 9 \times 10^{-2}$

Section P.3

For Thought: **1.** F **2.** T **3.** F **4.** T **5.** F **6.** F
7. F **8.** T **9.** T **10.** T

Exercises: **1.** -3 **3.** 8 **5.** -4 **7.** 81 **9.** 1/16
11. 1/2 **13.** 8/27 **15.** $|x|$ **17.** a^3
19. a^2 **21.** xy^2 **23.** y^3 **25.** $x^2 y^{1/2}$ **27.** $6a^{3/2}$
29. $3a^{1/6}$ **31.** $a^{7/3}$ **33.** $\dfrac{x^2 y}{z^3}$ **35.** 30 **37.** -2
39. -2 **41.** 2/3 **43.** 0.1 **45.** $-1/5$ **47.** 8

49. $\sqrt[3]{10^2}$ **51.** $\dfrac{3}{\sqrt[5]{y^3}}$ **53.** $x^{-1/2}$ **55.** $x^{3/5}$ **57.** $4x$
59. $2y^3$ **61.** $\dfrac{\sqrt{xy}}{10}$ **63.** $\dfrac{-2a}{b^5}$ **65.** $2\sqrt{7}$ **67.** $\dfrac{\sqrt{5}}{5}$
69. $\dfrac{\sqrt{2x}}{4}$ **71.** $2\sqrt[3]{5}$ **73.** $-5x\sqrt[3]{2x}$ **75.** $\dfrac{\sqrt[3]{4}}{2}$
77. $2\sqrt{2} + 2\sqrt{5} - 2\sqrt{3}$ **79.** $-30\sqrt{2}$ **81.** $60a$
83. 75 **85.** $\dfrac{3\sqrt{a}}{a^2}$ **87.** $\dfrac{5\sqrt{x}}{x}$ **89.** $5x\sqrt{5x}$ **91.** $\sqrt[6]{72}$
93. $\sqrt[12]{81x^3}$ **95.** $\sqrt[6]{4x^5 y^5}$ **97.** $\sqrt[6]{7}$ **99.** 46 **101.** 25.4
103. a. 50%, 30% **b.** 47.5%, 27.5% **105.** 14 in.
107. 19.0 ft^2 **109.** b **111.** No

Section P.4

For Thought: **1.** T **2.** T **3.** F **4.** T **5.** F **6.** F
7. T **8.** T **9.** T **10.** F

Exercises: **1.** Imaginary, $0 + 6i$ **3.** Imaginary, $\dfrac{1}{3} + \dfrac{1}{3}i$
5. Real, $\sqrt{7} + 0i$ **7.** Real, $\dfrac{\pi}{2} + 0i$ **9.** $7 + 2i$ **11.** $-2 - 3i$
13. $4 + i\sqrt{2}$ **15.** $\dfrac{9}{2} + \dfrac{5}{6}i$ **17.** $-12 - 18i$ **19.** 26
21. $34 - 22i$ **23.** 29 **25.** 4 **27.** $-7 + 24i$
29. $1 - 4i\sqrt{5}$ **31.** i **33.** -1 **35.** $-i$ **37.** i **39.** $-i$
41. -1 **43.** 90 **45.** 17/4 **47.** 1 **49.** 12 **51.** $\dfrac{2}{5} + \dfrac{1}{5}i$
53. $\dfrac{3}{2} - \dfrac{3}{2}i$ **55.** $3 + 3i$ **57.** $\dfrac{1}{13} - \dfrac{5}{13}i$ **59.** $\dfrac{1}{34} - \dfrac{13}{34}i$
61. $-i$ **63.** $-4 + 2i$ **65.** -6 **67.** -10 **69.** $-1 + i\sqrt{5}$
71. $-3 + i\sqrt{11}$ **73.** $-4 + 8i$ **75.** $-1 + 2i$ **77.** $\dfrac{-2 + i\sqrt{2}}{2}$
79. $-3 - 2i\sqrt{2}$ **81.** $\dfrac{3 + \sqrt{21}}{2}$ **83.** 34 **85.** 6
87. $-\dfrac{8}{17} - \dfrac{15}{17}i$ **89.** $-1 + i$ **91.** $3 - i$ **99.** $95 + 2i$

Section P.5

For Thought: **1.** F **2.** F **3.** F **4.** T **5.** F **6.** F
7. F **8.** F **9.** F **10.** T

Exercises: **1.** 3, 1, trinomial **3.** 2, -3, binomial
5. 0, 79, monomial **7.** 12 **9.** 77 **11.** $8x^2 + 3x - 1$
13. $-5x^2 + x - 3$ **15.** $(-5a^2 + 4a)x^3 + 2a^2x - 3$
17. $2x - 1$ **19.** $3x^2 - 3x - 6$ **21.** $-18a^5 + 15a^4 - 6a^3$
23. $3b^3 - 14b^2 + 17b - 6$ **25.** $8x^3 - 1$
27. $xz - 4z + 3x - 12$ **29.** $a^3 - b^3$ **31.** $a^2 + 7a - 18$
33. $2y^2 + 15y - 27$ **35.** $4x^2 - 81$ **37.** $4x^2 + 20x + 25$

39. $6x^4 + 22x^2 + 20$　**41.** $5 + 4\sqrt{2}$　**43.** $14 - 11\sqrt{2}$

45. $9 - \sqrt{6}$　**47.** $8 + 2\sqrt{15}$　**49.** $9x^2 + 30x + 25$

51. $x^{2n} - 9$　**53.** -23　**55.** $55 - 6\sqrt{6}$

57. $9x^6 - 24x^3 + 16$　**59.** $\dfrac{-5 - 5\sqrt{7}}{6}$　**61.** $5\sqrt{2} + 2\sqrt{10}$

63. $\dfrac{2\sqrt{6} - \sqrt{2}}{11}$　**65.** $2 + 2\sqrt{2} + \sqrt{3} + \sqrt{6}$

67. $\dfrac{2\sqrt{3} + 3\sqrt{2}}{6}$　**69.** $-9x^3$　**71.** $-x + 2$　**73.** $x + 3$

75. $a^2 + a + 1$　**77.** $x + 5, 13$　**79.** $2x - 6, 13$

81. $x^2 + x + 1, 0$　**83.** $x + 1 + \dfrac{-1}{x - 1}$　**85.** $2x - 3 + \dfrac{1}{x}$

87. $x + 1 + \dfrac{1}{x}$　**89.** $x + \dfrac{1}{x - 2}$　**91.** $x^2 + 2x - 24$

93. $2a^{10} - 3a^5 - 27$　**95.** $-y - 9$　**97.** $6a - 6$

99. $w^2 + 8w + 16$　**101.** $16x^2 - 81$　**103.** $3y^5 - 9xy^2$

105. $2b - 1$　**107.** $x^6 - 64$　**109.** $9w^4 - 12w^2n + 4n^2$

111. 1　**113.** $2x^2 + 5x - 3$　**115.** $x^2 + 5x + 6$

117. $4x^2 - 20x + 24$　**119. a.** $\approx \$0.20$　**b.** $\$0.21$　**121.** $\$5500$

123. 100 sq ft less

Section P.6

For Thought: **1.** F　**2.** T　**3.** T　**4.** F　**5.** F　**6.** F
7. F　**8.** F　**9.** T　**10.** T

Exercises: **1.** $6x^2(x - 2), -6x^2(-x + 2)$

3. $4a(1 - 2b), -4a(2b - 1)$

5. $ax(-x^2 + 5x - 5), -ax(x^2 - 5x + 5)$

7. $1(m - n), -1(n - m)$　**9.** $(x^2 + 5)(x + 2)$

11. $(y^2 - 3)(y - 1)$　**13.** $(d - w)(ay + 1)$

15. $(y^2 - b)(x^2 - a)$　**17.** $(x + 2)(x + 8)$　**19.** $(x - 6)(x + 2)$

21. $(m - 2)(m - 10)$　**23.** $(t - 7)(t + 12)$

25. $(2x + 1)(x - 4)$　**27.** $(4x + 1)(2x - 3)$

29. $(3y + 5)(2y - 1)$　**31.** $(t - u)(t + u)$　**33.** $(t + 1)^2$

35. $(2w - 1)^2$　**37.** $(y^{2t} - 5)(y^{2t} + 5)$　**39.** $(3zx + 4)^2$

41. $(t - u)(t^2 + tu + u^2)$　**43.** $(a - 2)(a^2 + 2a + 4)$

45. $(3y + 2)(9y^2 - 6y + 4)$

47. $(3xy^2 - 2z^3)(9x^2y^4 + 6xy^2z^3 + 4z^6)$

49. $(x^n - 2)(x^{2n} + 2x^n + 4)$　**51.** $(y^3 + 5)^2$

53. $(2a^2b^4 + 1)(2a^2b^4 - 5)$　**55.** $(2a + 7)(2a - 3)$

57. $(b^2 - 2)(b^2 + 1)$　**59.** $-3x(x - 3)(x + 3)$

61. $2t(2t + 3w)(4t^2 - 6tw + 9w^2)$

63. $(a - 2)(a + 2)(a + 1)$　**65.** $(x - 2)^2(x^2 + 2x + 4)$

67. $-2x(6x + 1)(3x - 2)$

69. $(a - 2)(a^2 + 2a + 4)(a + 2)(a^2 - 2a + 4)(a - 1)$

71. $-(3x + 5)(2x - 3)$　**73.** $(a - 1)(a + 1)(a^2 + 1)$

75. Yes　**77.** No　**79.** $(x - 1)(x + 2)(x + 3)$

81. $(x - 3)(x^2 + 2x + 2)$　**83.** $(x + 2)(x + 3)(x - 1)(x + 1)$

85. $x^2 + x + 1$　**87.** 15 in.^{3}, 20 in.^{3}, 12 in.^{3}, 1 in.　**89.** b　**91.** No

Section P.7

For Thought: **1.** F　**2.** T　**3.** F　**4.** F　**5.** F　**6.** T
7. T　**8.** T　**9.** T　**10.** T

Exercises: **1.** $\{x \mid x \neq -2\}$　**3.** $\{x \mid x \neq 4 \text{ and } x \neq -2\}$

5. $\{x \mid x \neq 3 \text{ and } x \neq -3\}$　**7.** All real numbers

9. $\dfrac{3}{x + 2}$　**11.** $-\dfrac{2}{3}$　**13.** $\dfrac{ab^4}{b - a^2}$　**15.** $\dfrac{y^2z}{x^3}$　**17.** $\dfrac{a^2 + ab + b^2}{a + b}$

19. $-\dfrac{3y - 1}{3y + 1}$　**21.** $\dfrac{3}{7ab}$　**23.** $\dfrac{42}{a^2}$　**25.** $\dfrac{a + 3}{3}$　**27.** $\dfrac{2x - 2y}{x + y}$

29. $x^2y^2 + xy^3$　**31.** $\dfrac{-2}{b + 4}$　**33.** $\dfrac{16a}{12a^2}$　**35.** $\dfrac{x^2 - 8x + 15}{x^2 - 9}$

37. $\dfrac{x^2 + x}{x^2 + 6x + 5}$　**39.** $12a^2b^3$　**41.** $6(a + b)$

43. $(x + 2)(x + 3)(x - 3)$　**45.** $\dfrac{9 + x}{6x}$　**47.** $\dfrac{x + 7}{(x - 1)(x + 1)}$

49. $\dfrac{3a + 1}{a}$　**51.** $\dfrac{t^2 - 2}{t + 1}$　**53.** $\dfrac{2x^2 + 3x - 1}{(x + 1)(x + 2)(x + 3)}$

55. $\dfrac{7}{2x - 6}$　**57.** $\dfrac{x^2}{x^3 - y^3}$　**59.** $\dfrac{x^2 + 2x - 1}{x(x^2 - 1)}$

61. $\dfrac{x^2 - 22x - 15}{x(x + 1)(x - 3)}$　**63.** $\dfrac{15}{8a^2}$　**65.** $\dfrac{4b^2 - 3ab}{b + 2a}$　**67.** $\dfrac{a^2 - a}{3b^2 + b}$

69. $\dfrac{a + 2}{a - 2}$　**71.** $\dfrac{6t - 3}{2t^2 + t - 4}$　**73.** $\dfrac{1 + x}{1 - x}$　**75.** $\dfrac{a^3b^3 + 1}{ab^3}$

77. $-x^2y - xy^2$　**79.** $\dfrac{m^2n^2}{n^2 - 2mn + m^2}$　**81.** 3/7　**83.** 3/506

85. 1/5　**87.** 1195/1191　**89.** 3.1087　**91.** 3.00001

93. a. Decreasing　**b.** $\$27.14, \$24.17, \$22.27$

95. a. $\$6$ million, $\$18$ million, $\$594$ million
b. Cost increases without bound and 100% cleanup is impossible.
c. $\{p \mid 0 \leq p < 100\}$　**97.** 5/12　**99.** 272.7 mph

103. a. 8/13　**b.** 3/2

Chapter P Review Exercises

1. F　**3.** F　**5.** F　**7.** F　**9.** F　**11.** F　**13.** $17x - 12$

15. $\dfrac{3x}{10}$　**17.** $\dfrac{x - 2}{3}$　**19.** $-\dfrac{3}{4}$　**21.** -2　**23.** 4　**25.** 7

27. 28.5　**29.** 625　**31.** 44　**33.** 3/2　**35.** $-16/3$

37. 1/4　**39.** $5x^2$　**41.** 11　**43.** $2s\sqrt{7s}$　**45.** $-10\sqrt[3]{2}$

47. $\dfrac{\sqrt{10a}}{2a}$　**49.** $\dfrac{\sqrt[3]{50}}{5}$　**51.** $8n\sqrt{2n}$　**53.** $3 + \sqrt{3}$

55. $\dfrac{\sqrt{3}}{5}$　**57.** 320,000,000　**59.** 0.000185　**61.** 5.6×10^{-5}

63. 2.34×10^6　**65.** 1.25×10^{20}　**67.** 4×10^6　**69.** $-1 - i$

71. $-9 - 40i$　**73.** 20　**75.** $-3 - 2i$　**77.** $\dfrac{1}{5} - \dfrac{3}{5}i$

79. $-\dfrac{1}{13} + \dfrac{5}{13}i$　**81.** $3 + i\sqrt{2}$　**83.** $\dfrac{3}{4} - \dfrac{\sqrt{5}}{4}i$　**85.** $-1 - i$

87. $2x^2 + x - 7$　**89.** $-5x^4 + 3x^3 + 3x$

91. $3a^3 - 8a^2 + 9a - 10$　**93.** $b^2 - 6by + 9y^2$

95. $3t^2 - 7t - 6$　**97.** $-5y^3$　**99.** 7　**101.** $4 + 2\sqrt{3}$

103. $23 + 4\sqrt{15}$　**105.** $x^2 + 4x - 1, 1$　**107.** $3x + 2, 4$

109. $x - 2 + \dfrac{1}{x + 2}$　**111.** $2 + \dfrac{13}{x - 5}$　**113.** $6x(x - 1)(x + 1)$

115. $(3h + 4t)^2$　**117.** $(t + y)(t^2 - ty + y^2)$　**119.** $(x - 3)(x + 3)^2$

121. $(t - 1)(t^2 + t + 1)(t + 1)(t^2 - t + 1)$

123. $(6x + 5)(3x - 4)$　**125.** $ab(a + 6)(a - 3)$

127. $(x - 1)(x + 1)(2x + y)$　**129.** 2　**131.** $\dfrac{2x + 2}{(x - 2)(x + 4)}$

133. $-\dfrac{1}{2}$ **135.** $\dfrac{bc^{17}}{a^8}$ **137.** $\dfrac{3x+7}{x^2-4}$ **139.** $\dfrac{5x-21}{30x^2}$

141. $\dfrac{a-1}{2}$ **143.** $\dfrac{-x^2+x-1}{2}$ **145.** $\dfrac{3a^2+8a+7}{(a+1)(a-1)(a+5)}$

147. $\dfrac{7}{2x-8}$ **149.** $\dfrac{-3y^2+7}{4y^2-3}$ **151.** $\dfrac{b^3-a^2}{ab^2}$ **153.** $\dfrac{q^3+p^2}{pq^3}$

155. -11 **157.** -9 **159.** $4/11$ **161.** $149/91$
163. a. 300 ft   **b.** 39.8 ft   **165.** 5.9×10^{26} **167.** 11.12
169. $5/6$

Chapter P Test

1. All   **2.** $\{-1.22, -1, 0, 2, 10/3\}$

3. $\{-\pi, -\sqrt{3}, \sqrt{5}, 6.020020002\ldots\}$ **4.** $\{0, 2\}$ **5.** 13   **6.** 2

7. $-1/9$ **8.** -1 **9.** $6x^5y^7$ **10.** 0   **11.** $a^2+2ab+b^2$

12. $-32a^3b^{10}$ **13.** $3\sqrt{3}+2\sqrt{2}$ **14.** $\sqrt{3}+1$ **15.** $\dfrac{\sqrt[3]{2x^2}}{2x}$

16. $2xy^4\sqrt{3xy}$ **17.** $7-24i$ **18.** $\dfrac{1}{2}-\dfrac{1}{2}i$ **19.** $-1+i$

20. $-4+4i\sqrt{3}$ **21.** $3x^3+3x^2-12x$ **22.** $-5x^2+9x-13$
23. x^3+x^2-7x-3 **24.** $4h^2+2h+1$ **25.** $x^2-6xy-27y^2$
26. x^2+2x+2 **27.** $9x^2-48x+64$ **28.** $4t^8-1$
29. $\dfrac{x^2-2x+4}{x}$ **30.** $\dfrac{2x^2+7x+21}{(x+4)(x-3)(x-1)}$ **31.** $\dfrac{2a^2-7}{4a^2-9}$
32. $\dfrac{6b^2-24a^3b^3}{3a+4a^2b^2}$ **33.** $a(x-9)(x-2)$
34. $m(m-1)(m+1)(m^2+1)$ **35.** $(3x-1)(x+5)$
36. $(bx+w)(x-3)$ **37.** \$33,075   **38.** 24.4%   **39.** 176 ft

CHAPTER 1

Section 1.1

For Thought: 1. T   **2.** T   **3.** F   **4.** T   **5.** F   **6.** T
7. F   **8.** F   **9.** F   **10.** F
Exercises: 1. No   **3.** Yes   **5.** $\{5/3\}$ **7.** $\{-2\}$ **9.** $\{1/2\}$
11. $\{11\}$ **13.** $\{-24\}$ **15.** $\{-6\}$ **17.** $\{-2/5\}$
19. R, identity   **21.** $\{0\}$, conditional   **23.** $\{9\}$, conditional
25. $\varnothing$, inconsistent   **27.** $\{x \mid x \neq 0\}$, identity
29. $\{w \mid w \neq 1\}$, identity   **31.** $\{x \mid x \neq 0\}$, identity
33. $\{9/8\}$, conditional   **35.** $\{x \mid x \neq 3 \text{ and } x \neq -3\}$, identity
37. $\varnothing$, inconsistent   **39.** $\{-2\}$, conditional   **41.** $\{-19.952\}$
43. $\{36.28\}$ **45.** $\{-2.562\}$ **47.** $\{1/3\}$ **49.** $\{0.199\}$
51. $\{0.425\}$ **53.** $\{-0.380\}$ **55.** $\{-8, 8\}$ **57.** $\{-4, 12\}$
59. $\{6\}$ **61.** $\varnothing$ **63.** $\{-2, 5\}$ **65.** $\{-23, 41\}$
67. $\{-10, 0\}$ **69.** $\{2/3\}$ **71.** $\varnothing$ **73.** $\{200\}$
75. $\{4\}$ **77.** $\{0\}$ **79.** $\{-1/2\}$ **81.** $\{-10\}$ **83.** $\{5\}$
85. $\{-8, -4\}$ **87.** $\{3/2\}$ **89.** $\varnothing$
91. $\{x \mid x \neq 2 \text{ and } x \neq -2\}$, identity
93. $\{x \mid x \neq -3 \text{ and } x \neq 2\}$, identity   **95.** $\varnothing$, inconsistent
97. a. About 1995   **b.** Increasing   **c.** 2015
99. \$18,260.87   **101.** 250,000

Section 1.2

For Thought: 1. F   **2.** F   **3.** F   **4.** T   **5.** T   **6.** F
7. F   **8.** T   **9.** F   **10.** F
Exercises: 1. $r = \dfrac{I}{Pt}$ **3.** $C = \dfrac{5}{9}(F-32)$ **5.** $b = \dfrac{2A}{h}$

7. $y = \dfrac{C-Ax}{B}$ **9.** $R_1 = \dfrac{RR_2R_3}{R_2R_3-RR_3-RR_2}$

11. $n = \dfrac{a_n-a_1+d}{d}$ **13.** $a_1 = \dfrac{S(1-r)}{1-r^n}$

15. $D = \dfrac{5.688-L+F\sqrt{S}}{2}$ **17.** 5.4%   **19.** 2.5 hr   **21.** $-5°C$

23. \$37,250   **25.** 33.9 in.   **27.** \$60,000   **29.** 16 ft, 7 ft, 7 ft
31. 6400 ft^2 **33.** 112.5 mph   **35.** 48 mph   **37.** \$106,000
39. Northside 600, Southside 900   **41.** 28.8 hr   **43.** 2 P.M.
45. 64.85 acres   **47.** 225 ft   **49.** 4.15 ft   **51.** 1.998 hectares
53. \$80,500   **55.** 8/3 liters   **57.** 4 lb apples, 16 lb apricots
59. 3 dimes, 5 nickels   **61.** 120 milliliters   **63.** 2.5 gal
65. 3 hr 5 min   **67. a.** About 1992   **b.** 1992
69. 1 hr, 3 min, 6 sec, no

Section 1.3

For Thought: 1. F   **2.** F   **3.** F   **4.** F   **5.** T   **6.** F
7. T   **8.** T   **9.** T   **10.** F
Exercises: 1. $(4, 1)$, I   **3.** $(1, 0)$, x-axis   **5.** $(5, -1)$, IV

7. $(-4, -2)$, III   **9.** $(-2, 4)$, II   **11.** $5, (2.5, 5)$

13. $2\sqrt{2}, (0, -1)$ **15.** $25, (17/2, 1)$ **17.** $6, \left(\dfrac{-2+3\sqrt{3}}{2}, \dfrac{5}{2}\right)$

19. $\sqrt{74}, (-1.3, 1.3)$ **21.** $|a-b|, \left(\dfrac{a+b}{2}, 0\right)$

23. $\dfrac{\sqrt{\pi^2+4}}{2}, \left(\dfrac{3\pi}{4}, \dfrac{1}{2}\right)$

25. $(0, 0), 4$ **27.** $(-6, 0), 6$

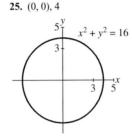

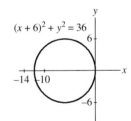

29. $(-1, 0), 5$ **31.** $(2, -2), 2\sqrt{2}$

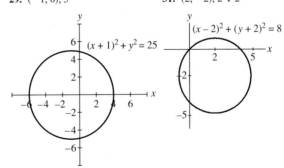

33. $x^2+y^2 = 49$ **35.** $(x+2)^2+(y-5)^2 = 1/4$
37. $(x-3)^2+(y-5)^2 = 34$ **39.** $(x-5)^2+(y+1)^2 = 32$

41. $(0, 0)$, 3

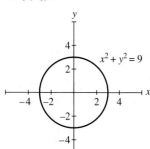

$x^2 + y^2 = 9$

43. $(0, -3)$, 3

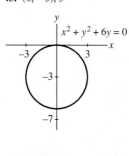

$x^2 + y^2 + 6y = 0$

59. $(0, -4)$, $(4/3, 0)$

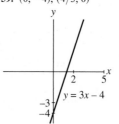

$y = 3x - 4$

61. $(0, -6)$, $(2, 0)$

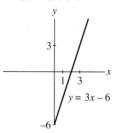

$y = 3x - 6$

63. $(0, 30)$, $(-90, 0)$

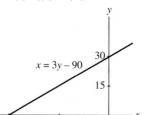

$x = 3y - 90$

65. $(0, 600)$, $(-800, 0)$

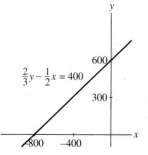

$\frac{2}{3}y - \frac{1}{2}x = 400$

45. $(-3, -4)$, 5

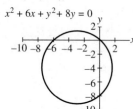

$x^2 + 6x + y^2 + 8y = 0$

47. $(3/2, -1)$, 2

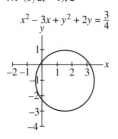

$x^2 - 3x + y^2 + 2y = \frac{3}{4}$

49. $(3, 4)$, 5

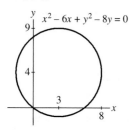

$x^2 - 6x + y^2 - 8y = 0$

51. $(2, 3/2)$, 5/2

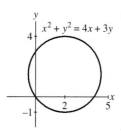

$x^2 + y^2 = 4x + 3y$

67. $(0, 0.0025)$, $(0.005, 0)$

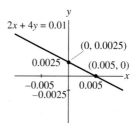

$2x + 4y = 0.01$
$(0, 0.0025)$
$(0.005, 0)$

69. $(0, 2500)$, $(5000, 0)$

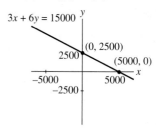

$3x + 6y = 15000$
$(0, 2500)$
$(5000, 0)$

71.

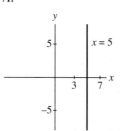

$x = 5$

73.

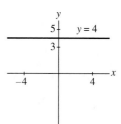

$y = 4$

53. $(1/4, -1/6)$, 1/6

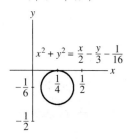

$x^2 + y^2 = \frac{x}{2} - \frac{y}{3} - \frac{1}{16}$

55. a. $x^2 + y^2 = 49$ **b.** $(x - 1)^2 + y^2 = 20$
   c. $(x - 1)^2 + (y - 2)^2 = 13$

57. a. $(x - 2)^2 + (y + 3)^2 = 4$ **b.** $(x + 2)^2 + (y - 1)^2 = 1$
   c. $(x - 3)^2 + (y + 1)^2 = 9$ **d.** $x^2 + y^2 = 1$

75.

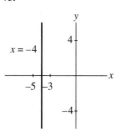

$x = -4$

77.

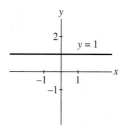

$y = 1$

79. $\{3.6\}$ **81.** $\{14\}$ **83.** $\{-2.83\}$ **85.** $\{558.54\}$

87. $\{116,566.67\}$ **89.** $\{4.91\}$

91. a. $(15, 22.95)$ The median of age at first marriage in 1985 was 22.95.
   b. 30.3, Because of the units, distance is meaningless.

93. $C = 1.8$

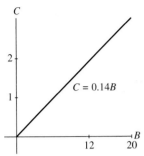

Section 1.4

For Thought: **1.** F   **2.** F   **3.** F   **4.** F   **5.** T   **6.** F
7. F   **8.** T   **9.** F   **10.** T

Exercises: **1.** $\dfrac{1}{3}$ **3.** -4 **5.** 0   **7.** 2   **9.** No slope

11. $y = \dfrac{5}{4}x + \dfrac{1}{4}$ **13.** $y = -\dfrac{7}{6}x + \dfrac{11}{3}$ **15.** $y = 5$

17. $x = 4$ **19.** $y = \dfrac{2}{3}x - 1$ **21.** $y = \dfrac{5}{2}x + \dfrac{3}{2}$

23. $y = -2x + 4$ **25.** $y = \dfrac{3}{2}x + \dfrac{5}{2}$

27. $y = \dfrac{3}{5}x - 2, \dfrac{3}{5}, (0, -2)$ **29.** $y = 2x - 5, 2, (0, -5)$

31. $y = \dfrac{1}{2}x + \dfrac{1}{2}, \dfrac{1}{2}, \left(0, \dfrac{1}{2}\right)$ **33.** $y = 4, 0, (0, 4)$

35. $y = 0.03x - 2.6, 0.03, (0, -2.6)$

37.   **39.**

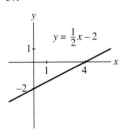   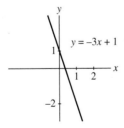

41.   **43.**

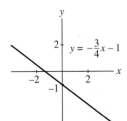   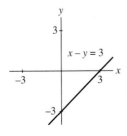

45.

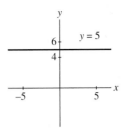

47. $4x - 3y = 12$ **49.** $4x - 5y = -7$ **51.** $x = -4$

53. $16x - 15y = 22$ **55.** $3x - 50y = -11$ **57.** 0.5   **59.** -1

61. 0   **63.** $2x - y = 4$ **65.** $3x + y = 7$ **67.** $5x - 7y = 23$

69. $3x + 2y = 0$ **71.** $2x + y = -5$ **73.** $y = 5$ **75.** -5

77. 8   **79.** T   **81.** T   **83.** F   **85.** Yes, no

87. $y = x - 2, x^3 - 8 = (x - 2)(x^2 + 2x + 4)$

89. $F = \dfrac{9}{5}C + 32, 302°F$ **91.** $c = 50 - n$, \$400

93. $S = -0.005D + 95$

95. $c = -\dfrac{3}{4}p + 30, -\dfrac{3}{4}$, If p increases by 4 then c decreases by 3.

Section 1.5

For Thought: **1.** T   **2.** T   **3.** F   **4.** T   **5.** T   **6.** T
7. F   **8.** F   **9.** F   **10.** F

Exercises: **1.** Linear   **3.** No relationship   **5.** Nonlinear

7. Linear   **9.** Linear   **11.** Linear   **13.** No relationship

15. $(3.3, 160), (4.0, 193)$ **17.** $(132, 34), (148, 25)$

19. a. $y = 331.5x + 4956.6$ where x is years since 2000
   b. 7,608,600   **c.** 1985

21. a. $y = 50.9x + 1083.8$ where x is years since 2000   **b.** 2018
   c. \$829 billion

23. $p = -0.069A + 0.403, 12.7\%$

Section 1.6

For Thought: **1.** F   **2.** F   **3.** F   **4.** F   **5.** F   **6.** T
7. F   **8.** F   **9.** T   **10.** T

Exercises: **1.** $\{\pm\sqrt{5}\}$ **3.** $\left\{\pm i\dfrac{\sqrt{6}}{3}\right\}$ **5.** $\{0, 6\}$ **7.** $\{1/3\}$

9. $\{-2, 3\}$ **11.** $\{-2 \pm 2i\}$ **13.** $\left\{0, \dfrac{4}{3}\right\}$ **15.** $\{-4, 5\}$

17. $\{-2, -1\}$ **19.** $\left\{-\dfrac{1}{2}, 3\right\}$ **21.** $\left\{\dfrac{2}{3}, \dfrac{1}{2}\right\}$ **23.** $\{-7, 6\}$

25. $x^2 - 12x + 36$ **27.** $r^2 + 3r + \dfrac{9}{4}$ **29.** $w^2 + \dfrac{1}{2}w + \dfrac{1}{16}$

31. $\{-3 \pm 2\sqrt{2}\}$ **33.** $\{1 \pm \sqrt{2}\}$ **35.** $\left\{\dfrac{-3 \pm \sqrt{13}}{2}\right\}$

37. $\left\{-4, \dfrac{3}{2}\right\}$ **39.** $\left\{\dfrac{-1 \pm i\sqrt{2}}{3}\right\}$ **41.** $\{-4, 1\}$

43. $\left\{-\dfrac{1}{2}, 3\right\}$ **45.** $\left\{-\dfrac{1}{3}\right\}$ **47.** $\left\{\pm\dfrac{\sqrt{6}}{2}\right\}$ **49.** $\{2 \pm i\}$

51. $\{1 \pm i\sqrt{3}\}$ **53.** $\left\{\dfrac{1}{2} \pm \dfrac{3}{2}i\right\}$ **55.** $\left\{1 \pm \dfrac{\sqrt{3}}{2}i\right\}$

57. $\{-3.24, 0.87\}$ **59.** $\{-1.99, 3.40\}$ **61.** $0, 1$ **63.** $-4, 0$

65. $172, 2$ **67.** $\left\{-\dfrac{2}{3}, \dfrac{1}{2}\right\}$ **69.** $\{-3, 5\}$ **71.** 1 **73.** 0

75. 2 **77.** $\left\{-\dfrac{1}{3}, \dfrac{5}{3}\right\}$ **79.** $\{\pm\sqrt[4]{2}\}$ **81.** $\left\{-\dfrac{\sqrt{6}}{6}, \dfrac{\sqrt{6}}{12}\right\}$

83. $\{-12, 6\}$ **85.** $\left\{\dfrac{1 \pm \sqrt{5}}{2}\right\}$ **87.** $\{2 \pm \sqrt{3}\}$ **89.** $\{-6, 8\}$

91. $\varnothing$ **93.** $\{1/2\}$ **95.** $r = \pm\sqrt{\dfrac{A}{\pi}}$ **97.** $x = -k \pm \sqrt{k^2 - 3}$

99. $y = x\left(-1 \pm \dfrac{\sqrt{6}}{2}\right)$ **101.** 5000 or $35{,}000$ **103.** 2.5 sec

105. 340 ft **107.** $2\sqrt{205} \approx 28.6$ yd **109.** $18{,}503.4$ pounds

111. $6 - \sqrt{10}$ ft **113.** 10 ft/hr **115.** 4.58 m/sec, 0.93 sec

117. a. $y = -0.067x^2 + 1.26x + 51.14$ where x is years since 1980
b. 2019

119. $\dfrac{9 + \sqrt{53}}{2} \approx 8.14$ days **121.** 40 lb or 20 lb

Section 1.7

For Thought: **1.** T **2.** F **3.** F **4.** T **5.** F **6.** F
7. F **8.** F **9.** F **10.** T

Exercises: **1.** $x < 12$ **3.** $x \geq -7$ **5.** $[-8, \infty)$

7. $(-\infty, \pi/2)$

9. $(5, \infty)$

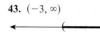

11. $[2, \infty)$

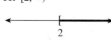

13. $(-\infty, 54)$

15. $(-\infty, 13/3)$

17. $(-\infty, 3/2]$

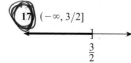

19. $(-\infty, 0]$

21. $(-\infty, -3.5)$ **23.** $(-\infty, 1.4]$ **25.** $(4, \infty)$ **27.** $(-\infty, -2]$
29. $(-3, \infty)$ **31.** $(-3, \infty)$ **33.** $(-5, -2)$ **35.** $\varnothing$
37. $(-\infty, 5]$
39. $(3, 6)$

41. $(1/2, \infty)$

43. $(-3, \infty)$

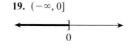

45. $(-\infty, \infty)$

47. $\varnothing$ **49.** $(2, 4)$ **51.** $(-3, 1]$
53. $(-1/3, 1)$

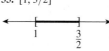

55. $[1, 3/2]$

57. $(-\infty, 0] \cup [2, \infty)$

59. $(-\infty, 2) \cup (8, \infty)$

61. $[-1, 9]$

63. $\varnothing$

65. $\varnothing$

67. $(-\infty, 1) \cup (3, \infty)$

69. $(-\infty, 1) \cup (5, \infty)$

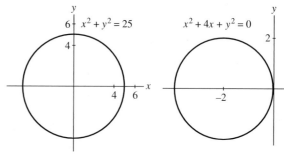

71. $|x| < 5$ **73.** $|x| > 3$ **75.** $|x - 6| < 2$
77. $|x - 4| > 1$ **79.** $|x| \geq 9$ **81.** $|x - 7| \leq 4$
83. $|x - 5| > 2$ **85.** $[2, \infty)$ **87.** $(-\infty, 2)$
89. $(-\infty, -3] \cup [3, \infty)$ **91.** $[\$0, \$7000]$ **93.** $(93, 115)$
95. $(86, 102.5)$ **97.** $(0$ in., 15 in.$)$ **99.** $96, 79, 68, 56, 47$, yes
101. $|x - 74{,}595| > 25{,}000, x > 99{,}595$ or $x < 49{,}595$

103. $\dfrac{|x - 35|}{35} < 0.01, (34.65°, 35.35°)$

105. $[2.26$ cm, 2.32 cm$]$ **107. a.** Florida, Hawaii, Alaska
b. Alabama, Maryland, New Jersey, Connecticut

Chapter 1 Review Exercises

1. $\{2/3\}$ **3.** $\left\{\dfrac{32}{15}\right\}$ **5.** $\{-2\}$ **7.** $\left\{-\dfrac{1}{3}\right\}$

9. $\sqrt{146}, (-1/2, -1/2)$ **11.** $\dfrac{\sqrt{73}}{12}, \left(\dfrac{3}{8}, \dfrac{2}{3}\right)$

13. $(0, 0), 5$ **15.** $(-2, 0), 2$

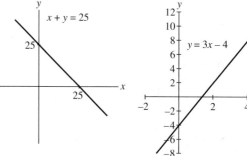

17. $(25, 0), (0, 25)$ **19.** $(4/3, 0), (0, -4)$

21. $(5, 0)$

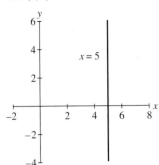

23. $(x + 3)^2 + (y - 5)^2 = 3$ **25.** $(4, 0), (0, -3)$ **27.** -2

29. $y = -\dfrac{4}{7}x + \dfrac{13}{7}$ **31.** $x - 3y = 14$ **33.** $y = \dfrac{2}{3}x - 2$

35. $y = \dfrac{1}{x - 3}$ **37.** $y = -\dfrac{a}{b}x + \dfrac{c}{b}$ **39.** 8, two real

41. 0, one real   **43.** $\{\pm\sqrt{5}\}$ **45.** $\{\pm 2i\sqrt{2}\}$ **47.** $\left\{\pm i\dfrac{\sqrt{2}}{2}\right\}$

49. $\{2 \pm \sqrt{17}\}$ **51.** $\{-3, 4\}$ **53.** $\{3 \pm i\}$ **55.** $\{2 \pm \sqrt{3}\}$

57. $\left\{\dfrac{1 \pm \sqrt{6}}{2}\right\}$ **59.** $\{1 \pm i\}$ **61.** $\left\{\dfrac{1}{3}, 2\right\}$ **63.** $\left\{\dfrac{2}{3}, 2\right\}$

65. $\{3/2\}$ **67.** No solutions

69. $(3, \infty)$

71. $(-\infty, 4)$

73. $(-\infty, -14/3)$

75. $(-1, 13]$

77. $(1/2, 1)$

79. $(-4, \infty)$

81. $(-\infty, 1) \cup (5, \infty)$

83. $\{7/2\}$

85. $(-\infty, \infty)$

87. $\{10\}$ **89.** $(-\infty, 8)$ **91.** $\dfrac{19 - \sqrt{209}}{4} \approx 1.14$ in.

93. 136.4 mi   **95.** 1600   **97.** 20 mi   **99.** $(0, \$12.50)$

101. (7 in., 11.5 in.)   **103.** 2.15×10^9 gal, 31.8 mpg

105. $y = 10.925x - 21,664.625$, 108.9 million

107. $p = 0.008a + 0.07$, 59%

109. a. $C = 263.4\,y - 521,811.4$ **b.** $7536

Chapter 1 Test

1. $\{-7\}$ **2.** $\{1\}$ **3.** $\left\{\pm\dfrac{\sqrt{6}}{3}\right\}$ **4.** $\{3 \pm 2\sqrt{2}\}$ **5.** $\{2, 7\}$

6. $\{0\}$ **7.** $\{1 \pm 2i\}$ **8.** $\{\pm i\}$

9.

10.

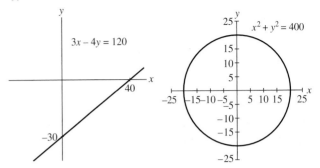

11.

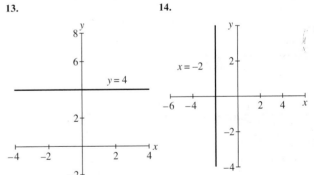

12.

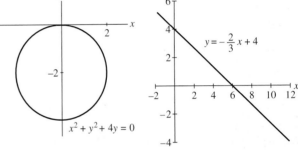

13.

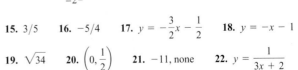

14.

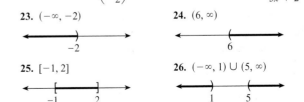

15. $3/5$ **16.** $-5/4$ **17.** $y = -\dfrac{3}{2}x - \dfrac{1}{2}$ **18.** $y = -x - 1$

19. $\sqrt{34}$ **20.** $\left(0, \dfrac{1}{2}\right)$ **21.** -11, none   **22.** $y = \dfrac{1}{3x + 2}$

23. $(-\infty, -2)$

24. $(6, \infty)$

25. $[-1, 2]$

26. $(-\infty, 1) \cup (5, \infty)$

27. 289 ft^2 **28.** 20 gal

29. $y = 3400x + 88,000$ where x is the years since 1994, $159,400

30. a. $y = -14.4x + 2189.8$, $y = -0.286x^2 - 13.257x + 2189.229$
where x is the years since 1999
b. 2,046,000, 2,028,000, 18,000 farms

Tying It All Together Chapters P–1

1. $7x$ **2.** $30x^2$ **3.** $\dfrac{3}{2x}$ **4.** $x^2 + 6x + 9$ **5.** $6x^2 + x - 2$

6. $2x + h$ **7.** $\dfrac{2x}{x^2 - 1}$ **8.** $x^2 + 3x + \dfrac{9}{4}$ **9.** R

10. $\left\{0, \dfrac{11}{30}\right\}$ **11.** $(-\infty, 0) \cup (0, \infty)$ **12.** $\{0\}$ **13.** $\left\{-\dfrac{2}{3}, \dfrac{1}{2}\right\}$

14. $\left\{\dfrac{8 \pm \sqrt{89}}{5}\right\}$ **15.** $\{0, 1\}$ **16.** $\{1\}$ **17.** 0 **18.** -2

19. $\dfrac{9}{8}$ **20.** $\dfrac{44}{27}$ **21.** -8 **22.** -2 **23.** -4 **24.** -2.75

CHAPTER 2

Section 2.1

For Thought: **1.** F **2.** F **3.** T **4.** F **5.** F **6.** F
7. T **8.** T **9.** T **10.** F
Exercises: **1.** Both **3.** a is a function of b **5.** b is a function of a
7. Neither **9.** Both **11.** No **13.** Yes **15.** Yes **17.** Yes
19. No **21.** No **23.** Yes **25.** Yes **27.** Yes **29.** No
31. Yes **33.** Yes **35.** No **37.** $\{-3, 4, 5\}, \{1, 2, 6\}$
39. $(-\infty, \infty), \{4\}$ **41.** $(-\infty, \infty), [5, \infty)$ **43.** $[-3, \infty), (-\infty, \infty)$
45. $[4, \infty), [0, \infty)$ **47.** $(-\infty, 0], (-\infty, \infty)$ **49.** 6 **51.** 11 **53.** 3
55. 7 **57.** 22 **59.** $3a^2 - a$ **61.** $4a + 6$ **63.** $3x^2 + 5x + 2$
65. $4x + 4h - 2$ **67.** $6xh + 3h^2 - h$ **69.** $-$2,400 per yr
71. $-32, -48, -62.4, -63.84,$ and -63.984 ft/sec

73. -10 million hectares per yr **75.** 4 **77.** 3 **79.** $2x + h + 1$

81. $-2x - h + 1$ **83.** $\dfrac{3}{\sqrt{x + h} + \sqrt{x}}$

85. $\dfrac{1}{\sqrt{x + h + 2} + \sqrt{x + 2}}$ **87.** $\dfrac{-1}{x(x + h)}$

89. $\dfrac{-3}{(x + 2)(x + h + 2)}$

91. a. $A = s^2$ **b.** $s = \sqrt{A}$ **c.** $s = \dfrac{d\sqrt{2}}{2}$ **d.** $d = s\sqrt{2}$

e. $P = 4s$ **f.** $s = P/4$ **g.** $A = \dfrac{P^2}{16}$ **h.** $d = \sqrt{2A}$

93. $C = 50 + 35n$

95. a. Amount spent in 2004, $9.6 billion **b.** 2010

97. $h = \left(2\sqrt{3} + 2\right)a$

99. At $18/ticket revenue is increasing at $1,950 per dollar change in ticket price. At $22/ticket revenue is decreasing at $2,050 per dollar change in ticket price.

Section 2.2

For Thought: **1.** T **2.** F **3.** T **4.** T **5.** F **6.** T
7. T **8.** T **9.** F **10.** T
Exercises:
1. $(-\infty, \infty), (-\infty, \infty)$, yes

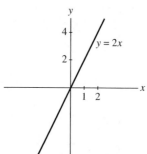

3. $(-\infty, \infty), (-\infty, \infty)$, yes

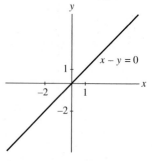

5. $(-\infty, \infty), \{5\}$, yes

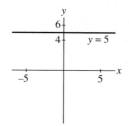

7. $(-\infty, \infty), [0, \infty)$, yes

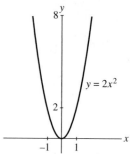

9. $(-\infty, \infty), (-\infty, 1]$, yes

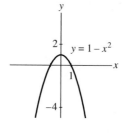

11. $[0, \infty), [1, \infty)$, yes

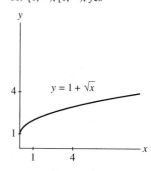

13. $[1, \infty), (-\infty, \infty)$, no

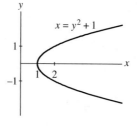

15. $[0, \infty), [0, \infty)$, yes

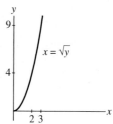

17. $(-\infty, \infty), (-\infty, \infty)$, yes

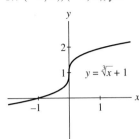

$y = \sqrt[3]{x} + 1$

19. $(-\infty, \infty), (-\infty, \infty)$, yes

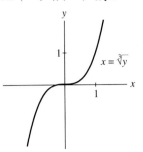

$x = \sqrt[3]{y}$

33. $(-\infty, \infty), \{-2, 2\}$

$$f(x) = \begin{cases} 2 & x < -1 \\ -2 & x \geq -1 \end{cases}$$

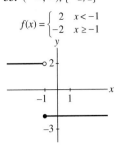

35. $(-\infty, \infty), (-\infty, -2] \cup (2, \infty)$

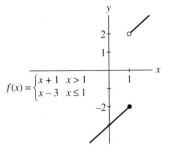

$$f(x) = \begin{cases} x+1 & x > 1 \\ x-3 & x \leq 1 \end{cases}$$

21. $[-1, 1], [-1, 1]$, no

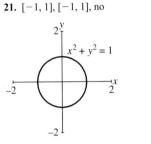

$x^2 + y^2 = 1$

23. $[-1, 1], [0, 1]$, yes

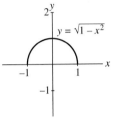

$y = \sqrt{1 - x^2}$

37. $[-2, \infty), (-\infty, 2]$

$$f(x) = \begin{cases} \sqrt{x+2} & -2 \leq x \leq 2 \\ 4 - x & x > 2 \end{cases}$$

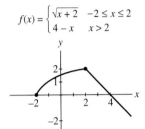

39. $(-\infty, \infty), [0, \infty)$

$$f(x) = \begin{cases} \sqrt{-x} & x < 0 \\ \sqrt{x} & x \geq 0 \end{cases}$$

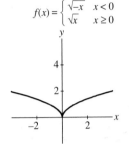

25. $(-\infty, \infty), (-\infty, \infty)$, yes

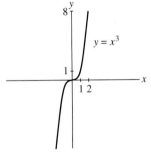

$y = x^3$

27. $(-\infty, \infty), [0, \infty)$, yes

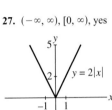

$y = 2|x|$

41. $(-\infty, \infty), (-\infty, \infty)$

$$f(x) = \begin{cases} x^2 & x < -1 \\ -x & x \geq -1 \end{cases}$$

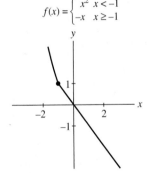

43. $(-\infty, \infty)$, integers

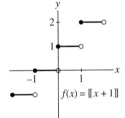

$f(x) = [\![x + 1]\!]$

29. $(-\infty, \infty), (-\infty, 0]$, yes

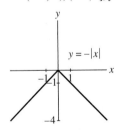

$y = -|x|$

31. $[0, \infty), (-\infty, \infty)$, no

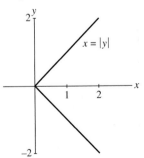

$x = |y|$

45. $[0, 4), \{2, 3, 4, 5\}$

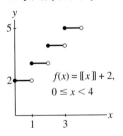

$f(x) = [\![x]\!] + 2,$
$0 \leq x < 4$

47. **a.** $D(-\infty, \infty), R(-\infty, \infty)$, dec $(-\infty, \infty)$
    b. $D(-\infty, \infty), R(-\infty, 4]$, inc $(-\infty, 0)$, dec $(0, \infty)$

49. **a.** $D[-2, 6], R[3, 7]$, inc $(-2, 2)$, dec $(2, 6)$
    b. $D(-\infty, 2], R(-\infty, 3]$, inc $(-\infty, -2)$, constant $(-2, 2)$

51. a. D$(-\infty, \infty)$, R $[0, \infty)$, dec $(-\infty, 0)$, inc $(0, \infty)$
   b. D$(-\infty, \infty)$, R$(-\infty, \infty)$, dec $(-\infty, -2)$ and $(-2/3, \infty)$, inc $(-2, -2/3)$
53. a. D$(-\infty, \infty)$, R$(-\infty, \infty)$, inc $(-\infty, \infty)$
   b. D$[-2, 5]$, R$[1, 4]$, dec $(-2, 1)$, inc $(1, 2)$, constant $(2, 5)$
55. $(-\infty, \infty)$, $(-\infty, \infty)$,   **57.** $(-\infty, \infty)$, $[0, \infty)$,
   inc $(-\infty, \infty)$ dec $(-\infty, 1)$, inc $(1, \infty)$

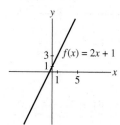

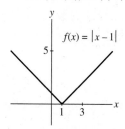

59. $(-\infty, 0) \cup (0, \infty)$, $\{-1, 1\}$,   **61.** $[-3, 3]$, $[0, 3]$,
   constant $(-\infty, 0)$, $(0, \infty)$ inc $(-3, 0)$, dec $(0, 3)$

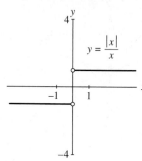

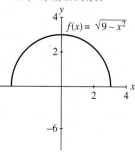

63. $(-\infty, \infty)$, $(-\infty, \infty)$, inc $(-\infty, 3)$, $(3, \infty)$

$$f(x) = \begin{cases} x+1 & x \ge 3 \\ x+2 & x < 3 \end{cases}$$

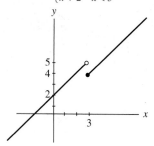

65. $(-\infty, \infty)$, $(-\infty, 2]$,
   inc $(-\infty, -2)$, $(-2, 0)$
   dec $(0, 2)$, $(2, \infty)$

$$f(x) = \begin{cases} x+3 & x \le -2 \\ \sqrt{4-x^2} & -2 < x < 2 \\ -x+3 & x \ge 2 \end{cases}$$

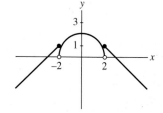

67. $f(x) = \begin{cases} 2 & \text{for } x > -1 \\ -1 & \text{for } x \le -1 \end{cases}$ **69.** $f(x) = \begin{cases} x-1 & \text{for } x \ge -1 \\ -x & \text{for } x < -1 \end{cases}$

71. $f(x) = \begin{cases} 2x-2 & \text{for } x \ge 0 \\ -x-2 & \text{for } x < 0 \end{cases}$

73. Dec $(-\infty, 0.83)$, inc $(0.83, \infty)$
75. Inc $(-\infty, -1)$, $(1, \infty)$, dec $(-1, 1)$
77. Dec $(-\infty, -1.73)$, $(0, 1.73)$, inc $(-1.73, 0)$, $(1.73, \infty)$
79. Inc $(30, 50)$, $(70, \infty)$, dec $(-\infty, 30)$, $(50, 70)$
81. c   **83.** d
85. Inc $(0, 3)$, $(6, 15)$, dec $(3, 6)$, $(30, 39)$, constant $(15, 30)$

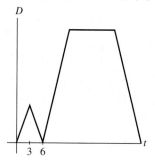

87. Inc $(0, 2)$, dec $(2.5, 4.5)$, constant $(2, 2.5)$

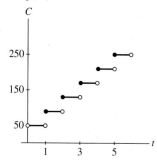

89. 565 million, 800 million, 14.5 million/yr   **91.** $(0, 10^4)$, $(10^4, \infty)$
93. $[5, \infty)$

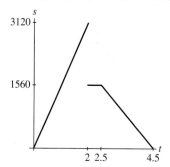

95. $f(x) = \begin{cases} 150 & 0 < x < 3 \\ 50x & 3 \le x \le 10 \end{cases}$

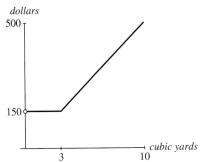

Section 2.3

For Thought: **1.** F   **2.** T   **3.** F   **4.** T   **5.** T   **6.** F
7. T   **8.** F   **9.** T   **10.** T

Exercises:

1.

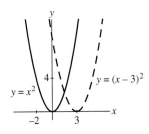

3.

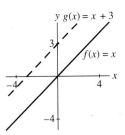

5.

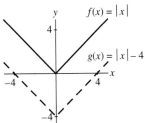

7.

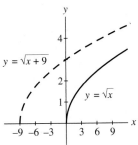

9.

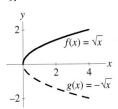

11.

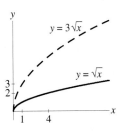

13.

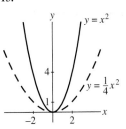

15.

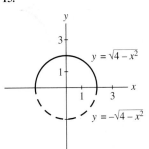

17. g   **19.** b   **21.** c   **23.** f   **25.** $y = \sqrt{x} + 2$
27. $y = (x - 5)^2$ **29.** $y = (x - 10)^2 + 4$ **31.** $y = -3\sqrt{x} - 5$
33. $y = -3|x - 7| + 9$
35. $(-\infty, \infty), [2, \infty)$

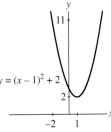

37. $(-\infty, \infty), [3, \infty)$

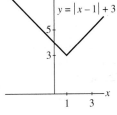

39. $(-\infty, \infty), (-\infty, \infty)$

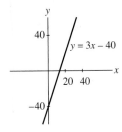

41. $(-\infty, \infty), (-\infty, \infty)$

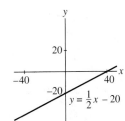

43. $(-\infty, \infty), (-\infty, 40]$

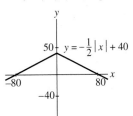

45. $(-\infty, \infty), (-\infty, 0]$ **47.** $[3, \infty), (-\infty, 1]$

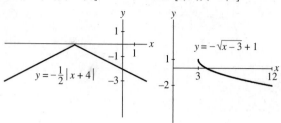

$y = -\frac{1}{2}|x + 4|$

$y = -\sqrt{x - 3} + 1$

49. $[-3, \infty), (-\infty, 2]$

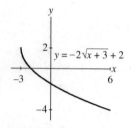

$y = -2\sqrt{x + 3} + 2$

51. y-axis, even   **53.** No symmetry, neither   **55.** $x = -3$, neither
57. $x = 2$, neither   **59.** Origin, odd   **61.** No symmetry, neither
63. No symmetry, neither   **65.** y-axis, even
67. No symmetry, neither   **69.** y-axis, even   **71.** e   **73.** g
75. b   **77.** c   **79.** $(-\infty, -1] \cup [1, \infty)$ **81.** $(-\infty, -1) \cup (5, \infty)$
83. $(-2, 4)$ **85.** $[0, 25]$ **87.** $\left(-\infty, 2 - \sqrt{3}\right) \cup \left(2 + \sqrt{3}, \infty\right)$
89. $(-5, 5)$ **91.** $(-3.36, 1.55)$
93. a.

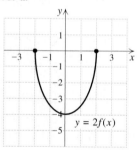

$y = 2f(x)$

b.

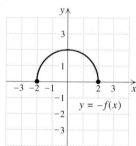

$y = -f(x)$

c.

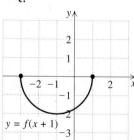

$y = f(x + 1)$

d.

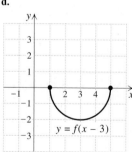

$y = f(x - 3)$

e.

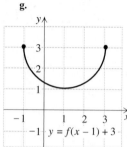

$y = -3f(x)$

f.

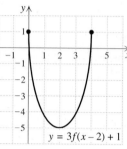

$y = f(x + 2) - 1$

g.

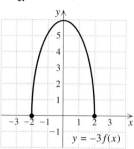

$y = f(x - 1) + 3$

h.

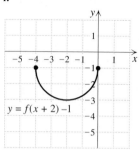

$y = 3f(x - 2) + 1$

95. $N(x) = x + 2000$ **97.** $x > 25\%$
99. a. Even function
 b. Odd function   **c.** Translation one unit to left
 d. Translation two units to right and three units up

Section 2.4

For Thought: **1.** F   **2.** T   **3.** T   **4.** T   **5.** T   **6.** T   **7.** F
8. T   **9.** F   **10.** T
Exercises: **1.** 1   **3.** -11 **5.** -8 **7.** $1/12$ **9.** $a^2 - 3$
11. $a^3 - 4a^2 + 3a$ **13.** $\{(-3, 3), (2, 6)\}, \{-3, 2\}$
15. $\{(-3, -1), (2, -6)\}, \{-3, 2\}$ **17.** $\{(-3, 2), (2, 0)\}, \{-3, 2\}$
19. $\{(-3, 2)\}, \{-3\}$ **21.** $(f + g)(x) = \sqrt{x} + x - 4, [0, \infty)$
23. $(f - h)(x) = \sqrt{x} - \dfrac{1}{x - 2}, [0, 2) \cup (2, \infty)$
25. $(g \cdot h)(x) = \dfrac{x - 4}{x - 2}, (-\infty, 2) \cup (2, \infty)$
27. $(g/f)(x) = \dfrac{x - 4}{\sqrt{x}}, (0, \infty)$
29. $\{(-3, 0), (1, 0), (4, 4)\}$ **31.** $\{(1, 4)\}$ **33.** $\{(-3, 4), (1, 4)\}$
35. 5   **37.** 5   **39.** 59.816   **41.** 5   **43.** 5   **45.** a
47. $3t^2 + 2$ **49.** $(f \circ g)(x) = \sqrt{x} - 2, [0, \infty)$
51. $(f \circ h)(x) = \dfrac{1}{x} - 2, (-\infty, 0) \cup (0, \infty)$
53. $(h \circ g)(x) = \dfrac{1}{\sqrt{x}}, (0, \infty)$
55. $(f \circ f)(x) = x - 4, (-\infty, \infty)$

57. $(h \circ g \circ f)(x) = \dfrac{1}{\sqrt{x-2}}, (2, \infty)$

59. $(h \circ f \circ g)(x) = \dfrac{1}{\sqrt{x-2}}, (0, 4) \cup (4, \infty)$

61. $F = g \circ h$ **63.** $H = h \circ g$ **65.** $N = h \circ g \circ f$

67. $P = g \circ f \circ g$ **69.** $S = g \circ g$ **71.** $g(x) = x^3$ and $h(x) = x - 2$

73. $g(x) = x + 5$ and $h(x) = \sqrt{x}$

75. $g(x) = 3x - 1$ and $h(x) = \sqrt{x},\ g(x) = 3x$ and $h(x) = \sqrt{x-1}$

77. $g(x) = |x|$ and $h(x) = 4x + 5,\ g(x) = 4|x|$ and $h(x) = x + 5$

79. $y = 6x - 1$ **81.** $y = x^2 + 6x + 7$ **83.** $y = x$

85. $y = (n - 4)^2$ **87.** $y = \sqrt{x + 16/8}$ **89.** $y = -x$, no

91. $[-1, \infty), [-7, \infty)$ **93.** $[1, \infty), [0, \infty)$ **95.** $[0, \infty), [4, \infty)$

97. $P(x) = 28x - 200, x \geq 8$ **99.** $A = d^2/2$

101. $(f \circ f)(x) = 0.899x, (f \circ f \circ f)(x) = 0.852x$ **103.** $T(x) = 1.26x$

105. $D = \dfrac{1.16 \times 10^7}{L^3}$ **107.** $W = \dfrac{(8 + \pi)s^2}{8}$ **109.** $s = \dfrac{d\sqrt{2}}{2}$

111. No, composition

Section 2.5

For Thought: **1.** F   **2.** F   **3.** F   **4.** T   **5.** F   **6.** F   **7.** F
8. F   **9.** F   **10.** T

Exercises: **1.** Not invertible   **3.** Invertible   **5.** Invertible

7. Invertible, $\{(3, 9), (2, 2)\}$ **9.** Not invertible

11. Invertible, $\{(3, 3), (2, 2), (4, 4), (7, 7)\}$ **13.** Not invertible

15. $\{(1, 2), (5, 3)\}, 3, 2$ **17.** $\{(-3, -3), (5, 0), (-7, 2)\}, 0, 2$

19. Not one-to-one   **21.** One-to-one   **23.** Not one-to-one

25. One-to-one   **27.** One-to-one   **29.** Not one-to-one

31. Not one-to-one   **33.** One-to-one   **35.** Not invertible

37. Not invertible   **39.** $f^{-1}(x) = \dfrac{x+7}{3}$

41. $f^{-1}(x) = (x - 2)^2 + 3$ for $x \geq 2$ **43.** $f^{-1}(x) = -x - 9$

45. $f^{-1}(x) = \dfrac{5x + 3}{x - 1}$ **47.** $f^{-1}(x) = -\dfrac{1}{x}$

49. $f^{-1}(x) = (x - 5)^3 + 9$ **51.** $f^{-1}(x) = \sqrt{x + 2}$

53. $f(g(x)) = x, g(f(x)) = x$, yes   **55.** $f(g(x)) = x, g(f(x)) = |x|$, no

57. $f(g(x)) = x, g(f(x)) = x$, yes   **59.** $f(g(x)) = x, g(f(x)) = x$, yes
61. The functions y_1 and y_2 are inverses.

63. No   **65.** Yes

67.   **69.**

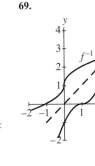

71. $f^{-1}(x) = \dfrac{x - 2}{3}$ **73.** $f^{-1}(x) = \sqrt{x + 4}$

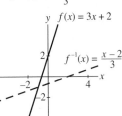   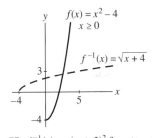

75. $f^{-1}(x) = \sqrt[3]{x}$ **77.** $f^{-1}(x) = (x + 3)^2$ for $x \geq -3$

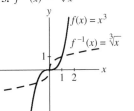   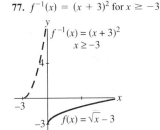

79. a. $f^{-1}(x) = \dfrac{x}{5}$ **b.** $f^{-1}(x) = x + 88$ **c.** $f^{-1}(x) = \dfrac{x + 7}{3}$

d. $f^{-1}(x) = \dfrac{x - 4}{-3}$ **e.** $f^{-1}(x) = 2x + 18$ **f.** $f^{-1}(x) = -x$

g. $f^{-1}(x) = (x + 9)^3$ **h.** $f^{-1}(x) = \sqrt[3]{\dfrac{x + 7}{3}}$

81. $C = 1.08P, P = \dfrac{C}{1.08}$ **83.** Yes, $r = \dfrac{7.89 - t}{0.39}, 6$

85. $w = \dfrac{V^2}{1.496}, 8840$ lb   **87. a.** 10.9%   **b.** $V = 50,000(1 - r)^5$

93. $1 + \dfrac{-5}{x + 2}$

Section 2.6

For Thought: **1.** F   **2.** F   **3.** T   **4.** T   **5.** T   **6.** F   **7.** T
8. T   **9.** T   **10.** F

Exercises: **1.** $G = kn$ **3.** $V = \dfrac{k}{P}$ **5.** $C = khr$ **7.** $Y = \dfrac{kx}{\sqrt{z}}$

9. A varies directly as the square of r.

11. The variable y varies inversely as x.   **13.** No variation

15. The variable a varies jointly as z and w.

17. H varies directly as the square root of t and inversely as s.

19. D varies jointly as L and J, and inversely as W.

21. $y = \dfrac{5}{9}x$ **23.** $T = \dfrac{-150}{y}$ **25.** $m = 3t^2$ **27.** $y = \dfrac{1.37x}{\sqrt{z}}$

29. $-27/2$ **31.** 1   **33.** $\sqrt{6}$ **35.** 7/4

37. Direct, $L_i = 12L_f$ **39.** Inverse, $P = 20/n$

41. Direct, $S_m = 0.6S_k$ **43.** Neither   **45.** Direct, $A = 30W$

47. Inverse, $n = \dfrac{5}{p}$ **49.** 2604 lb/in.^{2} **51.** 12.8 hr   **53.** \$50.70

55. \$19.84   **57.** 18.125 oz   **59.** 8 ft/yr   **61.** No   **63.** 38

65. 35.96 ft/sec

Chapter 2 Review Exercises

1. $\{-2, 0, 1\}, \{-2, 0, 1\}$, yes

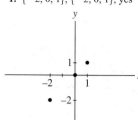

3. $(-\infty, \infty), (-\infty, \infty)$, yes

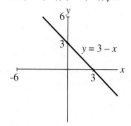

45.

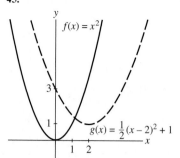

5. $\{2\}, (-\infty, \infty)$, no

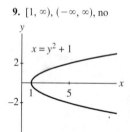

7. $[-0.1, 0.1], [-0.1, 0.1]$, no

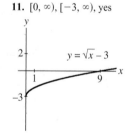

47.

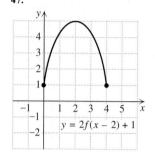

49.

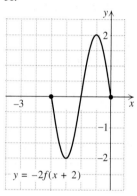

9. $[1, \infty), (-\infty, \infty)$, no

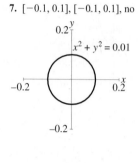

11. $[0, \infty), [-3, \infty)$, yes

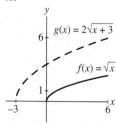

51.

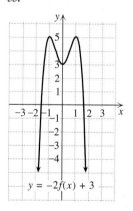

53.

13. 12   **15.** 17   **17.** ± 4 **19.** 17   **21.** 4   **23.** -36

25. 12   **27.** $4x^2 - 28x + 52$ **29.** $x^4 + 6x^2 + 12$

31. $a^2 + 2a + 4$ **33.** $6 + h$ **35.** $2x + h$ **37.** x

39. $\dfrac{x + 7}{2}$

41.

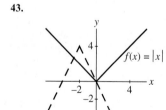

43.

55. $F = f \circ g$ **57.** $H = f \circ h \circ g \circ j$ **59.** $N = h \circ f \circ j$

61. $R = g \circ h \circ j$ **63.** -5 **65.** $\dfrac{-1}{2x(x + h)}$

67. $[-10, 10], [0, 10]$,
inc $(-10, 0)$, dec $(0, 10)$

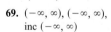

69. $(-\infty, \infty), (-\infty, \infty)$,
inc $(-\infty, \infty)$

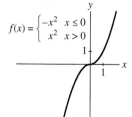

71. $(-\infty, \infty), [-2, \infty),$
inc $(-2, 0), (2, \infty),$
dec $(-\infty, -2)$ and $(0, 2)$

$$f(x) = \begin{cases} -x - 4 & x \le -2 \\ -|x| & -2 < x < 2 \\ x - 4 & x \ge 2 \end{cases}$$

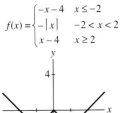

73. $y = |x| - 3, (-\infty, \infty), [-3, \infty)$
75. $y = -2|x| + 4, (-\infty, \infty), (-\infty, 4]$
77. $y = |x + 2| + 1, (-\infty, \infty), [1, \infty)$
79. *y*-axis   **81.** Origin   **83.** Neither symmetry   **85.** *y*-axis
87. $(-\infty, 2] \cup [4, \infty)$ **89.** $\left(-\sqrt{2}, \sqrt{2}\right)$ **91.** $\varnothing$

93. Inverse functions   **95.** Inverse functions

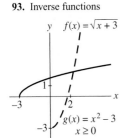   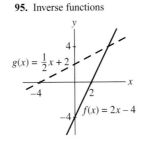

97. Not invertible   **99.** $f^{-1}(x) = \dfrac{x + 21}{3}, (-\infty, \infty), (-\infty, \infty)$

101. Not invertible
103. $f^{-1}(x) = x^2 + 9$ for $x \ge 0, [0, \infty), [9, \infty)$
105. $f^{-1}(x) = \dfrac{5x + 7}{1 - x}, (-\infty, 1) \cup (1, \infty), (-\infty, -5) \cup (-5, \infty)$
107. $f^{-1}(x) = -\sqrt{x - 1}, [1, \infty), (-\infty, 0]$
109. $C(x) = 1.20x + 40, R(x) = 2x, P(x) = 0.80x - 40$ where x is the number of roses, 51 or more roses

111. $t = \dfrac{\sqrt{64 - h}}{4}$, domain $[0, 64]$ **113.** $d = 2\sqrt{A/\pi}$

115. 0.5 in./lb   **117.** 36   **119.** 61 km/hr   **121.** \$30.72

Chapter 2 Test

1. No   **2.** Yes   **3.** No   **4.** Yes   **5.** $\{2, 5\}, \{-3, -4, 7\}$
6. $[9, \infty), [0, \infty)$ **7.** $[0, \infty), (-\infty, \infty)$
8.   **9.**

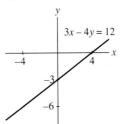   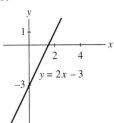

10.   **11.**

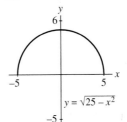   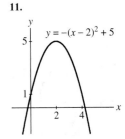

12.   **13.**

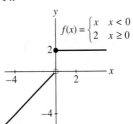   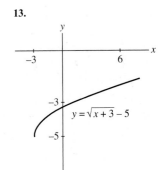

14.

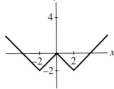

15. 3   **16.** $\sqrt{7}$ **17.** $\sqrt{3x + 1}$ **18.** $\dfrac{x + 1}{3}$

19. 45   **20.** 3   **21.** Inc $(3, \infty)$, dec $(-\infty, 3)$ **22.** *y*-axis

23. $(-2, 4)$ **24.** $g^{-1}(x) = (x - 3)^3 + 2$ **25.** \$0.125 per envelope

26. 12 candlepower   **27.** $V = \dfrac{\sqrt{2}\, d^3}{4}$

Tying It All Together Chapters 1–2

1. $\{3/2\}$ **2.** $\{3\}$ **3.** $\{\pm 100\}$ **4.** $\{-89.5, -90.5\}$

5. $\{-27.75\}$ **6.** $\varnothing$ **7.** $\left\{\dfrac{4 \pm \sqrt{2}}{2}\right\}$ **8.** $\{-5/2, -3/2\}$

9. $\left\{\pm\sqrt{5}\right\}$ **10.** $\varnothing$

11. $(-\infty, \infty), (-\infty, \infty), (3/2, 0)$

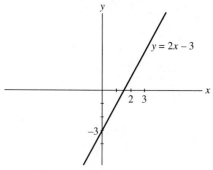

12. $(-\infty, \infty), (-\infty, \infty), (3, 0)$

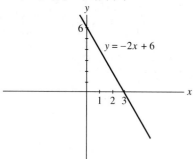

13. $(-\infty, \infty), [-100, \infty), (\pm 100, 0)$

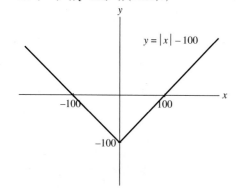

14. $(-\infty, \infty), (-\infty, 1], (-89.5, 0), (-90.5, 0)$

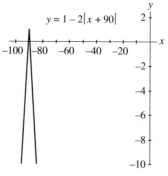

15. $[-30, \infty), (-\infty, 3], (-27.75, 0)$

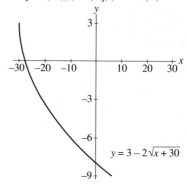

16. $[3, \infty), [15, \infty),$ no x-intercepts

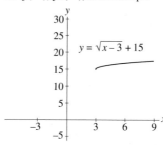

17. $(-\infty, \infty), (-\infty, 1], \left(\dfrac{4 \pm \sqrt{2}}{2}, 0\right)$

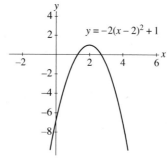

18. $(-\infty, \infty), [-1, \infty), (-5/2, 0), (-3/2, 0)$

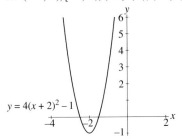

$y = 4(x + 2)^2 - 1$

19. $[-3, 3], [-2, 1], (\pm\sqrt{5}, 0)$

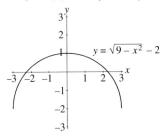

$y = \sqrt{9 - x^2} - 2$

20. $[-7, 7], [3, 10],$ no x-intercepts

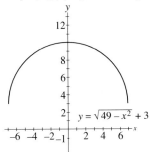

$y = \sqrt{49 - x^2} + 3$

21. $(3/2, \infty)$ **22.** $[3, \infty)$ **23.** $(-\infty, -100] \cup [100, \infty)$
24. $(-90.5, -89.5)$ **25.** $[-27.75, \infty)$ **26.** $(-\infty, \infty)$

27. $\left(-\infty, \dfrac{4 - \sqrt{2}}{2}\right) \cup \left(\dfrac{4 + \sqrt{2}}{2}, \infty\right)$

28. $(-\infty, -5/2] \cup [-3/2, \infty)$ **29.** $\left[-\sqrt{5}, \sqrt{5}\right]$
30. $\varnothing$ **31.** 57
32.

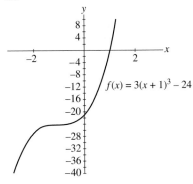

$f(x) = 3(x + 1)^3 - 24$

33. $f = K \circ F \circ H \circ G$ **34.** $\{1\}$ **35.** $[1, \infty)$

36. $x = \sqrt[3]{\dfrac{y + 24}{3}} - 1$ **37.** $f^{-1}(x) = \sqrt[3]{\dfrac{x + 24}{3}} - 1$

38.

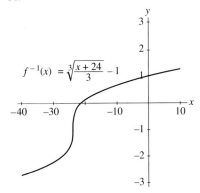

$f^{-1}(x) = \sqrt[3]{\dfrac{x + 24}{3}} - 1$

39. $(-21, \infty)$ **40.** $f^{-1} = G^{-1} \circ H^{-1} \circ F^{-1} \circ K^{-1}$

CHAPTER 3

Section 3.1

For Thought: **1.** F   **2.** F   **3.** T   **4.** T   **5.** T   **6.** T
7. T   **8.** T   **9.** T   **10.** F
Exercises:
 1. $y = (x + 2)^2 - 4$

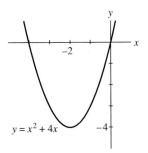

$y = x^2 + 4x$

3. $y = \left(x - \dfrac{3}{2}\right)^2 - \dfrac{9}{4}$

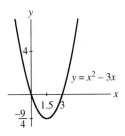

$y = x^2 - 3x$

5. $y = 2(x - 3)^2 + 4$

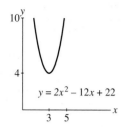

$y = 2x^2 - 12x + 22$

7. $y = -3(x - 1)^2$

$y = -3x^2 + 6x - 3$

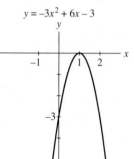

9. $y = \left(x + \dfrac{3}{2}\right)^2 + \dfrac{1}{4}$

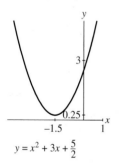

$y = x^2 + 3x + \dfrac{5}{2}$

11. $y = -2\left(x - \dfrac{3}{4}\right)^2 + \dfrac{1}{8}$

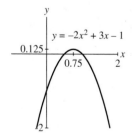

$y = -2x^2 + 3x - 1$

13. $(2, -11)$ **15.** $(4, 1)$ **17.** $(-1/3, 1/18)$
19. Up, $(1, -4), x = 1, [-4, \infty)$, min -4, dec $(-\infty, 1)$, inc $(1, \infty)$
21. $(-\infty, 3]$, max 3, inc $(-\infty, 0)$, dec $(0, \infty)$
23. $[-1, \infty)$, min -1, dec $(-\infty, 1)$, inc $(1, \infty)$
25. $[-18, \infty)$, min value -18, dec $(-\infty, -4)$, inc $(-4, \infty)$
27. $[4, \infty)$, min value 4, dec $(-\infty, 3)$, inc $(3, \infty)$
29. $(-\infty, 27/2]$, max 27/2, inc $(-\infty, 3/2)$, dec $(3/2, \infty)$
31. $(-\infty, 9]$, max 9, inc $(-\infty, 1/2)$, dec $(1/2, \infty)$

33. $(0, -3), x = 0, (0, -3),$ **35.** $\left(\dfrac{1}{2}, -\dfrac{1}{4}\right), x = \dfrac{1}{2},$
$\left(\pm\sqrt{3}, 0\right)$, up $(0, 0), (1, 0)$, up

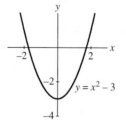

$y = x^2 - 3$

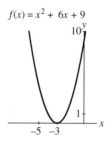

$y = x^2 - x$

37. $(-3, 0), x = -3, (0, 9),$ **39.** $(3, -4), x = 3, (0, 5), (1, 0),$
$(-3, 0)$, up $(5, 0)$, up

$f(x) = x^2 + 6x + 9$

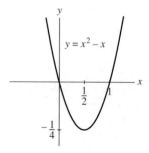

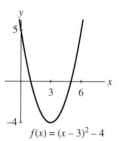

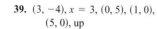

$f(x) = (x - 3)^2 - 4$

41. $(2, 12), x = 2, (0, 0),$ **43.** $(1, 3), x = 1, (0, 1),$
$(4, 0)$, down $\left(1 \pm \dfrac{\sqrt{6}}{2}, 0\right)$, down

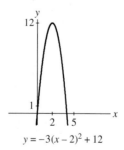

$y = -3(x - 2)^2 + 12$

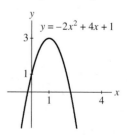

$y = -2x^2 + 4x + 1$

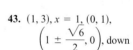

45. $(-1, 3/2)$

47. $(-\infty, -3) \cup (5, \infty)$

49. $(-\infty, -2] \cup [6, \infty)$

51. $[-4, 4]$

53. $\{-3\}$

55. $(-\infty, 3/2) \cup (3/2, \infty)$

57. $\left(2 - \sqrt{2}, 2 + \sqrt{2}\right)$

59. $\left(-\infty, -\sqrt{10}\right) \cup \left(\sqrt{10}, \infty\right)$

61. $\left(-\infty, 5 - \sqrt{7}\right) \cup$
$\left(5 + \sqrt{7}, \infty\right)$

63. $(-\infty, \infty)$

65. No solution
67. $(-\infty, \infty)$

69. $(-\infty, -1] \cup [3, \infty)$ **71.** $(-3, 1)$ **73.** $[-3, 1]$
75. a. $\{-2, 5\}$ **b.** $\{0, 3\}$ **c.** $(-\infty, -2) \cup (5, \infty)$ **d.** $[-2, 5]$
   e. $f(x) = \left(x - \frac{3}{2}\right)^2 - \frac{49}{4}$, Move $y = x^2$ to the right $\frac{3}{2}$ and down $\frac{49}{4}$ to obtain f.
   f. $(-\infty, \infty), \left[-\frac{49}{4}, \infty\right)$, minimum $-\frac{49}{4}$

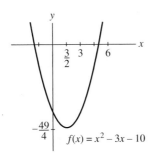

   g. The graph of f is above the x-axis when x is in $(-\infty, -2) \cup (5, \infty)$ and on or below the x-axis when x is in $[-2, 5]$.
   h. $(-2, 0), (5, 0), (0, -10), x = \frac{3}{2}, \left(\frac{3}{2}, -\frac{49}{4}\right)$, opens upward, dec on $\left(-\infty, \frac{3}{2}\right)$ and inc on $\left(\frac{3}{2}, \infty\right)$

77. 261 ft
79. a. 408 ft   **b.** $(10 + \sqrt{102})/2 \approx 10.05$ sec
81. a. Approximately 100 mph   **b.** 97.24 mph   **c.** 13.2 gal/hr
83. 50 yd by 50 yd   **85.** 20 ft by 30 ft   **87.** 7.5 ft by 15 ft
89. 5 in. wide, 2.5 in. high
91. a. $p = 50 - n$ **b.** $R = 50n - n^2$ **c.** \$625
93. 1/2
95. a.

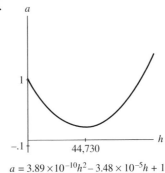

$$a = 3.89 \times 10^{-10} h^2 - 3.48 \times 10^{-5} h + 1$$

   b. Decreasing   **c.** Dec $(0, 44{,}730)$, inc $(44{,}730, \infty)$ **d.** No
   e. Approximately $(0, 30{,}000)$
97. a. $y = -2655x + 40{,}032, y = 45x^2 - 3105x + 40{,}858$
   b. Too close to call   **c.** \$10,827, \$12,149, quadratic

Section 3.2

For Thought: **1.** F   **2.** T   **3.** T   **4.** T   **5.** F   **6.** F
   7. F   **8.** T   **9.** T   **10.** F
Exercises: **1.** $x - 3, 1$ **3.** $-2x^2 + 6x - 14, 33$ **5.** $s^2 + 2, 16$
   7. $x + 6, 13$ **9.** $-x^2 + 4x - 16, 57$ **11.** $4x^2 + 2x - 4, 0$
   13. $2a^2 - 4a + 6, 0$ **15.** $x^3 + x^2 + x + 1, -2$
   17. $x^4 + 2x^3 - 2x^2 - 4x - 4, -13$
   19. 0   **21.** -33 **23.** 5   **25.** $\dfrac{55}{8}$ **27.** 0   **29.** 8
31. $(x + 3)(x + 2)(x - 1)$ **33.** $(x - 4)(x + 3)(x + 5)$ **35.** Yes
37. No   **39.** Yes   **41.** No   **43.** $\pm(1, 2, 3, 4, 6, 8, 12, 24)$
45. $\pm(1, 3, 5, 15)$
47. $\pm\left(1, 3, 5, 15, \dfrac{1}{2}, \dfrac{1}{4}, \dfrac{1}{8}, \dfrac{3}{2}, \dfrac{3}{4}, \dfrac{3}{8}, \dfrac{5}{2}, \dfrac{5}{4}, \dfrac{5}{8}, \dfrac{15}{2}, \dfrac{15}{4}, \dfrac{15}{8}\right)$
49. $\pm\left(1, 2, \dfrac{1}{2}, \dfrac{1}{3}, \dfrac{2}{3}, \dfrac{1}{6}, \dfrac{1}{9}, \dfrac{2}{9}, \dfrac{1}{18}\right)$
51. $2, 3, 4$ **53.** $-3, 2 \pm i$ **55.** $\dfrac{1}{2}, \dfrac{3}{2}, \dfrac{5}{2}$ **57.** $\dfrac{1}{2}, \dfrac{1 \pm i}{3}$
59. $\pm i, 1, -2$ **61.** $-1, \pm\sqrt{2}$ **63.** $\dfrac{1}{4}, \dfrac{1}{3}, \dfrac{1}{2}$ **65.** $\dfrac{1}{16}, 1 \pm 2i$
67. $-\dfrac{6}{7}, \dfrac{7}{3}, \pm i$ **69.** $-5, -2, 1, \pm 3i$ **71.** $1, 3, 5, 2 \pm \sqrt{3}$
73. $2 + \dfrac{5}{x - 2}$ **75.** $a + \dfrac{5}{a - 3}$ **77.** $1 + \dfrac{-3c}{c^2 - 4}$
79. $2 + \dfrac{-7}{2t + 1}$
81. a. 6 hr   **b.** ≈ 120 ppm   **c.** ≈ 3 hr   **d.** ≈ 4 hr
83. 5 in. by 9 in. by 14 in.   **85.** $c = 10$

Section 3.3

For Thought: **1.** F   **2.** T   **3.** T   **4.** F   **5.** F   **6.** T
7. F   **8.** F   **9.** T   **10.** T
Exercises: **1.** Degree 2, 5 with multiplicity 2
3. Degree 5, ± 3, 0 with multiplicity 3
5. Degree 4, 0, 1 each with multiplicity 2
7. Degree 4, $-\frac{4}{3}, \frac{3}{2}$ each with multiplicity 2

9. Degree 3, 0, $2 \pm \sqrt{10}$ **11.** $x^2 + 9$ **13.** $x^2 - 2x - 1$
15. $x^2 - 6x + 13$ **17.** $x^3 - 8x^2 + 37x - 50$
19. $x^2 - 2x - 15 = 0$ **21.** $x^2 + 16 = 0$ **23.** $x^2 - 6x + 10 = 0$
25. $x^3 + 2x^2 + x + 2 = 0$ **27.** $x^3 + 3x = 0$
29. $x^3 - 5x^2 + 8x - 6 = 0$ **31.** $x^3 - 6x^2 + 11x - 6 = 0$
33. $x^3 - 5x^2 + 17x - 13 = 0$ **35.** $24x^3 - 26x^2 + 9x - 1 = 0$
37. $x^4 - 2x^3 + 3x^2 - 2x + 2 = 0$ **39.** 3 neg; 1 neg, 2 imag
41. 1 pos, 2 neg; 1 pos, 2 imag   **43.** 4 imag
45. 4 pos; 2 pos, 2 imag; 4 imag   **47.** 4 imag and 0
49. $-1 < x < 3$ **51.** $-3 < x < 2$ **53.** $-1 < x \le 5$
55. $-1 < x < 3$ **57.** $-2, 1, 5$ **59.** $-3, \dfrac{3 \pm \sqrt{13}}{2}$
61. $\pm i, 2, -4$ **63.** $-5, \dfrac{1}{3}, \dfrac{1}{2}$
65. 1, -2 each with multiplicity 2   **67.** 0, 2 with multiplicity 3
69. 0, 1, ± 2, $\pm i\sqrt{3}$ **71.** $-2, -1, 1/4, 1, 3/2$
73. $-5 < x < 6, -5 < x < 6$
75. $-6 < x < 6, -5 < x < 5$ **77.** $-1 < x < 23, -1 < x < 23$
79. 4 hr and 5 hr   **81.** 3 in.   **87.** $f(x) = -\dfrac{1}{2}x^3 + 3x^2 - \dfrac{11}{2}x + 3$

Section 3.4

For Thought: **1.** F   **2.** F   **3.** F   **4.** F   **5.** T   **6.** F
7. F   **8.** T   **9.** T   **10.** F

Exercises: **1.** $\{\pm 2, -3\}$ **3.** $\left\{-500, \pm\dfrac{\sqrt{2}}{2}\right\}$

5. $\left\{0, \dfrac{15 \pm \sqrt{205}}{2}\right\}$ **7.** $\{0, \pm 2\}$ **9.** $\{\pm 2, \pm 2i\}$

11. $\{8\}$ **13.** $\{25\}$ **15.** $\left\{\dfrac{1}{4}\right\}$ **17.** $\left\{\dfrac{2 + \sqrt{13}}{9}\right\}$

19. $\{-4, 6\}$ **21.** $\{9\}$ **23.** $\{5\}$ **25.** $\{10\}$ **27.** $\{\pm 2\sqrt{2}\}$
29. $\left\{\pm\dfrac{1}{8}\right\}$ **31.** $\left\{\dfrac{1}{49}\right\}$ **33.** $\left\{\dfrac{5}{4}\right\}$ **35.** $\{\pm 3, \pm\sqrt{3}\}$

37. $\left\{-\dfrac{17}{2}, \dfrac{13}{2}\right\}$ **39.** $\left\{\dfrac{3}{20}, \dfrac{4}{15}\right\}$ **41.** $\{-2, -1, 5, 6\}$

43. $\{1, 9\}$ **45.** $\{9, 16\}$ **47.** $\{8, 125\}$ **49.** $\{\pm\sqrt{7}, \pm 1\}$

51. $\{0, 8\}$ **53.** $\{-3, 0, 1, 4\}$ **55.** $\{-2, 4\}$ **57.** $\left\{\dfrac{1}{2}\right\}$

59. $\{\pm 2, -1 \pm i\sqrt{3}, 1 \pm i\sqrt{3}\}$ **61.** $\{\sqrt{3}, 2\}$ **63.** $\{-2, \pm 1\}$
65. $\{5 \pm 9i\}$ **67.** $\left\{\dfrac{1 \pm 4\sqrt{2}}{3}\right\}$ **69.** $\{\pm 2\sqrt{6}, \pm\sqrt{35}\}$

71. $\{\pm 3, 2\}$ **73.** $\{-11\}$ **75.** $\{2\}$ **77.** 279.56 m^2 **79.** 23

81. $\dfrac{25}{4}$ and $\dfrac{49}{4}$ **83.** 5 in.   **85.** 1600 ft^2 **87.** 17,419.3 lbs

89. 462.89 in.^{3} **91.** 1 P.M.   **93. a.** \$2.49 billion   **b.** \$2.96 billion

95. 27.4 m

Section 3.5

For Thought: **1.** F   **2.** T   **3.** T   **4.** F   **5.** T   **6.** T
7. F   **8.** F   **9.** T   **10.** F
Exercises: **1.** Symmetric about y-axis   **3.** Symmetric about $x = 3/2$
5. Neither symmetry   **7.** Symmetric about origin
9. Symmetric about $x = 5$ **11.** Symmetric about origin
13. Does not cross at $(4, 0)$ **15.** Crosses at $(1/2, 0)$
17. Crosses at $(1/4, 0)$ **19.** No x-intercepts
21. Does not cross at $(0, 0)$, crosses at $(3, 0)$
23. Crosses at $(1/2, 0)$, does not cross at $(1, 0)$
25. Does not cross at $(-3, 0)$, crosses at $(2, 0)$ **27.** $y \to \infty$
29. $y \to -\infty$ **31.** $y \to -\infty$ **33.** $y \to \infty$ **35.** $y \to \infty$
37. Neither symmetry; crosses at $(-2, 0)$; does not cross at $(1, 0)$; $y \to \infty$
as $x \to \infty$; $y \to -\infty$ as $x \to -\infty$
39. Symmetric about y-axis; no x-intercepts; $y \to \infty$ as $x \to \infty$; $y \to \infty$ as
$x \to -\infty$

41.

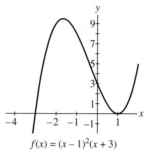

$f(x) = (x-1)^2(x+3)$

43.

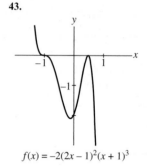

$f(x) = -2(2x-1)^2(x+1)^3$

45. (e)   **47.** (g)   **49.** (b)
51. (c)
53.

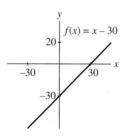

$f(x) = x - 30$

55.

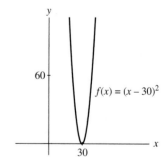

$f(x) = (x-30)^2$

57.

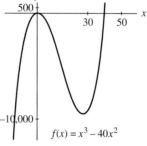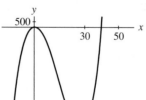

$f(x) = x^3 - 40x^2$

59.

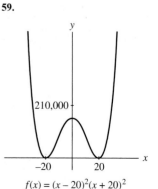

$f(x) = (x-20)^2(x+20)^2$

61.

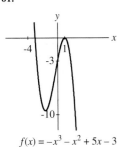

$f(x) = -x^3 - x^2 + 5x - 3$

63.

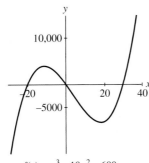

$f(x) = x^3 - 10x^2 - 600x$

87.

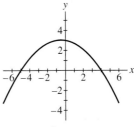

$y = -\dfrac{3}{20}(x + 5)(x - 4)$

89.

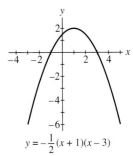

$y = -\dfrac{1}{2}(x + 1)(x - 3)$

65.

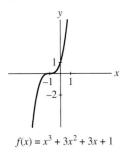

$f(x) = x^3 + 18x^2 - 37x + 60$

67.

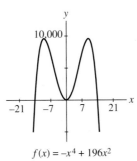

$f(x) = -x^4 + 196x^2$

91.

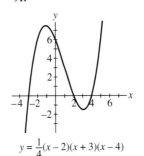

$y = \dfrac{1}{4}(x - 2)(x + 3)(x - 4)$

93.

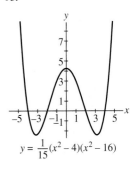

$y = \dfrac{1}{15}(x^2 - 4)(x^2 - 16)$

95. Loc max value 3.11, loc min value 0.37
97. Loc max value 23.74, loc min value -163.74
99. Loc max value 21.01, loc min value 13.99
101. \$3,400, \$2,600, $x > \$2,200$
103. Decreases to 0 at 30 stores, then increases.
105. $V = 3x^3 - 24x^2 + 48x$, 4/3 in. by 4 in. by 16/3 in.
107. All of the paint coats the inside of can. $r = 6.67 \times 10^{-4}$ ft and $h = 11,924.69$ ft, $r = 2.82$ ft and $h = 6.67 \times 10^{-4}$ ft

Section 3.6

For Thought: **1.** F   **2.** F   **3.** F   **4.** F   **5.** T   **6.** F
7. T   **8.** F   **9.** T   **10.** T
Exercises: **1.** $(-\infty, -2) \cup (-2, \infty)$
3. $(-\infty, -2) \cup (-2, 2) \cup (2, \infty)$
5. $(-\infty, 3) \cup (3, \infty)$ **7.** $(-\infty, 0) \cup (0, \infty)$
9. $(-\infty, -1) \cup (-1, 0) \cup (0, 1) \cup (1, \infty)$
11. $(-\infty, -3) \cup (-3, -2) \cup (-2, \infty)$
13. $(-\infty, 2) \cup (2, \infty), y = 0, x = 2$
15. $(-\infty, 0) \cup (0, \infty), y = x, x = 0$
17. $x = 2, y = 0$ **19.** $x = \pm 3, y = 0$ **21.** $x = 1, y = 2$
23. $x = 0, y = x - 2$ **25.** $x = -1, y = 3x - 3$
27. $x = -2, y = -x + 6$

69.

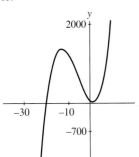

$f(x) = x^3 + 3x^2 + 3x + 1$

71.

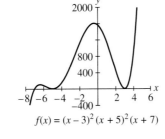

$f(x) = (x - 3)^2(x + 5)^2(x + 7)$

73. $\left(-\sqrt{3}, 0\right) \cup \left(\sqrt{3}, \infty\right)$ **75.** $\left(-\infty, -\sqrt{2}\right] \cup \{0\} \cup \left[\sqrt{2}, \infty\right)$
77. $(-4, -1) \cup (1, \infty)$ **79.** $[-4, 2] \cup [6, \infty)$ **81.** $(-\infty, 1)$
83. $\left[-\sqrt{10}, -3\right] \cup \left[3, \sqrt{10}\right]$ **85.** d

29. $x = 0, y = 0$

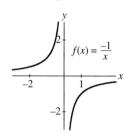

$f(x) = \dfrac{-1}{x}$

31. $x = 2, y = 0, (0, -1/2)$

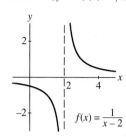

$f(x) = \dfrac{1}{x-2}$

41. $x = \pm 1, y = 0, (0, 0)$

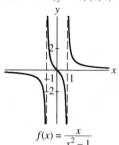

$f(x) = \dfrac{x}{x^2 - 1}$

43. $x = 1, y = 0, (0, 0)$

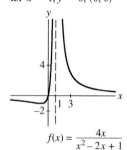

$f(x) = \dfrac{4x}{x^2 - 2x + 1}$

33. $x = \pm 2, y = 0, (0, -1/4)$

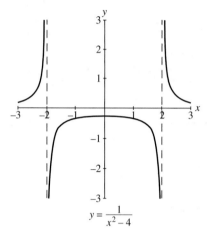

$y = \dfrac{1}{x^2 - 4}$

45. $x = \pm 3, y = -1, (0, -8/9),$
$(\pm \sqrt{8}, 0)$

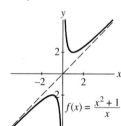

$f(x) = \dfrac{8 - x^2}{x^2 - 9}$

47. $x = -1, y = 2, (0, 2),$
$(-2 \pm \sqrt{3}, 0)$

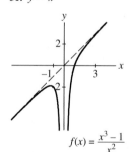

$f(x) = \dfrac{2x^2 + 8x + 2}{x^2 + 2x + 1}$

35. $x = -1, y = 0, (0, -1)$

$f(x) = \dfrac{-1}{(x+1)^2}$

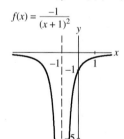

49. $y = x$

$f(x) = \dfrac{x^2 + 1}{x}$

51. $y = x$

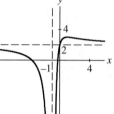

$f(x) = \dfrac{x^3 - 1}{x^2}$

37. $x = 1, y = 2, (0, -1),$
$\left(-\dfrac{1}{2}, 0\right)$

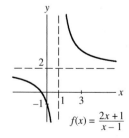

$f(x) = \dfrac{2x + 1}{x - 1}$

39. $x = -2, y = 1, (3, 0),$
$(0, -3/2)$

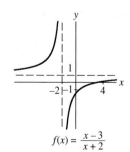

$f(x) = \dfrac{x - 3}{x + 2}$

53. $y = x - 1$

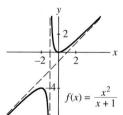

$f(x) = \dfrac{x^2}{x + 1}$

55. $y = 2x + 1$

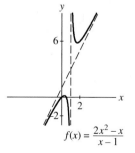

$f(x) = \dfrac{2x^2 - x}{x - 1}$

57. (e)   **59.** (a)   **61.** (b)   **63.** (c)
65. $x \neq \pm 1$ **67.** $x \neq 1$

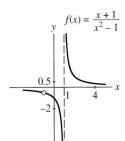

$$f(x) = \frac{x+1}{x^2-1}$$

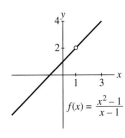

$$f(x) = \frac{x^2-1}{x-1}$$

99.

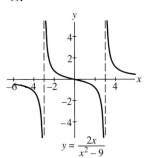

$$y = \frac{2x}{x^2-9}$$

101.

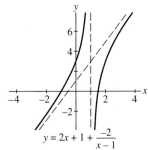

$$y = 2x + 1 + \frac{-2}{x-1}$$

103. $C = \dfrac{100+x}{x}, \$2, C \to \$1$ **105.** $S = \dfrac{100}{4-x}, S \to \infty$ as $x \to 4$

107. a. ≈ 15 min   **b.** ≈ 220 PPM
c. $t = 0, PPM = 0,$ Low concentration for long time or high concentration for short time will give permanent brain damage.

109. a. $h = 500/(\pi r^2)$
b. $S = 2\pi r^2 + 1000/r$

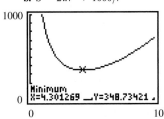

c. 4.3 ft   **d.** \$2789.87

69.

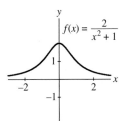

$$f(x) = \frac{2}{x^2+1}$$

71.

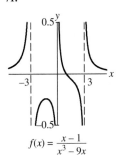

$$f(x) = \frac{x-1}{x^3-9x}$$

73.

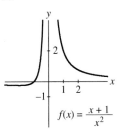

$$f(x) = \frac{x+1}{x^2}$$

75. $(-2, 4]$ **77.** $(-\infty, -8) \cup (-3, \infty)$ **79.** $[-2, 3] \cup (6, \infty)$
81. $(-2, 3)$ **83.** $(-\infty, -2) \cup (4, 5)$ **85.** $[-1, 3] \cup (5, \infty)$
87. $\left(-\infty, -\sqrt{7}\,\right] \cup \left(-\sqrt{2}, \sqrt{2}\,\right) \cup \left[\sqrt{7}, \infty\right)$
89. $(-\infty, -3) \cup \{-1\} \cup (5, \infty)$ **91.** $(1, \infty)$ **93.** $(3, \infty)$
95.   **97.**

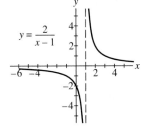
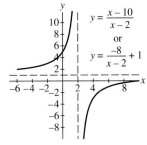

$$y = \frac{2}{x-1}$$

$$y = \frac{x-10}{x-2}$$
or
$$y = \frac{-8}{x-2} + 1$$

Chapter 3 Review Exercises

1. $f(x) = 3\left(x - \dfrac{1}{3}\right)^2 + \dfrac{2}{3}$

3. $(1, -3), x = 1, \left(\dfrac{2 \pm \sqrt{6}}{2}, 0\right), (0, -1)$

5. $y = -2x^2 + 4x + 6$ **7.** $1/3$ **9.** $\pm 2\sqrt{2}$ **11.** $\dfrac{1}{2}, \dfrac{-1 \pm i\sqrt{3}}{4}$

13. $\pm\sqrt{10}, \pm i\sqrt{10}$ **15.** $-\dfrac{1}{2}, \dfrac{1}{2}$ with multiplicity 2

17. $0, -1 \pm \sqrt{7}$ **19.** 83   **21.** 5   **23.** $\pm\left(1, 2, \dfrac{1}{3}, \dfrac{2}{3}\right)$

25. $\pm\left(1, 3, \dfrac{1}{2}, \dfrac{1}{3}, \dfrac{1}{6}, \dfrac{3}{2}\right)$ **27.** $2x^2 - 5x - 3 = 0$

29. $x^2 - 6x + 13 = 0$ **31.** $x^3 - 4x^2 + 9x - 10 = 0$
33. $x^2 - 4x + 1 = 0$ **35.** 0 with multiplicity 2, 6 imag
37. 1 pos, 2 imag; 3 pos   **39.** 3 neg; 1 neg, 2 imag
41. $-4 < x < 3$ **43.** $-1 < x < 8$ **45.** $-1 < x < 1$

47. $1, 2, 3$ **49.** $\dfrac{1}{2}, \dfrac{1}{3}, \pm i$ **51.** $3, 3 \pm i$ **53.** $2, 1 \pm i\sqrt{2}$

55. $0, \dfrac{1}{2}, 1 \pm \sqrt{3}$ **57.** $\{1/5\}$ **59.** $\{\pm\sqrt{2}\}$ **61.** $\{30\}$

63. $\{16\}$ **65.** $\{\pm 2\}$ **67.** $\{-7, 9\}$ **69.** No solution
71. $\{11/4\}$ **73.** $x = 3/4$ **75.** y-axis   **77.** Origin
79. $(-\infty, -2.5) \cup (-2.5, \infty)$ **81.** $(-\infty, \infty)$

83. $(-1, 0), (2, 0), (0, -2)$

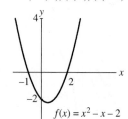

$f(x) = x^2 - x - 2$

85. $(-1, 0), (2, 0), (0, -2)$

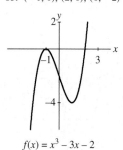

$f(x) = x^3 - 3x - 2$

99. $(-2, 0), (0, 2)$

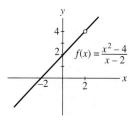

$f(x) = \dfrac{x^2 - 4}{x - 2}$

87. $(\pm 2, 0), (1, 0), (0, 2)$

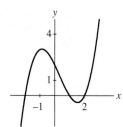

$f(x) = \frac{1}{2}x^3 - \frac{1}{2}x^2 - 2x + 2$

89. $(\pm 2, 0), (0, 4)$

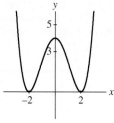

$f(x) = \frac{1}{4}x^4 - 2x^2 + 4$

101. $(1/4, 1/2)$ **103.** $[-5, 3]$ **105.** $[-1/2, 1/2] \cup [100, \infty)$
107. $(-\infty, -2) \cup (0, \infty)$ **109.** $(-\infty, 0) \cup (0, 3) \cup (4, \infty)$
111. $(-\infty, 1] \cup [2, 3) \cup (4, \infty)$ **113.** $x^2 - 3x, -15$
115. 380.25 ft   **117.** 24 ft wide, 7 ft high
119. a. 172.4 ft/sec   **b.** $V = 200$ **c.** 200 ft/sec

Chapter 3 Test

1. $y = 3(x - 2)^2 - 11$

2. $(2, -11), x = 2, (0, 1), \left(\dfrac{6 \pm \sqrt{33}}{3}, 0\right), [-11, \infty)$ **3.** -11

4. $2x^2 - 6x + 14, -37$ **5.** -14 **6.** $\pm\left(1, 2, 3, 6, \dfrac{1}{3}, \dfrac{2}{3}\right)$

7. $x^3 + 3x^2 + 16x + 48 = 0$ **8.** 2 pos, 1 neg; 1 neg, 2 imag

9. 256 ft   **10.** ± 3 **11.** $\pm 2, \pm 2i$ **12.** $1 \pm \sqrt{6}, 2$

13. $\pm i$ each with multiplicity 2

14. 2, 0 with multiplicity 2, $-\frac{3}{2}$ with multiplicity 3   **15.** $2 \pm i, \dfrac{1}{2}$

16.

$y = 2(x - 3)^2 + 1$

17.

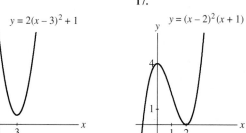

$y = (x - 2)^2(x + 1)$

91. $(0, 2/3), x = -3, y = 0$

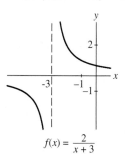

$f(x) = \dfrac{2}{x + 3}$

93. $(0, 0), x = \pm 2, y = 0$

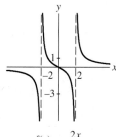

$f(x) = \dfrac{2x}{x^2 - 4}$

95. $(1, 0), \left(0, -\dfrac{1}{2}\right), x = 2,$

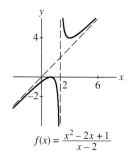

$f(x) = \dfrac{x^2 - 2x + 1}{x - 2}$

97. $\left(\dfrac{1}{2}, 0\right), \left(0, -\dfrac{1}{2}\right), x = 2,$
     $y = -2$

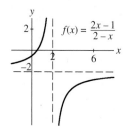

$f(x) = \dfrac{2x - 1}{2 - x}$

18.

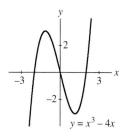

$y = x^3 - 4x$

19. $x = 2, y = 0$

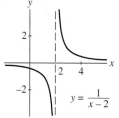

$y = \dfrac{1}{x - 2}$

20. $x = 2, y = 2$

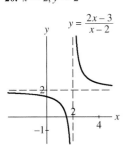

21. $x = 0, y = x$

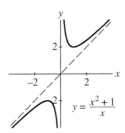

22. $x = \pm2, y = 0$

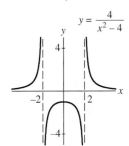

23.

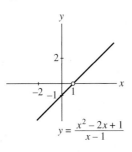

24. $(-2, 4)$ **25.** $(-\infty, 1/2) \cup (3, \infty)$ **26.** $(-\infty, -3] \cup (-1, 4)$
27. $\left(-\sqrt{7}, 0\right) \cup \left(\sqrt{7}, \infty\right)$ **28.** $\{3 \pm 3\sqrt{3}\}$ **29.** $\{16\}$

Tying It All Together Chapters 1–3

1. $3/2$ **2.** $0, 3/2$ **3.** -1 **4.** 1 **5.** ±27
6. $(-\infty, \infty), (-\infty, \infty), \left(\frac{3}{2}, 0\right)$, inc $(-\infty, \infty)$

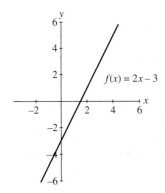

7. $(-\infty, \infty), \left[-\frac{9}{8}, \infty\right), (0, 0), \left(\frac{3}{2}, 0\right)$, dec $\left(-\infty, \frac{3}{4}\right)$, inc $\left(\frac{3}{4}, \infty\right)$

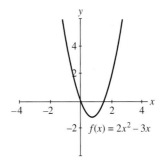

8. $(-\infty, \infty), (-\infty, \infty), (-1, 0)$, inc $(-\infty, \infty)$

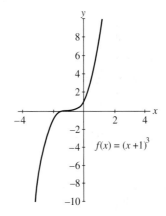

9. $(-\infty, 2) \cup (2, \infty), (-\infty, 1) \cup (1, \infty), (1, 0)$, dec $(-\infty, 2)$ and $(2, \infty)$

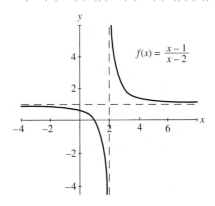

10. $(-\infty, \infty)$, $[-9, \infty)$, $(\pm 27, 0)$, $(0, -9)$, dec $(-\infty, 0)$, inc $(0, \infty)$

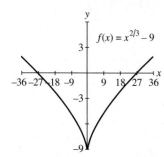

11. $\left(-\infty, \dfrac{3}{2}\right)$ **12.** $\left(0, \dfrac{3}{2}\right)$ **13.** $(-1, \infty)$ **14.** $(-\infty, 1) \cup (2, \infty)$

15. $(-27, 27)$ **16.** $f^{-1}(x) = \dfrac{x + 3}{2}$ **17.** Not invertible

18. $f^{-1}(x) = \sqrt[3]{x} - 1$ **19.** $f^{-1}(x) = \dfrac{2x - 1}{x - 1}$ **20.** Not invertible

CHAPTER 4

Section 4.1

For Thought: **1.** F   **2.** T   **3.** T   **4.** T   **5.** T   **6.** T
7. T   **8.** F   **9.** T   **10.** T
Exercises: **1.** 27   **3.** -1 **5.** 1/8   **7.** 16   **9.** 4
11. $-1/27$ **13.** 9   **15.** 1/9   **17.** 1/2   **19.** 8
21. 4   **23.** 2
25. $(-\infty, \infty)$, $(0, \infty)$, inc   **27.** $(-\infty, \infty)$, $(0, \infty)$, dec

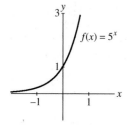   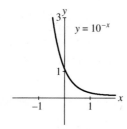

29. $(-\infty, \infty)$, $(0, \infty)$, dec   **31.** $(-\infty, \infty)$, $(-3, \infty)$,
                                                        $y = -3$, increasing

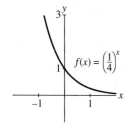   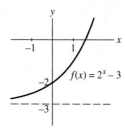

33. $(-\infty, \infty)$, $(-5, \infty)$,
         $y = -5$, increasing

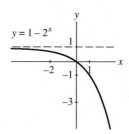

35. $(-\infty, \infty)$, $(-\infty, 0)$,
         $y = 0$, increasing

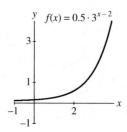

37. $(-\infty, \infty)$, $(-\infty, 1)$,
         $y = 1$, decreasing

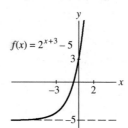

39. $(-\infty, \infty)$, $(0, \infty)$,
         $y = 0$, increasing

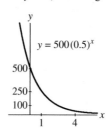

41. $(-\infty, \infty)$, $(0, \infty)$,
         $y = 0$, decreasing

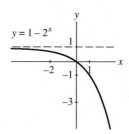

43. $y = 2^{x-5} - 2$ **45.** $y = -(1/4)^{x-1} - 2$ **47.** $\{6\}$
49. $\{-1\}$ **51.** $\{3\}$ **53.** $\{-2\}$ **55.** $\{1/3\}$ **57.** $\{-2\}$
59. $\{-3\}$ **61.** $\{-1\}$ **63.** 2   **65.** -1 **67.** 0   **69.** -3
71. 3   **73.** -1 **75.** 1   **77.** -1 **79.** 9, 1, 1/3, -2
81. 1, -2, 5, 1   **83.** -16, -2, $-1/2$, 5
85. a. \$7934.37, \$2934.37   **b.** \$8042.19, \$3042.19
   c. \$8067.51, \$3067.51   **d.** \$8079.95, \$3079.95
87. a. \$8080.37   **b.** \$9671.31   **c.** \$7694.93   **d.** \$26,570.30
89. \$2121.82   **91.** \$3934.30   **93. a.** \$6.85   **b.** \$6.87
95. 200 g, 121.3 g
97. a. $y = 5.68(1.31)^x$ where $x = 0$ corresponds to 1990   **b.** Yes
   c. 435.4 million
99. $P = 10\left(\dfrac{1}{2}\right)^n$ **101.** 6

Section 4.2

For Thought: **1.** T   **2.** F   **3.** T   **4.** T   **5.** F   **6.** T
7. T   **8.** F   **9.** T   **10.** T

Exercises: **1.** 6   **3.** -4 **5.** 1/4   **7.** -3 **9.** 6   **11.** -4
13. 1/4   **15.** -3 **17.** -1 **19.** 0   **21.** 1   **23.** -5
25. $(0, \infty), (-\infty, \infty)$ **27.** $(0, \infty), (-\infty, \infty)$

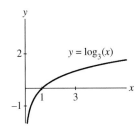

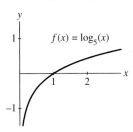

29. $(0, \infty), (-\infty, \infty)$ **31.** $(0, \infty), (-\infty, \infty)$

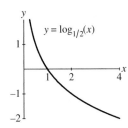

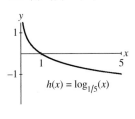

33. $(1, \infty), (-\infty, \infty)$ **35.** $(-2, \infty), (-\infty, \infty)$

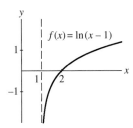

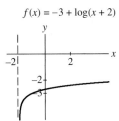

37. $(1, \infty), (-\infty, \infty)$

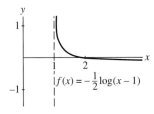

39. $y = \ln(x - 3) - 4$ **41.** $y = -\log_2(x - 5) - 1$
43. $2^5 = 32$ **45.** $5^y = x$ **47.** $10^z = 1000$ **49.** $e^x = 5$
51. $x = a^m$ **53.** $\log_5(125) = 3$ **55.** $\ln(y) = 3$
57. $\log(y) = m$ **59.** $z = \log_a(y)$ **61.** $\log_a(n) = x - 1$
63. $f^{-1}(x) = \log_2(x)$ **65.** $f^{-1}(x) = 7^x$ **67.** $f^{-1}(x) = e^x + 1$
69. $f^{-1}(x) = \log_3(x) - 2$ **71.** $f^{-1}(x) = \log(2x - 10) + 1$
73. 256   **75.** $\sqrt{3}$ **77.** 4   **79.** $\log_3(77)$ **81.** 6
83. $3\sqrt{2}$ **85.** $-1 + \log_3(7)$ **87.** 2   **89.** 27   **91.** 1/4
93. 6/5   **95.** 1.3979   **97.** 0.5493   **99.** -0.2231
101. -0.3010
103. a. 34.7 yr   **b.** 17.3 yr   **c.** 8.7 yr   **d.** 4.3 yr

105. a. 22.0%   **b.** 11.0%   **c.** 5.5%   **d.** 2.7%
107. 49 yr 125 days   **109.** $r = \ln(A/P)/t$, 23.1%
111. a. 6.9 yr   **113.** 3.5%   **115.** 9.8 yr   **117.** 1.87%, 7.3 billion
119. a. 99.2 billion gigabits/yr   **b.** 2005
121. a. $p = -45 \cdot \log(I) + 190$ **b.** 0%
123. 4.1   **125.** 3.7   **127.** $c = \ln(b)$, 3.54%

Section 4.3

For Thought: **1.** F   **2.** T   **3.** T   **4.** T   **5.** F   **6.** F
7. F   **8.** T   **9.** F   **10.** F

Exercises: **1.** $\log(15)$ **3.** $\log_2(x^2 - x)$ **5.** $\log_4(6)$ **7.** $\ln(x^5)$
9. $\log_2(3) + \log_2(x)$ **11.** $\log(x) - \log(2)$
13. $\log(x - 1) + \log(x + 1)$ **15.** $\ln(x - 1) - \ln(x)$
17. $3 \cdot \log_a(5)$ **19.** $\frac{1}{2} \cdot \log_a(5)$ **21.** $-1 \cdot \log_a(5)$ **23.** $\sqrt{y}$
25. $y + 1$ **27.** 999   **29.** $\log_a(2) + \log_a(5)$
31. $\log_a(5) - \log_a(2)$ **33.** $\log_a(2) + \frac{1}{2} \cdot \log_a(5)$
35. $2 \cdot \log_a(2) - 2 \cdot \log_a(5)$ **37.** $\log_3(5) + \log_3(x)$
39. $\log_2(5) - \log_2(2) - \log_2(y)$ **41.** $\log(3) + \frac{1}{2}\log(x)$
43. $\log(3) + (x - 1)\log(2)$ **45.** $\frac{1}{3}\ln(x) + \frac{1}{3}\ln(y) - \frac{4}{3}\ln(t)$
47. $\ln(6) + \frac{1}{2}\ln(x - 1) - \ln(5) - 3 \cdot \ln(x)$ **49.** $\log_2(5x^3)$
51. $\log_7(x^{-3})$ **53.** $\log\left(\frac{2xy}{z}\right)$ **55.** $\log\left(\frac{z\sqrt{x}}{y\sqrt[3]{w}}\right)$
57. $\log_4(x^{20})$ **59.** 3.1699   **61.** -3.5864 **63.** 11.8957
65. 2.2025   **67.** 1.5850   **69.** 0.3772   **71.** -3.5850
73. 13.8695   **75.** 34.3240   **77.** 0.3200   **79.** 0.0479, -24.0479
81. 2.0172   **83.** 1.5928   **85.** 11 yr 166 days   **87.** 44 quarters
89. 5.7%   **91.** 3.58%   **93.** $\log(I/I_0)$, 3
95. $t = \frac{1}{r}\ln(P) - \frac{1}{r}\ln(P_0)$
97. a. Decreasing   **b.** $n \le 4{,}892{,}961$
99. $MR(x) = \log\left[\left(\frac{x + 2}{x + 1}\right)^{500}\right]$, $MR(x) \to 0$
101. a. $y = 369.2(1.108)^x$ where $x = 0$ corresponds to 1996
b. $y = 369.2\,e^{0.102x}$ **c.** 10.2%   **d.** 2010   **e.** No
105. a. Same function   **b.** Same function   **c.** Inverse functions
d. Inverse functions

Section 4.4

For Thought: **1.** T   **2.** T   **3.** F   **4.** T   **5.** F   **6.** F
7. T   **8.** T   **9.** T   **10.** T

Exercises: **1.** 8   **3.** 80   **5.** ± 5 **7.** 3   **9.** 1/2   **11.** $\sqrt[3]{10}$
13. 1/4   **15.** 6   **17.** 20   **19.** 2   **21.** $\frac{\sqrt{5}}{2}$ **23.** $\frac{1}{2}$
25. $\frac{2}{\ln(6)}$ **27.** 3.8074   **29.** 3.5502   **31.** -3.0959
33. No solution   **35.** 1.5850   **37.** 0.7677   **39.** -0.2
41. 2.5850   **43.** 1/3   **45.** 1, 100   **47.** 29.4872   **49.** 2
51. $-5/4$ **53.** 0.194, 2.70   **55.** -49.73 **57.** $-0.767, 2, 4$

59. -6.93×10^{-5} **61.** 19,035 yr ago   **63.** 1507 yr
65. 24,850 yr   **67.** 30.5%   **69.** A.D. 34   **71.** 1 hr 11 min, forever
73. 5:20 A.M.   **75.** 10 yr 3 mo   **77.** 19,328 yr 307 days
79. a. 12,300   **b.** $\approx$ 15 yr   **c.** 13.9 yr
81. a. 2211   **b.** 99%
83. 1.32 parsecs
85. a. $P = 10^{-0.1826x+4.5}$ **b.** \$718.79   **c.** 2010
87. 10^{-3} watts/m^2 **89.** \$6791.91
91. a. $y = 114.8 - 12.8 \cdot \ln(x)$ in year $1960 + x$ **b.** 65%
   c. 2121   **d.** Quadratic or exponential
93. 1.105170833, 1.105170918

Chapter 4 Review Exercises

1. 64   **3.** 6   **5.** 0   **7.** 17   **9.** 32   **11.** 3   **13.** 9
15. 3   **17.** -3 **19.** 3   **21.** $\log(x^2 - 3x)$ **23.** $\ln(3x^2y)$
25. $\log(3) + 4 \cdot \log(x)$ **27.** $\log_3(5) + \dfrac{1}{2}\log_3(x) - 4 \cdot \log_3(y)$

29. $\ln(2) + \ln(5)$ **31.** $\ln(2) + 2 \cdot \ln(5)$ **33.** 10^{10} **35.** 3
37. -5 **39.** -4 **41.** $2 + \ln(9)$ **43.** -2 **45.** $100\sqrt{5}$
47. 8   **49.** 6   **51.** 3   **53.** $3, -3, \sqrt{3}, 0$ **55.** (c)   **57.** (b)
59. (d)   **61.** (e)
63. $(-\infty, \infty)$, $(0, \infty)$, inc, $y = 0$ **65.** $(-\infty, \infty)$, $(0, \infty)$, dec, $y = 0$

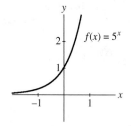

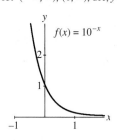

67. $(0, \infty)$, $(-\infty, \infty)$, inc, $x = 0$ **69.** $(-3, \infty)$, $(-\infty, \infty)$, inc, $x = -3$

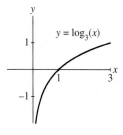

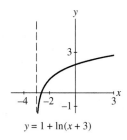

71. $(-\infty, \infty)$, $(1, \infty)$, inc, $y = 1$ **73.** $(-\infty, 2)$, $(-\infty, \infty)$, dec, $x = 2$

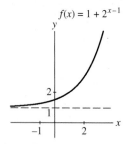

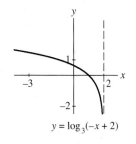

75. $f^{-1}(x) = \log_7(x)$ **77.** $f^{-1}(x) = 5^x$ **79.** $f^{-1}(x) = 10^{x/3} + 1$
81. $f^{-1}(x) = -2 + \ln(x + 3)$ **83.** 2.0959   **85.** 7.8538
87. -4.4243 **89.** T   **91.** F   **93.** T   **95.** F   **97.** F
99. F   **101.** T   **103.** The pH of A is one less than the pH of B.
105. \$122,296.01   **107.** 56 quarters   **109.** 25 g, 18.15 g, 2166 yr
111. 2,877 hr   **113.** $-2/3, -3, -1/3$

Chapter 4 Test

1. 3   **2.** -2 **3.** 6.47   **4.** $\sqrt{2}$ **5.** $f^{-1}(x) = e^x$
6. $f^{-1}(x) = -1 + \log_8(x + 3)$ **7.** $\log(xy^3)$ **8.** $\ln\!\left(\dfrac{\sqrt{x-1}}{33}\right)$
9. $2 \cdot \log_a(2) + \log_a(7)$ **10.** $\log_a(7) - \log_a(2)$ **11.** 4
12. 18   **13.** $\dfrac{\ln(5)}{\ln(5) - \ln(3)} \approx 3.1507$ **14.** $1 + 3^{5.46} \approx 403.7931$
15. $(-\infty, \infty)$, $(1, \infty)$, inc, $y = 1$ **16.** $(1, \infty)$, $(-\infty, \infty)$, dec, $x = 1$

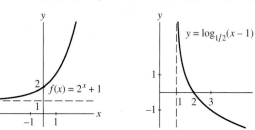

17. $(1, 0)$ **18.** \$9750.88, \$9906.06
19. 22.5 watts, 173.3 days, 428.7 days   **20.** 61.5 quarters
21. $p = 0.86$, no

Tying It All Together Chapters 1–4

1. 5, 1   **2.** 5   **3.** 19   **4.** 5   **5.** 19   **6.** 7, -1
7. $2 \pm \sqrt{2}$ **8.** -3 **9.** 130   **10.** $\log_2(3)$ **11.** 4
12. $-1, 2, 3$
13.   **14.**

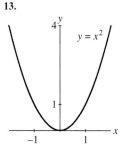

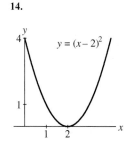

15.   **16.**

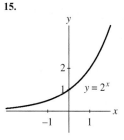

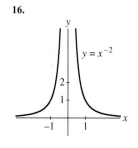

17.

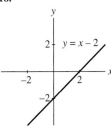

18.

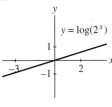

19.

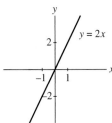

20.

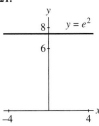

21.

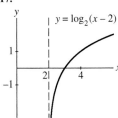

22.

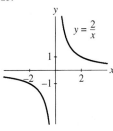

23.

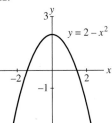

24.

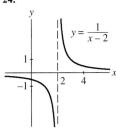

25. $f^{-1}(x) = 3x$ **26.** $f^{-1}(x) = -\log_3(x)$
27. $f^{-1}(x) = x^2 + 2$ for $x \geq 0$ **28.** $f^{-1}(x) = \sqrt[3]{x - 2} + 5$
29. $f^{-1}(x) = (10^x + 3)^2$ **30.** $f^{-1} = \{(1, 3), (4, 5)\}$
31. $f^{-1}(x) = \dfrac{1}{x - 3} + 5$ **32.** $f^{-1}(x) = (\ln(3 - x))^2$ for $x \leq 2$
33. $(p \circ m)(x) = e^{x+5}, (-\infty, \infty), (0, \infty)$
34. $(p \circ q)(x) = e^{\sqrt{x}}, [0, \infty), [1, \infty)$
35. $(q \circ p \circ m)(x) = \sqrt{e^{x+5}}, (-\infty, \infty), (0, \infty)$
36. $(m \circ r \circ q)(x) = \ln(\sqrt{x}) + 5, (0, \infty), (-\infty, \infty)$
37. $(p \circ r \circ m)(x) = e^{\ln(x+5)}, (-5, \infty), (0, \infty)$
38. $(r \circ q \circ p)(x) = \ln(\sqrt{e^x}), (-\infty, \infty), (-\infty, \infty)$ **39.** $F = f \circ g \circ h$
40. $H = g \circ h \circ f$ **41.** $G = h \circ g \circ f$ **42.** $M = h \circ f \circ g$

CHAPTER 5

Section 5.1

For Thought: **1.** T   **2.** F   **3.** F   **4.** F   **5.** T   **6.** F
7. F   **8.** T   **9.** F   **10.** T
Exercises: **1.** $420°, 780°, -300°, -660°$ **3.** $344°, 704°, -376°, -736°$
5. Yes   **7.** No   **9.** I   **11.** III   **13.** IV   **15.** I
17. $45°$ **19.** $60°$ **21.** $120°$ **23.** $40°$ **25.** $20°$ **27.** $340°$
29. $13.2°$ **31.** $-8.505°$ **33.** $28.0858°$ **35.** $75°30'$
37. $-17°19'48''$ **39.** $18°7'23''$ **41.** $\pi/6$ **43.** $\pi/10$
45. $-3\pi/8$ **47.** $7\pi/2$ **49.** 0.653 **51.** -0.241
53. -0.936 **55.** $75°$
57. $315°$ **59.** $-1080°$ **61.** $136.937°$
63. $7\pi/3, 13\pi/3, -5\pi/3, -11\pi/3$
65. $11\pi/6, 23\pi/6, -13\pi/6, -25\pi/6$
67. π **69.** $\pi/2$ **71.** $\pi/3$ **73.** $5\pi/3$ **75.** 2.04 **77.** No
79. Yes   **81.** I   **83.** III   **85.** IV   **87.** IV
89. $30° = \pi/6, 45° = \pi/4, 60° = \pi/3, 90° = \pi/2, 120° = 2\pi/3,$
$135° = 3\pi/4, 150° = 5\pi/6, 180° = \pi, 210° = 7\pi/6, 225° = 5\pi/4,$
$240° = 4\pi/3, 270° = 3\pi/2, 300° = 5\pi/3, 315° = 7\pi/4,$
$330° = 11\pi/6, 360° = 2\pi$
91. 3π ft   **93.** 209.4 mi   **95.** 1 mi   **97.** 3.18 km
99. 3102 mi   **101.** $29.0°$ **103.** 1256.6 in./min   **105.** 166.6 mph
107. 20.5 mph   **109.** 0.26 mph   **111.** 41,143 km, 40,074 km
113. 33.5 in.^{2} **115.** 10 ft/sec and 24 rad/sec, 10 ft/sec and 40 rad/sec
117. 12:16:22 and 12:49:05
119. a. 22.5 in.^{3} **b.** $V = \dfrac{\pi(360 - \alpha)^2 \sqrt{720\alpha - \alpha^2}}{2{,}187{,}000}$
c. $66.06°$ **d.** 25.8 in.^{3}

Section 5.2

For Thought: **1.** F   **2.** F   **3.** T   **4.** F   **5.** F   **6.** T
7. F   **8.** F   **9.** F   **10.** T
Exercises: **1.** $(1, 0), (\sqrt{2}/2, \sqrt{2}/2), (0, 1), (-\sqrt{2}/2, \sqrt{2}/2), (-1, 0),$
$(-\sqrt{2}/2, -\sqrt{2}/2), (0, -1), (\sqrt{2}/2, -\sqrt{2}/2)$
3. 0   **5.** 0   **7.** 0   **9.** 0   **11.** $\sqrt{2}/2$ **13.** $-\sqrt{2}/2$
15. $1/2$ **17.** $1/2$ **19.** $-\sqrt{3}/2$ **21.** $\sqrt{3}/2$ **23.** $1/2$
25. $1/2$ **27.** $\sqrt{3}/2$ **29.** $30°, \pi/6$ **31.** $60°, \pi/3$
33. $60°, \pi/3$ **35.** $30°, \pi/6$ **37.** $45°, \pi/4$ **39.** $45°, \pi/4$
41. $+$ **43.** $+$ **45.** $-$ **47.** $-$ **49.** $\sqrt{2}/2$ **51.** $1/2$
53. $-\sqrt{2}/2$ **55.** $-\sqrt{3}/2$ **57.** $-\sqrt{2}/2$ **59.** $-1/2$
61. $\sqrt{3}/3$ **63.** -1 **65.** 1   **67.** $2 + \sqrt{3}$ **69.** $\sqrt{2}$
71. 0.9999 **73.** 0.4035 **75.** -0.7438 **77.** 1.0000
79. -0.2588 **81.** 1   **83.** $1/2$ **85.** $\sqrt{2}/2$ **87.** $\sqrt{3}/2$
89. $-12/13$ **91.** $-4/5$ **93.** $2\sqrt{2}/3$
95. $x = 4 \sin t - 3 \cos t, 3.53$
97. 1.708 in., 1.714 in.
99. $\cos \alpha = \sqrt{1 - \sin^2 \alpha}$ for α in quadrants I or IV,
$\cos \alpha = -\sqrt{1 - \sin^2 \alpha}$ for α in quadrants II or III

Section 5.3

For Thought: **1.** F   **2.** F   **3.** F   **4.** T   **5.** T   **6.** T
7. F   **8.** T   **9.** T   **10.** T

Exercises: **1.** $y = -2\sin(x), 2$ **3.** $y = 3\cos(x), 3$

5. $2, 2\pi, 0$ **7.** $1, 2\pi, \pi/2$

9. $2, 2\pi, -\pi/3$

11. $1, 0, (0, 0), (\pi/2, -1), (\pi, 0),$
$(3\pi/2, 1), (2\pi, 0)$

13. $3, 0, (0, 0), (\pi/2, -3),$
$(\pi, 0), (3\pi/2, 3), (2\pi, 0)$

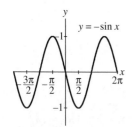

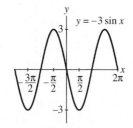

15. $1/2, 0, (0, 1/2), (\pi/2, 0),$
$(\pi, -1/2), (3\pi/2, 0),$
$(2\pi, 1/2)$

17. $1, -\pi, (0, 0), (\pi/2, -1),$
$(\pi, 0), (3\pi/2, 1), (2\pi, 0)$

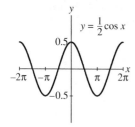

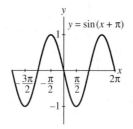

19. $1, \pi/3, (-2\pi/3, -1),$
$(-\pi/6, 0), (\pi/3, 1),$
$(5\pi/6, 0), (4\pi/3, -1)$

21. $1, 0, (0, 3), (\pi/2, 2), (\pi, 1),$
$(3\pi/2, 2), (2\pi, 3)$

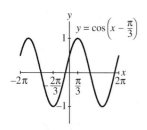

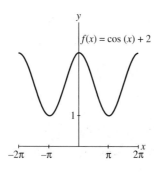

23. $1, 0, (0, -1), (\pi/2, -2), (\pi, -1), (3\pi/2, 0), (2\pi, -1)$

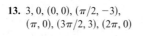

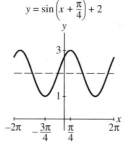

25. $1, -\pi/4, (-\pi/4, 2), (\pi/4, 3), (3\pi/4, 2), (5\pi/4, 1), (7\pi/4, 2)$

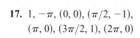

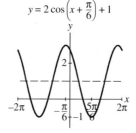

27. $2, -\pi/6, (-\pi/6, 3), (\pi/3, 1), (5\pi/6, -1), (4\pi/3, 1), (11\pi/6, 3)$

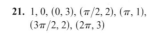

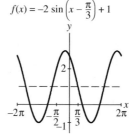

29. $2, \pi/3, (-\pi/6, 3), (\pi/3, 1), (5\pi/6, -1), (4\pi/3, 1), (11\pi/6, 3)$

$f(x) = -2\sin\left(x - \dfrac{\pi}{3}\right) + 1$

31. $3, \pi/2, 0$ **33.** $1, 4\pi, 0$ **35.** $2, 2\pi, \pi$ **37.** $2, \pi, -\pi/4$

39. $2, 4, -2$ **41.** $y = 2 \sin [2(x + \pi/2)] + 5$

43. $y = 5 \sin [\pi(x - 2)] + 4$ **45.** $y = 6 \sin [4\pi(x + \pi)] - 3$

47. $y = -\sin (x - \pi/4) + 1$ **49.** $y = -3 \cos (x - \pi) + 2$

51. $F(x) = \sin (3x - \pi/4)$ **53.** $F(x) = \sin (3x - 3\pi/4)$

55. $2\pi/3, 0, [-1, 1], (0, 0),$
$(\pi/6, 1), (\pi/3, 0),$
$(\pi/2, -1), (2\pi/3, 0)$

57. $\pi, 0, [-1, 1], (0, 0),$
$(\pi/4, -1), (\pi/2, 0),$
$(3\pi/4, 1), (\pi, 0)$

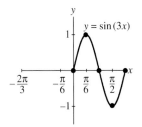

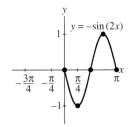

59. $\pi/2, 0, [1, 3], (0, 3), (\pi/8, 2),$
$(\pi/4, 1), (3\pi/8, 2), (\pi/2, 3)$

61. $8\pi, 0, [1, 3], (0, 2), (2\pi, 1),$
$(4\pi, 2), (6\pi, 3), (8\pi, 2)$

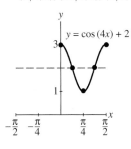

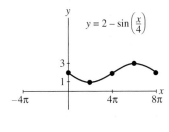

63. $6, 0, [-1, 1], (0, 0), (1.5, 1), (3, 0), (4.5, -1), (6, 0)$

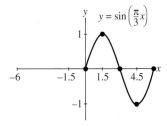

65. $\pi, \pi/2, [-1, 1], (\pi/2, 0),$
$(3\pi/4, 1), (\pi, 0),$
$(5\pi/4, -1), (3\pi/2, 0)$

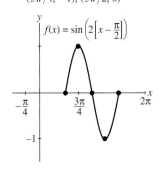

67. $4, -3, [-1, 1], (-3, 0), (-2, 1), (-1, 0), (0, -1), (1, 0)$

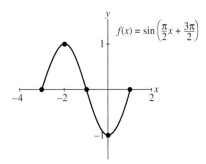

69. $\pi, -\pi/6, [-1, 3], (-\pi/6, 3),$
$(\pi/12, 1), (\pi/3, -1),$
$(7\pi/12, 1), (5\pi/6, 3)$

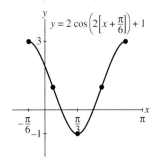

71. $2\pi/3, \pi/6, [-3/2, -1/2], (\pi/6, -1), (\pi/3, -3/2), (\pi/2, -1),$
$(2\pi/3, -1/2), (5\pi/6, -1)$

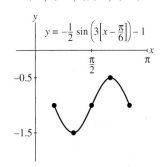

73. $y = 2 \sin \left(2\left[x - \dfrac{\pi}{4}\right]\right)$ **75.** $y = 3 \sin \left(\dfrac{3}{2}\left[x + \dfrac{\pi}{3}\right]\right) + 3$

77. 100 cycles/sec   **79.** 40 cycles/hr

81. $x = 3 \sin(2t), 3, \pi$

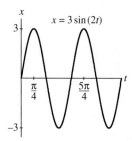

83. 11 yr **85. a.** 1300 cc, 500 cc **b.** 30
87. 12, 15,000, -3, 25,000, $y = 15{,}000 \sin(\pi x/6 + \pi/2) +$
25,000, $17,500
89. a. 40, 65, $y = 65 \sin(\pi x/20)$ **b.** 40 days **c.** -38.2 m/sec
d. Planet is between earth and Rho
91. $y = \sin(\pi x/10) + 1$
93. $y = 51.6 \sin(0.20x + 1.69) + 50.43, 31.7$ days, 87%

Section 5.4

For Thought: 1. T **2.** F **3.** T **4.** F **5.** F **6.** T
7. F **8.** T **9.** T **10.** T

Exercises:
1. $\tan(0) = 0, \tan(\pi/4) = 1, \tan(\pi/2)$ undefined, $\tan(3\pi/4) = -1$,
$\tan(\pi) = 0, \tan(5\pi/4) = 1, \tan(3\pi/2)$ undefined, $\tan(7\pi/4) = -1$
3. $\sqrt{3}$ **5.** -1 **7.** 0 **9.** $-\sqrt{3}/3$ **11.** $2\sqrt{3}/3$
13. Undefined **15.** Undefined **17.** $\sqrt{2}$ **19.** -1 **21.** $\sqrt{3}$
23. -2 **25.** $-\sqrt{2}$ **27.** 0 **29.** 48.0785 **31.** -2.8413
33. 500.0003 **35.** 1.0353 **37.** 636.6192 **39.** -1.4318
41. 71.6221 **43.** -0.9861 **45.** 4 **47.** $\sqrt{3}/3$ **49.** $-\sqrt{2}$
51. $\pi/3$ **53.** π

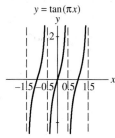

55. 2π **57.** 1

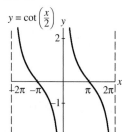

59. π **61.** π

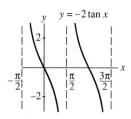 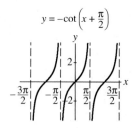

63. $\pi/2$ **65.** 2

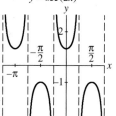 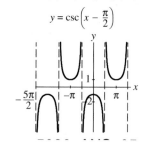

67. $\pi, (-\infty, -1] \cup [1, \infty)$ **69.** $2\pi, (-\infty, -1] \cup [1, \infty)$

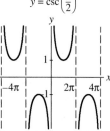 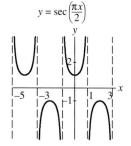

71. $4\pi, (-\infty, -1] \cup [1, \infty)$ **73.** $4, (-\infty, -1] \cup [1, \infty)$

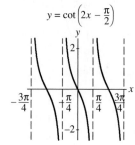 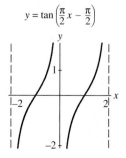

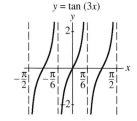

75. 2π, $(-\infty, -2] \cup [2, \infty)$ **77.** π, $(-\infty, -1] \cup [1, \infty)$

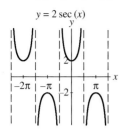

79. 4, $(-\infty, -1] \cup [1, \infty)$ **81.** π, $(-\infty, 0] \cup [4, \infty)$

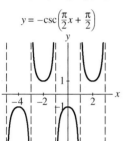

83. $\pi/2$, $(-\infty, \infty)$ **85.** 4π, $(-\infty, -3] \cup [1, \infty)$
87. π, $(-\infty, -7] \cup [-1, \infty)$
89. $y = 3\tan(x - \pi/4) + 2$ **91.** $y = -\sec(x + \pi) + 2$
93.

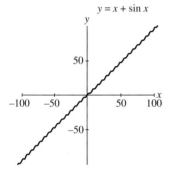

97. 2.3 yr, tangent

Section 5.5

For Thought: **1.** T   **2.** T   **3.** F   **4.** F   **5.** F   **6.** T
7. T   **8.** T   **9.** F   **10.** F
Exercises: **1.** $-\pi/6$ **3.** $\pi/6$ **5.** $\pi/4$ **7.** $-45°$ **9.** $30°$
11. $0°$ **13.** $-19.5°$ **15.** $34.6°$ **17.** $3\pi/4$ **19.** $\pi/3$
21. π **23.** $135°$ **25.** $180°$ **27.** $120°$ **29.** $173.2°$
31. $89.9°$ **33.** $-\pi/4$ **35.** $\pi/3$ **37.** $\pi/4$ **39.** $-\pi/6$
41. 0   **43.** $\pi/2$ **45.** $3\pi/4$ **47.** $2\pi/3$ **49.** 0.60   **51.** 3.02
53. -0.14 **55.** 1.87   **57.** 1.15   **59.** -0.36 **61.** 3.06
63. 0.06   **65.** $\sqrt{3}$ **67.** $-\pi/6$ **69.** $\pi/6$ **71.** $\pi/4$ **73.** 1
75. $\pi/2$ **77.** 0   **79.** $\pi/2$ **81.** 0.8930   **83.** Undefined
85. -0.9802 **87.** -0.4082 **89.** 3.4583   **91.** 1.0183
93. $f^{-1}(x) = 0.5\sin^{-1}(x)$, $[-1, 1]$
95. $f^{-1}(x) = \dfrac{1}{\pi}\tan^{-1}(x - 3)$, $(-\infty, \infty)$
97. $f^{-1}(x) = 2\sin(x - 3)$, $[3 - \pi/2, 3 + \pi/2]$ **99.** $67.1°$

Section 5.6

For Thought: **1.** F   **2.** T   **3.** F   **4.** T   **5.** T   **6.** F
7. F   **8.** T   **9.** T   **10.** F
Exercises: **1.** 4/5, 3/5, 4/3, 5/4, 5/3, 3/4
3. $3\sqrt{10}/10$, $-\sqrt{10}/10$, -3, $\sqrt{10}/3$, $-\sqrt{10}$, $-1/3$
5. $-\sqrt{3}/3$, $-\sqrt{6}/3$, $\sqrt{2}/2$, $-\sqrt{3}$, $-\sqrt{6}/2$, $\sqrt{2}$
7. $-1/2$, $\sqrt{3}/2$, $-\sqrt{3}/3$, -2, $2\sqrt{3}/3$, $-\sqrt{3}$
9. $\sqrt{5}/5$, $2\sqrt{5}/5$, $1/2$, $2\sqrt{5}/5$, $\sqrt{5}/5$, 2
11. $3\sqrt{34}/34$, $5\sqrt{34}/34$, 3/5, $5\sqrt{34}/34$, $3\sqrt{34}/34$, 5/3
13. 4/5, 3/5, 4/3, 3/5, 4/5, 3/4
15. $80.5°$ **17.** $60°$ **19.** 1.0   **21.** 0.4
23. $\beta = 30°$, $a = 10\sqrt{3}$, $b = 10$
25. $c = 10$, $\alpha = 36.9°$, $\beta = 53.1°$
27. $a = 5.7$, $\alpha = 43.7°$, $\beta = 46.3°$
29. $\beta = 74°$, $a = 5.5$, $b = 19.2$ **31.** $\beta = 50°51'$, $b = 11.1$, $c = 14.3$
33. 25   **35.** 0.831   **37.** 18.8   **39.** -289 **41.** 50 ft
43. 1.7 mi   **45.** 43.2 m   **47.** 25.1 ft   **49.** 22 m, $57.6°$
51. No, speed $\approx$ 11.2 mph   **53.** 153.1 m   **55.** 4.5 km
57. 1.9×10^{13} mi, ≈ 3.23 yr   **59.** 4391 mi
61. 41.60 ft by 90.14 ft   **63.** 78.1 ft   **65.** 2.768 ft

Chapter 5 Review Exercises

1. $28°$ **3.** $206°45'33''$ **5.** $180°$ **7.** $108°$ **9.** $300°$
11. $270°$ **13.** $11\pi/6$ **15.** $-5\pi/3$
17.

θ deg	0	30	45	60	90	120	135	150	180
θ rad	0	$\dfrac{\pi}{6}$	$\dfrac{\pi}{4}$	$\dfrac{\pi}{3}$	$\dfrac{\pi}{2}$	$\dfrac{2\pi}{3}$	$\dfrac{3\pi}{4}$	$\dfrac{5\pi}{6}$	π
$\sin\theta$	0	$\dfrac{1}{2}$	$\dfrac{\sqrt{2}}{2}$	$\dfrac{\sqrt{3}}{2}$	1	$\dfrac{\sqrt{3}}{2}$	$\dfrac{\sqrt{2}}{2}$	$\dfrac{1}{2}$	0
$\cos\theta$	1	$\dfrac{\sqrt{3}}{2}$	$\dfrac{\sqrt{2}}{2}$	$\dfrac{1}{2}$	0	$-\dfrac{1}{2}$	$-\dfrac{\sqrt{2}}{2}$	$-\dfrac{\sqrt{3}}{2}$	-1
$\tan\theta$	0	$\dfrac{\sqrt{3}}{3}$	1	$\sqrt{3}$		$-\sqrt{3}$	-1	$-\dfrac{\sqrt{3}}{3}$	0

19. $-\sqrt{2}/2$
21. $\sqrt{3}$ **23.** $-2\sqrt{3}/3$ **25.** 0   **27.** 0   **29.** -1
31. $\sqrt{3}/3$ **33.** $-\sqrt{2}/2$ **35.** -2 **37.** $-\sqrt{3}/3$
39. 5/13, 12/13, 5/12, 13/5, 13/12, 12/5
41. 0.6947   **43.** -0.0923 **45.** 0.1869   **47.** 1.0356
49. $-\pi/6$ **51.** $-\pi/4$ **53.** $\pi/4$ **55.** $\pi/6$ **57.** $90°$
59. $135°$ **61.** $30°$ **63.** $90°$
65. $c = \sqrt{13}$, $\alpha = 33.7°$, $\beta = 56.3°$
67. $\beta = 68.7°$, $c = 8.8$, $b = 8.2$
69. $2\pi/3$, $[-2, 2]$

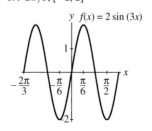

71. $\pi/2$, $(-\infty, \infty)$

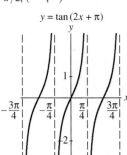

$y = \tan(2x + \pi)$

73. 4π, $(-\infty, -1] \cup [1, \infty)$

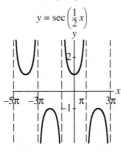

$y = \sec\left(\frac{1}{2}x\right)$

75. π, $[-1/2, 1/2]$

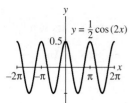

$y = \frac{1}{2}\cos(2x)$

77. $\pi/2$, $(-\infty, \infty)$

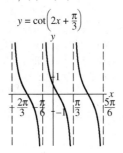

$y = \cot\left(2x + \frac{\pi}{3}\right)$

79. π, $(-\infty, -1/3] \cup [1/3, \infty)$

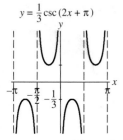

$y = \frac{1}{3}\csc(2x + \pi)$

81. $y = 2\sin\left(\frac{\pi}{2}[x - 2]\right)$ **83.** $y = 20\sin(x) + 40$ **85.** $150°$

87. $-2\sqrt{6}/5$ **89.** 6.6 ft   **91.** $53.1°$ **93.** 1.08×10^{-8} sec

95. 94.2 ft/hr   **97.** 6.9813 ft

99.

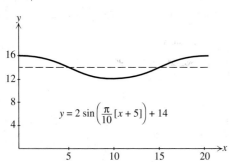

$y = 2\sin\left(\frac{\pi}{10}[x + 5]\right) + 14$

101. No   **103.** 0.33 rad/sec   **105.** 11.6 mi

Chapter 5 Test

1. $1/2$ **2.** $-1/2$ **3.** -1 **4.** 2 **5.** -2 **6.** $\sqrt{3}/3$

7. $-\pi/6$ **8.** $2\pi/3$ **9.** $-\pi/4$ **10.** Undefined

11. Undefined   **12.** $2\sqrt{2}/3$

13. $2\pi/3$, $[-3, -1]$, 1

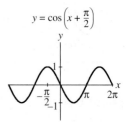

$y = \sin(3x) - 2$

14. 2π, $[-1, 1]$, 1

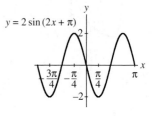

$y = \cos\left(x + \frac{\pi}{2}\right)$

15. 2, $(-\infty, \infty)$

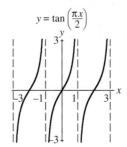

$y = \tan\left(\frac{\pi x}{2}\right)$

16. π, $[-2, 2]$, 2

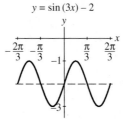

$y = 2\sin(2x + \pi)$

17. 2π, $(-\infty, -2] \cup [2, \infty)$

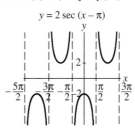

$y = 2\sec(x - \pi)$

18. 2π, $(-\infty, -1] \cup [1, \infty)$

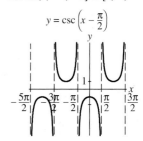

$y = \csc\left(x - \frac{\pi}{2}\right)$

19. $\pi/2$, $(-\infty, \infty)$

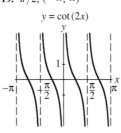

$y = \cot(2x)$

20. 2π, $[-1, 1]$, 1

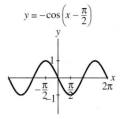

$y = -\cos\left(x - \frac{\pi}{2}\right)$

21. 28.85 m   **22.** $134.1°$ **23.** $-\sqrt{15}/4$

24. $\sin\alpha = -2/\sqrt{29}$, $\cos\alpha = 5/\sqrt{29}$, $\tan\alpha = -2/5$,
     $\csc\alpha = -\sqrt{29}/2$, $\sec\alpha = \sqrt{29}/5$, $\cot\alpha = -5/2$

25. 647.2 rad/min   **26.** 7.97 mph   **27.** 12.2 m   **28.** 1117 ft

29.

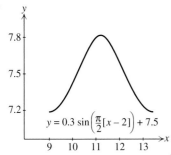

$$y = 0.3 \sin\left(\frac{\pi}{2}[x - 2]\right) + 7.5$$

Tying It All Together Chapters 1–5

1. $(-\infty, \infty), (2, \infty)$

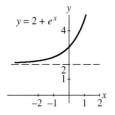

$y = 2 + e^x$

2. $(-\infty, \infty), [2, \infty)$

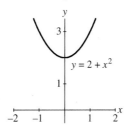

$y = 2 + x^2$

3. $(-\infty, \infty), [1, 3]$

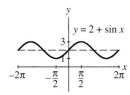

$y = 2 + \sin x$

4. $(0, \infty), (-\infty, \infty)$

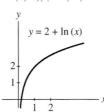

$y = 2 + \ln(x)$

5. $(\pi/4, \infty), (-\infty, \infty)$

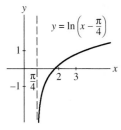

$y = \ln\left(x - \frac{\pi}{4}\right)$

6. $(-\infty, \infty), [-1, 1]$

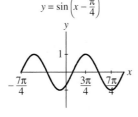

$y = \sin\left(x - \frac{\pi}{4}\right)$

7. $(0, \infty), (-\infty, \infty)$

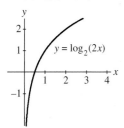

$y = \log_2(2x)$

8. $(-\infty, \infty), [-1, 1]$

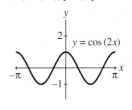

$y = \cos(2x)$

---

9. Neither   10. Odd   11. Even   12. Even   13. Odd
14. Even   15. Neither   16. Odd   17. Increasing
18. Increasing   19. Increasing   20. Decreasing
21. Decreasing   22. Increasing

CHAPTER 6

Section 6.1

For Thought: 1. F   2. T   3. T   4. F   5. T   6. F
7. F   8. T   9. F   10. T

Exercises: 1. $\sin x$ 3. 1   5. $\csc x$ 7. 1   9. 1
11. $\cos^2 \alpha$ 13. $-\cos^2 \beta$ 15. $2 \sin \alpha$
17. $\cot(x) = \pm\sqrt{\csc^2(x) - 1}$ 19. $\sin(x) = \dfrac{\pm 1}{\sqrt{1 + \cot^2(x)}}$
21. $\tan(x) = \dfrac{\pm 1}{\sqrt{\csc^2(x) - 1}}$
23. $\sin \alpha = 1/\sqrt{5}$, $\cos \alpha = 2/\sqrt{5}$, $\csc \alpha = \sqrt{5}$, $\sec \alpha = \sqrt{5}/2$, $\cot \alpha = 2$
25. $\sin \alpha = -\sqrt{22}/5$, $\tan \alpha = \sqrt{22}/\sqrt{3}$, $\csc \alpha = -5/\sqrt{22}$, $\sec \alpha = -5/\sqrt{3}$, $\cot \alpha = \sqrt{3}/\sqrt{22}$
27. $\sin \alpha = -3/\sqrt{10}$, $\cos \alpha = 1/\sqrt{10}$, $\tan \alpha = -3$, $\csc \alpha = -\sqrt{10}/3$, $\sec \alpha = \sqrt{10}$
29. $\sqrt{1 - x^2}$ 31. $1/\sqrt{1 + x^2}$ 33. $x/\sqrt{1 - x^2}$
35. $\sqrt{1 + x^2}$ 37. $\cos x$ 39. 0   41. 0   43. $\cos^2 \alpha$
45. $-\cos \beta$ 47. Odd   49. Neither   51. Even   53. Even
55. Even   57. Odd   59. h   61. n   63. m   65. k
67. l   69. g   71. b   73. f   75. d   87. $-\tan^2 x$
89. -1 91. 1   93. $\tan^2 x$ 95. $\csc x$ 97. $\sin^2 x - \cos^2 x$
99. $\pm 2\sqrt{2}/3$ 101. $\sqrt{1 - u^2}$
103. $x = \dfrac{\pi}{2} + k\pi$, where k is any integer

Section 6.2

For Thought: 1. T   2. F   3. T   4. T   5. T   6. F
7. T   8. T   9. F   10. F
Exercises: 1. D   3. A   5. B   7. H   9. G
11. $2 \cos^2 \beta - \cos \beta - 1$ 13. $\csc^2(x) + 2 + \sin^2(x)$
15. $4 \sin^2(\theta) - 1$ 17. $9 \sin^2 \theta + 12 \sin \theta + 4$
19. $4 \sin^4 y - 4 + \csc^4 y$ 21. $\cos^2 \alpha$ 23. $\cot^2 \alpha$ 25. -1
27. $(2 \sin \gamma + 1)(\sin \gamma - 3)$ 29. $(\tan \alpha - 4)(\tan \alpha - 2)$
31. $(2 \sec \beta + 1)^2$ 33. $(\tan \alpha - \sec \beta)(\tan \alpha + \sec \beta)$
35. $(\cos \beta)(\sin \beta + 2)(\sin \beta - 1)$ 37. $(2 \sec^2 x - 1)^2$
39. $(\cos \alpha + 1)(\sin \alpha + 1)$ 41. $\dfrac{\sin^2 x}{a}$ 43. $\dfrac{3 \sin(2x)}{2}$
45. $\dfrac{5 \tan x}{6}$ 47. $1 - \sin x$ 49. $\sin x + \cos x$
51. $\dfrac{\sin x + 1}{\sin x + 2}$ 53. $\sin x$
95. Identity   97. Not identity   99. Identity
101. Not identity   103. Identity

Section 6.3

For Thought: 1. F   2. T   3. T   4. F   5. T   6. F
7. T   8. T   9. F   10. T

Exercises: 1. $\dfrac{\sqrt{2} - \sqrt{6}}{4}$ 3. $\dfrac{\sqrt{2} + \sqrt{6}}{4}$
5. 70°   7. $\pi/3$ 9. 6°   11. $\dfrac{\sqrt{2} + \sqrt{6}}{4}$

13. $\dfrac{\sqrt{6} - \sqrt{2}}{4}$ **15.** $2 - \sqrt{3}$ **17.** $\sqrt{3} - 2$ **19.** $\dfrac{7\pi}{12}$

21. $\dfrac{13\pi}{12}$ **23.** 1   **25.** 0   **27.** 1   **29.** $\sin(3k)$

31. $30° + 45°$ **33.** $120° + 45°$ **35.** $\dfrac{\sqrt{6} - \sqrt{2}}{4}$

37. $\dfrac{\sqrt{6} + \sqrt{2}}{4}$ **39.** $2 + \sqrt{3}$ **41.** $\dfrac{\sqrt{2} - \sqrt{6}}{4}$

43. $\dfrac{-\sqrt{6} - \sqrt{2}}{4}$ **45.** $-2 + \sqrt{3}$ **47.** 1   **49.** $\cos(3\pi/10)$

51. $\tan(13\pi/42)$ **53.** $\sin 49°$ **55.** G   **57.** H   **59.** F

61. A   **63.** 16/65   **65.** $\dfrac{2 - \sqrt{15}}{6}$ **67.** 297/425

69. $-416/425$ **71.** $-\sin\alpha$ **73.** $-\cos\alpha$ **75.** $-\sin\alpha$
77. $\cos\alpha$ **103.** 91/89

Section 6.4

For Thought: 1. T   **2.** T   **3.** F   **4.** T   **5.** F   **6.** T
7. F   **8.** F   **9.** T   **10.** T
Exercises: 1. 1   **3.** $\sqrt{3}$ **5.** -1 **7.** $\sqrt{3}$ **9.** $\dfrac{\sqrt{2} + \sqrt{3}}{2}$

11. $\dfrac{\sqrt{2} - \sqrt{3}}{2}$ **13.** $2 - \sqrt{3}$ **15.** $\dfrac{\sqrt{2} - \sqrt{2}}{2}$ **17.** Positive

19. Negative   **21.** Negative   **23.** $\sin 26°$ **25.** $\sqrt{2}/2$
27. $\sqrt{3}/6$ **29.** $\tan 6°$ **31.** $-\sin(7\pi/9)$ **33.** $\cos(2\pi/9)$
35. c   **37.** g   **39.** a   **41.** h   **43.** f
45. $1/\sqrt{5}, 2/\sqrt{5}, 1/2, \sqrt{5}, \sqrt{5}/2, 2$
47. $\sqrt{26}/26, 5\sqrt{26}/26, 1/5, \sqrt{26}, \sqrt{26}/5, 5$
49. $-\sqrt{15}/8, -7/8, \sqrt{15}/7, -8/\sqrt{15}, -8/7, 7/\sqrt{15}$
51. $-24/25, -7/25, 24/7, -25/24, -25/7, 7/24$
67. Not identity   **69.** Not identity   **71.** Not identity   **73.** Identity

75. $-24/25$ **77.** 161/289   **79.** $\dfrac{5\sqrt{34} - 25}{3}$

81. $A = \dfrac{d^2}{2}\sin(2\alpha)$

Section 6.5

For Thought: 1. T   **2.** F   **3.** T   **4.** F   **5.** T   **6.** F
7. T   **8.** T   **9.** T   **10.** T
Exercises: 1. $0.5(\cos 4° - \cos 22°)$ **3.** $0.5(\sin 36° - \sin 4°)$
5. $0.5(\sin 15° - \sin 5°)$ **7.** $0.5(\cos(\pi/30) + \cos(11\pi/30))$
9. $0.5(\cos 2y^2 + \cos 12y^2)$ **11.** $0.5(\sin(3s) + \sin(s - 2))$

13. $\dfrac{\sqrt{2} - 1}{4}$ **15.** $\dfrac{\sqrt{2} + \sqrt{3}}{4}$ **17.** $2\cos 10° \sin 2°$

19. $2\sin 83.5° \sin 3.5°$ **21.** $-2\cos(4.2)\sin(0.6)$
23. $-2\sin(4\pi/15)\sin(\pi/15)$ **25.** $-2\sin(4y + 3)\sin(y - 6)$
27. $-2\cos(6.5\alpha)\sin(1.5\alpha)$ **29.** $\sqrt{6}/2$ **31.** $\dfrac{\sqrt{2} - \sqrt{2}}{2}$
33. $\sqrt{2}\sin(x - \pi/4)$ **35.** $\sin(x + 2\pi/3)$ **37.** $\sin(x - \pi/6)$
39. $y = \sqrt{2}\sin(x + 3\pi/4), \sqrt{2}, 2\pi, -3\pi/4$

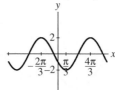

41. $y = 2\sin(x - \pi/4), 2, 2\pi, \pi/4$

$y = 2\sin\left(x - \dfrac{\pi}{4}\right)$

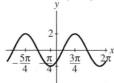

43. $y = 2\sin(x + 7\pi/6), 2, 2\pi, -7\pi/6$

$y = 2\sin\left(x + \dfrac{7\pi}{6}\right)$

45. $5, -0.9$ **47.** $\sqrt{37}, -3.0$ **49.** $\sqrt{34}, -4.2$
61. $x = 2\sin(t + \pi/6), 2$ m

Section 6.6

For Thought: 1. F   **2.** F   **3.** T   **4.** T   **5.** T   **6.** F
7. F   **8.** T   **9.** F   **10.** F
Exercises: 1. $\{x \mid x = \pi + 2k\pi\}$ **3.** $\{x \mid x = k\pi\}$

5. $\left\{x \mid x = \dfrac{3\pi}{2} + 2k\pi\right\}$

7. $\left\{x \mid x = \dfrac{\pi}{3} + 2k\pi \text{ or } x = \dfrac{5\pi}{3} + 2k\pi\right\}$

9. $\left\{x \mid x = \dfrac{\pi}{4} + 2k\pi \text{ or } x = \dfrac{3\pi}{4} + 2k\pi\right\}$

11. $\left\{x \mid x = \dfrac{\pi}{4} + k\pi\right\}$

13. $\left\{x \mid x = \dfrac{5\pi}{6} + 2k\pi \text{ or } x = \dfrac{7\pi}{6} + 2k\pi\right\}$

15. $\left\{x \mid x = \dfrac{5\pi}{4} + 2k\pi \text{ or } x = \dfrac{7\pi}{4} + 2k\pi\right\}$

17. $\left\{x \mid x = \dfrac{3\pi}{4} + k\pi\right\}$ **19.** $\{x \mid x = 90° + k180°\}$

21. $\{x \mid x = 90° + k360°\}$ **23.** $\{x \mid x = k180°\}$
25. $\{x \mid x = 29.2° + k360° \text{ or } x = 330.8° + k360°\}$
27. $\{x \mid x = 345.9° + k360° \text{ or } x = 194.1° + k360°\}$
29. $\{x \mid x = 79.5° + k180°\}$

31. $\left\{x \mid x = \dfrac{2\pi}{3} + 4k\pi \text{ or } x = \dfrac{10\pi}{3} + 4k\pi\right\}$

33. $\left\{x \mid x = \dfrac{2k\pi}{3}\right\}$

35. $\left\{x \mid x = \dfrac{\pi}{3} + 4k\pi \text{ or } x = \dfrac{5\pi}{3} + 4k\pi\right\}$

37. $\left\{x \mid x = \dfrac{5\pi}{8} + k\pi \text{ or } x = \dfrac{7\pi}{8} + k\pi\right\}$ **39.** $\left\{x \mid x = \dfrac{\pi}{6} + \dfrac{k\pi}{2}\right\}$

41. $\left\{x \mid x = \dfrac{k\pi}{4}\right\}$ **43.** $\left\{x \mid x = \dfrac{1}{6} + 2k \text{ or } x = \dfrac{5}{6} + 2k\right\}$

45. $\left\{x \mid x = \dfrac{1}{4} + \dfrac{k}{2}\right\}$ **47.** $\{240°, 300°\}$

49. $\{22.5°, 157.5°, 202.5°, 337.5°\}$

51. $\{45°, 75°, 165°, 195°, 285°, 315°\}$
53. $\{60°\}$
55. $\{\alpha \mid \alpha = 6.6° + k120° \text{ or } \alpha = 53.4° + k120°\}$
57. $\{\alpha \mid \alpha = 72.3° + k120° \text{ or } \alpha = 107.7° + k120°\}$
59. $\{\alpha \mid \alpha = 38.6° + k180° \text{ or } \alpha = 141.4° + k180°\}$
61. $\{\alpha \mid \alpha = 668.5° + k720° \text{ or } \alpha = 411.5° + k720°\}$
63. $\{0, 0.3, 2.8, \pi\}$ **65.** $\{\pi, 2\pi/3, 4\pi/3\}$
67. $\{0.7, 2.5, 3.4, 6.0\}$ **69.** $\{11\pi/6\}$ **71.** $\{1.0, 4.9\}$
73. $\{0, \pi\}$ **75.** $\{\pi/2, 3\pi/2\}$
77. $\{7\pi/12, 23\pi/12\}$ **79.** $\{7\pi/6, 11\pi/6\}$ **81.** $\{0°\}$
83. $\{26.6°, 206.6°\}$ **85.** $\{30°, 90°, 150°, 210°, 270°, 330°\}$
87. $\{67.5°, 157.5°, 247.5°, 337.5°\}$ **89.** $\{221.8°, 318.2°\}$
91. $\{120°, 300°\}$ **93.** $\{30°, 45°, 135°, 150°, 210°, 225°, 315°, 330°\}$
95. $\{0°, 60°, 120°, 180°, 240°, 300°\}$ **97.** 10.5   **99.** 56.4°
101. 19.6°   **103.** $\{0.4, 1.9, 2.2, 4.0, 4.4, 5.8\}$
105. $\{\pi/3\}$ **107.** $\dfrac{5\pi}{12} + \dfrac{k\pi}{2}$ for k a nonnegative integer
109. $1 + 6k$ and $2 + 6k$ for k a nonnegative integer
111. 44.4° or 45.6°   **113.** 12.5° or 77.5°, 6.3 sec

Chapter 6 Review Exercises

1. $\cos^2 \alpha$ **3.** $-\cot^2 x$ **5.** $\sec^2 \alpha$ **7.** $\tan 4s$ **9.** $-\sin 3\theta$
11. $\tan z$ **13.** e   **15.** c   **17.** a   **19.** g
21. $\sin \alpha = 12/13$, $\tan \alpha = -12/5$, $\csc \alpha = 13/12$, $\sec \alpha = -13/5$, $\cot \alpha = -5/12$
23. $\sin \alpha = -4/5$, $\cos \alpha = -3/5$, $\tan \alpha = 4/3$, $\csc \alpha = -5/4$, $\sec \alpha = -5/3$, $\cot \alpha = 3/4$
25. $\sin \alpha = -24/25$, $\cos \alpha = 7/25$, $\tan \alpha = -24/7$, $\csc \alpha = -25/24$, $\sec \alpha = 25/7$, $\cot \alpha = -7/24$
27. Identity   **29.** Not identity   **31.** Odd   **33.** Neither
35. Even   **37.** f   **39.** e   **41.** b   **43.** h   **45.** c
61. $\sqrt{3} - 2$ **63.** $-\dfrac{\sqrt{2 + \sqrt{3}}}{2}$
65. $y = 4\sqrt{2} \sin(x + \pi/4)$, $4\sqrt{2}$, $-\pi/4$

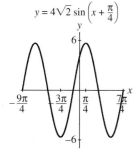

$$y = 4\sqrt{2} \sin\left(x + \frac{\pi}{4}\right)$$

67. $y = \sqrt{5} \sin(x + 2.68)$, $\sqrt{5}$, -2.68

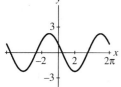

$$y = \sqrt{5} \sin(x + 2.68)$$

69. $\left\{x \mid x = \dfrac{\pi}{3} + k\pi \text{ or } x = \dfrac{2\pi}{3} + k\pi\right\}$
71. $\left\{x \mid x = \dfrac{\pi}{3} + 2k\pi, \dfrac{2\pi}{3} + 2k\pi, \dfrac{\pi}{6} + 2k\pi, \dfrac{5\pi}{6} + 2k\pi\right\}$
73. $\left\{x \mid x = \dfrac{\pi}{6} + 2k\pi, \dfrac{5\pi}{6} + 2k\pi, \dfrac{\pi}{2} + 2k\pi\right\}$
75. $\left\{x \mid x = \dfrac{2\pi}{3} + 4k\pi \text{ or } x = \dfrac{4\pi}{3} + 4k\pi\right\}$
77. $\left\{x \mid x = \pi + 2k\pi, \dfrac{\pi}{3} + 4k\pi, \dfrac{5\pi}{3} + 4k\pi\right\}$
79. $\left\{x \mid x = \dfrac{\pi}{2} + k\pi\right\}$
81. $\left\{x \mid x = \pi + 2k\pi \text{ or } x = \dfrac{3\pi}{2} + 2k\pi\right\}$
83. $\{45°, 225°\}$ **85.** $\{90°, 180°\}$ **87.** $\varnothing$
89. $\{22.5°, 67.5°, 112.5°, 157.5°, 202.5°, 247.5°, 292.5°, 337.5°\}$
91. $\{0°, 180°\}$ **93.** $\{15°, 75°, 135°, 195°, 255°, 315°\}$
95. $2 \cos 17° \cos 2°$ **97.** $2 \cos \dfrac{\pi}{16} \sin \dfrac{3\pi}{16}$
99. $\sin 24° - \sin 2°$ **101.** $\cos \dfrac{x}{12} + \cos \dfrac{7x}{12}$
103. $\dfrac{\sqrt{6} + \sqrt{2}}{4}$ **105.** $\dfrac{\sqrt{6} - \sqrt{2}}{4}$ **107.** 1.28, 2.85

Chapter 6 Test

1. $2 \cos x$ **2.** $\sin 7t$ **3.** $2 \csc^2 y$ **4.** $\tan(3\pi/10)$
9. $\left\{\theta \mid \theta = \dfrac{3\pi}{2} + 2k\pi\right\}$ **10.** $\left\{s \mid s = \dfrac{\pi}{9} + \dfrac{2k\pi}{3} \text{ or } s = \dfrac{5\pi}{9} + \dfrac{2k\pi}{3}\right\}$
11. $\left\{t \mid t = \dfrac{\pi}{3} + \dfrac{k\pi}{2}\right\}$
12. $\left\{\theta \mid \theta = \dfrac{\pi}{2} + k\pi \text{ or } \theta = \dfrac{\pi}{6} + 2k\pi \text{ or } \dfrac{5\pi}{6} + 2k\pi\right\}$
13. $\{19.5°, 90°, 160.5°\}$
14. $\{27°, 63°, 99°, 171°, 207°, 243°, 279°, 351°\}$
15. $y = 2 \sin(x + 5\pi/3)$, 2π, 2, $-5\pi/3$

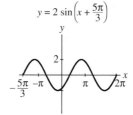

$$y = 2 \sin\left(x + \frac{5\pi}{3}\right)$$

16. $\sin \alpha = 1/2$, $\cos \alpha = -\sqrt{3}/2$, $\tan \alpha = -1/\sqrt{3}$, $\sec \alpha = -2/\sqrt{3}$, $\cot \alpha = -\sqrt{3}$
17. Even   **18.** $-\dfrac{\sqrt{2 - \sqrt{3}}}{2}$
20. 0.4 sec, 1.4 sec, 2.5 sec, 3.5 sec

Tying It All Together Chapters 1–6

1. Odd   **2.** Even   **3.** Odd   **4.** Even   **5.** Even   **6.** Odd
7. Even   **8.** Even   **9.** Not identity   **10.** Not identity
11. Identity   **12.** Identity   **13.** Not identity   **14.** Not identity
15. $\beta = 60°$, $b = 4\sqrt{3}$, $c = 8$ **16.** $\alpha = 60°$, $\beta = 30°$, $c = 2$
17. $\alpha = 17.5°$, $\beta = 72.5°$, $a = 1.6$, $c = 5.2$

18. $\alpha = 36.9°, \beta = 53.1°, b = 2.7, c = 3.3$

19. $\dfrac{1}{2} + \dfrac{\sqrt{3}}{2}\,i$ **20.** $-1 + \sqrt{3}\,i$ **21.** i **22.** 1

23. $2 + 11i$ **24.** -16

CHAPTER 7

Section 7.1

For Thought: **1.** T   **2.** F   **3.** F   **4.** T   **5.** T   **6.** F
7. T   **8.** F   **9.** T   **10.** T
Exercises: **1.** $\gamma = 44°, b = 14.4, c = 10.5$
3. $\beta = 134.2°, a = 5.2, c = 13.6$ **5.** $\beta = 26°, a = 14.6, b = 35.8$
7. $\alpha = 45.7°, b = 587.9, c = 160.8$
9. None   **11.** One: $\alpha = 30°, \beta = 90°, a = 10$
13. One: $\alpha = 26.3°, \gamma = 15.6°, a = 10.3$
15. Two: $\alpha_1 = 134.9°, \gamma_1 = 12.4°, c_1 = 11.4$; or $\alpha_2 = 45.1°,$
$\gamma_2 = 102.2°, c_2 = 51.7$
17. One: $\alpha = 25.4°, \beta = 55.0°, a = 5.4$
19. 9.8   **21.** 83.4   **23.** 66.3   **25.** 37.7   **27.** $22\sqrt{3} + 6$
29. 18.4 mi   **31.** 97,535 ft^2 **33.** 159.4 ft   **35.** 28.9 ft, 15.7 ft
37. 277.7 in.^{2} **39. a.** 19.9 mi   **b.** 1.0 sec before 4 P.M.   **c.** 281 mi
41. 62.2°

Section 7.2

For Thought: **1.** T   **2.** F   **3.** F   **4.** T   **5.** F   **6.** T
7. T   **8.** T   **9.** T   **10.** F
Exercises: **1.** $\alpha = 30.4°, \beta = 28.3°, c = 5.2$
3. $\alpha = 26.4°, \beta = 131.3°, \gamma = 22.3°$
5. $\alpha = 163.9°, \gamma = 5.6°, b = 4.5$
7. $\alpha = 130.3°, \beta = 30.2°, \gamma = 19.5°$
9. $\beta = 48.3°, \gamma = 101.7°, a = 6.2$
11. $\alpha = 53.9°, \beta = 65.5°, \gamma = 60.6°$
13. $\alpha = 120°, b = 3.5, c = 4.8$ **15.** 0   **17.** 1   **19.** 0   **21.** 1
23. 0   **25.** 40.9   **27.** 13.9   **29.** 40,471.9   **31.** 100
33. 20.8   **35.** 20.4   **37.** 9.90 ft   **39.** 783.45 ft   **41.** 20.6 mi
43. 19.2°   **45.** $\angle A = 78.5°, \angle B = 57.1°, \angle C = 44.4°$
47. 3.8 mi, 40.4°, 111.6°   **49.** 11.76 m   **51.** $\theta_1 = 13.3°, \theta_2 = 90.6°$
53. a. 0.52° to 0.54°   **b.** 0.49° to 0.55°   **c.** yes   **55.** 603 ft

Section 7.3

For Thought: **1.** T   **2.** F   **3.** T   **4.** T   **5.** F   **6.** F
7. T   **8.** T   **9.** T   **10.** T
Exercises:
1.

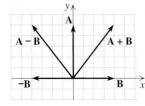

3.

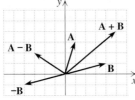

5.

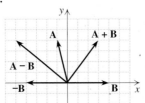

7. D   **9.** E   **11.** B   **13.** $|\mathbf{v}_x| = 1.9, |\mathbf{v}_y| = 4.1$
15. $|\mathbf{v}_x| = 7256.4, |\mathbf{v}_y| = 3368.3$ **17.** $|\mathbf{v}_x| = 87.7, |\mathbf{v}_y| = 217.0$
19. $2, 30°$ **21.** $2, 135°$ **23.** $16, 300°$ **25.** $5, 0°$
27. $\sqrt{13}, 146.3°$ **29.** $\sqrt{10}, 341.6°$ **31.** $\langle 4\sqrt{2}, 4\sqrt{2}\rangle$
33. $\langle -237.6, 166.3\rangle$ **35.** $\langle 17.5, -4.0\rangle$ **37.** $\langle 15, -10\rangle$
39. $\langle 18, -22\rangle$ **41.** $\langle 11, -13\rangle$ **43.** $\langle 0, -1\rangle$ **45.** -13
47. 32.5°   **49.** 27.3°   **51.** 90°   **53.** Perpendicular
55. Parallel   **57.** Neither   **59.** $2\mathbf{i} + \mathbf{j}$ **61.** $-3\mathbf{i} + \sqrt{2}\mathbf{j}$
63. $-9\mathbf{j}$ **65.** $-7\mathbf{i} - \mathbf{j}$ **67.** $\sqrt{17}, 76.0°$ **69.** $3\sqrt{10}, 198.4°$
71. $\sqrt{29}, 158.2°$ **73.** $\sqrt{65}/2, 172.9°$ **75.** $\sqrt{89}, 32.0°$
77. $2\sqrt{10}, 108.4°$ **79.** $\sqrt{5}, 243.4°$ **81.** $\sqrt{73}$ lb, 69.4°, 20.6°
83. 8.25 newtons, 22.9°, 107.1°   **85.** 5.1 lb, 75.2°
87. 127.0 lb   **89.** 450.3 mph, 260 mph   **91.** 1368.1 lb
93. 1026.1 lb   **95.** 78.5°   **97.** 97.1°, 241.9 mph
99. 66.5°, 38.5 mph   **101.** 108.2°, 451.0 mph
103. 198.4°, 6000 ft   **105. a.** $d = 0.2|\sec \beta|$ **b.** $r = 3\cos(\alpha)|\sec(\beta)|$
107. 1420.0 lb, E 13.5° N assuming Mama is pulling east

Section 7.4

For Thought: **1.** T   **2.** F   **3.** T   **4.** F   **5.** F   **6.** T
7. T   **8.** T   **9.** T   **10.** F
Exercises: **1.** 8   **3.** 9   **5.** $2\sqrt{10}$ **7.** 4   **9.** 1   **11.** $3\sqrt{2}$
13. $8(\cos 0° + i \sin 0°)$ **15.** $\sqrt{3}(\cos 90° + i \sin 90°)$
17. $3\sqrt{2}(\cos 135° + i \sin 135°)$ **19.** $3(\cos 135° + i \sin 135°)$
21. $2(\cos 150° + i \sin 150°)$ **23.** $5(\cos 53.1° + i \sin 53.1°)$
25. $\sqrt{34}(\cos 121.0° + i \sin 121.0°)$
27. $3\sqrt{5}(\cos 296.6° + i \sin 296.6°)$
29. $1 + i$ **31.** $-\dfrac{3}{4} + \dfrac{\sqrt{3}}{4}i$ **33.** $-0.42 - 0.26i$ **35.** $3i$
37. $-i\sqrt{3}$ **39.** $\dfrac{\sqrt{6}}{2} + \dfrac{3\sqrt{2}}{2}i$ **41.** $6i$ **43.** $\dfrac{3\sqrt{2}}{2} + \dfrac{\sqrt{6}}{2}i$
45. $9i$ **47.** $\sqrt{3} + i$ **49.** $0.34 - 0.37i$ **51.** $-40i, -0.8$
53. $8i, \dfrac{\sqrt{3}}{4} - \dfrac{1}{4}i$ **55.** $4\sqrt{2}, i\sqrt{2}$ **57.** $-7 - 26i, -\dfrac{23}{29} - \dfrac{14}{29}i$
59. $-18 + 14i, \dfrac{6}{13} + \dfrac{22}{13}i$ **61.** $-3 + 3i, 1.5 + 1.5i$
63. $3\left(\cos\left(-\dfrac{\pi}{4}\right) + i \sin\left(-\dfrac{\pi}{4}\right)\right)$
65. $2\sqrt{3}(\cos 20° + i \sin 20°)$ **67.** 9   **69.** 4   **71.** $32i$
73. $-54 + 54i$
79. $6\cos(9°) + 3\cos(5°) + i(6\sin(9°) + 3\sin(5°)), 6 - 4i$, standard
form

Section 7.5

For Thought: 1. F   **2.** F   **3.** T   **4.** F   **5.** F   **6.** F
7. T   **8.** T   **9.** T   **10.** F
Exercises: 1. $27i$ **3.** $-2 + 2i\sqrt{3}$ **5.** $-\dfrac{1}{2} + \dfrac{\sqrt{3}}{2}i$

7. $-18 + 18i\sqrt{3}$ **9.** $701.5 + 1291.9i$ **11.** $-16 + 16i$

13. $-8 - 8i\sqrt{3}$ **15.** $-3888 + 3888i\sqrt{3}$ **17.** $-119 - 120i$
19. $-7 - 24i$ **21.** $-44.928 - 31.104i$
23. $2(\cos 45° + i \sin 45°), 2(\cos 225° + i \sin 225°)$
25. $\cos \alpha + i \sin \alpha$ for $\alpha = 30°, 120°, 210°, 300°$

27. $2(\cos \alpha + i \sin \alpha)$ for $\alpha = \dfrac{\pi}{6}, \dfrac{\pi}{2}, \dfrac{5\pi}{6}, \dfrac{7\pi}{6}, \dfrac{3\pi}{2}, \dfrac{11\pi}{6}$

29. $1, -\dfrac{1}{2} \pm \dfrac{\sqrt{3}}{2}i$ **31.** $\pm 2, \pm 2i$

33. $\dfrac{\sqrt{2}}{2} \pm \dfrac{\sqrt{2}}{2}i, -\dfrac{\sqrt{2}}{2} \pm \dfrac{\sqrt{2}}{2}i$

35. $\pm\dfrac{\sqrt{3}}{2} + \dfrac{1}{2}i, -i$ **37.** $\pm(1 + i\sqrt{3})$
39. $1.272 + 0.786i, -1.272 - 0.786i$

41. $\dfrac{1}{2} \pm \dfrac{\sqrt{3}}{2}i, -1$ **43.** $\pm 3, \pm 3i$ **45.** $-1 + i, 1 - i$
47. $0, \pm 2, 1 \pm i\sqrt{3}, -1 \pm i\sqrt{3}$ **49.** $-2, \pm i\sqrt{5}, 1 \pm i\sqrt{3}$
51. $2^{1/5}(\cos \alpha + i \sin \alpha)$ for $\alpha = 0°, 72°, 144°, 216°, 288°$
53. $10^{1/8}(\cos \alpha + i \sin \alpha)$ for $\alpha = 40.4°, 130.4°, 220.4°, 310.4°$

55. $-\dfrac{1}{4} + \dfrac{1}{4}i$ **57.** $1, -i$

Section 7.6

For Thought: 1. T   **2.** F   **3.** F   **4.** F   **5.** T   **6.** T
7. T   **8.** F   **9.** F   **10.** F
Exercises: 1. $(3, \pi/2)$
3. $(3\sqrt{2}, \pi/4)$
5–19 odd

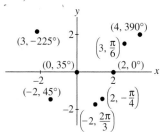

21. $(4, 0)$ **23.** $(0, 0)$ **25.** $\left(\dfrac{\sqrt{3}}{2}, \dfrac{1}{2}\right)$ **27.** $(0, 3)$ **29.** $(-1, 1)$

31. $\left(-\dfrac{\sqrt{6}}{2}, \dfrac{3\sqrt{2}}{2}\right)$ **33.** $(2\sqrt{3}, 60°)$ **35.** $(2\sqrt{2}, 135°)$

37. $(2, 90°)$ **39.** $(3\sqrt{2}, 225°)$ **41.** $(\sqrt{17}, 75.96°)$
43. $(\sqrt{6}, -54.7°)$ **45.** $(3.60, 1.75)$ **47.** $(1.80, 0.87)$
49. $(-0.91, -1.78)$ **51.** $(6.4, 51.3°)$ **53.** $(7.3, 254.1°)$

55.

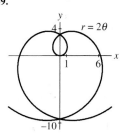

57.

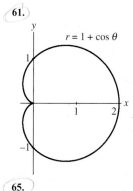

59.

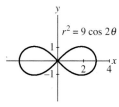

61.

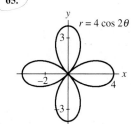

63.

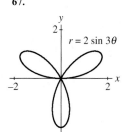

65.

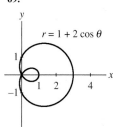

67.

69.

71.

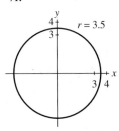

73.

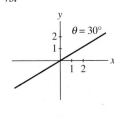

75. $x^2 - 4x + y^2 = 0$ **77.** $y = 3$ **79.** $x = 3$
81. $x^2 + y^2 = 25$ **83.** $y = x$ **85.** $x^2 - 4y = 4$
87. $r \cos \theta = 4$ **89.** $\theta = -\pi/4$ **91.** $r = 4 \tan \theta \sec \theta$
93. $r = 2$ **95.** $r = \dfrac{1}{2 \cos \theta - \sin \theta}$ **96.** $r = \dfrac{5}{\sin \theta + 3 \cos \theta}$
97. $r = 2 \sin \theta$
99. $(1, 0.17), (1, 0.87), (1, 2.27), (1, 2.97), (1, 4.36), (1, 5.06)$
101. $(0, 0), (0.9, 1.4), (1.2, 1.8), (1.9, 2.8), (1.9, 3.5), (1.2, 4.5), (0.8, 4.9)$

Section 7.7

For Thought: **1.** F   **2.** T   **3.** T   **4.** F   **5.** T   **6.** F
7. T   **8.** T   **9.** F   **10.** T

Exercises:

1.

t	x	y
0	1	-2
1	5	-1
1.5	7	-0.5
3	13	1

3.

t	x	y
1	1	2
2.5	6.25	6.5
$\sqrt{5}$	5	$3\sqrt{5} - 1$
4	16	11
5	25	14

5. $[-2, 10], [3, 7]$

7. $(-\infty, \infty), [0, \infty)$

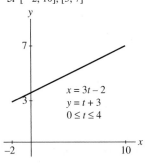

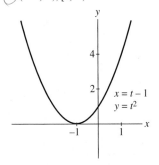

9. $(0, 1), (0, 1)$

11. $[-1, 1], [-1, 1]$

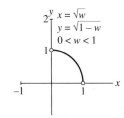

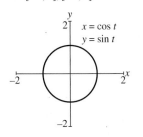

13. $x + y = -2$, line, $(-\infty, \infty), (-\infty, \infty)$
15. $x^2 + y^2 = 16$, circle, $[-4, 4], [-4, 4]$
17. $y = e^{4x}$, exponential, $(-\infty, \infty), (0, \infty)$
19. $y = 2x + 3$, line, $(-\infty, \infty), (-\infty, \infty)$
21. $x = \dfrac{3}{2}t + 2, y = 3t + 3, 0 \le t \le 2$
23. $x = 2 \cos t, y = 2 \sin t, \pi < t < 3\pi/2$
25. $x = 3, y = t, -\infty < t < \infty$
27. $x = \sin(2t), y = 2 \sin^2(t)$

29.

31.

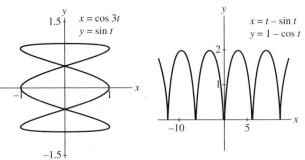

33.

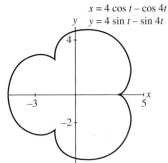

35.

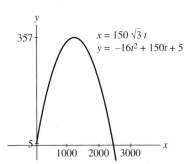

37. 9.4 sec

Chapter 7 Review Exercises

1. $\alpha = 82.7°, \beta = 49.3°, c = 2.5$ **3.** $\gamma = 103°, a = 4.6, b = 18.4$
5. No triangle   **7.** $\alpha = 107.7°, \beta = 23.7°, \gamma = 48.6°$
9. $\alpha_1 = 110.8°, \gamma_1 = 47.2°, a_1 = 6.2; \alpha_2 = 25.2°, \gamma_2 = 132.8°,$
   $a_2 = 2.8$
11. 92.4 ft^2 **13.** 23.0 km^2 **15.** $5.5, 2.4$ **17.** $2.0, 2.5$
19. $\sqrt{13}, 56.3°$ **21.** $6.0, 237.9°$ **23.** $\langle 1, 1 \rangle$ **25.** $\langle -3.0, 8.6 \rangle$
27. $\langle -6, 8 \rangle$ **29.** $\langle 0, -17 \rangle$ **31.** 6   **33.** $-4\mathbf{i} + 8\mathbf{j}$

35. $3.6\sqrt{3}\mathbf{i} + 3.6\mathbf{j}$ **37.** $\sqrt{34}$ **39.** $2\sqrt{2}$

41. $5.94(\cos 135° + i\sin 135°)$

43. $7.6(\cos 252.3° + i\sin 252.3°)$

45. $-\dfrac{3}{2} + \dfrac{\sqrt{3}}{2}i$ **47.** $5.4 + 3.5i$ **49.** $-15i, -\dfrac{5}{6}$

51. $8 - i, \dfrac{4}{13} + \dfrac{7}{13}i$ **53.** $-4\sqrt{2} + 4i\sqrt{2}$

55. $-128 + 128i$ **57.** $\dfrac{\sqrt{2}}{2} + \dfrac{\sqrt{2}}{2}i, -\dfrac{\sqrt{2}}{2} - \dfrac{\sqrt{2}}{2}i$

59. $2^{1/3}(\cos \alpha + i \sin \alpha)$ for $\alpha = 10°, 130°, 250°$

61. $5^{1/6}(\cos \alpha + i \sin \alpha)$ for $\alpha = 8.9°, 128.9°, 248.9°$

63. $5(\cos \alpha + i \sin \alpha)$ for $\alpha = 22.5°, 112.5°, 202.5°, 292.5°$

65. $\left(2.5, 5\sqrt{3}/2\right)$ **67.** $(-0.3, 1.7)$ **69.** $(4, 4\pi/3)$

71. $\left(\sqrt{13}, -0.98\right)$

73.

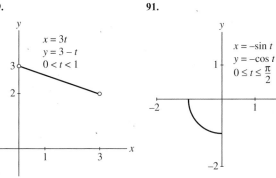

$r = -2\sin\theta$

75.

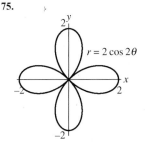

$r = 2\cos 2\theta$

77.

$r = 500 + \cos\theta$

79.

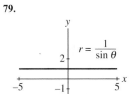

$r = \dfrac{1}{\sin\theta}$

81. $x + y = 1$ **83.** $x^2 + y^2 = 25$

85. $r = \dfrac{3}{\sin\theta}$ **87.** $r = 7$

89.

$x = 3t$
$y = 3 - t$
$0 < t < 1$

91.

$x = -\sin t$
$y = -\cos t$
$0 \le t \le \dfrac{\pi}{2}$

93. 18.4 lb, $11.0°$, $19.0°$ **95.** Seth, 42,785 ft^2 **97.** \$2639

99. b. 10.3 in.

Chapter 7 Test

1. One: $\beta = 90°, \gamma = 60°, c = 2\sqrt{3}$

2. Two: $\beta_1 = 111.1°, \gamma_1 = 8.9°, c_1 = 0.7; \beta_2 = 68.9°,$
$\gamma_2 = 51.1°, c_2 = 3.5$

3. One: $\gamma = 145.6°, b = 2.5, c = 5.9$

4. One: $\alpha = 33.8°, \beta = 129.2°, c = 1.5$

5. $\alpha = 28.4°, \beta = 93.7°, \gamma = 57.9°$ **6.** $2\sqrt{10}, 108.4°$

7. $2\sqrt{5}, 206.6°$ **8.** $3\sqrt{17}, 76.0°$ **9.** $3\sqrt{2}(\cos 45° + i\sin 45°)$

10. $2(\cos 120° + i\sin 120°)$ **11.** $2\sqrt{5}(\cos 206.6° + i\sin 206.6°)$

12. $3\sqrt{2} + 3i\sqrt{2}$ **13.** $512i$ **14.** $\dfrac{3\sqrt{2}}{4} + \dfrac{3\sqrt{2}}{4}i$

15. $\left(\dfrac{5\sqrt{3}}{2}, \dfrac{5}{2}\right)$ **16.** $\left(-\dfrac{3\sqrt{2}}{2}, \dfrac{3\sqrt{2}}{2}\right)$ **17.** $(-26.4, -19.9)$

18.

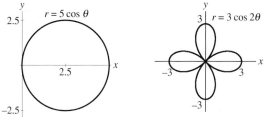

$r = 5\cos\theta$

19.

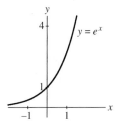

$r = 3\cos 2\theta$

20. 12.2 m^2 **21.** $\mathbf{v} = 3.66\mathbf{i} + 2.78\mathbf{j}$

22. $\dfrac{3\sqrt{2}}{2} \pm \dfrac{3\sqrt{2}}{2}i, -\dfrac{3\sqrt{2}}{2} \pm \dfrac{3\sqrt{2}}{2}i$

23. $r = -5\sin\theta$ **24.** $(x^2 + y^2)^{3/2} = 10xy$

25. $x = 6t - 2, y = 8t - 3, 0 \le t \le 1$ **26.** $47.2°, 239.3$ mph

Tying It All Together Chapters 1–7

1. $0, 1, -\dfrac{1}{2} \pm \dfrac{\sqrt{3}}{2}i$ **2.** $-2, 1, 3$

3. $-2, 1, \cos\theta + i\sin\theta$ for $\theta = 72°, 144°, 216°, 288°$

4. $2^{-1/4} \pm i2^{-1/4}, -2^{-1/4} \pm i2^{-1/4}, 1, -\dfrac{1}{2} \pm \dfrac{\sqrt{3}}{2}i$

5. $\dfrac{\pi}{6} + 2k\pi, \dfrac{5\pi}{6} + 2k\pi, \dfrac{2\pi}{3} + 2k\pi, \dfrac{4\pi}{3} + 2k\pi$

6. $-\dfrac{1}{2}, \dfrac{\pi}{6} + 2k\pi, \dfrac{5\pi}{6} + 2k\pi$ **7.** $k\pi$

8. $\ln\left(\dfrac{\pi}{6} + 2k\pi\right), \ln\left(\dfrac{5\pi}{6} + 2k\pi\right)$ for k a nonnegative integer

9. 4   **10.** $\varnothing$

11.

$y = \sin x$

12.

$y = e^x$

13.

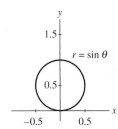

14.

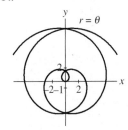

15.

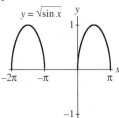

16.

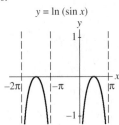

17.

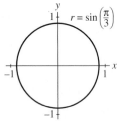

18.

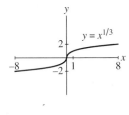

19. 0   **20.** 0   **21.** -1 **22.** 0   **23.** $\pi/6$ **24.** $\pi/3$

25. $-\pi/4$ **26.** $\pi/4$ **27.** $\dfrac{2x}{x^2-4}$ **28.** $2\sec^2(x)$ **29.** 4

30. $-\log(x^2-4)$ **31.** 2   **32.** $1/4$ **33.** 2   **34.** 0

35. $\dfrac{-2x}{x-3}$ **36.** 1

CHAPTER 8

Section 8.1

For Thought: **1.** T   **2.** F   **3.** F   **4.** T   **5.** T   **6.** T
7. F   **8.** F   **9.** F   **10.** T
Exercises: **1.** Yes   **3.** No   **5.** $\{(1,2)\}$ **7.** $\varnothing$ **9.** $\{(3,2)\}$
11. $\{(3,1)\}$ **13.** $\varnothing$ **15.** $\{(x,y)\,|\,x-2y=6\}$
17. $\{(-1,-1)\}$, independent   **19.** $\{(11/5,-6/5)\}$, independent
21. $\{(x,y)\,|\,y=3x+5\}$, dependent   **23.** $\varnothing$, inconsistent
25. $\{(150,50)\}$, independent   **27.** $\varnothing$, inconsistent
29. $\{(34,15)\}$, independent   **31.** $\{(13,7)\}$, independent
33. $\{(4,-1)\}$, independent   **35.** $\varnothing$, inconsistent
37. $\{(-1,1)\}$, independent   **39.** $\{(x,y)\,|\,x+2y=12\}$, dependent
41. $\{(4,6)\}$, independent   **43.** $\{(1.5,3.48)\}$, independent
45. Independent   **47.** Dependent   **49.** $(-1000,-497)$
51. $(6.18,-0.54)$ **53.** Althea $49,000, Vaughn $33,000
55. $10,000 at 10%, $15,000 at 8%   **57.** $6.50 adult, $4 child
59. Dependent, $m =$ number of male and $12 - m =$ number of female
  memberships, $0 \le m \le 12$

61. 23 ostriches, 19 cows   **63.** Dependent, lots of solutions
65. Inconsistent, no solution   **67.** 600 students
69. 65 pennies, 22 nickels   **71.** $x = 6$ oz, $y = 10$ oz
73. Plan A, 12 mo   **75.** 1.1 yr   **77.** $25/33$ sec ≈ 0.76 sec
79. $y = -2x + 3$ **81.** $y = -\dfrac{5}{3}x - \dfrac{1}{3}$

Section 8.2

For Thought: **1.** T   **2.** F   **3.** T   **4.** F   **5.** T   **6.** T
7. T   **8.** T   **9.** T   **10.** F
Exercises:
  1.   **3.**

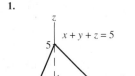
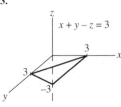

5. Yes   **7.** No   **9.** $\{(1,2,3)\}$ **11.** $\{(-1,1,2)\}$
13. $\{(0,-5,5)\}$ **15.** $(1,4,-4), (2,5,-3), (3,6,-2)$
17. $(2,1,-6), (4,2,-5), (6,3,-4)$ **19.** $y-3, y-8$
21. $z+1, z+2$ **23.** $\dfrac{y-1}{2}, \dfrac{3y-5}{2}$
25. $\{(x, x-1, x+5)\,|\,x \text{ is any real number}\}$
27. $\{(x,y,z)\,|\,x+2y-3z=5\}$ **29.** $\varnothing$
31. $\{(2,y,y)\,|\,y \text{ is any real number}\}$
33. $\{(x, 5-x, 3-x)\,|\,x \text{ is any real number}\}$
35. $\{(5,7,9)\}$ **37.** $\{(1.2,1.5,2.4)\}$ **39.** $\{(1000,2000,6000)\}$
41. $\{(15/11, 13/11, -3/11)\}$ **43.** $y = x^2 - 3$
45. $y = -2x^2 + 5x$ **47.** $y = x^2 + 4x + 4$ **49.** $x + y + z = 1$
51. $2x + y - z = 2$ **53.** 8, 12, 20   **55.** 5, 6, 10
57. $4000 in stocks, $7000 in bonds, $14,000 in mutual fund
59. $0.60, $0.80, $0.50
61. Weight in pounds

LR	LF	RR	RF
280	332	296	292
285	327	291	297
290	322	286	302

63. 116 pennies, 48 nickels, 68 dimes   **65.** $3.95
67. $x = 36$ lb, $y = 24$ lb, $z = 51$ lb
69. a. $y = -\dfrac{1}{20}x^2 + \dfrac{9}{2}x$ **b.** 101.25 m   **c.** 90 m

Section 8.3

For Thought: **1.** T   **2.** F   **3.** T   **4.** F   **5.** F   **6.** F
7. T   **8.** T   **9.** T   **10.** F
Exercises: **1.** Yes   **3.** No   **5.** $\{(0,0),(1,1)\}$ **7.** $\{(2,4),(3,9)\}$
9. $\left\{\left(-\dfrac{3}{2},\dfrac{3}{2}\right)\right\}$ **11.** $\{(-1,1),(0,0),(1,1)\}$
13. $\left\{(0,0),\left(\dfrac{1}{4},\dfrac{1}{2}\right)\right\}$ **15.** $\{(0,0),(2,8),(-2,-8)\}$
17. $\left\{(0,0),\left(\sqrt{2},\sqrt{2}\right),\left(-\sqrt{2},-\sqrt{2}\right)\right\}$

19. $\left\{\left(\dfrac{\sqrt{2}}{2}, \dfrac{\sqrt{2}}{2}\right)\left(-\dfrac{\sqrt{2}}{2}, -\dfrac{\sqrt{2}}{2}\right)\right\}$

21. $\{(-2 + \sqrt{3}, -2 - \sqrt{3}), (-2 - \sqrt{3}, -2 + \sqrt{3})\}$

23. $\{(1, \pm 1), (-1, \pm 1)\}$ **25.** $\left\{\left(-\dfrac{1}{2}, -2\right), (2, 3)\right\}$

27. $\{(2, 5)\}$ **29.** $\{(4, -3), (-1, 2)\}$ **31.** $\{(1, -2), (-1, 2)\}$

33. $\left\{\left(-\dfrac{1}{3}, 2^{2/3}\right)\right\}$ **35.** $\{(2, 1)\}$ **37.** $\{(2, 2)\}$ **39.** $\{(0, 1)\}$

41. $\{(2, 1), (0.3, -1.8)\}$ **43.** $\{(1.9, 0.6), (0.1, -2.0)\}$

45. $\{(-0.8, 0.6), (2, 4), (4, 16)\}$ **47.** $-2, 8$ **49.** 9 m and 12 m

51. $6 - 2\sqrt{3}$ ft, $6\sqrt{3} - 6$ ft, $12 - 4\sqrt{3}$ ft

53. $x = 8$ in. and $y = 8$ oz   **55.** A 9.6 min, B 48 min

57. $3 + i$ and $3 - i$ **59.** 60 ft and 30 ft

61. 0 and 9.65 yrs, 29.5 yr   **63.** 6 A.M.

65. $80 + 4\sqrt{29} \approx 101.54$ ft

Section 8.4

For Thought: 1. T   **2.** T   **3.** F   **4.** T   **5.** F   **6.** F
7. T   **8.** F   **9.** T   **10.** T

Exercises: 1. $\dfrac{7x - 5}{(x - 2)(x + 1)}$ **3.** $\dfrac{x^2 - 3x + 5}{(x - 1)(x^2 + 2)}$

5. $\dfrac{3x^3 + x^2 + 8x + 5}{(x^2 + 3)^2}$ **7.** $\dfrac{4x^2 - x - 1}{(x - 1)^3}$ **9.** $A = 2, B = -2$

11. $\dfrac{2}{x + 1} + \dfrac{3}{x - 2}$ **13.** $\dfrac{1/2}{x + 2} + \dfrac{3/2}{x + 4}$ **15.** $\dfrac{1/3}{x - 3} + \dfrac{-1/3}{x + 3}$

17. $\dfrac{-1}{x} + \dfrac{1}{x - 1}$ **19.** $A = 2, B = 5, C = -1$

21. $\dfrac{1}{x - 1} + \dfrac{1}{(x - 1)^2} + \dfrac{-1}{x + 2}$ **23.** $\dfrac{1}{x + 4} + \dfrac{-1}{x - 2} + \dfrac{2}{(x - 2)^2}$

25. $\dfrac{1}{x - 1} + \dfrac{2}{x + 1} + \dfrac{1}{(x + 1)^2}$ **27.** $A = -1, B = 2, C = -3$

29. $\dfrac{2}{x + 2} + \dfrac{3x - 1}{x^2 + 1}$ **31.** $\dfrac{-1}{x + 1} + \dfrac{2x - 1}{x^2 + x + 1}$

33. $\dfrac{-3}{(x + 2)^2} + \dfrac{-2}{x + 2}$ **35.** $\dfrac{4}{x + 1} + \dfrac{2x - 3}{x^2 + 1}$

37. $\dfrac{-8x}{(x^2 + 9)^2} + \dfrac{3x - 1}{x^2 + 9}$ **39.** $\dfrac{5}{x - 2} + \dfrac{-2x + 3}{x^2 + 2x + 4}$

41. $2x + 1 + \dfrac{2}{x - 1} + \dfrac{3}{x + 1}$ **43.** $\dfrac{3x + 3}{(x^2 + x + 1)^2} + \dfrac{3x - 5}{x^2 + x + 1}$

45. $\dfrac{3x}{x^2 + 4} + \dfrac{1}{x - 2} + \dfrac{-1}{x + 2}$ **47.** $\dfrac{1}{x} + \dfrac{2}{x^2} + \dfrac{3}{x^3} + \dfrac{4}{x - 1}$

49. $\dfrac{3}{x - 2} + \dfrac{-1}{(x - 2)^2} + \dfrac{3}{x - 3}$ **51.** $\dfrac{4}{x + 5} + \dfrac{2}{x + 2} + \dfrac{3}{x - 3}$

53. $\dfrac{-1}{(x - 1)^3} + \dfrac{2}{(x - 1)^2} + \dfrac{1}{x - 1}$ **55.** $\dfrac{-b/a}{(ax + b)^2} + \dfrac{1/a}{ax + b}$

57. $\dfrac{c/b}{x} + \dfrac{1 - ac/b}{ax + b}$ **59.** $\dfrac{1/b}{x^2} + \dfrac{-a/b^2}{x} + \dfrac{a^2/b^2}{ax + b}$

Section 8.5

For Thought: 1. F   **2.** F   **3.** T   **4.** F   **5.** T   **6.** T
7. F   **8.** F   **9.** T   **10.** T
Exercises: 1. (c)   **3.** (d)

5.
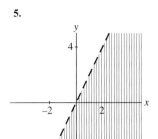

7.
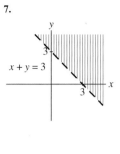

9.
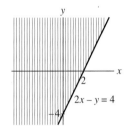

11.
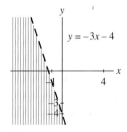

13.
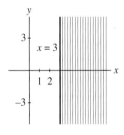

15.
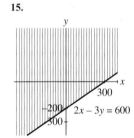

17.
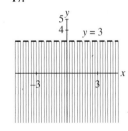

19.
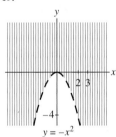

21.
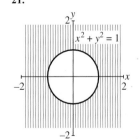

23.
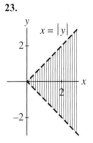

25.

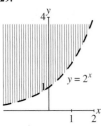

27.

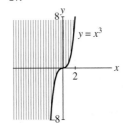

47.

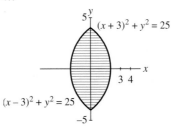

29.

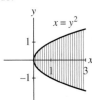

31.

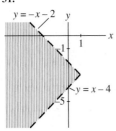

49.

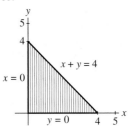

51.

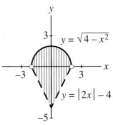

33.

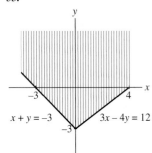

35.

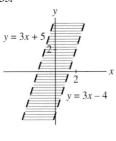

53.

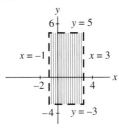

55.

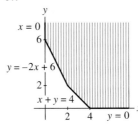

37. No solution

39.

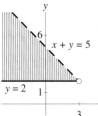

41.

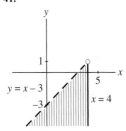

57.

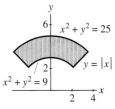

59.

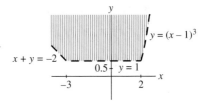

61.

43.

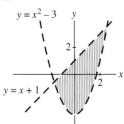

45.

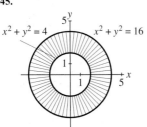

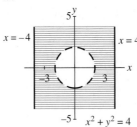

63.

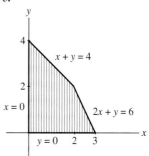

65. $x \geq 0, y \geq 0, y \leq -\dfrac{2}{3}x + 5, y \leq -3x + 12$

67. $y \geq 0, x \geq 0, y \geq -\dfrac{1}{2}x + 3, y \geq -\dfrac{3}{2}x + 5$

69. $|x| < 2, |y| < 2$ **71.** $x > 0, y > 0, x^2 + y^2 < 81$

73. $(-1.17, 1.84)$ **75.** $(150, 22.4)$

77. $w + h \leq 40$ **79.**

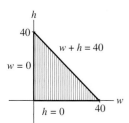

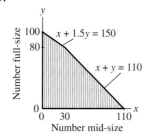

81.

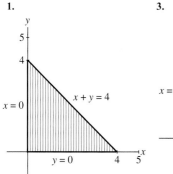

Section 8.6

For Thought: **1.** F   **2.** F   **3.** F   **4.** F   **5.** T   **6.** T
7. F   **8.** T   **9.** F   **10.** T

Exercises:

1.

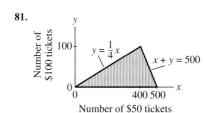

3.

5.

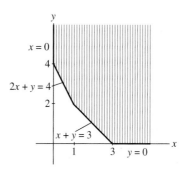

7.

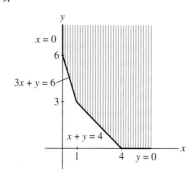

9.

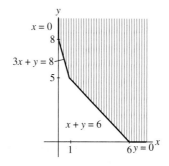

11.

13. 15   **15.** 8   **17.** 30   **19.** 8   **21.** 100
23. 8 bird houses, 6 mailboxes   **25.** 16 bird houses, 0 mailboxes
27. 0 small, 8 large   **29.** 5 small, 3 large

Chapter 8 Review Exercises

1. $\{(3, 5)\}$ **3.** $\{(-1, 3)\}$ **5.** $\{(-19/2, -19/2)\}$, independent
7. $\{(39/17, 18/17)\}$, independent
9. $\{(x, y) \mid y = -3x + 1\}$, dependent
11. $\varnothing$, inconsistent   **13.** $\{(1, 3, -4)\}$
15. $\{(x, 0, 1 - x) \mid x \text{ is any real number}\}$ **17.** $\varnothing$

19. $\left\{\left(\dfrac{-1 + \sqrt{17}}{2},\ \pm\sqrt{\dfrac{-1 + \sqrt{17}}{2}}\right)\right\}$

21. $\{(0, 0), (1, 1), (-1, 1)\}$ **23.** $\dfrac{2}{x - 3} + \dfrac{5}{x + 4}$

25. $\dfrac{2x - 1}{x^2 + 4} + \dfrac{5}{x - 3}$

27.

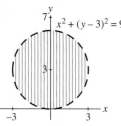

29.

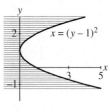

31.

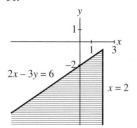

33.

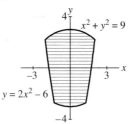

35.

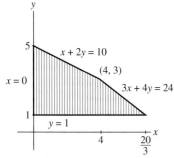

37.

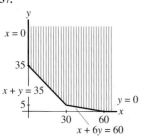

39. $y = -\dfrac{2}{3}x + \dfrac{5}{3}$ **41.** $y = 3x^2 - 4x + 5$

43. 57 tacos, 62 burritos   **45.** \$0.95   **47.** 16.8
49. 6 million barrels per day from each source

Chapter 8 Test

1. $\{(-3, 4)\}$ **2.** $\{(25/11, -6/11)\}$ **3.** $\{(4, 6)\}$
4. Inconsistent   **5.** Dependent   **6.** Independent
7. Inconsistent   **8.** $\{(x, 10 - x, 14 - 3x) \mid x \text{ is any real number}\}$
9. $\{(1, -2, 3)\}$ **10.** $\varnothing$ **11.** $\{(4, 0), (-4, 0)\}$
12. $\left\{\left(2 + \sqrt{2}, -4 - \sqrt{2}\right), \left(2 - \sqrt{2}, -4 + \sqrt{2}\right)\right\}$

13. $\dfrac{3}{x - 4} + \dfrac{-1}{x + 2}$ **14.** $\dfrac{2}{x^2} + \dfrac{1}{x} + \dfrac{3}{x - 1}$

15.

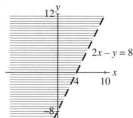

16.

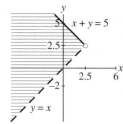

17.

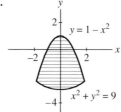

18. 24 males, 28 females   **19.** 5 television, 18 newspaper

Tying It All Together Chapters 1–8

1. $\left\{\dfrac{103}{13}\right\}$ **2.** $\left\{\dfrac{40}{13}, 3\right\}$ **3.** $\left\{-\dfrac{4}{5}\right\}$ **4.** $\{-1, 4\}$

5. $\{-11\}$ **6.** $\left\{\pm\dfrac{2\sqrt{3}}{3}\right\}$ **7.** $\{1\}$ **8.** $\{1 + \log_2(9)\}$

9. $\{1\}$ **10.** $\{\pm 8\}$ **11.** $\left\{\dfrac{3 \pm \sqrt{33}}{2}\right\}$ **12.** $\left\{\dfrac{6 \pm \sqrt{2}}{2}\right\}$

13. $(-\infty, 3/2)$ **14.** $(-\infty, 1.5) \cup (1.5, \infty)$

15. $(-\infty, -3] \cup [3, \infty)$ **16.** $[-7, 5]$

17.

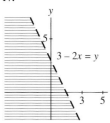

18.

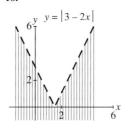

19.

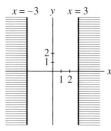

20.

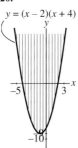

CHAPTER 9

Section 9.1

For Thought: **1.** F   **2.** F   **3.** T   **4.** T   **5.** T   **6.** T
7. F   **8.** T   **9.** F   **10.** T
Exercises: **1.** 1×3 **3.** 1×1 **5.** 3×2 **7.** $\begin{bmatrix} 1 & -2 & | & 4 \\ 3 & 2 & | & -5 \end{bmatrix}$

9. $\begin{bmatrix} 1 & -1 & -1 & | & 4 \\ 1 & 3 & -1 & | & 1 \\ 0 & 2 & -5 & | & -6 \end{bmatrix}$ **11.** $\begin{bmatrix} 1 & 3 & -1 & | & 5 \\ 1 & 0 & 1 & | & 0 \end{bmatrix}$

13. $3x + 4y = -2$ **15.** $5x = 6$
$\quad\;\; 3x - 5y = 0$ $\qquad\;\; -4x + 2z = -1$
$\qquad\qquad\qquad\qquad\qquad\; 4x + 4y = 7$

17. $x - y + 2z = 1$
$\qquad\;\; y + 4z = 3$

19. $\begin{bmatrix} 1 & 2 & | & 0 \\ -2 & 4 & | & 1 \end{bmatrix}$ **21.** $\begin{bmatrix} 1 & 4 & | & 1 \\ 0 & 3 & | & 6 \end{bmatrix}$ **23.** $\begin{bmatrix} 1 & -2 & | & 1 \\ 0 & -1 & | & 3 \end{bmatrix}$

25. $2x + 4y = 14, 5x + 4y = 5, \{(-3, 5)\}, \frac{1}{2}R_1 \rightarrow R_1,$
$\quad -5R_1 + R_2 \rightarrow R_2, -\frac{1}{6}R_2 \rightarrow R_2, -2R_2 + R_1 \rightarrow R_1$

27. $\{(2, 3)\}$, independent   **29.** $\{(1, 3)\}$, independent
31. $\{(3, 3)\}$, independent   **33.** $\{(0.5, 1)\}$, independent
35. $\varnothing$, inconsistent   **37.** $\{(u, v)\,|\,u + 3v = 4\}$, dependent
39. $\{(4, -4)\}$, independent   **41.** $\{(3, 2, 1)\}$, independent
43. $\{(1, 1, 1)\}$, independent   **45.** $\{(1, 2, 0)\}$, independent
47. $\{(1, 0, -1)\}$, independent
49. $\{(x, y, z)\,|\,x - 2y + 3z = 1\}$, dependent   **51.** $\varnothing$, inconsistent
53. $\{(x, 5 - 2x, 2 - x)\,|\,x$ is any real number$\}$, dependent
55. $\left\{\left(x, \dfrac{13 - 5x}{2}, \dfrac{5 - 3x}{2}\right)\,\middle|\,x$ is any real number$\right\}$, dependent
57. $\{(4, 3, 2, 1)\}$, independent
59. 38 hrs at Burgers-R-Us, 22 hrs at Soap Opera
61. \$20,000 mutual fund, \$11,333.33 treasury bills, \$8,666.67 bonds
63. $a = 1, b = -2, c = 3$

65. $y = 750 - x, z = x - 250$, and $250 \le x \le 750$. If $z = 50$,
then $x = 300$ and $y = 450$.

Section 9.2

For Thought: **1.** T   **2.** F   **3.** F   **4.** T   **5.** T   **6.** F
7. F   **8.** F   **9.** F   **10.** F
Exercises: **1.** $x = 2, y = 5$ **3.** $x = 3, y = 4, z = 2$

5. $\begin{bmatrix} 5 \\ 6 \end{bmatrix}$ **7.** $\begin{bmatrix} 1.5 & 0.97 \\ 1.95 & 0.67 \end{bmatrix}$ **9.** $\begin{bmatrix} 3 & -4 & 5 \\ 4 & -5 & 7 \\ 6 & -3 & 2 \end{bmatrix}$

11. $\begin{bmatrix} -1 & 4 \\ 5 & -6 \end{bmatrix}, \begin{bmatrix} 0 & 0 \\ 0 & 0 \end{bmatrix}$ **13.** $\begin{bmatrix} -3 & 0 & 1 \\ -8 & 2 & -1 \\ 3 & -6 & -3 \end{bmatrix}, \begin{bmatrix} 0 & 0 & 0 \\ 0 & 0 & 0 \\ 0 & 0 & 0 \end{bmatrix}$

15. $\begin{bmatrix} 3 & -3 \\ 4 & 4 \end{bmatrix}$ **17.** $\begin{bmatrix} 2 & 2 \\ 5 & 9 \end{bmatrix}$ **19.** Undefined

21. $\begin{bmatrix} -12 & 3 \\ 9 & 0 \end{bmatrix}$ **23.** $\begin{bmatrix} 4 \\ -5 \end{bmatrix}$ **25.** $\begin{bmatrix} -21 & -9 \\ 15 & -15 \end{bmatrix}$

27. $\begin{bmatrix} -7 & 4 \\ -1 & -4 \end{bmatrix}$ **29.** $\begin{bmatrix} -5 \\ 4 \end{bmatrix}$ **31.** Undefined   **33.** $\begin{bmatrix} -8 & -5 \\ 12 & -1 \end{bmatrix}$

35. $\begin{bmatrix} 0.4 & 0.15 \\ 0.7 & 1.1 \end{bmatrix}$ **37.** $\begin{bmatrix} 1/2 & 3/2 \\ 3 & -12 \end{bmatrix}$ **39.** $\begin{bmatrix} 10 & 0 \\ -2 & 16 \end{bmatrix}$

41. Undefined   **43.** $\begin{bmatrix} -1 & 13 \\ -9 & 3 \\ 6 & -2 \end{bmatrix}$ **45.** $\begin{bmatrix} 3\sqrt{2} & 2 & 3\sqrt{3} \end{bmatrix}$

47. $\begin{bmatrix} 13a \\ -b \end{bmatrix}$ **49.** $\begin{bmatrix} -x & -0.5y \\ -0.7x & 3.5y \end{bmatrix}$ **51.** $\begin{bmatrix} 3x & 2y & -z \\ -6x & 3y & 7z \\ 0 & -7y & -7z \end{bmatrix}$

53. $\{(3, 2)\}$ **55.** $\{(-1, 3)\}$ **57.** $\{(0.5, 3, 4.5)\}$ **59.** $\begin{bmatrix} \$390 \\ \$160 \\ \$135 \end{bmatrix}$

61. $\begin{bmatrix} 40 & 80 \\ 30 & 90 \\ 80 & 200 \end{bmatrix}, \begin{bmatrix} 60 & 120 \\ 45 & 135 \\ 120 & 300 \end{bmatrix}$ **63.** Yes, yes   **65.** Yes, yes

67. Yes, yes   **69.** $\begin{bmatrix} 0 & 0 \\ 0 & 0 \end{bmatrix}$

Section 9.3

For Thought: **1.** T   **2.** T   **3.** F   **4.** F   **5.** T   **6.** T
7. T   **8.** T   **9.** T   **10.** F
Exercises: **1.** 3×5 **3.** 1×1 **5.** 5×5 **7.** 3×3

9. AB is undefined   **11.** $[-10]$ **13.** $\begin{bmatrix} 10 \\ -10 \end{bmatrix}$ **15.** $\begin{bmatrix} 8 & 11 \\ 5 & 5 \end{bmatrix}$

17. $\begin{bmatrix} 15 & 18 \\ 5 & 6 \end{bmatrix}$ **19.** $\begin{bmatrix} -2 & 3 & 1 \\ -2 & 4 & 2 \\ -1 & 6 & 5 \end{bmatrix}, \begin{bmatrix} 6 & 9 \\ 1 & 1 \end{bmatrix}$

21. $\begin{bmatrix} 1 & 3 & 6 \\ 2 & 3 & 6 \\ 3 & 5 & 6 \end{bmatrix}, \begin{bmatrix} 6 & 5 & 7 \\ 5 & 3 & 4 \\ 3 & 2 & 1 \end{bmatrix}$ **23.** $\begin{bmatrix} 4 & 6 & 8 \\ -6 & -9 & -12 \\ 2 & 3 & 4 \end{bmatrix}$ **25.** $[2 \; 5 \; 9]$

27. Undefined   **29.** $\begin{bmatrix} 7 & 8 \\ 5 & 5 \\ 1 & 0 \end{bmatrix}$ **31.** $\begin{bmatrix} 1 & 1 \\ 14 & 15 \end{bmatrix}$

33. Undefined   **35.** $\begin{bmatrix} 0 \\ -2 \\ 1 \end{bmatrix}$ **37.** Undefined   **39.** $\begin{bmatrix} 6 & 8 & 10 \\ -6 & -7 & -10 \\ 2 & 3 & 6 \end{bmatrix}$

41. $\begin{bmatrix} 1 & 0 \\ 2 & 1 \end{bmatrix}$ **43.** $\begin{bmatrix} 1 & 0 \\ 4 & 1 \end{bmatrix}$ **45.** $\begin{bmatrix} 2 & 2 \\ 3 & 4 \end{bmatrix}$ **47.** $\begin{bmatrix} 1 & 0 \\ 0 & 1 \end{bmatrix}$

49. $\begin{bmatrix} -0.5 & 4 \\ 9 & 0.7 \end{bmatrix}$ **51.** $[4a \quad -3b]$ **53.** $\begin{bmatrix} -2a & 5a & 3a \\ b & 4b & 6b \end{bmatrix}$

55. $[4 \quad 2 \quad 6]$ **57.** $[17]$ **59.** $\begin{bmatrix} x^2 & xy \\ xy & y^2 \end{bmatrix}$ **61.** $\begin{bmatrix} 2\sqrt{2} \\ 7\sqrt{2} \end{bmatrix}$

63. $\begin{bmatrix} -17/3 & 11 \\ -3 & 6 \end{bmatrix}$ **65.** Undefined   **67.** $\begin{bmatrix} 2 & 0 & 6 \\ 3 & 5 & 2 \\ 19 & 21 & 16 \end{bmatrix}$

69. $\begin{bmatrix} -0.2 & 1.7 \\ 3 & 3.4 \\ -1.4 & -4.6 \end{bmatrix}$ **71.** $\{(3, 2)\}$ **73.** $\varnothing$ **75.** $\{(-1, -1, 6)\}$

77. $\begin{bmatrix} 2 & 3 \\ 4 & -1 \end{bmatrix} \begin{bmatrix} x \\ y \end{bmatrix} = \begin{bmatrix} 9 \\ 6 \end{bmatrix}$ **79.** $\begin{bmatrix} 1 & 2 & -1 \\ 3 & -1 & 3 \\ 2 & 1 & -4 \end{bmatrix} \begin{bmatrix} x \\ y \\ z \end{bmatrix} = \begin{bmatrix} 3 \\ 1 \\ 0 \end{bmatrix}$

81. $AQ = \begin{bmatrix} \$376,000 \\ \$642,000 \end{bmatrix}$. The entries of AQ are the total cost of labor and total cost of material for four economy and seven deluxe models.

83. F   **85.** T   **87.** T

Section 9.4

For Thought: **1.** T   **2.** T   **3.** T   **4.** F   **5.** T   **6.** F
7. F   **8.** T   **9.** F   **10.** F

Exercises:

1. $\begin{bmatrix} 1 & 3 \\ 4 & 6 \end{bmatrix} \begin{bmatrix} 1 & 0 \\ 0 & 1 \end{bmatrix} = \begin{bmatrix} 1 & 3 \\ 4 & 6 \end{bmatrix}, \begin{bmatrix} 1 & 0 \\ 0 & 1 \end{bmatrix} \begin{bmatrix} 1 & 3 \\ 4 & 6 \end{bmatrix} = \begin{bmatrix} 1 & 3 \\ 4 & 6 \end{bmatrix}$

3. $\begin{bmatrix} 3 & 2 & 1 \\ 5 & 6 & 2 \\ 7 & 8 & 3 \end{bmatrix} \begin{bmatrix} 1 & 0 & 0 \\ 0 & 1 & 0 \\ 0 & 0 & 1 \end{bmatrix} = \begin{bmatrix} 3 & 2 & 1 \\ 5 & 6 & 2 \\ 7 & 8 & 3 \end{bmatrix}$,

$\begin{bmatrix} 1 & 0 & 0 \\ 0 & 1 & 0 \\ 0 & 0 & 1 \end{bmatrix} \begin{bmatrix} 3 & 2 & 1 \\ 5 & 6 & 2 \\ 7 & 8 & 3 \end{bmatrix} = \begin{bmatrix} 3 & 2 & 1 \\ 5 & 6 & 2 \\ 7 & 8 & 3 \end{bmatrix}$

5. $\begin{bmatrix} -3 & 5 \\ 12 & 6 \end{bmatrix}$ **7.** $\begin{bmatrix} 1 & 0 \\ 0 & 1 \end{bmatrix}$ **9.** $\begin{bmatrix} 1 & 0 \\ 0 & 1 \end{bmatrix}$ **11.** $\begin{bmatrix} 3 & 5 & 1 \\ 4 & 5 & 7 \\ 4 & 9 & 2 \end{bmatrix}$

13. $\begin{bmatrix} 1 & 0 & 0 \\ 0 & 1 & 0 \\ 0 & 0 & 1 \end{bmatrix}$ **15.** $\begin{bmatrix} 1 & 0 & 0 \\ 0 & 1 & 0 \\ 0 & 0 & 1 \end{bmatrix}$ **17.** Yes

19. No   **21.** No   **23.** $\begin{bmatrix} 1 & -2 \\ 0 & 0.5 \end{bmatrix}$ **25.** $\begin{bmatrix} 3 & -2 \\ -1/3 & 1/3 \end{bmatrix}$

27. $\begin{bmatrix} 4 & 3 \\ -3 & -2 \end{bmatrix}$ **29.** $\begin{bmatrix} -1.5 & -2.5 \\ -0.5 & -0.5 \end{bmatrix}$ **31.** No inverse

33. No inverse   **35.** $\begin{bmatrix} 0.5 & 0.5 & 0 \\ 0.5 & 0 & -0.5 \\ 0 & -0.5 & 0.5 \end{bmatrix}$ **37.** $\begin{bmatrix} -3.5 & -1 & 2 \\ 0.5 & 0 & 0 \\ 4.5 & 2 & -3 \end{bmatrix}$

39. $\begin{bmatrix} 1/3 & -1/6 & 1/3 \\ -2/3 & 1/3 & 1/3 \\ 2/3 & 1/6 & -1/3 \end{bmatrix}$ **41.** $\begin{bmatrix} 2.5 & -3 & 1 \\ 1 & -1 & 0 \\ -1.5 & 2 & 0 \end{bmatrix}$

43. $\begin{bmatrix} 1 & -2 & 1 & 0 \\ 0 & 1 & -2 & 1 \\ 0 & 0 & 1 & -2 \\ 0 & 0 & 0 & 1 \end{bmatrix}$ **45.** $\{(3, -1)\}$ **47.** $\{(2, 1/3)\}$

49. $\{(1, -1)\}$ **51.** $\{(5, 2)\}$ **53.** $\{(1, -1, 3)\}$ **55.** $\{(1, 3, 2)\}$
57. $\{(11.3, -3.9)\}$ **59.** $\{(-2z - 4, -z - 9, z) \mid z \text{ is any real number}\}$
61. $\{(1/2, 1/6, 1/3)\}$ **63.** $\{(-165, 97.5, 240)\}$
65. $\{(-1.6842, -9.2632, -9.2632)\}$

67. $\begin{bmatrix} 1 & 7 \\ 3 & 20 \end{bmatrix}, \begin{bmatrix} 2 & 7 \\ 3 & 10 \end{bmatrix}, \begin{bmatrix} 4 & 7 \\ 3 & 5 \end{bmatrix}, \begin{bmatrix} 20 & 7 \\ 3 & 1 \end{bmatrix}, \begin{bmatrix} 10 & 7 \\ 3 & 2 \end{bmatrix}, \begin{bmatrix} 5 & 7 \\ 3 & 4 \end{bmatrix}$

69. \$0.75 eggs, \$1.80 magazine   **71.** \$400 plywood, \$150 insulation

73. Good luck   **75.** \$17,142.86 Asset Manager; \$42,857.14 Magellan
77. \$4.20 animal totems, \$2.75 necklaces, \$0.89 tribal masks

Section 9.5

For Thought: **1.** F   **2.** T   **3.** T   **4.** F   **5.** T   **6.** T
7. F   **8.** T   **9.** F   **10.** F
Exercises: **1.** 2   **3.** 19   **5.** -0.41 **7.** 23/32   **9.** 0
11. 0   **13.** 4   **15.** ± 4 **17.** $\{(1, 3)\}$ **19.** $\{(1, -3)\}$
21. $\{(-3, 5)\}$ **23.** $\{(11/2, -1/2)\}$ **25.** $\{(12, 6)\}$
27. $\{(500, 400)\}$ **29.** $\{(x, y) \mid 3x + y = 6\}$ **31.** $\varnothing$
33. $\{(-6, -9)\}$ **35.** $\{(\sqrt{2}/2, \sqrt{3})\}$ **37.** $\{(0, -5), (\pm 3, 4)\}$
39. $\{(8, 2), (0, -2)\}$ **41.** Invertible   **43.** Not invertible
45. $\{(146, 237)\}$ **47.** 175 boys, 440 girls
49. 89/3 degrees and 181/3 degrees
51. Vice-president \$150,000, president \$250,000   **53.** 2, 3, 6, yes
57. 8, no

Section 9.6

For Thought: **1.** F   **2.** F   **3.** T   **4.** T   **5.** F   **6.** F
7. F   **8.** T   **9.** F   **10.** F
Exercises: **1.** 14   **3.** 1   **5.** -23 **7.** -16 **9.** -33
11. -56 **13.** -115 **15.** 14   **17.** -54 **19.** 0   **21.** -3
23. -72 **25.** -137 **27.** -293 **29.** $\{(1, 2, 3)\}$
31. $\{(2, 3, 5)\}$ **33.** $\{(7/16, -5/16, -13/16)\}$
35. $\{(-2/7, 11/7, 5/7)\}$
37. $\{(x, x - 5/3, x - 4/3) \mid x \text{ is any real number}\}$ **39.** $\varnothing$
41. Jackie 27, Alisha 11, Rochelle 39   **43.** 30, 50, 100
45. $\{(16.8, 12.3, 11.2)\}$ **47.** \$1.799 regular, \$1.899 plus, \$1.999 supreme
49. $\{(1, -2, 3, 2)\}$

Chapter 9 Review Exercises

1. $\begin{bmatrix} 5 & 4 \\ -1 & 6 \end{bmatrix}$ **3.** $\begin{bmatrix} 1 & -13 \\ -5 & 6 \end{bmatrix}$ **5.** $\begin{bmatrix} 3 & 8 \\ -2 & -6 \end{bmatrix}$

7. Undefined   **9.** $\begin{bmatrix} -11 \\ 14 \end{bmatrix}$ **11.** $\begin{bmatrix} 3 & 2 & -1 \\ -12 & -8 & 4 \\ 9 & 6 & -3 \end{bmatrix}$

13. $[1 \quad -1 \quad -1]$ **15.** Undefined   **17.** $\begin{bmatrix} 2 & 1.5 \\ 1 & 1 \end{bmatrix}$

19. $\begin{bmatrix} -1 & 0 & 0 \\ 1 & 1 & 0 \\ -5 & -3 & 1 \end{bmatrix}$ **21.** $\begin{bmatrix} 3 & 4 \\ -1 & -1.5 \end{bmatrix}$ **23.** $\begin{bmatrix} 1 & 0 \\ 0 & 1 \end{bmatrix}$ **25.** 2

27. -1 **29.** $\{(10/3, 17/3)\}$ **31.** $\{(-2, 3)\}$

33. $\{(x, y) \mid x - 5y = 9\}$ **35.** $\varnothing$ **37.** $\{(1, 2, 3)\}$

39. $\{(-3, 4, 1)\}$ **41.** $\left\{ \left(\dfrac{3y + 3}{2}, y, \dfrac{1 - y}{2} \right) \mid y \text{ is any real number} \right\}$

43. $\varnothing$ **45.** $\{(9, -12)\}$ **47.** $\{(2, 4)\}$ **49.** $\{(0, -3)\}$

51. $\{(0.5, 0.5, 0.5)\}$ **53.** $\{(2, -3, 5)\}$

55. 225.56 gal A, 300.74 gal B

57. $22.88 water, $55.65 gas, $111.30 electric

Chapter 9 Test

1. $\{(1, 1/3)\}$ **2.** $\{(3, 2, 1)\}$

3. $\left\{ \left(x, \dfrac{-x - 1}{2}, \dfrac{3x - 1}{2} \right) \mid x \text{ is any real number} \right\}$

4. $\begin{bmatrix} 3 & -4 \\ -6 & 10 \end{bmatrix}$ **5.** $\begin{bmatrix} 0 & 1 \\ 0 & 2 \end{bmatrix}$ **6.** $\begin{bmatrix} 6 & -9 \\ -20 & 30 \end{bmatrix}$ **7.** $\begin{bmatrix} -3 \\ 8 \end{bmatrix}$

8. Undefined   **9.** $[-2 \; 1 \; 2]$ **10.** $\begin{bmatrix} 2 & 0 & -2 \\ 3 & 0 & -3 \\ -1 & 0 & 1 \end{bmatrix}$

11. $\begin{bmatrix} 2 & 0.5 \\ 1 & 0.5 \end{bmatrix}$ **12.** $\begin{bmatrix} 7 & -5 & -8 \\ 3 & -2 & -3 \\ 6 & -4 & -7 \end{bmatrix}$ **13.** 2   **14.** 0   **15.** -1

16. $\{(5, 3)\}$ **17.** $\varnothing$ **18.** $\{(6, 2, 4)\}$ **19.** $\{(-2, -3)\}$

20. $\{(15, 6, 13)\}$ **21.** Bought 46, sold 34

22. $y = 0.5x^2 - 4\sqrt{x} + 3$

Tying It All Together Chapters 1–9

1. $\{-1/3\}$ **2.** $\{29/15\}$ **3.** $\{1/3, 2\}$ **4.** $\{-3/2\}$

5. $\{(4, -2)\}$ **6.** $\{(2/5, 1/5)\}$ **7.** $\{(7, -100)\}$

8. $\{(-2, -3)\}$ **9.** $\{(1, 2)\}$ **10.** $\{(2, 1)\}$ **11.** $\varnothing$

12. $\{(x, y) \mid x - 2y = 3\}$ **13.** $\{(-4, -3), (3, 4)\}$

14. $\{(-2, 3), (1, 0)\}$ **15.** $\left\{ \left(\sqrt{2}, \pm 1 \right), \left(-\sqrt{2}, \pm 1 \right) \right\}$

16. $\{(\pm 1, 0)\}$

CHAPTER 10

Section 10.1

For Thought: **1.** F   **2.** T   **3.** T   **4.** F   **5.** T   **6.** F

7. F   **8.** T   **9.** F   **10.** F

Exercises: **1.** $(0, 0), (0, 1), y = -1$ **3.** $(1, 2), (1, 3/2), y = 5/2$

5. $(3, 1), (15/4, 1), x = 9/4$ **7.** $y = \dfrac{1}{8}x^2$ **9.** $y = -\dfrac{1}{12}x^2$

11. $y = \dfrac{1}{6}(x - 3)^2 + \dfrac{7}{2}$ **13.** $y = -\dfrac{1}{10}(x - 1)^2 - \dfrac{1}{2}$

15. $y = 1.25(x + 2)^2 + 1$ **17.** $y = \dfrac{1}{4}x^2$ **19.** $y = -x^2$

21. $(1, 0), (1, 1/4), y = -1/4$ **23.** $(3, 0), (3, 1), y = -1$

25. $(3, 4), (3, 31/8), y = 33/8$

27. $y = (x - 4)^2 - 13, (4, -13), (4, -51/4), y = -53/4$

29. $y = 2(x + 3)^2 - 13, (-3, -13), (-3, -103/8), y = -105/8$

31. $y = -2(x - 3/2)^2 + 11/2, (3/2, 11/2), (3/2, 43/8), y = 45/8$

33. $y = 5(x + 3)^2 - 45, (-3, -45), (-3, -44.95), y = -45.05$

35. $y = \dfrac{1}{8}(x - 2)^2 + 4, (2, 4), (2, 6), y = 2$

37. $(2, -1), (2, -3/4), y = -5/4,$ up

39. $(1, -4), (1, -17/4), y = -15/4,$ down

41. $(1, -2), (1, -23/12), y = -25/12,$ up

43. $(-3, 13/2), (-3, 6), y = 7,$ down

45. $(0, 5), (0, 6), y = 4,$ up

47. $(1/2, -9/4), x = 1/2, (-1, 0), (2, 0), (0, -2)$

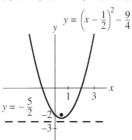

49. $(-1/2, 25/4), x = -1/2, (-3, 0), (2, 0), (0, 6)$

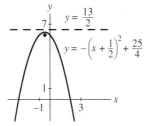

51. $(-2, 2), x = -2, (0, 4), (-2, 5/2), y = 3/2$

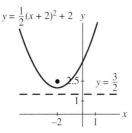

53. $(-4, 2), x = -4, \left(-4 \pm 2\sqrt{2}, 0 \right), (0, -2), (-4, 1), y = 3$

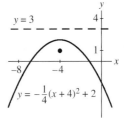

55. $(0, -2), x = 0, (\pm 2, 0), (0, -2), (0, -3/2), y = -5/2$

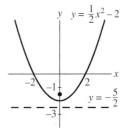

57. $(2, 0), x = 2, (2, 0), (0, 4), (2, 1/4), y = -1/4$

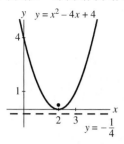

59. $(3/2, -3/4), x = 3/2, (0, 0), (3, 0), (0, 0), (3/2, 0), y = -3/2$

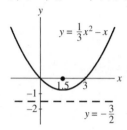

61. $(0, 0), y = 0, (0, 0), (0, 0), (-1/4, 0), x = 1/4$

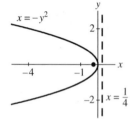

63. $(1, 0), y = 0, (1, 0), (0, \pm 2), (0, 0), x = 2$

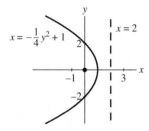

65. $(-25/4, -1/2), y = -1/2, (-6, 0), (0, -3), (0, 2), (-6, -1/2),$
  $x = -13/2$

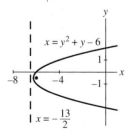

67. $(-7/2, -1), y = -1, (-4, 0), (-4, -1), x = -3$

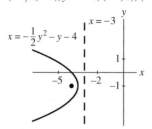

69. $(3, 1), y = 1, (5, 0), (25/8, 1), x = 23/8$

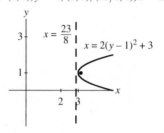

71. $(1, -2), y = -2, (-1, 0), (0, -2 \pm \sqrt{2}), (1/2, -2), x = 3/2$

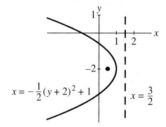

73. $y = \dfrac{1}{4}(x - 1)^2 + 4$ **75.** $x = \dfrac{1}{8}y^2$ **77.** $y = \dfrac{1}{2640}x^2$, 26.8 in.

79. Look alike
81. $y = \pm\sqrt{-x}$

Section 10.2

For Thought: **1.** F   **2.** T   **3.** T   **4.** T   **5.** T   **6.** F
7. T   **8.** F   **9.** F   **10.** F
Exercises: **1.** Foci $(\pm\sqrt{5}, 0)$, vertices $(\pm 3, 0)$, center $(0, 0)$
  3. Foci $(2, 1 \pm \sqrt{5})$, vertices $(2, 4)$ and $(2, -2)$, center $(2, 1)$
  5.   **7.**

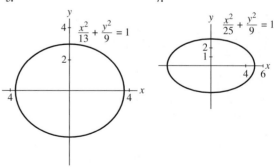

9.

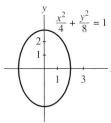
$$\frac{x^2}{4} + \frac{y^2}{8} = 1$$

11.

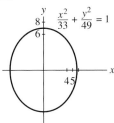
$$\frac{x^2}{33} + \frac{y^2}{49} = 1$$

29. $(-4 \pm \sqrt{35}, -3)$

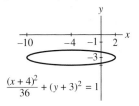
$$\frac{(x+4)^2}{36} + (y+3)^2 = 1$$

31. $(1, -2 \pm \sqrt{5})$

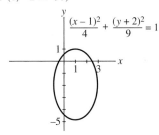
$$\frac{(x-1)^2}{4} + \frac{(y+2)^2}{9} = 1$$

13. $(\pm 2\sqrt{3}, 0)$

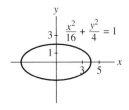
$$\frac{x^2}{16} + \frac{y^2}{4} = 1$$

15. $(0, \pm 3\sqrt{3})$

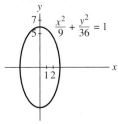
$$\frac{x^2}{9} + \frac{y^2}{36} = 1$$

33. $(3, -2 \pm \sqrt{5})$

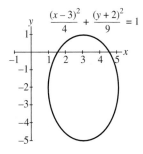
$$\frac{(x-3)^2}{4} + \frac{(y+2)^2}{9} = 1$$

17. $(\pm 2\sqrt{6}, 0)$

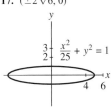
$$\frac{x^2}{25} + y^2 = 1$$

19. $(0, \pm 4)$

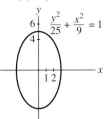
$$\frac{y^2}{25} + \frac{x^2}{9} = 1$$

35. $\frac{x^2}{16} + \frac{y^2}{4} = 1, (\pm 2\sqrt{3}, 0)$

37. $\frac{(x+1)^2}{4} + \frac{(y+2)^2}{16} = 1, (-1, -2 \pm 2\sqrt{3})$

39. $x^2 + y^2 = 4$ **41.** $x^2 + y^2 = 41$

43. $(x-2)^2 + (y+3)^2 = 20$ **45.** $(x-1)^2 + (y-3)^2 = 5$

47. $(0, 0), 10$ **49.** $(1, 2), 2$

21. $(0, \pm 2\sqrt{2})$

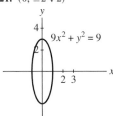
$9x^2 + y^2 = 9$

23. $(\pm \sqrt{5}, 0)$

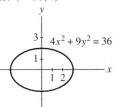
$4x^2 + 9y^2 = 36$

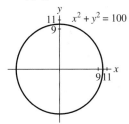
$x^2 + y^2 = 100$

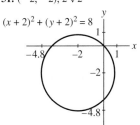
$(x-1)^2 + (y-2)^2 = 4$

25. $(1 \pm \sqrt{7}, -3)$

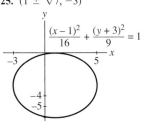
$$\frac{(x-1)^2}{16} + \frac{(y+3)^2}{9} = 1$$

27. $(3, 2), (3, -6)$

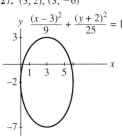
$$\frac{(x-3)^2}{9} + \frac{(y+2)^2}{25} = 1$$

51. $(-2, -2), 2\sqrt{2}$

$(x+2)^2 + (y+2)^2 = 8$

53. $(0, -1), 3$ **55.** $(-4, 5), \sqrt{41}$ **57.** $(-2, 0), 3$
59. $(0.5, -0.5), 1$ **61.** $(-1/3, -1/6), 1/2$ **63.** $(-1, 0), \sqrt{6}/2$
65. Circle   **67.** Ellipse   **69.** Parabola   **71.** Parabola

73. Ellipse   **75.** Circle   **77.** $\dfrac{x^2}{16} + \dfrac{y^2}{12} = 1$ **79.** $(12, 0)$

81. $\dfrac{x^2}{260.5^2} + \dfrac{y^2}{520} = 1, 0.996$ **83.** 5.25×10^9 km

85. $y = \pm\sqrt{6360^2 - x^2}$

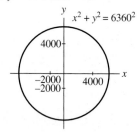

87. a. $y = \dfrac{4}{5}x + 5$ **b.** $(-4, 9/5)$ **89.** Parabolic

Section 10.3

For Thought: **1.** F   **2.** F   **3.** T   **4.** T   **5.** F   **6.** T
7. T   **8.** T   **9.** F   **10.** F
Exercises: **1.** Vertices $(\pm 1, 0)$, foci $(\pm\sqrt{2}, 0)$, asymptotes $y = \pm x$
3. Vertices $(1, \pm 3)$, foci $(1, \pm\sqrt{10})$, asymptotes $y = 3x - 3$ and $y = -3x + 3$

5. $(\pm\sqrt{13}, 0), y = \pm\dfrac{3}{2}x$ **7.** $(0, \pm\sqrt{29}), y = \pm\dfrac{2}{5}x$

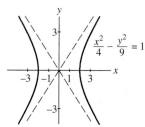   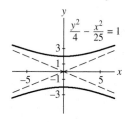

9. $(\pm\sqrt{5}, 0), y = \pm\dfrac{1}{2}x$ **11.** $(\pm\sqrt{10}, 0), y = \pm 3x$

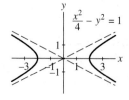   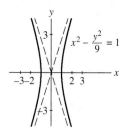

13. $(\pm 5, 0), y = \pm\dfrac{4}{3}x$ **15.** $(\pm\sqrt{2}, 0), y = \pm x$

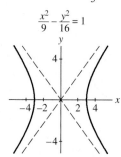   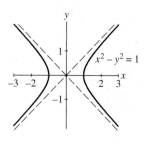

17. $(-1 \pm \sqrt{13}, 2), y = \dfrac{3}{2}x + \dfrac{7}{2}, y = -\dfrac{3}{2}x + \dfrac{1}{2}$

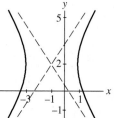

19. $(-2, 1 \pm \sqrt{5}), y = 2x + 5, y = -2x - 3$

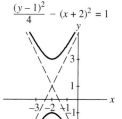

21. $(3, 3), (-7, 3), y = \dfrac{3}{4}x + \dfrac{9}{2}, y = -\dfrac{3}{4}x + \dfrac{3}{2}$

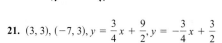
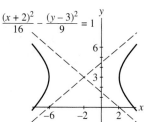

23. $(3, 3 \pm \sqrt{2}), y = x, y = -x + 6$

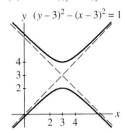
y $(y-3)^2 - (x-3)^2 = 1$

25. $\dfrac{x^2}{36} - \dfrac{y^2}{9} = 1$ **27.** $\dfrac{x^2}{9} - \dfrac{y^2}{16} = 1$ **29.** $\dfrac{x^2}{9} - \dfrac{y^2}{25} = 1$

31. $\dfrac{x^2}{9} - \dfrac{y^2}{16} = 1$ **33.** $\dfrac{y^2}{9} - \dfrac{(x-1)^2}{9} = 1$

35. $y^2 - (x-1)^2 = 1$, hyperbola   **37.** $y = x^2 + 2x$, parabola

39. $x^2 + y^2 = 100$, circle   **41.** $x = -4y^2 + 100$, parabola

43. $(x-1)^2 + (y-2)^2 = \dfrac{1}{2}$, circle

45. $\dfrac{(x+1)^2}{2} + \dfrac{(y+3)^2}{4} = 1$, ellipse

47. $\dfrac{(x-3)^2}{4} - \dfrac{(y+1)^2}{25} = 1$, hyperbola   **49.** $\dfrac{y^2}{64} - \dfrac{x^2}{36} = 1$

51. $\left(\dfrac{4\sqrt{14}}{7}, \dfrac{3\sqrt{7}}{7}\right)$ **53.** $\left(\pm\dfrac{3\sqrt{2}}{2}, \dfrac{\sqrt{14}}{2}\right), \left(\pm\dfrac{3\sqrt{2}}{2}, -\dfrac{\sqrt{14}}{2}\right)$

55. 0.01

Chapter 10 Review Exercises

1. $(-6, 0), (2, 0), (0, -12), (-2, -16), x = -2, (-2, -63/4),$
$y = -65/4$

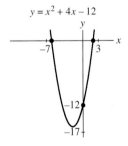
$y = x^2 + 4x - 12$

3. $(0, 0), (3, 0), (3/2, 9/2), x = 3/2,$
$(3/2, 35/8), y = 37/8$

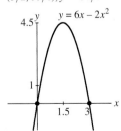
$y = 6x - 2x^2$

5. $(-6, 0), (0, -2 \pm \sqrt{10}), (-10, -2), y = -2,$
$(-39/4, -2), x = -41/4$

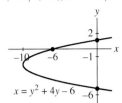
$x = y^2 + 4y - 6$

7. $(0, \pm 2\sqrt{5})$

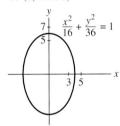
$\dfrac{x^2}{16} + \dfrac{y^2}{36} = 1$

9. $(1, 5), (1, -3)$

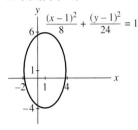
$\dfrac{(x-1)^2}{8} + \dfrac{(y-1)^2}{24} = 1$

11. $(1, -3 \pm \sqrt{2})$

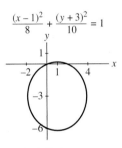
$\dfrac{(x-1)^2}{8} + \dfrac{(y+3)^2}{10} = 1$

13. $(0, 0), 9$

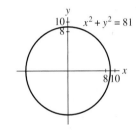
$x^2 + y^2 = 81$

15. $(-1, 0), 2$

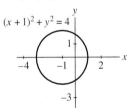
$(x+1)^2 + y^2 = 4$

17. $(-5/2, 0), \sqrt{6}$

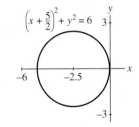
$\left(x + \dfrac{5}{2}\right)^2 + y^2 = 6$

19. $x^2 + (y + 4)^2 = 9$ **21.** $(x + 2)^2 + (y + 7)^2 = 6$

23. $(\pm 10, 0), y = \pm\dfrac{3}{4}x$

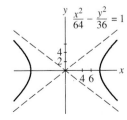
$\dfrac{x^2}{64} - \dfrac{y^2}{36} = 1$

25. $(4, 2 \pm 4\sqrt{5}), y = 2x - 6, y = -2x + 10$

$$\frac{(y-2)^2}{64} - \frac{(x-4)^2}{16} = 1$$

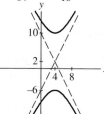

27. $(2 \pm \sqrt{5}, 4), y = \frac{1}{2}x + 3, y = -\frac{1}{2}x + 5$

$$\frac{(x-2)^2}{4} - (y-4)^2 = 1$$

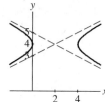

29. Hyperbola   **31.** Ellipse   **33.** Parabola   **35.** Hyperbola
37.   **39.**

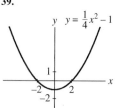

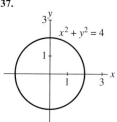

41.   **43.**

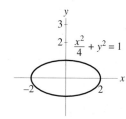

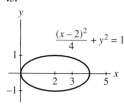

45. $x = (y - 3)^2 + \frac{3}{4}$ **47.** $\frac{x^2}{36} + \frac{y^2}{20} = 1$

49. $(x - 1)^2 + (y - 3)^2 = 20$ **51.** $\frac{x^2}{4} - \frac{y^2}{5} = 1$

53. $(x + 2)^2 + (y - 3)^2 = 9$ **55.** $\frac{(x+2)^2}{9} + (y - 1)^2 = 1$

57. $\frac{(y-1)^2}{9} - \frac{(x-2)^2}{4} = 1$ **59.** $\frac{x^2}{100^2} - \frac{y^2}{120^2} = 1$

61. $\frac{x^2}{34^2} + \frac{y^2}{16^2} = 1$, 10.81 ft

Chapter 10 Test

1.   **2.**

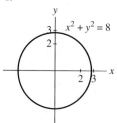

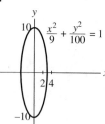

3.   **4.**

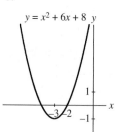

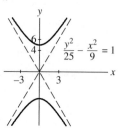

5.   **6.**

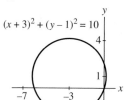

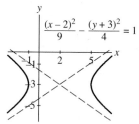

7. Hyperbola   **8.** Parabola   **9.** Circle   **10.** Ellipse

11. $(x + 3)^2 + (y - 4)^2 = 12$ **12.** $x = \frac{1}{8}y^2$ **13.** $\frac{x^2}{4} + \frac{y^2}{10} = 1$

14. $\frac{x^2}{36} - \frac{y^2}{28} = 1$ **15.** $(2, -15/4), y = -17/4, (2, -4), x = 2$

16. $(\pm 2\sqrt{3}, 0), 8, 4$ **17.** $(0, \pm\sqrt{17}), (0, \pm1), y = \pm\frac{1}{4}x, 2, 8$

18. $\left(-\frac{1}{2}, \frac{3}{2}\right), \frac{3}{2}$

19. 24 cm

Tying It All Together Chapters 1–10

1.   **2.**

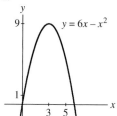

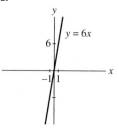

3.

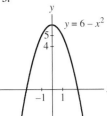

$y = 6 - x^2$

4.

$x^2 + y^2 = 6$

5.

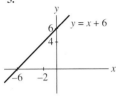

$y = x + 6$

6.

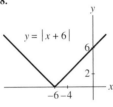

$x = 6 - y^2$

7.

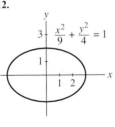

$y = (6 - x)^2$

8.

$y = |x + 6|$

9.

$y = 6^x$

10.

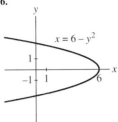

$y = \log_6(x)$

11.

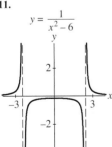

$y = \dfrac{1}{x^2 - 6}$

12.

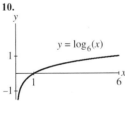

$\dfrac{x^2}{9} + \dfrac{y^2}{4} = 1$

13.

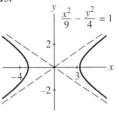

$\dfrac{x^2}{9} - \dfrac{y^2}{4} = 1$

14.

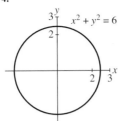

$y = 6x - x^3$

15.

$2x + 3y = 6$

16.

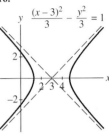

$\dfrac{(x - 3)^2}{3} - \dfrac{y^2}{3} = 1$

17. $\{-5/3\}$ **18.** $\{-5/3\}$ **19.** $\{11/15\}$ **20.** $\{0\}$
21. $\{22/3\}$ **22.** $\{50\}$ **23.** $\{-17, 3/2\}$ **24.** $\{-31/2, 0\}$
25. $\left\{17 \pm \sqrt{3}\right\}$ **26.** $\{17 \pm i\}$

CHAPTER 11

Section 11.1

For Thought: **1.** T **2.** F **3.** T **4.** F **5.** F **6.** F
7. F **8.** F **9.** F **10.** T

Exercises: 1. 1, 4, 9, 16, 25, 36, 49 **3.** $\dfrac{1}{2}, -\dfrac{1}{3}, \dfrac{1}{4}, -\dfrac{1}{5}, \dfrac{1}{6}, -\dfrac{1}{7}, \dfrac{1}{8}, -\dfrac{1}{9}$

5. 1, -2, 4, -8, 16, -32 **7.** 2, 1, $\dfrac{1}{2}, \dfrac{1}{4}, \dfrac{1}{8}$

9. -6, -10, -14, -18, -22 **11.** 5, 5.5, 6, 6.5, 7, 7.5, 8

13. 8.9, 8.8, 8.7, 8.6; 8 **15.** 8, 5, 2, -1; -19 **17.** $\dfrac{4}{3}, \dfrac{4}{5}, \dfrac{4}{7}, \dfrac{4}{9}; \dfrac{4}{21}$

19. $-\dfrac{1}{6}, \dfrac{1}{12}, -\dfrac{1}{20}, \dfrac{1}{30}; \dfrac{1}{132}$ **21.** 2, 24, 720, 40,320, 3,628,800

23. 2, 2, $\dfrac{4}{3}, \dfrac{2}{3}, \dfrac{4}{15}$ **25.** 1, 2, 3, 4, 5 **27.** $-\dfrac{1}{2}, \dfrac{1}{6}, -\dfrac{1}{24}, \dfrac{1}{120}, -\dfrac{1}{720}$

29. $\dfrac{1}{6e}, \dfrac{1}{24e^2}, \dfrac{1}{120e^3}, \dfrac{1}{720e^4}, \dfrac{1}{5040e^5}$ **31.** $a_n = 2n$

33. $a_n = 2n + 7$ **35.** $a_n = (-1)^{n+1}$ **37.** $a_n = n^3$

39. $a_n = e^n$ **41.** $a_n = \dfrac{1}{2^{n-1}}$ **43.** -4, -10, -28, -82; -6562

45. 2, 1, -2, 1; 1 **47.** -15, -8, -1, 6; 34 **49.** 6, 3, 0, -3; -21

51. 1, 0.9, 0.8, 0.7; 0.1 **53.** $\dfrac{14}{3}, \dfrac{13}{3}, \dfrac{12}{3}, \dfrac{11}{3}; \dfrac{5}{3}$ **55.** Yes

57. No **59.** No **61.** Yes **63.** $a_n = 5n - 4$

65. $a_n = 2n - 2$ **67.** $a_n = -4n + 9$ **69.** $a_n = 0.1n + 0.9$

71. $a_n = \dfrac{\pi}{6}n$ **73.** $a_n = 15n + 5$ **75.** 32 **77.** 27 **79.** 4.2

81. $a_n = 2.5n + 2.5$ **83.** $a_n = a_{n-1} + 9, a_1 = 3$
85. $a_n = 3a_{n-1}, a_1 = 1/3$ **87.** $a_n = \sqrt{a_{n-1}}, a_1 = 16$

89. $37,381, $39,624, $42,001, $44,521, $47,193;
  $a_n = 35,265(1.06)^n$ for $n = 1, 2, 3, 4, 5$
91. $a_n = 3n + 2, 92$ **93.** $67,200
95. $C_n = 4, E_n = 8n - 8, I_n = 4n^2 - 8n + 4$ for $n = 1, 2, 3, \ldots$

Section 11.2

For Thought: **1.** T   **2.** T   **3.** T   **4.** T   **5.** F   **6.** T
7. T   **8.** T   **9.** F   **10.** F

Exercises: **1.** 55   **3.** 15   **5.** 34   **7.** 30   **9.** 91   **11.** 0

13. 90   **15.** $\sum_{i=1}^{6} i$ **17.** $\sum_{i=1}^{5}(-1)^i(2i-1)$ **19.** $\sum_{i=1}^{5} i^2$

21. $\sum_{i=1}^{\infty}\left(-\frac{1}{2}\right)^{i-1}$ **23.** $\sum \ln(x_i)$ **25.** $\sum_{i=1}^{11} ar^{i-1}$ **27.** True

29. False   **31.** True   **33.** $\sum_{j=0}^{31}(-1)^{j+1}$ **35.** $\sum_{j=1}^{10}(2j+7)$

37. $\sum_{j=0}^{8}\frac{10!}{(j+2)!(8-j)!}$ **39.** $\sum_{j=5}^{\infty}\frac{5^{j-3}e^{-5}}{(j-3)!}$

41. $0.5 + 0.5r + 0.5r^2 + 0.5r^3 + 0.5r^4 + 0.5r^5$
43. $a^4 + a^3b + a^2b^2 + ab^3 + b^4$ **45.** $a^2 + 2ab + b^2$ **47.** 74/3
49. 16   **51.** 14.151   **53.** 360   **55.** 1.3   **57.** 1128
59. 950   **61.** $-28,386$ **63.** -36 **65.** 406   **67.** 52.25
69. 9   **71.** 72   **73.** $1,335,000, $44,500   **75.** $\sum_{i=1}^{9} 12i = 540$

77. $\sum_{i=1}^{10} 1000(1.05)^i$ **79.** 455 mg   **81.** 228.5   **83.** $-1/2$

Section 11.3

For Thought: **1.** F   **2.** T   **3.** T   **4.** F   **5.** T   **6.** F
7. F   **8.** T   **9.** T   **10.** T

Exercises: **1.** 3, 6, 12, 24; $r = 2$ **3.** 400, 200, 100, 50; $r = 1/2$
5. $1, -2/3, 4/9, -8/27; r = -2/3$ **7.** 1/2   **9.** 10   **11.** -2

13. -1 **15.** $a_n = \frac{1}{6}2^{n-1}$ **17.** $a_n = 0.9(0.1)^{n-1}$

19. $a_n = 4(-3)^{n-1}$ **21.** Arithmetic   **23.** Neither

25. Geometric   **27.** Arithmetic   **29.** Geometric   **31.** Neither
33. 2, 4, 6, 8; arithmetic   **35.** 1, 4, 9, 16; neither

37. $\frac{1}{2}, \frac{1}{4}, \frac{1}{8}, \frac{1}{16}$; geometric   **39.** 8, 32, 128, 512; geometric

41. 3, $-9, 27, -81$; geometric   **43.** 11   **45.** 3   **47.** ± 3
49. $a_n = -3(-2)^{n-1}$ **51.** 31   **53.** 126   **55.** 364/27
57. 242/27   **59.** -127.5 **61.** 31.8343   **63.** 1657.1341

65. $\sum_{n=1}^{5} 3\left(-\frac{1}{3}\right)^{n-1}$ **67.** $\sum_{n=1}^{\infty} 0.6(0.1)^{n-1}$ **69.** $\sum_{n=1}^{\infty} -4.5\left(-\frac{1}{3}\right)^{n-1}$

71. 9/4   **73.** 1   **75.** $-297/40$ **77.** 34/99   **79.** No sum
81. 2/3   **83.** 20   **85.** 2/45   **87.** 8172/990 or 454/55
89. 322.55 mg   **91.** 1,073,741,823   **93.** $4000(1.02)^n$, $8322.74
95. $2561.87   **97.** $183,074.35   **99.** 45 ft   **101.** $8,000,000

Section 11.4

For Thought: **1.** F   **2.** F   **3.** F   **4.** T   **5.** T   **6.** F
7. T   **8.** T   **9.** F   **10.** F
Exercises: **1.** 4   **3.** 8   **5.** 8   **7.** 6   **9.** 1024   **11.** 8
13. 210   **15.** 120   **17.** 78   **19.** 12   **21.** 60   **23.** 210
25. 15,120   **27.** 120   **29.** 990   **31.** 1   **33.** 10,920
35. 479,001,600   **37.** 2730   **39.** 6.3×10^{10} **41.** 4096
43. 72   **45.** 15,600   **47.** 30,000   **49.** 24   **51.** 131,072
53. 210

Section 11.5

For Thought: **1.** T   **2.** F   **3.** T   **4.** F   **5.** T   **6.** T
7. F   **8.** T   **9.** F   **10.** T
Exercises: **1.** 10   **3.** 7   **5.** 5   **7.** 70   **9.** 35   **11.** 1
13. 1   **15.** 6: $\{A, B\}, \{A, C\}, \{A, D\}, \{B, C\}, \{B, D\}, \{C, D\}$
17. 10   **19.** 10   **21.** 13,983,816   **23.** 2,598,960
25. 2520   **27.** 210   **29.** 3, 6, 10, C$(n, 2)$ **31.** 2520   **33.** 27
35. 56   **37.** 1260   **39.** 479,001,600   **41.** 36   **43.** 144
45. $x^2 + 2xy + y^2$ **47.** $4a^2 - 12a + 9$ **49.** $a^3 - 6a^2 + 12a - 8$
51. $8a^3 + 12a^2b^2 + 6ab^4 + b^6$
53. $x^4 - 8x^3y + 24x^2y^2 - 32xy^3 + 16y^4$
55. $x^8 + 4x^6 + 6x^4 + 4x^2 + 1$
57. $a^5 - 15a^4 + 90a^3 - 270a^2 + 405a - 243$
59. $x^6 + 12ax^5 + 60a^2x^4 + 160a^3x^3 + 240a^4x^2 + 192a^5x + 64a^6$
61. $x^9 + 9x^8y + 36x^7y^2$ **63.** $4096x^{12} - 24,576x^{11}y + 67,584x^{10}y^2$
65. $256s^8 - 512s^7t + 448s^6t^2$ **67.** $m^{18} - 18m^{16}w^3 + 144m^{14}w^6$
69. 56   **71.** $-41,184$ **73.** 13,860   **75.** 120

Section 11.6

For Thought: **1.** F   **2.** F   **3.** T   **4.** T   **5.** F   **6.** F
7. T   **8.** F   **9.** F   **10.** F
Exercises: **1.** $\{(H, H), (H, T), (T, H), (T, T)\}$
3. $\{(H, 1), (H, 2), (H, 3), (H, 4), (H, 5), (H, 6),$
  $(T, 1), (T, 2), (T, 3), (T, 4), (T, 5), (T, 6)\}$
5. 1/2   **7.** 1/36   **9.** 1/12   **11.** 1/8   **13.** 2/3, 1, 5/6, 0, 1/6
15. 1/4, 3/4, 1/4, 3/4   **17.** 1/36, 11/36, 5/36, 35/36, 1/36
19. 1/3, 2/3, 0   **21.** 3/13, 9/13, 9/13, 4/13   **23.** 1/72, 1/36, 1/18
25. 1/2,598,960, 1024/2,598,960   **27.** 0.9   **29.** 0.1   **31.** 0.5
33. Yes   **35.** No   **37.** Yes   **39.** 78%   **41.** 1/4   **43.** 4/13
45. 34%, 40%, 74%   **47.** 0.999   **49.** 1/36, 35/36, 35/36
51. 4 to 1   **53.** 4 to 1   **55.** 1 to 3, 3 to 1   **57.** 80%
59. 1 to 9, 9/10   **61.** 3 to 5   **63.** 1 to 5   **65.** 1 to 1,999,999
67. 1/32

Section 11.7

For Thought: **1.** T   **2.** F   **3.** F   **4.** F   **5.** T   **6.** F
7. F   **8.** F   **9.** T   **10.** F
Exercises: **1.** 1, 2, 3   **3.** 1, 2, 3   **5.** 1, 2   **7.** 2, 3

9. $S_1: \sum_{i=1}^{1} i = 1, S_2: \sum_{i=1}^{2} i = 3$,

  $S_3: \sum_{i=1}^{3} i = 6, S_4: \sum_{i=1}^{4} i = 10$

11. $S_1: \sum_{i=1}^{1} i^3 = 1, S_2: \sum_{i=1}^{2} i^3 = 9$,

  $S_3: \sum_{i=1}^{3} i^3 = 36, S_4: \sum_{i=1}^{4} i^3 = 100$

13. $S_1: 7^1 - 1$ is divisible by 6,
  $S_2: 7^2 - 1$ is divisible by 6,
  $S_3: 7^3 - 1$ is divisible by 6,
  $S_4: 7^4 - 1$ is divisible by 6

15. $S_1: 2(1) = 1(1 + 1), S_k: \sum_{i=1}^{k} 2i = k(k + 1)$,

  $S_{k+1}: \sum_{i=1}^{k+1} 2i = (k + 1)(k + 2)$

17. $S_1: 2 = 2(1)^2, S_k: 2 + 6 + \cdots + (4k - 2) = 2k^2$,
  $S_{k+1}: 2 + 6 + \cdots + (4k + 2) = 2(k + 1)^2$

19. $S_1: 2 = 2^2 - 2, S_k: \sum\limits_{i=1}^{k} 2^i = 2^{k+1} - 2, S_{k+1}: \sum\limits_{i=1}^{k+1} 2^i = 2^{k+2} - 2$

21. $S_1: (ab)^1 = a^1 b^1, S_k: (ab)^k = a^k b^k, S_{k+1}: (ab)^{k+1} = a^{k+1} b^{k+1}$

23. $S_1:$ If $0 < a < 1$, then $0 < a^1 < 1$,
$S_k:$ If $0 < a < 1$, then $0 < a^k < 1$,
$S_{k+1}:$ If $0 < a < 1$, then $0 < a^{k+1} < 1$

Chapter 11 Review Exercises

1. $1, 2, 4, 8, 16$ **3.** $-1, \dfrac{1}{2}, -\dfrac{1}{6}, \dfrac{1}{24}$ **5.** $3, 1.5, 0.75$

7. $3, 0, -3$ **9.** 0.9375 **11.** 5450 **13.** $1/3$

15. $33,065.9541$ **17.** $a_n = \dfrac{(-1)^n}{n + 2}$

19. $a_n = 6\left(\dfrac{1}{6}\right)^{n-1}$ **21.** $\sum\limits_{i=1}^{\infty} \dfrac{(-1)^{i+1}}{i + 1}$

23. $\sum\limits_{i=1}^{14} 2i$ **25.** ± 2 **27.** $\$243.52$ **29.** $\$13,971.64$

31. 1 million   **33.** $a^4 + 8a^3 b + 24a^2 b^2 + 32ab^3 + 16b^4$

35. $32a^5 - 80a^4 b + 80a^3 b^2 - 40a^2 b^3 + 10ab^4 - b^5$

37. $a^{10} + 10a^9 b + 45a^8 b^2$ **39.** $256x^8 + 512x^7 y + 448x^6 y^2$

41. 715   **43.** 36,960   **45.** 24   **47.** 1,953,125   **49.** 210
51. 60   **53.** 56   **55.** 70, 5, 30   **57.** 180, 120   **59.** 128
61. 1/1024, 1/1024   **63.** 5/13, 8/13, 0, 1
65. 1 to 7   **67.** 9 to 1   **69.** 90%   **71.** 40,320   **73.** 20
75. 1680   **77.** 28   **79.** 1680   **81.** 8

Chapter 11 Test

1. 2.3, 2.8, 3.3, 3.8   **2.** 20, 10, 5, 2.5   **3.** $a_n = (-1)^{n-1}(n - 1)^2$

4. $a_n = 3n + 4$ **5.** $a_n = \dfrac{1}{3}\left(-\dfrac{1}{2}\right)^{n-1}$ **6.** 4023

7. 13,050.5997   **8.** 49   **9.** $a_n = 1.5n - 4.5$ **10.** 3780
11. $\$445$ **12.** 10,000, 3/10,000   **13.** $\$487,521.25, \$515,024.49,$ no
14. $a^5 - 10a^4 x + 40a^3 x^2 - 80a^2 x^3 + 80ax^4 - 32x^5$
15. $x^{24} + 24x^{23} y^2 + 276x^{22} y^4$

16. $\sum\limits_{i=0}^{30} \binom{30}{i} m^{30-i} y^i$

17. 1/6, 1 to 5
18. 220
19. 2970
20. 1/336

Credits

CHAPTER P

Page 1, © Jean-Paul Pelisser/Reuters/Corbis; Page 13, © Sam Ogden/ Science Photo Library; Page 23, (Exercise 80) © Mike Segar/Reuters/ Corbis, (Exercise 82) © The Purcell Team/Corbis; Page 25, © Paul Barton/Corbis; Page 37, © Corbis; Page 55, © Dick Luria/Science Source/Photo Researchers; Page 56, © Stone/Getty Images; Page 77, © PhotoDisc; Page 78, © Corbis; Page 84, © AFP/Corbis; Page 85, © PhotoDisc.

CHAPTER 1

Page 87, © AP/WideWorld Photos; Page 97, (Exercises 97 and 98) © Getty, (Exercise 101) © Bob Krist/Corbis, (Exercise 102) © Corbis; Page 98, © Warren Morgan/Corbis; Page 110, © Catherine Karnow/Corbis; Page 111, © PhotoDisc; Page 123, (Exercise 91) © Dennis Degnan/Corbis, (Exercise 92) © Wartenburg/Picture Press/Corbis; Page 124, © Ed Young/Science Source/Photo Researchers; Page 132, © Walter Hodges/Corbis; Page 136, © Corbis; Page 138, © PhotoDisc; Page 140, © PhotoDisc; Page 141, © Reuters NewMedia Inc./Corbis; Page 144, (Exercise 19) © Beth Anderson, (Exercise 20) © Annie Griffiths Belt/Corbis, (Exercise 21) © Comstock; Page 145, (Exercise 22) © PhotoDisc, (Exercise 23) © AFP/Corbis; Page 154, © AFP/Corbis; Page 156, © Vince Streano/Corbis; Page 160, (Exercise 115) © AP/ WideWorld Photos, (Exercise 116) ©PhotoDisc, (Exercise 117) © Brand X Pictures; Page 161, © AP/WideWorld Photos; Page 173, (Exercise 98) © Reuters NewMedia Inc./Corbis, (Exercise 99) © PhotoDisc; Page 174, © PhotoDisc; Page 175, © PhotoDisc; Page 180, © PhotoDisc; Page 181, (Exercise 108) © AFP/Corbis, (Exercise 109) © Weststock/Image State, (Exercise 110) ©PhotoDisc; Page 182, © PhotoDisc.

CHAPTER 2

Page 184, © PhotoDisc; Page 195, © Chuck Savage/Corbis Stock Market; Page 196, (Exercise 70) © PhotoDisc, (Exercises 73 and 74) © Catherine Karnow/Corbis; Page 197, (Exercise 94) © PhotoDisc, (Exercises 95 and 96) © DigitalVision; Page 198, © PhotoDisc; Page 210, © PhotoDisc; Page 212, © Ed Bock/Corbis Stock Market; Page 226, (Exercise 97) © Bob Krist/Corbis, (Exercise 98) Courtesy Andersen Windows, Inc.; Page 236, © Corbis; Page 239 (Table 2.1) © PhotoDisc, (Table 2.2) © PhotoDisc; Page 250, © PhotoDisc; Page 251, (Exercise 87) © Mark Jenkinson/Corbis, (Exercise 88) © PhotoDisc; Page 258, © Reuters NewMedia Inc./Corbis; Page 259, © PhotoDisc; Page 260, NASA.

CHAPTER 3

Page 267, © Richard T. Nowitz/Corbis; Page 278, © PhotoDisc; Page 279, © Waren Morgan/Corbis; Page 280, © Corbis; Page 291, © Scott Camenzine/Photo Researchers, Inc.; Page 311, © Jan Butchofsky-House/Corbis; Page 314, © Beth Anderson; Page 345, © Jon Feingersh/The Stock Market; Page 350, © PHOTRI/The Stock Market.

CHAPTER 4

Page 353, © (PAL) Photo Researchers, Inc.; Page 357, © Adam Woolfit/ Corbis; Page 367, (Exercise 97) © PhotoDisc Red, (Exercise 98) © Corbis; Page 368, © Laura Dwight/Corbis; Page 380, (Exercise 119) © PhotoDisc, (Exercise 121) © Alan Towse, Ecoscene/Corbis; Page 382, © Chuck Savage/Corbis Stock Market; Page 394, (Exercise 99) © PhotoDisc, (Exercise 100) © Corbis, (Exercise 101) © R. W. Jones/Corbis; Page 395, © PhotoDisc; Page 401, © PhotoDisc; Page 404, © PhotoDisc; Page 405, © Kennan Ward/Corbis; Page 406, (Exercise 81) © PhotoDisc, (Exercises 85 and 86) © PhotoDisc; Page 407, (Exercise 91) © PhotoDisc, (Exercise 92) By permission of Betatherm Ireland Ltd.; Page 408, © Paul A. Souders/ Corbis; Page 412, © PhotoDisc; Page 413, © Miguel Gandert/Corbis.

CHAPTER 5

Page 416, Boston Redevelopment Authority; Page 445, © Digital Vision; Page 458, © PhotoDisc; Page 461, (Exercise 83) © PhotoDisc, (Exercise 84) © PhotoDisc; Page 462, © PhotoDisc; Page 475, PhotoDisc; Page 476, © PhotoDisc Red.

CHAPTER 6

Page 508, © PhotoDisc; Page 555, Christmas Store, © Corbis RF.

CHAPTER 7

Page 576, © Digital Vision; Page 597, © Corbis.

CHAPTER 8

Page 654, © Brand X Pictures; Page 664, © PhotoDisc; Page 665, ©PhotoDisc; Page 666, © Reed Kaestner/Corbis; Page 678, (Exercise 69) © Lindsay Hebberd/Corbis; Page 678, © PhotoDisc; Page 696, ©Macduff Everton/Corbis; Page 712, (Exercise 27) © PhotoDisc, (Exercise 28) © PhotoDisc Red; Page 716, © Kelly-Mooney/Corbis.

CHAPTER 9

Page 719, © DigitalVision; Page 734, © PhotoDisc; Page 750, © Corbis; Page 751, © Corbis; Page 779, © Comstock; Page 783, © DigitalVision.

CHAPTER 10

Page 785, NASA; Page 795, © Roger Ressmayer/Corbis.

CHAPTER 11

Page 830, © Medio Images; Page 850, © PhotoDisc.

Index of Applications

Index

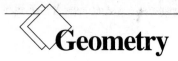 # Geometry

Rectangle

Area = LW

Perimeter = $2L + 2W$

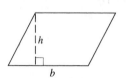

Square

Area = s^2

Perimeter = $4s$

Triangle

Area = $\frac{1}{2}bh$

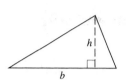

Right Triangle

Area = $\frac{1}{2}ab$

Pythagorean theorem:
$c^2 = a^2 + b^2$

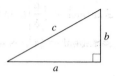

Parallelogram

Area = bh

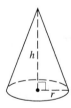

Trapezoid

Area = $\frac{1}{2}h(b_1 + b_2)$

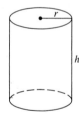

Circle

Area = πr^2

Circumference = $2\pi r$

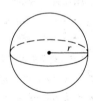

Right Circular Cone

Volume = $\frac{1}{3}\pi r^2 h$

Lateral surface area = $\pi r \sqrt{r^2 + h^2}$

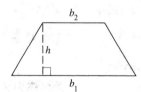

Right Circular Cylinder

Volume = $\pi r^2 h$

Lateral surface area = $2\pi rh$

Sphere

Volume = $\frac{4}{3}\pi r^3$

Surface area = $4\pi r^2$

Metric Abbreviations

Length		Volume		Weight	
mm	millimeter	mL	milliliter	mg	milligram
cm	centimeter	cL	centiliter	cg	centigram
dm	decimeter	dL	deciliter	dg	decigram
m	meter	L	liter	g	gram
dam	dekameter	daL	dekaliter	dag	dekagram
hm	hectometer	hL	hectoliter	hg	hectogram
km	kilometer	kL	kiloliter	kg	kilogram

English-Metric Conversion

Length	Volume (U.S.)	Weight
1 in. = 2.540 cm	1 pt = 0.4732 L	1 oz = 28.35 g
1 ft = 30.48 cm	1 qt = 0.9464 L	1 lb = 453.6 g
1 yd = 0.9144 m	1 gal = 3.785 L	1 lb = 0.4536 kg
1 mi = 1.609 km		

Length	Volume (U.S.)	Weight
1 cm = 0.3937 in.	1 L = 2.2233 pt	1 g = 0.0353 oz
1 cm = 0.03281 ft	1 L = 1.0567 qt	1 g = 0.002205 lb
1 m = 1.0936 yd	1 L = 0.2642 gal	1 kg = 2.205 lb
1 km = 0.6215 mi		

Algebra

Subsets of the Real Numbers

Natural numbers = $\{1, 2, 3, \ldots\}$
Whole numbers = $\{0, 1, 2, 3, \ldots\}$
Integers = $\{\ldots -3, -2, -1, 0, 1, 2, 3, \ldots\}$
Rational = $\left\{ \dfrac{a}{b} \,\middle|\, a \text{ and } b \text{ are integers with } b \neq 0 \right\}$
Irrational = $\{x \mid x \text{ is not rational}\}$

Properties of the Real Numbers

For all real numbers a, b, and c

$a + b$ and ab are real numbers.	Closure
$a + b = b + a; \; a \cdot b = b \cdot a$	Commutative
$(a + b) + c = a + (b + c);$ $(ab)c = a(bc)$	Associative
$a(b + c) = ab + ac;$ $a(b - c) = ab - ac$	Distributive
$a + 0 = a; \; 1 \cdot a = a$	Identity
$a + (-a) = 0; \; a \cdot \dfrac{1}{a} = 1 \quad (a \neq 0)$	Inverse
$a \cdot 0 = 0$	Multiplication property of 0

Absolute Value

$$|a| = \begin{cases} a & \text{for } a \geq 0 \\ -a & \text{for } a < 0 \end{cases}$$

$\sqrt{x^2} = |x|$ for any real x

$|x| = k \Leftrightarrow x = k \text{ or } x = -k \qquad (k > 0)$

$|x| < k \Leftrightarrow -k < x < k \qquad (k > 0)$

$|x| > k \Leftrightarrow x < -k \text{ or } x > k \qquad (k > 0)$

(The symbol $\Leftrightarrow$ means "if and only if.")

Interval Notation

$(a, b) = \{x \mid a < x < b\}$ $[a, b] = \{x \mid a \leq x \leq b\}$

$(a, b] = \{x \mid a < x \leq b\}$ $[a, b) = \{x \mid a \leq x < b\}$

$(-\infty, a) = \{x \mid x < a\}$ $(a, \infty) = \{x \mid x > a\}$

$(-\infty, a] = \{x \mid x \leq a\}$ $[a, \infty) = \{x \mid x \geq a\}$

Exponents

$a^n = a \cdot a \cdots \cdot a \; (n \text{ factors of } a)$

$a^0 = 1 \qquad\qquad a^{-n} = \dfrac{1}{a^n}$

$a^r a^s = a^{r+s} \qquad \dfrac{a^r}{a^s} = a^{r-s}$

$(a^r)^s = a^{rs} \qquad (ab)^r = a^r b^r$

$\left(\dfrac{a}{b}\right)^r = \dfrac{a^r}{b^r} \qquad \left(\dfrac{a}{b}\right)^{-r} = \left(\dfrac{b}{a}\right)^r$

Radicals

$a^{1/n} = \sqrt[n]{a} \qquad\qquad a^{m/n} = \left(\sqrt[n]{a}\right)^m = \sqrt[n]{a^m}$

$\sqrt[n]{ab} = \sqrt[n]{a} \cdot \sqrt[n]{b} \qquad \sqrt[n]{\dfrac{a}{b}} = \dfrac{\sqrt[n]{a}}{\sqrt[n]{b}}$

Factoring

$a^2 + 2ab + b^2 = (a + b)^2$

$a^2 - 2ab + b^2 = (a - b)^2$

$a^2 - b^2 = (a + b)(a - b)$

$a^3 - b^3 = (a - b)(a^2 + ab + b^2)$

$a^3 + b^3 = (a + b)(a^2 - ab + b^2)$

Rational Expressions

$\dfrac{ac}{bc} = \dfrac{a}{b} \qquad\qquad \dfrac{a}{b} + \dfrac{c}{d} = \dfrac{ad + bc}{bd}$

$\dfrac{a}{b} \cdot \dfrac{c}{d} = \dfrac{ac}{bd} \qquad \dfrac{a}{b} \div \dfrac{c}{d} = \dfrac{a}{b} \cdot \dfrac{d}{c}$

Quadratic Formula

The solutions to $ax^2 + bx + c = 0$ with $a \neq 0$ a

$$x = \frac{-b \pm \sqrt{b^2 - 4ac}}{2a}.$$

Distance Formula

The distance from (x_1, y_1) to (x_2, y_2), is

$$\sqrt{(x_2 - x_1)^2 + (y_2 - y}$$

Algebra

Midpoint Formula

The midpoint of the line segment with endpoints (x_1, y_1) and (x_2, y_2) is

$$\left(\frac{x_1 + x_2}{2}, \frac{y_1 + y_2}{2}\right).$$

Slope Formula

The slope of the line through (x_1, y_1) and (x_2, y_2) is

$$\frac{y_2 - y_1}{x_2 - x_1} \quad \text{(for } x_1 \neq x_2\text{)}.$$

Linear Function

$f(x) = mx + b$ with $m \neq 0$

Graph is a line with slope m.

Quadratic Function

$f(x) = ax^2 + bx + c$ with $a \neq 0$

Graph is a parabola.

Polynomial Function

$f(x) = a_n x^n + a_{n-1} x^{n-1} + \cdots + a_1 x + a_0$ for
n a nonnegative integer

ational Function

$\dfrac{p(x)}{q(x)}$, where p and q are polynomial functions
$\neq 0$

Expon   and Logarithmic Functions

$f(x) = a^x$
$f(x) = \log_a($ > 0 and $a \neq 1$
$a > 0$ and $a \neq 1$

Properties of Logarithms

Base-a logarithm: $y = \log_a(x) \Leftrightarrow a^y = x$

Natural logarithm: $y = \ln(x) \Leftrightarrow e^y = x$

Common logarithm: $y = \log(x) \Leftrightarrow 10^y = x$

One-to-one: $a^{x_1} = a^{x_2} \Leftrightarrow x_1 = x_2$

   $\log_a(x_1) = \log_a(x_2) \Leftrightarrow x_1 = x_2$

$\log_a(a) = 1$ $\log_a(1) = 0$

$\log_a(a^x) = x$ $a^{\log_a(N)} = N$

$\log_a(MN) = \log_a(M) + \log_a(N)$

$\log_a(M/N) = \log_a(M) - \log_a(N)$

$\log_a(M^x) = x \cdot \log_a(M)$

$\log_a(1/N) = -\log_a(N)$

$$\log_a(M) = \frac{\log_b(M)}{\log_b(a)} = \frac{\ln(M)}{\ln(a)} = \frac{\log(M)}{\log(a)}$$

Compound Interest

P = principal, t = time in years, r = annual interest rate, and A = amount:

$$A = P\left(1 + \frac{r}{n}\right)^{nt} \text{ (compounded } n \text{ times/year)}$$

$A = Pe^{rt}$ (compounded continuously)

Variation

Direct: $y = kx$ $(k \neq 0)$

Inverse: $y = k/x$ $(k \neq 0)$

Joint: $y = kxz$ $(k \neq 0)$

Straight Line

Slope-intercept form: $y = mx + b$

Slope: m y-intercept: $(0, b)$

Point-slope form: $y - y_1 = m(x - x_1)$

Standard form: $Ax + By = C$

Horizontal: $y = k$ Vertical: $x = k$

 # Algebra

Parabola

$y = a(x - h)^2 + k \quad (a \neq 0)$

Vertex: (h, k)

Axis of symmetry: $x = h$

Focus: $(h, k + p)$, where $a = \dfrac{1}{4p}$

Directrix: $y = k - p$

Circle

$(x - h)^2 + (y - k)^2 = r^2 \quad (r > 0)$

Center: (h, k) Radius: r

$x^2 + y^2 = r^2$

Center $(0, 0)$ Radius: r

Ellipse

$\dfrac{x^2}{a^2} + \dfrac{y^2}{b^2} = 1 \quad (a > b > 0)$

Center: $(0, 0)$ Major axis: horizontal

Foci: $(\pm c, 0)$, where $c^2 = a^2 - b^2$

$\dfrac{x^2}{b^2} + \dfrac{y^2}{a^2} = 1 \quad (a > b > 0)$

Center: $(0, 0)$ Major axis: vertical

Foci: $(0, \pm c)$, where $c^2 = a^2 - b^2$

Hyperbola

$\dfrac{x^2}{a^2} - \dfrac{y^2}{b^2} = 1$

Center: $(0, 0)$ Vertices: $(\pm a, 0)$

Foci: $(\pm c, 0)$, where $c^2 = a^2 + b^2$

Asymptotes: $y = \pm\dfrac{b}{a} x$

$\dfrac{y^2}{a^2} - \dfrac{x^2}{b^2} = 1$

Center: $(0, 0)$ Vertices: $(0, \pm a)$

Foci: $(0, \pm c)$, where $c^2 = a^2 + b^2$

Asymptotes: $y = \pm\dfrac{a}{b} x$

Arithmetic Sequence

$a_1, a_1 + d, a_1 + 2d, a_1 + 3d, \ldots$

Formula for nth term: $a_n = a_1 + (n - 1)d$

Sum of n terms:

$$S_n = \sum_{i=1}^{n} [a_1 + (i - 1)d] = \frac{n}{2}(a_1 + a_n)$$

Geometric Sequence

$a_1, a_1 r, a_1 r^2, a_1 r^3, \ldots$

Formula for nth term: $a_n = a_1 r^{n-1}$

Sum of n terms when $r \neq 1$:

$$S_n = \sum_{i=1}^{n} a_1 r^{i-1} = \frac{a_1 - a_1 r^n}{1 - r}$$

Sum of all terms when $|r| < 1$:

$$S = \sum_{i=1}^{\infty} a_1 r^{i-1} = \frac{a_1}{1 - r}$$

Counting Formulas

Factorial notation: $n! = 1 \cdot 2 \cdot 3 \cdot \cdots \cdot (n - 1) \cdot n$

Permutation: $P(n, r) = \dfrac{n!}{(n - r)!}$ for $0 \leq r \leq n$

Combination: $C(n, r) = \dbinom{n}{r} = \dfrac{n!}{(n - r)!r!}$ for $0 \leq r \leq n$

Binomial Expansion

$(a + b)^2 = a^2 + 2ab + b^2$

$(a + b)^3 = a^3 + 3a^2 b + 3ab^2 + b^3$

$(a + b)^4 = a^4 + 4a^3 b + 6a^2 b^2 + 4ab^3 + b^4$

$(a + b)^n = \sum_{r=0}^{n} \dbinom{n}{r} a^{n-r} b^r$, where $\dbinom{n}{r} = \dfrac{n!}{(n - r)!r!}$

Trigonometry

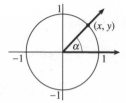

Trigonometric Functions

If the angle α
(in standard position)
intersects the unit circle
at (x, y), then

$$\sin \alpha = y \qquad \cos \alpha = x \qquad \tan \alpha = \frac{y}{x}$$

$$\csc \alpha = \frac{1}{y} \qquad \sec \alpha = \frac{1}{x} \qquad \cot \alpha = \frac{x}{y}$$

Trigonometric Ratios

If (x, y) is any point
other than the origin
on the terminal side of α
and $r = \sqrt{x^2 + y^2}$, then

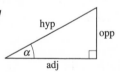

$$\sin \alpha = \frac{y}{r} \qquad \cos \alpha = \frac{x}{r} \qquad \tan \alpha = \frac{y}{x}$$

$$\csc \alpha = \frac{r}{y} \qquad \sec \alpha = \frac{r}{x} \qquad \cot \alpha = \frac{x}{y}$$

Right Triangle Trigonometry

If α is an acute angle
of a right triangle, then

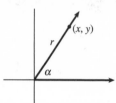

$$\sin \alpha = \frac{\text{opp}}{\text{hyp}} \qquad \cos \alpha = \frac{\text{adj}}{\text{hyp}} \qquad \tan \alpha = \frac{\text{opp}}{\text{adj}}$$

$$\csc \alpha = \frac{\text{hyp}}{\text{opp}} \qquad \sec \alpha = \frac{\text{hyp}}{\text{adj}} \qquad \cot \alpha = \frac{\text{adj}}{\text{opp}}$$

Special Right Triangles

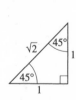
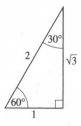

Exact Values of Trigonometric Functions

x degrees	x radians	$\sin x$	$\cos x$	$\tan x$
0°	0	0	1	0
30°	$\frac{\pi}{6}$	$\frac{1}{2}$	$\frac{\sqrt{3}}{2}$	$\frac{\sqrt{3}}{3}$
45°	$\frac{\pi}{4}$	$\frac{\sqrt{2}}{2}$	$\frac{\sqrt{2}}{2}$	1
60°	$\frac{\pi}{3}$	$\frac{\sqrt{3}}{2}$	$\frac{1}{2}$	$\sqrt{3}$
90°	$\frac{\pi}{2}$	1	0	—

Basic Identities

$$\tan x = \frac{\sin x}{\cos x} = \frac{1}{\cot x} \qquad \cot x = \frac{\cos x}{\sin x} = \frac{1}{\tan x}$$

$$\sin x = \frac{1}{\csc x} \qquad \csc x = \frac{1}{\sin x}$$

$$\cos x = \frac{1}{\sec x} \qquad \sec x = \frac{1}{\cos x}$$

Pythagorean Identities

$$\sin^2 x + \cos^2 x = 1 \qquad\qquad 1 + \cot^2 x = \csc^2 x$$

$$\tan^2 x + 1 = \sec^2 x$$

Odd Identities

$$\sin(-x) = -\sin(x) \qquad\qquad \csc(-x) = -\csc(x)$$

$$\tan(-x) = -\tan(x) \qquad\qquad \cot(-x) = -\cot(x)$$

Even Identities

$$\cos(-x) = \cos(x) \qquad\qquad \sec(-x) = \sec(x)$$